Lecture Notes in Computer Science 6302

Commenced Publication in 1973
Founding and Former Series Editors:
Gerhard Goos, Juris Hartmanis, and Jan van Leeuwen

Maria Serna   Ronen Shaltiel
Klaus Jansen   José Rolim (Eds.)

Approximation, Randomization, and Combinatorial Optimization

Algorithms and Techniques

13th International Workshop, APPROX 2010
and 14th International Workshop, RANDOM 2010
Barcelona, Spain, September 1-3, 2010
Proceedings

 Springer

Volume Editors

Maria Serna
Universitat Politècnica de Catalunya
Dept. de Llenguatges i Sistemes Informàtics
Jordi Girona Salgado, 1-3, 08034 Barcelona, Spain
E-mail: mjserna@lsi.upc.edu

Ronen Shaltiel
University of Haifa, Department of Computer Science
Mount Carmel, Haifa 31905, Israel
E-mail: ronen@cs.haifa.ac.il

Klaus Jansen
University of Kiel, Department of Computer Science
Olshausenstrasse 40, 24098 Kiel, Germany
E-mail: kj@informatik.uni-kiel.de

José Rolim
University of Geneva, Centre Universitaire d'Informatique
Battelle Bat. A, 7 rte de Drize, 1227 Carouge, Switzerland
E-mail: jose.rolim@unige.ch

Library of Congress Control Number: 2010932674

CR Subject Classification (1998): F.2, E.1, G.2, I.3.5, F.1, C.2

LNCS Sublibrary: SL 1 – Theoretical Computer Science and General Issues

ISSN   0302-9743
ISBN-10   3-642-15368-2 Springer Berlin Heidelberg New York
ISBN-13   978-3-642-15368-6 Springer Berlin Heidelberg New York

springer.com

© Springer-Verlag Berlin Heidelberg 2010
Printed in Germany

Typesetting: Camera-ready by author, data conversion by Scientific Publishing Services, Chennai, India
Printed on acid-free paper   06/3180

Preface

This volume contains the papers presented at the 13th International Workshop on Approximation Algorithms for Combinatorial Optimization Problems (APPROX 2010) and the 14th International Workshop on Randomization and Computation (RANDOM 2010), which took place concurrently in Universitat Politècnica de Catalunya (UPC) Barcelona, Spain, during September 1–3, 2010. APPROX focuses on algorithmic and complexity issues surrounding the development of efficient approximate solutions to computationally difficult problems, and was the 13th in the series after Aalborg (1998), Berkeley (1999), Saarbrücken (2000), Berkeley (2001), Rome (2002), Princeton (2003), Cambridge (2004), Berkeley (2005), Barcelona (2006), Princeton (2007), Boston (2008) and Berkeley (2009). RANDOM is concerned with applications of randomness to computational and combinatorial problems, and was the 14th workshop in the series following Bologna (1997), Barcelona (1998), Berkeley (1999), Geneva (2000), Berkeley (2001), Harvard (2002), Princeton (2003), Cambridge (2004), Berkeley (2005), Barcelona (2006), Princeton (2007), Boston (2008), and Berkeley (2009).

Topics of interest for APPROX and RANDOM are: design and analysis of approximation algorithms, hardness of approximation, small space algorithms, sub-linear time algorithms, streaming algorithms, embeddings and metric space methods, mathematical programming methods, combinatorial problems in graphs and networks, game theory, markets and economic applications, geometric problems, packing, covering, scheduling, approximate learning, design and analysis of randomized algorithms, randomized complexity theory, pseudorandomness and derandomization, random combinatorial structures, random walks/Markov chains, expander graphs and randomness extractors, probabilistic proof systems, random projections and embeddings, error-correcting codes, average-case analysis, property testing, computational learning theory, and other applications of approximation and randomness.

The volume contains 28 contributed papers, selected by the APPROX Program Committee out of 66 submissions, and 29 contributed papers, selected by the RANDOM Program Committee out of 61 submissions.

We would like to thank all of the authors who submitted papers and the members of the Program Committees, and the external Reviewers.

We gratefully acknowledge the support from the Software Department the Universitat Politecnica de Catalunya in Barcelona, the Department of Computer Science at the University of Haifa in Israel, the Institute of Computer Science of the Christian-Albrechts-Universität zu Kiel and the Department of Computer Science of the University of Geneva.

We also thank the support of the Technical University of Catalonia and the Spanish Ministry of Science and Innovation.

Finally, many thanks to Parvaneh Karimi-Massouleh for editing the proceedings.

September 2010   Maria Serna
                                                                             Ronen Shaltiel
                                                                             Klaus Jansen
                                                                           José D.P. Rolim

Organization

Program Committees

APPROX 2010

Chandra Chekuri	University of Illinois
Uriel Feige	Weizmann Institute of Science
Pierre Fraigniaud	Université Paris Diderot
Magnús M. Halldórsson	Reykjavik University
Christos Kaklamanis	University of Patras
Anna Karlin	University of Washington
Samir Khuller	University of Maryland
Guy Kortsarz	Rutgers University
Monaldo Mastrolilli	IDSIA
Claire Mathieu	Brown University
Zeev Nutov	The Open University of Israel
Giuseppe Persiano	Università di Salerno
Maria Serna	Universitat Politècnica de Catalunya (Chair)
Martin Skutella	Technische Universität Berlin
Maxim Sviridenko	IBM T. J. Watson Research Center
David P. Williamson	Cornell University

RANDOM 2010

Dimitris Achlioptas	UC Santa Cruz and University of Athens
Alexandr Andoni	Princeton University
Anna Gal	University of Texas at Austin
Valentine Kabanets	Institute for Advanced Study and Simon Fraser University
Swastik Kopparty	MIT
Michael Krivelevich	Tel Aviv University
Sofya Raskhodnikova	Pennsylvania State University
Ran Raz	The Weizmann Institute of Science
Atri Rudra	University at Buffalo, State University of New York
Rocco Servedio	Columbia University
Ronen Shaltiel	University of Haifa (Chair)
Angelika Steger	ETH Zurich
Christopher Umans	California Institute of Technology
Eric Vigoda	Georgia Institute of Technology
Sergey Yekhanin	Microsoft Research Silicon Valley

External Referees

Ittai Abraham, Ernst Althaus, Estie Arkin, Arash Asadpour, Stavros Athanassopoulos, Nikhil Bansal, Cristina Bazgan, Paul Beame, Luca Becchetti, Nayantara Bhatnagar, Arnab Bhattacharyya, Peter Biro, Punya Biswal, Eric Blais, Andreas Bley, Johannes Blömer, Andrej Bogdanov, Allen Borodin, Mark Braverman, Jaroslaw Byrka, Ioannis Caragiannis, Moses Charikar, Joseph Cheriyan, Janka Chlebikova, Edith Cohen, Amin Coja-Oghlan, Graham Cormode, Artur Czumaj, Dana Dachman-Soled, Anindya De, Ilias Diakonikolas, Josep Diaz, Feodor Dragan, Andrew Drucker, Yuval Emek, David Eppstein, Thomas Erlebach, Guy Even, Moran Feldman, Eldar Fischer, Alan Frieze, Ariel Gabizon, Hal Gabow, Naveen Garg, Dmitry Gavinsky, Cyril Gavoille, Parikshit Gopalan, Luca Gugelmann, Venkatesan Guruswami, Mohammad Taghi Hajiaghayi, Sariel Har-Peled, Avinatan Hassidim, Frederic Havet, Russell Impagliazzo, Sandy Irani, Madhav Jha, Riko Jacob, Lior Kamma, Panagiotis Kanellopoulos, Tali Kaufman, Ken-ichi Kawarabayashi, Rohit Khandekar, Sanjeev Khanna, Guy Kindler, Jon Kleinberg, Yusuke Kobayashi, Ronald Koch, Jochen Koenemann , Charalampos Konstantopoulos, Nitish Korula, Michal Koucky, Robi Krauthgamer, Michael Lampis, Homin Lee, James Lee, Asaf Levin, Yi-Kai Liu, Hamid Mahini, Konstantin Makarychev, Yury Makarychev, Dániel Marx, Fabien Mathieu, Kevin Matulef, Jannik Matuschke, Andrew McGregor, Manor Mendel, Julian Mestre, Adam Meyerson, Vahab Mirrokni, Joseph Mitchell, Pradipta Mitra, Dieter Mitsche, Mohammad Moharammi, Gianpiero Monaco, Ravi Montenegro, Koyel Mukherjee, S Muthukrishnan, Nikolaus Mutsanas, Torsten Mütze, Viswanath Nagarajan, Assaf Naor, Ilan Newman, Huy L. Nguyen, Thach Nguyen, Tim Nonner, Krzysztof Onak, Shayan Oveis Gharan, Igor Pak, Alessandro Panconesi, Evi Papaioannou, Preyas Popat, Christopher Peikert, Yuri Rabinovich, Rajiv Raman, Anup Rao, Thomas Rast, Dror Rawitz, Daniel Reichman, Renato Renner, Sebastien Roch, Dana Ron, Adi Rosen, Thomas Rothvoss, Alex Russell, Rishi Saket, Mike Saks, Mohammad Salavatipour, Alex Samorodnitsky, Rahul Santhanam, Nicolas Schabanel, Grant Schoenebeck, Oded Schwartz, Uwe Schwiegelshohn, Danny Segev, Comandur Seshadhri, Hadas Shachnai, Bruce Shepherd, Amir Shpilka, Anastasios Sidiropoulos, Alistair Sinclair, Mohit Singh, Rene Sitters, Adam Smith, Christian Sohler, Georgios Stamoulis, David Steurer, Sebastian Stiller, Ola Svensson, Zoya Svitkina, Inbal Talgam, Li–Yang Tan, Kunal Talwar, Thomas Thierauf, Henning Thomas, Ryuhei Uehara, Rob van Stee, Paul Valiant, Virginia Vassileska Williams, Juan Vera, Jose Verschae, Laurent Viennot, Aravindan Vijayraghavan, Emanuele Viola, Jan Vondrak, Andrew Wan, Andreas Wiese, Philipp Woelfel, David Woodruff, Yi Wu, Justin Yip, Raphy Yuster, Morteza Zadimoghaddam, Alex Zelikovsky, Pawel Zielinski, and David Zuckerman.

Table of Contents

Contributed Talks of APPROX

Contributed Talks of RANDOM

Approximation Algorithms for the Bottleneck Asymmetric Traveling Salesman Problem

Hyung-Chan An[1,*], Robert D. Kleinberg[1,**], and David B. Shmoys[2,***]

[1] Dept. of Computer Science
Cornell University, Ithaca, NY 14853
{anhc,rdk}@cs.cornell.edu
[2] School of ORIE and Dept. of Computer Science
Cornell University, Ithaca, NY 14853
shmoys@cs.cornell.edu

Abstract. We present the first nontrivial approximation algorithm for the *bottleneck asymmetric traveling salesman problem*. Given an asymmetric metric cost between n vertices, the problem is to find a Hamiltonian cycle that minimizes its *bottleneck* (or maximum-length edge) cost. We achieve an $O(\log n / \log \log n)$ approximation performance guarantee by giving a novel algorithmic technique to shortcut Eulerian circuits while bounding the lengths of the shortcuts needed. This allows us to build on the recent result of Asadpour, Goemans, Mądry, Oveis Gharan, and Saberi to obtain this guarantee. Furthermore, we show how our technique yields stronger approximation bounds in some cases, such as the bounded orientable genus case studied by Oveis Gharan and Saberi.

Keywords: Approximation algorithms, traveling salesman problem, bottleneck optimization.

1 Introduction

In this paper, we study the *bottleneck asymmetric traveling salesman problem*; that is, in contrast to the variant of traveling salesman problem most commonly studied, the objective is to minimize the maximum edge cost in the tour, rather than the sum of the edge costs. Furthermore, while the edge costs satisfy the triangle inequality, we do not require that they be symmetric, in that the distance from point a to point b might differ from the distance from b to a. The triangle inequality is naturally satisfied by many cost functions; for example, minimizing the longest interval between job completions in the no-wait flow-shop reduces

* Research supported in part by NSF under grants no. CCR-0635121, DMS-0732196, CCF-0832782, CCF-0729102 and the Korea Foundation for Advanced Studies.
** Supported by NSF grants no. CCF-0643934 and CCF-0729102, a grant from the Air Force Office of Scientific Research, a Microsoft Research New Faculty Fellowship, and an Alfred P. Sloan Foundation Fellowship.
*** Research supported in part by NSF under grants no. CCR-0635121, DMS-0732196, CCF-0832782.

M. Serna et al. (Eds.): APPROX and RANDOM 2010, LNCS 6302, pp. 1–11, 2010.

to the bottleneck asymmetric traveling salesman problem under a metric cost. The bottleneck asymmetric traveling salesman problem cannot be approximated within a reasonable factor without assuming the triangle inequality. Surprisingly, no approximation algorithm was previously known to deliver solutions within an $o(n)$ factor of optimal, where n denotes the number of nodes in the input. We present the first nontrivial approximation algorithm for the bottleneck asymmetric traveling salesman problem, by giving an $O(\log n / \log \log n)$-approximation algorithm. At the heart of our result is a new algorithmic technique for converting Eulerian circuits into tours while introducing "shortcuts" that are of bounded length.

For any optimization problem defined in terms of pairwise distances between nodes, it is natural to consider both the symmetric case and the asymmetric one, as well as the min-sum variant and the bottleneck one. The standard (min-sum symmetric) traveling salesman problem (TSP) has been studied extensively [18], and for approximation algorithms, Christofides' 3/2-approximation algorithm [5] remains the best known guarantee, and yet the strongest NP-hardness result, due to Papadimitriou and Vempala [22], states that the existence of a ρ-approximation algorithm with $\rho < 220/219$, implies that P=NP. In contrast, for the bottleneck symmetric TSP, Lau [17], and Parker & Rardin [23], building on structural results of Fleischner [7], give a 2-approximation algorithm, and based on the metric in which all costs are either 1 or 2, it is easy to show that, for any $\rho < 2$, the existence of a ρ-approximation algorithm implies that P=NP. For the asymmetric min-sum problem, Frieze, Galbiati, and Maffioli [8] gave the first $O(\log n)$-approximation algorithm, which is a guarantee that was subsequently matched by work of Kleinberg and Williamson [16], and only recently improved upon by work of Asadpour, Goemans, Mądry, Oveis Gharan, and Saberi [3].

This cross-section of results is mirrored in other optimization settings. For example, for the min-sum symmetric k-median problem in which k points are chosen as "medians" and each point is assigned to its nearest median, Arya, Garg, Khandekar, Meyerson, Munagala, and Pandit [2] give a ρ-approximation algorithm for each $\rho > 3$, whereas Jain, Mahdian, Markakis, Saberi and Vazirani prove hardness results for $\rho < 1 + 2/e$ [14]. In contrast, for the bottleneck symmetric version, the k-center problem, Hochbaum and Shmoys [12] gave a 2-approximation algorithm, whereas Hsu and Nemhauser [13] showed the NP-hardness of a performance guarantee of $\rho < 2$. For the asymmetric k-center, a matching upper and lower bound of $\Theta(\log^* n)$ for the best performance guarantee was shown by Panigrahy & Vishwanathan [21] and Chuzhoy, Guha, Halperin, Khanna, Kortsarz, Krauthgamer & Naor [6], respectively. In contrast, for the asymmetric k-median problem, a bicriterion result which allowed a constant factor increase in cost with a logarithmic increase in the number of medians was shown by Lin and Vitter [19], and a hardness tradeoff matching this (up to constant factors) was proved by Archer [1].

In considering these comparative results, there is a mixed message as to whether a bottleneck problem is easier or harder to approximate than its min-sum counterpart. On the one hand, for any bottleneck problem, one can immediately reduce the optimization problem with cost data to a more combinatorially defined question, since there is the trivial relationship that the optimal bottleneck solution is of objective function value at most T if and only if there exists a feasible solution that uses only those edges of cost at most T. Furthermore, there are only a polynomial number of potential thresholds T, and so a polynomial-time algorithm that answers this purely combinatorial decision question leads to a polynomial-time optimization algorithm.

For a ρ-approximation algorithm, it is sufficient for the algorithm to solve a "relaxed" decision question: either provide some certificate that no feasible solution exists, or produce a solution in which each edge used is of cost at most ρT. If G denotes the graph of all edges of cost at most T, then the triangle inequality implies that it is sufficient to find feasible solutions within G^ρ, the ρth power of G, in which we include an edge (u, v) whenever G contains a path from u to v with at most ρ edges. In the context of the TSP, this means that we either want to prove that G is not Hamiltonian, or else to produce a Hamiltonian cycle within, for example, the square of G (to yield a 2-approximation algorithm as in [17,23]).

Unfortunately, the techniques invented in the context of the min-sum problem do not seem to be amenable to bottleneck objective function. For example, the analysis of the $O(\log n)$-approximation algorithm for the min-sum asymmetric TSP due to Kleinberg and Williamson [16] depends crucially on the monotonicity of the optimal value over the vertex-induced subgraphs, and the fact that shortcutting a circuit does not increase the objective. That fact clearly is not true in the bottleneck setting: shortcutting arbitrary subpaths of a circuit may result in a tour that is valid only in a higher-order power graph. The aforementioned monotonicity is also lost as it relies on this fact as well.

In order to resolve this difficulty, we devise a condition on Eulerian circuits under which we can limit the lengths of the paths that are shortcut to obtain a Hamiltonian cycle. We will present a polynomial-time constructive proof of this condition using Hall's Transversal Theorem [10]; this proof is directly used in the algorithm. One of the special cases of the condition particularly worth mentioning is a degree-bounded spanning circuit (equivalently, an Eulerian spanning subgraph of bounded degree). If there exists a bound k on the number of occurrences of any vertex in a spanning circuit, our theorem provides a bound of $2k - 1$ on the length of the shortcut paths.

We will then show how thin trees defined in Asadpour et al. [3] can be used to compute these degree-bounded spanning circuits. An α-thin tree with respect to a weighted graph G is a unit-weighted spanning tree of G whose cut weights are no more than α times the corresponding cut weights of G. The min-sum algorithm due to Asadpour et al. [3] augments an $O(\log n / \log \log n)$-thin tree with respect to a (scaled) Held-Karp solution (Held and Karp [11]) into a spanning Eulerian graph by solving a circulation problem. The Held-Karp relaxation is a linear program consisting of the equality constraints on the in- and out-degree of each vertex and the inequality constraints on the directed cut weights: the equality constraints set the degrees to one, and the inequality constraints ensure that the total weight of edges leaving S is at least 1 for each subset S. We introduce vertex capacities to the circulation problem to impose the desired degree bound without breaking the feasibility of the circulation problem. This leads to an algorithm that computes degree-bounded spanning circuits with an $O(\log n / \log \log n)$ bound.

Recently, Oveis Gharan and Saberi [20] gave an $O(1)$-approximation algorithm for the min-sum asymmetric TSP when the support of the Held-Karp solution can be embedded on an orientable surface with a bounded genus. They achieved this by showing how to extract an $O(1)$-thin tree in this special case. Our result can be combined with this to yield an $O(1)$-approximation algorithm for the bottleneck asymmetric TSP when the support of the Held-Karp solution has a bounded orientable genus. Chekuri, Vondrák, and Zenklusen [4] showed that an alternative sampling procedure can be used to find the thin tree in Asadpour et al. [3].

Section 2 of this paper reviews some notation and previous results, and Section 3 describes the $O(\log n / \log \log n)$-approximation algorithm to the bottleneck asymmetric traveling salesman problem. Section 4 examines the special case when the support of the Held-Karp solution can be embedded on an orientable surface with a bounded genus. Some open questions are discussed in Section 5.

2   Preliminaries

We introduce some notation and review previous results in this section. Some notation was adopted from Asadpour et al. [3]

Let $G = (V, A)$ be a digraph and E be the underlying undirected edge set: $\{u, v\} \in E$ if and only if $\langle u, v \rangle \in A$ or $\langle v, u \rangle \in A$. For $S \subset V$, let

$$\delta^+(S) := \{\langle u, v \rangle \in A \mid u \in S, v \notin S\},$$
$$\delta^-(S) := \delta^+(V \setminus S),$$
$$\delta(S) := \{\{u, v\} \in E \mid |E \cap S| = 1\};$$

for $v \in V$,

$$\delta^+(v) := \delta^+(\{v\}),$$
$$\delta^-(v) := \delta^-(\{v\}),$$
$$\delta(v) := \delta(\{v\});$$

for $B \subset A$ and $x \in \mathbb{R}^A$,

$$x(B) := \sum_{b \in B} x_b;$$

similarly, for $F \subset E$ and $z \in \mathbb{R}^E$,

$$z(F) := \sum_{f \in F} z_f.$$

We need a notion of the non-Hamiltonicity certificate to solve the "relaxed" decision problem. We establish this certificate by solving the Held-Karp relaxation ([11]) in our algorithm. The Held-Karp relaxation to the asymmetric traveling salesman problem is the following linear program (we do not define an objective here):

$$\begin{cases} x(\delta^+(v)) = x(\delta^-(v)) = 1 & \forall v \in V \\ x(\delta^+(S)) \geq 1 & \forall S \subsetneq V, S \neq \emptyset \\ x \geq 0. \end{cases} \tag{1}$$

A graph is non-Hamiltonian if (1) is infeasible. This linear program can be solved in polynomial time [9].

A thin tree is defined as follows in Asadpour et al. [3].

Definition 1. *A spanning tree T is α-thin with respect to $z^* \in \mathbb{R}^E$ if $|T \cap \delta(U)| \leq \alpha z^*(\delta(U))$ for all $U \subset V$.*

Asadpour et al. [3] then prove Theorem 1: they show the thinness for $z^*_{uv} := \frac{n-1}{n}(x^*_{uv} + x^*_{vu})$ where $n = |V|$, and Theorem 1 is only weaker.

Theorem 1. *There exists a probabilistic algorithm that, given an extreme point solution $x^* \in \mathbb{R}^A$ to the Held-Karp relaxation, produces an α-thin tree T with respect to $z^*_{uv} := x^*_{uv} + x^*_{vu}$ with high probability, for $\alpha = \frac{4 \ln n}{\ln \ln n}$.*

Let $T_\rightarrow$ be a directed version of T, obtained by choosing the arcs in the support of x^*. If arcs exist in both directions, an arbitrary choice can be made. Consider a circulation problem on G: recall that the circulation problem requires, given a lower and upper bound on each arc, a set of flow values on arcs such that the sum of the incoming flows at every vertex matches the sum of outgoing, while honoring both bounds imposed on each arc. When all of the bounds are integers, an integral solution can be found in polynomial time unless the problem is infeasible [24]. Here we consider an instance where the lower bounds l and upper bounds u on the arcs are given as follows:

$$\begin{aligned} l(a) &= \begin{cases} 1 & \text{if } a \in T_\rightarrow \\ 0 & \text{otherwise} \end{cases} \\ u(a) &= \begin{cases} 1 + 2\alpha x^*_a & \text{if } a \in T_\rightarrow \\ 2\alpha x^*_a & \text{otherwise.} \end{cases} \end{aligned} \tag{2}$$

Asadpour et al. [3] show that this problem is feasible; the existence of an integral circulation under the rounded-up bounds follows from that.

Lemma 1. *The circulation problem defined by (2) is feasible.*

3   Algorithm

This section gives the $O(\frac{\log n}{\log \log n})$-approximation algorithm to the bottleneck asymmetric traveling salesman problem and its analysis. We present the lemmas to bound the lengths of the paths that are shortcut in the process of transforming a spanning circuit into a Hamiltonian cycle; we also show how a degree-bounded spanning circuit can be constructed.

Lemma 2. *Let $v_1, \ldots, v_m, v_1$ be a (non-simple) circuit that visits every vertex at least once. Partition $v_1, \ldots, v_m$ into contiguous subsequences of length k, except for the final subsequence whose length may be less than k. Denote the pieces of this partition by $P_1, \ldots, P_\ell$. If, for all t, the union of any t sets in $\{P_1, \ldots, P_\ell\}$ contains at least t distinct vertices, G^{2k-1} is Hamiltonian.*

Proof. From Hall's Transversal Theorem [10], if the given condition holds, $\{P_1, \ldots, P_\ell\}$ has a transversal: i.e., we can choose one vertex from each piece P_i such that no vertex is chosen more than once. If we take any subsequence of $v_1, \ldots, v_m$ that contains every vertex exactly once and includes all of the vertices in the transversal, this subsequence is a Hamiltonian cycle in G^{2k-1}. This is because any two contiguous vertices chosen in the transversal are at most $2k - 1$ arcs away. Since a transversal can be found in polynomial time (see Kleinberg and Tardos [15]), a Hamiltonian cycle can be constructed in polynomial time as well. □

Lemma 3 shows that a degree-bounded spanning circuit forms a special case of Lemma 2.

Lemma 3. *Given a circuit on G that visits every vertex at least once and at most k times, a Hamiltonian cycle on G^{2k-1} can be found in polynomial time.*

Proof. Consider $\{P_1, \ldots, P_\ell\}$ as defined in Lemma 2. For any t sets in $\{P_1, \ldots, P_\ell\}$, the sum of their cardinalities is strictly greater than $(t - 1)k$. If their union contained only $t - 1$ distinct vertices, then by the pigeonhole principle there would be some vertex that occurs at least $k + 1$ times, violating the upper bound on the number of occurrences of any vertex in the circuit.

Thus, by Lemma 2, there exists a Hamiltonian cycle in G^{2k-1}, and this can be found in polynomial time. □

Now we show how to construct a degree-bounded spanning circuit.

Lemma 4. *Let x^* be a feasible solution to the Held-Karp relaxation. Given an α-thin tree T with respect to $z_{uv}^* := x_{uv}^* + x_{vu}^*$, a circuit on G with every vertex visited at least once and at most $\lceil 4\alpha \rceil$ times can be found in polynomial time.*

Proof. We modify the circulation problem defined in (2) by introducing vertex capacities to the vertices: every vertex v is split into two vertices v_i and v_o, where all the incoming edges are connected to v_i and the outgoing edges are from v_o. We set the vertex capacity $u(\langle v_i, v_o \rangle)$ as $\sum_{a:\text{tail}(a)=v} u(a)$. (See Fig. 1.) It is easy to see that this modification does not change the feasibility; thus, from Lemma 1, this new circulation problem instance is also feasible.

Rounding up all u values of this instance preserves the feasibility and guarantees the existence of an integral solution. By contracting split vertices back in the

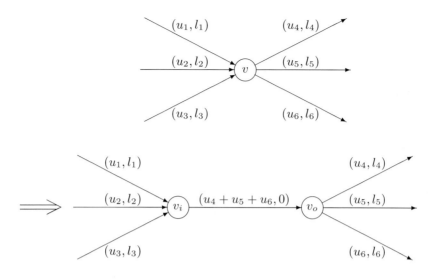

Fig. 1. Introducing vertex capacities

integral solution, we obtain a spanning Eulerian subgraph of $G = (V, A)$ (with arcs duplicated) whose maximum indegree is at most $\max_{v \in V} \lceil \sum_{a:\text{tail}(a)=v} u(a) \rceil$. Observe that, for any $v \in V$,

$$\sum_{a:\text{tail}(a)=v} u(a) = |\{a \in T_\rightarrow \mid \text{tail}(a) = v\}| + \sum_{a:\text{tail}(a)=v} 2\alpha x_a^*$$

$$\leq \alpha z^*(\delta(v)) + 2\alpha x^*(\delta^+(v))$$

$$= 4\alpha.$$

Thus, we can find a spanning Eulerian subgraph of $G = (V, A)$ whose maximum degree is at most $\lceil 4\alpha \rceil$, given the α-thin tree T. Any Eulerian circuit of this graph will satisfy the desired property. □

Theorem 1 and Lemmas 3 and 4 yield the algorithm.

Theorem 2. *There exists a probabilistic $O(\frac{\log n}{\log \log n})$-approximation algorithm for the bottleneck asymmetric traveling salesman problem under a metric cost.*

Proof. Let $A_{\leq \tau} := \{\langle u, v \rangle \mid c(u, v) \leq \tau\}$ and $G_{\leq \tau} := (V, A_{\leq \tau})$. The algorithm first determines the minimum τ such that the Held-Karp relaxation for $G_{\leq \tau}$ is feasible. Let τ^* be this minimum. If $\tau_1 \leq \tau_2$ and the Held-Karp relaxation for $G_{\leq \tau_2}$ is infeasible, the relaxation for $G_{\leq \tau_1}$ is also infeasible; therefore, τ^* can be discovered by binary search. Note that τ^* can serve as a lower bound on the optimal solution value.

Once τ^* is determined, we compute an extreme point solution x^* to the Held-Karp relaxation for $G_{\leq \tau^*}$. Then we sample an α-thin tree T with respect to $z_{uv}^* := x_{uv}^* + x_{vu}^*$ for $\alpha = \frac{4 \ln n}{\ln \ln n}$. By Theorem 1, this can be performed in polynomial time with high probability.

Then the algorithm constructs the circulation problem instance described in the proof of Lemma 4 and finds an integral solution. Lemma 4 shows that any Eulerian circuit of this integral solution is a spanning circuit where no vertex appears more than $\lceil 4\alpha \rceil$ times.

Let $\{P_1, \ldots, P_\ell\}$ be the partition of this spanning circuit as defined in Lemma 2 for $k = \lceil 4\alpha \rceil$. The algorithm computes a transversal of $\{P_1, \ldots, P_\ell\}$ and augments it into a Hamiltonian cycle C in $G^{2\lceil 4\alpha \rceil - 1}$. By the triangle inequality, the cost of C is at most $(2\lceil 4\alpha \rceil - 1) \cdot \tau^*$; thus, C is a $(2\lceil 4\alpha \rceil - 1)$-approximate solution to the given input. Note that $2\lceil 4\alpha \rceil - 1 = 2\lceil \frac{16 \ln n}{\ln \ln n} \rceil - 1 = O(\frac{\log n}{\log \log n})$.

The foregoing is a probabilistic $O(\frac{\log n}{\log \log n})$-approximation algorithm for the bottleneck asymmetric traveling salesman problem under a metric cost.   □

4   Special Case

In this section, we illustrate how our framework can be used together with other results to yield a stronger approximation guarantee in certain special cases. Lemmas 3 and 4 imply the following theorem.

Theorem 3. *If an $f(n)$-thin tree can be found in polynomial time for a certain class of metric, an $O(f(n))$-approximation algorithm exists for the bottleneck asymmetric traveling salesman problem under the same class of metric.*

In particular, Oveis Gharan and Saberi [20] investigate the case when the Held-Karp solution can be embedded on an orientable surface with a bounded genus; Oveis Gharan and Saberi [20], in addition to an $O(1)$-approximation algorithm for the min-sum problem, show the following:

Theorem 4. *Given a feasible solution $x^* \in \mathbb{R}^A$ to the Held-Karp relaxation, let $z_{uv}^* := x_{uv}^* + x_{vu}^*$. If the support of z^* can be embedded on an orientable surface with a bounded genus, an α-thin tree with respect to z^* can be found in polynomial time, where α is a constant that depends on the bound on the genus.*

Theorems 3 and 4 together imply the following.

Corollary 1. *There exists an $O(1)$-approximation algorithm for the bottleneck asymmetric traveling salesman problem when the support of the Held-Karp solution can be embedded on an orientable surface with a bounded genus.*

5   Open Questions

Given that the bottleneck symmetric TSP is 2-approximable [7,17,23], a naturally following question is if the asymmetric version also admits a 2-approximation algorithm. The algorithms for the symmetric case are based on the fact that the square of a 2-connected graph is Hamiltonian. One could regard the analogue of 2-connectedness of an undirected graph in a digraph as the following property: for any two vertices, there exists a simple directed cycle that includes both vertices. However, unfortunately, there exists such a graph that is non-Hamiltonian. In fact, for any constant k and p, the following can be shown:

Theorem 5. *For any constant $k, p \in \mathbb{N}$, there exists a digraph $G = (V, A)$ such that:*

(i) *for all $u, v \in V$, there exist k paths $P_1, \ldots, P_k$ from u to v and k paths $Q_1, \ldots, Q_k$ from v to u such that $P_1, \ldots, P_k, Q_1, \ldots, Q_k$ are internally vertex-disjoint;*

(ii) *G^p is non-Hamiltonian.*

As this approach appears unpromising, one could instead ask if some constant-order power of a graph whose Held-Karp relaxation is feasible is Hamiltonian.

Question 1. Does there exist a constant p such that the pth power of any digraph with a feasible Held-Karp relaxation is Hamiltonian?

One plausible way to affirmatively answer Question 1 is by proving that a graph whose Held-Karp relaxation is feasible contains a spanning circuit that satisfies the property of Lemma 2; Lemma 3 might be helpful in this. In particular, if there exists an efficient procedure that computes an $O(1)$-thin tree with respect to the Held-Karp solution, that would affirm Question 1.

Considering the undirected case, we can show that the set of graphs whose Held-Karp relaxation is feasible is a proper subset of the set of 2-connected graphs (see Theorem 6 for one direction); therefore, it is conceivable that one could attain a simpler proof that the square of a graph whose Held-Karp relaxation is feasible is Hamiltonian. Such a proof may provide inspiration for the asymmetric case.

Theorem 6. *For an undirected graph $G = (V, E)$, if the linear system*

$$\begin{cases} z(\delta(v)) = 2 & \forall v \in V \\ z(\delta(S)) \geq 2 & \forall S \subsetneq V, S \neq \emptyset \\ z \geq 0. \end{cases} \tag{3}$$

has a feasible solution $z^* \in \mathbb{R}^E$, G is 2-connected.

Proof. This proof borrows some idea from the proof of Lemma 4.

Let $G' = (V, A)$ be the digraph obtained from G by replacing each edge with two arcs in both directions. Consider a flow network on G', where the arc capacity is given as the z^* value of the underlying edge.

For any $u, v \in V$, a flow of 2 can be routed from u to v on this network. Let $f \in \mathbb{R}^A$ be this flow. Without loss of generality, we can assume that

$$\forall \{x, y\} \in E \quad f(x, y) = 0 \text{ or } f(y, x) = 0. \tag{4}$$

We drop the arcs on which the flow is zero from the network.

Let x be an arbitrary vertex other than u or v. Note that, from (4), the sum of the capacities of the arcs incident to/from x is at most 2. From the flow conservation, the incoming flow into x is at most 1; thus, introducing the vertex capacity of 1 to every vertex other than u and v does not break the feasibility of f.

Now we round up all of the capacities, and there exists an integral flow of value 2 from u to v on this flow network. This proves the existence of two vertex-disjoint paths from u to v. □

Acknowledgments. The authors thank the anonymous reviewers for their helpful comments.

References

1. Archer, A.: Inapproximability of the asymmetric facility location and k-median problems (2000),
   http://www2.research.att.com/~aarcher/Research/asym-hard.ps
2. Arya, V., Garg, N., Khandekar, R., Meyerson, A., Munagala, K., Pandit, V.: Local search heuristics for k-median and facility location problems. SIAM J. Comput. 33(3), 544–562 (2004)
3. Asadpour, A., Goemans, M.X., Mądry, A., Oveis Gharan, S., Saberi, A.: An $O(\log n/\log\log n)$-approximation algorithm for the asymmetric traveling salesman problem. In: SODA 2010: Proceedings of the 21st Annual ACM-SIAM Symposium on Discrete Algorithms, pp. 379–389 (2010)
4. Chekuri, C., Vondrák, J., Zenklusen, R.: Dependent randomized rounding for matroid polytopes and applications. CoRR abs/0909.4348 (2009),
   http://arxiv.org/abs/0909.4348
5. Christofides, N.: Worst-case analysis of a new heuristic for the travelling salesman problem. Tech. Rep. 388, Graduate School of Industrial Administration, CMU (1976)
6. Chuzhoy, J., Guha, S., Halperin, E., Khanna, S., Kortsarz, G., Krauthgamer, R., Naor, J.: Asymmetric k-center is $\log^* n$-hard to approximate. J. ACM 52(4), 538–551 (2005)
7. Fleischner, H.: The square of every two-connected graph is Hamiltonian. Journal of Combinatorial Theory, Series B 16(1), 29–34 (1974)
8. Frieze, A.M., Galbiati, G., Maffioli, F.: On the worst-case performance of some algorithms for the asymmetric traveling salesman problem. Networks 12, 23–39 (1982)
9. Grötschel, M., Lovász, L., Schrijver, A.: The ellipsoid method and its consequences in combinatorial optimization. Combinatorica 1(2), 169–197 (1981)
10. Hall, P.: On Representatives of Subsets. J. London Math. Soc. 10, 26–30 (1935)
11. Held, M., Karp, R.M.: The traveling-salesman problem and minimum spanning trees. Operations Research 18(6), 1138–1162 (1970)
12. Hochbaum, D.S., Shmoys, D.B.: A unified approach to approximation algorithms for bottleneck problems. J. ACM 33(3), 533–550 (1986)
13. Hsu, W.L., Nemhauser, G.L.: Easy and hard bottleneck location problems. Discrete Applied Mathematics 1(3), 209–215 (1979)
14. Jain, K., Mahdian, M., Markakis, E., Saberi, A., Vazirani, V.V.: Greedy facility location algorithms analyzed using dual fitting with factor-revealing LP. J. ACM 50(6), 795–824 (2003)

15. Kleinberg, J., Tardos, É.: Algorithm Design. Addison-Wesley Longman Publishing Co., Inc., Boston (2005)
16. Kleinberg, J., Williamson, D.P.: Unpublished manuscript, pp. 124–126 (1998), http://legacy.orie.cornell.edu/~dpw/cornell.ps
17. Lau, H.T.: Finding EPS-Graphs. Monatshefte für Mathematik 92(1), 37–40 (1981)
18. Lawler, E.L., Lenstra, J.K., Rinnooy Kan, A.H.G., Shmoys, D.B. (eds.): The Traveling Salesman Problem: A Guided Tour of Combinatorial Optimization. Wiley, Chichester (1985)
19. Lin, J.H., Vitter, J.S.: ϵ-approximations with minimum packing constraint violation (extended abstract). In: STOC 1992: Proceedings of the 24th Annual ACM Symposium on Theory of Computing, pp. 771–782. ACM, New York (1992)
20. Oveis Gharan, S., Saberi, A.: The asymmetric traveling salesman problem on graphs with bounded genus. CoRR abs/0909.2849 (2009), http://arxiv.org/abs/0909.2849
21. Panigrahy, R., Vishwanathan, S.: An $O(\log^* n)$ approximation algorithm for the asymmetric p-center problem. J. Algorithms 27(2), 259–268 (1998)
22. Papadimitriou, C.H., Vempala, S.: On the approximability of the traveling salesman problem. Combinatorica 26(1), 101–120 (2006)
23. Parker, R.G., Rardin, R.L.: Guaranteed performance heuristics for the bottleneck traveling salesman problem. Operations Research Letters 2(6), 269–272 (1984)
24. Schrijver, A.: Combinatorial Optimization: Polyhedra and Efficiency. Springer, Heidelberg (2003)

Improved Inapproximability for Submodular Maximization

Per Austrin[*]

Courant Institute of Mathematical Sciences
New York University
austrin@cims.nyu.edu

Abstract. We show that it is Unique Games-hard to approximate the maximum of a submodular function to within a factor 0.695, and that it is Unique Games-hard to approximate the maximum of a symmetric submodular function to within a factor 0.739. These results slightly improve previous results by Feige, Mirrokni and Vondrák (FOCS 2007) who showed that these problems are NP-hard to approximate to within $3/4 + \varepsilon \approx 0.750$ and $5/6 + \varepsilon \approx 0.833$, respectively.

1 Introduction

Given a ground set U, consider the problem of finding a set $S \subseteq U$ which maximizes some function $f : 2^U \to \mathbb{R}^+$ which is *submodular*, i.e., satisfies

$$f(S \cup T) + f(S \cap T) \le f(S) + f(T).$$

for every $S, T \subseteq U$. The submodularity property is also known as the property of *diminishing returns*, since it is equivalent with requiring that, for every $S \subset T \subseteq U$ and $i \in U \setminus T$, it holds that

$$f(T \cup \{i\}) - f(T) \le f(S \cup \{i\}) - f(S).$$

There has been a lot of attention on various submodular optimization problems throughout the years (e.g., [8,7,3], see also the first chapter of [10] for a more thorough introduction). Many natural problems can be cast in this general form – examples include natural graph problems such as maximum cut, and many types of combinatorial auctions and allocation problems.

A further restriction which is also very natural to study is *symmetric* submodular functions. These are functions which satisfy $f(S) = f(\overline{S})$ for every $S \subseteq U$, i.e., a set and its complement always have the same value. A well-studied example of a symmetric submodular maximization problem is the problem to find a maximum cut in a graph.

Since it includes familiar NP-hard problems such as maximum cut as a special case, submodular maximization is in general NP-hard, even in the symmetric

---

[*] Supported by NSF Expeditions grant CCF-0832795.

M. Serna et al. (Eds.): APPROX and RANDOM 2010, LNCS 6302, pp. 12–24, 2010.

case. As a side note, a fundamental and somewhat surprising result is that submodular *minimization* has a polynomial time algorithm [5].

To cope with this hardness, there has been much focus on efficiently finding good approximate solutions. We say that an algorithm is an α-approximation algorithm if it is guaranteed to output a set S for which $f(S) \geq \alpha \cdot f(S_{\mathrm{OPT}})$ where S_{OPT} is an optimal set. We also allow randomized algorithms in which case we only require that the expectation of $f(S)$ (over the random choices of the algorithm) is at least $\alpha \cdot f(S_{\mathrm{OPT}})$.

In many special cases such as the maximum cut problem, it is very easy to design a constant factor approximation (in the case of maximum cut it is easy to see that a random cut is a $1/2$-approximation). For the general case of an arbitrary submodular functions, Feige et al. [3] gave a $(2/5 - o(1))$-approximation algorithm based on local search, and proved that a uniformly random set is a $1/2$-approximation for the symmetric case. The $(2/5 - o(1))$-approximation has been slightly improved by Vondrák [9] who achieved a 0.41-approximation algorithm, which is currently the best algorithm we are aware of.

Furthermore, [3] proved that in the *(value) oracle model* (where the submodular function to be maximized is given as a black box), no algorithm can achieve a ratio better than $1/2 + \varepsilon$, even in the symmetric case. However, this result says nothing about the case when one is given an *explicit representation* of the submodular function – say, a graph in which one wants to find a maximum cut. Indeed, in the case of maximum cut there is in fact a 0.878-approximation algorithm, as given by a famous result of Goemans and Williamson [4]. In the explicit representation model, the best current hardness results, also given by [3], are that it is NP-hard to approximate the maximum of a submodular function to within $3/4 + \varepsilon$ in the general case and $5/6 + \varepsilon$ in the symmetric case.

1.1   Our Results

In this paper we slightly improve the inapproximability results of [3]. However, as opposed to [3] we do not obtain NP-hardness but only hardness assuming Khot's *Unique Games Conjecture* (UGC) [6]. The conjecture asserts that a problem known as Unique Games, or Unique Label Cover, is very hard to approximate. See e.g. [6] for more details. While the status of the UGC is quite open, our results still imply that obtaining efficient algorithms that beat our bounds would require a fundamental breakthrough.

For general submodular functions we prove the following theorem.

Theorem 1. *It is UG-hard to approximate the maximum of a submodular function to within a factor 0.695.*

In the case of symmetric functions we obtain the following bound.

Theorem 2. *For every $\varepsilon > 0$ it is UG-hard to approximate the maximum of a symmetric submodular function to within a factor $709/960 + \varepsilon < 0.739$.*

These improved inapproximability results still fall short of coming close to the
1/2-barrier in the oracle model. Unfortunately, while marginal improvments of
our results may be possible, we do not believe that our approach can come
close to a factor 1/2. It remains a challenging and interesting open question
to determine the exact approximability of explicitly represented submodular
functions.

1.2   Our Approach

As in [3], the starting point of our approach is hardness of approximation for
constraint satisfaction problems (CSPs), an area which, due to much progress
during the last 15 years, is today quite well understood. Here it is useful to take a
slightly different viewpoint. Instead of thinking of the family of subsets 2^U of U,
we consider the set of binary strings $\{0,1\}^n$ of length $n = |U|$, indentified with
2^U in the obvious way. These views are of course equivalent and throughout the
paper we shift between them depending on which view is the most convenient.

For a string $x \in \{0,1\}^n$ and a k-tuple $C \in [n]^k$ of indices, let $x_C \in \{0,1\}^k$
denote the string of length k which, in position $j \in [k]$ has the bit x_{C_j}. Now,
given a function $f : \{0,1\}^k \to \mathbb{R}^+$, we define the problem MAX CSP$^+(f)$ as
follows. An instance of MAX CSP$^+(f)$ consists of a list of k-tuples of variables
$C_1, \ldots, C_m \in [n]^k$. These specify a function $F : \{0,1\}^n \to \mathbb{R}^+$ by

$$F(x) = \frac{1}{m} \sum_{i=1}^m f(x_{C_i})$$

and the problem is to find an $x \in \{0,1\}^n$ to maximize x.

Note that if f is submodular then every instance F of MAX CSP$^+(f)$ is submodular and MAX CSP$^+(f)$ is a special case of the submodular maximization
problem.

Next, we use a variation of a result by the author and Mossel [2]. The result
of [2] is for CSPs where one allows *negated literals*[1], which can not be allowed
in the context of submodular maximization. However, in Theorem 3 we give a
simple analogue of the result of [2] for the MAX CSP$^+(f)$ setting.

Roughly speaking the hardness result says the following. Suppose that there
is a *pairwise independent* distribution μ such that the expectation of f under μ
is at least c, but that the expectation of f under the *uniform* distribution is at
most s. Then MAX CSP$^+(f)$ is UG-hard to approximate to within a factor of
s/c.

The hardness result suggests the following natural approach: take a pairwise
independent distribution μ with small support, and let $\mathbf{1}_\mu : \{0,1\}^k \to \{0,1\}$
be the indicator function of the support of μ. Then take f to be a "minimum
submodular upper bound" to $\mathbf{1}_\mu$, by which we mean a submodular function
satisfying $f(x) \geq \mathbf{1}_\mu(x)$ for every x while having small expectation under the
uniform distribution.

---

[1] Where each "constraint" $f(x_{C_i})$ of F is of the more general form $f(x_{C_i} + l_i)$ for
  some $l_i \in \{0,1\}^k$, where $+$ is interpreted as addition over $GF(2)^k$.

To make this plan work, there are a few small technical complications (hidden in the "roughly speaking" part of the description of the hardness result above) that we need to overcome, making the final construction slightly more complicated. Unfortunately, understanding the "minimum submodular upper bound" of the families of indicator functions that we use appears difficult, and to obtain our results, we resort to explicitly computing the resulting submodular functions for small k.

Let us compare our approach with that of [3]. As mentioned above, their starting point is also hardness of approximation for constraint satisfaction. However, here their approach diverges from ours: they construct a gadget reduction from the k-LIN problem (linear equations mod 2 where each equation involves only k variables). This gadget introduces two variables x_i^0 and x_i^1 for every variable x_i in the k-LIN instance, where x_i^j is intended to be an indicator of the event that $x_i = j$. Each equation $x_{i_1} \oplus \ldots \oplus x_{i_k} = b$ is replaced by some submodular function f on the $2k$ new variables corresponding to the x_{i_j}'s. The analysis then has to make sure that there is always an optimal assignment where for each i exactly one of x_i^0 and x_i^1 equals 1, which for the inapproximability of $3/4$ becomes quite delicate. In our approach, which we feel is more natural and direct, we don't run into any such issues.

1.3   Organization

In Section 2 we set up some more notation that we use throughout the paper and give some additional background. In Section 3 we describe the hardness result that is our starting point. In Section 4 we describe in more detail the construction outlined above, and finally, in Section 5, we describe how to obtain the concrete bounds given in Theorems 1 and 2.

2   Notation and Background

Throughout the paper, we identify binary strings in $\{0,1\}^n$ and subsets of $[n]$ in the obvious way. Analogously to the notation $|S|$ and $\overline{S}$ for the cardinality and complement of a subset $S \subseteq [n]$ we use $|x|$ and $\overline{x}$ for the Hamming weight and coordinatewise complement of a string $x \in \{0,1\}^n$.

2.1   Submodularity

Apart from the two definitions in the introduction, a third characterization of submodularity is that a function $f : 2^X \to \mathbb{R}^+$ is submodular if and only if

$$f(S) - f(S \cup \{i\}) - f(S \cup \{j\}) + f(S \cup \{i\} \cup \{j\}) \leq 0 \tag{1}$$

for every $S \subseteq X$, and $i, j \in X \setminus S$, $i \neq j$. It is straightforward to check that this condition is equivalent to the diminishing returns property mentioned in the introduction.

2.2   Probability

For $p \in [0,1]$, we use $\{0,1\}^k_{(p)}$ to denote the k-dimensional boolean hypercube with the p-biased product distribution, i.e., if x is a sample from $\{0,1\}^k_{(p)}$ then the probability that the i'th coordinate $x_i = 1$ is p, independently for each $i \in [k]$.

We abuse notation somewhat by making no distinction between probability distribution functions $\mu : \{0,1\}^k \to [0,1]$ and the probability space $(\{0,1\}^k, \mu)$ for such μ. Hence we write, e.g., $\mu(x)$ for the probability of $x \in \{0,1\}^k$ under μ and $\mathbb{E}_{x \sim \mu}[f(x)]$ for the expectation of a function $f : \{0,1\}^k \to \mathbb{R}$ under μ.

A distribution μ over $\{0,1\}^k$ is *balanced pairwise independent* if every two-dimensional marginal distribution of μ is the uniform distribution, or formally, if for every $1 \le i < j \le n$ and $b_1, b_2 \in \{0,1\}$, it holds that

$$\Pr_{x \sim \mu}[x_i = b_1 \wedge x_j = b_2] = 1/4.$$

Recall that the support $\mathrm{Supp}(\mu)$ of a distribution μ over $\{0,1\}^k$ is the set of strings with non-zero probability under μ, i.e., $\mathrm{Supp}(\mu) = \{\, x \in \{0,1\}^k \,:\, \mu(x) > 0 \,\}$.

We conclude this section with a lemma that will be useful to us.

Lemma 1. *Let $f : \{0,1\}^k \to \mathbb{R}^+$ be a symmetric set function. For $t \in \{0,\ldots,k\}$ let $a(t)$ denote the average of f on strings of weight t, $a(t) = \frac{1}{\binom{k}{t}}\sum_{|x|=t} f(x)$. If a is monotonely nondecreasing in $\{0,\ldots,k/2\}$, then the maximum average of f under any p-biased distribution is achieved by the uniform distribution. I.e.,*

$$\max_{p \in [0,1]} \mathbb{E}_{x \sim \{0,1\}^k_{(p)}}[f(x)] = 2^{-x} \sum_{x \in \{0,1\}} f(x)$$

This intuitively obvious lemma is probably well known but as we do not know a reference we give a proof in the full version of this paper [1].

3   Hardness from Pairwise Independence

In this section we state formally the variation of the hardness result of [2] that we use. We first define the parameters which control the inapproximability ratio that we obtain.

Definition 1. *Let $f : \{0,1\}^k \to \mathbb{R}^+$ be a submodular function.*

We define the completeness $c_\mu(f)$ *of f with respect to a distribution μ over $\{0,1\}^k$ by the expected value of f under μ, i.e.,*

$$c_\mu(f) := \mathbb{E}_{x \sim \mu}[f(x)]$$

We define the soundness $s_p(f)$ *of f with respect to bias p by the expected value of f under the p-biased distribution, i.e.,*

$$s_p(f) := \mathbb{E}_{x \sim \{0,1\}^k_{(p)}}[f(x)].$$

Finally, we define the soundness $s(f)$ *of f by its maximum soundness with respect to any bias, i.e.,*

$$s(f) := \max_{p \in [0,1]} s_p(f)$$

We can now state the hardness result.

Theorem 3. *Let μ be a balanced pairwise independent distribution over $\{0,1\}^k$. Then for every function $f : \{0,1\}^k \to \mathbb{R}^+$ and $\varepsilon > 0$, given a MAX CSP$^+(f)$ instance $F : \{0,1\}^n \to \mathbb{R}^+$ it is UG-hard to distinguish between the cases:*

Yes: *There is an $S \subseteq X$ such that $F(S) \geq c_\mu(f) - \varepsilon$.*
No: *For every $S \subseteq X$ it holds that $F(S) \leq s(f) + \varepsilon$.*

The proof of Theorem 3 follows the proof of [2] almost exactly. A proof can be found in the full version of this paper [1].

   Consequently, for any submodular function f and pairwise independent distribution μ with all marginals equal, it is UG-hard to approximate MAX CSP$^+(f)$ to within a factor $s(f)/c_\mu(f)+\varepsilon$ for every $\varepsilon > 0$. Note also that the **No** case is the best possible: there is a trivial algorithm which finds a set such that $F(S) \geq s(f)$ for every F, by simply letting each input be 1 with probability p for the p that maximizes $s_p(f)$.

   As a somewhat technical remark, we mention that Theorem 3 still holds if μ is not required to be balanced – it suffices that all the one-dimensional marginal probabilities $\Pr_{x \sim \mu}[x_i = 1]$ are identical, not necessarily equal to $1/2$ as in the balanced case. We state the somewhat simpler form since that is sufficient to obtain our results for submodular functions and since that makes it more similar to the result of [2], which requires the distribution μ to be balanced.

   Let us then briefly discuss the difference between Theorem 3 and the main result of [2]. First, the result of [2] only applies in the more general setting when one allows negated literals, which is why it can not be used to obtain inapproximability for submodular functions. On the other hand, this more general setting allows for a stronger conclusion: in the **No** case, [2] achieves a soundness of $s_{1/2}(f) + \varepsilon$ which in general can be much smaller than $s(f)$. As an example, consider the case when $f : \{0,1\}^3 \to \{0,1\}$ is the logical OR function on 3 bits. In this case the MAX CSP$^+(f)$ problem is of course trivial – the all-ones assignment satisfies all constraints – and $s(f) = 1$, whereas $s_{1/2}(f) = 7/8$. Letting μ be the uniform distribution on strings of odd parity (it is readily verified that this is a balanced pairwise independent distribution) one gets $c_\mu(f) = 1$, showing that the MAX 3-SAT problem is hard to approximate to within $7/8 + \varepsilon$.

4   The Construction

In this section we make formal the construction outlined in Section 1.2.

   Theorem 3 suggests the following natural approach: pick a pairwise independent distribution μ over $\{0,1\}^k$ and let $\mathbf{1}_\mu : \{0,1\}^k \to \{0,1\}$ be the indicator function of the support of μ. Then take f to be a "minimum submodular upper bound" to $\mathbf{1}_\mu$, by which we mean a submodular function satisfying $f(x) \geq \mathbf{1}_\mu(x)$ for every x while having $s(f)$ as small as possible (whereas $c_\mu(f)$ is clearly at least 1). Note that the smaller the support of μ, the less constrained f is, meaning that there should be more room to make $s(f)$ small.

To this end, let us make the following definition.

Definition 2. *For a subset $\mathcal{C} \subseteq \{0,1\}^k$, we denote by $\mathsf{SM}(\mathcal{C})$ the optimum function $f : \{0,1\}^k \to \mathbb{R}^+$ of the following program[2]:*

$$\begin{aligned}&\textit{Minimize } s(f)\\&\textit{Subject to } f(x) \geq 1 \textit{ for every } x \in \mathcal{C}\\&\qquad f \textit{ is submodular}\end{aligned}$$

In addition, we write $\mathsf{SM}_p(\mathcal{C})$ for the optimal f when the objective to be minimized is changed to $s_p(f)$ instead of $s(f)$. Analogously, we define $\mathsf{SM}^{sym}(\mathcal{C})$ and $\mathsf{SM}_p^{sym}(\mathcal{C})$ as the optimal f with the additional restriction that f is symmetric.

While the objective function $s(f)$ is not linear (or even convex), it turns out that for the $\mathcal{C}$'s that we are interested in, $\mathsf{SM}(\mathcal{C})$ is actually quite well approximated by $\mathsf{SM}_{1/2}(\mathcal{C})$, i.e., we simply minimize $\sum_x f(x)$ (in fact, we even believe that for our $\mathcal{C}$'s $\mathsf{SM}_{1/2}(\mathcal{C})$ gives the exact optimum for $\mathsf{SM}(\mathcal{C})$, though we have not attempted to prove it). The advantage of considering $\mathsf{SM}_{1/2}(\mathcal{C})$ is of course that it is given by a linear program, which gives us a reasonably efficient way of finding it. Armed with this definition, let us now describe the constructions we use.

4.1   The Asymmetric Case

The family of pairwise independent distributions μ that we consider is a standard construction based on the Hadamard code. Fix a parameter $l > 0$ and let $k = 2^l - 1$. We identify the set of coordinates $[k]$ with the set of non-empty subsets of $[l]$, in some arbitrary way. A string x from the distribution μ is sampled as follows: pick a uniformly random string $y \in \{0,1\}^l$ and defining, for each $\emptyset \neq T \subseteq [l]$, the coordinate $x_T = \bigoplus_{i \in T} y_i$.

This construction already has an issue: since the all-zeros string $\mathbf{0}$ is in the support of the distribution, any submodular upper bound to $\mathbf{1}_\mu$ must have $f(\mathbf{0}) \geq 1$, implying that $s_0(f) = 1$. To fix this, we simply ignore $\mathbf{0}$ when constructing f. Formally, let $\mathcal{C}_l = \mathrm{Supp}(\mu) \setminus \{\mathbf{0}\} \subseteq \{0,1\}^k$ be the $2^l - 1$ strings in the support of μ except $\mathbf{0}$. Now we would like to take our submodular function f to be $\mathsf{SM}(\mathcal{C})$, but we instead take it to be $\mathsf{SM}_{1/2}(\mathcal{C})$, as this function is much more easily computed.

Definition 3. *For a parameter $l > 0$, let $k = 2^l - 1$ and take $\mathcal{C}_l \subseteq \{0,1\}^k$ as above. We define $f_l = \mathsf{SM}_{1/2}(\mathcal{C}_l)$.*

---

[2] In the case when the optimum is not unique, we choose an arbitrary optimal f as $\mathsf{SM}(\mathcal{C})$.

Note that using only $\mathcal{C}_l$ instead of the entire support costs us a little in that the completeness is now reduced from 1 to $c_\mu(f_l) \geq 1 - 2^{-l}$, but one can hope (and it indeed turns out that this is the case) that this loss is compensated by a greater improvement in soundness.

Also, we stress that $s(f_l)$ is typically *not* given by the average $s_{1/2}(f_l)$ (which is the quantity actually minimized by f_l). Indeed, the points in $\mathcal{C}_l$ all have Hamming weight $(k+1)/2$ and this is also where f_l is typically the largest. This causes $s(f)$ to be achieved by the p-biased distribution for some p slightly larger than $1/2$.

An obvious question to ask is whether using $\mathsf{SM}(\mathcal{C}_l)$ would give a better result than using $\mathsf{SM}_{1/2}(\mathcal{C}_l)$. For the values of l that we have been able to handle, it appears that the answer to this question is negative: computing $\mathsf{SM}_p(\mathcal{C}_l)$ for a p that approximately maximizes $s_p(f_l)$ gives f_l, indicating that we in fact have $f_l = \mathsf{SM}(\mathcal{C}_l)$.

4.2   Symmetric Functions

One way of constructing symmetric functions would be to use the exact same construction as above but taking $\mathsf{SM}^{\mathsf{sym}}(\mathcal{C}_l)$ rather than $\mathsf{SM}(\mathcal{C}_l)$. However, that is somewhat wasteful, and we achieve better results by also taking symmetry into account when constructing the family of strings $\mathcal{C}$.

Thus, we alter the above construction as follows: rather than identifying the coordinates with all non-empty subsets of $[l]$, we identify them with all subsets of $[l]$ of odd cardinality. In other words, we take $k = 2^{l-1}$ and associate $[k]$ with all $T \subseteq [l]$ such that $|T|$ is odd. The resulting distribution μ is symmetric in the sense that if x is in the support then so is $\overline{x}$.

In this case, both the all-zeros string $\mathbf{0}$ and the all-ones string $\mathbf{1}$ are in the support which is not acceptable for the same reason as above. Hence, we construct a submodular function by taking $\mathcal{C}_l^{\mathsf{sym}} = \mathrm{Supp}(\mu) \setminus \{\mathbf{0}, \mathbf{1}\}$ (note that $|\mathcal{C}_l^{\mathsf{sym}}| = 2^l - 2$).

Definition 4. *For a parameter $l > 0$, let $k = 2^{l-1}$ and take $\mathcal{C}_l^{\mathsf{sym}} \subseteq \{0,1\}^k$ as above. We define $f_l^{\mathsf{sym}} = \mathsf{SM}_{1/2}^{\mathsf{sym}}(\mathcal{C}_l^{\mathsf{sym}})$.*

In this case, since we removed 2 out of the 2^l points of the support of μ to construct $\mathcal{C}_l^{\mathsf{sym}}$, we have that $c_\mu(f_l^{\mathsf{sym}}) \geq 1 - 2^{1-l}$.

An salient feature of f_l^{sym} is that all strings of $\mathcal{C}_l^{\mathsf{sym}}$ have Hamming weight exactly $k/2$. By Lemma 1, this causes $s_p(f_l^{\mathsf{sym}})$ to be maximized by $p = 1/2$ (the monotonicity of the function a in Lemma 1 is not immediately clear). This means that in the symmetric case, using $\mathsf{SM}_{1/2}^{\mathsf{sym}}(\mathcal{C}_l^{\mathsf{sym}})$ rather than $\mathsf{SM}^{\mathsf{sym}}(\mathcal{C}_l^{\mathsf{sym}})$ is provably without loss of generality.

5   Concrete Bounds

Unfortunately, understanding the behaviour of the two families of functions f_l and f_l^{sym} (or even just their soundnesses) for large l appears difficult. There seems to be two conflicting forces at work: on the one hand, $\mathcal{C}_l$ only has $2^l - 1 = k$ points so even though f_l is forced to be large on these there may still be plenty of room to make it small elsewhere. But on the other hand, since $\mathcal{C}_l$ is a good code the elements of $\mathcal{C}_l$ are very pread out (their pairwise Hamming distances are roughly $k/2$), which together with the submodularity condition appears to force f_l to be large.

In this section we study f_l for small l, obtaining our hardness results. As discussed towards the end of the section, there are indications that the inapproximability given by f_l actually becomes worse for large l and that our results are the best possible for this family of functions, but we do not yet know whether these indications are correct.

5.1   Symmetric Functions

We start with the symmetric functions, as these are somewhat nicer than the asymmetric ones in that their symmetry turn out to cause $s(f_l^{\mathsf{sym}})$ to be achieved by $p = 1/2$, i.e., $s(f_l^{\mathsf{sym}})$ simply equals the average of f_l^{sym}. Table 1 gives a summary of the completeness, soundness, and inapproximability obtained by f_l^{sym} for $l \in \{3, 4, 5\}$. We now describe these functions in a more detail.

Table 1. Behaviour of f_l^{sym} for small l

l	c	$s(f_l^{\mathsf{sym}})$	Inapproximability s/c	
3	3/4	5/8	5/6	< 0.8334
4	7/8	43/64	43/56	< 0.7679
5	15/16	709/1024	709/960	< 0.7386

As a warmup, let us first describe the quite simple function $f_4^{\mathsf{sym}} : 2^{[8]} \to [0, 1]$ (we leave the even easier function f_3^{sym} to the interested reader). Its definition is as follows:

$$f_4^{\mathsf{sym}}(S) = \begin{cases} f(\overline{S}) & \text{if } |S| > 4 \\ |S|/4 & \text{if } |S| < 4 \\ 1 & \text{if } |S| = 4 \text{ and } S \text{ is in } \mathcal{C}_4^{\mathsf{sym}} \\ 3/4 & \text{otherwise} \end{cases}.$$

That $f_4^{\mathsf{sym}}(S)$ is submodular is easily verified. It is also easy to check that Lemma 1 applies and therefore we have that $s(f_4^{\mathsf{sym}}) = s_{1/2}(f_4^{\mathsf{sym}})$, which is straightforward to compute (note that $|\mathcal{C}_4^{\mathsf{sym}}| = 14$):

$$s_{1/2}(f_4^{\mathsf{sym}}) =$$
$$2^{-8} \left(2\binom{8}{1} \cdot \frac{1}{4} + 2\binom{8}{2} \cdot \frac{2}{4} + 2\binom{8}{3} \cdot \frac{3}{4} + 14 \cdot 1 + \left(\binom{8}{4} - 14 \right) \cdot \frac{3}{4} \right) = \frac{43}{64}$$

Let us then move on to the next function $f_5^{\text{sym}} : 2^{[16]} \to [0, 1]$, giving an inapproximability of 0.7386. It turns out that one can take $f_5^{\text{sym}}(S)$ to be a function of two simple properties of S, namely its cardinality $|S|$, and the distance from S to $\mathcal{C}_5^{\text{sym}}$. Specifically, for $|S| \leq 8$ let us define the *number of errors* $e(S)$ as the minimum number of elements that must be removed from S to get a subset of some set in $\mathcal{C}_5^{\text{sym}}$. Formally

$$e(S) = \min_{C \in \mathcal{C}_5^{\text{sym}}} |S \setminus C|,$$

or equivalently, $d(S, \mathcal{C}_5^{\text{sym}}) = 8 - |S| + 2e(S)$, where $d(S, \mathcal{C}_5^{\text{sym}})$ is the Hamming distance from the binary string corresponding to S to the nearest element in $\mathcal{C}_5^{\text{sym}}$. Table 2 gives the values of f_5^{sym} for all $|S| \leq 8$, and for $|S| > 8$ the value of $f_5^{\text{sym}}(S)$ is given by $f_5^{\text{sym}}(\overline{S})$. Note that, for sets with $e(S) = 0$, i.e., no errors, $f_5^{\text{sym}}(S)$ is simply $|S|/8$, which is what one would expect. However, for sets with errors, $f_5^{\text{sym}}(S)$ has a more complicated behaviour and it is far from clear how this generalizes to larger l.

Table 2. Description of $f_5^{\text{sym}}(S)$ as a function of $|S|$ and $e(S)$ for $|S| \leq 8$

$e(S)$	0	1	2	3	4	5	6	7	8
0	0	1/8	2/8	3/8	4/8	5/8	6/8	7/8	1
1	–	–	–	–	–	19/32	22/32	24/32	26/32
2	–	–	–	–	–	–	20/32	23/32	24/32

Veryfing that f_5^{sym} is indeed submodular is not as straightforward as with f_4^{sym}. We have not attempted to construct a shorter proof of this than simply checking condition (1) for every S, i and j, a task which is of course best suited for a computer program (which is straightforward to write and runs in a few seconds).

A computer program is also the best way to compute the soundness $s(f_5^{\text{sym}})$. It is almost obvious from inspection of Table 2 that f_5^{sym} satisfies the monotonicity condition of Lemma 1 (the only possible source of failure is that the table only implies that the average of f_5^{sym} on sets of size 6 is between 20/32 and 24/32, and that the average on sets of size 7 is between 23/32 and 28/32). It turns out that the conditions of Lemma 1 are indeed satisfied and that the average of f_5^{sym} is $s_{1/2}(f_5^{\text{sym}}) = 709/1024$.

Concluding this discussion on f_l^{sym}, it is tempting to speculate on its behaviour for larger l. We have made a computation of $f_6^{\text{sym}} : 2^{[32]} \to [0, 1]$, under the assumption that $f_6^{\text{sym}}(S)$ only depends on $|S|$ and the multiset of distances to every point of the support of $\mathcal{C}_6^{\text{sym}}$. Under this assumption, our computations indicate that $s(f_6^{\text{sym}}) \approx 0.7031$ giving an inapproximatibility of $s(f_6^{\text{sym}})/(31/32) \approx 0.7258$, improving upon f_5^{sym}. However, as these computations took a few days they are quite cumbersome to verify (and we have not even made a careful verification of them ourselves) and therefore we do not claim this stronger hardness as a theorem.

5.2 Asymmetric Functions

We now return our focus to the asymmetric case. Table 3 describes the hardness ratios obtained from f_l for the cases $l = 3$ and $l = 4$.

Table 3. Behaviour of f_l for small l

l	c	$s(f_l)$	Inapproximability s/c
3	7/8	< 0.6275	< 0.7172
4	15/16	< 0.6508	< 0.6942

We begin with the description of the function $f_3 : 2^{[7]} \to [0, 1]$. Similarly to the definition $e(S)$ used in the description of f_5^{sym}, let us say that $S \subseteq [7]$ *has no errors* if it is a subset or a superset of some $C \in \mathcal{C}_3$. In other words, if $|S| < 4$ it has no errors if it can be transformed to a set in $\mathcal{C}_3$ by adding some elements, and if $|S| > 4$ it is has no errors if it can be transformed to a codeword by removing some elements. The function f_3 is as follows:

$$f_3(S) = \begin{cases} |S|/4 & \text{if } |S| \leq 4 \text{ and has no errors} \\ (7 - |S|)/3 & \text{if } |S| > 4 \text{ and has no errors} \\ 11/24 & \text{if } |S| = 3 \text{ and has errors} \\ 17/24 & \text{if } |S| = 4 \text{ and has errors} \end{cases}$$

As with f_5^{sym}, it is not completely obvious that f_3 satisfies the submodularity condition and there are a few cases to verify, best left to a computer program.

The average of f_3 is $637/1024 \approx 0.622$. However, since f_3 takes on its largest values at sets of size $(k + 1)/2 = 4$, the p-biased average is larger than this for some $p > 1/2$. It turns out that $s(f_4)$ is obtained by the p-biased distribution for $p \approx 0.542404$, giving $s(f_4) \approx 0.627434 < 0.6275$.

We are left with the description of $f_4 : 2^{[15]} \to [0, 1]$, which is also the most complicated function yet. One might hope that f_4 shares the simple structure of the previous functions – that it depends only on $|S|$ and the distance of S to the nearest $C \in \mathcal{C}_4$. However, the best function under this assumption turns out to give a worse result than f_3. Instead, f_4 depends on $|S|$ and the multiset of distances to all elements of $\mathcal{C}_4$.

To describe f_4, define for $S \subseteq [15]$ the multiset $\mathcal{D}(S)$ as the multiset of distances to all the 15 strings in $\mathcal{C}_4$. For instance, for $S = \emptyset$, $\mathcal{D}(S)$ consists of the number 8 repeated 15 times, reflecting the fact that all strings of $\mathcal{C}_4$ have weight 8, and for $S \in \mathcal{C}_4$ we have that $\mathcal{D}(S)$ consists of the number 8 repeated 14 times,

Table 4. Description of f_4

$|S|$	$\mathcal{D}(\mathcal{S})$	$\#S$	$448 \cdot f_4(S)$	$|S|$	$\mathcal{D}(\mathcal{S})$	$\#S$	$448 \cdot f_4(S)$
0	8^{15}	1	0	8	$0^1 8^{14}$	15	448
1	$7^8 9^7$	15	56	8	$2^1 6^4 8^7 10^3$	840	358
2	$6^4 8^8 10^3$	105	112	8	$4^3 8^{11} 12^1$	420	328
3	$5^2 7^6 9^6 11^1$	420	168	8	$4^2 6^4 8^5 10^4$	2520	328
3	$7^{12} 11^3$	35	138	8	$4^1 6^6 8^5 10^2 12^1$	2520	298
4	$4^2 8^{12} 12^1$	105	224	8	$6^7 8^7 14^1$	120	253
4	$4^1 6^4 8^6 10^4$	840	224	9	$1^1 7^8 9^6$	105	384
4	$6^6 8^6 10^2 12^1$	420	194	9	$3^1 5^2 7^6 9^5 11^1$	2520	324
5	$3^1 5^1 7^6 9^6 11^1$	840	280	9	$5^6 9^9$	280	324
5	$5^5 9^{10}$	168	280	9	$5^4 7^6 9^3 11^2$	1680	294
5	$5^3 7^6 9^4 11^2$	1680	250	9	$5^3 7^8 9^3 13^1$	420	279
5	$5^2 7^8 9^4 13^1$	315	220	10	$2^1 6^4 8^8 10^2$	315	320
6	$2^1 6^3 8^8 10^3$	420	336	10	$4^2 6^4 8^6 10^3$	1680	290
6	$4^2 6^3 8^6 10^4$	1680	306	10	$4^1 6^6 8^6 10^1 12^1$	840	275
6	$4^1 6^5 8^6 10^2 12^1$	2520	276	10	$6^{10} 10^5$	168	260
6	$6^9 10^6$	280	276	11	$3^1 5^2 7^6 9^6$	420	256
6	$6^6 8^8 14^1$	105	216	11	$3^1 7^{12} 11^2$	105	256
7	$1^1 7^7 9^7$	120	392	11	$5^4 7^6 9^4 11^1$	840	241
7	$3^1 5^2 7^5 9^6 11^1$	2520	332	12	$4^3 8^{12}$	35	192
7	$3^1 7^{11} 11^3$	420	302	12	$4^1 6^6 8^6 10^2$	420	192
7	$5^4 7^5 9^4 11^2$	2520	302	13	$5^3 7^8 9^4$	105	128
7	$5^3 7^7 9^4 13^1$	840	272	14	$6^7 8^8$	15	64
7	$7^{14} 15^1$	15	197	15	7^{15}	1	0

together with a single 0, because the distance between any pair of strings in $\mathcal{C}_4$ is 8.

Table 4 describes the behaviour of $f_4(S)$ as a function of $|S|$ and $\mathcal{D}(S)$.[3] In the table $\mathcal{D}(S)$ is described by a string of the form $d_1^{m_1} d_2^{m_2} \ldots$, with $d_1 < d_2 < \ldots$ and $\sum m_i = 15$, indicating that m_1 strings of $\mathcal{C}_4$ are at distance d_1 from S, that m_2 strings are at distance d_2, and so on. Thus, for $S = \emptyset$ the description of $\mathcal{D}(S)$ is "8^{15}", and for $S \in \mathcal{C}_4$ the description of $\mathcal{D}(S)$ is "$0^1 8^{14}$".

The $\#S$ column of Table 4 gives the total number of $S \subseteq [15]$ having this particular value of $(|S|, \mathcal{D}(S))$, and the last column gives the actual value of f_4, multiplied by 448 to make all values integers.

---

[3] It is not necessary to include $|S|$ as it is uniquely determined by $\mathcal{D}(S)$, but we find that explicitly including $|S|$ makes the table somewhat less obscure.

Again, checking that f_4 is submodular is a tedious task best suited for a computer. The average of f_4 is $9519345/(448 \cdot 2^{15}) \approx 0.6485$, but, as with f_3, $s(f_4)$ is somewhat larger than this. It turns out that the p maximizing $s_p(f_4)$ is roughly $p \approx 0.526613$, and that $s(f_4) \approx 0.650754 < 0.6508$.

Finally, we mention that as in the symmetric case, we have made a computation of the next function, f_5, again under the assumption that it depends only on the multiset of distances to the codewords. Under this assumption it turns out that $s_{1/2}(f_5) \approx 0.6743$, meaning that the inapproximability obtained can not be better than $s_{1/2}(f_5)/(31/32) \approx 0.6961$ which is worse than the inapproximability obtained from f_4.

Acknowledgments

We are grateful to Jan Vondrák for stimulating discussions.

References

1. Austrin, P.: Improved Inapproximability For Submodular Maximization. arXiv:1004.3777v1 [cs.CC] (2010)
2. Austrin, P., Mossel, E.: Approximation Resistant Predicates from Pairwise Independence. Computational Complexity 18(2), 249–271 (2009)
3. Feige, U., Mirrokni, V.S., Vondrák, J.: Maximizing non-monotone submodular functions. In: IEEE Symposium on Foundations of Computer Science (FOCS), pp. 461–471 (2007)
4. Goemans, M.X., Williamson, D.P.: Improved Approximation Algorithms for Maximum Cut and Satisfiability Problems Using Semidefinite Programming. Journal of the ACM 42, 1115–1145 (1995)
5. Grötschel, M., Lovász, L., Schrijver, A.: The ellipsoid method and its consequences in combinatorial optimization. Combinatorica 1(2), 169–197 (1981)
6. Khot, S.: On the Power of Unique 2-prover 1-round Games. In: ACM Symposium on Theory of Computing (STOC), pp. 767–775 (2002)
7. Lovász, L.: Submodular functions and convexity. In: Grötschel, M., Bachem, A., Korte, B. (eds.) Mathematical Programming: The State of the Art - Bonn 1982, pp. 235–257. Springer, Heidelberg (1983)
8. Nemhauser, G.L., Wolsey, L.A., Fisher, M.L.: An analysis of approximations for maximizing submodular set functions–I. Mathematical Programming 14, 265–294 (1978)
9. Vondrák, J.: Submodular maximization by simulated annealing. Unpublished manuscript
10. Vondrák, J.: Submodularity in Combinatorial Optimization. PhD thesis, Charles University (2007)

Approximation Algorithms for
the Directed k-Tour and k-Stroll Problems

MohammadHossein Bateni[1,*] and Julia Chuzhoy[2,**]

[1] Princeton University, Princeton NJ 08540, USA
mbateni@cs.princeton.edu
[2] Toyota Technological Institute, Chicago, IL 60637, USA
cjulia@ttic.edu

Abstract. We consider two natural generalizations of the Asymmetric Traveling Salesman problem: the k-Stroll and the k-Tour problems. The input to the k-Stroll problem is a directed n-vertex graph with nonnegative edge lengths, an integer k, and two special vertices s and t. The goal is to find a minimum-length s-t walk, containing at least k distinct vertices. The k-Tour problem can be viewed as a special case of k-Stroll, where $s = t$. That is, the walk is required to be a tour, containing some pre-specified vertex s. When $k = n$, the k-Stroll problem becomes equivalent to Asymmetric Traveling Salesman Path, and k-Tour to Asymmetric Traveling Salesman.

Our main result is a polylogarithmic approximation algorithm for the k-Stroll problem. Prior to our work, only bicriteria $(O(\log^2 k), 3)$-approximation algorithms have been known, producing walks whose length is bounded by 3OPT, while the number of vertices visited is $\Omega(k/\log^2 k)$. We also show a simple $O(\log^2 n/ \log \log n)$-approximation algorithm for the k-Tour problem. The best previously known approximation algorithms achieved $\min(O(\log^3 k), O(\log^2 n \cdot \log k/ \log \log n))$-approximation in polynomial time, and $O(\log^2 k)$-approximation in quasipolynomial time.

1 Introduction

In the Asymmetric Traveling Salesman Problem (ATSP), the input is a directed n-vertex graph $G = (V, E)$ with nonnegative edge lengths, and the goal is to find a minimum-length tour, visiting each vertex at least once. ATSP, along with its undirected counterpart, the Traveling Salesman problem, is a classical combinatorial optimization problem, that has been studied extensively. In a recent breakthrough, Asadpour et al. [1] have shown an $O(\log n/ \log \log n)$-approximation algorithm for ATSP, breaking the long-standing barrier of $O(\log n)$ on its approximation ratio [11,3,5,10,12]. With only APX-hardness known on the negative side, this remains one of the central open problems in the area of approximation. A closely related problem is Asymmetric Traveling Salesman Path (ATSPP), defined exactly like ATSP, except that the input also contains

---

* The author was supported by a Gordon Wu fellowship as well as NSF ITR grants CCF-0205594, CCF-0426582 and NSF CCF 0832797, NSF CAREER award CCF-0237113, MSPA-MCS award 0528414, NSF expeditions award 0832797.
** Supported in part by NSF CAREER award CCF-0844872.

M. Serna et al. (Eds.): APPROX and RANDOM 2010, LNCS 6302, pp. 25–38, 2010.
© Springer-Verlag Berlin Heidelberg 2010

two vertices s and t, and instead of a tour, we are required to find a minimum length s-t walk, visiting every vertex at least once. While ATSPP appears to be very similar to ATSP, an $O(\log n)$-approximation algorithm has only been discovered recently by Chekuri and Pál [9], and required new nontrivial ideas.

In this paper we focus on two natural and well-studied generalizations of ATSP: the k-Stroll and the k-Tour problems[1]. In the k-Stroll problem, in addition to the edge-weighted graph G, we are also given a parameter k, and two special vertices s and t. The goal is to find a minimum-length walk from s to t, containing at least k distinct vertices. The k-Tour problem is defined similarly, except that instead of the vertices s and t, the input contains one root vertex r, and we are required to find a minimum-length tour containing r, that visits at least k distinct vertices. Therefore, k-Tour can be viewed as a special case of k-Stroll, where $s = t$. When the input graph is undirected, we get the undirected k-Tour and undirected k-Stroll problems, respectively[2]. For the special case where $k = n$, k-Tour becomes equivalent to ATSP, and k-Stroll becomes equivalent to ATSPP.

A bicriteria (α, β)-approximation algorithm for the k-Stroll problem is an algorithm that returns a walk of length at most $\beta \cdot$ OPT, containing at least k/α distinct vertices. Chekuri, Korula and Pál [7] and Nagarajan and Ravi [15] have independently shown, using different methods, $(O(\log^2 k), 3)$ bicriteria approximation algorithms for the k-Stroll problem. To the best of our knowledge, these are the only known approximation algorithms for the problem. The main result of our paper is a polylogarithmic approximation algorithm for the k-Stroll problem. We note that undirected k-Stroll has a $(2 + \epsilon)$-approximation algorithm, due to Chaudhuri et al. [6].

The first nontrivial approximation algorithm for the k-Tour problem, due to Chekuri and Pál [8], achieved an $O(\log^2 k)$-approximation in quasi-polynomial time. Chekuri, Korula and Pál [7] and Nagarajan and Ravi [15] have later independently shown polynomial time algorithms achieving $O(\log^3 k)$ and $O(\log^2 n \cdot \log k)$ approximation, respectively. Using the recent result of [1] for ATSP, the latter approximation factor improves to $O(\log^2 n \cdot \log k / \log \log n)$. We show a simple $O(\log^2 n / \log \log n)$-approximation algorithm for the problem.

Related Work. There is a large body of research on ATSP and its variants. We only mention here results most closely related to the problems we study. The Orienteering problem is defined as follows: given an edge-weighted graph, two vertices s and t and a budget B, find an s-t walk of length at most B, maximizing the number of distinct vertices visited. The problem is closely related to the k-Stroll problem, and this relationship has been made formal by Blum et al. [4], who showed that an α-approximation algorithm for k-Stroll gives an $O(\alpha)$-approximation for Orienteering, in both the directed and the undirected settings. This result was later generalized by Chekuri, Korula and Pál [7] and Nagarajan and Ravi [15], who proved that an (α, β)-bicriteria approximation for k-Stroll implies an $O(\alpha\beta)$-approximation for Orienteering, in both directed and

---

[1] k-Tour is sometimes referred to as k-ATSP in the literature. Similarly, k-Stroll is sometimes called k-ATSPP.

[2] Since we will be focusing on directed graphs, the names k-Tour and k-Stroll will refer to the directed versions of the problems throughout the paper, unless stated otherwise.

undirected graphs. Chekuri and Pál [8] showed that for any fixed integer h, the directed Orienteering problem has an $O(\log \text{OPT} / \log h)$-approximation algorithm, whose running time is $(n \log B)^{O(h \log n)}$. In particular, they obtain $O(\log \text{OPT})$-approximation in quasipolynomial time, and sublogarithmic approximation in subexponential time. Chekuri, Korula and Pál [7] and Nagarajan and Ravi [15] have later independently obtained a polynomial-time $O(\log^2 \text{OPT})$ approximation algorithm for directed Orienteering. The results of [8] also hold for generalizations of the directed Orienteering problem: directed Submodular Orienteering, where instead of maximizing the number of distinct vertices contained in the tour, the goal is to maximize the value of some given submodular function over the set of vertices the tour visits, and directed Submodular Orienteering with time windows, where each vertex is associated with a time window, and a vertex is covered by the tour only if it is visited during its time window. The undirected version of the Orienteering problem has also been studied extensively. The first constant factor approximation algorithm, due to Blum et al. [4], achieved a factor 4 approximation, and was later improved by Bansal et al. [2] to factor 3. The best currently known approximation algorithm, due to Chekuri, Korula and Pál [7], gives a factor $(2 + \epsilon)$ approximation. On the negative side, the basic Orienteering problem is known to be APX-hard for both directed and undirected graphs [4]. Chekuri and Pál [8] have shown that an α-approximation for undirected Submodular Orienteering implies an $O(\alpha \log k)$-approximation for the Group Steiner tree problem, and therefore undirected Submodular Orienteering is hard to approximate to within factor $\Omega(\log^{1-\epsilon} n)$ unless $\mathsf{NP} \subseteq \mathsf{ZTIME}\left(n^{\text{poly} \log(n)}\right)$ [13].

Problem definitions, our results and techniques. The input to the k-Stroll problem is a **complete** directed n-vertex graph $G = (V, E)$ with lengths $c_e \geq 0$ on edges, satisfying the triangle inequalities. Additionally, we are given two special vertices s and t and an integer k. The goal is to find an s-t walk of minimum length that visits at least k distinct vertices.

The input to the k-Tour problem is a **complete** directed n-vertex graph $G = (V, E)$ with edge lengths $c_e \geq 0$, satisfying the triangle inequality, an integer k and a root vertex r. The objective is to find a minimum-length tour $\mathcal{T}$, containing at least k distinct vertices, including r. Let β denote the best approximation factor efficiently achievable for the k-Tour problem. Our result for the k-Stroll problem is summarized in the following theorem:

Theorem 1. *There is an efficient $O(\log k) \cdot \beta$-approximation algorithm for the k-Stroll problem.*

The algorithm is somewhat similar to the quasipolynomial time algorithm of Chekuri and Pál [8] for the Orienteering problem, in the following sense: the algorithm also guesses the middle point v of the walk, partitioning the problem into two subproblems, and then solves the two subproblems separately. This is done by means of dynamic programming, and the main challenge is to keep the size of the dynamic programming table polynomial in n. To demonstrate this difficulty, consider the top-most level of the recursion, and let v be the guessed vertex that appears in the middle of the tour. Our algorithm partitions all the vertices into three subsets L_v, R_v, and C_v, with the following properties: All vertices of L_v that are covered by the optimal walk, must appear before v on it, and similarly all vertices of R_v belonging to the optimal walk appear after v on it. The vertices of C_v may appear either before or after v, and we can solve the problem induced by these vertices using the algorithm for the k-Tour problem. The main challenge is that when we continue to recursively solve the problem induced by, say, L_v, we need to ensure that the vertices of R_v are not used in its solution, so we do not over-count the vertices we cover. Therefore, for each subproblem that we solve, we need to find a way to concisely represent the vertices that have been removed in previous recursive levels. Equivalently, we need to keep the number of entries in the dynamic programming table polynomial in the input size, while ensuring that we do not over-count vertices that the solution visits.

We now turn to the k-Tour problem. Let β_{HK} be the best approximation factor achievable for the ATSP problem, via LP rounding of the Held-Karp LP relaxation [14] (see Section 3.1 for formal definitions). From the work of Asadpour et al. [1], $\beta_{HK} \leq O(\log n / \log \log n)$. We obtain the following result for the k-Tour problem.

Theorem 2. *There is an efficient $O(\log n) \cdot \beta_{HK}$ approximation algorithm for the k-Tour problem. In particular, the problem is approximable to within factor $O(\log^2 n / \log \log n)$.*

From the work of Chekuri, Korula and Pál [7], and from Theorem 2, the approximation factor β for the k-Tour problem is therefore bounded by $\min(O(\log^2 n / \log \log n), O(\log^3 k))$. Therefore, we establish the following result for the k-Stroll problem:

Corollary 1. *The k-Stroll problem has an efficient $\min(O(\log^2 n \cdot \log k / \log \log n), O(\log^4 k))$ approximation algorithm.*

Our algorithm for the k-Tour problem is simple, and it is very similar to the $O(\log^2 n)$-approximation algorithm of Nagarajan and Ravi [15] for the minimum ratio ATSP problem. Nagarajan and Ravi then use this algorithm as a subroutine to obtain an $O(\log^2 n \cdot \log k)$-approximation for k-Tour. We bypass this step by solving the k-Tour problem directly, and this allows us to save the $O(\log k)$ factor in the approximation ratio. We note that following the work of Asadpour et al. [1], the approximation factors in [15] improve to $O(\log^2 n / \log \log n)$ for minimum ratio ATSP, and to $O(\log^2 n \cdot \log k / \log \log n)$ for the k-Tour problem.

Our algorithm starts by solving a linear programming relaxation of the k-Tour problem, which can be seen as an extension of the Held-Karp LP relaxation for ATSP. Each vertex v is associated with an indicator variable z_v, for covering v by the solution. We then partition all vertices geometrically into $O(\log n)$ buckets, according to their values z_v, with bucket B_i containing vertices v with $2^{-i} < z_v \leq 2^{-i+1}$. Next, using the LP-rounding algorithm for ATSP, we find, for each bucket B_i, a tour $\mathcal{T}_i$ of length $O(\beta_{HK} \cdot 2^i \cdot \text{OPT})$, containing all vertices of B_i. This tour is then partitioned into $\Theta(2^i)$ segments, containing $\lceil |B_i|/2^i \rceil$ vertices each, and the cheapest such segment, $\mathcal{T}_i^*$, is selected. We then connect together the selected segments $\mathcal{T}_i^*$, for all buckets B_i to obtain the final tour $\mathcal{T}$.

Organization. Section 2 is devoted to the polylogarithmic approximation algorithm for the k-Stroll problem, and the algorithm for the k-Tour problem appears in Section 3. All proofs omitted from this version can be found in the full version of the paper available on authors' web pages.

2 Approximation Algorithm for the k-Stroll Problem

2.1 Preliminaries

We assume that we are given a **complete** directed n-vertex graph $G = (V, E)$ with nonnegative lengths c_e on edges, satisfying the triangle inequality. Additionally, we are given two special vertices s and t, called the source and the sink, and an integer k. The goal is to find an s-t walk of minimum length, visiting at least k distinct vertices. For any instance $\mathcal{I}$ of the problem, we denote by $\mathsf{OPT}(\mathcal{I})$ the cost of the optimal solution for this instance, and when the instance is clear from context, we use the short-hand OPT. For each pair u, v of vertices, we denote by $d(u, v)$ the length of the shortest path connecting u to v in G.

Let α denote the desired approximation factor. We assume throughout the algorithm that we know the value L^* of the optimal solution. This can be assumed w.l.o.g. using standard techniques: we can perform a binary search on the value L^*, and run our approximation algorithm for each such guessed value L. If the algorithm produces a solution whose cost is bounded by αL, then $L^* \leq L$, and otherwise $L^* > L$, so we can adjust our guessed value L accordingly. Therefore, from now on we assume that we have a value $L^* \geq \mathsf{OPT}$, and our goal is to produce a solution of cost at most αL^*. Our first step is to make the edge lengths polynomially bounded. The proof of the next claim uses standard techniques and is omitted.

Claim 1. *We can assume, at the cost of losing a constant factor in the approximation ratio, that all edge lengths c_e are integers in $\{0, \ldots, N\}$, where $N = \mathrm{poly}(n)$.*

We use the following notation in describing the algorithm. For a vertex $v \in V$ and a parameter D, let $B(v, D) = \{u \in V \mid d(v, u) \leq D, d(u, v) \leq D\}$. For a pair x, y of vertices and a parameter D, let $S(x, y, D) = \{u \in V \mid d(x, u) + d(u, y) \leq D\}$. Therefore, $S(x, y, D)$ is the set of all vertices that may appear on a path of length D connecting x to y.

For technical reasons that will be apparent later, we need to ensure that $B(s, L^*) = \{s\}$ and $B(t, L^*) = \{t\}$. We can do so, w.l.o.g., by adding a new source vertex s' and a new sink vertex t', and setting the lengths of edges (s', s) and (t, t') to 0, and the lengths of all other edges incident to s' and t' to $n^2 \cdot L^*$. (Recall that the graph is required to be complete). This does not affect the solution cost or the approximation factor. So from now on we assume that in the input instance $\mathcal{I}$, $B(s, L^*) = \{s\}$ and $B(t, L^*) = \{t\}$.

Throughout the algorithm, we will be solving instances of the k-Tour problem on subgraphs of G. Let $\mathsf{Alg}_{k\text{-tour}}$ be a β-approximation algorithm for the k-Tour problem. An instance $I(V', r, k')$ of the k-Tour problem, where $V' \subseteq V$, $r \in V$, $k' \in \mathbb{Z}^+$, is an instance defined on the subgraph of G induced by $V' \cup \{r\}$, with the root vertex r, and the parameter k' denoting the number of vertices that need to be covered. We denote by $\mathsf{Alg}_{k\text{-tour}}(V', r, k')$ the output of $\mathsf{Alg}_{k\text{-tour}}$ on instance $I(V', r, k')$.

2.2   Algorithm Overview

Let $\theta = 3/2$ and $\alpha(k') = 9\log_\theta k' + 3$, for $k' > 1$. Our final approximation factor is $O(\beta \cdot \alpha(k)) = O(\beta \cdot \log k)$, as required.

We solve the problem using dynamic programming. Each entry of the dynamic programming table is parametrized by $T(x, y, k', D, \Delta_1, \Delta_2)$, where $x, y \in V$, k' is an integer, $1 \le k' \le k$, and D, Δ_1, Δ_2 are integers between 0 and L^*. Let

$$V(x, y, D, \Delta_1, \Delta_2) = S(x, y, D) \setminus (B(x, \Delta_1) \cup B(y, \Delta_2)).$$

Entry $T(x, y, k', D, \Delta_1, \Delta_2)$ is associated with an instance of the k-Stroll problem denoted by $\pi(x, y, k', D, \Delta_1, \Delta_2)$. The instance is defined on the subgraph of G induced by $V(x, y, D, \Delta_1, \Delta_2) \cup \{x, y\}$. The number of vertices to be covered by the stroll is k', and the endpoints of the stroll are x and y.

We say that entry $T(x, y, k', D, \Delta_1, \Delta_2)$ is *feasible* iff $\Delta_1, \Delta_2 \ge D$, $d(x, y) \le D$, and the value of the optimal solution of instance $\pi(x, y, k', D, \Delta_1, \Delta_2)$ is at most D. A feasible entry $T(x, y, k', D, \Delta_1, \Delta_2)$ must contain a feasible solution for problem $\pi(x, y, k', D, \Delta_1, \Delta_2)$, whose length is at most $3\beta(\Delta_1 + \Delta_2) + \beta \cdot \alpha(k') \cdot D$. Notice that since we have ensured that $B(s, L^*) = \{s\}$ and $B(t, L^*) = \{t\}$, entry $T(s, t, k, L^*, L^*, L^*)$ is feasible, with $V(s, t, L^*, L^*, L^*) = V$. So if the entries of the dynamic programming table are computed correctly, it must contain a solution to $\mathcal{I}$ of cost $O(\beta\alpha(k))L^* = O(\log k)\beta L^*$, as desired. The entries of the dynamic programming table are filled in from smaller to larger values k'. After entry $T = T(x, y, k', D, \Delta_1, \Delta_2)$ is processed, it either contains a feasible solution to problem $\pi(x, y, k', D, \Delta_1, \Delta_2)$ of cost at most $3\beta(\Delta_1 + \Delta_2) + \beta \cdot \alpha(k') \cdot D$, in which case we say that T is *good*, or the value of T is undefined, and we say that it is *bad*. The latter will only happen if T is infeasible.

2.3   Computing the Entries of the Dynamic Programming Table

Let $T = T(x, y, k', D, \Delta_1, \Delta_2)$ be a feasible entry of the dynamic programming table that needs to be processed. Recall that $\Delta_1, \Delta_2 \ge D$, and we can assume that the cost of the optimal solution for instance $\pi = \pi(x, y, k', D, \Delta_1, \Delta_2)$ is bounded by D. For simplicity, we denote $V' = V(x, y, D, \Delta_1, \Delta_2)$. We say that the problem instance π is *easy* iff one of the following happens—in fact, these are the base cases of the dynamic programming.

1. $k' \le 4$, or
2. $d(y, x) \le 3(\Delta_1 + \Delta_2) + D$, or
3. none of the above holds, and there are two integers k_1, k_2, with $k_1 + k_2 \ge k'$, such that the tours $\mathcal{T}_1 = \text{Alg}_{k\text{-tour}}(B(x, 3\Delta_1) \cap V', x, k_1)$ and $\mathcal{T}_2 = \text{Alg}_{k\text{-tour}}(B(y, 3\Delta_2) \cap V', y, k_2)$ have total length at most $3\beta(\Delta_1 + \Delta_2) + 2\beta D$. In other words, we can find two tours: $\mathcal{T}_1$ rooted at x inside the subgraph induced by $B(x, 3\Delta_1) \cap V'$, and $\mathcal{T}_2$ rooted at y inside the subgraph induced by $B(y, 3\Delta_2) \cap V'$, that together cover k' vertices (we show below that the two tours are disjoint), and their total length is at most $3\beta(\Delta_1 + \Delta_2) + 2\beta D$.

Notice that we can check if π is easy in polynomial time.

Claim 2. *If $T = T(x, y, k', D, \Delta_1, \Delta_2)$ is feasible, and $\pi = \pi(x, y, k', D, \Delta_1, \Delta_2)$ is easy, then we can find a solution for π of cost at most $3\beta(D + \Delta_1 + \Delta_2) \leq 3\beta(\Delta_1 + \Delta_2) + \beta \cdot \alpha(k') \cdot D$.*

Proof. If $k' \leq 4$, an optimal solution of cost at most D can be found by exhaustive search. Otherwise, if $d(y, x) \leq 3(\Delta_1 + \Delta_2) + D$, then there is a solution to instance $I(V', x, k')$ of the k-**Tour** problem of cost at most $2D + 3\Delta_1 + 3\Delta_2$. We obtain a solution to π by concatenating $\mathsf{Alg}_{k\text{-tour}}(V', x, k')$ with edge (x, y). The cost of the solution is bounded by $D + \beta(2D + 3\Delta_1 + 3\Delta_2) \leq 3\beta(D + \Delta_1 + \Delta_2)$.

Finally, if none of the above happens, the sets $B(x, 3\Delta_1)$ and $B(y, 3\Delta_2)$ are completely disjoint. So if the third condition holds, the two tours T_1, T_2 are completely disjoint, covering together k' distinct vertices. We can connect them to each other by adding the edge (x, y), obtaining a solution of cost at most $3\beta(\Delta_1 + \Delta_2 + D)$ to π. $\quad\square$

From now on we assume that the instance π is not easy. We also assume that for all $k'' < k'$, all entries $T(x', y', k'', D', \Delta_1', \Delta_2')$ have been computed correctly. That is, if $T(x', y', k'', D', \Delta_1', \Delta_2')$ is a feasible entry, then it is good.

Our high-level idea is to subdivide π into two subinstances, and then look the corresponding values up in the dynamic programming table. Let $\mathcal{P}$ denote the optimal solution for π. Roughly speaking, we would like to find a pivot vertex v that lies "in the middle" of $\mathcal{P}$, with roughly half the vertices appearing before and after v on $\mathcal{P}$, and then obtain two subproblems: one that appears "to the left" and one that appears "to the right" of v on $\mathcal{P}$. Let v be the guessed "middle" vertex, and let D_L, D_R be the guessed values of the lengths of the segments of $\mathcal{P}$ before and after it visits v (since we have a complete graph, v is visited at most once). We require that $D_L + D_R = D$, $d(x, v) \leq D_L$, and $d(v, y) \leq D_R$. We now define the following three sets of vertices:

- $C_v = B(v, D) \cap V'$.
- $L_v \;=\; \{u \in V' \setminus C_v \mid d(x, u) + d(u, v) \leq D_L\}$. Equivalently, $L_v = (S(x, v, D_L) \setminus B(v, D)) \cap V'$. Notice that if $u \in L_v$, then $d(v, u) + d(u, y) > D_R$ (otherwise u must belong to C_v). Therefore, if $u \in \mathcal{P}$, then it has to appear before v on $\mathcal{P}$.
- $R_v \;=\; \{u \in V' \setminus C_v \mid d(v, u) + d(u, y) \leq D_R\}$. Equivalently, $R_v = (S(v, y, D_R) \setminus B(v, D)) \cap V'$. Notice that if $u \in R_v$, then $d(x, u) + d(u, v) > D_L$ (otherwise $u \in C_v$). Therefore, if $u \in \mathcal{P}$, then u has to appear after v on $\mathcal{P}$.

Clearly, the three sets C_v, L_v and R_v are completely disjoint. It is easy to see that we can transform $\mathcal{P}$ into another x-y walk $\mathcal{P}'$, that visits the same vertices as $\mathcal{P}$, and it consists of three segments: the first segment connects x to v and only contains vertices of $L_v \cup \{x, v\}$; the second segment is a tour containing only vertices of C_v, including v; and the third segment connects v to y and only contains vertices of $R_v \cup \{v, y\}$. The lengths of these segments are bounded by D_L, $D_L + 2D + D_R \leq 3D$ and D_R, respectively. Let k_L, k_C, k_R be the numbers of distinct vertices contained in each of the segments, respectively, $k_L + k_C + k_R = k' + 2$. (Notice that vertex v appears on all three segments).

Observe that if value k_C has been guessed correctly, then $\mathrm{Alg}_{k\text{-tour}}(C_v, v, k_C)$ returns a tour $\mathcal{P}_C$, containing k_C vertices from C_v, including v, of length at most $3\beta D$. We would like now to look the remaining two segments up in the dynamic programming table. The first segment should appear in $T(x, v, k_L, D_L, \Delta_1, D)$, and the second segment in $T(v, y, k_R, D_R, D, \Delta_2)$. Indeed, this approach works if we can ensure that $V(x, v, D_L, \Delta_1, D) = L_v$ and $V(v, y, D_R, D, \Delta_2) = R_v$. Unfortunately this is not necessarily true. To overcome this issue, we proceed as follows. First, we define a set of *admissible* pivots. We then show that if v is an admissible pivot, then indeed $V(x, v, D_L, \Delta_1, D) = L_v$ and $V(v, y, D_R, D, \Delta_2) = R_v$. Finally, we show how to take care of the case where no pivot is admissible.

Definition 1. We say that v is an *admissible pivot* iff $v \notin B(x, 2\Delta_1)$ and $v \notin B(y, 2\Delta_2)$.

Claim 3. *If v is admissible, then $V(x, v, D_L, \Delta_1, D) = L_v$ and $V(v, y, D_R, D, \Delta_2) = R_v$.*

Proof. Let $V^* = V(x, v, D_L, \Delta_1, D)$. We show that $V^* = L_v$; the other case is symmetric. Assume first that $u \in L_v$. We show that $u \in V^*$ as well. By the definition of L_v, $u \in (S(x, v, D_L) \setminus B(v, D)) \cap V'$. On the other hand, $V' = S(x, y, D) \setminus (B(x, \Delta_1) \cup B(y, \Delta_2))$. In particular, $u \notin B(x, \Delta_1)$. Therefore, $u \in S(x, v, D_L) \setminus (B(v, D) \cup B(x, \Delta_1))$ and so $u \in V^*$.

Assume now that $u \in V^*$. From the definition of V^*, this means that $u \in S(x, v, D_L)$, $u \notin B(x, \Delta_1)$ and $u \notin B(v, D)$. Assume for contradiction that $u \notin L_v$. Since L_v contains all vertices in $S(x, v, D_L) \setminus B(v, D)$ that participate in V', it means that $u \notin V'$. But since $u \in S(x, v, D_L)$, we have that $d(x, u) + d(u, v) \leq D_L$, which together with $d(v, y) \leq D_R$ implies that $d(x, u) + d(u, y) \leq D_L + D_R \leq D$, and hence $u \in S(x, y, D)$. Therefore, the only possibility for $u \notin V'$ is that $u \in B(y, \Delta_2)$. Then we have that $d(y, u) \leq \Delta_2$; $d(u, v) \leq D_L$ and $d(v, y) \leq D_R$. Since $D_L + D_R = D \leq \Delta_2$, we get that $v \in B(y, 2\Delta_2)$, a contradiction for v being an admissible pivot. $\square$

We now proceed as follows. First, we define the notion of *good* admissible pivots. Intuitively, an admissible pivot is good iff it lies "in the middle" of the optimal solution. More precisely, we use the following definition.

Definition 2. An admissible pivot v is *good* iff there are integers k_L, k_R, k_C, D_L, D_R, with $k_L + k_R + k_C = k' + 2$, $D_L + D_R = D$, $k_L, k_R \leq 2k'/3$, such that both $T(x, v, k_L, D_L, \Delta_1, D)$ and $T(v, y, k_R, D_R, D, \Delta_2)$ are good entries, and the length of the tour $\mathrm{Alg}_{k\text{-tour}}(C_v, v, k_C)$ is at most $3\beta D$.

Observe that we can check whether a pivot v is good and admissible in polynomial time. The next claim shows that if a good admissible pivot exists, then we can find the required solution to the instance π. After that we show how to handle the case where no admissible pivot exists. In this case, we show that we can decompose the problem into two subproblems, one of which is easy, while the other is "small," in the sense that the number of vertices that we need to cover in the second subproblem is significantly smaller than k'.

Claim 4. *If there is a good admissible pivot, then we can find a solution to π of cost at most $3\beta(\Delta_1 + \Delta_2) + \beta \cdot \alpha(k') \cdot D$.*

Proof. Since pivot v is admissible, from Claim 3, $V(x, v, D_L, \Delta_1, D) = L_v$ and $V(v, y, D_R, D, \Delta_2) = R_v$. Consider the three paths $\mathcal{P}_L = T(x, v, k_L, D_L, \Delta_1, D)$, $\mathcal{P}_R = T(v, y, k_R, D_R, D, \Delta_2)$, and $\mathcal{P}_C = \text{Alg}_{k\text{-tour}}(C_v, v, k_C)$. Since the sets L_v, C_v and R_v are completely disjoint, the three paths are also completely disjoint, except for the vertex v, that appears on each one of them. (Notice that since v is admissible, $x, y \notin C_v$.) So altogether these paths cover $k_L + k_R + k_C - 2 = k'$ distinct vertices of $V' \cup \{x, y\}$. Let $\mathcal{P}$ be the path obtained by concatenating $\mathcal{P}_L, \mathcal{P}_C$ and $\mathcal{P}_R$. It now only remains to bound the length of $\mathcal{P}$. The lengths of $\mathcal{P}_L$ and $\mathcal{P}_R$ are bounded by $3\beta(\Delta_1 + D) + \beta(9\log_\theta k_L + 3)D_L$ and $3\beta(D + \Delta_2) + \beta(9\log_\theta k_R + 3)D_R$, respectively. Since $k_L, k_R \leq 2k'/3$ and $\theta = 3/2$, $\log_\theta k_L \leq \log_\theta k' - 1$ and $\log_\theta k_R \leq \log_\theta k' - 1$. Therefore, the total solution cost is bounded by

$$
\begin{aligned}
3\beta D &+ [3\beta(\Delta_1 + D) + \beta(9\log_\theta k_L + 3)D_L] \\
&+ [3\beta(D + \Delta_2) + \beta(9\log_\theta k_R + 3)D_R] \\
\leq 3\beta D &+ 3\beta(\Delta_1 + D) + \beta(9(\log_\theta k' - 1) + 3)D_L \\
&+ 3\beta(D + \Delta_2) + \beta(9(\log_\theta k' - 1) + 3)D_R \\
= 3\beta D &+ 3\beta(\Delta_1 + \Delta_2 + 2D) + \beta(9(\log_\theta k' - 1) + 3)(D_L + D_R) \\
\leq 3\beta(\Delta_1 &+ \Delta_2) + \beta(9\log_\theta k' + 3)D \\
= 3\beta(\Delta_1 &+ \Delta_2) + \beta \cdot \alpha(k') \cdot D.
\end{aligned}
$$
□

It now only remains to take care of the case where no good admissible pivots exist. This is done in the following claim.

Claim 5. *If T is a feasible entry, π is not easy, and no good admissible pivot exists, then there is an admissible (non-good) pivot v, integers k_L, k_R, k_C, D_L, D_R, with $k_L + k_R + k_C = k' + 2$, $D_L + D_R = D$, such that the length of the tour $\text{Alg}_{k\text{-tour}}(C_v, v, k_C)$ is at most $3\beta D$, and the entries $T(x, v, k_L, D_L, \Delta_1, D)$ and $T(v, y, k_R, D_R, D, \Delta_2)$ are good. Moreover, either $k_R \leq 2k'/3$, and problem $\pi(x, v, k_L, D_L, \Delta_1, D)$ is easy, or $k_L \leq 2k'/3$, and problem $\pi(v, y, k_R, D_R, D, \Delta_2)$ is easy. In either case, we can find a solution to π of cost at most $3\beta(\Delta_1 + \Delta_2) + \beta \cdot \alpha(k') \cdot D$.*

Proof. For simplicity, we call vertices of V' that belong to $B(x, 2\Delta_1)$ "red", and vertices of V' that belong to $B(y, 2\Delta_2)$ "blue". Consider the optimal solution $\mathcal{P}$ to the problem π. First, it is easy to see that all red vertices appear before all blue vertices on $\mathcal{P}$: otherwise, if some blue vertex b appears before some red vertex r on $\mathcal{P}$, then $d(y, x) \leq d(y, b) + d(b, r) + d(r, x) \leq D + 2\Delta_1 + 2\Delta_2$, so π is an easy problem (case 2). Similarly, no vertex can be blue and red simultaneously.

Let r be the last red vertex and b the first blue vertex on path $\mathcal{P}$. Observe that all vertices lying before r on $\mathcal{P}$ belong to $B(x, 3\Delta_1)$, and all vertices appearing after b on $\mathcal{P}$ belong to $B(y, 3\Delta_2)$, since $\Delta_1, \Delta_2 \geq D$. Thus, if no vertex lies between r and b on $\mathcal{P}$, there are two integers k_1 and k_2, $k_1 + k_2 = k'$, such that the two instances $I(B(x, 3\Delta_1) \cap V', x, k_1)$ and $I(B(y, 3\Delta_2) \cap V', y, k_2)$ of the k-Tour problem have solutions of total cost at most $3(\Delta_1 + \Delta_2) + 2D$, so problem π is easy (case 3).

Let Q be the set of all vertices lying between r and b on $\mathcal{P}$. Then all vertices in Q are admissible pivots. Let $\mathcal{P}_1$ be the portion of $\mathcal{P}$ lying between x and r, and let $\mathcal{P}_2$ be the portion of $\mathcal{P}$ lying between b and y. If both $\mathcal{P}_1$ and $\mathcal{P}_2$ contain less than $2k'/3$ distinct vertices, then one of the pivots in Q must be good. Since we have assumed that there are no good admissible pivots, either $\mathcal{P}_1$ or $\mathcal{P}_2$ contains more than $2k'/3$ vertices. For simplicity, assume the former; the other case is symmetric. Let v be the vertex appearing on $\mathcal{P}$ right after r. We choose v as the pivot. Observe that v is an admissible pivot. Consider the corresponding sets L_v, C_v and R_v. As before, we can replace $\mathcal{P}$ by a path $\mathcal{P}'$ that consists of three segments. The first segment connects x to v and only visits vertices of $L_v \cup \{x, v\}$; the second segment is a tour containing v and only visiting vertices of C_v, and the third segment connects v to y and only contains vertices of $R_v \cup \{v, y\}$. The lengths of these segments are bounded by $D_L, 3D$ and D_R, respectively, where $D_L + D_R = D$. Let k_L, k_C and k_R denote the number of distinct vertices appearing on each one of the three segments, respectively, $k_L + k_R + k_C = k' + 2$. Observe that only vertices that appear after v on $\mathcal{P}$ belong to R_v, so $k_R \leq 2k'/3$. Finally, we need to show that problem $\pi' = \pi(x, v, k_L, D_L, \Delta_1, D)$ is easy. First, if $d(v, x) \leq 3(\Delta_1 + D) + D_L$, problem π' is easy (case 2). So assume this is not the case. Recall that since v is an admissible pivot, $V(x, v, D_L, \Delta_1, D) = L_v$, and since all vertices of L_v appear before v on $\mathcal{P}$, $L_v \subseteq B(x, 3\Delta_1) \cap V'$. Therefore, there is a solution to the k-tour problem instance $I(B(x, 3\Delta_1) \cap V', x, k_L - 1)$ of cost at most $D_L + 3\Delta_1$, and solution to $I(B(v, 3D) \cap V', v, 1))$ of cost 0, and so π' is easy (case 3). Since the two entries $T(x, v, k_L, D_L, \Delta_1, D)$ and $T(v, y, k_R, D_R, D, \Delta_2)$ are feasible, they must also be good.

It now only remains to bound the solution cost. We assume again w.l.o.g. that the first case happens, that is, $k_R \leq 2k'/3$, and problem $\pi(x, v, k_L, D_L, \Delta_1, D)$ is easy. Using Claim 2, we can find a solution $\mathcal{T}_L$ to instance $\pi(x, v, k_L, D_L, \Delta_1, D)$ of cost at most $3\beta(D_L + \Delta_1 + D)$. We let $\mathcal{T}_C = \text{Alg}_{k\text{-tour}}(C_v, v, k_C)$ be the tour of cost at most $3\beta D$, and recall that the entry $T(v, y, k_R, D_R, D, \Delta_2)$ is good, so it contains a path, denoted by $\mathcal{T}_R$, of length at most $3\beta(D + \Delta_2) + \beta(9 \log_\theta k_R + 3)D_R$.

Then the total cost is bounded by

$$
\begin{aligned}
& 3\beta D + 3\beta(D_L + \Delta_1 + D) + [3\beta(D + \Delta_2) + \beta(9 \log_\theta k_R + 3)D_R] \\
\leq\ & 3\beta D + 3\beta(D_L + \Delta_1 + D) + 3\beta(D + \Delta_2) + \beta(9(\log_\theta k' - 1) + 3)D_R \\
\leq\ & 3\beta D + 3\beta(\Delta_1 + 2D + \Delta_2) + \beta(9(\log_\theta k' - 1) + 3)(D_L + D_R) \\
\leq\ & 3\beta(\Delta_1 + \Delta_2) + \beta(9(\log_\theta k' + 3)D \\
\leq\ & 3\beta(\Delta_1 + \Delta_2) + \beta \cdot \alpha(k') \cdot D.
\end{aligned}
$$
$\qquad\qquad\qquad\qquad\qquad\qquad\qquad\qquad\qquad\qquad\qquad\qquad\qquad\qquad\qquad\qquad\square$

We now summarize our algorithm for computing entry $T(x, y, k', D, \Delta_1, \Delta_2)$:

- If instance $\pi(x, y, k', D, \Delta_1, \Delta_2)$ is easy, return the solution of cost at most $3\beta(D + \Delta_1 + \Delta_2) \leq 3\beta(\Delta_1 + \Delta_2) + \beta \cdot \alpha(k') \cdot D$, guaranteed by Claim 2.
- Otherwise, if there is a good admissible pivot v, return the solution of cost at most $3\beta(\Delta_1 + \Delta_2) + \beta \cdot \alpha(k') \cdot D$, guaranteed by Claim 4.
- Otherwise, if there is an admissible pivot v, and integers k_L, k_R, k_C, D_L, D_R, with $k_L + k_R + k_C = k' + 2$, $D_L + D_R = D$, such that the length of the tour $\text{Alg}_{k\text{-tour}}(C_v, v, k_C)$ is at most $3\beta D$, the entries $T(x, v, k_L, D_L, \Delta_1, D)$

and $T(v, y, k_R, D_R, D, \Delta_2)$ are good, and either (1) $k_R \leq 2k'/3$, and $\pi(x, v, k_L, D_L, \Delta_1, D)$ is easy, or (2) $k_L \leq 2k'/3$ and $\pi(v, y, k_R, D_R, D, \Delta_2)$ is easy: return a solution of cost at most $3\beta(\Delta_1 + \Delta_2) + \beta \cdot \alpha(k') \cdot D$, guaranteed by Claim 5.

- Otherwise, the entry $T(x, y, k', D, \Delta_1, \Delta_2)$ is undefined.

From the above discussion, if $T(x, y, k', D, \Delta_1, \Delta_2)$ is feasible, and all entries $T(x', y', k'', D', \Delta'_1, \Delta'_2)$ for $k'' < k'$ have been computed correctly, the algorithm finds a solution to the k-Stroll instance $\pi(x, y, k', D, \Delta_1, \Delta_2)$ of cost at most $3\beta(\Delta_1 + \Delta_2) + \beta\alpha(k')D$. In particular, the entry $T(s, t, k, L^*, L^*, L^*)$ will contain an s-t walk covering k vertices, of length at most $O(\beta \cdot \alpha(k) \cdot L^*) = O(\log k) \cdot \beta \cdot L^*$.

3 Approximation Algorithm for the k-Tour Problem

3.1 Preliminaries and Notation

We assume that we are given a directed graph $G = (V, E)$ with nonnegative lengths c_e for all edges $e \in E$. For each vertex $v \in V$, we denote by $\delta^-(v)$ and $\delta^+(v)$ the sets of the incoming and the outgoing edges of v, respectively. Similarly, for a subset $U \subseteq V$ of vertices, $\delta^-(U) = \{(v, u) \in E \mid v \in V \setminus U, u \in U\}$ and $\delta^+(U) = \{(u, v) \in E \mid u \in U, v \in V \setminus U\}$. Given a pair u, v of vertices, the distance $d(u, v)$ is the length of the shortest path from u to v in G, where the length of each edge e is c_e.

Held-Karp LP: We will use the famous Held-Karp LP relaxation for the ATSP problem [14], defined as follows:

$$(\text{LP-HK}) \quad \text{minimize } \sum_{e \in E} c_e x_e$$

$$\text{s.t.}$$

$$\sum_{e \in \delta^-(v)} x_e = \sum_{e \in \delta^+(v)} x_e \quad \forall v \in V \tag{1}$$

$$\sum_{e \in \delta^+(U)} x_e \geq 1 \qquad \forall U \subset V \tag{2}$$

$$x_e \geq 0 \qquad \forall e \in E$$

For each edge $e \in E$, the LP relaxation contains an indicator variable x_e for including e in the solution. The objective is to minimize the total length of edges in the solution. An integral solution to LP-HK induces a subgraph of G, and the set (1) of constraints ensures that the in-degree of every vertex equals its out-degree, while the set (2) of constraints requires each subset $U \subset V$ of vertices to have at least one edge leaving the set in this subgraph. Although (LP-HK) has an exponential number of constraints, it can be solved in polynomial time, either by the Ellipsoid algorithm with a separation oracle, or by writing an equivalent LP relaxation with a polynomial number of variables and constraints.

Let β_{HK} denote the best approximation factor achievable by any LP-rounding algorithm based on (LP-HK). More precisely, β_{HK} is the smallest approximation factor, for which there is an efficient algorithm $\mathcal{A}$, that for any instance $\mathcal{I}$ of the

ATSP problem, produces a solution whose cost is at most $\beta_{HK} \cdot \text{OPT}_{HK}(\mathcal{I})$, where $\text{OPT}_{HK}(\mathcal{I})$ is the value of the optimal solution of (LP-HK) for $\mathcal{I}$. From the recent result of Asadpour et al. [1], $\beta_{HK} \leq O(\log n / \log \log n)$. The goal of this section is to show an $O(\log n)\beta_{HK}$-approximation algorithm for the k-Tour problem. Let $\alpha = O(\log n)\beta_{HK}$ denote the desired approximation factor.

LP relaxation for k-Tour. Throughout the algorithm, we assume that we know the value L^* of the optimal solution to the k-Tour problem. This is done using standard techniques. We now perform the following simple transformation to our input graph G: first, we discard all vertices v, for which $d(r, v) > L^*$ or $d(v, r) > L^*$. Next, we discard all edges e with $c_e > L^*$. Since the discarded edges and vertices do not participate in the optimal tour, the value of the optimal tour in the new graph does not change. For simplicity, we will use G to denote the new graph. Clearly, a tour of length αL^* in the new graph translates to a tour of the same length in the old graph. We are now ready to define the linear programming relaxation, extending (LP-HK) to the k-Tour problem. In addition to variables x_e for all $e \in E$, the LP relaxation contains, for each vertex $v \in V$, a variable z_v, indicating whether v belongs to the tour.

(LP-k-Tour)   minimize $\sum_{e \in E} c_e x_e$

s.t.

$$\sum_{e \in \delta^-(v)} x_e = \sum_{e \in \delta^+(v)} x_e \quad \forall v \in V \tag{3}$$

$$\sum_{e \in \delta^+(U)} x_e \geq z_v \qquad \forall U \subseteq V \setminus \{r\}, \forall v \in U \tag{4}$$

$$z_v \leq 1 \qquad \forall v \in V \tag{5}$$

$$z_r = 1 \tag{6}$$

$$\sum_{v \in V} z_v \geq k \tag{7}$$

$$z_v, x_e \geq 0 \qquad \forall v \in V, \forall e \in E$$

The set (3) of constraints is identical to constraints (1) of (LP-HK). The second set of constraints, (4), corresponds to constraints (2) of (LP-HK), and it requires that whenever a vertex v belongs to the solution, every cut U containing v but not r, has an edge $e \in \delta^+(U)$ in the solution. The next three constraints (5)–(7) ensure that each vertex is covered at most once, the root vertex r belongs to the solution, and the total number of vertices covered is k, respectively.

The LP relaxation has exponentially many constraints, but similarly to (LP-HK), it can be solved efficiently. Let OPT_{LP} denote the optimal solution value of (LP-k-Tour). Notice that we can assume that $\text{OPT}_{LP} \leq L^*$, the guessed value of the optimal solution cost.

3.2   LP Rounding

We start with initial rounding of the LP solution.

Lemma 1. *We can efficiently find a feasible solution (x', z') to (LP-k-Tour) of cost at most $4 \cdot \text{OPT}_{LP}$, such that all nonzero values z'_v belong to the set $\{1/2^i \mid 0 \leq i \leq \lceil 3 \log n \rceil\}$.*

Proof. Let (x, z) be the optimal feasible solution to (LP-k-Tour), whose cost is OPT_{LP}. We transform it to solution (x', z') as follows: for each edge $e \in E$, set $x'_e = 4x_e$. For each $v \in V$, if $1/2^i < z_v \leq 1/2^{i-1}$, then if $i > \lceil 3 \log n \rceil$, set $z'_v = 0$; otherwise, $z'_v = \min(1, 1/2^{i-2})$.

It is immediately seen that the cost of the new solution (x', z') is bounded by 4OPT_{LP}. We now only need to verify that it is a feasible solution. First, since all values x_e were multiplied by the same factor, constraints (3) continue to hold. It is also easy to see that for each vertex v, $z'_v \leq 1$, and $z'_r = 1$, and therefore constraints (5) and (6) still hold. Consider now constraint (4) for some $v \in V$, $U \subseteq V \setminus \{r\}$ with $v \in U$. The value of z_v has increased by at most a factor 4, while the values x_e for all $e \in \delta^+(U)$ have increased by a factor 4. Therefore, the constraint continues to hold.

Finally, it remains to show that $\sum_{v \in V} z'_v \geq k$. Let Z_0 contain the set of vertices v, for which $z_v \leq 1/2^{\lceil 3 \log n \rceil} \leq \frac{1}{n^3}$. These are the only vertices whose LP values have decreased. The total value $\sum_{v \in Z_0} z_v \leq 1/n^2$. Let Z_1 denote the set of vertices v for which $z'_v = 1$. If $|Z_1| \geq k$, then clearly constraint (7) holds. Otherwise, $\sum_{v \notin Z_1} z_v \geq 1$ must hold in the original solution, and therefore $\sum_{v \notin Z_1 \cup Z_0} z_v \geq 1 - 1/n^2 \geq \sum_{v \in Z_0} z_v$. For each vertex $v \notin Z_1 \cup Z_0$, we have that $z'_v \geq 2z_v$. So overall $\sum_{v \notin Z_1} z'_v \geq 2 \sum_{v \notin Z_1 \cup Z_0} z_v \geq \sum_{v \notin Z_1} z_v$. Since $\sum_{v \in Z_1} z'_v \geq \sum_{v \in Z_1} z_v$, constraint (7) continues to hold. $\square$

For each $i : 0 \leq i \leq \lceil 3 \log n \rceil$, we denote by B_i the set of vertices v with $z'_v = 1/2^i$, and set $k_i = |B_i|$. Recall that $\sum_{i=0}^{\lceil 3 \log n \rceil} k_i/2^i \geq k$.

Theorem 3. *For each $i : 0 \leq i \leq \lceil 3 \log n \rceil$, we can efficiently find a tour $\mathcal{T}_i$ of cost at most $\beta_{HK} \cdot 2^{i+5} \cdot L^*$, visiting all vertices in B_i.*

The proof of the theorem is omitted due to lack of space. We now show that Theorem 2 follows from it. We first show that for each $i : 0 \leq i \leq \lceil 3 \log n \rceil$, there is a path $\mathcal{T}_i^*$, containing at least $\lceil k/2^i \rceil$ vertices of B_i, of length at most $O(\beta_{HK}) \cdot L^*$. Since we have discarded all vertices v with $d(v, r) > L^*$ or $d(r, v) > L^*$, we can turn $\mathcal{T}_i^*$ into a tour containing the vertex r, at the additional cost of $2L^*$. Therefore, for each $i : 0 \leq i \leq \lceil 3 \log n \rceil$, we obtain a tour containing the vertex r, and additional $\lceil k/2^i \rceil$ vertices of B_i, of length at most $O(\beta_{HK}) \cdot L^*$. Connecting all these tours together gives a tour of length at most $O(\beta_{HK} \cdot \log n) \cdot L^*$, containing at least $\sum_{i=0}^{\lceil 3 \log n \rceil} k_i/2^i \geq k$ vertices.

It now only remains to show how to find the paths $\mathcal{T}_i^*$. Fix some $i : 0 \leq i \leq \lceil 3 \log n \rceil$. If $k_i/2^i \leq 1$, then choose any vertex $v \in B_i$, and the path $\mathcal{T}_i^*$ then only consists of the vertex v. Otherwise, consider the tour $\mathcal{T}_i$. This tour contains all k_i vertices of B_i, and its length is at most $\beta_{HK} \cdot 2^{i+5} \cdot L^*$. We partition $\mathcal{T}_i$ into at least 2^{i-2} disjoint consecutive segments, each containing $\lceil k_i/2^i \rceil$ vertices of B_i. We let $\mathcal{T}_i^*$ be the segment of minimum length, so the length of $\mathcal{T}_i^*$ is bounded by $O(\beta_{HK} \cdot L^*)$.

Acknowledgment. We would like to thank Chandra Chekuri for suggesting the problems, and for sharing with us his survey on open problems related to Orienteering.

References

1. Asadpour, A., Goemans, M.X., Madry, A., Oveis Gharan, S., Saberi, A.: An $O(\log n/\log \log n)$-approximation algorithm for the asymmetric traveling salesman problem. In: SODA (2010)
2. Bansal, N., Blum, A., Chawla, S., Meyerson, A.: Approximation algorithms for deadline-TSP and vehicle routing with time-windows. In: STOC, pp. 166–174 (2004)
3. Blaser, M.: A new approximation algorithm for the asymmetric TSP with triangle inequality. In: SODA, pp. 638–645 (2003)
4. Blum, A., Chawla, S., Karger, D.R., Lane, T., Meyerson, A., Minkoff, M.: Approximation algorithms for orienteering and discounted-reward TSP. In: FOCS, pp. 46–55 (2003)
5. Charikar, M., Goemans, M.X., Karloff, H.: On the integrality ratio for the asymmetric traveling salesman problem. Math. Oper. Res. 31, 245–252 (2006)
6. Chaudhuri, K., Godfrey, B., Rao, S., Talwar, K.: Paths, trees, and minimum latency tours. In: FOCS, pp. 36–45 (2003)
7. Chekuri, C., Korula, N., Pál, M.: Improved algorithms for orienteering and related problems. In: SODA, pp. 661–670 (2008)
8. Chekuri, C., Pál, M.: A recursive greedy algorithm for walks in directed graphs. In: FOCS, pp. 245–253 (2005)
9. Chekuri, C., Pál, M.: An O($log\ n$) approximation ratio for the asymmetric traveling salesman path problem. Theory of Computing 3, 197–209 (2007)
10. Feige, U., Singh, M.: Improved approximation ratios for traveling salesperson tours and paths in directed graphs. In: Charikar, M., Jansen, K., Reingold, O., Rolim, J.D.P. (eds.) RANDOM 2007 and APPROX 2007. LNCS, vol. 4627, pp. 104–118. Springer, Heidelberg (2007)
11. Frieze, A., Galbiati, G., Maffioli, F.: On the worst-case performance of some algorithms for the asymmetric traveling salesman problem. Networks 12, 23–39 (1982)
12. Goemans, M.X., Harvey, N.J.A., Jain, K., Singh, M.: A randomized rounding algorithm for the asymmetric traveling salesman problem, CoRR, abs/0909.0941 (2009)
13. Halperin, E., Krauthgamer, R.: Polylogarithmic inapproximability. In: STOC, pp. 585–594 (2003)
14. Held, M., Karp, R.M.: The traveling-salesman problem and minimum spanning trees. Oper. Res. 18, 1138–1162 (1970)
15. Nagarajan, V., Ravi, R.: Poly-logarithmic approximation algorithms for directed vehicle routing problems. In: Charikar, M., Jansen, K., Reingold, O., Rolim, J.D.P. (eds.) RANDOM 2007 and APPROX 2007. LNCS, vol. 4627, pp. 257–270. Springer, Heidelberg (2007)

Submodular Secretary Problem and Extensions*

MohammadHossein Bateni[1,**], MohammadTaghi Hajiaghayi[2,***],
and Morteza Zadimoghaddam[3,†]

[1] Princeton University, Princeton NJ 08540, USA
mbateni@cs.princeton.edu
[2] AT&T Labs–Research, Florham Park, NJ 07932, USA
hajiagha@research.att.com
[3] MIT, CSAIL, Cambridge, MA 02139, USA
morteza@mit.edu

Abstract. Online auction is the essence of many modern markets, particularly networked markets, in which information about goods, agents, and outcomes is revealed over a period of time, and the agents must make irrevocable decisions without knowing future information. Optimal stopping theory, especially the classic *secretary problem*, is a powerful tool for analyzing such online scenarios which generally require optimizing an objective function over the input. The secretary problem and its generalization the *multiple-choice secretary problem* were under a thorough study in the literature. In this paper, we consider a very general setting of the latter problem called the *submodular secretary problem*, in which the goal is to select k secretaries so as to maximize the expectation of a (not necessarily monotone) submodular function which defines efficiency of the selected secretarial group based on their overlapping skills. We present the first constant-competitive algorithm for this case. In a more general setting in which selected secretaries should form an independent (feasible) set in each of l given matroids as well, we obtain an $O(l \log^2 r)$-competitive algorithm generalizing several previous results, where r is the maximum rank of the matroids. Another generalization is to consider l knapsack constraints (i.e., a knapsack constraint assigns a nonnegative cost to each secretary, and requires that the total cost of all the secretaries employed be no more than a budget value) instead of the matroid constraints, for which we present an $O(l)$-competitive algorithm. In a sharp contrast, we show for a more general setting of *subadditive secretary problem*, there is no $\tilde{o}(\sqrt{n})$-competitive algorithm and thus submodular functions are the most general functions to consider for constant-competitiveness in our setting. We complement this result by giving a matching $O(\sqrt{n})$-competitive algorithm for the subadditive case.

* An early version of this paper was publicly released as an AT&T technical report [7] in July 2009. Several proofs and further discussion are omitted from this extended abstract. The reader is referred to the full version [8].
** The author was supported by a Gordon Wu fellowship as well as NSF ITR grants CCF-0205594, CCF-0426582 and NSF CCF 0832797, NSF CAREER award CCF-0237113, MSPA-MCS award 0528414, NSF expeditions award 0832797. He is also with the Center for Computational Intractability, Princeton, NJ, USA.
*** He is also with University of Maryland, Department of Computer Science, College Park, MD, USA.
† Part of the work was done while the author was visiting EPFL, Lausanne, Switzerland.

M. Serna et al. (Eds.): APPROX and RANDOM 2010, LNCS 6302, pp. 39–52, 2010.

1   Introduction

Online auction is the essence of many modern markets, particularly networked markets, in which information about goods, agents, and outcomes is revealed over a period of time, and the agents must make irrevocable decisions without knowing future information. Optimal stopping theory is a powerful tool for analyzing such scenarios which generally require optimizing an objective function over the space of stopping rules for an allocation process under uncertainty. Combining optimal stopping theory with game theory allows us to model the actions of rational agents applying competing stopping rules in an online market. This first has been done by Hajiaghayi et al. [24] who considered the well-known *secretary problem* in online settings and initiated several follow-up papers (see e.g. [4,5,6,25,29,33]).

   Perhaps the most classic problem of stopping theory is the secretary problem. Imagine that you manage a company, and you want to hire a secretary from a pool of n applicants. You are very keen on hiring only the best and brightest. Unfortunately, you cannot tell how good a secretary is until you interview him, and you must make an irrevocable decision whether or not to make an offer at the time of the interview. The problem is to design a strategy which maximizes the probability of hiring the most qualified secretary. It is well-known since 1963 [12] that the optimal policy is to interview the first $t - 1$ applicants, then hire the next one whose quality exceeds that of the first $t - 1$ applicants, where t is defined by $\sum_{j=t+1}^{n} \frac{1}{j-1} \leq 1 < \sum_{j=t}^{n} \frac{1}{j-1}$; as $n \to \infty$, the probability of hiring the best applicant approaches $1/e$, as does the ratio t/n. Note that a solution to the secretary problem immediately yields an algorithm for a slightly different objective function optimizing the expected value of the chosen element. Subsequent papers have extended the problem by varying the objective function, varying the information available to the decision-maker, and so on, see e.g., [2,21,40,42].

   An important generalization of the secretary problem with several applications (see e.g., a survey by Babaioff et al. [5]) is called the *multiple-choice secretary problem* in which the interviewer is allowed to hire up to $k \geq 1$ applicants in order to maximize performance of the secretarial group based on their overlapping skills (or the joint utility of selected items in a more general setting). More formally, assuming applicants of a set $S = \{a_1, a_2, \cdots, a_n\}$ (applicant pool) arriving in a uniformly random order, the goal is to select a set of at most k applicants in order to maximize a profit function $f : 2^S \mapsto \mathbb{R}$. We assume f is non-negative throughout this paper. For example, when $f(T)$ is the maximum individual value [19,20], or when $f(T)$ is the sum of the individual values in T [33], the problem has been considered thoroughly in the literature. Indeed, both of these cases are special monotone non-negative submodular functions that we consider in this paper. A function $f : 2^S \mapsto \mathbb{R}$ is called *submodular* if and only if $\forall A, B \subseteq S : f(A) + f(B) \geq f(A \cup B) + f(A \cap B)$. An equivalent characterization is that the marginal profit of each item should be non-increasing, i.e., $f(A \cup \{a\}) - f(A) \leq f(B \cup \{a\}) - f(B)$ if $B \subseteq A \subseteq S$ and $a \in S \setminus B$. A function $f : 2^S \mapsto \mathbb{R}$ is *monotone* if and only if $f(A) \leq f(B)$ for $A \subseteq B \subseteq S$; it is *non-monotone* if is not necessarily the case. Since the number of sets is exponential, we assume a value oracle access to the submodular function; i.e., for a given set T, an algorithm can query an oracle to find its value $f(T)$. As we discuss below, maximizing a (monotone or non-monotone) submodular function which demonstrates economy of scale is a central and very general problem in combinatorial optimization and has been subject of a thorough study in the literature.

The closest setting to our submodular multiple-choice secretary problem is the *matroid secretary problem* considered by Babaioff et al. [6]. In this problem, we are given a matroid by a ground set U of elements and a collection of independent (feasible) subsets $I \subseteq 2^U$ describing the sets of elements which can be simultaneously accepted. We recall that a matroid has three properties: 1) the empty set is independent; 2) every subset of an independent set is independent (closed under containment)[1]; and finally 3) if A and B are two independent sets and A has more elements than B, then there exists an element in A which is not in B and when added to B still gives an independent set[2]. The goal is to design online algorithms in which the structure of U and I is known at the outset (assume we have an oracle to answer whether a subset of U belongs to I or not), while the elements and their values are revealed one at a time in random order. As each element is presented, the algorithm must make an irrevocable decision to select or reject it such that the set of selected elements belongs to I at all times. Babaioff et al. present an $O(\log r)$-competitive algorithm for general matroids, where r is the rank of the matroid (the size of the maximal independent set), and constant-competitive algorithms for several special cases arising in practical scenarios including graphic matroids, truncated partition matroids, and bounded degree transversal matroids. However, they leave as a main open question the existence of constant-competitive algorithms for general matroids. Our constant-competitive algorithms for the submodular secretary problem in this paper can be considered in parallel with this open question. To generalize both results of Babaioff et al. and ours, we also consider the *submodular matroid secretary problem* in which we want to maximize a submodular function over all independent (feasible) subsets I of the given matroid. Moreover, we extend our approach to the case in which l matroids are given and the goal is to find the set of maximum value which is independent with respect to all the given matroids. We present an $O(l \log^2 r)$-competitive algorithm for the submodular matroid secretary problem generalizing previous results.

Prior to our work, there was no polynomial-time algorithm with a nontrivial guarantee for the case of l matroids—even in the offline setting—when l is not a fixed constant. Lee et al. [34] give a local-search procedure for the offline setting that runs in time $O(n^l)$ and achieves approximation ratio $l + \varepsilon$. Even the simpler case of having a linear function cannot be approximated to within a factor better than $\Omega(l/\log l)$ [28]. Our results imply an algorithm with guarantees $O(l \log r)$ and $O(l \log^2 r)$ for the offline and (online) secretary settings, respectively. Both these algorithms run in time polynomial in l. In case of the knapsack constraints, the only previous relevant work that we are aware of is that of Lee et al. [34] which gives a $(5 + \varepsilon)$ approximation in the offline setting if the number of constraints is a constant. In contrast, our results work for arbitrary number of knapsack constraints, albeit with a loss in the guarantee; see Theorem 3.

Our competitive ratio for the submodular secretary problem is $\frac{7}{1-1/e}$. Though our algorithm is relatively simple, it has several phases and its analysis is relatively involved. As we point out below, we cannot obtain any approximation factor better than $1 - 1/e$

---

[1] This is sometimes called the *hereditary property*.
[2] This is sometimes called the *augmentation property* or the *independent set exchange property*.

even for offline special cases of our setting unless $\mathbf{P} = \mathbf{NP}$. A natural generalization of a submodular function while still preserving economy of scale is a subadditive function $f : 2^S \mapsto \mathbb{R}$ in which $\forall A, B \subseteq S : f(A) + f(B) \geq f(A \cup B)$. In this paper, we show that if we consider the subadditive secretary problem instead of the submodular secretary problem, there is no algorithm with competitive ratio $\tilde{o}(\sqrt{n})$. We complement this result by giving an $O(\sqrt{n})$-competitive algorithm for the subadditive secretary problem.

Background on submodular maximization. Submodularity, a discrete analog of convexity, has played a central role in combinatorial optimization [35]. It appears in many important settings including cuts in graphs [30,22,37], plant location problems [11,10], rank function of matroids [13], and set covering problems [14].

The problem of maximizing a submodular function is of essential importance, with special cases including Max Cut [22], Max Directed Cut [26], hypergraph cut problems, maximum facility location [1,11,10], and certain restricted satisfiability problems [27,16]. While the Min Cut problem in graphs is a classical polynomial-time solvable problem, and more generally it has been shown that any submodular function can be minimized in polynomial time [30,38], maximization turns out to be more difficult and indeed all the aforementioned special cases are NP-hard.

Max-k-Cover, where the goal is to choose k sets whose union is as large as possible, is another related problem. It is shown that a greedy algorithm provides a $(1 - 1/e)$ approximation for Max-k-Cover [32] and this is optimal unless $\mathbf{P} = \mathbf{NP}$ [14]. More generally, we can view this problem as maximization of a monotone submodular function under a cardinality constraint, that is, we seek a set S of size k maximizing $f(S)$. The greedy algorithm again provides a $(1 - 1/e)$ approximation for this problem [36]. A $1/2$ approximation has been developed for maximizing monotone submodular functions under a matroid constraint [18]. A $(1 - 1/e)$ approximation has been also obtained for a knapsack constraint [39], and for a special class of submodular functions under a matroid constraint [9].

Recently constant factor $(\frac{3}{4} + \varepsilon)$-approximation algorithms for maximizing non-negative non-monotone submodular functions has also been obtained [17]. Typical examples of such a problem are max cut and max directed cut. Here, the best approximation factors are 0.878 for max cut [22] and 0.859 for max directed cut [16]. The approximation factor for max cut has been proved optimal, assuming the Unique Games Conjecture [31]. Generalizing these results, Vondrák very recently obtains a constant factor approximation algorithm for maximizing non-monotone submodular functions under a matroid constraint [41]. Subadditive maximization has been also considered recently (e.g. in the context of maximizing welfare [15]).

Submodular maximization also plays a role in maximizing the difference of a monotone submodular function and a modular function. A typical example of this type is the maximum facility location problem in which we want to open a subset of facilities and maximize the total profit from clients minus the opening cost of facilities. Approximation algorithms have been developed for a variant of this problem which is a special case of maximizing nonnegative submodular functions [1,11,10]. The current best approximation factor known for this problem is 0.828 [1]. Asadpour et al. [3] study the problem of maximizing a submodular function in a stochastic setting, and obtain constant-factor approximation algorithms.

Our results and techniques. The main theorem in this paper is as follows.

Theorem 1. *There exists a $\frac{7}{1-1/e}$-competitive algorithm for the monotone submodular secretary problem. More generally there exists a $8e^2$-competitive algorithm for the non-monotone submodular secretary problem.*

We prove Theorem 1 in Section 2. We first present our simple algorithms for the problem. Since our algorithm for the general non-monotone case uses that of monotone case, we first present the analysis for the latter case and then extend it for the former case. We divide the input stream into equal-sized segments, and show that restricting the algorithm to pick only one item from each segment decreases the value of the optimum by at most a constant factor. Then in each segment, we use a standard secretary algorithm to pick the best item conditioned on our previous choices. We next prove that these local optimization steps lead to a global near-optimal solution.

The argument breaks for the non-monotone case since the algorithm actually approximates a set which is larger than the optimal solution. The trick is to invoke a new structural property of (non-monotone) submodular functions which allows us to divide the input into two equal portions, and randomly solve the problem on one.

Indeed Theorem 1 can be extended for the submodular matroid secretary problem as follows.

Theorem 2. *There exists an $O(l \log^2 r)$ competitive algorithm for the (non-monotone) matroid submodular secretary problem, where r is the maximum rank of the given l matroids.*

We prove theorem 2 in Section 3. We note that in the submodular matroid secretary problem, selecting (bad) elements early in the process might prevent us from selecting (good) elements later since there are matroid independence (feasibility) constraints. To overcome this issue, we only work with the first half of the input. This guarantees that at each point in expectation there is a large portion of the optimal solution that can be added to our current solution without violating the matroid constraint. However, this set may not have a high value. As a remedy we prove there is a near-optimal solution all of whose large subsets have a high value. This novel argument may be of its own interest.

We shortly mention in Section 4 our results for maximizing a submodular secretary problem with respect to l knapsack constraints. In this setting, there are l knapsack capacities $C_i : 1 \leq i \leq l$, and each item j has different weights w_{ij} associated with each knapsack. A set T of items is feasible if and only if for *each* knapsack i, we have $\sum_{j \in T} w_{ij} \leq C_i$.

Theorem 3. *There exists an $O(l)$-competitive algorithm for the (non-monotone) multiple knapsack submodular secretary problem, where l denotes the number of given knapsack constraints.*

Lee et al. [34] gives a better $(5 + \varepsilon)$ approximation in the offline setting if l is a fixed constant.

We next show that indeed submodular secretary problems are the most general cases that we can hope for constant competitiveness.

Theorem 4. *For the subadditive secretary problem, there is no algorithm with competitive ratio in $\tilde{o}(\sqrt{n})$. However there is an algorithm with almost tight $O(\sqrt{n})$ competitive ratio in this case.*

We prove Theorem 4 in Section 5. The algorithm for the matching upper bound is very simple, however the lower bound uses clever ideas and indeed works in a more general setting. We construct a subadditive function, which interestingly is almost submodular, and has a "hidden good set". Roughly speaking, the value of any query to the oracle is proportional to the intersection of the query and the hidden good set. However, the oracle's response does not change unless the query has considerable intersection with the good set which is hidden. Hence, the oracle does not give much information about the hidden good set.

Remark. Subsequent to our study of online submodular maximization [7], Gupta et al. [23] consider similar problems. By reducing the case of non-monotone submodular functions to several runs of the greedy algorithm for monotone submodular functions, they present $O(p)$-approximation algorithms for maximizing submodular functions (in the offline setting) subject to p-independence systems (which include the intersection of p matroids), and constant factor approximation algorithms when the maximization is subject to a knapsack constraint. In the online secretary setting, they provide $O(1)$-competitive results for maximizing a submodular function subject to cardinality or partition matroid constraints. They also obtain an $O(\log r)$ competitive ratio for maximization subject to a general matroid of rank r. The latter result improves our Theorem 2 when $l = 1$.

2   The Submodular Secretary Problem

2.1   Algorithms

In this sections, we present the algorithms used to prove Theorem 1. In the classic secretary problem, the efficiency value of each secretary is known only after she arrives. In order to marry this with the value oracle model, we say that the oracle answers the query regarding the efficiency of a set $S' \subseteq S$ only if all the secretaries in S' have already arrived and been interviewed.

Our algorithm for the monotone submodular case is relatively simple though its analysis is relatively involved. First we assume that n is a multiple of k, since otherwise we could virtually insert $n - k\lfloor \frac{n}{k} \rfloor$ dummy secretaries in the input: for any subset A of dummy secretaries and a set $B \subseteq S$, we have that $f(A \cup B) = f(B)$. In other words, there is no profit in employing the dummy secretaries. To be more precise, we simulate the augmented input in such a way that these secretaries are arriving uniformly at random similarly to the real ones. Thus, we say that n is a multiple of k without loss of generality.

We partition the input stream into k equally-sized segments, and, roughly speaking, try to employ the *best* secretary in each segment. Let $l := \frac{n}{k}$ denote the length of each segment. Let $a_1, a_2, \cdots, a_n$ be the actual ordering in which the secretaries are interviewed. Break the input into k segments such that $S_j = \{a_{(j-1)l+1}, a_{(j-1)l+2}, \ldots, a_{jl}\}$

for $1 \le j < k$, and $S_k = \{a_{(k-1)l+1}, a_{(k-1)l+2}, \ldots, a_n\}$. We employ at most one secretary from each segment S_i. Note that this way of having several phases of (almost) equal length for the secretary problem seems novel to this paper, since in previous works there are usually only two phases (see e.g. [24]). The phase i of our algorithm corresponds to the time interval when the secretaries in S_i arrive. Let T_i be the set of secretaries that we have employed from $\bigcup_{j=1}^{i} S_j$. Define $T_0 := \emptyset$ for convenience. In phase i, we try to employ a secretary e from S_i that maximizes $f(T_{i-1} \cup \{e\}) - f(T_{i-1})$. For each $e \in S_i$, we define $g_i(e) = f(T_{i-1} \cup \{e\}) - f(T_{i-1})$. Then, we are trying to employ a secretary $x \in S_i$ that has the maximum value for $g_i(e)$. Using a classic algorithm for the *secretary problem* (see [12] for instance) for employing the single secretary, we can solve this problem with constant probability $1/e$. Hence, with constant probability, we pick the secretary that maximizes our local profit in each phase. It leaves us to prove that this local optimization leads to a reasonable global guarantee.

The previous algorithm fails in the non-monotone case. Observe that the first **if** statement is never true for a monotone function, however, for a non-monotone function this guarantees the values of sets T_i are non-decreasing. Algorithm 2.1 first divides the input stream into two equal-sized parts: U_1 and U_2. Then, with probability $1/2$, it calls Algorithm 2.1 on U_1, whereas with the same probability, it skips over the first half of the input, and runs Algorithm 2.1 on U_2.

2.2 Analysis

In this section, we prove Theorem 1. Since the algorithm for the non-monotone submodular secretary problem uses that for the monotone submodular secretary problem, first we start with the monotone case.

Monotone Submodular. We prove in this section that for Algorithm 2.1, the expected value of $f(T_k)$ is within a constant factor of the optimal solution. Let $R = \{a_{i_1}, a_{i_2}, \cdots, a_{i_k}\}$ be the opti-

Algorithm 2.1. Monotone Submodular Secretary Algorithm

Input: A monotone submodular function $f : 2^S \mapsto \mathbb{R}$, and a randomly permuted stream of secretaries, denoted by $(a_1, a_2, \ldots, a_n)$, where n is an integer multiple of k.

Output: A subset of at most k secretaries.

Let $T_0 \leftarrow \emptyset$
Let $l \leftarrow n/k$
for $i \leftarrow 1$ **to** k **do** {phase i}
  Let $u_i \leftarrow (i-1)l + l/e$
  Let $\alpha_i \leftarrow \max_{(i-1)l \le j < u_i} f(T_{i-1} \cup \{a_j\})$
  if $\alpha_i < f(T_{i-1})$ **then**
    $\alpha_i \leftarrow f(T_{i-1})$
  end if
  Pick an index $p_i : u_i \le p_i < il$ such that $f(T_{i-1} \cup \{a_{p_i}\}) \ge \alpha_i$
  if such an index p_i exists **then**
    Let $T_i \leftarrow T_{i-1} \cup \{a_{p_i}\}$
  else
    Let $T_i \leftarrow T_{i-1}$
  end if
end for
Output T_k as the solution

mal solution. Note that the set $\{i_1, i_2, \cdots, i_k\}$ is a uniformly random subset of $\{1, 2, \cdots, n\}$ with size k. It is also important to note that the permutation of the elements of the optimal solution on these k places is also uniformly random, and is independent from the set $\{i_1, i_2, \cdots, i_k\}$. For example, any of the k elements of the optimum can appear as a_{i_1}. These are two key facts used in the analysis.

Before starting the analysis, we present a simple property of submodular functions which will prove useful in the analysis. The proof of the lemma is standard, and is included in the appendix for the sake of completeness.

Lemma 1. *If* $f : 2^S \mapsto \mathbb{R}$ *is a submodular function, we have* $f(B) - f(A) \leq \sum_{a \in B \setminus A} [f(A \cup \{a\}) - f(A)]$ *for any* $A \subseteq B \subseteq S$.

Algorithm 2.2. Submodular Secretary Algorithm

Input: A (possibly non-monotone) submodular function $f : 2^S \mapsto \mathbb{R}$, and a randomly permuted stream of secretaries, denoted by $(a_1, a_2, \ldots, a_n)$, where n is an integer multiple of $2k$.

Output: A subset of at most k secretaries.

Let $U_1 := \{a_1, a_2, \ldots, a_{n/2}\}$
Let $U_2 := \{a_{n/2} + 1, \ldots, a_{n-1}, a_n\}$
Let $0 \leq X \leq 1$ be a uniformly random value.
if $X \leq 1/2$ **then**
    Run Algorithm 2.1 on U_1 to get S_1
    Output S_1 as the solution
else
    Run Algorithm 2.1 on U_2 to get S_2
    Output S_2 as the solution
end if

Define $\mathcal{X} := \{S_i : |S_i \cap R| \neq \emptyset\}$. For each $S_i \in \mathcal{X}$, we pick one element, say s_i, of $S_i \cap R$ randomly. These selected items form a set called $R' = \{s_1, s_2, \cdots, s_{|\mathcal{X}|}\} \subseteq R$ of size $|\mathcal{X}|$. Since our algorithm approximates such a set, we study the value of such random samples of R in the following lemmas. We first show that restricting ourselves to picking at most one element from each segment does not prevent us from picking many elements from the optimal solution (i.e., R).

Lemma 2. *The expected value of the number of items in* R' *is at least* $k(1 - 1/e)$.

The next lemma materializes the proof of an intuitive statement: if you randomly sample elements of the set R, you expect to obtain a profit proportional to the size of your sample. An analog of this is proved in [15] for the case when $|R|/|A|$ is an integer.

Lemma 3. *For a random subset* A *of* R, *the expected value of* $f(A)$ *is at least* $\frac{|A|}{k} \cdot f(R)$.

Here comes the crux of our analysis where we prove that the local optimization steps (i.e., trying to make the best move in each segment) indeed lead to a globally approximate solution.

Lemma 4. *The expected value of* $f(T_k)$ *is at least* $\frac{|R'|}{7k} \cdot f(R)$.

The following theorem wraps up the analysis of the algorithm.

Theorem 5. *The expected value of the output of our algorithm is at least* $\frac{1-1/e}{7} f(R)$.

Non-monotone Submodular. Before starting the analysis of Algorithm 2.1 for non-monotone functions, we show an interesting property of Algorithm 2.1. Consistently with the notation of Section 2.2, we use R to refer to some optimal solution. Recall that we partition the input stream into (almost) equal-sized segments $S_i : 1 \leq i \leq k$, and pick one item from each. Then T_i denotes the set of items we have picked at the completion of segment i. We show that $f(T_k) \geq \frac{1}{2e} f(R \cup T_i)$ for some integer i, even when f is not monotone. Roughly speaking, the proof mainly follows from the submodularity property and Lemma 1.

Lemma 5. *If we run the monotone algorithm on a (possibly non-monotone) submodular function f, we obtain $f(T_k) \geq \frac{1}{2e^2} f(R \cup T_i)$ for some i.*

Unlike the case of monotone functions, we cannot say that $f(R \cup T_i) \geq f(R)$, and conclude that our algorithm is constant-competitive. Instead, we need to use other techniques to cover the cases that $f(R \cup T_i) < f(R)$. The following lemma presents an upper bound on the value of the optimum.

Lemma 6. *For any pair of disjoint sets Z and Z', and a submodular function f, we have $f(R) \leq f(R \cup Z) + f(R \cup Z')$.*

We are now at a position to prove the performance guarantee of our main algorithm.

Theorem 6. *Algorithm 2.1 has competitive ratio $8e^2$.*

3   The Submodular Matroid Secretary Problem

In this section, we prove Theorem 2. We first design an $O(\log^2 r)$-competitive algorithm for maximizing a monotone submodular function, when there are matroid constraints for the set of selected items. Here we are allowed to choose a subset of items only if it is an independent set in the given matroid.

The matroid $(\mathcal{U}, \mathcal{I})$ is given by an oracle access to $\mathcal{I}$. Let n denote the number of items, i.e., $n := |\mathcal{U}|$, and r denotes the rank of the matroid. Let $S \in \mathcal{I}$ denote an optimal solution that maximizes the function f. We focus our analysis on a refined set $S^* \subseteq S$ that has certain nice properties: 1) $f(S^*) \geq (1-1/e)f(S)$, and 2) $f(T) \geq f(S^*)/\log r$ for any $T \subseteq S^*$ such that $|T| = \lfloor |S^*|/2 \rfloor$. We cannot necessarily find S^*, but we prove that such a set exists.

Start by letting $S^* = S$. As long as there is a set T violating the second property above, remove T from S^*, and continue. The second property clearly holds at the termination of the procedure. In order to prove the first property, consider one iteration. By submodularity (subadditivity to be more precise) we have $f(S^* \setminus T) \geq f(S^*) - f(T) \geq (1 - 1/\log r)f(S^*)$. Since each iteration halves the set S^*, there are at most $\log r$ iterations. Therefore, $f(S^*) \geq (1 - 1/\log r)^{\log r} \cdot f(S) \geq (1 - 1/e)f(S)$.

We analyze the algorithm assuming the parameter $|S^*|$ is given, and achieve a competitive ratio $O(\log r)$. If $|S^*|$ is unknown, though, we can guess its value (from a pool of $\log r$ different choices) and continue with Lemma 7. This gives an $O(\log^2 r)$ competitive ratio.

Lemma 7. *Given $|S^*|$, the above algorithm picks an independent subset of items with size $|S^*|/2$ whose expected value is at least $f(S^*)/4e \log r$.*

Finally, it is straightforward (and hence the details are omitted) to combine the algorithm in this section with Algorithm 2.1 for the non-monotone submodular secretary problem, to obtain an $O(\log^2 r)$-competitive algorithm for the non-monotone submodular secretary problem subject to a matroid constraint.

Here we show the same algorithm works when there are $l \geq 1$ matroid constraints and achieves a competitive ratio of $O(l \log^2 r)$. We just need to respect all matroid constraints in the algorithm. This finishes the proof of Theorem 2.

Lemma 8. *Given $|S^*|$, the above algorithm picks an independent subset of items (i.e., independent with respect to all matroids) with expected value at least $f(S^*)/4el \log r$.*

4   Knapsack Constraints

In this section, we prove Theorem 3. We first outline how to reduce an instance with multiple knapsacks to an instance with only one knapsack, and then we show how to solve the single knapsack instance.

Without loss of generality, we can assume that all knapsack capacities are equal to one. Let I be the given instance with the value function f, and item weights w_{ij} for $1 \leq i \leq l$ and $1 \leq j \leq n$. Define a new instance I' with one knapsack of capacity one in which the weight of the item j is $w'_j := \max_i w_{ij}$. We first prove that this reduction loses no more than a factor $4l$ in the total value. Take note that both the scaling and the weight transformation can be carried in an online manner as the items arrive. Hence, the results of this section hold for the online as well as the offline setting.

Lemma 9. *With instance I' defined above, we have $\frac{1}{4l} \text{OPT}(I) \leq \text{OPT}(I') \leq \text{OPT}(I)$.*

Here we show how to achieve a constant competitive ratio when there is only one knapsack constraint. Let w_j denote the weight of item $j : 1 \leq j \leq n$, and assume without loss of generality that the capacity of the knapsack is 1. Moreover, let f be the value function which is a non-monotone submodular function. Let T be the optimal solution, and define $\text{OPT} := f(T)$. The value of the parameter $\lambda \geq 1$ will be fixed below. Define T_1 and T_2 as the subsets of T that appears in the first and second half of the input stream, respectively. We first show the this solution is broken into two *balanced* portions.

Lemma 10. *If the value of each item is at most OPT/λ, for sufficiently large λ, the random variable $|f(T_1) - f(T_2)|$ is bounded by $\text{OPT}/2$ with a constant probability.*

The algorithm is as follows. Without loss of generality assume that all items are feasible, i.e., any one item fits into the knapsack. We flip a coin, and if it turns up "heads," we simply try to pick the one item with the maximum value. We do the following if the coin turns up "tails." We do not pick any items from the first half of the stream. Instead, we compute the maximum value set in the first half with respect to the knapsack constraint; Lee et al. give a constant factor approximation for this task. From the above argument, we know that $f(T_1)$ is at least $OPT/4$ since all the items have limited value in this case (i.e., at most OPT/λ). Therefore, we obtain a constant factor estimation of OPT by looking at the first half of the stream: i.e., if the estimate is $\hat{\text{OPT}}$, we get $\text{OPT}/c \leq \hat{\text{OPT}} \leq \text{OPT}$. After obtaining this estimate, we go over the second half of the input, and pick an item j if and only if it is feasible to pick this item, and moreover, the ratio of its marginal value to w_j is at least $\hat{\text{OPT}}/6$.

Lemma 11. *The above algorithm is a constant competitive algorithm for the non-monotone submodular secretary problem with one knapsack constraint.*

5   The Subadditive Secretary Problem

In this section, we prove Theorem 4 by presenting first a hardness result for approximation subadditive functions in general. The result applies in particular to our online setting. Surprisingly, the monotone subadditive function that we use here is *almost submodular*; see Proposition 1 below. Hence, our constant competitive ratio for submodular functions is nearly the most general we can achieve.

Definition (Subadditive function maximization). Given a nonnegative subadditive function f on a ground set U, and a positive integer $k \leq |U|$, the goal is to find a subset S of U of size at most k so as to maximize $f(S)$. The function f is accessible through a value oracle.

5.1   Hardness Result

In the following discussion, we assume that there is an upper bound of m on the size of sets given to the oracle. We believe this restriction can be lifted. If the function f is not required to be monotone, this is quite easy to have: simply let the value of the function f be zero for queries of size larger than m. Furthermore, depending on how we define the online setting, this may not be an *additional* restriction here. For example, we may not be able to query the oracle with secretaries that have already been rejected.

The main result of the section is the following theorem. It shows the subadditive function maximization is difficult to approximate, even in the offline setting.

Theorem 7. *There is no polynomial time algorithm to approximate an instance of subadditive function maximization within $\tilde{O}(\sqrt{n})$ of the optimum. Furthermore, no algorithm with exponential time 2^t can achieve an approximation ratio better than $\tilde{O}(\sqrt{n/t})$.*

First, we are going to define our *hard* function. Afterwards, we continue with proving certain properties of the function which finally lead to the proof of Theorem 7.

Let n denote the size of the universe, i.e., $n := |U|$. Pick a random subset $S^* \subseteq U$ by sampling each element of U with probability k/n. Thus, the expected size of S^* is k.

Define the function $g : U \mapsto \mathbb{N}$ as $g(S) := |S \cap S^*|$ for any $S \subseteq U$. One can easily verify that g is submodular. We have a positive r whose value will be fixed below. Define the final function $f : U \mapsto \mathbb{N}$ as

$$f(S) := \begin{cases} 1 & \text{if } g(S) = 0 \\ \lceil g(S)/r \rceil & \text{otherwise.} \end{cases}$$

It is not difficult to verify the subadditivity of f; it is also clearly monotone.

In order to prove the core of the hardness result in Lemma 12, we now let $r := \lambda \cdot \frac{mk}{n}$, where $\lambda \geq 1 + \sqrt{\frac{3tn}{mk}}$ and $t = \Omega(\log n)$ will be determined later.

Lemma 12. *An algorithm making at most 2^t queries to the value oracle cannot solve the subadditive maximization problem to within k/r approximation factor.*

Now we can prove the main theorem of the section.

Proof (Theorem 7). We just need to set $k = m = \sqrt{n}$. Then, $\lambda = \sqrt{3t}$, and the inapproximability ratio is $\Omega(\sqrt{\frac{n}{t}})$. Restricting to polynomial algorithms, we obtain $t := O(\log^{1+\varepsilon} n)$, and considering exponential algorithms with running time $O(2^{t'})$, we have $t = O(t')$, giving the desired results.   □

In case the query size is not bounded, we can define $f(X) := 0$ for large sets X, and pull through the same result; however, the function f is no longer monotone in this case.

We now show that the function f is almost submodular. Recall that a function g is submodular if and only if $g(A) + g(B) \geq g(A \cup B) + g(A \cap B)$.

Proposition 1. *For the hard function f defined above, $f(A) + f(B) \geq f(A \cup B) + f(A \cap B) - 2$ always holds; moreover, $f(X)$ is always positive and attains a maximum value of $\tilde{\Theta}(\sqrt{n})$ for the parameters fixed in the proof of Theorem 7.*

5.2 Algorithm

An algorithm that only picks the best item clearly gives a k competitive ratio. We now show how to achieve an $O(n/k)$ competitive ratio, and thus by combining the two, we obtain an $O(\sqrt{n})$-competitive algorithm for the monotone subadditive secretary problem. This result complements our negative result nicely.

Partition the input stream S into $\ell := n/k$ (almost) equal-sized segments, each of size at most k. Randomly pick all the elements in one of these segments. Let the segments be denoted by $S_1, S_2, \ldots, S_\ell$. Subadditivity of f implies $f(S) \leq \sum_i f(S_i)$. Hence, the expected value of our solution is $\sum_i \frac{1}{\ell} f(S_i) \geq \frac{1}{\ell} f(S) \geq \frac{1}{\ell} \text{OPT}$, where the two inequalities follow from subadditivity and monotonicity, respectively.

6 Conclusions and Further Results

In this paper, we consider the (non-monotone) submodular secretary problem for which we give a constant-competitive algorithm. The result can be generalized when we have a matroid constraint on the set that we pick; in this case we obtain an $O(\log^2 r)$-competitive algorithm where r is the rank of the matroid. However, we show that it is very hard to compete with the optimum if we consider subadditive functions instead of submodular functions. This hardness holds even for "almost submodular" functions; see Proposition 1.

Acknowledgments

The second author wishes to thank Bobby Kleinberg for useful discussions, and the anonymous reviewer for the simpler proof of Lemma 3.

References

1. Ageev, A.A., Sviridenko, M.I.: An 0.828-approximation algorithm for the uncapacitated facility location problem. Discrete Appl. Math. 93, 149–156 (1999)
2. Ajtai, M., Megiddo, N., Waarts, O.: Improved algorithms and analysis for secretary problems and generalizations. SIAM J. Discrete Math. 14, 1–27 (2001)
3. Asadpour, A., Nazerzadeh, H., Saberi, A.: Stochastic submodular maximization. In: Papadimitriou, C., Zhang, S. (eds.) WINE 2008. LNCS, vol. 5385, pp. 477–489. Springer, Heidelberg (2008)
4. Babaioff, M., Immorlica, N., Kempe, D., Kleinberg, R.: A knapsack secretary problem with applications. In: Charikar, M., Jansen, K., Reingold, O., Rolim, J.D.P. (eds.) RANDOM 2007 and APPROX 2007. LNCS, vol. 4627, pp. 16–28. Springer, Heidelberg (2007)
5. Babaioff, M., Immorlica, N., Kempe, D., Kleinberg, R.: Online auctions and generalized secretary problems. SIGecom Exch. 7, 1–11 (2008)
6. Babaioff, M., Immorlica, N., Kleinberg, R.: Matroids, secretary problems, and online mechanisms. In: SODA, pp. 434–443 (2007)
7. Bateni, M., Hajiaghayi, M., Zadimoghaddam, M.: The submodular secretary problem, Tech. Report TD-7UEP26, AT&T Labs–Research (July 2009)
8. Bateni, M., Hajiaghayi, M., Zadimoghaddam, M.: Submodular secretary problem and extensions, Tech. Report 2010-002, CSAIL, MIT (February 2010)
9. Calinescu, G., Chekuri, C., Pál, M., Vondrák, J.: Maximizing a submodular set function subject to a matroid constraint (extended abstract). In: Fischetti, M., Williamson, D.P. (eds.) IPCO 2007. LNCS, vol. 4513, pp. 182–196. Springer, Heidelberg (2007)
10. Cornuejols, G., Fisher, M., Nemhauser, G.L.: On the uncapacitated location problem. In: Studies in Integer Programming (Proc. Workshop, Bonn. 1975). Ann. of Discrete Math., vol. 1, pp. 163–177. North-Holland, Amsterdam (1977)
11. Cornuejols, G., Fisher, M.L., Nemhauser, G.L.: Location of bank accounts to optimize float: an analytic study of exact and approximate algorithms. Manage. Sci. 23, 789–810 (1976/1977)
12. Dynkin, E.B.: The optimum choice of the instant for stopping a markov process. Sov. Math. Dokl. 4, 627–629 (1963)
13. Edmonds, J.: Submodular functions, matroids, and certain polyhedra. In: Combinatorial Structures and their Applications (Proc. Calgary Internat. Conf., Calgary, Alta., 1969), pp. 69–87. Gordon and Breach, New York (1970)
14. Feige, U.: A threshold of ln n for approximating set cover. J. ACM 45, 634–652 (1998)
15. Feige, U.: On maximizing welfare when utility functions are subadditive. In: STOC, pp. 41–50 (2006)
16. Feige, U., Goemans, M.X.: Approximating the value of two power proof systems, with applications to MAX 2SAT and MAX DICUT. In: ISTCS, p. 182 (1995)
17. Feige, U., Mirrokni, V.S., Vondrák, J.: Maximizing non-monotone submodular functions. In: FOCS, pp. 461–471 (2007)
18. Fisher, M.L., Nemhauser, G.L., Wolsey, L.A.: An analysis of approximations for maximizing submodular set functions. II. Math. Prog. Stud., 73–87 (1978), Polyhedral combinatorics

19. Freeman, P.R.: The secretary problem and its extensions: a review. Internat. Statist. Rev. 51, 189–206 (1983)
20. Gilbert, J.P., Mosteller, F.: Recognizing the maximum of a sequence. J. Amer. Statist. Assoc. 61, 35–73 (1966)
21. Glasser, K.S., Holzsager, R., Barron, A.: The d choice secretary problem. Comm. Statist. C—Sequential Anal. 2, 177–199 (1983)
22. Goemans, M.X., Williamson, D.P.: Improved approximation algorithms for maximum cut and satisfiability problems using semidefinite programming. J. Assoc. Comput. Mach. 42, 1115–1145 (1995)
23. Gupta, A., Roth, A., Schoenebeck, G., Talwar, K.: Constrained non-monotone submodular maximization: offline and secretary algorithms (2010), http://www.cs.cmu.edu/alroth/submodularsecretaries.html
24. Hajiaghayi, M.T., Kleinberg, R., Parkes, D.C.: Adaptive limited-supply online auctions. In: EC, pp. 71–80 (2004)
25. Hajiaghayi, M.T., Kleinberg, R., Sandholm, T.: Automated online mechanism design and prophet inequalities. In: AAAI, pp. 58–65 (2007)
26. Halperin, E., Zwick, U.: Combinatorial approximation algorithms for the maximum directed cut problem. In: SODA, pp. 1–7 (2001)
27. Håstad, J.: Some optimal inapproximability results. J. ACM 48, 798–859 (2001)
28. Hazan, E., Safra, S., Schwartz, O.: On the complexity of approximating k-set packing. Computational Complexity 15, 20–39 (2006)
29. Immorlica, N., Kleinberg, R.D., Mahdian, M.: Secretary problems with competing employers. In: Spirakis, P.G., Mavronicolas, M., Kontogiannis, S.C. (eds.) WINE 2006. LNCS, vol. 4286, pp. 389–400. Springer, Heidelberg (2006)
30. Iwata, S., Fleischer, L., Fujishige, S.: A combinatorial strongly polynomial algorithm for minimizing submodular functions. J. ACM 48, 761–777 (2001)
31. Khot, S., Kindler, G., Mossel, E., O'Donnell, R.: Optimal inapproximability results for max-cut and other 2-variable csps? In: FOCS, pp. 146–154 (2004)
32. Khuller, S., Moss, A., Naor, J.: The budgeted maximum coverage problem. Inf. Process. Lett. 70, 39–45 (1999)
33. Kleinberg, R.: A multiple-choice secretary algorithm with applications to online auctions. In: SODA, pp. 630–631 (2005)
34. Lee, J., Mirrokni, V., Nagarajan, V., Sviridenko, M.: Maximizing non-monotone submodular functions under matroid and knapsack constraints. In: STOC, pp. 323–332 (2009)
35. Lovász, L.: Submodular functions and convexity. In: Mathematical programming: the state of the art (Bonn, 1982), pp. 235–257. Springer, Berlin (1982)
36. Nemhauser, G.L., Wolsey, L.A., Fisher, M.L.: An analysis of approximations for maximizing submodular set functions. I. Math. Program. 14, 265–294 (1978)
37. Queyranne, M.: A combinatorial algorithm for minimizing symmetric submodular functions. In: SODA, pp. 98–101 (1995)
38. Schrijver, A.: A combinatorial algorithm minimizing submodular functions in strongly polynomial time. J. Combin. Theory Ser. B 80, 346–355 (2000)
39. Sviridenko, M.: A note on maximizing a submodular set function subject to a knapsack constraint. Oper. Res. Lett. 32, 41–43 (2004)
40. Vanderbei, R.J.: The optimal choice of a subset of a population. Math. Oper. Res. 5, 481–486 (1980)
41. Vondrák, J.: Symmetry and approximability of submodular maximization problems. In: FOCS (2009)
42. Wilson, J.G.: Optimal choice and assignment of the best m of n randomly arriving items. Stochastic Process. Appl. 39, 325–343 (1991)

Approximation Algorithms
for Min-Max Generalization Problems

Piotr Berman and Sofya Raskhodnikova*

Pennsylvania State University
{berman,sofya}@cse.psu.edu

Abstract. We provide improved approximation algorithms for the *min-max generalization problems* considered by Du, Eppstein, Goodrich, and Lueker [1]. In min-max generalization problems, the input consists of data items with weights and a lower bound w_{lb}, and the goal is to partition individual items into groups of weight at least w_{lb}, while minimizing the maximum weight of a group. The rules of legal partitioning are specific to a problem. Du *et al.* consider several problems in this vein: (1) partitioning a graph into connected subgraphs, (2) partitioning unstructured data into arbitrary classes and (3) partitioning a 2-dimensional array into non-overlapping contiguous rectangles (subarrays) that satisfy the above size requirements.

We significantly improve approximation ratios for all the problems considered by Du *et al.*, and provide additional motivation for these problems. Moreover, for the first problem, while Du *et al.* give approximation algorithms for specific graph families, namely, 3-connected and 4-connected planar graphs, no approximation algorithm that works for all graphs was known prior to this work.

1 Introduction

We provide improved approximation algorithms for the *min-max generalization problems* considered by Du, Eppstein, Goodrich, and Lueker [1]. In min-max generalization problems, the input consists of data items with weights and a lower bound w_{lb}, and the goal is to partition individual items into groups of weight at least w_{lb}, while minimizing the maximum weight of a group. The rules of legal partitioning are specific to a problem. Du *et al.* consider several problems in this vein: (1) partitioning a graph into connected subgraphs, (2) partitioning unstructured data into arbitrary classes and (3) partitioning a 2-dimensional array into non-overlapping contiguous rectangles (subarrays) that satisfy the above size requirements. We call these problems (1) Min-Max Graph Partition, (2) Min-Max Bin Covering and (3) Min-Max Rectangle Tiling.

Du *et al.* motivate the min-max generalization problems by applications to privacy-preserving data mining. Generalization is widely used in the data mining

---

* S.R. was supported by National Science Foundation (NSF/CCF award 0729171 and NSF/CCF CAREER award 0845701).

M. Serna et al. (Eds.): APPROX and RANDOM 2010, LNCS 6302, pp. 53–66, 2010.

community as means for achieving k-anonymity (see [2] for a survey). Generalization involves replacing a value with a less specific value. To achieve k-anonymity each record should be generalized to the same value as at least $k - 1$ other records. For example, if the records contain geographic information (e.g., GPS coordinates), and the plane is partitioned into axis-parallel rectangles each containing locations of at least k records, to achieve k-anonymity, the coordinates of each record can be replaced with the corresponding rectangle. Generalization can also be viewed as a natural way of compressing a dataset.

We briefly discuss several other applications of generalization. Geographic Information Systems contain very large data sets that are organized either according to the (almost) planar graph of the road network, or according to geographic coordinates (see, e.g., [3]). These sets have to be partitioned into *pages* that can be transmitted to a mobile device or retrieved from secondary storage. Because of the high overhead of a single transmission/retrieval operation, we want to assure a minimum size of a single part (page), while controlling the maximum size. When the process that is exploring a graph needs to investigate a node whose information it has not retrieved yet, it has to request a new page. Therefore, pages are more useful if they contain information about connected subgraphs. Min-Max Graph Partition captures the problem of distributing information about the graph among pages.

Min-Max Bin Covering is a variant of the classical Bin Covering problem. In the classical version, the input is a set of items with positive weights and the goal is to pack items into bins, so that the number of bins that receive items of total weight at least 1 is maximized (see [4,5,6] and references therein). Both variants are natural. For example, when Grandfather Frost[1] partitions presents into bundles for kids, he clearly wants to ensure that each bundle has items of at least a certain value to make kids happy. Grandfather Frost could try to minimize the value of the maximum bundle, to avoid jealousy (Min-Max Bin Covering), or to maximize the number of kids who get presents (classical Bin Covering). Min-Max Bin Covering can also be viewed as a variant of scheduling on parallel identical machines where, given n jobs and their processing times, the goal is to schedule them on m identical parallel machines while minimizing *makespan*, that is, the maximum time used by any machine [8]. In our variant, the number of machines is not given in advance, but instead, there is a lower bound on the processing time. This requirement is natural, e.g., when "machines" represent workers that must be hired for at least a certain number of hours.

Rectangle tiling problems with various optimization criteria arise in applications ranging from databases and data mining to video compression and manufacturing, and have been extensively studied [9,10,11,12,13,14,15,16]. The min-max version can be used to design a Geographic Information System, described above. If the data is a set of coordinates specifying object positions, as opposed to a road network, we would like to partition it into pages that correspond to

---

[1] Grandfather Frost is a secular character that played the role of Santa Claus for Soviet children. The Santa Claus problem [7] is not directly related to our problem.

rectangles on the plane. As before, we would like to ensure that pages have at least the minimum size while controlling the maximum size.

1.1   Problems

In each of the problems we consider, the input is an item set $\mathcal{I}$, non-negative weights w_i for all $i \in \mathcal{I}$ and a non-negative bound w_{lb}. For $\mathcal{I}' \subseteq \mathcal{I}$, we use $w(\mathcal{I}')$ to denote $\sum_{i \in \mathcal{I}'} w_i$. Each problem below specifies a class of *allowed* subsets of $\mathcal{I}$. A valid solution is a partition P of $\mathcal{I}$ into allowed subsets such that $w(\mathcal{I}') \geq w_{\mathrm{lb}}$ for each $\mathcal{I}' \in P$. The goal is to minimize the cost of P, defined as $\max_{\mathcal{I}' \in P} w(\mathcal{I}')$.

In *Min-Max Graph Partition*, $\mathcal{I}$ is the vertex set V of an (undirected) graph (V, E), and a subset of V is allowed if it induces a connected subgraph. In *Min-Max Bin Covering*, every subset of $\mathcal{I}$ is allowed. A partition of $\mathcal{I}$ is called a *packing*, and the parts of a partition are called *bins*. In *Min-Max Rectangle Tiling*, $\mathcal{I} = \{1, \ldots, m\} \times \{1, \ldots, n\}$, and the allowed sets are rectangles, i.e., sets of the form $\{a, \ldots, b\} \times \{c, \ldots, d\}$. A partition of $\mathcal{I}$ is called a *tiling*, and the parts of a partition are called *tiles*.

All three min-max problems above are NP-complete. Moreover, if P$\neq$NP no polynomial time algorithm can achieve an approximation ratio better than 2 for Bin Covering (and hence for Graph Partition) or better than 1.33 for Graph Partition on 3-connected planar graphs and Rectangle Tiling [1].

1.2   Our Results and Techniques

Our main technical contribution is a 3-approximation algorithm for Min-Max Graph Partition. The remaining algorithms are very simple, even though the analysis is non-trivial.

Min-Max Graph Partition. We present the first polynomial time approximation algorithm for Min-Max Graph Partition. Du *et al.* gave approximation algorithms for specific graph families, *i.e.*, a 4-approximation for 3-connected and a 3-approximation for 4-connected planar graphs. We give a 3-approximation algorithm for the general case, simultaneously improving the approximation ratio

Table 1. Approximation Ratios for Min-Max Generalization Problems. (Note: Graph Partition generalizes Bin Covering, and hence inherits its inapproximability.)

Min-Max Problem	Hardness [1]	Ratio in [1]	Our ratio
Graph Partition	2	—	3
on 3-connected planar graphs	1.33	4	
on 4-connected planar graphs	—	3	2.5
Bin Covering	2	$2 + \varepsilon$ in time exp in ε^{-1}	2
Rectangle Tiling	1.33	5	4
with 0-1 entries	—	—	3

and applicability of the algorithm. We also improve the approximation ratio for 4-connected planar graphs from 3 to 2.5.

Our 3-approximation algorithm for Min-Max Graph Partition constructs a 2-tier partition where nodes are partitioned into *groups*, and groups are partitioned into *supergroups*. Intuitively, supergroups represent parts in a legal partition, while groups represent (nearly) indivisible subparts. The initial 2-tier partition is obtained greedily and then transformed using 4 carefully designed transformations until all supergroups of large weight have well-defined central nodes, and almost all non-central nodes in those supergroups are only connected to central nodes (possibly of multiple supergroups). Supergroups of small weight are used as parts in the final solution. The remaining supergroups are more tricky to deal with. We create one part in the final solution for each supergroup or, more precisely, for each group with a central node. We redistribute other groups among supergroups using a scheduling algorithm of Lenstra, Shmoys and Tardos [17], while ! leaving all central nodes in separate parts. Roughly, central nodes play a role of the machines and the groups that we need to redistribute play a role of jobs to be scheduled on these machines. The final part of the algorithm repairs parts of insufficient weight to obtain the final partition.

Our use of the scheduling algorithm of Lenstra *et al.* is *gray-box* in the following sense: our algorithm runs the scheduling algorithm in a black-box manner. However, in the analysis, we look inside the black box. Namely, we apply the Rounding Theorem of Lenstra *et al.* to show that the LP used by their algorithm yields a good solution for our problem.

For partitioning 4-connected planar graphs, following Du *et al.*, we use the fact that such graphs have Hamiltonian cycles [18] which can be found in linear time [19]. Our algorithm is simple and efficient: It goes around the Hamiltonian cycle and greedily partitions the nodes, starting from the lightest contiguous part of the cycle that satisfies the weight lower bound. If the last part is too light, it is combined with the first part. Thus, the algorithm runs in linear time. Our algorithm and analysis apply to any graph that contains a Hamiltonian cycle which can be computed efficiently or is given as part of the input.

Min-Max Bin Covering. We present a simple 2-approximation algorithm that runs in linear time. Du *et al.* gave a schema with approximation ratio $2 + \varepsilon$, and time complexity exponential in ε^{-1}. They also showed that approximation ratio better than 2 cannot be achieved in polynomial time unless P=NP. Thus, we completely resolve the approximability of this problem.

Our algorithm greedily places items in the bins in the order of decreasing weights, and then redistributes items in the first and the last three bins.

Min-Max Rectangle Tiling. We improve the approximation ratio for this problem from 5 to 4. We can get a better ratio of 3 when the entries in the matrix are restricted to be 0 or 1. This case covers the scenarios where each entry indicates the presence or absence of some object, as in applications with geographic data, such as GPS coordinate data originally considered by Du *et al.*

Our algorithm builds on the *slicing and dicing* method introduced by Berman *et al.* [15]. The idea is to first partition the rectangle horizontally into *slices*, and then partition slices vertically. The straightforward application of *slicing and dicing* gives ratio 5. We improve it by doing simple preprocessing. For the case of 0-1 entries, the preprocessing step is more sophisticated.

Summary and Organization. We summarize our results in Table. 1. The results on Graph Partition are stated in Theorems 2.1 and 2.2 in Sect. 2, on Bin Covering, in Theorem 3.1 in Sect. 3, and on Rectangle Tiling, in Theorems 4 and 4.2 in Sect. 4. All omitted proofs are deferred to the full version.

Terminology and Notation. Here we describe terminology and notation common to all technical sections. We use *opt* as the cost of an optimal solution.

Definition 1.1. *An item (or a set of items) is* fat *if it has weight at least* w_{lb}, *and* lean *otherwise. We apply this terminology to nodes and sets of nodes in an instance of Graph Partition, and to elements and rectangles in Rectangle Tiling.*

A solution is *legal* if it obeys the *minimum weight constraint*, i.e., all parts are fat.

2   Min-Max Graph Partition

We present two approximation algorithms for Min-Max Graph Partition whose performance is summarized in Theorems 2.1 and 2.2.

Theorem 2.1. *Min-Max Graph Partition can be approximated with ratio 3 in polynomial time.*

Theorem 2.2. *Min-Max Graph Partition on 4-connected planar graphs can be approximated with ratio 2.5 in linear time.* (The proof is omitted.)

The rest of this section is devoted to the proof of Theorem 2.1.

Recall that an input to Min-Max Graph Partition is a graph (V, E) with node weights $w : V \to \mathbb{R}^+$ and a weight lower bound w_{lb}. W.l.o.g. assume that $w_{lb} = 1$. (All weights can be divided by w_{lb} to obtain an equivalent instance with $w_{lb} = 1$.) For now, we will also assume that all nodes in the graph are *lean*. (Recall Definition 1.1 of *fat* and *lean*.) We remove this assumption in Sect. 2.4.

As described in Sect. 1.2, our algorithm first constructs a 2-tier partition into groups and supergroups (Sect. 2.1), then transforms it until all supergroups of large weight have well-defined central nodes, and nearly all non-central nodes in those supergroups are only connected to central nodes (Sect. 2.2) and finally solves an instance of Scheduling on Unrelated Parallel Machines (Sect. 2.5), interprets it as a partition and adjusts it to get the final solution.

2.1   A Preliminary 2-Tier Partition

We start by defining a 2-tier partition. (See illustration in Fig. 1.)

Definition 2.1 (2-tier partition). *A 2-tier partition of a graph (V, E, w) containing only lean nodes is a partition of V into lean sets, called* groups, *together with a partition of the groups into fat sets, called* supergroups. *The set of nodes in a group, or in a supergroup, should induce a connected graph. The set of groups contained in a supergroup S is denoted by $\mathcal{G}(S)$.*

Since groups are lean and supergroups are fat, each supergroup contains at least two groups. We assign names to some types of groups and supergroups.

Definition 2.2. Group-pair, triangle, star supergroups; central group

- *A supergroup is a* group-pair *if it consists of two groups.*
- *A supergroup is a* triangle *if it consists of three groups, pairwise connected by an edge.*
- *A supergroup S with 3 or more groups is a* star *if it forms a star graph on groups, i.e., it contains a group G, called* central, *such that groups in $\mathcal{G}(S) - \{G\}$ form connected components of $S - G$.*

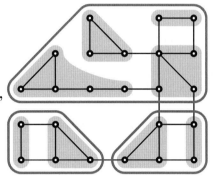

Fig. 1. An example of a 2-tier partition. Shaded background indicates groups; curved lines indicate supergroups. The top supergroup has a 4-node central group and 3 mobile groups. The two bottom supergroups are group-pairs.

Lemma 2.1 (Initial partition). *Given a connected graph on lean nodes, in polynomial time we can compute a 2-tier partition where* (a) *each supergroup is a group-pair, a triangle or a star and* (b) *$w(G) + w(H) \geq 1$ for all adjacent groups G and H.*

Proof. First, form the groups greedily: Make each node a group. While there are two groups G, H such that $G \cup H$ is lean and connected, merge G and H.

Second, form group-pairs greedily: While there are two adjacent groups G, H that are not included in a supergroup, form a supergroup $G \cup H$.

Next, insert remaining groups into supergroups: For each group G still not included in a supergroup, pick an adjacent group H. Since the second step halted, H is in some group-pair created in that step. Insert G into H's supergroup.

Finally, break down supergroups that are not stars: Consider a group-pair P created in the second step from groups G and H, and let S be the supergroup that was formed from P. Suppose S has 4 or more groups, but is not a star. Since groups in $S - P$ are not connected, and neither G nor H can become the center of S, there are two different groups G' and H' in S that are adjacent to G and H, respectively. Let S_1 be the the union of G, G' and all other groups in S that are not adjacent to H. Replace S with S_1 and $S - S_1$. In the resulting 2-tier partition, all supergroups with 4 or more groups are stars, so item (a) of the lemma holds. Item (b) is guaranteed by the first step of the construction. □

2.2   Improving the Initial 2-Tier Partition

In this section, we modify the initial 2-tier partition, while maintaining property (a) and a weaker version of property (b) of Lemma 2.1. As we are working on our 2-tier partition, we will rearrange groups and supergroups. A group G is called *mobile* if it can be removed from its supergroup S while keeping property (a) of Lemma 2.1. Namely, the modified S has to be a group-pair or a star.

Definition 2.3 (Mobile group). *A group is* mobile *if it is not in a group-pair and it is not a central group.*

The goal of this phase of the algorithm is to separate supergroups into two *types*: (i) the ones that will be repartitioned by the scheduling algorithm and (ii) the ones that will be used in the final partition as they are. Supergroups of *type* (i) will be well structured: in such a supergroup, the central group will have a unique *central* node, and mobile groups will be connected only to central nodes (possibly in multiple central groups). Supergroups of *type* (ii) will have at most 3 groups, and thus weight at most 3— sufficiently light to form parts in a 3-approximate solution. Central groups of supergroups of *type* (i) will be allocated their own parts in the final partition. Mobile groups will be distributed among these parts by the scheduling algorithm. To guarantee that the optimal distribution of central nodes and mobile groups into parts provides a sufficiently good solution, we require that mobile groups are connected only to central nodes of the supergroups of *type* (i). (Non-central nodes of central groups will join the parts of their central nodes after the scheduling algorithm produces a 2-approximate solution. Since, by definition, each group is lean, even after adding central groups, we will still be able to guarantee a 3-approximation.).

We explain this phase of the algorithm by specifying several transformations of a 2-tier partition (see Figs. 2 and 3). The algorithm applies these transformations to the initial 2-tier partition from Lemma 2.1. Each transformation is defined by the *trigger* and the *action*. The algorithm performs the action for the first transformation for which the trigger condition is satisfied for some group(s) in the current 2-tier partition. This phase terminates when no transformation can be applied.

The purpose of the first transformation, **CombG**, is to ensure that $w(G) + w(H) \geq 1$ for all adjacent groups G and H, where one of the groups is mobile. Even though an even stronger condition, property (b) of Lemma 2.1, holds for the initial 2-tier partition, it might be violated by other transformations. The second transformation, **ConP**, is getting rid of edges between mobile group. The third transformation, **SplitC**, is ensuring that each central group has a unique central node to which mobile groups connect. To accomplish this, while there is a central group G that violates this condition, **SplitC** splits G into two parts, each containing a node to which mobile groups connect. Later, it rearranges resulting groups and supergroups to ensure that all previously achieved properties of our 2-tier partition are preserved (in some cases, relying on **CombG** and **ConP** to reinstate these properties).

• **CombG = Combine groups.**

Trigger: Groups G and H are connected by an edge, $G \cup H$ is lean and H is mobile.

Action: Remove H from its supergroup and merge the two groups.

• **ConP = Connect group-pairs.**

Trigger: Two mobile groups are connected with an edge, and they belong either to two different supergroups or to a supergroup with more than three groups.

Action: Remove them from their supergroup(s) and combine them into a group-pair.

• **SplitC = Split the center.**

Trigger: G is the central group of a supergroup S, u and v are two different nodes in G, and two mobile groups H_u, H_v (not necessarily from $\mathcal{G}(S)$) have edges to u and v, respectively.

Action: Split G into two connected sets, G_u and G_v, containing u and v, respectively. Split S into S_u and S_v, by attaching each non-central group to G_u or G_v. If $H_u \in \mathcal{G}(S)$ attach H_u to G_u. Similarly, if $H_v \in \mathcal{G}(S)$ attach H_v to G_v.

[LeanLean case]: If both S_u and S_v are lean, we make them groups, and S becomes a group-pair.

[FatFat case]: If both S_u and S_v are fat, they become new supergroups.

Now assume that S_u is fat and S_v is lean.

[FatLean-IN case]: If $H_v \in \mathcal{G}(S)$ then change the partition of S by replacing G and H_v with G_u and S_v. If S is not a star, but has 4 or more groups, apply **CombG** or **ConP**.

[FatLean-OUT case]: If $H_v \notin \mathcal{G}(S)$ then remove S_v from G and S and treat it like a mobile group in contact with H_v, which triggers **CombG** or **ConP**.

• **ChainR = Chain Reconnect.**

Trigger: An unstructured supergroup S has 4 or more groups.

Action: Since **CombG**, **ConP** cannot be applied, we have a chain of supergroups $S = S_1, \ldots, S_k$ where S_k is a group-pair, a mobile group of S_{k-1} is adjacent to S_k and for $i = 1, \ldots, k - 2$ a mobile group of S_i is adjacent to the central group of S_{i+1}. Then for $i = 1, \ldots, k - 1$, move a mobile group from S_i to S_{i+1}.

Fig. 2. Transformations. (Perform the first one that applies.)

If the previously described transformations cannot be applied, star supergroups in the current 2-tier partition are well structured: they have unique central nodes, and all mobile groups connect only to these central nodes, with one

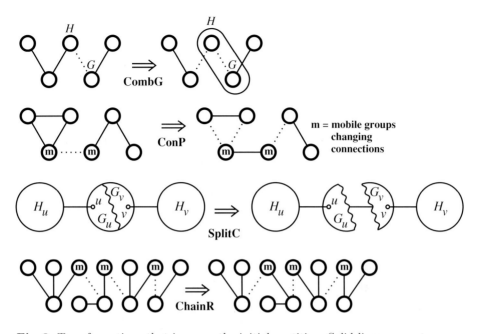

Fig. 3. Transformations that improve the initial partition. Solid lines connect groups of a supergroup, circles indicate groups, unless they are within ovals—then ovals are groups and circles are fragments of groups. **SplitC** transformation has four cases: all split a central group G into two parts and combine them with groups H_u and H_v to form new groups or supergroups (depending on the weight of the resulting pieces).

exception—they could still connect to group-pairs. Group-pairs are not guaranteed to have any structure. But they are light enough to be used as parts in the final partition. The same applies to triangles and stars with 3 groups. All group-pairs and triangles will be used as parts in the final partition, and thus will be of *type* (ii), according to the description after Definition 2.3. Stars with 3 groups could be of *type* (i) or (ii). As we already explained, it is important for the success of the next (scheduling) phase of the algorithm that mobile groups of supergroups of *type* (i) are adjacent only to the central nodes of supergroups of *type* (i). Next, we define *structured* and *unstructure! d* supergroups. After this phase completes, *structured* supergroups of the resulting 2-tier partition will be assigned *type* (i) and *unstructured* supergroups will be assigned *type* (ii). We call all group-pairs *unstructured*. Each star whose mobile group is adjacent to an unstructured supergroup is not ready to become a group of *type* (i) and is also called *unstructured*.

Definition 2.4 (Structured and unstructured supergroups). *An* unstructured *supergroup is either a group-pair or (recursively) a star that has a mobile group adjacent to an unstructured supergroup. A* structured *supergroup is a star that is not* unstructured.

The purpose of **ChainR** is to ensure that each remaining unstructured super-group has at most 3 groups. **ChainR** is triggered if there is an unstructured supergroup S with 4 or more groups. This can happen only if S is connected by a chain of unstructured supergroups to a group-pair. The mobile nodes along this chain are reconnected, as explained in Fig. 2 and illustrated in Fig. 3. This completes the description of transformations and this phase of the algorithm.

2.3   Analysis of Transformations

We analyze the properties of a 2-tier partition to which our transformations cannot be applied in Lemma 2.2 and bound the running time of this stage of the algorithm in Lemma 2.3. (The proofs of these lemmas are omitted.)

Lemma 2.2. *When transformations* **CombG**, **ConP**, **SplitC** *and* **ChainR** *cannot be applied, the resulting 2-tier partition satisfies the following:*

a. *If G is a center group and H is a mobile group of the same supergroup then $w(G) + w(H) \geq 1$.*
b. *No edges exist between mobile groups except for groups in the same triangle.*
c. *Each supergroup S with a central group G also has a central node $c(S)$ such that all edges between G and mobile groups include node $c(S)$.*
d. *Each supergroup with 4 or more groups is structured.*

Lemma 2.3. *An algorithm performing transformations defined in Fig. 2 on a 2-tier partition until none of them are applicable runs in polynomial time.*

2.4   A 2-Tier Partition on Graphs with Arbitrary Weights

In this section we remove the assumption that all nodes in our input graph are lean. To obtain a 2-tier partition of a graph with arbitrary node weights, first allocate a separate supergroup for each fat node. Let V_{lean} be the set of lean nodes. Form *isolated groups* from lean connected components of V_{lean}. For fat connected components of V_{lean}, compute the 2-tier partition using the method from Sections 2.1 and 2.2.

The next lemma summarizes the main outcome of improving the 2-tier partition using transformations in Fig. 2. It follows directly from Lemma 2.2.

Lemma 2.4 (Main). *Consider a 2-tier partition of a graph $G = (V, E, w)$ obtained by our method. Let C be the set consisting of fat nodes and central nodes of structured supergroups in that 2-tier partition. Then mobile groups of structured supergroups are connected components of $V - C$.*

Proof. By definition, each group is connected. It remains to show that a node in a mobile group cannot be adjacent to nodes of $V - C$ which are in different groups. Recall that all groups are either central, mobile or in a group-pair. A node in a mobile group cannot be adjacent to a node in a different mobile group by Lemma 2.2(b). It cannot be adjacent to a non-central node in a central group by Lemma 2.2(c). Finally, it cannot be adjacent to a node in a group-pair by Definition 2.4 and Lemma 2.2(d). □

2.5   Reduction to Scheduling and the Final Partition

We reduce Min-Max Graph Partition to Scheduling Unrelated Parallel Machines (SUPM), and use a 2-approximation algorithm of Lenstra *et al.* for SUPM to get a 3-approximation for graph partition.

The number of parts in the final partition will be equal to the number of super-groups in the 2-tier partition of Sect. 2.4. We use all unstructured supergroups and triangles as parts in the final partition. By Lemma 2.2(d), the weight of these supergroups is below 3. We use central groups of structured supergroups and fat nodes as *seeds* of the remaining parts, that is, in the final partition, we create a part for each central group and each fat node, and partition the remaining groups among these parts using a reduction to SUMP.

Now we explain our reduction. In SUPM, the input is m parallel machines, n jobs and processing times p_{ji} of job j on machine i. For each job j, we can also specify a set $M(j)$ of machines on which it can be scheduled. (This is equivalent to setting p_{ji} to infinity for $i \notin M(j)$). The starting point of the reduction is the 2-tier partition from Sect. 2.4. We create a machine for every node in C, where C is the set consisting of fat nodes and central nodes of structured supergroups, as defined in Lemma 2.4. We create a job for every node in C, and for every mobile and isolated group. To simplify the notation, we identify the names of the machines and jobs with the names of the corresponding nodes and groups. A job corresponding to a node i in C can be scheduled only on machine i, that is, $M(i) = \{i\}$, and we set $p_{ii} = w(i)$. A job corresponding to a mobile or isolated group j can be scheduled on machine m_i iff group j is connected to C-node i. This defines $M(j)$. We set $p_{ji} = w(j)$.

We run the algorithm of [17] for SUPM on the instance defined above. The solution returned by the algorithm is interpreted as a partition of the nodes of the original graph as follows. If job j is scheduled on machine i then node (group) j is assigned to part i of the partition. Each central group is assigned to the same part as the central node of the group.

The final part of the algorithm repairs lean parts in the resulting partition. While there is a lean part P in the partition, reassign a group as follows. Let S be the supergroup in the 2-tier partition whose center was a seed for P. (A lean part cannot have a fat node as a seed.) Let C be the central group of S. Then, by construction, P contains C. Remove a mobile group of S, say H, from its current part and insert it into P. Now, by Lemma 2.2(a), $w(P) \geq w(C) + w(H) \geq 1$ because P contains C and H.

This repair process will terminate because each part is repaired at most once. Since we repair P using a mobile group from the supergroup corresponding to P (that is, the supergroup from the 2-tier partition whose center is C), the future repairs of other parts will not remove H from part P. Later, even if P looses a mobile group when we repair some other part P', the weight of P will still satisfy: $w(P) \geq w(C) + w(H) \geq 1$. Thus, after a number of steps which is at most the number of parts, all parts will be fat.

Theorem 2.1 follows from the following lemma whose proof is omitted.

Lemma 2.5. *The final partition returned by the algorithm above has parts of weight at most* $opt + 2$.

3   Min-Max Bin Covering

In this section, we present our algorithm for Min-Max Bin Covering.

Theorem 3.1. *Min-Max Bin Covering can be approximated with ratio 2 in time* $O(n)$.

Proof. W.l.o.g. assume that $w_{lb} = 1$, $\mathcal{I} = \{1, \ldots, n\}$ and $w_1 \geq w_2 \geq \ldots \geq w_n$. We also assume that $w_i < 1$ for all items i, since items of larger weight can be placed in their own bins without affecting the quality of the solution. (Each such bin has weight at least 1 and at most opt.)

If $w(\mathcal{I}) < 3$, a legal packing consists of ≤ 2 bins. Therefore, $opt \geq w(\mathcal{I})/2$. Thus, $w(\mathcal{I}) \leq 2opt$, and we get a 2-approximation by returning one bin $B_1 = \mathcal{I}$. Theorem 3.1 follows from Lemma 3.1, dealing with instances with $w(\mathcal{I}) \geq 3$.   □

Lemma 3.1. *Given a Min-Max Bin Covering instance* $\mathcal{I}$ *with* n *items and* $w(\mathcal{I}) \geq 3$, *a solution with cost at most* $opt + 1$ *can be found in time* $O(n)$.

Proof. We compute a preliminary packing greedily, filling successive bins with items in order (of decreasing weights), and moving to a new bin when the weight of the current bin reaches or exceeds 1. Let $B_1, \ldots, B_k$ be the resulting bins.

Definition 3.1. *A bin* B *is* good *if* $w(B) \in [1, 2]$. *A packing where all bins are good is called* good.

All bins in the preliminary packing, excluding B_k, are good. If $w(B_k) \geq 1$, the preliminary packing is good. However, B_k can have weight less than 1. If $w(B_{k-1}) + w(B_k) \leq 2$, we obtain a good packing by combining B_{k-1} and B_k. In the remainder of the proof, we show how to rearrange items in B_k when

$$w(B_k) < 1; \tag{1}$$
$$w(B_{k-1}) + w(B_k) > 2 \tag{2}$$

to obtain a legal packing with cost at most $opt + 1$.

Observation 3.2 *If* $i \in B_j$ *then* $w(B_j) < 1 + w_i$. *Thus,* $w(B_j) - w_i < 1$.

Definition 3.2. *An item* i *is called* small *if* $w_i \leq 1/2$, *and* large *otherwise.*

Since $w(B_k) < 1$, $w(B_{k-1}) < 2$ and $w(\mathcal{I}) \geq 3$, the number of bins $k \geq 3$. We repack bins B_1, B_{k-2}, B_{k-1} and B_k to ensure that the last bin satisfies the weight lower bound. The remaining proof (omitted) is broken down into cases, depending on how many bins contain small items.   □

4   Min-Max Rectangle Tiling

We present two approximation algorithms for Min-Max Rectangle Tiling whose performance is summarized in Theorems 4.1 and 4.2.

Theorem 4.1. *Min-Max Rectangle Tiling can be approximated with ratio 4 in time $O(mn)$.*

Proof. Our algorithm first preprocesses the array to ensure that the last row is fat. (Recall that *fat* and *lean* were defined in Definition 1.1.) Then it greedily *slices* the array, that is, partitions it using horizontal lines. The resulting groups of consecutive rows are called *slices*. Finally, each slice is greedily *diced* using vertical lines into sub-rectangles, called *chunks*.

Let R_i denote the ith row of A. While R_m is thin, we perform a step of preprocessing that replaces the last two rows, R_{m-1} and R_m, with row $R_{m-1} + R_m$ (and decrements m by 1). When R_m is thin, every subset of R_m is thin, and cannot be a valid tile. Thus, every element of R_m has to be in the same tile as the element directly above it. Therefore, a preprocessing step does not change the set of valid tilings of A.

In a step of slicing, we start at the top (that is, go through the rows in the increasing order of indices). Let j be the smallest index such that remaining (not yet sliced) top rows up to row R_j form a fat rectangle. Then we cut horizontally between rows R_j and R_{j+1}, and call the top set of rows a slice. Continue on the matrix formed by the bottom rows. Since the preprocessing ensured that the last row is fat, all resulting slices are fat.

In a step of dicing, analogously to the slicing step, we cut up a slice vertically, dicing away *chunks*, minimal fat sets of leftmost columns, unless the remaining columns form a lean rectangle.

Consider a tile/chunk produced by our algorithm. The rectangle formed by all rows of the tile, ex-

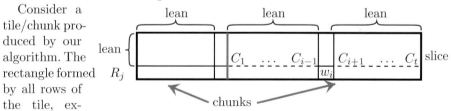

cluding the bottom row, is lean because it is obtained by partitioning a valid slice. Thus, the weight of this rectangle is less than w_{lb}, and consequently, less than opt. Let $C_1, \ldots, C_t$ be the columns of the tile (partial columns of the original matrix), and $w_1, \ldots, w_t$ be the entries in the bottom row of the slice. Let i be the smallest index such that $C_1, \cdots, C_i$ form a fat rectangle. (If this tile is the last chunk in its slice, then i might be less than t.) By the choice of i, the rectangle formed by $C_1, \ldots, C_{i-1}$ is lean, and so is the rectangle formed by $C_{i+1}, \ldots, C_t$. C_i without w_i is also lean, because it is a subset of the lean part of the slice. Finally, since w_i has to participate in a tile, $w_i \geq opt$. Consequently, the weight of the tile is smaller than $opt + 3w_{\mathrm{lb}} \leq 4opt$.

It is easy to implement the algorithm so that each step performs a constant number of operations per matrix entry, and the algorithm takes time $O(mn)$. $\square$

We can get a better approximation ratio when the entries in the matrix are restricted to be 0 or 1. This case covers the scenarios where each entry indicates the presence or absence of some object.

Theorem 4.2. *Min-Max Rectangle Tiling with 0-1 entries can be approximated with ratio 3 in time $O(mn)$.* (The proof is omitted.).

References

1. Du, W., Eppstein, D., Goodrich, M.T., Lueker, G.S.: On the approximability of geometric and geographic generalization and the min-max bin covering problem. In: WADS, pp. 242–253 (2009)
2. Ciriani, V., di Vimercati, S.D.C., Foresti, S., Samarati, P.: k-anonymous data mining: A survey. In: Aggarwal, C.C., Yu, P.S. (eds.) Privacy-Preserving Data Mining: Models and Algorithms, Springer, Heidelberg (2008)
3. Garcia, Y.J., Lopez, M.A., Leutenegger, S.T.: A greedy algorithm for bulk loading r-trees. In: GIS 1998: Proceedings of the 1998 ACM Int. Symp. on Advances in Geographic Information Systems, pp. 163–164. ACM, New York (1998)
4. Assmann, S.F., Johnson, D.S., Kleitman, D.J., Leung, J.Y.T.: On a dual version of the one-dimensional bin packing problem. J. Algorithms 5(4), 502–525 (1984)
5. Csirik, J., Johnson, D.S., Kenyon, C.: Better approximation algorithms for bin covering. In: SODA, pp. 557–566 (2001)
6. Jansen, K., Solis-Oba, R.: An asymptotic fully polynomial time approximation scheme for bin covering. Theor. Comput. Sci. 306(1-3), 543–551 (2003)
7. Bansal, N., Sviridenko, M.: The santa claus problem. In: STOC 2006: Proceedings of the Thirty-Eighth Annual ACM Symposium on Theory of Computing, pp. 31–40. ACM, New York (2006)
8. Graham, R.L., Lawler, E.L., Lenstra, J.K., Kan, A.H.G.R.: Optimization and approximation in deterministic sequencing and scheduling: A survey. Annals of Discrete Mathematics 5, 287–326 (1979)
9. Manne, F.: Load Balancing in Parallel Sparse Matrix Computation. PhD thesis, University of Bergen, Norway (1993)
10. Khanna, S., Muthukrishnan, S., Paterson, M.: On approximating rectangle tiling and packing. In: SODA, pp. 384–393 (1998)
11. Sharp, J.P.: Tiling multi-dimensional arrays. In: Ciobanu, G., Păun, G. (eds.) FCT 1999. LNCS, vol. 1684, pp. 500–511. Springer, Heidelberg (1999)
12. Smith, A., Suri, S.: Rectangular tiling in multi-dimensional arrays. In: SODA, pp. 786–794 (1999)
13. Muthukrishnan, S., Poosala, V., Suel, T.: On rectangular partitionings in two dimensions: Algorithms, complexity, and applications. In: Beeri, C., Bruneman, P. (eds.) ICDT 1999. LNCS, vol. 1540, pp. 236–256. Springer, Heidelberg (1998)
14. Berman, P., DasGupta, B., Muthukrishnan, S., Ramaswami, S.: Improved approximation algorithms for rectangle tiling and packing. In: SODA, pp. 427–436 (2001)
15. Berman, P., DasGupta, B., Muthukrishnan, S.: Slice and dice: A simple, improved approximate tiling recipe. In: SODA, pp. 455–464 (2002)
16. Berman, P., DasGupta, B., Muthukrishnan, S.: Approximation algorithms for max-min tiling. J. Algorithms 47(2), 122–134 (2003)
17. Lenstra, J.K., Shmoys, D.B., Tardos, E.: Approximation algorithms for scheduling unrelated parallel machines. Math. Program. 46(3), 259–271 (1990)
18. Tutte, W.T.: A theorem on planar graphs. Trans. Amer. Math. Soc. 82, 99–116 (1956)
19. Chiba, N., Nishizeki, T.: The hamiltonian cycle problem is linear-time solvable for 4-connected planar graphs. J. Algorithms 10(2), 187–211 (1989)

Min-Power Strong Connectivity

Gruia Calinescu[*]

Abstract. Given a directed simple graph $G = (V, E)$ and a cost function $c : E \to R_+$, the *power* of a vertex u in a directed spanning subgraph H is given by $p_H(u) = \max_{uv \in E(H)} c(uv)$, and corresponds to the energy consumption required for wireless node u to transmit to all nodes v with $uv \in E(H)$. The *power* of H is given by $p(H) = \sum_{u \in V} p_H(u)$.

Power Assignment seeks to minimize $p(H)$ while H satisfies some connectivity constraint. In this paper, we assume E is bidirected (for every directed edge $e \in E$, the opposite edge exists and has the same cost), while H is required to be strongly connected. This is the original power assignment problem introduce in 1989 and since then the best known approximation ratio is 2 and is achieved by a bidirected minimum spanning tree. We improve this to $2 - \epsilon$ for a small $\epsilon > 0$. We do this by combining techniques from Robins-Zelikovsky (2000) for Steiner Tree, Christofides (1976) for Metric Travelling Salesman, and Caragiannis, Flammini, and Moscardelli (2007) for the broadcast version of Power Assignment, together with a novel property on T-joins in certain two-edge-connected hypergraphs. With the restriction that $c : E \to \{A, B\}$, where $0 \le A < B$, we improve the best known approximation ratio from 1.8 to $\pi^2/6 - 1/36 + \epsilon \le 1.61$ using an adaptation of the algorithm developed by Khuller, Raghavachari, and Young (1995,1996) for (unweighted) Minimum Strongly Connected Subgraph.

1 Introduction

There has been a surge of research in Power Assignment problems since 2000 (among the earlier papers are [23,29,14]) This class of problems take as input a directed simple graph $G = (V, E)$ and a cost function $c : E \to R_+$. The *power* of a vertex u in a directed spanning subgraph H of G is given by $p_H(u) = \max_{uv \in E(H)} c(uv)$, and corresponds to the energy consumption required for wireless node u to transmit to all nodes v with $uv \in E(H)$. The *power* (or total power) of H is given by $p(H) = \sum_{u \in V} p_H(u)$.

The study of the min-power power assignment was started by Chen and Huang [6], which consider, as we do, the case when E is bidirected (the case is sometimes called "symmetric" or "undirected" in the literature) while H is required to be strongly connected. We call this problem *Min-Power Strong Connectivity*. We use with the same name both the (bi)directed and the undirected version of G. [6] prove that the bidirected version of a minimum (cost) spanning tree (MST) of the input graph G has power at most twice the optimum, and therefore the

---

[*] Department of Computer Science, Illinois Institute of Technology, Chicago, IL 60616, USA. `calinescu@iit.edu`. Research supported in part by NSF grant CNS-0916743.

M. Serna et al. (Eds.): APPROX and RANDOM 2010, LNCS 6302, pp. 67–80, 2010.

MST algorithm has approximation ratio at most 2. This is known to be tight (see Section 2).

We improve this to $2 - \epsilon$ for a small $\epsilon > 0$. We do this by combining techniques from Robins-Zelikovsky [24,25] for Steiner Tree, Christofides [7] for Metric Travelling Salesman, Caragiannis, Flammini, and Moscardelli [4] for the broadcast version of symmetric Power Assignment (assuming a bidirected $G = (V, E, c)$ and a "root" $u \in V$ is given, H must contain a path from u to every other vertex of G), together with what we believe to be a new property on T-joins in certain two-edge-connected hypergraphs. Familiarity with the fundamental NP-Hard optimization problems Steiner Tree and Travelling Salesman will help the reader make sense of this extended abstract; however all our proofs are self-contained and only refer to previous papers for intuition.

Carmi and Katz [5] consider the case $c : E \to \{A, B\}$, where $0 \leq A < B$. It is easy to see that for minimizing total power, the hardest case is $A = 0$ and $B = 1$. Then we have to minimize the number of nodes with power B. [5] gives a 9/5 approximation and a faster 11/6 approximation algorithm. We adapt the algorithms of Khuller, Raghavachari, and Young [17,18] for (unweighted) Minimum Strongly Connected Subgraph (given a strongly connected directed graph $G = (V, E)$, find minimum-size $F \subseteq E$ with (V, F) strongly connected) to obtain a $\pi^2/6 - 1/36 + \epsilon \leq 1.61$ approximation in this case. We also show that the algorithms of Carmi and Katz have approximation ratio exactly 7/4.

Very restricted versions of Min-Power Strong Connectivity have been proven NP-Hard [19,8,5]. We are not aware of better than a factor of 2 approximation except for [5], [2] (where c is assumed to be a metric), and the exact (dynamic programming) algorithms [19] for the specific case where each vertex of G maps to a point on a line, and $c(uv)$ is an increasing function of the Euclidean distance between the images of u and v, A related version, also NP-Hard, asks for H to be bidirected (also known as "undirected" or "symmetric"). This problem is called Min-Power Symmetric Connectivity, and the best known ratio of $5/3 + \epsilon$ [1] is obtained with techniques first applied to Steiner Tree; when $c : E \to \{A, B\}$ one gets 3/2 with the same method [22]. In fact, many but not all power assignment algorithms use techniques from Steiner Tree variants (or direct reduction to Steiner Tree variants; these connections to Steiner Tree are not obvious and cannot be easily explained), and in particular Caragiannis et al [4] uses the relative greedy heuristic of Zelikovsky [30]. New interesting techniques were also developed for power assignment problem, as in [20], an improvement over [15].

The existing lower bound of the optimum, which we use, is the cost of the minimum spanning tree of G. Indeed, the optimum solution OPT contains an in-arborescence rooted at v, for some $v \in V$, and then, for all $u \in V \setminus \{v\}$, $p_{OPT}(u)$ is at least the cost of the directed edge connecting u to its parent in the arborescence.

We also use a relative greedy method as in [30,24]; Robins-Zelikovsky [24] is rarely used as a technique, and not by only citing the ratio (improved by now in [3]). We use the natural structures of [4] to improve over the minimum spanning tree. We are at a disadvantage as our new lower bound is not far from optimum (and the algorithm more than adds the two lower bounds); still the greedy Robins-Zelikovsky method allows an improvement over the factor of 2.

The new lower bound resembles the T-joins implicit in Christofides' approximation algorithm for Traveling Salesman Problem, however we have hypergraphs rather than graphs. Precisely, we are given an edge-weighted hypergraph $K = (V_K, E_K)$ and an even-sized set of vertices $S \subseteq V_K$, and we need a minimum weight set M of hyperedges such that every T-cut is covered by M; that is, for any $Q \subset V$ with $|Q \cap S|$ odd, there is a hyperedge $e \in M$ intersecting both Q and $V \setminus Q$. Such a set M is called a T-join, and a minimum-weight one can be computed in polynomial-time if K is a graph (Chapter 29 of [26]) . Our generalization of Minimum Weight Graph T-join, which we call *Hypergraph T-join*, is however NP-hard (Section 2) and we cannot directly use the Christofides approach; instead we resort to Robins-Zelikovsky. For our new lower bound, we also need to know the maximum, over two-edge-connected instances of Hypergraph T-join, of the minimum weight T-join divided by the weight of (all edges in) the hypergraph. For graphs, this *T-ratio* is known (and not too hard to prove) to be $1/2$ [9]. For hypergraphs, we discovered a sequence of examples where the T-ratio converges to $2/3$, and prove it is at most $7/8$ for the particular hypergraphs we need.

Min-Power Strong Connectivity has a flavor similar to two more fundamental problems for which the best known ratio is 2: Min-Cost Strong Connectivity and Min-Cost Two Edge Connectivity, and we hope this extended abstract will renew the interest in those two problems.

2 Preliminaries

In directed graphs, we use *arc* to denote a directed edge. In a directed graph K, an *incoming arborescence* rooted at $x \in V(K)$ is a subgraph T of K such that the underlying undirected graph of T is a tree and every vertex of T other than x has exactly one outgoing arc in T. The vertices of T with no incoming arcs in T are called *leafs*.

Given an arc xy, its undirected version is the undirected edge with endpoints x and y. Arcs xy and yx are antiparallel, and the antiparallel arcs resulting from undirected edge uv are uv and vu; if undirected edge uv has cost then each of the two antiparallel arcs resulting from undirected edge uv have this cost.

An alternative definition of our problem is: we are given a simple undirected graph $G = (V, E)$ and a cost function $c : E \to R_+$. A power assignment is a function $p : V \to R_+$, and it induces a simple directed graph $H(p)$ on vertex set V given by xy being an arc of $H(p)$ if and only if $\{x, y\} \in E$ and $p(x) \geq c(\{x, y\})$. The problem is to minimize $\sum_{u \in V} p(u)$ subject to $H(p)$ being strongly connected. To see the equivalence of the definition, given directed spanning subgraph H, define for each $u \in V$ the power assignment $p(u) = p_H(u)$.

The following example shows that the ratio of 2 for the MST algorithm is tight. Consider $2n$ points located on a single line such that the distance between consecutive points alternates between 1 and $\epsilon < 1$, and c be the square of the

Euclidean distance (see Figure 1). Then the minimum spanning tree MST connects consecutive neighbors and has power $p(MST) = 2n$. On the other hand, the tree T with edges connecting each other node (see Figure 1(b)) has power equal $p(T) = n(1 + \epsilon)^2 + (n-1)\epsilon^2 + 1$. When $n \to \infty$ and $\epsilon \to 0$, we obtain that $p(MST)/p(T) \to 2$.

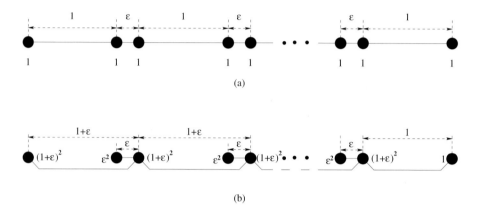

(a)

(b)

Fig. 1. Tight example for the performance ratio of the MST algorithm. (a) The MST-based power assignment (b) Optimum power assignment.

The example above may give intuition on how Power Assignment (and even more specifically, Min-Power Symmetric Connectivity) relates to the k-restricted Steiner trees, with stars (trees of height 1) taking the place of restricted components. Another example from [1], below, shows how Min-Power Strong Connectivity differs from Min-Power Symmetric Connectivity, and may give intuition how Min-Power Strong Connectivity relates to Travelling Salesman, and also Min-Cost Strong Connectivity and Min-Cost Two-Edge Connectivity (a two edge connected graph has an edge orientation that makes it strongly connected - see for example Chapter 2, written by A. Frank, of [12]). However we cannot think of direct reductions either way, and, as we mention in conclusions, the methods we use (more precisely, the Christofides algorithm) only applies to certain instances of Min-Cost Strong Connectivity and Min-Cost Two-Edge Connectivity.

The power of a Min-Power Strong Connectivity optimum solution can be almost half the power of an Min-Power Symmetric Connectivity optimum solution for the same instance: we present a series of examples illustrated in Figure 2. The $n(n + 1)$ vertices are embedded in the plane in n groups of $n + 1$ points each. Each group has two "terminals" (represented as thick circles in Figure 2), and the $2n$ terminals are the corners of a regular $2n$-gon with sides of length 1. Each group has another $n - 1$ equally spaced points (dashes in Figure 2) on the line segment between the two terminals. The cost function c is the square of the Euclidean distance. It is easy to see that a minimum power assignment ensuring strong connectivity assigns a power of 1 to one thick terminal in each group and a power of $\epsilon^2 = (1/n)^2$ to all other points in the group - the arcs going clockwise. The total power then equals $n + 1$. For symmetric connectivity it is necessary to assign power of 1 to all but two of the thick points, and of ϵ^2 to the remaining points, which results in total power of $2n - 1 - 1/n + 2/n^2$. Also, keep in mind that the minimum spanning tree solution is a symmetric solution.

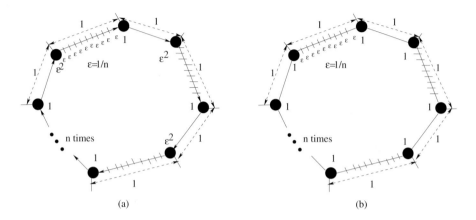

Fig. 2. Total power for the Min-Power Strong Connectivity can be half the total power for Min-Power Symmetric Connectivity. (a) Minimum power assignment ensuring strong connectivity. (b) Minimum power assignment ensuring symmetric connectivity.

While it may be already known, it is easy to check that Hypergraph T-join is indeed NP-hard, by a reduction from 4-D Matching (which asks if a 4-regular hypergraph K with $V(K)$ a multiple of 4, contains a perfect matching, that is, a set of disjoint hyperedges containing every vertex of the input hypergraph; see Garey and Johnson [11] problem SP1). It is easy to check that for $S = V(K)$, a T-join of size at most $|V(K)|/4$ must be a perfect matching.

3   (Weighted) Min-Power Strong Connectivity

Theorem 1. *There exists an $\epsilon > 0$ and a polynomial time algorithm for Min-Power Strong Connectivity with approximation ratio $2 - \epsilon$.*

Our algorithm uses a greedy approach similar to [30,24,4]. Let T be the undirected minimum spanning tree of G. We will not use that T is minimum except to note $opt \geq c(T)$, where $opt = p(OPT)$ for an optimum solution OPT. Let $\hat{T}$ be the bidirected version of T.

For $u \in V$ and $r \in \{c(uv) \mid uv \in E\}$, let $S(u, r)$ be the directed star with center u containing all the arcs uv with $c(uv) \leq r$; note that r is the power of S. For a directed star S, let $E(S)$ be its set of arcs and $V(S)$ be its set of vertices.

For given $S(u, r)$, let $Q(u, r)$ be the set of edges of T on a path in T between some $x, y \in V(S(u, r))$. Let $\hat{Q}(u, r)$ be the set of arcs of $\hat{T}$ on a directed path from u to some $x \in V(S(u, r))$; it is easy to verify that the undirected version of $\hat{Q}(u, r)$ is $Q(u, r)$.

For a collection $\mathcal{A}$ of directed stars $S(u_i, r_i)$, define $Q(\mathcal{A}) = \cup_{S(u_i, r_i) \in \mathcal{A}} Q(u_i, r_i)$ and $f(\mathcal{A}) = \sum_{e \in Q(\mathcal{A})} c(e)$. We will use later that this function is increasing and submodular, a well-known fact which appears as an example in Subchapter 44.1.a of [26]. Also define $w(\mathcal{A}) = \sum_{S(u_1, r_i) \in \mathcal{A}} r_i$, the total power used by the stars in $\mathcal{A}$. For $S = S(u, r)$, define $f_{\mathcal{A}}(S) = f(\mathcal{A} \cup \{S\}) - f(\mathcal{A}) = \sum_{e \in Q(u, r) \setminus Q(\mathcal{A})} c(e) = \sum_{e \in I_{\mathcal{A}}(S)} c(e)$, where $I_{\mathcal{A}}(S)$ is defined to be those arcs of $\hat{Q}(u, r)$ for which the undirected version is not in $Q(\mathcal{A})$.

The algorithm starts with $M = \hat{T}$ as the set of arcs, and greedily adds directed stars to collection $\mathcal{A}$ (initially empty) while removing arcs from M to improve the following quantity which is an upper bound on our output: the cost the arcs in M plus the total power of the stars in $\mathcal{A}$. To simplify later proofs, we make changes even if our quantity stays the same. To be precise:

Algorithm **Greedy:**
    $\mathcal{A} \leftarrow \emptyset$, $M \leftarrow \hat{T}$
    While ($f(\mathcal{A}) < c(T)$) do
    $(u, r) \leftarrow \mathrm{argmax}_{(u', r')} f_{\mathcal{A}}(S(u', r'))/r'$
    $M \leftarrow M \setminus I_{\mathcal{A}}(S(u, r))$
    $\mathcal{A} \leftarrow \mathcal{A} \cup \{S(u, r)\}$
    Output $\cup_{S \in \mathcal{A}} E(S) \cup M$

Note that a star $S(u, r)$ always exists for which $f_{\mathcal{A}}(S(u, r)) > 0$ and $f_{\mathcal{A}}(S(u, r))/r \geq 1$. Indeed, as long as a pair of antiparallel arcs e' and e'' are in M, we can pick as next star $S(u, r)$ the one given by u being the tail of e' and $r = c(e')$; this star will be added to $\mathcal{A}$ while e' is removed from M.

Lemma 1. *The output is a spanning strongly connected subgraph of G.*

Proof Sketch. We prove the following invariant: $X := \cup_{S \in \mathcal{A}} E(S) \cup M$ gives a spanning strongly connected subgraph whenever the **while** condition is checked by the algorithm. Moreover, suppose we remove from T all edges e for which both antiparallel arcs appear in M, splitting T in components with vertex sets T_j, for some range of j. We prove that for every j and every $x, y \in T_j$, there exists a directed path P from x to y using only vertices of T_j and arcs from X.

The invariant is true before the first iteration, when each T_j has just one vertex, and keeping it is rather straightforward, based on the following idea: when we add an arc uz and remove arcs $x_i y_i$ on the simple path in T from u to z, then arcs $y_i x_i$ are kept and can be used together with paths inside components T_j as above, and arc uz, to produce a directed cycle C containing none of the arcs $x_i y_i$ and passing through u, z and all the y_i, x_i. ∎

Lemma 2. *There exists a collection of stars $\mathcal{B}$ with $f(\mathcal{B}) = c(T)$ and $w(\mathcal{B}) \leq (7/8)opt$, where opt is the power of the optimum solution.*

The factor of $7/8$ above cannot be improved to a constant better than $1/2$, as we can see by looking at the example in Figure 1. To "cover" the whole minimum spanning tree (that is, have $f(\mathcal{B}) = c(T)$) one needs to select every second star of optimum (that is, use with power $(1 + \epsilon)^2$ nodes $3, 7, 11 \ldots$). Our proof is based on T-joins and the $7/8$ above cannot be improved to more than $2/3$ with the same method. Due to page limitations, we omit the long, technical part of the proof.

Proof Idea. Let $(S_v)_{v \in V}$, be the directed stars of OPT, with S_v centered at v, and let $\mathcal{A}$ be collection of these stars. Let $K = (V_K, E_K)$ be the (undirected) hypergraph defined by $V_K = V$ and $E_K = \{V(S) \mid S \in \mathcal{A}\}$. Define the weight of an hyperedge to be the power of the corresponding directed star. A path in a hypergraph consists of an alternating sequence of vertices and hyperedges for which each hyperedge contains the two vertices which precede and follow it in the sequence. A hypergraph is two-edge-connected if there exist two hyperedge-disjoint paths between any two vertices.

Recall from the introduction that, with given $R \subseteq V$ with $|R|$ even, a T-cut is a partition of V into two parts Q and $\bar{Q} := V \setminus Q$ such that $|Q \cap R|$ is odd. A T-join in K for R is a set of hyperedges $M \subseteq E_K$ such that for every T-cut $(Q, \bar{Q})$ has a hyperedge $e \in M$ intersecting both Q and $\bar{Q}$.

The following is the equivalent of Christofides' method:

Claim. Let R be the vertices of the tree T of odd degree (note that $|R|$ is even). Let $\mathcal{D}$ be a set of stars such that the corresponding hyperedges form a T-join in K for R. Then $f(\mathcal{D}) = c(T)$.

Proof. Note that when it comes to computing f, for each star S only $V(S)$ counts, and which vertex is the center is not relevant. For star $S = S(u, r)$, let $Q(S) := Q(u, r)$, defined previously as be the set of edges of T on a path in T between some $x, y \in V(S)$.

We need to show that for every $e \in T$, there is a star $S \in \mathcal{D}$ with $e \in Q(S)$. Indeed, if we remove e from T, we create two subtrees T_u and T_v, where u and v are the endpoints of e. Then $|R \cap V(T_u)|$ is odd, since if we take T_u and add the vertex v and the edge uv, we have an even number of vertices of odd degree, of which one is v. Thus $(V(T_u), V(T_v))$ is a T-cut for R and the T-join given by $\mathcal{D}$ must have an hyperedge intersecting both $V(T_u)$ and $V(T_v)$, and thus a star $S \in \mathcal{D}$ with $e \in Q(S)$. ∎

Based on the claim above, Lemma 2 follows from:

Claim. For any arbitrary R, there exists a T-join in K with weight at most $(7/8)w(K)$.

Due to page limitations, we must omit the very long and technical proof of the above claim. We discuss it nevertheless. While K is indeed two-edge-connected (which we do not prove here) and it may very well be the case that the $7/8$

ratio holds for any two-edge-connected hypergraph, our current proof relies on the structure of K (how it was obtained from OPT).

The series of example showing that $2/3$ is the best we can hope instead of $7/8$ also requires too much space. As mentioned before, if we were dealing with graphs rather than hypergraphs the ratio would be $1/2$. We only have space for a $3/5$ example: nine vertices $x, y_0, y_1, y_2, z_0, z_1, z_2, u_1, u_2$, edges of cost 2: xy_0 and xz_0, edges of cost 1: y_1y_2, z_1z_2, and u_1u_2, and edges of cost 0: y_0y_1, z_0z_1, y_2u_1, z_2u_1, and u_2x. OPT has cost 5: x has power 2 and y_1, z_1, and u_1 each have power 1. The minimum spanning tree has the four edges of cost 0 and the three edges of cost 1. R consists of the vertices x, y_0, z_0, u_1, and one can check by inspection that any T-join has weight 3 (here one can "cover" the minimum spanning tree with just the star of power 2 rooted at x, but our proof method relies on T-joins!). ∎

Now we need the following lemma, whose proof is obtained from Robins-Zelikovsky as presented in [13] by changing what quantities represent and some parameters.

Lemma 3. *Assuming that there exists a collection of stars $\mathcal{B}$ with $f(\mathcal{B}) = c(T)$ and $w(\mathcal{B}) \leq \alpha opt$, where opt is the power of the optimum solution and $\alpha < 1$, the algorithms' output has power at most β where $\beta = 1 + \alpha + \alpha \ln(1/\alpha)$.*

Proof. First, if $c(T) \leq \alpha opt$, then before any improvement we have a solution of cost at most $2\alpha opt$ and $2\alpha < \beta$. Thus in the following we assume $c(T) > \alpha opt$.

Note that at the end of the algorithm, M contains one of the two antiparallel arcs for each edge of T. Then the power of the output cannot exceed

$$c(T) + w(\mathcal{A}) \tag{1}$$

for the final $\mathcal{A}$, as a $f(\mathcal{A}) = c(T)$, and for every H and u, $p_H(u) = max_{uv \in H} c(uv) \leq \sum_{uv \in H} c(uv)$.

Let $S_1, S_2, \ldots, S_q$ be the stars picked by our algorithm and let $\mathcal{A}_i$, for $1 \leq i \leq q$ be the collection of the first i stars; also let for convenience $\mathcal{A}_0$ be the empty collection. For $1 \leq i \leq q$, let $p_i = p(S_i) = r_i$, where r_i comes from $S_i = S(u_i, r_i)$, and let $f_i = f_{\mathcal{A}_{i-1}}(S_i)$.

The greedy choice of the algorithm and the submodularity of f gives:

$$\frac{p_i}{f_i} \leq \frac{w(\mathcal{B})}{f_{\mathcal{A}_{i-1}}(\mathcal{B})} = \frac{w(B)}{c(T) - f(\mathcal{A}_{i-1})} = \frac{w(B)}{c(T) - \sum_{j=1}^{i-1} f_j}.$$

Rewriting and replacing $w(\mathcal{B})$ with αopt, we obtain

$$p_i \leq f_i \frac{\alpha opt}{c(T) - \sum_{j=1}^{i-1} f_j}. \tag{2}$$

Define the function $g : [0..c(T)] \to [0..1]$ by $g(x) = \alpha opt/(c(T) - x)$ for $x \leq c(T) - \alpha opt$, and $g(x) = 1$ for $x > c(T) - \alpha opt$. Then from Equation 2 and the observation (made right after the algorithm) that $\frac{p_i}{f_i} \leq 1$, we obtain:

$$\sum_{i=1}^{q} p_i \leq \int_0^{c(T)} g(x)dx = \int_0^{c(T)-\alpha opt} \frac{\alpha opt}{c(T) - x} dx + \int_{c(T)-\alpha opt}^{c(T)} 1 dx$$

$$= (-\alpha opt)\ln(c(T) - x)\big|_0^{c(T)-\alpha opt} + (c(T) - (c(T) - \alpha opt))$$

$$= \alpha opt \left(1 + \ln \frac{c(T)}{\alpha opt}\right)$$

Using this and $c(T) \leq opt$ and Equation 1 (recall that $w(\mathcal{A}) = \sum_{i=1}^q p_i$), we obtain that the power of the output is at most

$$c(T) + \alpha opt \left(1 + \ln \frac{c(T)}{\alpha opt}\right) \leq opt \left(1 + \alpha + \alpha \ln(1/\alpha)\right)$$

finishing the proof.   ∎

Based on Lemmas 1 and 2, Theorem 1 follows immediately from the fact that $\alpha < 1$ implies $\beta < 2$, which follows from $\alpha(1 + \ln(1/\alpha)) < 1$, which is equivalent to $\ln(1/\alpha) < 1/\alpha - 1$, a fact that holds for all $\alpha > 1$. For $\alpha = 7/8$, we obtain $\beta \leq 1.992$. If one were to prove $\alpha = 1/2$ or $\alpha = 2/3$ in Lemma 2, the resulting approximation ratio would be less than 1.85 or 1.94 respectively.

4 Min-Power Strong Connectivity with Two Power Levels

In this section, it is more convenient to to work with the alternative definition of power assignment described in the preliminary section.

Theorem 2. *For every integer $k \geq 4$, there exists a $O(n^{k+3})$ algorithm with approximation ratio $\pi^2/6 - 1/36 + 1/(k(k-1))$.*

Given set of vertices $S \subseteq V$, we define the function $p^S : V \rightarrow \{0, 1\}$ as follows: $p^S(u) = 1$ if $u \in S$ and $p^S(u) = 0$ if $u \in V \setminus S$.

We say that $S \subseteq V$ is *impeccable* if no arc of $H(p^S)$ has endpoints in distinct strongly connected components of $H(p^S)$. See Figures 3 and 4 for illustrations, where we use the following conventions: In our pictures, we use full small circles to denote nodes with power 1, that is, S. Empty little circles are nodes with power 0, continuous thin line segments are edges of E of cost $c = 0$; both antiparallel corresponding arcs are in H. Thicker segments with arrows are for arcs of $H(p^S)$ coming from edges of E of cost 1, and thicker dashed segments for edges of E of cost 1 for which none of the two antiparallel corresponding arcs are in H. Larger ellipses represent strongly connected components of $H(p^S)$.

We say that $Q \subseteq V$ is *quasi-perfect* with respect to impeccable $S \subseteq V$ if the following holds: Let $X_1, X_2, \ldots, X_s$ be the strongly connected components of $H(p^S)$. Then for all $i = 1, 2, \ldots, q$, $|Q \cap X_i| \leq 1$, and all vertices of Q are in the same strongly connected component of $H(p^{Q \cup S})$. See Figure 5 for illustrations. If Q is quasi-perfect w.r.t impeccable S and $Q \cup S$ is impeccable, then Q is called *perfect* w.r.t. S.

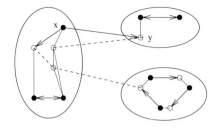

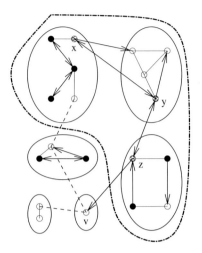

Fig. 3. This S is not impeccable as $H(p^S)$ has the arc xy crossing between its strongly connected component

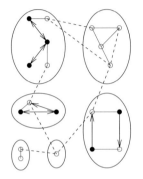

Fig. 4. An impeccable S

Fig. 5. The set $Q = \{x, y, z\}$ is quasiperfect w.r.t the given S as the thick dotted closed curve gives the strongly connected component of $H(p^{Q \cup S})$ containing Q. Q is not perfect since the arc zv has endpoints in different strongly connected component of $H(p^{Q \cup S})$

Our algorithm grows an impeccable set of vertices S (S_i, below) by adding to S a perfect set Q. Perfect sets of vertices play for us the role of directed cycles in the Khuller et al. algorithms [17,18]. In [17,18] strongly connected components are contracted, for simplicity, to a single vertex; we have to keep them during the algorithm. Below is our modified version of the Khuller et al. algorithm. Both running time and approximation ratio depend on integer parameter $k \geq 4$.

1 Set $i = 0$, $j = k$, $S_0 = \emptyset$.
2 **while** $j \geq 4$
3   **if** there exists $Q \subseteq V$ with $|Q| \geq j$ and perfect w.r.t. S_i
4          $S_{i+1} = S_i \cup Q$; $i = i + 1$
5   **else**
6          $j = j - 1$
7 **endwhile**
8 Find minimum $R \subseteq V$ that makes $H(p^{S_i \cup R})$ strongly connected
9 Return $p^{S_i \cup R}$

We need to elaborate for step 3 and step 8, which differ from [17,18]. Step 3 adapts the [17,18] method; this cannot be done for Step 8. Due to space

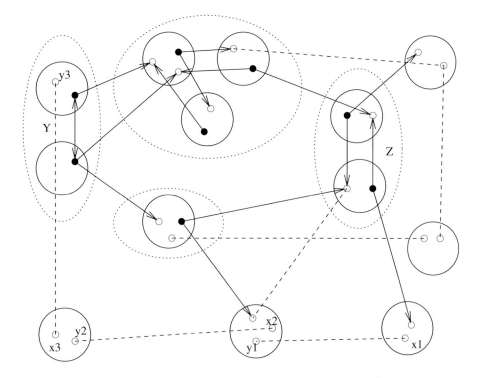

Fig. 6. Here circles are the strongly connected components of $H(p^S)$; members of S are not depicted and the filled small circles give the set J. Empty small circles are in neither S nor J. The dashed ellipses give the strongly connected components of L (we kept parallel arcs in the picture). In this example, Y and Z exist and are depicted. M is also implicit in the figure: the solid circles not in dashed ellipse, and here one such path P can be $x_1 y_1$, $x_2 y_2$, and $x_3 y_3$.

limitations we omit our solution (based on Graphical Matroid Parity [21,10,27]), but mention that we believe the algorithms of Carmi and Katz [5] gives the optimum indeed for instances where every perfect set has at most three vertices.

Step 3. First, Step 3 can be implemented in $O(n^{k+2})$, as described in the following procedure, whose input is S_i (from now on we use S instead of S_i). Compute $X_1, X_2, \ldots, X_s$, the strongly connected components of $H(p^S)$. Try all combinations of sets $J \subseteq S \setminus S_i$ of exactly j elements. If there exists a $k \in \{1, 2, \ldots, s\}$ with $|J \cap X_k| > 1$, reject J (and try another combination).

Otherwise, construct the directed graph $L = L(S, J)$ with vertex set $\{k \mid |J \cap X_k| = 1\}$ and put an arc from k to q if there exists an edge $e \in E$ endpoints in $J \cap X_k$ and X_q. See Figure 6 for an illustration of this paragraph. Compute the strongly connected components of L. If L is strongly connected proceed to the extension phase, described in the next paragraph. If L has more than one strongly connected component with no incoming arc, or more than one strongly connected

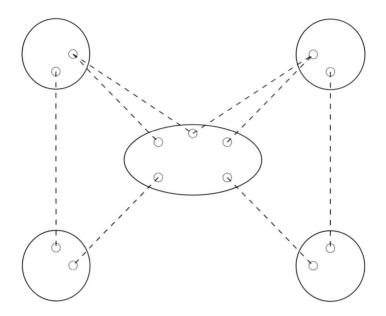

Fig. 7. Splitting the graph in two by the middle cut-vertex results in a loss in approximation ratio as some vertices of G - like the middle one here - can contribute to both sides of the split

component with no outgoing arc, reject J. Else, let Y be the vertex set of the strongly connected component of L with no incoming arc, and Z be the vertex set of the strongly connected component of L with no outgoing arc. Construct an undirected simple graph M with vertex set $\{v_k \mid |J \cap X_k| = 0 \text{ or } k \in Y\}$ and put an edge from v_k to v_q if there exists an edge $e \in E$ with endpoints in X_q and X_k. Let $W = \{v_k \in V(M) \mid \exists x \in J \cap (\cup_{r \in Z} X_r) \; \exists y \in X_k \; \exists e \in E \; (e = \{x, y\})\}$. If M does not have a path from some v_k with $k \in W$ to some v_q with $q \in Y$, then reject J. Else, let P be such a path with minimum number of edges. Each edge f of P comes from an edge $\{x_f, y_f\} \in E$, and permute x_f and y_f if necessary such that, if $x_f \in X_k$ and $y_f \in X_q$, then k is before q on P when starting from W. Put all such x_f in J and proceed.

Now we have the extension phase: as long as there exists an edge $e \in E$ with one endpoint in J and the other in some X_k with $|J \cap X_k| = 0$, we add to J the endpoint of e in X_k.

Due to space limitations, the correctness and analysis of this algorithm is left for a full (journal) version of this extended abstract.

5   Conclusions

Based on the discussion in Section 3, we also have: if we are looking for a minimum weight strongly connected spanning subgraph in a directed graph, and a bidirected tree exists of cost at most twice optimum, then Christofides' method gives a 1.5 approximation. The same holds for Minimum Weight Two-Edge-Connected Spanning Subgraph if there always exists a "double tree" (that is, where each edge has a twin) of weight at most twice optimum. These facts are probably known.

The approximation ratio of Khuller, Raghavachari, and Young [18] has been improved to 1.5 by Vetta [28]. However we see no way of adapting [28] (or [16]) to power assignment with two power levels: the first step of [28] (Lemma 3.1) is to split the graph until no cut vertex exists. For us, the equivalent of a cut vertex would be a strongly connected component of p^0 - and what we choose in this strongly connected component can affect all the graph as in Figure 7. Thus we cannot split the input graph, and with cut vertices the lower bound of [28] is not 1.5 but 2.

It is possible that the very recent breakthrough by Byrka et. al [3] for Steiner Tree gives techniques to improve our approximation ratio for Min-Power Strong Connectivity.

We leave open the correct T-ratio in two-edge-connected hypergraphs.

References

1. Althaus, E., Calinescu, G., Mandoiu, I., Prasad, S., Tchervenski, N., Zelikovsky, A.: Power efficient range assignment for symmetric connectivity in static ad hoc wireless networks. Wireless Networks 12(3), 287–299 (2006)
2. Ambühl, C., Clementi, A.E.F., Penna, P., Rossi, G., Silvestri, R.: On the approximability of the range assignment problem on radio networks in presence of selfish agents. Theor. Comput. Sci. 343(1-2), 27–41 (2005)
3. Byrka, J., Grandoni, F., Rothvoß, T., Sanità, L.: An improved LP-based approximation for Steiner tree. In: STOC 2010: Proceedings of the 42nd ACM Symposium on Theory of Computing, pp. 583–592. ACM, New York (2010)
4. Caragiannis, I., Flammini, M., Moscardelli, L.: An Exponential Improvement on the MST Heuristic for Minimum Energy Broadcasting in Ad Hoc Wireless Networks. In: Arge, L., Cachin, C., Jurdziński, T., Tarlecki, A. (eds.) ICALP 2007. LNCS, vol. 4596, pp. 447–458. Springer, Heidelberg (2007)
5. Carmi, P., Katz, M.J.: Power assignment in radio networks with two power levels. Algorithmica 47(2), 183–201 (2007)
6. Chen, W.T., Huang, N.F.: The strongly connecting problem on multihop packet radio networks. IEEE Transactions on Communications 37(3), 293–295 (1989)
7. Christofides, N.: Worst-case analysis of a new heuristic for the traveling salesman problem. Technical Report 388, Carnegie Mellon University (1976)
8. Clementi, A.E.F., Penna, P., Silvestri, R.: On the power assignment problem in radio networks. In: Electronic Colloquium on Computational Complexity (ECCC) (054) (2000)
9. Edmonds, J., Johnson, E.L.: Matching, Euler tours and the Chinese postman. Mathematical Programming 5(1), 88–124 (1973)
10. Gabow, H.N., Stallmann, M.: Efficient algorithms for graphic matroid intersection and parity. In: 12th Colloq. on Automata, Language and Programming, pp. 210–220 (1985)
11. Garey, M.R., Johnson, D.S.: Computers and Intractability. W.H. Feeman and Co., NY (1979)

12. Graham, R.L., Grötschel, M., Lovász, L. (eds.): Handbook of combinatorics, vol. 1, 2. Elsevier Science B.V., Amsterdam (1995)
13. Gröpl, C., Hougardy, S., Nierhoff, T., Prömel, H.J.: Approximation algorithms for the Steiner tree problem in graphs. In: Du, D.-Z., Cheng, X. (eds.) Steiner Trees in Industries, pp. 235–279. Kluwer Academic Publishers, Dordrecht (2001)
14. Hajiaghayi, M.T., Immorlica, N., Mirrokni, V.S.: Power optimization in fault-tolerant topology control algorithms for wireless multi-hop networks. In: MOBI-COM, pp. 300–312 (2003)
15. Hajiaghayi, M.T., Kortsarz, G., Mirrokni, V.S., Nutov, Z.: Power optimization for connectivity problems. Math. Program. 110(1), 195–208 (2007)
16. Jothi, R., Raghavachari, B., Varadarajan, S.: A 5/4-approximation algorithm for minimum 2-edge-connectivity. In: SODA 2003: Proceedings of the Fourteenth Annual ACM-SIAM Symposium on Discrete Algorithms, pp. 725–734. Society for Industrial and Applied Mathematics, Philadelphia (2003)
17. Khuller, S., Raghavachari, B., Young, N.: Approximating the minimum equivalent digraph. SIAM J. Comput. 24(4), 859–872 (1995)
18. Khuller, S., Raghavachari, B., Young, N.: On strongly connected digraphs with bounded cycle length. Discrete Appl. Math. 69(3), 281–289 (1996)
19. Kirousis, L.M., Kranakis, E., Krizanc, D., Pelc, A.: Power consumption in packet radio networks. Theoretical Computer Science 243, 289–305 (2000)
20. Kortsarz, G., Mirrokni, V.S., Nutov, Z., Tsanko, E.: Approximating minimum-power degree and connectivity problems. In: Laber, E.S., Bornstein, C., Nogueira, L.T., Faria, L. (eds.) LATIN 2008. LNCS, vol. 4957, pp. 423–435. Springer, Heidelberg (2008)
21. Lovász, L., Plummer, M.D.: Matching Theory. Elsevier, Amsterdam (1986)
22. Nutov, Z., Yaroshevitch, A.: Wireless network design via 3-decompositions. Inf. Process. Lett. 109(19), 1136–1140 (2009)
23. Ramanathan, R., Rosales-Hain, R.: Topology control of multihop wireless networks using transmit power adjustment. In: INFOCOM (2), pp. 404–413 (2000)
24. Robins, G., Zelikovsky, A.: Improved Steiner tree approximation in graphs. In: ACM-SIAM Symposium on Discrete Algorithms, pp. 770–779 (2000)
25. Robins, G., Zelikovsky, A.: Tighter bounds for graph Steiner tree approximation. SIAM Journal of Discrete Mathematics 19(1), 122–134 (2005)
26. Schrijver, A.: Combinatorial Optimization. Springer, Heidelberg (2003)
27. Szigeti, Z.: On the graphic matroid parity problem. J. Combin. Theory Ser. B 88, 247–260 (2003)
28. Vetta, A.: Approximating the minimum strongly connected subgraph via a matching lower bound. In: SODA 2001: Proceedings of the Twelfth Annual ACM-SIAM Symposium on Discrete Algorithms, pp. 417–426. Society for Industrial and Applied Mathematics, Philadelphia (2001)
29. Wattenhofer, R., Li, L., Bahl, P., Wang, Y.-M.: Distributed topology control for wireless multihop ad-hoc networks. In: IEEE INFOCOM 2001 (2001)
30. Zelikovsky, A.: Better approximation bounds for the network and Euclidean Steiner tree problems. Technical Report CS-96-06, Department of Computer Science, University of Virginia (1996)

The Complexity of Approximately Counting Stable Matchings

Prasad Chebolu*, Leslie Ann Goldberg, and Russell Martin*

Department of Computer Science, University of Liverpool

Abstract. We investigate the complexity of *approximately counting* stable matchings in the k-attribute model, where the preference lists are determined by dot products of "preference vectors" with "attribute vectors", or by Euclidean distances between "preference points" and "attribute points". Irving and Leather [16] proved that counting the number of stable matchings in the general case is $\#P$-complete. Counting the number of stable matchings is reducible to counting the number of downsets in a (related) partial order [16] and is interreducible, in an approximation-preserving sense, to a class of problems that includes counting the number of independent sets in a bipartite graph ($\#BIS$) [7]. It is conjectured that no FPRAS exists for this class of problems. We show this approximation-preserving interreducibilty remains even in the restricted k-attribute setting when $k \geq 3$ (dot products) or $k \geq 2$ (Euclidean distances). Finally, we show it is easy to count the number of stable matchings in the 1-attribute dot-product setting.

1 Introduction

1.1 Stable Matchings

The *stable matching problem* (or *stable marriage problem*) is a classical combinatorics problem. An instance of this problem consists of n men and n women, where each man has his own preference list (a total ordering) of the women, and, similarly, each woman has her own preference list of the men. A one-to-one pairing of the men with the women is called a *matching* (or *marriage*). Given a matching, if there exists a man M and a woman w in the matching who prefer each other over their partners in the matching, then the matching is considered *unstable* and the man-woman pair (M, w) is called a *blocking pair*. (M and w would prefer to drop their current partners and pair up with each other.) If a matching has no blocking pairs, then we call it a *stable matching*. In 1962, Gale and Shapley proved that every stable matching instance has a stable matching, and described an $O(n^2)$ algorithm for finding one [8].

The stable matching problem has many variants, where ties in the preference lists could be allowed, where people might have partial preference lists (i.e. someone might prefer to remain single rather than be paired with certain members of the opposite sex), generalizations to men/women/pets, universities and

* Research supported in part by EPSRC Grant EP/F020651/1.

M. Serna et al. (Eds.): APPROX and RANDOM 2010, LNCS 6302, pp. 81–94, 2010.
© Springer-Verlag Berlin Heidelberg 2010

applicants, students and projects, etc. Some of these generalizations have also been well-studied and, indeed, algorithms for finding stable matchings are used for assigning residents to hospitals in Scotland, Canada, and the USA [4,20,22].

In this paper, we concentrate solely on the classical problem, so the term "matching instance" will refer to one where the number of men is equal to the number of women, and each man or women has their own full totally-ordered (i.e. no ties allowed) preference list for the opposite sex.

Irving and Leather [16] demonstrated that counting the *number* of stable matchings for a given instance is #*P*-complete. This completeness result relies on the connection between stable marriages and *downsets* in a related partial order (explained in more detail in Section 2), as counting the number of downsets in a partial order is another classical #*P*-complete problem [21].

Knowing that exactly counting stable matchings is difficult (under standard complexity-theoretic assumptions), one might turn to methods for *approximately* counting this number. In particular, we would like to find a *fully-polynomial randomized approximation scheme* (an *FPRAS*) for this task, i.e. an algorithm that provides an arbitrarily close approximation in time polynomial in the input size and the desired error. One method that has proven successful for other counting problems is the Markov Chain Monte Carlo (MCMC) method. This technique exploits a relationship between counting and sampling described by Jerrum, Valiant, and Vazirani [17], namely, for *self-reducible* combinatorial structures, the existence of an FPRAS is computationally equivalent to a polynomial-time algorithm for approximate sampling from the set of structures. Although the set of stable matchings for an instance does not obviously fit into the class of self-reducible problems, an efficient algorithm for (approximately) sampling a random stable matching can be transformed into a method for (approximately) counting this number.

Bhatnagar, Greenberg, and Randall [1] considered this problem of sampling a random stable matching using the MCMC method. They examined a natural Markov chain that uses "male-improving" and "female-improving" *rotations* (see Section 2.3) to define a random walk on the state space of stable matchings for a given instance. In the most general setting, matching instances can be exhibited for which the *mixing time* of the random walk has an exponential lower bound, meaning that it will take an exponential amount of time to (approximately) sample a random stable matching. This exponential mixing time is due to the existence of a "bad cut" in the state space. Bhatnagar, et al. considered several restricted settings for matching instances and were still able to show instances for which such a bad cut exists in the state space, implying an exponential mixing time in these restricted settings.

Of particular interest to us in this paper, Bhatnagar et al. examined the so-called *k-attribute model*. In this setting each man and woman has two *k*-dimensional vectors associated with them, a "preference" vector and a "position" (or "attribute") vector. A man M_i has a preference vector denoted by $\hat{M}_i$, and a position vector denoted by $\bar{M}_i$ (similarly denoted for the woman w_j). Then, M_i prefers w_j over w_k (i.e. w_j appears higher on his preference list than w_k) if

and only if $\hat{M}_i \cdot \bar{w}_j > \hat{M}_i \cdot \bar{w}_k$, where $\hat{M}_i \cdot \bar{w}_j$ denotes the usual k-dimensional dot product of vectors. Since we assume that each man has a total order over the women (and vice-versa), we note that $\hat{M}_i \cdot \bar{w}_j \neq \hat{M}_i \cdot \bar{w}_k$ whenever $j \neq k$ (and analogously for the women's preference vectors/men's position vectors).

Even in this restricted k-attribute setting (not every matching instance can be represented in this manner if k is small [3]), Bhatnagar, Greenberg, and Randall were still able to demonstrate examples of matching instances having a "bad cut" where the Markov chain has an exponential mixing time. Bhatnagar et al. also considered two other restricted settings, the so-called k-range and k-list models, but we will not be considering those cases here. (Again, they gave instances having an exponential mixing time for the Markov chain.)

It must be noted that even though the male-improving/female-improving Markov chain might have an exponential mixing time, this does not necessarily imply the non-existence of an FPRAS for the corresponding counting problems. However, Dyer et al. [7] give evidence suggesting that even *approximately* counting the number of stable matchings is itself difficult, i.e. suggesting that an FPRAS may not exist. They do this by demonstrating *approximation-preserving reductions* amongst several counting problems, one being that of counting downsets in a partial order (once again, the connection to stable matchings is outlined in Section 2). The main point is that the existence of an FPRAS for one problem would imply the existence of an FPRAS for this entire class of counting problems. Currently, the existence of such an FPRAS remains an open question.

It is precisely the goal of this paper to consider the complexity of the approximate counting problem for the k-attribute model.

Before we continue, let us formally define some counting problems. Two counting problems relevant to us are $\#SM$, the problem of computing the number of stable matchings, given a stable matchings instance, and $\#SM(k\text{-attribute})$, the problem of computing the number of stable matchings, given a stable matching instance in the k-attribute setting, where preference lists are determined using dot products between k-dimensional preference and position vectors.

If k is small (relative to n), there exist preference lists that are not realizable in the k-attribute setting [3]. On the other hand, if $k = n$ then we can clearly represent any set of n preference lists by simply using a separate coordinate for each person to rank the members of the opposite sex. Another counting problem we consider in this paper is $\#SM(k\text{-Euclidean})$, which is the problem of computing the number of stable matchings given an instance in the k-dimensional Euclidian setting. In this setting, men and women each have a "preference point" and "position point". Preference lists are determined using Euclidean distances between preference points and position points. In other words, for a k-Euclidean stable matching instance man M_i prefers woman w_j to woman w_k if and only if $d(\hat{M}_i, \bar{w}_j) < d(\hat{M}_i, \bar{w}_k)$, where $d(x, y)$ is the Euclidean distance between points x and y. Once again, ties are not allowed in the preference lists. Before we describe our results, let us give a brief introduction to approximation-preserving (AP) reductions and AP-reducibility.

1.2 AP-Reducibility (A Brief Introduction)

A *randomized approximation scheme* is an algorithm for approximately computing the value of a function $f : \Sigma^* \to \mathbb{R}$. The approximation scheme has a parameter $\varepsilon > 0$ which specifies the error tolerance. A *randomized approximation scheme* for f is a randomized algorithm that takes as input an instance $x \in \Sigma^*$ (e.g., for the problem $\#SM$, the input would be an encoding of a stable matching instance) and a rational error tolerance $\varepsilon > 0$, and outputs a rational number z (a random variable of the "coin tosses" made by the algorithm) such that, for every instance x, $\Pr\left[e^{-\epsilon} f(x) \le z \le e^{\epsilon} f(x)\right] \ge \frac{3}{4}$. The randomized approximation scheme is said to be a *fully polynomial randomized approximation scheme*, or *FPRAS*, if it runs in time bounded by a polynomial in $|x|$ and ϵ^{-1}. Note that the quantity $3/4$ in the definition could be changed to any value in the open interval $(\frac{1}{2}, 1)$ without changing the set of problems that have randomized approximation schemes [17, Lemma 6.1].

We now define the notion of an approximation-preserving (AP) reduction. Suppose that f and g are functions from Σ^* to $\mathbb{R}$. Informally speaking, an AP-reduction from f to g gives a way to turn an FPRAS for g into an FPRAS for f. Here is the formal definition. An *approximation-preserving reduction* from f to g is a randomized algorithm $\mathcal{A}$ for computing f using an oracle for g. The algorithm $\mathcal{A}$ takes as input a pair $(x, \varepsilon) \in \Sigma^* \times (0, 1)$, and satisfies the following three conditions: (i) every oracle call made by $\mathcal{A}$ is of the form (w, δ), where $w \in \Sigma^*$ is an instance of g, and $0 < \delta < 1$ is an error bound satisfying $\delta^{-1} \le \text{poly}(|x|, \varepsilon^{-1})$; (ii) the algorithm $\mathcal{A}$ meets the specification for being a randomized approximation scheme for f (as described above) whenever the oracle meets the specification for being a randomized approximation scheme for g; and (iii) the run-time of $\mathcal{A}$ is polynomial in $|x|$ and ε^{-1}. We write $f \le_{AP} g$ to mean that f has an AP-reduction to g. Similarly, we write $f \equiv_{AP} g$ to mean that $f \le_{AP} g$ and $g \le_{AP} f$, or that f and g are AP-interreducible.

The complexity class $\#\text{RH}\Pi_1$ of counting problems was introduced by Dyer, Goldberg, Greenhill and Jerrum [7] as a means to classify a wide class of approximate counting problems that were previously of indeterminate computational complexity. The problems in $\#\text{RH}\Pi_1$ are those that can be expressed in terms of counting the number of models of a logical formula from a certain syntactically restricted class. Although the authors were not aware of it at the time, this syntactically restricted class had already been studied under the title "restricted Krom SNP" [5]. The complexity class $\#\text{RH}\Pi_1$ has a completeness class (with respect to AP-reductions) which includes a wide and ever-increasing range of natural counting problems, including: independent sets in a bipartite graph, downsets in a partial order, configurations in the Widom-Rowlinson model (all [7]) and the partition function of the ferromagnetic Ising model with mixed external field [10]. Either all of these problems have an FPRAS, or none do. No FPRAS is currently known for any of them, despite much effort having been expended on finding one.

All the problems in the completeness class mentioned above are inter-reducible via AP-reductions, so any of them could be said to exemplify the completeness class. However, mainly for historical reasons, the problem $\#BIS$, which is the problem of computing the number of independent sets in a given bipartite graph, tends to be taken as a key example in the class, much in the same way that SAT has a privileged status in the theory on NP-completeness.

Ge and Štefankovič [9] recently proposed an interesting new MCMC algorithm for sampling independent sets in bipartite graphs. Unfortunately, however, the relevant Markov chain mixes slowly [12] so even this interesting new idea does not give an FPRAS for $\#BIS$. In fact, Goldberg and Jerrum [11] conjecture that no FPRAS exists for $\#BIS$ (or for the other problems in the completeness class). We make this conjecture on empirical grounds, namely that the problem has survived its first decade despite considerable efforts to find an FPRAS and the collection of known $\#BIS$-equivalent problems is growing.

Since Dyer et al. show that $\#BIS$ and counting downsets are both complete in this class, and it is known that counting downsets is equivalent to counting stable matchings, the result of Dyer et al. implies $\#BIS \equiv_{AP} \#SM$. The goal of this paper is to demonstrate AP-interreducbility of $\#BIS$ with the two restricted stable matching problems defined in Section 1.1.

1.3 Our Results

In this paper we outline the following results:

Theorem 1. $\#BIS \equiv_{AP} \#SM(k-\text{attribute})$ *when* $k \geq 3$.

In other words, $\#BIS$ is AP-interreducible with counting stable matchings in the k-attribute setting when $k \geq 3$, so this problem is equivalent in terms of approximability to the complete problems in the complexity class $\#\text{RH}\mathit{\Pi}_1$.

Theorem 2. $\#SM(1-\text{attribute})$ *is solvable in polynomial time.*

We can also prove AP-interreducibility with $\#BIS$ in the k-Euclidean setting (when $k \geq 2$) in a similar manner. Recall that in the k-Euclidean setting, preference lists are determined by (closest) Euclidean distances between the "preference points" and "position points".

Theorem 3. $\#BIS \equiv_{AP} \#SM(k-\text{Euclidean})$ *when* $k \geq 2$.

Section 2 reviews some combinatorics of the stable matching problem that is relevant for our purposes in this paper. Section 3 gives the construction that can be used to demonstrate Theorem 1. Theorems 2 and 3 are not proven in this extended abstract.

2   Combinatorics of the Stable Matching Problem

The (classical) stable matching problem has a rich combinatorial structure which has been widely studied. We relate some aspects of this structure that we will need in this paper. Many of the definitions and results that follow can be found, for example, in [18,16,14,13].

2.1   The Gale-Shapley Algorithm

In their seminal paper on the stable matching problem, Gale and Shapley [8] gave a polynomial-time algorithm for constructing a stable matching. This is generally referred to as the "proposal algorithm" and bears the names of Gale and Shapley in all of the literature on stable matchings. One sex (typically the men) make proposals to members of the other, forming "engagements". Once all the "proposers" are engaged, the algorithm terminates with a stable matching. As noted by Gale and Shapley (and others), their algorithm computes the male-optimal stable matching, which is optimal in the very strong sense that every man likes his partner in this matching at least as much as his partner in any other stable matching. Given an instance with n men and n women, the algorithm computes the male-optimal stable matching in time $O(n^2)$.

During the algorithm, after a woman becomes "engaged" she never becomes free, though she might be engaged to different men at different times during the execution of the algorithm. On the other hand, a man could oscillate between being free and being engaged. It is well-known (see, e.g. [8,18]) that the male-optimal matching may be obtained by taking *any* ordering of the men and have them make proposals in that order, i.e. when "a free man M proposes..." we can take the highest free man in our ordering of the men to perform the next proposal. By reversing the roles of men and women (i.e. the women are the "proposers"), we can obtain the female-optimal stable matching.

2.2   Stable Matching Lattice

Given a matching instance and two stable matchings $\mathcal{M}$ and $\mathcal{M}'$ where

$$\mathcal{M} = \quad \{(M_1, w_1), \cdots, (M_n, w_n)\}, \quad \mathcal{M}' = \quad \{(M_1', w_1), \cdots, (M_n', w_n)\},$$

we define $\max\{M_i, M_i'\}$, $\min\{M_i, M_i'\}$, $\max\{\mathcal{M}, \mathcal{M}'\}$ and $\min\{\mathcal{M}, \mathcal{M}'\}$ as follows:

$$\max\{M_i, M_i'\} = \text{ favorite choice of woman } w_i \text{ between men } M_i \text{ and } M_i'$$
$$\min\{M_i, M_i'\} = \text{ least preferred choice of woman } w_i \text{ between men } M_i \text{ and } M_i'$$
$$\max\{\mathcal{M}, \mathcal{M}'\} = \{(\max\{M_1, M_1'\}, w_1), \cdots, (\max\{M_n, M_n'\}, w_n)\}$$
$$\min\{\mathcal{M}, \mathcal{M}'\} = \{(\min\{M_1, M_1'\}, w_1), \cdots, (\min\{M_n, M_n'\}, w_n)\}$$

Note that in the expression $\max\{M_i, M_i'\}$, the woman w_i can deduced from the arguments since she is the only woman married to M_i in $\mathcal{M}$ and in M_i' in $\mathcal{M}'$. From [18], we have that $\max\{\mathcal{M}, \mathcal{M}'\}$ and $\min\{\mathcal{M}, \mathcal{M}'\}$ are themselves stable matchings. Further, we define the relation $\mathcal{M} \leq \mathcal{M}'$ if and only if $\mathcal{M}' = \max\{\mathcal{M}, \mathcal{M}'\}$. It is clear that the relation $\leq$ is reflexive, antisymmetric, and transitive. Hence, the stable matchings of a stable matching instance form a lattice under the $\leq$ relation.

In fact, this lattice is a *distributive lattice* under the "max" and "min" operations defined above [18]. The male-optimal matching is the minimum element

in this lattice (under the $\leq$ relation), while the female-optimal matching is the maximum element.

It is well-known (see, for instance, [6]) that a finite distributive lattice is isomorphic to the lattice of *downsets* of another partial order (ordered by subset inclusion). We shall shortly see how this other downset lattice arises in the context of stable matchings, and its connection to the stable matching lattice.

2.3 Stable Pairs and Rotations

Definition 1. *A pair (M, w) is called* stable *if and only if (M, w) is a pair in some stable matching $\mathcal{M}$. A pair (M, w) that is not stable is called an* unstable pair.

Definition 2. *Let $\mathcal{M}$ be a stable matching. For any man M (woman w), let $sp_{\mathcal{M}}(M)$ ($sp_{\mathcal{M}}(w)$) denote the spouse of man M (woman w) in the matching $\mathcal{M}$.*

Definition 3. *[1] Let $\mathcal{M}$ be a stable matching. The* suitor *of a man M is defined to be the first woman w' on M's preference list such that (i) M prefers $sp_{\mathcal{M}}(M)$ over w' and (ii) w' prefers M over $sp_{\mathcal{M}}(w)$. The suitor of man M is denoted by $S_{\mathcal{M}}(M)$.*

We note that $S_{\mathcal{M}}(M)$ may not exist for every man. For instance, if $\mathcal{M}$ is the female-optimal stable matching, then $S_{\mathcal{M}}(M)$ would not exist.

Definition 4. *[16] Let $\mathcal{M}$ be a stable matching. Let $R = \{(M_0, w_0), (M_1, w_1), \cdots, (M_{k-1}, w_{k-1})\}$ be an ordered list of pairs from $\mathcal{M}$ such that for every i, $0 \leq i \leq k-1$, $S_{\mathcal{M}}(M_i)$ is $w_{i+1 (\bmod k)}$. Then R is a* rotation *(exposed in the matching $\mathcal{M}$).*

A stable matching may have many or no exposed rotations. Applying an exposed rotation to a stable matching (i.e. breaking the pairs (M_i, w_i) and forming the new pairs (M_i, w_{i+1})) gives a new stable matching in which the women are "happier" and the men are less happy. In other words, after a rotation, every woman (respectively, man) involved in the rotation is married to someone higher (resp. lower) on her (resp. his) preference list than her (resp. his) partner in the rotation.

We can similarly define suitors for the women, given some stable matching $\mathcal{M}$. We do not need to do so for the purposes of this paper, but the Markov chain that Bhatnagar, et al. examine in [1] consists of moves that are "male-improving" and "female-improving" rotations. Starting from any stable matching, it is possible to obtain any other stable matching using some (appropriately chosen) sequence of male-improving and/or female-improving rotations [16].

Definition 5. *[13] A pair (M, w), not necessarily stable, is said to be* eliminated *by the rotation R if R moves w from M or below on her preference list to a man strictly above M.*

Note that if a stable pair (M, w) in a rotation R is eliminated by R, and if (M, w') is any other pair eliminated by R, then man M prefers w over w', for otherwise no matching that has R exposed in it could be stable.

Lemma 1. *[16] No pair is eliminated by more than one rotation, and for any pair (M, w), at most one rotation moves M to w.*

We can now define a relation on rotations.

Definition 6. *[16] Let R and R' be two distinct rotations. Rotation R is said to explicitly precede R' if and only if R eliminates a pair (M, w), and R' moves M to a woman w' such that M (strictly) prefers w to w'. The relation "precedes" is defined as the transitive closure of the "explicitly precedes" relation.*

If a rotation R explicitly precedes R' then there is no stable matching with R' exposed such that applying R' results in a stable matching with R exposed — the intermediate matching would have a blocking pair (hence would not be stable). The relation *precedes* ($\leq$) defines a partial order on the set of rotations of the stable matching instance. We call the partial order on the set of rotations the *rotation poset* of the instance and denote it $(P, \leq)$. The following theorem relates the rotations in the rotation poset to the stable matchings of the instance via the downsets of P.

Theorem 4. *[16, Theorem 4.1] For any stable matching instance, there is a one-to-one correspondence between the stable matchings of that instance and the downsets of its rotation poset.*

Every stable matching of the instance can be obtained by starting with the male-optimal stable matching and performing the rotations in the corresponding downset (ensuring that a rotation is performed before any rotation that succeeds it is performed). Note that the downsets corresponding to the male-optimal stable matching and the female-optimal stable matching are $\emptyset$ and P, respectively.

To construct the rotation poset, we need (i) the rotations and (ii) the precedence relations between them. We note that once we have all the rotations in the poset, we can establish the precedence relations using the "explicitly precedes" relation, i.e. by determining which (stable or unstable) pairs are eliminated by each rotation.

Gusfield [13] gave an algorithm that runs in $O(n^2)$ time for finding all rotations of a stable matching instance. His algorithm is a refinement of successive applications of the "*breakmarriage*" procedure of McVitie and Wilson [19]. For the sake of presentation, we can use a slower variant of his algorithm (the variant still runs in polynomial time, which suffices for our purposes).

2.4   #BIS, Independent Sets, and Stable Matchings

The rotation poset for a matching instance plays a key role in what follows. To prove Theorem 1, we take a #BIS instance $G = (V_1 \cup V_2, E)$ and view this as

the rotation poset of a matching instance. In particular, G is the Hasse diagram of the poset when we draw G with the set V_2 "above" V_1.

Each independent set in the bipartite graph naturally corresponds to a downset in the partial order, and vice-versa. See Figure 1 for an example. An independent set, namely $\{d, f, g\}$, is shown in the left of that figure. The corresponding downset is shown on the right. This downset is obtained by taking the set $\{d, f, g\}$ and adding the two elements a and b, as $a < f$ and $b < f$ (and $b < g$) in the Hasse diagram. Conversely, given a downset, such as the one on the right of the diagram, we can find the corresponding independent set in G by taking the set of maximal elements of the downset. So given G, we then construct a matching instance (using 3-dimensional preference and attribute vectors) whose rotation poset is (isomorphic to) G, giving a 1-1 correspondence for our AP-reduction from $\#BIS$ to $\#SM(k\text{-attribute})$, showing that $\#BIS \leq_{AP} \#SM(k\text{-attribute})$. The reverse implication $\#SM(k\text{-attribute}) \leq_{AP} \#BIS$ follows from the two results that $\#SM \leq_{AP} \#Downsets$ (Theorem 4, quoted here from [16]) and $\#Downsets \leq \#BIS$ [7, Lemma 9], where $\#Downsets$ is the problem of counting the number of downsets in a partial order.

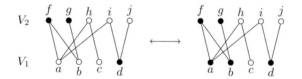

Fig. 1. The correspondence between independent sets and downsets

3   Stable Matchings in the k-attribute Model ($k \geq 3$)

In this section we give our construction to show AP-reducibility from $\#BIS$ to the k-attribute stable matching model when $k \geq 3$. Given our previous remarks about the relation between $\#BIS$, independent sets, and stable matchings, our procedure is as follows:

1. Let $G = (V_1 \cup V_2, E)$ denote a bipartite graph where $|E| = n$. Our goal will be to construct a k-attribute stable matching instance for which we can show that the Hasse diagram of its rotation poset is G. This will give a bijection between stable matchings and downsets of G, hence a bijection between stable matchings and independent sets of G.

2. Using G, in the manner to be specified in Section 3.1, we construct preference lists for a 3-attribute stable matching instance with $3n$ men and $3n$ women.

3. Given this matching instance, we find the male-optimal and female-optimal matchings.

4. Using the **Find-All-Rotations** algorithm, we extract the rotations from our stable matching instance.

5. Having these rotations, we construct the partial order, P, on these rotations (specified by the transitive closure of the "explicitly precedes" relation).

6. We finally show that P is isomorphic to G (when G is viewed as a partial order), thereby showing our construction is an approximation-preserving reduction from $\#BIS$ to $\#SM(3-\text{attribute})$.

Due to space constraints, we are only able to specify the position and preference vectors of the men and women. The proof that these vectors give rise to a matching instance whose rotation poset is isomorphic to G will appear in the full version of the paper.

3.1 Construction of the Stable Matching Instance

BIS and Permutations. Let $G = (V_1 \cup V_2, E)$ denote our BIS instance, where $E \subseteq V_1 \times V_2$ and $|E| = n$. Using G we will construct a 3-attribute stable matching instance with $3n$ men and $3n$ women. The men and women of the instance are denoted $\{A_1, \ldots, A_n, B_1, \ldots, B_n, C_1, \ldots, C_n\}$ and $\{a_1, \ldots, a_n, b_1, \ldots, b_n, c_1, \ldots, c_n\}$, respectively. To describe our construction, we label the edges of G B_1 through B_n from "left-to-right" with respect to the vertices (V_1) on the bottom. This becomes more clear from the example in Figure 2. We refer to edge B_i as man B_i, and this will be clear from the context.

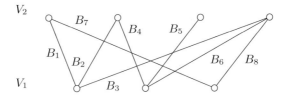

Fig. 2. A BIS instance and our labeling of its edges

For our construction we associate two permutations, ρ and σ, of $[n] = \{1, \ldots, n\}$ with the BIS instance. The cycles of ρ correspond to vertices in V_1 and those of σ correspond to vertices in V_2. In other words, if the edges incident to a vertex in V_2 are $B_{i_1}, B_{i_2}, \ldots, B_{i_d}$, then $(i_1, i_2, \ldots, i_d)$ is a σ-cycle. We define ρ-cycles in a similar fashion. If G has $k = |V_1|$ vertices on the bottom and $l = |V_2|$ vertices on the top, then the permutations ρ and σ have k and l cycles, respectively. Since the graph G will turn out to be isomorphic to a rotation poset, every vertex in G will represent a rotation in the stable matching instance. The rotations of the stable matching instance will be governed by the ρ- and σ-cycles in a manner to be specified. The rotations corresponding to the ρ-cycles will be called ρ-rotations and those corresponding to the σ-cycles will be called σ-rotations.

In the example of Figure 2, the three ρ-cycles are $\rho_1 = (1, 2, 3), \rho_2 = (4, 5, 6)$, and $\rho_3 = (7, 8)$. The four σ-cycles are $\sigma_1 = (1, 7), \sigma_2 = (2, 4), \sigma_3 = (5)$, and $\sigma_4 = (3, 6, 8)$.

Here is a brief overview of how we go about constructing a stable matching instance from a given bipartite graph. First of all, the male-optimal stable matching that we construct in our matching instance will consist of the pairs $(A_i, a_i), (B_i, b_i), (C_i, c_i)$ for all $i \in [n]$. (We must show later this is indeed the case for the construction we describe.) A ρ-cycle of the form $(i_1, i_1 + 1, \ldots, i_2)$ will correspond to the ρ-rotation, R, of the form $\{(B_{i_1}, b_{i_1}), (A_{i_1}, a_{i_1}), (B_{i_1+1}, b_{i_1+1}),$ $(A_{i_1+1}, a_{i_1+1}), \ldots, (B_{i_2}, b_{i_2}), (A_{i_2}, a_{i_2})\}$. This rotation R arises from a vertex $v \in V_1$ with edges $B_{i_1}, B_{i_1+1}, \ldots, B_{i_2}$ incident to it. We can show that a σ-rotation R' is of the form $\{(B_{i_1}, a_{i_1}), (C_{i_1}, c_{i_1}), (B_{i_2}, a_{i_2}), (C_{i_2}, c_{i_2}), \ldots, (B_{i_p}, a_{i_p}), (C_{i_p}, c_{i_p})\}$, where $(i_1, i_2, \ldots, i_p)$ is the corresponding σ-cycle, and that the rotation R' corresponds to the vertex $v' \in V_2$ with edges $B_{i_1}, B_{i_2}, \ldots, B_{i_p}$ incident to it. In this manner, every rotation in the rotation poset is defined in terms of the men involved in them, the women being the (then-current) partners of the men that are in the rotation. Assuming that the above two claims regarding rotations are valid (as we will show below), we make the following observation.

Observation 5. *A ρ-cycle and a σ-cycle can have at most one element in common. (This is because G is a graph and not a multi-graph.) This means that a ρ-rotation and a σ-rotation can have at most one man in common. This similarly holds for the women.*

Assigning Preference and Position Vectors. Suppose $D_1, \ldots, D_l$ are the l cycles of σ of lengths $p_1, \ldots, p_l$, respectively. Let e_i be a representative element of cycle D_i. In other words, we can represent the σ-cycle D_i as $D_i = \left(e_i, \sigma(e_i), \ldots, \sigma^{p_i-1}(e_i)\right)$. (We may, for example, select e_i to be the smallest number in the cycle, and we will do so here). In what follows we will often abbreviate $\sigma x = \sigma(x), \sigma^2 x = \sigma^2(x), \sigma^{-1}x = \sigma^{-1}(x)$, etc, and, similarly, $\rho x = \rho(x)$, etc. Let $Rep(\sigma) = \{e_1, e_2 \cdots, e_l\}$ be the set of representative elements we choose for the cycles of σ. Let $W_i = \{a_x : x \in D_i\} \cup \{b_{\rho x} : x \in D_i\} \cup \{c_{\sigma^{-1}x} : x \in D_i\}$. Let $T(x) = \{c_{\sigma^{-1}x}, a_x, b_{\rho x}\}$ where $x \in D_i$. It follows that $W_i = \cup_{x \in D_i} T(x)$ and $T(i) \cap T(j) = \emptyset$ for $i \neq j$.

Using the definitions above, first we fix the position vectors of the women. The z-coordinate of women a_i and c_i is set to 0 for $1 \leq i \leq n$. The z-coordinate of woman b_i is set to 4^i for $1 \leq i \leq n$. The x- and y-coordinates of a_i, b_i, and c_i are such that the projection of each women's position vector onto the x-y plane lies on the unit circle $x^2 + y^2 = 1$. Furthermore, we group the projections according to the sets W_i. In other words, all women in W_i are embedded in an angle of ϵ on the unit circle, where $\epsilon = 2\pi/n^2$. These groups are embedded around the circle in the order W_1 through W_l, and the angle between two adjacent groups is $(2\pi - l\epsilon)/l$. Note that W_l is adjacent to W_{l-1} and W_1. Group W_i starts at angle $2\pi(i-1)/l$ and ends at $2\pi(i-1)/l + \epsilon$. Within the group W_i, the women are further sub-grouped into triplets $T(e_i), T(\sigma(e_i)), \ldots, T(\sigma^{p_i-1}(e_i))$.

Within the angle of size ϵ, the sub-groups are embedded in the order $T(e_i)$ through $T(\sigma^{p_i-1}(e_i))$, with each $T(\cdot)$ spanning an angle of $6\theta_i$. The angle between two adjacent $T(\cdot)$'s is θ_i, where $\theta_i = \epsilon/(7p_i - 1)$. Within each $T(x)$, the women appear in the order $c_{\sigma^{-1}x}, a_x$, and $b_{\rho x}$, and the angle between $\bar{c}_{\sigma^{-1}x}$ and $\bar{a}_x$ is $4\theta_i$, and the angle between $\bar{a}_x$ and $\bar{b}_{\rho x}$ is $2\theta_i$. We summarize the above description by giving the exact coordinates for the position vector for the women.

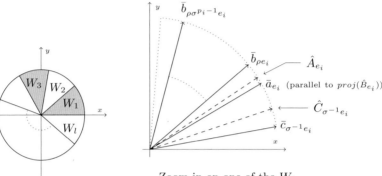

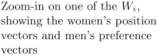

Arrange (projections of) position vectors for each W_i inside the sectors shown

Zoom-in on one of the W_i, showing the women's position vectors and men's preference vectors

Fig. 3. Placement of the women's position vectors and men's preference vectors

Let $\epsilon = \dfrac{2\pi}{n^2}$. For $e_i \in Rep(\sigma)$, let $\theta_i = \dfrac{\epsilon}{7p_i - 1}$. Then for $0 \le m \le p_i - 1$ define

$$\bar{a}_{\sigma^m e_i} = \left(\cos(2\pi(i-1)/l + 7m\theta_i + 4\theta_i),\ \sin(2\pi(i-1)/l + 7m\theta_i + 4\theta_i),\ 0\right),$$
$$\bar{c}_{\sigma^{m-1} e_i} = \left(\cos(2\pi(i-1)/l + 7m\theta_i),\ \sin(2\pi(i-1)/l + 7m\theta_i),\ 0\right),\ \text{and}$$
$$\bar{b}_{\rho\sigma^m e_i} = \left(\cos(2\pi(i-1)/l + 7m\theta_i + 6\theta_i),\ \sin(2\pi(i-1)/l + 7m\theta_i + 6\theta_i),\ 4^{\rho\sigma^m e_i}\right).$$

Next we define the preference vectors of the men. The z-coordinates of all $\hat{A}_i$ and $\hat{C}_i$ are set to 0. We place $\hat{A}_i$ between $\bar{a}_i$ and the projection onto the x-y plane of $\bar{b}_{\rho i}$. If the angle between $\bar{a}_i$ and (the projection of) $\bar{b}_{\rho i}$ is α, then the angle between $\bar{a}_i$ and $\hat{A}_i$ is $\frac{1}{3}\alpha$, and the angle between $\hat{A}_i$ and (the projection of) $\bar{b}_{\rho i}$ is $\frac{2}{3}\alpha$. This will ensure that A_i prefers a_i over $b_{\rho i}$. We will later show that the preference list of A_i starts with $a_i b_{\rho i}$. We place $\hat{C}_i$ between $\bar{c}_i$ and $\bar{a}_{\sigma i}$ such that if the angle between $\bar{c}_i$ and $\bar{a}_{\sigma i}$ is β, then the angle between $\bar{c}_i$ and $\hat{C}_i$ is $\frac{2}{5}\beta$ and the angle between $\hat{C}_i$ and $\bar{a}_{\sigma i}$ is $\frac{3}{5}\beta$. This will ensure that C_i prefers c_i over $a_{\sigma i}$. We will later show that the preference list of C_i starts with $c_i a_{\sigma i}$. Finally, we place $\hat{B}_i$, which is of unit length, such that $\hat{B}_i$ makes an angle of $\phi = 2\pi/100$ with the vertical axis (z-axis) and its projection on the x-y plane is parallel to $\bar{a}_i$. In other words, the projection of $\hat{B}_i$ on the $z = 0$ plane is $\sin\phi\,\bar{a}_i$. We summarize the above discussion by providing the exact coordinates of $\hat{A}_i$, $\hat{B}_i$, and $\hat{C}_i$.

Let $\epsilon = \dfrac{2\pi}{n^2}$, $\phi = \dfrac{2\pi}{100}$ and $\gamma_i = \dfrac{2\pi(i-1)}{l}$. For $e_i \in Rep(\sigma)$, let $\theta_i = \dfrac{\epsilon}{7p_i - 1}$.

Then for $0 \le m \le p_i - 1$ define

$$\hat{A}_{\sigma^m e_i} = (\cos(\gamma_i + 7m\theta_i + (14/3)\theta_i),\ \sin(\gamma_i + 7m\theta_i + (14/3)\theta_i),\ 0),$$

$$\hat{B}_{\sigma^m e_i} = (\sin\phi\cos(\gamma_i + 7m\theta_i + 4\theta_i),\ \sin\phi\sin(\gamma_i + 7m\theta_i + 4\theta_i),\ \cos\phi), \text{ and}$$

$$\hat{C}_{\sigma^{m-1} e_i} = (\cos(\gamma_i + 7m\theta_i + (8/5)\theta_i),\ \sin(\gamma_i + 7m\theta_i + (8/5)\theta_i),\ 0).$$

We can similarly define the position vectors of the men and preference vectors of the women. Because of space constraints, we simply give these vectors below.

Let $\epsilon = \dfrac{2\pi}{n^2}$, $\phi = \dfrac{2\pi}{100}$ and $\eta_i = \dfrac{2\pi(i-1)}{k}$.   For $f_i \in Rep(\rho)$, let $\omega_i = \dfrac{\epsilon}{7q_i - 1}$.

Then, for $0 \le m \le q_i - 1$, we define

$$\bar{A}_{\rho^{m-1} f_i} = (\cos(\eta_i + 7m\omega_i),\ \sin(\eta_i + 7m\omega_i),\ 0),$$

$$\bar{B}_{\rho^m f_i} = (\cos(\eta_i + 7m\omega_i + 4\omega_i),\ \sin(\eta_i + 7m\omega_i + 4\omega_i),\ 0), \text{ and}$$

$$\bar{C}_{\rho^m f_i} = (\cos(\eta_i + 7m\omega_i + 6\omega_i),\ \sin(\eta_i + 7m\omega_i + 6\omega_i),\ 4^{\rho^m f_i}). \quad \text{Also,}$$

$$\hat{a}_{\rho^m f_i} = (\sin\phi\cos(\eta_i + 7m\omega_i + 4\omega_i),\ \sin\phi\sin(\eta_i + 7m\omega_i + 4\omega_i),\ \cos\phi),$$

$$\hat{b}_{\rho^m f_i} = (\cos(\eta_i + 7m\omega_i + (8/5)\omega_i),\ \sin(\eta_i + 7m\omega_i + 8/5\omega_i),\ 0), \text{ and}$$

$$\hat{c}_{\rho^m f_i} = (\cos(\eta_i + 7m\omega_i + 14/3\omega_i),\ \sin(\eta_i + 7m\omega_i + (14/3)\omega_i),\ 0).$$

Using the vectors we define above, we can show that the preference lists of men A_i, C_i and B_i and women b_i, c_i and a_i begin as follows:

for $e_i \in Rep(\sigma)$ and $f_i \in Rep(\rho)$,

$$A_{\sigma^m e_i} : \quad a_{\sigma^m e_i} b_{\rho\sigma^m e_i} \ , \quad 0 \le m \le p_i - 1,$$

$$C_{\sigma^{(m-1)} e_i} : \quad c_{\sigma^{(m-1)} e_i} a_{\sigma^m e_i} \ , \quad 0 \le m \le p_i - 1,$$

$$B_{\sigma^m e_i} : \quad b_n \cdots b_1 a_{\sigma^m e_i} c_{\sigma^m e_i} \ , \quad 0 \le m \le p_i - 2, \text{ and}$$

$$B_{\sigma^{(p_i-1)} e_i} : \quad b_n \cdots b_1 a_{\sigma^{(p_i-1)} e_i} c_{\sigma^{(p_i-2)} e_i} a_{\sigma^{(p_i-2)} e_i} \cdots a_{\sigma e_i} c_{e_i} a_{e_i} c_{\sigma^{(p_i-1)} e_i}. \quad \text{Also,}$$

$$b_{\rho^m f_i} : \quad A_{\rho^{(m-1)} f_i} B_{\rho^m f_i} \ , \quad 0 \le m \le q_i - 1,$$

$$c_{\rho^m f_i} : \quad B_{\rho^m f_i} C_{\rho^m f_i} \ , \quad 0 \le m \le q_i - 1,$$

$$a_{\rho^m f_i} : \quad C_n \cdots C_1 B_{\rho^m f_i} A_{\rho^m f_i} \ , \quad 0 \le m \le q_i - 2, \text{ and}$$

$$a_{\rho^{(q_i-1)} f_i} : \quad C_n \cdots C_1 B_{\rho^{(q_i-1)} f_i} A_{\rho^{(q_i-2)} f_i} \cdots B_{\rho^2 f_i} A_{\rho f_i} B_{\rho f_i} A_{f_i} B_{f_i} A_{\rho^{(q_i-1)} f_i}.$$

We have not specified the entire preference lists for the men and women. The remaining portion of each preference list appears *after* the part that we have given above, and there will never be any stable pairs involving a man/woman pair that is not shown on the partial preference lists given. The partial lists we have given are sufficient to find the male- and female-optimal matchings, and they contain the necessary information to generate *all* of the stable matchings for our constructed instance, or equivalently, to find all of the rotations for this instance.

References

1. Bhatnagar, N., Greenberg, S., Randall, D.: Sampling stable marriages: why spouse-swapping won't work. In: Proc. 19th Annual ACM-SIAM Symposium on Discrete Algorithms (SODA 2008), pp. 1223–1232 (2008)
2. Blair, C.: Every finite distributive lattice is a set of stable matchings. J. Combinatorial Theory (A) 37, 353–356 (1984)
3. Bogomolnaia, A., Laslier, J.-F.: Euclidean preferences. J. Mathematical Economics 43, 87–98 (2007)
4. Canadian Resident Matching Service,
   `http://www.carms.ca/eng/operations_algorithm_e.shtml`
5. Dalmau, V.: Linear datalog and bounded path duality of relational structures. Logical Methods in Computer Science 1, 1–32 (2005)
6. Davey, B.A., Priestley, H.A.: Introduction to Lattices and Order. Cambridge University Press, Cambridge (1990)
7. Dyer, M., Goldberg, L.A., Greenhill, C., Jerrum, M.: The relative complexity of approximate counting problems. Algorithmica 38, 471–500 (2004)
8. Gale, D., Shapley, L.S.: College admissions and the stability of marriage. American Mathematical Monthly 69, 9–15 (1962)
9. Ge, Q., Štefankovič, D.: A graph polynomial for independent sets of bipartite graphs (2009), `http://arxiv.org/abs/0911.4732`
10. Goldberg, L.A., Jerrum, M.: The complexity of ferromagnetic Ising with local fields. Combinatorics, Probability & Computing 16, 43–61 (2007)
11. Goldberg, L.A., Jerrum, M.: Approximating the partition function of the ferromagnetic Potts model (2010), `http://arxiv.org/abs/1002.0986`
12. Goldberg, L.A., Jerrum, M.: Counterexample to rapid mixing of the GS Process, technical note (March 2010)
13. Gusfield, D.: Three fast algorithms for four problems in stable marriage. SIAM J. Computing 16, 111–128 (1987)
14. Gusfield, D., Irving, R.W.: The Stable Marriage Problem: Structure and Algortihms. MIT Press, Boston (1989)
15. Gusfield, D., Irving, R.W., Leather, P., Saks, M.: Every finite distributive lattice is a set of stable matchings for a small stable marriage instance. J. Combinatorial Theory (A) 44, 304–309 (1987)
16. Irving, R.W., Leather, P.: The complexity of counting stable marriages. SIAM J. Computing 15, 655–667 (1986)
17. Jerrum, M.R., Valiant, L.G., Vazirani, V.V.: Random generation of combinatorial structures from a uniform distribution. Threoretical Computer Science 43, 169–188 (1986)
18. Knuth, D.E.: Stable Marriage and its Relation to Other Combinatorial Problems. American Mathematical Society, Providence (1997) (English edition)
19. McVitie, D., Wilson, L.: The stable marriage problem. Communications of the ACM 14, 486–490 (1971)
20. National Resident Matching Program,
    `http://www.nrmp.org/res_match/about_res/algorithms.html`
21. Provan, J.S., Ball, M.O.: The complexity of counting cuts and of computing the probability that a graph is connected. SIAM J. Computing 12, 777–788 (1983)
22. Scottish Foundation Allocation Scheme,
    `http://www.nes.scot.nhs.uk/sfas/About/default.asp`

Constant Approximation Algorithms for Embedding Graph Metrics into Trees and Outerplanar Graphs[*]

Victor Chepoi[1], Feodor F. Dragan[2], Ilan Newman[3],
Yuri Rabinovich[3], and Yann Vaxès[1]

[1] LIF, Universitée d'Aix-Marseille,
13288 Marseille, France
{chepoi,vaxes}@lif.univ-mrs.fr
[2] Computer Science Department, Kent State University,
Kent, OH 44242, USA
dragan@cs.kent.edu
[3] Department of Computer Science, University of Haifa,
Mount Carmel, Haifa 31905, Israel
{ilan,yuri}@cs.haifa.ac.il

Abstract. We present a simple factor 6 algorithm for approximating the optimal multiplicative distortion of embedding (unweighted) graph metrics into tree metrics (thus improving and simplifying the factor 100 and 27 algorithms of Bădoiu et al. (2007) and Bădoiu et al. (2008)). We also present a constant factor algorithm for approximating the optimal distortion of embedding graph metrics into outerplanar metrics. For this, we introduce a notion of metric relaxed minor and show that if G contains an α-metric relaxed H-minor, then the distortion of any embedding of G into any metric induced by a H-minor free graph is $\geq \alpha$. Then, for $H = K_{2,3}$, we present an algorithm which either finds an α-relaxed minor, or produces an $O(\alpha)$-embedding into an outerplanar metric.

1 Introduction

1.1 Avant-Propos

The structure of the shortest-path metrics of special classes of graphs, in particular, graphs families defined by forbidden minors (e.g., line metrics, tree metrics, planar metrics) is one of the main areas in the theory of metric spaces. From the algorithmic point of view, such metrics have more structure than general metrics, and this structure can often be exploited algorithmically. Thus, if the input metric can be well approximated by a special metric, this usually leads to an algorithmic advantage; see, e.g., [13] for a survey of algorithmic applications of embeddings. One way of understanding this structure is to study the low distortion embeddings from one metric class to˙another. To do this successfully,

---

[*] This research was partly supported by the ANR grant BLANC GGAA.

M. Serna et al. (Eds.): APPROX and RANDOM 2010, LNCS 6302, pp. 95–109, 2010.

one needs to develop tools allowing a decomposition of the host space consistent with the embedded space. If this is impossible, one usually learns much about the limitations of the host space and the richness of the embedded space. In this paper, we pursue this direction and study the embeddings into tree metrics and the metrics of $K_{2,3}$-minor free graphs (essentially outerplanar metrics).

The study of tree metrics can be traced back to the beginning of the 20th century, when it was first realized that weighted trees can in some cases serve as an (approximate) model for the description of evolving systems. More recently, as indicated in [16], it was observed that certain Internet originated metrics display tree-like properties. It is well known [17] that tree metrics have a simple structure: d is a tree metric iff all submetrics of d of size 4 are such. Moreover, the underlying tree is unique, easily reconstructible, and has rigid local structure corresponding to the local structure of d. But what about the structure of *approximately* tree metrics? We have only partial answers for this question, and yet what we already know seems to indicate that a rich theory might well be hiding there. The strongest results were obtained, so far, for the *additive* distortion. A research on the algorithmical aspects of finding a tree metric of least additive distortion has culminated in the paper [1] (see also [8]), where a 6-approximation algorithm was established (in the notation of [1], their algorithm is a 3 approximation, however, in our more restrictive definition, this is a 6-approximation), together with a (rather close) hardness result. Relaxing the local condition on d by allowing its size-4 submetrics to be δ-close to a tree metric, one gets precisely Gromov's δ-hyperbolic geometry. For study of algorithmic and other aspects of such geometries, see e.g. [7,14]. The situation with the *multiplicative* distortion is less satisfactory. The best result for general metrics is obtained in [4]: the approximation factor is exponential in $\sqrt{\log \Delta}/\log \log n$, where Δ is the aspect ratio. Judging from the parallel results of [2] for line metrics, it is conceivable that any constant factor approximation for the general metric is NP-hard. For some small constant γ, the hardness result of [1] implies that it is NP-hard to approximate the multiplicative distortion better than γ even for metrics that come from unit-weighted graphs. For a special interesting case of shortest path metrics of *unit-weighted* graphs, [4] gets a large (around 100) constant approximation factor (which was improved in [3] to a factor 27). The proof introduces a certain metric-topological obstacle for getting embeddings of distortion better than α, and then algorithmically either produces an $O(\alpha)$-embedding, or an α-obstacle (such an obstacle was used also in [11], and, essentially, in [15]).

1.2   Our Results

In this paper, we simplify and improve the construction of [4], using a decomposition procedure developed earlier in [5,6]. The improved constant is 6 and the running time of the algorithm is linear once the distance matrix is computed. We also introduce the notion of metric relaxed minor and show that if G contains an α-metric relaxed H-minor, then the distortion of any embedding of G into any metric induced by a H-minor free graph is at least α. This generalizes the obstacle of [4]. Using this newly defined H-obstacle, we show that it is an essential obstacle not only for trees, but also for graphs without $H = K_{2,3}$ minors. We further develop an efficient algorithm which either embeds the input metric induced by a unit-weighted graph G into an outerplanar metric with distortion $O(\alpha)$, or finds an α-metric relaxed $K_{2,3}$-minor in G. This is a first result of this kind for any H different from a C_4 (which is the α-metric relaxed minor corresponding to the four-point condition used for embedding into tree-metrics).

1.3   Preliminaries

A metric space (X, d) is *isometrically embeddable* into a host metric space (Y, d') if there exists a map $\varphi : X \mapsto Y$ such that $d'(\varphi(x), \varphi(y)) = d(x, y)$ for all $x, y \in X$. More generally, $\varphi : X \mapsto Y$ is an *embedding with (multiplicative) distortion* $\lambda \geq 1$ if $d(x, y) \leq d'(\varphi(x), \varphi(y)) \leq \lambda \cdot d(x, y)$ for all $x, y \in X$. Given a metric space (X, d) and a class $\mathcal{M}$ of host metric spaces, we denote by $\lambda^* := \lambda^*(X, \mathcal{M})$ the minimum distortion of an embedding of (X, d) into a member of $\mathcal{M}$. Analogously, $\varphi : X \mapsto Y$ is an *embedding with additive distortion* $\lambda \geq 0$ if $d(x, y) \leq d'(\varphi(x), \varphi(y)) \leq d(x, y) + \lambda$ for all $x, y \in X$ and, in a similar way, we can define the minimum additive distortion. In this paper, we consider unweighted graphs as input metric spaces and *tree metrics* (trees) or *outerplanar metrics* as the class of host metric spaces. If not specified, all our results concern embeddings with multiplicative distortion. For a connected unweighted graph $G = (V, E)$, we denote by $d_G(u, v)$ the shortest-path distance between u and v. A finite metric space (X, d) is called a *tree metric* [17] if it isometrically embeds into a tree, i.e., there exists an weighted tree $T = (X', E')$ such that $X \subseteq X'$ and $d(u, v) = d_T(u, v)$ for any two points $u, v \in X$, where $d_T(u, v)$ is the length of the unique path connecting u and v in T. Analogously, an *outerplanar metric* is a metric space isometrically embeddable into an outerplanar weighted graph. Denote by $\mathcal{T}$ the class of tree metric spaces and by $\mathcal{O}$ the class of outerplanar metric spaces.

2   Preliminary Results

In this section, we establish some properties of layering partitions and of embeddings with distortion λ of graph metrics into weighted graphs.

2.1   Layering Partitions

The layering partitions have been introduced in [5,6] and recently used in a slightly more general forms in both approximation algorithms of [3,4] and in other similar contexts [7,9,10]. Let $G = (V, E)$ be a graph with a distinguished vertex s and let $r := \max\{d_G(s, x) : x \in V\}$. A *layering* of G with respect to s is the decomposition of V into the *spheres* $L^i = \{u \in V : d(s, u) = i\}$, $i = 0, 1, 2, \ldots, r$. A *layering partition* $\mathcal{LP}(s) = \{L_1^i, \ldots, L_{p_i}^i : i = 0, 1, 2, \ldots, r\}$ of G is a partition of each L^i into *clusters* $L_1^i, \ldots, L_{p_i}^i$ such that two vertices $u, v \in L^i$ belong to the same cluster L_j^i iff they can be connected by a path outside the ball $B_{i-1}(s)$ of radius $i-1$ centered at s. Let Γ be a graph whose vertex set is the set of all clusters L_j^i in a layering partition $\mathcal{LP}$ and $C = L_j^i$ and $C' = L_{j'}^{i'}$ are adjacent in Γ iff there exist $u \in L_j^i$ and $v \in L_{j'}^{i'}$ such that u and v are adjacent in G. Γ is a tree [6], called the *layering tree* of G. $\mathcal{LP}$ and Γ are computable in linear time [6]. We can construct a new tree $H = (V, F)$ (closely reproducing the global structure of Γ) by identifying for each cluster $C = L_j^i$ an arbitrary vertex $x_C \in L^{i-1}$ (the *support vertex* for cluster C) which has a neighbor in C and by making x_C adjacent in H with all vertices $v \in C$. In what follows, we assume that Γ and H are rooted at s. Let D be the largest diameter of a cluster in $\mathcal{LP}$, i.e., $D := \max_{C \in \mathcal{LP}} \max_{v,u \in C} \{d_G(u,v)\}$. The following result (also implicitly used in [5,6,7]) shows that the additive distortion of the embedding of G into H is essentially D:

Proposition 1. *If $x, y \in V$, then $d_H(x,y) - 2 \le d_G(x,y) \le d_H(x,y) + D$.*

Proof. Let C_x and C_y be the clusters containing x and y. Let C be the nearest common ancestor of C_x and C_y in Γ. For $C \ne C_x$, let $x', y' \in C$ be the ancestors of x and y in a $BFS(G, s)$-tree. Then $d_\Gamma(C_x, C) = d_G(x, x')$ and $d_\Gamma(C_y, C) = d_G(y, y')$. By construction of H, $d_H(x, y)$ is equal to $d_\Gamma(C_x, C) + d_\Gamma(C_y, C)$ or to $d_\Gamma(C_x, C) + d_\Gamma(C_y, C) + 2$. By the triangle inequality, $d_G(x, y) \le d_G(x, x') + d_G(x', y') + d_G(y, y') \le d_\Gamma(C_x, C) + d_\Gamma(C_y, C) + D \le d_H(x, y) + D$. By definition of clusters, $d_G(x, y) \ge d_G(x, x') + d_G(y, y') \ge d_H(x, y) - 2$. $\square$

The BFS-tree H preserves the distances between the root s and any other vertex of G. We can locally modify H by assigning uniform weights to its edges or by adding Steiner points to obtain a number of other desired properties. Assigning length $w := D + 1$ to each edge of H, we will get a tree $H_w = (V, F, w)$ in which G embeds with distortion essentially equal to $D + 1$: $d_G(u, v) \le d_{H_w}(u, v) \le (D+1)(d_G(u, v) + 2) \ \forall u, v \in V$. Adding Steiner points and using edge lengths 0 and 1, H can be transformed into a tree H' which has the same additive distortion and satisfies the non-expansive property. For this, for each cluster $C := L_j^i$ we introduce a Steiner point p_C, and add an edge of length 0 between any vertex of C and p_C and an edge of length 1 between p_C and the support vertex x_C for C: $d_{H'}(u, v) \le d_G(u, v) \le d_{H'}(u, v) + D \ \forall u, v \in V$. Finally, by replacing each edge in H' with edge of length $w := \frac{D+1}{2}$, we obtain a tree H'_w so that $d_G(u, v) \le d_{H'_w}(u, v) \le (D+1)(d_G(u, v) + 1) \ \forall u, v \in V$.

2.2 Embeddings with Distortion λ of Graph Metrics

We continue with two auxiliary standard results about embeddings.

Lemma 1. *If $G = (V, E), G' = (V', E')$ are two graphs, one unweighted and second weighted, and $\varphi : V \mapsto V'$ is a map so that $d_{G'}(\varphi(u), \varphi(v)) \le \lambda \ \forall uv \in E$, then $d_{G'}(\varphi(x), \varphi(y)) \le \lambda d_G(x, y) \ \forall x, y \in V$.*

Lemma 2. *If $G = (V, E), G' = (V', E')$ are two graphs, one unweighted and second weighted, and $\varphi : V \mapsto V'$ is a map so that $d_{G'}(\varphi(u), \varphi(v)) \ge d_G(u, v)$ $\forall \varphi(u)\varphi(v) \in E'$, then $d_{G'}(\varphi(x), \varphi(y)) \ge d_G(x, y) \ \forall x, y \in V$.*

3   Embedding into Trees

We describe now a simple factor 6 algorithm for approximating the optimal distortion $\lambda^* = \lambda^*(G, \mathcal{T})$ of embedding finite unweighted graphs G into trees. For this, we first investigate the properties of layering partitions of graphs which λ-embed into trees, i.e., for each such graph $G = (V, E)$ there exists a tree $T = (V', E')$ with $V \subseteq V'$ such that (1) $d_G(x, y) \leq d_T(x, y)$ (non-contractibility) and (2) $d_T(x, y) \leq \lambda \cdot d_G(x, y)$ (bounded expansion) for every $x, y \in V$. Denote by $P_T(x, y)$ the path connecting the vertices x, y in T. For $x \in V'$ and $A \subseteq V'$, we denote by $d_T(x, A) = \min\{d_T(x, v) : v \in A\}$ the distance from x to A. First we show that the diameters of clusters in a layering partition of such a graph G are at most 3λ, allowing already to build a tree with distortion $9\lambda^*$. Refining this property of layering partitions, we construct in $O(|V||E|)$ time a tree into which G embeds with distortion $\leq 6\lambda^*$.

Lemma 3. *If G λ-embeds into a tree, then for any $x, y \in V$, any (x, y)-path $P_G(x, y)$ of G and any vertex $c \in P_T(x, y)$, $d_T(c, P_G(x, y)) \leq \lambda/2$.*

Proof. Removing c from T, we separate x from y. Let T_y be the subtree of $T \setminus \{c\}$ containing y. Since $x \notin T_y$, we can find an edge ab of $P_G(x, y)$ with $a \in T_y$ and $b \notin T_y$. Therefore, the path $P_T(a, b)$ must go via c. If $d_T(c, a) > \lambda/2$ and $d_T(c, b) > \lambda/2$, then $d_T(a, b) = d_T(a, c) + d_T(c, b) > \lambda$ and since $d_G(a, b) = 1$, we obtain a contradiction with the assumption that the embedding of G in T has distortion λ (condition (2)). Hence $d_T(c, P_G(x, y)) \leq \min\{d_T(c, a), d_T(c, b)\} \leq \lambda/2$, concluding the proof. $\quad\square$

Lemma 4. *If G λ-embeds into a tree T, then the diameter in G of any cluster C of a layering partition of G is $\leq 3\lambda$, i.e., $d_G(x, y) \leq 3\lambda$ for any $x, y \in C$. In particular, $\lambda^*(G, \mathcal{T}) \geq D/3$.*

Proof. Let $P_G(x, y)$ be a (x, y)-path of G outside the ball $B_k(s)$, where $k = d_G(s, x) - 1$. Let $P_G(x, s)$ and $P_G(y, s)$ be two shortest paths of G connecting x, s and y, s, respectively. Let $c \in V(T)$ be the unique vertex of T in $P_T(x, y) \cap P_T(x, s), \cap P_T(y, s)$. Since c belongs to each of the paths $P_T(x, y), P_T(x, s)$, and $P_T(y, s)$, applying Lemma 3 three times, we infer that $d_T(c, P_G(x, y))$, $d_T(c, P_G(x, s))$, and $d_T(c, P_G(y, s))$ are $\leq \lambda/2$. Let a be a closest to c vertex of $P_G(x, s)$ in the tree T, i.e., $d_T(a, c) = d_T(c, P_G(x, s)) \leq \lambda/2$. Let z be a closest to a vertex of $P_G(x, y)$ in T. From (1) and previous inequalities we conclude that $d_G(a, z) \leq d_T(a, z) = d_T(a, P_G(x, y)) \leq d_T(a, c) + d_T(c, P_G(x, y)) \leq \lambda$. Since $z \in P_G(x, y)$ and $P_G(x, y) \cap B_k(s) = \emptyset$, necessarily $d_G(s, z) \geq d_G(s, y) = d_G(s, a) + d_G(a, x)$, yielding $d_G(a, x) \leq d_G(a, z) \leq \lambda$. Analogously, if b is a closest to c vertex of $P_G(y, s)$ in T, then $d_G(b, y) \leq \lambda$ and $d_T(b, c) \leq \lambda/2$. By non-contractibility condition (1) and triangle condition, $d_G(a, b) \leq d_T(a, b) \leq d_T(a, c) + d_T(b, c) \leq \lambda$. Summarizing, we obtain the desired inequality $d_G(x, y) \leq d_G(x, a) + d_G(a, b) + d_G(b, y) \leq 3\lambda$. $\quad\square$

Lemma 1 and the properties of H' imply that one can construct in linear time an unweighted tree $H = (V, F)$ (without Steiner points) and a $\{0, 1\}$-weighted tree $H' = (V \cup S', F')$ (with Steiner points), so that $d_H(x, y) - 2 \leq d_G(x, y) \leq d_H(x, y) + 3\lambda$ and $d_{H'}(x, y) \leq d_G(x, y) \leq d_{H'}(x, y) + 3\lambda \; \forall x, y \in V$. Hence, for any graph G, it is possible to turn its non-contractive multiplicative distortion embedding into a weighted tree to a non-expanding additive distortion embedding into a $\{0, 1\}$-weighted tree. ¿From properties of the trees H_w and H'_w, we obtain:

Corollary 1. *If $G = (V, E)$ λ-embeds into a tree, then there exists uniformly weighted trees $H_w = (V, F, w)$ and $H'_w = (V \cup S', F', w)$ (without and with Steiner points), both constructible in $O(|V||E|)$ time, such that $d_G(u, v) \leq d_{H_w}(u, v) \leq (3\lambda+1)(d_G(u, v)+2)$ and $d_G(u, v) \leq d_{H'_w}(u, v) \leq (3\lambda+1)(d_G(u, v)+1) \; \forall u, v \in V$.*

Corollary 1 implies already that there exists a factor 12 (factor 8 if Steiner points are used) approximation algorithm for considered problem. We will show now that, by strengthening Lemma 4, one can improve the approximation ratio from 12 to 9 and from 8 to 6.

Lemma 5. *If $G = (V, E)$ λ-embeds into a tree T, $C = L^i_j \in \mathcal{LP}$ is a cluster of a layering partition of G and v is a vertex of C, then $d_G(v', u) \leq \max\{3\lambda-1, 2\lambda+1\}$ for any neighbor $v' \in L^{i-1}$ of v and any $u \in C$.*

Proof. Let $c \in V(T)$ be the nearest common ancestor in the tree T (rooted at s) of all vertices of cluster $C = L^i_j$. Let x and y be two vertices of C separated by c. Let $P_G(x, y)$ be a path of G connecting vertices x and y outside the ball $B_{i-1}(s)$. Then, as in the proof of Lemma 4, we have $d_T(c, P_G(x, y)) \leq \lambda/2$. Pick an arbitrary vertex $v \in C$ and a shortest path $P_G(v, s)$ connecting v with s in G. Since c separates v from s in T, by Lemma 3, $d_T(c, P_G(v, s)) \leq \lambda/2$ holds. Let a_v be a closest to c vertex of $P_G(v, s)$ in the tree T. Then, $d_T(a_v, P_G(x, y)) \leq d_T(a_v, c) + d_T(c, P_G(x, y)) \leq \lambda$. The choice of the path $P_G(x, y)$ and inequality (1) imply that $d_G(a_v, v) \leq d_G(a_v, P_G(x, y)) \leq d_T(a_v, P_G(x, y)) \leq \lambda$.

Consider an arbitrary vertex $u \in C$, $u \neq v$. By the triangle inequality and (1), we have $d_G(a_v, a_u) \leq d_T(a_v, a_u) \leq d_T(a_v, c) + d_T(a_u, c) \leq \lambda$, thus $d_G(a_v, u) \leq d_G(a_v, a_u) + d_G(a_u, u) \leq 2\lambda$. Let $v' \in L^{i-1}$ be a neighbor of v in $P_G(v, s)$. If $a_v = v$, then $d_G(v, u) = d_G(a_v, u) \leq 2\lambda$, i.e., $d_G(v', u) \leq d_G(v, u) + 1 \leq 2\lambda + 1$. Otherwise, if $a_v \neq v$, then $d_G(v', u) \leq d_G(v', a_v) + d_G(a_v, u) \leq \lambda - 1 + 2\lambda = 3\lambda - 1$, $d_G(v', u) \leq \max\{3\lambda - 1, 2\lambda + 1\}$. □

To make the embedding non-contractive, it suffices to assign the length $\ell := \max\{3\lambda - 1, 2\lambda + 1\}$ to each edge of H and get a uniformly weighted tree $H_\ell = (V, F, \ell)$. Then $d_G(u, v) \leq d_{H_\ell}(u, v) \leq \max\{3\lambda-1, 2\lambda+1\}(d_G(u, v)+2)$. The tree H_ℓ (without Steiner points) provides a 9-approximation to our problem. If we allow Steiner points and assign the length $\ell := \frac{3\lambda}{2}$ to each edge of H', then get a uniformly weighted tree H'_ℓ such that $d_G(u, v) \leq d_{H'_\ell}(u, v) \leq 3\lambda(d_G(u, v)+1)$.

For a graph $G = (V, E)$, we do not know λ in advance, however we know from Lemma 4 that $\lambda^*(G, \mathcal{T}) \geq D/3$. Therefore, the length ℓ to be assigned to the edges of the tree H (which is defined independently of the value of λ), can be found as follows: $\ell = \max\{d_G(u, v) : uv \text{ is an edge of } H\}$. The length

ℓ, which needs to be assigned to each edge of H', can be found as follows: $\ell = \frac{1}{2}\max\{D, \max\{d_G(u,v) : uv \text{ is an edge of } H\}\}$. Hence, ℓ can be computed in $O(|V||E|)$ time. Our main result of this section is the following theorem.

Theorem 1. *There exists a factor 6 approximation algorithm for the optimal distortion of embedding an unweighted graph G into a tree.*

The approximation ratio 6 of our algorithm holds only for adjacent vertices of G. It decreases when distances in G increase. Our tree H_ℓ does not have any Steiner points and the edges of both trees H_ℓ and H'_ℓ are uniformly weighted. The tree H'_ℓ, with Steiner points, is better than the tree H_ℓ only for small graph distances. So, the Steiner points do not really help, confirming A. Gupta's claim [12].

4   Minors, Relaxed Minors, and Metric Minors

We define metric relaxed minors, which, together with layering partitions, are used for approximate embedding of graphs into outerplanar metrics.

4.1   Minors and Relaxed Minors

A graph H is a *minor* of a graph G if a graph isomorphic to H can be obtained from G by contracting or delating some edges and some isolated vertices. To adapt the concept of minor to our embedding purposes, note that $H = (V', E')$ is a minor of $G = (V, E)$ if there exists a map $\mu : V' \cup E' \mapsto 2^V$, such that

(i)   for any vertex v of H, $G(\mu(v))$ is connected;
(ii) for any vertices $v \neq v'$ of H, $G(\mu(v)) \cap G(\mu(v')) = \emptyset$;
(iii) for any edge $e = uv$ of H, $G(\mu(e))$ is a path P_e of G with ends in $G(\mu(u))$ and $G(\mu(v))$;
(iv) for any vertex v and any edge e of H with $v \notin e$, $P_e \cap G(\mu(v)) = \emptyset$;
(v') for any edges $e = (x, y), e' = (u, v)$ of H, P_e and $P_{e'}$ intersect iff $\{x, y\} \cap \{u, v\} \neq \emptyset$ and if $e = (x, y), e' = (x, v)$, then $P_e \cap P_{e'} = \mu(x)$.

Indeed, if μ exists, then contracting each $\mu(v), v \in V'$, to a single vertex v and each P_e to an edge e, (ii),(iii), and (v') ensure that the resulting graph will be isomorphic to H. Note that if in (v') two paths P_e and $P_{e'}$ intersect, then they intersect in $G(\mu(u))$, where u is the common end of e and e'. In particular, if e, e' are non-incident, then P_e and $P_{e'}$ are disjoint. For our metric purposes we need a weaker notion of minor by allowing intersecting paths to intersect anywhere. A graph $H = (V', E')$ is a *relaxed minor* of a graph $G = (V, E)$ if there exists a map $\mu : V' \cup E' \mapsto 2^V$ satisfying (i)-(iv) and the following relaxation of (v'):

(v)   for any two non-incident edges e, e' of H, the paths $P_e \cap P_{e'} = \emptyset$.

The concept of relaxed minor is weaker than that of minor: the triangle C_3 is not a minor of any tree, but it is a relaxed minor of the star $K_{1,3}$: μ maps the three vertices of C_3 to the three leaves of $K_{1,3}$ and maps each edge uv of C_3 to the path of $K_{1,3}$ between the leaves $\mu(u)$ and $\mu(v)$. The map μ satisfies (i)-(v) but does not satisfy (v'). Relaxed and α-metric relaxed minors (see Subsection are crucial because their existence corresponds to a witness that G *cannot* be embedded into H-relaxed-minor-free graphs with small distortion (see Proposition 3). Thus it seems important to relate this notion to standard minors. We conjecture that *if the graph H is triangle-free, then the notion of relaxed minor is not weaker than that of minor.* We established a weaker statement which is enough to deal with H of special form: H will be bipartite $H = (V, F; E)$ with every vertex $f \in F$ of degree two. Such *subdivided* graphs H can be seen as a subdivision of an arbitrary graph $H' = (V, E')$ where $(u, v) \in H'$ iff there is a member $f \in F$ such that $(u, f), (v, f) \in E$.

Proposition 2. *If a graph $G = (V, E)$ has a subdivided graph $H = (V', E')$ as a relaxed minor, then G has H as a minor.*

4.2 α-Metric Relaxed Minors

Two sets A, B are α-*far* if $\min\{d_G(a, b) : a \in A, b \in B\} > \alpha$. For $\alpha \geq 1$, we call a graph $H = (V', E')$ an α-*metric relaxed minor* of $G = (V, E)$ if there exists a map $\mu : V' \cup E' \mapsto 2^V$ satisfying (i)-(v) and the following stronger version of condition (v):

(v$^+$) for any non-incident edges $e = uv$ and $e' = u'v'$ of H, the sets $\mu(u) \cup P_e \cup \mu(v)$ and $\mu(u') \cup P_{e'} \cup \mu(v')$ are α-far in G.

Let φ be an embedding of a graph $G = (V, E)$ into a graph $G' = (V', E')$ with distortion $\leq \alpha$. For $S \subseteq V$ inducing a connected subgraph $G(S)$ of G, we denote by $[\varphi(S)]$ a union of shortest paths of G' running between each pair of vertices of $\varphi(S)$ which are images of adjacent vertices of $G(S)$, one shortest path per pair.

Lemma 6. *If G α-embeds into G' and two sets of vertices A, B inducing connected subgraphs of G are α-far, then $[\varphi(A)] \cap [\varphi(B)] = \emptyset$.*

Proposition 3. *If a subdivided 2-connected graph $H = (V', E')$ is an α-metric relaxed minor of $G = (V, E)$, then any embedding of G into an H-minor free graph has distortion $> \alpha$.*

Proof. Suppose G has an embedding φ with distortion $\leq \alpha$ into an H-minor free graph G'. Let $\mu : V' \cup E' \mapsto 2^V$ be a map showing that H is an α-metric relaxed minor of G. Extend φ from V to the edge-set E by associating with each edge e of G the shortest path $P_e := [\varphi(e)]$ of G'. Pick any vertex v of H. Then, $\varphi(\mu(v))$ is a connected subgraph of G' because μ and φ map connected subgraphs to connected subgraphs. From Lemma 6 we know that φ maps two α-far connected subgraphs of G to two disjoint subgraphs of G'. As to the map

μ, we assert that *for any distinct vertices* v, v' *of* H, $\mu(v)$ *and* $\mu(v')$ *are* α-*far and for any vertex* v *and any edge* e *of* H *with* $v \notin e$, $\mu(v)$ *and* $\mu(e) = P_e$ *are* α-*far.* We will prove the first part. Since H is 2-connected, any two vertices v, v' belong to a common cycle C of H. Since H is triangle-free, v and v' belong to non-incident edges e, e' of C. Applying (v^+) to e and e', we conclude that $\mu(v)$ and $\mu(v')$ are α-far. Now, we define the following map $\nu : V' \cup E' \mapsto 2^{V(G')}$ from H to G'. For each $v \in V'$, set $\nu(v) = \varphi(\mu(v))$. For each edge $e = uv$ of H, $\mu(e) = P_e$ is a path of G with end-vertices $u^* \in \mu(u)$ and $v^* \in \mu(v)$. Each edge f of P_e is mapped by φ to a path $\varphi(f)$ of G'. Let $\nu(e)$ be any path of G' between $u' = \varphi(u^*)$ and $v' = \varphi(v^*)$ contained in the set $\bigcup \{\varphi(f) : f$ is an edge of $P_e\}$. From definition of ν and properties of μ and φ it follows that ν satisfies (i) and (iii). We will show that ν satisfies (ii), (iv), and (v) as well. To verify (ii), pick two vertices u, v of H. The sets $\mu(u)$ and $\mu(v)$ are α-far, thus Lemma 6 implies that $\nu(u) = \varphi(\mu(u))$ and $\nu(v) = \varphi(\mu(v))$ are disjoint, showing (ii). Analogously, if v is a vertex and e is an edge of H with $v \notin e$, then, since the sets $\mu(v)$ and $P_e = \mu(e)$ are α-far, thus, by Lemma 6, $\nu(v) = \varphi(\mu(v))$ and $\varphi(P_e)$ are disjoint. Since $\nu(e) \subseteq \varphi(P_e)$, $\nu(v)$ and $\nu(e)$ are disjoint as well, establishing (iv). The last condition (v) can be derived in a similar way by using (v^+) and Lemma 6. Hence, ν satisfies (i)-(v), i.e., H is a relaxed minor of G'. By Proposition 2, H is a minor of G', contradicting that G' is H-minor free.   □

4.3   Lower Bounds for α-Embeddings into $K_{2,r}$-Minor Free Graphs

We use the previous results to give lower bounds for the distortion of embedding a graph $G = (V, E)$ into $K_{2,r}$-minor free graphs.

Proposition 4. *If a cluster* C *of a layering partition* $\mathcal{LP}$ *of* G *contains* $r \geq 3$ *vertices* $v_1^*, \ldots, v_r^*$ *that are pairwise* $(4\alpha + 2)$-*far, then any embedding* φ *of* G *into a* $K_{2,r}$-*minor free graph has distortion* $> \alpha$.

Proof. Let $\mathcal{LP}$ be defined with respect to s and let T be a BFS tree rooted at s. Let k be the distance from s to C. Since C contains $(4\alpha + 2)$-far vertices $v_1^*, \ldots, v_r^*$, $k \geq 2\alpha + 2$. We will define a mapping μ from $K_{2,r}$ to G allowing to conclude that $K_{2,r}$ is an α-metric relaxed minor of G. Since $K_{2,r}$ is a subdivided graph, Proposition 3 will show that any embedding of G into a $K_{2,r}$-minor free graph has distortion $> \alpha$.

   Let $u_1, \ldots, u_r, v, w$ be the vertices of $K_{2,r}$, where v, w have degree r. Denote by e_i the edge vu_i and by f_i the edge wu_i, $i = 1, \ldots, r$. Let $P_1, \ldots, P_r$ be the paths of T of length $\alpha + 1$ from $v_1^*, \ldots, v_r^*$ towards the root s. Denote by $u_1^*, \ldots, u_r^*$ the other end vertices of the paths $P_1, \ldots, P_r$. Let $R_1, \ldots, R_r$ be the paths of T of length $\alpha + 1$ from $u_1^*, \ldots, u_r^*$ towards s. Denote by $w_1^*, \ldots, w_r^*$ the other end vertices of the paths $R_1, \ldots, R_r$. Set $\mu(u_i) := u_i^*$, $\mu(e_i) := P_i$ and $\mu(f_i) := R_i$ for $i = 1, \ldots, r$. Let $\mu(v)$ be the connected subgraph of G induced by all (or some) paths connecting the vertices $v_1^*, \ldots, v_r^*$ outside the ball $B_{k-1}(s)$. Finally, let $\mu(w) := B_{k-2\alpha-2}(s)$ (clearly, $w_1^*, \ldots, w_r^*$ belong to $\mu(w)$). From the definitions of μ and $\mathcal{LP}$, we conclude that μ satisfies (i) and (iii). Since

$\mu(v) \subseteq \cup_{j \geq k} L^j, \mu(w) = B_{k-2\alpha-2}(s)$, and the vertices $u_1^* = \mu(u_1), \ldots, u_r^* = \mu(u_r)$ belong to $L^{k-\alpha-1}$, the μ-images of the vertices of $K_{2,r}$ are pairwise α-far in G. Analogously, any vertex of $\mu(v)$ is at distance $> \alpha$ from any path $R_i = \mu(f_i)$ and any vertex of $\mu(w)$ is at distance $> \alpha$ from any path $P_i = \mu(e_i)$. If a vertex u_i^* is at distance $\leq \alpha$ from $x \in P_j \cup R_j$ for $j \neq i$, then, by triangle inequality, we obtain $d_G(v_i^*, v_j^*) \leq d_G(v_i^*, u_i^*) + d_G(u_i^*, x) + d_G(x, v_j^*) \leq \alpha + 1 + \alpha + d_G(v_j^*, x)$. Since $x \neq w_j^*, d_G(v_j^*, x) \leq 2\alpha + 1$, yielding $d_G(v_i^*, v_j^*) \leq \alpha + 1 + \alpha + 2\alpha + 1 = 4\alpha + 2$, contrary to the assumption that v_i^* and v_j^* are $(4\alpha + 2)$-far. This contradiction shows that the μ-images of any vertex and any non-incident edge of $K_{2,r}$ are α-far. It remains to show that any two paths P_i and R_j with $i \neq j$ are α-far. If $d_G(x, y) \leq \alpha$ for $x \in P_i \setminus \{v_i^*, u_i^*\}$ and $y \in R_j \setminus \{u_j^*, w_j^*\}$, then $d_G(v_i^*, v_j^*) \leq d_G(v_i^*, x) + d_G(x, y) + d_G(y, v_j^*) \leq \alpha + \alpha + 2\alpha + 1 \leq 4\alpha + 1$, contrary to the assumption that v_i^* and v_j^* are α-far. This contradiction shows that $K_{2,r}$ is an α-metric relaxed minor of G. $\qquad\square$

5   Embedding into Outerplanar Graphs

We present now the algorithm for approximate embedding of graph metrics into into outerplanar metrics.

5.1   The Algorithm

Let $G = (V, E)$ be the input graph and let $\mathcal{LP}$ be a layering partition of G. We assume that $\lambda \geq 1$ is so that each cluster C of $\mathcal{LP}$ contains at most two $(4\lambda + 2)$-far vertices (otherwise, by Proposition 4, the optimal distortion is larger than λ). Set $\Lambda := 4\lambda + 2$. We call a cluster C *bifocal* if it has two Λ-far vertices c_1 and c_2. In addition, for such cluster C let $C_1 = \{x \in C : d_G(x, c_1) \leq d_G(x, c_2)\}$ and $C_2 = \{x \in C : d_G(x, c_2) \leq d_G(x, c_1)\}$, and call C_1 and C_2 the *cells* of C centered at c_1 and c_2 (we will suppose below that c_1 and c_2 form a diametral pair of C). If $\operatorname{diam}(C) \leq \Lambda$ (i.e., C is not bifocal), then the cluster C is called *small*. Then C has a unique cell centered at an arbitrary vertex of C. A bifocal cluster C is called *big* if $\operatorname{diam}(C) > 16\lambda + 12$, otherwise, if $\Lambda < \operatorname{diam}(C) \leq 16\lambda + 12$, then C is a *medium* cluster. An *almost big cluster* is a medium cluster C such that $\operatorname{diam}(C) > 16\lambda + 10$. A cluster C is Δ-*separated* if C is bifocal with cells C_1 and C_2 and $d_G(u, v) > \Delta$ for any $u \in C_1$ and $v \in C_2$. Further, we will set $\Delta := 8\lambda + 6$. A bifocal cluster C' is *spread* if both cells C_1, C_2 of its father C are adjacent to C'. Given a cluster C at distance k from s and its son C', we call the union of C with the connected component of $G(V \setminus B_k(s))$ containing C' the CC'-*fiber* of G and denote it by $\mathcal{F}(C, C')$. We now ready to describe the algorithm.

5.2   Small, Medium, and Big Clusters

We present here without proof several simple properties of clusters of $\mathcal{LP}$.

Lemma 7. *If C is bifocal, then the diameter of each of its cells is $\leq 2\Lambda$.*

Algorithm APPROXIMATION BY OUTERPLANAR METRIC

Input: A graph $G = (V, E)$, a layering partition $\mathcal{L}P$ of G, and λ

Output: An outerplanar graph $G' = (V, E')$ or an answer "not"

1. **For** each cluster C of the layering partition $\mathcal{L}P$ **do**
2.   **If** C has two big sons or C is big and has two spread sons, **then** **return** "not".
3.   **Else for** each son C' of C **do**
4.   Case 1: **If** C' is small, **then** pick the center c of a cell of C adjacent to C' and in G' make c adjacent to all vertices of C'.
5.   Case 2: **If** C' is medium and C is not big, or C' is medium and not spread and C is big, **then** pick the center c of a cell of C adjacent to C' and in G' make c adjacent to all vertices of C'.
6.   Case 3: **If** C' is medium, C is big, and C' is the (unique) spread son of C, **then** in G' make the center c_1 of cell C_1 of C adjacent to all vertices of C'. Additionally, make the center c_2 of cell C_2 of C adjacent to all vertices of C'.
7.   Case 4: **If** $C' = C'_1 \cup C'_2$, such that C'_1 is adjacent to C_1 and C'_2 is adjacent to C_2, where C_1 and C_2 are the cells of C with centers c_1 and c_2, **then** in G' make c_1 adjacent to all vertices of C'_1 and c_2 adjacent to all vertices of C'_2.

Lemma 8. *If C is bifocal and* $\mathrm{diam}(C) = d_G(c_1, c_2) > 12\lambda + 6$, *then (i) C is* $(\mathrm{diam}(C) - 2\Lambda - 1)$-*separated, in particular $C_1 \cap C_2 = \emptyset$ and (ii)* $\mathrm{diam}(C_1) \leq \Lambda$ *and* $\mathrm{diam}(C_2) \leq \Lambda$.

If C is big, then C is $(8\lambda + 8)$-*separated and if C is almost big, then C is* $(8\lambda + 6)$-*separated, whence big and almost big clusters are Λ-separated. If C is big or almost big, then* $\mathrm{diam}(C_1) \leq \Lambda$ *and* $\mathrm{diam}(C_2) \leq \Lambda$.

Lemma 9. *If C is big, then C has a bifocal spread son C' such that contracting the four cells of C and C' (but preserving the inter-cell edges), we will obtain a* $2K_2$.

Lemma 10. *If C' is big or almost big, then its father C is bifocal and the neighbors in C of the centers c'_1 and c'_2 of the cells C'_1 and C'_2 of C' belong to different cells of C. Big and almost big clusters are spread.*

Lemma 11. *If C is big, no son of C has a cell adjacent to both cells of C. No big cluster C has a small son adjacent to both cells of C.*

5.3   Correctness of the Algorithm

The following results establish the correctness and the approximation ratio of our algorithm.

Theorem 2. *Let $G = (V, E)$ be a graph and $\lambda \geq 1$. If the algorithm returns the answer "not", then any embedding of G into a $K_{2,3}$-minor free graph has distortion $> \lambda$. If the algorithm returns the outerplanar graph $G' = (V, E')$, then assigning to its edges weight $w := 20\lambda + 15$, we obtain an embedding of G to G'*

such that $d_G(x,y) \leq d_{G'}(x,y) \leq 5wd_G(x,y) \ \forall x, y \in V$. As a result, we obtain a factor $100\lambda + 75$ approximation of the optimal distortion of embedding a graph into an outerplanar metric.

The proof of this theorem is subdivided into two propositions. We start with a technical result, essentially showing that in both cases when our algorithm returns the answer "not", any embedding of G into an outerplanar metric requires distortion $> \lambda$:

Proposition 5. *Let C be a big or an almost big cluster having two sons C', C'' such that the two cells of C can be connected in both CC'- and CC''-fibers of G. Then, any embedding of G in a $K_{2,3}$-minor free graph has distortion $> \lambda$. These conditions hold in the following two cases: (i) C is big and has two spread sons; (ii) C has two big sons C', C''. In particular, if the algorithm returns the answer "not", then any embedding of G in a $K_{2,3}$-minor free graph requires distortion $> \lambda$.*

Now suppose that the algorithm returns the graph G'. By construction, G' is outerplanar. Let $d_{G'}(x,y)$ be the distance in G' between x and y, where each edge of G' has length $w := 20\lambda + 15$. We continue with the basic property of G' allowing to analyze the approximation ratio.

Proposition 6. *For each edge xy of G, x and y can be connected in the graph G' by a path consisting of at most 5 edges, i.e. $d_{G'}(x,y) \leq 5w$. Conversely, for each edge xy of G', $d_G(x,y) \leq 20\lambda + 15$.*

5.4 Proof of Proposition 6

We start with first assertion. First suppose that $d_G(s,x) = d_G(s,y)$. Let C be the cluster of G containing xy. Then, either C is not big or C is big and x, y belong to the same cell of C. In both cases, by construction of G', we deduce that x and y will be adjacent in G' to the same vertex from the father C_0 of C, implying $d_{G'}(x,y) = 2w$. Now suppose that $x \in C, y \in C'$ and C' is a son of C. Let C_0 be the father of C. Let z be a vertex of C to which y is adjacent in G'. If C is small, medium, or C is big but x and z belong to the same cell, then in G' the vertices z and x will be adjacent to the same vertex x_{C_0} of C_0, yielding $d_{G'}(x,y) \leq 3w$. So, suppose that C is big and the vertices z and x belong to different cells C_1 and C_2 of C, say $z \in C_1$ and $x \in C_2$. By Lemma 11, C' is not small. According to the algorithm, z is the center of the cell C_1, i.e., $z = c_1$. Note also that x and the center c_2 of its cell are both adjacent in G' to a vertex $x_{C_0} \in C_0$, whence $d_{G'}(x, c_2) = 2w$. If C' is big and say $y \in C_1'$, then since y is adjacent to z in G', from the algorithm we conclude that a vertex of C_1' is adjacent in G to a vertex of C_1. On the other hand, $y \in C_1'$ is adjacent in G to $x \in C_2$. As a consequence, the cell C_1' is adjacent in G to both cells C_1 and C_2 of C, which is impossible by Lemma 11. So, the cluster C' must be medium. If C has a big son C''', then since both cells of C are adjacent in G to the medium son C', we obtain a contradiction with Proposition 5(i). Hence, C cannot have big sons. Moreover, by Proposition 5, C' is the unique spread son of C. According to the algorithm (see **Case 3**), the centers $z = c_1$ and c_2 of the cells of C are adjacent in G' to a common vertex u from C', yielding $d_{G'}(z, c_2) = 2w$. As a result, we obtain a path with at most 5 edges connecting the vertices y and x in $G : (y, z = c_1, u, c_2, x_{C_0}, x)$.

We continue with second assertion. Any edge xy of G' runs between two clusters lying in consecutive layers of G (and G'); let $x \in C$ and $y \in C'$, where C is the father of C'. In G, y has a neighbor $x' \in C$. Let $x' \neq x$, otherwise we are done. If C is not big, then $d_G(x, x') \leq 16\lambda + 12$, whence $d_G(x, y) \leq 16\lambda + 13$. So, suppose C is big. If x, x' belong to the same cell of C, then Lemma 7 implies that $d_G(x, x') \leq 2\Lambda = 8\lambda + 4$, yielding $d_G(x, y) \leq 8\lambda + 5$. Now, let $x \in C_1$ and $x' \in C_2$. By Lemma 11, C' is medium or big. If C' is big and $y \in C_1'$, since x and y are adjacent in G', according to the algorithm, C_1' contains a vertex that is adjacent in G to a vertex of C_1. Since $y \in C_1'$ is adjacent in G to $x' \in C_2$, we obtain a contradiction with Lemma 11. Hence C' is a medium cluster. According to the algorithm, x is the center of the cell C_1 and C_1 contains a vertex z adjacent in G to a vertex $v \in C'$. Since $x, z \in C_1$ implies $d_G(x, z) \leq 4\lambda + 2$ and $y, v \in C'$ implies $d_G(y, v) \leq 16\lambda + 12$, we obtain $d_G(x, y) \leq 20\lambda + 15$.

5.5   Proof of Proposition 5

By Proposition 3, it suffices to show that G contains $K_{2,3}$ as a λ-metric relaxed minor. Indeed, suppose that C is a big or an almost big cluster with cells C_1 and C_2 having two sons C', C'', such that C_1 and C_2 can be connected by a path in each of the CC'- and CC''-fibers of G. Let $k = d_G(s, C)$. Denote by P' and P'' the shortest such paths connecting two vertices of C, one in C_1 and another in C_2, in $\mathcal{F}(C, C')$ and $\mathcal{F}(C, C'')$, respectively. Denote by $x' \in C_1$ and $y' \in C_2$ the end-vertices of P' and by $x'' \in C_1$ and $y'' \in C_2$ the end-vertices of P''. The choice of P' implies $P' \cap C = \{x', y'\}$ and the choice of P'' implies $P'' \cap C = \{x'', y''\}$. Let w' and w'' be middle vertices of P' and P'', respectively. Let a', b' be the vertices of P' at distance $\lambda + 1$ (measured in P') from w', where a' is between w' and x' and b' is between w' and y'. Let L' be the subpath of P' between a' and w' and R' the subpath of P' between w' and b'. Analogously, for P'' we can define the vertices a'', b'' and the paths L'', R'' of length $\lambda + 1$ each. Finally, denote by P_1', P_2' the subpaths of P' between a' and x' and between b' and y'. Analogously, define the supbaths P_1'' and P_2'' of P''. Pick any shortest path M' in G between the vertices x', x'' and any shortest path M'' between y', y''. Let F' be a subpath of a shortest path $P(x', s)$ from x' to the root s starting with x' and having length 3λ. Analogously, let F'' be a subpath of a shortest path $P(y'', s)$ from y'' to s starting with y'' and having length 3λ. Let J' and J'' be the subpaths of length $\lambda + 1$ of $P(x', s)$ and $P(y', s)$, which continue F' and F'', respectively, towards s.

Now we define a mapping $\mu : V(K_{2,3}) \cup E(K_{2,3}) \mapsto V(G)$ certifying that $K_{2,3}$ is a λ-metric relaxed minor of G. Denote the vertices of $K_{2,3}$ by a, b, c, q', q'', where the vertices q' and q'' are assumed to be adjacent to each of the vertices a, b, c. We set $\mu(a) := \{w'\}, \mu(b) := \{w''\}, \mu(q') := P_1' \cup P_1'' \cup M' \cup F' =:$

$Q', \mu(q'') := P_2' \cup P_2'' \cup M'' \cup F'' := Q''$, and $\mu(c) := B_{k'} =: S$, where $k' = k - 4\lambda - 1$. Additionally, for each edge of $K_{2,3}$, we set $\mu(aq') := L', \mu(aq'') := R', \mu(bq') := L'', \mu(bq'') := R'', \mu(q's) := J', \mu(q''s) := J''$. We will call the paths $L', L'', R', R'', P_1', P_2', P_1'', P_2'', F', F'', J', J'', M', M''$, the vertices w', w'', and the set S the *elements* of the map μ. Notice first that each vertex of $K_{2,3}$ is mapped to a connected subgraph of G and each edge of $K_{2,3}$ is mapped to a path of G, thus μ satisfies the conditions (i) and (iii) of a metric relaxed minor. It remains to show that μ satisfies the remaining conditions of a λ-metric relaxed minor. The proof of this is subdivided into several results: (1) $d_G(w', C) \geq 4\lambda + 3$ and $d_G(w'', C) \geq 4\lambda + 3$, (2) S is λ-far from all elements of μ except J', J'' (3) w' is λ-far from all elements of μ except L', R' and w'' is λ-far from all elements of μ except L'', R'', (4) L', R' are λ-far from $L'', R'', P_1'', P_2'', J', J''$ and L'', R'' are λ-far from P_1', P_2', J', J'', (5) Q' is λ-far from the R', R'', J'' and Q'' is λ-far L', L'', J', and (6) Q' and Q'' are λ-far.

To prove the second assertion of Proposition 5, first suppose that the cluster C is big and C has a big and a medium sons C', C'' such that both cells C_1 and C_2 are adjacent to C'' or that C has two medium sons C', C'' adjacent to both cells of C. By definition of the layering, each vertex of $C' \cup C''$ is adjacent to a vertex of C. If all vertices of C' are adjacent to vertices from the same cell of C, say C_1, then for any $x', y' \in C'$ we have $d_G(x', y') \leq 2 + 4\lambda + 2$, contrary to the assumption that C' is big. Hence, both cells of C are adjacent to C', say $x \in C_1$ is adjacent to $x' \in C'$ and $y \in C_2$ is adjacent to $y' \in C'$. By Lemma 11, x' and y' belong to different cells of C', say $x' \in C_1'$ and $y' \in C_2'$. Let $k := d_G(s, C)$. Since $x', y' \in C'$, the vertices x' and y' are adjacent in $G(V \setminus B_k(s))$ by a path $P(x', y')$. Then $P(x, y) := xx' \cup P(x', y') \cup y'y$ is a path between x and y in the CC'-fiber $\mathcal{F}(C, C')$. Analogously, since both cells C_1 and C_2 are adjacent to C'', we conclude that two vertices from different cells of C can be connected by a path belonging to the CC''-fiber, showing that the first condition of Proposition 5 is fulfilled. This establishes (i). Now suppose that C has two big sons C' and C''. Then C is either a big or an almost big cluster. By Lemma 9, each of the clusters C', C'' is $(8\lambda + 8)$-separated while the cluster C is $(8\lambda + 6)$-separated and that its cells C_1 and C_2 have diameters at most Λ. As in previous cases, one can deduce that C_1 is adjacent to one cell of each of the clusters C' and C'', while C_2 is adjacent to the second cell of these clusters, establishing (ii).

5.6   Proof of Theorem 2

The algorithm returns the answer "not" when a cluster C has two big sons or a big cluster C has two spread sons. In this case, by Proposition 5 any embedding of G into a $K_{2,3}$-minor free graph requires distortion $> \lambda$, whence $\lambda^*(G, \mathcal{O}) > \lambda$. Now suppose that the algorithm returns the outerplanar graph G' weighted uniformly with $w = 20\lambda + 15$. Notice that in Case 4 of the algorithm, the required matching between the four cells of the big clusters C and C' exists by Lemma 9 and because C' is the unique spread son of C. By Proposition 6 we have $d_G(x, y) \leq 20\lambda + 15 = d_{G'}(x, y)$ for each edge xy of the graph G'. By Lemma 2 we conclude that $d_G(x, y) \leq d_{G'}(x, y)$ for any pair $x, y \in V$. By

Proposition 6, for any edge xy of G, the vertices x and y can be connected in G' by a path with at most 5 edges, i.e., $d_{G'}(x, y) \leq 5w = 100\lambda + 75$. By Lemma 1 we conclude that $d_{G'}(x, y) \leq (100\lambda + 75)d_G(x, y)$ for any pair x, y of V. Hence $d_G \leq d_{G'} \leq (100\lambda + 75)d_G$.

References

1. Agarwala, R., Bafna, V., Farach, M., Narayanan, B., Paterson, M., Thorup, M.: On the approximability of numerical taxonomy (fitting distances by tree metrics). SIAM J. Comput. 28 (1999)
2. Bădoiu, M., Chuzhoy, J., Indyk, P., Sidiropoulos, A.: Low-distortion embeddings of general metrics into the line. In: STOC 2005 (2005)
3. Bădoiu, M., Demaine, E.D., Hajiaghayi, M.T., Sidiropoulos, A., Zadimoghaddam, M.: Ordinal embedding: approximation algorithms and dimensionality reduction. In: APPROX-RANDOM 2008 (2008)
4. Bădoiu, M., Indyk, P., Sidiropoulos, A.: Approximation algorithms for embedding general metrics into trees. In: SODA 2007 (2007)
5. Brandstädt, A., Chepoi, V., Dragan, F.: Distance approximating trees for chordal and dually chordal graphs. J. Algorithms 30 (1999)
6. Chepoi, V., Dragan, F.: A note on distance approximating trees in graphs. Europ. J. Combin. 21 (2000)
7. Chepoi, V., Dragan, F., Estellon, B., Habib, M., Vaxès, Y.: Diameters, centers, and approximating trees of δ-hyperbolic geodesic spaces and graphs. In: SoCG 2008 (2008)
8. Chepoi, V., Fichet, B.: l_∞-Approximation via subdominants. J. Math. Psychol. 44 (2000)
9. Dourisboure, Y., Dragan, F.F., Gavoille, C., Yan, C.: Spanners for bounded tree-length graphs. Theor. Comput. Sci. 383 (2007)
10. Dourisboure, Y., Gavoille, C.: Tree-decompositions with bags of small diameter. Discr. Math. 307 (2007)
11. Emek, Y., Peleg, D.: Approximating minimum max-stretch spanning trees on unweighted graphs. In: SODA 2004 (2004)
12. Gupta, A.: Steiner points in tree metrics don't (really) help. In: SODA 2001 (2001)
13. Indyk, P., Matousek, J.: Low-distortion embeddings of finite metric spaces. In: Handbook of Discrete and Computational Geometry. CRC Press, LLC (2004)
14. Krauthgamer, R., Lee, J.R.: Algorithms on negatively curved spaces. In: FOCS 2006 (2006)
15. Rabinovich, Y., Raz, R.: Lower bounds on the distortion of embedding finite metric spaces in graphs. Discr. Comput. Geom. 19 (1998)
16. Shavitt, Y., Tankel, T.: On internet embedding in hyperbolic spaces for overlay construction and distance estimation. In: INFOCOM 2004 (2004)
17. Semple, C., Steel, M.: Phylogenetics. Oxford University Press, Oxford (2003)

Approximating Linear Threshold Predicates*

Mahdi Cheraghchi[1], Johan Håstad[2], Marcus Isaksson[3], and Ola Svensson[2]

[1] School of Computer and Communication Sciences, EPFL, Lausanne, Switzerland
mahdi.cheraghchi@epfl.ch
[2] KTH - Royal Institute of Technology, Stockholm, Sweden
{johanh,osven}@kth.se
[3] Chalmers University of Technology and University of Gothenburg,
Gothenburg, Sweden
maris@chalmers.se

Abstract. We study constraint satisfaction problems on the domain
$\{-1, 1\}$, where the given constraints are homogeneous linear threshold
predicates. That is, predicates of the form $\mathrm{sgn}(w_1 x_1 + \cdots + w_n x_n)$ for
some positive integer weights $w_1, \ldots, w_n$. Despite their simplicity, current
techniques fall short of providing a classification of these predicates in
terms of approximability. In fact, it is not easy to guess whether there
exists a homogeneous linear threshold predicate that is approximation
resistant or not.

The focus of this paper is to identify and study the approximation
curve of a class of threshold predicates that allow for non-trivial approximation. Arguably the simplest such predicate is the majority predicate
$\mathrm{sgn}(x_1 + \cdots + x_n)$, for which we obtain an almost complete understanding of the asymptotic approximation curve, assuming the Unique Games
Conjecture. Our techniques extend to a more general class of "majority-
like" predicates and we obtain parallel results for them. In order to classify these predicates, we introduce the notion of *Chow-robustness* that
might be of independent interest.

Keywords: Approximability, constraint satisfaction problems, linear
threshold predicates.

1 Introduction

Constraint satisfaction problems or more succinctly CSPs are at the heart of
theoretical computer science. In a CSP we are given a set of constraints, each
putting some restriction on a constant size set of variables. The variables can
take values in many different domains but in this paper we focus on the case of
variables taking Boolean values. This is the most fundamental case and it has
also attracted the most attention over the years. We also focus on the case where
each condition is given by the same predicate, P, applied to a sequence of literals.
The role of this predicate P is key in this paper and as it is more important for

* This research is supported by the ERC Advanced investigator grant 226203.
M. Cheraghchi is supported by the ERC Advanced investigator grant 228021.

M. Serna et al. (Eds.): APPROX and RANDOM 2010, LNCS 6302, pp. 110–123, 2010.

us than the number of variables, we reserve the letter n for the arity of this predicate while using N to be the number of variables in the instance. We also reserve m to denote the number of constraints.

Traditionally we ask for an assignment that satisfies all constraints and in this case it turns out that all Boolean CSPs are either NP-complete or belong to P and this classification was completed already in 1978 by Schaefer [15]. In this paper we study Max-CSPs which are optimization problems where we want to satisfy as many constraints as possible. Almost all Max-CSPs of interest turn out to be NP-hard and the main focus is that of efficient approximability.

The standard measure of approximability is given by a single number C and an algorithm is a C-approximation algorithm if it, on each input, finds an assignment with an objective value that is at least C times the optimal value. Here we might allow randomization and be content if the assignment found satisfies these many constraints on average. A more refined question is to study the approximation curve where for each constant c, assuming that the optimal assignment satisfies cm constraints, we want to determine the maximal number of constraints that we can satisfy efficiently.

To get a starting point to discuss the quality of approximation algorithms it is useful to first consider the most simple algorithm that chooses the values of the variables randomly and uniformly from all values in $\{0,1\}^N$. If the predicate P is satisfied by t inputs in $\{0,1\}^n$ it is easy to see that this algorithm, on the average, satisfies $mt2^{-n}$ constraints. By using the method of conditional expectations it is also easy to deterministically find an assignment that satisfies this number of constraints.

A very strong type of hardness result possible for a Max-CSP is to prove that, even for instances where the optimal assignment satisfies all constraints, it is NP-hard to find an assignment that does significantly better (by a constant factor independent of N) than the above trivial algorithm. We call such a predicate "approximation resistant on satisfiable instances". A somewhat weaker, but still strong, negative result is to establish that the approximation ratio given by the trivial algorithm, namely $t2^{-n}$, is the best approximation ratio that can be obtained by an efficient algorithm. This is equivalent to saying that we cannot satisfy significantly more than $mt2^{-n}$ constraints when given an almost satisfiable instance. We call such a predicate "approximation resistant". It is well known that, unless P=NP, Max-3-Sat (i.e. when P is the disjunction of the three literals) is approximation resistant on satisfiable instances and Max-3-Lin (i.e. when P is the exclusive-or of three literals) is approximation resistant [8].

When it comes to positive results on approximability the most powerful technique is semi-definite programming introduced in this context in the classical paper by Goemans and Williamson [6] studying the approximability of Max-Cut, establishing the approximability constant $\alpha_{GW} \approx .878$. In particular, this result implies that Max-Cut is not approximation resistant. Somewhat surprisingly as proved by Khot et al. [12], this constant has turned out, assuming the Unique Games Conjecture, to be best possible. We note that these results have been extended in great generality and O'Donnell and Wu [14] determined the complete approximation curve of Max-Cut.

The general problem of determining which predicates are approximation resistant is still not resolved but as this is not the main theme of this paper let us cut this discussion short by mentioning a general result by Austrin and Mossel [2]. This paper relies on the Unique Games Conjecture by Khot [11] and proves that, under this conjecture, any predicate such that the set $P^{-1}(1)$ supports a pairwise independent measure is approximation resistant.

On the algorithmic side there is a general result by Hast, [7], that is somewhat complementary to the result of Austrin and Mossel. Hast considers the real valued function $P^{\leq 2}$ which is the sum of the linear and quadratic parts of the Fourier expansion of P. Oversimplifying slightly, the result by Hast says that if $P^{\leq 2}$ is positive on all inputs accepted by P then we can derive a non-trivial approximation algorithm and hence P is not approximation resistant.

To see the relationship between the results of Austrin and Mossel, and Hast, note that the condition of Austrin and Mossel is equivalent to saying that there is a probability distribution on inputs accepted by P such that the average of any unbiased quadratic function[1] is 0. In contrast, Hast needs that a particular unbiased quadratic function is positive on all inputs accepted by P. It is not difficult to come up with predicates that satisfies neither of these two conditions and hence we do not have a complete classification, even if we are willing to assume the Unique Games Conjecture. The combination of the two results, however, points to the class of predicates that can be written on the form

$$P(x) = \text{sgn}(Q(x))$$

for a quadratic function Q as an interesting class of predicates to study and this finally brings us to the topic of this paper. We study this scenario in the simplest form by assuming that Q is in fact an unbiased linear function, L. In other words we have

$$P(x) = \text{sgn}(L(x)) = \text{sgn}\left(\sum_{i=1}^{n} w_i x_i\right),$$

for some, without loss of generality, positive integral weights $(w_i)_{i=1}^{n}$. Note that if we allow a constant term in L the situations is drastically different as for instance 3-Sat is the sign of linear form if we allow a non-zero constant term. One key difference is that a probability distribution supported on the set "$L(x) > 0$" cannot have even unbiased variables in the case when L is without constant term and thus hardness results such as the result by Austrin and Mossel do not apply.

To make life even simpler we make sure that L never takes the value 0 and as $L(-x) = -L(x)$, P accepts precisely half of the inputs and thus the number of constraints satisfied by a random assignment is, on the average, $m/2$.

---

[1] Throughout this work, we find it more convenient to represent Boolean values by $\{-1,+1\}$ rather than $\{0,1\}$.

The simplest such predicate is majority of an odd number of inputs. For this predicate it easy to see that Hast's condition is fulfilled and hence, for any odd value of n, his results imply that majority is not approximation resistant. This result generalizes to "majority-like" functions as follows. For a linear threshold functions, the Chow parameters, $\bar{P} = (\hat{P}(i))_{i=0}^{n}$, [3] are for, $i > 0$, defined to be the correlations between the output of the function and inputs x_i. We have that $\hat{P}(0)$ is the bias of the function and thus in our case this parameter is always equal to 0 and hence ignored.

Now if we order the weights $(w_i)_{i=1}^{n}$ in nondecreasing order then also the $\hat{P}(i)$'s are nondecreasing but in general quite different from the weights. It is well known that the Chow parameters determine the threshold function uniquely [3] but the computational problem of given $\bar{P}$, how to recover the weights, or even to compute P efficiently is an interesting problem and several heuristics have been proposed [10,17,9,4] together with an empirical study that compares various methods [18]. More recently, the problem of finding an approximation of P given the Chow parameters has received increased attention, see e.g. [13] and [5]. The most naive method is to use $\bar{P}$ as weights. This does not work very well in general but this is a case of special interest to us as it is precisely when this method gives us back the original function that we can apply Hast's results directly. We call such a threshold function "Chow-robust" and we have not been able to find the characterization of this class of functions in the literature. If we ignore some error terms and technical conditions a sufficient condition to be Chow-robust is roughly that

$$\sum_{i=1}^{n}(w_i^3 - w_i) \le 3 \sum_{i=1}^{n} w_i^2 \tag{1}$$

and thus it applies to functions with rather modest weights. We believe that this condition is not very far from necessary but we have not investigated this in detail.

Having established non-approximation resistance for such predicates we turn to study the full curve of approximability and, in an asymptotic sense as a function of n, we get almost tight answers establishing both approximability results and hardness results. Our results do apply with degrading constants to more general threshold functions but let us here state them for majority. We have the following theorem.

Theorem 1. *(Informal) Given an instance of Max-Maj-n with n odd and m constraints and assume that the optimal assignment satisfies $(1 - \frac{\delta}{n+1})m$ constraints, for some $\delta < 1$. Then it is possible to efficiently find an assignment that satisfies*

$$\left(\frac{1}{2} + \Omega\left(\frac{(1-\delta)^{3/2}}{n^{1/2}}\right) - \mathcal{O}\left(\frac{\log^4 n}{n^{5/6}}\right)\right)m$$

constraints.

Thus for large n we need almost satisfiable instances to get above the threshold $\frac{1}{2}$ obtained by a a random assignment. This might seem weak but we prove that this is probably the correct threshold.

Theorem 2. *(Informal) Assume the Unique Games Conjecture and let $\epsilon > 0$ be arbitrary. Then it is NP-hard to distinguish instances of Max-Maj-n where the optimal value is $(1 - \frac{1}{n+1} - \epsilon)m$, from those where the optimal value is $(\frac{1}{2} + \epsilon)m$.*

This proves that the range of instances to which Theorem 1 applies is essentially the correct one. A drawback is that the error term $\mathcal{O}\left(\frac{\log^4 n}{n^{5/6}}\right)$ in Theorem 1 dominates the systematic contribution of $(1 - \delta)^{3/2}n^{-1/2}$ for δ very close to 1 and hence the threshold is not sharp. We are, however, able to sharply locate the threshold where something nontrivial can be done by combining our result with the general results by Hast. For details, see Section 3.

To see that the advantage obtained by the algorithm is also the correct order of magnitude we have the following theorem.

Theorem 3. *(Informal) Assume the Unique Games Conjecture and let $\epsilon > 0$ be arbitrary. Then there is an absolute constant c such that it is NP-hard to distinguish instances of Max-Maj-n where the optimal value is $(1 - \epsilon)m$, from those where the optimal value is $(\frac{1}{2} + \frac{c}{\sqrt{n}} + \epsilon)m$.*

In summary, we get an almost complete understanding of the approximability curve of majority, at least in an asymptotic sense as a function of n. This complements the results for majority on three variables, for which there is a 2/3-approximation algorithm [19] and it is NP-hard to do substantially better [8].

The idea of the algorithm behind Theorem 1 is quite straightforward while its analysis gets rather involved. We set up a natural linear program which we solve and then use the obtained solution as biases in a randomized rounding. The key problem that arises is to carefully analyze the probability that a sum of biased Boolean variables is positive. In the case of majority-like variables we have the additional complication of the different weights. This problem is handled by writing the probability in question as a complex integral and then estimating this integral by the saddle-point method. The resulting proof is quite long and does not fit within the page limit of the current abstract. This proof and several other proofs are hence omitted and can be found in the full version of the paper.

The hardness results given in Theorem 2 and Theorem 3 resort to the techniques of Austrin and Mossel [2]. The key to these results is to find suitable pairwise independent distributions relating to our predicate. In the case of majority it is easy to find such distributions explicitly, while in the case of more general weights the construction gets more involved. In particular, we need to answer the following question: What is the minimal value of $\Pr[L(x) < 0]$ when x is chosen according to a pairwise independent distribution. This is a nice combinatorial question of independent interest.

An outline of the paper is as follows. Notation and conventions used throughout the paper are presented in Section 2. This is followed by the adaptation of Hast's algorithm for odd Chow-robust predicates and the result that (essentially) the condition $\sum_{j=1}^{n} w_j^3 - w_j \leq 3 \sum_{j=1}^{n} w_j^2$ on the weights is sufficient for a predicate to be Chow-robust. In Section 4, we present our main algorithm for Chow-robust predicates which establishes Theorem 1 in the special case of majority. These positive results are then complemented in Section 5 where we show essentially tight hardness results assuming the Unique Games Conjecture. Finally, we discuss the obtained results together with interesting future directions (Section 6). As already stated, the current abstract only contains some of our shorter proofs and a reader interested in the full proofs must turn to the full version of the paper.

2   Preliminaries

We consider the optimization problem Max-CSP(P) for homogeneous linear threshold predicates $P : \{-1,1\}^n \rightarrow \{-1,1\}$ of the form

$$P(x) = \mathrm{sgn}(w_1 x_1 + \cdots + w_n x_n),$$

where we assume that the weights are non-decreasing positive integers $1 \leq w_1 \leq \ldots \leq w_n$ such that $\sum_{j=1}^{n} w_j$ is odd and $w_{\max} := \max_j w_j = w_n$. The special case of equal weights, which requires n to be odd, is denoted by Maj_n, and we also write Max-Maj-n for Max-CSP(Maj_n). Using Fourier expansion, any such function can be written uniquely as

$$P(x) = \sum_{S \subseteq [n]} \hat{P}(S) \prod_{j \in S} x_j.$$

The Fourier coefficients are given by $\hat{P}(S) = \mathbb{E}[P(X)\prod_{j \in S} X_j]$, where X is uniform on $\{-1,1\}^n$. Since all homogeneous linear threshold predicates are odd we have $\hat{P}(S) = 0$ when $|S|$ is even. We will also write $\hat{P}(j) = \hat{P}(\{j\})$ for the first level Fourier coefficients (i.e. the Chow parameters) and let $P^{-1}(1)$ denote the set of assignments that satisfy P, i.e. $P^{-1}(1) = \{x : P(x) = 1\}$.

For an instance $\mathcal{I} = (m, N, l, s)$ of Max-CSP(P) consisting of m constraints, N variables and matrices $l \in N^{m \times n}, s \in \{-1,1\}^{m \times n}$, the objective is to maximize the number of satisfied constraints or, equivalently since $P(-x) = -P(x)$ and thus $\mathbb{E}[P(x)] = 0$, the average advantage

$$\mathsf{Adv}(x) := \frac{1}{m} \sum_{i=1}^{m} P(s_{i,1} x_{l_{i,1}}, \ldots, s_{i,n} x_{l_{i,n}})$$

subject to $x \in \{-1,1\}^N$.

3   Adaptation of the Algorithm by Hast

Using Fourier expansion we may write the advantage of an assignment to a Max-CSP(P) instance as

$$\mathsf{Adv}(x) = \frac{1}{m} \sum_{i=1}^{m} \mathrm{sgn}\left(\sum_{j=1}^{n} w_j s_{i,j} x_{l_{i,j}} \right) = \sum_{S \subseteq [N]:|S| \leq n} c_S \prod_{k \in S} x_k. \qquad (2)$$

Hast [7] gives a general approximation algorithm for Max-CSP(P) that achieves a non-trivial approximation ratio whenever the linear part of the instance's objective function is large enough. We use his algorithm, but as our basic predicates are odd we have that $c_S = 0$ for any S of even size and we get slightly better bounds.

Theorem 4. *For any $\delta > 0$, there is a probabilistic polynomial time algorithm which given an instance of Max-CSP(P) with objective function*

$$\mathsf{Adv}(x_1, \ldots, x_N) = \sum_{S \subseteq [N], |S| \leq n} c_S \prod_{k \in S} x_k$$

satisfying $\sum_{k=1}^{N} |c_{\{k\}}| \geq \delta$ and $c_S = 0$ for any set S of even cardinality, achieves $\mathbb{E}[\mathsf{Adv}(x)] \geq \frac{\delta^{3/2}}{8n^{3/4}}$.

Proof. Let $\epsilon > 0$ be a parameter to be determined. We set each x_i randomly and independently to one with probability $(1 + \mathrm{sgn}(c_{\{i\}})\epsilon)/2$. Clearly this implies that $\mathbb{E}[c_{\{i\}} x_i] = \epsilon |c_{\{i\}}|$ and that $|\mathbb{E}[\prod_{k \in S} x_k]| = \epsilon^{|S|}$.

By Cauchy Schwarz inequality and Parseval's identity we have that

$$\sum_{|T|=k} |\hat{P}(T)| \leq \binom{n}{k}^{1/2} \left(\sum_{|T|=k} \hat{P}^2(T) \right)^{1/2} \leq \binom{n}{k}^{1/2}$$

and hence

$$\sum_{|S|=k} |c_S| \leq \binom{n}{k}^{1/2}. \tag{3}$$

We conclude that the advantage of the given algorithm is, given that $c_S = 0$ for even cardinality S, at least

$$\epsilon \sum_{i=1}^{n} |c_i| - \sum_{|S| \geq 3} \epsilon^k |c_S| \geq \epsilon\delta - \sum_{k=3}^{n} \epsilon^k \binom{n}{k}^{1/2}. \tag{4}$$

The sum in (4) is, provided $\epsilon \leq (2\sqrt{n})^{-1}$, and using Cauchy-Schwarz bounded by

$$\left(\sum_{k=3}^{n} \left(\frac{1}{n}\right)^k \binom{n}{k} \right)^{1/2} \left(\sum_{k=3}^{n} (\epsilon^2 n)^k \right)^{1/2} \leq \left(1 + \frac{1}{n} \right)^{n/2} (2\epsilon^6 n^3)^{1/2} \leq 3\epsilon^3 n^{3/2},$$

where we used $\sum_{k=0}^{n} \left(\frac{1}{n}\right)^k \binom{n}{k} = \left(1 + \frac{1}{n}\right)^n$ and $\sum_{k=3}^{n} (\epsilon^2 n)^k \leq \epsilon^6 n^3 \sum_{k=0}^{\infty} \frac{1}{2^k}$ for the first inequality. Setting $\epsilon = \delta^{1/2}(2n^{3/4})^{-1}$, which is at most $(2\sqrt{n})^{-1}$ by (3) with $k = 1$, we see that the advantage of the algorithm is

$$\epsilon\delta - 3\epsilon^3 n^{3/2} = \frac{\delta^{3/2}}{8n^{3/4}}.$$

and the proof is complete.

Let us see how to apply Theorem 4 in the case when P is majority of n variables. Suppose we are given an instance that is $1 - \frac{\delta}{n+1}$ satisfiable and let us consider

$$\sum_{i=1}^{N} c_{\{i\}} \alpha_i \tag{5}$$

where $x_i = \alpha_i$ is the optimal solution and prove that this is large. Any lower bound for this is clearly a lower bound for $\sum_{i=1}^{N} |c_{\{i\}}|$.

Let $\hat{P}_1$ be the value of any Fourier coefficient of a unit size set. Then any satisfied constraint contributes at least $\hat{P}_1$ to (5) while any other constraint contributes at least $-n\hat{P}_1$. We conclude that (5) is at least

$$\left(1 - \frac{\delta}{n+1}\right) \hat{P}_1 - \frac{\delta}{n+1} n\hat{P}_1 = (1 - \delta)\hat{P}_1.$$

Using Theorem 4 and the fact that $\hat{P}_1 = \Theta(n^{-1/2})$ we get the following corollary.

Theorem 5. *Suppose we are given an instance of Max-Maj-n which is $(1 - \frac{\delta}{n+1})$-satisfiable. Then it is possible, in probabilistic polynomial time, to find an assignment that satisfies a fraction*

$$\frac{1}{2} + \Omega((1 - \delta)^{3/2} n^{-3/2})$$

of the constraints.

Let us sketch how to generalize this theorem to predicates other than majority. Clearly the key property is to establish that the sum (5) is large when most constraints can be simultaneously satisfied. In order to have any possibility for this to be true it must be that whenever a constraint is satisfied, then the contribution to (5) is positive and this is exactly being "Chow-robust" as discussed in the introduction. Furthermore, to get a quantitative result we must also make sure that it is positive by some fixed amount. Let us turn to a formal definition.

Recall that the Chow parameters of a predicate P are given by its degree-0 and degree-1 Fourier coefficients, i.e., $\hat{P}(0), \hat{P}(1), \ldots, \hat{P}(n)$ for $i = 1, 2, \ldots, n$. As we are here dealing with an odd predicate, $\hat{P}(0) = 0$. If it holds that

$$P(x) = \mathrm{sgn}(\hat{P}(1)x_1 + \hat{P}(2)x_2 + \cdots + \hat{P}(n)x_n) \qquad \text{for all } x \in \{-1, 1\}^n$$

we say that such a predicate is *Chow-robust* and it is *γ-Chow-robust* iff

$$0 < \gamma \le \min_{x:P(x)=1} \left(\sum_{j=1}^{n} \hat{P}(j)x_j\right).$$

Note that $\gamma \le \hat{P}(1)$ and in fact $\gamma = \Theta\left(\frac{1}{\sqrt{n}}\right)$ for majority. Let us state our extension of Theorem 5 in the present context.

Theorem 6. *Let $P(x) = \operatorname{sgn}(w_1 x_1 + w_2 x_2 + \cdots + w_n x_n)$ be a γ-Chow-robust predicate and suppose that $\mathcal{I}$ is a $1 - \frac{\delta\gamma}{\gamma + \sum_{j=1}^{n} \hat{P}(j)}$ satisfiable instance of Max-CSP(P) where $\delta < 1$. Then there is a probabilistic polynomial time algorithm that achieves $\mathbb{E}[\operatorname{Adv}(x)] = \frac{(1-\delta)^{3/2} \gamma^{3/2}}{8 n^{3/4}}$.*

The proof of this theorem is given in the full version of the paper.

Given Theorem 6 it is interesting to discuss sufficient conditions for P to be Chow-robust and we have the following theorem.

Theorem 7. *Suppose we are given positive integers $(w_j)_{j=1}^{n}$ such that*

$$\beta(w) := 1 - \frac{\sum_{j=1}^{n} (w_j^3 - w_j)}{3 \sum_{j=1}^{n} w_j^2} > 0.$$

Further, suppose that for at least $400 \log n$ different values of j, say $1, 2, \ldots, n_1$, we have $w_j = 1$. Then the predicate $P(x) = \operatorname{sgn}(x_1 + \cdots + x_{n_1} + w_{n_1+1} x_{n_1+1} + \cdots + w_n x_n)$ is γ-Chow-robust with $\gamma = \left(\beta(w) - \mathcal{O}\left(\frac{w_{\max}^2}{n}\right)\right) \hat{P}(1)$.

Note that we need n sufficiently large to make γ positive.

Also this proof is postponed to the full version. Let us comment on the condition on the $\Omega(\log n)$ weights that we require to be one. This should be viewed as a technical condition and we could have chosen other similar conditions. In particular, we have made no effort to optimize the constant 400. In our calculations this condition is used to bound the integrand of a complex integral on the unit circle when we are not close to the point $z = 1$ and this could be done in many ways. We would like to point out that although there are choices for the technical condition, some condition is needed. The condition should imply some mathematical form of "when z on the unit circle is far from 1 then many numbers of the form z^{w_j} are not close to 1". Sets of weights violating such conditions are cases when almost all weights have a common factor. An interesting example is the function which, for odd n, has $n - 4$ weights equal to 3 and 4 weights equal to 1. This function is not Chow-robust for any value of n. The above example shows that there are functions with weights of at most 3 that are not Chow-robust. This is a tight bound as the techniques used in the proof of Theorem 7 can be used to show that a function with weights equal to 1 or 2 is Chow-robust.

4   Our Main Algorithm

We now give an improved algorithm for Max-CSP(P) for homogeneous linear threshold predicates. On almost satisfiable instances, this algorithm achieves an advantage $\Omega\left(\frac{1}{\sqrt{n}}\right)$ over a random assignment in comparison to the $\Omega\left(\frac{1}{n^{3/2}}\right)$ advantage achieved by the adaptation of Hast's algorithm presented in the previous section. However, a drawback of the more advanced algorithm is that we are unable to analyze its advantage on instances that are close to the threshold where Hast's algorithm still achieves a non-trivial advantage. Thus, in order to fully understand the approximability curve, a combination of the algorithm presented below and Hast's algorithm is needed. We now proceed by describing the algorithm. Recall that we write the i'th constraint as

$$P(s_{i,1}x_{l_{i,1}}, \ldots, s_{i,n}x_{l_{i,n}}) = \text{sgn}(L_i(x)),$$

where $L_i(x) = \sum_{j=1}^{n} w_j s_{i,j} x_{l_{i,j}}$, and let $W := \sum_{j=1}^{n} w_j$. The algorithm which is parameterized by a noise parameter $0 < \epsilon < 1$ is described as follows:

Algorithm $A_{\text{LP},\epsilon}$

1. Let x^*, Δ^* be the optimal solution to the following linear program

$$\begin{aligned} &\text{maximize } \tfrac{1}{m} \sum_{i=1}^{m} \Delta_i \\ &\text{subject to } L_i(x) \geq \Delta_i, \forall i \in [m] \\ &\qquad\quad x \in [-1,1]^N, \Delta \in [-W,1]^m \end{aligned}$$

2. Pick $X_1, \ldots, X_N \in \{-1,1\}$ independently with bias $\mathbb{E}[X_i] = \epsilon x_i^*$ and return this assignment.

As in Theorem 7 we now define $\beta(w)$ for a set of weights $w = (w_1, \ldots, w_n)$ as

$$\beta(w) = 1 - \frac{\sum_{j=1}^{n}(w_j^3 - w_j)}{3 \sum_{j=1}^{n} w_j^2}.$$

Note that $\beta \leq 1$ for any set of weights, while for majority $\beta = 1$. Further, if $\beta(w) > 0$, then Theorem 7 shows that P is γ-Chow-robust provided that n is large enough.

We have the following theorem whose proof will appear in the full version.

Theorem 8. *Fix any homogeneous threshold predicate $P(x) = \text{sgn}(w_1 x_1 + \cdots + w_n x_n)$ having $w_j = 1$ for at least $200 \log n$ different values of j and satisfying $\beta := \beta(w) > 0$. Then, for any $1 - \frac{\delta}{1+W}$ satisfiable instance $\mathcal{I}$ of Max-CSP(P), where $\delta < \beta$, we have*

$$\mathbb{E}[\text{Adv}(A_{LP,\epsilon}(\mathcal{I}))] = (\beta - \delta)^{3/2} \Omega\left(\frac{1}{\sqrt{n}}\right) - \mathcal{O}\left(\frac{\log^4 n}{n^{5/6}}\right), \qquad (6)$$

where $\epsilon = (\beta - \delta)^{1/2}\epsilon_0$ and $\epsilon_0 > 0$ is an absolute constant.

Thus, for δ bounded away from β, and large enough n, this algorithm is an improvement over the algorithm of Theorem 6. We may also note that both the algorithm $A_{\text{LP},\epsilon}$ and the algorithm of Theorem 6 can be de-randomized using the method of conditional expectation.

As $\beta = 1$ for Maj_n the following result follows directly from Theorem 8:

Corollary 1. *For all $1 - \frac{\delta}{n+1}$ satisfiable instances $\mathcal{I}$ of Max-Maj-n, where $\delta < 1$, we have*

$$\mathbb{E}[\text{Adv}(A_{LP,\epsilon}(\mathcal{I}))] = (1 - \delta)^{3/2}\Omega\left(\frac{1}{\sqrt{n}}\right) - \mathcal{O}\left(\frac{\log^4 n}{n^{5/6}}\right),$$

where $\epsilon = (1 - \delta)^{1/2}\epsilon_0$ and $\epsilon_0 > 0$ is an absolute constant.

5   Unique Games Hardness

The hardness results in this section are under the increasingly prevalent assumption that the Unique Games Conjecture (UGC) holds. The conjecture was made by Khot [11] and states that a specific combinatorial problem known as Unique Games, or Unique Label Cover, is very hard to approximate (see e.g. [11] for more details). The basic tool that we use is the result by Austrin and Mossel [2], which states that the UGC implies that a predicate is approximation resistant if it supports a uniform pairwise independent distribution, and hard to approximate if it "almost" supports a uniform pairwise independent distribution. We now state their result in a simplified form tailored for the application at hand:

Theorem 9 ([2]). *Let $P \colon \{-1, 1\}^n \to \{-1, 1\}$ be a n-ary predicate and let μ be a balanced pairwise independent distribution over $\{-1, 1\}^n$. Then, for any $\epsilon > 0$, the UGC implies that it is NP-hard to distinguish between those instances of Max-CSP(P).*

- *that have an assignment satisfying at least a fraction $\mathrm{Pr}_{x \in (\{-1,1\}^n, \mu)}[P(x) = 1] - \epsilon$ of the constraints;*
- *and those for which any assignment satisfies at most a fraction $|P^{-1}(1)|/2^n + \epsilon$ of the constraints.*

We first give a fairly easy application of the above theorem to the predicate Maj_n. We then generalize this approach to more general homogeneous linear threshold predicates.

Theorem 10. *For any $\epsilon > 0$ the UGC implies that it is NP-hard to distinguish between those instances of Max-Maj-n*

- *that have an assignment satisfying at least a fraction $1 - \frac{1}{n+1} - \epsilon$ of the constraints;*
- *and those for which any assignment satisfies at most a fraction $1/2 + \epsilon$ of the constraints.*

Proof. Consider the following distribution μ over $\{-1, +1\}^n$: with probability $\frac{1}{n+1}$, all the bits in μ are fixed to -1, and with probability $\frac{n}{n+1}$, μ samples a vector with $(n + 1)/2$ ones, chosen uniformly at random among all possibilities. To see that this gives a pairwise independent distribution let $X = (X_1, \ldots, X_n)$ be drawn from μ. Then $\mathbb{E}\left[\sum_{i=1}^n X_i\right] = \frac{1}{n+1} \cdot (-n) + \frac{n}{n+1} \cdot 1 = 0$ and

$$\mathbb{E}\left[\sum_{\substack{i,j=1 \\ i \neq j}}^n X_i X_j\right] = \mathbb{E}\left[\left(\sum_{i=1}^n X_i\right)^2\right] - n = \frac{1}{n+1} \cdot (n^2) + \frac{n}{n+1} \cdot 1 - n = 0.$$ Because

of the symmetry of the coordinates, it follows that for all i, $\mathbb{E}[X_i] = 0$ and for every $i \neq j$, $\mathbb{E}[X_i X_j] = 0$. Therefore, the distribution μ is balanced pairwise independent. Theorem 9 now gives the result.

For predicate Maj_n, we can also obtain a hardness result for almost satisfiable instances:

Theorem 11. *For any $\epsilon > 0$ the UGC implies that it is NP-hard to distinguish between those instances of Max-Maj-n*

- *that have an assignment satisfying at least a fraction $1 - \epsilon$ of the constraints;*
- *and those for which any assignment satisfies at most a fraction $\frac{1}{2} + c_n \frac{1}{\sqrt{n}} + \epsilon$ of the constraints, where*

$$c_n = \frac{\sqrt{n}}{2^{n-2}} \binom{n-2}{\frac{n-1}{2}} \approx \sqrt{\frac{2}{\pi}}.$$

Proof. Let $k = n - 2$ and consider the predicate $P \colon \{-1,1\}^k \to \{-1,1\}$ defined as $P(x) = \operatorname{sgn}(x_1 + \cdots + x_k + 2)$. Our interest in P stems from the fact that Max-Maj-n is at least as hard to approximate as Max-CSP(P). Indeed, given an instance of Max-CSP(P), we can construct an instance of Max-Maj-n by letting each constraint $P(l_1, \ldots l_k)$ equal $\operatorname{Maj}_n(y_1, y_2, l_1, \ldots, l_k)$ for two new variables y_1 and y_2, that are the same in all constraints and always appear in the positive form. As any good solution to the instance of Max-Maj-n sets both y_1 and y_2 to one, we can conclude that any optimal assignments to the two instances satisfy the same fraction of constraints.

Now consider the following distribution μ over $\{-1,1\}^k$: with probability $\frac{1}{k+1}$, all the bits in μ are fixed to ones, and with probability $\frac{k}{k+1}$, μ samples a vector with $(k+1)/2$ minus ones, chosen uniformly at random among all possibilities. The same argument as in the proof of Theorem 10 shows that the distribution μ is uniform and pairwise independent. Theorem 9 now gives that for any $\epsilon > 0$ the UGC implies that it is NP-hard to distinguish between those instances of Max-CSP(P) that have an assignment satisfying a fraction $1 - \epsilon$ of the constraints; and those for which any assignment satisfies at most a fraction

$$\frac{|P^{-1}(1)|}{2^k} + \epsilon = \frac{1}{2^k} \sum_{j=0}^{\frac{k+1}{2}} \binom{k}{j} + \epsilon = \frac{1}{2} + \frac{\binom{k}{\frac{k+1}{2}}}{2^k} + \epsilon = \frac{1}{2} + \sqrt{\frac{2}{\pi k}} + o(1/k) + \epsilon.$$

The result now follows from the observation above that we can construct an instance of Max-Maj-n from an instance of Max-CSP(P) such that optimal assignments to the two instances satisfy the same fraction of the constraints.

Taking the convex combination of the results in Theorems 10 and 11 yields:

Corollary 2. *For any $\delta : 0 \le \delta \le 1$ and any $\epsilon > 0$, the UGC implies that it is NP-hard to find an assignment x to a given $1 - \frac{\delta}{n+1} - \epsilon$ satisfiable instance of Max-Maj-n achieving*

$$\operatorname{Adv}(x) \ge (1 - \delta) c_n \frac{1}{\sqrt{n}} + \epsilon,$$

where c_n is the constant defined in Theorem 11.

The above techniques also extend to general weights and we have the following theorem.

Theorem 12. *Suppose we are given positive integers $(w_j)_{j=1}^n$ such that $\sum_{j=1}^n w_j^3 < 100n$ and $\sum_{j=1}^n w_j$ is odd. Further, suppose that for at least $400 \log n$ different* *values of j we have $w_j = 1$. Let $P(x) = \text{sgn}(w_1 x_1 + \cdots + w_n x_n)$, then, for any $\epsilon > 0$, the UGC implies that it is NP-hard to distinguish between those instances of Max-CSP(P)*

- *that have an assignment satisfying at least a fraction $1 - \mathcal{O}\left(\frac{w_{\max}^4}{n}\right) - \epsilon$ of the constraints;*
- *and those for which any assignment satisfies at most a fraction $1/2 + \epsilon$ of the constraints.*

Of course the key to this theorem is to study suitable pairwise independent distributions. In particular, we prove that similar ideas as used in the proof of Theorem 10 can be used to construct *almost* pairwise distributions for more general "majority-like" threshold predicates. As we allow predicates with different weights, the analysis gets more involved and again the problem reduces to estimating complex integrals using the saddle point method. For this reason we need the technical conditions on the weights that were previously discussed after Theorem 7. We then show that such distributions can be slightly adjusted to obtain perfect balanced pairwise distributions and the final result follows by applying Theorem 9. The details will appear in the full version of the paper.

6 Conclusions

We have studied, and obtained rather tight bounds for the approximability curve of "majority-like" predicates. There are still many questions to be addressed and let us mention a few.

This work has been in the context of predicates given by Chow-robust threshold functions. Within this class we already knew, by the results of Hast [7], that no such predicate can be approximation resistant and our contribution is to obtain sharp bounds on the nature of how approximable these predicates are. It is a very nice open question whether there are any approximation resistant predicates given as thresholds of balanced linear functions. It is not easy to guess the answer to this question.

Looking at our results from a different angle one has to agree that the approximation algorithm we obtain is rather weak. For large values of n we only manage to do something useful on almost satisfiable instances and in this case we beat the random assignment by a rather slim margin. On the other hand we also prove that this is the best we can do. One could ask the question whether there is any other predicate that genuinely depends on n variables, accepts about half the inputs and which is easier to approximate than majority. It is not easy to guess what such a predicate would be but there is also very little information to support the guess that majority is the easiest predicate to approximate.

Using the results of Austrin and Mossel, Austrin and Håstad [1] proved that almost all predicates are approximation resistant. One way to interpret the results of this paper is that for the few predicates of large arity where we can get some nontrivial approximation, we should not hope for too strong positive results.

References

1. Austrin, P., Håstad, J.: Randomly supported independence and resistance. In: 38th Annual ACM Symposium on Theory of Computation, pp. 483–492 (2009)
2. Austrin, P., Mossel, E.: Approximation resistant predicates from pairwise independence. Computational Complexity 18(2), 249–271 (2009)
3. Chow, C.K.: On the characterization of threshold functions. In: Proceedings of the 2nd Annual Symposium on Switching Circuit Theory and Logical Design, pp. 34–38 (1961)
4. Dertouzos, M.: Threshold logic: a synthesis approach. MIT Press, Cambridge (1965)
5. Diakonikolas, I., Servedio, R.A.: Improved approximation of linear threshold functions. In: IEEE Conference on Computational Complexity, pp. 161–172 (2009)
6. Goemans, M., Williamson, D.: Improved approximation algorithms for maximum cut and satisfiability problems using semidefinite programming. Journal of the ACM 42, 1115–1145 (1995)
7. Hast, G.: Beating a random assignment. In: APPROX-RANDOM, pp. 134–145 (2005)
8. Håstad, J.: Some optimal inapproximability results. Journal of ACM 48, 798–859 (2001)
9. Kaplan, K.R., Winder, R.O.: Chebyshev approximation and threshold functions. IEEE Transactions on Electronic Computers EC -14(2), 250–252 (1965)
10. Kaszerman, P.: A geometric test-synthesis procedure for a threshold device. Information and Control 6(4), 381–398 (1963)
11. Khot, S.: On the power of unique 2-prover 1-round games. In: Proceedings of 34th ACM Symposium on Theory of Computating, pp. 767–775 (2002)
12. Khot, S., Mossel, E., Kindler, G., O'Donnell, R.: Optimal inapproximability results for Max-Cut and other 2-variable CSPs? In: Proceedings of 45th Annual IEEE Symposium of Foundations of Computer Science, pp. 146–154 (2004)
13. O'Donnell, R., Servedio, R.A.: The Chow parameters problem. In: Proceedings of the 40th Annual ACM Symposium on Theory of Computing, pp. 517–526 (2008)
14. O'Donnell, R., Wu, Y.: An optimal SDP algorithm for Max-Cut and equally optimal long code tests. In: Proceedings of 40th ACM Symposium on Theory of Computating, pp. 335–344 (2008)
15. Schaefer, T.: The complexity of satisfiability problems. In: Conference Record of the Tenth Annual ACM Symposium on Theory of Computing, pp. 216–226 (1978)
16. Shiganov, I.S.: Refinement of the upper bound of a constant in the remainder term of the central limit theorem. Journal of Soviet mathematics 35, 109–115 (1986)
17. Winder, R.O.: Threshold logic in artificial intelligence. Artificial Intelligence S-142, 107–128 (1963)
18. Winder, R.O.: Threshold gate approximations based on Chow parameters. IEEE Transactions on Computers 18(4), 372–375 (1969)
19. Zwick, U.: Approximation algorithms for constraint satisfaction problems involving at most three variables per constraint. In: Proceedings of the Ninth Annual ACM-SIAM Symposium on Discrete Algorithms, pp. 201–210 (1998)

Approximating Sparsest Cut in Graphs of Bounded Treewidth[*]

Eden Chlamtac[1,**], Robert Krauthgamer[1,***], and Prasad Raghavendra[2]

[1] Weizmann Institute of Science, Rehovot, Israel
{eden.chlamtac,robert.krauthgamer}@weizmann.ac.il
[2] Microsoft Research New England, Cambridge, MA, USA
pnagaraj@microsoft.com

Abstract. We give the first constant-factor approximation algorithm for Sparsest-Cut with general demands in bounded treewidth graphs. In contrast to previous algorithms, which rely on the flow-cut gap and/or metric embeddings, our approach exploits the Sherali-Adams hierarchy of linear programming relaxations.

1   Introduction

The Sparsest-Cut problem is one of the most famous graph optimization problems. The problem has been studied extensively due to the central role it plays in several respects. First, it represents a basic graph partitioning task that arises in several contexts, such as divide-and-conquer graph algorithms (see e.g. [29, 40] and [42, Chapter 21]). Second, it is intimately related to other graph parameters, such as flows, edge-expansion, conductance, spectral gap and bisection-width. Third, there are several deep technical links between Sparsest-Cut and two seemingly unrelated concepts, the Unique Games Conjecture and Metric Embeddings.

Given that Sparsest-Cut is known to be NP-hard [34], the problem has been studied extensively from the perspective of polynomial-time approximation algorithms. Despite significant efforts and progress in the last two decades, we are still quite far from determining the approximability of Sparsest-Cut. This is true not only for general graphs, but also for several important graph families, such as planar graphs or bounded treewidth graphs. The latter family is the focus of this paper; we shall return to it after setting up some notation and defining the problem formally.

Problem definition. For a graph $G = (V, E)$ we let $n = |V|$. For $S \subset V$, the cutset $(S, \bar{S}) \subset V \times V$ is the set of unordered pairs with exactly one endpoint in S, i.e. $\{\{u, v\} \in V \times V : u \in S, v \notin S\}$. In the Sparsest-Cut problem (with general

---

[*] A full version appears at http://arxiv.org/abs/1006.3970
[**] Supported in part by a Sir Charles Clore postdoctoral fellowship.
[***] Supported in part by The Israel Science Foundation (grant #452/08), and by a Minerva grant.

M. Serna et al. (Eds.): APPROX and RANDOM 2010, LNCS 6302, pp. 124–137, 2010.
© Springer-Verlag Berlin Heidelberg 2010

demands), the input is a graph $G = (V, E)$ with edge capacities $\mathsf{cap} : E \to \mathbb{R}_{\geq 0}$ and a set of *demand pairs*, $D = (\{s_1, t_1\}, \ldots, \{s_k, t_k\})$ with a demand function $\mathsf{dem} : D \to \mathbb{R}_{\geq 0}$. The goal is to find $S \subset V$ (a cut of G) that minimizes the ratio

$$\Phi(S) = \frac{\sum_{(u,v)\in(S,\bar{S})\cap E} \mathsf{cap}(u, v)}{\sum_{(u,v)\in(S,\bar{S})\cap D} \mathsf{dem}(u, v)}.$$

The demand function dem is often set to $\mathsf{dem}(s, t) = 1$ for all $(s, t) \in D$. The special case where, in addition to this, the demand set D includes all vertex pairs is referred to as *uniform demands*.

Treewidth. Let $G = (V, E)$ be a graph. A *tree decomposition* of $G = (V, E)$ is a pair $(\mathcal{B}, T)$ where $\mathcal{B} = \{B_1, \ldots, B_m\}$ is a family of subsets $B_i \subseteq V$ called *bags*, and T is a tree whose nodes are the bags B_i, satisfying the following properties: (i) $V = \bigcup_i B_i$; (ii) For every edge $(u, v) \in E$, there is a bag B_j that contains both u, v; and (iii) For each $v \in V$, all the bags B_i containing v form a connected subtree of T. The *width* of the tree decomposition is $\max_i |B_i| - 1$. The *treewidth* of G, denoted $\mathsf{tw}(G)$, is the smallest width among all tree decompositions of G. The *pathwidth* of G is defined similarly, except that T is restricted to be a path; thus, it is at least $\mathsf{tw}(G)$. It is straightforward to see that every graph G excludes as a minor the complete graph on $\mathsf{tw}(G) + 2$ vertices. Thus, the family of graphs of tree width r contains the family of graphs with pathwidth r, and is contained in the family of graphs excluding K_{r+2} as a minor (here K_{r+2} refers to the complete graph on $r + 2$ vertices).

1.1   Results

We present the first algorithm for general demand Sparsest-Cut that achieves a constant factor approximation for graphs of bounded treewidth r (the restriction is only on the structure of the graph, not the demands). Such an algorithm is conjectured to exist by [20] (they actually make a stronger conjecture, see Section 1.3 for details). However, previously such an algorithm was not known even for $r = 3$, although several algorithms are known for $r = 2$ [20, 7, 14] and for bounded-pathwidth graphs [28] (which is a subfamily of bounded-treewidth graphs).

Theorem 1. *There is an algorithm for* Sparsest-Cut *(general demands) on graphs of treewidth r, that runs in time $(2^r n)^{O(1)}$ and achieves approximation factor $C = C(r)$ (independently of n, the size of the graph).*

Table 1 lists the best approximation algorithms known for various special cases of Sparsest-Cut. We remark that the problem (with general demands) is NP-hard even for pathwidth 2 (see the full version for details).

Table 1. Approximation algorithms for Sparsest-Cut

Demands	Graphs	Approximation	Based on	Reference
general	arbitrary	$\tilde{O}(\sqrt{\log|D|})$	SDP	[2]
	treewidth 2	2	LP (flow)	[20, 7]
	fixed outerplanarity	$O(1)$	LP (integer flow)	[12, 14]
	excluding W_4-minor	$O(1)$	LP (flow)	[7]
	fixed pathwidth	$O(1)$	LP (flow)	[28]
	fixed treewidth	$O(1)$	LP (lifted)	This work
uniform	arbitrary	$O(\sqrt{\log n})$	SDP	[3]
	excluding fixed-minor	$O(1)$	LP (flow)	[24, 18]
	fixed treewidth	$O(1)$	LP (flow)	[35, 13]
	fixed treewidth	1	dynamic programming	

Techniques. Similarly to almost all previous work, our algorithm is based on rounding a linear programming (LP) relaxation of the problem. A unique feature of our algorithm is that it employs an LP relaxation derived from the hierarchy of (increasingly stronger) LPs, designed by Sherali and Adams [39]. Specifically, we use level $r + O(1)$ of this hierarchy. In contrast, all prior work on Sparsest-Cut uses either the standard LP (that arises as the dual of the concurrent-flow problem, see e.g. [29]), or its straightforward strengthening to a semidefinite program (SDP). Consequently, the entire setup changes significantly (e.g. the known connections to embeddings and flow, see Section 1.2), and we face the distinctive challenges of exploiting the complex structure of these relaxations (see Section 1.3).

While bounding the integrality gap of the standard LP (the flow-cut gap) for various graph families remains an important open problem with implications in metric embeddings (see Section 1.2), our focus is on directly approximating Sparsest-Cut. Accordingly, our LP is larger and (possibly much) stronger than the standard flow LP, and hence our rounding does not imply a bound on the flow-cut gap (akin to rounding of the SDP relaxation in [3, 10, 2]).

Finally, note that the running time stated in Theorem 1 is much better than the $n^{O(r)}$ running time typically needed to solve the $r + O(1)$ level of Sherali-Adams (or any other hierarchy). The reason is that only $O(3^r n|D|)$ of the Sherali-Adams variables and constraints are really needed for our analysis to go through (see Remark 1), thus greatly improving the time needed to solve the LP. As the rounding algorithm we use is a simple variant of the standard method of randomized rounding for LP's (adapted for Sherali-Adams relaxations on bounded-treewidth graphs), the entire algorithm is both efficient and easily implementable.

1.2   The GNRS Excluded-Minor Conjecture

Gupta, Newman, Rabinovich and Sinclair (GNRS) conjectured in [20] that metrics supported on graphs excluding a fixed minor embed into ℓ_1 with distortion

$O(1)$ (i.e. independent of the graph size). By the results of [30, 4, 20], this conjecture is equivalent to saying that in all such graphs (regardless of the capacities and demands), the ratio between the sparsest-cut and the concurrent-flow, called the *flow-cut gap*, is bounded by $O(1)$. Since the concurrent-flow problem is polynomial-time solvable (e.g. by linear programming), the conjecture would immediately imply that Sparsest-Cut admits $O(1)$ approximation (in polynomial-time) on these graphs.

Despite extensive research, the GNRS conjecture is still open, even in the special cases of planar graphs and of graphs of treewidth 3. The list of special cases that have been resolved includes graphs of treewidth 2, $O(1)$-outerplanar graphs, graphs excluding a 4-wheel minor, and bounded-pathwidth graphs; see Table 1, where the flow LP is mentioned.

Our approximation algorithm may be interpreted as evidence supporting the GNRS conjecture (for graphs of bounded treewidth), since by the foregoing discussion, the conjecture being true would imply the existence of such approximation algorithms, and moreover that our LP's integrality gap is bounded. In fact, one consequence of our algorithm and its analysis can be directly phrased in the language of metric embeddings:

Corollary 1. *For every r there is some constant $C = C(r)$ such that every shortest-path metric on a graph of treewidth $\leq r$, for which every set of size $r + 3$ is isometrically embeddable into L_1 in a locally consistent way (i.e. the embeddings of two such sets, when viewed as probability distributions over cuts, are consistent on the intersection of the sets), can be embedded into L_1 with distortion at most C.*

If, on the other hand, the GNRS conjecture is false, then our algorithm (and its stronger LP) gives a substantial improvement over techniques using the flow LP, and may have surprising implications for the Sherali-Adams hierarchy (see Section 1.3). Either way, our result opens up several interesting questions, which we discuss in Section 1.4.

1.3 Related Work

Relaxation hierarchies and approximation algorithms. A research plan that has attracted a lot of attention in recent years is the use of lift-and-project methods to design improved approximation algorithms for NP-hard optimization problems. These methods, such as Sherali-Adams [39], Lovász-Schrijver [31], and Lasserre [26] (see [27] for a comparison), systematically generate, for a given $\{0, 1\}$ program (which can capture many combinatorial optimization problems, e.g. Vertex-Cover), a sequence (aka *hierarchy*) of increasingly stronger relaxations. The first relaxation in this sequence is often a commonly-used LP relaxation for that combinatorial problem. After n steps (which are often called *rounds* or *levels*), the sequence converges to the convex hull of the integral solutions, and the k-th relaxation in the sequence is a convex program (LP or SDP) that can be solved in time $n^{O(k)}$. Therefore, the first few, say $O(1)$, relaxations in the sequence offer a great promise to approximation algorithms — they could be much stronger than the commonly-used LP relaxation, yet are polynomial-time computable. This is particularly promising for problems for which there is a gap between known approximations and proven hardness of approximation (or when the hardness relies on weaker assumptions than $P \neq NP$).

Unfortunately, since the work of Arora, Bollobás, Lovász, and Tourlakis [1] on Vertex-Cover, there has been a long line of work showing that for various problems, even after a large (super-constant) number of rounds, various hierarchies do not yield smaller integrality gaps than a basic LP/SDP relaxation (see, e.g. [38, 19, 37, 41, 8]). In particular, Raghavendra and Steurer [36] have recently shown that a superconstant number of rounds of certain SDP hierarchies does not improve the integrality gap for any constraint satisfaction problem (MAX-CSP).

In contrast, only few of the known results are positive, i.e. show that certain hierarchies give a sequence of improvements in the integrality gap in their first $O(1)$ levels — this has been shown for Vertex-Cover in planar graphs [32], Max-Cut in dense graphs [43], Knapsack [21, 6], and Maximum Matching [33]. There are even fewer results where the improved approximation is the state-of-the-art for the respective problem — such results include recent work on Chromatic Number [15], Hypergraph Independent Set [16], and MaxMin Allocation [5].

In the context of bounded-treewidth graphs, a bounded number of rounds in the Sherali-Adams hierarchy is known to be tight (i.e. give exact solutions) for many problems that are tractable on this graph family, such as CSPs [44]. This is only partially true for Sparsest-Cut — due to the exact same reason, we easily find in the graph a cut whose edge capacity exactly matches the corresponding expression in the LP. However, the demands are arbitrary (and in particular do not have a bounded-treewidth structure), and analyzing them requires considerably more work.

Hardness and integrality gaps for sparsest-cut. As mentioned earlier, Sparsest-Cut is known to be NP-hard [34], and we further show in the full version that it is even NP-hard on graphs of pathwidth 2. Two results [23, 9] independently proved that under Khot's unique games conjecture [22], the Sparsest-Cut problem is NP-hard to approximate within any constant factor. However, the graphs produced by the reductions in these two results have large treewidth.

The standard flow LP relaxation for Sparsest-Cut was shown in [29] to have integrality gap $\Omega(\log n)$ in expander graphs, even for uniform demands. Its standard strengthening to an SDP relaxation (the SDP used by the known approximation algorithms of [3, 2]) was shown in [23, 25, 17] to have integrality gap $\Omega(\log \log n)$, even for uniform demands. For the case of general demands, a stronger bound $(\log n)^{\Omega(1)}$ was recently shown in [11]. Some of these results were extended in [8, 36] to certain hierarchies and a nontrivial number of rounds, even for uniform demands. Again, the graphs used in these results have large treewidth.

Integrality gaps for graphs of treewidth r (or excluding a fixed minor of size r) follow from the above in the obvious way of replacing n with r (or so), for instance, the standard flow LP has integrality gap $\Omega(\log r)$. However, no stronger gaps are known for these families; in particular, it is possible that the integrality gap approaches 1 with sufficiently many rounds (depending on r, but not on n).

1.4 Discussion and Further Questions

We show that for the Sparsest-Cut problem, the Sherali-Adams (SA) LP hierarchy can yield algorithms with better approximation ratio than previously known. Moreover, our analysis exhibits a strong (but rather involved) connection between the input graph's treewidth and the SA hierarchy level. Several interesting questions arise immediately:

1. Can this approach be generalized to excluded-minor graphs?
2. Can the approximation factor be improved to an absolute constant (independent of the treewidth)?

A particularly intriguing and more fundamental question is whether this hierarchy (or a related one, or for a different input family) is strictly stronger than the standard LP (or SDP) relaxation. One possibility is that our relaxation can actually yield an absolute constant factor approximation (as in Question 2). Such an approximation factor is shown in [8] to require at least $\Omega(\log r)$ rounds of Sherali-Adams, and we would conclude that hierarchies yield strict improvement — higher (yet constant) levels of the Sherali-Adams hierarchy do give improved approximation factors, for an increasing sequence of graph families. We note, however, that this would require a different rounding algorithm (see Remark 2). Another possibility is that the GNRS conjecture does not hold even for bounded treewidth graphs, in which case the integrality gap of the standard LP exhibits a dependence on n, while, as we prove here, the stronger LP does not.

2 Technical Overview

Relaxations arising from the Sherali-Adams (SA) hierarchy, and lift-and-project techniques in general, are known to give LP (or SDP) solutions which satisfy the following property: for every subset of variables of bounded size (bounded by the level in the hierarchy used), the LP/SDP solution restricted to these variables is a convex combination of valid $\{0, 1\}$ assignments. Such a convex combination can naturally be viewed as a distribution on local assignments. In our case, for example, in an induced subgraph on $r + 1$ vertices S, an $(r + 1)$-level relaxation gives a local distribution on assignments $f : S \to \{0, 1\}$ such that for every edge (i, j) within S, the probability that $f(i) \neq f(j)$ is exactly the contribution of edge (i, j) to the objective function (which we also call the *LP-distance* of this pair). Our algorithm makes explicit use of this property, which is very useful for treewidth r graphs.

Given an $(r+3)$-level Sherali-Adams relaxation, for every demand pair there is some distribution which (within every bag) matches the local distributions suggested by the LP, and also *cuts/separates* this demand pair (i.e. assigns different values to its endpoints) with the correct probability (the LP distance). Unfortunately, there might not be any single distribution which is consistent with all demand pairs, so instead our algorithm assigns $\{0, 1\}$ values at random to the vertices of the graph G in a stochastic process which matches the local distributions suggested by the LP solution (per bag), but is oblivious to the structure of the demands D.

Intuition. To achieve a good approximation ratio, it suffices to ensure that every demand pair is cut with probability not much smaller than the its LP distance. To achieve this, the algorithm fixes an arbitrary bag as the root, and traverses the tree decomposition one bag at a time, from the root towards the leaves, and samples the assignment to currently unassigned vertices in the current bag. This assignment is sampled in a way that ignores all previous assignments to vertices outside the current bag, but achieves the correct distribution on assignments to the current bag. Essentially, the algorithm finds locally correct distributions while maximizing the entropy of the overall distribution. Intuitively, this should only "distort" the distribution suggested by the LP (for a given demand pair) only by introducing noise, which (if the noise is truly unstructured) mixes the correct global distribution with a completely random one in which every two vertices are separated with probability $\frac{1}{2}$. In this case, the probability of separating any demand pair would decrease by at most a factor 2. Unfortunately, we are not able to translate this intuition into a formal proof (and on some level, it is not accurate – see Remark 2). Thus we are forced to adopt a different strategy in analyzing the performance of the rounding algorithm. Let us see one illustrative special case.

Example: Simple Paths. Consider, for concreteness, the case of a single simple path $v_1, v_2, \ldots, v_n$. For every edge in the path (v_{i-1}, v_i), the LP suggests cutting it (assigning different values) with some probability p_i. Our algorithm will perform the following Markov process: pick some assignment $f(v_1) \in \{0, 1\}$ at random according to the LP, and then, at step i (for $i = 2, \ldots, n$) look only at the assignment $f(v_{i-1})$ and let $f(v_i) = 1 - f(v_{i-1})$ with probability p_i, and $f(v_i) = f(v_{i-1})$ otherwise. Each edge has now been cut with exactly the probability corresponding to its LP distance. However, for (v_1, v_n), which could be a demand pair, the LP distance between them might be much greater than the probability $q_n = \Pr[f(v_1) \neq f(v_n)]$. Let us see that the LP distance can only be a constant factor more.

First, if the above probability satisfies $q_n \geq \frac{1}{3}$, then clearly we are done, as all LP distances will be at most 1. Thus we may assume that $q_n \leq \frac{1}{3}$. Let us examine what happens at a single step. Suppose the algorithm has separated v_1 from v_{i-1} with some probability $q_{i-1} \leq \frac{1}{3}$ (assuming that all $q_i \leq \frac{1}{3}$ is a somewhat stronger assumption than $q_n \leq \frac{1}{3}$, but a more careful analysis shows it is also valid). After the current step (flipping sides with probability p_i), the probability that v_i is separated from v_1 is exactly $(1 - q_{i-1})p_i + q_{i-1}(1 - p_i)$. This is an increase over the previous value q_{i-1} of at least

$$[(1 - q_{i-1})p_i + q_{i-1}(1 - p_i)] - q_{i-1} = (1 - 2q_{i-1})p_i \geq p_i/3.$$

However, the LP distance from v_1 can increase by at most p_i (by triangle inequality). Thus, we can show inductively that we never lose more than a factor 3.

In general, our analysis will consider paths of bags of size $r + 1$. Even though we can still express the distribution on assignments chosen by the rounding algorithm as a Markov process (where the possible states at every step will be assignments to some set of at most r vertices), it will be less straightforward to relate the LP values to this process. It turns out that we can get a handle on the LP distances by modeling the Markov process as a layered digraph H with edges capacities representing the transitions (this is only in the analysis, or in the derandomization of our algorithm). In this case the LP distance we wish to bound becomes the value of a certain (s, t)-flow in H. We then bound the flow-value from above by finding a small cut in H. Constructing and bounding the capacity of such a cut in H constitutes the technical core of this work.

3   The Algorithm

3.1   An LP Relaxation Using the Sherali Adams Hierarchy

Let us start with an informal overview of the Sherali-Adams (SA) hierarchy. In an LP relaxation for a 0–1 program, the linear variables $\{y_i \mid i \in [n]\}$ represent linear relaxations of integer variables $x_i \in \{0, 1\}$. We can extend such a relaxation to include variables $\{y_I\}$ for larger subsets $I \subseteq [n]$ (usually, up to some bounded cardinality). These should be interpreted as representing the products $\prod_{i \in I} x_i$ in the intended (integer) solution. Now, for any pair of sets $I, J \subseteq [n]$, we will denote by $y_{I,J}$ the linear relaxation for the polynomial $\prod_{i \in I}(1 - x_i) \prod_{j \in J} x_j$. These can be derived from the variables y_I by the inclusion-exclusion principle. That is, we define

$$y_{I,J} = \sum_{I' \subseteq I}(-1)^{|I'|} y_{I' \cup J}.$$

The constraints defined by the polytope $\mathbf{SA}_t(n)$, that is, level t of the Sherali-Adams hierarchy starting from the trivial n-dimensional LP, are simply the inclusion-exclusion constraints:

$$\forall I, J \subseteq [n] \text{ s.t. } |I \cup J| \leq t \ : \ y_{I,J} \geq 0 \tag{1}$$

For every solution other than the trivial (all-zero) solution, we can define a normalized solution $\{\tilde{y}_I\}$ as follows:

$$\tilde{y}_I = y_I/y_\emptyset,$$

and the normalized derived variables $\tilde{y}_{I,J}$ can be similarly defined.

As is well-known, in a non-trivial level t Sherali-Adams solution, for every set of (at most) t vertices, constraints (1) imply a distribution on $\{0, 1\}$ assignments to these vertices matching the LP values:

Lemma 1. *Let $\{y_I\}$ be a non-zero vector in the polytope $\boldsymbol{SA}_t(n)$. Then for every set $L \subseteq [n]$ of cardinality $|L| \leq t$, there is a distribution μ_L on assignments $f : L \to \{0, 1\}$ such that for all $I, J \subseteq L$,*

$$\Pr_{\mu_L}\left[(\forall i \in I : f(i) = 0) \wedge (\forall j \in J : f(j) = 1)\right] = \tilde{y}_{I,J}.$$

In a Sparsest Cut relaxation, we are interested in the event in which a pair of vertices is cut (i.e. assigned different values). This is captured by the following linear variable:

$$y_{i \neq j} = y_{\{i\},\{j\}} + y_{\{j\},\{i\}}.$$

We can now define our relaxation for Sparsest Cut, $\mathbf{SC}_r(G)$:

$$\min \quad \sum_{(i,j) \in E} \mathsf{cap}(i,j) y_{i \neq j} \tag{2}$$

$$\text{s.t.} \quad \sum_{i,j \in D} \mathsf{dem}(i,j) y_{i \neq j} = 1 \tag{3}$$

$$\{y_I\} \in \mathbf{SA}_{r+3}(n) \tag{4}$$

$$y_{I,J} = y_{J,I} \qquad\qquad \forall I, J \text{ s.t. } |I \cup J| \leq r + 3 \tag{5}$$

Note that constraint (3) is simply a normalization ensuring that the objective function is really a relaxation for the ratio of the two sums. Also note that constraint (5), which ensures that the LP solution is fully symmetric, does not strengthen the LP, in the following sense: For any solution $\{y_I'\}$ to the above LP without constraint (5), a new solution to the symmetric LP (with the same value in the objective function) can be achieved by taking $y_I = (y_I' + y_{I,\emptyset}')/2$ without violating any of the other constraints. In particular, for every vertex $i \in V$ this gives $\tilde{y}_i = 1 - \tilde{y}_i = \frac{1}{2}$. While our results hold true without imposing this constraint, we will retain it as it simplifies our analysis.

Remark 1. The size of this LP (and the time needed to solve it) is $n^{O(r)}$. Specifically for bounded-treewidth graphs, we could also formulate a much smaller LP, where constraint (4) would be replaced with the condition $\{y_I \mid I \subseteq B \cup \{i,j\}\} \in \mathbf{SA}_{r+3}(r + 3)$ for every bag B and demand pair $(i, j) \in D$. This would reduce the size of the LP to (and time needed to solve it) to at most $\mathrm{poly}(2^r n)$, and our rounding algorithm and analysis would still hold.

3.2   Rounding the LP

Before we present the rounding algorithm, let us introduce some notation which will be useful in describing the algorithm. This notation will allow us to easily go back-and-forth between the LP solution and the local distributions on assignments described in Lemma 1. For ease of notation, whenever two functions f_1, f_2 have disjoint domains, we will denote by $f_1 \cup f_2$ the unique function from the union of the domains which is an extension of both f_1 and f_2.

- For every set of vectors $\{y_I\}$ and subset $L \subseteq [n]$ as in Lemma 1, we will denote by $\mu_L^{\{y_I\}}$ the distribution on random assignments to L guaranteed by the lemma. We will omit the superscript $\{y_I\}$, and simply write μ_L, when it is clear from the context.

- Conversely, for any fixed assignment $f' : L \to \{0,1\}$, we will write $\tilde{y}_{f'} = \tilde{y}_{L_0, L_1}$, where $L_b = \{i \in L \mid f'(i) = b\}$ for $b = 0, 1$. Thus, for a random assignment $f : L \to \{0,1\}$ distributed according to μ_L, we have $\Pr[f = f'] = \tilde{y}_{f'}$.
- For any nonempty subset $L' \subseteq L$, and a given assignment $f_0 : L \setminus L' \to \{0,1\}$ in the support of $\mu_{L \setminus L'}$, we will denote by μ_{L', f_0} the distribution on random assignments $f \sim \mu_L$ conditioned on the partial assignment f_0. Formally, a random assignment $f' : L' \to \{0,1\}$, distributed according to μ_{L', f_0} satisfies $\Pr_{f'}[f' = f_1] = \tilde{y}_{f_0 \cup f_1} / \tilde{y}_{f_0}$ for every choice of $f_1 : L' \to \{0,1\}$.

Let G be an graph with treewidth r for some integer $r > 0$, and let $(\mathcal{B}, T)$ be the corresponding tree decomposition. Let $\{y_I\}$ be a vector satisfying $\mathbf{SC}_r(G)$. We now present the rounding algorithm:

Algorithm SC-Round$(G, (\mathcal{B}, T), \{y_I\})$ [Constructs a random assignment f]

1. Pick an arbitrary $B_0 \in \mathcal{B}$ as the root of T, and sample $f|_{B_0} \sim \mu_{B_0}$.
2. Traverse the rest of the tree T in any order from the root towards the leaves. For each bag B traversed, do the following:
   (a) Let B^+ be the set of vertices in B for which f is already defined, and let $B^- = B \setminus B^+$. Let f_0 be the existing assignment $f_0 = f|_{B^+}$.
   (b) If B^- is non-empty, sample $f|_{B^-}$ at random according to μ_{B^-, f_0}.

Let us first see that every edge $(i, j) \in E$ is cut with probability exactly $\tilde{y}_{i \neq j}$. Since every edge is contained in at least one bag, it suffices to show that within every bag B, the assignment $f|_B$ is distributed according to μ_B. This is shown by the following lemma, whose straightforward proof appears in the full version.

Lemma 2. *For every bag B, the assignment $f|_B$ produced by running algorithm SC-Round$(G, (\mathcal{B}, T), \{y_I\})$ is distributed according to μ_B.*

This lemma shows that the expected value of the cut is $\sum_{(i,j) \in E} \mathsf{cap}(i, j) \tilde{y}_{i \neq j}$, which is exactly the value of the objective function (2) scaled by $1/y_\emptyset$. In particular, for a host of other problems where the objective function and constraints depend only on the edges (e.g. Minimum Vertex Cover, Chromatic Number), this type of LP relaxation (normalized by setting $y_\emptyset = 1$), along with the above rounding, always produces an optimal solution for bounded-treewidth graphs. Thus, in some sense, we consider this to be a "natural" rounding algorithm.

Before we analyze the expected value of the cut demands (or specifically, the probability that each demand is cut), let us show that the order in which the tree T is traversed has no effect on the distribution of cuts produced (it will suffice to show a slightly weaker claim – that the joint distribution of cuts in any two bags is not affected). This is shown in the following lemma, whose proof appears in the appendix.

Lemma 3. *Let $B_1, B_2 \in \mathcal{B}$ be two arbitrary bags. Then the distribution on assignments $f|_{B_1 \cup B_2}$ is invariant under any connected traversal of T.*

4 Markov Flow Graphs

We show the following lemma, whose proof appears in the full version, which together with Lemma 2 implies Theorem 1 (see Remark 3).

Lemma 4. *For every integer $r > 0$ there exists a constant $c_r > 0$ such that for any treewidth-r graph G with tree decomposition $(\mathcal{B}, T)$, and vectors $\{y_I\}$ satisfying $SC_r(G)$, algorithm SC-Round$(G, (\mathcal{B}, T), \{y_I\})$ outputs a random $f : V \to \{0, 1\}$ s.t. for every $i, j \in V$,*

$$\Pr[f(i) \neq f(j)] \geq c_r \tilde{y}_{i \neq j}. \tag{6}$$

Remark 2. The constant c_r arising in our analysis is quite small (roughly 2^{-r2^r}). While we believe this can be improved, we cannot eliminate the dependence on r, as a lower bound on the performance of our rounding algorithm (which appears in the full version) shows that c_r cannot be more than $2^{-r/2}$.

Remark 3. In fact, Lemmas 2 and 4 taken together show the following: Given any solution to $SC_r(G)$ with objective function value $\alpha > 0$, algorithm SC-Round produces a random assignment f satisfying

$$\mathbb{E}\left[\sum_{(i,j) \in E} \mathsf{cap}(i,j)\,|f(i) - f(j)| - \tfrac{\alpha}{c_r} \sum_{(i,j) \in D} \mathsf{dem}(i,j)\,|f(i) - f(j)| \right] \leq 0.$$

This means the algorithm produces a $1/c_r$-approximation with positive probability, but does not immediately imply a lower bound on that probability. Fortunately, following the analysis in this section, the algorithm can be derandomized by the method of conditional expectations, since, at each step, finding the probability of separating each demand pair reduces to calculating the probability of reaching a certain state at a certain phase in some Markov process, which simply involves multiplying $O(n)$ transition matrices of size at most $2^r \times 2^r$ (in fact, these can be consolidated so that every step of the algorithm involves a total of $O(n|T|)$ small matrix multiplications for all demands combined, where T is the set of vertices participating in demand pairs).

For vertices $i, j \in V$ belonging to (at least) one common bag, Lemma 2 implies equality in (6) for $c_r = 1$. For $i, j \in V$ which do not lie in the same bag, consider the path of bags $B_1, \ldots, B_N$ in tree T from the (connected) component of bags containing i to the component of bags containing j. By Lemma 3, we may assume that the algorithm traverses the path in order from B_1 to B_N.

    To understand the event that vertices i and j are separated, it suffices to consider the following incomplete (but consistent) description of the stochastic process involved: Let $S_0 = \{i\}$ and $S_N = \{j\}$, and let $S_l = B_l \cap B_{l+1}$ for $l = 1, \ldots, N - 1$. The algorithm assigns $f(i)$ a value in $\{0, 1\}$ uniformly at random, and then for $l = 1, \ldots, N$, samples $f|_{S_l}$ from the distribution $\mu_{S_l, f|_{S_{l-1}}}$ (we extend the definition of $\mu_{S, f'}$ in the natural way to include the case where S may intersect the domain of f').

This is a Markov process, and can be viewed as a Markov flow graph. That is, a layered graph, where each layer consists of nodes representing the different states (in this case, assignments to S_l), with exactly one unit of flow going from the first to the last layer, with all edges having flow at full capacity. Since all edges in the flow graph represent pairs of assignments within the same bag, Lemma 2 implies that the capacity of an edge (transition) (f_1, f_2) is exactly $\tilde{y}_{f_1 \cup f_2}$, and the amount of flow going through each node f_0 is $\tilde{y}_{f_0}$.

We now would like to analyze the contribution of a demand pair to the LP. By constraint (5), this contribution (up to a factor $\mathsf{dem}(i,j)$) is $\tilde{y}_{i \neq j} = 2\tilde{y}_{\{i\},\{j\}} = 2\tilde{y}_{f^*}$, where $f^* : \{i, j\} \to \{0, 1\}$ is the function assigning 0 to i and 1 to j. Now consider a layer graph as above where each edge (f_1, f_2) has flow $\tilde{y}_{f^* \cup f_1 \cup f_2}$. To see that this is indeed a flow, note that two consecutive layers along with i and j only involve at most $r+3$ vertices in G, and so by Lemma 1 for any $l > 0$ and function $f_2 : S_l \to \{0, 1\}$ the incoming flow at f_2 must be $\displaystyle\sum_{f_1 \in S_{l-1}} \tilde{y}_{f^* \cup f_1 \cup f_2} = \tilde{y}_{f^* \cup f_2}$, and so is the outgoing flow. The total flow in this graph is exactly $\tilde{y}_{f^*}$ (half the LP contribution $\tilde{y}_{i \neq j}$). Moreover, for each such edge (transition) we also have $\tilde{y}_{f^* \cup f_1 \cup f_2} \leq \tilde{y}_{f_1 \cup f_2}$. Hence, the flow with values $\{\tilde{y}_{f^* \cup f_1 \cup f_2}\}$ is a legal flow respecting the capacities $\{\tilde{y}_{f_1 \cup f_2}\}$ in the Markov flow graph which represents the rounding algorithm.

Thus it suffices to show the following theorem (proved in the full version):

Theorem 2. *For every integer $k > 1$, there is a constant $C = C(k) > 0$ such that for any symmetric Markov flow graph $G = (L_0, \ldots, L_N, E)$ representing a Markov process $X_0, \ldots, X_N$ with sources $L_0 = \{s_0, s_1\}$ and sinks $L_N = \{t_0, t_1\}$ and at most k nodes per layer, the total amount of capacity-respecting flow in G from s_0 to t_1 can be at most $C \cdot \Pr[X_0 = s_0 \wedge X_N = t_1]$.*

Applying this theorem to the Markov flow graph described above with $k = 2^r$ immediately implies Lemma 4. As usual, to bound the amount of flow in a graph from above, it suffices to find a suitable cut. See the full version for details.

Acknowledgments. We would like to thank Claire Mathieu for a series of helpful conversations.

References

[1] Arora, S., Bollobás, B., Lovász, L., Tourlakis, I.: Proving integrality gaps without knowing the linear program. Theory of Computing 2(1), 19–51 (2006)

[2] Arora, S., Lee, J.R., Naor, A.: Euclidean distortion and the sparsest cut. J. Amer. Math. Soc. 21(1), 1–21 (2008)

[3] Arora, S., Rao, S., Vazirani, U.: Expander flows, geometric embeddings and graph partitioning. J. ACM 56(2), 1–37 (2009)

[4] Aumann, Y., Rabani, Y.: An $O(\log k)$ approximate min-cut max-flow theorem and approximation algorithm. SIAM J. Comput. 27(1), 291–301 (1998)

[5] Bateni, M., Charikar, M., Guruswami, V.: Maxmin allocation via degree lower-bounded arborescences. In: 41st Annual ACM Symposium on Theory of Computing, pp. 543–552. ACM, New York (2009)

[6] Bienstock, D.: Approximate formulations for 0-1 knapsack sets. Oper. Res. Lett. 36(3), 317–320 (2008)

[7] Chakrabarti, A., Jaffe, A., Lee, J.R., Vincent, J.: Embeddings of topological graphs: Lossy invariants, linearization, and 2-sums. In: 49th Annual IEEE Symposium on Foundations of Computer Science, pp. 761–770 (2008)

[8] Charikar, M., Makarychev, K., Makarychev, Y.: Integrality gaps for Sherali-Adams relaxations. In: 41st ACM Symposium on Theory of Computing, pp. 283–292 (2009)

[9] Chawla, S., Krauthgamer, R., Kumar, R., Rabani, Y., Sivakumar, D.: On the hardness of approximating multicut and sparsest-cut. Computational Complexity 15(2), 94–114 (2006)

[10] Chawla, S., Gupta, A., Räcke, H.: Embeddings of negative-type metrics and an improved approximation to generalized sparsest cut. ACM Transactions on Algorithms 4(2) (2008)

[11] Cheeger, J., Kleiner, B., Naor, A.: A $(\log n)^{\Omega(1)}$ integrality gap for the sparsest cut SDP. In: 50th Annual IEEE Symposium on Foundations of Computer Science, pp. 555–564. IEEE, Los Alamitos (2009)

[12] Chekuri, C., Gupta, A., Newman, I., Rabinovich, Y., Sinclair, A.: Embedding k-outerplanar graphs into ℓ_1. SIAM J. Discret. Math. 20(1), 119–136 (2006)

[13] Chekuri, C., Khanna, S., Shepherd, F.B.: A note on multiflows and treewidth. Algorithmica 54(3), 400–412 (2009)

[14] Chekuri, C., Shepherd, B., Weibel, C.: Flow-cut gaps for integer and fractional multiflows. In: 21st ACM-SIAM Symposium on Discrete Algorithms (2010)

[15] Chlamtac, E.: Approximation algorithms using hierarchies of semidefinite programming relaxations. In: 48th Annual IEEE Symposium on Foundations of Computer Science, pp. 691–701 (2007)

[16] Chlamtac, E., Singh, G.: Improved approximation guarantees through higher levels of SDP hierarchies. In: Goel, A., Jansen, K., Rolim, J.D.P., Rubinfeld, R. (eds.) APPROX and RANDOM 2008. LNCS, vol. 5171, pp. 49–62. Springer, Heidelberg (2008)

[17] Devanur, N.R., Khot, S.A., Saket, R., Vishnoi, N.K.: Integrality gaps for sparsest cut and minimum linear arrangement problems. In: 38th Annual ACM Symposium on Theory of Computing, pp. 537–546 (2006)

[18] Fakcharoenphol, J., Talwar, K.: Improved decompositions of graphs with forbidden minors. In: 6th International Workshop on Approximation Algorithms for Combinatorial Optimization, pp. 36–46 (2003)

[19] Georgiou, K., Magen, A., Pitassi, T., Tourlakis, I.: Integrality gaps of 2 - o(1) for vertex cover SDPs in the Lovász-Schrijver hierarchy. In: 48th Annual IEEE Symposium on Foundations of Computer Science, pp. 702–712 (2007)

[20] Gupta, A., Newman, I., Rabinovich, Y., Sinclair, A.: Cuts, trees and ℓ_1-embeddings of graphs. Combinatorica 24(2), 233–269 (2004)

[21] Karlin, A.R., Mathieu, C., Thach Nguyen, C.: Integrality Gaps of Linear and Semidefinite Programming Relaxations for Knapsack. CoRR abs/1007.1283 (2010), http://arxiv.org/abs/1007.1283

[22] Khot, S.: On the power of unique 2-prover 1-round games. In: 34th Annual ACM Symposium on the Theory of Computing, pp. 767–775 (July 2002)

[23] Khot, S., Vishnoi, N.K.: The unique games conjecture, integrality gap for cut problems and the embeddability of negative type metrics into ℓ_1. In: 46th IEEE Annual Symposium on Foundations of Computer Science, pp. 53–62 (2005)

[24] Klein, P., Plotkin, S.A., Rao, S.: Excluded minors, network decomposition, and multicommodity flow. In: 25th Annual ACM Symposium on Theory of Computing, pp. 682–690 (May 1993)

[25] Krauthgamer, R., Rabani, Y.: Improved lower bounds for embeddings into L_1. SIAM J. Comput. 38(6), 2487–2498 (2009)

[26] Lasserre, J.B.: Semidefinite programming vs. LP relaxations for polynomial programming. Math. Oper. Res. 27(2), 347–360 (2002)

[27] Laurent, M.: A comparison of the Sherali-Adams, Lovász-Schrijver, and Lasserre relaxations for 0–1 programming. Math. Oper. Res. 28(3), 470–496 (2003)

[28] Lee, J.R., Sidiropoulos, A.: Pathwidth, trees, and random embeddings. CoRR abs/0910.1409 (2009)

[29] Leighton, T., Rao, S.: Multicommodity max-flow min-cut theorems and their use in designing approximation algorithms. J. ACM 46(6), 787–832 (1999)

[30] Linial, N., London, E., Rabinovich, Y.: The geometry of graphs and some of its algorithmic applications. Combinatorica 15(2), 215–245 (1995)

[31] Lovász, L., Schrijver, A.: Cones of matrices and set-functions and 0-1 optimization. SIAM J. Optim. 1(2), 166–190 (1991)

[32] Magen, A., Moharrami, M.: Robust algorithms for maximum independent set on minor-free graphs based on the Sherali-Adams hierarchy. In: Dinur, I., Jansen, K., Naor, J., Rolim, J. (eds.) APPROX 2009. LNCS, vol. 5687, pp. 258–271. Springer, Heidelberg (2009)

[33] Mathieu, C., Sinclair, A.: Sherali-Adams relaxations of the matching polytope. In: 41st Annual ACM Symposium on Theory of Computing. pp. 293–302 (2009)

[34] Milman, V.D., Schechtman, G.: Asymptotic theory of finite-dimensional normed spaces. Springer, Berlin (1986)

[35] Rabinovich, Y.: On average distortion of embedding metrics into the line and into L_1. In: 35th Annual ACM Symposium on Theory of Computing, pp. 456–462 (2003)

[36] Raghavendra, P., Steurer, D.: Integrality gaps for strong SDP relaxations of unique games. In: 50th IEEE Symposium on Foundations of Computer Science, pp. 575–585 (2009)

[37] Schoenebeck, G.: Linear level Lasserre lower bounds for certain k-CSPs. In: 49th IEEE Symposium on Foundations of Computer Science, pp. 593–602 (2008)

[38] Schoenebeck, G., Trevisan, L., Tulsiani, M.: A linear round lower bound for Lovász-Schrijver SDP relaxations of vertex cover. In: 22nd Annual IEEE Conference on Computational Complexity, pp. 205–216 (2007)

[39] Sherali, H.D., Adams, W.P.: A hierarchy of relaxation between the continuous and convex hull representations. SIAM J. Discret. Math. 3(3), 411–430 (1990)

[40] Shmoys, D.: Cut problems and their applications to divide-and-conquer. In: Hochbaum, D. (ed.) Approximation Algorithms for NP-Hard Problems. PWS Publishing Company (1997)

[41] Tulsiani, M.: CSP gaps and reductions in the Lasserre hierarchy. In: 41st Annual ACM Symposium on Theory of Computing, pp. 303–312 (2009)

[42] Vazirani, V.V.: Approximation algorithms. Springer, Berlin (2001)

[43] de la Vega, W.F., Kenyon-Mathieu, C.: Linear programming relaxations of max-cut. In: 18th ACM-SIAM Symposium on Discrete Algorithms, pp. 53–61 (2007)

[44] Wainwright, M.J., Jordan, M.I.: Treewidth-based conditions for exactness of the Sherali-Adams and Lasserre relaxations. Tech. Rep. 671, University of California, Berkeley, Department of Statistics (September 2004)

On the Conditional Hardness of Coloring a 4-Colorable Graph with Super-Constant Number of Colors

Irit Dinur and Igor Shinkar[*]

Department of Computer Science and Applied Mathematics
The Weizmann Institute of Science, Rehovot, 76100 Israel
irit.dinur@weizmann.ac.il, igor.shinkar@weizmann.ac.il

Abstract. For $3 \leq q < Q$ we consider the APPROXCOLORING(q, Q) problem of deciding whether $\chi(G) \leq q$ or $\chi(G) \geq Q$ for a given graph G. Hardness of this problem was shown in [7] for $q = 3, 4$ and arbitrary large constant Q under variants of the Unique Games Conjecture [10].

We extend this result to values of Q that depend on the size of a given graph. The extension depends on the parameters of the conjectures we consider. Following the approach of [7], we find that a careful calculation of the parameters gives hardness of coloring a 4-colorable graph with $\lg^c(\lg(n))$ colors for some constant $c > 0$. By improving the analysis of the reduction we show that under related conjectures it is hard to color a 4-colorable graph with $\lg^c(n)$ colors for some constant $c > 0$.

The main technical contribution of the paper is a variant of the Majority is Stablest Theorem, which says that among all balanced functions whose each coordinate has $o(1)$ influence, the *Majority* function has the largest noise stability. We adapt the theorem for our applications to get a better dependency between the parameters required for the reduction.

Keywords: Hardness of Approximation, Graph Coloring, Majority is Stablest.

1   Introduction

Graph Coloring is one of the most fundamental problems in combinatorics and computer science. A graph G on n vertices is said to be q-colorable if there is an assignment of labels $\{1, \ldots, q\}$ to the vertices of G, so that every two neighboring vertices receive different colors. The chromatic number of G, denoted by $\chi(G)$, is the minimal number q such that G is q-colorable. Due to the self reducibility of the coloring problem, APPROXCOLORING(q, Q) is computationally equivalent to coloring a q-colorable graph with $Q - 1$ colors.

For $q < Q$ we consider the problem APPROXCOLORING(q, Q): Given a graph G, decide whether $\chi(G) \leq q$ or $\chi(G) \geq Q$. It is well known that for any constant

---

[*] Both authors' work supported by ISF grant 1179/09, BSF grant 2008293, and ERC starting grant 239985.

M. Serna et al. (Eds.): APPROX and RANDOM 2010, LNCS 6302, pp. 138–151, 2010.

$q \geq 3$ the problem APPROXCOLORING$(q, q+1)$ is NP-hard [9]. If we consider q to be some small fixed number (e.g. 3 or 4), there is a huge gap between the values of Q for which an efficient algorithm for the problem is known, and that for which hardness results exist. For example for $q = 3$ the best known polynomial time algorithm is due to Chlamtac [5]. The semi-definite programming based algorithm solves the problem for $Q = O(n^{0.2072})$ colors (see also [1], [2], [12]). On the other hand, the strongest known hardness result shows that the problem is NP-hard for $Q = 5$ (see [8], [11]). So for $q = 3$ the problem is open for all $5 < Q < O(n^{0.2072})$.

Many inapproximability results are shown by a reduction from the PCP theorem, formulated in terms of the hardness of gap Label Cover. An instance of the Label Cover problem is a bipartite graph $G = (V \cup W, E)$, a number R, and a constraint $\pi_e \subset [R] \times [R]$ per edge e. The goal is to find a labeling that maximizes the fraction of satisfied constraints, i.e. of constraints that are satisfied by the labels on the relevant vertices. The value of instance Φ, denoted by $val(\Phi)$, is the fraction of satisfied constraints under such assignment. A Label Cover instance has the "d-to-1 property" if there are subsets $R_V = \{1, \ldots, R\}$, $R_W = \{1, \ldots, dR\}$, and the constraints are projections $\pi_e : R_W \to R_V$, such that for every $a \in R_V$ there are at most d values $b \in R_W$ that $(a, b) \in \pi_e$. Khot [10] has made following conjecture.

Conjecture 1 (d-to-1 Conjecture[10]). For any $\epsilon > 0$ there is $R = R(\epsilon)$ s.t. the following problem is NP-hard. Given a d-to-1 Label Cover instance $\Phi = (V \cup W, E)$ with label sets $R_V = \{1, \ldots, R\}$ and $R_W = \{1, \ldots, dR\}$, distinguish between the case where $val(\Phi) = 1$ and the case where $val(\Phi) < \epsilon$.

1.1   Our Result

Assuming Khot's 2-to-1 conjecture it is shown in [7] that the problem of coloring a 4-colorable graph with any constant number of colors is NP-hard. We give a quantitative version of this result. Specifically, we analyze the dependency between the inapproximability factor of the 2-to-1 Label Cover problem and the number Q of colors with which it is still hard to color a 4-colorable graph. Our main result is the following theorem:

Theorem 1. *Assume that given a 2-to-1 Label Cover instance Φ with the label set of size $R = O(\lg(n))$, it is NP-hard to distinguish between the case where $val(\Phi) = 1$ and the case where $val(\Phi) < \frac{1}{f(n)}$, for some $f(n)$.*

Then it is NP-hard to color a 4-colorable graph with $f^c(n)$ colors for some constant $c > 0$. For example if $f(n) = \lg^\delta(n)$, then it is NP-hard to color a 4-colorable graph with $\lg^{c\delta}(n)$ colors.

The theorem improves the dependency between the inapproximability factor of 2-to-1 Label Cover ($\frac{1}{f(n)}$) and the hardness of the graph coloring problem. For comparison, the (implicit) dependency in [7] is logarithmic, i.e. the soundness of $1/f(n)$ in the Label Cover is translated into hardness of coloring a 4-colorable graph with $\Omega(\lg(f(n)))$ colors.

The main technical contribution of the paper is the following theorem. It follows from a variation of the *Majority is Stablest* Theorem, which has been developed in the paper of Mossel et al. [13].

Theorem 2. *Let q be a fixed integer and let T be a symmetric Markov operator on $[q]$ with spectral radius $\rho = r(T) < 1$. Then for any $\epsilon > 0$ there exist $\delta = \epsilon^{O(1)}$ and $k = O(\lg(\frac{1}{\epsilon}))$, where the constants in the O notation depend only on ρ and q, such that the following holds: For any $f, g : [q]^n \to [0,1]$, if $E[f] > \epsilon$, $E[g] > \epsilon$ and $\langle f, T^{\otimes n} g \rangle = 0$, then*

$$\exists i \in \{1, \ldots, n\} \quad s.t. \quad \mathrm{Inf}_i^{\leq k}(f) \geq \delta \quad and \quad \mathrm{Inf}_i^{\leq k}(g) \geq \delta$$

In the analogous theorem in [7, Corollary 4.12], the (implicit) dependence between δ and ϵ is exponential, e.g. $\delta = \exp(-\frac{1}{\epsilon})$. Our contribution is a new analysis that gives a polynomial dependence between the parameters, which in turn allows us to improve our inapproximability factor to be polynomial rather that logarithmic in the assumed gap of the 2-to-1 Label Cover problem.

In Section 3 we prove a variant of the Majority is Stablest Theorem with adjustments for our purposes and conclude Theorem 2. For the sake of completeness we present the reduction of [7] from 2-to-1 Label Cover problem to the APPROXCOLORING problem and work out the parameters of the reduction, proving Theorem 1.

2   Preliminaries

2.1   Functions on the q-Ary Hypercube

Let q be a fixed integer. Let $[q]$ denote the set $\{0, \ldots, q-1\}$. For an element $x \in [q]^n$ denote by $|x|$ the number of nonzero coordinates of x. Consider the space of real valued function with domain $[q]$ or, equivalently, a vector space $\mathbb{R}^q$ with inner product defined as

$$\langle v, w \rangle = \mathbb{E}[vw] = \frac{1}{q} \sum_{i=1}^{q} v_i w_i$$

and norm of a vector defined as

$$\|v\| = \sqrt{\langle v, v \rangle}$$

Let $\alpha_0 = \mathbf{1}, \alpha_1, \ldots, \alpha_{q-1}$ be some orthonormal basis of $\mathbb{R}^q$. It defines naturally an orthonormal basis of $\mathbb{R}^{q^n}$ by applying the n-fold tensor product. It is easy to see that the set $\{\alpha_x = \alpha_{x_1} \otimes \alpha_{x_2} \otimes \cdots \otimes \alpha_{x_n} \in \mathbb{R}^{q^n} : x \in [q]^n\}$ is indeed an orthonormal basis of $\mathbb{R}^{q^n}$. Equivalently, we may think of α_x as a function from $[q]^n$ to $\mathbb{R}$ defined by $\alpha_x(y) = \prod_{i=1}^n \alpha_{x_i}(y_i)$. Thus any function $f : [q]^n \to \mathbb{R}$ can be written as

$$f = \sum_{x \in [q]^n} \hat{f}(\alpha_x) \alpha_x \tag{1}$$

Next we define the notion of influence of a variable on a function, introduced to computer science by Ben-Or and Linial in [3].

Definition 1. *Let* $f : [q]^n \to \mathbb{R}$ *be a function on a q-ary hypercube. The influence of the i'th variable on* f*, is defined as*

$$\mathrm{Inf}_i(f) = \mathbb{E}_{x \setminus i}[\mathrm{Var}_{x_i}[f(x)|x_1, \ldots, x_{i-1}, x_{i+1}, \ldots, x_n]]$$

where $x_1, \ldots, x_n$ *are uniformly distributed in* $[q]$*.*

Some standard formulas are easily checkable using independence and orthonormality.

Proposition 1. *Let* $f : [q]^n \to \mathbb{R}$ *be as in (1). Then*

$$\mathbb{E}[f] = \hat{f}(\alpha_0) \qquad \mathbb{E}[f^2] = \sum_x \hat{f}(\alpha_x)^2$$

$$\mathrm{Var}[f] = \sum_{|x|>0} \hat{f}(\alpha_x)^2 \qquad \mathrm{Inf}_i(f) = \sum_{x:x_i \neq 0} \hat{f}(\alpha_x)^2$$

Analogously we can define "low-degree influence", a notion useful in PCPs due to the fact that a bounded function cannot have too many coordinates with non-negligible low-degree influences.

Definition 2. *The d-low-degree influence of* $f : [q]^n \to \mathbb{R}$ *is*

$$\mathrm{Inf}_i^{\leq d}(f) = \sum_{x:x_i \neq 0, |x| \leq d} \hat{f}(\alpha_x)^2$$

The remark above follows from the following easy proposition

Proposition 2. *Let* $f : [q]^n \to \mathbb{R}$ *be as in (1). Then*

$$\sum_i \mathrm{Inf}_i^{\leq d}(f) \leq d \cdot \mathrm{Var}[f]$$

In particular for $f : [q]^n \to [-1, 1]$ *holds*

$$\sum_i \mathrm{Inf}_i^{\leq d}(f) \leq d$$

and thus there are at most d/ϵ *variables* i *with* $\mathrm{Inf}_i^{\leq d}(f) \geq \epsilon$*.*

Instead of picking x at random, changing one coordinate, and seeing how it changes the value of f, we can change a constant fraction (in expectation) of the coordinates.

Definition 3. *Let* $f : [q]^n \to \mathbb{R}$*, and let* $\rho \in [0, 1]$*. Suppose the string* x *is picked uniformly at random and each coordinate* y_i *is independently chosen to be* x_i *with probability* ρ *and is a uniformly random element of* $[q]$ *otherwise. We define the noise stability of* f *to be*

$$\mathbb{S}_\rho(f) = \mathbb{E}[f(x)f(y)]$$

Analogously we generalize the notion of stability with respect to two functions:

$$\mathbb{S}_\rho(f, g) = \mathbb{E}[f(x)g(y)]$$

The notion above can be also considered as following: For any $\rho \in [0,1]$ define the following Markov operator on $[q]$ (called the Bonami-Beckner operator)

$$T_\rho = \begin{pmatrix} \rho + \frac{1-\rho}{q} & \frac{1-\rho}{q} & \cdots & \frac{1-\rho}{q} \\ \vdots & \ddots & & \\ \vdots & & \ddots & \\ \frac{1-\rho}{q} & \cdots & \frac{1-\rho}{q} & \rho + \frac{1-\rho}{q} \end{pmatrix}$$

Clearly $T_\rho \mathbf{1} = \mathbf{1}$ and $T_\rho v = \rho \cdot v$ for any vector $v \perp \mathbf{1}$. In particular holds $T_1(f) = f$ and $T_0(f) = \mathbb{E}[f]$. The following formulas are standard and easily checkable

Proposition 3. *Let $f, g : [q]^n \to \mathbb{R}$ be as in (1) w.r.t. some orthonormal basis $\{\alpha_i\}$. Then*

$$T_\rho^{\otimes n}(f) = \sum_x \rho^{|x|} \hat{f}(\alpha_x) \alpha_x$$

and by orthonormality

$$\mathbb{S}_\rho(f, g) = \langle f, T_\rho^{\otimes n} g \rangle = \sum_x \rho^{|x|} \hat{f}(\alpha_x) \hat{g}(\alpha_x)$$

By applying T_ρ on a function $f : [q]^n \to [0,1]$, the weight of f on higher levels reduces exponentially. More precisely if $g = T_\rho f$, then $\sum_{x:|x| \geq k} \hat{g}(\alpha_x)^2 \leq \rho^{2k} \sum_x \hat{f}(\alpha_x)^2 \leq \rho^{2k}$. We think of T_ρ as a smoothing operator.

Definition 4. *Let $g : [q]^n \to \mathbb{R}$, and let $\eta \in (0,1)$. We say that g is η-smooth if $\sum_{x:|x| \geq k} \hat{g}(\alpha_x)^2 \leq \eta^k$ for all $k \geq 0$.*

2.2   Functions in Gaussian Space

Before we continue, we need to define some basic notions in $L^2(\mathbb{R}^n, \gamma)$, the space of real valued functions with domain $\mathbb{R}^n$ equipped with the standard Gaussian measure. The density function of the standard normal distribution is denoted by $\gamma(x) = \frac{1}{(2\pi)^{n/2}} \exp(-\frac{\|x\|^2}{2})$. The inner product is defined as

$$\langle f, g \rangle = \mathbb{E}_\gamma[fg] = \int_{R^n} f(x)g(x)\gamma(x)dx$$

For $\rho \in [-1, 1]$ denote by U_ρ the Ornstein-Uhlenbeck operator

$$(U_\rho f)(x) = \mathbb{E}_{y \sim \gamma}[f(\rho x + \sqrt{1 - \rho^2} y)]$$

For $\mu \in (0, 1)$ define an indicator of half space function $L^2(\mathbb{R}, \gamma)$ as

$$F_\mu(x) = \mathbf{1}_{x < \Phi^{-1}(\mu)}(x)$$

where $\Phi(t) = \int_{-\infty}^{t} \gamma(x)dx$ is the cumulative distribution function.

A useful quantity that will appear later is $\langle F_\epsilon, U_\rho(1 - F_{1-\epsilon}) \rangle = \langle F_\epsilon, U_{-\rho}(F_\epsilon) \rangle$ for $\rho \in (0,1)$. Observe that $\langle F_\epsilon, U_{-\rho}F_\epsilon \rangle = \Pr[X < \Phi^{-1}(\epsilon), Y < \Phi^{-1}(\epsilon)]$, where X and Y are $-\rho$ correlated normal random variables with mean 0, variance 1. That is $X \sim N(0,1)$ and for $Z \sim N(0,1)$ the r.v. Y is $-\rho X + \sqrt{1 - \rho^2}Z$.

It can be found in the literature, (see e.g. in [14], [6]) that as $\epsilon \to 0$

$$\langle F_\epsilon, U_{-\rho}F_\epsilon \rangle \sim \epsilon^{2/(1-\rho)}(4\pi \ln(1/\epsilon))^{\rho/(1-\rho)}\frac{(1-\rho)^{3/2}}{(1+\rho)^{1/2}}$$

In particular if ρ is a constant bounded below 1, then

$$\langle F_\epsilon, U_\rho(1 - F_{1-\epsilon}) \rangle_\gamma = \text{poly}(\epsilon) \tag{2}$$

2.3 The Majority Is Stablest Theorem

The Majority is Stablest Theorem [13] roughly says that for all functions $f : [q]^n \to [0,1]$ in which each coordinate has $o(1)$ influence, the noise stability of f is bounded by some function of $\mathbb{E}[f]$. More specifically.

Theorem 3 ([13, Theorem 4.4]). *Fix $q \geq 2$ and $\rho \in [0,1]$. Then for any $\epsilon > 0$ there is a small enough $\delta = \delta(\epsilon, \rho, q)$ such that for any function $f : [q]^n \to [0,1]$ such that*

$$\text{Inf}_i(f) \leq \delta \quad \forall i \in \{1, \ldots, n\}$$

holds

$$\mathbb{S}_\rho(f) \leq \langle F_{\mathbb{E}[f]}, U_\rho F_{\mathbb{E}[f]} \rangle_\gamma + \epsilon$$

In particular case of $q = 2$ and a balanced functions $f : \{0,1\} \to \{0,1\}$ the theorem states that if

$$\mathbb{S}_\rho(f) > \langle F_{\mathbb{E}[f]}, U_\rho F_{\mathbb{E}[f]} \rangle_\gamma + \epsilon = \frac{1}{4} + \frac{1}{2\pi}\arcsin\rho + \epsilon = \mathbb{S}_\rho(Maj) + \epsilon$$

then f has some influential coordinate. That is among all balanced boolean functions in which each coordinate has $o(1)$ influence, the *Majority* function has the largest noise stability.

This theorem is generalized in [7] in two directions: the stability is defined with respect to two functions and for any Markov operator T on $[q]$ (not only for T_ρ). The idea is that given a symmetric Markov operator T with eigenvalues $1 = \lambda_0 > \lambda_1 \geq \cdots \geq \lambda_{q-1}$, it is enough to bound its spectral radius $\rho = r(T) = \max\{|\lambda_1|, |\lambda_{q-1}|\}$ below 1. Suppose we are given a symmetric Markov operator T on $[q]$ with spectral radius $\rho < 1$, and two functions $f, g : [q]^n \to [0,1]$ that satisfy the inequality

$$\langle f, T^{\otimes n} g \rangle > \langle F_{\mathbb{E}[f]}, U_\rho F_{\mathbb{E}[g]} \rangle_\gamma + \epsilon$$

The main technical result in [7, Theorem 3.1] says that in such case f and g have a common coordinate with non-negligible influence. In our setup, however, we consider functions f and g with small expectation and ρ some fixed constant and thus we allow ourselves to consider the case of $\langle f, T^{\otimes n} g \rangle > \langle F_{\mathbb{E}[f]}, U_{\rho'} F_{\mathbb{E}[g]} \rangle_\gamma + \epsilon$, for some $\rho' > \rho$. In return we conclude that f and g have a common coordinate with relatively large influence on both functions. The exact formulation and the proof appear in the next section.

3   A Variant of the Majority Is Stablest Theorem

In this section we prove our main technical theorem, which is used in the soundness of the reduction. This is a variant of [7, Theorem 3.1] that we adjust for our purposes. The main new idea is in Lemma 2 (analogue of Lemma 3.9 in [7]).

Let q be a fixed integer, and let T be a symmetric Markov operator on $[q]$ with eigenvalues $1 = \lambda_0 > \lambda_1 \geq \cdots \geq \lambda_{q-1} > -1$, and let $\alpha_0 = \mathbf{1}, \alpha_1, \ldots, \alpha_{q-1}$ be the corresponding eigenvectors. Denote the spectral radius of T by $\rho = r(T) = \max\{|\lambda_1|, |\lambda_{q-1}|\} < 1$.

Now suppose we are given two functions $f, g : [q]^n \to [0,1]$ that do not have any common influential coordinates. We show the following bound on the quantity $\langle f, T^{\otimes n} g \rangle$:

Theorem 4. *Let q be a fixed integer, and let T be a symmetric Markov operator on $[q]$ such that $\rho = r(T) < 1$ and let $\rho' \in (\rho, 1)$. Then for any $\epsilon > 0$ there are $\delta = \epsilon^{O(1)}$ and $k = O(\lg(\frac{1}{\epsilon}))$, where the constants in the O notation depend only on $\frac{\rho}{\rho'}$ and q, such that the following holds: If $f, g : [q]^n \to [0,1]$ are two functions with $\mu = \mathbb{E}[f]$, $\nu = \mathbb{E}[g]$ satisfying*

$$\forall i \quad \min\left(\mathrm{Inf}_i^{\leq k}(f), \mathrm{Inf}_i^{\leq k}(g)\right) < \delta$$

then

$$\langle f, T^{\otimes n} g \rangle \geq \langle F_\mu, U_{\rho'}(1 - F_{1-\nu}) \rangle_\gamma - \epsilon \tag{3}$$

and

$$\langle f, T^{\otimes n} g \rangle \leq \langle F_\mu, U_{\rho'} F_\nu \rangle_\gamma + \epsilon \tag{4}$$

Observe that compared to the analogous theorem of [7, Theorem 3.1], we gain a better tradeoff between ϵ and δ. We allow δ to be poly(ϵ), i.e. not too small, (instead of $\delta = \exp(-\frac{1}{\epsilon})$ implicitly appearing in [7]). On the other hand, we get a bound on $\langle f, T^{\otimes n} g \rangle$ as a function of $\rho' \in (\rho, 1)$ instead of ρ.

For our application of the theorem, we think of ρ and ρ' as constants smaller than 1, and of μ and ν as small quantities compared to ϵ. In this setup, the polynomial dependency between ϵ and δ improves the dependency of [13] and [7]. The following corollary proves Theorem 2.

Corollary 1. *Let q be a fixed integer and T be a symmetric Markov operator on $[q]$ with spectral radius $\rho = r(T) < 1$. Then for any $\epsilon > 0$ there exist $\delta = \epsilon^{O(1)}$ and $k = O(\lg(\frac{1}{\epsilon}))$, where the constants in the O notation depend only on ρ and q, such that the following holds: For any $f, g : [q]^n \to [0,1]$, if*

$$E[f] > \epsilon \quad E[g] > \epsilon \quad \text{and} \quad \langle f, T^{\otimes n} g \rangle = 0$$

then

$$\exists i \in \{1,\ldots,n\} \quad s.t. \quad \mathrm{Inf}_i^{\leq k}(f) \geq \delta \quad and \quad \mathrm{Inf}_i^{\leq k}(g) \geq \delta$$

Proof. Let $\rho' = \sqrt{\rho}$ (note that $\rho < \rho' < 1$), and let $\epsilon_0 = \langle F_\epsilon, U_{\rho'}(1 - F_{1-\epsilon})\rangle_\gamma$. Then

$$\langle f, T^{\otimes n}g\rangle < \langle F_{\mathbb{E}[f]}, U_{\rho'}(1 - F_{1-\mathbb{E}[g]})\rangle_\gamma - \epsilon_0$$

We apply Theorem 4 to get $\delta = \mathrm{poly}(\epsilon_0)$ and $k = O(\lg(\frac{1}{\epsilon_0}))$, and some $i \in \{1,\ldots,n\}$ s.t. $\mathrm{Inf}_i^{\leq k}(f) > \delta$ and $\mathrm{Inf}_i^{\leq k}(g) > \delta$. The corollary follows from equation (2), since $\epsilon_0 = \langle F_\epsilon, U_{\rho'}(1 - F_{1-\epsilon})\rangle_\gamma = \mathrm{poly}(\epsilon)$, where the degree of the polynomial depends only on ρ'.

3.1   Proof of Theorem 4

Note that (3) follows from (4). Indeed, apply (4) to $1 - g$ to obtain

$$\langle f, T^{\otimes n}(1 - g)\rangle \leq \langle F_\mu, U_{\rho'}F_{1-\nu}\rangle_\gamma + \epsilon$$

and then use the equalities

$$\langle f, T^{\otimes n}(1 - g)\rangle = \langle f, 1\rangle - \langle f, T^{\otimes n}g\rangle = \mu - \langle f, T^{\otimes n}g\rangle = \langle F_\mu, U_{\rho'}1\rangle_\gamma - \langle f, T^{\otimes n}g\rangle.$$

So our goal in this section is to prove (4).

The following lemma is the first step in the proof of Theorem 4. It is proven in [7] and essentially follows from the Invariance Principle [13] and Borell's inequality [4].

Let T be some fixed Markov operator on $[q]$ with eigenvalues $1 = \lambda_0 > \lambda_1 \geq \cdots \geq \lambda_{q-1} > -1$ and let $\alpha_0 = \mathbf{1}, \alpha_1, \ldots, \alpha_{q-1}$ be the corresponding orthonormal eigenbasis. Assume that $\rho = r(T) < 1$.

Lemma 1 ([7, Lemma 3.9]). *Let T be a symmetric linear operator on $\mathbb{R}^q$ with spectral radius $\rho = \max\{|\lambda_1|, |\lambda_{q-1}|\} < 1$. Then for any $\eta < 1$, $\epsilon > 0$ there is $\delta = \epsilon^{O(\frac{\lg(q)}{1-\eta})}$ s.t. the following holds: Let $f, g : [q]^n \to [0,1]$ be two functions with $\mathbb{E}[f] = \mu$, $\mathbb{E}[g] = \nu$ and decomposition as in (1). If both functions are η-smooth, i.e.*

$$\forall k \sum_{x:|x|\geq k} \hat{f}(\alpha_x)^2 \leq \eta^k \quad and \quad \forall k \sum_{x:|x|\geq k} \hat{g}(\alpha_x)^2 \leq \eta^k$$

and all influences in both of them are bounded by δ, i.e.

$$\forall i \ \mathrm{Inf}_i(f) < \delta \quad and \quad \forall i \ \mathrm{Inf}_i(g) < \delta$$

then

$$\langle f, T^{\otimes n}g\rangle \leq \langle F_\mu, U_\rho F_\nu\rangle + \epsilon$$

We complete the proof of Theorem 4 in the following lemma.. Recall that for any $\gamma > 0$ the linear operator T_γ on $\mathbb{R}^q$ is defined by: $T_\gamma \mathbf{1} = \mathbf{1}$ and $T_\gamma v = \gamma v$ for $v \perp \mathbf{1}$. It is easy to see that the operator $S = TT_\gamma$ has the same eigenvectors as T and the corresponding eigenvalues are $1 = \lambda_0 > \lambda_1\gamma \geq \cdots \geq \lambda_{q-1}\gamma > -1$ (as long as $\gamma < 1/\rho$).

Lemma 2. *Fix T as above and let $\rho' \in (\rho, 1)$. For any $\epsilon > 0$ there are $\delta = \epsilon^{O(1)}$ and $k = O(\lg(\frac{1}{\epsilon}))$, where the constants in the O notation depend only on $\frac{\rho}{\rho'}$ and q, such that the following holds: If $f, g : [q]^n \to [0, 1]$ are two functions with $\mu = \mathbb{E}[f]$, $\nu = \mathbb{E}[g]$ satisfying*

$$\forall i \quad \min\left(\operatorname{Inf}_i^{\leq k}(f), \operatorname{Inf}_i^{\leq k}(g)\right) < \delta$$

then

$$\langle f, T^{\otimes n} g \rangle \leq \langle F_\mu, U_{\rho'} F_\nu \rangle_\gamma + \epsilon$$

In order to apply Lemma 1 we need to make sure that all variables of f and g have small influence and that the given functions are smooth. The first part is achieved by observing that coordinates that have large influence either on f or on g, make small contribution to $\langle f, T^{\otimes n} g \rangle$.

One approach to the second part is to smooth f and g a little, that is to define $f_1 = T_{1 - \epsilon \lg(\frac{1}{\epsilon})} f$ and $g_1 = T_{1 - \epsilon \lg(\frac{1}{\epsilon})} g$ and show that $|\langle f, T^{\otimes n} g \rangle - \langle f_1, T^{\otimes n} g_1 \rangle| < \epsilon$. But then f_1 and g_1 are only $1 - \epsilon \lg(\frac{1}{\epsilon})$-smooth and δ that we get from Lemma 1 is exponential in ϵ.

A different approach is to use the fact that ρ is some constant smaller than 1 and to define $f_1 = T_\eta f$ and $g_1 = T_\eta g$ for some constant $\eta \in (\rho, 1)$. Then f_1 and g_1 are η-smooth and $\langle f, T^{\otimes n} g \rangle = \langle f_1, S^{\otimes n} g_1 \rangle$ for some operator S whose spectral radius is larger than $r(T)$, but still constant smaller than 1. By applying Lemma 1 on f_1 and g_1 with the operator S we get $\delta = \operatorname{poly}(\epsilon)$.

Proof. Set $\eta = \frac{\rho}{\rho'} < 1$ and denote $S = TT_{\frac{1}{\eta}}$. Then S has the same eigenvectors as T, largest eigenvalue 1 and $r(S) = \frac{\rho}{\eta} = \rho' < 1$. We also denote

$$f_1 = T_{\sqrt{\eta}} f = \sum \hat{f}(\alpha_x) \eta^{\frac{|x|}{2}} \alpha_x \quad and \quad g_1 = T_{\sqrt{\eta}} g = \sum \hat{g}(\alpha_x) \eta^{\frac{|x|}{2}} \alpha_x$$

Using this notation it is easy to see that we can express $\langle f, T^{\otimes n} g \rangle$ as

$$\langle f, T^{\otimes n} g \rangle = \sum_x \hat{f}(\alpha_x) \lambda_x \hat{g}(\alpha_x) = \langle f_1, S^{\otimes n} g_1 \rangle \tag{5}$$

We apply Lemma 1 with operator S and parameters η and $\epsilon/2$ to get $\delta' = \delta_1(S, \eta, \frac{\epsilon}{2})/2 = \epsilon^{O(\frac{1}{1-\eta})} = \operatorname{poly}(\epsilon)$, where the degree of the polynomial depends only on η and q. Let $k = O(\lg(\frac{1}{\epsilon}))$ be such that $\eta^k < \min(\delta', \epsilon/4)$, and let $\delta = (\frac{\epsilon \delta'}{8k})^2 = \operatorname{poly}(\epsilon)$. We show that these δ and k satisfy the requirements of the lemma.

Take two functions $f, g : [q]^n \to [0, 1]$ such that $\forall i \min\left(\operatorname{Inf}_i^{\leq k}(f), \operatorname{Inf}_i^{\leq k}(g)\right) < \delta$. Then f_1 and g_1 are η-smooth and satisfy the same assumption. However, we cannot apply Lemma 1 on them with the operator S, as the requirement is that all influences in both of them are small. In order overcome this problem we define two functions f_2 and g_2 with small influences such that $\langle f_1, S^{\otimes n} g_1 \rangle \approx \langle f_2, S^{\otimes n} g_2 \rangle$. Define

$$B_f = \{i : \operatorname{Inf}_i^{\leq k}(f) \geq \delta'\} \qquad B_g = \{i : \operatorname{Inf}_i^{\leq k}(g) \geq \delta'\}$$

Then $|B_f|, |B_g| \leq k/\delta'$. Moreover $B_f \cap B_g = \emptyset$ as $\delta < \delta'$ and $\forall i \ \min(\mathrm{Inf}_i^{\leq k}(f),$ $\mathrm{Inf}_i^{\leq k}(g)) < \delta$. We define $f_2(y)$ and $g_2(y)$ as the average over the coordinates in B_f and B_g respectively, namely

$$f_2(y) = \mathop{\mathbb{E}}_{y_i : i \in B_f} [f_1(y)] = \sum_{x:x_{B_f}=0} \hat{f}(\alpha_x) \eta^{\frac{|x|}{2}} \alpha_x(y)$$

$$g_2(y) = \mathop{\mathbb{E}}_{y_i : i \in B_g} [g_1(y)] = \sum_{x:x_{B_g}=0} \hat{g}(\alpha_x) \eta^{\frac{|x|}{2}} \alpha_x(y)$$

Clearly $\mathbb{E}[f_2] = \mathbb{E}[f] = \mu$, $\mathbb{E}[g_2] = \mathbb{E}[g] = \nu$. We have $\mathrm{Inf}_i(f_2) = 0$ for $i \in B_f$ and $\mathrm{Inf}_i(f_2) \leq \mathrm{Inf}_i^{\leq k}(f) + \eta^k < 2\delta'$ otherwise. Same holds for g_2. Their smoothness follows from smoothness of f_1, g_1 and we can apply Lemma 1 with the operator S to get

$$\langle f_2, S^{\otimes n} g_2 \rangle \leq \langle F_\mu, U_{\rho'} F_\nu \rangle + \epsilon/2 \tag{6}$$

It is only left to show that

$$|\langle f_1, S^{\otimes n} g_1 \rangle - \langle f_2, S^{\otimes n} g_2 \rangle| \leq \epsilon/2 \tag{7}$$

Here we use the assumption that a coordinate cannot have a significant influence on both functions.

$$|\langle f_1, S^{\otimes n} g_1 \rangle - \langle f_2, S^{\otimes n} g_2 \rangle| = \left| \sum_{x:x_{B_f \cup B_g} \neq 0} \left(\prod_{i:x_i \neq 0} \frac{\lambda_{x_i}}{\eta} \right) \eta^{|x|} \hat{f}(\alpha_x) \hat{g}(\alpha_x) \right|$$

$$\leq \sum_{\substack{x:|x| \leq k \\ x:x_{B_f \cup B_g} \neq 0}} |\hat{f}(\alpha_x) \hat{g}(\alpha_x)| + \sum_{x:|x|>k} \left| \rho^{|x|} |\hat{f}(\alpha_x) \hat{g}(\alpha_x)| \right|$$

$$[\rho < \eta] \leq \sum_{i \in B_f \cup B_g} \sum_{\substack{x:|x| \leq k \\ x_i \neq 0}} |\hat{f}(\alpha_x) \hat{g}(\alpha_x)| + \eta^k$$

$$[\text{Cauchy Schwartz}] \leq \sum_{i \in B_f \cup B_g} \sqrt{\mathrm{Inf}_i^{\leq k}(f)} \sqrt{\mathrm{Inf}_i^{\leq k}(g)} + \eta^k$$

$$[i \in B_f \Rightarrow \mathrm{Inf}_i(g) < \delta] \leq (|B_f| + |B_g|)\sqrt{\delta} + \eta^k$$

$$[|B_f|, |B_g| \leq k/\delta', \eta^k \leq \epsilon/4] \leq \frac{2k}{\delta'} \frac{\epsilon \delta'}{8k} + \epsilon/4$$

$$= \epsilon/2$$

Combining (5), (6) and (7) we get the required result $\langle f, T^{\otimes n} g \rangle \leq \langle F_\mu, U_{\rho'} F_\nu \rangle + \epsilon$ and complete the proof of theorem.

Theorem 1 is proven by following the reduction of [7], applying Theorem 2 and calculating the exact parameters of the reduction. For the sake of completeness we include the reduction in the appendix.

Theorem 5. *There is a reduction from 2-to-1 Label Cover problem to* Approx-Coloring *problem with the following properties: Given an instance of 2-to-1 Label Cover $\Phi = (V \cup W, E, R, \Pi)$ it produces a graph G on $|W| \cdot 4^{2R}$ vertices.*

- *If $val(\Phi) = 1$, then G is 4-colorable.*
- *If G contains an independent set of size ϵ, then $val(\Phi) \geq \Omega\left(\frac{\epsilon\delta^2}{k^2}\right) = \text{poly}(\epsilon)$,*
  where δ and k are as in Theorem 2. In other words, if $val(\Phi) \leq \frac{1}{f(n)}$, then
  $\chi(G) \geq f^c(n)$ for some constant $c > 0$.

The running time of the reduction is linear in the size of the output.

References

1. Arora, S., Chlamtac, E.: New approximation guarantee for chromatic number. In: STOC 2006: Proceedings of the Thirty-Eighth Annual ACM Symposium on Theory of Computing, pp. 215–224 (2006)
2. Blum, A., Karger, D.: An $\tilde{O}(n^{3/14})$-coloring algorithm for 3-colorable graphs. Information Processing Letters 61, 49–53 (1997)
3. Ben-Or, M., Linial, N.: Collective Coin Flipping. Randomness and Computation, pp. 91–115 (1989)
4. Borell, C.: Geometric Bounds on the Ornstein-Uhlenbeck Velocity Process. Z. Wahrsch. Verw. Gebiete 70(1), 1–13 (1985)
5. Chlamtac, E.: Approximation Algorithms Using Hierarchies of Semidefinite Programming Relaxations. In: FOCS 2007: Proceedings of the 48th Annual IEEE Symposium on Foundations of Computer Science, pp. 691–701 (2007)
6. de Klerk, E., Pasechnik, D.V., Warners, J.P.: Approximate Graph Colouring and MAX-k-CUT Algorithms Based on the ϑ-Function. Journal of Combinatorial Optimization 8(3), 267–294 (2004)
7. Dinur, I., Mossel, E., Regev, O.: Conditional Hardness for Approximate Coloring. SIAM Journal on Computing 39(3), 843–873 (2009)
8. Guruswami, V., Khanna, S.: On the Hardness of 4-coloring a 3-colorable Graph. SIAM J. Discret. Math. (2004)
9. Karp, R.: Reducibility Among Combinatorial Problems. Complexity of Computer Computations 43, 85–103 (1972)
10. Khot, S.: On the Power of Unique 2-Prover 1-Round Games. In: Proceedings of the 34th Annual ACM Symposium on Theory of Computing, pp. 767–775 (2002)
11. Khanna, S., Linial, N., Safra, S.: On the Hardness of Approximating the Chromatic Number. Combinatorica 20(3), 393–415 (2000)
12. Karger, D., Motwani, R., Sudan, M.: Approximate Graph Coloring by Semidefinite Programming. J. ACM 45(2), 246–265 (1998)
13. Mossel, E., O'Donnell, R., Oleszkiewicz, K.: Noise Stability of Functions with Low Influences: Invariance and Optimality. Annals of Mathematics 171(1), 295–341 (2010)
14. Rinott, Y., Rotar', V.: A Remark on Quadrant Normal Probabilities in High Dimensions. Statistics and Probability Letters 51(1), 47–51 (2001)

A   Reduction

In this section we present the reduction of [7], which proves our main Theorem 1. Our starting point is the following conjecture [10]:

Conjecture 2 (2-to-1 Conjecture). For any $\epsilon > 0$ there is $R = R(\epsilon)$ s.t. the following problem is NP-hard. Given a bipartite 2-to-1 LC instance $\Phi = (V \cup W, E)$ with label sets $\{1, \ldots, R\}$ for V and $\{1, \ldots, 2R\}$ for W, distinguish between the case where $val(\Phi) = 1$ and the case where $val(\Phi) < \epsilon$.

In our case we make a stronger conjecture. We allow the label set to be of size $O(\lg(n))$. The best possible soundness we could hope for is $\lg(n)^{-c}$ for some constant $c > 0$. In addition, as a small technicality, we assume that all vertices of W have the same degree.

The proof of Theorem 1 is by showing a reduction with the following properties: given an instance of 2-to-1 Label Cover $\Phi = (V \cup W, E, R, \Pi)$ it produces a graph G on $|W| \cdot 4^{2R}$ vertices. In the completeness part it is shown that if $val(\Phi) = 1$, then G is 4-colorable. The soundness of the reduction says that if G contains an independent set of size ϵ, then $val(\Phi) \geq \Omega\left(\frac{\epsilon\delta^2}{k^2}\right) = \text{poly}(\epsilon)$, where δ and k are as in Theorem 2. In other words if $val(\Phi) \leq \frac{1}{f(n)}$, then $\chi(G) \geq f^c(n)$ for some constant $c > 0$.

Definition 5. *We define a symmetric Markov operator T on $\{0, 1, 2, 3\}^2$ such that $r(T) < 1$ and such that $T((x_1, x_2) \leftrightarrow (y_1, y_2)) > 0$ if and only if $\{x_1, x_2\} \cap \{y_1, y_2\} = \emptyset$.*

Our operator has three types of transitions, with transitions probabilities β_1, β_2, and β_3.

- *With probability β_1 we have $(x, x) \leftrightarrow (y, y)$ where $x \neq y$.*
- *With probability β_2 we have $(x, x) \leftrightarrow (y, z)$ where x, y, z are all different.*
- *With probability β_3 we have $(x, y) \leftrightarrow (z, w)$ where x, y, z, w are all different.*

For T to be a symmetric Markov operator, we need that β_1, β_2 and β_3 are non-negative and

$$3\beta_1 + 6\beta_2 = 1, \quad 2\beta_2 + 2\beta_3 = 1.$$

For example for $\beta_1 = \frac{1}{12}$, $\beta_2 = \frac{1}{8}$, and $\beta_3 = \frac{3}{8}$ we have $\rho = r(T) = 5/6$

Reduction. We start with a 2-to-1 Label Cover instance $\Phi = (V \cup W, E)$. Each $(v, w) \in E$ is associated with a constraint π_{vw} s.t. for each $b \in 2R$ there is a unique a s.t. $(a, b) \in \pi_{vw}$ (we denote $a = \pi_{vw}(b)$) and for each $a \in R$ there are exactly two $b_1, b_2 \in 2R$ s.t. $(a, b_i) \in \pi_{vw}$ (denote $(b_1, b_2) = \pi_{vw}^{-1}(a)$). We construct $G' = (V', E')$ as follows:

- Each vertex $w \in W$ is replaced by a copy of $\{0, 1, 2, 3\}^{2R}$ (denote by $[w]$). The set of vertices in G' is $V' = \bigcup_{w \in W}[w] = W \times \{0, 1, 2, 3\}^{2R}$.

- Let T be as in Definition 5. For every $w_1, w_2 \in W$ that have a common neighbor $v \in V$ let π_1, π_2 be the corresponding constraints. We set an edge between (w_1, x) and (w_2, y) if $T(x_{\pi_1^{-1}(k)}, y_{\pi_2^{-1}(k)}) \neq 0$ for all $k \in R$, or equivalently $\{x_{i_1}, x_{j_1}\} \cap \{x_{i_2}, x_{j_2}\} = \emptyset$ where $\pi_1^{-1}(k) = (i_1, j_1)$ and $\pi_2^{-1}(k) = (i_2, j_2)$.

Completeness. Assume there is a labeling L such that $w_L(\Phi) = 1$. Let $c(w, x) = x_{L(w)}$ for all $w \in W$. We show that this is a legal coloring of G'. Pick an edge $((w_1, x), (w_2, y)) \in E'$. Then w_1, w_2 have a common neighbor $v \in V$. Let π_1 and π_2 be the corresponding constraints, and let $k = L(v)$. Then $\pi_1(L(w_1)) = k = \pi_2(L(w_2))$, as L satisfies all the constraints.

Since $((w_1, x), (w_2, y)) \in E'$, the sets $x_{\pi_1^{-1}(k)}$ and $y_{\pi_2^{-1}(k)}$ are disjoint and hence $c(w_1, x) \neq c(w_2, y)$ as $c(w_1, x) = x_{L(w_1)} \in x_{\pi_1^{-1}(k)}$ and $c(w_2, y) = y_{L(w_2)} \in y_{\pi_2^{-1}(k)}$. $\qquad\square$

Soundness. Assume that $\chi(G') \leq Q$. Then G' contains an independent set $S \subseteq V'$ s.t. $\frac{|S|}{|V'|} \geq \frac{1}{Q} = 2\epsilon$. Our goal is to show that is such case $val(\Phi) > \text{poly}(\epsilon)$. Let J be a subset of W that make a non-negligible contribution to S

$$J = \{w \in W : \frac{[w] \cap S}{[w]} > \epsilon\}$$

Markov inequality implies $|J| \geq \epsilon|W|$.

For each $w \in J$ let $f_w : \{0, 1, 2, 3\}^{2R} \to \{0, 1\}$ be the indicator function of S, i.e. $f_w(x) = 1$ iff $(w, x) \in S$. Then $\mathbb{E}[f_w] > \epsilon$ for such w's. Let δ and k be as in Theorem 2 applied on the operator T from Definition 5 with parameter ϵ. We define a small set of labels for w.

$$L(w) = \{i : \text{Inf}_i^{\leq 2k} > \delta/2\}$$

Observe that $|L(w)| < \frac{4k}{\delta}$. Next we give labels to neighbors of J in Φ.

Claim. Let $v \in N(J)$ and let $w_1, w_2 \in N(v) \cap J$. Let π_1, π_2 be the corresponding constraints. Then there are $i \in L_{w_1}, j \in L_{w_2}$ s.t. $\pi_1(i) = \pi_2(j)$.

Proof. Recall that f_w's are indicators of an independent set. Thus $f_{w_1}(x) = 1 = f_{w_2}(y)$ implies that $((w_1, x), (w_2, y)) \notin E'$. Therefore $T\left(x_{\pi_1^{-1}(k)}, y_{\pi_2^{-1}(k)}\right) = 0$ for some $k \in R$ and thus

$$T^{\otimes R}\left((x_{\pi_1^{-1}(1)}, \ldots, x_{\pi_1^{-1}(R)}), (y_{\pi_2^{-1}(1)}, \ldots, y_{\pi_2^{-1}(R)})\right) = 0$$

Define

$$\overline{f}(x_{\pi_1^{-1}(1)}, \ldots, x_{\pi_1^{-1}(R)}) = f_{w_1}(x_1, \ldots, x_{2R})$$

$$\overline{g}(y_{\pi_2^{-1}(1)}, \ldots, y_{\pi_2^{-1}(R)}) = f_{w_2}(y_1, \ldots, y_{2R})$$

where we think of $\overline{f}, \overline{g}$ as functions in R variables, each taking values in $\{0, 1, 2, 3\}^2$. We show that $\langle \overline{f}, T^{\otimes R}\overline{g} \rangle = 0$. Then using Theorem 2 we conclude that there is $\ell \in R$ s.t. $\mathrm{Inf}_\ell^{\leq k}(\overline{f}) > \delta$ and $\mathrm{Inf}_\ell^{\leq k}(\overline{g}) > \delta$. Using the relation between $\overline{f}$ and f_{w_1}, we conclude that there is some $i \in \pi_1^{-1}(\ell)$ such that $\mathrm{Inf}_i^{\leq 2k}(f_{w_1}) > \delta/2$. Similarly for g there is some $j \in \pi_2^{-1}(\ell)$ such that $\mathrm{Inf}_j^{\leq 2k}(f_{w_2}) > \delta/2$. Therefore there are $i \in L_{w_1}, j \in L_{w_2}$ s.t. $\pi_1(i) = \pi_2(j)$.

So it is left to show that $\langle \overline{f}, T^{\otimes R}\overline{g} \rangle = 0$. And indeed:

$$
\begin{aligned}
\langle \overline{f}, T^{\otimes R}\overline{g} \rangle &= \frac{1}{4^{2R}} \sum_{x \in (\{0,1,2,3\}^2)^R} \overline{f}(x) \sum_{y \in (\{0,1,2,3\}^2)^R} T^{\otimes R}(x, y)\overline{g}(y) \\
&= \frac{1}{4^{2R}} \sum_x f_{w_1}(x) \sum_y T^{\otimes R}(x_{\pi_1^{-1}}, y_{\pi_2^{-1}}) f_{w_2}(y) \\
&= \frac{1}{4^{2R}} \sum_{\substack{x: f_{w_1}(x)=1 \\ y: f_{w_2}(y)=1}} T^{\otimes R}(x_{\pi_1^{-1}}, y_{\pi_2^{-1}}) \\
&= \frac{1}{4^{2R}} \sum_{\substack{x: f_{w_1}(x)=1 \\ y: f_{w_2}(y)=1}} 0 \\
&= 0
\end{aligned}
$$

From the claim above we get that for all $v \in N(J)$ and any $w_1, w_2 \in N(v) \cap J$

$$
\Pr_{\substack{i \in L(w_1) \\ j \in L(w_2)}} [\pi_1(i) = \pi_2(j)] \geq \frac{1}{|L(w_1)||L(w_2)|} \geq \left(\frac{\delta}{4k}\right)^2
$$

By averaging there is $L_0 : V \cup W \to 2R$ such that

$$
\Pr_{v \in N(w)} [L_0(v) = \pi(L_0(w)) | w \in J] \geq \left(\frac{\delta}{4k}\right)^2
$$

Hence, if we assume regularity on the vertices of W, we get

$$
\Pr_{vw}[L(v) = \pi(L(w))] \geq \Pr_{w \in W}[w \in J] \Pr_{v \in N(w)}[L_0(v) = \pi(L_0(w)) | w \in J]
$$

$$
\geq \epsilon \left(\frac{\delta}{4k}\right)^2
$$

$$
= \mathrm{poly}(\epsilon)
$$

We conclude that $val(\Phi) > \mathrm{poly}(\epsilon)$, which completes the soundness analysis of the reduction. $\qquad \square$

Vertex Sparsifiers: New Results from Old Techniques[*]

Matthias Englert[1,**], Anupam Gupta[2,***], Robert Krauthgamer[3,†], Harald Räcke[1,‡],
Inbal Talgam-Cohen[3], and Kunal Talwar[4]

[1] Department of Computer Science and DIMAP, University of Warwick, Coventry, UK
[2] Computer Science Department, Carnegie Mellon University, Pittsburgh, PA, USA
[3] Weizmann Institute of Science, Rehovot, Israel
[4] Microsoft Research Silicon Valley. Mountain View, CA, USA

Abstract. Given a capacitated graph $G = (V, E)$ and a set of terminals $K \subseteq V$,
how should we produce a graph H only on the terminals K so that every (multicommodity) flow between the terminals in G could be supported in H with low
congestion, and vice versa? (Such a graph H is called a *flow-sparsifier* for G.)
What if we want H to be a "simple" graph? What if we allow H to be a convex
combination of simple graphs?

Improving on results of Moitra [FOCS 2009] and Leighton and Moitra [STOC
2010], we give efficient algorithms for constructing: (a) a flow-sparsifier H that
maintains congestion up to a factor of $O(\frac{\log k}{\log \log k})$, where $k = |K|$. (b) a convex
combination of trees over the terminals K that maintains congestion up to a factor
of $O(\log k)$. (c) for a planar graph G, a convex combination of planar graphs
that maintains congestion up to a constant factor. This requires us to give a new
algorithm for the 0-extension problem, the first one in which the preimages of
each terminal are connected in G. Moreover, this result extends to minor-closed
families of graphs.

Our bounds immediately imply improved approximation guarantees for several terminal-based cut and ordering problems.

1 Introduction

Given an undirected capacitated graph $G = (V, E)$ and a set of terminal nodes $K \subseteq V$,
we consider the question of producing a graph H only on the terminals K so that the
congestion incurred on G and H for any multicommodity flow routed between terminal
nodes is similar. Often, we will want the graph H to be structurally "simpler" than G
as well. Such a graph H will be called a *flow-sparsifier* for G; the *loss* (also known as
quality) of the flow-sparsifier is the factor by which the congestions in the graphs G and
H differ. For instance, when $K = V$, the results of [Räc08] give a convex combination

---

[*] A full version appears at http://arxiv.org/abs/1006.4586

[**] Supported by EPSRC grant EP/F043333/1 and DIMAP (the Centre for Discrete Mathematics
and its Applications).

[***] Research was partly supported by the NSF award CCF-0729022, and an Alfred P. Sloan
Fellowship. Research done when visiting Microsoft Research SVC.

[†] Supported in part by The Israel Science Foundation (grant #452/08), and by a Minerva grant.

[‡] Supported by DIMAP (the Centre for Discrete Mathematics and its Applications).

M. Serna et al. (Eds.): APPROX and RANDOM 2010, LNCS 6302, pp. 152–165, 2010.
© Springer-Verlag Berlin Heidelberg 2010

of trees H with a loss of $O(\log n)$. We call this a *tree-based flow-sparsifier*—it uses a convex combination of trees.[1] Here and throughout, $k = |K|$ denotes the number of terminals, and $n = |V|$ the size of the graph.

For the case where $K \neq V$, it was shown by Moitra [Moi09] and by Leighton and Moitra [LM10] that for every G and K, there exists a flow-sparsifier $H = (K, E_H)$ whose loss is $O(\frac{\log k}{\log \log k})$, and moreover, one can efficiently find an $H' = (K, E_{H'})$ whose loss is $O(\frac{\log^2 k}{\log \log k})$. They used these to give approximation algorithms for several terminal-based problems, where the approximation factor depended poly-logarithmically on the number of terminals k, and not on n. We note that they construct an arbitrary graph on K, and do not attempt to directly obtain "simple" graphs; e.g., to get tree-based flow-sparsifiers on K, they apply [Räc08] to H', and increase the loss by an $O(\log k)$ factor.

In this paper, we simplify and unify some of these results: we show that using the general framework of interchanging distance-preserving mappings and capacity-preserving mappings from [Räc08] (which was reinterpreted in an abstract setting by Andersen and Feige [AF09]), we obtain the following improvements over the results of [Moi09, LM10].[2]

1. We show that using the 0-extension results [CKR04, FHRT03] in the framework of [Räc08, AF09] almost immediately gives us *efficent* constructions of flow-sparsifiers with loss $O(\frac{\log k}{\log \log k})$. While the existential result of [LM10] also used the connection between 0-extensions and flow sparsifiers, the algorithmically-efficient version of the result was done *ab initio*, increasing the loss by another $O(\log k)$ factor. We use existing machinery, thereby simplifying the exposition somewhat, and avoiding the increased loss.

2. We then use a randomized tree-embedding due to [GNR10], which is a variant of the so-called FRT tree-embedding [FRT04] where the expected stretch is reduced to $O(\log k)$ by requiring the non-contraction condition only for terminal pairs. Using this refined embedding in the framework of [Räc08, AF09], we obtain efficient constructions of *tree-based flow-sparsifiers* with loss $O(\log k)$.

3. We then turn to special families of graphs. For planar graphs, we give a new 0-extension algorithm that outputs a convex combination of 0-extensions $f : V \to K$ (with $f(x) = x$ for all $x \in K$), such that all the corresponding 0-extension graphs $H_f = (K, E_f)$ (namely, $E_f = \{(f(u), f(v)) : (u, v) \in E\}$) are *planar graphs*, and its expected stretch $\max_{u,v \in V} \mathbb{E}[d_{H_f}(f(u), f(v))]/d_G(u,v) \leq O(1)$. In particular, the planar graphs H_f produced are graph-theoretic minors of G. We remark that the known 0-extension algorithms [CKR04, AFH+04, LN05] do not ensure planarity of H_f.

   It follows that planar graphs admit a *planar-based flow-sparsifier* (i.e., which is a convex combination of capacitated planar graphs on vertex-set K) with loss $O(1)$, and that we can find these efficiently. The fact that flow-sparsifiers with this loss *exist* was shown by [LM10], but their sparsifiers are not planar-based.

---

[1] Given a class $\mathcal{F}$ of graphs, we define an $\mathcal{F}$-flow-sparsifier to be a sparsifier that uses a single graph from $\mathcal{F}$ and an $\mathcal{F}$-based flow sparsifier to be a sparsifier that uses a convex combination of graphs from $\mathcal{F}$.

[2] Recently, it has come to our attention that, independent of and concurrent to our work, Charikar, Leighton, Li, and Moitra, and independently Makarychev and Makarychev, obtained results similar to the first two below, as well as related lower bounds.

Moreover, the 0-extension algorithm itself can be viewed as a randomized version of Steiner point removal in metrics: previously, it was only known how to remove Steiner points from tree metrics with $O(1)$ distortion [Gup01]. We believe this randomized procedure is of independent interest; e.g., combined with an embedding of [GNRS04], this gives an alternate proof of the fact that the metric induced on the vertices of a single face of a planar graph can be embedded into a distribution over trees [LS09].

4. The results for planar graphs are in fact much more general. Suppose G is a β_G-decomposable graph (see definition in Section 1.1). Then we can efficiently output a distribution over graphs $H_f = (K, E_f)$ such that these are all minors of G, and the expected stretch $\max_{u,v \in V} \mathbb{E}[d_{H_f}(f(u), f(v))]/d_G(u,v)$ is bounded by $O(\beta_G \log \beta_G)$. Now applying the same ideas of interchanging distance and capacity preservation, given any G and K, we can find *minor-based flow sparsifiers* with loss $O(\beta_G \log \beta_G)$.

5. Finally, we show lower bounds on flow-sparsifiers: we show that flow-sparsifiers that are 0-extensions of the original graph must have loss at least $\Omega(\sqrt{\log k})$ in the worst-case. For this class of possible flow sparsifiers, this improves on the $\Omega(\log \log k)$ lower bound for sparsifiers proved in [LM10]. We also show that any flow-sparsifier that only uses edge capacities which are bounded from below by a constant, must suffer a loss of $\Omega(\sqrt{\log k}/ \log \log k)$ in the worst-case.

We can use these results to improve the approximation ratios of several application problems. In many cases, constructions based on trees allow us to use better algorithms. Our results are summarized in Table 1. Note that apart from the two linear-arrangement problems, our results smoothly approach the best known results for the case $k = n$.

Table 1. Summary of our results. Previous results marked with † from [Räc08], all others from [Moi09, LM10].

	Previous Best Result	Our Result	Best Result when $k = n$
Flow Sparsifiers (efficient)	$O(\frac{\log^2 k}{\log \log k})$	$O(\frac{\log k}{\log \log k})$	—
Tree-Based Flow Sparsifiers	$O(\log n)^\dagger$, $O(\frac{\log^3 k}{\log \log k})$	$O(\log k)$	$\Theta(\log n)$
Minor-based Flow Sparsifiers	—	$O(\beta_G \log \beta_G)$	—
Steiner Oblivious Routing	$\widetilde{O}(\log^2 k)$	$O(\log k)$	$\Theta(\log n)$
ℓ-Multicut	$\widetilde{O}(\log^3 k)$	$O(\log k)$	$O(\log n)$
Steiner Minimum Linear Arrangement (SMLA)	$\widetilde{O}(\log^{2.5} k)$	$O(\log k \log \log k)$	$O(\sqrt{\log n} \log \log n)$
SMLA in planar graphs	$\widetilde{O}(\log^{1.5} k)$	$O(\log \log k)$	$O(\log \log n)$
Steiner Min-Cut Linear Arrangement	$\widetilde{O}(\log^4 k)$	$O(\log^2 k)$	$O(\log^{1.5} n)$
Steiner Graph Bisection	$O(\log n)^\dagger$, $O(\frac{\log^3 k}{\log \log k})$	$O(\log k)$	$O(\log n)$

Many of these applications further improve when the graph comes from a minor-closed family (and hence has good β-decompositions): e.g., for the Steiner Minimum

Linear Arrangement problem on planar graphs, we can get an $O(\log \log k)$-approximation by using our minor-based flow-sparsifiers to reduce the problem to planar instances on the k terminals. Finally, in the full version we show how to get better approximations for the Steiner linear arrangement problems above using direct LP/SDP approaches.

1.1 Notation

Our graphs will have edge lengths or capacities; all edge-lengths will be denoted by $\ell : E \rightarrow \mathbb{R}_{\geq 0}$, and edge costs/capacities will be denoted by $c : E \rightarrow \mathbb{R}_{\geq 0}$. When we refer to a graph (G, ℓ), we mean a graph G with edge-lengths $\ell(\cdot)$; similarly (H, c) denotes one with capacities $c(\cdot)$. When there is potential for confusion, we will add subscripts (e.g., $c_H(\cdot)$ or $\ell_G(\cdot)$) for disambiguation. Given a graph (G, ℓ), the shortest-path distances under the edge lengths ℓ is denoted by $d_G : V \times V \rightarrow \mathbb{R}_{\geq 0}$.

Given a graph $G = (V, E)$ and a subset of vertices $K \subseteq V$ designated as *terminals*, a *retraction* is a map $f : V \rightarrow K$ such that $f(x) = x$ for all $x \in K$. For (G, c) and terminals $K \subseteq V$, a *K-flow in G* is a multicommodity flow whose sources and sinks lie in K.

Decomposition of Metrics. Let (X, d) be a metric space with terminals $K \subset X$. A partition (i.e., a set of disjoint "clusters") P of X is called Δ-*bounded* if every cluster $S \in P$ satisfies $max_{u,v \in S} d(u, v) \leq \Delta$. The metric (X, d) with terminals K is called β-*decomposable* if for every $\Delta > 0$ there is polynomial time algorithm to sample from a probability distribution μ over partitions of X, with the following properties:

- *Diameter bound:* Every partition $P \in \text{supp}(\mu)$ is Δ-bounded.
- *Separation event:* For all $u, v \in X$, $\text{Pr}_{P \in \mu}[\exists S \in P \text{ such that } u \in S \text{ but } v \notin S] \leq \beta \cdot d(u, v)/\Delta$.

β-decompositions of metrics have become standard tools with many applications; for more information see, e.g., [LN05].

We say that *a graph* $G = (V, E)$ is β-decomposable if for every nonnegative edge-lengths ℓ_G, the resulting shortest-path metric d_G is β-decomposable. Additionally, we assume that each cluster S in any partition P induces a *connected* subgraph of G; if not, break such a cluster into its connected components. The diameter bound and separation probabilities for edges remain unchanged by this operation; the separation probability for non-adjacent pairs (u, v) can be bounded by $\beta \cdot d(u, v)/\Delta$ by noting that some edge on the u-v shortest path must be separated for (u, v) to be separated, and applying the union bound.

2   0-Extensions

In this section we provide a definition of 0-extension which is somewhat different than the standard definition, and review some known results for 0-extensions. We also derive in Corollary 1 a variation of a known result on tree embeddings, which will be applied in Section 3.

A 0-*extension* of graph $(G = (V, E), \ell_G)$ with terminals $K \subseteq V$ is usually defined as a retraction $f : V \rightarrow K$. We define a 0-extension to be a retraction $f : V \rightarrow K$

along with another graph $(H = (K, E_H), \ell_H)$; here, the length function $\ell_H : E_H \to \mathbb{R}_+$ is defined as $\ell_H(x, y) = d_G(x, y)$ for every edge $(x, y) \in E_H$. Note that this immediately implies $d_H(x, y) \geq d_G(x, y)$ for all $x, y \in K$. Note also that H_f defined in Section 1 is a special case of H in which $E_H = \{(f(u), f(v)) : (u, v) \in E\}$, whereas, in general, H is allowed more flexibility (e.g., H can be a tree). This flexibility is precisely the reason we are interested both in the retraction f and in the graph H—we will often want H to be structurally simpler than G (just like we want a flow-sparsifier to be simpler than the original graph).

For a (randomized) algorithm $\mathcal{A}$ that takes as input (G, ℓ_G) and outputs a (random) 0-extension (H, ℓ_H), the *stretch factor* of algorithm $\mathcal{A}$ is the minimum $\alpha \geq 1$ such that

$$\mathbb{E}_H[\, d_H(f(x), f(y))\,] \leq \alpha\, d_G(x, y) \qquad \text{for all } x, y \in V.$$

The following are well-known results for 0-extension.

Theorem 1 ([FHRT03]). *There is an algorithm $\mathcal{A}_{FHRT}$ for 0-extension with stretch* $\alpha = \alpha_{FHRT} := O(\frac{\log k}{\log \log k})$.

Theorem 2 ([CKR04], see also [LN05]). *If the graph is β-decomposable, there is an algorithm $\mathcal{A}_{CKR}$ for 0-extensions with stretch $\alpha = \alpha_{CKR} := O(\beta)$.*

In particular, if the graph G belongs to a non-trivial family of graphs that is minor-closed, it follows from [KPR93, FT03] that $\alpha = O(1)$.

2.1   0-Extension with Trees

The following result is a direct corollary of [GNR10, Theorem 7] (which in turn is an extension of the tree-embedding theorem of [FRT04]). Details omitted from this version.

Corollary 1 (Tree 0-extension). *There is a randomized polynomial-time algorithm $\mathcal{A}_{GNR}$ for 0-extension that has $\alpha_{GNR} = O(\log k)$; furthermore, the graphs output by the algorithm are trees on the vertex set K.*

As an aside, a weaker version of Corollary 1 with $O(\frac{\log^2 k}{\log \log k})$ can be proved as follows. First use Theorem 1 to obtain a random 0-extension H from G such that $\mathbb{E}_H[d_H(x, y)] \leq O(\frac{\log k}{\log \log k})\, d_G(x, y)$ for all $x, y \in K$. Then use the result of [FRT04] to get a random tree $H' = (K, E_{H'})$ such that $\mathbb{E}_{H'}[d_{H'}(x, y)] \leq O(\log k)\, d_H(x, y)$ for all $x, y \in V(H)$. Combining these two results proves the weaker claim.

3   Flow-Sparsifiers

Recall that given an edge-capacitated graph (G, c) and a set $K \subseteq V$ of terminals, a *flow-sparsifier with quality ρ* is another capacitated graph $(H = (K, E_H), c_H)$ such that (a) any feasible K-flow in G can be feasibly routed in H, and (b) any feasible K-flow in H can be routed in G with congestion ρ.

3.1 Interchanging Distance and Capacity

We use the framework of Räcke [Räc08], as interpreted by Andersen and Feige [AF09]. Given a graph $G = (V, E)$, let $\mathcal{P}$ be a collection of multisets of E, which will henceforth be called *paths*. A mapping $M : E \rightarrow \mathcal{P}$ maps each edge e to a path $M(e)$ in $\mathcal{P}$. Such a map can be represented as a matrix $\mathbf{M}$ in $\mathbb{Z}^{|E| \times |E|}$ where $\mathbf{M}_{e,e'}$ is the number of times the edge e' appears in the path (multiset) $M(e)$. Given a collection $\mathcal{M}$ of mappings (which we call the *admissible* mappings), a *probabilistic mapping* is a probability distribution over (or, convex combination of) admissible mappings; i.e., define $\lambda_M \geq 0$ for each $M \in \mathcal{M}$ such that $\sum_{M \in \mathcal{M}} \lambda_M = 1$.

Distance Mappings. Given $G = (V, E)$ and lengths $\ell : E \rightarrow \mathbb{R}_{>0}$,
- The *stretch* of an edge $e \in E$ under a mapping M is $\sum_{e'} \mathbf{M}_{e,e'} \ell(e')/\ell(e)$.
- The *average stretch* of e under a probabilistic mapping $\{\lambda\}$ is $\sum_M \lambda_M (\sum_{e'} \mathbf{M}_{e,e'} \frac{\ell(e')}{\ell(e)})$.
- The *stretch of a probabilistic mapping* is the maximum over all edges of their average stretch.

Capacity Mappings. Given a graph G with edge capacities $c : E \rightarrow \mathbb{R}_{>0}$,
- The *load* of an edge $e' \in E$ under a mapping M is $\sum_e \mathbf{M}_{e,e'} c(e)/c(e')$.
- The *expected load* of e' under a probabilistic mapping $\{\lambda\}$ is $\sum_M \lambda_M (\sum_e \mathbf{M}_{e,e'} \frac{c(e)}{c(e')})$.
- The *congestion of a probabilistic mapping* is the maximum over all edges of their expected loads.

The Transfer Theorem. Andersen and Feige [AF09] distilled ideas from Räcke [Räc08] to state:

Theorem 3 (Theorem 6 in [AF09]). *Fix a graph G and a collection $\mathcal{M}$ of admissible mappings. For every $\rho \geq 1$, the following are equivalent:*

1. *For every collection of edge lengths ℓ_e, there is a probabilistic mapping with stretch at most ρ.*
2. *For every collection of edge capacities c_e, there is a probabilistic mapping with congestion at most ρ.*

In our settings, the techniques of Räcke [Räc08] can be used to make the result algorithmic: if one can efficiently sample from the probabilistic mapping with stretch ρ (which is true for the settings in this paper), one can efficiently sample from a probabilistic mapping with congestion $O(\rho)$ (and *vice versa*). In fact, one can obtain an explicit distribution on polynomially many admissible mappings. We defer further discussion of efficiency issues to the full version of the paper.

3.2 Tree-Based Flow Sparsifiers

The distance mappings we will consider will be similar to Räcke's application. Let us first fix for each $u, v \in K$ a canonical shortest-path S_{uv} between u, v in G. Now, consider a tree 0-extension (T, f) where $T = (K, E_T)$ and $f : V \rightarrow K$ is a retraction. For each edge $e = (w, x) \in E(G)$, consider the (unique) $f(w)$-$f(x)$-path $P_T(f(w), f(x))$ in the tree T. Define the mapping $M_T : E \rightarrow \mathcal{P}$ corresponding to the 0-extension (T, f) by

$$M_T((w, x)) = \biguplus_{(u,v) \in P_T(f(w), f(x))} S_{uv}. \tag{3.1}$$

In other words, this maps each tree edge (w, x) to its canonical path; for each non-tree edge (w, x), it considers the edges on the tree-path between the images of w and x in the tree, and maps (w, x) to the disjoint union of the canonical paths of these edges. Recall that $M_T((w, x))$ is a multiset. In the corresponding matrix representation, $\mathbf{M}_{e,e'}$ is the multiplicity of e' in the set $\biguplus_{(u,v) \in P_T(f(w), f(x))} S_{uv}$. Corollary 1 now implies the following:

Theorem 4. *Given a graph (G, ℓ) with terminals $K \subseteq V(G)$, there is a polynomial-time procedure to sample from a probabilistic mapping (which is a distribution over tree 0-extensions) with stretch $\rho_{dist} = O(\log k)$. Moreover, $\rho_{dist} \geq 1$ if $K \neq \emptyset$.*

Now we can apply the Transfer Theorem. Recall that in a K-flow, all source-sink pairs belong to set K.

Theorem 5 (Tree-Based Flow-Sparsifiers). *Given an edge-capacitated graph (G, c), and a set of terminals $K \subseteq V$, there is a polynomial-time algorithm that outputs a graph $H = (K, E_H)$ that is a convex combination of edge capacitated trees such that:*

(a) *every K-flow that can be routed in G, can also be routed in H; and*
(b) *every K-flow that can be feasibly routed in H, can be routed with congestion $O(\log k)$ in G.*

In other words, if we were to scale up the capacities in G to route all feasible flows in H, then the factor by which we would have to scale up capacities would only be $O(\log k)$.

Proof. We apply Theorem 3 and Theorem 4 to $G = (V, E)$ to get a convex combination $\{\lambda_{T,f}\}$ of maps $(T = (K, E_T), f)$ such that each edge in E has an average load of $O(\rho_{dist})$. Let us see how this implies (a) and (b) above: this is essentially a matter of unraveling the definitions. For each such (T, f), we define capacities on the edges $e_T \in E_T$ thus: let (A, B) be node sets of the two connected components of T formed by deleting the edge e_T, where $A \cup B = K$. Let $A' = \{v \in V \mid f(v) \in A\}$, and $B' = V \setminus A'$. Define

$$c_{T,f}(e_T) := \sum_{e \in E \cap (A' \times B')} c(e). \tag{3.2}$$

We claim that this convex combination $\{\lambda_{T,f}\}$ of capacitated trees satisfies (a) and (b). For (a), the definition of the capacities $c_{T,f}$ ensures that each edge of G can be concurrently routed feasibly in each T using capacities $c_{T,f}(\cdot)$, hence so can any K-flow feasible in G. Since this holds for each (T, f) pair, it holds for the convex combination.

To prove (b), we want to route edges in the convex combination of trees in the graph G, where we scale the capacities $c_{T,f}$ of edges from (T, f) by its convex multiplier $\lambda_{T,f}$. Consider any edge $e_T = (u, v) \in E_T$ with capacity $c_{T,f}(e_T)$ defined in (3.2): we can use the canonical shortest path S_{uv} to route this flow. Hence the load on any edge $e' = (w', x') \in E$ due to the convex combination of trees is at most

$$\frac{1}{c(e')} \sum_{T,f} \lambda_{T,f} \sum_{e_T \in E_T : e' \in S_{uv}} c_{T,f}(e_T). \tag{3.3}$$

Since $c_{T,f}(e_T)$ is the sum of the capacity of all edges $e = (w, x)$ such that e_T lies on the unique tree-path between $f(w), f(x)$, we rewrite (3.3) as

$$\frac{1}{c(e')} \sum_{T,f} \lambda_{T,f} \sum_{e_T=(u,v)\in E_T : e' \in S_{uv}} \sum_{(w,x)\in E : e_T \in P_T(f(w),f(x))} c(wx) \qquad (3.4)$$

$$= \frac{1}{c(e')} \sum_{T,f} \lambda_{T,f} \sum_{(w,x)\in E} c(wx) \times (\text{multiplicity of } e' \text{ in } \uplus_{(u,v)\in P_T(f(w),f(x))} S_{uv}). \qquad (3.5)$$

However, this is exactly the *expected load* for e' under the notion of admissible maps defined in (3.1); hence this is bounded by the congestion (the maximum expected load over all edges), which is at most ρ_{dist} by Theorem 3. This proves condition (b) above, that the congestion to route any K-flow in the convex combination H in the graph G is at most ρ_{dist}.   □

3.3   General Flow Sparsifiers

Theorem 6 (Flow-Sparsifiers). *Given any graph G and terminals K, there is a randomized polynomial-time algorithm to output a flow-sparsifier H with loss $O(\frac{\log k}{\log \log k})$.*

Proof. Suppose we use Theorem 1 instead of using the tree 0-extension result (Corollary 1), we use the constructive version of the Transfer Theorem to get a polynomial number of graphs $H_1, H_2, \ldots$ on the vertex set K such that a convex combination of these graphs is a flow-sparsifier for the original graph G where the load is $O(\frac{\log k}{\log \log k})$. We can then construct a single graph H by setting the capacity of an edge to be the appropriate weighted combination of capacities of those edges in H_i; all feasible K-flows in G can be routed in H, and all feasible K-flows in H can be routed in G with congestion $O(\frac{\log k}{\log \log k})$.   □

The same idea using 0-extension results for β-decomposable graphs (Theorem 2) gives us the following:

Theorem 7 (Flow-Sparsifiers for Minor-Closed Families). *For any graph G that is β-decomposable and any K, there is a randomized polynomial-time algorithm to construct a flow-sparsifier with loss $O(\beta)$.*

Note that the decomposability holds if G belongs to a non-trivial minor-closed-family $\mathcal{G}$ (e.g., if G is planar). However, Theorem 7 does not claim that the flow-sparsifier for G also belongs to the family $\mathcal{G}$; this is the question we resolve in the next section.

4   Connected 0-Extensions and Minor-Based Flow-Sparsifiers

The results in this section apply to β-decomposable graphs. A prominent example of such graphs are planar graphs, which (along with every family of graphs excluding a fixed minor) are $O(1)$-decomposable [KPR93, FT03]. Thus, Theorem 8, Corollary 5 and Theorem 9 below all apply to planar graphs (and more generally to excluded-minor graphs) with $\beta = O(1)$. We now state our results for β-decomposable graphs in general.

In Section 4.2 we define a related notion called *terminal-decomposability*, and show analogous results for $\hat{\beta}$-terminal-decomposable graphs.

In what follows we use the definition of 0-extension from Section 2 with $H = H_f$, i.e., $E_H = \{(f(u), f(v)) : (u, v) \in E\}$, hence the 0-extension is completely defined by the retraction f. We say that a 0-extension f is *connected* if for every x, $f^{-1}(x)$ induces a connected component in G. Our main result shows that we get connected 0-extensions with stretch $O(\beta \log \beta)$ for β-decomposable metrics.

Theorem 8 (Connected 0-Extension). *There is a randomized polynomial-time algorithm that, given $(G = (V, E), \ell_G)$ with terminals K such that d_G is β-decomposable, produces a connected 0-extension $f : V \to K$ such that for all $u, v \in V$, we have*

$$\mathbb{E}[d_H(f(u), f(v))] \leq O(\beta \log \beta) \cdot d_G(u, v).$$

Note that if f is a connected 0-extension, the graph H_f is a minor of G. Applying Theorem 3 to interchange the distance preservation with capacity preservation, we get the following analogue of Theorem 5.

Corollary 2 (Minor-Based Flow-Sparsifiers). *For every β-decomposable graph $G = (V, E)$ with edge capacities c_G and a subset $K \subset V$ of k terminals, there is a minor-based flow-sparsifier with quality $O(\beta \log \beta)$. Moreover, a minor-based flow-sparsifier for G, c_G, K can be computed efficiently in randomized poly-time.*

Since planar graphs are $O(1)$-decomposable and since their minors are planar, by Corollary 2 they have an efficiently constructable planar-based flow-sparsifier with quality $O(1)$. By Theorem 8, they always have a connected 0-extension with stretch at most $O(1)$. An interesting consequence of the latter result is that given any planar graph (G, ℓ_G), and a set K of terminals, we can "remove" the non-terminals and get a related planar graph on K while preserving inter-terminal distances in expectation. This generalizes a result of Gupta [Gup01] who showed a similar result for trees. (Obviously, this extends to every family of graphs excluding a fixed minor.)

Theorem 9 (Steiner Points Removal). *There is a randomized polynomial-time algorithm that, given $(G = (V, E), \ell_G)$ and K such that d_G is β-decomposable, outputs minors $H = (K, E_H)$ of G such that $1 \leq \frac{\mathbb{E}[d_H(x,y)]}{d_G(x,y)} \leq O(\beta \log \beta)$ for all $x, y \in K$.*

Note that these results only give us an $O(\log n \log \log n)$-approximation for connected 0-extension on arbitrary graphs (or an $O(\log^2 k \log \log k)$-approximation using results of Section 4.2). We can improve that to $O(\log k)$; details in the full version.

Theorem 10 (Connected CKR). *There is a randomized polynomial-time algorithm that on input $(G = (V, E), \ell_G)$ and K, produces a connected 0-extension f with stretch factor $\mathbb{E}[d_H(f(u), f(v))] \leq O(\log k) \cdot d_G(u, v)$ for all $u, v \in V$.*

Using the semi-metric relaxation for 0-extension, we get a connected 0-extension whose cost is at most $O(\log k)$ times the optimal (possibly disconnected) 0-extension. To our knowledge, this is the first approximation algorithm for connected 0-extension, and in fact shows that the gap between the optimum connected 0-extension and the optimum 0-extension is bounded by $O(\log k)$. The same is true with an $O(1)$ bound for planar graphs. We remark that the connected 0-extension problem is a special case of the connected metric labeling problem, which has recently received attention in the vision community [VKR08, NL09].

4.1 The Algorithm for Decomposable Metrics

We now give the algorithm behind Theorem 8. Assume that edge lengths ℓ_G are integral and scaled such that the shortest edge is of length 1. Let the diameter of the metric be at most 2^δ. For each vertex $v \in V$, define $A_v = \min_{x \in K} d_G(v, x)$ to be the distance to the closest terminal. The algorithm maintains a partial mapping f at each point in time—some of the $f(v)$'s may be undefined (denoted by $f(v) = \bot$) during the run, but f is a well-defined 0-extension when the algorithm terminates. We say a vertex $v \in V$ is *mapped* if $f(v) \neq \bot$. The algorithm appears as Algorithm 1.

Algorithm 1. Algorithm for Connected 0-extension

 1: **input:** $(G, \ell_G), K$.
 2: **let** $i \leftarrow 0$, $f(x) = x$ for all $x \in K$, $f(v) = \bot$ for all $v \in V \setminus K$.
 3: **while** there is a v such that $f(v) = \bot$ **do**
 4:     **let** $i \leftarrow i + 1$, $r_i \leftarrow 2^i$
 5:     sample a β-decomposition of d_G with diameter bound r_i to get a partition P
 6:     **for all** clusters C_s in the partition P that contains both mapped and unmapped vertices **do**
 7:         delete all vertices u in C_s with $f(u) \neq \bot$
 8:         **for** each connected component C from C_s **do**
 9:             choose a vertex $w_C \in C_s$ that was deleted and had an edge to C
10:             reset $f(u) = f(w_C)$ for all $u \in C$.
11:         **end for**
12:     **end for**
13: **end while**

We can assume that in round $\delta = \log \text{diam}(G)$, the partitioning algorithm returns a single cluster, in which case all vertices are mapped and the algorithm terminates. Let f_i be the mapping at the end of iteration i. For $x \in K$, let V_i^x denote $f_i^{-1}(x)$, the set of nodes colored x. The following claim follows inductively:

Lemma 1. *For every i and $x \in K$, the set V_i^x induces a connected component in G.*

Proof. We prove the claim inductively. For $i = 0$, there is nothing to prove since $V_i^x = \{x\}$. Suppose that in iteration i, we map vertex u to x so that $u \in V_i^x$. Thus for some component C containing u, the mapped neighbor w_C chosen by the algorithm was in V_{i-1}^x. Since we map all of C to x, there is a path connecting v to w_C in V_i^x. Inductively, w_C is connected to x in $V_{i-1}^x \subseteq V_i^x$, and the claim follows. $\quad\square$

The following lemma will be useful in the analysis of the stretch; it says that any node mapped in iteration i is mapped to a terminal at distance $O(2^i)$.

Lemma 2. *For every iteration i and $x \in K$, and every $u \in V_i^x$, $d_G(x, u) \leq 2r_i$.*

Proof. The proof is inductive. For $i = 0$, the claim is immediate. Suppose that in iteration i, we map vertex u to x so that $u \in V_i^x$. Thus for some component C containing u, the mapped neighbor w_C chosen by the algorithm was in V_{i-1}^x. Moreover, u and

w_C were in the same cluster in the decomposition so that $d(u, w_C) \leq r_i$. Inductively, $d(w_C, x) \leq 2r_{i-1}$ and the claim follows by triangle inequality. □

In the rest of the section, we bound the stretch of the 0-extension; for every edge $e = (u, v)$ of G, we show that

$$\mathbf{E}[d_G(f(u), f(v))] \leq O(\beta \log \beta) \, d_G(u, v).$$

Note that for $e = (u, v)$, $d_G((f(u), f(v)) = d_H((f(u), f(v))$, and so it's enough to prove the claim for d_G. The analogous claim for non-adjacent pairs will follow by triangle inequality, but here with d_H. We say that the edge $e = (u, v)$ is *settled in round j* if the later of its endpoints gets mapped in this round; e is *untouched after round j* if both u and v are unmapped at the end of round j. Let $d_G(u, K) \leq d_G(v, K)$ and let A_e denote the distance $d_G(u, K)$. Let $j_e := \lfloor \log(A_e) \rfloor - 1$.

Lemma 3. *For edge $e = (u, v)$,*
   (a) edge e is untouched after round $j_e - 1$,
   (b) if edge e is settled in round j then $d_G(f(u), f(v)) = O(2^j + d_G(u, v))$.

Proof. For (a), if one of the end points of e is mapped before round j_e, then $2 \cdot 2^{j_e} \leq A_e = d_G(e, K)$, which contradicts Lemma 2. For (b), both $d_G(u, f(u)), d_G(v, f(v)) \leq 2^{j+1}$ by Lemma 2; the triangle inequality completes the proof. □

Let $\mathcal{B}_j$ denote the "bad" event that the edge is settled in round j and that both end-points are mapped to different terminals. Let $z := \max\{A_e, d_G(u, v)\}$. We want to use

$$\mathbf{E}[d(f(u), f(v))] = \sum_j \mathbf{Pr}[\mathcal{B}_j] \cdot \mathbf{E}[d(f(u), f(v)) \mid \mathcal{B}_j].$$

Claim. $\mathbf{Pr}[\mathcal{B}_j] \leq \min\{4\beta \frac{z}{2^j}, 1\} \cdot 5\beta \frac{d_G(u,v)}{2^j}$.

Proof. Recall that an edge is untouched after round j' if neither of its endpoints is mapped at the end of this round. For this to happen, u must be separated from its closest terminal in the clustering in round j', which happens with probability at most $\min\{\beta \frac{A_e}{2^{j'}}, 1\}$. Also recall that the probability that an edge $e = (u, v)$ is cut in a round j' is at most $\beta \frac{d_G(u,v)}{2^{j'}}$. Let i denote the round in which the edge is first touched. We upper bound the probability of the event $\mathcal{B}_j$ separately depending on how i and j compare. Note that for $j \leq 2$, the right hand side is at least 1 so the claim holds trivially.

- $i \leq j - 2$. For $\mathcal{B}_j$ to occur, the edge e must be cut in round $j - 2$ and $j - 1$, as otherwise it would already be settled in one of these rounds. The probability of this is at most $\min\{\beta \frac{d_G(u,v)}{2^{j-2}}, 1\} \cdot \beta \frac{d_G(u,v)}{2^{j-1}} \leq \min\{4\beta \frac{z}{2^j}, 1\} \cdot 2\beta \frac{d_G(u,v)}{2^j}$.
- $i = j - 1$. For $\mathcal{B}_j$ to occur, the edge e must be cut in round $j - 1$ and must be untouched after round $j-2$. The probability of this is at most $\min\{\beta \frac{A_e}{2^{j-2}}, 1\} \cdot \beta \frac{d_G(u,v)}{2^{j-1}} \leq \min\{4\beta \frac{z}{2^j}, 1\} \cdot 2\beta \frac{d_G(u,v)}{2^j}$.

- $i = j$. For $\mathcal{B}_j$ to occur, e must be cut in round j and must be untouched after round $j-1$. The probability of this is at most $\min\{\beta\frac{A_e}{2^{j-1}}, 1\}\cdot\beta\frac{d_G(u,v)}{2^j} \le \min\{4\beta\frac{z}{2^j}, 1\}\cdot\beta\frac{d_G(u,v)}{2^j}$.

Since $\mathbf{Pr}[\mathcal{B}_j] = \mathbf{Pr}[\mathcal{B}_j \wedge (i \le j-2)] + \mathbf{Pr}[\mathcal{B}_j \wedge (i = j-1)] + \mathbf{Pr}[\mathcal{B}_j \wedge (i = j)]$, the claim follows.   □

Lemma 3(b) implies that if the edge is settled before round $j_d := \lfloor\log(d_G(u,v))\rfloor$, the conditional expectation $\mathbf{E}[d_G(f(u), f(v)) \mid \mathcal{B}_j]$ is $O(d_G(u,v))$. Moreover the edge e cannot be settled before round $j_e = \lfloor\log(A_e)\rfloor - 1$ by Lemma 3(a). Let $j_m := \max\{j_d, j_e\}$. It therefore suffices to to show that

$$\sum_{j \ge j_m} \mathbf{Pr}[\mathcal{B}_j] \cdot O(2^j) \le O(\beta\log\beta)\, d_G(u, v) \ .$$

Plugging in the upper bound for $\mathbf{Pr}[\mathcal{B}_j]$ into the left hand side, we get

$$\sum_{j \ge j_m} \mathbf{Pr}[\mathcal{B}_j] \cdot O(2^j) \le \sum_{j \ge j_m} \min\{4\beta\tfrac{z}{2^j}, 1\} \cdot 5\beta\frac{d_G(u,v)}{2^j} \cdot O(2^j)$$
$$\le \sum_{j \ge j_m} \min\{4\beta\tfrac{z}{2^j}, 1\} \cdot \beta \cdot O(d_G(u,v)) \quad \le O(\beta\log\beta)\, d_G(u, v) \ .$$

In the last step, we used that $z = \max\{A_e, d_G(u, v)\} \le \max\{2^{j_e+2}, 2^{j_d+1}\} \le 2^{j_m+2}$, so the first $O(\log\beta)$ terms contribute $O(\beta\, d_G(u, v))$, while the remaining terms form a geometric series and sum to $O(d_G(u, v))$. This completes the proof of Theorem 8.

4.2   Terminal Decompositions

The general theorem for connected 0-extensions gives a guarantee in terms of its decomposition parameter β, and in general this quantity may depend on n. This seems wasteful, since we decompose the entire metric while we mostly care about separating the terminals.

To this end, we define *terminal decompositions* (the reader might find it useful to contrast it with definition of decompositions in Section 1.1). A *partial partition* of a set X is a collection of disjoint subsets (called "clusters" of X). A metric (X, d) with terminals K is called $\hat{\beta}$-*terminal-decomposable* if for every $\varDelta > 0$ there is probability distribution μ over partial partitions of X, with the following properties:

- *Diameter bound:* Every partial partition $\widehat{P} \in \mathrm{supp}(\mu)$ is connected and $\varDelta$-bounded.
- *Separation event:* For all $u, v \in X$, $\mathrm{Pr}_{\widehat{P}\in\mu}[\exists S \in \widehat{P}$ such that $u \in S$ but $v \notin S] \le \hat{\beta} \cdot d(u, v)/\varDelta$.
- *Terminal partition:* For all $x \in K$, every partial partition $\widehat{P} \in \mathrm{supp}(\mu)$ has a cluster containing x.
- *Terminal-centered clusters:* For every partial partition $\widehat{P} \in \mathrm{supp}(\mu)$, every cluster $S \in \widehat{P}$ contains a terminal.

A graph $G = (V, E)$ with terminals K is $\hat{\beta}$-terminal-decomposable if for every nonnegative lengths ℓ_G assigned to its edges, the resulting shortest-path metric d_G with terminals K is $\hat{\beta}$-terminal-decomposable. Throughout, we assume that there is a polynomial time algorithm that, given the metric, terminals and $\varDelta$ as input, samples a partial partition $\widehat{P} \in \mu$. Note that if $K = V$, the above definitions coincide with the definitions of β-decomposable metrics and graphs.

Our main theorem for terminal decomposable metrics is the following:

Theorem 11. *Given $(G = (V, E), \ell_G)$, suppose d_G is $\hat{\beta}$-terminal-decomposable with respect to terminals K. There is a randomized polynomial-time algorithm that produces a connected 0-extension $f : V \to K$ such that for all $u, v \in V$, we have $\mathbb{E}[d_G(f(u), f(v))] \leq O(\hat{\beta}^2 \log \hat{\beta}) \cdot d_G(u, v)$.*

This theorem is interesting when $\hat{\beta}$ is much less than β, the decomposability of the metric itself. E.g., one can alter the CKR decomposition scheme to get $\hat{\beta}(k, n) = O(\log k)$, while $\beta = O(\log n)$.

The Modified Algorithm. Algorithm 2 for the terminal-decomposable case is very similar to Algorithm 1: the main difference is that in each iteration we only obtain a partial partition of the vertices, we color only the nodes that lie in clusters of this partial partition.

A few words about the algorithm: recall that a partial partition returns a set of connected diameter-bounded clusters such that each cluster contains at least one terminal, and each terminal is in exactly one cluster— we use V^x to denote the cluster containing $x \in K$. (Hence either $V^x = V^y$ or $V^x \cap V^y = \emptyset$.) Now when we delete all the vertices in some cluster V^x that are already mapped, this includes the terminal x—and hence there is at least one candidate for w_C in Line 9. Eventually, there will be only one cluster, in which case all vertices are mapped and the algorithm terminates.

Algorithm 2. Algorithm for Connected 0-extension: the terminal-decomposable case

---

1: **input:** $(G, \ell_G), K$.
2: **let** $i \leftarrow 0$, $f(x) = x$ for all $x \in K$, $f(v) = \perp$ for all $v \in V \setminus K$.
3: **while** there is a v such that $f(v) = \perp$ **do**
4:     **let** $i \leftarrow i + 1$, $r_i \leftarrow 2^i$
5:     find a $\hat{\beta}$-terminal-decomposition of d_G with diameter bound r_i; let V^x be the cluster containing terminal x.
6:     **for all** clusters V^x in the partial partition **do**
7:         delete all vertices u in V^x with $f(u) \neq \perp$
8:         **for** each connected component C from V^x thus formed **do**
9:             choose a vertex $w_C \in V^x$ that was deleted and had a neighbor in C
10:             reset $f(u) = f(w_C)$ for all $u \in C$.
11:         **end for**
12:     **end for**
13: **end while**

The analysis for Theorem 11 is almost the same as for Theorem 8; the only difference is that Claim 4.1 is replaced by the following weaker claim (proof omitted from this version), which immediately gives the $O(\hat{\beta}^2 \log \hat{\beta})$ bound.

Claim. $\mathbf{Pr}[\mathcal{B}_j] \leq \min\{8\hat{\beta}\frac{z}{2^j}, 1\} \cdot 23\hat{\beta}^2 \frac{d(u,v)}{2^j}$.

5   Future Directions

We gave a set of results on and around the idea of flow-sparsifiers and 0-extensions. Some of these results are not tight, and it would be interesting to obtain better bounds for these problems. Another interesting direction for future work is this: define an ℓ-*sparse-extension* of graph $G = (V, E)$ with terminals K to be any graph $H = (Z, E_H)$ with $|Z| = \ell$, $K \subseteq Z \subseteq V$, along with a retraction $f : V \rightarrow Z$ that satisfies $d_H(x, y) \geq d_G(x, y)$ for all $x, y \in Z$. (Note that a $|K|$-sparse-extension is just a 0-extension; one possible $|V|$-sparse-extension is G itself.) What if we consider ℓ-sparse-extensions (H, f) with

$$E[\, d_H(f(x), f(y))\,] \leq \alpha \, d_G(x, y) \qquad \text{for all } x, y \in V,$$

where ideally $\ell = \text{poly}(k)$, and $\alpha = O(1)$ (or just $\alpha \ll \frac{\log k}{\log \log k}$)? In other words, if we are willing to retain a small number of non-terminals, can we achieve better stretch bounds? Note that standard lower bounds for 0-extension have the property that $|V| = \text{poly}(k)$—hence the entire graph G is a "good" solution (poly(k)-sparse-extension with $\alpha = 1$).

References

[AF09]   Andersen, R., Feige, U.: Interchanging distance and capacity in probabilistic map-pings. CoRR, abs/0907.3631 (2009)

[AFH+04]   Archer, A., Fakcharoenphol, J., Harrelson, C., Krauthgamer, R., Talwar, K., Tardos, É.: Approximate classification via earthmover metrics. In: Proc. 15th SODA, pp. 1079–1087 (2004)

[CKR04]   Calinescu, G., Karloff, H.J., Rabani, Y.: Approximation algorithms for the 0-extension problem. SIAM J. Comput. 34(2), 358–372 (2004)

[FHRT03]   Fakcharoenphol, J., Harrelson, C., Rao, S., Talwar, K.: An improved approximation algorithm for the 0-extension problem. In: Proc. 14th SODA, pp. 257–265 (2003)

[FRT04]   Fakcharoenphol, J., Rao, S., Talwar, K.: A tight bound on approximating arbitrary metrics by tree metrics. J. Comput. System Sci. 69(3), 485–497 (2004)

[FT03]   Fakcharoenphol, J., Talwar, K.: Improved decompositions of graphs with forbid-den minors. In: Arora, S., Jansen, K., Rolim, J.D.P., Sahai, A. (eds.) 6th APPROX. LNCS, vol. 2764, pp. 36–46. Springer, Heidelberg (2003)

[GNR10]   Gupta, A., Nagarajan, V., Ravi, R.: Improved approximation algorithms for require-ment cut. Operations Research Letters 38(4), 322–325 (2010)

[GNRS04]   Gupta, A., Newman, I., Rabinovich, Y., Sinclair, A.: Cuts, trees and ℓ_1-embeddings of graphs. Combinatorica 24(2), 233–269 (2004)

[Gup01]   Gupta, A.: Steiner points in tree metrics don't (really) help. In: Proc. 12th SODA, pp. 220–227 (2001)

[KPR93]   Klein, P., Plotkin, S.A., Rao, S.B.: Excluded minors, network decomposition, and multicommodity flow. In: Proc. 25th STOC, pp. 682–690 (1993)

[LM10]   Leighton, T., Moitra, A.: Extensions and limits to vertex sparsification. In: Proc. 42th STOC, pp. 47–56 (2010)

[LN05]   Lee, J.R., Naor, A.: Extending Lipschitz functions via random metric partitions. In-vent. Math. 160(1), 59–95 (2005)

[LS09]   Lee, J.R., Sidiropoulos, A.: On the geometry of graphs with a forbidden minor. In: Proc. 41st STOC, pp. 245–254 (2009)

[Moi09]   Moitra, A.: Approximation algorithms for multicommodity-type problems with guarantees independent of the graph size. In: Proc. 50th FOCS, pp. 3–12 (2009)

[NL09]   Nowozin, S., Lampert, C.H.: Global connectivity potentials for random field models. In: Proc. 22nd CVPR, pp. 818–825 (2009)

[Räc08]   Räcke, H.: Optimal hierarchical decompositions for congestion minimization in net works. In: Proc. 40th STOC, pp. 255–264 (2008)

[VKR08]   Vicente, S., Kolmogorov, V., Rother, C.: Graph cut based image segmentation with connectivity priors. In: Proc. 21st CVPR (2008)

PTAS for Weighted Set Cover on Unit Squares

Thomas Erlebach[1] and Erik Jan van Leeuwen[2]

[1] Department of Computer Science, University of Leicester, University Road,
Leicester, LE1 7RH, UK
t.erlebach@mcs.le.ac.uk
[2] Department of Informatics, University of Bergen, P.O. Box 7803,
N-5020 Bergen, Norway
E.J.van.Leeuwen@ii.uib.no

Abstract. We study the planar version of Minimum-Weight Set Cover, where one has to cover a given set of points with a minimum-weight subset of a given set of planar objects. For the unit-weight case, one PTAS (on disks) is known. For arbitrary weights however, the problem appears much harder, and in particular no PTASs are known. We present the first PTAS for Weighted Geometric Set Cover on planar objects, namely on axis-parallel unit squares. By extending the algorithm, we also obtain a PTAS for Minimum-Weight Dominating Set on intersection graphs of unit squares and Geometric Budgeted Maximum Coverage on unit squares. The running time of the developed algorithms is optimal under the exponential time hypothesis. We also show inapproximability results for Geometric Set Cover on various object shapes that are more general than unit squares.

1 Introduction

One of the most fundamental and best-known optimization problems is Minimum Set Cover. Given a universe $\mathbb{U}$, a set of elements $\mathcal{P} \subseteq \mathbb{U}$, and a set $\mathcal{S}$ of subsets of $\mathbb{U}$, one should find a minimum set $S \subseteq \mathcal{S}$ such that each element of $\mathcal{P}$ is contained in (covered by) a set in S. If $\mathbb{U} = \mathbb{R}^d$ for some $d > 0$, we talk about *Geometric Set Cover*. In particular, we are interested in the case where $d = 2$ and the sets in $\mathcal{S}$ are induced by simple geometric shapes, such as disks or squares. Geometric Set Cover can be better approximated than general Minimum Set Cover [2,5,23,29], but for many object shapes the approximability has not been settled yet, particularly in the weighted case. In this paper, we consider the approximability of Geometric Set Cover and several of its variants, with emphasis on weighted cases.

Motivation. Minimum Set Cover is known to be approximable within $1 + \ln |\mathcal{P}|$, even in the weighted case [20,25,4]. This algorithm is also optimal. That is, Minimum Set Cover has no polynomial-time algorithm attaining an approximation ratio of $(1 - \epsilon) \ln |\mathcal{P}|$ for any $\epsilon > 0$, unless $\mathrm{NP} \subset \mathrm{DTIME}(n^{O(\log \log n)})$ [11]. Because of its applicability in the design of (wireless) networks, Geometric Set Cover has recently received a lot of attention. Geometric Set Cover is NP-hard on unit

M. Serna et al. (Eds.): APPROX and RANDOM 2010, LNCS 6302, pp. 166–177, 2010.
© Springer-Verlag Berlin Heidelberg 2010

squares and on unit disks [12,21], even if the point set $\mathcal{P}$ corresponds to the centers of the squares or disks. This has led to researchers studying approximation algorithms for variations of the problem.

The biggest focus of approximation algorithms for Geometric Set Cover has been on (unit) disks. Several constant-factor approximation algorithms were proposed on unweighted unit disks [2,6,30,3]. This line of research recently culminated in the discovery of a PTAS for Geometric Set Cover on general disks [29] using a transformation into the geometric version of Minimum Hitting Set on three-dimensional half-spaces.

The above algorithms are not known to be applicable to the weighted case and only recently have algorithms approximating the problem started to appear. After a few iterations [1,19,7], a $(4+\epsilon)$-approximation algorithm on unit disks, independently proposed by Zou et al. [34] and Erlebach and Mihalák [9], currently is the best known result. Varadarajan [32] gives a $2^{O(\log^* n)}$-approximation on general disks. On unit squares, a 2-approximation algorithm exists [28]. It seems however that past approaches are insufficient to reach a PTAS, except when the disk centers have a constant minimum distance from each other [13,24].

We also consider the geometric version of the Budgeted Maximum Coverage problem. Here each element u of $\mathcal{P}$ has a profit $p(u)$, each set $\mathcal{S}_i$ of $\mathcal{S}$ a cost $c(\mathcal{S}_i)$, and we aim to maximize the total profit of the points covered by some $S \subseteq \mathcal{S}$, while the total cost of S is no more than a given budget B. Budgeted Maximum Coverage has a $(1 - \frac{1}{e})$-approximation algorithm in both the unit cost [33,18,16] and the general case [22]. Khuller, Moss, and Naor [22] proved that no polynomial-time algorithm can obtain an approximation ratio better than $1 - \frac{1}{e}$, unless NP $\subset$ DTIME$(n^{O(\log \log n)})$. As far as we know, Geometric Budgeted Maximum Coverage has not been studied yet. The problem can be shown to be NP-hard on unit squares by reduction from Geometric Set Cover.

Observe that Geometric Set Cover differs significantly from the Geometric Covering problem, where the position of the objects may be chosen freely. This problem has a well-known PTAS both on unit disks [17] and on unit squares [14].

Our Results. In this paper, we present a PTAS for Geometric Set Cover on any set of axis-parallel unit squares. Using a novel dynamic programming idea, refining the classic sweep-line technique, we are able to solve this problem optimally in $n^{O(k)}$ time when the given sets of points lie within a horizontal strip of height k. Combining this with the well-known shifting technique then yields the PTAS. We also observe that it follows from Marx [26,27] that the scheme has essentially optimal running time (up to constants), unless the exponential time hypothesis is false.

The presented scheme extends to the weighted case of Geometric Set Cover and Minimum Dominating Set on intersection graphs of unit squares and in fact to the more general Geometric Budgeted Maximum Coverage problem. We note that the optimality result for our PTAS continues to hold.

Beside these positive algorithmic results, we also give several negative results. In particular, we show that Geometric Set Cover is APX-hard on arbitrary four-sided convex polygons. We also obtain APX-hardness results on axis-parallel rectangles and ellipses. Finally, we show that on convex polygons, translated copies of a single polygon, rotated copies of a single convex polygon, and α-fat objects, Geometric Set Cover is as hard as Minimum Set Cover.

2   A PTAS on Unit Squares

We consider Geometric Set Cover on unit squares and show that it has a PTAS by applying the shifting technique. So let $\mathcal{P}$ be a set of points and $\mathcal{S}$ a set of axis-aligned unit squares. For sake of notation, when referring to the (x, y)-coordinates of a square, we mean the coordinates of the bottom left corner of that square. For a square s, the x-coordinate of s is denoted by $x(s)$, while the y-coordinate is denoted by $y(s)$. We can assume that no horizontal (vertical) boundary of a square is on the same line as the horizontal (vertical) boundary of another square, that no point lies on the boundary of a square, and that none of the square or point coordinates are integers.

Consider the horizontal lines $y = h$ ($h \in \mathbb{Z}$). They partition the plane into horizontal slabs of height 1. Any point is contained in a slab and every square intersects precisely one line. Let $k \geq 1$ be an integer determined later. Using the shifting technique, it suffices to prove that we can optimally solve Geometric Set Cover on unit squares if we restrict to k consecutive slabs and the $k + 1$ lines defining them.

Theorem 1. *For any instance of Geometric Set Cover on a set of unit squares $\mathcal{S}$ where all points of $\mathcal{P}$ are inside $k \geq 1$ consecutive height 1 horizontal slabs, one can find an optimal solution in $O((3|\mathcal{S}|)^{4k+4} |\mathcal{P}|)$ time.*

The idea of the proof of this theorem will be to apply a sweep-line algorithm. To this end, consider the subset of squares of an optimum solution intersecting a horizontal line $y = h$ for some $h \in \mathbb{Z}$. Any such square must appear on the lower or upper envelope of this subset, or all points it covers would be covered by other squares. Following this observation, for each position of the sweep-line and for each of the $k+1$ integer horizontal lines, we should consider at most two squares intersecting the sweep-line: one that will appear on the upper envelope and one that will appear on the lower envelope of the final solution.

However, a square might appear on the lower envelope for some position of the sweep-line and on the upper envelope for a later position. This makes it difficult to avoid counting certain squares twice. To circumvent this, we split the sweep-line into k parts, one part per slab. We move these parts at different speeds, but always in such a way that if a square appears both on the lower and the upper envelope, then the split sweep-line is positioned such that it intersects the square both at the point where the square appears on the lower and on the upper envelope. We formalize this intuition below. We remark that the basic idea of having a split sweep-line was also used by Erlebach and Mihalák [9], but the details of how the split sweep-line is then handled by a dynamic programming approach are very different in our case.

Just as in any sweep-line algorithm, we maintain a data structure (the *front*) containing the squares that are 'active' at a given position of the sweep-lines and allow only a limited number of operations on it.

Let $\mathcal{S}^l$ and $\mathcal{S}^r$ be two dummy sets of $k+1$ squares each, such that the squares in $\mathcal{S}^l$ ($\mathcal{S}^r$) are to the left (right) of all squares in $\mathcal{S}$ and each integer horizontal line intersects precisely one square of $\mathcal{S}^l$ and one square of $\mathcal{S}^r$. Let $\overline{\mathcal{S}} = \mathcal{S} \cup \mathcal{S}^l \cup \mathcal{S}^r$. Given some set $S \subseteq \overline{\mathcal{S}}$, let S_i denote the set of squares in S intersecting line i. Let $R_i \subseteq S_i$ be the set containing precisely the rightmost square of S_i (denote it by s_i) and those squares s that overlap part of the left boundary of s_i and whose right boundary is not fully covered by squares of S_i.

We now define a front. For a better understanding of the definition, imagine that the squares are being inserted in order of increasing x-coordinate and that we want to keep track of the upper and lower envelope of each line i.

Definition 1. *Let S be the union of $\mathcal{S}^l$ and some subset of $\overline{\mathcal{S}}$. Then a* front *$F = \{u_1, \ldots, u_{k+1}, l_1, \ldots, l_{k+1}, b_1, \ldots, b_{k+1}, x_1, \ldots, x_k\}$ for S has the following:*

- *$u_i, l_i \in R_i$ with $u_i = s_i$ or $l_i = s_i$, $y(s) \leq y(u_i)$ for any $s \in S_i$ to the right of u_i (i.e. with $x(s) > x(u_i)$) and $y(s) \geq y(l_i)$ for any $s \in S_i$ to the right of l_i (i.e. with $x(s) > x(l_i)$),*
- *b_i equals the lowest square of S_i to the right of l_i if $x(u_i) > x(l_i)$, the highest square of S_i to the right of u_i if $x(l_i) > x(u_i)$, and s_i if $x(u_i) = x(l_i)$,*
- *x_i equals the larger x-coordinate from which l_{i+1} appears on the lower envelope of S_{i+1} and from which u_i appears on the upper envelope of S_i.*

Fronts are the representative of the current state of the sweep-line algorithm. The squares u_i and l_i track the current square on respectively the upper and the lower envelope of line i. The value of x_i is the x-coordinate of the part of the sweep-line between lines i and $i+1$. The square b_i is used in checking if a certain square may be inserted into the front or not. An example is depicted in Figure 1.

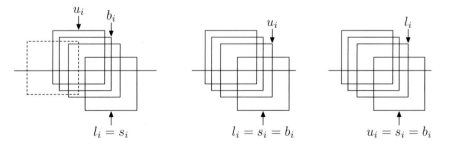

Fig. 1. The left figure shows a set S_i. The solid squares are in R_i, the dashed square is not. By Definition 1, the labeling of the left figure is correct. The middle figure shows the same set R_i, with a different and still correct labeling. The labeling in the right figure is incorrect.

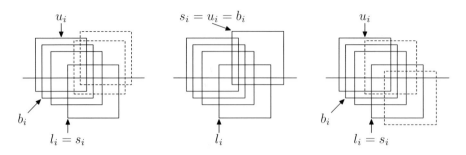

Fig. 2. The left figure shows two (dashed) squares that are upper-insertable. The middle figure shows the resulting front after upper-inserting the rightmost square. The right figure shows two (dashed) non-upper-insertable squares.

We make two observations about fronts. Firstly, $y(u_i) \geq y(l_i)$ and as $u_i, l_i \in R_i$, $|x(u_i) - x(l_i)| < 1$ for any $i = 1, \ldots, k+1$. Secondly, $y(l_i) \leq y(b_i) \leq y(u_i)$.

For a given front, we distinguish four types of insertions that are possible: an upper-insertion for squares that will appear only on the upper envelope for some line, a lower-insertion for squares appearing only on the lower envelope, and a middle-insertion and a skip-insertion for squares appearing on both envelopes. We define these four insertions, describe when they may be applied, and prove that any geometric set cover can be obtained using these insertions.

From now on, S will denote the union of $\mathcal{S}^l$ and some subset of $\overline{\mathcal{S}}$.

Definition 2. *Let F be a front for some S and let $s \notin S$ be a square intersecting line $i \in \{1, \ldots, k\}$. We say that s is upper-insertable into F if all of the following hold: 1) $y(s) > y(l_i)$ and if $x(l_i) > x(u_i)$, then $y(s) > y(b_i)$, 2) $x(s) \in (x(l_i), x(l_i) + 1]$ and $x(s) \in (x(u_i), x(u_i) + 1]$, 3) $x_i' > x_i$, 4) any point of $\mathcal{P}$ in $[x_i, x_i'] \times [i, i+1]$ is covered by u_i or l_{i+1}, where x_i' is the x-coordinate from which s is on the upper envelope of $(S \cup \{s\})_i$.*

Condition 1 ensures that s lies above l_i and all squares between u_i and l_i (represented by b_i), Condition 2 ensures that s appears on the upper envelope of $(S \cup \{s\})_i$, Condition 3 ensures that this appearance happens after u_i appears on the upper envelope, and Condition 4 ensures that we cover all points between two consecutive sweep-line positions. An example of upper-insertable squares and squares that are not upper-insertable is given in Figure 2.

Lemma 1. *Let F be a front for some S and let $s \notin S$ be a square intersecting line $i \in \{1, \ldots, k\}$ that is upper-insertable into F. Then between the appearance of u_i and the appearance of s on the upper envelope of $(S \cup \{s\})_i$ no other squares appear on the upper envelope of $(S \cup \{s\})_i$.*

Proof. If $u_i = s_i$, this follows from $x(s) > x(u_i) = x(s_i)$ and $x_i' > x_i$. So assume that $u_i \neq s_i$. Then $l_i = s_i$ and $x(l_i) > x(u_i)$. Recall the definition of a front and observe that b_i is the highest square of S_i to the right of u_i. As $x(l_i) - x(u_i) < 1$ and $y(b_i) < y(u_i)$, it suffices for s to lie above b_i (i.e. $y(s) > y(b_i)$) and for s to cover the x-range $[x(u_i) + 1, x(l_i) + 1]$ (i.e. $x(l_i) < x(s) < x(u_i) + 1$). This holds from the definition of upper-insertable.   □

Lemma 2. *Let F be a front for some S and let $s \notin S$ be a square intersecting line $i \in \{1, \ldots, k\}$ that is upper-insertable into F. Then $S \cup \{s\}$ has a front F' equal to F, except u_i is replaced by s, x_i is set to x_i', where x_i' is equal to $x(s)$ if $y(s) > y(u_i)$ and to $x(u_i) + 1$ otherwise, and if $x(u_i) \leq x(l_i)$ or $y(s) \leq y(b_i)$, b_i is set to s.*

Proof. Since $x(s) > \max\{x(u_i), x(l_i)\} = x(s_i)$ by Condition 2 of Definition 2, we can replace u_i by s. Note that l_i can remain the same by Condition 1 and 2. By Lemma 1, x_i' is indeed the x-coordinate from which s appears on the upper envelope of $(S \cup \{s\})_i$. From Condition 3, x_i should be set to x_i'. If $x(u_i) \leq x(l_i)$, then as $x(s) > x(l_i)$, b_i should be set to s. If $x(u_i) > x(l_i)$, then b_i must be changed if s lies below b_i, i.e. if $y(s) \leq y(b_i)$. Then F' is indeed a front for $S \cup \{s\}$.   □

Constructing the front F' from F as prescribed in the lemma statement is called the *upper-insertion* of s into F.

We can define the notions of lower-/middle-/skip-insertable and lower-/middle-/skip-insertions in similar ways. The definitions of upper- and lower-insertable/-insertion are similar, except that we check if the square we want to insert will appear on the lower envelope directly after l_i appears on the lower envelope. The definition of middle-insertable/-insertion combines the definitions of upper- and lower-insertable/-insertion. Skip-insertions are used when the square we want to insert does not intersect u_i or l_i, i.e. when $x(s) > 1 + \max\{x(u_i), x(l_i)\}$. Full definitions are in [31].

In general, we call an upper-/lower-/middle-/skip-insertion an *insertion* and we say s is *insertable* if it is upper-/lower-/middle-/skip-insertable. A *valid insertion* is the upper- (respectively lower-/middle-/skip-) insertion of a square that is upper- (respectively lower-/middle-/skip-) insertable.

Denote by F^l and F^r the fronts for $\mathcal{S}^l$ and $\overline{\mathcal{S}}$ respectively.

Lemma 3. *Assume $\mathcal{P} = \emptyset$. Let S be some set such that $S = \mathcal{S}^l \cup S_i \cup \mathcal{S}^r$ for some $i \in \{1, \ldots, k+1\}$ and any square in S_i appears on the lower or the upper envelope of S_i. Then there is a sequence of $|S_i| + k - 1$ valid insertions starting from F^l, leading to fronts $F^l = F_0, F_1, \ldots, F_{|S_i|+k-1} = F^r$ such that for any square $s \in S_i$, there is a front F_j containing s.*

Proof (Sketch). We assume that if $i = 1$, then no squares of S_i appear only on the lower envelope of S_i. Similarly, if $i = k + 1$, assume that no squares of S_i appear only on the upper envelope of S_i. Order the squares in $S_i \setminus \mathcal{S}^l$ by increasing x-coordinate, i.e. $s_1, \ldots, s_{|S_i|-1}$. Note that the squares appearing on the upper envelope form an increasing subsequence of S_i. Similarly, the squares appearing on the lower envelope form an increasing subsequence. We claim that one can obtain the requested sequence of valid insertions by inserting s_j into F_{j-1} for all $j = 1, \ldots, |S_i| - 1$ as follows. If s_j appears only on the upper envelope of S_i, then s_j is upper-insertable and will be upper-inserted. If s_j appears only on the lower envelope of S_i, then s_j is lower-insertable and will be lower-inserted. If s_j appears on the upper and lower envelope of S_i and a square of S_i covers part of its left boundary, then s_j is middle-insertable and will be middle-inserted. If s_j appears on the upper and lower envelope of S_i and no square of S_i covers part of its left boundary, then s_j is skip-insertable and will be skip-inserted. Now apply induction on the number of inserted squares.   □

Lemma 4. *Assume $\mathcal{P} = \emptyset$. Let S be some subset of $\overline{S}$ containing $\mathcal{S}^l \cup \mathcal{S}^r$, such that for the set S_i of squares in S intersecting line i for $i \in \{1, \ldots, k+1\}$, any square in S_i appears on the upper or lower envelope of S_i. Then there is a sequence of $|S| - k - 1$ valid insertions starting from $F_0 = F^l$, leading to $F_1, \ldots, F_{|S|-k-1} = F^r$ such that for any square $s \in S$, there is a front F_j containing s.*

Proof (Sketch). By the previous lemma, we can insert the squares intersecting each horizontal line in order of increasing x-coordinate. However, we should interleave the sequences of the different lines. For any $i = 1, \ldots, k$, consider the squares appearing on the upper envelope of S_i and the lower envelope of S_{i+1}. Order these squares according to the x-coordinate from which they appear on the upper envelope of S_i or on the lower envelope of S_{i+1} respectively. Combining these two orders, we can extend this to an order by which to insert the squares of S. We can then prove that the j-th square s_j according to this order is insertable into F_{j-1} and that after inserting s_j, all squares $s_{j'}$ with $j' > j$ are still insertable.   □

The next lemmas follow from the coverage constraints on valid insertions.

Lemma 5. *Let S be any smallest subset of $\overline{S}$ containing $\mathcal{S}^l \cup \mathcal{S}^r$ and covering all points in $\mathcal{P}$. Then there is a sequence of $|S| - k - 1$ valid insertions starting from F^l, leading to $F_1, \ldots, F_{|S|-k-1} = F^r$ such that for any square $s \in S$, there is a front F_j containing s.*

Lemma 6. *Let $l \geq 0$. Then any sequence of $l + k + 1$ valid insertions starting from F^l and resulting in F^r corresponds to a set $S \subseteq \mathcal{S}$ of cardinality l covering all points in $\mathcal{P}$.*

Proof (of Theorem 1). Construct a directed graph G with $V(G)$ equal to the set of all fronts and a directed edge from front F to front F' if F' can be obtained from F by a single valid insertion. From the definition of a front, $|V(G)| = O(|\overline{S}|^{4k+3})$. As each front allows for at most $4|\overline{S}|$ valid insertions, $|E(G)| = O(|\overline{S}|^{4k+4})$. Because the validity of an insertion can be checked in $O(|\mathcal{P}|)$ time, G can be constructed in $O(|\overline{S}|^{4k+4}|\mathcal{P}|)$ time.

From Lemma 5 and 6, a shortest path in G from F^l to F^r corresponds to a minimum subset of $\mathcal{S}$ covering all points in $\mathcal{P}$. Then a shortest path can be found in $O(|E(G)|) = O(|\overline{S}|^{4k+4})$ time. Observe that $|\overline{S}| = |\mathcal{S}| + |\mathcal{S}^l| + |\mathcal{S}^r| \leq 3|\mathcal{S}|$, because if no square intersects a certain line, we may ignore it. The running time of the algorithm is $O((3|\mathcal{S}|)^{4k+4}|\mathcal{P}|)$.   □

Using Theorem 1 with the shifting technique, we get a PTAS for Geometric Set Cover on unit squares. The proof of this theorem can be found in [31].

Theorem 2. *There is a PTAS for Geometric Set Cover on unit squares.*

3 Geometric Budgeted Maximum Coverage

The above PTAS easily extends to the weighted case of Geometric Set Cover, by weighting the graph constructed in the proof of Theorem 1. We can however extend to the more general budgeted case as well.

Let $\mathcal{S}$ be a set of unit squares, $\mathcal{P}$ a set of points, c a cost function over $\mathcal{S}$, p a nonnegative profit function over $\mathcal{P}$, and B a budget. Let $p_{\max}$ denote the maximum profit of any single point. We define the function $cov(s)$ as the set of points in $\mathcal{P}$ covered by a square $s \in \mathcal{S}$. This notation extends to $cov(S)$ for a set $S \subseteq \mathcal{S}$. Abusing notation, we will use $p(S)$ to denote $p(cov(S))$.

Theorem 3. *In Geometric Budgeted Maximum Coverage on a set of unit squares $\mathcal{S}$ where all points are inside $k-1$ consecutive height 1 horizontal slabs and all profits are positive integers, one can find a cheapest set of profit at least r (if one exists) for all $0 \leq r \leq |\mathcal{P}| \cdot p_{\max}$ in time $O((3|\mathcal{S}|)^{4k}(|\mathcal{P}| \cdot p_{\max}))$.*

Proof. We modify the algorithm described above. Assume the cost of squares in $\mathcal{S}^l \cup \mathcal{S}^r$ to be zero. Remove the coverage constraints from the four definitions of insertable. Then, as in the proof of Theorem 1, we construct a directed graph G with $V(G)$ equal to the set of all fronts and an edge from F to F' if F' can be obtained from F by a single valid insertion.

We assign two weights, a *cost* and a *profit*, to each edge of this graph G. For any edge in $E(G)$ from some front F to a front F' that represents the insertion of a square s, the cost of the edge is the cost $c(s)$ of s and the profit of the edge is the total profit of the points covered by the insertion of s. For example, for an upper-insertion of a square s intersecting line i, the profit of the edge is the total profit of the points covered by u_i or l_{i+1} in $[x_i, x_i'] \times [i, i+1]$.

Now the sum of the profits of the edges on a F^l–F^r path equals the profit of the solution corresponding to this path. Moreover, the sum of the costs of the edges of the path equals the cost of that solution. Hence we aim to find for any $0 \leq r \leq |\mathcal{P}| \cdot p_{\max}$ a lightest path (with respect to edge costs) of total edge profit at least r. A straightforward dynamic programming algorithm for this problem takes $O(|E(G)| \cdot |\mathcal{P}| \cdot p_{\max}) = O((3|\mathcal{S}|)^{4k}(|\mathcal{P}| \cdot p_{\max}))$ time. $\qquad\square$

We now apply the shifting technique and scaling to obtain a PTAS. First assume integer profits. For each integer $0 \leq a \leq k-1$, let N_a denote the set of points between lines $y = bk + a$ and $y = bk + a + 1$ for any $b \in \mathbb{Z}$. Moreover, for any $b \in \mathbb{Z}$, let $\mathcal{P}_a^b$ be the set of points between lines $y = bk + a + 1$ and $y = (b+1)k + a$.

For any $0 \leq r \leq |\mathcal{P}| \cdot p_{\max}$, let $C_a^b(r)$ denote the set returned by the algorithm of Theorem 3, applied on $\mathcal{S}$ and $\mathcal{P}_a^b$, attaining profit at least r. We assume that $c(C_a^b(r)) = \infty$ if profit at least r cannot be attained.

Let the nonempty sets $\mathcal{P}_a^b$ be numbered $\mathcal{P}_a^0, \ldots, \mathcal{P}_a^{l_a}$ in an arbitrary way, and let $C_a^0, \ldots, C_a^{l_a}$ be the corresponding solutions. Define

$$\mathsf{s}_a(0, r) = c(C_a^0(r))$$
$$\mathsf{s}_a(b, r) = \min_{0 \le r' \le r} \{c(C_a^b(r')) + \mathsf{s}_a(b - 1, r - r')\}$$

for $1 \le b \le l_a$ and $0 \le r \le |\mathcal{P}| \cdot p_{\max}$. Observe that computing s_a can be done in $O(|\mathcal{P}| \cdot (|\mathcal{P}| \cdot p_{\max})^2)$ time.

Let C_a denote a set attaining $\max_{0 \le r \le |\mathcal{P}| \cdot p_{\max}} \{r \mid \mathsf{s}_a(l_a, r) \le B\}$ and let $C_{\max}$ denote a most profitable such set. By definition, $c(C_{\max}) \le B$.

Lemma 7. $p(C_{\max}) \ge (1 - 1/k) \cdot p(OPT)$, where OPT is an optimal solution.

Proof. Let $\mathcal{S}_a^b$ denote the set of squares in $\mathcal{S}$ covering at least one point in $\mathcal{P}_a^b$. Then it can be easily seen that $c\left(C_a^b\left(p\left(cov\left(OPT \cap \mathcal{S}_a^b\right) \cap \mathcal{P}_a^b\right)\right)\right) \le c(OPT \cap \mathcal{S}_a^b)$ for any $0 \le a \le k - 1$ and $0 \le b \le l_a$. Because for fixed a the sets $\mathcal{S}_a^b$ are pairwise disjoint, $\sum_{b=0}^{l_a} c(OPT \cap \mathcal{S}_a^b) \le B$. Then it follows from the definition of s and by induction that $p(C_a) \ge \sum_{b=0}^{l_a} p(cov(OPT \cap \mathcal{S}_a^b) \cap \mathcal{P}_a^b)$. Since we can show that $\sum_{b=0}^{l_a} p(cov(OPT \cap \mathcal{S}_a^b) \cap \mathcal{P}_a^b) = p(OPT) - p(cov(OPT) \cap N_a)$ and any point is in N_a for precisely one value of a, $k \cdot p(C_{\max}) \ge \sum_{a=0}^{k-1} p(C_a) \ge (k - 1) \cdot p(OPT)$. It follows immediately that $p(C_{\max}) \ge (1 - 1/k) \cdot p(OPT)$. $\square$

Using scaling for noninteger profits, we obtain a PTAS. Details can be found in [31].

Theorem 4. *There is a PTAS for Geometric Budgeted Maximum Coverage on unit squares.*

4   Optimality and Relation to Domination

Geometric Set Cover and the geometric version of Minimum Dominating Set are closely related. An instance of Minimum Dominating Set on an intersection graph of unit squares can be easily transformed into an instance of Geometric Set Cover on unit squares [28]. Then the following is immediate from Theorem 2 and the remarks at the beginning of Section 3.

Theorem 5. *There is a PTAS for Minimum-Weight Dominating Set on intersection graphs of unit squares.*

Theorem 5 is the first PTAS for Minimum-Weight Dominating Set on intersection graphs of two-dimensional objects. Another consequence of the above reduction from Minimum Dominating Set on unit square graphs to Geometric Set Cover on unit squares is the following. Recall that the exponential time hypothesis (ETH) states that n-variable 3SAT cannot be decided in $2^{o(n)}$ time.

Theorem 6. *If there exist constants $\delta \ge 1$, $0 < \beta < 1$ such that Geometric Set Cover or Geometric Budgeted Maximum Coverage on n unit squares has a PTAS with running time $2^{O(1/\epsilon)^\delta} n^{O(1/\epsilon)^{1-\beta}}$, then ETH is false.*

This holds, as Marx [27] showed that Minimum Dominating Set on intersection graphs of unit squares cannot have such a PTAS. Similarly, one can show from Marx [26] that Geometric Set Cover and Geometric Budgeted Maximum Coverage on unit squares have no EPTAS.

Theorem 7. *Geometric Set Cover and Geometric Budgeted Maximum Coverage on unit squares cannot have an EPTAS, unless FPT=W[1].*

This is an indication that one cannot hope to improve the running time of the algorithms of Theorems 2 and 4.

5 Hardness of Approximation

Not many explicit inapproximability results for Geometric Set Cover problems can be found in the literature. Our approximation scheme settles the approximability of Geometric Set Cover on unit squares. In this section, we adapt known results for related problems to give several hardness results for more general shapes. A convex subset s of $\mathbb{R}^2$ is α-*fat* for some $\alpha \geq 1$ if the ratio between the radii of the smallest disk enclosing s and the largest disk inscribed in s is at most α [8].

Theorem 8. *Geometric Set Cover is not approximable within $(1 - \epsilon)\ln n$ for any $\epsilon > 0$, unless $NP \subset DTIME(n^{O(\log \log n)})$, on convex polygons, translated copies of a single polygon, rotated copies of a single convex polygon, and α-fat objects for any $\alpha > 1$, where n is the number of points,*

Theorem 9. *Geometric Set Cover is APX-hard on convex polygons with $r \geq 4$ corners, α-fat objects of constant description complexity for any $\alpha > 1$, axis-parallel rectangles, and ellipses.*

Theorem 8 and Theorem 9 can be proved using constructions where points are arranged on a line or circle and the objects can cover arbitrary subsets (of bounded size, in case of Theorem 9) of these points. On axis-parallel rectangles and ellipses we need a more elaborate construction. The ideas are similar to ones used for the geometric version of Minimum Dominating Set [10]. See [31] for details. We remark that Har-Peled [15] recently showed that Geometric Set Cover is even APX-hard on fat convex polygons with $r \geq 3$ corners.

Using ideas from Khuller et al. [22], one can obtain the following.

Corollary 1. *Geometric Budgeted Maximum Coverage is not approximable with ratio better than $(1 - 1/e)$, unless $NP \subset DTIME(n^{O(\log \log n)})$, on convex polygons, translated copies of a single polygon, rotated copies of a single convex polygon, and α-fat objects for any $\alpha > 1$.*

6 Conclusions

We have given the first PTAS for Weighted Geometric Set Cover, in the case of axis-parallel unit squares. The scheme extends to Geometric Budgeted Maximum

Coverage. Moreover, we presented evidence that one cannot hope to improve on the running time of these algorithms. This settles the approximability of Geometric Set Cover on unit squares.

Many problems surrounding Weighted Geometric Set Cover remain open however. In particular, the question of a PTAS on (unit) disks or arbitrary squares is very interesting. The techniques in this paper seem insufficient to deal with these problems and probably completely different insight is required. In general, it is an interesting question for which objects (Weighted) Geometric Set Cover can still be approximated well. The hardness results of this paper however set clear limits to its approximability.

References

1. Ambühl, C., Erlebach, T., Mihalák, M., Nunkesser, M.: Constant-Factor Approximation for Minimum-Weight (Connected) Dominating Sets in Unit Disk Graphs. In: Díaz, J., Jansen, K., Rolim, J.D.P., Zwick, U. (eds.) APPROX 2006 and RANDOM 2006. LNCS, vol. 4110, pp. 3–14. Springer, Heidelberg (2006)
2. Brönnimann, H., Goodrich, M.T.: Almost Optimal Set Covers in Finite VC-Dimension. Discrete Comput. Geom. 14(1), 463–479 (1995)
3. Carmi, P., Katz, M.J., Lev-Tov, N.: Covering Points by Unit Disks of Fixed Location. In: Tokuyama, T. (ed.) ISAAC 2007. LNCS, vol. 4835, pp. 644–655. Springer, Heidelberg (2007)
4. Chvátal, V.: A Greedy Heuristic for the Set-Covering Problem. Math. Oper. Research 4(3), 233–235 (1979)
5. Clarkson, K.L., Varadarajan, K.R.: Improved Approximation Algorithms for Geometric Set Cover. Discrete Comput. Geom. 37(1), 43–58 (2007)
6. Călinescu, G., Mandoiu, I.I., Wan, P.J., Zelikovsky, A.: Selecting forwarding neighbors in wireless ad hoc networks. In: Proc. DIAL-M 2001, pp. 34–43. ACM, New York (2001)
7. Dai, D., Yu, C.: A $5+\epsilon$-approximation algorithm for minimum weighted dominating set in unit disk graph. Theoret. Comput. Sci. 410(8-10), 756–765 (2009)
8. Efrat, A., Sharir, M.: The Complexity of the Union of Fat Objects in the Plane. Discrete Comput. Geom. 23(2), 171–189 (2000)
9. Erlebach, T., Mihalák, M.: A (4+epsilon)-Approximation for the Minimum-Weight Dominating Set Problem in Unit Disk Graphs. In: Bampis, E., Jansen, K. (eds.) WAOA 2009. LNCS, vol. 5893, pp. 135–146. Springer, Heidelberg (2010)
10. Erlebach, T., van Leeuwen, E.J.: Domination in Geometric Intersection Graphs. In: Laber, E.S., Bornstein, C.F., Nogueira, L.T., Faria, L. (eds.) LATIN 2008. LNCS, vol. 4957, pp. 747–758. Springer, Heidelberg (2008)
11. Feige, U.: A Threshold of $\ln n$ for Approximating Set Cover. J. ACM 45(4), 634–652 (1998)
12. Fowler, R.J., Paterson, M.S., Tanimoto, S.L.: Optimal Packing and Covering in the Plane are NP-Complete. Inform. Process. Lett. 12(3), 133–137 (1981)
13. Glaßer, C., Reitwießner, C., Schmitz, H.: Multiobjective Disk Cover Admits a PTAS. In: Hong, S.-H., Nagamochi, H., Fukunaga, T. (eds.) ISAAC 2008. LNCS, vol. 5369, pp. 40–51. Springer, Heidelberg (2008)
14. Gonzalez, T.F.: Covering a Set of Points in Multidimensional Space. Inform. Process. Lett. 40(4), 181–188 (1991)

15. Har-Peled, S.: Being Fat and Friendly is Not Enough. arXiv:0908.2369v1 [cs.CG] (2009)
16. Hochbaum, D.S.: Approximating Covering and Packing Problems: Set Cover, Vertex Cover, Independent Set, and Related Problems. In: Hochbaum, D.S. (ed.) Approximation Algorithms for NP-hard Problems, pp. 46–93. PWS Publishing Company, Boston (1997)
17. Hochbaum, D.S., Maass, W.: Approximation Schemes for Covering and Packing Problems in Image Processing and VLSI. J. ACM 32(1), 130–136 (1985)
18. Hochbaum, D.S., Pathria, A.: Analysis of the greedy approach in covering problems (1994) (unpublished manuscript)
19. Huang, Y., Gao, X., Zhang, Z., Wu, W.: A Better Constant-Factor Approximation for Weighted Dominating Set in Unit Disk Graph. J. Combin. Optim. 18(2), 179–194 (2009)
20. Johnson, D.S.: Approximation algorithms for combinatorial problems. J. Comput. System Sci. 9(3), 256–278 (1974)
21. Johnson, D.S.: The NP-Completeness Column: An Ongoing Guide. J. Algorithms 3(2), 182–195 (1982)
22. Khuller, S., Moss, A., Naor, J.S.: The Budgeted Maximum Coverage Problem. Inform. Process. Lett. 70(1), 39–45 (1999)
23. Laue, S.: Geometric Set Cover and Hitting Sets for Polytopes in $\mathbb{R}^3$. In: Albers, S., Weil, P. (eds.) Proc. STACS 2008, LIPIcs, vol. 1, pp. 479–490. Schloss Dagstuhl – Leibniz-Zentrum fuer Informatik, Dagstuhl (2008)
24. Liao, C., Hu, S.: Polynomial time approximation schemes for minimum disk cover problems. J. Global Optim. (2009), doi: 10.1007/s10878-009-9216-y
25. Lovász, L.: On the ratio of optimal integral and fractional covers. Discrete Math. 13(4), 383–390 (1975)
26. Marx, D.: Parameterized Complexity of Independence and Domination Problems on Geometric Graphs. In: Bodlaender, H.L., Langston, M.A. (eds.) IWPEC 2006. LNCS, vol. 4169, pp. 154–165. Springer, Heidelberg (2006)
27. Marx, D.: On the Optimality of Planar and Geometric Approximation Schemes. In: Proc. FOCS 2007, pp. 338–348. IEEE, Los Alamitos (2007)
28. Mihalák, M.: Optimization Problems in Communication Networks, PhD thesis, Univ. Leicester (2006)
29. Mustafa, N., Ray, S.: PTAS for Geometric Hitting Set Problems via Local Search. In: Proc. SoCG 2009, pp. 17–22. ACM, New York (2009)
30. Narayanappa, S., Vojtěchovský, P.: An Improved Approximation Factor For The Unit Disk Covering Problem. In: Proc. CCCG 2006 (2006)
31. van Leeuwen, E.J.: Optimization and Approximation on Systems of Geometric Objects, PhD thesis, University of Amsterdam (2009)
32. Varadarajan, K.: Weighted Geometric Set Cover via Quasi-Uniform Sampling. In: Proc. STOC 2010, pp. 641–648. ACM, New York (2010)
33. Vohra, R., Hall, N.G.: A Probabilistic Analysis of the Maximal Covering Location Problem. Discrete Appl. Math. 43(2), 175–183 (1993)
34. Zou, F., Wang, Y., Xu, X.-H., Li, X., Du, H., Wan, P., Wu, W.: New approximations for minimum-weighted dominating sets and minimum-weighted connected dominating sets on unit disk graphs. Theoret. Comput. Sci. (to appear)

Improved Lower Bounds for the Universal and *a priori* TSP

Igor Gorodezky[1,*], Robert D. Kleinberg[2,**],
David B. Shmoys[3,***], and Gwen Spencer[4,†]

[1] Center for Applied Mathematics, Cornell University, Ithaca, NY 14853, and
Palantir Technologies, Palo Alto, CA 94301
igor@cam.cornell.edu
[2] Dept. of Computer Science, Cornell University, Ithaca, NY 14853
rdk@cs.cornell.edu
[3] School of ORIE and Dept. of Computer Science, Cornell University,
Ithaca, NY 14853
shmoys@cs.cornell.edu
[4] School of ORIE, Cornell University, Ithaca, NY 14853
gms39@cornell.edu

Abstract. We consider two partial-information generalizations of the metric traveling salesman problem (TSP) in which the task is to produce a total ordering of a given metric space that performs well for a subset of the space that is not known in advance. In the universal TSP, the subset is chosen adversarially, and in the *a priori* TSP it is chosen probabilistically. Both the universal and *a priori* TSP have been studied since the mid-80's, starting with the work of Bartholdi & Platzman and Jaillet, respectively. We prove a lower bound of $\Omega(\log n)$ for the universal TSP by bounding the competitive ratio of shortest-path metrics on Ramanujan graphs, which improves on the previous best bound of Hajiaghayi, Kleinberg & Leighton, who showed that the competitive ratio of the $n \times n$ grid is $\Omega(\sqrt[6]{\log n / \log \log n})$. Furthermore, we show that for a large class of combinatorial optimization problems that includes TSP, a bound for the universal problem implies a matching bound on the approximation ratio achievable by deterministic algorithms for the corresponding black-box *a priori* problem. As a consequence, our lower bound of $\Omega(\log n)$ for the universal TSP implies a matching lower bound for the black-box *a priori* TSP.

* Supported by an NSF Graduate Research Fellowship.
** Supported by NSF grants no. CCF-0643934 and CCF-0729102, a grant from the Air Force Office of Scientific Research, a Microsoft Research New Faculty Fellowship, and an Alfred P. Sloan Foundation Fellowship.
*** Supported in part by NSF under grants no. CCR-0635121, DMS-0732196, CCF-0832782.
† Supported in part by NSF under a Graduate Research Fellowship and grants no. CCR-0635121, DMS-0732196, CCF-0832782.

M. Serna et al. (Eds.): APPROX and RANDOM 2010, LNCS 6302, pp. 178–191, 2010.

1   Introduction

A delivery person has a fixed list of potential clients that is known in advance, but on each day only a subset of the clients must be served. Rather than re-optimizing the delivery route every day, the delivery person might seek to minimize distance traveled using the following heuristic: decide in advance on a master tour of the entire client list, and on each particular day visit the set of active clients in the order that they would be served by the master tour. How well does this heuristic perform? That is, given a list of clients residing in an arbitrary finite metric space, is there a master tour of the list such that the tour induced on an active subset of clients by shortcutting the master tour tends to be close to optimal? Naturally, the answer depends on how active subsets are generated; the two canonical models are the *adversarial* and the *probabilistic*.

In the adversarial model, the master tour is announced and an adversary chooses an active subset S that maximizes the ratio of the length of the induced tour to the length of the optimal tour of S. In this case the delivery person faces the *universal traveling salesman problem* (TSP). If for a given master tour there is a ρ such that for every active subset, the length of the induced tour is guaranteed to be no more than ρ times the length of the optimal tour, we say that this master tour achieves *competitive ratio ρ*.

Alternately, in the probabilistic model there is a fixed probability distribution over subsets of clients, and the delivery person seeks a master tour that minimizes the expected cost of the induced tour. This is known as the *a priori* TSP. In this paper we will focus on the *black-box a priori* model in which information about the distribution on active client sets is available only through calls to a black-box oracle. Note that both the universal and *a priori* variants of TSP are NP-hard since they are both generalizations of metric TSP.

In this paper we give improved lower bounds for the universal and *a priori* TSP. In particular, we exhibit a family of metric spaces on which the competitive ratio of any solution to the universal TSP is $\Omega(\log n)$, improving on the previous bound of $\Omega(\sqrt[6]{\log n / \log \log n})$ (proved for a different family of metric spaces; details below). We then extend this bound to the *a priori* TSP by showing how to translate a universal lower bound into a matching lower bound on approximation for a large class of *a priori* optimization problems (that includes TSP).

Prior work. The universal TSP was originally motivated by the need for an easy and efficient vehicle routing system for the "Meals on Wheels" program in Atlanta, GA. This problem was considered in Bartholdi et al. [3], who phrased it as the universal TSP on a finite subset of the plane and suggested a heuristic based on space-filling curves. Bartholdi & Platzman proved in [2] that this heuristic yields master tours with a competitive ratio of $O(\log n)$ (not only on the plane but in Euclidean space in general; see also [18]).

Bertsimas & Grigni subsequently showed in [4] that $O(\log n)$ is asymptotically tight for this particular method, and made the general conjecture that $\Omega(\log n)$ is a lower bound for the universal TSP on the plane. Working towards this conjecture, Hajiaghayi, Kleinberg & Leighton showed in [12] that any master tour of the vertices of the $n \times n$ grid has its competitive ratio bounded by $\Omega(\sqrt[6]{\log n / \log \log n})$. The main result of this paper (Theorem 1) is an improved lower bound of $\Omega(\log n)$ on the competitive ratio for the universal TSP, though for a different, non-planar family of metrics (see the discussion below).

Recently, there has been significant progress on algorithms for the universal TSP; inspired by [2], Jia et al. formulate in [14] the universal TSP for general finite metric spaces and give an algorithm that produces, given a metric space on n points, a master tour with competitive ratio $O(\log^4 n / \log \log n)$. Their algorithm achieves a competitive ratio of $O(\log n)$ on doubling metrics (which include constant-dimensional Euclidean metrics). At present, the best universal TSP algorithm is due to Gupta et al. [10] (using ideas of [7]) and produces a master tour with competitive ratio $O(\log^2 n)$ on n-point metric spaces.

There has been considerably less work on the *a priori* TSP. The general notion of *a priori* optimization was introduced in Jaillet's PhD thesis [13] (see also [5]). The current best algorithmic results for the black-box *a priori* TSP are a 4-approximation due to Shmoys & Talwar [20] for the case of *independent activation* (in which points appear in the active set independently), and a randomized $O(\log n)$-approximation algorithm due to Shmoys & Schalekamp [19] for the general problem. The former algorithm depends deterministically on a single black-box sample while the latter is entirely distribution-free (that is, it requires no samples). We observe that any algorithm for the universal TSP with competitive ratio ρ can be made into a distribution-free algorithm for the *a priori* TSP that achieves a ρ-approximation. Save for the trivial observation that an inapproximability result for the classical metric TSP translates to a matching lower bound for the *a priori* TSP, no lower bounds for the *a priori* TSP were known prior to this paper: a corollary to our second main result (Corollary 1) is a lower bound of $\Omega(\log n)$ on the approximation ratio of any deterministic algorithm for the black-box *a priori* TSP.

A great deal of recent work in approximation algorithms has been devoted to the design and analysis of optimization algorithms that must produce solutions when the input is only partially known, or is determined stochastically. Examples include online algorithms, stochastic optimization (see [11]), the work of Jia et al. [14] on universal optimization, as well as recent results on oblivious routing (see [10] and the references within). Our work on universal and *a priori* variants of the classical metric TSP fits squarely within this conceptual framework. Moreover, though we focus on the TSP, there are many other classical optimization problems which yield natural and interesting universal and *a priori* variants. The second main result of the paper provides a direct connection between worst-case, universal lower bounds and average-case, *a priori* lower bounds, elucidating the connection between these two problem variants not only for the TSP but for a large class of natural combinatorial optimization problems.

Main results and techniques. Recall that given a master tour of a metric space X, its competitive ratio is the maximum ratio, over all $S \subseteq X$, of the length of the induced tour of S to the length of the optimum tour of S. The competitive ratio of the universal TSP on X is the minimum of the competitive ratio over all master tours of X. In our first main result (Theorem 1) we exhibit a family of metric spaces (in particular, shortest-path metrics on Ramanujan graphs) for which the competitive ratio of the universal TSP is $\Omega(\log n)$. This proves a conjecture of Hajiaghayi, Kleinberg & Leighton from [12].

Our proof of this result is probabilistic and proceeds as follows. Let X be the shortest-path metric on a Ramanujan graph, and fix some master tour of X. We give a simple procedure for producing $S \subseteq X$ such that the length of the induced tour of S is at least $\Omega(\log n)$ times the length of the optimal tour: take a random walk on the graph and let S be the set of distinct vertices encountered by the walk. We show that with positive probability a constant fraction of the vertices in S are $\Omega(\log n)$-far from their successors with respect to the tour of S induced by the master tour. Since the length of an optimal tour of S is $O(|S|)$, this proves the desired property. Our proof makes essential use of the spectral properties of Ramanujan graphs.

Our second main result (Theorem 3) shows that for a large class of natural optimization problems that includes TSP, a lower bound on the competitive ratio of the universal version of the problem implies a matching bound on the approximation ratio of a deterministic algorithm for the black-box *a priori* version. Thus our lower bound implies that no deterministic approximation algorithm for the black-box *a priori* TSP has approximation ratio $o(\log n)$. This limit on the power of deterministic algorithms provides an interesting complement to positive results in the existing literature ([19],[20],[1]). For example, if randomization is allowed then the distribution-free algorithm of Shmoys & Schalekamp [19] (using ideas from [7]) achieves an approximation of $O(\log n)$ for the *a priori* TSP, but no deterministic algorithm that is allowed even a polynomial number of samples from the distribution can achieve a better asymptotic approximation. For the full version of this paper, see [9].

2   Universal and *a priori* TSP

Given a metric space (X, δ) with $|X| = n$, a *master tour* is a total ordering of the points of X, which we write as a bijection $\tau : [n] \to X$. Given $x \in X$ we will refer to $\tau^{-1}(x)$, the preimage of x under τ, as its *index*. The total length of a master tour τ will be written $\delta(\tau)$: $\delta(\tau) = \sum_{i=1}^{n} \delta(\tau(i), \tau(i+1))$, where we identify element $\tau(n+1)$ with element $\tau(1)$. In words, $\delta(\tau)$ is the total distance traveled when traversing X in the order given by τ (including returning, at the end of the tour, from $\tau(n)$ back to $\tau(1)$).

If $S \subseteq X$ then we write $\tau|_S$ for the tour of S obtained by *shortcutting* τ; this tour starts at the element of S of smallest index, travels to the element of S of second smallest index, etc., and at the end returns to the element of smallest index. We will write $\delta(\tau|_S)$ for the total length of this tour. Given $S \subseteq X$ let $\text{opt}(S)$ denote some optimal tour of (S, δ), i.e. a tour of S that minimizes total distance traveled. The length of an optimal tour of S will be written $\delta(\text{opt}(S))$. The object $\text{opt}(S)$ is clearly not well-defined as there could exist multiple optimal tours, but we will ignore this technicality since our analysis is only concerned with the quantity $\delta(\text{opt}(S))$.

We define $\rho(X, \delta, \tau)$ to be the competitive ratio of a master tour τ of (X, δ), $\rho(X, \delta, \tau) = \max_{S \subseteq X}(\delta(\tau|_S)/\delta(\mathrm{opt}(S)))$, and define $\rho(X)$ to be the competitive ratio of the universal TSP on (X, δ), $\rho(X) = \min_\tau \rho(X, \delta, \tau) = \min_\tau \max_{S \subseteq X} \frac{\delta(\tau|_S)}{\delta(\mathrm{opt}(S))}$.

Notation: In this paper, (X, δ) will usually be the metric space associated with an undirected graph G; that is, X is the vertex set of G and δ is the shortest-path metric on G (the edges of G will always be unit-weighted). In this case, we write $\rho(G)$ instead of $\rho(X)$. The universal TSP has as its input a finite metric space (X, δ) and asks to produce a master tour τ that minimizes $\rho(X, \delta, \tau)$. In the *a priori* TSP, we are given (X, δ) with a probability distribution D over 2^X. The problem asks to produce a master tour τ that minimizes $\mathbb{E}_D[\delta(\tau|_S)]$, the expected length of the induced tour of S with respect to D. In this paper we consider the black-box model of the *a priori* TSP in which no information about D is part of the input, but an algorithm may sample subsets of X according to D from a black box in polynomial time.

3   A Lower Bound for the Universal TSP

In this section we prove a lower bound for the universal TSP on Ramanujan graphs. In order to define these precisely and formally state our result, we must make a few additional definitions. Given an $n \times n$ symmetric matrix M, we denote its (real) eigenvalues, in decreasing order, by $\lambda_i(M)$ for $i = 1, \ldots, n$. Recall that if G is an undirected d-regular graph, A is its adjacency matrix, and $M = \frac{1}{d}A$ is the transition matrix for a random walk on G, then $\lambda_1(M) = 1$. Let us refer to the eigenvalues of the transition matrix M as the eigenvalues of the graph G, written $1 = \lambda_1(G) \geq \lambda_2(G) \geq \cdots \geq \lambda_n(G)$. We say that a d-regular graph G is a *Ramanujan graph* if

$$|\lambda_i(G)| \leq 2\sqrt{d-1}/d \qquad (3.1)$$

for $i \geq 2$ (i.e., if the Alon-Boppana bound [17] is asymptotically tight for G).

Now we may state our main result. Recall that the *girth* of G is the length of its shortest cycle.

Theorem 1. *For all sufficiently large d, $\exists n_0$ such that if G is a d-regular Ramanujan graph on $n \geq n_0$ vertices whose girth is $\geq \frac{2}{3}\log_{d-1} n$, then $\rho(G) = \Omega(\log n)$.*

Before moving on to the proof of this theorem, we note that its hypotheses are an explicit reference to the Ramanujan graphs constructed by Lubotzky, Phillips & Sarnak in [15]: for every odd prime p, there exists an infinite family of $(p+1)$-regular Ramanujan graphs whose girth is at least $\frac{2}{3}\log_p n$ (this construction was later extended by Morgenstern in [16] to allow prime powers). It is well-known that if G is a d-regular graph on n vertices then its girth can be at most $2\log_{d-1} n$ (see, for instance, Section IV.1 of [6]), so that the girth of the Lubotzky-Phillips-Sarnak graphs is very nearly as large as possible. As we will see, this large-girth property will be central to the proof of Theorem 1.

The strategy: a probabilistic proof. We will prove Theorem 1 by using the probabilistic method. Namely, in this section we will reduce the theorem to a claim about random walks on Ramanujan graphs (then prove this claim in subsequent sections). We require several lemmas analyzing random walks on graphs of large girth; these lemmas and their proofs can be found in the full version of the paper [9].

Let G be a d-regular Ramanujan graph on the vertex set V, with $|V| = n$, and let δ be the shortest-path metric on G. Given a master tour τ of the metric space (V, δ), let us use the term *bad set* for a set $S \subset V$ such that $\delta(\tau|_S)/\operatorname{opt}(S) = \Omega(\log n)$. We would like to prove that a bad set exists for every master tour of G, which would imply $\rho(G) = \Omega(\log n)$. Our strategy is as follows.

Fix a master tour τ of G, let g be the girth of G (recall that one of the assumptions in Theorem 1 is that g is large), and consider uniformly sampling a random walk of $L = 70g$ steps; that is, choose a starting vertex uniformly at random and perform a random walk for L steps. We regard an L-step random walk as a map $\mathsf{W} : [L] \to [n]$ where $\mathsf{W}(i)$ is the index of the ith vertex in the walk. Note that $\mathsf{W}(i)$ refers to an index rather than a vertex; the ith vertex of W would be written $\tau(\mathsf{W}(i))$.

Define V_W to be the set of distinct vertices encountered by W: that is, $V_\mathsf{W} = \{\tau(\mathsf{W}(i)) \mid i = 1, \dots, L\}$. It is easy to see that $\delta(\operatorname{opt}(V_\mathsf{W})) = O(L)$, since walking along W and then back again produces a tour of V_W. If we can show that there exists a V_W for which $\delta(\tau|_{V_\mathsf{W}}) = \Omega(L \log n)$ then we are done, for such a V_W is a bad set. We will prove the existence of such a V_W using probabilistic arguments. To this end, given a walk W and an index $\mathsf{W}(i)$ define the *successor* of $\mathsf{W}(i)$ (with respect to τ) to be $\mathsf{W}_{succ}(i) = \min\{\mathsf{W}(j) \mid \mathsf{W}(i) < \mathsf{W}(j)\}$. That is, $\mathsf{W}_{succ}(i)$ is the index of the vertex in V_W that follows $\mathsf{W}(i)$ when V_W is traversed by shortcutting τ (as usual, the successor of the vertex in V_W with largest index is the vertex with the smallest index).

Theorem 2. *Let G be a d-regular Ramanujan graph on n vertices with girth $g \geq \frac{2}{3} \log_{d-1} n$, fix a master tour of G, and set $L = 70g$. For all sufficiently large d there exists n_0 such that if $n \geq n_0$ and W is a uniformly sampled L-step random walk on G, then*

$$\Pr\left[\delta\big(\tau(\mathsf{W}(i)), \tau(\mathsf{W}_{succ}(i))\big) > \frac{2}{3d} \log_{d-1} n\right] = \Omega(1)$$

for every $i = 1, 2, \dots, g/2$, where δ is the shortest-path metric on G.

Before proving this theorem, let us see how it implies the existence of a V_W that is a bad set. Given an L-step walk W, let $X(\mathsf{W})$ be the number of events of the form

$$\delta\big(\tau(\mathsf{W}(i)), \tau(\mathsf{W}_{succ}(i))\big) > \frac{2}{3d} \log_{d-1} n, \text{ and}$$

$$\text{the vertex } \tau(\mathsf{W}(i)) \text{ is visited at most 140 times by } \mathsf{W}. \qquad (3.2)$$

The key observation is that we can bound $\delta(\tau|_{V_\mathsf{W}})$ from below by

$$\delta(\tau|_{V_\mathsf{W}}) \geq \left(\frac{X(\mathsf{W})}{140}\right)\left(\frac{2}{3d}\log_{d-1} n\right) = \left(\frac{1}{210d\log(d-1)}\right)X(\mathsf{W})\log n$$

because traversing V_W according to τ means making at least $X(\mathsf{W})/140$ trips of length at least $\frac{2}{3d}\log_{d-1} n$. We will use Theorem 2 to prove the existence of a W such that $X(\mathsf{W}) = \Omega(L)$, which would prove the existence of a bad V_W.

For brevity, let us write the events "$\delta\big(\tau(\mathsf{W}(i)), \tau(\mathsf{W}_{succ}(i))\big) > \frac{2}{3d}\log_{d-1} n$" and "$\tau(\mathsf{W}(i))$ is visited at most 140 times by W" as E_i and E_i', respectively. Now, $\mathbb{E}[X(\mathsf{W})]$ is the sum of the probabilities of $E_i \cap E_i'$:

$$\mathbb{E}[X(\mathsf{W})] = \sum_{i=1}^{L} \Pr[E_i \cap E_i'] \geq \sum_{i=1}^{g/2} \Pr[E_i \cap E_i']. \tag{3.3}$$

First note that $\Pr[E_i \cap E_i'] \geq \Pr[E_i] - \Pr[\overline{E_i'}]$. Theorem 2 says that there exists a constant A such that $\Pr[E_i] \geq A$ when $i \leq g/2$. It is proved in the full paper that $\Pr[\overline{E_i'}]$ can be made arbitrarily small by taking d and then n large enough (hence this requirement in the statement of Theorem 1). Let us take d and n large enough so that $\Pr[\overline{E_i'}] \leq A/2$. In light of all this, equation (3.3) now gives

$$\mathbb{E}[X(\mathsf{W})] \geq \sum_{i=1}^{g/2}\left(\Pr[E_i] - \Pr[\overline{E_i'}]\right) \geq \frac{A}{4}g = \frac{A}{280}L.$$

Since $\mathbb{E}[X(\mathsf{W})] = \Omega(L)$, there must exist a walk $\overline{\mathsf{W}}$ for which $X(\overline{\mathsf{W}}) = \Omega(L)$, meaning the event (3.2) holds for a constant fraction of the L steps of the walk. As noted in the discussion above, this implies $\delta(\tau|_{V_{\overline{\mathsf{W}}}}) = \Omega(L\log n)$, which shows that $V_{\overline{\mathsf{W}}}$ is a bad set. Thus we find that Theorem 2 implies Theorem 1, so it remains to prove the former.

Proving Theorem 2. This section is dominated by technical arguments concerning random walks on graphs. In order to present these as clearly as possible, we introduce some notation. Recall that Theorem 2 is a statement about a random walk W on a Ramanujan graph; we shall reserve this notation for this particular random walk, owing to its central role in the proof of Theorem 1 (as discussed in Section 3). Our proof of Theorem 2, however, will require general lemmas about random walks on regular graphs, which we will later apply to W, or more often sub-walks of W. In these lemmas we will use the notation w for an arbitrary random walk. As before, $w(i)$ is the index of the vertex at step i of w, and $\tau(w(i))$, where τ will always be some fixed master tour, is the vertex at step i.

One last piece of notation: given $j \in [n]$ and a positive integer $\ell \leq n$, define

$$[j : \ell]_n = \{(j+1) \bmod n, (j+2) \bmod n, \ldots, (j+\ell) \bmod n\}.$$

This notation will appear in the following context: fix a master tour τ of a graph G and consider a random walk w on G. Then $w(i)$ is the index of the vertex at step i of w, and $[w(i) : \ell]_n$ consists of the ℓ indices that follow $w(i)$ in τ, with wraparound modulo n.

The proof of Theorem 2 requires the following technical lemmas.

Lemma 1. *Let G be a d-regular, n-vertex graph such that $|\lambda_i(G)| \leq \eta$ for $i \geq 2$, say $\eta = .05$. Fix a master tour of $V(G)$, some integer k, and set $\ell = \lceil n/(32k) \rceil$. If w is a uniformly sampled k-step random walk on G, then the probability that w encounters a vertex whose index is in $[w(1) : \ell]_n$ is less than $1/4$.*

Lemma 2. *Let G be a d-regular graph, $d \geq 7$, with n vertices and girth g, and consider an L'-step random walk, with $g/2 \leq L' \leq kg/2$, for some integer $k \geq 2$, that begins at a fixed vertex v. Then the probability that*
  (i) in the first $g/2$ steps the walk leaves the g/d-ball around v, and
  (ii) the walk never returns to this g/d-ball in subsequent steps
is at least $1 - 6ke^{-\Omega(g)}$. In particular, if $g = \Omega(\log_{d-1} n)$ then the probability that (i) and (ii) both hold is at least $1 - 6kn^{-\Omega(1)/\log(d-1)}$.

The proofs of these lemmas can be found in the full version of the paper [9]. We proceed to the proof of Theorem 2.

Proof of Theorem 2. Recall the setting of Theorem 2: we have a d-regular Ramanujan graph G on n vertices whose girth is $g \geq \frac{2}{3}\log_{d-1} n$. We fix a master tour of G and uniformly sample an L-step random walk W, where $L = 70g$. We want to show that for every $i = 1, 2, \ldots, g/2$ we have

$$\Pr\left[\delta\big(\tau(W(i)), \tau(W_{succ}(i))\big) > \frac{2}{3d}\log_{d-1} n\right] = \Omega(1).$$

Fix, then, some $i \in \{1, \ldots, g/2\}$. We can regard W as two independent random walks w_1, w_2 that both begin at $\tau(W(i))$, i.e. $w_1(1) = w_2(1) = W(i)$. See Fig. 1. Note that $\tau(W(i))$ is uniformly sampled from V since W itself is a uniformly sampled walk. Let L_1, L_2 be the number of steps of w_1 and w_2, respectively. Since $i \leq g/2$ we know that $L_1 \leq g/2$ and $\frac{139}{2}g \leq L_2 \leq 70g$.

First we apply Lemma 1 to w_1 (with $k = g/2$) and then once again to the first $g/2$ steps of w_2. Observe that this requires choosing d large enough so that

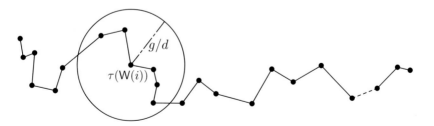

Fig. 1. The g/d-ball centered at $\tau(W(i))$. The walk w_1 is drawn to the left of $\tau(W(i))$ and the much longer walk w_2 is drawn to the right. Note that $W(i) = w_1(1) = w_2(1)$.

$|\lambda_i(G)| \leq .05$ for $i \geq 2$, which is possible since G is a Ramanujan graph (recall inequality (3.1)). Now, combining these two applications of Lemma 1 with the union bound, we find that with probability at least $1/2$ both of the following events hold:

> (i) w_1 doesn't encounter any of the $\lceil n/(16g) \rceil$ indices that follow $w_1(1)$;
> (ii) in its first $g/2$ steps, w_2 doesn't encounter these indices either.

Now, we apply Lemma 2 to the walk w_2 (use $k = 140$), and choose n large enough so that the probability of the two events

> (iii) in its first $g/2$ steps, w_2 leaves the g/d-ball around $\tau(W(i))$, and
> (iv) w_2 never returns to this g/d-ball in subsequent steps

is at least $9/10$. Putting it all together via the union bound, we find that all four of the events $(i) - (iv)$ hold with probability at least $1 - 1/2 - 1/10 = 2/5$. Next, note that there are at least $69g$ steps left in w_2 after the first $g/2$. We claim that with probability $4/5$,

> (v) one of the last $64g$ steps of w_2 contains one of the $\lceil n/(16g) \rceil$ indices that follow $w_2(1) = W(i)$.

This claim will complete the proof: if event (v) holds along with events $(i) - (iv)$ (which happens with probability at least $1 - 3/5 - 1/5 = 1/5$ by the union bound) then $W_{succ}(i)$, which is the next-largest (modulo n) index in W after $W(i)$, must be one of the $\lceil n/(16g) \rceil$ indices that follow $W(i)$. But none of these indices are encountered by W inside the g/d-ball around $\tau(W(i))$, which implies $\delta\big(\tau(W(i)), \tau(W_{succ}(i))\big) > \frac{2}{3d} \log_{d-1} n$.

The claim about (v) is motivated by the following intuition. Because our graph is a strong expander, w_2 rapidly mixes in its first $\frac{11}{2}g$ steps. Therefore the last $64g$ steps approximate a stationary random walk on G. Since the indices that follow $w_2(1)$ are a $(16g)^{-1}$-fraction of the vertices and stationary random walks resemble independent sampling, we would expect a $64g$-step, almost stationary random walk to hit this index set with large probability. This argument is formalized in the full version of the paper [9] and relies on a lemma from [8]. $\square$

4   Universal TSP Bound Implies *a priori* TSP Bound

Using a framework for abstract optimization (similar to [14], [11] for, respectively, universal and stochastic optimization) we show how to convert universal lower bounds into matching *a priori* lower bounds on deterministic algorithms for a large and natural class of optimization problems that contains TSP. In order to be applicable to the TSP, our definitions describe minimization problems, but analogous definitions and results hold for maximization problems.

Abstract optimization problems. An abstract combinatorial optimization problem is a triple $\Pi = (X, c, M)$ where X is a universe of *clients* and c, M are functions defined as follows.[1] With each $S \subseteq X$ we associate a collection of

---

[1] Regard c and M as polynomial-time algorithms that are part of the description of Π in the form of a Turing machine encoding.

feasible solutions that we denote sol(S). The function c is a *cost function* that maps, for every $S \subseteq X$, solutions in sol(S) to $\mathbb{R}$. The function M is a *restriction map* that given $S' \subseteq S$ and $F \in$ sol(S) outputs a feasible solution for S': $M(F, S') \in$ sol(S'). When $S = X$ with $F \in$ sol(X), we will use the notation $F|_{S'}$ for $M(F, S')$, and say that $F|_{S'}$ is the master solution F *restricted* to S', or that $F|_{S'}$ is the solution for S' *induced* by F.

Given $S \subseteq X$ we use opt(S) to denote a feasible solution for S of minimum cost. The *classical* version of Π specifies $S \subseteq X$ and asks to produce $F \in$ sol(S) of minimum cost. The *universal* version of Π asks for $F \in$ sol(X) that minimizes the *competitive ratio*

$$\rho(X, F) = \max_{S \subseteq X} \frac{c(F|_S)}{c(\mathrm{opt}(S))}.$$

We write $\rho(\Pi)$ for the competitive ratio of Π

$$\rho(\Pi) = \min_{F \in \mathrm{sol}(X)} \left(\max_{S \subseteq X} \frac{c(F|_S)}{c(\mathrm{opt}(S))} \right)$$

and use the notation $\mathrm{opt}_u(X)$ for a solution that is optimal with respect to this universal objective, i.e. for which $\rho(X, \mathrm{opt}_u(X)) = \rho(\Pi)$.

In the *a priori* version of Π, there is a distribution D on subsets of X. The *a priori* objective is to return $F \in$ sol(X) that minimizes $\mathbb{E}_D[F|_S]$. We denote a solution optimal with respect to this objective by $\mathrm{opt}_{ap}(X, D)$, so that if $p(S)$ is the probability given to $S \subseteq X$ by D, then

$$c(\mathrm{opt}_{ap}(X, D)) = \min_{F \in \mathrm{sol}(X)} \sum_{S \subseteq X} c(F|_S) p(S).$$

We will focus on the *black-box* model of *a priori* optimization in which no information about D is part of the input describing the *a priori* version of Π, but an algorithm may sample subsets of X according to D from a black box in polynomial time.

Notice that the definitions in this section generalize the content of Section 2: in the case of TSP, X is a set of points, feasible solutions to $S \subseteq X$ are total orderings of S, the cost function c encodes the metric on X, and M restricts tours by shortcutting.

Approximate solutions to a priori *problems.* There are two key definitions of approximate *a priori* solutions that feature in the literature. Given the *a priori* optimization problem (Π, D), an algorithm A is a *hard β-approximation* if A runs in polynomial time (in the description of Π), makes polynomially-many black-box queries, and produces a solution $F \in$ sol(X) such that

$$\Pr\left[\frac{\mathbb{E}_D[F|_S]}{\mathbb{E}_D[\mathrm{opt}_{ap}(X, D)|_S]} \leq \beta \right] > \frac{1}{2} \tag{4.1}$$

where the probability is over the black-box output as well random choices made by A.

We say that an algorithm A is a *soft β-approximation* if A runs in polynomial time, makes polynomially-many black-box queries, and produces a solution $F \in \mathrm{sol}(X)$ such that

$$\mathbb{E}\left[\frac{\mathbb{E}_D[F|_S]}{\mathbb{E}_D[\mathrm{opt}_{ap}(X, D)|_S]}\right] \leq \beta. \tag{4.2}$$

The expectation is over the black-box output as well as choices made by A.

Notice that, via Markov's inequality, any soft β-approximation to an *a priori* minimization problem is a hard 2β-approximation.

We say that A is a *deterministic* hard or soft β-approximation if it is deterministic in the black-box output. In this case the outermost probability in inequality (4.1) and the outermost expectation in inequality (4.2) depend only on the black-box output. On the other hand, we may explicitly say that A is randomized when we want to stress this fact.

Worst-case bounds to average-case bounds. In this section we show how to convert universal lower bounds into *a priori* lower bounds on deterministic hard approximations for a certain class of optimization problems.

Our result applies to the class of optimization problems $\Pi = (X, c, M)$ that satisfy three properties:

- *Non-negative costs*: For all $S \subseteq X$, $c(F) \geq 0$ for every $F \in \mathrm{sol}(S)$.
- *Trivial set*: There exists a (possibly empty) *trivial* $T \subseteq X$ such that $c(F|_T) = 0$ for all $F \in \mathrm{sol}(X)$.
- *Optimal sub-solutions are recoverable*: For any $S \subseteq X$ there exists some $F \in \mathrm{sol}(X)$ such that M maps F to $F' \in \mathrm{sol}(S)$ with $c(F') = c(\mathrm{opt}(S))$.

Problems satisfying these axioms will be called *regular*. The reader can verify that TSP is a regular problem. It is easy to see that most of the examples given in [5] are regular, including Steiner Tree and Max Cut.

Theorem 3. *If Π is a regular minimization problem with competitive ratio $\rho(\Pi) \geq \gamma$ then no deterministic hard approximation for the a priori version of Π can have approximation ratio better than γ.*

Proof. We prove the contrapositive: if there exists a deterministic hard β-approximation algorithm A for the *a priori* version of Π then $\rho(\Pi) \leq \beta$. Since A is a hard β-approximation, there exists $\alpha < 1/2$ such that

$$\Pr\left[\frac{\mathbb{E}_D[F|_S]}{\mathbb{E}_D[\mathrm{opt}_{ap}(X, D)|_S]} \leq \beta\right] \geq 1 - \alpha. \tag{4.3}$$

Let $f(|\Pi|)$ be the number of black-box samples queried by A. Choose $\delta > 0$ sufficiently small so that $1 - \alpha - \delta > 0$ and set $k = \lceil \log\left(f(|\Pi|)/\delta\right)\rceil$ so that

$$\left(1 - \frac{1}{2^k}\right)^{f(|\Pi|)} \geq 1 - \frac{f(|\Pi|)}{2^k} \geq 1 - \delta. \tag{4.4}$$

Let T be a trivial subset (as in the definition of a regular optimization problem). Since A is deterministic, when it observes problem instance Π and $f(|\Pi|)$ samples which are each T, it must return some specific $\hat{F} \in \mathrm{sol}(X)$. We claim that with probability at least $1 - \delta - \alpha > 0$, $\hat{F}$ has competitive ratio $\rho(X, \hat{F}) \le \beta$. Proving this claim will imply $\rho(X) \le \beta$ and complete the proof.

To prove the claim, construct an instance of *a priori* Π by defining a distribution D as follows: choose $R \subseteq X$ that maximizes $\frac{c(\hat{F}|_R)}{c(\mathrm{opt}(R))}$, i.e. that achieves $\rho(X, \hat{F})$, and set

$$\Pr{}_D[S] = \begin{cases} 1 - \frac{1}{2^k} & \text{if } S = T \\ \frac{1}{2^k} & \text{if } S = R. \end{cases}$$

We have chosen k so that when we run A on (Π, D) the probability that the black-box queries made by A return a non-T subset is at most δ (see inequality (4.4)). Therefore, when we run A on (Π, D) it follows from the union bound (and the fact that A is a hard β-approximation) that the following happens with probability at least $1 - \delta - \alpha > 0$: the collection of black-box samples contains no non-T subset, and A produces a β-approximate solution.

What are the consequences of this event? Since each sample observed is T, and the problem instance is Π, A must return $\hat{F}$ (since A is deterministic). Let $\mathrm{opt}_{ap}(X, D)$ denote the optimal *a priori* solution in the instance (Π, D). Since A gave a β-approximate solution in this case, we have (cf. inequality 4.3)

$$\mathbb{E}_D[c(\hat{F}|_S)] \le \beta\, \mathbb{E}_D[c(\mathrm{opt}_{ap}(X, D)|_S)]$$
$$\Longrightarrow \Pr(S = T) \cdot 0 + \Pr(S = R)c(\hat{F}|_R) \le$$
$$\beta\big(\Pr(S = T) \cdot 0 + \Pr(S = R)c(\mathrm{opt}_{ap}(X, D)|_R)\big)$$
$$\Longrightarrow c(\hat{F}|_R) \le \beta c(\mathrm{opt}_{ap}(X, D)|_R).$$

Consider $\mathrm{opt}_{ap}(X, D)|_R$: all tours perform the same on the active set T (by the second regularity axiom), so $\mathrm{opt}_{ap}(X, D)$ is any solution in $\mathrm{sol}(X)$ which has minimal cost when restricted to a solution on R via M, i.e. $c(\mathrm{opt}_{ap}(X, D)|_R) = c(\mathrm{opt}(R))$. Note that such a master solution must exist since Π is regular, hence optimal sub-solutions are recoverable.

We conclude that $c(\hat{F}|_R) \le \beta c(\mathrm{opt}(R))$, so that $c(\hat{F}|_R)/c(\mathrm{opt}(R)) \le \beta$. Since we chose R to maximize this ratio, the bound holds for every $Q \subseteq X$:

$$\frac{c(\hat{F}|_Q)}{c(\mathrm{opt}(Q))} \le \beta.$$

Thus, with probability at least $1 - \delta - \alpha > 0$, the master solution $\hat{F}$ has competitive ratio β for the universal version of Π. In particular, there exists a master solution for universal Π with this competitive ratio, hence $\rho(\Pi) \le \beta$. □

Combining Theorem 3 with our lower bound on the competitive ratio for universal TSP (Theorem 1), we immediately get the following corollary.

Corollary 1. *No deterministic (hard or soft) approximation algorithm for a priori TSP has approximation ratio $o(\log n)$.*

Deterministic vs. randomized approximations. We can use Theorem 3 to translate universal lower bounds into lower bounds on deterministic *a priori* approximations for other regular optimization problems. Applying the theorem to the Steiner Tree problem illustrates that randomized *a priori* approximations can sometimes perform dramatically better than deterministic ones.

In the *a priori* Steiner Tree problem, we have a graph $G = (V, E)$ and a distribution D over 2^V. A feasible solution for $U \subseteq V$ is a subtree of G that spans U. The solution induced by $U' \subseteq U$ and some $T \in \text{sol}(U)$ is the smallest connected subtree of T that spans U'. The *a priori* objective is to minimize the expected cost of the induced subtree. To bound the competitive ratio of universal Steiner tree, let G be the n-cycle with all edges of unit length. A master solution T is a spanning tree of G, hence is simply G with some edge (u, v) omitted. Setting $U = \{u, v\}$, the sub-solution on U induced by T has length $n - 1$, but the optimal solution on U has length 1. Thus the competitive ratio of any master solution is $\Omega(n)$.

A randomized algorithm can do much better: simply sample a tree-metric approximation of the shortest-path metric on G using the algorithm of Fakcharoenphol, Rao & Talwar [7]. Since the expected distortion is $O(\log n)$, this randomized (distribution-free!) algorithm achieves a soft approximation of $O(\log n)$ (the analysis is identical to the proof of Corollary 4 in [19]).

Corollary 2. *No deterministic (hard or soft) approximation algorithm for a priori Steiner Tree has approximation ratio $o(n)$, whereas there exists a randomized, distribution-free soft $O(\log n)$-approximation.*

Returning briefly to TSP, Schalekamp & Shmoys give in [19] a randomized soft $O(\log)$-approximation algorithm for *a priori* TSP that is distribution-free. Corollary 1 shows that no polynomial number of black-box samples can help a deterministic algorithm asymptotically beat this randomized approximation guarantee. In the independent activation case of *a priori* TSP, Shmoys & Talwar give in [20] a deterministic soft 4-approximation which relies on a single black-box sample. Our results show that without the assumption of independence, no polynomial number of samples is enough to help a deterministic algorithm achieve a $o(\log n)$ approximation ratio.

5   Open Problems

Though we have proved a general lower bound of $\Omega(\log n)$ for the universal TSP, the conjecture of Bertsimas & Grigni [4] that this bound holds even for finite subsets of the plane remains open. The best known algorithm for the universal TSP (Gupta, Hajiaghayi & Räcke [10]) produces, for an arbitrary metric space (X, δ) on n points, a master tour τ with $\rho(X, \delta, \tau) = O(\log^2 n)$. Is either bound tight? In [14], Jia et al. give an algorithm that produces τ with $\rho(X, \delta, \tau) = O(\log n)$ when (X, δ) is a constant-dimension Euclidean or bounded-growth metric. Schalekamp & Shmoys observe in [19] that a straightforward application of the tree-metric embedding scheme of Fakcharoenphol et al. [7] gives a randomized algorithm that produces a master tour whose expected competitive ratio for any fixed $S \subseteq X$ is $O(\log n)$. We conjecture that this expected guarantee for a fixed S can be matched by a deterministic guarantee for all S.

Conjecture 1. There is a deterministic, polynomial-time algorithm that, given a metric space (X, δ) with $|X| = n$, produces τ with $\rho(X, \delta, \tau) = O(\log n)$.

References

1. Arora, S.: Polynomial time approximation schemes for euclidean traveling salesman and other geometric problems. J. ACM 45(5), 753–782 (1998)
2. Bartholdi, J., Platzman, L.: An $o(n \log n)$ planar traveling salesman heuristic based on spacefilling curves. Operations Research Letters 1(4), 121–125 (1981)
3. Bartholdi, J., Platzman, L., Collins, R.L., Warden III:, W.H.: A minimal technology routing system for meals on wheels. Interfaces 13(3), 1–8 (1983)
4. Bertsimas, D., Grigni, M.: On the spacefilling curve heuristic for the euclidean traveling salesman problem. Operations Research Letters 8(5), 241–244 (1989)
5. Bertsimas, D., Jaillet, P., Odoni, A.: On the spacefilling curve heuristic for the euclidean traveling salesman problem. Operations Research Letters 38(6), 1019–1033 (1990)
6. Bollobás, B.: Modern Graph Theory. Springer, Heidelberg (2001)
7. Fakcharoenphol, J., Rao, S., Talwar, K.: A tight bound on approximating arbitrary metrics by tree metrics. J. Comput. Syst. Sci. 69(3), 485–497 (2004)
8. Goldreich, O., Impagliazzo, R., Levin, L.A., Venkatesan, R., Zuckerman, D.: Security preserving amplification of hardness. In: FOCS, pp. 318–326 (1990)
9. Gorodezky, I., Kleinberg, R., Shmoys, D., Spencer, G.: Improved lower bounds for the universal and *a priori* TSP. In: TSP (2010) (Preprint),
   http://people.orie.cornell.edu/~shmoys/GKSS10.pdf
10. Gupta, A., Hajiaghayi, M., Räcke, H.: Oblivious network design. In: SODA, pp. 970–979 (2006)
11. Gupta, A., Pál, M., Ravi, R., Sinha, A.: Boosted sampling: approximation algorithms for stochastic optimization. In: STOC, pp. 417–426 (2004)
12. Hajiaghayi, M., Kleinberg, R., Leighton, F.: Improved lower and upper bounds for universal tsp in planar metrics. In: SODA, pp. 649–658 (2006)
13. Jaillet, P.: Probabilistic traveling salesman problems. Technical Report, MIT Operations Research Center, 185 (1985)
14. Jia, L., Lin, G., Noubir, G., Rajaraman, R., Sundaram, R.: Universal approximations for tsp, steiner tree, and set cover. In: STOC, pp. 386–395 (2005)
15. Lubotzky, A., Phillips, R., Sarnak, P.: Ramanujan graphs. Combinatorica 8(3), 261–277 (1988)
16. Morgenstern, M.: Existence and explicit constructions of + 1 regular ramanujan graphs for every prime power. J. Comb. Theory, Ser. B 62(1), 44–62 (1994)
17. Nilli, A.: On the second eigenvalue of a graph. Discrete Mathematics 91(2), 207–210 (1991)
18. Platzman, L., Bartholdi III, J.: Spacefilling curves and the planar travelling salesman problem. J. ACM 36(4), 719–737 (1989)
19. Schalekamp, F., Shmoys, D.: Algorithms for the universal and *a priori* tsp. Oper. Res. Lett. 36(1), 1–3 (2008)
20. Shmoys, D., Talwar, K.: A constant approximation algorithm for the *a priori* traveling salesman problem. In: Lodi, A., Panconesi, A., Rinaldi, G. (eds.) IPCO 2008. LNCS, vol. 5035, pp. 331–343. Springer, Heidelberg (2008)

Proximity Algorithms for Nearly-Doubling Spaces[*]

Lee-Ad Gottlieb and Robert Krauthgamer

Weizmann Institute of Science, Rehovot, Israel
{lee-ad.gottlieb,robert.krauthgamer}@weizmann.ac.il

Abstract. We introduce a new problem in the study of doubling spaces: Given a point set S and a target dimension d^*, remove from S the fewest number of points so that the remaining set has doubling dimension at most d^*. We present a bicriteria approximation for this problem, and extend this algorithm to solve a group of proximity problems.

1   Introduction

In the last few years, researchers have increasingly made use of the doubling dimension in the design of algorithms. Analyzing algorithmic tasks via the doubling dimension is natural for proximity problems such as nearest neighbor search [KL04, BKL06, CG06b] and clustering [Tal04, ABS08, FM10], and for graph problems such as spanner construction [GGN06, CG06a, DPP06, GR08a, GR08b], the traveling salesman problem [Tal04], and routing [KSW04, Sli05, AGGM06, KRXY07, KRX08]. The doubling dimension has proved to be a powerful tool in embeddings [Ass83, GKL03, ABN07, ABN08, CGT08, BRS07, GK09] and has found applications in fields such as machine learning [BLL09, GKK10]. Interestingly, the problem of computing the exact doubling dimension of a point set is NP-hard. (This result seems to be folklore.) Yet this fact has not deterred the development of algorithms that are based on the doubling dimension, partly because it can be approximated within a constant factor, and partly because many of these algorithms function without explicit knowledge of the doubling dimension – it appears only in the analysis.

However, a host of algorithms previously developed for doubling dimension – perhaps even the majority of them – suffer from a more serious problem: They are not robust to severe yet infrequent irregularities in the space. The guarantees provided by these algorithms are markedly degraded even if only a small subset of the working set possesses high doubling dimension. This problem was noted for example by [CG08] who instead defined a global notion of dimension (which can be thought of as the average doubling dimension over the set) and developed an algorithm under this new definition.

We pursue a different approach. We introduce the following key problem: Given an n-point set S and a target dimension d^*, remove from S the fewest

---

[*] This work was supported in part by The Israel Science Foundation (grant #452/08), and by a Minerva grant.

M. Serna et al. (Eds.): APPROX and RANDOM 2010, LNCS 6302, pp. 192–204, 2010.

number of points so that the remaining set has doubling dimension at most d^* (or equivalently, target doubling constant $\lambda^* = 2^{d^*}$). We thus call a data set *nearly-doubling* if all but a negligible fraction of the points have bounded doubling dimension.

A solution to this point removal problem yields a contribution in two related areas. The first paradigm, broadly speaking, is outlier detection. In this scenario, the removed points are ignored and only the remaining points are processed. A direct motivation for this model stems from the dimension induced clustering framework of [GHPT05], which given a point set seeks a subset with low intrinsic dimension. Further motivation stems from algorithms which have "slack"; that is, they give guarantees for most but not all of the point set [KRXY07, FM10]. These algorithm can be extended to nearly-doubling data sets by simply ignoring the removed points (i.e. throwing them into the slack). The second paradigm is an original one: Here, both the removed points and the remaining ones are processed, albeit by separate algorithms tailored to the properties of the two point sets.

Results. The point removal problem is NP-hard, and it is not difficult to show that the problem does not admit even an approximate multiplicative-factor solution (see Lemma 1). However, we develop a framework that yields a bicriteria approximation for this problem. In Section 3, we present bicriteria algorithms that achieve the following bounds:

1. In time $2^{O(d^*)}n^3$, we remove a number of points arbitrarily close to optimal, while achieving doubling dimension $4d^* + O(1)$ (Corollary 1).
2. In time $2^{O(d^*)}n \log \alpha$ (where α is the aspect ratio of S), we remove a number of points arbitrarily close to optimal, while achieving doubling dimension $10d^* + O(1)$ (Corollary 2).
3. In time $2^{O(d^*)}n \log^3 n$, we remove a number of points arbitrarily close to optimal, while achieving doubling dimension $12d^* + O(1)$ (Corollary 2).

Returning to the first paradigm presented above, our algorithms solve the clustering problem posed by [GHPT05]. (They provided heuristic solutions to this question.) In Section 4, we present algorithms that function under the second paradigm delineated above: These algorithms process the removed points and the remaining ones with separate techniques tailored to the properties of the two point sets. When the data set is nearly-doubling, or more precisely, when all but at most square root of the points have bounded doubling dimension, we give near-linear time algorithms for constructing $(1 + \varepsilon)$-stretch spanners, approximate minimum spanning trees, $O(1)$-query time distance oracles, and calculating approximate all points nearest neighbor.

2   Preliminaries

In this section we define doubling dimension, and present some basic hardness results. We then review point hierarchies for doubling spaces.

Doubling dimension. For a metric (X, d), let λ be the infimum value such that every closed ball in X can be covered by λ closed balls of half the radius, where a ball is centered at a point of the metric. λ is the *doubling constant* of X, and the *doubling dimension* of X is $\dim(X) = \log_2 \lambda$. A metric is *doubling* when its doubling dimension is finite. It is a folklore result that determining the doubling constant (and dimension) of a point set is an NP-hard problem. We formalize this result below.

Lemma 1. *Given a metric (S, d), computing the doubling constant of S is NP-hard.*

Proof. The proof is a reduction from vertex cover with bounded degree Δ [PY91]. Let $G = (V, E)$ be an input instance of vertex cover with degree $\Delta < \sqrt{|V|}$. Note that the size of any vertex cover of G must be greater than $\sqrt{|V|}$. Create a set S containing $|V|$ points, each corresponding to a vertex in V. Let $d(u, v) = \frac{1}{2}$ for $u, v \in S$ if the corresponding vertices have an edge in E, and let $d(u, v) = 1$ otherwise. The radius of S is 1.

Now, any subset of S found in a closed ball of radius $\frac{1}{2}$ contains fewer than $\sqrt{|V|}$ points (since the degree of V is less than $\sqrt{|V|}$), so the doubling constant of the subset is less than $\sqrt{|V|}$. However, the minimum covering of all of S by closed balls of radius $\frac{1}{2}$ is equivalent to the minimum vertex cover of V, which is necessarily greater than $\sqrt{|V|}$. It follows that determining the doubling constant of S is equivalent to determining the minimum vertex cover of V.

Note that the above reduction preserves hardness of approximation: It is NP-hard to determine the doubling constant of a metric within a factor $\frac{16}{15} - \varepsilon$ (see [Cle99]). This problem does admit an approximation – for example, a 2-approximation to the doubling dimension (equivalently, the square of the doubling constant) can be determined by the algorithm of Lemma 3 (see also [HM05, Theorem 9.1]).

A further consequence of Lemma 1 is that the problem of removing the minimum number of points from a set S in order to obtain a set S' with some target doubling constant does not admit a multiplicative-factor approximation algorithm: That is, it is NP-hard to distinguish the case where no points need be removed, from the case that one point must be removed.

Point hierarchies. Here, we define hierarchical partitions and describe three different partitions that have appeared in the literature and will be utilized in this paper.

Similar to what was described in [GGN06, KL04], a subset of points $X \subseteq Y$ is an (r, s)-discrete center set (or *net* in the terminology of [KL04]) of Y ($r \leq s$) if it satisfies the following properties:

(i) Packing: For every $x, y \in X$, $d(x, y) > r$.
(ii) Covering: Every point $y \in Y$ is strictly within distance s of some point $x \in X$: $d(x, y) \leq s$.

We say that x *covers* y if $x \in X$, $y \in Y$ and $d(x, y) \leq s$. The previous conditions require that the points of X be spaced out, yet nevertheless cover all points of Y. A hierarchical partition for a set S is a hierarchy of discrete center sets, where each level of the hierarchy is a discrete center set of the level beneath it. The bottom level contains all points, and the top level contains only a single point. (For ease of presentation, we assume that the minimum inter-point distance in S is 1.)

The first hierarchy we describe is that of [KL04]. The hierarchy is composed of levels H_{2^i} (for integer $i = 0, \ldots$), where each level H_{2^i} ($i > 0$) is a $(2^i, 2^i)$-discrete center set for the previous level $H_{2^{i-1}}$. (The subscript in the notation of the level indicates that the packing and covering properties of subsequent levels grow by a factor of 2.) The bottom level of the hierarchy is the set $Y_{2^0=1} = S$, and the top level is the set $Y_{2^{\lceil \log \alpha \rceil}}$ that contains only a single point. The construction supports insertions and deletions to the hierarchy in time $2^{O(\log \lambda)} \log \alpha$. (Recall that α is the aspect ratio of S.)

The second hierarchy is that of [GR08a]. This hierarchy is similar to that of [KL04], but level H_{2^i} is a $(\frac{1}{2} 2^i, 2^i)$-discrete center set for $H_{2^{i-1}}$. This hierarchy supports insertions and deletions in $2^{O(\log \lambda)} \log n$ amortized time. Hence, a series of n insertions and deletions can be done deterministically in $2^{O(\log \lambda)} n \log n$ time.

The third hierarchy is that of [CG06b]. In this hierarchy, level H_{5^i} is a $(\frac{1}{5} 5^i, \frac{3}{5} 5^i)$-discrete center set for $H_{5^{i-1}}$. (The packing and covering properties of subsequent levels grow by a factor of 5.) The hierarchy supports insertions in time $2^{O(\log \lambda)} \log n$, though points cannot be removed from within the hierarchy. (A static hierarchy with similar construction time was also presented in [HM06].)

On top of these hierarchies, we define a parent-child relationship: Point $y \in H_{2^i}$ (or H_{5^i}) is the child of one of the points in $H_{2^{i+1}}$ (or H_{5^i}) that covers y. This immediately defines an ancestral relationship as well.

3   Point Removal Algorithm

In this section, we present the bicriteria algorithm for the problem of removing points to obtain a target doubling constant. The construction, presented below, proceeds roughly as follows: We formulate the notion of a "bad" witness set, which can be found efficiently and exists if and only if the doubling constant is too large (to within some constant factors). Given this setup, the algorithms is greedy: Repeatedly find such a witness set and remove it entirely.

We first define the density constant (in Section 3.1), and explain the existence of witness sets for the density constant. We show that it is NP-hard to locate a maximum witness set, but we are able to give an approximation algorithm for locating witness sets. In Section 3.2, we use this approximation algorithm for witness sets to develop a bicriteria point removal algorithm for achieving a target density constant. This bicriteria algorithm in turn yields a bicriteria point removal algorithm for achieving a target doubling constant. Finally, in Section 3.3, we show how to improve the runtime of the two bicriteria algorithms.

3.1 Density Constant and Witness Sets

Let a closed ball $B(x, r) \in S$ be centered at point x and include all points of set S within distance r of x. We define the *density constant* $\mu(S)$ of point set S as follows: $\mu(S)$ is the smallest number such that every open r-radius ball of S (for every r) contains at most $\mu(S)$ points of mutual inter-point distance greater than $r/2$. Clearly the doubling constant cannot be greater than the density constant. Further, the density constant is not greater than the square of the doubling constant (since $\mu(S)$ balls of radius $\frac{r}{4}$ are required to cover these points). It follows that

$$\sqrt{\mu(S)} \leq \lambda(S) \leq \mu(S).$$

Now, we consider the following point removal problem: Given a point set S and a target density constant $\mu^* \leq \mu(S)$, remove the minimum number of points from S to obtain a set S^* with density constant μ^*. (This problem can serve as a proxy for the problem of removing points to obtain a target doubling constant.) However, we demonstrate in Lemma 2 below that the problem of determining the density constant of a point set S is NP-hard. An immediate consequence of Lemma 2 is that the point removal problem to achieve a target density constant is NP-hard.

Lemma 2. *Given a point set S, the problem of determining the density constant of S is NP-hard.*

Proof. The proof is a reduction from the maximum independent set problem with bounded degree Δ [PY91]. Let $G = (V, E)$ be an input instance of the max independent set problem with degree $\Delta < \sqrt{|V|}$. Note that the size of any maximal independent set for G is greater than $\sqrt{|V|}$. Create a set S containing $|V|$ points, each corresponding to a vertex in V. Let $d(u, v) = \frac{1}{2} + \varepsilon$ for $u, v \in S$ (and an infinitely small ε) if the corresponding vertices have an edge in E, and let $d(u, v) = 1$ otherwise. The radius of S is 1.

Now, any subset of S found in a closed ball of radius $\frac{1}{2} + \varepsilon$ contains fewer than $\sqrt{|V|}$ points (since the degree of V is less than $\sqrt{|V|}$), so the density constant of any subset of points of S that all fall in a ball of radius $\frac{1}{2} + \varepsilon$ and have inter-point distance greater than $\frac{1}{4} + \frac{\varepsilon}{2}$ is less than $\sqrt{|V|}$. However, the maximum number of points in all of S with inter-point distance greater than $\frac{1}{2}$ is necessarily greater than $\sqrt{|V|}$. It follows that determining the density constant of S is equivalent to determining the maximum independent set in V.

As an aside, note that the reduction preserves hardness of approximation: It is NP-hard to approximate the density constant of a point set S within a factor of $|S|^{\frac{1}{2} - \varepsilon}$ (this follows easily from [Has96]).

It follows from Lemma 2 that the point removal problem to achieve a target density constant is NP-hard. Further, this problem does not even admit a multiplicative-factor approximation algorithm: It is NP-hard to distinguish the case where no points need be removed, from the case that one point must be removed. However, we can approximate the density constant of a point set, as in Lemma 3 below. We will first require a definition.

Definition 1. *Given a point set S, a witness set $S' \subset S$ is a set of points contained in a closed ball of radius r with mutual inter-point distance greater than $\frac{r}{2}$.*

Comment. Note that the existence of a witness set $S' \subset S$ implies that $\mu(S) \geq |S'|$. The notion of a witness set exists for the density constant, but a similar notion does not exist for the doubling constant. That is, the addition of points to a set S with doubling constant $\lambda(S)$ may in fact result in a set with somewhat lower doubling constant than $\lambda(S)$. This motivates our decision to define the density constant.

Lemma 3. *Given an n-point set S with minimum inter-point distance 1, there exists an $O(2^{O(\log \mu(S))} n^3)$ time algorithm that locates a witness set of size $\lceil \sqrt{\mu(S)} \rceil$.*

Proof. Note that there are $O(n^2)$ inter-point distances in S, so there exist $O(n^2)$ distinct balls of S, each of size $O(n)$. For each ball $B(x \in S, r)$, we greedily build the point hierarchy of [KL04] consisting of four radii levels $\{r, \frac{r}{2}, \frac{r}{4}, 1\}$, where level r contains only one point, and level 1 contains all points. This can be done in time $2^{O(\log \mu)} n$ per ball (where $\mu = \mu(S)$), yielding a total runtime of $O(2^{O(\log \mu)} n^3)$.

Now there must exist in S a point set S' of size exactly μ with radius r and minimum inter-point distance greater than $r/2$, for some r. In the hierarchy for the ball that contains S' (and possibly contains other points as well), one of the following must hold:

(i) Level $\frac{r}{2}$ contains at least $\lceil \sqrt{\mu} \rceil$ points; it follows that these points are contained in a ball of radius r and have minimum inter-point distance greater than $\frac{r}{2}$, so that they are a witness set. Or,

(ii) Level $\frac{r}{2}$ contains fewer than $\lceil \sqrt{\mu} \rceil$ points. Now, since the μ points of S' have minimum distance $\frac{r}{2}$, they must be covered by distinct points of level $\frac{r}{4}$, so there must exist more than μ points in level $\frac{r}{4}$. It follows that some point of level $\frac{r}{2}$ covers more than $\sqrt{\mu}$ points of level $\frac{r}{4}$. These points have minimum inter-point distance greater than $\frac{r}{4}$ and are found in a set of radius less than $\frac{r}{2}$, so they are a witness set.

Comment. As an aside, note that the algorithm of Lemma 3 yields a 2-approximation to the doubling dimension of S.

Lemma 3 shows that the density constant can be approximated. In the next section, we will use this tool to develop a bicriteria algorithm for the problem of removing points to obtain a target density constant. This will in turn allow us to develop a bicriteria algorithm for the problem of removing points to obtain a target doubling constant. However, for the purposes of efficient algorithmic runtime, we need to introduce a slightly stronger variant of Lemma 3, as follows:

Lemma 4. *Given an n-point set S with minimum inter-point distance 1 and a parameter $\mu' \leq \mu(S)$, there exists an $O(2^{O(\log \mu')} n^3)$ time algorithm that locates a maximal collection of distinct witness sets each of size $\lceil \sqrt{\mu'} \rceil$.*

Proof. The construction is similar to the one presented in the proof of Lemma 3. We identify all $O(n^2)$ balls, and for each ball $B(x, r)$ we build its hierarchy one point at a time. If the insertion of a point into the hierarchy of $B(x, r)$ implies a witness set of size $\lceil \sqrt{\mu'} \rceil$ – that is, either level $\frac{r}{2}$ contains $\lceil \sqrt{\mu'} \rceil$ points, or a point of level $\frac{r}{2}$ covers $\lceil \sqrt{\mu'} \rceil$ points – then we output the witness set as an element of the collection, and delete the points of this witness set from all ball hierarchies. We then repair the hierarchies (as usual after a deletion, see [KL04]) and resume the hierarchy construction. The removal of the witness set points and subsequent repair of the hierarchies do not increase the runtime.

3.2   Bicriteria Algorithm

Given Lemma 4, we prove the following theorem, which is a bicriteria algorithm for the problem of removing points to achieve a target density constant. A corollary of this theorem gives a bicriteria algorithm for the problem of removing points to achieve a target doubling constant.

Theorem 1. *Given a point set S with density constant $\mu(S)$ and a target density constant μ^*, let $k^* = k^*(S, \mu^*)$ be the minimum number of points that must be removed from S to obtain a set S^* with density constant μ^*. Then there exists a $2^{O(\log \mu^*)} n^3$ time algorithm that removes $k' \leq \frac{c\mu^* + 1}{(c-1)\mu^* + 1} \cdot k^*$ points from S (for any desired $c \geq 1$), and yields a point set S' with density constant $\mu(S') \leq (c\mu^*)^2$.*

Proof. We first prove the theorem for $c = 1$. We run the algorithm of Lemma 3 to find a collection of distinct witness sets of size exactly $\mu^* + 1$. Remove these sets from S. It follows that the resulting set S' has density constant at most $(\mu^*)^2$. Now, in the optimal solution S^*, at least one of the points in each witness set must be removed. Hence, the algorithm removes $k' \leq (\mu^* + 1)k^*$ points.

Turning to $c > 1$, we run the algorithm of Lemma 3 to find a collection of distinct witness sets of size exactly $c\mu^* + 1$. Remove these sets from S. It follows that the resulting set S' has density constant at most $(c\mu^*)^2$. Now, if our algorithm has removed a witness set of size m, then in the optimal solution at least $m - \mu^*$ of these points must have been removed. It follows that the algorithm removes $k' \leq \frac{c\mu^* + 1}{c\mu^* + 1 - \mu^*} k^* = \frac{c\mu^* + 1}{(c-1)\mu^* + 1} \cdot k^*$ points from S.

This algorithm for the density constant implies a similar one for the doubling constant:

Corollary 1. *Given a point set S with doubling constant $\lambda(S)$ and a target doubling constant λ^*, let $m^* = m^*(S, \lambda^*)$ be the minimum number of points that must be removed from S to obtain a set S^* with doubling constant λ^*. Then there exists an $2^{O(\log \lambda^*)} n^3$ time algorithm that removes $m' \leq \frac{c(\lambda^*)^2 + 1}{(c-1)(\lambda^*)^2 + 1} \cdot m^*$ points from S (for any desired $c \geq 1$), and yields a point set S' with doubling constant $\lambda(S') \leq (c(\lambda^*)^2)^2$.*

Proof. The proof follows from Theorem 1 with $\mu^* = (\lambda^*)^2$.

3.3   Improved Run Time

While the bicriteria algorithms implied by Theorem 1 and Corollary 1 provide a powerful tradeoff for the two point removal problems, the algorithmic runtime may be undesirable for some applications. Here we present bicriteria algorithms that feature near-linear runtime at the expense of slightly higher dimension.

Theorem 2. *Given a point set S with density constant $\mu(S)$ and a target density constant μ^*, let $k^* = k^*(S, \mu^*)$ be the minimum number of points that must be removed from S to obtain a set S^* with density constant μ^*. Then there exists*

(i) *An algorithm that runs in $2^{O(\log \mu^*)} n \log \alpha$ time that removes $k' \leq \frac{c\mu^*+1}{(c-1)\mu^*+1}$. k^* points from S (for any desired $c \geq 1$), and yields a point set S' with density constant $\mu(S') \leq (c\mu^*)^5$.*

(ii) *An algorithm that runs in $2^{O(\log \mu^*)} n \log^3 n$ time that removes $k' \leq \frac{c\mu^*+1}{(c-1)\mu^*+1}$. k^* points from S (for any desired $c \geq 1$), and yields a point set S' with density constant $\mu(S') \leq (c\mu^*)^6$.*

Proof. We begin by building the hierarchy of [KL04] for S, inserting one point at a time. Now, if a point insertion causes a point of level H_{2^i} to possess more than $(c\mu^*)^5$ neighbors in H_{2^i} within distance $32 \cdot 2^i$, then we can find a witness set: By building a hierarchy for just the neighbor set on distances $\{32 \cdot 2^i, 16 \cdot 2^i, 8 \cdot 2^i, 4 \cdot 2^i, 2 \cdot 2^i, 2^i\}$, we locate in the neighbor set some witness set of size at least $c\mu^*$. (That is, some point in the neighbor set hierarchy must cover $c\mu^* + 1$ points one level down, and these points form a witness set for S.) As before, the points of the witness set are then deleted from the hierarchy of S. The algorithm terminates with set S' when no more witness sets can be found. This can all be be done in $O(2^{O(\log \mu^*)} n \log \alpha)$ time. The analysis for optimality of removed points is the same as above.

It is only left to show that the resulting set cannot have density constant greater than $(c\mu^*)^5$: Suppose in contradiction that S' contained a witness set of size greater than $(c\mu^*)^5$ with diameter r and minimum inter-point distance $\frac{r}{2}$. Now, each point of the witness set appears in H_1, the bottom level of the full hierarchy, and a geometric series gives that the distance between each point and its ancestor in level H_{2^i} is less than 2^{i+1}. Hence, the distance between the level H_i ancestors of two different points of the witness set is greater than $\frac{r}{2} - 2^{i+2}$ and less than $r + 2^{i+2}$. Now let j be the index for which $r \geq 2^j > \frac{r}{2}$. Let $i = j - 3$, so that the distance between the ancestors of two points is greater than $\frac{r}{2} - 2^{j-1} \geq 0$ (and so at least 2^i) and less than $r + 2^{j-1} \geq 20 \cdot 2^i$. This contradicts the assumption that there does not exist in the hierarchy a set of more than $(c\mu)^5$ points within radius $32 \cdot 2^i$ and minimum inter-point distance 2^i.

The runtime of $2^{O(\log \mu^*)} n \log^3 n$ can be achieved by using the hierarchy of [GR08a] instead of the hierarchy of [KL04]. (Note however that the semi-dynamic hierarchy of [CG06b] or the static hierarchy of [HM05] are not sufficient for our purposes.) The analysis is similar.

This above point removal algorithm for the density constant implies a similar one for the doubling constant:

Corollary 2. *Given a point set S with doubling constant $\lambda(S)$ and a target doubling constant λ^*, let $m^* = m^*(S, \lambda^*)$ be the minimum number of points that must be removed from S to obtain a set S^* with doubling constant λ^*. Then there exists*

(i) *An algorithm that runs in $2^{O(\log \lambda)} n \log \alpha$ time that removes $m' \leq \frac{c(\lambda^*)^2 + 1}{(c-1)(\lambda^*)^2 + 1} \cdot$ m^* points from S (for any desired $c \geq 1$), and yields a point set S' with doubling constant $\lambda(S') \leq (c(\lambda^*)^2)^5$.*

(ii) *An algorithm that runs in $2^{O(\log \lambda)} n \log^3 n$ time that removes $m' \leq \frac{c(\lambda^*)^2 + 1}{(c-1)(\lambda^*)^2 + 1} \cdot$ m^* points from S (for any desired $c \geq 1$), and yields a point set S' with doubling constant $\lambda(S') \leq (c(\lambda^*)^2)^6$.*

4   Applications

The algorithms of Section 3 are given a point set S, and remove from S a set R, resulting in a set $S' = S - R$ of low doubling constant $(\lambda^*)^{O(1)}$. If $|R| = O(n^{1/2})$ (that is, S is nearly-doubling), we can use techniques from [GR08b, BGK+10] to construct near-linear runtime algorithms for spanners and fast distance oracles.

We first review the spanner of [GR08b] in Section 4.1, and then present the near-linear algorithms in Section 4.2

4.1   Spanner Review

We review the $(1 + \varepsilon)$-stretch spanner presented in [GR08b] (which itself draws on the work of [GGN06]). This spanner is constructed as follows: Given a point set S, the point hierarchy of [CG06b] is constructed for S. First, all parent-child pairs in the hierarchy are connected by edges in the spanner; these are the parent-child edges. Next, we add edges to connect all point pairs $p, q \in H_{5^i}$ (for all i) if p and q are c-neighbors, that is if $d(p, q) \leq c5^i$ for some fixed constant $c = \Theta(1/\varepsilon)$. These are the lateral edges. Notice that the lateral edges of level H_{5^i} are much longer than the parent-child edges of that level (by a factor of $\theta(1/\varepsilon)$). The entire construction can be done in time $2^{O(\log \lambda(S))} \log n + \varepsilon^{-O(\log \lambda(S))}$.

It was shown in [GR08b] that given two points $p, q \in S$, there exists a simple spanner path that connects p and q and has stretch at most $(1 + \varepsilon)$. Let $p', q' \in H_{5^j}$ be the lowest ancestral c-neighbors of p and q. (That is, j is the smallest index for which p' and q', the respective ancestors of p and q in H_{5^j}, are c-neighbors.) The low stretch spanner path is the path that begins at p, follows a series of parent-child edges up to p', a single lateral edge to q', and a series of parent-child edges down to q. The length of the path is dominated by the length of the single lateral edge: The length of the lateral edge is $\Theta(5^j/\varepsilon)$, while the length of all other edges in the path are bounded by two geometric series that each sum to $O(5^j)$. This implies a $(1 + \varepsilon)$-stretch spanner path for the pair p, q.

4.2   Near Linear Algorithms

In this section we present near-linear algorithms for nearly-doubling spaces. We have the following theorem:

Theorem 3. *There exists an algorithm that, given point sets S' and R ($|R| = O(\sqrt{|S'|})$), builds a $(1 + \varepsilon)$-stretch spanner for $S = S' \cup R$ ($n = |S|$) with $\varepsilon^{-O(\log \lambda^*)}n$ edges in $2^{O(\log \lambda^*)}n \log n + \varepsilon^{-O(\log \lambda^*)}n$ time.*

Proof. We first construct the full graph for R, which is a 1-stretch spanner of $O(n)$ edges for these points. We then construct a $(1 + \varepsilon)$ spanner for S' in the manner described above. It is left only to guarantee $(1 + \varepsilon)$ stretch between the points of S' and R. To this end, for each point $p \in R$ we locate the lowest hierarchical level H_{2^i} of S' in which p is covered, and connect p to its covering point with a parent-child edge, and to all points of levels H_{2^i} and below within distance $c2^i$ using lateral edges. As in [GR08b], there are $\varepsilon^{-O(\log \lambda)}$ edges incident on p, and this construction mimics an actual insertion of p into the hierarchy. It follows that there exists low stretch paths connecting $p \in R$ to all points of S'. ∎

The following corollary is a consequence of the spanner construction of Theorem 3.

Corollary 3. *There exists an algorithm that, given point sets S' and R ($|R| = O(\sqrt{|S'|})$), computes*

(i) A $(1+\varepsilon)$ approximation to the minimum spanning tree(MST) for S, in time $2^{O(\log \lambda^*)}n \log n + \varepsilon^{-O(\log \lambda^*)}n$.

(ii) A $(1 + \varepsilon)$ approximation to all pairs nearest neighbor in time $2^{O(\log \lambda^*)}n \log n + \varepsilon^{-O(\log \lambda^*)}n$.

(iii) A $(1+\varepsilon)$-approximate distance oracle that supports $O(1)$-time distance queries with storage $2^{O(\log \lambda^* \log \log \lambda^*)}n + \varepsilon^{-O(\log \lambda^*)}n$, with construction time $2^{O(\log \lambda^*)}n \log n + 2^{O(\log \lambda^* \log \log \lambda^*)}n + \varepsilon^{-O(\log \lambda^*)}n$.

Proof

(i) Given the above $(1+\varepsilon)$-stretch spanner, a $(1+\varepsilon)$-approximate MST for S can be construction by a simple breadth first search algorithm (such as Dijkstra's algorithm) on the points and edges of the spanner.

(ii) A $(1 + \varepsilon)$-approximate nearest neighbor for each point p may be found by consulting the spanner edges incident on p, and choosing the closest incident point. This can be maintained in $O(1)$ time per edge insertion.

(iii) A $(1 + \varepsilon)$-approximate $O(1)$ query time distance oracle for doubling spaces was presented in [BGK+10]. This oracle was built on the spanner of [GR08b] described above: The structure records the exact distance between any pair of points that are connected in the spanner. For query points p and q, the algorithm simply locates the lowest ancestral c-neighbors p', q' of p, q, and returns their distance. (Recall that p', q' are connected by a lateral edge, so their true distance is recorded in the spanner.) It follows from the discussion above that the distance between p' and q' is a $(1 + \varepsilon)$-approximation to the distance between p and q.

We extend the construction of [GR08b] to the spanner in the proof of Theorem 3: We record the distance between any pairs of points that are connected in the spanner of the proof of Theorem 3. It follows that the distance between any pair $p, q \in R$ is recorded explicitly. For all pairs $p, q \in S'$, a search for the lowest ancestral c-neighbors of p and q returns an approximation for the distance between p and q. For points $p \in R$ and $q \in S'$, their distance can similarly be derived via a search for the lowest ancestral c-neighbors, making use of the edges added to the spanner in the construction for the proof of Theorem 3.

Acknowledgments. The authors thank Uri Feige and Liam Roditty for useful discussions.

References

[ABN07]   Abraham, I., Bartal, Y., Neiman, O.: Local embeddings of metric spaces. In: Proceedings of 39th Annual ACM Symposium on Theory of Computing, pp. 631–640 (2007)

[ABN08]   Abraham, I., Bartal, Y., Neiman, O.: Embedding metric spaces in their intrinsic dimension. In: 19th Annual ACM-SIAM Symposium on Discrete Algorithms, pp. 363–372 (2008)

[ABS08]   Ackermann, M.R., Blömer, J., Sohler, C.: Clustering for metric and non-metric distance measures. In: SODA 2008: Proceedings of the Nineteenth Annual ACM-SIAM Symposium on Discrete Algorithms, pp. 799–808 (2008)

[AGGM06]   Abraham, I., Gavoille, C., Goldberg, A.V., Malkhi, D.: Routing in networks with low doubling dimension. In: 26th IEEE International Conference on Distributed Computing Systems, p. 75 (2006)

[Ass83]   Assouad, P.: Plongements lipschitziens dans $\mathbf{R}^n$. Bull. Soc. Math. France 111(4), 429–448 (1983)

[BGK+10]   Bartal, Y., Gottlieb, L., Kopelowitz, T., Lewenstein, M., Roditty, L.: Fast and precise distance queries (2010) (manuscript)

[BKL06]   Beygelzimer, A., Kakade, S., Langford, J.: Cover trees for nearest neighbor. In: 23rd International Conference on Machine Learning, pp. 97–104 (2006)

[BLL09]   Bshouty, N.H., Li, Y., Long, P.M.: Using the doubling dimension to analyze the generalization of learning algorithms. Journal of Computer and System Sciences 75(6), 323–335 (2009)

[BRS07]   Bartal, Y., Recht, B., Schulman, L.: A Nash-type dimensionality reduction for discrete subsets of L_2. Manuscript (2007), http://www.ist.caltech.edu/~brecht/publications.html

[CG06a]   Chan, T.-H., Gupta, A.: Small hop-diameter sparse spanners for doubling metrics. In: 17th Annual ACM-SIAM Symposium on Discrete Algorithm, pp. 70–78 (2006)

[CG06b]   Cole, R., Gottlieb, L.: Searching dynamic point sets in spaces with bounded doubling dimension. In: 38th Annual ACM Symposium on Theory of Computing, pp. 574–583 (2006)

[CG08]   Chan, T.-H.H., Gupta, A.: Approximating tsp on metrics with bounded global growth. In: SODA 2008: Proceedings of the Nineteenth Annual ACM-SIAM Symposium on Discrete Algorithms, pp. 690–699 (2008)

[CGT08]   Chan, H., Gupta, A., Talwar, K.: Ultra-low-dimensional embeddings for
           doubling metrics. In: 19th Annual ACM-SIAM Symposium on Discrete
           Algorithms, pp. 333–342 (2008)

[Cle99]   Clementi, A.E.F.: Improved non-approximability results for minimum
           vertex cover with density constraints. Theor. Comput. Sci. 225(1-2), 113–
           128 (1999)

[DPP06]   Damian, M., Pandit, S., Pemmaraju, S.V.: Distributed spanner construc-
           tion in doubling metric spaces. In: Shvartsman, M.M.A.A. (ed.) OPODIS
           2006. LNCS, vol. 4305, pp. 157–171. Springer, Heidelberg (2006)

[FM10]   Friedler, S.A., Mount, D.M.: Approximation algorithm for the kinetic
           robust k-center problem. Comput. Geom. Theory Appl. 43(6-7), 572–
           586 (2010)

[GGN06]   Gao, J., Guibas, L.J., Nguyen, A.: Deformable spanners and applications.
           Comput. Geom. Theory Appl. 35(1), 2–19 (2006)

[GHPT05]   Gionis, A., Hinneburg, A., Papadimitriou, S., Tsaparas, P.: Dimension
           induced clustering. In: KDD 2005: Proceedings of the Eleventh ACM
           SIGKDD International Conference on Knowledge Discovery in Data
           Mining, pp. 51–60 (2005)

[GK09]   Gottlieb, L., Krauthgamer, R.: A nonlinear approach to dimension re-
           duction. In: CoRR, abs/0907.5477 (2009)

[GKK10]   Gottlieb, L., Kontorovich, A., Krauthgamer, R.: Efficient classification
           for metric data. In: COLT (2010)

[GKL03]   Gupta, A., Krauthgamer, R., Lee, J.R.: Bounded geometries, fractals,
           and low-distortion embeddings. In: 44th Annual IEEE Symposium on
           Foundations of Computer Science, pp. 534–543 (October 2003)

[GR08a]   Gottlieb, L., Roditty, L.: Improved algorithms for fully dynamic geomet-
           ric spanners and geometric routing. In: SODA 2008: Proceedings of the
           Nineteenth Annual ACM-SIAM Symposium on Discrete Algorithms, pp.
           591–600 (2008)

[GR08b]   Gottlieb, L., Roditty, L.: An optimal dynamic spanner for doubling
           metric spaces. In: Halperin, D., Mehlhorn, K. (eds.) ESA 2008. LNCS,
           vol. 5193, pp. 478–489. Springer, Heidelberg (2008)

[Has96]   Hastad, J.: Clique is hard to approximate within $n^{1-\epsilon}$. Acta Mathemat-
           ica, 627–636 (1996)

[HM05]   Har-Peled, S., Mendel, M.: Fast construction of nets in low dimensional
           metrics, and their applications. In: 21st Annual Symposium on Compu-
           tational Geometry, pp. 150–158. ACM, New York (2005)

[HM06]   Har-Peled, S., Mendel, M.: Fast construction of nets in low-dimensional
           metrics and their applications. SIAM Journal on Computing 35(5), 1148–
           1184 (2006)

[KL04]   Krauthgamer, R., Lee, J.R.: Navigating nets: Simple algorithms for prox-
           imity search. In: 15th Annual ACM-SIAM Symposium on Discrete Al-
           gorithms, pp. 791–801 (January 2004)

[KRX08]   Konjevod, G., Richa, A.W., Xia, D.: Dynamic routing and location ser-
           vices in metrics of low doubling dimension. In: Taubenfeld, G. (ed.) DISC
           2008. LNCS, vol. 5218, pp. 379–393. Springer, Heidelberg (2008)

[KRXY07]   Konjevod, G., Richa, A.W., Xia, D., Yu, H.: Compact routing with slack
           in low doubling dimension. In: 26th Annual ACM Symposium on Prin-
           ciples of Distributed Computing, pp. 71–80. ACM, New York (2007)

[KSW04]   Kleinberg, J.M., Slivkins, A., Wexler, T.: Triangulation and embedding using small sets of beacons. In: 45th Annual IEEE Symposium on Foundations of Computer Science, pp. 444–453 (2004)

[PY91]   Papadimitriou, C.H., Yannakakis, M.: Optimization, approximation, and complexity classes. J. Comput. System Sci. 43(3), 425–440 (1991)

[Sli05]   Slivkins, A.: Distance estimation and object location via rings of neighbors. In: Proceedings of the 24th Annual ACM Symposium on Principles of Distributed Computing, pp. 41–50 (2005)

[Tal04]   Talwar, K.: Bypassing the embedding: Algorithms for low dimensional metrics. In: Proceedings of the 36th Annual ACM Symposium on Theory of Computing, pp. 281–290 (2004)

Matrix Sparsification and the Sparse Null Space Problem

Lee-Ad Gottlieb[1] and Tyler Neylon[2]

[1] Weizmann Institute of Science
lee-ad.gottlieb@weizmann.ac.il
[2] Bynomial, Inc.
tylerneylon@gmail.com

Abstract. We revisit the matrix problems sparse null space and matrix sparsification, and show that they are equivalent. We then proceed to seek algorithms for these problems: We prove the hardness of approximation of these problems, and also give a powerful tool to extend algorithms and heuristics for sparse approximation theory to these problems.

1 Introduction

In this paper, we revisit the matrix problems sparse null space and matrix sparsification.

The sparse null space problem was first considered by Pothen in 1984 [27]. The problem asks, given a matrix A, to find a matrix N that is a full null matrix for A – that is, N is full rank and the columns of N span the null space of A. Further, N should be sparse, i.e. contain as few nonzero values as possible. The sparse null space problem is motivated by its use to solve Linear Equality Problems (LEPs) [9]. LEPs arise in the solution of constrained optimization problems via generalized gradient descent, segmented Lagrangian, and projected Lagrangian methods. Berry *et al.* [4] consider the sparse null space problem in the context of the dual variable method for the Navier-Stokes equations, or more generally in the context of null space methods for quadratic programming. Gilbert and Heath [16] noted that among the numerous applications of the sparse null space problem arising in solutions of underdetermined system of linear equations, is the efficient solution to the force method (or flexibility method) for structural analysis, which uses the null space to create multiple linear systems. Finding a sparse null space will decrease the run time and memory required for solving these systems. More recently, it was shown [36,26] that the sparse null space problem can be used to find correlations between small numbers of times series, such as financial stocks. The decision version of the sparse null space problem is known to be NP-Complete [9], and only heuristic solutions have been suggested for the minimization problem [9,16,4].

The matrix sparsification problem is of the same flavor as sparse null space. One is given a full rank matrix A, and the task is to find another matrix B that is equivalent to A under elementary column operations, and contains as

M. Serna et al. (Eds.): APPROX and RANDOM 2010, LNCS 6302, pp. 205–218, 2010.
© Springer-Verlag Berlin Heidelberg 2010

few nonzero values as possible. Many fundamental matrix operations are greatly simplified by first sparsifying a matrix (see [12]) and the problem has applications in areas such as machine learning [30] and in discovering cycle bases of graphs [20]. But there seem to be only a small number of heuristics for matrix sparsification ([7] for example), or algorithms under limiting assumptions ([17] considers matrices that satisfy the Haar condition), but no general approximation algorithms. McCormick [22] established that the decision version of this problem is NP-Complete.

For these two classic problems, we wish to investigate potentials and limits of approximation algorithms both for the general problems and for some variants under simplifying assumptions. To this end, we will need to consider the well-known vector problems min unsatisfy and exact dictionary representation (elsewhere called the sparse approximation or highly nonlinear approximation problem [32]).

The min unsatisfy problem is an intuitive problem on linear equations. Given a system $Ax = b$ of linear equations (where A is an integer $m \times n$ matrix and b is an integer m-vector), the problem is to provide a rational n-vector x; the measure to be minimized is the number of equations not satisfied by $Ax = b$. The term "min unsatisfy" was first coined by Arora et al. [2] in a seminal paper on the hardness of approximation, but they claim that the the NP-Completeness of the decision version of this problem is implicit in a 1978 paper of Johnson and Preparata [18]. Arora et al. demonstrated that it is hard to approximate min unsatisfy to within a factor $2^{\log^{.5-o(1)} n}$ of optimal (under the assumption that NP does not admit a quasi-polynomial time deterministic algorithm). This hardness result holds over $\mathbb{Q}$, and stronger results are known for finite fields [10]. For this problem, Berman and Karpinski [3] gave a randomized $\frac{m}{c \log m}$-approximation algorithm (where c is a constant). We know of no heuristics studied for this problem.

The exact dictionary representation problem is the fundamental problem in sparse approximation theory (see [23]). In this problem, we are given a matrix of dictionary vectors D and a target vector s, and the task is to find the smallest set $D' \subset D$ such that a linear combination of the vectors of D' is equal to s. This problem and its variants have been well studied. According to Temlyakov [31], a variant of this problem may be found as early as 1907, in a paper of Schmidt [28]. The decision version of this problem was shown to be NP-Complete by Natarajan [24]. (See [21] for further discussion.)

The field of sparse approximation theory has become exceedingly popular: For example, SPAR05 was largely devoted to it, as was the SparseLand 2006 workshop at Princeton, and a mini-symposium at NYU's Courant Institute in 2007. The applications of sparse approximation theory include signal representation and recovery [8,25], amplitude optimization [29] and function approximation [24]. When the dictionary vectors are Fourier coefficients, this problem is a classic problem in Fourier analysis, with applications in data compression, feature extraction, locating approximate periods and similar data mining problems [37,14,15,6]. There is a host of results for this problem, though all are heuristics or approximations under some qualifying assumptions. In fact, Amaldi and Kann [1] showed that this problem (they called it RVLS – 'relevant variables in the linear system') is as hard to approximate as min unsatisfy, though their result seems to have escaped the notice of the sparse approximation theory community.

Our contribution. As a first step, we note that the matrix problems sparse null space and matrix sparsification are equivalent, and that the vector problems min unsatisfy and exact dictionary representation are equivalent as well. (Due to space constraints, proofs of equivalence are deferred to the full version of this paper. Note that although these equivalences are straightforward, they seem to have escaped researchers in this field. For example, [5] claimed that the sparse null space problem is computationally more difficult than matrix sparsification.)

We proceed to show that matrix sparsification is hard to approximate, via a reduction from min unsatisfy. We will thereby show that the two matrix problems are hard to approximate within a factor $2^{\log^{.5-o(1)} n}$ of optimal (assuming NP does not admit quasi-polynomial time deterministic algorithms).

This hardness result for matrix sparsification is important in its own right, but it further leads us to ask what *can* be done for this problem. Specifically, what restrictions or simplifying assumptions may be made upon the input matrix to make matrix sparsification problem tractable? In addressing this question, we provide the major contribution of this paper and show how to adapt the vast number of heuristics and algorithms for exact dictionary representation to solve matrix sparsification (and hence sparse null space as well). This allows us to conclude, for example, that matrix sparsification admits a randomized $\frac{m}{c \log m}$-approximation algorithm, and also to give limiting conditions under which a known ℓ_1 relaxation scheme for exact dictionary matching solves matrix sparsification exactly. All of our results assume that the vector variables are over $\mathbb{Q}$.

An outline of our paper follows: In Section 2 we review some linear algebra and introduce notation. In Section 3 we prove that matrix sparsification is hard to approximate, and in Section 4 we show how to adapt algorithms for exact dictionary representation to solve matrix sparsification.

2   Preliminaries

In this section we review some linear algebra, introduce notation and definitions, and formally state our four problems.

2.1   Linear Algebra and Notation

Matrix and vector properties. Given a set V of n m-dimensional column vectors, an m-vector $v \notin V$ is *independent* of the vectors of V if there is no linear combination of vectors in V that equals v. A set of vectors is independent if each vector in the set is independent of the rest.

Now let the vectors of V be arranged as columns of an $m \times n$ matrix A; we refer to a column of A as a_i, and to a position in A as a_{ij}. We define #col(A) to be the number of columns of A. The column *span* of A (col(A)) is the (infinite) set of column vectors that can be produced by a linear combination of the columns of A. The *column rank* of A is the dimension of the column space of A (rank(A) = dim(col(A))); it is the size of the maximal independent subset in the columns of A. If the column rank of A is equal to n, then the columns of A are independent, and A is said to be *full rank*.

Other matrices may be produced from A using *elementary column operations*. These include multiplying columns by a nonzero factor, interchanging columns, and adding a multiple of one column to another. These operations produce a matrix A' which has the same column span as A; we say A and A' are *column equivalent*. It can be shown that A, A' are column equivalent iff $A' = AX$ for some invertible matrix X.

Let R be a set of rows of A, and C be a set of columns. $A(R,C)$ is the submatrix of A restricted to R and C. Let $A(:,C)$ ($A(R,:)$) be the submatrix of A restricted to all rows of A and to columns in C (restricted to the rows of R and all columns in A). A *square matrix* is an $m \times m$ matrix. A square matrix is *nonsingular* if it is invertible.

Null space. The *null space* (or *kernel*) of A (null(A)) is the set of all nonzero n-length vectors b for which $Ab = 0$. The rank of A's null space is called the *corank* of A. The rank-nullity theorem states that for any matrix A, rank(A)+ corank(A) = n. Let N be a matrix consisting of column vectors in the null space of A; we have that $AN = 0$. If the rank of N is equal to the corank of A then N is a *full null matrix* for A.

Given matrix A, a full null matrix for A can be constructed in polynomial time. Similarly, given a full rank matrix N, polynomial time is required to construct a matrix A for which N is a full null matrix [26].

Notation. Throughout this paper, we will be interested in the number of zero and nonzero entries in a matrix A. Let nnz(A) denote the number of nonzero entries in A. For a vector x, let $||x||_0$ denote the number of nonzero entries in x. This notation refers to the quasi-norm ℓ_0, which is not a true norm since $\lambda ||x||_0 \neq ||\lambda x||_0$, although it does honor the triangle inequality.

For vector x, let x_i be the value of the ith position in x. The *support* of x (supp(x)) is the set of indices in x which correspond to nonzero values, $i \in supp(x) \Leftrightarrow x_i \neq 0$.

The notation $A|B$ indicates that the rows of matrix B are concatenated to the rows of matrix A. The notation $\binom{A}{B}$ indicates that the columns of B are appended to the columns of A. $M = A \otimes B$ denotes the Kronecker product of two matrices, where M is formed by multiplying each individual entry in A by the entire matrix B. (If A is $m \times n$, B is $p \times q$, then M is $mp \times nq$.)

By *equivalent* problems, we mean that reductions between them preserve approximation factors. A formal definition of approximation equivalence is deferred to the full version of this paper.

2.2   Minimization Problems

In this section, we formally state the four major minimization problems discussed in this paper. The first two problems have vector solutions, and the second two problems have matrix solutions. Our results hold when the variables are over $\mathbb{Q}$, although these problems can be defined over $\mathbb{R}$. $\mathcal{I}_F$ is the set of input instances, $S_F(x)$ is the solution space for $x \in \mathcal{I}_F$, $M_F(x, y)$ is the objective metric for $x \in \mathcal{I}_F$ and $y \in S_F(x)$.

Problem 1. exact dictionary representation (EDR)
$\mathcal{I}_{\mathsf{EDR}} = \langle D, s \rangle$, $m \times n$ matrix D, vector s with $s \in \mathsf{col}(D)$
$S_{\mathsf{EDR}}(D, s) = \{v \in \mathbb{Q}^n : Dv = s\}$
$m_{\mathsf{EDR}}(\langle D, s \rangle, v) = ||v||_0$

Problem 2. min unsatisfy (MU)
$\mathcal{I}_{\mathsf{MU}} = \langle A, y \rangle$, $m \times n$ matrix A, vector $y \in \mathbb{Q}^m$
$S_{\mathsf{MU}}(A, y) = \{x : x \in \mathbb{Q}^n\}$
$m_{\mathsf{MU}}(\langle A, y \rangle, x) = ||y - Ax||_0$

Problem 3. sparse null space (SNS)
$\mathcal{I}_{\mathsf{SNS}} = $ matrix A
$S_{\mathsf{SNS}}(A) = \{N : N$ is a full null matrix for $A\}$
$m_{\mathsf{SNS}}(A, N) = \mathsf{nnz}(N)$

Problem 4. matrix sparsification (MS)
$\mathcal{I}_{\mathsf{MS}} = $ full rank $m \times n$ matrix B
$S_{\mathsf{MS}}(B) = \{$matrix $N : N = BX$ for some invertible matrix $X\}$
$m_{\mathsf{MS}}(B, N) = \mathsf{nnz}(N)$

3   Hardness of Approximation for Matrix Problems

In this section, we prove the hardness of approximation of matrix sparsification (and therefore sparse null space). This motivates the search for heuristics or algorithms under simplifying assumptions for matrix sparsification, which we undertake in the next section. For the reduction, we will need a relatively dense matrix which we know cannot be further sparsified. We will prove the existence of such a matrix in the first subsection.

3.1   Unsparsifiable Matrices

Any $m \times n$ matrix A may be column reduced to contain at most $(m - r + 1)r$ nonzeros, where $r = \mathsf{rank}(A)$. For example, Gaussian elimination on the columns of the matrix will accomplish this sparsification. We will say that a rank r, $m \times n$ matrix A is *completely unsparsifiable* if and only if, for any invertible matrix X, $\mathsf{nnz}(AX) \geq (m - r + 1)r$. A matrix A is *optimally sparse* if, for any invertible X, $\mathsf{nnz}(AX) \geq \mathsf{nnz}(A)$. The main result of this section follows.

Theorem 1. *Let A be an $m \times n$ matrix with $m \geq n$. If every square submatrix of A is nonsingular, then A has rank n and is completely unsparsifiable. Moreover, in such case the matrix $\binom{I}{A}$ is optimally sparse, where I is the $n \times n$ identity matrix.*

Before attempting a proof of the theorem, we need a few intermediate results.

Lemma 1. *Matrix A is optimally sparse if and only if, for any vector $x \neq 0$, $||Ax||_0 \geq \max_{i \in supp(x)} ||a_i||_0$.*

Proof.   Suppose that there exists an x that, for some $i \in \text{supp}(x)$, $||Ax||_0 < ||a_i||_0$. Then we may replace the matrix column a_i by Ax, and create a matrix with the same rank as A which is sparser than A; a contradiction. Similarly, suppose that A is not optimally sparse, so that there exists $B = AX$ with $\text{nnz}(B) < \text{nnz}(A)$, for some invertible X. Assume without loss of generality that the diagonal of X is full, $x_{ii} \neq 0$ (otherwise just permute the columns of X to make it so). Then there must exist an index $j \in [n]$ with $||b_j||_0 < ||a_j||_0$, and we have $||Ax_j||_0 = ||b_j||_0 < ||a_j||_0 \leq \max_{i \in supp(x_j)} ||a_i||_0$, since $x_{jj} \neq 0$.   □

A submatrix $A(R, C)$ is *row-inclusive* iff $r \notin R$ implies that $A(r, C)$ is not in the row span of $A(R, C)$. In other words, $A(R, C)$ includes all the rows of $A(:, C)$ which are in the row span of this submatrix. A submatrix $A(R, C)$ is a *candidate submatrix* of A (written $A(R, C) \triangleleft A$) if and only if $A(R, C)$ is both row-inclusive and $\text{rank}(A(R, C)) = |C| - 1$. This last property is equivalent to stating that the columns of $A(R, C)$ form a *circuit* – they are minimally linearly dependent. We can potentially zero out $|R|$ entries of A by using the column dependency of $A(R, C)$; being row-inclusive means there would be exactly $|R|$ zeros in the modified column of A.

The next lemma demonstrates the close relationship between candidate submatrices and vectors x which may sparsify A as in Lemma 1.

Lemma 2. *For any $m \times n$ matrix A: (1) For any $x \neq 0$ and $i \in supp(x)$, there exists $A(R, C) \triangleleft A$ for which $|R| \geq m - ||Ax||_0$, and $i \in C \subset supp(x)$. (2) For any $A(R, C) \triangleleft A$ there exists a vector x for which $supp(x) = C$ and $||Ax||_0 = m - |R|$.*

Proof.   Part 1: Let $R' = [m] - \text{supp}(Ax)$ (where $[m] = \{1, 2, ..., m\}$), and choose C so that $i \in C \subset \text{supp}(x)$, and the columns of $A(R', C)$ form a circuit. (Note that the columns $A(R', \text{supp}(x))$ are dependent since $A(R', :)x = 0$). Now expand R' to R so that $A(R, C)$ is row-inclusive. Then $\text{rank}(A(R, C)) = \text{rank}(A(R', C)) = |C| - 1$, so that $A(R, C) \triangleleft A$.

Part 2: Since the columns of $A(R, C)$ form a circuit, there is an $\tilde{x}$ with $\tilde{x}_i \neq 0 \, \forall i$ and $A(R, C)\tilde{x} = 0$. Then $\dim(\text{col}(A(R, C)^T)) = |C| - 1 = \dim(\text{null}(\tilde{x}^T))$ and also $\text{col}(A(R, C)^T) \subset \text{null}(\tilde{x}^T)$, which together imply $\text{col}(A(R, C)^T) = \text{null}(\tilde{x}^T)$. So $A(r, C)\tilde{x} = 0$ is true iff $r \in R$ (using the fact that $A(R, C)$ is row-inclusive). Now choose x so that $x(C) = \tilde{x}$ and all other coordinates are zero; then $\text{supp}(Ax) = [m] - R$.   □

The following is an immediate consequence of the lemma, and is crucial to our proof of Theorem 1.

Corollary 1. *The $m \times n$ matrix A is optimally sparse if and only if there is no candidate submatrix $A(R, C) \lhd A$ with $m - |R| < ||a_i||_0$ for some $i \in C$.*

We are now ready to prove the theorem.

Proof of Theorem 1.   Let $B = \binom{I}{A}$. We prove that B is optimally sparse. Suppose $B(R, C) \lhd B$. Let $R_I = R \cap [n]$ and $R_A = R - [n]$. Now $B(R_I, C)$ is a submatrix of I with dependent columns, so $B(R_I, C) = 0$. By row-inclusiveness, R_I must include all zero rows in $B([n], C)$, so $|R_I| = n - |C|$. Since $B(R_I, C) = 0$, it follows that $\mathsf{rank}(B(R_A, C)) = \mathsf{rank}(B(R, C)) = |C| - 1$, and $|R_A| \geq |C| - 1$. Any $|C| \times |C|$ subsquare of $B(R_A, C)$ would make the rank at least $|C|$, so we must have $|R_A| < |C|$; thus $|R_A| = |C| - 1$. Combined with $|R_I| = n - |C|$, this implies that $|R| = n - 1$. Then $m + n - |R| = m + 1 = ||b_i||_0$ for any column b_i of B, proving that B is optimally sparse by corollary 1.

Recall that Gaussian elimination on matrix $A \to G$ yields $\mathsf{nnz}(G) = (m - n + 1)n$. Now suppose there is an invertible matrix X with $\mathsf{nnz}(AX) < (m - n + 1)n$. Then $\mathsf{nnz}(BX) = \mathsf{nnz}(\binom{X}{AX})) < n^2 + (m - n + 1)n = (m + 1)n$, contradicting the optimal sparsity of B. Hence no such X exists and A is completely unsparsifiable. □

3.2   Efficiently Building an Unsparsifiable Matrix

The next lemma establishes that we can easily construct an unsparsifiable matrix with a given column, a useful fact for the reductions to follow.

Lemma 3. *If $n \times n$ matrix $M = (M_{ij})$ has entries $m_{ij} = i^{p_j}$ for distinct positive reals $p_1, p_2, \ldots, p_n$, then every subsquare of M is nonsingular.*

Proof.   Let f be a *signomial* (a polynomial allowed to have nonintegral exponents). We define $\mathsf{positive\_zeros}(f) := \{x : x > 0 \; \& \; f(x) = 0\}$ and $\#\mathsf{sign\_changes}$ $(f) := \#\{i : \mu_i \mu_{i+1} < 0\}$, where $f = \sum_i \mu_i x^{p_i}$, and no $\mu_i = 0$. A slight generalization of Descartes' rule of signs [35] states that

$$\#\mathsf{positive\_zeros}(f) \leq \#\mathsf{sign\_changes}(f) \qquad (1)$$

Consider any $k \times k$ subsquare $M(R, C)$ given by $R = \{r_1, \ldots, r_k\}, C = \{c_1, \ldots, c_k\} \subset [n]$, and any nonzero vector $\mu \in \mathbb{R}^k$. Then $M(R, C) \cdot \mu$ matches the signomial $f(x) = \sum \mu_i x^{p_{c_i}}$ evaluated at $x = r_1, \ldots, r_k$. Using (1), $\#\{i : f(r_i) = 0\} \leq \#\mathsf{sign\_changes}(f) < k$, so that some $f(r_i) \neq 0$, and $M(R, C)\mu \neq 0$. Hence the subsquare has a trivial kernel, and is nonsingular. □

To avoid problems of precision, we will choose powers of p_j to be consecutive integers beginning at 0. This yields the Vandermonde matrix over $\mathbb{Q}$. It can also be shown, by an elementary cardinality argument, that a random matrix (using a non-atomic distribution) is unsparsifiable with probability 1 over infinite fields. The above lemma avoids any probability and allows us to construct such a matrix as quickly as we can iterate over the entries.

3.3   Reduction for Matrix Problems

After proving the existence of an unsparsifiable matrix in the last section, we can now prove the hardness of approximation of matrix sparsification. We reduce min unsatisfy to matrix sparsification. Given an instance $\langle A, y \rangle$ of min unsatisfy, we create a matrix M such that matrix sparsification on M solves the instance of min unsatisfy.

Before describing the reduction, we outline the intuition behind it. We wish to create a matrix M with many copies of y and some copies of A. The number of copies of y should greatly outnumber the number of copies of A. The desired approximation bounds will be achieved by guaranteeing that M is composed mostly of zero entries and of copies of y. It follows that minimizing the number of nonzero entries in the matrix (solving matrix sparsification) will reduce to minimizing the number of nonzero entries in the copies of y by finding a sparse linear combination of y with some other dictionary vectors (solving min unsatisfy).

The construction is as follows: Given an instance $\langle A, y \rangle$ of min unsatisfy (where A is an $m \times n$ matrix, $y \in R^m$, and $q \geq p$ are free parameters), take an optimally sparse $(p + q) \times p$ matrix $\binom{I_p}{X}$ as given by Lemma 3 and Theorem 1 (where I_p is a $p \times p$ identity matrix), and create matrix $M_l = \binom{I_p}{X} \otimes y = \binom{I_p \otimes y}{X \otimes y}$ (of size $(p+q)m \times p$). Further create matrix $I_q \otimes A$ (of size $qm \times qn$), and take matrix 0 (of size $pm \times qn$) and form matrix $M_r = \binom{0}{I_q \otimes A}$ (of size $(p+q)m \times qn$). Append M_r to the right of M_l to create matrix $M = M_l | M_r$ of size $(p+q)m \times (p + qn)$. We can summarize this construction as $M = \begin{pmatrix} I_p \otimes y & 0 \\ X \otimes y & I_q \otimes A \end{pmatrix}$.

M_l is composed of $p + pq$ m-length vectors, all corresponding to copies of y. M_r is composed of qn m-length vectors, all corresponding copies of vectors in A. By choosing $p = q = n^2$, we ensure that the term pq is larger than qn by a factor of n. Note that M now contains $O(n^3)$ columns.

It follows that the number of zeros in M depends mostly on the number of zeros induced by a linear combination of dictionary vectors that include y. Because M_l is unsparsifiable, vectors in the rows of M_l will not contribute to sparsifying other vectors in these rows; only vectors in M_r (which are copies of the vectors of A) may sparsify vectors in M_l (which are copies of the vectors in y). It follows that an approximation to matrix sparsification will yield a similar approximation – within a factor of $1 + n^{-\frac{1}{3}}$ – to min unsatisfy, and that matrix sparsification is hard to approximate within a factor $2^{\log^{.5-o(1)} n^{1/3}} = 2^{\log^{.5-o(1)} n}$ of optimal (assuming NP does not admit quasi-polynomial time deterministic algorithms).

4   Solving matrix sparsification through min unsatisfy

In the previous section we showed that matrix sparsification is hard to approximate. This motivates the search for heuristics and algorithms under simplifying assumptions for matrix sparsification. In this section we show how to extend algorithms and heuristics for min unsatisfy to apply to matrix sparsification – and hence sparse null space – while preserving approximation guarantees. (Note that this result is distinct from the hardness result; neither one implies the other.)

We first present an algorithm for matrix sparsification which is in essence identical to the one given by Coleman and Pothen [9] for sparse null space. The algorithm assumes the existence of an oracle for a problem we will call the sparsest independent vector problem. The algorithm makes a polynomial number of queries to this oracle, and yields an optimal solution to matrix sparsification.

The sparsest independent vector problem takes full-rank input matrices A and B, where the columns of B are a contiguous set of right-most columns from A (informally, one could say that B is a suffix of A, in terms of columns). The output is the sparsest vector in the span of A but not in the span of B. For convenience, we add an extra output parameter — a column of $A \setminus B$ which can be replaced by the sparsest independent vector while preserving the span of A. More formally, sparsest independent vector is defined as follows.

Problem 5. sparsest independent vector (SIV)
$\mathcal{I}_{\mathsf{SIV}} = \langle A, B \rangle$; A is an $m \times n$ full rank matrix with $A = (C|B)$ for some non-empty matrix C.
$S_{\mathsf{SIV}}(A, B) = \{a : a \in \mathsf{col}(A), a \notin \mathsf{col}(B)\}$
$m_{\mathsf{SIV}}(\langle A, B \rangle, a) = \mathsf{nnz}(a)$

The following algorithm reduces matrix sparsification on an $m \times n$ input matrix A to making a polynomial number of queries to an oracle for sparsest independent vector:

Algorithm Matrix_Sparsification(A)
$B \leftarrow$ null
for $i = n$ to 1:
    $\langle b_i, a_j \rangle = \mathsf{SIV}(A, B)$
    $A \leftarrow (A \setminus \{a_j\}|b_i)$
    $B \leftarrow (b_i|B)$
return B

This greedy algorithm sparsifies the matrix A by generating a new matrix B one column at a time. The first-added column (b_n) is the sparsest possible, and each subsequent column is the next sparsest. It is decidedly non-obvious why such a greedy algorithm would actually succeed; we refer the reader to [9] where it is proven that greedy algorithms yield an optimal result on matroids such as the set of vectors in $\mathsf{col}(A)$. Our first contribution is in expanding the result of [9] as follows.

Lemma 4. *Let subroutine* SIV *in algorithm Matrix_Sparsification be a λ-approximation oracle for* sparse independent vector. *Then the algorithm yields a λ-approximation to* matrix sparsification.

Proof. Given $m \times n$ matrix A, suppose $\tilde{C}$ exactly solves $\mathsf{MS}(A)$, and that the columns $\tilde{c}_1, \ldots, \tilde{c}_n$ of $\tilde{C}$ are sorted in decreasing order by number of nonzeros. Let $s_i = ||\tilde{c}_i||_0$; then $s_1 \geq s_2 \geq \ldots \geq s_n$. As already mentioned, given a true oracle to sparsest independent vector, algorithm Matrix_Sparsification would first discover a column with s_n nonzeros, then a column with s_{n-1} nonzeros, etc.

Now suppose algorithm Matrix_Sparsification made calls to a $\lambda-$approximation oracle for sparse independent vector. The first column generated by the algorithm, call it b_n, will have at most λs_n nonzeros, since the optimal solution has s_n nonzeros. The second column generated will have at most λs_{n-1} nonzeros, since the optimal solution to the call to SIV has no more than s_{n-1} nonzeros: even if b_n is suboptimal, it is true that at least one of $\tilde{c}_n$ or $\tilde{c}_{n-1}$ is an optimal solution to $SIV(A, b_n)$.

More generally, the ith column found by the algorithm has no more then λs_i nonzeros, since at least one of $\{\tilde{c}_n, \ldots, \tilde{c}_i\}$ is an optimal solution to the ith query to SIV. Thus we have $\text{nnz}(B) = \sum_i ||b_i||_0 \leq \sum \lambda ||\tilde{c}_i||_0 = \lambda \ \text{nnz}(\tilde{C})$, and may conclude that the algorithm yields a $\lambda-$approximation to matrix sparsification. □

It follows that in order to utilize the aforementioned algorithm for matrix sparsification, we need some algorithm for sparsest independent vector. This is in itself problematic, as the sparsest independent vector problem is hard to approximate – in fact, we will demonstrate later that sparsest independent vector is as hard to approximate as min unsatisfy. Hence, although we have extended the algorithm of [9] to make use of an approximation oracle for sparsest independent vector, the benefit of this algorithm remains unclear.

To this end, we will show how to solve sparsest independent vector while making queries to an approximate oracle for min unsatisfy. This algorithm preserves the approximation ratio of the oracle. This implies that *all* algorithms for min unsatisfy immediately carry over to sparsest independent vector, and further that they carry over to matrix sparsification as well. This also implies a useful tool for applying heuristics for min satisfy to the other problems.

The problem sparsest independent vector on input $\langle A, B \rangle$ asks to find the sparsest vector in the span of A but not in the span of B. It is not difficult to see that min unsatisfy solves a similar problem: Given a matrix A and target vector y not in the span of A, find the sparsest vector in the span of $(A|y)$ but not in the span of A. Hence, if we query the oracle for min unsatisfy once for each vector $a_j \notin col(B)$, one of these queries must return the solution for the sparsest independent vector problem. This discussion implies the following algorithm:

Algorithm Sparse_Independent_Vector(A, B)
$s \leftarrow m + 1$
for $j = 1$ to n:
    if $a_j \notin col(B)$:
        $A_j \leftarrow A\backslash\{a_j\}$
        $x \leftarrow \text{MU}(A_j, a_j)$
        $c' \leftarrow A_j x - a_j$
        if $||c'||_0 < s$
            $c \leftarrow c'; \ s \leftarrow ||c||_0; \ \alpha \leftarrow a_j$
    return $\langle c, \alpha \rangle$

Note that when this algorithm is given a λ-approximate oracle for min unsatisfy, it yields a λ-approximate algorithm for sparsest independent vector. (In this case, the approximation algorithm is valid over the field for which the oracle is valid.)

We conclude this section by giving hardness results for sparsest independent vector by reduction from min unsatisfy; we show that any instance $\langle A, b \rangle$ of min unsatisfy may be modeled as an instance $\langle A', B' \rangle$ of sparsest independent vector: Let $A' = A|y$, and $B' = A$. This suffices to force the linear combination to include y. It follows that sparsest independent vector is as hard to approximate as min unsatisfy, and in fact that the two problems are approximation equivalent.

4.1   Approximation Algorithms

We have presented a tool for extending algorithms and heuristics for exact dictionary representation to min unsatisfy and then directly to the matrix problems. When these algorithms make assumptions on the dictionary of EDR, it is necessary to investigate how these assumptions carry over to the other problems.

To this end, we consider here one of the most popular heuristic for EDR – ℓ_1-minimization – and the case where it is guaranteed to provide the optimal result. The heuristic is to find a vector v that satisfies $Dv = s$, while minimizing $||v||_1$ instead of $||v||_0$. (See [34,33,11] for more details.) In [13], Fuchs shows that under the following relatively simple condition ℓ_1-minimization provides the optimal answer to EDR.

In the following, we write $\mathsf{sgn}(x)$ to indicate $\frac{x}{|x|}$, or zero if $x = 0$. Given a matrix D whose columns are divided into two submatrices D_0 and D_1, we may write $D = (D_0 \ D_1)$, even though D_0 and D_1 may not be contiguous portions of the full matrix. (The reader may view this as permuting the columns of D before splitting into D_0 and D_1.)

Theorem 2 (Fuchs). *Suppose that* $s = Dv$, *and that* $||v||_0$ *is minimal (so that this* v *solves* EDR(D, s)*). Split* $D = (D_0 \ D_1)$ *so that* D_0 *contains all the columns in the support of* v*. Accordingly, we split the vector* $v = \binom{v_0}{0}$*, in which all coordinates of* v_0 *are nonzero.*

If there exists a vector h *so that* $D_0^T h = \mathsf{sgn}(v_0)$*, and* $||D_1^T h||_\infty < 1$*, then* $||v||_1 < ||w||_1$ *for all vectors* $w \neq v$ *with* $Dw = s$*.*

We extend this result to each of our major problems.

Theorem 3.   *min unsatisfy. Suppose, for a given* A, y *pair, that* x *minimizes* $||y - Ax||_0$*. Split* $y = \binom{y_0}{y_1}$ *and* $A = \binom{A_0}{A_1}$ *so that* A_1 *is maximal such that* $y_1 = A_1 x$*, and let* $v = y_0 - A_0 x$*. If there is a matrix* u *with* $||u||_\infty < 1$ *and* $A_1^T u = -A_0^T \mathsf{sgn}(v)$*, then our reduction of* MU(A, y) *to an* ℓ_1 *approximation of* EDR(D, s) *gives the truly optimal answer.*

matrix sparsification. For a given $m \times n$ *matrix* B*, suppose* C *minimizes* $\mathsf{nnz}(C)$ *such that* $C = BX$ *for invertible* X*. For any* $i \in [n]$*, split column* $c_i = \binom{c_{i,0}}{0}$ *so that* $c_{i,0}$ *is completely nonzero, and, respectively,* $B = \binom{B_{i,0}}{B_{i,1}}$*, so that* $c_{i,0} =$

$B_{i,0}x_i$. If, for all $i \in [n]$, there exists vector u_i with $||u_i||_\infty < 1$ and $B_{i,1}^T u_i = -B_{i,0}^T \mathsf{sgn}(c_{i,0})$, then our reduction algorithm to an ℓ_1 approximation of EDR via min unsatisfy will give a truly optimal answer to this MS instance.

sparse null space For a given matrix A with corank c, suppose matrix V solves SNS(A). For each $i \in [c]$, split column $v_i = \binom{v_{i,0}}{\mathbf{0}}$ so that $v_{i,0}$ is completely nonzero and, respectively, $A = (A_{i,0}\ A_{i,1})$ so that $A_{i,0}v_{i,0} = 0$. If, for all $i \in [c]$, there exists vector h_i with $||A_{i,1}^T h_i||_\infty < 1$ and $A_{i,0}^T h_i = \mathsf{sgn}(v_{i,0})$, then our reduction to an ℓ_1 approximation of EDR via matrix sparsification and min unsatisfy gives a truly optimal answer to this SNS instance.

Proof.   min unsatisfy. As in our reduction from MU to EDR, we find matrix D with $DA = 0$ and vector $s = Dy$. Then

$$\binom{\mathsf{sgn}(v_0)}{u} \in \mathsf{null}(A^T) = \mathsf{col}(D^T) \implies \exists h : D^T h = \binom{\mathsf{sgn}(v)}{u}.$$

Splitting $D = (D_0\ D_1)$, we see that $D_0^T h = \mathsf{sgn}(v)$ and $||D_1^T h||_\infty < 1$, exactly what is required for theorem 2, showing that ℓ_1 minimization gives the answer $D_0 v_0$. Since $D_0 v_0 = (D_0\ D_1)\binom{v}{0} = D(y - Ax) = s$, this completes the proof.

matrix sparsification. We write $A \setminus i$ to denote matrix A with the ith column removed. In our reduction of MS to MU, we need to solve instances of MU over equations of the form $(B \setminus i)x = b_i$. According to the MU portion of this theorem, it suffices to show that $(B_{i,1} \setminus i)^T u_i = -(B_{i,0} \setminus i)^T \mathsf{sgn}(c_{i,0})$. The condition for this portion of the theorem implies this, since removing any corresponding rows from a matrix equation of the form $Ax = By$ still preserves the equality.

sparse null space. As in our reduction from SNS to MS, we find a matrix B such that A is a full null matrix for B. For any i, let $u_i = A_{i,1}^T h_i$ so that $A^T h_i = \binom{\mathsf{sgn}(v_{i,0})}{u_i}$. Then $\binom{\mathsf{sgn}(v_{i,0})}{u_i} \in \mathsf{col}(A^T) = \mathsf{null}(B^T)$, and $B_{i,1}^T u_i = -B_{i,0}^T \mathsf{sgn}(v_{i,0})$, which is exactly what is necessary for matrix sparsification to function through ℓ_1 approximation. □

The following intuitive conditions give insight into which matrices are amenable to ℓ_1 approximations. A^+ denotes $(A^T A)^{-1} A^T$, the pseudoinverse of A.

Corollary 2. min unsatisfy. *Suppose matrix* $A = \binom{A_0}{A_1}$ *is split by an optimal answer as in theorem 3. If* $\mathsf{row}(A_0) \subset \mathsf{row}(A_1)$ *and* $||(A_1^T)^+ A_0^T||_{1,1} < 1$, *our* ℓ_1 *approximation scheme will give a truly optimal answer.*

matrix sparsification. *Suppose matrix* $B = \binom{B_{i,0}}{B_{i,1}}$ *is split by the columns of an optimal answer* $C = BX$ *as in theorem 3. If, for any i,* $\mathsf{row}(B_{i,0}) \subset \mathsf{row}(B_{i,1})$ *and* $||(B_{i,1}^T)^+ B_{i,0}^T||_{1,1} < 1$, *then our* ℓ_1 *approximation will give the optimal answer.*

sparse null space. *Suppose matrix* $A = (A_{i,0}\ A_{i,1})$ *is split by the columns of an optimal answer* V *with* $AV = 0$ *as in theorem 3. If* $\mathsf{col}(A_{i,0}) \subset \mathsf{col}(A_{i,1})$ *and* $||A_{i,0}^+ A_{i,1}||_{1,1} < 1\ \forall i$, *then our* ℓ_1 *approximation will give an optimal answer.*

Acknowledgements

We thank Daniel Cohen for finding an error in an earlier version of this paper.

References

1. Amaldi, E., Kann, V.: The complexity and approximability of finding maximum feasible subsystems of linear relations. Theoretical computer science 147(1-2), 181–210 (1995)
2. Arora, S., Babai, L., Stern, J., Sweedyk, Z.: The hardness of approximate optima in lattices, codes and linear equations. JCSS 54(2) (1997)
3. Berman, P., Karpinski, M.: Approximating minimum unsatisfiability of linear equations. Electronic Colloquium on Computational Complexity (ECCC) 8(25) (2001)
4. Berry, M., Heath, M., Kaneko, I., Lawo, M., Plemmons, R., Ward, R.: An algorithm to compute a sparse basis of the null space. Numer. Math. 47, 483–504 (1985)
5. Brualdi, R.A., Friedland, S., Pothen, A.: The sparse basis problem and multilinear algebra. SIAM Journal on Matrix Analysis and Applications 16(1), 1–20 (1995)
6. Candès, E., Romberg, J., Tao, T.: Robust uncertainty principles: Exact signal reconstruction from highly incomplete frequency information. IEEE Trans. Inform. Theory 52, 489–509 (2006)
7. Frank Chang, S., Thomas McCormick, S.: A hierarchical algorithm for making sparse matrices sparser. Mathematical Programming 56(1-3), 1–30 (1992)
8. Coifman, R., Wickerhauser, M.: Entropy-based algorithms for best basis selection. IEEE Transactions on Information Theory 38(2), 713–718 (1992)
9. Coleman, T.F., Pothen, A.: The null space problem I. complexity. SIAM Journal on Algebraic and Discrete Methods 7(4), 527–537 (1986)
10. Dinur, I., Kindler, G., Raz, R., Safra, S.: Approximating CVP to within almost-polynomial factors is NP-hard. Combinatorica 23(2), 205–243 (2003)
11. Donoho, D.L.: For most large underdetermined systems of linear equations, the minimal l1-norm solution is also the sparsest (2004),
    http://www-stat.stanford.edu/ donoho/Reports/2004/
    l1l0EquivCorrected.pdf
12. Duff, I.S., Erisman, A.M., Reid, J.K.: Direct Methods for Sparse Matrices. Oxford University Press, Oxford (1986)
13. Fuchs, J.J.: On sparse representations in arbitrary redundant bases. IEEE Trans. Inf. Th. 50(6), 1341–1344 (2004)
14. Gilbert, A.C., Guha, S., Indyk, P., Muthukrishnan, S., Strauss, M.: Near-optimal sparse fourier representations via sampling. In: STOC (2002)
15. Gilbert, A.C., Muthukrishnan, S., Strauss, M.: Improved time bounds for near-optimal sparse fourier representations. In: Proc. SPIE Wavelets XI (2005)
16. Gilbert, J.R., Heath, M.T.: Computing a sparse basis for the null space. SIAM Journal on Algebraic and Discrete Methods 8(3), 446–459 (1987)
17. Hoffman, A.J., McCormick, S.T.: A fast algorithm that makes matrices optimally sparse. In: Progress in Combinatorial Optimization. Academic Press, London (1984)
18. Johnson, D.S., Preparata, F.P.: The densent hemisphere problem. Theoret. Comput. Sci. 6, 93–107 (1978)
19. Kann, V.: Polynomially bounded minimization problems that are hard to approximate. Nordic Journal of Computing 1(3), 317–331 (Fall 1994)
20. Kavitha, T., Mehlhorn, K., Michail, D., Paluch, K.: A faster algorithm for minimum cycle basis of graphs. In: Díaz, J., Karhumäki, J., Lepistö, A., Sannella, D. (eds.) ICALP 2004. LNCS, vol. 3142, pp. 846–857. Springer, Heidelberg (2004)
21. Matoušek, J., Gartner, B.: Understanding and Using Linear Programming. Springer, Heidelberg (2007)

22. McCormick, S.T.: A combinatorial approach to some sparse matrix problems. PhD thesis, Stanford Univ., Stanford, California, Technical Report 83-5, Stanford Optimization Lab. (1983)
23. Muthukrishnan, S.: Nonuniform sparse approximation with haar wavelet basis. DIMACS TR:2004-42 (2004)
24. Natarajan, B.K.: Sparse approximate solutions to linear systems. SIAM J. on Comput. 24(2), 227–234 (1995)
25. Needell, D., Tropp, J.: Cosamp: Iterative signal recovery from incomplete and inaccurate samples. Applied and Computational Harmonic Analysis 26(3), 301–322 (2009)
26. Neylon, T.: Sparse solutions for linear prediction problems. PhD thesis, New York University, New York, New York (2006)
27. Pothen, A.: Sparse null bases and marriage theorems. PhD thesis, Cornell University, Ithica, New York (1984)
28. Schmidt, E.: Zur thoerie der linearen und nichtlinearen integralgleichungen, i. Math. Annalen. 64(4), 433–476 (1906-1907)
29. Singhal, S.: Amplitude optimization and pitch predictin in multi-pulse coders. IEEE Transactions on Acoustics, Speech and Signal Processing 37(3) (March 1989)
30. Smola, A., Schölkopf, B.: Sparse greedy matrix approximation for machine learning. In: Proc. 17th International Conf. on Machine Learning, pp. 911–918. Morgan Kaufmann, San Francisco (2000)
31. Temlyakov, V.: Nonlinear methods of approximation. Foundations of Comp. Math. 3(1), 33–107 (2003)
32. Tropp, J.A.: Greed is good: algorithmic results for sparse approximation. IEEE Transactions on Information Theory 50(10), 2231–2242 (2004)
33. Tropp, J.A.: Just relax: convex programming methods for subset selection and sparse approximation. IEEE Transactions on Information Theory 50(3), 1030–1051 (2006)
34. Tropp, J.A.: Recovery of short, complex linear combinations via ℓ_1 minimization. IEEE Trans. Inform. Theory 51(1), 188–209 (2006)
35. Wang, X.: A simple proof of descartes' rule of signs. Amer. Math. Monthly 111, 525–526 (2004)
36. Zhao, X., Zhang, X., Neylon, T., Shasha, D.: Incremental methods for simple problems in time series: algorithms and experiments. In: Ninth International Database Engineering and Applications Symposium (IDEAS 2005), pp. 3–14 (2005)
37. Zhou, J., Gilbert, A., Strauss, M., Daubechies, I.: Theoretical and experimental analysis of a randomized algorithm for sparse fourier transform analysis. Journal of Computational Physics 211, 572–595 (2006)

The Checkpoint Problem

MohammadTaghi Hajiaghayi[1], Rohit Khandekar[2],
Guy Kortsarz[3,*], and Julián Mestre[4]

[1] AT&T Labs– Research & University of Maryland
hajiagha@research.att.com
[2] IBM T.J. Watson Research Center
rohitk@us.ibm.com
[3] Rutgers University-Camden
guyk@camden.rutgers.edu
[4] Max-Planck-Institut für Informatik
jmestre@mpi-inf.mpg.de

Abstract. In this paper we consider the *checkpoint problem*. The input consists of an undirected graph G, a set of source-destination pairs $\{(s_1, t_1), \ldots, (s_k, t_k)\}$, and a collection $\mathcal{P}$ of paths connecting the (s_i, t_i) pairs. A feasible solution is a multicut E'; namely, a set of edges whose removal disconnects every source-destination pair. For each $p \in \mathcal{P}$ we define $\mathsf{cp}_{E'}(p) = |p \cap E'|$. In the *sum checkpoint (SCP)* problem the goal is to minimize $\sum_{p \in \mathcal{P}} \mathsf{cp}_{E'}(p)$, while in the *maximum checkpoint (MCP)* problem the goal is to minimize $\max_{p \in \mathcal{P}} \mathsf{cp}_{E'}(p)$. These problem have several natural applications, e.g., in urban transportation and network security. In a sense, they combine the *multicut problem* and the *minimum membership set cover problem*.

For the sum objective we show that weighted *SCP* is equivalent, with respect to approximability, to undirected multicut. Thus there exists an $O(\log n)$ approximation for *SCP* in general graphs.

Our current approximability results for the max objective have a wide gap: we provide an approximation factor of $O(\sqrt{n \log n / \mathsf{opt}})$ for *MCP* and a hardness of 2 under the assumption P $\neq$ NP. The hardness holds for trees, in which case we can obtain an asymptotic approximation factor of 2.

Finally we show strong hardness for the well-known problem of *finding a path with minimum forbidden pairs*, which in a sense can be considered the dual to the checkpoint problem. Despite various works on this problem, hardness of approximation was not known prior to this work. We show that the problem cannot be approximated within cn for some constant $c > 0$, unless P $=$ NP. This is the strongest type of hardness possible. It carries over to directed acyclic graphs and is a huge improvement over the plain NP-hardness of Gabow (*SIAM J. Comp 2007, pages 1648–1671*).

1   Introduction

In many countries, trains and urban transport operate largely on the honor system with enforcement by roving inspectors or conductors. The typical transaction consists of a

* Partially supported by NSF grant number 0829959.

M. Serna et al. (Eds.): APPROX and RANDOM 2010, LNCS 6302, pp. 219–231, 2010.

user buying a ticket from a vending machine or a salesperson in advance and then time stamping it with a validating machine at the station just before use. Inspectors check the tickets at certain stations (or indeed on the train in between stations) called *checkpoints* that might vary from day to day and fine people without validated tickets. In these scenarios, the transportation companies generally want to make sure a ticket is checked at least once, (in all routes, even those that carry small number of people) but avoid many checkpoints at popular source-destination travel paths. Due to the inconvenience of checking tickets for passengers many times, potential delays, and lack of resources, we consider the problem of placing checkpoints to minimize the average or maximum checks of tickets for some popular source-destination paths.

This problem can be modeled as follows. We are given an undirected graph $G(V, E)$ corresponding to stations and their connections via the transit system. We are also given a set of source-destinations $\{(s_1, t_1), (s_2, t_2), \ldots, (s_k, t_k)\}$ and a set of fixed paths $\mathcal{P}$ between them. The goal is to find a set of *checkpoint* edges E' that forms a *multicut*, i.e., for every i, s_i and t_i are in different connected components in $G(V, E \setminus E')$ and minimizes the average (equivalently sum) or minimizes the maximum intersection with each path $p \in \mathcal{P}$. In this paper, we consider this problem, which we call the *checkpoint problem*. We note that the problem has other potential applications beyond our motivating example in transportation networks; for instance, in network security, we may want to check certain malicious source-destinations pair without incurring too much delay along certain critical paths.

1.1   Related Work

Closely related to the checkpoint problem are the more common multicut problems in which given an edge-weighted (undirected or directed) graph and a collection of pairs $\{(s_i, t_i)\}_{i=1}^{k}$, the goal is to find a subset $E' \subseteq E$ of minimum cost so that in $E \setminus E'$ there is no s_i-t_i path for any i. The only difference between our problems and these is the objective function.

The literature on undirected multicut (*UM*) problems is extensive. Garg *et al.* [7] showed that *UM* is at least as hard to approximate as the well-known vertex-cover problem even if the underlying graph is a star. This implies that unless P = NP, it is hard to approximate the undirected multicut problem on stars within a factor better than $10\sqrt{5} - 21$ [4]. Garg *et al.* [7] also gave a 2 approximation for *UM* in trees via the primal dual approach. The best known approximation for *UM* in general undirected graphs is $O(\log n)$ [8]. Conditional on the Unique Game Conjecture [11], Chawla *et al.* [1] proved that the *UM* problem admits no constant approximation ratio for any constant. A stronger version of the conjecture implies that the *UM* problem can not be approximated within a factor $\Omega(\sqrt{\log \log n})$.

The multicut problem can also be viewed as a set cover problem in which we want to cover all paths between specific source-destination pairs (as elements) by a minimum number of edges (as sets). Set cover problems in which we are restricted to cover elements are also considered. In the *minimum membership set-cover problem* of Kuhn *et al.* [16], we want to cover all elements while minimizing the maximum number of sets covering an element. In the *unique coverage problem* of Demaine *et al.* [3], we want to maximize elements which are covered exactly once. Both these problems have applications concerning interference reduction in cellular networks. Roughly speaking, our checkpoint problem combines multicut and minimum membership set cover.

Multicut problems are associated with a dual multicommodity flow problem, where instead of disconnecting pairs, the objective is to connect them. In our setting, these flow problems are quite hard even if the input contains a single pair. We consider the well-known problem of *finding a path with minimum forbidden pairs (PAFP)*, a problem that has been studied since the seventies [15]. The input consists of a (directed or undirected) graph $G(V, E)$, a pair (s, t) of vertices, and a collection of forbidden pairs $\mathcal{F} = \{b_i b_i'\}_{i=1}^{\ell}$ where the forbidden pairs are the pairs of vertices *that may not appear simultaneously on the solution path*. A vertex may appear in many forbidden pairs. The goal is to find an *s-t* path with the minimum number of pairs $b_i b_i' \in \mathcal{F}$ such that *both* $b_i \in p$ and $b_i' \in p$.

The *PAFP* problem is particularly important for its relation to automatic software testing and validation [15,19], and its applications in bioinformatics [2]. In [5] it is proved that *PAFP* is NP-complete on directed acyclic graphs. Yinnone [22] studied the problem in directed graphs under the so called skew-symmetry condition constraining the set of edges and the set of forbidden pairs. Yinnone gives a polynomial algorithm for the problem under that restriction. Chen *et al.* [2] study a special case of the problem coming from protein identification via tandem mass spectronomy. Kolman *et al.* [12] study *PAFP* under the so-called halving structure for which they prove the problem remains NP-complete, and also under the hierarchical structures condition for which they give a polynomial-time algorithm.

Notation and problem definitions. In this section, we define useful notations and formally define the problems considered in this paper. Let OPT be the optimum solution and opt be its value for the problem and instance at hand. For the rest of the paper we fix a collection $\mathcal{H} = \{(s_i, t_i)\}_{i=1}^{k}$ of k source-destination pairs. The given set of s_i-t_i paths will throughout be denoted by $\mathcal{P}$. We require that every (s_i, t_i) pair has at least one path in $\mathcal{P}$. Generally we assume that $|\mathcal{P}|$ is polynomial in n, unless stated otherwise. When working with trees, $\mathcal{P}$ is uniquely defined by $\mathcal{H}$ because there is a single path connecting every source-sink pair.

Let p be a path in $\mathcal{P}$ and e be an edge in p. We will say that e *stabs* or *covers* p. For a set $E' \subseteq E$ we denote by $\mathsf{cp}_{E'}(p) = |p \cap E'|$ the number of edges in E' that stab p.

Definition 1. *The sum checkpoint (SCP) problem is to find a multicut $E' \subseteq E$ minimizing $\sum_{p \in \mathcal{P}} \mathsf{cp}_{E'}(p)$. The max checkpoint (MCP) problems is to find a multicut $E' \subseteq E$ minimizing $\max_{p \in \mathcal{P}} \mathsf{cp}_{E'}(p)$.*

The checkpoint value cp treats all edges uniformly since it simply counts the *number* of checkpoints in the multicut. In some cases, though, edges may be endowed with weights. In these cases, cp can be defined as the weight of edges chosen in the multicut that are in the path. We explore this variant for *SCP*.

Definition 2. *Given a (directed or undirected) graph G, a pair st, and a collection $\mathcal{F} = \{b_i b_i'\}_{i=1}^{\ell}$ of forbidden pairs of vertices, the path with minimum forbidden pairs (PAFP) problem is to find a path p from s to t minimizing the number of pairs $b_i b_i' \in \mathcal{F}$ such that both $b_i \in p$ and $b_i' \in p$.*

Note that a forbidden pair $b_i b_i' \in \mathcal{F}$ such that *at most* one of b_i or b_i' lies on p does not contribute towards the *PAFP* objective function.

Our results. First, we study *MCP* in trees. A tree input for *MCP* is said to have *ascending paths* if for all $(s_i, t_i) \in \mathcal{H}$ either s_i is an ancestor of t_i or vise-versa. *MCP* on ascending path tree inputs can be solved in polynomial time by linear programming. This follows from the well-known fact [21] that the edge-path incident matrix is totally unimodular. However, such a solution would have a very large running time. Even if T is a path, it is non-trivial to come up with purely combinatorial algorithms. We develop a linear-time algorithm for *MCP* in trees with ascending paths, which gives a solution with cost opt $+ 1$. Then we build upon this to obtain a combinatorial polynomial-time exact algorithm, which runs orders of magnitude faster than the obvious linear-programming based algorithm.

Beyond this special case, the problem becomes hard. We prove that unless P = NP, *MCP* in trees does not admit an approximation ratio better than 2. This solves an open problem of [17]. On the positive side, using standard techniques one can show a nearly matching approximation ratio.

For general graphs, we design an $O\left(\sqrt{\frac{n \log n}{\mathrm{opt}}}\right)$-approximation algorithm for *MCP* using a more sophisticated approach. Our algorithm is based on a somewhat unusual application of sphere growing. First the sphere growing is combinatorial, that is, we grow spheres on the graph itself rather than on the LP solution à la Garg *et al.* [8]. Second, we use an LP solution to remove some edges in order to ensure that the every source-sink pair is "far apart". Combining these two ingredients, we guarantee that when the neighborhood of a set S is removed to disconnect a source-sink pair, the set S contains no "uncut" pairs.

Then we focus our attention on the weighted version of *SCP*. We show that weighted *SCP* is equivalent to *UM* from the point of view of approximability. In particular, *SCP* admits an $O(\log n)$ approximation ratio in general graphs and a 2 approximation ratio in trees.

Finally, we give a strong hardness of approximation for *PAFP* for undirected graphs. We show that unless P $\neq$ NP, *PAFP* admits no $c \cdot n$ approximation ratio for some $c > 0$. Moreover, our construction can be easily modified to give the same hardness of approximation on directed acyclic graphs. This represents a huge improvement over the plain NP-hardness result of Gabow [6]. In fact, such a linear lower bound is one of the largest that can be found in the literature.

We close the section by mentioning that, independently, Nelson [17] also studied *MCP*. He designed an exact algorithm for paths, an asymptotic 2 approximation for trees, and showed 1.5-hardness for general graphs. The algorithm in Section 2 is a generalization of Nelson's algorithm for paths. We thank him for letting us include his result here. Our 2-approximation for general trees is slightly different. Our hardness result improves the one of Nelson in two aspects. First, our hardness ratio is slightly better; second, our proof is for trees and Nelson's is for general graphs. In fact establishing whether the problem is hard on trees is stated as an open problem in [17].

2   The *MCP* Problem in Trees with Ascending Paths

In this subsection we consider *MCP* in trees with ascending paths. That is, we look at instances where G is a rooted tree and for each pair (s_i, t_i) we have a unique path connecting them where s_i is an ancestor of t_i. For a given path $p \in \mathcal{P}$ we denote with $s(p)$ the *starting point* (closest vertex to the root) of p and with $f(p)$ the *finishing point* (furthest vertex from the root) of p. We call the edge $e \in p$ that is adjacent to $f(p)$ the *furthest* edge in p. For a given set X of paths, we define

$$F(X) = \cup_{p \in X} \{\text{the furthest edge in } p\}. \tag{1}$$

For a path $p \in \mathcal{P}$ and a set of paths A, we define $I_A(p)$ to be the number of paths in A that are contained in p.

Additive one approximation. Our main algorithm builds upon the following greedy procedure for computing a set paths. First, we show that the set of paths found by GREEDY can be used to produce a solution for *MCP* that

Algorithm GREEDY($\mathcal{P}$)
1. $A \leftarrow \emptyset$
2. **for** $p \in \mathcal{P}$ in increasing depth of $f(p)$ **do**
3.   **if** $p \cap F(A) = \emptyset$
4.   **then** $A \leftarrow A \cup \{p\}$
5. **return** A

is close to optimum. Later, we show how this algorithm can be used to find an optimal solution.

Let A be the set returned by GREEDY. Notice that taking the furthest edge of each path in A yields a feasible solution: For any path $p \in \mathcal{P}$, if $p \in A$ then it is clear that $F(A)$ stabs p; otherwise, by Line 3, we know that $F(A)$ stabs p. The next lemma shows that the set $F(A)$ is a good approximation of the optimum.

Lemma 1. *Let A be the set returned by* GREEDY *and p be an arbitrary path in* $\mathcal{P}$. *Then every feasible solution stabs p at least* $I_A(p)$ *times and* $F(A)$ *stabs p at most* $I_A(p) + 1$ *times.*

Proof. We claim that the intervals in A contained in p are pairwise disjoint. Suppose, for the sake of contradiction, that there are paths $a, a' \in A$ contained in p that share an edge. Assume without loss of generality that a was added to A before a'. Thus we have that either $f(a) = f(a')$ or $f(a)$ is a proper ancestor of $f(a')$. Now because both paths lie in p and they intersect, it must be the case that the furthest edge of a stabs a'—here we use the property that all paths are ascending. Thus, we reach the contradiction that a' was not added to A. We conclude that the paths in A are pairwise disjoint.

Let a be a path in A whose furthest edge stabs p. Because the paths are ascending, either $s(a)$ is a proper ancestor of $s(p)$, or $s(a)$ belongs to p (that is, a lies inside p); let us call these paths of type 1 and 2, respectively. The key observation is that there is at most one path of type 1 (otherwise the furthest edge of one would stab the other and hence they could not be disjoint). Also, all paths of type 2 are disjoint and lie inside p;

that is, $I_A(p)$ equals the number of type 2 paths. Therefore, furthest edges of paths of type 2 stab p exactly $I_A(p)$ times and the type 1 path, if any, can stab p one more time.

Notice that any solution must stab the type 2 paths using different edges. Therefore, any solution must stab p at least $I_A(p)$ times. □

It follows that $F(A)$ is a feasible solution that uses at most one extra checkpoint than the optimal solution. In addition, GREEDY can be implemented to run in linear time. The proof of this fact is omitted due to lack of space.

Lemma 2. *There is an $O(n + k)$ time additive-1 approximation for MCP in trees with ascending paths.*

From approximate to optimal. Our exact algorithm is based on the idea of trying to weed out the structure that forces the previous algorithm to use an extra checkpoint. We call $a \in A$, $p \in \mathcal{P}$ a *bad pair* if $I_A(p) = M$ and $s(a)$ is proper ancestor of

---

Algorithm ITERATIVE-REFINEMENT($\mathcal{P}$)

1. $A \leftarrow B \leftarrow$ GREEDY($\mathcal{P}$)
2. $M \leftarrow \max_{p \in \mathcal{P}} I_A(p)$
3. **while** $\max_{p \in \mathcal{P}} I_A(p) = M$ **do**
4.   **if** $\exists$ bad pair $a \in A, p \in \mathcal{P}$
5.   **then** $f(a) \leftarrow s(p)$ and $A \leftarrow$ GREEDY($\mathcal{P}$)
6.   **else** return $F(A)$ // opt $= M$
7. **return** $F(B)$ // opt $= M + 1$

$s(p)$, and $s(p)$ is a proper ancestor of $f(a)$, and $f(a)$ is a proper ancestor of $f(p)$; notice that in this case the furthest edge of a stabs p. From the proof of Lemma 1, it immediately follows that the solution $F(A)$ has cost $M + 1$ if and only if there is a bad pair because if p is involved in a bad pair, then it will be stabbed by the furthest edge of a.

Lemma 3. *If $a \in A$, $p \in \mathcal{P}$ is a bad pair and there is a feasible solution with cost M then the solution is also feasible for the modified instance where $f(a) \leftarrow s(p)$. Also, any feasible solution to the modified instance is feasible for the original instance.*

Proof. Recall that there are $I_A(p) = M$ disjoint paths in A inside of p. These paths together with a form a set of disjoint paths. Suppose X is solution with cost M. Since p is stabbed only M times in X then it must be that a is stabbed in $a \setminus p$. Therefore, X remains feasible after we set $f(a) \leftarrow s(p)$.

The second part is trivial since after the modification, a is a subset of its original self. □

With this observation in hand, an algorithm follows suit. Compute A and iteratively try to find a bad pair. If we cannot find a bad pair then $F(A)$ has cost M and this is optimal. Otherwise, we modify the instance as described in Lemma 3 and recompute A. If $\max_{p \in \mathcal{P}} I_A(p)$ becomes $M + 1$ then the new instance cannot have a solution with cost M and hence our implicit assumption that the original instance admitted a solution with cost M must have been wrong.

Theorem 1. *There is a polynomial-time algorithm for MCP in trees with ascending paths.*

Proof. As mentioned above the correctness follows directly from repeatedly applying Lemma 3. To bound the running time we note that each iteration runs in $O(n + k)$ time and that there could be at most k^2 iterations since once a bad pair (a, p) is fixed, it never again becomes a bad pair. We note that the number of iterations can be brought down to $\min\{n, k^2\}$ if we are more aggressive when handling a bad pair (a, p). $\square$

3 Hardness

In this section we show hardness of approximation for *MCP* in trees via a gap-inducing reduction from *1-in-3-SAT*. Recall that a 3-CNF formula belongs to *1-in-3-SAT* if there exists a satisfying assignment where each clause has exactly one true literal. Schaefer [18] proved that *1-in-3-SAT* is NP-complete. Our reduction maps yes (no) instances of *1-in-3-SAT* to instances of *MCP* with cost two (one). Due to lack of space we state our result without proof.

Theorem 2. *Unless* $P = NP$, *MCP in trees admits no better than ratio 2 approximation.*

4 Approximations for *MCP*

LP formulation. In this section we present our approximation results for *MCP*. Our algorithms are based on the following linear programming relaxation. Let $\mathcal{Q}$ be the full set of paths connecting the source-sink pairs. (Recall that $\mathcal{P}$ is just a subset of $\mathcal{Q}$.)

$$\text{minimize } z \qquad\qquad\qquad \text{(LP1)}$$

subject to

$$\sum_{e \in q} x_e \geq 1 \qquad\qquad \text{for all } q \in \mathcal{Q} \qquad (2)$$

$$\sum_{e \in p} x_e \leq z \qquad\qquad \text{for all } p \in \mathcal{P} \qquad (3)$$

$$x_e \geq 0 \qquad\qquad e \in E$$

Variable x_e indicates whether edge e is chosen in the multicut. Constraint (2) enforces that the set of edges chosen indeed forms a multicut. The objective is to minimize z, the maximum number of edges any one path sees (3). For general graph, the set $\mathcal{Q}$ can be exponentially large. The program (LP1) can be solved in polynomial time by running the Ellipsoid algorithm on its dual.

Due to lack of space the presentation of our approximation algorithm for *MCP* in trees is deferred for the journal version.

Theorem 3. *There is a polynomial-time algorithm for MCP in trees that returns a solution with cost no more than* $2 \cdot \mathsf{opt} + 2$.

4.1 *MCP* in General Graphs

Throughout this section, *Sol* will denote the partial solution accumulated by our algorithm. We say that a source s_i is *uncut*, if $G(V, E \setminus Sol)$ contains an s_i-t_i path. For simplicity, we assume that opt, the value of the optimal solution, is known. This value can easily be guessed, or, alternatively, we can use the value of the optimal fractional solution instead.

Algorithm APPROXIMATING-MCP$(G, \mathcal{H})$

1. $x \leftarrow$ fraction optimal solution for (LP1)
2. $Sol \leftarrow \left\{ e \in E : x_e \geq \frac{1}{2} \sqrt{\frac{\text{opt}}{n \cdot (\ln n + 1)}} \right\}$
3. remove the edges Sol from G
4. **while** Sol is not a multicut **do**
5. $\quad S \leftarrow \{s\}$, for some arbitrary uncut source s
6. $\quad$ **while** $|N(S)| \geq \left(1 + \sqrt{\frac{\text{opt} \cdot (\ln n + 1)}{n}}\right) |S|$ **do**
7. $\quad\quad S \leftarrow N(S)$
8. $\quad\quad Sol \leftarrow Sol \cup E(N(S) \setminus S)$
9. $\quad\quad$ remove $E(N(S) \setminus S)$ from G
10. **return** Sol

Along the way, we prove the following result: If the minimum distance between every (s_i, t_i) pair is ℓ, then there exists a vertex cut of size at most $\tilde{O}(n/\ell)$ whose deletion disconnects all pairs. We believe this fact is known, but are not aware of any specific reference. Some results along these lines are known for the directed case; for example, it was shown independently in [20] and [13] that if every pair in a *directed graph* has distance at least ℓ, then there is an *edge cut* separating all pairs whose size is at most $\tilde{O}(n^2/\ell^2)$.

Given a fractional solution x to (LP1), we denote the *fractional checkpoint value* of a path p by $\text{cp}_x(p) = \sum_{e \in p} x(e)$. Let $dist(u, v)$ denote the length of the shortest path in G between u and v measured by the number of edges. The following operators are used by our algorithm:

$$N(X) = X \cup \{v \in V : \exists u \in X \text{ s.t. } (u, v) \in E)\},$$

and

$$E(X) = \{(u, v) \in E : u \in X \vee v \in X\}.$$

In other words, $N(X)$ equals X and its all neighbors, while $E(X)$ equals the set of edges with at least one endpoint in X. We note that both operators are defined with respect to the graph $G(V, E)$. As the algorithm progresses and removes edges from G, these operators change accordingly.

The algorithm can be thought off as having two main parts: a *filtering step* and a *region-growing step*. The next two lemmas, which we state without proof, establish some important properties of the first step.

Lemma 4. *Consider the value of Sol and G right after Line 3. Then $dist(s_i, t_i) > 2\sqrt{\frac{n \cdot (\ln n + 1)}{\text{opt}}}$ for all $(s_i, t_i) \in \mathcal{H}$.*

Lemma 5. *Consider Sol right after Line 2. Then $\text{cp}_{Sol}(p) = \text{cp}_x(p) \cdot O\left(\sqrt{\frac{n \log n}{\text{opt}}}\right)$ for all $p \in \mathcal{P}$.*

After the initial filtering step (after the initial *Sol* is computed in Line 2), the algorithm iteratively finds sets $S_1, S_2, \ldots$, using a region-growing procedure out of uncut sources $s_1, s_2, \ldots$, respectively. We note that our approach is related to that of Garg *et al.*[8]. There are, however, two major differences. First, instead of "growing our regions on the LP solution", we do so in the input graph itself. Second, instead of using edge cuts, we use vertex cuts—indeed, the edges removed in Line 9 correspond to removing the vertices $N(X) \setminus S$.

Lemma 6. *The sets $S_1, S_2, \ldots$ are pair-wise disjoint.*

Proof. Consider an arbitrary set S_i. Upon existing the while loop in Line 6, the algorithm adds $E(N(S_i) \setminus S_i)$ to *Sol*. This effectively disconnects S_i from the rest of the graph defined by $E \setminus Sol$.

We claim that the number of iterations of the while loop in Line 6 needed to compute S_i is at most $\sqrt{\frac{n \cdot (\ln n + 1)}{\mathsf{opt}}}$. Indeed, since the size of $|S|$ increases by $1 + \sqrt{\frac{\mathsf{opt} \cdot (\ln n + 1)}{n}}$ factor in each iteration, if the while loop were to run for $\sqrt{\frac{n \cdot (\ln n + 1)}{\mathsf{opt}}}$ iterations then we would reach the contradiction that

$$|S| > \left(1 + \sqrt{\frac{\mathsf{opt} \cdot (\ln n + 1)}{n}}\right)^{\sqrt{\frac{n \cdot (\ln n + 1)}{\mathsf{opt}}}} \geq n \geq |S|.$$

A corollary of this, is that the diameter of the graph induced by S_i is most $2\sqrt{\frac{n \cdot (\ln n + 1)}{\mathsf{opt}}}$.

Now consider a set S_j constructed in some subsequent iteration. If $s_j \notin S_i$ then clearly S_j and S_i must be disjoint. We claim that this is the only option. Indeed, if $s_j \in S_i$ then, since the diameter of S_i is at most $2\sqrt{\frac{n \cdot (\ln n + 1)}{\mathsf{opt}}}$, it follows that right after S_i is created $dist(s_j, t_j) \leq 2\sqrt{\frac{n \cdot (\ln n + 1)}{\mathsf{opt}}}$, which contradicts Lemma 4.   □

Everything is in place to prove the main result of this section.

Theorem 4. *The MCP problem admits a polynomial-time $O\left(\sqrt{\frac{n \log n}{\mathsf{opt}}}\right)$ approximation algorithm.*

Proof. Let p be an arbitrary path in $\mathcal{P}$. Notice that when $E(N(S_i) \setminus S_i)$ is added to *Sol*, since p is simple, the value of $\mathsf{cp}_{Sol}(p)$ increases by at most $|N(S_i) \setminus S_i|$. Therefore, in order to bound total increase in $\mathsf{cp}_{Sol}(p)$ due to edges to *Sol* after Line 2, we need to bound $\sum_i |N(S_i) \setminus S_i|$:

$$\sum_i |N(S_i) \setminus S| < \sum_i \sqrt{\frac{\mathsf{opt} \cdot (\ln n + 1)}{n}} \cdot |S_i| \leq \sqrt{n \cdot \mathsf{opt} \cdot (\ln n + 1)}, \qquad (4)$$

where the first inequality follows from the exit condition of the while loop in Line 6, and the second, from Lemma 6. Putting (4) and Lemma 4 together, we conclude that

$$\mathsf{cp}_p(Sol) = \mathsf{cp}_x(Sol) \cdot O\left(\sqrt{\frac{n \ln n}{\mathsf{opt}}}\right) + \sqrt{n \cdot \mathsf{opt} \cdot (\ln n + 1)} = \mathsf{opt} \cdot O\left(\sqrt{\frac{n \ln n}{\mathsf{opt}}}\right).$$

□

5   Approximation for *SCP*

For this problem we allow the graph to be weighted, in which case $\mathsf{cp}_{E'}(p)$ is the combined weight of edges in $E' \cap p$. Recall that $w(e)$ denotes the edge e.

Theorem 5. *Any ρ approximation for UM gives a ρ approximation for weighted SCP, and vise-versa*

Proof. We first show the forward direction. We construct edge capacities c as follows: For every edge e let $p(e)$ be the number of paths $p \in \mathcal{P}$ that use e; notice that if for a fixed pair (s_i, t_i) the set $\mathcal{P}_i$ contains many paths going trough e, each one will contribute towards $p(e)$. We give edge e capacity $c(e) = p(e)w(e)$. We show that capacity of a multicut E' equals the min-sum checkpoint value, that is, $\sum_{e \in E'} c(e) = \sum_{p \in \mathcal{P}} \mathsf{cp}_{E'}(p)$. Given an edge $e \in E'$, we charge $w(e)$ to each of the $p(e)$ paths containing that edge; this exhausts the $c(e)$ term in the cost of the *UM* objective. Therefore, $\sum_{e \in E'} c(e) = \sum_{p \in \mathcal{P}} \mathsf{cp}_{E'}(p)$ and every ratio ρ that applies to *UM* also applies to *SCP*.

In the other direction, assume we have a ρ approximation for *SCP*. We approximate *UM* by a reduction to *SCP* as follows. Create a *SCP* instance with every e having capacity $w(e)/p(e)$. For any multicut E', $w(e)$ will be counted $p(e)$ times thus the checkpoint cost of E' is $c(E')$. Thus the best solution is the minimum capacity multicut. Thus, *SCP* and *UM* are equivalent with respect to approximation. $\square$

Corollary 1. *SCP admits an $O(\log n)$ approximation in general graphs and a 2 approximation in trees.*

6   A Lower Bound for *PAFP*

Recall that in *PAFP* we are given a pair (s, t) to connect and a collection of forbidden pairs $\{(b_i, b_i')\}_{i=1}^{\ell} \subseteq V \times V$. The goal is to find an s-t path minimizing the number of pairs (b_i, b_i') such that both belong to the path. To disallow a zero cost solution we may arbitrarily define (s, t) as a forbidden pair or we can define the cost of a solution as the maximum between the number of bad pairs in the path and 1.

Background. The *LABELCOVER-MAX* problem is introduced in [10, Chapter 10] for presenting one-round two-provers systems. Here we use an alternative formulation, called *MAX-REP*, defined in [14]. In *MAX-REP*, we are given a bipartite graph $G(V_1, V_2, E)$. The sets V_1 and V_2 are partitioned into a disjoint union of q sets: $V_1 = \bigcup_{i=1}^q A_i$ and $V_2 = \bigcup_{j=1}^q B_j$. The bipartite graph and the partition of V_1 and V_2 induce a super-graph $\mathcal{H}$ in the following way: The vertices in $\mathcal{H}$ are the sets A_i and B_j. Two sets A_i and B_j are connected by a super-edge in $\mathcal{H}$ if and only if there exist $a_i \in A_i$ and $b_j \in B_j$ which are adjacent in G. In *MAX-REP* we are to select a unique *representative* vertex $a_i \in A_i$ from each subset A_i, and a unique *representative* vertex $b_j \in B_j$ from each B_j. We say that a super-edge (A_i, B_j) is *covered* if the two corresponding representatives are neighbors in G; that is, $(a_i, b_j) \in E$. The goal is to select unique representatives so as to maximize the number of super-edges covered. Håstad's breakthrough hardness for *3-SAT-5* [9] translates into the following hardness for *MAX-REP*.

Theorem 6 ([9]). *There is a polynomial time reduction that maps each instance ϕ of SAT into an instance G of MAX-REP with n' vertices and $h = \Theta(n')$ super-edges. If ϕ is satisfiable then there exists a set of unique representatives of G that covers all h super-edges. If ϕ is not satisfiable then every set of unique representatives of G covers at most $\frac{23}{24}h$ super-edges.*

Reduction. The reduction from *MAX-REP* to *PAFP* is relatively simple. Arbitrarily order the super-vertices from left to right: $X_1, X_2, \ldots, X_{2q}$. Join X_i to X_{i+1} with a complete bipartite graph for every $1 \leq i \leq 2q - 1$. Let (A, B) be a super-edge in our *MAX-REP* instance. For each $a \in A$ and $b \in B$ such that $(a, b) \notin E$, we create a forbidden pair (a, b). Thus, forbidden pairs correspond to vertices that *are not* connected in the *MAX-REP* graph and whose corresponding super-nodes *are* connected in the super-graph. Finally, join a vertex s to all the vertices of X_1 and join a vertex t to all the vertices of X_h. This defines the *PAFP* instance.

Theorem 7. *Unless* $P = NP$, *PAFP on undirected graphs admits no c·n approximation ratio, where n is the number of vertices and $c > 0$ is some constant. The same holds for directed acyclic graphs.*

Proof. Consider the reduction above. We show that a solution for the *PAFP* instance with t forbidden pairs translates into solution for the *MAX-REP* instance covering $h - t$ super-edges, and vice-versa. Without loss of generality we restrict our attention to *PAFP* solutions that use a single vertex from each super-vertex X_i. Under this restriction, there is a clear one-to-one correspondence between solutions for the *PAFP* instance (s-t paths) and solutions for the *MAX-REP* instance (unique representative choices). Let X be a unique representative choice and p its corresponding s-t path. Let (A, B) be an arbitrary super-edge, and let a and b be the representatives of A and B respectively. If (A, B) is covered by X then none of the forbidden pairs induced by (A, B) appear in p. Otherwise, if (A, B) is not covered by X, we know that (a, b) is a forbidden pair. It follows that the number of super-edges covered by X is h minus the number of forbidden pairs in p.

In Theorem 6, satisfiable formulas map to instances of *MAX-REP* that have a perfect cover, which in turn our reduction maps to instances of *PAFP* that have a path with no forbidden pairs, which have value 1 (recall that the cost of a path is the maximum of 1 and the number of forbidden pairs.) On the other hand, unsatisfiable formulas map to instances of *MAX-REP* with value at most $\frac{23}{24}h$, which in turn map to instances of *PAFP* having value at least $\frac{h}{24}$. In addition, Theorem 6 tells us that the *MAX-REP* instance has $h = \Theta(n)$, where n is the number of vertices in the *PAFP* instance. This finishes the proof for undirected order. In order to get the result on directed acyclic graphs, just direct all the edges from s to t. □

7   Discussion and Open Problems

Can the approximation for *SCP* be used to approximate *MCP*? If the optimum for *SCP* is opt_s then the optimum for *MCP* is at least $\text{opt}_s / |\mathcal{P}|$. Therefore, by approximating the *SCP* objective, we obtain a lower bound for the *MCP* objective. The multicut for the *SCP* problem, however, cannot be used directly as a solution for the *MCP* problem since the path with the largest checkpoint value may be well above the average checkpoint value. One could try to deal with these "expensive" paths in a later stage, but this may increase the checkpoint value of paths previously having low checkpoint value in the *SCP* solution. Indeed, the *MCP* problem seems highly non-separable.

Finally, it would be interesting to know whether *MCP* admits a $\text{polylog}(n)$ approximation, or whether it has a hardness is similar to *MAX-REP*.

Acknowledgement. The first author thanks Erik Demaine and Jelani Nelson for several fruitful discussions especially on initiating the problem.

References

1. Chawla, S., Krauthgamer, R., Kumar, R., Rabani, Y., Sivakumar, D.: On the hardness of approximating multicut and sparsest-cut. Computational Complexity 15(2), 94–114 (2006)
2. Chen, T., Kao, M.Y., Tepel, M., Rush, J., Church, G.: A dynamic programming approach to de novo peptide sequencing via tandem mass spectrometry. Journal of Computational Biology 8(3), 325–337 (2001)
3. Demaine, E.D., Feige, U., Hajiaghayi, M.T., Salavatipour, M.: Combination can be hard: approximability of the unique coverage problem. SIAM J. Comp. 38(4), 1464–1483 (2008)
4. Dinur, I., Safra, S.: The importance of being biased. In: STOC, pp. 33–42 (2002)
5. Gabow, H., Maheswari, S., Osterweil, L.: On two problems in the generation of program test paths. IEEE Trans. Software Eng. 2(3), 227–231 (1976)
6. Gabow, H.N.: Finding paths and cycles of superpolylogarithmic length. SIAM J. Comp. 36(6), 1648–1671 (2007)
7. Garg, N., Vazirani, V., Yannakakis, M.: Primal-dual approximation algorithms for integral flow and multicut in trees. Algorithmica 18(1), 3–20 (1997)
8. Garg, N., Vazirani, V.V., Yannakakis, M.: Approximate max-flow min-(multi) cut theorems and their applications. SIAM Journal on Computing 25(2), 235–251 (1996)
9. Håstad, J.: Some optimal inapproximability results. J. of the ACM 48(4), 798–859 (2001)
10. Hochbaum, D.: Approximation algorithms for NP-hard problems. PWS Publishing Co. (1997)
11. Khot, S.: On the unique games conjecture. In: FOCS, p. 3 (2005)
12. Kolman, P., Pangrac, O.: On the complexity of paths avoiding forbidden pairs. Discrete applied math. 157, 2871–2877 (2009)
13. Kortsarts, Y., Kortsarz, G., Nutov, Z.: Greedy approximation algorithms for directed multicuts. Networks 45(4), 214–217 (2005)
14. Kortsarz, G.: On the hardness of approximating spanners. Algorithmica 30(3), 432–450 (2001)
15. Krause, K., Smith, R., Goodwin, M.: Optimal software test planning through authomated search analysis. In: IEEE Symp. Computer Software Reliability, pp. 18–22 (1973)
16. Kuhn, F., von Rickenbach, P., Wattenhofer, R., Welzl, E., Zollinger, A.: Interference in cellular networks: The minimum membership set cover problem. In: Wang, L. (ed.) COCOON 2005. LNCS, vol. 3595, pp. 188–198. Springer, Heidelberg (2005)
17. Nelson, J.: Notes on min-max multicommodity cut on paths and trees. Manuscript (2009)
18. Schaefer, T.J.: The complexity of satisfiability problems. In: Proc. of the 10th of the Tenth Annual ACM Symposium on Theory of Computing, pp. 216–226 (1978)

19. Strimani, P., Sinha, B.: Impossible pair-constrained test path generation in a program. Information Sciences 28, 87–103 (1982)
20. Varadarajan, K., Venkataraman, G.: Graph decomposition and a greedy algorithm for edge-disjoint paths. In: SODA, pp. 379–380 (2004)
21. Yannakakis, M.: On a class of totally unimodular matrices. In: FOCS, pp. 10–16 (1980)
22. Yinnone, H.: On paths avoiding forbidden pairs of vertices in a graph. Discrete Appl. Math. 74(1), 85–92 (1997)

The Euclidean Distortion of Flat Tori

Ishay Haviv[*] and Oded Regev[**]

The Blavatnik School of Computer Science, Tel Aviv University, Israel

Abstract. We show that for every n-dimensional lattice $\mathcal{L}$ the torus $\mathbb{R}^n/\mathcal{L}$ can be embedded with distortion $O(n \cdot \sqrt{\log n})$ into a Hilbert space. This improves the exponential upper bound of $O(n^{3n/2})$ due to Khot and Naor (FOCS 2005, Math. Annal. 2006) and gets close to their lower bound of $\Omega(\sqrt{n})$. We also obtain tight bounds for certain families of lattices.

Our main new ingredient is an embedding that maps any point $u \in \mathbb{R}^n/\mathcal{L}$ to a Gaussian function centered at u in the Hilbert space $L_2(\mathbb{R}^n/\mathcal{L})$. The proofs involve Gaussian measures on lattices, the smoothing parameter of lattices and Korkine-Zolotarev bases.

Keywords: Lattice, Embedding, Torus.

1 Introduction

An n-dimensional full-rank lattice $\mathcal{L} \subseteq \mathbb{R}^n$ is the set of all integer combinations of n linearly independent vectors. Such a lattice defines the *torus* $\mathbb{R}^n/\mathcal{L}$, i.e., the space $\mathbb{R}^n$ where two points are identified if and only if the difference between them is a lattice vector. For $u, v \in \mathbb{R}^n/\mathcal{L}$ the distance $\mathrm{dist}_{\mathbb{R}^n/\mathcal{L}}(u, v)$ in the torus $\mathbb{R}^n/\mathcal{L}$ is defined as the distance between a representative of $u - v$ in $\mathbb{R}^n$ from the lattice $\mathcal{L}$.

In this paper we study the ability to embed a torus $\mathbb{R}^n/\mathcal{L}$ into a Hilbert space in a distance-preserving manner. For a lattice $\mathcal{L}$ we are interested in a Hilbert space L_2, an embedding $H : \mathbb{R}^n/\mathcal{L} \to L_2$ and a number $c_2 > 0$ such that for any $u, v \in \mathbb{R}^n/\mathcal{L}$, $\mathrm{dist}_{\mathbb{R}^n/\mathcal{L}}(u, v) \leq \mathrm{dist}_{L_2}(H(u), H(v)) \leq c_2 \cdot \mathrm{dist}_{\mathbb{R}^n/\mathcal{L}}(u, v)$. The *distortion* of an embedding H is the least c_2 for which the above holds. The least distortion that one can get over all the embeddings H is known as the *Euclidean distortion* of $\mathbb{R}^n/\mathcal{L}$ and is denoted by $c_2(\mathbb{R}^n/\mathcal{L})$.

For example, consider the n-dimensional lattice $\mathbb{Z}^n$. The torus $\mathbb{R}^n/\mathbb{Z}^n$ can be embedded into the Euclidean space $\mathbb{R}^{2n}$ by the embedding $H : \mathbb{R}^n/\mathbb{Z}^n \to \mathbb{R}^{2n}$ defined by $H(x_1, \ldots, x_n) = (\cos 2\pi x_1, \sin 2\pi x_1, \ldots, \cos 2\pi x_n, \sin 2\pi x_n)$. It is easy to see that H has a constant distortion independent of n. It is not difficult to

[*] Supported by the Adams Fellowship Program of the Israel Academy of Sciences and Humanities.

[**] Supported by the Binational Science Foundation, by the Israel Science Foundation, by the European Commission under the Integrated Project QAP funded by the IST directorate as Contract Number 015848, by the Wolfson Family Charitable Trust, and by a European Research Council (ERC) Starting Grant.

M. Serna et al. (Eds.): APPROX and RANDOM 2010, LNCS 6302, pp. 232–245, 2010.
© Springer-Verlag Berlin Heidelberg 2010

extend this example and to achieve an embedding with constant distortion for every lattice generated by n orthogonal vectors.

Metric embeddings have been extensively investigated in the last few years by the theoretical computer science community. One of the main motivations for research on embedding metric spaces comes from applications to designing geometric approximation algorithms. Indeed, in order to approximate the distance between two points in a certain metric space one can apply an efficient low distortion embedding and then compute (or approximate) the distance between the corresponding embedded points. Studying the Euclidean distortion of flat tori might have applications to the complexity of lattice problems, and might also lead to more efficient algorithms for lattice problems through the use of our metric embeddings. For example, consider the Closest Vector Problem with Preprocessing (CVPP). In this problem a (not necessarily efficient) preprocessing step is applied to the lattice. Then, given a target point, we are supposed to efficiently approximate its distance from the lattice. Embedding flat tori suggests a special type of algorithms for CVPP, in which the data performed in the preprocessing step enables to approximate distances in the embedded space efficiently. A recent result by Micciancio and Voulgaris [8] demonstrates how CVPP can lead to breakthroughs for standard lattice problems. For further information on CVPP we refer the reader to [3].

In this work we study the distortion required to embed an n-dimensional torus into a Hilbert space. This question was introduced by Khot and Naor in [4] who provided a partial answer as stated below. The following theorem provides a lower bound on $c_2(\mathbb{R}^n/\mathcal{L})$ in terms of $\lambda_1(\mathcal{L}^*)$ and $\mu(\mathcal{L}^*)$, which are, respectively, the length of a shortest nonzero vector and the covering radius of $\mathcal{L}^*$, the dual lattice of $\mathcal{L}$.

Theorem 1 ([4]). *For any $n \geq 1$ and an n-dimensional lattice $\mathcal{L}$, $c_2(\mathbb{R}^n/\mathcal{L}) = \Omega\left(\frac{\lambda_1(\mathcal{L}^*)}{\mu(\mathcal{L}^*)} \cdot \sqrt{n}\right)$.*

It is known that for every large enough n there exists an n-dimensional self-dual lattice $\mathcal{L}$ (i.e., $\mathcal{L} = \mathcal{L}^*$) such that $\lambda_1(\mathcal{L}) = \Theta(\mu(\mathcal{L}))$. This fact is due to Conway and Thompson; see [9, Page 46] for details. Theorem 1 and this family of lattices imply that for any large enough n there exists an n-dimensional lattice $\mathcal{L}$ for which $c_2(\mathbb{R}^n/\mathcal{L}) = \Omega(\sqrt{n})$. We note that in [4] it was shown that the bound in Theorem 1 holds even for embeddings into the space L_1. The next theorem shows an upper bound on $c_2(\mathbb{R}^n/\mathcal{L})$ for n-dimensional lattices and in particular implies that the supremum of $c_2(\mathbb{R}^n/\mathcal{L})$ over all n-dimensional lattices $\mathcal{L}$ is finite.

Theorem 2 ([4]). *For any $n \geq 1$ and an n-dimensional lattice $\mathcal{L}$, $c_2(\mathbb{R}^n/\mathcal{L}) = O(n^{3n/2})$.*

We note that the true performance of the embedding of Khot and Naor used in the proof of Theorem 2 is not clear. Yet, it can be shown that there are lattices for which the distortion achieved by their embedding is super-polynomial. We discuss this issue in the full version of the paper.

1.1 Our Results

The gap between the above lower and upper bounds on $c_2(\mathbb{R}^n/\mathcal{L})$ is huge. In this work we significantly reduce this gap. Our main result is that for every lattice the torus $\mathbb{R}^n/\mathcal{L}$ can be embedded into a Hilbert space with distortion slightly higher than linear in n.

Theorem 3. *For any $n \geq 1$ and an n-dimensional lattice $\mathcal{L}$, $c_2(\mathbb{R}^n/\mathcal{L}) = O(n \cdot \sqrt{\log n})$.*

For n-dimensional lattices $\mathcal{L}$ with ratio $\frac{\mu(\mathcal{L})}{\lambda_1(\mathcal{L})} \leq n^{o(n)}$ we provide the following better bound.

Theorem 4. *For any $n \geq 1$ and an n-dimensional lattice $\mathcal{L}$, $c_2(\mathbb{R}^n/\mathcal{L}) = O\left(\sqrt{n \cdot \log\left(\frac{4\mu(\mathcal{L})}{\lambda_1(\mathcal{L})}\right)}\right)$.*

Notice that Theorem 1 yields that the bound in Theorem 4 is tight up to a multiplicative constant for the self-dual lattices that were mentioned above (see Corollary 1).

Finally, we observe that Theorem 1 can be slightly improved to the following.

Theorem 5. *For any $n \geq 1$ and an n-dimensional lattice $\mathcal{L}$, $c_2(\mathbb{R}^n/\mathcal{L}) \geq \frac{\lambda_1(\mathcal{L}^*) \cdot \mu(\mathcal{L})}{4\sqrt{n}}$.*

It can be shown that $\mu(\mathcal{L}) \cdot \mu(\mathcal{L}^*) \geq \Omega(n)$ holds for any n-dimensional lattice and hence Theorem 5 improves Theorem 1.

1.2 Intuitive Overview of Proofs and Techniques

Our goal is to construct, given a lattice $\mathcal{L}$, a function H from the torus $\mathbb{R}^n/\mathcal{L}$ to a Hilbert space such that H preserves distances up to a multiplicative factor that is as small as possible. Our basic idea is to map any $u \in \mathbb{R}^n$ to the Gaussian function defined on $\mathbb{R}^n$ centered at u with parameter s, i.e., the function mapping $x \in \mathbb{R}^n$ to $e^{-\pi\|(x-u)/s\|^2}$. It is not difficult to see that the L_2 distance between $H(u)$ and $H(v)$ depends more or less linearly on the distance between u and v as long as the latter is at most s, beyond which the distance between $H(u)$ and $H(v)$ is saturated and no longer increases linearly. This is illustrated in the left side of Figure 1.

However, the embedding defined above is not an embedding of $\mathbb{R}^n/\mathcal{L}$ because it is not $\mathcal{L}$-periodic. We therefore replace the Gaussian function centered at u with the sum of all Gaussian functions centered at points in $u + \mathcal{L}$, i.e., all the shifts of u by vectors of $\mathcal{L}$. See the right side of Figure 1.

An important role in the performance of our basic embedding is played by the choice of the parameter s. Notice that we cannot take s to be significantly smaller than the covering radius of $\mathcal{L}$ (the maximum distance between two elements in $\mathbb{R}^n/\mathcal{L}$). Indeed, as mentioned above, the distance between the embedded functions is saturated beyond distance s, thereby leading to a distortion of at least

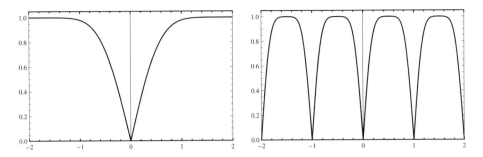

Fig. 1. The left plot shows the L_2 distance between the (one-dimensional) Gaussian function centered at 0 and the Gaussian function centered at $u \in \mathbb{R}$ (as a function of u; $s = 1$). The right plot shows the L_2 distance between the sum of all Gaussian functions centered at points in $\mathbb{Z}$ and the sum of all Gaussian functions centered at points in $u + \mathbb{Z}$ (as a function of u; $s = 0.3$).

$\mu(\mathcal{L})/s$. On the other hand, s cannot be larger than $\lambda_1(\mathcal{L})$: for such s, small shifts in the direction of a shortest vector of $\mathcal{L}$ are much less noticeable than shifts in directions orthogonal to it, and this creates a huge distortion. By choosing s to be slightly smaller than $\lambda_1(\mathcal{L})$ our basic embedding achieves distortion proportional to $\frac{\mu(\mathcal{L})}{\lambda_1(\mathcal{L})}$ (see Theorem 7).

In order to improve the distortion we need two more ideas. First, we combine several basic embeddings for various choices of the parameter s in the range $[\lambda_1(\mathcal{L}), \mu(\mathcal{L})]$. The idea is that every distance in $\mathbb{R}^n/\mathcal{L}$ is handled by at least one of these choices. This proves Theorem 4. The second idea which is used in the proof of Theorem 3 is to use our basic embedding on projected lattices using Korkine-Zolotarev bases. In our analysis of the basic embedding we employ and extend techniques originating in a paper by Banaszczyk [2] that were found useful in several recent papers on the complexity of lattice problems (see, e.g., [1]).

1.3   Open Question

As mentioned before, we show in this paper that any n-dimensional lattice $\mathcal{L}$ satisfies $c_2(\mathbb{R}^n/\mathcal{L}) = O(n \cdot \sqrt{\log n})$, and it was shown in [4] that there are lattices for which $c_2(\mathbb{R}^n/\mathcal{L}) = \Omega(\sqrt{n})$. The main open question raised by our work is the following.

Question 1. Is it true that for any n-dimensional lattice $\mathcal{L}$, $c_2(\mathbb{R}^n/\mathcal{L}) = O(\sqrt{n})$?

We observe that a positive answer to this question using Theorem 5 immediately implies that any n-dimensional lattice $\mathcal{L}$ satisfies $\lambda_1(\mathcal{L}^*) \cdot \mu(\mathcal{L}) \leq O(n)$. The only proof we are aware of for this tight bound is the one of Banaszczyk [2] whose tools and techniques are the heart of the current paper. This might hint that our approach to the embedding question is natural and that it has not been pushed to its limit yet.

A more ambiguous open question is to obtain tight bounds on $c_2(\mathbb{R}^n/\mathcal{L})$ for every lattice $\mathcal{L}$ in terms of geometrical parameters of $\mathcal{L}$.

1.4   Outline

The paper is organized as follows. In Section 2 we gather all the definitions on embeddings, lattices, Gaussian measures and Korkine-Zolotarev bases that we need in this paper. In Section 3 we prove properties of Gaussian distributions on lattices. In Section 4 we prove Theorems 4 and 3. The proof of Theorem 5 and a discussion on the performance of the embedding of Khot and Naor used in the proof of Theorem 2 can be found in the full version of the paper.

2   Preliminaries

For a real x, $\lceil x \rfloor$ stands for the integer that satisfies $-0.5 < x - \lceil x \rfloor \leq 0.5$. The ℓ_2 norm of $u \in \mathbb{C}^n$ is defined as $\|u\| = (\sum_{i=1}^{n} |u_i|^2)^{1/2}$ where u_i is the ith coordinate of u. The inner product of $u, v \in \mathbb{C}^n$ is defined as $\langle u, v \rangle = \sum_{i=1}^{n} u_i \overline{v_i}$. For a point $u \in \mathbb{C}^n$ and a set $S \subseteq \mathbb{C}^n$, denote $u + S = \{u + x \mid x \in S\}$ and $\mathrm{dist}(u, S) = \inf_{x \in S} \|u - x\|$. The open unit ball is defined as $\mathcal{B} = \{w \in \mathbb{R}^n \mid \|w\| < 1\}$. For a scalar function f and a subset A of its domain, we use the notation $f(A) = \sum_{x \in A} f(x)$.

We will need the following simple fact, in which we do not make any attempt to optimize the constants.

Fact 6. *For any $a \geq 0$ and $0 \leq b < \frac{1}{\sqrt{2}}$, $\cosh(2\pi ab) - 1 \leq 230 \cdot b^2 e^{\frac{3\pi}{4} a^2}$.*

Proof: We separate the proof into two cases as follows. If $\pi ab \leq 1$ then use the fact that any $\alpha \in [-2, 2]$ satisfies $\cosh(\alpha) - 1 \leq \alpha^2$ and $\alpha \leq e^{\alpha}$ to obtain $\cosh(2\pi ab) - 1 \leq 4\pi^2 \cdot b^2 e^{a^2}$. Otherwise, use the fact that any $\alpha \geq 0$ satisfies $\cosh(\alpha) \leq e^{\alpha}$ and $\alpha^2 \leq e^{\alpha}$ and the assumption $b \leq \frac{1}{\sqrt{2}}$ to obtain

$$\cosh(2\pi ab) - 1 \leq e^{2\pi ab} \leq (\pi ab)^2 \cdot e^{\sqrt{2}\pi a} \leq \pi^2 \cdot b^2 e^{(\sqrt{2}\pi + 1)a} \leq 230 \cdot b^2 e^{\frac{3\pi}{4} a^2},$$

where the last inequality is easy to prove by taking the logarithm on both sides. ∎

2.1   Embeddings

For two metric spaces (X, dist_X) and (Y, dist_Y) and a function $f : X \to Y$ we define the *Lipschitz constant* of f as

$$\|f\|_{\mathtt{Lip}} = \sup_{x \neq y \in X} \frac{\mathrm{dist}_Y(f(x), f(y))}{\mathrm{dist}_X(x, y)}.$$

If f is injective we define its *distortion* as $\mathrm{distortion}(f) = \|f\|_{\mathtt{Lip}} \cdot \|f^{-1}\|_{\mathtt{Lip}}$, and otherwise $\mathrm{distortion}(f) = \infty$. By $c_Y(X)$ we denote the least distortion with which X can be embedded into Y, i.e.,

$$c_Y(X) = \inf \{\mathrm{distortion}(f) \mid f : X \to Y\}.$$

We use $c_p(X)$ to denote $c_{L_p}(X)$. Of special interest are embeddings into Hilbert spaces and in this case the parameter $c_2(X)$ is called the *Euclidean distortion* of X.

2.2 Lattices

An n-dimensional *lattice* $\mathcal{L} \subseteq \mathbb{R}^n$ is the set of all integer combinations of a set of linearly independent vectors $\{b_1, \ldots, b_m\} \subseteq \mathbb{R}^n$, i.e., $\mathcal{L} = \{\sum_{i=1}^m a_i b_i \mid \forall i.\ a_i \in \mathbb{Z}\}$. The set $\{b_1, \ldots, b_m\}$ is called a *basis* of $\mathcal{L}$ and m, the number of vectors in it, is the *rank* of $\mathcal{L}$. Let B be the n by m matrix whose ih column is b_i. We identify the matrix and the basis that it represents and denote by $\mathcal{L}(B)$ the lattice that B generates. The determinant of $\mathcal{L}$ is defined by $\det(\mathcal{L}) = \sqrt{\det(B^T B)}$. It is not difficult to verify that $\det(\mathcal{L})$ is independent of the choice of the basis. The *dual lattice*, denoted by $\mathcal{L}^*$, is defined as the set of all vectors in $\mathbb{R}^n$ that have integer inner product with all the lattice vectors of $\mathcal{L}$, that is $\mathcal{L}^* = \{u \in \mathbb{R}^n \mid \forall v \in \mathcal{L}.\ \langle u, v \rangle \in \mathbb{Z}\}$, and a *self-dual* lattice is one that satisfies $\mathcal{L} = \mathcal{L}^*$. The length of a shortest nonzero vector in $\mathcal{L}$ is denoted by $\lambda_1(\mathcal{L}) = \min\{\|u\| \mid u \in \mathcal{L} \setminus \{0\}\}$. This definition is naturally extended to the *successive minima* $\lambda_1, \ldots, \lambda_m$ defined as follows:

$$\lambda_i(\mathcal{L}) = \inf\{r > 0 \mid \operatorname{rank}(\operatorname{span}(\mathcal{L} \cap r \cdot \mathcal{B})) \geq i\}.$$

It will be convenient to define also $\lambda_0(\mathcal{L}) = 0$. For a full-rank lattice $\mathcal{L}$ (that is, $m = n$) the *covering radius* $\mu(\mathcal{L})$ is defined as the smallest r such that balls of radius r centered at all lattice points cover the entire space, or equivalently $\mu(\mathcal{L}) = \max\{\operatorname{dist}(x, \mathcal{L}) \mid x \in \mathbb{R}^n\}$. It is well known that $\frac{1}{2} \cdot \lambda_n(\mathcal{L}) \leq \mu(\mathcal{L}) \leq \frac{\sqrt{n}}{2} \cdot \lambda_n(\mathcal{L})$ (see, e.g., [6, Page 138]).

The space $\mathbb{R}^n/\mathcal{L}$ is the quotient space defined by a lattice $\mathcal{L}$. Let $u, v \in \mathbb{R}^n/\mathcal{L}$ be two points. By abuse of notation we sometimes identify between points in $\mathbb{R}^n/\mathcal{L}$ and their representatives in $\mathbb{R}^n$. For example, $\operatorname{dist}_{\mathbb{R}^n/\mathcal{L}}(u, v)$ is defined as $\operatorname{dist}(u, \mathcal{L} + v)$, i.e., the distance between representatives of u and v modulo the lattice. A function $f : \mathbb{R}^n \to \mathbb{C}$ is $\mathcal{L}$-*periodic* if $f(x) = f(x+y)$ for all $x \in \mathbb{R}^n$ and $y \in \mathcal{L}$. The Hilbert space $L_2(\mathbb{R}^n/\mathcal{L})$ is a space of scalar functions with domain $\mathbb{R}^n/\mathcal{L}$. We sometimes identify a function in $L_2(\mathbb{R}^n/\mathcal{L})$ with its corresponding $\mathcal{L}$-periodic function with domain $\mathbb{R}^n$. For $f, g \in L_2(\mathbb{R}^n/\mathcal{L})$, the distance between them is defined as

$$\operatorname{dist}_{L_2(\mathbb{R}^n/\mathcal{L})}(f, g) = \left(\int_{\mathbb{R}^n/\mathcal{L}} |f(x) - g(x)|^2 \mathrm{d}x \right)^{1/2}.$$

2.3 Gaussian Measures and the Smoothing Parameter

For $n \in \mathbb{N}$ and $s > 0$ let $\rho_s : \mathbb{R}^n \to (0, 1]$ be the Gaussian function centered at the origin scaled by a factor of s defined by

$$\forall x \in \mathbb{R}^n.\ \rho_s(x) = e^{-\pi \|x/s\|^2}.$$

We omit the subscript when $s = 1$. We define the *discrete Gaussian distribution* with parameter s on a lattice $\mathcal{L}$ by its probability function

$$\forall x \in \mathcal{L}.\ D_{\mathcal{L},s}(x) = \frac{\rho_s(x)}{\rho_s(\mathcal{L})}.$$

Notice that the sum $\rho_s(\mathcal{L})$ over all lattice vectors is finite, as follows from the fact that $\int_{\mathbb{R}^n} \rho_s(x) \mathrm{d}x = s^n$. It can be shown that a vector sampled from $D_{\mathcal{L},s}$ has the zeros vector as expectation and has expected squared norm close to $s^2 n/2\pi$ if s is large enough. Micciancio and Regev [7] defined a lattice parameter that measures how big s should be for the distribution $D_{\mathcal{L},s}$ to "behave like" a continuous Gaussian distribution in $\mathbb{R}^n$ (and in particular to have expected squared norm close to $s^2 n/2\pi$). This parameter is called the *smoothing parameter* and is defined as follows.

Definition 1. *For a lattice $\mathcal{L}$ and a positive $\varepsilon > 0$ the* smoothing parameter $\eta_\varepsilon(\mathcal{L})$ *is defined as the smallest $s > 0$ such that $\rho_{1/s}(\mathcal{L}^* \setminus \{0\}) \leq \varepsilon$.*

A main property of the smoothing parameter is that, roughly speaking, the distribution of a uniformly chosen random lattice point from $\mathcal{L}$ perturbed by a Gaussian with $s = \eta_\varepsilon(\mathcal{L})$ is $\varepsilon/2$-close to a uniform distribution on the entire space. For more details on the smoothing parameter the reader is referred to [7].

We state below a lemma due to Banaszczyk [2] and a simple bound on the smoothing parameter that it yields (whose proof can be found in the full version of the paper).

Lemma 1 ([2]). *For any $n \geq 1$, an n-dimensional lattice $\mathcal{L}$ and a vector $u \in \mathbb{R}^n$,*

$$\rho((\mathcal{L} - u) \setminus 2\sqrt{n}\mathcal{B}) \leq 2^{-11n} \cdot \rho(\mathcal{L}).$$

Lemma 2. *For any $n \geq 1$ and an n-dimensional lattice $\mathcal{L}$, $\eta_\varepsilon(\mathcal{L}) \leq \frac{2\sqrt{n}}{\lambda_1(\mathcal{L}^*)}$ where $\varepsilon = 2^{-10n}$.*

2.4   Korkine-Zolotarev Bases

The question of specifying a basis of a lattice with valuable properties is known as *reduction theory*. In 1873, Korkine and Zolotarev [5] defined and studied a notion of a reduced basis whose vectors are in some sense close to orthogonal. These bases are known as Korkine-Zolotarev bases.

Before defining Korkine-Zolotarev bases we need to define the *Gram-Schmidt orthogonalization process*. For a sequence of vectors $b_1, \ldots, b_n$ define the corresponding Gram-Schmidt orthogonalized vectors $\tilde{b}_1, \ldots, \tilde{b}_n$ by

$$\tilde{b}_i = b_i - \sum_{j=1}^{i-1} \mu_{i,j} \tilde{b}_j, \quad \mu_{i,j} = \frac{\langle b_i, \tilde{b}_j \rangle}{\langle \tilde{b}_j, \tilde{b}_j \rangle}.$$

In words, $\tilde{b}_i$ is the component of b_i orthogonal to $b_1, \ldots, b_{i-1}$. A Korkine-Zolotarev basis is defined as follows.

Definition 2. *Let B be a basis of an n-dimensional lattice $\mathcal{L}$ and let $\tilde{B}$ be the corresponding Gram-Schmidt orthogonalized basis. For $1 \leq i \leq n$ define the projection function $\pi_i^{(B)}(x) = \sum_{j=i}^n \langle x, \tilde{b}_j \rangle \cdot \tilde{b}_j / \|\tilde{b}_j\|^2$ that maps x to its projection on $\mathrm{span}(\tilde{b}_i, \ldots, \tilde{b}_n)$. A basis B is a Korkine-Zolotarev basis if for all $1 \leq i \leq n$,*

- $\tilde{b}_i$ is a shortest nonzero vector in $\pi_i^{(B)}(\mathcal{L}) = \{\pi_i^{(B)}(u) \mid u \in \mathcal{L}\}$,
- and for all $j < i$, the Gram-Schmidt coefficients $\mu_{i,j}$ of B satisfy $|\mu_{i,j}| \leq \frac{1}{2}$.

We now state two simple lemmas on Korkine-Zolotarev bases whose proofs can be found in the full version of the paper. For an n-dimensional lattice $\mathcal{L}$ and a Korkine-Zolotarev basis B that generates it, let $\mathcal{L}_i = \pi_i^{(B)}(\mathcal{L})$ be the projection of $\mathcal{L}$ on $\mathrm{span}(\tilde{b}_i, \ldots, \tilde{b}_n)$. Notice that $\mathcal{L}_i$ is a lattice for every $1 \leq i \leq n$. Intuitively speaking, since the vectors of B are close to being orthogonal, we expect a shortest nonzero vector in $\mathcal{L}_i$ to have length similar to $\lambda_i(\mathcal{L})$. This is stated formally in the following lemma. Notice that the lower bound is meaningful only when there is a gap between $\lambda_{i-1}(\mathcal{L})$ and $\lambda_i(\mathcal{L})$.

Lemma 3. *Let B be a Korkine-Zolotarev basis of an n-dimensional lattice $\mathcal{L}$ and denote $\mathcal{L}_i = \pi_i^{(B)}(\mathcal{L})$. Then for all $1 \leq i \leq n$,*

$$\frac{4}{i+3} \cdot \lambda_i(\mathcal{L})^2 - \frac{i-1}{4} \cdot \lambda_{i-1}(\mathcal{L})^2 \leq \lambda_1(\mathcal{L}_i)^2 \leq \lambda_i(\mathcal{L})^2.$$

The next lemma says that if the distance of a vector $u \in \mathbb{R}^n$ from $\mathcal{L}$ is somewhat higher than $\lambda_{i-1}(\mathcal{L})$, then it is close to the distance between $\mathcal{L}_i$ and the projected vector $\pi_i(u)$.

Lemma 4. *Let B be a Korkine-Zolotarev basis of an n-dimensional lattice $\mathcal{L}$ and denote $\mathcal{L}_i = \pi_i^{(B)}(\mathcal{L})$. Then for any $u \in \mathbb{R}^n$ and $1 \leq i \leq n$,*

$$\mathrm{dist}(u, \mathcal{L})^2 - \frac{i-1}{4} \cdot \lambda_{i-1}(\mathcal{L})^2 \leq \mathrm{dist}(\pi_i(u), \mathcal{L}_i)^2 \leq \mathrm{dist}(u, \mathcal{L})^2.$$

3   Properties of Gaussian Distributions

For an n-dimensional lattice $\mathcal{L} \subseteq \mathbb{R}^n$ and a positive number $s > 0$ we define the function $h_{\mathcal{L},s} : \mathbb{R}^n \to [0, 1)$ by

$$\forall u \in \mathbb{R}^n. \quad h_{\mathcal{L},s}(u) = 1 - \frac{\rho_s(\mathcal{L} - u)}{\rho_s(\mathcal{L})}.$$

It can be shown that the function $h_{\mathcal{L},s}$ is nonnegative.[1] Notice that if $u \in \mathcal{L}$ then $h_{\mathcal{L},s}(u) = 0$.

In this section we gather and prove several results on $h_{\mathcal{L},s}$ that, roughly speaking, show that for certain choices of s, $h_{\mathcal{L},s}(u)$ is closely related to the distance of u from $\mathcal{L}$. The following lemma provides upper and lower bounds on $h_{\mathcal{L},s}(u)$. Its first item is due to [2] and we include its proof for completeness. We remark that the lemma can also be proven using Fourier transform.

Lemma 5. *For any $n \geq 1$, an n-dimensional lattice $\mathcal{L}$, a vector $u \in \mathbb{R}^n$ and $s > 0$,*

---

[1] For example, this follows from Proposition 1.

1. $h_{\mathcal{L},s}(u) \leq \frac{\pi}{s^2} \cdot \mathrm{dist}(u, \mathcal{L})^2$.
2. If $0 < \varepsilon \leq \frac{1}{1000}$, $s \leq \frac{1}{2\eta_\varepsilon(\mathcal{L}^*)}$ and $\mathrm{dist}(u, \mathcal{L}) \leq \frac{s}{\sqrt{2}}$ then $h_{\mathcal{L},s}(u) \geq \frac{c}{s^2} \cdot$ $\mathrm{dist}(u, \mathcal{L})^2$, where c is an absolute constant.

Proof: Assume without loss of generality that $\mathrm{dist}(u, \mathcal{L}) = \|u\|$ and observe that

$$h_{\mathcal{L},s}(u) = 1 - \frac{1}{\rho_s(\mathcal{L})} \cdot \sum_{x \in \mathcal{L}} e^{-\frac{\pi\|x-u\|^2}{s^2}} = 1 - \frac{1}{2\rho_s(\mathcal{L})} \cdot \sum_{x \in \mathcal{L}} \left(e^{-\frac{\pi\|x-u\|^2}{s^2}} + e^{-\frac{\pi\|x+u\|^2}{s^2}} \right)$$

$$= 1 - \frac{e^{-\frac{\pi\|u\|^2}{s^2}}}{\rho_s(\mathcal{L})} \cdot \sum_{x \in \mathcal{L}} \left(e^{-\frac{\pi\|x\|^2}{s^2}} \cdot \cosh\left(\frac{2\pi\langle x, u\rangle}{s^2} \right) \right)$$

$$= 1 - e^{-\frac{\pi\|u\|^2}{s^2}} - \frac{e^{-\frac{\pi\|u\|^2}{s^2}}}{\rho_s(\mathcal{L})} \cdot \sum_{x \in \mathcal{L}\setminus\{0\}} \left(e^{-\frac{\pi\|x\|^2}{s^2}} \cdot \left(\cosh\left(\frac{2\pi\langle x, u\rangle}{s^2} \right) - 1 \right) \right).$$

For Item 1, use the fact that for all $\alpha \in \mathbb{R}$, $\cosh(\alpha) \geq 1$ and $1 - e^{-\alpha} \leq \alpha$ to get that

$$h_{\mathcal{L},s}(u) \leq 1 - e^{-\frac{\pi\|u\|^2}{s^2}} \leq \frac{\pi\|u\|^2}{s^2} = \frac{\pi}{s^2} \cdot \mathrm{dist}(u, \mathcal{L})^2.$$

For Item 2, use the Cauchy-Schwarz inequality and Fact 6 to get that any $x \in \mathcal{L} \setminus \{0\}$ satisfies

$$\cosh\left(\frac{2\pi\langle x, u\rangle}{s^2} \right) - 1 \leq \cosh\left(\frac{2\pi\|x\| \cdot \|u\|}{s^2} \right) - 1 \leq 230 \cdot \frac{\|u\|^2}{s^2} \cdot e^{\frac{3\pi\|x\|^2}{4s^2}}.$$

This implies that

$$h_{\mathcal{L},s}(u) \geq 1 - e^{-\frac{\pi\|u\|^2}{s^2}} - \frac{230\|u\|^2}{s^2} \cdot \sum_{x \in \mathcal{L}\setminus\{0\}} \left(e^{-\frac{\pi\|x\|^2}{s^2}} \cdot e^{\frac{3\pi\|x\|^2}{4s^2}} \right)$$

$$= 1 - e^{-\frac{\pi\|u\|^2}{s^2}} - \frac{230\|u\|^2}{s^2} \cdot \rho_{2s}(\mathcal{L}\setminus\{0\}) \geq \frac{\|u\|^2}{s^2}\left(\frac{\pi}{4} - 230\varepsilon \right),$$

where the last inequality follows from the inequality $1 - e^{-\alpha} \geq \frac{\alpha}{4}$ that holds for any $\alpha \leq 2$ and the assumptions $\|u\| \leq \frac{s}{\sqrt{2}}$ and $\eta_\varepsilon(\mathcal{L}^*) \leq \frac{1}{2s}$. This completes the proof by our assumption on ε. ∎

We turn to deal with lower bounds on $h_{\mathcal{L},s}(u)$ for vectors u that are far from the lattice.

Lemma 6. *For any $n \geq 1$, an n-dimensional lattice $\mathcal{L}$, $s > 0$ and $u \in \mathbb{R}^n$,*

1. *If $\mathrm{dist}(u, \mathcal{L}) > 2s \cdot \sqrt{n}$ then $h_{\mathcal{L},s}(u) \geq 1 - 2^{-11n}$.*
2. *If $\lambda_1(\mathcal{L}) \geq 4s\sqrt{n}$ then $h_{\mathcal{L},s}(u) \geq 1 - e^{-\pi\,\mathrm{dist}(u,\mathcal{L})^2/s^2} - 2^{-11n}$.*

Proof: First, apply Lemma 1 to $\frac{1}{s} \cdot \mathcal{L}$ to get that

$$h_{\mathcal{L},s}(u) \geq 1 - 2^{-11n} - \frac{\rho_s((\mathcal{L} - u) \cap 2s \cdot \sqrt{n}\mathcal{B})}{\rho_s(\mathcal{L})}.$$

If $\text{dist}(u, \mathcal{L}) > 2s \cdot \sqrt{n}$ then the intersection $(\mathcal{L} - u) \cap (2s \cdot \sqrt{n}\mathcal{B})$ is empty and we are done. For Item 2, notice that there is at most one point of $\mathcal{L} - u$ inside the (open) ball of radius $2s \cdot \sqrt{n}$. ∎

4   The Embedding

In this section we prove the main results of the paper. We define an embedding from a torus $\mathbb{R}^n/\mathcal{L}$ into the Hilbert space $L_2(\mathbb{R}^n/\mathcal{L})$ and relate the distortion that it achieves to the function $h_{\mathcal{L},s}$ defined in Section 3.

For an n-dimensional lattice $\mathcal{L}$ and $s > 0$ we define the embedding $H_{\mathcal{L},s} : \mathbb{R}^n/\mathcal{L} \to L_2(\mathbb{R}^n/\mathcal{L})$ that maps any vector $u \in \mathbb{R}^n/\mathcal{L}$ to the function that maps any $x \in \mathbb{R}^n$ to

$$\frac{s}{\sqrt{2\rho_s(\mathcal{L})}} \cdot \left(\frac{2}{s}\right)^{n/2} \cdot \rho_{\frac{s}{\sqrt{2}}}(\mathcal{L} + x - u).$$

In words, $H_{\mathcal{L},s}(u)$ is the function that maps any $x \in \mathbb{R}^n$ to the mass of the Gaussian function centered at u with parameter $\frac{s}{\sqrt{2}}$ on all the shifts of x by lattice vectors (up to some normalization factor).

The following proposition relates the distance between two embedded points and the function $h_{\mathcal{L},s}$ from Section 3. This enables us to use the lemmas from Section 3 to bound the distortion achieved by our embedding.

Proposition 1. *For any $n \geq 1$, an n-dimensional lattice $\mathcal{L}$, a real $s > 0$ and $u, v \in \mathbb{R}^n/\mathcal{L}$,*

$$\text{dist}_{L_2(\mathbb{R}^n/\mathcal{L})}(H_{\mathcal{L},s}(u), H_{\mathcal{L},s}(v))^2 = s^2 \cdot h_{\mathcal{L},s}(u - v).$$

Proof: We start by calculating the integral $\int_{\mathbb{R}^n/\mathcal{L}} \rho_{\frac{s}{\sqrt{2}}}(\mathcal{L} + z - u)\rho_{\frac{s}{\sqrt{2}}}(\mathcal{L} + z - v)$ dz for general $u, v \in \mathbb{R}^n$. Notice that $\mathcal{L} = x + \mathcal{L}$ for every $x \in \mathcal{L}$ and hence the integral equals to

$$\sum_{x \in \mathcal{L}} \int_{\mathbb{R}^n/\mathcal{L}} \rho_{\frac{s}{\sqrt{2}}}(x + z - u)\rho_{\frac{s}{\sqrt{2}}}(\mathcal{L} + x + z - v)dz = \int_{\mathbb{R}^n} \rho_{\frac{s}{\sqrt{2}}}(w)\rho_{\frac{s}{\sqrt{2}}}(\mathcal{L} + w + u - v)dw$$

$$= \sum_{y \in \mathcal{L}} \int_{\mathbb{R}^n} \rho_s(2w + y + u - v)\rho_s(y + u - v)dw = \left(\frac{s}{2}\right)^n \cdot \rho_s(\mathcal{L} + v - u),$$

where for the second equality we use the parallelogram law.

Now we prove the lemma using the integral from above. The squared distance $\text{dist}_{L_2(\mathbb{R}^n/\mathcal{L})}(H_{\mathcal{L},s}(u), H_{\mathcal{L},s}(v))^2$ equals to

$$\frac{s^2}{2\rho_s(\mathcal{L})} \cdot \left(\frac{2}{s}\right)^n \cdot \int_{\mathbb{R}^n/\mathcal{L}} (\rho_{\frac{s}{\sqrt{2}}}(\mathcal{L} + z - u) - \rho_{\frac{s}{\sqrt{2}}}(\mathcal{L} + z - v))^2 dz$$

$$= \frac{s^2}{2\rho_s(\mathcal{L})} \cdot \left(\frac{2}{s}\right)^n \cdot \left(2 \cdot \left(\frac{s}{2}\right)^n \cdot \rho_s(\mathcal{L}) - 2 \cdot \left(\frac{s}{2}\right)^n \cdot \rho_s(\mathcal{L} + v - u)\right)$$

$$= s^2 \cdot \left(1 - \frac{\rho_s(\mathcal{L} + v - u)}{\rho_s(\mathcal{L})}\right) = s^2 \cdot h_{\mathcal{L},s}(u - v).$$

∎

4.1 Upper Bounds in Terms of Lattice Parameters

In this section we prove an upper bound on $c_2(\mathbb{R}^n/\mathcal{L})$ in terms of $\lambda_1(\mathcal{L})$ and $\mu(\mathcal{L})$. We start with the following theorem for didactical reasons and then prove its strengthening Theorem 4.

Theorem 7. *For any $n \geq 1$ and an n-dimensional lattice $\mathcal{L}$, $c_2(\mathbb{R}^n/\mathcal{L}) = O\left(\frac{\mu(\mathcal{L})}{\lambda_1(\mathcal{L})} \cdot \sqrt{n}\right)$.*

Proof: Let $\mathcal{L}$ be an n-dimensional lattice, consider the embedding $H_{\mathcal{L},s}$ for $s = \frac{\lambda_1(\mathcal{L})}{4\sqrt{n}}$, and fix distinct $u, v \in \mathbb{R}^n/\mathcal{L}$. By Proposition 1 our goal is to bound

$$A := \frac{\text{dist}_{L_2(\mathbb{R}^n/\mathcal{L})}(H_{\mathcal{L},s}(u), H_{\mathcal{L},s}(v))^2}{\text{dist}_{\mathbb{R}^n/\mathcal{L}}(u, v)^2} = \frac{s^2 \cdot h_{\mathcal{L},s}(u - v)}{\text{dist}_{\mathbb{R}^n/\mathcal{L}}(u, v)^2}$$

from above and from below. For the upper bound use Item 1 of Lemma 5 to obtain $A \leq s^2 \cdot \frac{\pi}{s^2} = \pi$. For the lower bound consider the following two cases. If $\text{dist}_{\mathbb{R}^n/\mathcal{L}}(u, v) \leq \frac{s}{\sqrt{2}}$ then by Item 2 of Lemma 5 applied to $u - v$ we get $A \geq s^2 \cdot \frac{c}{s^2} = c$, using Lemma 2 that yields $2\eta_\varepsilon(\mathcal{L}^*) \leq \frac{4\sqrt{n}}{\lambda_1(\mathcal{L})} = \frac{1}{s}$ for $\varepsilon = 2^{-10n} \leq \frac{1}{1000}$. Otherwise, if $\text{dist}_{\mathbb{R}^n/\mathcal{L}}(u, v) > \frac{s}{\sqrt{2}}$, by Item 2 of Lemma 6 applied to $u - v$ we have $A \geq s^2 \cdot \frac{1 - e^{-\pi/2} - 2^{-11n}}{\mu(\mathcal{L})^2}$, using the fact that $\lambda_1(\mathcal{L}) = 4s \cdot \sqrt{n}$. Hence, our embedding achieves distortion $O\left(\frac{\mu(\mathcal{L})}{s}\right) = O\left(\frac{\mu(\mathcal{L})}{\lambda_1(\mathcal{L})} \cdot \sqrt{n}\right)$.

∎

For the proof of Theorem 4 we extend the embedding $H_{\mathcal{L},s}$ as follows. For an n-dimensional lattice $\mathcal{L}$, $s > 0$ and $k \geq 1$, we define the embedding $H_{\mathcal{L},s}^{(k)} : \mathbb{R}^n/\mathcal{L} \to L_2(\mathbb{R}^n/\mathcal{L})^k$ by $H_{\mathcal{L},s}^{(k)} = (H_{\mathcal{L},s_1}, H_{\mathcal{L},s_2}, \ldots, H_{\mathcal{L},s_k})$, where $s_i = 2^{i-1} \cdot s$.

Proof of Theorem 4: Let $\mathcal{L}$ be an n-dimensional lattice and consider the embedding $H_{\mathcal{L},s}^{(k)}$ for $s = \frac{\lambda_1(\mathcal{L})}{4\sqrt{n}}$ and $k = \left\lceil \log\left(\frac{4\mu(\mathcal{L})}{\lambda_1(\mathcal{L})}\right) \right\rceil$. This embedding maps any point $u \in \mathbb{R}^n/\mathcal{L}$ to a vector of Gaussian functions with various radii in the interval between the length of a shortest nonzero vector in $\mathcal{L}$ and its covering radius. Intuitively, in this way for every possible distance between two points in $\mathbb{R}^n/\mathcal{L}$ we have a Gaussian function sensitive to it.

Fix distinct $u, v \in \mathbb{R}^n/\mathcal{L}$ and use Proposition 1 to observe that

$$\frac{\text{dist}_{L_2(\mathbb{R}^n/\mathcal{L})^k}(H_{\mathcal{L},s}^{(k)}(u), H_{\mathcal{L},s}^{(k)}(v))^2}{\text{dist}_{\mathbb{R}^n/\mathcal{L}}(u, v)^2} = \sum_{i=1}^{k} \frac{\text{dist}_{L_2(\mathbb{R}^n/\mathcal{L})}(H_{\mathcal{L},s_i}(u), H_{\mathcal{L},s_i}(v))^2}{\text{dist}_{\mathbb{R}^n/\mathcal{L}}(u, v)^2},$$

where $s_i = 2^{i-1} \cdot s$. Denoting $A_i(u,v) = \frac{s_i^2 \cdot h_{\mathcal{L},s_i}(u-v)}{\text{dist}_{\mathbb{R}^n/\mathcal{L}}(u,v)^2}$ for $1 \leq i \leq k$, the above equals to $\sum_{i=1}^{k} A_i(u,v)$. We will show that

$$\Omega\left(\frac{1}{n}\right) \leq \sum_{i=1}^{k} A_i(u,v) \leq O(k), \tag{1}$$

which implies that our embedding has distortion $O(\sqrt{nk})$, as required.

By Item 1 of Lemma 5 we have $A_i(u,v) \leq s_i^2 \cdot \frac{\pi}{s_i^2} = \pi$ for every $1 \leq i \leq k$, which proves the upper bound in (1). In order to prove the lower bound in (1) we now show that there exists an i such that $A_i(u,v) \geq \Omega(\frac{1}{n})$. Consider the following three cases:

- Case 1: $\text{dist}_{\mathbb{R}^n/\mathcal{L}}(u,v) \leq \frac{1}{4\sqrt{2n}} \cdot \lambda_1(\mathcal{L})$.

  Notice that by Lemma 2 we have $2\eta_\varepsilon(\mathcal{L}^*) \leq \frac{4\sqrt{n}}{\lambda_1(\mathcal{L})} = \frac{1}{s_1}$ for $\varepsilon = 2^{-10n} \leq \frac{1}{1000}$. Hence, by Item 2 of Lemma 5, $A_1(u,v) \geq s_1^2 \cdot \frac{c}{s_1^2} = c$.

- Case 2: $\frac{1}{4\sqrt{2n}} \cdot \lambda_1(\mathcal{L}) < \text{dist}_{\mathbb{R}^n/\mathcal{L}}(u,v) \leq \lambda_1(\mathcal{L})$.

  Since $\lambda_1(\mathcal{L}) = 4s_1\sqrt{n}$, by Item 2 of Lemma 6 we get $h_{\mathcal{L},s_1}(u-v) \geq 1 - e^{-\pi/2} - 2^{-11n}$ and hence

  $$A_1(u,v) \geq (1 - e^{-\pi/2} - 2^{-11n}) \cdot \frac{s_1^2}{\text{dist}_{\mathbb{R}^n/\mathcal{L}}^2(u,v)} \geq \frac{1 - e^{-\pi/2} - 2^{-11n}}{16n}.$$

- Case 3: $\lambda_1(\mathcal{L}) < \text{dist}_{\mathbb{R}^n/\mathcal{L}}(u,v) \leq \mu(\mathcal{L})$.

  Let $1 \leq i \leq k$ be the index that satisfies $s_i < \frac{\text{dist}_{\mathbb{R}^n/\mathcal{L}}(u,v)}{2\sqrt{n}} \leq 2 \cdot s_i = s_{i+1}$. This index exists due to our choice of k. So $\text{dist}_{\mathbb{R}^n/\mathcal{L}}(u,v) = \text{dist}(u-v,\mathcal{L}) > 2s_i \cdot \sqrt{n}$. Hence, by Item 1 of Lemma 6, we have $h_{\mathcal{L},s_i}(u-v) \geq 1 - 2^{-11n}$ and we get that $A_i(u,v) \geq (1 - 2^{-11n}) \cdot \frac{s_i^2}{\text{dist}_{\mathbb{R}^n/\mathcal{L}}^2(u,v)} \geq \frac{1-2^{-11n}}{16n}$. ∎

In the following two corollaries we observe that our bounds are nearly tight for certain families of lattices. The first follows immediately by combining Theorems 5 and 7 and the second requires a transference theorem by Banaszczyk [2].

Corollary 1. *Let $\mathcal{L}$ be an n-dimensional lattice such that $\lambda_1(\mathcal{L}), \mu(\mathcal{L})$ are equal up to a multiplicative constant and $\lambda_1(\mathcal{L}^*) \cdot \mu(\mathcal{L}) \geq \Omega(n)$. Then, $c_2(\mathbb{R}^n/\mathcal{L}) = \Theta(\sqrt{n})$.*

Corollary 2. *Let $\mathcal{L}$ be an n-dimensional lattice such that $\lambda_1(\mathcal{L}^*)$ and $\mu(\mathcal{L}^*)$ are equal up to a multiplicative constant. Then, $\Omega(\sqrt{n}) \leq c_2(\mathbb{R}^n/\mathcal{L}) \leq O(\sqrt{n \log n})$.*

4.2 General Upper Bound

In this section we prove an upper bound on $c_2(\mathbb{R}^n/\mathcal{L})$ that depends only on n and is almost linear. Before presenting the proof let us start with some intuition. Notice that using the tools presented in Section 3 we have an embedding that works well for distances at most $\lambda_1(\mathcal{L})$ (Item 2 of Lemma 5) and an embedding that works for specific distances (Lemma 6). Consider a Korkine-Zolotarev basis and the projections that it defines: the lattice $\mathcal{L}_i = \pi_i(\mathcal{L})$ is the lattice $\mathcal{L}$ projected to span$(\tilde{b}_i, \ldots, \tilde{b}_n)$. We think of $\mathcal{L}_i$ as a full-rank lattice inside an $(n - i + 1)$-dimensional space. Our embedding consists of n Gaussian functions where the ith function corresponds to the lattice $\mathcal{L}_i$. Due to the use of a Korkine-Zolotarev basis using Item 2 of Lemma 5 we can show that the ith Gaussian function handles distances that are both somewhat larger than $\lambda_{i-1}(\mathcal{L})$ and somewhat smaller than $\lambda_i(\mathcal{L})$. In order to treat distances around the $\lambda_i(\mathcal{L})$'s we add additional $O(\log n)$ Gaussian functions for every i and use Lemma 6 to prove correctness.

Proof of Theorem 3: Let $\mathcal{L}$ be an n-dimensional lattice generated by a Korkine-Zolotarev basis B. For $1 \le i \le n$ let $\pi_i = \pi_i^{(B)}$ be the corresponding orthogonal projection function that maps vectors to the orthogonal complement of span$(b_1, \ldots, b_{i-1})$. Denote $\mathcal{L}_i = \pi_i(\mathcal{L})$, $n_i = n - i + 1$, $s_i = \frac{\lambda_i(\mathcal{L})}{4n}$, $k = \lceil \frac{1}{2} \cdot (5 + 3 \log n) \rceil$ and $r_{i,j} = 2^{j-1} \cdot \frac{\lambda_i(\mathcal{L})}{8n\sqrt{2n}}$ for $1 \le j \le k$. Consider the embedding $H_\mathcal{L}$ that maps $u \in \mathbb{R}^n / \mathcal{L}$ to the vector of $n + nk = O(n \log n)$ functions

$$(H_{\mathcal{L}_1, s_1}(\pi_1(u)), \ldots, H_{\mathcal{L}_n, s_n}(\pi_n(u)), H_{\mathcal{L}, r_{1,1}}^{(k)}(u), \ldots, H_{\mathcal{L}, r_{n,1}}^{(k)}(u)).$$

This is an element in the space $L_2 = L_2(\mathbb{R}^{n_1} / \mathcal{L}_1) \oplus \cdots \oplus L_2(\mathbb{R}^{n_n} / \mathcal{L}_n) \oplus L_2(\mathbb{R}^n / \mathcal{L})^{nk}$.

Fix distinct $u, v \in \mathbb{R}^n / \mathcal{L}$. By Proposition 1,

$$\frac{\text{dist}_{L_2}(H_\mathcal{L}(u), H_\mathcal{L}(v))^2}{\text{dist}_{\mathbb{R}^n / \mathcal{L}}(u, v)^2} = \sum_{i=1}^n A_i(u, v) + \sum_{1 \le i \le n, 1 \le j \le k} B_{i,j}(u, v),$$

where

$$A_i(u, v) = \frac{s_i^2 \cdot h_{\mathcal{L}_i, s_i}(\pi_i(u) - \pi_i(v))}{\text{dist}_{\mathbb{R}^n / \mathcal{L}}(u, v)^2}, \quad B_{i,j}(u, v) = \frac{r_{i,j}^2 \cdot h_{\mathcal{L}, r_{i,j}}(u - v)}{\text{dist}_{\mathbb{R}^n / \mathcal{L}}(u, v)^2}.$$

We will show that

$$\Omega\left(\frac{1}{n}\right) \le \sum_{i=1}^n A_i(u, v) + \sum_{1 \le i \le n, 1 \le j \le k} B_{i,j}(u, v) \le O(n \log n), \qquad (2)$$

which implies that our embedding has distortion $O(n \cdot \sqrt{\log n})$, as required.

By Item 1 of Lemma 5, for every $1 \le i \le n$ and $1 \le j \le k$ we have

$$A_i(u, v) \le \frac{\text{dist}_{\mathbb{R}^{n_i} / \mathcal{L}_i}(\pi_i(u), \pi_i(v))^2}{\text{dist}_{\mathbb{R}^n / \mathcal{L}}(u, v)^2} \cdot s_i^2 \cdot \frac{\pi}{s_i^2} \le \pi, \quad B_{i,j}(u, v) \le r_{i,j}^2 \cdot \frac{\pi}{r_{i,j}^2} = \pi,$$

where for the bound on $A_i(u, v)$ we use the upper bound from Lemma 4 applied to $u - v$. This yields the upper bound in (2). In order to prove the lower bound in (2) we now show that there exists an i such that $A_i(u, v) \ge \Omega(\frac{1}{n})$ or there exist i, j such that $B_{i,j}(u, v) \ge \Omega(\frac{1}{n})$. Since $\text{dist}_{\mathbb{R}^n / \mathcal{L}}(u, v) = \text{dist}(u - v, \mathcal{L}) \le \mu(\mathcal{L}) \le \frac{\sqrt{n}}{2} \cdot \lambda_n(\mathcal{L})$ the vectors u and v correspond to one of the following two cases:

– Case 1: There exists an i s.t. $\frac{\sqrt{n}}{2} \cdot \lambda_{i-1}(\mathcal{L}) < \mathrm{dist}_{\mathbb{R}^n/\mathcal{L}}(u,v) \le \frac{1}{4\sqrt{2}\cdot n} \cdot \lambda_i(\mathcal{L})$.
Think of $\mathcal{L}_i$ as a full-rank n_i-dimensional lattice, and use Lemma 2 to obtain for $\varepsilon_i = 2^{-10n_i} \le \frac{1}{1000}$ that

$$\eta_{\varepsilon_i}(\mathcal{L}_i^*)^2 \le \frac{4n_i}{\lambda_1(\mathcal{L}_i)^2} \le \frac{4n_i}{\frac{4}{i+3} \cdot \lambda_i(\mathcal{L})^2 - \frac{i-1}{4} \cdot \lambda_{i-1}(\mathcal{L})^2} \le \frac{4n^2}{\lambda_i(\mathcal{L})^2},$$

where the second inequality follows from Lemma 3 and the third from a straightforward calculation. This yields that $2 \cdot \eta_{\varepsilon_i}(\mathcal{L}_i^*) \le \frac{4n}{\lambda_i(\mathcal{L})} = \frac{1}{s_i}$, so we get that $A_i(u,v) \ge s_i^2 \cdot \frac{\mathrm{dist}_{\mathbb{R}^{n_i}/\mathcal{L}_i}(\pi_i(u),\pi_i(v))^2}{\mathrm{dist}_{\mathbb{R}^n/\mathcal{L}}(u,v)^2} \cdot \frac{c}{s_i^2} \ge c \cdot \left(1 - \frac{i-1}{4} \cdot \frac{\lambda_{i-1}(\mathcal{L})^2}{\mathrm{dist}_{\mathbb{R}^n/\mathcal{L}}(u,v)^2}\right) \ge \frac{c}{n}$, where the first inequality follows from Item 2 of Lemma 5, the second follows from Lemma 4 applied to $u - v$, and the third from the assumption that $\frac{\sqrt{n}}{2} \cdot \lambda_{i-1}(\mathcal{L}) < \mathrm{dist}_{\mathbb{R}^n/\mathcal{L}}(u,v)$.

– Case 2: There exists an i s.t. $\frac{1}{4\sqrt{2}\cdot n} \cdot \lambda_i(\mathcal{L}) < \mathrm{dist}_{\mathbb{R}^n/\mathcal{L}}(u,v) \le \frac{\sqrt{n}}{2} \cdot \lambda_i(\mathcal{L})$.
Let $1 \le j \le k$ be the index that satisfies $r_{i,j} < \frac{\mathrm{dist}_{\mathbb{R}^n/\mathcal{L}}(u,v)}{2\sqrt{n}} \le 2 \cdot r_{i,j} = r_{i,j+1}$. This index exists due to our choice of k. So $\mathrm{dist}_{\mathbb{R}^n/\mathcal{L}}(u,v) = \mathrm{dist}(u-v,\mathcal{L}) > 2r_{i,j} \cdot \sqrt{n}$. Hence, by Item 1 of Lemma 6, we have $h_{\mathcal{L},r_{i,j}}(u-v) \ge 1 - 2^{-11n}$ and we get that $B_{i,j}(u,v) \ge (1 - 2^{-11n}) \cdot \frac{r_{i,j}^2}{\mathrm{dist}_{\mathbb{R}^n/\mathcal{L}}^2(u,v)} \ge \frac{1 - 2^{-11n}}{16n}$. ∎

References

1. Aharonov, D., Regev, O.: Lattice problems in NP intersect coNP. Journal of the ACM 52(5), 749–765 (2005); Preliminary version in FOCS 2004
2. Banaszczyk, W.: New bounds in some transference theorems in the geometry of numbers. Mathematische Annalen 296(4), 625–635 (1993)
3. Feige, U., Micciancio, D.: The inapproximability of lattice and coding problems with preprocessing. J. Comput. System Sci. 69(1), 45–67 (2004)
4. Khot, S., Naor, A.: Nonembeddability theorems via Fourier analysis. Mathematische Annalen 334(4), 821–852 (2006); Preliminary version in FOCS 2005
5. Korkine, A., Zolotareff, G.: Sur les formes quadratiques. Mathematische Annalen 6, 366–389 (1873)
6. Micciancio, D., Goldwasser, S.: Complexity of Lattice Problems: A Cryptographic Perspective. The Kluwer International Series in Engineering and Computer Science, vol. 671. Kluwer Academic Publishers, Boston (2002)
7. Micciancio, D., Regev, O.: Worst-case to average-case reductions based on Gaussian measures. SIAM Journal on Computing 37(1), 267–302 (2007)
8. Micciancio, D., Voulgaris, P.: A deterministic single exponential time algorithm for most lattice problems based on voronoi cell computations. In: Proc. 42nd ACM Symposium on Theory of Computing (STOC), pp. 351–358 (2010)
9. Milnor, J., Husemoller, D.: Symmetric bilinear forms. Springer, Berlin (1973)

Online Embeddings

Piotr Indyk[1], Avner Magen[2],
Anastasios Sidiropoulos[3], and Anastasios Zouzias[2]

[1] MIT
indyk@theory.csail.mit.edu
[2] University of Toronto
{avner,zouzias}@cs.toronto.edu
[3] Toyota Technological Institute
tasos@tti-c.org

Abstract. We initiate the study of *on-line metric embeddings*. In such an embedding we are given a sequence of n points $X = x_1, \ldots, x_n$ one by one, from a metric space $M = (X, D)$. Our goal is to compute a low-distortion embedding of M into some host space, which has to be constructed in an on-line fashion, so that the image of each x_i depends only on $x_1, \ldots, x_i$. We prove several results translating existing embeddings to the on-line setting, for the case of embedding into ℓ_p spaces, and into distributions over ultrametrics.

Keywords: Computational geometry, online algorithms, metric embeddings.

1 Introduction

A low-distortion (or bi-Lipschitz) embedding between two metric spaces $M = (X, D)$ and $M' = (X', D')$ is a mapping f such that for any pair of points $p, q \in X$ we have $D(p, q) \leq D'(f(p), f(q)) \leq c \cdot D(p, q)$; the factor c is called the *distortion* of f. In recent years, low-distortion embeddings found numerous applications in computer science [17,12]. This can be, in part, attributed to the fact that embeddings provide a general method for designing approximation algorithms for problems defined over a "hard" metric, by embedding the input into an "easy" metric and solving the problem in there.

For some problems, however, applying this paradigm encounters difficulties. Consider for example the nearest neighbor problem: given a set P of n points in some metric (X, D), the goal is to build a data structure that finds the nearest point in P to a query point $q \in X$. A fundamental theorem of Bourgain [4] shows that it is possible to embed P and the query point q into an "easy" metric space, such as ℓ_2 with distortion $\log n$. This, however, does not translate to an efficient approximation algorithm for the problem for the simple reason that the query point q is *not known* at the preprocessing stage, so it cannot be embedded together with the set P. More specifically, for the approach to work in this scenario we must require that we can *extend* the embeddings $f : P \to \ell_2$

M. Serna et al. (Eds.): APPROX and RANDOM 2010, LNCS 6302, pp. 246–259, 2010.

to $g : P \cup \{q\} \to \ell_2$. We note that the aforementioned Bourgain's theorem [4] does not have such an extendability property.

An even more straightforward setting in which the standard notion of embeddings is not quite the right notion comes up in the design of on-line algorithms. Often, the input considered is metric space; at each step the algorithm receives an input point and needs to make decisions about it instantly. In order to use the embedding method, we must require that the embedding would observe the inputs sequentially, so that a point is mapped based only on the *distance information* of the points observed so far. Here is a precise definition of the desired object.

Definition 1. *An on-line embedding of an n-point metric space $M = (X, D)$ where $X = \{x_1, \ldots x_n\}$ into some host metric space M' is a sequence of functions f_k for $k = 1, \ldots, n$ (possibly randomized) such that*

– *f_k depends only on M_k, the restriction of M on $\{x_1, \ldots, x_k\}$.*
– *f_k extends f_{k-1}: for each $x \in \{x_1, \ldots, x_{k-1}\}$, $f_k(x) = f_{k-1}(x)$. If the functions are randomized, the extendability property means that the random bits used for f_{k-1} are a subset of the random bits for f_k, and when these bits between f_{k-1} and f_k coincide the (deterministic) image of $x \in \{x_1, \ldots, x_{k-1}\}$ is the same for these functions.*

The associated distortion of the above $f_1, \ldots, f_n$ is the distortion of f_n. If f_i can be obtained algorithmically, then we say that we have an on-line algorithm for the embedding problem. We also consider on-line embeddings into shortest-path metrics of graphs. In this case, we require that M_k is mapped into a graph G_k, and that every G_k is subgraph of G_{k+1}.

In this work we investigate fundamental embedding questions in the on-line context. Can we hope, for example, to embed a general metric space in Euclidean space in an on-line fashion? Not surprisingly, the use of randomization is almost always essential in the design of such embeddings. It is interesting to relate the above notion to "oblivious embeddings". An embedding is said to be oblivious, if the image of a point does not depend on other points. In the usual (off-line) embeddings, the image of a point may depend on all other points. In this language, on-line embedding is some type of middle-ground between these two types of embeddings. In particular, oblivious embeddings are a special, very restricted case of on-line embedding. Oblivious embeddings play an important role in the design of algorithms, for example in the context of streaming algorithms [11] or in the design of near linear algorithms that rely on embeddings [1]. Indeed, some of our results use oblivious embeddings as a building block, most notably, random projections and construction of random decompositions.

1.1   Results and Motivation

Embedding into ℓ_p spaces, and into distributions over ultrametrics. We start our investigation by considering embeddings into ℓ_p spaces, and into distributions over ultrametrics. These target spaces have been studied extensively in the embedding literature.

We observe that Bartal's embedding [2] can be easily modified to work in the on-line setting. We remark that this observation was also made by Englert, Räcke, and Westermann [6]. As a consequence, we obtain an on-line analog of Bourgain's theorem [4]. More specifically, we deduce that any n-point metric space with spread[1] Δ can be embedded on-line into ℓ_p with distortion $O((\log \Delta)^{1/p} \log n)$. Similarly, we also obtain an analog of a theorem due to Bartal [2] for embedding into ultrametrics. More precisely, we give an on-line probabilistic embedding of an input metric into a distribution over ultrametrics with distortion $O(\log n \cdot \log \Delta)$.

Doubling metrics. For the special case when the input space is doubling, we give an improved on-line embedding into ultrametrics with distortion $O(\log \Delta)$. We complement this upper bound by exhibiting a distribution $\mathcal{F}$ over doubling metrics (in fact, subsets of $\mathbb{R}^1$) such that any on-line embedding of a metric chosen from $\mathcal{F}$ into ultrametrics has distortion $\Omega(\min\{n, \log \Delta\})$.

Embedding into ℓ_∞. We also consider on-line analogs of another embedding theorem, due to Fréchet, which states that any n-point metric can be embedded into ℓ_∞ with distortion 1. We show that this theorem extends to the on-line setting with the same distortion, albeit larger dimension. By composing our on-line embedding into ℓ_2, with a random projection, we obtain for any $\alpha > \sqrt{2}$, an on-line embedding into ℓ_∞ with distortion $O(\alpha \cdot \log n \sqrt{\log \Delta})$, and dimension $\Omega(\max\{(\log n)^{2/(1-1/e)}, n^{4/(\alpha^2-2)}\})$.

On-line embedding when an (off-line) isometry or near-isometry is possible. Finally, we consider the case of embedding into constant-dimensional ℓ_p spaces. It is well known ([18]) that for any constant dimension there are spaces that require polynomial distortion (e.g. via a simple volume argument). It is therefore natural to study the embedding question for instances that do embed with small distortion. When a metric embeds isometrically into ℓ_2 or ℓ_2^d, it is clear that this isometry can be found on-line. We exhibit a sharp contrast with this simple fact for the case when there is only a near-isometry guaranteed. Using a topological argument, we prove that there exists a distribution $\mathcal{D}$ over metric spaces that $(1 + \varepsilon)$-embed into ℓ_2^d, yet any on-line algorithm with input drawn from $\mathcal{D}$ computes an embedding with distortion $n^{\Omega(1/d)}$. In light of our positive results about embedding into ℓ_2 and a result of Matoušek [18], this bound can be shown to be a near-optimal for on-line embeddings.

Remark 1. For simplicity of the exposition, we will assume that n is given to the on-line algorithm in advance. We remark however that with the single exception of embedding into ℓ_∞, all of our algorithms can be modified to work without this knowledge.

---

[1] The ratio between the largest and the smallest non-zero distances in the metric space.

Table 1. Summary of results

Input Space	Host Space	Distortion	Section	Comments
General	ℓ_p	$O(\log n (\log \Delta)^{1/p})$	2	$p \in [1, \infty]$
General	Ultrametrics	$O(\log n \log \Delta)$	2	
Doubling	ℓ_2	$O(\log \Delta)$	3	
Doubling	Ultrametrics	$O(\log \Delta)$	3	
Doubling	Ultrametrics	$\Omega(\min\{n, \log \Delta\})$	4	
$(1,2)$-metric	ℓ_∞^n	1	5	
General	ℓ_∞	1	5	The input is drawn from a fixed finite set of metrics.
ℓ_2	ℓ_∞^d	D	5.1	$d \approx \Omega(n^{4/(D^2-2)})$
ℓ_∞	ℓ_∞	> 1	[21]	
$(1+\varepsilon)$-embeddable into ℓ_2^d	ℓ_2^d	$\Omega(n^{1/(d-1)})$	6	

1.2 Related Work

The notion of low-distortion on-line embeddings is related to the well-studied notion of Lipschitz extensions. A prototypical question in the latter area is: for spaces Y and Z, is it true that for every $X \subset Y$, and every C-Lipschitz[2] mapping $f : X \to Z$ it is possible to extend f to $f' : Y \to Z$ which is C'-Lipschitz, for C' not much greater than C? For many classes of metric spaces the answer to this question is positive (e.g., see the overview in [16]).

One could ask if analogous theorems hold for low-distortion (i.e., bi-Lipschitz) mapping. If so, we could try to construct on-line embeddings by repeatedly constructing bi-Lipschitz extensions to points $p_1, p_2, \ldots$. Unfortunately, bi-Lipschitz extension theorems are more rare, since the constraints are much more stringent.

In the context of the aforementioned work, the on-line embeddings can be viewed as "weak" bi-Lipschitz extension theorems, which hold for only *some* mappings $f : X \to Z$, $X \subset Y$.

1.3 Notation and Definitions

For a point $y \in \mathbb{R}^d$, we denote by y_i the i-th coordinate of y. That is, $y = (y_1, \ldots, y_d)$. Similarly, for a function $f : A \to \mathbb{R}^d$, and for $a \in A$, we use the notation $f(a) = (f_1(a), \ldots, f_d(a))$. Also, we denote by ℓ_p the space of sequences with finite p-norm, i.e., $\|x\|_p = \left(\sum_{i=1}^\infty |x_i|^p\right)^{1/p} < \infty$.

Consider a finite metric space (X, D) and let $n = |X|$. For any point $x \in X$ and $r \geq 0$, the ball with radius r around x is defined as $B_X(x, r) = \{z \in X \mid D(x, z) \leq r\}$. We omit the subscript when it is clear from the context. A metric space (X, D) is called Λ-*doubling* if for any $x \in X$, $r \geq 0$ the ball $B(x, r)$ can be covered by Λ balls of radius $r/2$. The *doubling constant* of X is the infimum Λ so that X is Λ-doubling. The *doubling dimension* of X is $\dim(X) = \log_2 \Lambda$. A metric space with $\dim(X) = O(1)$ is called *doubling*. A γ-net for a metric space (X, D) is a set $N \subseteq X$ such that for any $x, y \in N$,

---

[2] I.e., a mapping which expands the distances by a factor at most C.

$D_X(x, y) \geq \gamma$ and $X \subseteq \bigcup_{x \in N} B_X(x, \gamma)$. Let $M_1 = (X, D_1)$ and $M_2 = (X, D_2)$ be two metric spaces. We say that M_1 *dominates* M_2 if for every $i, j \in X$, $D_1(i, j) \geq D_2(i, j)$. Let (X, D_1) and (Y, D_2) be two metric space and an embedding $f : X \to Y$. We say that f is *non-expanding* if f doesn't expand distances between every pair $x_1, x_2 \in X$, i.e., $D_2(f(x_1), f(x_2)) \leq D_1(x_1, x_2)$. Similarly, f is *non-contracting* if it doesn't contract pair-wise distances. Also we say that f is α-bi-Lipschitz if there exists $\beta > 0$ such that for every $x_1, x_2 \in X$, $\beta D_1(x_1, x_2) \leq D_2(f(x_1), f(x_2)) \leq \alpha \beta D_1(x_1, x_2)$.

2   Embedding General Metrics into Ultrametrics and into ℓ_p

In this section we will describe an on-line algorithm for embedding arbitrary metrics into ℓ_p, with distortion $O(\log n \cdot (\log \Delta)^{1/p})$, for any $p \in [1, \infty]$. We also give an on-line probabilistic embedding into a distribution over ultrametrics with distortion $O(\log n \cdot \log \Delta)$. Both algorithms are on-line versions of the algorithm of Bartal [2], for embedding metrics into a distribution of dominating HSTs, with distortion $O(\log^2 n)$. Before we describe the algorithm we need to introduce some notation.

Definition 2 ([2]). *An l-partition of a metric* $M = (X, D)$ *is a partition* $Y_1, \ldots, Y_k$ *of* X, *such that the diameter of each* Y_i *is at most* l.

For a distribution $\mathcal{F}$ over l-partitions of a metric $M = (X, D)$, and for $u, v \in X$, let $p_{\mathcal{F}}(u, v)$ denote the probability that in an l-partition chosen from $\mathcal{F}$, u and v belong to different clusters.

Definition 3 ([2]). *An* (r, ρ, λ)-*probabilistic partition of a metric* $M = (X, D)$ *is a probability distribution* $\mathcal{F}$ *over* $r\rho$-*partitions of* M, *such that for each* $u, v \in X$, $p_{\mathcal{F}}(u, v) \leq \lambda \frac{D(u,v)}{r}$. *Moreover,* $\mathcal{F}$ *is* ε-*forcing if for any* $u, v \in X$, *with* $D(u, v) \leq \varepsilon \cdot r$, *we have* $p_{\mathcal{F}}(u, v) = 0$.

We observe that Bartal's algorithm [2] can be interpreted as an on-line algorithm for constructing probabilistic partitions. The input to the problem is a metric $M = (X, D)$, and a parameter r. In the first step of Bartal's algorithm, every edge of length less than r/n is contracted. This step cannot be directly performed in an on-line setting, and this is the reason that the parameters of our probabilistic partition will depend on Δ. More precisely, our partition will be $1/\Delta$-forcing, while the one obtained by Bartal's off-line algorithm is $1/n$-forcing.

The algorithm proceeds as follows. We begin with an empty partition P. At every step j, each $Y_t \in P$ will correspond to a ball of some fixed radius r_t around a point $y_t \in X_j$. Once we have picked y_t, and r_t, they will remain fixed until the end of the algorithm. Assume that we have partitioned all the points $x_1, \ldots, x_{i-1}$, and that we receive x_i. Let $P = \{Y_1, \ldots, Y_k\}$. If $x_i \notin \bigcup_{j \in [k]} B(y_j, r_j)$, then we add a new cluster Y_{k+1} in P, with center $y_{k+1} = x_i$, and we pick the radius $r_{k+1} \in [0, r \log n)$, according to the probability distribution $p(r_{k+1}) = \left(\frac{n}{n-1}\right) \frac{1}{r} e^{-r_{k+1}/r}$.

Otherwise, let Y_s be the minimum-index cluster in P, such that $x_i \in B(y_s, r_s)$, and add x_i to Y_s.

By Bartal's analysis on the above procedure, we obtain the following lemma.

Lemma 1. *Let M be a metric, and $r \in [1, \Delta]$. There exists an $1/\Delta$-forcing, $(r, O(\log n), O(1))$-probabilistic partition $\mathcal{F}$ of M, and a randomized on-line algorithm that against any non-adaptive adversary, given M computes a partition P distributed according to $\mathcal{F}$. Moreover, after each step i, the algorithm computes the restriction of P on X_i.*

By the above discussion it follows that for any $r > 0$ we can compute an $(r, O(\log n), O(1))$-probabilistic partition of the input space $M = (X, D)$. It is well known that this implies an embedding into ℓ_p for any $p \in [1, \infty]$. Since the construction is folklore (see e.g. [8,9,22]), we will only give a brief overview, demonstrating that the embedding can be indeed computed in an on-line fashion.

For each $i \in \{1, \ldots, \log \Delta\}$, and for each $j \in \{1, \ldots, O(\log n)\}$ we sample a probabilistic partition $P_{i,j}$ of M with clusters of radius 2^i. Each such cluster corresponds to a subset of a ball of radius 2^i centered at some point of M. For every i, j we compute a mapping $f_{i,j} : X \to \mathbb{R}$ as follows. For each cluster $C \in P_{i,j}$ we chose $s_{i,j} \in \{-1, 1\}$ uniformly at random. Next, for each point $x \in X$ we need to compute its distance $h_{i,j}(x)$ to the "boundary" of the union of all clusters. For every $C \in P_{i,j}$ let $a(C), r(C)$ be the center and radius of C, respectively. We can order the clusters in $P_{i,j} = (C_1, \ldots, C_k)$, so that C_t is created by the on-line algorithm before C_l for every $t < l$. For a point $x \in X$ let $C(x)$ be the cluster containing x. Suppose $C(x) = C_t$. We set $h_{i,j}(x) = \min_{l \in \{1, \ldots, t\}} |r(C_l) - D(x, a(C_l))|$. Note that $h_{i,j}(x)$ can be computed in an on-line fashion. We set $f_{i,j}(x) = s_{i,j} \cdot h_{i,j}(x)$. The resulting embedding is $\varphi(x) = \bigoplus_{i,j} f_{i,j}(x)$. It is now straightforward to verify that with high probability, for all $x, y \in X$ we have $D(x, y) \cdot \Omega((\log n)^{1/p} / \log n) \geq \|\varphi(x) - \varphi(y)\|_p \geq D(x, y) \cdot \Omega((\log n)^{1/p} / \log n)$, implying the following result.

Theorem 1. *There exists an on-line algorithm that for any $p \in [1, \infty]$, against a non-adaptive adversary, computes an embedding of a given metric into $\ell_p^{O(\log n \log \Delta)}$ with distortion $O(\log n \cdot (\log \Delta)^{1/p})$. Note that for $p = \infty$ the distortion is $O(\log n)$.*

Following the analysis of Bartal [2], we also obtain the following result.

Theorem 2. *There exists an on-line algorithm that against a non-adaptive adversary, computes a probabilistic embedding of a given metric into a distribution over ultrametrics with distortion $O(\log n \cdot \log \Delta)$.*

We remark that in the off-line probabilistic embedding into ultrametrics of [2] the distortion is $O(\log^2 n)$. In this bound there is no dependence on Δ due to a preprocessing step that contracts all sufficiently small edges. This step however cannot be implemented in an on-line fashion, so the distortion bound in Theorem 2 is slightly weaker. Interestingly, Theorem 5 implies that removing the dependence on Δ is impossible, unless the distortion becomes polynomially large.

3   Embedding Doubling Metrics into Ultrametrics and into ℓ_2

In this section we give an embedding of doubling metrics into ℓ_2 with distortion $O(\log \Delta)$. We proceed by first giving a probabilistic embedding into ultrametrics. Let $M = (X, D)$ be a doubling metric, with doubling dimension $\lambda = \log_2 \Lambda$.

We begin with an informal description of our approach. Our algorithm proceeds by incrementally constructing an HST[3], and embedding the points of the input space M into its leaves. The algorithm constructs an HST incrementally, embedding X into its leaves. The construction is essentially greedy: assume a good HST was constructed to the points so far, then when a new point p arrives it is necessary to "go down the right branch" of the tree so as to be at a small tree-distance away from points close to p. This is done by letting each internal vertex of the HST of height i correspond to a subset of M of (appropriately randomized) radius about 2^i. When p is too far from the previous centers of the balls it will branch out. The only issue that can arise (and in general, the only reason for randomness) is that while p is too far from the centre of a ball, it is in fact close to some of its members, and so a large expansion may occur when it is not placed in that part of the tree. Randomness allows to deal with this, but when decisions are made online and cannot be changed as in our case, it is not guaranteed to work. What saves the day is the fact that when a metric has bounded doubling dimension the obtained tree has *bounded* degree. This is crucial when bounding the probability of the bad event described above to happen, as at every level of the tree there could be only constant number of possible conflicts, each with low probability.

We now give a formal argument. Let $\delta = \Lambda^3$. Let $T = (V, E)$ be a complete δ-ary tree of depth $\log \Delta$, rooted at a vertex r. For each $v \in V(T)$, let $l(v)$ be the number of edges on the path from r to v in T. We set the length of an edge $\{u, v\} \in E(T)$ to $\Delta \cdot 2^{-\min\{l(u), l(v)\}}$. That is, the length of the edges along a branch from r to a leaf, are $\Delta, \Delta/2, \Delta/4, \ldots, 1$. Fix a left-to-right orientation of the children of each vertex in T. For a vertex $v \in V(T)$, let T_v denote the sub-tree of T rooted at v, and let $c(v)$ denote the left-most leaf of T_v. We refer to the point mapped to $c(v)$ as the *center* of T_v. Let $B(x, r)$ denote the ball centered at x with radius r.

We will describe an on-line embedding f of M into T, against a non-adaptive adversary. We will inductively define mappings $f_1, f_2, \ldots, f_n = f$, with $f_i : \{x_1, \ldots, x_i\} \to V(T)$, such that f_{i+1} is an extension of f_i. We pick a value $\alpha \in [1, 2]$, uniformly at random.

We inductively maintain the following three invariants.

(I1) For any $v \in V(T)$, if a point of X_i is mapped to the subtree T_v, then there is a point of X_i that is mapped to $c(v)$. In other words, the first point of X that is mapped to a subtree T_v has image $c(v)$, and is therefore the center of T_v. Formally, if $f_i(X_i) \cap V(T_v) \neq \emptyset$, then $c(v) \in f_i(X_i)$.

---

[3] See [3, Definition 8] for a definition of HST.

(I2) For any $v \in V(T)$, all the points in X_i that are mapped to T_v are contained inside a ball of radius $\Delta/2^{l(v)-1}$ around the center of T_v in M. Formally, $f_i^{-1}(V(T_v)) \subset B(f_i^{-1}(c(v)), \Delta/2^{l(v)-1})$.

(I3) For any $v \in V(T)$, and for any children $u_1 \neq u_2$ of v, the centers of T_{u_1} and T_{u_2} are at distance at least $\Delta/2^{l(v)+1}$ in M. Formally, $D(f_i^{-1}(c(u_1)), f_i^{-1}(c(u_2))) > \Delta/2^{l(v)+1}$.

We begin by setting $f_1(x_1) = c(r)$. This choice clearly satisfies invariants (I1)–(I3). Upon receiving a point x_i, we will show how to extend f_{i-1} to f_i. Let $P = p_0, \ldots, p_t$ be the following path in T. We have $p_0 = r$. For each $j \geq 0$, if there exists a child q of p_j such that $V(T_q) \cap f_{i-1}(X_{i-1}) \neq \emptyset$, and $D(f_{i-1}^{-1}(c(q)), x_i) < \alpha \cdot \Delta/2^j$, we set p_{j+1} to be the left-most such child of p_j. Otherwise, we terminate P at p_j.

Claim. There exists a child u of p_t, such that $c(u) \notin f_{i-1}(\{x_1, \ldots, x_{i-1}\})$.

Proof. Suppose that the assertion is not true. Let $y = f_{i-1}^{-1}(p_t)$. Let $v_1, \ldots, v_\delta$ be the children of p_t. By the inductive invariants (I1) and (I2), it follows that for each $i \in [\delta]$, $D(f_{i-1}^{-1}(c(v_i)), y) \leq \Delta/2^{t-1}$. Moreover, by the choice of p_t, $D(y, x_i) \leq \alpha \cdot \Delta/2^{t-1} \leq \Delta/2^{t-2}$. Therefore, the ball of radius $\Delta/2^{t-2}$ around z in M, contains the $\delta+1 = \Lambda^3 + 1$ points $x_i, f^{-1}(c(v_1)), \ldots, f^{-1}(c(v_\delta))$. However, by the choice of p_t, and by the inductive invariant (I3), it follows that the balls in M of radius $\Delta/2^{t+1}$ around each one of these points are pairwise disjoint, contradicting the fact that the doubling constant of M is Λ.

By Claim 3, we can find a sub-tree rooted at a child q of p_t such that none of the points in X_{i-1} has its image in T_q. We extend f_{i-1} to f_i by setting $f_i(x_i) = c(q)$. It is straight-forward to verify that f_i satisfies the invariants (I1)–(I3). This concludes the description of the embedding. It remains to bound the distortion of f.

Lemma 2. *For any $x, y \in X$, $D_T(f(x), f(y)) \geq \frac{1}{3}D(x, y)$.*

Proof. Let v be the nearest-common ancestor of $f(x)$ and $f(y)$ in T. By invariant (I2) we have $D(x, y) \leq D(x, f^{-1}(c(v))) + D(y, f^{-1}(c(v))) \leq \Delta \cdot 2^{-l(v)+2}$. Moreover, $D_T(f(x), f(y)) = 2 \cdot \Delta \sum_{i=l(v)}^{\log \Delta} 2^{-i} = \Delta \cdot 2^{-l(v)+2} - 2$. The lemma follows since the minimum distance in M is 1.

Lemma 3. *For any $x, y \in X$, $\mathbb{E}[D_T(f(x), f(y))] \leq O(\Lambda^3 \cdot \log \Delta) \cdot D(x, y)$.*

Proof (Proof sketch (full details in the full version):). The distance between x and y is about 2^i when they are separated at level i. For this to happen, y must be assigned to a sibling of x at level $i - 1$. The probability of assigning to any particular such sibling is $O(D(x, y)/2^i)$. It is here that we utilize the bounded-degree property. By a union bound over all siblings at this level we get a contribution of $O(\Lambda^3)$ on the expected expansion. Summing up over all $\log \Delta$ levels we get the desired bound.

Theorem 3. *There exists an on-line algorithm that against any non-adaptive adversary, given a metric $M = (X, D)$ of doubling dimension λ, computes a probabilistic embedding of M into a distribution over ultrametrics with distortion $2^{O(\lambda)} \cdot \log \Delta$.*

It is well known that ultrametrics embed isometrically into ℓ_2, and it is easy to see that such an embedding can be computed in an on-line fashion for the HSTs constructed above. We therefore also obtain the following result.

Theorem 4. *There exists an on-line algorithm that against any non-adaptive adversary, given a doubling metric $M = (X, D)$ of doubling dimension λ, computes a probabilistic embedding of M into ℓ_2 with distortion $2^{O(\lambda)} \cdot \log \Delta$.*

Remark 2. In the off-line setting, Krauthgamer et al. [15] have obtained embeddings of doubling metrics into Hilbert space with distortion $O(\sqrt{\log n})$. Their approach however is based on the random partitioning scheme of Calinescu, Karloff, and Rabani [5], and it is not known how to perform this step in the on-line setting.

4 Lower Bound for Probabilistic Embeddings into Ultrametrics

In this section we present a lower bound for on-line probabilistic embeddings into ultrametrics. Consider the following distribution $\mathcal{F}$ over metric spaces. Each space $M = (X, D)$ in the support of $\mathcal{F}$ is induced by an n-point subset of $\mathbb{R}^1$, with $X = \{x_1, \ldots, x_n\}$, and $D(x_i, x_j) = |x_i - x_j|$. We have $x_1 = 0$, $x_2 = 1$, $x_3 = 1/2$. For each $i \geq 4$, we set $x_i = x_{i-1} + b_i 2^{-i+2}$, where $b_i \in \{-1, 1\}$ is chosen uniformly at random.

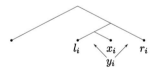

Fig. 1. The evolution of the construction of the ultrametric

It is easy to see that for each $i \geq 3$, there exist points $l_i, r_i \in X$ such that $l_i = x_i - 2^{-i+2}$, $r_i = x_i + 2^{-i+2}$, and $\{x_1, \ldots, x_i\} \cap [l_i, r_i] = \{l_i, x_i, r_i\}$. Moreover, for each $i \in \{3, \ldots, n-1\}$, there uniquely exists $y_i \in \{l_i, r_i\}$, such that $\{x_{i+1}, \ldots, x_n\} \subset [\min\{x_i, y_i\}, \max\{x_i, y_i\}]$.

Claim. Let $M = (X, D)$ be a metric from the support of $\mathcal{F}$. Let f be an embedding of M into an ultrametric $M' = (X, D')$. Then, for each $i \geq 3$, there exists $z_i \in \{l_i, r_i\}$, such that $D'(x_i, z_i) \geq D'(l_i, r_i)$.

Proof. It follows immediately by the fact that M' is an ultrametric, since for any $x_i, l_i, r_i \in X$, $D'(l_i, r_i) \leq \max\{D'(l_i, x_i), D'(x_i, r_i)\}$.

In order to simplify notation, we define for any $i \geq 4$, $\delta_i = D(x_i, y_i)$, and $\delta_i' = D'(x_i, y_i)$.

Claim. Let $M = (X, D)$ be a metric from the support of $\mathcal{F}$. Let f be an on-line embedding of M into an ultrametric $M' = (X, D')$. Then, for any $i \geq 3$, $\mathbf{Pr}[\delta_i' \geq \delta_{i-1}' | \forall j \in \{4, \ldots, i-1\}, \delta_j' \geq \delta_{j-1}'] \geq 1/2$.

Proof. Assume without loss of generality that $z_i = l_i$, since the case $z_i = r_i$ is symmetric. By the construction of M, we have that $\mathbf{Pr}[y_i = z_i | \forall j \in \{4, \ldots, i-1\}, \delta_j' \geq \delta_{j-1}'] = 1/2$. If $y_i = z_i$, then $\delta_i' = D'(x_i, z_i) \geq D'(l_i, r_i) = \delta_{i-1}'$, concluding the proof.

Lemma 4. *Let f be an on-line, non-contracting embedding of M into an ultrametric M'. Then, $\mathbb{E}[\delta_{n-1}'/\delta_{n-1}] = \Omega(n)$.*

Proof. Let $i \geq 4$, and $1 \leq t \leq i-1$. By Claim 4 we have $\mathbf{Pr}[\delta_i' \geq \delta_{i-t}] \geq \mathbf{Pr}[\delta_i' \geq \delta_{i-t}'] \geq \mathbf{Pr}[\forall j \in \{1, \ldots, t\}, \delta_{i-j+1}' \geq \delta_{i-j}'] = \prod_{j=1}^{t} \mathbf{Pr}[\delta_{i-j+1}' \geq \delta_{i-j}' | \forall s \in \{1, \ldots, j-1\}, \delta_{i-s+1}' \geq \delta_{i-s}'] \geq 2^{-t}$. Therefore $\mathbb{E}[\delta_{n-1}'] \geq \sum_{i=3}^{n-1} \delta_i \cdot 2^{-n+i+1} = \sum_{i=3}^{n-1} 2^{-i+2} \cdot 2^{-n+i+1} = \Omega(n \cdot 2^{-n}) = \Omega(n) \cdot \delta_{n-1}$.

Since the aspect ratio (spread) is $\Delta = \Theta(2^n)$, we obtain the following result.

Theorem 5. *There exists a non-adaptive adversary against which any on-line probabilistic embedding into a distribution over ultrametrics has distortion $\Omega(\min\{n, \log \Delta\})$.*

We remark that the above bound is essentially tight, since the input space is a subset of the line, and therefore doubling. By Theorem 3, every doubling metric space probabilistically embeds into ultrametrics with distortion $O(\log \Delta)$.

5   Embedding into ℓ_∞

In the off-line setting, it is well-known that any n-point metric space isometrically embeds into n-dimensional ℓ_∞. Moreover, there is an explicit construction of the embedding due to Fréchet. Let $M = (X, D)$ be an arbitrary metric space. The embedding $f : (X, D) \to \ell_\infty^d$ is simply $f(x_i) = (D(x_i, x_1), D(x_i, x_2), \ldots, D(x_i, x_n))$. It is clear that the Fréchet embedding does not fit in the on-line setting, since the image of any point x depends on the distances between x and all points of the metric space, in particular the future points.

A similar question regarding the existence of on-line embeddings can be posed: does there exist a bi-Lipschitz extension for *any* embedding into ℓ_∞. The connection with the on-line setting is immediate; it is well-known (see e.g. [16]) that for any metric space $M = (X, D)$, for any $Y \subset X$, and for any a-Lipschitz function $f : Y \to \ell_\infty$, there exists an a-Lipschitz extension $\tilde{f}$ of f, with $\tilde{f} : X \to \ell_\infty$.

It seems natural to ask whether this is also true when f and $\tilde{f}$ are required to be a-bi-Lipschitz. Combined with the fact that any metric embeds isometrically into ℓ_∞, this would immediately imply an on-line algorithm for embedding isometrically into ℓ_∞: start with an arbitrary isometry, and extend it at each step to include a new point. Unfortunately, as the next proposition explains, this is not always possible, even for the special case of $(1, 2)$-metrics (the proof appears in the full version). We need some new ideas to obtain such an embedding.

Proposition 1. *There exists a finite metric space $M = (X, D)$, $Y \subset X$, and an isometry $f : Y \to \ell_\infty$, such that any extension $\tilde{f} : X \to \ell_\infty$ of f is not an isometry.*

Although it is not possible to extend *any* 1-bi-Lipschitz mapping into ℓ_∞, there exists a *specific* mapping that is extendable, provided that the input space is drawn from a fixed finite family of metrics. We will briefly sketch the proof of this fact, and defer the formal analysis to the full version. Consider a metric space M' obtained from M by adding a point p. Suppose that we have an isometry $f : M \to \ell_\infty$. As explained above, f might not be isometrically extendable to M'. The key step is proving that f is always Lipschitz-extendable to M'. We can therefore get an on-line embedding as follows: We maintain a concatenation of embeddings for all metrics in the family of input spaces. When we receive a new point x_i, we isometrically extend all embeddings of spaces that agree with our input on $\{x_1, \ldots, x_i\}$, and Lipschitz-extend the rest.

Theorem 6. *Let $\mathcal{F}$ be a finite collection of n-point metric spaces. There exists an on-line embedding algorithm that given a metric $M \in \mathcal{F}$, computes an isometric embedding of M into ℓ_∞.*

5.1   Low-Distortion Embeddings into Low-Dimensional ℓ_∞

In the pursuit of a good embedding of a general metric space into low dimensional ℓ_∞ space we demonstrate the usefulness (and feasibility) of concatenation of two on-line embeddings. In fact one of these embeddings is oblivious, which in particular makes it on-line. Why the concatenation of two on-line embeddings results in yet another on-line embeddings is fairly clear when the embeddings are deterministic; in the case of probabilistic embeddings it suffices to simply concatenate the embeddings in an independent way. In both cases the distortion is the product of the distortions of the individual embeddings. Recall that Section 2 provides us with an on-line embedding of a metric space into Euclidean space. The rest of the section shows that the classical method of projection of points in Euclidean space onto a small number of dimensions supplies low distortion embedding when the host space is taken to be ℓ_∞. To put things in perspective, the classical Johnson-Lindenstrauss lemma [13] considers the case where the image space is equipped with the ℓ_2 norm, and it is well-known that a similar result can be achieved with ℓ_1 as the image space [10, p. 92]. As we will see, ℓ_∞ metric spaces behave quite differently than ℓ_2 and ℓ_1 spaces in this respect, and while a dimension reduction is possible, it is far more limited than the first two spaces.

The main technical ingredient we need is the following concentration result. See also [23] for a similar analysis. The proof is given in the full version.

Lemma 5. *Let $u \in \mathbb{R}^n$ be a nonzero vector and let $\alpha > 1$ and $d \geq e^2$. Let y be the normalized projection of u onto d dimensions by a Gaussian matrix as follows: $y = (2/m)Ru$ where R is a $d \times n$ Gaussian random matrix, i.e., a matrix with i.i.d. normal entries and $m = 2\sqrt{\ln d}$. Then $\mathbf{Pr}\left[\|y\|_\infty/\|u\|_2 \leq 1\right] \leq \exp(-\frac{1}{4}\sqrt{d/\ln d})$, and $\mathbf{Pr}\left[\|y\|_\infty/\|u\|_2 \geq \alpha\right] \leq (2/\alpha)d^{1-\alpha^2/2}$.*

With the concentration bound of Lemma 5 it is not hard to derive a good embedding for any n-point set, as is done, say, in the Johnson Lindenstrauss Lemma [13], and we get

Lemma 6. *Let $X \subset \mathbb{R}^n$ an n-point set and let $\alpha > \sqrt{2}$. If $d = \Omega(\max\{(\log n)^{2/(1-1/e)}, n^{4/(\alpha^2-2)}\})$, then the above mapping $f : X \to \ell_\infty^d$ satisfies $\forall x, y \in X$, $\|x - y\|_2 \leq \|f(x) - f(y)\|_\infty \leq \alpha\|x - y\|_2$ with high probability.*

By a straightforward composition of the embeddings in Theorem 1 and Lemma 6, we get

Theorem 7. *There exists an on-line algorithm against any non-adaptive adversary that for any $\alpha > \sqrt{2}$, given a metric $M = (X, D_M)$, computes an embedding of M into ℓ_∞^d with distortion $O(\alpha \cdot \log n \cdot \sqrt{\log \Delta})$ and $d = \Omega(\max\{(\log n)^{2/(1-1/e)}, n^{4/(\alpha^2-2)}\})$.*

Remark 3. The embeddings into ℓ_∞ given in Theorems 1 and 7 are incomparable: the distortion in Theorem 1 is smaller, but the dimension is larger than the one in Theorem 7 for large values of Δ.

6 On-Line Embedding When an Off-Line (Near-)Isometry Is Possible

It is not hard to see that given an n-point ℓ_2^d metric M, one can compute an online isometric embedding of M into ℓ_2^d. This is simply because there is essentially (up to translations and rotations) a unique isometry, and so keeping extending the isometry online is always possible. However, as soon as we deal with near isometries this uniqueness is lost, and the situation changes dramatically as we next show: even when the input space embeds into ℓ_2^d with distortion $1 + \varepsilon$, the best online embedding we can guarantee in general will have distortion that is polynomial in n. We use the following topological lemma from [20].

Lemma 7 ([20]). *Let $\delta < \frac{1}{4}$ and let $f_1, f_2 : S^{d-1} \to \mathbb{R}^d$ be continuous maps satisfying*

- *$\|f_i(x) - f_i(y)\|_2 \geq \|x - y\|_2 - \delta$ for all $x, y \in S^{d-1}$ and all $i \in \{1, 2\}$,*
- *$\|f_1(x) - f_2(x)\|_2 \leq \frac{1}{4}$ for all $x \in S^{d-1}$, and*
- *$\Sigma_1 \cap \Sigma_2 = \emptyset$, where $\Sigma_i = f_i(S^{d-1})$.*

Let U_i denote the unbounded component of $\mathbb{R}^d \setminus \Sigma_i$. Then, either $U_1 \subset U_2$, or $U_2 \subset U_1$.

Theorem 8. *For any $d \geq 2$, for any $\varepsilon > 0$, and for sufficiently large $n > 0$, there exists a distribution $\mathcal{F}$ over n-point metric spaces that embed into ℓ_2^d with distortion $1 + \varepsilon$, such that any on-line algorithm on input a metric space chosen from $\mathcal{F}$ outputs an embedding into ℓ_2^d with distortion $\Omega(n^{1/(d-1)})$, and with probability at least $1/2$.*

Proof. The proof will appear in the full version of this paper.

Acknowledgements. We thank the anonymous referee for pointing out Theorem 4.

References

1. Andoni, A., Onak, K.: Approximating Edit Distance in Near-Linear Time. In: Proceedings of the Symposium on Theory of Computing (STOC), pp. 199–204 (2009)
2. Bartal, Y.: Probabilistic approximation of metric spaces and its algorithmic applications. In: Proceedings of the Symposium on Foundations of Computer Science (FOCS), pp. 184–193 (1996)
3. Bartal, Y.: On approximating arbitrary metrics by tree metrics. In: Proceedings of the Symposium on Theory of Computing (STOC), pp. 161–168 (1998)
4. Bourgain, J.: On Lipschitz Embedding of Finite Metric Spaces in Hilbert space. Israel J. Math. 52(1-2), 46–52 (1985)
5. Calinescu, G., Karloff, H., Rabani, Y.: Approximation Algorithms for the 0-extension Problem. In: Proceedings of the ACM-SIAM Symposium on Discrete Algorithms (SODA), pp. 8–16 (2001)
6. Englert, M., Räcke, H., Westermann, M.: Reordering Buffers for General Metric Spaces. In: Proceedings of the Symposium on Theory of Computing (STOC), pp. 556–564 (2007)
7. Feller, W.: An Introduction to Probability Theory and Its Applications, vol. 1. John Wiley & Sons, Chichester (January 1968)
8. Gupta, A., Krauthgamer, R., Lee, J.R.: Bounded Geometries, Fractals, and Low-Distortion Embeddings. In: Proceedings of the Symposium on Foundations of Computer Science (FOCS), pp. 534–543 (2003)
9. Gupta, A., Ravi, R.: Lecture Notes of Metric Embeddings and Methods (2003), http://www.cs.cmu.edu/~anupamg/metrics/lectures/lec10.ps
10. Indyk, P.: High-dimensional Computational Geometry. PhD thesis, Stanford University (2000)
11. Indyk, P.: Stable Distributions, Pseudorandom Generators, Embeddings and Data Stream Computation. In: Proceedings of the Symposium on Foundations of Computer Science (FOCS), pp. 189–197 (2000)
12. Indyk, P.: Tutorial: Algorithmic Applications of Low-distortion Geometric Embeddings. In: Proceedings of the Symposium on Foundations of Computer Science (FOCS), pp. 10–33 (2001)
13. Johnson, W.B., Lindenstrauss, J.: Extensions of Lipschitz Mappings into a Hilbert Space. In: Amer.Math. Soc. (ed.) Conference in Modern Analysis and Probability, pp. 189–206 (1984)

14. Kirszbraun, M.D.: Uber die Zusammenziehenden und Lipschitzchen Transformationen. Fund. Math. 22, 77–108 (1934)
15. Krauthgamer, R., Lee, J.R., Mendel, M., Naor, A.: Measured Descent: A new Embedding Method for Finite Metrics. Geometric Aspects of Functional Analysis, 839–858 (2005)
16. Lee, J.R., Naor, A.: Extending Lipschitz Functions via Random Metric Partitions. Invent. Math. 160(1), 59–95 (2005)
17. Linial, N., London, E., Rabinovich, Y.: The Geometry of Graphs and some of its Algorithmic Applications. Combinatorica 15(2), 215–245 (1995)
18. Matoušek, J.: Bi-Lipschitz Embeddings into Low-dimensional Euclidean Spaces. Comment. Math. Univ. Carolinae 31, 589–600 (1990)
19. Matoušek, J.: Lectures on Discrete Geometry. Springer, New York (2002)
20. Matoušek, J., Sidiropoulos, A.: Inapproximability for Metric Embeddings into $\mathbb{R}^d$. In: Proceedings of the Symposium on Foundations of Computer Science (FOCS), pp. 405–413 (2008)
21. Newman, I., Rabinovich, Y.: Personal communication (2008)
22. Rao, S.: Small Distortion and Volume Preserving Embeddings for Planar and Euclidean Metrics. In: Proceedings of the ACM Symposium on Computational Geometry (SoCG), pp. 300–306 (1999)
23. Schechtman, G.: The Random Version of Dvoretzky's Theorem in ℓ_∞. In: Zighed, D.A., Komorowski, J., Żytkow, J.M. (eds.) PKDD 2000. LNCS (LNAI), vol. 1910, pp. 265–270. Springer, Heidelberg (2000)

Approximation Algorithms
for Intersection Graphs

Frank Kammer, Torsten Tholey, and Heiko Voepel

Institut für Informatik, Universität Augsburg, D-86135 Augsburg, Germany
{kammer,tholey,voepel}@informatik.uni-augsburg.de

Abstract. We study three complexity parameters that in some sense measure how chordal-like a graph is. The similarity to chordal graphs is used to construct simple polynomial-time approximation algorithms with constant approximation ratio for many $\mathcal{NP}$-hard problems, when restricted to graphs for which at least one of the three complexity parameters is bounded by a constant. As applications we present approximation algorithms with constant approximation ratio for maximum weighted independent set, minimum (independent) dominating set, minimum vertex coloring, maximum weighted clique, and minimum clique partition for large classes of intersection graphs.

1   Introduction

Complexity parameters can help to solve many NP-hard problems of theoretical or practical importance on a subclass of instances for which the chosen parameter is very small. Treewidth is one of the classical complexity parameters studied in graph theory. Graphs of bounded treewidth have a tree-like structure that allows a generalization of efficient algorithms for hard problems on trees to graphs of bounded treewidth. In particular, all decision problems that can be expressed in monadic second-order logic can be solved in polynomial time on graphs of bounded treewidth [3,8].We study three complexity parameters that all generalize in some kind another class of graphs, namely chordal graphs. One of them is new, whereas the others also appear in [34] and [25], but were not analyzed in detail in these papers. See Section 2 for a detailed definition of the complexity parameters. Like trees, chordal graphs have a simple structure that facilitates the solution of a large number of NP-hard problems. For example, there are linear time algorithms on chordal graphs for maximum clique (MC), for minimum clique partition (MCP) [15], for maximum weighted independent set (MWIS) [13], and for minimum vertex coloring (MVC).Thus, it seems natural to search for a generalization of chordal graphs. In doing so, we obtain new approximation algorithms for the problems above on big graph classes containing many intersection graph classes such as t-interval graphs, circular-arc graphs, (unit) disk graphs, and intersection graphs of regular polygons or of arbitrary polygons of so-called bounded fatness. In general, intersection graphs are useful subclasses of graphs with several practical applications. See [17] or [18] for an overview of applications on these graphs. It is not surprising that, for small

M. Serna et al. (Eds.): APPROX and RANDOM 2010, LNCS 6302, pp. 260–273, 2010.

graph classes such as unit disk graphs, one can achieve better results than by our new algorithms designed for bigger classes of graphs. Nevertheless, also on small graph classes such as disk graphs we obtain new results for some of the problems above as well as for minimum dominating set (MDS) and minimum independent dominating set (MIDS).

Table 1 summarizes the best previously known and new approximation results for the intersection graphs of disks, regular polygons, fat objects, t-intervals, and t-fat-objects. MIS denotes the unweighted version of MWIS and MWC the weighted version of MC. By an r-*regular polygon* we mean a polygon with r corners placed on a cycle such that all pairs of consecutive corners of the polygon have the same distance. We assume that $r \in O(1)$. We define a set $\mathcal{C}$ of geometric objects in $\mathbb{R}^d$—i.e., a set of points in $\mathbb{R}^d$—to be a set of *fat objects* if the following holds: First of all, let us call the radius of a smallest d-dimensional ball containing the closure of a geometric object S in $\mathbb{R}^d$ the *size* of S. Moreover, let R be the size of the largest object in $\mathcal{C}$. $\mathcal{C}$ is called *fat* if there is a constant c such that, for each d-dimensional ball B of radius r with $0 < r \le R$, there exist c points (possibly also outside B) such that every B-intersecting object $S \in \mathcal{C}$ of size at least r contains at least one of the c points. We also say that $\mathcal{C}$ has *fatness* c. $\mathcal{C}$ is called a *(c-)restricted set of fat objects* if in the condition above every B-intersecting object in $\mathcal{C}$ (with arbitrary size) contains at least one of the c points. By a *unit* set of objects—in opposite to *arbitrary*—we mean that each object must be a copy of each other object, i.e., it has to be of the same size and shape. However, unit and arbitrary objects may be rotated and moved to any position. An intersection graph G of t-intervals is an intersection graph, where each vertex represents a t-*interval*, i.e., the union of t intervals taken from a set S of intervals. By the intersection graph G of t-fat-objects we mean an intersection graph, where each vertex represents a t-*fat-object*, i.e., the union of t objects taken from a fat set S of objects. In both cases S is the *universe* of G.

As usual, disks and regular polygons should be defined in the plane $\mathbb{R}^2$, intervals in $\mathbb{R}$ and fat objects in $\mathbb{R}^d$, where we assume that $d = O(1)$. Concerning the results in table 1 including the hardness results, we assume that—beside an intersection graph itself—a representation of the intersection graph is given. More precisely, for the intersection graph of a set S of (1) disks, (2) r-regular polygons, (3) t-intervals, (4) fat objects, or (5) t-fat-objects, we are given for each element in S its radius and the coordinates of its center in case 1, the coordinates of the center and at least one corner in case 2, the start and end point of each interval in case 3. In case 4, we should be given a representation that, for each pair X, Y of objects, each point $p \in \mathbb{R}^d$, and each d-dimensional ball B represented by the coordinates of its center and its radius $r \le R$, supports the following computations in polynomial time: Decide whether X and Y intersect, whether X and B intersect, and whether p is contained in X. Moreover, determine the size s of X as well as the center of the ball with a radius s containing the closure of X, and find c points that are contained in every object of size $\ge r$ intersecting B. In case 5, each t-fat-object has a representation of its objects as described in case 4. The representations described are given explicitly in many applications.

Table 1. Approximation results. We use PA. and NP-h. as abbreviation for polynomial-time approximation algorithm and NP-hard, respectively. By n we denote the number of vertices of the intersection graph. [*] denotes a new result shown in this paper.

	disk	r-reg. polygon	fat objects	t-interval	t-fat-objects

MIS	arbitrary:
	  PTAS [6,10]
	unit:
	  PTAS [22]
	  NP-h. [12]
		arbitrary:
		  PTAS [6,10]
		unit:
		  PTAS [22]
		  NP-h. [12]
			fatness c:
			  PTAS [6,10]
			unit:
			
			  NP-h. [12]
				$2t$-PA. [2]
				
				$t \geq 3$:
				  $\mathcal{APX}$-h.
				  [21,31]
					fatness c:
					  $2tc$-PA. [*]
					
					  NP-h. [12]

MWIS	arbitrary:
	  PTAS [10]
		arbitrary:
		  PTAS [10]
			fatness c:
			  PTAS [10]
				$2t$-PA. [2]	fatness c:
					  $2tc$-PA. [*]

MDS	arbitrary:
	  PTAS [16]
	unit:
	  PTAS [22]
	  NP-h. [7]
		c-restricted:
		  c-PA. [*]
		unit:
		  PTAS [22]
		  NP-h. [7]
			c-restricted:
			  c-PA. [*]
			unit:
			
			  NP-h. [7]
				t^2-PA. [4]
				
				$t \geq 2$:
				  $\mathcal{APX}$-h.
				  [21,31]
					c-restricted:
					  tc-PA. [*]
					$t \geq 2$:
					  $\mathcal{APX}$-h.
					  [21,31]

MIDS	c-restricted:
	  c-PA. [*]
	unit:
	  PTAS [23]
	  NP-h. [7]
		c-restricted:
		  c-PA. [*]
		unit:
		  PTAS [23]
		  NP-h. [7]
			c-restricted:
			  c-PA. [*]
			unit:
			
			  NP-h. [7]
					c-restricted:
					  tc-PA. [*]
					$t \geq 2$:
					
					  NP-h. [7]

MVC	arbitrary:
	  5-PA.
	  [19,28,29]
	unit: 3-PA. [29]
	  4/3-PA. is NP
	  -h. [7,14,24]
		arbitrary:
		  $O(1)$-PA.
		  [27,34]
		unit:
		  $O(1)$-PA. [29]
		  NP-h. [14,24]
			fatness c:
			  c-PA. [34]
			
			unit:
			
			  NP-h. [14,24]
				$2t$-PA. [2]	fatness c:
					  $2tc$-PA. [*]

MC	arbitrary:
	  8-PA. [*]
	unit:
	  $\in$ P [7]
		arbitrary:
		  $O(1)$-PA. [*]
			fatness c:
			  c-PA. [*]
				$\frac{t^2-t+1}{2}$-PA. [4]
				$4t$-PA. [*]
				$t \geq 3$:
				  NP-h. [4]
					fatness c:
					  $2tc$-PA. [*]
					$t \geq 3$:
					  NP-h. [4]

MWC	arbitrary:
	  8-PA. [*]
		arbitrary:
		  $O(1)$-PA. [*]
			fatness c:
			  c-PA. [*]
				$\frac{t^2-t+1}{2}$-PA. [4]
				$4t$-PA. [*]
					fatness c:
					  $2tc$-PA. [*]

MCP	arbitrary:
	  8-PA. [*]
	unit: 3-PA. [5]
	  PTAS [9,32]
		arbitrary:
		  $O(1)$-PA. [*]
			fatness c:
			  c-PA. [*]
				$O(\log^2 n/$
				  $\log(1+1/t))$-
				  PA. [*]
					c-restricted:
					  tc-PA. [*]

Very related to the graph classes considered in this paper is the so-called class of *sequentially k-independent graphs* introduced by Akcoglu, Aspnes, Das-Gupta, and Kao [1] and studied more extensively in a recent paper by Ye and Borodin [34]. We omit an exact definition of this graph class, but want to remark that even though the results of Ye et al. and our results are achieved completely independently, there are similarities between the papers. This indicates that our generalizations of chordal graphs are quite natural, but surprisingly have not been studied more extensively before. Other generalized classes of graphs including the intersection graphs of unit disks or r-regular polygons of unit size are graph classes of so-called *polynomially bounded growth* studied by Nieberg, Hurink and Kern [23,30]. Nieberg et al. presented a PTAS for MWIS, MDS and MIDS for these classes of graphs. However, graphs of polynomially bounded growth do not include the intersection graphs of arbitrary disks, arbitrary r-regular polygons, t-interval graphs, etc.

Our results include the first polynomial-time approximation algorithms with constant approximation ratio for maximum clique and minimum clique partition on disk graphs and on intersection graphs of r-regular polygons. We also present a polynomial-time approximation algorithm with constant approximation ratio for minimum dominating set on the intersection graphs of a restricted set of r-regular polygons. Recently, Erlebach and van Leeuwen [11] presented an approximation algorithm with constant approximation ratio for the same problem on an arbitrary set of r-regular polygons, however, they do not allow to rotate the polygons in contrast to this paper. Our results also imply an approximation algorithm with constant approximation ratio for minimum dominating set on intersection graphs of an arbitrary set of *non-rotated* r-regular polygons. With the introduction of the completely new graph class of k-perfect orientable graphs, we also can solve an open question posted by Butman et al. [4], namely to improve their approximation bound of $(t^2 - t + 1)/2$ for maximum clique on t-interval graphs. Our results lead to a $4t$-approximation. In general, our results also extend to intersection graphs of a restricted set of t-fat objects and further classes of graphs not discussed in this paper.

2   New Complexity Parameters

In this section, the following definitions introduce three complexity parameters. For each complexity parameter, we present examples of classes of intersection graphs for which the complexity parameter is bounded by a constant. For a set S of vertices in a graph G, let $G[S]$ be the subgraph of G induced by S.

Definition 1 (k-perfectly groupable). *A graph is k-perfectly groupable if the neighbors of each vertex v can be partitioned into k sets $S_1, \ldots, S_k$ such that $G[S_i \cup \{v\}]$ is a clique for each $i \in \{1, \ldots, k\}$.*

For each object S of a k-restricted set $\mathcal{C}$ of fat objects and a smallest ball B containing S, there exists a set P_S of k points such that every object in $\mathcal{C}$ intersecting B covers a point in P_S. For each S-intersecting and hence also B-intersecting object $S' \in \mathcal{C}$, choose one of the points in $S' \cap P_S$ as a representative. Then all S-intersecting objects having the same representative in P_S induce a clique in the intersection graph. Hence, the intersection graph of a k-restricted set of fat objects is k-perfectly groupable. Note that graphs of maximum degree k are also k-perfectly groupable. The set of the well studied $(k+1)$-clawfree graphs contains all k-groupable graphs. As we see in this section, unit disk graphs and unit square graphs are k-perfectly groupable for a suitable constant k.

Definition 2 (k-simplicial, k-simplicial elimination order, successor). *A graph G is k-simplicial if there is an order $v_1, \ldots, v_n$ of the vertices of G such that, for each vertex v_i ($1 \leq i \leq n$), the subset of neighbors of v_i contained in $\{v_j \mid j > i\}$ can be partitioned into k sets $S_1, \ldots, S_k$ such that $G[S_j \cup \{v_i\}]$ is a clique for each $j \in \{1, \ldots, k\}$. The vertices in $\{v_j \mid j > i, \{v_i, v_j\} \in E(G)\}$ are called the successors of v_i and the order above of the vertices in G is called a k-simplicial elimination order.*

The k-simplicial graphs are already defined in [25] and [34], whereas in the latter paper they are called $\tilde{G}(VCC_k)$. Let $\mathcal{C}$ be a set of fat objects $S_1, \ldots, S_n$ ordered by non-decreasing size. Let k be the fatness of $\mathcal{C}$. Then, for each object S_i with $i \in \{1, \ldots, n\}$, we can find k points such that every S_i-intersecting object in $\{S_{i+1}, \ldots, S_n\}$ contains one of the k points. Let v_i be the vertex representing S_i in the intersection graph G of $\mathcal{C}$. Then $v_1, \ldots, v_n$ defines a k-simplicial elimination order. Therefore, G is k-simplicial. Also note that disk graphs and square graphs are k-simplicial for a suitable constant k. Chordal graphs are exactly the 1-simplicial graphs. Moreover, every planar graph is 5-simplicial and every k-simplicial graph is sequentially k-independent (see [34]).

Definition 3 (k-perfectly orientable). *A graph G is called k-perfectly orientable if each edge $\{u_1, u_2\}$ of G can be assigned to exactly one of its endpoints u_1 and u_2 such that, for each vertex v, the vertices connected to v by edges assigned to v can be partitioned into k sets $S_1, \ldots, S_k$ such that $G[S_i \cup \{v\}]$ is a clique for each $i \in \{1, \ldots, k\}$. We write $a(\{u_1, u_2\}) = u_1$ if $\{u_1, u_2\}$ is assigned to u_1.*

We now show that the intersection graph $G = (V, E)$ of a set of t-fat-objects $\mathcal{C}$ with a universe of fatness c is $(t \cdot c)$-perfectly orientable. Let $V = \{v_1, \ldots, v_n\}$ and, for each $i \in \{1, \ldots, n\}$, let $\mathcal{S}_i$ be the union of t objects $S_{i,1}, \ldots, S_{i,t}$ represented by v_i. Choose, for each edge $\{v_i, v_j\}$ in G with $i < j$, a pair $\{p, q\}$ of indices such that $S_{i,p}$ and $S_{j,q}$ intersect. Assign $\{v_i, v_j\}$ to v_i if the size of $S_{i,p}$ is smaller than the size of $S_{j,q}$ and to v_j otherwise. Then, for each vertex v_i, one can find $t \cdot c$ points such that each $\mathcal{S}_i$-intersecting t-fat-object $\mathcal{S}_j$ with $\{v_i, v_j\}$ being assigned to v_i must intersect S_i in at least one of the $t \cdot c$ points. Therefore, the set of vertices being endpoints of edges assigned to v_i can be partitioned into $\leq t \cdot c$ cliques. This proves that G is $(t \cdot c)$-perfectly orientable. Note also that the intersection graphs of t-intervals are $2t$-perfectly orientable. For these graphs, an edge $\{v_i, v_j\}$ with $i < j$ is assigned to v_i if the t-interval represented by v_j intersects one of $2t$ endpoints of the intervals whose union is represented by v_i. Otherwise, $\{v_i, v_j\}$ is assigned to v_j.

We next present explicit upper bounds for the three complexity parameters on some special intersection graphs. Before that let us define the *inball* and the *outball* of a geometric object S to be a ball with largest radius contained in the closure of S and the ball with smallest radius containing the closure of S, respectively. The *center* of S is the center of its outball.

Theorem 4. *An intersection graph of t-squares, i.e., of unions of t (not necessarily axis-parallel) squares, is*

1. *10-perfectly groupable if $t = 1$ and if the squares are of unit size,*
2. *10-simplicial if $t = 1$, and*
3. *$10t$-perfectly orientable.*

Proof. For proving the first two cases, let G be the intersection graph of a set S of squares. It remains to show that, for a square Q of minimal side length ℓ, there are 10 points—called the *barriers* of Q—such that every Q-intersecting square Q' of length $\geq \ell$ must cover at least one of them. This fact also proves case 3 since the universe of a set of t-squares then has fatness 10.

We first describe our choice of the 10 barriers of Q. See also the left side of Fig. 1 for the following construction. Let b_1 and b_2 be the two perpendicular bisectors of the sides of Q. Choose two barriers x and y of Q as points on b_1 such that the part of b_1 inside Q is divided into three parts of equal length. We call these two points the *inner barriers* of Q. Let C be the curve surrounding Q that consists of all points having a distance of exactly ℓ to one of the inner barriers and a distance of at least ℓ to the remaining inner barrier. The remaining 8 barriers, called *outer barriers*, are almost equidistant points on C. More exactly, 4 outer barriers of Q are placed on the $2 + 2$ intersection points of C with b_1 and b_2. Choosing the other 4 outer barriers of Q is more sophisticated. Let x' and y' be the two points on b_1 having the same distance to the center of Q as to x and y, respectively. In addition, let $r_1, \ldots, r_4$ be the 4 rays starting from x' and y', respectively, and intersecting a corner of Q but neither b_1 nor b_2. The four remaining outer barriers are placed on the intersection points of C with the rays $r_1, \ldots, r_4$.

By a simple mathematical analysis one can show that the distance between any two consecutive outer barriers on C is strictly smaller than ℓ. It remains to show that each square of side length at least ℓ intersecting Q also covers one of the barriers of Q. Assume for a contradiction that we can find a square Q' of side length at least ℓ intersecting Q but none of the barriers of Q. W.l.o.g. we can assume that Q' has side length exactly ℓ since otherwise Q' also contains a smaller square intersecting Q. Let $\mathcal{H}$ be the convex hull of the outer barriers and let B be the largest circle contained in $\mathcal{H}$ such that B has the same center as Q. B and thus also $\mathcal{H}$ contain at least one corner of Q' since Q' intersects Q and B, and since a simple mathematical analysis shows that each chord of B with length at most l does not intersect Q. We now distinguish two cases.

Case 1: No side of Q' is completely contained in the convex hull $\mathcal{H}$ of the outer barriers. For each pair of consecutive outer barriers p and q on C, let us define $C_{p,q}$ to be the semi-circle inside $\mathcal{H}$ with endpoints p, q and hence having a diameter equal to the distance between p and q. See again the left side of Fig. 1. Let z be the corner of Q' inside B with the smallest distance to a point in Q. Note that the two sides of Q' ending in z are not completely contained in $\mathcal{H}$. Consequently, by Thales' theorem and the fact that Q' does not contain any

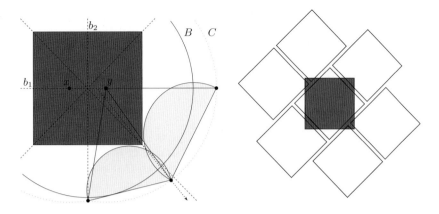

Fig. 1. The left side shows a square with some barriers, and on the right side, we see a square intersecting 7 disjoint squares

barriers there must be two consecutive outer barriers p and q on C such that z is contained in the face enclosed by $C_{p,q}$ and $\overline{pq}$. Again a simple mathematical analysis shows that none of our semi-circles intersects Q. Thus, neither z nor any other point of Q' is covered by Q. Contradiction.

Case 2: At least one side of Q' is completely contained in $\mathcal{H}$. Since each pair of consecutive outer barriers on C has a distance smaller than ℓ, the center q of Q' is inside $\mathcal{H}$.

By symmetry, w.l.o.g. we can assume that the distance between q and y is smaller or equal than the distance between q and x. Let $\mathcal{H}'$ be the convex hull of x and the outer barriers having a distance of at most ℓ to y. On the one hand, for each pair of consecutive barriers q_1 and q_2 on $\mathcal{H}'$, there is at most one corner in the face bounded by $\overline{q_1q_2}$ and the semi-circle outside $\mathcal{H}'$ with endpoints q_1 and q_2. On the other hand, at least one corner of Q' is outside $\mathcal{H}'$ since the inball of Q', which does not contain y, must intersect the border of $\mathcal{H}'$. Consequently, there are two sides s_1 and s_2 of Q' that have a common corner p outside $\mathcal{H}'$ and that intersect $\mathcal{H}'$ between to outer barriers, say q_1 and q_2.

Let T be the triangle with corners y, q_1 and q_2. Since Q' is a square of side length ℓ, since p is not covered by T, and since T is a triangle with two sides of length ℓ and with an s_1-intersecting side of length at most ℓ, y has to be inside Q'. Contradiction. $\square$

Observation 5. *Some square graphs are not 6-perfectly groupable as shown on the right side of Fig. 1.*

Lemma 6. *The intersection graph of a set of rectangles, all having aspect ratio of α, is $10\lceil\alpha\rceil$-simplicial.*

Proof. Consider each rectangle as a set of $\lceil\alpha\rceil$ squares. For each rectangle r_1 replaced by squares of a size s_1, one can find $10\lceil\alpha\rceil$ points such that every r_1-intersecting square of size $s_2 \geq s_1$ replacing another rectangle r_2 must cover one of these points. Here we use the fact that each rectangle can be replaced by $\lceil \alpha \rceil$ unit squares.

Theorem 7. *Let c be a fixed constant and G be an intersection graph, where each vertex represents a union of t polygons taken from a universe of non-rotated c-regular polygons. Then G is $(t \cdot c)$-perfectly orientable.*

Proof. The intersection of two non-rotated c-regular polygons must contain at least one of the corners of the two polygons. Note that this does not hold for general rotated polygons. Let $\{v_1, \ldots, v_n\}$ be the vertices of G. We assign an edge $\{v_i, v_j\}$ in G with $i < j$ to v_i if and only if one of the polygons in the union of polygons represented by v_i has a corner contained in the union of polygons represented by v_j. Otherwise, we assign it to v_j. The edges assigned to a vertex v can be partitioned into $t \cdot c$ sets such that the endpoints of the edges of each set induce a clique in G since we have one clique for each corner of the t polygons. $\square$

Theorem 8. *Let G be the intersection graph of some geometric objects in $\mathbb{R}^d$. If the objects are convex and if, additionally, there is a constant k such that, for each object, the ratio between its size and the radius of its inball is bounded by k, then G is $(\frac{3}{2}\sqrt{d\pi}(k+1))^d/\Gamma(d/2+1)$-simplicial, where Γ should denote the Gamma function. If the ratio between the largest size of the objects and the radius of a smallest inball of the objects is bounded by a constant k', G is $(\frac{3}{2}\sqrt{d\pi}(k'+1))^d/\Gamma(d/2+1)$-perfectly groupable (even in the case of non-convex objects).*

Proof. For proving the lemma we first show how to find, for a given ball B with radius $\leq R'$ and a real number $r > 0$, a set of points such that every ball b with radius at least r intersecting B must cover at least one of these points. Therefore, let us consider the d-dimensional space, paved with d-dimensional cubes of edge length $s = 2r/\sqrt{d}$ and volume $s^d = 2^d r^d d^{-\frac{d}{2}}$. Then, every ball b of radius at least r must contain at least one of their midpoints, as the cubes' diagonals have length $2r$. Furthermore, the distance between the center of a ball b of radius $\geq r$ intersecting B and B's center is at most $R' + r$. Hence it suffices to pave a ball of radius $R' + 2r$. To do this, we do not need more cubes than completely fit in a ball of radius $R' + 3r$. A ball of radius $R' + 3r$ has volume $(\sqrt{\pi}(R' + 3r))^d/\Gamma(\frac{d}{2} + 1)$ and hence the following number of cubes are enough:

$$\left\lfloor \frac{(\sqrt{\pi}(R' + 3r))^d}{\Gamma(\frac{d}{2} + 1)} \cdot \frac{1}{2^d r^d d^{-\frac{d}{2}}} \right\rfloor = \left\lfloor \left(\frac{\sqrt{d\pi}}{2} \left(\frac{R'}{r} + 3 \right) \right)^d / \Gamma\left(\frac{d}{2} + 1 \right) \right\rfloor$$

Let $\mathcal{S}$ be a set of geometric objects such that G is the intersection graph of $\mathcal{S}$. We first consider the case, where all objects are convex and where there is a k such that, for each object, the ratio between its size and the radius of its inball is bounded by k. Let S_1 be an object of $\mathcal{S}$ with smallest size R and let S_2 be an S_1-intersecting object in $\mathcal{S}$ with size $s_2 \geq R$. Choose S_2' as the image of a dilation of S_2 with an arbitrary point $p \in S_1 \cap S_2$ as center and scaling factor $\lambda = R/s_2 > 0$. Then—as S_2 is convex—every point covered by S_2' is also covered by S_2. Furthermore, the inball of S_2' having radius $r \geq R/k$ must be completely contained in the ball of radius $R' := 3R$ around the center of S_1. Now the considerations above imply that S_2'—and hence S_2—must cover the midpoint of at least one cube of edge length $s = 2r/\sqrt{d}$ completely contained in a ball of radius $R' + 3r$. If we number the vertices of G in an order such that the sizes of objects represented by the vertices do not decrease, we obtain a $(\frac{3}{2}\sqrt{d}\pi(k+1))^d/\Gamma(d/2+1))$-simplicial elimination order proving the claim.

Finally, let us consider the case, where the objects of $\mathcal{S}$ are not necessarily convex, but the ratio between the largest size of the objects and the radius of a smallest inball of the objects is bounded by a constant k'. Consider intersecting geometric objects S_1 (with size R_1) and S_2 (with size R_2 and inball radius r_2) in $\mathcal{S}$. Then the considerations above imply, that the inball of S_2 must completely lie inside the ball of radius $R' := R_1 + 2R_2$ around the center of S_1. With $\frac{R'}{r_2} = \frac{R_1 + 2R_2}{r_2} \leq 3k'$ the second part of the lemma follows immediately. $\square$

Theorem 9. *An intersection graph of t-disks, i.e., of unions of t disks, is*

1. *8-perfectly groupable if $t = 1$ and if the squares are of unit size,*
2. *8-simplicial if $t = 1$, and*
3. *8t-perfectly orientable.*

The theorem above can be shown with a proof similar to Theorem 4. Due to space limitations, we only want to remark that one can choose barriers—defined as in the proof of Theorem 4—of a disk with radius r as follows: One barrier is placed on the center of the disk and the remaining 7 barriers are placed equidistant on a circle of radius $3/2r$ with the same center than the disk.

3   Relations and Recognition

In the following we study the relations between the complexity parameters defined in the last section to each other and the NP-hardness of determining their minimal possible value.

Observation 10. *Each k-perfectly groupable graph is k-simplicial since any ordering of the vertices defines a k-simplicial elimination order. Conversely, an n-vertex star, i.e., an n-vertex tree with $n - 1$ leaves, is not k-perfectly groupable for all $k < n - 1$, but it is 1-simplicial.*

Lemma 11. *A k-simplicial graph is also k-perfectly orientable, but for every $n \in \mathbb{N}$ with $n \geq 12$, there exists a 2-perfectly orientable graph with n vertices that is not ℓ-simplicial for all $\ell < \lfloor \sqrt{n/3} \rfloor$.*

Proof. Let G be a k-simplicial graph having a k-simplicial elimination order $v_1, \ldots, v_n$. If all edges incident to a vertex v and one of its successors are assigned to v, the endpoints $u \neq v$ of the edges assigned to v can be partitioned into k

sets $S_1, \ldots, S_k$ such that $G[S_i \cup \{v\}]$ is a clique for every $i \in \{1, \ldots, k\}$. In other words, G is k-perfectly orientable.

Let us choose an arbitrary $n \in \mathbb{N}$ with $n \geq 12$ and let $k = \lfloor \sqrt{n/3} \rfloor$. We now construct a 2-perfectly orientable graph $G = (V, E)$ with n vertices that is not ℓ-simplicial for any $\ell < k$. The vertices of this graph are divided into three disjoint sets S_0, S_1 and S_2 of size k^2 and, if $n - 3k^2 > 0$, a further set $R = V \setminus (S_0 \cup S_1 \cup S_1)$ of isolated vertices. Each set S_i ($i \in \{0, 1, 2\}$) is divided into k subsets $S_{i,1}, \ldots, S_{i,k}$ of size k. For each $i \in \{0, 1, 2\}$ and each $j \in \{1, \ldots, k\}$, we introduce edges between each pair of vertices contained in the same subset $S_{i,j}$ and assign each of these edges arbitrarily to one of its endpoints. Let us define a numbering on the vertices of $S_{i,j}$ such that we can refer to the h-th vertex of $S_{i,j}$. For each $i \in \{0, 1, 2\}$ and each $h, j \in \{1, \ldots, k\}$, we additionally introduce edges between the h-th vertex u of $S_{i,j}$ and all vertices of $S_{(i+1) \bmod 3, h}$. We assign them to u. The constructed graph G is 2-perfectly orientable since the endpoints of an edge assigned to a vertex u being the h-th vertex of a subset $S_{i,j}$ belong to one of the two cliques induced by the vertices of $S_{i,j}$ and $S_{(i+1) \bmod 3, h}$. However, u is also adjacent to k vertices in $S_{(i-1) \bmod 3}$. Since there is no edge between a vertex in $S_{(i-1) \bmod 3, j_1}$ and a vertex in $S_{(i-1) \bmod 3, j_2}$ for $j_1 \neq j_2$, G cannot be ℓ-simplicial for any $\ell < k$. $\qquad\square$

A graph has *inductive degree* k if it can be obtained from a single vertex by repeatedly adding a new vertex with k edges. Then we can easily conclude:

Lemma 12. *All graphs of inductive degree k are k-simplicial and therefore also k-perfectly orientable.*

Note that an important subclass of the graphs of inductive degree k is the exentsively studied class of graphs of treewidth k (not defined in this paper).

Observation 13. *The n-vertex clique is an example for a 1-perfectly groupable graph G that does not have treewidth $n - 2$. Conversely, the n-vertex star is a graph with treewidth 1 that is not $(n-2)$-perfectly groupable.*

Lemma 14. *It is NP-hard to decide, for a tuple (G, k) of graph G and an integer k, whether G is k-perfectly groupable, k-simplicial, or k-perfectly orientable.*

Proof. In this version of the paper we only proof the result for k-perfectly orientable graphs. The proofs for the other graph classes are based on similiar reductions. Given an n-vertex graph $G = (V, E)$ as an instance of the minimum clique partition problem, we add a set V' of $nk+1$ new vertices to G and connect each new vertex to each vertex in V. Let G' be the graph obtained. We next show that G' is k-perfectly orientable if G has clique partition of size at most k. For this purpose, assign all incident edges of a vertex $v' \in V'$ to v' and edges $e \in E$ to an arbitrary endpoint of e. Then a vertex v together with the endpoints of edges assigned to $v \in V \cup V'$ induce k cliques, i.e., G' is k-perfectly orientable.

Conversely, let us assume that G' is k-perfectly orientable and let $a : E \to V \cup V'$ be a suitable assignment of the edges to their endpoints. For each vertex $v \in V$ at most k of the $nk+1$ new edges incident to v can be assigned by a to v

since there are no edges between two vertices of V'. Thus, there is at least one $v' \in V'$ with all its edges assigned to itself. Thus, G must have a clique partition of size at most k.

For each constant k, one can use a fixed parameter algorithm for the MCP, e.g., see [20], to decide in polynomial time whether a graph G is k-groupable.

4   Algorithms

We present now polynomial time approximation algorithms for several NP-hard problems on graph classes with one of the three complexity parameters bounded by a constant. We implicitly assume that we are given an explicit *representation* of a graph as a k-perfectly groupable, k-simplicial, or k-perfectly orientable graph G. By that we mean that we are given, for each vertex v, a partition of its neighbors, of its successors, and of the vertices connected to v by edges assigned to v, respectively, into k sets $S_1, \ldots, S_k$ such that $G[S_i \cup \{v\}]$ is a clique for all $i \in \{1, \ldots, k\}$. In addition, we are given a k-simplicial elimination order in the case of a k-simplicial graph and, for each vertex of G, the edges assigned to it in the case of a k-perfectly orientable graph. These representations are sufficient even for intersection graphs. We do not need the explicit representations as intersection graphs described in Section 1, but we can use them to construct our new representations in polynomial time (see also the Theorems 4 and 9).

Theorem 15. *On k-perfectly groupable graphs, minimum dominating set and minimum independent dominating set can be k-approximated in polynomial time.*

Proof. As a k-approximative solution on a k-perfectly groupable graph G we output a maximal—not necessarily maximum—independent set S of G. To prove correctness, let us consider a minimum (independent) dominating set S_{opt} of G. For all $v \in S \setminus S_{\mathrm{opt}}$, there must be a neighbor of v in S_{opt}. However, each such neighbor cannot cover more than k vertices of S, since G is k-perfectly groupable. Consequently, S is an independent dominating set of size at most $k|S_{\mathrm{opt}}|$.   □

Theorem 16. *Minimum clique partition, maximum weighted independent set, and maximum weighted clique, are k-approximable on k-simplicial and on k-perfectly groupable graphs in polynomial time.*

Proof (minimum clique partition). Given a graph G and a k-simplicial elimination order $v_1, \ldots, v_n$ for G, we first compute the graph G' obtained by removing v_1 and its neighbors from G. We then solve the problem recursively on G'. Let S' be the collection of vertex sets obtained as a solution for G'. Note that the graph induced by the removed vertices can be partitioned into a set Z of at most k cliques. We output $S = S' \cup Z$ as a solution for G. Note that v_1 is not incident to any vertex of G'. This guarantees that the difference between the size of a clique partition for G and for G' is at least 1. Thus, the clique partition obtained uses at most k times as many cliques as an optimal clique partition for G.   □

Proof (maximum weighted independent set). See [1], [27], or [34].   □

Proof (maximum weighted clique). Given a k-simplicial graph, choose, for each vertex v, a clique C_v of maximal weight among the cliques obtained from one of the k cliques induced by v and the successors of v. Return the clique with maximal weight among the cliques in $\{C_v \mid v \in V\}$. This solution has approximation ratio k since a maximum weighted clique C_{opt} must also contain a vertex v with C_{opt} consisting only of v and a subset of its successors. □

Theorem 17. *On k-perfectly orientable n-vertex graphs, there are polynomial-time algorithms with approximation ratio*

1. *$2k$ for maximum weighted independent set, minimum vertex coloring and maximum weighted clique.*
2. *$O(\log^2 n / \log(1 + 1/k))$ for minimum clique partition.*

For the following proofs let $G = (V, E)$ be a k-perfectly orientable n-vertex graph, and for each $u \in V$, let $V_{u,1}, \ldots, V_{u,k}$ be k pairwise disjoint vertex sets such that their union are the neighbors of u and such that $C_{u,i} = G[V_{u,i} \cup \{u\}]$ is a clique for all $1 \le i \le k$. Moreover, define $\mathcal{C} = \{C_{u,i} \mid u \in V, 1 \le i \le k\}$.

The proof for maximum weighted independent set bases on the ideas including the local ratio technique of [2] and is omitted here.

Proof (minimum vertex coloring). Construct an order $v_1, \ldots, v_n$ of the vertices of G such that, for each vertex v_i ($i \in \{1, \ldots, n\}$), at least half of the edges in $G[\{v_i, \ldots, v_n\}]$ being adjacent to v_i are assigned to v_i. We now want to color the vertices $v_n, \ldots, v_1$ in this order with numbers in $\{1, \ldots, n\}$. We color each vertex $v \in V$ with the smallest number different from the colors of all already colored neighbors of v. Concerning the approximation ratio, let us define, for each vertex v, D_v to be a set of vertices of maximal weight such that D_v consists only of successors of v with respect to the order above and such that $G[D_v]$ is a clique. Then, each vertex v of G obtains a color smaller or equal $2k|D_v| + 1$, whereas an optimal coloring must color v and its neighbors with at least $|D_v| + 1$ colors. Therefore, the coloring obtained is a $2k$-approximation.

Proof (maximum weighted clique). As a $2k$-approximative solution, return the clique $C \in \mathcal{C}$ of maximal weight. Let us compare the weight of C with the weight of a maximal clique C_{OPT} of G. The subgraph of G induced by the vertices of C_{OPT} contains at least one vertex u for which the sum of the weights of the neighbors not being endpoints of edges assigned to u does not exceed the sum of the weights of the neighbors being endpoints of edges assigned to u. Thus, the weight of C is at most a factor $2k$ smaller than the weight of C_{OPT}. □

Proof (minimum clique partition). As part of our computation, we want to find a minimal number of cliques in $\mathcal{C}$ in polynomial time such that the union of their vertex sets is V. Unfortunately, this is an instance of the NP-hard set cover problem. However, using the Johnson's algorithm [26] we can find a subset of the cliques in $\mathcal{C}$ that covers V and that is at most a factor $O(\log |V|)$ larger than the minimal number of cliques in $\mathcal{C}$. We return this subset as an approximative solution. We achieve the approximation ratio $O(\log^2 |V| / \log \frac{2k}{2k-1}) =$

$O(\log^2 |V|/\log(1 + \frac{1}{k}))$ since there is a clique partition of V using only cliques in $\mathcal{C}$ that uses $O(\log |V|/\log \frac{2k}{2k-1})$ as many cliques as a minimum clique partition C_{OPT} of $q \leq n$ arbitrary cliques $C_1, \ldots, C_q$ of G: Choose a vertex v of C_1 such that in the subgraph of G induced by the vertices of C_1 at least half of the edges adjacent to v are assigned to v. Remove the clique among $C_{v,1}, \ldots, C_{v,k}$ containing the largest number of not already deleted vertices in C_1. This decreases the number of vertices of C_1 by a factor of at least $1 - \frac{1}{2k} = \frac{2k-1}{2k}$. Repeat this step recursively until, after $O(\log |V|/\log \frac{2k}{2k-1})$ steps, C_1 contains no vertices any more. More precisely, when choosing a vertex v for which at least half of the adjacent edges are assigned to v, only count the edges not already being deleted. If we do the same for the remaining cliques, we obtain a clique partition with $O(q \log |V|/\log \frac{2k}{2k-1})$ cliques part of $\mathcal{C}$.   □

References

1. Akcoglu, K., Aspnes, J., DasGupta, B., Kao, M.-Y.: Opportunity cost algorithms for combinatorial auctions. In: Kontoghiorghes, E.J., Rustem, B., Siokos, S. (eds.) Applied Optimization: Computational Methods in Decision Making, Economics and Finance, vol. 74, pp. 455–479 (2002)
2. Bar-Yehuda, R., Halldórsson, M.M., Naor, J., Shachnai, H., Shapira, I.: Scheduling split intervals. SIAM J. Comput. 36, 1–15 (2006)
3. Bodlaender, H.L.: A linear-time algorithm for finding tree-decompositions of small treewidth. SIAM J. Comput. 25, 1305–1317 (1996)
4. Butman, A., Hermelin, D., Lewenstein, M., Rawitz, D.: Optimization problems in multiple-interval graphs. In: Proc.18th Annual ACM-SIAM Symposium on Discrete Algorithms (SODA 2007), pp. 268–277 (2007)
5. Cerioli, M.R., Faria, L., Ferreira, T.O., Protti, F.: On minimum clique partition and maximum independent set on unit disk graphs and penny graphs: complexity and approximation. Electronic Notes in Discrete Mathematics 18, 73–79 (2004)
6. Chan, T.M.: Polynomial-time approximation schemes for packing and piercing fat objects. J. Algorithms 46, 178–189 (2003)
7. Clark, B.N., Colbourn, C.J., Johnson, D.S.: Unit disk graphs. Discrete Math. 86, 165–177 (1990)
8. Courcelle, B.: The monadic second-order logic of graphs. I. Recognizable sets of finite graphs. Inform. and Comput. 85, 12–75 (1990)
9. Dumitrescu, A., Pach, J.: Minimum clique partition in unit disk graphs, arXiv:0909.1552v1
10. Erlebach, T., Jansen, K., Seidel, E.: Polynomial-time approximation schemes for geometric intersection graphs. In: Proc. 12th Annual ACM-SIAM Symposium on Discrete Algorithms (SODA 2001), pp. 671–679 (2001)
11. Erlebach, T., van Leeuwen, E.J.: Domination in geometric intersection graphs. In: Laber, E.S., Bornstein, C., Nogueira, L.T., Faria, L. (eds.) LATIN 2008. LNCS, vol. 4957, pp. 747–758. Springer, Heidelberg (2008)
12. Fowler, R.J., Paterson, M.S., Tanimoto, S.L.: Optimal packing and covering in the plane are NP-complete. Inform. Process. Lett. 12, 133–137 (1981)
13. Frank, A.: Some polynomial algorithms for certain graphs and hypergraphs. In: Proc. 5th British Combinatorial Conference (Aberdeen 1975), Congr. Numer. XV., pp. 211–226 (1976)

14. Garey, M.R., Johnson, D.S., Stockmeyer, L.: Some simplified NP-complete graph problems. Theoret. Comput. Sci. 1, 237–267 (1976)
15. Gavril, F.: Algorithms for minimum coloring, maximum clique, minimum covering by cliques, and maximum independent set of a chordal graph. SIAM J. Comput. 1, 180–187 (1972)
16. Gibson, T., Pirwani, I.A.: Approximation algorithms for dominating set in disk graphs. In: Proc. 18th Annual European Symposium on Algorithms (ESA 2010). LNCS. Springer, Heidelberg (2010) (to appear)
17. Golumbic, M.C.: Algorithmic Graph Theory and Perfect Graphs. Academic Press, New York (1980)
18. Golumbic, M.C.: Algorithmic aspects of intersection graphs and representation hypergraphs. Graphs and Combinatorics 4, 307–321 (1988)
19. Gräf, A.: Coloring and recognizing special graph classes, Technical Report Musik-informatik und Medientechnik Bericht 20/95, Johannes Gutenberg-Universität Mainz (1995)
20. Gramm, J., Guo, J., Hüffner, F., Niedermeier, R.: Data reduction and exact algorithms for clique cover. Journal of Experimental Algorithmics 13, Article No. 2 (2009)
21. Griggs, J.R., West, D.B.: Extremal values of the interval number of a graph. SIAM Journal on algebraic and discrete methods 1, 1–7 (1980)
22. Hunt III, H.B., Marathe, M.V., Radhakrishnan, V., Ravi, S.S., Rosenkrantz, D.J., Stearns, R.E.: NC-pproximation schemes for NP- and PSPACE-hard problems for geometric graphs. J. Algorithms 26, 238–274 (1998)
23. Hurink, J.L., Nieberg, T.: Approximating minimum independent dominating sets in wireless networks. Inform. Process. Lett. 109, 155–160 (2008)
24. Imai, H., Asano, T.: Finding the connected components and a maximum clique of an intersection graph of rectangles in the plane. J. Algorithms 4, 310–323 (1983)
25. Jamison, R.E., Mulder, H.M.: Tolerance intersection graphs on binary trees trees with constant tolerance 3. Discrete Math. 215, 115–131 (2000)
26. Johnson, D.S.: Approximation algorithms for combinatorial problems. J. Comput. System Sci. 9, 256–278 (1974)
27. Kammer, F., Tholey, T., Voepel, H.: Approximation algorithms for intersection graphs, Report 2009-6, Institut für Informatik, Universität Augsburg (2009)
28. Malesińska, E.: Graph-theoretical models for frequency assignment problems, PhD thesis, University of Berlin (1997)
29. Marathe, M.V., Breu, H., Hunt III, H.B., Ravi, S.S., Rosenkrantz, D.J.: Simple heuristics for unit disk graphs. Networks 25, 59–68 (1995)
30. Nieberg, T., Hurink, J., Kern, W.: Approximation Schemes for Wireless Networks. ACM Transactions on Algorithms 4, Article No. 49 (2008)
31. Papadimitriou, C.H., Yannakakis, M.: Optimization, approximation, and complexity classes. J. Comput. System Sci. 43, 425–440 (1991)
32. Pirwani, I.A., Salavatipour, M.R.: A weakly-robust PTAS for minimum clique partition on unit disk graphs. In: Kaplan, H. (ed.) SWAT 2010. LNCS, vol. 6139, pp. 188–199. Springer, Heidelberg (2010)
33. Tardos, É.: A strongly polynomial algorithm to solve combinatorial linear programs. Operations Research 34, 250–256 (1986)
34. Ye, Y., Borodin, A.: Elimination graphs. In: Albers, S., Marchetti-Spaccamela, A., Matias, Y., Nikoletseas, S., Thomas, W. (eds.) ICALP 2009. LNCS, vol. 5556, pp. 774–785. Springer, Heidelberg (2009)

An $O(\log n)$-Approximation Algorithm for the Disjoint Paths Problem in Eulerian Planar Graphs and 4-Edge-Connected Planar Graphs

Ken-ichi Kawarabayashi[*] and Yusuke Kobayashi[**]

National Institute of Informatics, 2-1-2, Hitotsubashi, Chiyoda-ku, Tokyo, Japan.
Department of Mathematical Informatics, Graduate School of Information Science
and Technology, University of Tokyo, Tokyo, 113-8656, Japan
k_keniti@nii.ac.jp, kobayashi@mist.i.u-tokyo.ac.jp

Abstract. In this paper, we study an approximation algorithm for the maximum edge-disjoint paths problem. In the maximum edge-disjoint paths problem, we are given a graph and a collection of pairs of vertices, and the objective is to find the maximum number of pairs that can be connected by edge-disjoint paths. We give an $O(\log n)$-approximation algorithm for the maximum edge-disjoint paths problem when an input graph is either 4-edge-connected planar or Eulerian planar. This improves an $O(\log^2 n)$-approximation algorithm given by Kleinberg [10] for Eulerian planar graphs. Our result also generalizes the result by Chekuri, Khanna and Shepherd [2,3] who gave an $O(\log n)$-approximation algorithm for the edge-disjoint paths problem with congestion 2 when an input graph is planar.

1 Introduction

1.1 Background and Our Main Result

In the edge- (vertex-) disjoint paths problem, we are given a graph G and a set of k pairs of vertices in G, and we have to decide whether or not G has k edge-(vertex-) disjoint paths connecting given pairs of terminals. This is certainly a central problem in algorithmic graph theory and combinatorial optimization. See the surveys [5,19]. It has attracted attention in the contexts of transportation networks, VLSI layout and virtual circuit routing in high-speed networks or Internet. A basic technical problem is to interconnect certain prescribed "channels" on the chip such that wires belonging to different pins do not touch each other. In this simplest form, the problem mathematically amounts to finding disjoint trees in a graph or disjoint paths in a graph, each connecting a given set of vertices.

[*] Research partly supported by Japan Society for the Promotion of Science, Grant-in-Aid for Scientific Research, by C & C Foundation, by Kayamori Foundation and by Inoue Research Award for Young Scientists.
[**] Supported by the Global COE Program "The research and training center for new development in mathematics", MEXT, Japan.

M. Serna et al. (Eds.): APPROX and RANDOM 2010, LNCS 6302, pp. 274–286, 2010.

Let us give previous known results on the edge-disjoint paths problem. If k is a part of the input of the problem, then this is one of Karp's NP-complete problems [7], and it remains NP-complete even if G is constrained to be planar [13]. In fact, even for series-parallel graphs (allowing multiple edges), it remains NP-complete [15]. This is one of the few problems that are known to be NP-complete for series-parallel graphs or bounded tree-width graphs. Let us observe that the vertex-disjoint paths problem is solvable for bounded tree-width graphs (and hence for series-parallel graphs), see [17]. On the positive side, the seminal work of Robertson and Seymour says that there is a polynomial-time algorithm for the edge-disjoint paths problem when the number of terminals, k, is fixed [18]. The running time of their algorithm is $O(m^3)$ and it is improved to $O(n^2)$ in [9], where m, n are the number of edges and vertices, respectively. Robertson and Seymour's algorithm is one of the spin-offs of their groundbreaking work on graph minor project, spanning 23 papers, and giving several deep and profound results and techniques in discrete mathematics.

We can also consider the maximization problem of the edge-disjoint paths problem. In the *maximum edge-disjoint paths problem*, we are given a graph and a collection of pairs of vertices, and the objective is to find the maximum number of pairs that can be connected by edge-disjoint paths. The maximum edge-disjoint paths problem receives considerable attention in view of approximation algorithms and hardness results. A *c-approximation algorithm* for this problem is a polynomial-time algorithm that finds at least $1/c$ of the maximum possible number of the edge-disjoint paths. Although a significant amount of research in this area is done, there are wide gaps in understanding of the approximability of the maximum edge-disjoint paths problem. It is known that no polynomial-time algorithm can achieve an approximation guarantee of $O(m^{1/2-\varepsilon})$ for any $\varepsilon > 0$ on directed graphs with m edges unless P = NP [6], but the result is intrinsically based on intractability for the directed case (specifically, hardness with just two terminals pair) that does not have analogues in the undirected case. The current strongest hardness of approximation bound for the undirected case is due to Andrews, Chuzhoy, Khanna and Zhang [1], leading to a lower bound of $\Omega((\log m)^{1/2-\varepsilon})$ for any $\varepsilon > 0$. This may be the right upper bound — though there may be a far reaching approximation algorithm that achieves a corresponding poly-logarithmic upper bound. Currently constant factor and poly-logarithmic factor approximation algorithms are known for restricted classes of graphs such as trees, meshes, and highly connected graphs such as expanders.

Concerning planar graphs, there is also an $O(\log n)$-approximation algorithm by Chekuri, Khanna and Shepherd [2,3] if a congestion two is allowed, that is, we allow up to two paths to share an edge. But on the other hand, a poly-logarithmic approximation algorithm for the maximum edge-disjoint paths problem for planar graphs seems to be very hard, despite Kleinberg and Tardos [11,12] have obtained poly-logarithmic approximation algorithms for some special class of planar graphs. They posed an open question that there is a poly-logarithmic approximation algorithm for the maximum edge-disjoint paths problem when a given graph is planar and Eulerian. This conjecture was solved by Kleinberg [10] who gave an

$O(\log^2 n)$-approximation algorithm. In this paper, we improve Kleinberg's result by showing that there is an $O(\log n)$-approximation algorithm for the maximum edge-disjoint paths problem when an input graph is planar and either Eulerian or 4-edge-connected. Thus we give a logarithmic approximation algorithm for bigger class of planar graphs (i.e, Eulerian planar graphs and 4-edge-connected planar graphs).

Theorem 1. *Suppose G is either a 4-edge-connected planar graph or an Eulerian planar graph. Then there is an $O(\log n)$-approximation algorithm for the maximum edge-disjoint paths problem.*

Let us observe that if G is a planar graph, and we replace each edge by parallel edges, then the resulting graph is clearly Eulerian planar. Thus a c-approximation algorithm for the maximum edge-disjoint paths problem for Eulerian planar graphs would imply a c-approximation algorithm for the maximum edge-disjoint paths problem for planar graphs with congestion two. Hence, our result generalizes the above mentioned result by Chekuri, Khanna and Shepherd [2,3] who gave an $O(\log n)$-approximation algorithm for the edge-disjoint paths problem if an input graph is planar and a congestion is two.

Let us point out that obtaining a poly-logarithmic approximation algorithm for the maximum edge-disjoint paths problem when an input graph is even a wall (for the definition of the wall, we refer the reader to the next section) is actually an open problem that appears to be quite difficult. In fact, this problem is very close in nature to the comparably hard vertex-disjoint paths problem for grids. Understanding the tractability of the disjoint paths problem is a fundamental issue in graph algorithms and combinatorial optimization, and progress in seeking for the boundary between approximability and inapproximability is certainly a crucial issue in this light. We believe that Theorem 1 makes some contribution on these issues. We, however, remark that it may be possible to obtain an $O(1)$-approximation algorithm for Theorem 1, because Chekuri, Khanna and Shepherd [4] proved that there is an $O(1)$-approximation algorithm for the maximum edge-disjoint paths problem for planar graphs with congestion four, and they conjectured that the same conclusion would hold with congestion two.

1.2   4-Edge-Connected Graphs and Eulerian Graphs

In this paper, we are interested in 4-edge-connected graphs and Eulerian graphs. 4-edge-connected graphs and Eulerian graphs appear very often in the context of the edge-disjoint paths problem. First, let us mention that there are many exactly solvable special cases for the edge-disjoint paths problem in planar graphs [16,20], but almost all require some type of Eulerian assumption. For more details, see [5,19]. But on the other hand, if k is a part of input, then the edge-disjoint paths problem is NP-complete even for 4-edge-connected graphs or Eulerian graphs [14].

We now mention how 4-edge-connected graphs play a role in the edge-disjoint paths problem. It helps to consider the following point. Consider the edge-disjoint paths problem in planar graphs with all terminal pairs on the outer face boundary. If every vertex has degree at most 3, then the edge-disjoint paths problem and the vertex-disjoint problem are essentially same (not only for planar graphs, but also for general graphs), thus there is a topological obstruction. On the other hand, if a given graph is 4-edge-connected, we can produce large sets of mutually crossing paths, which can act as "switching" structures for connecting up terminal pairs. Thus the edge-disjoint paths problem seems more tractable if an input graph is 4-edge-connected. In fact, in [8], we significantly simplify the algorithm of Robertson and Seymour for the edge-disjoint paths problem when an input graph G is either 4-edge-connected or Eulerian.

It is natural to ask at this point why we do not consider the weaker condition that the minimum degree being at least four. This assumption plays the same role to create a cross for sure, but in fact this weaker restriction would not gain us anything. Consider an instance of the edge-disjoint paths problem on an arbitrary graph G that may have degree three vertices. Then attach by two edges to each vertex in G a constant-sized graph of high minimum degree. This new graph G' has minimum degree high, but the resulting instance of the edge-disjoint paths problem is clearly equivalent to the original one in G. This example shows that 4-edge-connectivity is necessary. Thus we really need to stick the 4-edge-connectivity in our proof.

1.3 Overview

We now give a sketch for Theorem 1. At a high level, we follow the approach adapted by Kleinberg [10], and Kleinberg follows the approach by Chekuri, Khanna and Shepherd [2,3]. In fact, we first adapt very interesting recent work by them [2,3]. The crucial ingredient of their work is the following. For a vertex set X in a graph G, we say that X is *well-linked* if for every vertex set Z of G containing at most half of the vertices of X, $|\delta(Z)| \geq |X \cap Z|$ holds, where $\delta(Z)$ denotes the set of all edges with one end in Z and the other end in $G - Z$.

Kleinberg [10] proved that, by adapting the idea of Chekuri et al., for an input graph G that is Eulerian planar with the set of terminal pairs $\mathcal{T}$, one can compute vertex disjoint subgraphs $G_1, \ldots, G_r$ of G and their corresponding disjoint sets of vertex pairs $\mathcal{T}_1, \ldots, \mathcal{T}_r$ of $\mathcal{T}$ such that

(a) each vertex in G_i, except for the outer face boundary of G_i, has even degree,
(b) each $\mathcal{T}_i$ consists of the pairs of terminals,
(c) each terminal pair in $\mathcal{T}_i$ belongs to G_i for $i = 1, \ldots, r$,
(d) the members of the terminal pairs in $\mathcal{T}_i$ are well-linked in G_i, and
(e) the total size of the sets $\mathcal{T}_i$ is at least $\text{OPT}/f(n)$, where $f(n)$ is bounded by $O(\log^2 n)$.

We will refer to each instance $(G_i, \mathcal{T}_i)$ as an *even well-linked instance*. Here, OPT is the optimum value of the maximum edge-disjoint paths problem. Note that the maximum value of the LP-relaxation of the maximum edge-disjoint paths problem can be used as an upper bound of OPT.

For each of the even well-linked instances $(G_i, \mathcal{T}_i)$, Kleinberg found a "crossbar". More precisely, he found a wall W of height $\Omega(|\mathcal{T}_i|)$ in G_i, and he used W as a "crossbar" to route the paths with terminals of $\mathcal{T}_i$ in G_i. In fact, one can find, in polynomial time, such a wall W which is "attached" to the terminals in $\mathcal{T}_i$, i.e, roughly there are $\Omega(|\mathcal{T}_i|)$ edge-disjoint paths from the terminals of $\mathcal{T}_i$ to W in G_i. This, together with the famous result of Okamura and Seymour [16], allows Kleinberg to route $\Omega(|\mathcal{T}_i|)$ pairs of terminals in $\mathcal{T}_i$ in G_i. Since the total size of the sets $\mathcal{T}_i$ is at least $\Omega(\text{OPT}/\log^2 n)$, thus he can get an $O(\log^2 n)$-approximation algorithm for the maximum edge-disjoint paths problem when an input graph is Eulerian planar.

We now sketch how to improve to an $O(\log n)$-approximation algorithm. Let us first observe that if we do not impose the condition (a), then the result by Chekuri, Khanna and Shepherd [2] implies that the condition (e) can be replaced by the following: the total size of the sets $\mathcal{T}_i$ is at least $\text{OPT}/f(n)$, where $f(n)$ is bounded by $O(\log n)$. This is our first key observation.

Our second new technique is the following: We first construct a wall W of height $\Omega(|\mathcal{T}_i|)$ which is "attached" to the terminals in $\mathcal{T}_i$ in G_i. As discussed above, this can be done in polynomial time. We then divide this wall W into $\Omega(|\mathcal{T}_i|^2)$ parts $W_1, \ldots, W_l$, where $l = \Omega(|\mathcal{T}_i|^2)$ and each W_j is a proper subwall of W of height 2. The outer face boundary of W_j bounds a disk D_j. Let W_j' be the induced subgraph of G_i embedded into the disk D_j. We now try to find two edge-disjoint paths connecting two opposite corners of W_j in W_j'. This may not be possible if, after $\leq$ 3-edge-cuts reduction, W_j' has no vertex of degree 4 or more. But since the original graph G is either 4-edge-connected or Eulerian and W_j' contains at least one vertex of degree at least 3 in W_j', so if we add all the well-linked instances $G_{i'}$ that are contained in the disk D_j in G, then we can always find such desired two paths. By applying this argument to each W_j', we can construct a clique minor of order $\Omega(|\mathcal{T}_i|)$ in the line graph of G, which allows us to route $\Omega(|\mathcal{T}_i|)$ edge-disjoint paths.

The problem here is that the above constructed paths may go through some other well-linked instances. In order to avoid this problem, we adapt the idea of Kleinberg [10] who constructs a tree-representation of the well-linked instances. Roughly, if G_i is contained in a face of G_j, we say that G_i is a descendant of G_j. This allows us to construct a tree-representation of the well-linked instances with the corresponding tree T. Kleinberg picked up disjoint subtrees of T that are pairwise far apart from any other in T. This allows Kleinberg to focus on each subtree which gives rise to an "almost" Eulerian planar graph, i.e, each vertex, except for the outer face boundary, has even degree. Then using Okamura and Seymour's result [16], he can reroute $\Omega(|\mathcal{T}_i|)$ edge-disjoint paths for each well-linked instance G_i. But his method has to loose $\log n$ factor.

Our new idea is the following: Instead of keeping "almost evenness" in each well-linked instance (which was the case for Kleinberg's method), for each well-linked instance G_i, we also take all the descendants of G_i too. Let $\overline{G_i}$ be the resulting graph. Then $\overline{G_i}$ is 4-edge-connected or Eulerian (more or less, except for some part in the boundary of $\overline{G_i}$). As remarked above, we can certainly find

$\Omega(|\mathcal{T}_i|)$ desired edge-disjoint paths with terminals in $\mathcal{T}_i$ in $\overline{G_i}$, but we also impose the following important condition:

> Each descendant G_j of G_i can intersect at most one subpath of the constructed edge-disjoint paths in G_i.

This important property allows us to show that, in each well-linked instance G_i, we can always find $\Omega(|\mathcal{T}_i|) - |C(G_i)|$ edge-disjoint paths, where $C(G_i)$ is the set of "children" of G_i. More precisely, when constructing the edge-disjoint paths for G_i, we sacrifice one path for each child of G_i. As a result, we get $\Omega(|\mathcal{T}_i|) - |C(G_i)|$ edge-disjoint paths in G_i. Then, we can get $\Omega(\sum_i |\mathcal{T}_i|)$ edge-disjoint paths in total. This allows us to show an $O(\log n)$-approximation algorithm for the edge-disjoint paths problem when an input graph is Eulerian planar or 4-edge-connected planar.

This paper is organized as follows. In Section 2, we give some notations needed in this paper. In Section 3, we introduce Chekuri, Khanna and Shepherd's theorem [2] and give our key theorem finding a wall attached to the well-linked sets. In Section 4, we present a result concerning tree-representation of the well-linked instances. In Section 5, we give an algorithm for finding edge-disjoint paths in each graph G_i. This is one of the key ideas in our paper. Finally in Section 6, we give a complete description of our algorithm.

2   Preliminary

In this paper, n and m always mean the numbers of vertices and edges of a given graph, respectively. For a vertex set X in a graph $G = (V, E)$, let $\delta(X)$ be the set of edges between X and $V \setminus X$, and such an edge set is called an *edge-cut*.

An *elementary wall* of height eight is depicted in Fig. 1. An elementary wall of height h for $h \geq 2$ is similar. It consists of h levels each containing h bricks, where a brick is a cycle of length six. A *wall* of height h is obtained from an elementary wall of height h by subdividing some of the edges, i.e. replacing the edges with internally vertex disjoint paths with the same endpoints. The *nails* of a wall are the vertices of degree three within it. Any wall has a unique planar

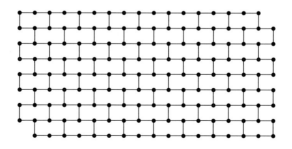

Fig. 1. An elementary wall of height 8

embedding. The *perimeter* of a wall W, denoted per(W) is the boundary of the unique face in this embedding which contains $4(h-1)$ nails. Nails on per(W) are called *perimeter-nails*, and the set of perimeter-nails is denoted by pn(W). For any wall W in a given graph G, there is a unique component U of $G - \text{per}(W)$ containing $W - \text{per}(W)$. The *compass* of W in G, denoted $\text{comp}_G(W)$, consists of the graph with vertex set $V(U) \cup V(\text{per}(W))$ and edge set

$$E(U) \cup E(\text{per}(W)) \cup \{xy \mid x \in V(U), y \in V(\text{per}(W))\}.$$

A *subwall* of a wall W is a wall which is a subgraph of W. A subwall of W of height h is *proper* if it consists of h consecutive bricks from each of h consecutive rows of W.

3   Well-Linked Instance

Our algorithm builds on very useful work of Chekuri, Khanna and Shepherd (CKS) [2,3], together with apparently new techniques. Here is the key CKS theorem [2].

Theorem 2 (Chekuri, Khanna and Shepherd [2]). *Let T be the set of k terminal pairs of the edge-disjoint paths problem (k is not fixed). Then there is a polynomial-time algorithm to compute vertex-disjoint induced connected subgraphs $G_1, \ldots, G_r$, and corresponding disjoint terminal pairs $T_1, \ldots, T_r$ such that the following holds:*

(1) each T_i consists of the pairs of terminals in T, and belongs to G_i;
(2) the members of the terminal pairs in T_i are well-linked in G_i;
(3) the total size of the sets T_i is at least $\text{OPT}/f(n)$ for a poly-logarithmic function $f(n)$.

Furthermore, $f(n)$ is bounded by $O(\log n)$ if a given graph G is planar.

We will refer to each instance (G_i, T_i) as a *well-linked instance*.

Chekuri, Khanna and Shepherd [2,3] also use the fact that we may assume that the original graph G has maximum degree 5 (In fact, maximum degree at most 4). This is because if G has a vertex v of degree more than 5, then we replace v by a grid as in Fig. 2. One can see that this reduction does not affect the existence of the edge-disjoint paths. We also note that if the original graph is 4-edge-connected planar (resp. Eulerian planar), then the obtained graph is 4-edge-connected planar (resp. Eulerian planar). Note that if we only consider Eulerian planar graphs, our construction implies that maximum degree is at most 4 (as Kleinberg did in [10]), because the size of a grid we will attach to each vertex of degree ≥ 5 has even order. But on the other hand, if we consider 4-edge-connected planar graphs, we may need one vertex of degree 5 in the grid when we apply our operation to the odd degree vertex of degree at least five (because the size of a grid we will attach has odd order).

The key for the proof of our main result is the following, which allows us to find a "crossbar" for routings. Due to the space constraint, the proof of this theorem is omitted.

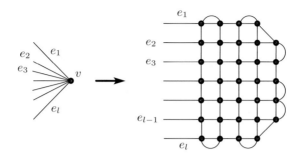

Fig. 2. Vertex of degree more than 5

Theorem 3. *Let $(G, \mathcal{T})$ be a well-linked instance, where G is a planar graph of maximum degree 5. Then, there exist an integer $t = \Omega(|\mathcal{T}|)$, a wall W of size $10t$, and a set of t terminal pairs $\mathcal{T}' \subseteq \mathcal{T}$ satisfying the following. There are $2t$ edge-disjoint paths from the terminals in $\mathcal{T}'$ to the perimeter-nails of W such that their end vertices are distinct and they do not intersect with $\mathrm{comp}_G(W)$ except for their end vertices.*

4 Rooted Forest Representation

In this section, we shall borrow some tools from [10].

We fix an embedding of an input planar graph G in the plane; we will also use G to denote the drawing of G too, if there is no fear of confusion. Since we fix an embedding of G, we can uniquely determine the *outer face* of G. All other faces are called *internal faces* of G.

Suppose that the graph G is partitioned into r graphs $G_1, \dots, G_r$ as in Theorem 2. From the drawing of G, we can define a drawing of each G_i. Now every internal face of G_i is either also a face of G, or else it is the result of deletion of some other well-linked instance. In the second case, we call it an *exceptional face*. Note that we will not refer to the outer face of each graph as exceptional.

Given two graphs G_i and G_j, since they are connected and disjoint, either one is drawn inside an exceptional face of the other, or each is drawn in the outer face of the other. We define a partial order on the graphs G_i, G_j, writing $G_i \prec G_j$ if G_i is drawn inside an exceptional face of G_j. For each G_i, let $\overline{G_i}$ be the subgraph of G induced by the vertex set $\bigcup_{G_j \prec G_i} V(G_j)$.

The following is the key observation in [10].

Lemma 1. *The graphs $G_1, \dots, G_r$ obtained in Theorem 2 are tree-representable (see Fig. 3). More precisely, there are rooted trees $R_1, \dots, R_l$ such that each vertex v_i in R_l corresponds to one graph G_i, and $G_i \prec G_j$ if and only if v_i and v_j are in the same tree, and v_i is a descendant of v_j.*

Kleinberg [10] takes each of the rooted tree in Lemma 1, and then partitions it into components. This decomposition has to pay $O(\log n)$ factor. Instead of paying it, we shall directly consider each rooted tree.

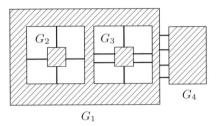

Fig. 3. Example of the decomposition

5   Disjoint Paths in Planar Graphs

In this section, we prove a key lemma for Theorem 1. Let $(G_d, \mathcal{T}_d)$ be a well-linked instance obtained in Theorem 2. Note that an input graph G is 4-edge-connected planar or Eulerian planar, and recall that for each G_i, $\overline{G_i}$ is the subgraph of G induced by the vertex set $\bigcup_{G_j \prec G_i} V(G_j)$.

Lemma 2. *Let W be a wall in G_d and W' be its proper subwall of height 2. Then, $\overline{G_d}$ has two edge-disjoint paths inside W' such that*

1. *each path connects the diagonally opposite corners of W', and*
2. *for each $G_j \prec G_d$, at most one of the two paths intersects with G_j.*

Proof. Let v be a nail of W' that is not on the perimeter of W'. So v has degree three in W', and W' contains three edge-disjoint paths P_1, P_2, P_3 from v to $\mathrm{per}(W')$. Let u_1, u_2, u_3 be the end vertices of P_1, P_2, P_3 that are on the perimeter of W' (see Fig. 4).

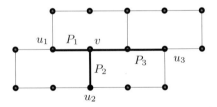

Fig. 4. Three disjoint paths from v to $\mathrm{per}(W')$

We now contract each of the G_j that is in an exceptional face of G_d, into a single vertex, and let G' be the resulting graph. Since G is 4-edge-connected or Eulerian, there are at least four edge-disjoint paths from v to $\mathrm{per}(W')$ in G, and hence there are at least four edge-disjoint paths from v to $\mathrm{per}(W')$ in G', too.

Since we have three edge-disjoint paths P_1, P_2, P_3 from v to $\mathrm{per}(W')$, we can find four edge-disjoint paths in G' from v to $\mathrm{per}(W')$ by finding one augmenting path P from v to $\mathrm{per}(W')$. More precisely, let G'' be the digraph obtained from G' by replacing every edge with two directed edges in opposite directions to each other, and then removing directed edges along P_i from v to u_i for $i = 1, 2, 3$.

Then we find a simple directed path P from v to per(W') in G'', and hence we can modify P_1, P_2, P_3 along P to obtain four edge-disjoint paths P'_1, P'_2, P'_3, P'_4 from v to per(W').

Suppose that the end vertices of P'_1, P'_2, P'_3, P'_4 are on per(W') in this order. Then, G' contains two edge-disjoint paths $P'_1 \cup P'_3$ and $P'_2 \cup P'_4$, which, together with the outer face boundary of W', give rise to two edge-disjoint paths P''_1, P''_2 connecting the diagonally opposite corners of W' inside W'.

Let v_j be the vertex in G' corresponding to $G_j \prec G_d$. Since P_1, P_2, and P_3 do not go through v_j and P is simple, at most one of P'_1, P'_2, P'_3, and P'_4 contains v_j. This means that at most one of P''_1 and P''_2 intersects with G_j.

We are now ready to give an algorithm that finds edge-disjoint paths in $\overline{G_i}$.

Theorem 4. *Let $(G_1, \mathcal{T}_1), \ldots, (G_r, \mathcal{T}_r)$ be well-linked instances obtained in Theorem 2. Then, for each $d = 1, \ldots, r$, $\overline{G_d}$ has $\Omega(|\mathcal{T}_d|)$ edge-disjoint paths such that*

1. *each path connects a pair of terminals in $\mathcal{T}_d$, and*
2. *for each $G_j \prec G_d$, at most one path intersects with G_j.*

Proof. By Theorem 3, we can find an integer $t = \Omega(|\mathcal{T}_d|)$, a wall W_0 of size $10t$, and a set of t terminal pairs $\mathcal{T}' \subseteq \mathcal{T}$ satisfying the following. There are $2t$ edge-disjoint paths from the terminals in $\mathcal{T}'$ to the perimeter-nails of W_0 such that they do not intersect with comp$_G(W_0)$ except for their end vertices. Let $v_1, v_2, \ldots, v_{2t}$ be the end vertices of these $2t$ paths on the perimeter-nails of W_0. We show that $v_1, v_2, \ldots, v_{2t}$ can be linked in W_0 in any desired way.

Let W be the proper subwall of W_0 of height $6t$ that is at the center of W_0. Let P_i be the horizontal path in W separating the i-th level of W and the $(i+1)$-st level of W for each i. Similarly, let P'_i be the vertical path in W separating the i-th column of W and the $(i+1)$-st column of W for each i. Let $Q_i = P_{3i-2} \cup P'_{3i-2}$ for $i = 1, 2, \ldots, 2t$.

We first observe that the intersection of Q_i and Q_j consists of two paths, and each of them is the "center" of a subwall W' of height 2 (Fig. 5). Then we shall make a "cross" in the line graph of comp$_{\overline{G_d}}(W')$. More precisely, let Q'_i, Q'_j be the subpaths of Q_i, Q_j, respectively, in W'. Then by Lemma 2, the paths Q'_i, Q'_j can be modified in comp$_{\overline{G_d}}(W')$ so that they are edge-disjoint and their endpoints are still same. Moreover, by the second condition in Lemma 2, each graph G_l contained in an exceptional face in G_d intersects with at most one of Q'_i and Q'_j. Note that clearly G_l would not hit any other subpaths of $Q_{i'}$ for $i' \neq i, j$. We now perform the above operation for each intersection of Q_i and Q_j for $i \neq j$. Then clearly we can modify Q_i in comp$_{\overline{G_i}}(W)$ so that each Q_i is edge-disjoint from any other Q_j, and any pair Q_i and Q_j share a common vertex.

We find $2t$ edge-disjoint paths in $W_0 - W$ connecting $\{v_1, v_2, \ldots, v_{2t}\}$ and $\{u_1, u_2, \ldots, u_{2t}\}$, where $u_i \in V(Q_i) \cap$ pn(W) for $i = 1, 2, \ldots, 2t$. Such $2t$ paths can be easily found, since we still have enough spaces in $W_0 - W$. Since any pair Q_i and Q_j share a common vertex, we can link up $\{u_1, u_2, \ldots, u_{2t}\}$ by edge-disjoint paths in W.

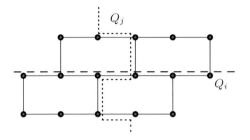

Fig. 5. Intersection of Q_i and Q_j

Thus, we can link up the terminals in $\mathcal{T}'$ by edge-disjoint paths in any desired way. The second condition in this theorem follows from the construction of Q_i, which completes the proof.

6   The Full Routing Algorithm

We are now ready to describe our algorithm for Theorem 1. Suppose an input graph G that is either Eulerian planar or 4-edge-connected planar is given. Suppose furthermore that the set of terminal pairs $\mathcal{T}$ is given. By using the reduction in Section 3, we can work on a graph G of maximum degree at most 5.

We begin with the well-linked decomposition of G as in Section 3. Suppose that G is decomposed into r vertex-disjoint induced connected subgraphs $G_1, \ldots, G_r$, as in Theorem 2. By Lemma 1, the subgraphs $G_1, \ldots, G_r$ are tree-representable. Then, the subgraphs $G_1, \ldots, G_r$ are tree-representable. Thus there are rooted trees $R_1, \ldots, R_l$ such that each vertex v_i in R_j corresponds to one subgraph G_i. If $G_i \prec G_j$ and there exists no subgraph G_q with $G_i \prec G_q \prec G_j$, then G_i is called a *child* of G_j. Let $C(G_j)$ be the set of children of G_j.

By Theorem 4, there exists a constant $\alpha > 0$ such that for each G_i, we can connect at least $\alpha|\mathcal{T}_i|$ pairs of terminals in $\mathcal{T}_i$ by edge-disjoint paths in $\overline{G_i}$. By ignoring the paths through $C(G_i)$, G_i contains at least $\alpha|\mathcal{T}_i| - |C(G_i)|$ edge-disjoint paths. Thus, G contains at least

$$\sum_i \left(\alpha|\mathcal{T}_i| - |C(G_i)| \right) \geq \sum_i \alpha|\mathcal{T}_i| - r$$

edge-disjoint paths, where r is the number of subgraphs.

On the other hand, since each G_i is connected, we can easily find one path connecting a terminal pair in G_i, and hence we can find r edge-disjoint paths in G.

Thus, we can find

$$\max \left\{ \sum_i \alpha|\mathcal{T}_i| - r, r \right\} \geq \frac{1}{2} \sum_i \alpha|\mathcal{T}_i|$$

edge-disjoint paths, which is $\Omega(\text{OPT}/\log n)$ by Theorem 2.

References

1. Andrews, M., Chuzhoy, J., Khanna, S., Zhang, L.: Hardness of the undirected edge-disjoint paths problem with congestion. In: Proc. 46th IEEE Symposium on Foundations of Computer Science (FOCS), 226–244 (2005)
2. Chekuri, C., Khanna, S., Shepherd, B.: Edge-disjoint paths in planar graphs. In: Proc. 45th IEEE Symposium on Foundations of Computer Science (FOCS), pp. 71–80 (2004)
3. Chekuri, C., Khanna, S., Shepherd, B.: Multicommodity flow, well-linked terminals, and routing problems. In: Proc. 37th ACM Symposium on Theory of Computing (STOC), pp. 183–192 (2005)
4. Chekuri, C., Khanna, S., Shepherd, B.: Edge-disjoint paths in planar graphs with constant congestion. In: Proc. 38th ACM Symposium on Theory of Computing (STOC), pp. 757–766 (2006)
5. Frank, A.: Packing paths, cuts and circuits – a survey. In: Korte, B., Lovász, L., Promel, H.J., Schrijver, A. (eds.) Paths, Flows and VLSI-Layout, pp. 49–100. Springer, Berlin (1990)
6. Guruswami, V., Khanne, S., Rajaraman, R., Shephard, B., Yannakakis, M.: Near-optimal hardness results and apprixmaiton algorithms for edge-disjoint paths and related problems. J. Comp. Styst. Science 67, 473–496 (2003); Also Proc. 31st ACM Symposium on Theory of Computing (STOC 1999), pp. 19–28 (1999)
7. Karp, R.M.: On the computational complexity of combinatorial problems. Networks 5, 45–68 (1975)
8. Kawarabayashi, K., Kobayashi, Y.: The edge-disjoint paths problem for 4-edge-connected graphs and Eulerian graphs. To appear in ACM-SIAM Symposium on Discrete Algorithms, SODA (2010)
9. Kawarabayashi, K., Kobayashi, Y., Reed, B.: The disjoint paths problem in quadratic time, manuscript
10. Kleinberg, J.: An approximation algorithm for the disjoint paths problem in even-degree planar graphs. In: Proc. 46th IEEE Symposium on Foundations of Computer Science (FOCS), pp. 627–636 (2005)
11. Kleinberg, J., Tardos, E.: Disjoint paths in densely embedded graphs. In: Proc. 36th IEEE Symposium on Foundations of Computer Science (FOCS), pp. 52–61 (1995)
12. Kleinberg, J., Tardos, E.: Approximations for the disjoint paths problem in high-diameter planar networks. In: Proc. 27th ACM Symposium on Theory of Computing (STOC), pp. 26–35 (1995)
13. Kramer, M.R., van Leeuwen, J.: The complexity of wirerouting and finding minimum area layouts for arbitrary VLSI circuits. Adv. Comput. Res. 2, 129–146 (1984)
14. Middendorf, M., Pfeiffer, F.: On the complexity of the disjoint paths problem. Combinatorica 13, 97–107 (1993)
15. Nishizeki, T., Vygen, J., Zhou, X.: The edge-disjoint paths problem is NP-complete for series-parallel graphs. Discrete Applied Math. 115, 177–186 (2001)
16. Okamura, H., Seymour, P.D.: Multicommodity flows in planar graphs. J. Combin. Theory Ser. B 31, 75–81 (1981)
17. Reed, B.: Tree width and tangles: a new connectivity measure and some applications. In: Surveys in Combinatorics. London Math. Soc. Lecture Note Ser., vol. 241, pp. 87–162. Cambridge Univ. Press, Cambridge (1997)

18. Robertson, N., Seymour, P.D.: Graph minors. XIII. The disjoint paths problem. J. Combin. Theory Ser. B 63, 65–110 (1995)
19. Schrijver, A.: Combinatorial Optimization: Polyhedra and Efficiency, Algorithm and Combinatorics, vol. 24. Springer, Heidelberg (2003)
20. Seymour, P.D.: On odd cuts and plane multicommodity flows. Proceedings of the London Mathematical Society 42, 178–192 (1981)

Improved Algorithm for the Half-Disjoint Paths Problem

Ken-ichi Kawarabayashi[*] and Yusuke Kobayashi[**]

National Institute of Informatics, 2-1-2, Hitotsubashi, Chiyoda-ku, Tokyo, Japan
Department of Mathematical Informatics, Graduate School of Information Science
and Technology, University of Tokyo, Tokyo, 113-8656, Japan
k_keniti@nii.ac.jp, kobayashi@mist.i.u-tokyo.ac.jp

Abstract. In this paper, we consider the half integral disjoint paths packing. For a graph G and k pairs of vertices $(s_1, t_1), (s_2, t_2), \ldots, (s_k, t_k)$ in G, the objective is to find paths $P_1, \ldots, P_k$ in G such that P_i joins s_i and t_i for $i = 1, 2, \ldots, k$, and in addition, each vertex is on at most two of these paths. We give a polynomial-time algorithm to decide the feasibility of this problem with $k = O((\log n / \log \log n)^{1/12})$. This improves a result by Kleinberg [12] who proved the same conclusion when $k = O((\log \log n)^{2/15})$. Our algorithm still works for several problems related to the bounded unsplittable flow. These results can all carry over to problems involving edge capacities. Our main technical contribution is to give a "crossbar" of a polynomial size of the tree-width of the graph.

1 Introduction

1.1 Background of the Disjoint Paths Problem

In the vertex- (edge-) disjoint paths problem, we are given a graph G and a set of k pairs of vertices in G, and we have to decide whether or not G has k vertex- (edge-) disjoint paths connecting given pairs of terminals. This is certainly a central problem in algorithmic graph theory and combinatorial optimization. See the surveys [6,22]. It has attracted attention in the contexts of transportation networks, VLSI layout and virtual circuit routing in high-speed networks or internet. A basic technical problem here is to interconnect certain prescribed "channels" on the chip such that wires belonging to different pins do not touch each other. In this simplest form, the problem mathematically amounts to finding disjoint trees in a graph or disjoint paths in a graph, each connecting a given set of vertices.

Let us give previous known results on the vertex-disjoint paths problem. If k is a part of the input of the problem, then this is one of Karp's NP-complete

---

[*] Research partly supported by Japan Society for the Promotion of Science, Grant-in-Aid for Scientific Research, by C & C Foundation, by Kayamori Foundation and by Inoue Research Award for Young Scientists.
[**] Supported by the Global COE Program "The research and training center for new development in mathematics", MEXT, Japan.

M. Serna et al. (Eds.): APPROX and RANDOM 2010, LNCS 6302, pp. 287–297, 2010.

problems [9], and it remains NP-complete even if an input graph G is constrained to be planar [14]. The seminal work of Robertson and Seymour says that there is a polynomial-time algorithm (actually $O(n^3)$ time algorithm, where n is the number of vertices of an input graph G) for the disjoint paths problem when the number of terminals, k, is fixed [20]. Actually, this algorithm is one of the spin-offs of their groundbreaking work on graph minor project, spanning 23 papers, and giving several deep and profound results and techniques in discrete mathematics.

In this multicommodity flow question, the commodities at the sources s_1 through s_k are different and the demand at each t_i is for a specific commodity. This is the type of question we need to resolve when sending information through the information highway network and so has become increasingly of interest to computer scientists (see, for example the work of Chekuri et al. [3,4] and of Kleinberg [11,12]). The unsplittable flow problem, which generalizes the disjoint paths problem is also motivated by these practical issues. In one basic version of this problem, we are given a graph, a set of k pairs of vertices in G and a nonnegative demand d_i associated with each given pair of terminals. We now have to decide whether or not it is possible to choose a single path for each pair of terminals so that the cumulative demand sent by these paths through any vertex is at most 1. So the vertex disjoint paths problem is one of the special cases of this problem. The other special case which is of great interest is that all demands are at most $1/2$. This is the bounded unsplittable flow problem, which often behaves very different from the disjoint paths problem. For example, the natural multicommodity flow relaxation of the problem provides very strong information for designing approximation algorithms when all demands are at most $1/2$, but the relaxation seems to be very weak when the demands are as large as 1 (see [7]). Thus by imposing this mild relaxation, one can get fairly dramatic changes on the global structure of the routing problem.

Indeed there are many such flow type problems for which the half integral version can be at least approximately solved although the integral version is intractable ([13,16]). A similar situation holds with respect to the k-disjoint path problem. The proof of correctness of Robertson and Seymour's algorithm requires almost all of the graph minors project spanning 23 papers and more than 500 pages. Its running time has the form $f(k)n^3$, where f is an extremely rapidly growing function. Actually this function f depends on the size of grid minor in Robertson-Seymour structure theorem, and it is believed to have very large bounds (see [8]).

We now contrast this with the solution to the following problem:

Half-integral Disjoint Paths Packing

Input: A graph G and k pair of vertices $(s_1, t_1), (s_2, t_2), \ldots, (s_k, t_k)$ in G (which are sometimes called *terminals*).

Problem: Determine whether or not there exist paths $P_1, \ldots, P_k$ in G such that P_i joins s_i and t_i for $i = 1, 2, \ldots, k$, and in addition, each vertex is on at most two of these paths.

It is known that the half-integral disjoint paths packing is still NP-complete, see [16]. When the number of terminals k is fixed, a polynomial time algorithm for this problem is obtained from Robertson and Seymour's algorithm for the integral version [20]. Kleinberg [12] gave a simpler polynomial-time algorithm for the half-integral disjoint paths packing. The correctness of his algorithm is much simpler than that of Robertson and Seymour's. This algorithm was generalized by Kawarabayashi and Reed [10] who gave a nearly linear time algorithm. Furthermore, the constants in both algorithms [10,12] are much smaller, and actually both polynomial-time algorithms work provided $k = O((\log \log n)^{2/15})$.

1.2 Our Main Results

The main purpose of this paper is to prove the following.

Theorem 1. *There is a polynomial-time algorithm to solve the half-integral disjoint paths packing with $k = O((\log n / \log \log n)^{1/12})$ terminals, where n is the number of vertices of an input graph.*

Thus this improves the above mentioned result by Kleinberg [12] who proved the same conclusion of Theorem 1 when $k = O((\log \log n)^{2/15})$. Our proof method for Theorem 1 can be used to design a polynomial-time algorithm for the following problems when the number of terminals $k = O((\log n / \log \log n)^{1/12})$.

1. Not only half-integral, but also that each vertex is on at most c paths, where $c \geq 2$.
2. Not only vertex capacities problem, but also edge capacities problem.

We note that our method can be also applied to the *bounded unsplittable flow problem* (see [12]). In the bounded unsplittable flow problem, each terminal pair (s_i, t_i) has a real-valued demand d_i with $0 \leq d_i \leq 1/2$, and the objective is to find a single path for each terminal pair so that the cumulative demand sent by these paths thorough any vertex is at most 1. This problem can be solved in $(w + k)^{O((w+k)k)} n^{O(1)}$ time when the tree-width of the graph is bounded by w. By using this running time instead of Theorem 5, we can show that the bounded unsplittable flow problem can be solved in polynomial time when $k = O((\log n / \log \log n)^{1/24})$.

Since our proof can be easily modified for these problems, so we omit proofs.

2 Overview

We now give an overview of our algorithm, but since we are going to improve Kleinberg's algorithm, let us sketch his algorithm first, and then clarify our improvement.

Kleinberg's algorithm differentiates between two different kind of inputs which it treats differently: either a given graph G has bounded tree-width or else it has a large tree-width. In the first case, one can apply dynamic programming [1,2,17] to a tree-decomposition of bounded tree-width. In the second case, there is a huge grid minor by the result in [5,17,19,21]. Now, we shall use this grid minor as a "crossbar structure". Let us see how to use the grid minor.

We can think of a $k \times k$ grid minor as a union of k^2 disjoint trees $T_{i,j}$ for $1 \leq i, j \leq k$ such that when we contract each tree $T_{i,j}$, then it becomes a $k \times k$ grid such that the image $t_{i,j}$ of $T_{i,j}$ (that is, the vertex obtained from $T_{i,j}$ by contracting it into a single point) is adjacent to $t_{i-1,j}, t_{i+1,j}, t_{i,j-1}, t_{i,j+1}$. Suppose we have a $4k \times 4k$ grid minor and $2k$ disjoint paths from the terminals to images of the vertices on the top and bottom rows which are pairwise nonadjacent and internally disjoint from the image of the grid. Within the grid minor, we can find paths between the endpoints of the paths using no vertex more than twice and using the endpoint of each path only once. Combining these with the paths from the terminals to the grid minor gives the desired half-integral disjoint paths packing. Actually it turns out that given a $2k^{3/2} \times 2k^{3/2}$ grid minor for which there is no small cut separating the terminals from this grid minor, we can still find the desired paths.

Even if there is a small cutset that separates the $2k$ terminals and this grid minor, we will be able to reduce the problem. So once we have a huge grid minor, either we can make a smaller graph or we can conclude that the desired half-integral disjoint paths packing exists. This was Kleinberg's key idea, and his approach is a simplification of Robertson and Seymour's algorithm for the integral case.

Our algorithm of Theorem 1 also follows this approach, but there is one big problem. Unfortunately, the tree-width results that guarantee the existence of a gird minor are too weak to obtain our main result by this method. The current best known result is the following: if G has tree-width at least 20^{2r^5}, then G has a $r \times r$ grid minor [5,17,21]. So, we cannot use a grid minor as a "crossbar" at the moment. We would need a result saying that if G has tree-width poly(r), then G would have a $r \times r$ grid minor. This has been conjectured for more than 20 years, but it is still wide open at the moment.

Instead, we shall use a "grid-like minor" as a crossbar, which is introduced by Reed and Wood [18]. Here we say that G has a grid-like minor of order r if the graph $G^* = G \square K_2$, i.e, Cartesian product of G and K_2, contains a clique minor of order r with some additional condition (see Section 4). Note that a $r \times r$ grid contains a grid like minor of order $r + 1$. Reed and Wood [18] proved that if tree-width is at least $\Omega(r^4 \sqrt{\log r})$, then G has a grid-like minor of order r. Very recently, Kruetzer and Tazari [15] provided a polynomial-time algorithm to construct such a grid-like minor.

An outline of our algorithm is as follows:

We first take Cartesian product $G^* = G \square K_2$. By the definition of the grid-like minor, there is a clique minor M^* of order r in G^*. By using the result of Robertson and Seymour [20], we show that we can use M^* as a "crossbar" to find vertex disjoint paths connecting the specified terminals in G^*. Let us observe that if there are desired paths in G^*, clearly they correspond to a half-integral disjoint paths packing in G.

Moreover, if there is a small separation that separates the terminals and the clique minor M^*, then there is an irrelevant vertex v in G, i.e, G has a feasible solution if and only if $G - v$ has.

Note that finding the above irrelevant vertex v is not trivial compared to the disjoint paths case, in the sense that we have to work on the original graph G but not on the graph G^* (the image creates a little problem).

This allows us to prove Theorem 1. Our main technical contribution can be regarded as constructing a "crossbar" of order $\text{poly}(r)$, where r is tree-width of the input graph. Note that if the tree-width is w, then by using dynamic programming approach [1,2,17], we can solve the half-integral disjoint paths packing in $(k + w)^{O(k+w)} n^{O(1)}$ time, where n is the number of vertices. It seems that this dynamic programming step is unavoidable, and the running time is polynomial of n when $w = O(\log n/ \log \log n)$. Thus we believe that it is hopeless to solve the half-integral disjoint paths packing when the number of terminals, k, is bigger than $O(\log n/ \log \log n)$, say.

This paper is organized as follows: In Section 3, we give some notation. In Section 4, we introduce the key concept of the paper, which is called "grid-like minor", and show how to construct a grid-like minor. Finally, in Section 5, we will complete the proof of Theorem 1.

3   Preliminary

In this paper, n and m always mean the number of vertices of a given graph and the number of edges of a given graph, respectively. A pair of subgraphs (A, B) of a graph G is a *separation* of G if $G = A \cup B$ and there are no edges between $A - B$ and $B - A$. The *order* of the separation (A, B) is $|V(A \cap B)|$. An $r \times r$ *grid* is a graph with vertex set $\{(i, j) \mid 1 \le i \le r,\ 1 \le j \le r\}$ in which two vertices (i, j) and (i', j') are adjacent if and only if $|i - i'| + |j - j'| = 1$. A complete graph (or a clique) with n vertices is denoted by K_n.

A *tree-decomposition* of a graph G is a pair $(T, \mathcal{W})$, where T is a tree and $\mathcal{W}$ is a family $\{W_t \mid t \in V(T)\}$ of vertex sets $W_t \subseteq V(G)$, such that the following two properties hold:

(W1) $\bigcup_{t \in V(T)} W_t = V(G)$, and every edge of G has both ends in some W_t.
(W2) If $t, t', t'' \in V(T)$ and t' lies on the path in T between t and t'', then $W_t \cap W_{t''} \subseteq W_{t'}$.

For two graphs $G_1 = (V_1, E_1)$ and $G_2 = (V_2, E_2)$, their *Cartesian product* $G_1 \square G_2 = (V^*, E^*)$ is the graph defined as follows:

- the vertex set V^* is $\{(v_1, v_2) \mid v_1 \in V_1,\ v_2 \in V_2\}$, and
- there exists an edge between $(v_1, v_2) \in V^*$ and $(v_1', v_2') \in V^*$ if and only if either $v_1 = v_1'$ and $v_2 v_2' \in E_2$, or $v_2 = v_2'$ and $v_1 v_1' \in E_1$.

For example, an $r \times r$ grid is a Cartesian product of two paths of length $r - 1$.

A clique minor of order r can be thought of r disjoint trees $T_1, \ldots, T_r$ such that there is an edge between T_i and T_j for any i, j with $i \neq j$. Sometimes, one tree T_i is called a *node* of the clique minor. A *topological clique minor* is a subgraph obtained from a clique by subdividing edges, and a *node* of the topological clique minor means a node of the original clique.

4   Grid-Like Minor

Let $\mathcal{P}_1$ and $\mathcal{P}_2$ be sets of disjoint connected subgraphs in a given graph G. We denote by $I(\mathcal{P}_1, \mathcal{P}_2)$ the *intersection graph* of $\mathcal{P}_1$ and $\mathcal{P}_2$ defined as follows: $I(\mathcal{P}_1, \mathcal{P}_2)$ is the bipartite graph with partite sets $\mathcal{P}_1$ and $\mathcal{P}_2$, which has one vertex for each element of $\mathcal{P}_1$ and $\mathcal{P}_2$, and an edge between two vertices exists if the corresponding subgraphs in $\mathcal{P}_1$ and in $\mathcal{P}_2$, respectively, intersect. Thus there are $|\mathcal{P}_1|$ vertices in one partite set of the bipartite graph, and $|\mathcal{P}_2|$ vertices in the other partite set.

We say that G contains a *grid-like minor* of order r if G has sets of disjoint paths $\mathcal{P}_1$ and $\mathcal{P}_2$ such that $I(\mathcal{P}_1, \mathcal{P}_2)$ contains the complete graph K_r as a minor. If the K_r-minor is, in fact, a topological minor, we call the structure a *topological grid-like minor* of order r. Note that a grid-like minor of order r consists of two sets $\mathcal{P}_1, \mathcal{P}_2$, each consists of disjoint paths, with $|\mathcal{P}_1| \geq r - 1$ and $|\mathcal{P}_2| \geq r - 1$ (for otherwise, we cannot construct a K_r-minor in the intersection graph $I(\mathcal{P}_1, \mathcal{P}_2)$). Let us observe that the intersection graph of the rows and columns of the $r \times r$ grid is the complete bipartite graph $K_{r,r}$ which clearly contains a K_{r+1}-minor (formed by contracting a matching of $r - 1$ edges). Hence the $r \times r$ grid minor contains a grid-like minor of order $r + 1$.

The motivation to introduce the grid-like minor is the following: The current best known result for the existence of the $r \times r$ grid minor is the following: if a given graph G has tree-width at least 20^{2r^5}, then G has a $r \times r$ grid minor [5,17,21]. An outstanding open problem in this area is whether polynomial tree-width forces a large grid minor. This question seems to be out of reach at the moment, but polynomial tree-width does force a large "grid-like minor". In [18], Reed and Wood proved the following useful result.

Theorem 2 (Reed and Wood [18]). *Every graph with tree-width at least $cr^4\sqrt{\log r}$ contains a grid-like minor of order r, for some constant c.*

The proof given in [18] can be converted into polynomial-time algorithm. This was done by Kreutzer and Tazari [15] as follows.

Theorem 3 (Kreutzer and Tazari [15, Theorem 4.2]). *There is a constant c such that if tree-width of a given graph G is cr^{12}, then G has either a K_r-minor or a topological gird-like minor of order r. Furthermore, given such a graph, there is a polynomial-time algorithm to construct the corresponding object.*

5   Main Proof

In this section, we give our main proof. To do so, we observe that if G contains a grid-like minor of order r, then $G \square K_2$ contains a topological clique minor of order r (Lemma 3.4 in [18]).

Let $\mathcal{P} = \{P_1, \ldots P_t\}$ be a set of paths in a graph H and r be an integer with $r \leq t$. We say that H contains a $\mathcal{P}$-*contracted topological K_r-minor* if there exists a subset $\mathcal{P}'$ of $\mathcal{P}$ such that the graph obtained from H by contracting all paths in $\mathcal{P}'$ contains a topological K_r-minor whose nodes correspond to paths in $\mathcal{P}'$.

For a graph $G = (V, E)$, let $G^* = G \square K_2$, i.e., G^* consists of G, its copy $G' = (V', E')$, and $|V|$ edges each connecting one vertex in V and its corresponding vertex in V'.

Lemma 1. *Suppose G has sets of disjoint paths $\mathcal{P}_1 = \{P_1, \ldots P_t\}$ and $\mathcal{P}_2 = \{P_1', \ldots P_t'\}$ such that $I(\mathcal{P}_1, \mathcal{P}_2)$ contains a topological K_{2r}-minor, where $t \geq 2r - 1$. Let G' be a copy of G as in the definition of G^*, and $\mathcal{P}_2'$ be a set of paths in G' corresponding to $\mathcal{P}_2$. Then, G^* contains a $\mathcal{P}_1$-contracted topological clique minor of order r, or a $\mathcal{P}_2'$-contracted topological clique minor of order r.*

Proof. One can see that $I(\mathcal{P}_1, \mathcal{P}_2)$ can be obtained from a subgraph of G^* by contracting all paths in $\mathcal{P}_1$ and $\mathcal{P}_2'$ (See Figure 1). Since the intersection graph $I(\mathcal{P}_1, \mathcal{P}_2)$ contains a topological K_{2r}-minor, thus G^* contains a topological K_{2r}-minor as well. Since $\mathcal{P}_1$ or $\mathcal{P}_2'$ contains at least r nodes of the K_{2r}-minor, we obtain a $\mathcal{P}_1$-contracted topological K_r-minor, or a $\mathcal{P}_2'$-contracted topological K_r-minor. □

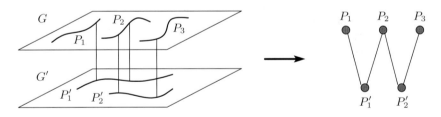

Fig. 1. Construction of $I(\mathcal{P}_1, \mathcal{P}_2)$

Our plan is to consider the clique minor as in Lemma 1 in G^*, and use it as a "crossbar". Thus let us give a theorem concerning a graph with a huge clique minor.

Theorem 4 (Robertson and Seymour [20, Theorem (5.4)])
Let $s_1, \ldots, s_k, t_1, \ldots, t_k$ be the terminals in a given G. If there is a clique minor of order at least $3k$ in G, and there is no separation (A, B) of order at most $2k - 1$ in G such that A contains all the terminals and $B - A$ contains at least one node of the clique minor, then there are disjoint paths P_i with two ends in s_i, t_i for $i = 1, \ldots, k$.

Furthermore, given the above clique minor, desired disjoint paths can be found in $O(km)$ time.

Even if there is a separation (A, B) of order at most $2k - 1$, we can reduce the problem to a smaller problem as follows.

Corollary 1. *Let $G, s_1, \ldots, s_k, t_1, \ldots, t_k$ be as in Theorem 4. If there is a clique minor of order at least $3k$ in G, then either there are desired k disjoint paths P_i with two ends in s_i, t_i for $i = 1, \ldots, k$, or there is a separation (A, B) of order at most $2k - 1$ such that A contains all the terminals and $B - A$ contains at least one node of the clique minor. Moreover, if the second happens, we can replace B by $A \cap B$ with $A \cap B$ becoming a clique, so that the resulting graph has the feasible solution if and only if G has. Furthermore, given the above clique minor, desired disjoint paths or such a separation (A, B) can be found in $O(km)$ time.*

Proof. If there is no separation (A, B) of order at most $2k - 1$ such that A contains all the terminals and $B - A$ contains at least one node of the clique minor, then we obtain the first result by Theorem 4.

If such a separation exists. We take a separation (A, B) with minimum $|V(A \cap B)|$. Furthermore, we assume that $V(B)$ is minimal among such separations. Note that such a separation can be found by a standard flow algorithm in $O(km)$ time. Then, by applying Theorem 4 to B (in place of G) with the terminals in $A \cap B$ (in place of $s_1, \ldots, s_k, t_1, \ldots, t_k$), we can link up $V(A \cap B)$ in any desired way in B. Thus, replacing B with a clique does not affect the solution of the disjoint paths problem. □

We are now ready to describe our algorithm for Theorem 1.

Proof of Theorem 1. Let T be the set of $2k$ terminals, where k satisfies that $k = O((\log n / \log \log n)^{1/12})$. By Theorem 3, in which $r = 12k$, we can find one of the following.

- a tree-decomposition of G of width $O(\log n / \log \log n)$,
- sets of disjoint paths $\mathcal{P}_1 = \{P_1, \ldots P_t\}$ and $\mathcal{P}_2 = \{P'_1, \ldots P'_t\}$ such that $t \geq 6k$ and $I(\mathcal{P}_1, \mathcal{P}_2)$ contains a topological K_{12k}-minor, or
- a K_{12k}-minor.

Case 1. Suppose that we have a tree-decomposition of G of width $O(\log n / \log \log n)$.

In this case, we can apply dynamic programming to solve problems on graphs of bounded tree-width, in the same way that we apply it to trees. In fact, the k-vertex-disjoint paths problem (and also half-integral k-disjoint paths packing) can be solved efficiently (see e.g. [1,2,17]).

Theorem 5. *For integers w and k, there exists a $(w + k)^{O(w+k)} n^{O(1)}$ time algorithm for the half-integral k-vertex-disjoint paths packing problem in graphs of tree-width w.*

Thus if the tree-width of G is $O(k)$ and $k = O(\log n/\log\log n)$, then the half-integral k-vertex-disjoint paths packing problem in G is solvable in polynomial time of n.

Case 2. Suppose that we obtain a topological K_{12k}-minor in $I(\mathcal{P}_1, \mathcal{P}_2)$.

We now take Cartesian product of G and K_2, and let $G^* = G\square K_2$. We assume $V(G^*) = V(G) \cup V(G')$, where G' is a copy of G. By symmetry, we may assume that G^* contains a $\mathcal{P}_1$-contracted topological clique minor of order $6k$ by Lemma 1, which we call M^*. Note that every node of M^* is contained in the original graph G. Let T' be the corresponding vertex set of the terminal set T in the copy G' of G. By Theorem 4, if there is no separation (A^*, B^*) of order at most $4k - 1$ in G' such that A^* contains all the terminals in $T \cup T'$ and $B^* - A^*$ contains at least one node of the clique minor, then there are $2k$ disjoint paths with one terminal in T and the other pair of the terminals in T', i.e, each path joins $s_i \in T$ and $t'_i \in T'$, or $s'_i \in T'$ and $t_i \in T$. This clearly gives rise to desired paths P_i in G with two ends in s_i, t_i in T for $i = 1, \ldots, k$.

On the other hand, suppose such a separation (A^*, B^*) exists in G^*. We take such a separation with minimum order. Let Q_1 be the node of the clique minor M^* that is contained in $B^* - A^*$. By the definition of $\mathcal{P}_1$-contracted topological clique minor, Q_1 is a path in $\mathcal{P}_1$.

Let (A_1, B_1) be the corresponding separation of (A^*, B^*) restricted to G, and let (A'_2, B'_2) be the corresponding separation of (A^*, B^*) restricted to G'. Then Q_1 is in $B_1 - A_1$.

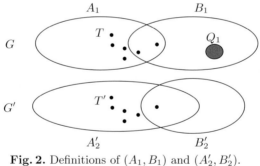

Fig. 2. Definitions of (A_1, B_1) and (A'_2, B'_2).

We need to consider two cases:

Case 2.1. $|V(A_1) \cap V(B_1)| \leq |V(A'_2) \cap V(B'_2)|$.

Let (A'_1, B'_1) be the separation of G' corresponding to (A_1, B_1). Then, $(A_1 \cup A'_1, B_1 \cup B'_1)$ is a separation of G^*, $A_1 \cup A'_1$ contains all terminals, $(B_1 \cup B'_1) - (A_1 \cup A'_1)$ contains Q_1, and $|V(A_1 \cup A'_1) \cap V(B_1 \cup B'_1)| \leq |V(A^*) \cap V(B^*)|$. Since we take (A^*, B^*) with the minimum order, we can apply Corollary 1 to replace $(B_1 \cup B'_1)$ by $(A_1 \cup A'_1) \cap (B_1 \cup B'_1)$ in such a way that $(A_1 \cup A'_1) \cap (B_1 \cup B'_1)$ becomes a clique, and the resulting graph has the same solution to find $2k$ disjoint paths, each connects one terminal in T and the other pair of the terminals in T', if and only if G^* has.

Thus after replacing B_1 by $A_1 \cap B_1$ in such a way that $A_1 \cap B_1$ is a clique, the resulting graph has a feasible solution if and only if G has. By Corollary 1, such a replacement can be done in polynomial time. We just recurse the algorithm to the resulting graph which has smaller vertices. This completes the description of Case 2.1.

Case 2.2. $|V(A_1) \cap V(B_1)| > |V(A_2') \cap V(B_2')|$. So $|V(A_2') \cap V(B_2')| \leq 2k - 1$.

Let (A_2, B_2) be the separation of G corresponding to the separation (A_2', B_2') of G'. Since M^* is a $\mathcal{P}_1$-contracted topological clique minor of order $6k$ and its node Q_1 is contained in $B_1 - A_1$, there are at least $6k - |V(A^*) \cap V(B^*)| \geq 2k + 1$ nodes of M^* in $B_1 - A_1$. Hence at least one node Q_2 of M^* in $B_1 - A_1$ does not intersect with $V(A_2) \cap V(B_2)$, which implies that Q_2 is contained in $B_2 - A_2$ by the construction of G^*. Then, $(A_2 \cup A_2', B_2 \cup B_2')$ is a separation of G^*, $A_2 \cup A_2'$ contains all terminals, $(B_2 \cup B_2') - (A_2 \cup A_2')$ contains Q_2, and $|V(A_2 \cup A_2') \cap V(B_2 \cup B_2')| \leq |V(A^*) \cap V(B^*)|$. Hence, in the same way as Case 2.1, we can replace B_2 by $A_2 \cap B_2$ in such a way that $A_2 \cap B_2$ is a clique. Then, the resulting graph has a feasible solution if and only if G has.

Case 3. Suppose that we obtain a K_{12k}-minor in G.

In this case, we can reduce the problem in the same way as Corollary 1. Then, we recurse the algorithm to the resulting graph which has smaller vertices.

With this observation, Theorem 1 follows.   □

References

1. Arnborg, S., Proskurowski, A.: Linear time algorithms for NP-hard problems restricted to partial k-trees. Discrete Appl. Math. 23, 11–24 (1989)
2. Bodlaender, H.L.: A linear-time algorithm for finding tree-decomposition of small treewidth. SIAM J. Comput. 25, 1305–1317 (1996)
3. Chekuri, C., Khanna, S., Shepherd, B.: The all-or-nothing multicommodity flow problem. In: Proc. 36th ACM Symposium on Theory of Computing (STOC), pp. 156–165 (2004)
4. Chekuri, C., Khanna, S., Shepherd, B.: Multicommodity flow, well-linked terminals, and routing problems. In: Proc. 37th ACM Symposium on Theory of Computing (STOC), pp. 183–192 (2005)
5. Diestel, R., Gorbunov, K.Y., Jensen, T.R., Thomassen, C.: Highly connected sets and the excluded grid theorem. J. Combin. Theory Ser. B 75, 61–73 (1999)
6. Frank, A.: Packing paths, cuts and circuits – a survey. In: Korte, B., Lovász, L., Promel, H.J., Schrijver, A. (eds.) Paths, Flows and VLSI-Layout, pp. 49–100. Springer, Berlin (1990)
7. Garg, N., Varzirani, V., Yannakakis, M.: Primal-dual approximation algorithms for integral flow and multicut in trees with applications to matching and set cover. In: Lingas, A., Carlsson, S., Karlsson, R. (eds.) ICALP 1993. LNCS, vol. 700, pp. 64–75. Springer, Heidelberg (1993)
8. Johnson, D.: The many faces of polynomial time. J. Algorithms 8, 285–303 (1987)
9. Karp, R.M.: On the computational complexity of combinatorial problems. Networks 5, 45–68 (1975)

10. Kawarabayashi, K., Reed, B.: A nearly linear time algorithm for the half disjoint paths packing. In: ACM-SIAM Symposium on Discrete Algorithms (SODA), pp. 446–454 (2008)
11. Kleinberg, J.: Single-source unsplittable flow. In: Proc. 37th Ann. IEEE Symp. Found. Comp. Sci. (FOCS), pp. 68–77 (1996)
12. Kleinberg, J.: Decision algorithms for unsplittable flows and the half-disjoint paths problem. In: Proc. 30th ACM Symposium on Theory of Computing (STOC), pp. 530–539 (1998)
13. Kolliopoulos, S., Stein, C.: Improved approximation algorithm for unsplittable flow problems. In: Proc. 38th Ann. IEEE Symp. Found. Comp. Sci. (FOCS), pp. 426–435 (1997)
14. Kramer, M.R., van Leeuwen, J.: The complexity of wirerouting and finding minimum area layouts for arbitrary VLSI circuits. Adv. Comput. Res. 2, 129–146 (1984)
15. Kreutzer, S., Tazari, S.: On brambles, grid-like minors and parameterized intractability of monadic second-order logic. In: ACM-SIAM Symposium on Discrete Algorithms (SODA), pp. 354–362 (2010)
16. Middendorf, M., Pfeiffer, F.: On the complexity of the disjoint paths problem. Combinatorica 13, 97–107 (1993)
17. Reed, B.: Tree width and tangles: a new connectivity measure and some applications. In: Surveys in Combinatorics. London Math. Soc. Lecture Note Ser., vol. 241, pp. 87–162. Cambridge Univ. Press, Cambridge (1997)
18. Reed, B., Wood, D.: Polynomial treewidth forces a large grid-like minor, available at arXiv:0809.0724v3[math.CO] (2008)
19. Robertson, N., Seymour, P.D.: Graph minors. V. Excluding a planar graph. J. Combin. Theory Ser. B 41, 92–114 (1986)
20. Robertson, N., Seymour, P.D.: Graph minors. XIII. The disjoint paths problem. J. Combin. Theory Ser. B 63, 65–110 (1995)
21. Robertson, N., Seymour, P.D., Thomas, R.: Quickly excluding a planar graph. J. Combin. Theory Ser. B 62, 323–348 (1994)
22. Schrijver, A.: Combinatorial Optimization: Polyhedra and Efficiency. Algorithm and Combinatorics, vol. 24. Springer, Heidelberg (2003)

Approximate Lasserre Integrality Gap for Unique Games

Subhash Khot[1], Preyas Popat[1], and Rishi Saket[2]

[1] Computer Science Department
New York University
[2] Computer Science Department
Carnegie Mellon University

Abstract. In this paper, we investigate whether a constant round Lasserre Semi-definite Programming (SDP) relaxation might give a good approximation to the UNIQUE GAMES problem. We show that the answer is negative if the relaxation is insensitive to a sufficiently small perturbation of the constraints. Specifically, we construct an instance of UNIQUE GAMES with k labels along with an approximate vector solution to t rounds of the Lasserre SDP relaxation. The SDP objective is at least $1 - \varepsilon$ whereas the integral optimum is at most γ, and all SDP constraints are satisfied up to an accuracy of $\delta > 0$. Here $\varepsilon, \gamma > 0$ and $t \in \mathbb{Z}^+$ are arbitrary constants and $k = k(\varepsilon, \gamma) \in \mathbb{Z}^+$. The accuracy parameter δ can be made sufficiently small independent of parameters $\varepsilon, \gamma, t, k$ (but the size of the instance grows as δ gets smaller).

1 Introduction

In recent years the UNIQUE GAMES problem and the Unique Games Conjecture (UGC) stating that the problem is hard to approximate [16] have received considerable attention thanks to their connection to inapproximability results and Semi-definite Programming based algorithms for a wide range of optimization problems. An inapproximability (a.k.a. hardness of approximation) result, under a widely believed hypothesis such as P $\neq$ NP, shows that there is no polynomial time algorithm achieving a good approximation. On the other hand, existence of an *integrality gap* instance is taken as evidence that an algorithm based on LP/SDP relaxation is unlikely to give a good approximation. An integrality gap instance is a specific instance (or a family of instances) where the optimum of the LP/SDP relaxation differs significantly from the integral (i.e. true) optimum. In the following, we review (a subset of) the known results for three problems: the MAXIMUM CUT, the SPARSEST CUT, and the UNIQUE GAMES.

For the MAXIMUM CUT problem, Goemans and Williamson [11] showed that a basic SDP relaxation combined with a random hyperplane rounding achieves an approximation guarantee of $\alpha_{GW}^{-1} \approx 1.13$ where α_{GW} is a certain trigonometric constant. Based on the Unique Games Conjecture [16] and the Majority is Stablest Theorem [23], a matching inapproximability result was shown in [17]. However, as the hardness is based on a conjecture, it remained an open question

M. Serna et al. (Eds.): APPROX and RANDOM 2010, LNCS 6302, pp. 298–311, 2010.

(in addition to resolving the Unique Games Conjecture itself) whether introducing additional SDP constraints such as the *triangle-inequality constraints* improves the approximation guarantee. Khot and Vishnoi [19] were able to construct an integrality gap instance to show that adding the triangle-inequality constraints does not help.

For the SPARSEST CUT problem, adding triangle inequality constraints to the basic SDP relaxation does indeed help. In a breakthrough work, Arora, Rao, and Vazirani [3] gave an upper bound of $O(\sqrt{\log n})$ on the integrality gap of the SDP relaxation equipped with triangle inequality constraints. This was subsequently extended by Arora, Lee, and Naor [2] to the *non-uniform* version of the problem. On the other hand, good lower bounds are known on the integrality gap as well: $(\log \log n)^{\Omega(1)}$ by Khot and Vishnoi [19], $\Omega(\log \log n)$ by Krauthgamer and Rabani [20] as well as by Devanur, Khot, Saket and Vishnoi [9]. In recent work, Cheeger, Kleiner, and Naor [7] have shown an integrality gap of $(\log n)^{\Omega(1)}$ based on earlier works of Lee and Naor [22] and Cheeger and Kleiner [6]. These lower bounds are for the non-uniform version except [9] that holds for the uniform version.

For the UNIQUE GAMES problem itself several approximation algorithms have been developed, see [16,28,13,4]. All these algorithms are based on LP or SDP relaxation and find a near satisfying assignment to a UNIQUE GAMES instance if there exists one. However their performance deteriorates as the number of labels and/or the size of the instance grows, and therefore they fall short of disproving the UGC. On the other hand, Khot and Vishnoi [19] give a strong integrality gap for a basic SDP relaxation of the UNIQUE GAMES problem (the algorithmic result of Charikar, Makarychev, and Makarychev [4] essentially matches this integrality gap).

Given the above mentioned works, it is worthwhile to investigate whether stronger LP/SDP relaxations help for problems like UNIQUE GAMES, MAXIMUM CUT or SPARSEST CUT. One can obtain stronger relaxations by adding (say polynomially many) natural constraints that an integral solution must satisfy.

Natural families of constraints considered in literature include the Lovász-Schrijver LP and SDP heirarchies, the Sherali-Adams LP heirarchy, and Lasserre SDP heirarchy. Instead of attempting a complete survey of known results, we refer the reader to the relevant papers [1,27,26,10,25,5,24,18]. and focus on the results pertaining to the Sherali-Adams and Lasserre heirarchies. The t-round Sherali-Adams LP hierarchy enforces the existence of local distributions over integral solutions. Specifically, a solution to such an LP gives a distribution over assignments to every set of at most t variables and the distributions over pairwise intersecting sets are consistent on the intersection. Strong lower bounds have been obtained by Charikar, Makarychev, and Makarychev [5] for up to n^δ rounds of Sherali-Adams relaxation for the MAXIMUM CUT problem. Their result shows $2 - \varepsilon$ gap for MAXIMUM CUT, and since the gap of the basic SDP relaxation is at most α_{GW}^{-1}, their result shows that even a *large* number of rounds of the Sherali-Adams heirarchy fail to capture the power of the basic SDP. In recent work, Raghavendra and Steurer [24] have obtained integrality gaps for a combination of a basic SDP and $(\log \log n)^{\Omega(1)}$ rounds of the Sherali-Adams LP: they obtain a strong gap for UNIQUE GAMES, $\alpha_{GW}^{-1} - \varepsilon$ for MAXIMUM CUT and $(\log \log n)^{\Omega(1)}$ for SPARSEST CUT. Simultaneously, Khot and Saket [18] also obtained similar but quantitatively weaker results.

One may also consider the t-round Lasserre SDP hierarchy [21] which introduces a SDP vector for every subset of variables of size at most t and each integral assignment to that subset. Appropriate consistency and orthogonality constraints are also added. As it turns out, a vector solution to the t-round Lasserre SDP also yields a solution to the t-round Sherali-Adams LP, and therefore the Lasserre SDP is at least as powerful as the Sherali-Adams LP.

Currently, we know very few integrality gap results for the Lassere hierarchy. Schoenebeck [25] obtained Lasserre integrality gap for MAX-3-LIN and Tulsiani extended it to MAX-k-CSP, and also obtained a gap of 1.36 for VERTEX COVER. However, we already know corresponding NP-hardness results, e.g. Håstad's [14] hardness result for MAX-3-LIN and Dinur and Safra's 1.36 hardness result for VERTEX COVER. Indeed Tulsiani's integrality gap for VERTEX COVER follows by *simulating* the Dinur-Safra reduction. It would be very interesting to have Lasserre gaps where we only know UGC-based hardness results, e.g. $2 - \varepsilon$ for VERTEX COVER, $\alpha_{GW}^{-1} - \varepsilon$ for MAXIMUM CUT, and a superconstant gap for SPARSEST CUT. Currently, such gaps are not known even for the third level of Lassere hierarchy, leaving open the tantalizing possibility that a constant round Lassere SDP relaxation might give better approximations to these problems, and consequently disprove the UGC.

In this paper, we make a partial progress towards this question. We show that if the constraints of a t-round Lasserre SDP are allowed to have a tiny but non-zero error $\delta > 0$, then a strong integrality gap exists for the UNIQUE GAMES problem. Using standard reductions from UNIQUE GAMES, similar integrality gaps can be obtained for MAX-CUT, VERTEX COVER etc. (we omit the details in this extended abstract). In fact the error can be made as small as desired independent of other parameters (except the size of the instance). All recent integrality gap constructions involving Sherali-Adams LP (see [5,24,18]) first construct such approximate solutions followed by an error-correction step. However correcting Lasserre vector solution seems challenging (due to a global constraint of positive definiteness) and we leave this as an open problem. On the other hand, our result does demonstrate that a Lasserre SDP relaxation will not give good approximation if it is insensitive to a tiny perturabation of the vector solution. To the best of our knowledge, all SDP based algorithms known are indeed insensitive to tiny perturbations (usually because rounding is very local). Next we introduce the Lasserre SDP hierarchy, informally state our results and give an overview of the construction.

Lasserre Hierarchy of SDP Relaxations

For a CSP such as UNIQUE GAMES on n vertices with a label set $[k]$, a t-round Lasserre SDP relaxation introduces vectors $\mathbf{x}_{S,\sigma}$ for every subset S of vertices of size at most t and every assignment $\sigma : S \mapsto [k]$ of labels to the vertices in

S. The intention is that in an integral solution, $\mathbf{x}_{S,\sigma} = 1$ if σ is restriction of the global assignment and $\mathbf{x}_{S,\sigma} = 0$ otherwise. Therefore, for a fixed set S, one adds the SDP constraint that the vectors $\{\mathbf{x}_{S,\sigma}\}_\sigma$ are orthogonal and the sum of their squared Euclidean norms is 1. One may interpret the squared Euclidean norms of these vectors as a probability distribution over assignments to S (in an integral solution the distribution is concentrated on a single assignment).

Natural consistency constraints satisfied by an integral solution are added as well. Specifically, for two sets $T \subseteq S$, each of size at most t, and every assignment τ to T, the following natural constaint is added:

$$\sum_{\sigma:S\mapsto[k],\sigma|_T=\tau} \mathbf{x}_{S,\sigma} = \mathbf{x}_{T,\tau}, \qquad (1)$$

where $\sigma|_T$ denotes the restriction of σ to subset T. Note that in an integral solution, both sides of the above equation are 1 if τ is restriction of the global assignment to T and zero otherwise. The objective value of the relaxation can be written in terms of pairwise inner products of vectors on singleton sets. The t-round Lasserre SDP relaxation entails adding $O(n^t)$ constraints in the SDP relaxation.

We will be interested in *approximate* solutions to the Lasserre hierarchy. Towards this end, we call a vector solution δ-approximate if Equation (1) is satisfied with error δ, i.e.

$$\left\| \sum_{\sigma:\sigma|_T=\tau} \mathbf{x}_{S,\sigma} - \mathbf{x}_{T,\tau} \right\| \le \delta. \qquad (2)$$

We now state informally the main result of this paper.

Theorem 1. *(Informal) Let $\varepsilon > 0$ and $k, t \in \mathbb{Z}^+$ be arbitrary constants. Then for every constant $\delta > 0$, there is an instance $\mathcal{U}$ of* Unique Games *with label set $[k]$ that satisfies:*

1. *There exist vectors $\mathbf{x}_{S,\sigma}$ for every set S of vertices of $\mathcal{U}$ of size at most t, and every assignment of labels σ to the vertices in S such that it is a δ-approximate solution to the SDP relaxation with t-round Lasserre hierarchy.*
2. *The SDP objective value of the above approximate vector solution is at least $1 - \varepsilon$.*
3. *Any labeling to the vertices of $\mathcal{U}$ satisfies at most $k^{-\varepsilon/2}$ fraction of edges.*

Overview of Our Construction

Our construction relies in large part on the work of Khot and Vishnoi [19] who gave SDP integrality gap examples for Unique Games and cut-problems including Maximum Cut. We also borrow ideas from [18] and [24] who build upon the work of [19] to obtain stronger integrality gap results as mentioned earlier.

Our strategy is to first construct approximate Lasserre vectors for the Unique Games instance $\mathcal{U}$ presented in [19]. This construction is not good enough by itself as the number of labels $[N]$ is too large relative to the quality of the accuracy parameter. We therefore apply the reduction of [17] to the instance $\mathcal{U}$ to obtain a new instance $\tilde{\mathcal{U}}$ of UNIQUE GAMES with a much smaller label set $[k]$. This reduction preserves the low integral optimum, transforms the vectors corresponding to the instance $\mathcal{U}$ into corresponding vectors for the instance $\tilde{\mathcal{U}}$, and preserves the high SDP objective. These new vectors constitute the final δ-approximate Lasserre solution to $\tilde{\mathcal{U}}$. Below we describe the construction of Lasserre vectors for the instance $\mathcal{U}$. In the actual construction we present, we do no explicitly construct these vectors, but rather directly construct the instance $\tilde{\mathcal{U}}$ along with its approximate Lasserre solution. However, the description of the implicit intermediate step does illustrate the main ideas involved.

Lasserre Vectors for [19] UNIQUE GAMES Instance. We start with the UNIQUE GAMES instance $\mathcal{U}$ along with a basic SDP solution constructed in [19]. Let $G(V, E)$ be its constraint graph and $[N]$ be the label set. The SDP solution consists of (up to a normalization) an orthonormal tuple $\{\mathbf{T}_{u,j}\}_{j \in [N]}$ for every vertex $u \in V$. A useful property of this solution is that the sum of vectors in every tuple is the same, i.e. for some fixed unit vector $\mathbf{T}$,

$$\mathbf{T} = \frac{1}{\sqrt{N}} \sum_{j \in [N]} \mathbf{T}_{u,j} \quad \forall u \in V. \tag{3}$$

As observed in [18], one can define a single vector $\mathbf{T}_u := \frac{1}{\sqrt{N}} \sum_{j \in [N]} \mathbf{T}_{u,j}^{\otimes 4}$ for each tuple $\{\mathbf{T}_{u,j}\}$ such that the distance $\|\mathbf{T}_u - \mathbf{T}_v\|$ captures the closeness between the pairs of tuples $\{\mathbf{T}_{u,j}\}$ and $\{\mathbf{T}_{v,j}\}$. Roughly speaking, the edge (i.e. constraint) set E corresponds to all pairs (u, v) such that $\|\mathbf{T}_u - \mathbf{T}_v\| \leq \gamma$ for a sufficiently small $\gamma > 0$. For any such edge, it necessarily holds that $\forall j \in [N]$, $\|\mathbf{T}_{u,j} - T_{v,\pi(j)}\| \leq O(\gamma)$ for some bijection $\pi = \pi(u, v) : [N] \mapsto [N]$. This is precisely the bijection defining the UNIQUE GAMES constraint on edge (u, v) and also ensures that the SDP objective is high, i.e. $1 - O(\gamma^2)$.

Another key observation is that in the graph $G(V, E)$, any *local* neighborhood can be given a consistent labeling; in fact, once an arbitrary label for a vertex is fixed, it uniquely determines labels to all other vertices in a local neighborhood. Specifically, fix a small positive constant $p \leq 0.1$. A set $C \subseteq V$ is called *p-local* if $\|\mathbf{T}_u - \mathbf{T}_v\| \leq p \ \forall \ u, v \in C$. As observed in [18], for any p-local set C, there is a set $L(C)$ of N labelings, such that each labeling $\tau \in L(C)$ satisfies all the induced edges inside C. The jth labeling is obtained by fixing the label of one vertex in C to be $j \in [N]$ and then uniquely fixing labels to all other vertices in C.

This gives a natural way to define Lasserre vectors for all subsets $S \subseteq C$. Fix an arbitrary vertex $w \in C$. Consider any subset $S \subseteq C$, and a labeling σ to the vertices in S. We wish to construct a vector $\mathbf{y}_{S,\sigma}$. If σ is not consistent with any of the N labelings $\tau \in L(C)$ then set $\mathbf{y}_{S,\sigma} = 0$. Otherwise, let $\mathbf{y}_{S,\sigma} = \frac{1}{\sqrt{N}} \mathbf{T}_{w,j}$ where the labeling σ is consistent with a labeling $\tau \in L(C)$ which assigns j to w. It can be seen that this is a valid Lasserre SDP solution for all subsets of C. All edges that are inside C contribute well (i.e. $1 - O(\gamma^2)$) towards the SDP objective.

We now try to extend the above strategy to the whole set V. Even though the following naive approach does not work, it helps illustrate the main idea behind the construction. We partition V into local sets and construct Lasserre vectors that are a tensor product of vectors constructed for each local set. Towards this end, we think of the set of vectors $\{\mathbf{T}_u\}_{u \in V}$ as embedded on the unit sphere $\mathbb{S}^{|V|-1}$. Partition the unit sphere into clusters of diameter at most p. This naturally partitions the set of vertices V into disjoint p-local subsets $C_1, \ldots, C_m$. As before, fix w_i to be any arbitrary vertex in C_i for $i = 1, \ldots, m$. Now consider a subset $S \subseteq V$, and a labeling σ to the vertices in S, for which we wish to construct a vector $\mathbf{x}_{S,\sigma}$. Suppose that there is a subset C_i such that $\sigma|_{S \cap C_i}$ is not consistent with any labeling in $L(C_i)$; in this case set $\mathbf{x}_{S,\sigma} = 0$. Otherwise, construct vector $\mathbf{y}_{S,\sigma}^i$ as follows: if $|S \cap C_i| = \emptyset$, then let $\mathbf{y}_{S,\sigma}^i = \mathbf{T}$; else set $\mathbf{y}_{S,\sigma}^i = \frac{1}{\sqrt{N}} \mathbf{T}_{w_i,j}$, where $\sigma|_{S \cap C_i}$ is consistent with a labeling in $L(C_i)$ that assigns label j to w_i. Finally, let $\mathbf{x}_{S,\sigma} := \bigotimes_{i=1}^m \mathbf{y}_{S,\sigma}^i$. It can be seen that this construction is a valid SDP Lasserre solution. The tensor product is a vector analogue of assigning labeling to different clusters independently.

However, the above construction does not work because the unit sphere has dimension $|V|-1$ and partitioning such a high-dimensional sphere into local clusters necessarily means that almost all edges of $G(V, E)$ will have two endpoints in different clusters, and therefore the two endpoints get labels independently. This results in a very low SDP objective. A natural approach is to use dimensionality reduction that w.h.p. preserves the geometry of any set points that is not too large.

We therefore first randomly project the vectors $\{\mathbf{T}_u\}_{u \in V}$ onto $\mathbb{S}^{d-1}$ for an appropriate constant d. The Johnson-Lindenstrauss lemma implies that for a set $S \subseteq V$ of at most t vertices, w.h.p. the mapping approximately preserves all pairwise distances between the vectors $\{\mathbf{T}_u\}_{u \in S}$. This is followed, as before, by a (randomized) partition of $\mathbb{S}^{d-1}$ into low-diameter clusters that induces a partition of V into subsets $C_1, \ldots, C_m$. The dimension d is low enough to ensure that most of the edges in E fall inside some cluster. However, since the projection fails to preserve distances with some non-zero probability, the subsets C_i $(1 \leq i \leq m)$ are *not* guaranteed to be p-local. Nevertheless, for any set S of at most t vertices, if the projection preserves all distances between vectors $\{\mathbf{T}_u\}_{u \in S}$, then each of the sets $S \cap C_i$ for $i = 1, \ldots, m$ is a p-local set. For a fixed projection and a partition, a vector $\mathbf{x}_{S,\sigma}$ for the set S and its labeling σ can then be constructed as described earlier, except that there is no fixed representative vertex w_i for each C_i. Instead, an arbitrary vertex is chosen from the set $S \cap C_i$ to serve as the representative vertex w_i, and the set of labelings $L(S \cap C_i)$ is used. Since the projection and the partitioning are randomized, we implement the construction for each choice of random string and let the final vectors to be a (weighted) direct sum of the vectors constructed for each random string.

The above approach yields Lasserre vectors which have a good SDP objective value but only approximately satisfy the Lasserre constraints. There are two sources of error. One is that the random projection preserves distances within a set $S, |S| \leq t$, w.h.p. but not with probability 1. Secondly, since an arbitrary vertex from $S \cap C_i$ is chosen as a representative, for $T \subseteq S$, the representative for $S \cap C_i$ need not coincide with the representative for $T \cap C_i$. Still, since $S \cap C_i$ and $T \cap C_i$ are local sets (provided that the random projection has succeeded in preserving distances in S), their representative vectors are close enough.

Obtaining a δ-Approximate Lasserre Solution. As stated earlier, once we have the SDP vectors to instance of [19], we apply the reduction of [17] and obtain a new instance of UNIQUE GAMES with a constant label set $[k]$. We also obtain vectors which constitute the δ-approximate Lasserre solution to the new instance of UNIQUE GAMES. We ensure that the objective value of the vectors remains high.

Organization of the Paper

In Section 2 we formally define the UNIQUE GAMES problem and a formulation of the Lasserre hierarchy. In Section 2.3, we formally state our main theorem with quantitative parameters. In Section 3 we describe the basic UNIQUE GAMES instance from [19] along with the reduction from [17] to obtain a new UNIQUE GAMES instance with a constant label set $[k]$. Finally, in Section 4 we construct Lasserre vectors for the new UNIQUE GAMES instance.

2   Preliminaries

2.1   Unique Games

An instance of UNIQUE GAMES $\mathcal{U}(G(V, E), [k], \{\pi_e\}_{e \in E})$ is a constraint satisfaction problem. For every edge $e = (u, v)$ in the graph, there is a bijection $\pi_e^{uv} : [k] \mapsto [k]$ on the label set $[k]$, and a weight function $wt(e)$. For notational convenience we define $\pi_e^{vu} := (\pi_e^{uv})^{-1}$. A labeling $\sigma : V \mapsto [k]$ satisfies an edge $e = (u, v) \in E$ iff $\pi_e^{uv}(\sigma(u)) = \sigma(v)$. The goal is to find a labeling that satisfies the maximum fraction of edges.

Let $\mathcal{U}$ be an instance of UNIQUE GAMES. Figure 1 gives a natural SDP relaxation SDP-UG. The relaxation is over the vector variables $\mathbf{x}_{u,i}$ for every vertex u of the graph G and label $i \in [k]$.

2.2   Lasserre Relaxation

One can write a natural integer quadratic program for solving UNIQUE GAMES, where the set of variables is $\mathbf{x}_{S,\sigma}$ for every $S \subseteq V$ and every assignment $\sigma : S \mapsto [k]$ to vertices in S. The solution to this quadratic program would ensure $\mathbf{x}_{S,\sigma} = 1$ if the global labeling of V induces the assignment σ on S and $\mathbf{x}_{S,\sigma} = 0$ otherwise.

The Lasserre semi-definite relaxation of UNIQUE GAMES L'-UG(t) (see the full version of the paper) is obtained by relaxing the variables of this quadratic program to vectors instead of integers and replacing the multiplication of two

$$\max \sum_{e=(u,v)\in E} \sum_{i\in[k]} \left\langle \mathbf{x}_{u,i}, \mathbf{x}_{v,\pi_e^{uv}(i)} \right\rangle \; wt(e)$$

Subject to,

$$\forall u \in V \qquad \sum_{i\in[k]} \|\mathbf{x}_{u,i}\|^2 = 1 \quad \text{(I)}$$

$$\forall u \in V, \; i,j \in [k], \; i \neq j \; \left\langle \mathbf{x}_{u,i}, \mathbf{x}_{u,j} \right\rangle = 0 \quad \text{(II)}$$

$$\forall u,v \in V, \; i,j \in [k] \qquad \left\langle \mathbf{x}_{u,i}, \mathbf{x}_{v,j} \right\rangle \geq 0 \quad \text{(III)}$$

Fig. 1. Relaxation SDP-UG for UNIQUE GAMES

$$\max \sum_{e=(u,v)\in E} \sum_{i\in[k]} \left\langle \mathbf{x}_{u,i}, \mathbf{x}_{v,\pi_e^{uv}(i)} \right\rangle \; wt(e)$$

Subject to,

$$\|\mathbf{x}_\phi\|^2 = 1 \qquad\qquad\qquad \text{(IV)}$$

$$\forall \, S, |S| \leq t, \sigma \neq \sigma' \qquad \left\langle \mathbf{x}_{S,\sigma}, \mathbf{x}_{S,\sigma'} \right\rangle = 0 \qquad\qquad \text{(V)}$$

$$\forall \, T \subseteq S, \tau \in [k]^T \qquad \sum_{\sigma:\sigma|_T=\tau} \mathbf{x}_{S,\sigma} = \mathbf{x}_{T,\tau} \qquad\qquad \text{(VI)}$$

Fig. 2. Relaxation L-UG(t) for UNIQUE GAMES

numbers by dot products of the corresponding vectors. In the t-round Lasserre relaxation, we consider sets of size up to t. In this paper, we work with another relaxation L-UG(t) in Figure 2 which is essentially equivalent to L'-UG(t), but rephrases the constraints in terms of vector sums instead of dot-products. The two relaxations have the exact same objective function. In the full version of the paper, we show that the two relaxations are essentially equivalent.

We say $\sigma|_T$ to mean assignment σ restricted to set T. We say $(S, \sigma) \simeq (S', \sigma')$ to mean that the assignments σ and σ' are consistent i.e. $\sigma|_{S\cap S'} = \sigma'|_{S\cap S'}$. Otherwise, we say $(S, \sigma) \not\simeq (S', \sigma')$. Let $\mathbf{x}_{u,i} := \mathbf{x}_{S,\sigma}$ for $S = \{u\}$ and $\sigma(u) = i$.

Thus, we want to construct $k^{|S|}$ orthogonal vectors for each set S of size up to t, such that the vectors for different sets are consistent with each other in the sense of Equation (VI).

2.3   Main Theorem

Theorem 2. *Fix an arbitrarily small constant $\varepsilon > 0$ and integer $k \in \mathbb{Z}^+$. Then for all sufficiently large N (that is a power of 2), there is an instance $\mathcal{U}$ of* UNIQUE GAMES *on $\frac{2^N}{N} \cdot k^{N-1}$ vertices with label set $[k]$ such that,*

1. *There exist vectors $\mathbf{x}_{S,\sigma}$ for every set S of vertices of $\mathcal{U}$ of size at most t, and every assignment of labels $\sigma : S \mapsto [k]$ such that it is a $O(t \cdot \eta^{1/16})$-approximate solution for $\eta := (\log N)^{-0.99}$ to the SDP relaxation with t-round Lasserre hierarchy of constraints.*
2. *The SDP objective value of the above approximate vector solution is at least $1 - O(\varepsilon)$.*
3. *Any labeling to the vertices of $\mathcal{U}$ satisfies at most $k^{-\varepsilon/2}$ fraction of edges.*

Proof. The construction is presented in Section 4 and properties (1), (2) and (3) are proved in the full version of the paper.

3   The Instance

3.1   Basic Instance

The starting point of our reduction is a UNIQUE GAMES integrality gap instance $\mathcal{U}_\eta$ for SDP-UG constructed in [19]. Our presentation of the UNIQUE GAMES instance $\mathcal{U}_\eta$ follows that in [18].

For $\eta > 0$ and $N = 2^m$ for some $m \in \mathbb{Z}^+$, Khot and Vishnoi [19] construct the UNIQUE GAMES instance $\mathcal{U}_\eta(G'(V', E'), [N], \{\pi_e\}_{e \in E})$ where the number of vertices $|V'| = 2^N/N$. The instance has no good labeling, i.e. has low optimum.

Lemma 1. *Any labeling to the vertices of the* UNIQUE GAMES *instance $\mathcal{U}_\eta(G'(V', E'), [N], \{\pi_e\}_{e \in E})$ satisfies at most $\frac{1}{N^\eta}$ fraction of the edges.*

In the construction of [19] the elements of $[N]$ are identified with the additive group $(\mathbb{F}[2]^m, \oplus)$. The authors construct a vector solution that consists of unit vectors $\mathbf{T}_{u,i}$ for every vertex $u \in V'$ and label $i \in [N]$. These vectors (up to a normalization) form the solution to the UNIQUE GAMES SDP relxation SDP-UG. We highlight the important properties of the SDP solution below:

Properties of the Unique Games SDP Solution

– **(Orthonormal basis)** $\forall\, u \in V', \;\; \forall\, i \neq j \in [N]$,

$$\|\mathbf{T}_{u,i}\| = 1, \quad \langle \mathbf{T}_{u,i}, \mathbf{T}_{u,j} \rangle = 0. \tag{4}$$

– **(Non-negativity)** $\forall\, u, v \in V', \;\; \forall\, i, j \in [N]$,

$$\langle \mathbf{T}_{u,i}, \mathbf{T}_{v,j} \rangle \geq 0. \tag{5}$$

– **(Symmetry)** $\forall\, u, v \in V'$, $\forall\, i, j, s \in [N]$,

$$\langle \mathbf{T}_{u,i}, \mathbf{T}_{v,j} \rangle = \langle \mathbf{T}_{u,s \oplus i}, \mathbf{T}_{v,s \oplus j} \rangle \tag{6}$$

where '$\oplus$' is the group operation on $[N]$ as described above.
– **(High SDP Value)** For every edge $e = (u, v) \in E'$,

$$\forall\, i \in [N], \quad \langle \mathbf{T}_{u,i}, \mathbf{T}_{v,\pi_e^{uv}(i)} \rangle \geq 1 - 4\eta. \tag{7}$$

In fact, there is $s_e^{uv} \in [N]$ such that $\forall\, i \in [N]$, $\pi_e^{uv}(i) = s_e^{uv} \oplus i$.
– **(Sum to a Constant Vector)** For every vertex $u \in V'$,

$$\frac{1}{\sqrt{N}} \sum_{i=1}^{N} \mathbf{T}_{u,i} = \mathbf{T} \tag{8}$$

where $\mathbf{T}$ is a fixed unit vector.
– **(Local Consistency)** A set $W \subseteq V'$ of vertices is p-local if $\|\mathbf{T}_u - \mathbf{T}_v\| \leq p \leq 0.1$ for all $u, v \in W$.

Lemma 2 ([18]). *Suppose a set $W \subseteq V'$ is p-local. Then there is set $L(W)$ of N locally consistent assignments to vertices in W such that if $\mu : W \mapsto [N] \in L(W)$ then*

$$\forall u, v \in W : \langle \mathbf{T}_{u,\mu(u)}, \mathbf{T}_{v,\mu(v)} \rangle \geq 1 - O(p^2). \tag{9}$$

The assignments in $L(W)$ are disjoint i.e. if $\mu \neq \mu' \in L(W)$ then $\forall\, u \in W, \mu(u) \neq \mu'(u)$.

The authors in [18] define for every vertex $u \in V'$ a unit vector $\mathbf{T}_u$

$$\mathbf{T}_u := \frac{1}{\sqrt{N}} \sum_{i \in [N]} \mathbf{T}_{u,i}^{\otimes 4}. \tag{10}$$

and prove that that the Euclidean distances between the vectors $\{\mathbf{T}_u\}_{u \in V'}$ are a measure of the 'closeness' between the orthonormal tuples $\{\mathbf{T}_{u,i} \mid i \in [N]\}_{u \in V'}$.

Lemma 3 ([18]). *For every $u, v \in V'$,*

$$\min_{i,j \in [N]} \|\mathbf{T}_{u,i} - \mathbf{T}_{v,j}\| \;\leq\; \|\mathbf{T}_u - \mathbf{T}_v\| \;\leq\; 2 \cdot \min_{i,j \in [N]} \|\mathbf{T}_{u,i} - \mathbf{T}_{v,j}\| \tag{11}$$

3.2   Reduction to Constant Label Size

In this section we transform the instance $\mathcal{U}_\eta(G'(V', E'), [N], \{\pi_e\}_{e \in E'})$ described in the previous section to another UNIQUE GAMES instance $\mathcal{U}_\varepsilon(G(V, E), [k], \{\pi_e\}_{e \in E})$ using a reduction presented in [17]. Here $[k]$ is to be thought of as the set $\{0, 1, \ldots, k - 1\}$ with the group operation of addition modulo k.

We start with the UNIQUE GAMES instance $\mathcal{U}_\eta(G'(V', E'), [N], \{\pi_e\}_{e \in E'})$ and replace each vertex $v \in V'$ by a block of k^{N-1} vertices $(v, \mathbf{s})$ where $\mathbf{s} \in [k]^N$ and $\mathbf{s}_1 = 0$.

For every pair of edges $e = (v, w), e' = (v, w') \in E'$, there are (all possible) weighted edges between the blocks $(w, \cdot)$ and $(w', \cdot)$ in the instance $\mathcal{U}_\varepsilon(G(V, E), [k], \{\pi_e\}_{e \in E})$. The edge between $a := (w, \mathbf{s})$ and $b := (w', \mathbf{s}')$ is constructed as follows:-

1. Pick $\mathbf{p}$ uniformly at random from $[k]^N$ and $\mathbf{p}' \in [k]^N$ such that each co-ordinate $\mathbf{p}'_i$ is chosen to be $\mathbf{p}_i$ with probability $1 - \varepsilon$ and is chosen uniformly at random from $[k]$ with probability ε for all $i \in [N]$.
2. Define $\mathbf{q}, \mathbf{q}' \in [k]^N$ as $\mathbf{q} := \mathbf{p} \circ \pi_e^{wv}$, $\mathbf{q}' := \mathbf{p}' \circ \pi_{e'}^{w'v}$ where $\mathbf{p} \circ \pi := (\mathbf{p}_{\pi(1)}, \ldots, \mathbf{p}_{\pi(N)})$.
3. Define $\mathbf{r}, \mathbf{r}' \in [k]^N$ as $\mathbf{r}_i := \mathbf{q}_i - \mathbf{q}_1$ and $\mathbf{r}'_i := \mathbf{q}'_i - \mathbf{q}'_1$ for all i from 1 through N.
4. Add an edge e^* between $a = (w, \mathbf{s})$ and $b = (w', \mathbf{s}')$ such that $\pi_{e^*}^{ab}(i) := (i + \mathbf{q}'_1 - \mathbf{q}_1)$ for all $i \in [k]$ and $wt(e^*) := \Pr[\mathbf{s} = \mathbf{r}, \mathbf{s}' = \mathbf{r}']$.

The third step in the construction incorporates a PCP trick called folding. To prove that the instance constructed has low optimum, we need the property that any labelling to vertices in $\mathcal{U}_\varepsilon$ is balanced on every block of vertices arising out of some vertex in $\mathcal{U}_\eta$ i.e. it assigns each label in every block equally often.

We achieve this by reducing the number of vertices in each block by a factor of $\frac{1}{k}$, and then extend any labelling on the reduced vertex set to a balanced labelling on the original vertex set. In our case, we only consider strings $\mathbf{s}$ with $\mathbf{s}_1 = 0$ and as a mental exercise we extend any labeling σ to all strings as

$$\sigma(\mathbf{s}'_1, \mathbf{s}'_2, \ldots, \mathbf{s}'_N) := \sigma(0, \mathbf{s}'_2 - \mathbf{s}'_1, \ldots, \mathbf{s}'_N - \mathbf{s}'_1) + \mathbf{s}'_1$$

The following is a reformulation of Theorem 12 and Corollary 13 of [17].

Lemma 4. *Any labeling to the vertices of the* UNIQUE GAMES *instance* $\mathcal{U}_\varepsilon(G(V, E), [k], \{\pi_e\}_{e \in E})$ *satisfies at most* $k^{-\varepsilon/2}$ *fraction of the edges provided the optimum of the instance* $\mathcal{U}_\eta$ *(which is at most* $N^{-\eta}$*)) is sufficiently small as a function of* ε *and* k*.*

4   Approximate Vector Construction

In this section we construct Lasserre vectors for the UNIQUE GAMES instance $\mathcal{U}_\varepsilon(G(V, E), [k], \{\pi_e\}_{e \in E})$ described in the previous section. Our construction will be randomized, i.e. we first create vectors $\mathbf{y}^r_{S,\sigma}$ for every choice of random bits r and then set

$$\mathbf{x}_{S,\sigma} := \bigoplus_r \sqrt{\Pr[r]} \, \mathbf{y}^r_{S,\sigma} \tag{12}$$

where $\Pr[r]$ is the probability of choosing the random bit-sequence r (vectors for different choices of randomness live in independent, mutually orthogonal spaces).

4.1 Construction

We will use the following results.

Lemma 5. *(see full version for proof) There is a randomized mapping Γ : $\mathbb{S}^{n-1} \mapsto \mathbb{S}^{d-1}$ with $d = 8 \ln(2t^2/\eta)/p^2$, such that for any set $X \subseteq \mathbb{S}^{n-1}$, $|X| \leq t$, with probability $1 - \eta$, we have*

$$\forall x, y \in X, \quad \frac{1}{32} \|\Gamma(x) - \Gamma(y)\| \leq \|x - y\| \leq 4p + 2 \|\Gamma(x) - \Gamma(y)\|.$$

If this conclusion holds, we say that the randomized mapping (projection) succeeded.

Theorem 6 ([12]). *Let $\mathbb{S}^{d-1} = \{x \in \mathbb{R}^d : \|x\| = 1\}$ denote the $(d-1)$ dimensional unit sphere. For every choice of diameter $p > 0$ there is a randomized partition $\tilde{P}$ of $\mathbb{S}^{d-1}$ into disjoint clusters such that,*

1. *For every cluster $C \in \tilde{P}$, $C \subseteq \mathbb{S}^{d-1}$, $\mathtt{diam}(C) \leq p$.*
2. *For any pair of points $u, v \in \mathbb{S}^{d-1}$ such that $\|u - v\| = \beta \leq \frac{p}{4}$,*

$$\Pr_{\tilde{P}} \left[u \text{ and } v \text{ fall into different clusters} \right] \leq \frac{100\beta d}{p}.$$

We intend to construct vectors $\mathbf{x}_{S,\sigma}$ for every set $S \subseteq V$, $|S| \leq t$, and every assignment $\sigma : S \mapsto [k]$. Set $p = \eta^{1/16}$ and $d = 8 \ln(2t^2/\eta)/p^2$.

1. **Projection:**
   Use Lemma 5 to obtain a mapping $\mathbf{T}_u \mapsto \mathbf{T}'_u \in \mathbb{S}^{d-1} \ \forall \ u \in V'$.
2. **Partition:**
   Use Theorem 6 to randomly partition $\mathbb{S}^{d-1}$ with diameter p. Let $C_1, C_2, \ldots$, C_m denote this partition of $\mathbb{S}^{d-1}$ as well as the induced partition of V' (by a slight abuse of notation).
3. **Constructing vectors for a fixed set $S \subseteq V$, $|S| \leq t$:**
   Recall that every vertex of S is of the form $a = (v, \mathbf{s})$ for some $v \in V'$ and $\mathbf{s} \in [k]^N$, $\mathbf{s}_1 = 0$. Let $S = \cup_{l=1}^m S_\ell$ be a partition of S such that

$$S_\ell := \{a = (v, \mathbf{s}) \in S \mid v \in C_\ell\}.$$

Also define for the sake of notational ease,

$$S'_\ell := \{v \mid \exists \ a = (v, \mathbf{s}) \in S_\ell\} \subseteq C_\ell \quad \text{and} \quad S' := \cup_{l=1}^m S'_\ell.$$

Since $|S| \leq t$, at most t of the sets S_ℓ (and hence S'_ℓ) are non-empty. Let r be the randomness used in Steps (1) and (2). If the Projection succeeds for the entire set S' (see Lemma 5), go to Step 4.
   Otherwise set $\mathbf{y}^r_{S,\sigma} := 0$ for all $\sigma : S \mapsto [k]$ and go to Step 5.

4. Since $S = \cup_{\ell=1}^{m} S_\ell$ is a partition, an assignment $\sigma : S \mapsto [k]$ can be split into assignments $\sigma_\ell : S_\ell \mapsto [k]$ for $\ell = 1, \ldots, m$. The construction below is the vector analogue of choosing an assignment σ_ℓ for set S_ℓ from a certain distribution, but independently for all $\ell = 1, \ldots, m$.

For each ℓ such that $S_\ell = \emptyset$, let $\mathbf{y}_{S_\ell, \sigma_\ell}^{r,l} := \mathbf{T}$.

For each ℓ such that $S_\ell \neq \emptyset$, observe that the set S'_ℓ is $O(p)$-local since the projection succeeded for S' and since the diameter of C_ℓ is at most p. Let $L(S'_\ell)$ denote the set of N locally consistent assignments to S'_ℓ as in Lemma 2, Equation (9).

We partition the set $L(S'_\ell)$ of locally consistent assignments into different classes depending on how they behave w.r.t. assignments $\sigma_\ell : S_\ell \mapsto [k]$. Towards this end, let

$$L_{S_\ell, \sigma_\ell}^{r, \ell} := \left\{ \mu \mid \mu \in L(S'_\ell) \text{ such that } \forall\, a = (v, \mathbf{s}) \in S_\ell, \ \mathbf{s}_{\mu(v)} = \sigma_\ell(a) \right\}.$$

Now arbitrarily pick a *representative* element $u \in S'_\ell$ and set

$$\mathbf{y}_{S_\ell, \sigma_\ell}^{r, \ell} := \frac{1}{\sqrt{N}} \sum_{\mu \in L_{S_\ell, \sigma_\ell}^{r, \ell}} \mathbf{T}_{u, \mu(u)}.$$

Finally define,

$$\mathbf{y}_{S, \sigma}^{r} := \bigotimes_{l=1}^{m} \mathbf{y}_{S_\ell, \sigma_\ell}^{r, \ell} \tag{13}$$

5. Construct vectors $\mathbf{x}_{S, \sigma} := \bigoplus_r \sqrt{\Pr[r]} \, \mathbf{y}_{S, \sigma}^{r}$ as in Equation (12).

References

1. Arora, S., Bollobás, B., Lovász, L., Tourlakis, I.: Proving integrality gaps without knowing the linear program. Theory of Computing 2(1), 19–51 (2006)
2. Arora, S., Lee, J.R., Naor, A.: Euclidean distortion and the sparsest cut. J. AMS 21(1), 1–21 (2008)
3. Arora, S., Rao, S., Vazirani, U.: Expander flows, geometric embeddings and graph partitioning. In: Proc. 36th ACM STOC, pp. 222–231 (2004)
4. Charikar, M., Makarychev, K., Makarychev, Y.: Near-optimal algorithms for unique games. In: Proc. 38th ACM STOC, pp. 205–214 (2006)
5. Charikar, M., Makarychev, K., Makarychev, Y.: Integrality gaps for Sherali-Adams relaxations. In: Proc. 41st ACM STOC, pp. 283–292 (2009)
6. Cheeger, J., Kleiner, B.: Generalized differentiation and bi-Lipschitz nonembedding in L^1. Comptes Rendus Mathematique 343(5), 297–301 (2006)
7. Cheeger, J., Kleiner, B., Naor, A.: A $(\log n)^{\Omega(1)}$ integrality gap for the sparsest cut SDP. In: Proc. 50th IEEE FOCS (2009)
8. Dasgupta, S., Gupta, A.: An elementary proof of the johnson-lindenstrauss lemma. Tech. Rep. TR-99-006, U. C. Berkeley (1999)

9. Devanur, N., Khot, S., Saket, R., Vishnoi, N.: Integrality gaps for sparsest cut and minimum linear arrangement problems. In: Proc. 38th ACM STOC, pp. 537–546 (2006)

10. Georgiou, K., Magen, A., Pitassi, T., Tourlakis, I.: Integrality gaps of 2 - o(1) for vertex cover SDPs in the Lovész-Schrijver hierarchy. In: Proc. 48th IEEE FOCS, pp. 702–712 (2007)

11. Goemans, M.X., Williamson, D.P.: Improved approximation algorithms for maximum cut and satisfiability problems using semidefinite programming. J. ACM 42(6), 1115–1145 (1995)

12. Gupta, A., Krauthgamer, R., Lee, J.R.: Bounded geometries, fractals, and low-distortion embeddings. In: Proc. 44th IEEE FOCS (2003)

13. Gupta, A., Talwar, K.: Approximating unique games. In: SODA 2006: Proceedings of the Seventeenth Annual ACM-SIAM Symposium on Discrete Algorithm (2006)

14. Håstad, J.: Some optimal inapproximability results. J. ACM 48(4), 798–859 (2001)

15. Johnson, W., Lindenstrauss, J.: Extensions of lipschitz maps into a hilbert space. Contemporary Mathematics 26, 189–206 (1984)

16. Khot, S.: On the power of unique 2-prover 1-round games. In: Proc. 34th ACM STOC, pp. 767–775 (2002)

17. Khot, S., Kindler, G., Mossel, E., O'Donnell, R.: Optimal inapproximability results for MAX-CUT and other 2-variable CSPs? SIAM J. Comput. 37(1), 319–357 (2007)

18. Khot, S., Saket, R.: SDP integrality gaps with local ℓ_1-embeddability. In: Proc. 50th IEEE FOCS (2009)

19. Khot, S., Vishnoi, N.: The unique games conjecture, integrality gap for cut problems and embeddability of negative type metrics into l_1. In: Proc. 46th IEEE FOCS, pp. 53–62 (2005)

20. Krauthgamer, R., Rabani, Y.: Improved lower bounds for embeddings into l_1. In: ACM SODA, pp. 1010–1017 (2006)

21. Lasserre, J.B.: An explicit exact SDP relaxation for nonlinear 0-1 programs. In: Aardal, K., Gerards, B. (eds.) IPCO 2001. LNCS, vol. 2081, pp. 293–303. Springer, Heidelberg (2001)

22. Lee, J.R., Naor, A.: l_p metrics on the Heisenberg group and the Goemans-Linial conjecture. In: Proc. 47th IEEE FOCS, pp. 99–108 (2006)

23. Mossel, E., O'Donnell, R., Oleszkiewicz, K.: Noise stability of functions with low infuences invariance and optimality. In: Proc. 46th IEEE FOCS (2005)

24. Raghavendra, P., Steurer, D.: Integrality gaps for strong SDP relaxations of Unique Games. In: Proc. 50th IEEE FOCS (2009)

25. Schoenebeck, G.: Linear level Lasserre lower bounds for certain k-CSPs. In: Proc. 49th IEEE FOCS, pp. 593–602 (2008)

26. Schoenebeck, G., Trevisan, L., Tulsiani, M.: A linear round lower bound for Lovasz-Schrijver SDP relaxations of vertex cover. In: IEEE Conference on Computational Complexity, pp. 205–216 (2007)

27. Schoenebeck, G., Trevisan, L., Tulsiani, M.: Tight integrality gaps for Lovasz-Schrijver lp relaxations of vertex cover and max cut. In: Proc. 39th ACM STOC, pp. 302–310 (2007)

28. Trevisan, L.: Approximation algorithms for Unique Games. In: Proc. 46th IEEE FOCS (2005)

Exploiting Concavity in Bimatrix Games:[*]
New Polynomially Tractable Subclasses

Spyros Kontogiannis[1,2] and Paul Spirakis[2]

[1] Dept. of Computer Science, University of Ioannina, 45110 Ioannina, Greece
kontog@cs.uoi.gr
[2] R.A. Computer Technology Institute, Patras Univ. Campus,
26504 Rio-Patra, Greece
spirakis@cti.gr

Abstract. We study the fundamental problem of computing an arbitrary Nash equilibrium in bimatrix games. We start by proposing a novel *characterization* of the set of Nash equilibria, via a bijective map to the solution set of a (parameterized) quadratic program, whose feasible space is the (highly structured) set of correlated equilibria. We then proceed by proposing *new* subclasses of bimatrix games for which either an exact polynomial-time construction, or at least a FPTAS, is possible. In particular, we introduce the notion of *mutual (quasi-) concavity* of a bimatrix game, which assures (quasi-) convexity of our quadratic program, for at least one value of the parameter. For mutually concave bimatrix games, we provide a polynomial-time computation of a Nash equilibrium, based on the polynomial tractability of convex quadratic programming. For the mutually quasi-concave games, we provide (to our knowledge) the first FPTAS for the construction of a Nash equilibrium.

Of course, for these new polynomially tractable subclasses of bimatrix games to be useful, *polynomial-time certificates* are also necessary that will allow us to efficiently identify them. Towards this direction, we provide various characterizations of mutual concavity, which allow us to construct such a certificate. Interestingly, these characterizations also shed light to some structural properties of the bimatrix games satisfying mutual concavity. This subclass entirely contains the most popular subclass of polynomial-time solvable bimatrix games, namely, all the constant-sum games (rank-0 games). It is though incomparable to the subclass of games with fixed rank [16]: Even rank-1 games may not be mutually concave (eg, Prisoner's dilemma), but on the other hand, there exist mutually concave games of arbitrary (even full) rank. Finally, we prove closeness of mutual concavity under (Nash equilibrium preserving) positive affine transformations of bimatrix games having the *same scaling factor* for both payoff matrices. For different scaling factors the property is not necessarily preserved.

Keywords: Bimatrix games, Nash equilibrium, correlated equilibrium, quadratic optimization.

[*] This work has been partially supported by the ICT Programme of the EU under contract number FP7-215270 (FRONTS), and the ERC/StG Programme of the EU under the contract number 210743 (RIMACO).

M. Serna et al. (Eds.): APPROX and RANDOM 2010, LNCS 6302, pp. 312–325, 2010.

1   Introduction

One of the "holy grail quests" of theoretical computer science in the last decade, has been the characterization of the computational complexity of constructing an arbitrary Nash equilibrium (NE) in a finite normal form game. There has been a massive attack on various refinements of the problem (eg, a NE maximizing the payoff of some player, or its support size), that have lead to **NP**−hardness results (eg, [15,9]). Eventually the unconstraint problem of finding *any* NE proved in finite normal form games proved to be **PPAD**−complete [10,13], even for the bimatrix case [6]. Moreover, even the problem of computing an $\left(n^{-\Theta(1)}\right)$ −approximate NE for the bimatrix case is **PPAD**−complete [7], excluding even the existence of a FPTAS for the problem, unless **PPAD** = **P**. Additionally, it was proved that the celebrated combinatorial algorithm of Lemke and Howson [20] may take an exponential number of steps to terminate [26].

Given the apparent hardness in computing NE in bimatrix games, two main research trends emerged quite naturally: To discover polynomial-time, constant-approximation algorithms (or even a PTAS) for computing NE for the general case, or to identify general subclasses of games that admit a polynomial-time construction of *exact* NE, or at least a (F)PTAS. Even if one exchanges the "polynomiality" to "strict subexponentiality", there is still much room for research. Indeed, the first subexponential-time approximation scheme was provided in [21] (see also [2]), while a new one appeared only recently [28]. A sequence of papers have also provided polynomial-time algorithms for various notions of approximation (eg, [17,11,4,12,27,19]), the current winners being the gradient-based algorithm of [27] that provides 0.3393−approximation for the most common notion of $\varepsilon-Nash\ equilibria$, while [19] provides an LP-based 0.667−approximation for the stricter notion of $\varepsilon-well\ supported\ approximate\ NE$.

As for exact solutions (or even FPTAS) for general subclasses of bimatrix games, it is well known (due to von Neumann's minimax theorem [23]) that any constant-sum bimatrix game is solvable in polynomial time. Trivially, any bimatrix game with a pure Nash equilibrium is also solvable in polynomial time. Finally, for the particular case of win-lose bimatrix games, [8] provided a linear-time (exact) algorithm for games with very sparse payoff matrices and [1] provided a polynomial-time algorithm when the graph of the game is planar. [16] introduced a hierarchy of the bimatrix games, according to the rank of the matrix $R + C$ of the game $\langle R, C \rangle$, which was called the *rank of the game*. Then, for any *fixed* constant $k \geq 0$, they present a FPTAS for bimatrix games of rank k. It is worth mentioning at this point that the bimatrix games of zero rank are the zero-sum games, while the class of rank−1 games is already a rich class.

An alternative pathway to tractability is, rather than compute, to have the players cooperatively *learn* a Nash equilibrium of a game. For example, [24] initially proved that the fictitious play on behalf of both players leads to a Nash equilibrium, for every zero-sum bimatrix game. In [25] Rosen introduced the notion of *concave* strategic games, ie, strategic games in which every player's utility function (to be maximized) is concave in her own strategy. Then, for the special case of *strictly concave* games, he proved global asymptotic stability and provided a gradient method for constructing an equilibrium point. Of course, for the bimatrix case Rosen's concavity trivially holds, due to the bilinearity of the payoff functions. Very recently, [14] introduced the notion of *social concavity*, according to which (i) there must exist a strict convex combination of all the players' payoff functions, that is a concave function in its domain, and (ii) each player's payoff function is convex in the opponents' profile of strategies. It was then proved that for any strategic game possessing this property, any no-external regret protocol adopted by every player assures that the empirical distributions of the players converge to a NE point.

1.1   Our Contribution and Roadmap

In this work we first propose a novel quadratic program which characterizes the NE set of a bimatrix game (cf. Section 3). Of course, this is not the first quadratic program that characterizes this set. For example [22] already provided a quadratic program whose solution set is exactly the NE set of the bimatrix game at hand. Nevertheless, our approach is (to our knowledge) the first to make a direct connection between the set of correlated equilibria (and their marginal distributions) and the NE points of a bimatrix game. Indeed, the feasible space of our program is exactly the set of correlated equilibria of the game at hand. The profiles of (independent) strategies for the players that we consider, are provided as the marginal probability distributions of correlated strategies which are already correlated equilibria. We prove (cf. Theorem 1) that the solution set of our program has a bijective correspondence with the NE set of the game.

We then proceed to determine new subclasses of bimatrix games which admit polynomial-time algorithms, or at least FPTAS, for finding any NE. The natural thing to do, is to consider properties that assure (at least quasi-) convexity of our quadratic program, in order to be solvable in polynomial time, or at least admit a FPTAS. Therefore, we introduce the property of *mutual concavity* (cf. Definition 1), which demands the existence of some *strict* convex combination of the two payoff matrices, such that the corresponding convex combination of the payoff functions of the two players be a concave function. This is enough to guarantee convexity of our proposed quadratic program. Indeed, our quadratic program may be seen as a generalization of a parameterized version of the quadratic program of [22], where we have substituted the feasible space with correlated equilibria of the game, and we have changed their objective function (which was an unweighed sum of the players' regrets) to a properly weighed sum of the regrets of the profile of marginal distributions from the original correlated strategy. Therefore, mutual concavity assures also convexity of the variant of the (parameterized) quadratic program of [22]. If we now substitute in our property the demand for concavity with a demand for quasi-concavity, then it is easy to see that the *mutual quasi-concavity* property of bimatrix games (cf. Definition 1) is even broader than simple mutual concavity, and implies quasi-convexity for our quadratic program. This is still a nice thing, since instead of a polynomial-time exact algorithm, it provides a FPTAS of the quasi-convex program, implying an FPTAS for $\varepsilon-$NE in the original game. Observe that the social concavity of [14] boils down to mutual concavity for bimatrix games, since the second condition of the social concavity is trivial in our case. But rather than trying to *learn*, we choose to *compute* in polynomial-time a NE of any mutually concave bimatrix game, or at least provide a FPTAS for any mutually quasi-concave bimatrix game (cf. Theorem 2).

Nevertheless, we are still far from being able to claim tractability of a certain new subclass of bimatrix games (even if we are able either to compute or to learn a NE for any particular member of the class), unless we are able to solve the corresponding decision problem in polynomial time. For example, it is trivial to recognize in polynomial time either a constant-sum game, a game of fixed rank, or a game possessing pure NE. In order to explore this possibility also for the new subclasses, we proceed (cf. Section 4) with a series of characterizations of mutual concavity in bimatrix games (Propositions 4 and 8), which eventually allow us to solve the corresponding decision problem in polynomial time (cf. Theorem 3). Therefore, we conclude that the class of mutually concave bimatrix games is indeed solvable in polynomial time. Moreover, these characterizations allow us to understand in more depth the structure and the expressiveness of this subclass of games. For example, we prove that any constant-sum game belongs to this class (cf. Corollary 1), but also demonstrate that there is a continuum of mutually concave games which are non-constant-sum. Additionally, we observe that the subclass of mutually concave bimatrix games, is incomparable to the subclass of games of fixed rank [16]. A game $\langle R, C \rangle$ is of fixed rank when $rank(A + B)$ is some *constant*. The mutually concave games demand the existence of *some* strict convex combination of the payoff matrices which has rank of at most 2 (cf. Proposition 8). Nevertheless, even rank-1 games *may not be* mutually concave (eg, Prisoner's dilemma), but on the other hand, there exist mutually concave games of arbitrary (even full) rank. As for mutually concave games of fixed rank, rather than providing an *approximate* Nash equilibrium (eg, via the approximation technique of [29], as in [16]), we provide an *exact* Nash equilibrium via the solution to a convex quadratic optimization problem. We conclude by studying the closeness of mutual concavity of bimatrix games under Nash equilibrium preserving game transformations, such as the positive affine transformations.

Due to space limitations, the technical proofs are deferred to the full version [18] of the paper.

2   Preliminaries

Algebraic Notation. For any positive number $k \in \mathbb{N}$, $[k] \equiv \{1, 2, \ldots, k\}$. In a $k-$dimensional space, for any positive integer i, $\mathbf{e_i}$ is the vector having all its elements zero, except for its $i-$th element which is equal to 1. $\mathbf{1} = \sum_{i \in [k]} \mathbf{e_i}$ is the "all-ones" vector, $\mathbf{0}$ is the "all-zeroes" vector, and $E = \mathbf{1} \cdot \mathbf{1}^T$ is an "all-ones" square matrix. For any positive integer k, $\Delta_k = \{\mathbf{z} \in \mathbb{R}^k : \mathbf{1}^T \mathbf{z} = 1; \mathbf{z} \geq \mathbf{0}\}$ is the set of all probability distributions over a $k-$element set, and $O_k = \{\mathbf{z} \in \mathbb{R}^k : \mathbf{1}^T \mathbf{z} = 0\}$. For any vector $\mathbf{x} \in \mathbb{R}^k$ and any $i \in [k]$, $(\mathbf{x})_i = x_i$ is the $i-$th element of

x. For any matrix $A \in \mathbb{R}^{m \times n}$, $(A)_{i,j} = A_{i,j}$ is the value of the corresponding cell in the matrix, $A_{i,\star}$ is the i−th row of A (as a row vector) and $A_{\star,j}$ is the j−th column of A (as a column vector). A^T denotes the transpose matrix of A. For any pair of $m \times n$ real matrices $A, B \in \mathbb{R}^{m \times n}$, $A \bullet B \equiv \sum_{i \in [m]} \sum_{j \in [n]} A_{i,j} B_{i,j}$.

Game Theoretic Notation. For any $2 \leq m \leq n$, we denote by $\langle R, C \rangle$ an $m \times n$ **bimatrix game**, where the first player (aka the row player) has $R \in \mathbb{R}^{m \times n}$ as its payoff matrix and the second player (aka the column player) has $C \in \mathbb{R}^{m \times n}$ as its payoff matrix. If both the payoff matrices have exclusively rational entries, $R, C \in \mathbb{Q}^{m \times n}$, then we refer to a **rational bimatrix game**. These are mainly the games of concern in this work, for computational reasons. The row (column) player is assumed to choose as her action one of the rows (columns) of the payoff bimatrix $(R, C) = (R_{i,j}, C_{i,j})_{(i,j) \in [m] \times [n]}$. For any pair of choices, $(i, j) \in [m] \times [n]$, the payoff to the row (column) player is $R_{i,j}$ $(C_{i,j})$.

- A (mixed in general) **strategy** for the row (column) player is a probability distribution $\mathbf{x} \in \Delta_m$ ($\mathbf{y} \in \Delta_n$), according to which she determines her action, independently of the opponent's final choice of action. If all the probability mass of a mixed strategy is assigned to a particular action of the corresponding player, then we refer to a **pure strategy**.
- The utility of the row (column) player for the profile $(\mathbf{x}, \mathbf{y})$ is the expected payoff $\mathbf{x}^T R \mathbf{y}$ ($\mathbf{x}^T C \mathbf{y}$) that she gets.
- For any real number $\varepsilon \geq 0$, a profile of strategies $(\bar{\mathbf{x}}, \bar{\mathbf{y}}) \in \Delta_m \times \Delta_n$ is an ε−**Nash equilibrium** (ε−NE in short) of $\langle R, C \rangle$, iff each player's strategy is an approximate best response (within an additive term of ε) to the opponent's strategy: $\forall \mathbf{x} \in \Delta_m, \bar{\mathbf{x}}^T R \bar{\mathbf{y}} \geq \mathbf{x}^T R \bar{\mathbf{y}} - \varepsilon$ and $\forall \mathbf{y} \in \Delta_n, \bar{\mathbf{x}}^T C \bar{\mathbf{y}} \geq \bar{\mathbf{x}}^T C \mathbf{y} - \varepsilon$. We denote by $NE(R, C)$ the set of (exact) 0−NE of $\langle R, C \rangle$.
- A **correlated strategy** of $\langle R, C \rangle$ is a *joint* probability distribution $W \in \Delta_{m \times n}$ over the whole set of action profiles $[m] \times [n]$ for both players.
- A correlated strategy $W \in \Delta_{m \times n}$ is a **correlated equilibrium** (CE in short) of $\langle R, C \rangle$, iff it satisfies the following system of linear inequalities:

$$\begin{array}{|ll|} \hline \forall i, k \in [m], \; \sum_{j \in [n]} (R_{i,j} - R_{k,j}) W_{i,j} \geq 0 \\ \forall j, \ell \in [n], \; \sum_{i \in [m]} (C_{i,j} - C_{i,\ell}) W_{i,j} \geq 0 \\ \hspace{4em} \sum_{i \in [m]} \sum_{j \in [n]} W_{i,j} = 1 \\ \forall (i,j) \in [m] \times [n], \hspace{4em} W_{i,j} \geq 0 \\ \hline \end{array} \quad \text{[CE Property]}$$

We denote by $CE(R, C)$ the (polyhedral) set of correlated equilibria of $\langle R, C \rangle$.

3   A Quadratic Formulation of NE Points

In this section we provide a parameterized quadratic program that computes CE of a bimatrix game, whose optimal solutions are in a *bijective correspondence* with the NE of the game. For any correlated strategy $W \in \Delta_{m \times n}$, we consider the **marginal probabilities** $\mathbf{x}(W), \mathbf{y}(W)$ defined as follows:

$$\forall i \in [m], \ x_i(W) = \sum_{\ell \in [n]} W_{i,\ell} \qquad \forall j \in [n], \ y_j(W) = \sum_{k \in [m]} W_{k,j} \qquad (1)$$

Consequently, we define the row player's loss $g_R(W)$ when *both players* adopt the marginal distributions of a correlated strategy W, rather than abiding with the correlated strategy (the column player's loss $g_C(W)$ is analogous, we only substitute matrix R with matrix C):

$$g_R(W) \equiv \sum_{i \in [m]} \sum_{j \in [n]} R_{i,j} W_{i,j} - \mathbf{x}(W)^T R \mathbf{y}(W)$$

$$= \sum_{i \in [m]} \sum_{j \in [n]} R_{i,j} \left[W_{i,j} - \sum_{k \in [m]} \sum_{\ell \in [n]} W_{i,\ell} W_{k,j} \right]$$

We consider the following parameterized quadratic program, $\text{NEQP}(\lambda)$, for any constant $\lambda \in (0,1)$, bounded away from both its boundaries:

$$\boxed{\textbf{minimize} \left\{ \lambda \cdot g_R(W) + (1-\lambda) \cdot g_C(W) : W \in CE(R,C) \right\}} \quad [\text{NEQP}(\lambda)]$$

We denote with $opt(\text{NEQP}(\lambda))$ the set of optimal solutions for $\text{NEQP}(\lambda)$. We shall prove that there is a *bijective map* between $opt(\text{NEQP}(\lambda))$ and $NE(R,C)$. The proof proceeds in steps which we present as a sequence of propositions, whose correctness is provided in the full version of the paper. We start by showing that both players' losses for (mutually) adopting their marginal distributions, rather than abiding with the correlated strategy, are non-negative:

Proposition 1. $\forall R, C \in \mathbb{R}^{m \times n}, \forall W \in CE(R,C), \ g_R(W) \geq 0 \wedge g_C(W) \geq 0 .$

We continue by showing that any NE point of $\langle R, C \rangle$ induces an optimal solution of $\text{NEQP}(\lambda)$:

Proposition 2. $\forall \lambda \in (0,1), \forall (\bar{\mathbf{x}}, \bar{\mathbf{y}}) \in NE(R,C), \ \bar{W} \equiv \bar{\mathbf{x}} \cdot \bar{\mathbf{y}}^T \in opt(\text{NEQP}(\lambda)) .$

Our final step is to verify that every optimal solution $\bar{W} \in opt(\text{NEQP}(\lambda))$ induces a profile of marginal strategies which is a NE point of the game.

Proposition 3. $\forall \lambda \in (0,1), \forall \bar{W} \in opt(\text{NEQP}(\lambda)), \ (\mathbf{x}(\bar{W}), \mathbf{y}(\bar{W})) \in NE(R,C) .$

From the above discussion it is now obvious that for any $\lambda \in (0,1)$, $\text{NEQP}(\lambda)$ provides a characterization of the NE property in bimatrix games, as claimed in the following theorem:

Theorem 1. *For any real number* $\lambda \in (0,1)$, *any pair of payoff matrices,* $R, C \in \mathbb{R}^{m \times n}$, *and any profile of strategies* $(\bar{\mathbf{x}}, \bar{\mathbf{y}}) \in \Delta_m \times \Delta_n$, *the following holds:* $(\bar{\mathbf{x}}, \bar{\mathbf{y}}) \in NE(R,C)$ *if and only if* $\bar{W} \equiv \bar{\mathbf{x}} \cdot \bar{\mathbf{y}}^T \in opt(\text{NEQP}(\lambda))$.

It is now quite natural to investigate for which cases the characterization of Theorem 1 can help us in constructing an arbitrary Nash equilibrium of the game. For computational reasons we have to focus our attention to rational bimatrix games only. Observe that for any $\lambda \in (0,1)$, the program NEQP(λ) corresponding to such a game, has at least one *rational* solution: Any rational bimatrix game $\langle R, C \rangle$ has at least one solution $(\bar{\mathbf{x}}, \bar{\mathbf{y}}) \in NE(R, C)$ in which both players' strategies are rational vectors. The joint distribution $\bar{W} \equiv \bar{\mathbf{x}}\bar{\mathbf{y}}^T$ is also a rational optimum to NEQP(λ). Observe also that strict containment of λ in $(0,1)$ is crucial, in order for Proposition 3 to hold. We consider the following two properties of bimatrix games, which assure tractability (or at least FPTAS) of $NEQP(\lambda)$ for *some* value $\lambda \in (0,1)$, when all the payoff values are rational numbers:

Definition 1. *Consider any* $R, C \in \mathbb{Q}^{m \times n}$, $\lambda \in (0,1)$, $Z(\lambda) = \lambda R + (1 - \lambda)C$, *and* $H_\lambda(\mathbf{x}, \mathbf{y}) \equiv \mathbf{x}^T Z(\lambda)\mathbf{y}$.
1. $\langle R, C \rangle$ *is* **mutually concave** *(MC in short) game iff the following holds:*

[MC1] $\exists \lambda \in (0,1)$ *s.t.* $H_\lambda(\mathbf{x}, \mathbf{y})$ *is a concave function of* $(\mathbf{x}, \mathbf{y}) \in \Delta_m \times \Delta_n$: $\forall \mu \in (0,1)$, $\forall (\bar{\mathbf{x}}, \bar{\mathbf{y}}), (\hat{\mathbf{x}}, \hat{\mathbf{y}}) \in \Delta_m \times \Delta_n$, $H_\lambda(\mu(\bar{\mathbf{x}}, \bar{\mathbf{y}}) + (1 - \mu)(\hat{\mathbf{x}}, \hat{\mathbf{y}})) \geq \mu H_\lambda(\bar{\mathbf{x}}, \bar{\mathbf{y}}) + (1 - \mu)H_\lambda(\hat{\mathbf{x}}, \hat{\mathbf{y}})$.

2. $\langle R, C \rangle$ *is* **mutually quasi-concave** *(MQC in short) game, iff the following holds:*

[MQC] $\exists \lambda \in (0,1)$ *s.t.* $H_\lambda(\mathbf{x}, \mathbf{y})$ *is a quasi-concave function of* $(\mathbf{x}, \mathbf{y}) \in \Delta_m \times \Delta_n$: $\forall \mu \in (0,1)$, $\forall (\bar{\mathbf{x}}, \bar{\mathbf{y}}), (\hat{\mathbf{x}}, \hat{\mathbf{y}}) \in \Delta_m \times \Delta_n$, $H_\lambda(\mu(\bar{\mathbf{x}}, \bar{\mathbf{y}}) + (1 - \mu)(\hat{\mathbf{x}}, \hat{\mathbf{y}})) \geq \min\{H_\lambda(\bar{\mathbf{x}}, \bar{\mathbf{y}}), H_\lambda(\hat{\mathbf{x}}, \hat{\mathbf{y}})\}$.

Returning to our quadratic formulation for the NE set of a bimatrix game, observe that the objective function of NEQP(λ) is:

$$
\begin{aligned}
G_\lambda(W) &\equiv \lambda \cdot g_R(W) + (1 - \lambda) \cdot g_C(W) \\
&= \sum_{i \in [m]} \sum_{j \in [n]} [\lambda R_{i,j} + (1 - \lambda)C_{i,j}] W_{i,j} - \mathbf{x}(W)^T [\lambda R + (1 - \lambda)C]\mathbf{y}(W) \\
&= Z(\lambda) \bullet W - H_\lambda(\mathbf{x}(W), \mathbf{y}(W))
\end{aligned}
$$

Theorem 2. *Consider any pair of natural numbers* $2 \leq m \leq n$, *and any* $R, C \in \mathbb{Q}^{m \times n}$.

- *If* $\langle R, C \rangle$ *is mutually concave, then either there is a pure NE, or else mutual concavity holds for a* unique *rational number* $\lambda^* \in (0,1)$ *(of the same bit complexity as the input). In the latter case, an exact NE can be constructed in time* $\mathcal{O}(T(m,n))$, *where* $T(m,n)$ *is the time for solving the corresponding instance of* $NEQP(\lambda^*)$.
- *If* $\langle R, C \rangle$ *is mutually quasi-concave game for some fixed rational number* $\lambda^* \in (0,1)$ *bounded away from both boundaries, then for any* $\varepsilon > 0$ *an* $\frac{\varepsilon}{\min\{\lambda^*, 1-\lambda^*\}} - NE$ *can be constructed in time* $\mathcal{O}\left(\log_2\left(\frac{1}{\varepsilon}\right) \cdot T(m,n)\right)$, *where* $T(m,n)$ *is an upper bound on the time for solving any of the corresponding convex (feasibility) quadratic programs used in the bisection that is employed.*

Remark: The time complexity for solving an N−variable convex program

$$\min\{f_0(\mathbf{x}) : \forall i \in [M], f_i(\mathbf{x}) \le 0; A\mathbf{x} = \mathbf{b}\}$$

where $A \in \mathbb{Q}^{k \times N}$, $\mathbf{b} \in \mathbb{Q}^k$, and the functions f_i are all convex, is roughly proportional to $\sqrt{M} \cdot \max\{N^3, N^2 \cdot M, F\}$, where F is the worst-case cost for evaluating any of the f_i's, their first and second derivatives (eg, see [3,5]). For the particular case of NEQP(λ), for some rational number $\lambda \in (0, 1)$, $N = n \cdot m$ and $M = 2(m^2 + n^2) - (m + n)$ while $F = \mathcal{O}((m \cdot n)^2)$. This would therefore imply that the worst case cost for NEQP(λ) is $T(m, n) = \mathcal{O}(n^7)$ (assuming $m \le n$) which is quite large. This is because of the large number of variables and inequality constraints in NEQP(λ). Nevertheless, this heavy time-complexity can be very easily dropped down to $\mathcal{O}(n^{3.5})$, if we replace NEQP(λ) with a simple quadratic relaxation that only considers the marginal distributions (rather than the correlated strategy itself), and thus involves only $m+n$ variables and $m+n-2$ constraints, and the convexity of this latter program again depends on the mutual concavity of the game. But then we lose the direct connection to the correlated equilibria that NEQP(λ) returns, since the mapping from correlated strategies to their marginal distributions is a many-to-one relation. We leave the details of this construction for the full version of the paper.

4   Tractability of Mutual Concavity

Although we already know that for any mutually (quasi-) concave rational bimatrix game it is possible to construct in polynomial time (or provide a FPTAS for) a Nash equilibrium via convex optimization techniques, we still need to be able to recognize such games in polynomial time. For example, it is trivial to check whether a bimatrix game has the constant-sum property, possesses a pure NE, is a coordination game (ie, equal payoffs to both players, for any action profile), or is a game of fixed rank, and then solve it (in case of an affirmative answer). In this section we provide a characterization of the mutual-concavity property that: (i) demonstrates the generality of the subclass, and (ii) shows how to check it in polynomial time.

It should be reminded at this point that mutual concavity [MC1] matches condition [A1] of *social concavity* of [14, Definition 2.1], when we consider a bimatrix game. As for the second condition ([A2]) of social concavity, namely, the convexity of each player's payoff in the profile of the opponents, it trivially holds for the bimatrix case. That is, social concavity of strategic games boils down to mutual concavity in the bimatrix case. In our approach though, rather than having the two players *learn* a NE using no-external-regret algorithms in the corresponding infinitely repeated game, we choose to *solve* the proper (convex) quadratic optimization problem. But as said before, for any of the two approaches for constructing NE points to be really a polynomial-time method, mutual concavity should be checkable in polynomial time for rational bimatrix games. The following proposition provides a characterization of mutual concavity, which shall be quite useful (the proof is deferred to the full version):

Proposition 4. *For any* rational *bimatrix game* $\langle A, B \rangle$ *and* $\lambda \in (0, 1)$, *let* $Z(\lambda) = \lambda A + (1 - \lambda)B$, $H_\lambda(\mathbf{x}, \mathbf{y}) \equiv \mathbf{x}^T Z(\lambda)\mathbf{y}$. $\langle A, B \rangle$ *is mutually concave iff any the following conditions holds:*

[MC1] $\exists \lambda \in (0, 1)$ *s.t.* $H_\lambda(\mathbf{x}, \mathbf{y})$ *is a* concave *function of* $(\mathbf{x}, \mathbf{y}) \in \Delta_m \times \Delta_n$: $\forall \mu \in (0, 1)$, $\forall (\bar{\mathbf{x}}, \bar{\mathbf{y}}), (\hat{\mathbf{x}}, \hat{\mathbf{y}}) \in \Delta_m \times \Delta_n$, $H_\lambda(\mu(\bar{\mathbf{x}}, \bar{\mathbf{y}}) + (1 - \mu)(\hat{\mathbf{x}}, \hat{\mathbf{y}})) \geq \mu H_\lambda(\bar{\mathbf{x}}, \bar{\mathbf{y}}) + (1 - \mu)H_\lambda(\hat{\mathbf{x}}, \hat{\mathbf{y}})$.

[MC3] $\exists \lambda \in (0, 1) : \forall (\xi, \psi) \in O_m \times O_n$, $\xi^T Z(\lambda)\psi = 0$.

[MC4] $\exists \lambda \in (0, 1) : \forall \psi \in O_n$, $Z(\lambda)\psi = \mathbf{1} \cdot (Z(\lambda)\psi)_1$.

[MC5] $\exists \lambda \in (0, 1) : \forall \xi \in O_m$, $Z(\lambda)^T \xi = \mathbf{1} \cdot (Z(\lambda)^T \xi)_1$.

It is now easy to observe that any constant-sum bimatrix game is mutually concave:

Corollary 1. *For any* $A \in \mathbb{R}^{m \times n}$ *and* $c \in \mathbb{R}$, *the* $c-$sum $m \times n$ *bimatrix game* $\langle A, -A + c \cdot E \rangle$, *where* $E = \mathbf{1} \cdot \mathbf{1}^T$, *is a mutually concave game.*

Proof. If we set $\lambda = \frac{1}{2}$ then $Z(\lambda) = \frac{c}{2} \cdot E$ and now, for any $(\xi, \psi) \in O_m \times O_n$ it holds that $\xi^T Z(\lambda)\psi = \frac{c}{2} \cdot \xi^T \cdot E \cdot \psi = \frac{c}{2} \cdot \underbrace{\xi^T \cdot \mathbf{1}}_{=0} \cdot \underbrace{\mathbf{1}^T \cdot \psi}_{=0} = 0$. □

At this point we explore the mutual concavity of 2×2 bimatrix games. The following proposition provides a simple characterization of mutual concavity for this simple case:

Proposition 5. *For any* $A, B \in \mathbb{R}^{2 \times 2}$, *let* $\bar{a} = A_{1,1} + A_{2,2} - A_{1,2} - A_{2,1}$ *and* $\bar{b} = B_{1,1} + B_{2,2} - B_{1,2} - B_{2,1}$. *The bimatrix game* $\langle A, B \rangle$ *is mutually concave iff the following condition holds:*

[B1] $\bar{a} = \bar{b} = 0 \vee \min\{\bar{a}, \bar{b}\} < 0 < \max\{\bar{a}, \bar{b}\}$

For example, in figure 1.(a,b,c) we present 2×2 games both with and without the mutual concavity. Observe that example (c) actually presents a continuum of 2×2, non-constant MC-games, rather than just a single game.

Consider now, for integers $2 \leq m \leq n$, an arbitrary pair of payoff matrices $A, B \in \mathbb{R}^{m \times n}$. Let $\forall 1 \leq i < k \leq m, \forall 1 \leq j < \ell \leq n$, $\bar{a}_{ik,j\ell} = A_{i,j} + A_{k,\ell} - A_{i,\ell} - A_{k,j}$ and $\bar{b}_{ik,j\ell} = B_{i,j} + B_{k,\ell} - B_{i,\ell} - B_{k,j}$. An obvious *necessary condition* for the mutual concavity of $\langle A, B \rangle$ is the following:

Proposition 6. *If* $\langle A, B \rangle$ *is mutually concave, then the following condition holds:*

[B2] $\exists \lambda \in (0, 1) : \forall 1 \leq i < k \leq m, \forall 1 \leq j < \ell \leq n,$

$$[\bar{a}_{ik,j\ell} = \bar{b}_{ik,j\ell} = 0]$$
$$\vee$$
$$\left[\max\{\bar{a}_{ik,j\ell}, \bar{b}_{ik,j\ell}\} > 0 > \min\{\bar{a}_{ik,j\ell}, \bar{b}_{ik,j\ell}\} \quad \wedge \quad \lambda = \frac{-\bar{b}_{ik,j\ell}}{\bar{a}_{ik,j\ell} - \bar{b}_{ik,j\ell}} \right]$$

	silent	betray
silent	$(-1,-1)$	$(-10,0)$
betray	$(0,-10)$	$(-5,-5)$

(a) Prisoners' Dilemma.

	cinema	football
cinema	$(2,1)$	$(0,0)$
football	$(0,0)$	$(1,2)$

(b) Battle of Sexes.

	L	R
T	$(2,1)$	$(1,1+\gamma)$
B	$(1,1)$	$(3,0)$

(c) 2×2 MC game.

	L	R
T	$(2,1)$	$(1,1+\gamma)$
M	$(1,1)$	$(3,0)$
B	(δ_1,δ_2)	$(\varepsilon_1,\varepsilon_2)$

(d) 3×2 MC game.

Fig. 1. (a) The non-mutually concave rank-1 "Prisoner's Dilemma" game. (b) The non-mutually concave rank-2 game "Battle of Sexes". (c) A generic (non-constant-sum) 2×2 game that is mutually concave, $\forall \gamma > -1$. (d) A generic (non-constant-sum) 3×2 game that is a mutually concave $\forall \gamma > -1$, $\delta_1, \delta_2, \varepsilon_1 \in \mathbb{R}$, $\varepsilon_2 = \delta_2 + \frac{1}{3} \cdot [2\gamma - 1 - (1+\gamma) \cdot (\varepsilon_1 - \delta_1)]$. Observe that the *unique* value $\lambda = \frac{1+\gamma}{4+\gamma}$ makes $\mathbf{x}^T Z(\lambda)\mathbf{y}$ in this game a concave function.

Remark: [B2] is also a sufficient condition, for any $2 \times n$ bimatrix game, since if for any $\begin{bmatrix} \xi \\ -\xi \end{bmatrix} \in O_2$ it holds that $\forall j \in [n]$, $[\xi, -\xi] \cdot (Z(\lambda))_{\star,1} = [\xi, -\xi] \cdot (Z(\lambda))_{\star,j}$, then clearly $\forall \psi \in O_n$, $[\xi, -\xi] \cdot Z(\lambda) \cdot \psi = 0$. The unique parameter of mutual concavity is $\lambda^* = \frac{-\bar{b}_{12,12}}{\bar{a}_{12,12} - \bar{b}_{12,12}}$. The example in figure 1.(d) provides a continuum of non-constant 3×2 MC-games.

Next we prove that computing a NE point in a bimatrix game with all the $\bar{a}$ and $\bar{b}$ values equal to zero, is polynomial-time tractable (independently of the validity of the mutual concavity).

Proposition 7. *For any pair of $m \times n$ matrices $A, B \in \mathbb{Q}^{m \times n}$, for which $\forall \{i, k\} \subseteq [m], \forall j, \ell \subseteq [n], \bar{a}_{ik,j\ell} = \bar{b}_{ik,j\ell} = 0$, finding a (pure) Nash equilibrium of the rational bimatrix game $\langle A, B \rangle$ is tractable in polynomial time.*

We now provide a necessary and sufficient condition for mutual concavity of $\langle A, B \rangle$ that applies directly to the combined matrix $Z(\lambda) = \lambda A + (1 - \lambda)B$. This property also indicates how to construct non-trivial instances of mutually concave bimatrix games. It also indicates that $Z(\lambda)$ must have rank at most 2, but of course this is not a sufficient condition for mutual concavity, as was shown in previous examples (cf. figure 1). On the other hand, as we shall see shortly, a mutually concave game may have arbitrary (even full) rank.

Proposition 8. *For any $m, n \geq 2$ and real matrices $A, B \in \mathbb{R}^{m \times n}$, the bimatrix game $\langle A, B \rangle$ is mutually concave if and only if any of the following properties holds:*

[MC6] $\exists \lambda \in (0, 1), \exists\ \mathbf{a} \in \mathbb{R}^m, \exists\ \delta = [0, \delta_2, \ldots, \delta_n]^T \in \mathbb{R}^n : \forall j \in [n], Z(\lambda)_{\star,j} = -\delta_j \cdot \mathbf{1} + \mathbf{a}$.

[MC7] $\exists \lambda \in (0,1), \exists \; \mathbf{b} \in \mathbb{R}^n, \exists \; \varepsilon = [0, \varepsilon_2, \dots, \varepsilon_m]^T \in \mathbb{R}^m : \forall i \in [m], \; Z(\lambda)_{i,\star} = -\varepsilon_i \cdot \mathbf{1}^T + \mathbf{b}^T.$

It is not hard to construct $n \times n$ games of full rank, which are nevertheless mutually concave. Eg, for $n = 7$, consider the vectors $\mathbf{b} = [1, 2, 4, 8, 16, 32, 64]^T, \varepsilon = [0, -1, 1, -2, 2, -3, 3]^T$. The corresponding matrix Z that complies with [MC7] is the following:

$$Z = \begin{bmatrix} 1 & 2\,4 & 8\;16\;32\;64 \\ 0 & 1\,3 & 7\;15\;31\;63 \\ 2 & 3\,5 & 9\;17\;33\;65 \\ -1 & 0\,2 & 6\;14\;30\;62 \\ 3 & 4\,6 & 10\;18\;34\;66 \\ -2 & -1\,1 & 5\;13\;29\;61 \\ 4 & 5\,7 & 11\;19\;35\;67 \end{bmatrix}$$

with $rank(Z) = 2$. Nevertheless, for $A = I_7$ being the identity matrix and $\lambda = 1/4$, the unique choice of B so that $Z(\lambda) = \lambda A + (1 - \lambda)B = Z$ is $B = \frac{4}{3}Z - \frac{1}{3}A$. It is now easy to check that $rank(A) = rank(B) = rank(A + B) = 7$, and thus the game has full rank.

The above discussion allows us now to prove the efficiency of checking mutual concavity in non-trivial bimatrix games (eg, those having no pure Nash equilibria):

Theorem 3. *For any natural numbers $n \geq m \geq 2$, and real matrices $A, B \in \mathbb{R}^{m \times n}$, for the bimatrix game $\langle A, B \rangle$ either all the $\bar{a}-$ and $\bar{b}-$values of its 2×2 submatrices are zero (and then there is a pure Nash equilibrium), or there is at most one candidate parameter $\lambda^* \in (0, 1)$ for which mutual concavity must be checked, and this can be done in time $\mathcal{O}(n^2 \cdot m^2)$. Moreover, in the latter case, if the bimatrix game is rational, then (in case of non-trivial mutual concavity) the unique parameter λ^* is also a rational number.*

5   Closeness of Mutual Concavity

In game theory literature it is quite common to consider various *Nash equilibrium preserving* transformations of a game, and then try to work on the new game. One of the most typical transformations of this kind is defined as follows:

Definition 2. *Consider the matrices $R, C \in \mathbb{R}^{m \times n}$ and the corresponding bimatrix game $\langle R, C \rangle$. For any scalars $\gamma_I, \gamma_{II} \in \mathbb{R}_{>0}$ and vectors $\mathbf{r} \in \mathbb{R}^n$, $\mathbf{c} \in \mathbb{R}^m$, the game $\langle A, B \rangle$ where:*

$$A = \gamma_I \cdot R + \begin{bmatrix} \mathbf{r}^T \\ \vdots \\ \mathbf{r}^T \end{bmatrix}, \quad B = \gamma_{II} \cdot C + [\mathbf{c}, \; \mathbf{c}, \; \cdots, \mathbf{c}],$$

*is called a **positive affine transformation (PAT)** of $\langle R, C \rangle$. In case that $\gamma_I = \gamma_{II} \in \mathbb{R}_{>0}$, then we refer to a **uniform positive affine transformation***

(UPAT) *of the game* $\langle R, C \rangle$. *If additionally* $\gamma_I = \gamma_{II} = 1$, *then we refer to an*
additive transformation (AdT).

A well-known (and trivial to prove) fact is that any PAT of a bimatrix game
preserves the Nash equilibrium set:

Fact 1 *For any scalars* $\gamma_I, \gamma_{II} \in \mathbb{R}_{>0}$ *and any real vectors* $\mathbf{r} \in \mathbb{R}^n$, $\mathbf{c} \in \mathbb{R}^m$, *any*
bimatrix game $\langle R, C \rangle$ *and its corresponding PAT* $\langle A, B \rangle$ *with parameters* $(\gamma_I, \mathbf{r})$
for the row player and $(\gamma_{II}, \mathbf{c})$ *for the column player, have exactly the same Nash*
equilibria.

We shall prove that the mutual concavity of bimatrix games is closed under
UPAT, but unfortunately is not necessarily closed under non-uniform PAT.

Theorem 4. *For any scalar* $\gamma \in \mathbb{R}_{>0}$, *any real vectors* $\mathbf{r} \in \mathbb{R}^n$, $\mathbf{c} \in \mathbb{R}^m$ *and*
any bimatrix game $\langle R, C \rangle$, *let* $\langle A, B \rangle$ *be its UPAT with common scaling factor* γ
and additive vectors $\mathbf{r}, \mathbf{c}$. *Then,* $\langle R, C \rangle$ *is mutually concave if and only if* $\langle A, B \rangle$
is mutually concave.

Unfortunately, the MC property of bimatrix games is not necessarily preserved under general (in particular, non-uniform) PAT. Additionally, one might also argue
that there exist (non-affine) transformations of *particular* bimatrix games, which
are Nash equilibrium preserving, for which the MC property is not preserved.
For example, one might use the form for the prisoner's dilemma game given in
figure 2. It is trivial to see that this form of PD has the MC property, but there
is actually no affine transformation (let alone PAT) that can lead from the non-
mutually concave PD version in figure 1, to this mutually concave PD version.
In our opinion, it is crucial to focus our interest on *generic* Nash equilibrium
preserving transformations of games, such as PAT. It is nevertheless mentioned
that also other crucial properties, such as the ratio of *approximate* (rather than
exact) Nash equilibria is not preserved, indeed not even under AdT alone. This
of course does not lessen the importance of such properties, but rather indicates
an additional difficulty that we face when exploring them.

		silent	betray
silent		$(6, 6)$	$(0, 10)$
betray		$(10, 0)$	$(4, 4)$

Fig. 2. An alternative form of the PD game, which happens to have the MC property

6   Conclusions and Future Work

In this paper we have presented a novel characterization of the Nash equilibrium
set of a bimatrix game, via the solution set of a proper quadratic program, whose
feasible space contains all the correlated equilibria of the game. Consequently,

we explored two properties (mutual concavity and mutual quasi-concavity) of a bimatrix game that we can exploit in this quadratic formulation, in order to assure polynomial time construction, or at least a FPTAS, for a Nash equilibrium point. We then proceeded to characterize the first of the two classes, and proved that it entirely contains all constant-sum games but is incomparable to the subclass of games with fixed rank. Nevertheless, it should be noted that for the intersection of mutually concave games of fixed rank, rather than giving a FPTAS, we find in polynomial time an *exact* Nash equilibrium point via the optimal solution to our convex quadratic program.

For further research, we shall explore the expressiveness and decidability of the mutual quasi-concavity property, which (when it holds) allows for a FPTAS for the game. Of course, the main challenge still remains to improve the approximation ratio, or even find a PTAS, for the general case of rational bimatrix games.

References

1. Addario-Berry, L., Olver, N., Vetta, A.: A polynomial time algorithm for finding nash equilibria in planar win-lose games. Journal of Graph Algorithms and Applications 11(1), 309–319 (2007)
2. Althöfer, I.: On sparse approximations to randomized strategies and convex combinations. Linear Algebra and Applications 199, 339–355 (1994)
3. Bazaraa, M.S., Sherali, H.D., Shetty, C.: Nonlinear Programming: Theory and Algorithms, 2nd edn. John Wiley & Sons, Inc., Chichester (1993)
4. Bosse, H., Byrka, J., Markakis, E.: New algorithms for approximate nash equilibria in bimatrix games. In: Deng, X., Graham, F.C. (eds.) WINE 2007. LNCS, vol. 4858, pp. 17–29. Springer, Heidelberg (2007)
5. Boyd, S., Vandenberghe, L.: Convex Optimization, 7th edn. Cambridge University Press, Cambridge (2009)
6. Chen, X., Deng, X.: Settling the complexity of 2-player nash equilibrium. In: Proc. of 47th IEEE Symp. on Found. of Comp. Sci. (FOCS 2006), pp. 261–272. IEEE Comp. Soc. Press, Los Alamitos (2006)
7. Chen, X., Deng, X., Teng, S.-H.: Computing nash equilibria: Approximation and smoothed complexity. In: Proc. of 47th IEEE Symp. on Found. of Comp. Sci. (FOCS 2006), pp. 603–612. IEEE Comp. Soc. Press, Los Alamitos (2006)
8. Codenotti, B., Leoncini, M., Resta, G.: Efficient computation of nash equilibria for very sparse win-lose bimatrix games. In: Azar, Y., Erlebach, T. (eds.) ESA 2006. LNCS, vol. 4168, pp. 232–243. Springer, Heidelberg (2006)
9. Conitzer, V., Sandholm, T.: Complexity results about nash equilibria. In: Proc. of 18th Int. Joint Conf. on Art. Intel. (IJCAI 2003), pp. 765–771. Morgan Kaufmann, San Francisco (2003)
10. Daskalakis, C., Goldberg, P.W., Papadimitriou, C.H.: The complexity of computing a nash equilibrium. SIAM Journal on Computing 39(1), 195–259 (2009); Preliminary version in ACM STOC 2006
11. Daskalakis, C., Mehta, A., Papadimitriou, C.: A note on approximate equilibria. In: Spirakis, P.G., Mavronicolas, M., Kontogiannis, S.C. (eds.) WINE 2006. LNCS, vol. 4286, pp. 297–306. Springer, Heidelberg (2006)

12. Daskalakis, C., Mehta, A., Papadimitriou, C.: Progress in approximate nash equilibrium. In: Proc. of 8th ACM Conf. on El. Comm. (EC 2007), pp. 355–358 (2007)
13. Daskalakis, C., Papadimitriou, C.: Three player games are hard. Technical Report TR05-139, Electr. Coll. on Comp. Compl., ECCC (2005)
14. Even-Dar, E., Mansour, Y., Nadav, U.: On the convergence of regret minimization dynamics in concave games. In: Proc. of 41st ACM Symp. on Th. of Comp. (STOC 2009), pp. 523–532 (2009)
15. Gilboa, I., Zemel, E.: Nash and correlated equilibria: Some complexity considerations. Games & Econ. Behavior 1, 80–93 (1989)
16. Kannan, R., Theobald, T.: Games of fixed rank: A hierarchy of bimatrix games. Economic Theory 42, 157–173 (2010); Preliminary version appeared in ACM-SIAM SODA 2007
17. Kontogiannis, S., Panagopoulou, P., Spirakis, P.: Polynomial algorithms for approximating nash equilibria in bimatrix games. In: Spirakis, P.G., Mavronicolas, M., Kontogiannis, S.C. (eds.) WINE 2006. LNCS, vol. 4286, pp. 286–296. Springer, Heidelberg (2006)
18. Kontogiannis, S., Spirakis, P.: Exploiting concavity in bimatrix games: New polynomially tractable subclasses. In: Proc. of 13th W. on Appr. Alg. for Comb. Opt., APPROX'10 (2010),
    http://www.cs.uoi.gr/~kontog/pubs/approx10paper-full.pdf
19. Kontogiannis, S., Spirakis, P.: Well supported approximate equilibria in bimatrix games. ALGORITHMICA 57, 653–667 (2010)
20. Lemke, C., Howson, J.: Equilibrium points of bimatrix games. Journal of the Society for Industrial and Applied Mathematics 12, 413–423 (1964)
21. Lipton, R., Markakis, E., Mehta, A.: Playing large games using simple strategies. In: Proc. of 4th ACM Conf. on El. Comm (EC 2003), pp. 36–41. Assoc. of Comp. Mach. (ACM), New York (2003)
22. Mangasarian, O.L., Stone, H.: Two-person nonzero-sum games and quadratic programming. Journal of Mathematical Analysis and Applications 9(3), 348–355 (1964)
23. Morgenstern, O., von Neumann, J.: The Theory of Games and Economic Behavior. Princeton University Press, Princeton (1947)
24. Robinson, J.: An iterative method of solving a game. Annals of Mathematics 54, 296–301 (1951)
25. Rosen, J.: Existence and uniqueness of equilibrium points for concave $n-$person games. Econometrica 33(3), 520–534 (1965)
26. Savani, R., von Stengel, B.: Exponentially many steps for finding a nash equilibrium in a bimatrix game. In: Proc. of 45th IEEE Symp. on Found. of Comp. Sci. (FOCS 2004), pp. 258–267 (2004)
27. Tsaknakis, H., Spirakis, P.G.: An optimization approach for approximate nash equilibria. In: Deng, X., Graham, F.C. (eds.) WINE 2007. LNCS, vol. 4858, pp. 42–56. Springer, Heidelberg (2007)
28. Tsaknakis, H., Spirakis, P.G.: A graph spectral approach for computing approximate nash equilibria. Technical report, Electronic Colloquium on Computational Complexity, Report No. 96 (2009)
29. Vavasis, S.: Approximation algorithms for indefinite quadratic programming. Mathematical Programming 57, 279–311 (1992)

Maximum Flows on Disjoint Paths

Guyslain Naves[1], Nicolas Sonnerat[1], and Adrian Vetta[2]

[1] Department of Mathematics and Statistics, McGill University
naves@math.mcgill.ca, sonnerat@math.mcgill.ca
[2] Department of Mathematics and Statistics, and School of Computer Science,
McGill University
vetta@math.mcgill.ca

Abstract. We consider the question: What is the maximum flow achievable in a network if the flow must be decomposable into a collection of edge-disjoint paths? Equivalently, we wish to find a maximum weighted packing of disjoint paths, where the weight of a path is the minimum capacity of an edge on the path. Our main result is an $\Omega(\log n)$ lower bound on the approximability of the problem. We also show this bound is tight to within a constant factor. Surprisingly, the lower bound applies even for the simple case of undirected, planar graphs.

Our results extend to the case in which the flow must decompose into at most k disjoint paths. There we obtain $\Theta(\log k)$ upper and lower approximability bounds.

1 Introduction

Network flows have played a fundamental role in the advancement of combinatorial optimization [13] and are ubiquitous in applications [2]. In the standard single-commodity flow problem we have a capacitated graph $G = (V, E)$ and terminal vertices s and t.[1] The goal is to find a maximum valued flow from s to t that satisfies the capacity constraints on each edge. Equivalently, we are searching for a maximum packing of weighted $s-t$ paths; the packing constraints simply state that the total weight of all paths passing through an edge must not exceed the *capacity/weight*, w_e, of that edge. Viewed in this light, a special case is the classical problem of finding a maximum collection of disjoint paths.

Thus, there has been a long-standing and close relationship between network flows and the disjoint packing of *unweighted* paths. An immediate question arises: what about the weighted case, namely, what if we desire that our network flow decomposes into a disjoint collection of *weighted* paths? Surprisingly given the apparent simplicity of the question, as far as we are aware, this question has not previously been considered in the literature.

Consequently, this paper investigates how to find a maximum flow whose path decomposition consists only of disjoint paths. Specifically, take a collection of pairwise edge-disjoint $s - t$ paths $\mathcal{P}$. Then the maximum flow, $w(P)$, we can

---

[1] Of course, by incorporating a supersource and supersink, this framework also models the case of multiple sources and sinks, provided any source can route to any sink.

M. Serna et al. (Eds.): APPROX and RANDOM 2010, LNCS 6302, pp. 326–337, 2010.

send down a path $P \in \mathcal{P}$ is simply the minimum capacity of an edge in P. The value of the flow is then the sum of the flows along each path, $\sum_{P \in \mathcal{P}} w(P)$. Our goal is to obtain a flow $\mathcal{P}$ consisting of edge-disjoint paths that has maximum value. We call this the *Disjoint Weighted Flow Problem*.

Observe this problem does indeed correspond to a simple weighted path-packing problem. Specifically, let the weight of an $s - t$ path P be $w(P) := \min\{w(e) \,|\, e \in E(P)\}$. Then we are looking for a collection $\mathcal{P}$ of disjoint paths of maximum total weight, that is $\sum_{P \in \mathcal{P}} w(P)$ as before. As well as being an elegant combinatorial question, we remark that this requirement for disjoint paths is also a natural one in applications where flow paths can interfere with one another or where technological constraints at links and nodes compel disjointness.

In this paper, we examine approximation algorithms for the the disjoint weighted flow problem. We present $\Theta(\log n)$ lower and upper approximation bounds for the weighted disjoint paths problem, where n is the number of vertices. Standard reductions show that these bounds also apply in directed graphs and/or if we insist the paths be vertex-disjoint rather than edge-disjoint. Furthermore, our lower bound applies even for the special case of planar graphs.

1.1   Related Work

Given the applicability of network flows there is a vast literature optimizing flows given additional constraints. These side-constraints may arise from the application itself, but they can also arise due to restrictions induced by available technology or by the choice of routing protocol; see [12] for a survey illustrating some of these issues. The work most closely related to our own, though, concerns k-splittable flows introduced by Baier, Köhler and Skutella [7]. A k-splittable flow is a flow that can be routed along k paths - note that these paths are not required to be disjoint. Thus, Kleinberg's unsplittable flows [9] can be viewed as 1-splittable flows. Baier et al. present a 2-approximation algorithm for the k-splittable single-commodity flow problem.

Our results also extend to the case in which feasible solutions must be decomposable into at most k disjoint paths, for some k. Disjoint weighted flows are, however, are harder to deal with and approximate than k-splittable flows. In particular, for single-commodity flows, we obtain $\Theta(\log k)$ lower and upper approximations bounds when we are constrained to use at most k disjoint paths.

2   The Lower Bound

In this section we present our main result:

Theorem 1. *For undirected planar networks, the hardness of approximation for the maximum disjoint weighted flow problem is $\Omega(\log n)$, unless $P = NP$.*

Before proving Theorem 1 we outline the structure of the proof. First, we introduce a graph G_N that has a maximum disjoint weighted flow of value equal to the harmonic number $H_N \approx \log N$. But if we use a slightly modified weight function for the paths then G_N has a maximum disjoint weighted flow of value one.

We then build a new network $\mathcal{G}$ by replacing each node of G_N by an instance of an NP-hard routing problem. The routing problem will be chosen to have the following properties. If it is a YES-instance then path weightings for the disjoint weighted flow problem on $\mathcal{G}$ will correspond to the original weighting scheme on G_N. In contrast, if it is a NO-instance, then path weightings for the disjoint weighted flow problem on $\mathcal{G}$ will correspond to the modified weighting scheme on G_N.

It follows that an approximation algorithm with guarantee better than logarithmic would allow us to distinguish between YES- and NO-instances of our routing problem, giving a lower bound of $\Theta(\log N)$. We will see that this bound is equal to $\Theta(\log n)$.

Furthermore, at all stages we will show this reduction can be applied using only undirected, planar graphs. Theorem 1 will follow.

2.1   A Half-Grid Graph

Let's begin by defining the graph G_N. There are N rows (numbered from top to bottom) and N columns (numbered from left to right). All the edges in the ith row and all the edges in the ith column have weight $\frac{1}{i}$. The ith row extends as far as the ith column and vice versa; thus, we obtain a "half-grid" that is a weighted version of the network considered by Guruswami et al [8]. Finally we add a source s and a sink t. There are edges of weight $\frac{1}{i}$ from s to the first vertex in row i and from t to the last vertex in column i. The complete construction is shown in Figure 1.

Note that there is a unique $s - t$ path P_i consisting only of edges of weight $\frac{1}{i}$, that is, the L-shaped path that goes from s along the ith row and then down the ith column to t. Moreover, for $i \neq j$, the path P_i intersects P_j precisely once. Clearly each path P_i has weight $w(P_i) = \frac{1}{i}$, so the collection of edge-disjoint paths $\mathcal{P}^* = \{P_1, P_2, \ldots, P_N\}$ gives a flow of total value $H_N = 1 + \frac{1}{2} + \ldots \frac{1}{N}$. Since every edge incident to s is used in $\mathcal{P}^*$ with its maximum weight, this solution is optimal. Similarly, if we are constrained to use flows that decompose into at most k disjoint paths then the optimal flow has weight H_k.

Now consider what happens when we modify the weight function for the paths. Given a collection $\mathcal{P}$ of paths, let the *modified weight*, $\hat{w}_{\mathcal{P}}(P)$, of a path $P \in \mathcal{P}$ be the the minimum weight amongst its edges and those edges incident to a vertex at which P crosses another path $Q \in \mathcal{P}$. Formally,

$$\hat{w}_{\mathcal{P}}(P) = \min\{w_{uv} \mid v \in P, uv \in Q \text{ for some } Q \in \mathcal{P}\}$$

where we will omit the subscript if $\mathcal{P}$ is clear.

The maximum value of a flow is significantly reduced if we use these modified weights.

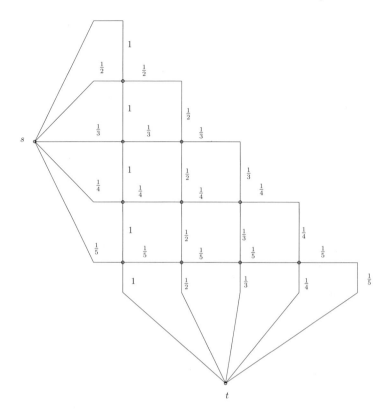

Fig. 1. Grid Graph G_N

Lemma 1. *The maximum modified value of a weighted flow in G_N is 1.*

Proof. Define the *rank* of a path P to be the index j for which this path uses the weight $\frac{1}{j}$ edge incident to t. Suppose $\frac{1}{i}$ is the maximum modified value of any path in a flow $\mathcal{P}$. Let j be the rank of some path $Q \in \mathcal{P}$ of modified weight $\frac{1}{i}$. Then set $\mathcal{P}^+$ to be the collection of paths in $\mathcal{P}$ with ranks greater than j, and $\mathcal{P}^-$ to be the paths with ranks less than j.

Observe that Q must contain as a sub-path all the edges in column j that lie below row i. Otherwise, Q would contain an edge in a row of lower weight than $\frac{1}{i}$, contradicting the fact that Q has modified weight $\frac{1}{i}$. Similarly, no other other path in $\mathcal{P}$ crosses Q on this sub-path, as this would reduce Q's modified weight. This implies that any path in $\mathcal{P}^+$ must use one edge of the columns j to $i + 1$ between row i and row $i + 1$. Consequently, $|\mathcal{P}^+| \leq i - j$. Obviously $|\mathcal{P}^-| \leq j - 1$ and so $|\mathcal{P}| \leq 1 + (i - j) + (j - 1) = i$. Since each path has modified weight at most $\frac{1}{i}$, this gives an upper bound of 1 on the modified value of the flow.

For $\mathcal{P}^* = \{P_1, P_2, \ldots, P_N\}$, we see that $\hat{w}(P_i) = \frac{1}{N}$, for all i. Thus, this collection of paths obtains the maximum modified value of one. $\square$

2.2   The 2-Edge-Disjoint Weighted Paths Problem

Recall the next step is to replace, in G_N, each vertex at the crossing of two paths P_i and P_j with an instance of an NP-hard routing problem. To define this routing problem, let H be an undirected graph whose edges have weight either a or b, where $b > a$. Given two pairs of vertices (s_1, t_1) and (s_2, t_2), we wish to find a path P_1 from s_1 to t_1 and a path P_2 from s_2 to t_2 with the properties that

(i) P_1 and P_2 are edge-disjoint.
(ii) P_2 may only use edges of weight b (P_1 may use either weight a or weight b edges).

We call this the *Two Edge-Disjoint Weighted Paths Problem*, or 2-EDWP.

Evidently, we will be most interested in the case where the graph H is planar. Then we have:

Theorem 2. PLANAR-2-EDWP *is NP-hard, even if the pairs of terminals lie on the outer face of H in the order s_1, s_2, t_1, t_2.*

We remark that in graphs which are directed and not planar, the hardness of 2-EDWP follows directly from the hardness of the 2-arc-disjoint paths problem ([6]).

Before embarking on the proof, observe that Theorem 2 immediately tells us that the maximum disjoint weighted flow problem is hard in planar graphs. Simply take an instance of PLANAR-2-EDWP and add a super-source s and a super-sink t. Then connect s to s_1 and s_2 with edges of weights a and b, respectively. Similarly, connect t to t_1 and t_2 with edges of weights a and b, respectively. Then there is a disjoint weighted $s - t$ flow of value $a + b$ if and only if there are paths P_1 and P_2 satisfying properties (i) and (ii). Of course, we desire a much stronger hardness result than this, but this observation will be useful in motivating the subsequent construction.

In order to prove Theorem 2, we will need the following geometric result.

Lemma 2. *Let $G = (V, E)$ be an embedded planar graph, and $\phi_e \subseteq \mathbb{R}^2$ the open curve corresponding to the embedding of e, for each edge $e \in E$. Then there is a simple closed curve in $\mathcal{R} = \mathbb{R}^2 \setminus \bigcup_{e \in E} \phi_e$ that intersects the image of every vertex.*

Proof. Let $\phi_u \in \mathbb{R}^2$ denote the image of $u \in V$. We prove the following stronger property by induction on the number of vertices: for every edge uv of G, there is a simple curve in $\mathcal{R}$ with endpoints ϕ_u and ϕ_v that intersects the image of every vertex. We remark that we may add an embedded edge to G or remove loops and parallel edges from G, without loss of generality.

If G has only two vertices or G is a cycle, the property is obvious. Otherwise, we can suppose G is two vertex connected - to see this, note that we can always add edges between the neighbours of a vertex such that its neighbourhood becomes connected. Now let F be a simple face containing uv. Let the ordered vertices on the cycle defined by the boundary of F be $\{u_1 = u, u_2, \ldots, u_k = v\}$, where $k \geq 3$.

Let $\Gamma(u)$ be the neighbourhood of u. If $\Gamma(u) = \{u_2, u_k\}$ then contract u along one of its two incident edges, say uu_2. Since u_2 and $v = u_k$ are now adjacent, we can find by induction a curve $\mathcal{D}$ from u_2 to u_k using every vertex except u. Extending the curve from u_2 to u in $\mathcal{R}$ gives the desired curve in G.

So assume that $|\Gamma(u)| \geq 3$. Let $F' \neq F$ be the other face containing u_1u_2, and let u_1w, $w \neq u_2$ be the second edge incident to u_1 in F'. Let G' be the graph obtained by contracting the edges contained in F and removing the loops and parallel edges, and call the new contracted vertex v_F. By the induction hypothesis, there is a curve between v_F and w going consecutively through the images of every vertex in G'. Let this curve be $\mathcal{D}_{G'} = \{x_1 = v_F, x_2, \ldots, x_l = w\}$. After uncontracting v_F, the curve $\mathcal{D}_{G'}$ begins at some vertex u_j in F.

Suppose $\mathcal{D}_{G'}$ enters x_2 through F'. Then we may assume that $u_j = u_2$. So we may extend the curve $\mathcal{D}_{G'}$ at one end from w to $u = u_1$ within F', and at the other end by a path from $x_j = u_2$ through $u_3, \ldots, u_k = v$ within F. Since $F \cup F' \subseteq \mathcal{R}$, we obtain our desired curve in $\mathcal{R}$ from u to v.

On the other hand, suppose that $x_2 \notin F'$. If $u_j = u_1$ then we first extend the curve $\mathcal{D}_{G'}$ from w to u_2 within F' and thence extend it from u_2 through $u_3, \ldots u_k = v$ within F. Again, we obtain our desired curve in $\mathcal{R}$ from u to v. Otherwise suppose $j \geq 2$. At one end, we first extend the curve $\mathcal{D}_{G'}$ from w to u within F' and at the other end extend it from u_j through $u_{j-1}, u_{j-2}, \ldots, u_2, u_{j+1}, u_{j+2}, \ldots, u_k = v$ within F. Again, we obtain our desired curve in $\mathcal{R}$ from u to v. $\qquad\square$

In the following, we identify vertices, edges and graphs with their respective images on the plane. For $\gamma \in \{a, b\}$, we call an edge of weight γ a γ-edge.

Proof of Theorem 2. We give a reduction from PLANAR-3-SAT to PLANAR-2-EDWP. Let $\mathcal{C}$ be a set of clauses over the variables $\mathcal{X}$, such that the bipartite graph $G = (\mathcal{X} \cup \mathcal{C}, \{xC : x \in \mathcal{X}, C \in \mathcal{C}, x \in C \vee \overline{x} \in C\})$ is planar. Without loss of generality, we can suppose that each variable appears at most three times. To see this, observe that if x appears in $k \geq 4$ clauses we may introduce k new variables, $x_1, \ldots, x_k$, and new clauses $\overline{x_1} \vee x_2$, $\overline{x_2} \vee x_3$, $\ldots$, $\overline{x_k} \vee x_1$, and replace each occurrence of x by an occurrence of one of the x_i. Without loss of generality, we can also suppose that each variable appears exactly once negatively. These transformations can clearly be implemented whilst preserving the planarity of G. Thus, we obtain a formula whose corresponding bipartite graph G has maximum degree 3.

Now take a planar embedding for G. By Lemma 2, we may find a closed curve $\mathcal{D}$ intersecting the embedding of G exactly on its vertices.

We will transform $(G, \mathcal{D})$ into an instance of PLANAR-2-EDWP in polynomial-time. To do this, we need to build an auxiliary edge-weighted planar graph G' for the routing problem. Towards this goal, we first take G and use $\mathcal{D}$ to induce an additional set of embedded a-edges whose endpoints are in $V(G)$.

Then we replace each edge $e = uv \in E(G)$ by a 4-cycle consisting of b-edges us_e, ut_e, vs_e, vt_e, where s_e and t_e are new vertices.

Next we replace each *variable* vertex $x \in X$ by a variable gadget and each *clause* vertex by a clause gadget. Each variable vertex x of degree three is replaced by one of four possible *variable gadgets*; the actual choice is dependent upon the relative position of $\mathcal{D}$ with respect to the edges incident to x and upon the sign of x in the adjacent clauses. These four gadgets are illustrated in Figure 2, where the edges corresponding to $\mathcal{D}$ and the other a-edges are dashed, the edges corresponding to $E(G)$ and the other b-edges are bold (recall there must be two edges out of the gadget for each edge in G as we initially replaced such edges by a 4-cycle). The $+$ and $-$ signs indicate whether the variable appears positively or negatively in the adjacent clause.

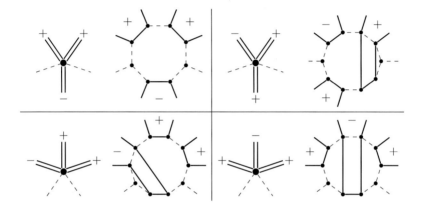

Fig. 2. Variable gadgets

Each clause vertex C of degree three is replaced by one of two possible *clause gadgets*; again, the actual choice is dependent upon the relative position of $\mathcal{D}$ with respect to the edges incident to C. These two gadgets are shown in Figure 3. The gadgets for clauses with two literals and for variables occuring only twice are similar to those presented, but simpler.

To complete the construction we need to specify the sources and the sinks. To do so, we first specify a multicommodity flow formulation with many source-sink pairs. Later we will show how to implement it as a flow with just two source-sink pairs. Towards the former goal, we will have a source-sink pair (s_e, t_e) for each edge $e \in G$. Furthermore, we will have one additional source-sink pair (s_a, t_a). To define this pair, arbitrarily choose one of the edges uv of $\mathcal{D}$. Then replace uv by two edges us_a and vt_a each with weight a. Observe that s_a and t_a are on the boundary of a common face of the resultant planar graph G'.

This multicommodity flow problem relates to the planar 3-SAT instance in the following manner.

Claim. The formula is satisfiable if and only if there are edge-disjoint paths $\{P_e\}_{e \in E(G)}$ and Q in G', with the following properties.

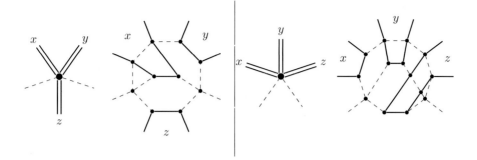

Fig. 3. Clause gadgets, with the same convention as in Figure 2

(i) P_e has endpoints s_e and t_e and uses only b-edges.
(ii) Q has endpoints s_a and t_a.

Proof. First, note the 4-cycles of b-edges that initially replaced each edge have become larger under the construction but are still b-cycles. Moreover, these b-cycles (call them H_e, for each $e \in G$) are edge disjoint and their union covers all of the b-edges in G'.

Now suppose that all the paths exist. There are only two possible routes in H_e between s_e and t_e that P_e can take; if $e = xC$ then one route passes through the variable gadget x and the other passes through the clause gadget C. Since s_e and t_e have degree two, it follows immediately that Q cannot use any of the edges incident to them. Consequently, Q must *follow* the curve $\mathcal{D}$. We will show how to obtain from Q a satisfying truth-assignment.

For any edge $e = xC$, we say that the cycle H_e is *positive* if x appears positively in C, *negative* otherwise. Then, for a variable gadget x, it is easy to see that if Q does not intersect the unique negative cycle going through the gadget then it must use at least one edge of each of the positive cycles H_e going through that gadget. If it intersects the negative cycle, set variable x to *true*, otherwise set it to *false*.

To see that this does produce a satisfying assignment, take any clause C, say over the variables x, y and z. Since Q follows $\mathcal{D}$ it must pass through each clause gadget. Consequently, Q intersects at least one of H_{xC}, H_{yC}, and H_{zC}. Without loss of generality, let it intersect H_{xC}. This means that P_{xC} cannot go through the clause gadget C and, hence, must go through the variable gadget x. But, again, as Q follows $\mathcal{D}$ it must pass through the variable gadget x too. Therefore, Q cannot intersect H_{xC} in the variable gadget x. This precisely means that x is *true* if x appears positively in C, and x is *false* if it appears negatively. So C is satisfied by x.

On the other hand, given a satisfying assignment, it is easy to find a collection of feasible paths. This is because, for each variable gadget, there is a sub-path that intersects only the positive cycles in that gadget and there is a sub-path that intersects only the negative cycle. Therefore, Q can always follow the appropriate sub-path.   □

To complete the proof of Theorem 2 we need to reduce the number of commodities in the flow to two. For this, we will keep the source-sink pair (s_a, t_a) but group into one all of the pairs (s_e, t_e) via the use of a new source sink pair (s_b, t_b). To accomplish this, we first need to position the new vertices s_b and t_b in G'. Let $\mathcal{B}$ be a closed curve that intersects G' on s_a and t_a only. Then add s_b arbitrarily on the "upper" path between s_a and t_a induced by $\mathcal{B}$. Similarly add t_b on the "lower" path between s_a and t_a induced by $\mathcal{B}$.

Our goal now is to force any path of b-edges between s_b and t_b to follow b-paths between s_e and t_e for every $e \in G$. To do this, let $e_1, e_2, \ldots, e_m$ be any ordering of the edges of G. For a cycle H_e, we define its *inside* as the connected component of $\mathbb{R}^2 \setminus H_e$ that does not contain any vertex of G'. Then set $\mathcal{R}$ to be the union of the inside of every cycle H_e plus $V(G')$ and the inside of $\mathcal{B}$. Observe that $\mathcal{R}$ is a union of disjoint balls, so its complement is connected. Let P be a path between s_b and s_{e_1} in this complement. Build a path of b-edges along P and add them to G', inserting new vertices whenever P crosses an a-edge (note that these are the only edges P can cross). Next add P to $\mathcal{R}$; this does not change the connectedness of its complementary set. In this manner, we may iteratively add paths of b-edges between t_i and s_{i+1}, for $1 \le i \le m-1$, and finally between between t_m and t_b. By construction, these paths are disjoint and cross only a-edges. We thus obtain a new planar graph G'' with four terminals on the same face, as desired.

Clearly this new instance of PLANAR-2-EDWP is equivalent to the previous multicommodity flow problem. To see this, simply note that the new b-edges are isthmi in the subgraph consisting of the b-edges. Consequently, the (s_b, t_b)-path must use each of these new b-edges and then, as before, in each H_e route through either the variable gadget or through the clause gadget. This completes the reduction. $\square$

2.3   The Hardness Result

We can now complete the proof of the approximation hardness. Observe that any vertex of degree four in G_N is incident to two edges of weight $\frac{1}{i}$ and to two edges of weight $\frac{1}{j}$, for some $i \ne j$. We construct a graph $\mathcal{G}$ by replacing each vertex of degree four with the routing graph H. We do this in such a way that the weight $\frac{1}{i}$ edges of G_N are incident to s_1 and t_1, and the weight $\frac{1}{j}$ edges are incident to s_2 and t_2. Moreover, for that copy of H placed at the intersection of P_i and P_j, we then let $a = \frac{1}{i}$ and $b = \frac{1}{j}$, where we may assume that $j < i$.

The hardness result will follow once we see how this construction relates to the original and modified weight functions.

Lemma 3. *If H is a YES-instance then the optimal disjoint weighted flow in $\mathcal{G}$ has value H_N. If H is a NO-instance then the optimal disjoint weighted flow in $\mathcal{G}$ has value at most 1.*

Proof. It is clear that if H is a YES-instance, then paths in $\mathcal{G}$ induce paths in G_N which are free to cross at any vertex without restrictions on their values. This means we obtain a flow of value H_N by using the canonical paths P_i, $1 \le i \le N$.

However, if H is a NO-instance, then it contains only an $s_1 - t_1$ path, or only an $s_2 - t_2$ path, or the $s_2 - t_2$ path is forced to use a lower weight a-edge. This implies that the induced paths $\mathcal{P}$ in G_N either do not cross at all, or if they cross then the weight of the path using the $\frac{1}{j}$-edge is forced down to a weight of $\frac{1}{i}$ (recall $j < i$). But this means that the weight of a path is upper bounded by the modified weight function $\hat{w}$. This allows us to apply Lemma 1, and hence the value of an optimal flow in this case is at most 1. $\square$

Proof of Theorem 1. It follows that if we could approximate the maximum disjoint weighted flow problem in $\mathcal{G}$ to a factor better than H_N, we could determine whether the optimal solution is 1 or H_N. This in turn would allow us to determine whether H is a YES- or a NO-instance.

Note that $\mathcal{G}$ has $n = \Theta(pN^2)$ edges, where $p = |V(H)|$. If we take $N = \Theta(p^{\frac{1}{2}(\frac{1}{\epsilon}-1)})$, where $\epsilon > 0$ is a small constant, then $\log n = \Theta(H_N) = \Theta(\log p)$. This gives our lower bound of $\Omega(\log n)$. $\square$

Similarly, if we are restricted to consider only flows that decompose into k disjoint paths then it is not hard to see that:

Theorem 3. *For undirected, planar networks, there is a $\Omega(\log k)$ hardness of approximation, unless $P = NP$, for the problem of finding a maximum flow that decomposes into at most k edge-disjoint paths.* $\square$

3   An Approximation Algorithm

Our lower bound is tight to within a constant factor - there is a simple approximation algorithm that gives an almost matching upper bound.

Theorem 4. *For any network, there is an $O(\log n)$ approximation algorithm for the maximum disjoint weighted flow problem.*

Proof. To begin, round each edge weight down to the nearest power of 2. This can only cost us a factor 2 in our approximation guarantee. Next, we claim that we may assume that every edge weight lies between $1 = 2^0$ and 2^t where $t = 1 + \lceil \log n \rceil$. To see this, first note that there can be at most n edge-disjoint $s - t$ paths in any flow. Therefore, for any j, the total contribution from all paths that contain an edge of weight 2^j or less is upper bounded by $n2^j$. Now, let 2^{j_0} be the highest edge weight such that there exists a path of weight 2^{j_0}. Deleting the edges of weight 2^j for all indices j where $2^j < \frac{1}{n}2^{j_0-1}$ loses us at most 2^{j_0-1} in weight, that is, half of the optimal flow value. The lowest remaining edge weight, 2^{j_1}, then satisfies $j_1 \geq j_0 - 1 - \lceil \log n \rceil$. Scaling down the edge weights by a factor 2^{j_1} gives the claim.

The approximation algorithm now proceeds as follows. For each i such that $0 \leq i \leq t = 1 + \lceil \log n \rceil$, let E_i be the edges of weight at least 2^i, and let $G_i = (V, E_i)$. Let ϕ_i be the maximum number of edge-disjoint $s - t$ paths in G_i. Clearly, these paths induce a weighted disjoint flow of value at least $2^i \phi_i$ in G. Furthermore the optimal weighted disjoint flow must have value at most

$\sum_{i=0}^{t} 2^i \phi_i$. To see this, note that the paths of weight 2^i in the optimal solution together form a feasible solution for the disjoint paths problem in G_i. Then, since $t = 1 + \lceil \log n \rceil$, one of the G_i produces a weighted disjoint flow whose value is at least a logarithmic fraction of the optimal flow value. As we can easily solve the maximum disjoint paths problem in G_i in polynomial time, this gives the claimed $O(\log n)$ approximation algorithm.   □

Corollary 1. *There is an $O(\log k)$ approximation algorithm for the problem of finding a maximum flow that decomposes into at most k edge-disjoint paths.*

Proof. This previous argument applies. The approximation guarantee, however, improves to $O(\log k)$ because now the paths of weight at most 2^j can only contribute a total value of at most $k2^j$.   □

4   Conclusion

We have given approximation guarantees for the maximum disjoint weighted flow problem in single-commodity networks. Therefore, a natural question would be to look at the multi-commodity case, where we wish to find weighted flows between s_i and t_i, for $i = 1, \ldots, k$, that are disjoint and maximize total weight. By the techniques of Section 3, we can easily obtain an upper bound of $O(\alpha \log n)$, where α is the approximation achievable in the unweighted case. Unfortunately, the unweighted version is extremely hard to approximate since it is the edge-disjoint paths problem studied by Guruswami et al. [8]. They show this problem is inapproximable to within $\alpha = m^{\frac{1}{2} - \epsilon}$, for any $\epsilon > 0$, in directed graphs and give an approximation algorithm that essentially matches this lower bound.

In addition, given that our lower bound is essentially tight, the search for bi-criteria results is of interest. Here we would relax the condition that the paths in a weighted flow be strictly disjoint; instead, one would allow a limited amount $c \geq 2$ of congestion on each edge. For multi-commodity flows, the unweighted version of the problem has recently been studied extensively; for ground-breaking results in this area, see Chekuri et al. [3] for upper bounds in planar graphs, and Andrews et al. [1] and Chuzhoy et al. [4] for lower bounds in general graphs.

References

1. Andrews, M., Chuzhoy, J., Guruswami, V., Khanna, S., Talwar, K., Zhang, L.: Inapproximability of edge-disjoint paths and low congestion routing on undirected graphs. Electronic Colloquium on Computational Complexity 14(113) (2007)
2. Ahuja, R., Magnanti, T., Orlin, J.: Network Flows: Theory, Algorithms, and Applications. Prentice-Hall, Englewood Cliffs (1993)
3. Chekuri, C., Khanna, S., Shepherd, B.: Edge-disjoint paths in planar graphs with constant congestion. SIAM J. Computing 39(1), 281–301 (2009)
4. Chuzhoy, J., Guruswami, V., Khanna, S., Talwar, K.: Hardness of routing with congestion in directed graphs. In: Proceedings of the 39th ACM Symposium on Theory of Computing, STOC (2007)

5. Dinitz, Y., Garg, N., Goemans, M.: On the single-source unsplittable flow problem. Combinatorica 19, 17–41 (1999)
6. Fortune, S., Hopcroft, J., Wyllie, J.: The directed subgraph homeomorphism problem. Theoretical Computer Science 10, 111–121 (1980)
7. Baier, G., Kohler, E., Skutella, M.: The k-splittable flow problem. Algorithmica 42, 231–248 (2005)
8. Guruswami, V., Khanna, S., Rajaraman, R., Shepherd, B., Yannakakis, M.: Near-optimal hardness results and approximation algorithms for edge-disjoint paths and related problems. Journal of Computer and System Sciences 67(3), 473–496 (2003)
9. Kleinberg, J.: Single-source unsplittable flow. In: Proceedings of the 37th on Foundations of Computer Science (FOCS), pp. 68–77 (1996)
10. Koch, R., Spenke, I.: Complexity and approximability of k-splittable flows. Theoretical Computer Science 369, 338–347 (2006)
11. Salazar, F., Skutella, M.: Single-source k-splittable min-cost flows. Operations Research Letters 37(2), 71–74 (2009)
12. Shepherd, B.: Single-sink multicommodity flow with side constraints. In: Cook, W., Lovasz, L., Vygen, J. (eds.) Research Trends in Combinatorial Optimization, pp. 429–450. Springer, Heidelberg (2009)
13. Schrijver, A.: Combinatorial Optimization: Polyhedra and Efficiency. Springer, Heidelberg (2003)

Approximation Algorithms for Reliable Stochastic Combinatorial Optimization

Evdokia Nikolova[*]

Massachusetts Institute of Technology
nikolova @mit.edu

Abstract. We consider optimization problems that can be formulated as minimizing the cost of a feasible solution $\mathbf{w}^T\mathbf{x}$ over an arbitrary combinatorial feasible set $\mathcal{F} \subset \{0,1\}^n$. For these problems we describe a broad class of corresponding stochastic problems where the cost vector $\mathbf{W}$ has independent random components, unknown at the time of solution. A natural and important objective that incorporates risk in this stochastic setting is to look for a feasible solution whose stochastic cost has a small tail or a small convex combination of mean and standard deviation. Our models can be equivalently reformulated as nonconvex programs for which no efficient algorithms are known. In this paper, we make progress on these hard problems.

Our results are several efficient general-purpose approximation schemes. They use as a black-box (exact or approximate) the solution to the underlying deterministic problem and thus immediately apply to arbitrary combinatorial problems. For example, from an available δ-approximation algorithm to the linear problem, we construct a $\delta(1 + \epsilon)$-approximation algorithm for the stochastic problem, which invokes the linear algorithm only a logarithmic number of times in the problem input (and polynomial in $\frac{1}{\epsilon}$), for any desired accuracy level $\epsilon > 0$. The algorithms are based on a geometric analysis of the curvature and approximability of the nonlinear level sets of the objective functions.

Keywords: Approximation algorithms, reliable optimization, stochastic optimization, risk, mean-risk, nonlinear programming, nonconvex optimization.

1 Introduction

In this paper, we consider generic combinatorial problems and ask what happens when their associated costs are stochastic. The most common approaches in stochastic optimization are to find the solution of minimum expected cost. However, in many applications reliability considerations are very important: risk-averse users need reassurance regarding the level of risk, and not just the expected cost of the provided solution. For example, the transportation community

---

[*] This work was supported in part by the National Science Foundation under grant 0931550.

M. Serna et al. (Eds.): APPROX and RANDOM 2010, LNCS 6302, pp. 338–351, 2010.

has recognized the importance of reliable route plans (*e.g.,* [7,27,24,36,9]), however the solutions offered are typically inefficient or heuristic with unknown approximation guarantee. Similarly, reliability is a key consideration in finance and
other *continuous* optimization settings [33]. It has been noted that incorporating
reliability [33,28] transforms the problems into nonconvex ones for which there
are no known efficient algorithms and rigorous approximative analysis is scarce.
In this paper, we provide a rigorous treatment of reliable combinatorial optimization, offering fully-polynomial approximation schemes for a rich framework
of reliability measures.

To illustrate our framework, consider an application such as driving to the
airport in uncertain traffic. Our goal is to find a route that gets us to the airport
on time. Clearly, the route which minimizes our expected travel time may not
be an appropriate choice. In fact, the natural objectives may vary depending on
when we are submitting the route query: ahead of time, when we are debating
how much time to budget for our trip, or at the start of our trip, when we
are optimizing our chance of ontime arrival. In the former setting, we would
typically want to allocate enough time to ensure some confidence of ontime
arrival, say 95%. In the latter, given a deadline to reach our destination, we
need to find the route which will most likely reach by the deadline. Another
natural objective, used for example by the Federal Highway Administration as
a travel time reliability criterion, is given by the mean plus standard deviation
of a route [10]. The latter reliability criterion has been considered in the context
of stochastic minimum spanning treess as well [2], and this model is sometimes
referred to as mean-risk optimization (*e.g.,* [2]).

We thus focus on a general framework for reliable stochastic combinatorial
optimization, which includes the following problem settings:

1. minimize (*mean* + *c · standard deviation*) for a non-negative constant c
   which parametrizes the level of risk-aversion. [Call this the *Mean-risk model*
   or objective.]
2. maximize $\Pr(solution\ cost \leq budget)$ for a given *budget*. [*Probability tail
   model* / objective.]
3. minimize *budget* such that $\Pr(solution\ cost \leq budget) \geq p$ for a given confidence probability p. [*Value-at-risk model.*]

In contrast with the diversity in model specifications above, we will show that the
same approximation algorithm design can simultaneously address all. Throughout, we assume that the cost distributions are independent, although our algorithms also extend to the case of correlations of neighboring edges for example
in shortest path problems (the graph with correlated edges is transformed into
a slightly larger graph with independent edges and thus all our results here
immediately carry through.)

Contributions. We start our discussion with the (relatively) simpler mean-risk
model, which is equivalent to minimizing $(mean + c \cdot \sqrt{variance})$. We provide
strong results that apply to *arbitrary* cost distributions with given means and

variances, and achieve essentially the same approximation factor as what is possible for the underlying deterministic problem. In particular, we provide general-purpose algorithms that use as a black-box an algorithm for the deterministic problem. We summarize our results for this setting below:

Theorem 1 (See Theorems 3, 4). *There is a fully-polynomial approximation scheme for the mean-risk stochastic model, when there is an exact or fully-polynomial approximation algorithm for the underlying deterministic problem.*

In addition, there is a $(1+\epsilon)\delta$-approximation for the stochastic model running in time polynomial in $\frac{1}{\epsilon}$, when there is an available δ-approximation for the deterministic problem.

A rigorous approximation-algorithmic analysis of the second and third models in the framework, which involve optimization of the probability tails, necessitates an assumption on the distribution: in the absence of any knowledge on the distributions, the best one can do is bound the tails, for example using Chernoff or Chebyshev bounds, and optimize those tail bounds instead—this will yield a conservative overestimate of the probability of exceeding the budget.

We provide strict approximation results under the commonly assumed Gaussian distributions; we then show how the same algorithmic techniques can apply to arbitrary distributions using tail bounds. In the former setting, minimizing the probability tail in the second model is equivalent to maximizing $\frac{budget-mean}{\sqrt{variance}}$ and we get the following approximations:

Theorem 2 (See Theorems 3, 5). *There is a fully-polynomial approximation scheme for the probability tail model, when there is an exact or fully-polynomial approximation algorithm for the underlying deterministic problem.*

In addition, when there is an available δ-approximation for the deterministic problem, there is a $\sqrt{1 - \left[\frac{\delta-(1-\epsilon^2/4)}{(2+\epsilon)\epsilon/4}\right]}$-approximation for the stochastic model running in time polynomial in $\frac{1}{\epsilon}$.

We remark that the above algorithms find the approximate solution, assuming there is a feasible solution with expected cost at most the budget, or $(1-\epsilon)$ times the budget in the exact and approximate deterministic settings respectively (in other words, the probability of exceeding the budget is at most $\frac{1}{2}$). Otherwise, if a given budget is so small that the probability of exceeding it is greater than $\frac{1}{2}$, we are in a risk-loving, rather than a risk-averse situation, which would be similar to minimizing a *(mean − standard deviation)*-type objective in model (1). In other words, we would prefer solutions with higher variances (for example, looking for longest paths).

The third (value-at-risk) model under Gaussian distributions is equivalent to the mean-risk model, with risk-aversion coefficient $c = \Phi^{-1}(p)$, where $\Phi^{-1}(\cdot)$ is the inverse cumulative distribution function of the standard normal $N(0,1)$.

For arbitrary distributions, the third model again reduces to the mean-risk model, but with a more conservative risk-aversion coefficient $c = \sqrt{\frac{p}{1-p}}$, as a

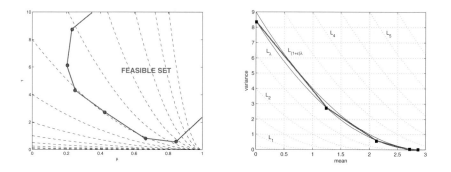

Fig. 1. *(a)* Level sets of the probability tail objective function and the convex hull of the projected feasible set on the mean-variance plane. *(b)* Level sets and approximate separation oracle for the mean-risk objective on the mean-variance plane.

result of which our algorithms provide an overestimate of the true error probability of exceeding the budget. Optimizing a tail bound in the second model similarly provides an overestimate of the true probability, which is again the best one can hope to achieve in the absence of other distributional information.

Background and Challenges. Our algorithms build on the fact that the model formulations in our framework are all instances of concave minimization, for which it is known that the optimal solution is attained at an extreme point of the feasible set (see, *e.g.,* [4]). In particular, our objective functions depend only on the means and variances of feasible solutions. Thus, we can project the feasible set on the plane spanned by the mean and variance vectors and only consider extreme points on the projection (see Figure 1(a)). This greatly restricts the number of relevant extreme points. For example, in the case of minimum spanning trees and matroids there are only polynomially many such extreme points, which can be efficiently enumerated, hence the corresponding reliable spanning trees and matroids in a stochastic environment can be found with a straightforward polynomial-time algorithm. However, an arbitrary combinatorial problem would most likely have too many extreme points even on a two-dimensional projection (for example, shortest paths have $n^{\log n}$ such points [29]), hence our focus on approximation in this paper.

We can geometrically visualize the objective function in terms of its level sets on the mean-variance plane. These form parabolas, corresponding to higher objective function values at greater mean and variance values. The optimal solution is obtained at the lowest parabola touching the projected feasible set. Figure 1(a) depicts these parabolas and the challenge that arises with concave minimization problems: along the convex hull boundary of the feasible set, the objective function fluctuates and, in particular, many extreme points may be local optima and thus local search algorithms would fail to find a good approximation. What we do instead is follow the objective function levels to guide us into the relevant portion of the feasible set, as explained below.

Overview of Algorithms and Techniques. [For the case of easy deterministic problems.] The algorithm constructs a (non-linear) separation oracle for telling us whether, for a given function level set,[1] there is a feasible solution below the level set (with value less than the given function value) or else, whether the entire feasible set is above the given level. Afterwords, a binary search on the optimum objective function value combined with the separation oracle finds the desired approximate solution.

The separation oracle approximates a given level set curve by inscribing a (partial) polygon in it. Each side of the polygon induces a linear objective over the feasible set, which we minimize via a black-box call to the algorithm for the deterministic problem. If the resulting solution is below the current level set (more precisely, its associated original objective function value is smaller than $(1 + \epsilon)$ times the given level), the separation oracle returns that solution. Else, if after minimizing with respect to all linear segments, we do not find any solutions below the level set, the separation oracle returns a negative answer that the entire feasible set is above it.

The subtlety arises in how to construct the polygonal segments to ensure a good and efficient approximation. To get an efficient algorithm, we need to approximate the level set curves with as few linear segments as possible. On the other hand, to get a good approximation factor, we need a finer polygon (with more and smaller sides), which is sandwiched between the desired level set with function value λ and the level set with function value $\lambda(1 + \epsilon)$ (See Figure 1(b)). In particular, in the worst case when the level sets touch, as is the case for the probability tail objective, a polygon sandwiched between the two level sets will have infinitely many sides. We resolve this problem by carefully bounding the optimal solution so that we do not need all infinitely many linear segments from the polygon, and we prove that it suffices to consider only polynomially many such segments.

[Hard deterministic problems.] We could use the same algorithm design as above, by appropriately modifying its analysis and approximation factors, when we have a δ-approximation rather than an exact algorithm for solving the underlying deterministic problem. It turns out that for this case, a cruder and simpler algorithm gives the same approximation factor. In particular, all we need to do here is apply the algorithm for the deterministic problem on a small sequence of linear cost functions of the form $mean + k \cdot variance$, for a geometric progression of coefficients k.

However, even if we know what single choice of k would find the optimal solution, the difficulty is to translate the approximation given by the deterministic black-box algorithm for the *linear function* into an approximation for the *original concave function*: the two functions have nothing in common (except that the former is a gradient of the latter at some point), and a priori it is not clear that an approximation of the former would at all yield a meaningful approximation factor for the original objective. Fortunately, all objective functions in our

---

[1] The level set of a function f for value λ is the subset of the domain on which the function equals λ, $L_\lambda = \{\mathbf{x} \mid f(\mathbf{x}) = \lambda\}$.

framework admit such an approximation (the probability tail objective is again more challenging due to the given budget and requires us to know that there is a feasible solution at least a small distance away from the budget).

Related Work. A rich body of work in stochastic combinatorial optimization focuses on two-stage and multistage optimization (*e.g.,* [35,17,21,16,18]). The models there typically look for solutions of minimum expected cost, and Swamy and Shmoys remark that "it would be interesting to explore stochastic models that incorporate risk" [38]. There are models that incorporate additional budget constraints [37] or threshold constraints for specific problems such as knapsack, load balancing and others [8,13,23].

At the other end of the spectrum is the paradigm of robust optimization (see survey [5]), which provides completely reliable (robust) solutions, though this is only possible when the uncertainty is bounded, namely the random variables have bounded support. Our framework for reliable optimization falls between stochastic optimization, which minimizes expected cost, and robust optimization, which minimizes the maximum cost. Interestingly, part of our framework (the mean-risk model) arises in robust discrete optimization under ellipsoidal uncertainty sets [6]. Bertsimas and Sim offer for it pseudopolynomial algorithms, assuming that the underlying deterministic problem can be solved exactly, in contrast with our fully polynomial approximation schemes that work with both exact and approximate algorithms for the deterministic problem.

Atamtürk and Narayanan [2] also consider mean-risk minimization in discrete optimization, giving a characterization in terms of submodular minimization. Our feasible set is an arbitrary subset of the hypercube vertices, on which it is not known how to do submodular minimization.

The probability tail objective was previously considered in the special context of stochastic shortest paths and an exact algorithm was given based on enumerating relevant extreme points from the path polytope [29]. The same type of algorithm extends to arbitrary combinatorial problems and its complexity is polynomial for minimum spanning trees and matroids. However, in general, it is superpolynomial or exponential, hence our focus on approximation algorithms in this paper.

A comprehensive survey of models that incorporate risk in *continuous* settings is provided by Rockafellar [33]. The solution concepts and continuous nature of the problems make this work very different from ours. Similarly, continuous optimization work with probability (chance) constraints (*e.g.,* [28]) applies for linear and *not* discrete optimization problems. Additional related work on the combinatorial optimization side includes research on multi-criteria optimization (*e.g.,* [31,1,34,39]) and combinatorial optimization with a ratio of linear objectives [26,32]. Our models can also be seen as instances of concave discrete minimization; however, the existing work in this area requires assumptions that do not hold in our framework, such as restrictive properties on the feasible set, strictly positive range of the objective function, or boundedness/positivity of the objective function gradient [30,3,22,14].

2 An FPTAS for the Reliable Versions of Easy Combinatorial Problems

In this section, we formally define the models in our reliable stochastic optimization framework and present a general-purpose FPTAS design for these problems. The FPTAS uses as a black-box an exact algorithm for the underlying deterministic problem and is based on a geometric analysis of the curvature and approximability of the level sets of the objective functions.

Suppose we have an arbitrary combinatorial set of feasible solutions $\mathcal{F} \subset \{0,1\}^n$, together with an oracle for optimizing linear objectives over the set. In addition, we are given nonnegative vectors of means $\boldsymbol{\mu} \in \mathbf{R}^n$ and variances $\boldsymbol{\tau} \in \mathbf{R}^n$ for the stochastic cost vector $\mathbf{W}$, coming from independent distributions so that the mean and variance of a solution $\mathbf{x} \in \mathcal{F}$ is $\boldsymbol{\mu}^T \mathbf{x}$ and $\boldsymbol{\tau}^T \mathbf{x} \geq 0$ respectively. We are interested in finding a feasible solution with optimal cost, where the notion of optimality incorporates risk.

1. *[Mean-risk model]* A family of objectives that has been analyzed in continuous optimization settings, mostly in the context of finance [11,25], is the family of convex combinations of mean and standard deviation. Formally, this problem is to:

$$\text{minimize} \quad \boldsymbol{\mu}^T \mathbf{x} + c\sqrt{\boldsymbol{\tau}^T \mathbf{x}} \qquad (1)$$
$$\text{subject to} \quad \mathbf{x} \in \mathcal{F},$$

   where the constant c parametrizes the degree of the user's risk aversion.

2. *[Probability tail model]* An alternative natural model maximizes the probability that the stochastic solution cost is within a desired budget or threshold t: maximize $\Pr\left(\mathbf{W}^T \mathbf{x} \leq t\right)$ subject to $\mathbf{x} \in \mathcal{F}$. When the stochastic costs $\mathbf{W}$ are Gaussian, subtracting the mean and dividing by the standard deviation transforms the problem into the following equivalent formulation (which is also approximation-preserving as we show in the extended version):

$$\text{maximize} \quad \frac{t - \boldsymbol{\mu}^T \mathbf{x}}{\sqrt{\boldsymbol{\tau}^T \mathbf{x}}} \qquad (2)$$
$$\text{subject to} \quad \mathbf{x} \in \mathcal{F}.$$

When the stochastic costs $\mathbf{W}$ come from arbitrary distributions, the maximum probability is lower-bounded by $\frac{(t-\boldsymbol{\mu}^T\mathbf{x})^2}{(t-\boldsymbol{\mu}^T\mathbf{x})^2+(\boldsymbol{\tau}^T\mathbf{x})}$ (by the one-sided Chebyshev bound, also known as Cantelli's inequality [15], $\Pr(X \leq E[X] + k\sqrt{Var(X)}) \geq 1 - \frac{1}{1+k^2}$, with $k = \frac{t-\boldsymbol{\mu}^T\mathbf{x}}{\sqrt{\boldsymbol{\tau}^T\mathbf{x}}}$). While maximizing a lower-bound will not yield a strict approximation of the probability tail objective, it is the best one can achieve in the absence of other distributional information—and our techniques can strictly approximate this bound as well:

$$\text{maximize} \quad \frac{(t - \boldsymbol{\mu}^T \mathbf{x})^2}{(t - \boldsymbol{\mu}^T \mathbf{x})^2 + \boldsymbol{\tau}^T \mathbf{x}} \qquad (3)$$
$$\text{subject to} \quad \mathbf{x} \in \mathcal{F}.$$

3. *[Value-at-risk model]* Finally, we may wish to minimize the budget t such that the probability of not exceeding it is at least a given confidence level p:

$$\text{minimize} \quad t \tag{4}$$
$$\text{subject to} \quad \Pr(\mathbf{W}^T\mathbf{x} \leq t) \geq p$$
$$\mathbf{x} \in \mathcal{F}.$$

Depending on whether we have Gaussian or arbitrary distributions, this problem is exactly equivalent to, or its solution can be upper-bounded using Chebyshev's bound by the mean-risk model (1) with $c = \Phi^{-1}(p)$ or $c = \sqrt{\frac{p}{1-p}}$ (See Ghaoui *et al.* [12]; more details are provided in the extended version of this paper).

We can obtain fully-polynomial approximation schemes (FPTAS) for all models above, with the same FPTAS template, which we explain below. All models are instances of concave minimization (equivalently, convex maximization) over $\mathbf{x} \in \mathcal{F}$. Our algorithms make black-box calls to an exact algorithm (sometimes referred to as the *linear oracle*) for solving the underlying deterministic (linear) problem:

$$\text{minimize} \quad \mathbf{w}^T\mathbf{x} \tag{5}$$
$$\text{subject to} \quad \mathbf{x} \in \mathcal{F},$$

for a carefully chosen *small* set of linear objectives $\mathbf{w} \geq 0$. We remark that in general such a set may not even exist; for example, the necessary number of linear objectives may be large or even infinite if the objective function has unbounded gradient (as is the case in the second model above). From a complexity perspective, minimizing a concave function over some feasible set may be hard to approximate even if minimizing a linear function over the same set can be done in polynomial time [22].

Thanks to the form of the objective functions, they can all be projected onto the mean-variance plane $span(\boldsymbol{\mu}, \boldsymbol{\tau})$ and can be thought of as functions on two dimensions. In that plane, the projected level sets of the objective functions are parabolas. We construct an approximate separation oracle, which tells us whether for a given function value λ there is a feasible solution below the $(1-\epsilon)\lambda$-level set or else if the entire feasible set is above the λ-level set. We do this by inscribing a (partial) polygon between these two level sets. Geometrically, the optimal polygon choice (with fewest sides) is such that its vertices are on one level set and its sides are tangent to the other, as shown in Figure 1(b).

Theorem 3. *There is an oracle fully-polynomial time approximation scheme for all problems in the reliable stochastic framework above, which uses as a black-box an exact algorithm for solving the underlying deterministic problem (5).*

In the rest of this section we prove this theorem. The crux of the proof is in establishing that the approximate separation oracle can be constructed from polynomially many linear segments, as described in the following main technical lemma. (The Lemma is stated for a stochastic maximization problem as in Eq. (2); the analogous statement holds for a stochastic minimization problem as in Eq. (1).) The argument for how the theorem follows from the Lemma is provided in the extended version.

Lemma 1 (Approximate Separation Oracle). *Suppose we have an exact algorithm for solving the deterministic problem (5). Then, we can construct an oracle which solves the following approximate separation problem: given a level λ and $\epsilon \in (0, 1)$, the oracle returns*

1. *A solution $\mathbf{x} \in \mathcal{F}$ with $f(\mathbf{x}) \geq (1 - \epsilon)\lambda$,   or*
2. *An answer that $f(\mathbf{x}) < \lambda$ for all $\mathbf{x} \in \mathcal{F}$,*

and the number of linear oracle calls it makes is polynomial in $\frac{1}{\epsilon}$ and the size of the input.

The proof-construction of the Approximate Separation Oracle from Lemma 1 follows from a series of lemmas about bounding the size and number of the linear segments that approximate a level set and comprise the separation oracle. Since the level sets and their position with respect to each other is different for the different objectives, the actual computations of the size and number of linear segments differs. For lack of space we provide the proof for the probability tail formulation (2), which is more subtle due to the budget threshold and the fact the level sets are tangent to each other. The proofs for the remaining objectives are analogous.

Consider the lower level sets $\underline{L}_\lambda = \{\mathbf{z} \mid f(\mathbf{z}) \leq \lambda\}$ of the objective function $f(m, s) = \frac{t-m}{\sqrt{s}}$, where $m, s \in \mathbf{R}$. Denote $L_\lambda = \{\mathbf{z} \mid f(\mathbf{z}) = \lambda\}$. We will prove that any level set boundary can be approximated by a small number of linear segments. The main work here involves deriving a condition for a linear segment with endpoints on L_λ, to have objective function values within $(1 - \epsilon)$ of λ.

Lemma 2. *Consider the points $(m_1, s_1), (m_2, s_2) \in L_\lambda$ with $s_1 > s_2 > 0$. The segment connecting these two points is contained in the level set region $\underline{L}_\lambda \backslash \underline{L}_{\lambda(1-\epsilon)}$ whenever $s_2 \geq (1 - \epsilon)^4 s_1$, for every $\epsilon \in (0, 1)$.*

Using Lemma 2, we show that any level set L_λ can be approximated within a multiplicative factor of $(1-\epsilon)$ via a small number of segments. Let s_{min} and s_{max} be a lower and upper bound respectively for the variance of the optimal solution. For example, take s_{min} to be the smallest positive coordinate of the variance vector, and s_{max} the variance of the feasible solution with smallest mean.

Lemma 3. *The level set $L_\lambda = \{(m, s) \in \mathbf{R}^2 \mid \frac{t-m}{\sqrt{s}} = \lambda\}$ can be approximated within a factor of $(1 - \epsilon)$ by $\left\lceil \frac{1}{4} \log \left(\frac{s_{max}}{s_{min}} \right) / \log \frac{1}{1-\epsilon} \right\rceil$ linear segments.*

The above lemma yields the approximate separation oracle for the level set L_λ and the feasible set $\mathcal{F}$, by applying the black-box algorithm for the deterministic problem to cost vectors $a\boldsymbol{\mu} + \boldsymbol{\tau}$, for all possible slopes $(-a)$ of the segments approximating the level set. This concludes the proof-construction for the separation oracle in Lemma 1.

3   Approximating the Reliable Versions of Hard Combinatorial Problems

In this section, we show that a δ-approximate oracle to the deterministic problem (5), also called the linear oracle, can be used to construct efficient approximation algorithms for the reliable stochastic models. As in the approximative analysis for easy combinatorial problems, we first check whether the optimal solution has zero variance and if not, proceed with the algorithm and analysis below.

We can use the same approximation algorithm template that constructs a separation oracle as in the previous section, but it turns out that a cruder algorithm which simply tests a geometric progression of mean-variance tradeoffs provides the same approximation guarantees. The main technical challenge in the algorithm analysis is that even if we know the optimal mean-variance tradeoff to query from the black-box algorithm for the deterministic problem, it is not obvious and not intuitive what approximation factor one can get for the reliable objectives from the δ-approximation factor for the deterministic one.

We obtain a very strong result for the relatively simpler mean-risk objective— we can get essentially the same approximation factor as the available one for the deterministic problem:

Theorem 4. *Suppose we have a δ-approximation oracle for solving the deterministic combinatorial problem (5). The mean-risk model (1) can be approximated to a multiplicative factor of $\delta(1+\epsilon)$ by calling the oracle for the deterministic problem polynomially many times in the input size and $\frac{1}{\epsilon}$.*

We can also get the following approximation for the probability tail formulation (2):

Theorem 5. *Suppose we have a δ-approximation oracle for solving the deterministic combinatorial problem (5). The probability tail model (2) has a $\sqrt{1 - \left[\frac{\delta-(1-\epsilon^2/4)}{(2+\epsilon)\epsilon/4}\right]}$ -approximation algorithm that calls the algorithm for the deterministic problem polynomially many times in $\frac{1}{\epsilon}$ and the input size, assuming the optimal solution to (2) satisfies $\boldsymbol{\mu}^T \mathbf{x}^* \leq (1-\epsilon)t$.*

The high-level analysis for these approximation algorithms is the same; it differs in the computation of the approximation factors. For lack of space, we only offer an overview of the proof of Theorem 5; the remaining details for both theorems are in the extended version.

We first prove several geometric lemmas that enable us to derive the approximation factor. The first lemma is key for the transition from approximating a linear objective (by the algorithm for the deterministic problem) to approximating the probability tail objective.

Lemma 4 (Geometric lemma). *Consider two objective function values $\lambda^* > \lambda$ and points $(m^*, s^*) \in L_{\lambda^*}$, $(m, s) \in L_\lambda$ with positive coordinates, such that* the tangents to the points at the corresponding level sets are parallel. Then, the y-intercepts b^*, b of the two tangent lines satisfy

$$b - b^* = s^* \left[1 - \left(\frac{\lambda}{\lambda^*} \right)^2 \right].$$

The next lemma shows that if we know the optimal linear objective to use with the available δ-approximate algorithm for the deterministic problem (5), then we can approximate the optimal solution well.

Lemma 5 (Optimal Linear Objective Lemma). *Suppose we have a δ-approximate linear oracle for optimizing over the feasible set $\mathcal{F}$ and suppose that the optimal solution satisfies $\boldsymbol{\mu}^T \mathbf{x}^* \leq (1 - \epsilon)t$. If we can guess the slope of the tangent to the corresponding level set at the optimal point $\mathbf{x}^*$, then we can find a $\sqrt{1 - \delta \frac{2-\epsilon}{\epsilon}}$-approximate solution to the nonconvex problem (2).*

In particular, setting $\epsilon = \sqrt{\delta}$ gives a $(1 - \sqrt{\delta})$-approximate solution.

Next, we give a geometric lemma that is needed to analyze the approximation factor we get when applying the linear oracle on an approximately optimal slope.

Lemma 6. *Consider the level set L_λ and points (m^*, s^*) and (m, s) on it, at which the tangents to L_λ have slopes $-a$ and $-a(1 + \xi)$ respectively. Let the y-intercepts of the tangent line at (m, s) and the line parallel to it through (m^*, s^*) be b_1 and b respectively. Then $\frac{b}{b_1} \leq \frac{1}{1 - \xi^2}$.*

We now show that we get a good approximation even when we use an approximately optimal linear objective with our linear oracle.

Lemma 7. *Suppose that we use an approximately optimal linear objective with a δ-approximate linear oracle for solving the probability tail model (2). In particular, suppose the linear objective (slope) that we use is within $(1 + \xi)$ of the slope of the tangent at the optimal solution. Then this will give a solution to the probability tail model (2) with value at least $\sqrt{1 - \left[\frac{\delta}{1 - \xi^2} - 1 \right] \frac{2-\epsilon}{\epsilon}}$ times the optimal, provided the optimal solution satisfies $\boldsymbol{\mu}^T \mathbf{x}^* \leq (1 - \epsilon)t$.*

Consequently, we can approximate the optimal solution by applying the approximate linear oracle on a small number of appropriately chosen linear functions and picking the best resulting solution, as explained in the proof of Theorem 5 in the extended version.

When $\delta = 1$, that is when we can solve the underlying linear problem exactly in polynomial time, the above algorithm gives an approximation factor of $\sqrt{\frac{1}{1+\epsilon/2}}$, or equivalently $1 - \epsilon'$, where $\epsilon = 2\left[\frac{1}{(1-\epsilon')^2} - 1 \right]$. While this algorithm is still an oracle-fully polynomial time approximation scheme, it gives a bi-criteria approximation: it requires that there is a small gap between the mean of the optimal solution and the budget t so it is weaker than our previous algorithm, which had no such requirement. This is expected since, of course, this algorithm is cruder, simply taking a geometric progression of linear functions rather than tailoring the black-box algorithm calls for the deterministic problem to the objective function value that it is searching for, as in the approximate separation oracle that the FPTAS from the previous section is based on.

4   Conclusion

We have presented a framework for reliable stochastic combinatorial optimization that includes mean-risk minimization and models involving the probability tail of the stochastic cost of a solution. Our algorithms are independent of the feasible set structure and use solutions for the underlying linear (deterministic) problems as oracles for solving the corresponding stochastic models. As such, they apply to very general combinatorial settings for which *exact* or *approximate* linear oracles are available.

Our primary motivation for this work was to design an approximation algorithm for finding the most reliable route in a network with uncertain edge delays (in the sense that the route maximizes the probability of arriving on time under a given deadline), which consequently extended to the rich class of problems and reliability models considered here. An implementation of our approximation algorithm in the context of reliable routes reveals that they are also very practical: for example, they achieve 99.9%-accuracy with only up to 6 iterations of an algorithm for the deterministic problem.

In future work, it would be interesting to extend our offline stochastic models to online models, as has previously been done with offline linear to online linear problems [20,19]. It would be also useful to consider adaptive stochastic reliability models, building on the framework of multistage stochastic optimization.

Acknowledgments. The author thanks Hari Balakrishnan, Dimitris Bertsimas, Shuchi Chawla, Costis Daskalakis, Nick Harvey, Michel Goemans, David Karger, David Kempe, Asu Ozdaglar, Christos Papadimitriou, Pablo Parrilo, Andreas Schulz, Nicolás Stier Moses, Madhu Sudan and John Tsitsiklis for valuable suggestions at various stages of this work.

References

1. Ackermann, H., Newman, A., Röglin, H., Vöcking, B.: Decision making based on approximate and smoothed pareto curves. In: Deng, X., Du, D.-Z. (eds.) ISAAC 2005. LNCS, vol. 3827, pp. 675–684. Springer, Heidelberg (2005)
2. Atamtürk, A., Narayanan, V.: Polymatroids and risk minimization in discrete optimization. Operations Research Letters 36, 618–622 (2008)
3. Berstein, Y., Lee, J., Onn, S., Weismantel, R.: Nonlinear optimization for matroid intersection and extensions. Manuscript at arXiv:0807.3907 (2008)
4. Bertsekas, D., Nedić, A., Ozdaglar, A.: Convex Analysis and Optimization. Athena Scientific, Belmont (2003)
5. Bertsimas, D., Brown, D., Caramanis, C.: Theory and applications of robust optimization (2007) (manuscript)

6. Bertsimas, D., Sim, M.: Robust discrete optimization and network flows (2004) (manuscript)
7. Chen, A., Ji, Z.: Path finding under uncertainty. Journal of advanced transportation 39(1), 19–37 (2005)
8. Dean, B., Goemans, M.X., Vondrák, J.: Approximating the stochastic knapsack: the benefit of adaptivity. In: Proceedings of the 45th Annual Symposium on Foundations of Computer Science, pp. 208–217 (2004)
9. Fan, Y., Kalaba, R., Moore, I.J.E.: Arriving on time. Journal of Optimization Theory and Applications 127(3), 497–513 (2005)
10. Federal Highway Administration: Traffic Congestion and Reliability: Trends and advanced strategies for congestion mitigation. Cambridge Systematics Inc., Texas Transportation Institute (2005)
11. Föllmer, H., Schied, A.: Stochastic Finance: An Introduction in Discrete Time. Walter de Gruyter, Berlin (2004)
12. Ghaoui, L.E., Oks, M., Oustry, F.: Worst-case value-at-risk and robust portfolio optimization: A conic programming approach. Oper. Res. 51(4), 543–556 (2003)
13. Goel, A., Indyk, P.: Stochastic load balancing and related problems. In: Proceedings of the 40th Symposium on Foundations of Computer Science (1999)
14. Goyal, V.: An FPTAS for minimizing a class of quasi-concave functions over a convex set. Technical Report Tepper WP 2008-E24, Carnegie Mellon University Tepper School of Business (2008)
15. Grimmett, G., Stirzaker, D.: Probability and Random Processes, 3rd edn. Oxford Univ. Press, Oxford (2001)
16. Gupta, A., Pál, M., Ravi, R., Sinha, A.: Boosted sampling: approximation algorithms for stochastic optimization. In: Proceedings of the 36th Annual ACM Symposium on Theory of Computing, Chicago, IL, USA, pp. 365–372 (2004)
17. Gupta, A., Pál, M., Ravi, R., Sinha, A.: What about Wednesday? Approximation algorithms for multistage stochastic optimization. In: Chekuri, C., Jansen, K., Rolim, J.D.P., Trevisan, L. (eds.) APPROX 2005 and RANDOM 2005. LNCS, vol. 3624, pp. 86–98. Springer, Heidelberg (2005)
18. Immorlica, N., Karger, D., Minkoff, M., Mirrokni, V.S.: On the costs and benefits of procrastination: approximation algorithms for stochastic combinatorial optimization problems. In: Proceedings of the Fifteenth Annual ACM-SIAM Symposium on Discrete Algorithms, pp. 691–700 (2004)
19. Kakade, S.M., Kalai, A.T., Ligett, K.: Playing games with approximation algorithms. In: STOC 2007: Proceedings of the Thirty-Ninth Annual ACM Symposium on Theory of Computing, pp. 546–555. ACM, New York (2007)
20. Kalai, A., Vempala, S.: Efficient algorithms for on-line optimization. Journal of Computer and System Sciences 71, 291–307 (2005)
21. Katriel, I., Kenyon-Mathieu, C., Upfal, E.: Commitment under uncertainty: Two-stage stochastic matching problems. Theor. Comput. Sci. 408(2-3), 213–223 (2008)
22. Kelner, J.A., Nikolova, E.: On the hardness and smoothed complexity of quasi-concave minimization. In: Proceedings of the 48th Annual Symposium on Foundations of Computer Science, Providence, RI, USA (2007)
23. Kleinberg, J., Rabani, Y., Tardos, É.: Allocating bandwidth for bursty connections. SIAM Journal on Computing 30(1), 191–217 (2000)
24. Nie, Y(M.), Xing, W.: Shortest path problem considering on-time arrival probability. Transportation Research Part B: Methodological 43(6), 597–613 (2009)
25. Markowitz, H.M.: Mean-Variance Analysis in Portfolio Choice and Capital Markets. Basil Blackwell, Cambridge (1987)

26. Megiddo, N.: Combinatorial optimization with rational objective functions. Mathematics of Operations Research 4, 414–424 (1979)
27. Miller-Hooks, E., Mahmassani, H.: Path comparisons for a priori and time-adaptive decisions in stochastic, time-varying networks. European Journal of Operational Research 146(1), 67–82 (2003)
28. Nemirovski, A., Shapiro, A.: Convex approximations of chance constrained programs. SIAM Journal on Optimization 17(4), 969–996 (2006)
29. Nikolova, E., Kelner, J.A., Brand, M., Mitzenmacher, M.: Stochastic shortest paths via quasi-convex maximization. In: Azar, Y., Erlebach, T. (eds.) ESA 2006. LNCS, vol. 4168, pp. 552–563. Springer, Heidelberg (2006)
30. Onn, S.: Convex discrete optimization. Encyclopedia of Optimization, 513–550 (2009)
31. Papadimitriou, C.H., Yannakakis, M.: On the approximability of trade-offs and optimal access of web sources. In: Proceedings of the 41st Annual Symposium on Foundations of Computer Science, Washington, DC, USA, pp. 86–92 (2000)
32. Radzik, T.: Newton's method for fractional combinatorial optimization. In: Proceedings of the 33rd Annual Symposium on Foundations of Computer Science, pp. 659–669 (1992)
33. Rockafellar, R.T.: Coherent approaches to risk in optimization under uncertainty. In: Tutorials in Operations Research INFORMS, pp. 38–61 (2007)
34. Safer, H., Orlin, J.B., Dror, M.: Fully polynomial approximation in multi-criteria combinatorial optimization. MIT Working Paper (February 2004)
35. Shmoys, D.B., Swamy, C.: An approximation scheme for stochastic linear programming and its application to stochastic integer programs. Journal of the ACM 53(6), 978–1012 (2006)
36. Sigal, C.E., Pritsker, A.A.B., Solberg, J.J.: The stochastic shortest route problem. Operations Research 28(5), 1122–1129 (1980)
37. Srinivasan, A.: Approximation algorithms for stochastic and risk-averse optimization. In: SODA 2007: Proceedings of the Eighteenth Annual ACM-SIAM Symposium on Discrete Algorithms, Philadelphia, PA, USA, pp. 1305–1313 (2007)
38. Swamy, C., Shmoys, D.B.: Approximation algorithms for 2-stage stochastic optimization problems. ACM SIGACT News 37(1), 33–46 (2006)
39. Warburton, A.: Approximation of pareto optima in multiple-objective, shortest-path problems. Oper. Res. 35(1), 70–79 (1987)

How to Schedule When You Have to Buy Your Energy

Kirk Pruhs[1,*] and Cliff Stein[2,**]

[1] Computer Science Department. University of Pittsburgh, Pittsburgh, PA, USA
kirk@cs.pitt.edu
[2] Department of IEOR, Columbia University, New York, NY
cliff@ieor.columbia.edu

Abstract. We consider a situation where jobs arrive over time at a data center, consisting of identical speed-scalable processors. For each job, the scheduler knows how much income is lost as a function of how long the job is delayed. The scheduler also knows the fixed cost of a unit of energy. The online scheduler determines which jobs to run on which processors, and at what speed to run the processors. The scheduler's objective is to maximize profit, which is the income obtained from jobs minus the energy costs. We give a $(1+\epsilon)$-speed $O(1)$-competitive algorithm, and show that resource augmentation is necessary to achieve $O(1)$-competitiveness.

1   Introduction

As the price of server hardware has remained relatively stable, energy cost becomes one of the primary components in the total cost of ownership for computer server systems in data centers [11]. In fact, according to Dr. Eric Schmidt, CEO of Google:

> "What matters most to the computer designers at Google is not speed, but power, low power, because data centers can consume as much electricity as a city." [24].

A commonly used power management technique is speed scaling, changing the speed of the processor. As the dynamic power used by a processor is approximately the cube of the speed of the processor (this is called the cube-root rule for CMOS based processors [12,25]), even a modest reduction in speed can have a dramatic impact on power. Researchers at Google reported an approximately twenty percent energy savings from implementing the following reactive strategy: When the workload of a processor was light, the speed was scaled down, and when most processors were at maximum speed, some less time critical tasks were suspended, to be restarted when the system was not so heavily loaded [19].

---

* Supported in part by NSF grants CNS-0325353, CCF-0514058, IIS-0534531, and CCF-0830558, and an IBM Faculty Award.
** Supported in part by NSF grants CCF-0728733 and CCF-0915681.

M. Serna et al. (Eds.): APPROX and RANDOM 2010, LNCS 6302, pp. 352–365, 2010.
© Springer-Verlag Berlin Heidelberg 2010

Scheduling problems related to speed scaling and power management naturally have competing dual objectives: some quality of service (QoS) objective, and some power related objective. By now there are many tens of papers on speed scaling in the algorithmic literature (and many more in the general computer science literature). Roughly speaking, all of the formal problems considered in the algorithmic speed scaling literature fall into one of two categories. The first type of problem turns one of the QoS or power objectives into a constraint, and optimizes the other objective. An example is minimizing the total flow time subject to the constraint that the energy used doesn't exceed an energy bound representing the energy stored in a battery. The second type of problem optimizes the sum of the QoS and power objectives. An example of this type of problem is minimizing the sum of energy used and total flow time.

In this paper, we introduce a new class of speed scaling problems, which makes the monetary cost of energy more explicit, and we provide algorithmic results for a particular problem in this class. We assume that the scheduler is aware of the income obtainable from finishing particular jobs by particular times, and is aware of the cost of energy. We then naturally assume that the scheduler's goal is to maximize profit, which is the aggregate income minus the aggregate energy cost. One can easily formulate many natural problems within this framework, depending on how one formalizes income and energy costs (and also, of course, depending on the processor and job environments). Here we consider a rather general model for the income of jobs: We assume that there is an non-negative non-increasing income function $I_i(t)$ associated with each job i that specifies the income that is obtained if the job is finished at time t. And we consider the most natural and simple model for energy costs: We assume a fixed cost per unit of energy.

We now explain the job and machine environments that we consider in this paper. Jobs arrive over time at the data center consisting of m identical speed-scalable processors. There is an arbitrary power function $P(s)$ that specifies the power when a processor is run at speed s. Job i arrives at time r_i, with known work/size w_i, and known income function $I_i(t)$. The online scheduler must decide, at each time, which job to run on each processor, and at what speed to run each processor. We allow preemption and migration, that is, jobs can be suspended at any time, and restarted from the point of suspension at a later time, possibly on a different machine. Recall that our objective is to maximize the income from the scheduled jobs minus the total energy costs.

The standard measure of goodness for an online algorithm is competitiveness, which in this setting is, roughly speaking, the worst-case, over all possible inputs, of the relative error between the optimal profit and the profit achieved by the online algorithm. One generally seeks algorithms that are competitive, that is, where this relative error is bounded. The motivation for seeking competitive algorithms is that if the online algorithm achieves very little profit, then it must be because great profit was not achievable, and not because the algorithm was at fault.

1.1   Our Results

The most obvious first concern that arises when seeking a competitive algorithm for this problem is that one can imagine a situation where the online algorithm does not achieve a positive profit, even though a positive profit is achievable, immediately killing any hope of a competitive algorithm. We start by observing, in Section 3 that this situation cannot occur, that is, that there is a simple online algorithm that achieves a positive profit if it is possible to do so. Unfortunately, we show that, in some sense, this result is the best positive result possible for the competitive ratio by showing that the competitive ratio can not be bounded by any function of the number of jobs. The intuition behind this lower bound is that the online algorithm can be forced to run, and then later abandon, a high-cost low-profit job, thus wasting a lot of energy and money on this job.

Reflecting on this lower bound instance, one notices that if the processors used by the online algorithm were only slightly more energy efficient, then the online algorithm could be competitive on this instance. We show in Section 4 that this phenomenon holds for all instances. More specifically, we assume that the online algorithm has $(1+\epsilon)$-speed augmentation, which in this setting means that if a processor can run at power P and speed s, then the online algorithm can run the processor at power P and speed $(1+\epsilon)s$. We then give an online scheduling algorithm that we show is $O(\frac{1}{\epsilon^3})$-competitive in terms of profit. Using standard terminology [20,26,28,27], one could say that this algorithm is a scalable scheduling algorithm, that is, it is $(1+\epsilon)$-speed $O(1)$-competitive. Intuitively, scalable algorithms can handle almost the same load as optimal. For elaboration see [28,27].

We now give an overview of the development of our scalable algorithm. The first key idea is that of a critical speed function $\hat{s}_i(t)$, which, for job i, specifies the fastest speed that the adversary can run job i and still obtain a non-negative profit if the job completes at time t. When a job i is released, the online algorithm determines whether to admit the job, and if the job is admitted, determines a deadline d_i for the job. Whenever an admitted job i is run by the online algorithm, it will be run at speed slightly faster than the critical speed for its deadline, $\hat{s}_i(d_i)$. Fixing the speed for a job defines a density for the job, which is roughly the profit that will be obtained by the job if it is completed at its deadline divided by the time that the job must be run to be completed. Intuitively, the online algorithm always picks the highest density jobs to run. Also intuitively, when a job is released, the online algorithm sets the deadline to be the time where it will obtain maximum profit from this job, assuming that in the future no more jobs arrive and that the highest density jobs will be run at their critical speeds.

To show that the online algorithm is scalable, we show that the profit obtained by the online algorithm is a constant fraction of the profit of the jobs that the online algorithms admits, and that the profit of these admitted jobs is a constant fraction of the optimal profit. In order to accomplish the latter goal, we show that there is a near optimal schedule OPT', that, with modest speed augmentation, is $O(1)$-competitive in terms of profit with the optimal schedule, and OPT' has the property that it runs each job i at speed approximately equal to the critical speed of the job for the completion time $\hat{s}_i(C_i^O)$. of that job in the optimal schedule OPT. Thus OPT' is still nearly optimal, but is structurally similar to the online schedule in that jobs are run at their critical speeds. A priori, it is not clear that such a schedule OPT' exists since a job i may be run at very different speeds in OPT' and in the online schedule. In other words, $\hat{s}_i(C_i^O)$ and $\hat{s}_i(d_i)$ may be very different, since there is no reason that the completion time in the optimal schedule, C_i^O, and the deadline set by the online algorithm, d_i, need be similar.

Note that our algorithm can be converted into one that constructs non-migratory schedules using the results in [18].

The income model in our paper was considered in [7], a scalable algorithm for maximizing income on a single fixed speed processor was given. Our algorithm and analysis necessarily generalize the results in [7] as we have multiple processors instead of a single processor, our processors are speed scalable instead of fixed speed, and we have profit as the objective instead of income. The fact that the processors are speed scalable creates complications because the algorithm and analysis in [7] use the fact that the processing time for a job is fixed. The objective of profit also creates complications because the algorithm and analysis in [7] use the fact that income is monotonic in time, which isn't true for profit.

1.2   Related Results

The first theoretical study of speed scaling algorithms was in the seminal paper [31], which introduced the deadline feasibility framework, and considered minimizing energy usage on a single processor. This problem is the most investigated speed scaling problem in the literature [31,5,14,4,23,22,2,9]. In [31], the authors showed that the optimal offline schedule can be efficiently computed by a greedy algorithm. Several online algorithms for this problem have been proposed and analyzed, including AVR [31,5], OA [31,4], BKP [4], and qOA[9]. The competitive ratios of all of these algorithms grow in an unbounded manner as the power function becomes steeper, but the competitive ratio is $O(1)$ if the power function is bounded by a fixed polynomial. These results have been extended in several ways including to parallel processors[2], analyzing BKP with respect to temperature minimization[4], a variant in which one minimizes the recharge rate from a solar cell[3], and scalable algorithms for throughput optimization on a single speed scalable processor with a polynomial power function and an upper bound on the maximum speed[15,6].

Another class of problems considers flow time and energy. [29] give an offline algorithm to minimize flow time subject to an energy budget for a single processor. For a single processor [1] gives a competitive online algorithm for unit work jobs for the objective of total flow plus energy assuming the power function is a polynomial. Their results were extended to arbitrary sized and arbitrary weighted jobs in [10], and to arbitrary power functions in [8]. An extension to nonclairvoyant algorithms on a uniprocessor is given in [16]. An extension to nonclairvoyant algorithms on a multiprocessor is given in [17].

There have been several papers in the literature on speed scaling with the makespan objective [13,30].

The scalable algorithm for income in [7] generalizes a scalable algorithm given in [20] for the special case of maximizing the profit of jobs completed before their deadline (there are many papers on this problem in the literature).

2   Definitions

Jobs arrive over time at the data center. Job i (also referred to as job i)arrives at time r_i, with known work/size w_i, and known income function $I_i(t)$. The function $I_i(t)$, defined for all $t > 0$, gives the income earned if job i is completed at time t. We assume that the income function $I_i(t)$ is non-negative and non-increasing. We assume that the income goes to zero at the completion time approaches infinity, that is, $\lim_{t \to \infty} I_i(t) = 0$. And we assume that if a job doesn't complete, the income is zero.

We allow preemption, that is, jobs can be suspended at any time, and restarted at a later time, possibly on a different machine. However a schedule can run a job on at most one machine at a time. To formally define a schedule, one needs to describe, for each time on each machine, which job is run on that machine and the speed at which it is run. Due to the convexity of the power function, we need only consider schedules where each job is only run at a fixed speed. With this in mind, a schedule can be given by describing, for each job, the speed at which the job runs, and, for each time, which (if any) machine it is running on. A job i completes at the first time C_i where the speed that the job is run, integrated over all the times the that job is run, equals the work w_i. If a job does not complete, $C_i = \infty$.

We are also given an arbitrary power function $P(s)$ that, for any non-negative s, specifies the power used while running at speed s on each machine. As observed in [8], one can without loss of generality assume that P is convex and increasing. The energy used is the sum over the processors, of the integral over time of the power of that machine. The income associated with a job is $I_i(C_i)$. If a job does not complete, the income is zero. One can also compute the energy cost associated with running a job i at speed s_i as $E_i = P(s_i)w_i/s_i$, since it runs for w_i/s_i units of time, at power $P(s_i)$. The profit associated with job i is $p_i = I_i(C_i) - E_i$. Our objective is the total profit $\sum_i p_i$. We will sometimes superscript these quantities by A for the online algorithm, or by O for the optimal/adversary.

An online algorithm A is c-competitive if for all inputs the total profit achieved by A is at least $\frac{1}{c}$ of the maximum achievable profit. An online algorithm A is $(1 + \epsilon)$-speed c-competitive if for all inputs the total profit achieved by A with power function $P(\frac{s}{1+\epsilon})$ is at least $\frac{1}{c}$ of the maximum achievable profit. More precisely, to have power function $P(\frac{s}{1+\epsilon})$ means that if the adversary can run at power P and speed s, then the online algorithm can run at power P and speed $(1 + \epsilon)s$.

We ignore any issues about the time to access and solve equations involving the income and power functions. Presumably in most applications these functions will be compactly described, and sufficiently simple, that manipulating these functions, as required by our algorithms, should not be a significant issue.

3   No Resource Augmentation

Our first results concern the situation where we do not allow resource augmentation, that is the adversary and the on-line algorithm both have the same power function. We first note that it is always possible for an on-line algorithm to obtain positive profit if the adversary can also receive positive profit. In contrast, we then show that the competitive ratio can be arbitrarily large.

Lemma 1. *At the release time r_i for job i, an online algorithm can compute whether it is possible to achieve positive profit for job i.*

Proof. Assume job i is completed at time t. To minimize energy, job i should be run at constant speed $w_i/(t - r_i)$ during the time period $[r_i, t]$. Job i will then have energy cost $E_i(t) = P(s(t))w_i/(t - r_i)$. So the online algorithm need only determine whether there exists t such that $I_i(t) - E_i(t) > 0$.

Lemma 2. *If the adversary can obtain a positive profit, then so can the online algorithm.*

Proof. For each job i, when it arrives, the online algorithm computes whether it is possible to make a positive profit by running the job, using Lemma 1. For the first such job, the online algorithm runs the job and obtains a positive profit. If no such job arrives, then the adversary cannot obtain positive profit either.

We now show that it is possible that the competitive ratio can be arbitrary large.

Lemma 3. *The competitive ratio of any deterministic algorithm can not be bounded by any function of n, even if jobs have unit work.*

Proof. We consider a two job instance and a power function for which $P(1) = 1$ and $P(1/\epsilon) = L/\epsilon$, where $\epsilon > 0$ is a small number and $L > 1$. (For intuition, think of L as large.) So each processor has only two possible speeds, 1 and $1/\epsilon$. Job 1 has $r_1 = 0$, $w_1 = 1$ and $I_1(t) = 1 + \epsilon$ if $t \le 1$ and $I_1(t) = 0$ for $t > 1$. When this job is released at time 0, by the reasoning of Lemma 2, the algorithm has to run this job immediately, or else the algorithm will have non-positive profit while the adversary could run the job for positive profit. Therefore, we assume that the algorithm runs job 1 at speed 1 (any other speed would incur a loss).

Job 2 has $r_2 = 1 - \epsilon$, $w_2 = 1$, and $I_2(t) = L + 1 - \epsilon$ if $t \le 1$ and $I_2(t) = 0$ for $t > 1$. When job 2 is released, the algorithm can either run job 2 or not. If the algorithm does not switch to job 2 and finishes job 1, then it obtains $p_1 = (1+\epsilon) - P(1) \cdot 1 = \epsilon$. If it switches, it obtains profit $p_2 = (L+1-\epsilon) - P(1/\epsilon)\epsilon = 1 - \epsilon$ from job 2 , but it also has to pay the energy cost of running job 1 which is $1 - \epsilon$. Thus, if it switches it actually obtains no profit. Therefore we can assume the algorithm does not switch and obtains a net profit of ϵ from running job 1. The adversary on the other hand, only runs job 2 and obtains a profit of $1 - \epsilon$. The competitive ratio is therefore at least $(1 - \epsilon)/\epsilon$ and by making ϵ small, we can make this ratio as large as we like.

4   The Online Algorithm and Its Analysis

Given the results in the previous section, we consider resource augmentation in the remainder of this paper. Our main result is the following theorem:

Theorem 1. *For any $\epsilon > 0$, there is an online algorithm A that is $(1+\epsilon)$-speed $O(\frac{1}{\epsilon^3})$-competitive for profit maximization.*

The purpose of this section is to prove Theorem 1.

   In subsection 4.1 we define the concept of critical speed function, which is required for both the definition and the analysis of our online algorithm. In subsection 4.2 we describe our online algorithm. In subsection 4.3 we prove the existence of a near-optimal schedule with nice structural properties that will facilitate the comparison with the online schedule. Finally, in subsection 4.4 we compare the online schedule to this structurally-nice near-optimal schedule.

4.1   Critical Speed Function

If i is completed at time t, then the minimum speed at job i is run is $s_i^{\min}(t) = w_i/(t - r_i)$. Recall that we can assume without loss of generality that each job runs at a fixed speed. Thus in order for the adversary to obtain positive profit from job i with power function $P(s)$, when completing the job at time t, it must be the case that:

$$I_i(t) - P(s_i^{\min}(t)) \frac{w_i}{s_i^{\min}(t)} > 0. \tag{1}$$

Alternatively, a feasible schedule complete job i at time t by running job i at a faster speed than $s_i^{\min}(t)$, and then no running job i for some times during the time interval $[r_i, t]$. As the speed that job i is run increases (with the completion time fixed at t), the energy cost increases. Thus there is a maximum speed at which job i can run, and complete at time t, while still having non-negative profit. We call this speed the *critical speed function* $\hat{s}_i(t)$. For time t, $\hat{s}_i(t)$ is by the unique solution to the equation:

$$I_i(t) - P(\hat{s}_i(t)) \frac{w_i}{\hat{s}_i(t)} = 0. \tag{2}$$

Dividing (2) through by $1 + \epsilon$ and regrouping terms, we can rewrite (2) as

$$I_i(t) - P(\hat{s}_i(t)) \frac{w_i}{(1+\epsilon)\hat{s}_i(t)} = \frac{\epsilon}{1+\epsilon} I_i(t) \tag{3}$$

Lemma 4. *In any schedule in which non-negative profit is earned from job i with power function $P(s)$, the speed s_i that job i runs is in the range $[s_i^{\min}(C_i), \hat{s}_i(C_i)]$.*

Proof. Immediate from equations 1 and 2.

4.2   The Description of the Online Algorithm

We break the description of the online algorithm into four parts: invariants that are maintained throughout the course of the algorithm, the policy for setting deadlines and assigning jobs, the policy for job selection, and the speed scaling policy.

At a high level, when a job arrives, we use the deadline setting and job assignment policy to set a deadline and assign the job to various time intervals on machines. This assignment is not a schedule, as we may assign multiple jobs to the same machine at the same time. We also set a speed via the speed scaling policy. We then use the job selection policy to take some jobs from the intervals and machines on which they are assigned and actually run them on machines.

Throughout this section, we use $\delta = \epsilon/2$.

Invariants: The online algorithm maintains a pool Q of *admitted* jobs. A job i remains in Q until it is completed or its deadline passes. Each job i in Q has several associated attributes:

- A deadline d_i assigned to job i when it was released.
- The critical speed $\hat{s}_i^A = \hat{s}_i^A(d_i)$ derived from the deadline d_i, and defined by (2). The online algorithm will run job i at speed $(1 + 2\delta)\hat{s}_i^A$.
- A collection of time intervals $J(i) = \{[t_1, t_1'], [t_2, t_2'], \ldots [t_h, t_h']\}$, where $r_i \le t_1 \le t_1' \le t_2 \le \ldots \le t_h \le t_h' = d_i$. This collection $J(i)$ is fixed when job i is released (but depends on previously scheduled jobs). The total length of the time intervals in $J(i)$ will be $\frac{(1+\delta)w_i}{(1+2\delta)\hat{s}_i^A}$.
- A processor $m_{i,k}$ associated with job i and each time interval $[t_k, t_k'] \in J(i)$ that was fixed at time r_i. Intuitively, at the time that job i was released, the online algorithm is tentatively planning on running job i on processor $m_{i,k}$ during the time period $[t_k, t_k']$. We say that job i is *assigned* to run on m_k during times $[t_k, t_k']$.

Deadline Setting and Job Assignment Policy: Consider a job i that is released at time r_i. Setting the deadline at some d_i will fix a critical speed $\hat{s}_i^A = \hat{s}_i(d_i)$ for job i, a job profit $p_i^A = I_i(d_i) - \frac{P(\hat{s}_i^A)w_i}{\hat{s}_i^A(1+2\delta)}$, and an online density $u_i^A = p_i^A \hat{s}_i^A / w_i$. The online algorithm considers the possible choices for deadlines by nonincreasing order of the resulting profit p_i^A. So assume that the online algorithm is considering setting the deadline d_i to be a time t. Let $c = 1 + \frac{2}{\delta}$. Let $X(\frac{u_i^A}{c})$ be the set of jobs in Q with density at least $\frac{u_i^A}{c}$. Consider the time interval $[r_i, t]$ and the associated intervals of jobs in $X(\frac{u_i^A}{c})$. Let A be the maximal subintervals of $[r_i, t]$ of times such that for each $[a, a'] \in A$, there is a processor m_k for which no job in $X(\frac{u_i^A}{c})$ is assigned to run on m_k during any time in $[a, a']$. We now consider two cases. In the first case assume that the total length of intervals in A is at least $\frac{(1+\delta)w_i}{(1+2\delta)\hat{s}_i^A}$. The deadline d_i is then set to be the time such that the measure of the portion A^* of A earlier than d_i is exactly $\frac{(1+\delta)w_i}{(1+2\delta)\hat{s}_i^A}$.

$J(i)$ is set to be A^*, and the processor associated with each interval in $J(i)$ is the processor m_k in the definition of the interval in A. If the job profit for the adversary with completion time $C_i^{OPT} = d_i$, $I_i(d_i) - P(\hat{s}_i^A)w_i/\hat{s}_i^A$, is positive, then job i is admitted to the pool Q, In the second case, when the total length of intervals in A is less than $\frac{(1+\delta)w_i}{(1+2\delta)\hat{s}_i^A}$, the online algorithm *rejects* this candidate deadline, and the next most profitable time t is considered for the deadline. A job is not admitted if there is no time t satisfying the stated conditions.

Speed Scaling Policy: Every job is run at its critical speed for its set deadline.

Job Selection Policy: At any time t, on any processor m_k, run the job i,assigned to m_k at time t, of maximum density.

4.3   Construction of a Structurally-Nice Near-Optimal Schedule OPT'

Using the results from [21], given an optimal schedule on m processors, one can create a *non-migratory* schedule on $6m$ processors that has objective value at least as large. A schedule is non-migratory if no job ever runs on more than one processor. Therefore, by taking the m processors with the largest total net profit, one can assume that the optimal schedule is non-migratory (modulo a factor of 6 in the profit objective). Thus for the rest of this subsection, we assume that the optimal schedule is non-migratory.

In order to facilitate the comparison of the online schedule to the optimal schedule, we assume that each job has a separate power function $P_i'(s)$ that is slightly smaller than $P(s)$ for speeds less than the critical speed $\hat{s}_i(C_i^O)$. More formally,

$$P_i'(s) = \begin{cases} P(s/(1+\epsilon)) & \text{if } s \in [s_i^{\min}(C_i), \hat{s}_i(C_i^O)] \\ P(s) & \text{otherwise} \end{cases}. \tag{4}$$

Notice that $P_i'(s) \leq P(s)$, so clearly the optimal schedule with respect to power function P' is at least as profitable as the optimal with respect to the original power function. We then show that with these modified per-job power functions, there is a near optimal schedule where each job runs at near this critical speed.

Lemma 5. *There is a schedule OPT' such that in OPT' each job i that runs does so at speed $(1 + \epsilon)\hat{s}_i(C_i^O)$, and the total profit obtained using the modified power functions is at least $\left(\frac{\epsilon}{1+\epsilon}\right)$ times the profit that OPT achieves using the power function P.*

Proof. For notation simplicity let $\hat{s}_i = \hat{s}_i(C_i^O)$, and $s_i^{\min} = s_i^{\min}(C_i^O)$. We modify the optimal schedule so that each job i is run at speed $\hat{s}_i$, and the profit p_i^O decreases by at most a factor of $\frac{\epsilon}{1+\epsilon}$. In OPT, by the definition of $\hat{s}_i$ in equation 2, we know that each job that runs is already running at speed at most $\hat{s}_i$. Combined with equation 1, we see that i is running at a speed s_i satisfying $s_i^{\min} \leq s_i \leq \hat{s}_i$. Thus, if we change the speed to $\hat{s}_i(1 + \epsilon)$, we are speeding the job up, which implies that the schedule will certainly be feasible, i.e. each job still completes by its completion time C_i^O. Now, the net profit associated with job i is at least

$$I_i(C_i^O) - P_i'(\hat{s}_i(1+\epsilon))\frac{w_i}{\hat{s}_i(1+\epsilon)} = I_i(C_i^O) - P(\hat{s}_i)\frac{w_i}{\hat{s}_i(1+\epsilon)}$$

$$= \frac{\epsilon}{1+\epsilon}I_i(C_i^O)$$

$$\geq \left(\frac{\epsilon}{1+\epsilon}\right)p_i^O.$$

The first equality follows from the definition of P', the second equality follows from (3), and the final inequality follows because the income must be greater than the profit.

4.4   Analysis of the Online Algorithm

In this section we compare the online algorithm with $(1+\epsilon)^2$ speed augmentation to OPT' in terms of profit. To simplify the analysis, we will generously assume that the power function for OPT' is $P(s/(1+\epsilon))$, and that the power function for the online algorithm is $P(s/(1+\epsilon)^2)$. In our analysis, it will be convenient to scale work or speed so that the power functions for OPT' and the online algorithm are $P(s)$ and $P(s/(1+\epsilon))$ respectively. We also generously assume that the online algorithm only gains income $I_i(d_i)$ from finishing a job i before its deadline, when in fact its real income is $I_i(C_i^A)$. Superscripting or subscripting by the variable O means that we are referring to the schedule OPT'.

We define the notion of *adversarial density* of a job i as $u_i^O = p_i^O s_i^O / w_i$, where p_i^O is the profit obtained from job i in the schedule OPT'. The *density of the online schedule at time t on a processor m_k is the density of the highest density admitted job assigned to processor m_k at that time t. The density of the online schedule at time t is the minimum density of any processor at that time.* Let C be the set of jobs completed by the online algorithm, and let R be the set admitted jobs. For any set X of jobs, let $||X||_A$ be the total profit of jobs in X if each job in X was run at its critical speed and finished at its deadline. Similarly, let $||X||_O$ be the total profit of jobs in X if each job in X was run at its critical speed and finished at the completion time of the job in OPT'.

Observation 2. *At any time t, let i and j be two admitted jobs where there is a time t and a processor m_k such that both jobs are assigned to m_k at time t. Then, either $u_i^A > c \cdot u_j^A$ or $u_j^A > c \cdot u_i^A$.*

Observation 3. *Consider any job i and a time t that the deadline setting policy of the online algorithm considered, but decided not to use as the deadline. Let v be what the density for job i would have been, if d_i were set to t. Let L be the amount of time during $[r_i, t]$ such that the density of the online schedule at that time is at least v/c. Then, $L \geq \frac{\delta}{1+2\delta}(t - r_i)$.*

Lemma 6. *For C and R as defined above, $||C||_A \geq (1 - \frac{1}{\delta(c-1)})||R||_A$, or equivalently, $||R||_A \leq \frac{\delta(c-1)}{\delta(c-1)-1}||C||_A$.*

Proof. We use a charging scheme to prove the lemma. We initially give p_i^A units of credit to each job $i \in C$. The jobs in $R - C$ are initially given 0 units of credit. We will describe a method to transfer the credits such that at the end, each job $i \in R$ has credit at least $(1 - \frac{1}{\delta(c-1)})p_i^A$, which completes the proof.

The method to transfer credit is as follows. At any time t, and any processor m_k, let S be the set of jobs assigned to m_k at time t. Let job i be the highest density job in S. Then, for each other job $j \in S$, at time t job i transfers credit to j at a rate of $(\frac{1+2\delta}{\delta})u_j^A$ units of credit per unit time.

We first show that every job $j \in R$ receives credit at least p_j^A either initially or transferred from other jobs. This clearly holds for jobs in C. For any job $j \in R - C$, as job j could not be completed during $J(j)$, it must have received credit for at least $\frac{\delta}{1+2\delta} \cdot \frac{w_j}{s_j^A}$ units of time. Thus, the total credit obtained is at least

$$\left(\frac{\delta}{1+2\delta}\right)\left(\frac{w_j}{s_j^A}\right)\left(\frac{1+2\delta}{\delta}\right)u_j^A = \frac{w_j u_j^A}{s_j^A} = p_j^A$$

We now show that the credit transferred out of each job i is at most $\frac{1}{\delta(c-1)}p_i^A$. When a job i is the highest density job in S, by observation 2 the remaining jobs in S have geometrically decreasing densities and hence their total density is at most $\frac{1}{c-1}u_i^A$. Therefore, the rate of credit transferring out of i is at most $(\frac{u_i^A}{c-1})(\frac{1+2\delta}{\delta})$. Since job i is the highest density job for at most $\frac{w_i}{\hat{s}_i^A(1+2\delta)}$ units of time, the total credit transferred out of job i is at most

$$\left(\frac{u_i^A}{c-1}\right)\left(\frac{1+2\delta}{\delta}\right)\left(\frac{w_i}{\hat{s}_i^A(1+2\delta)}\right) = \frac{1}{\delta(c-1)}p_i^A.$$

Next, we upper bound the profit obtained by the adversary. Let B be the set of jobs completed by the adversary. Let B_2 be the set of jobs in B for which the adversary's completion time is a deadline rejected by the online algorithm. Let $B_1 = B \setminus B_2$. For any $u > 0$, let $T(u)$ be the total length of time that the adversary is running a job in B_2 with adversarial density at least u. Let $L(\frac{u}{c})$ be the total length of time such that the density of the online schedule at that time is at least $\frac{u}{c}$.

Lemma 7. *For every $u > 0$, $T(u) \le \frac{2(1+2\delta)}{\delta}L(\frac{u}{c})$.*

Proof. For any job $i \in B_2$, let the span of i be the time interval $[r_i, C_i^O]$, where C_i^O is the completion time for the adversary. For any $u > 0$, let $B_2(u)$ be the set of jobs in B_2 with density at least u. Consider the union of spans of all jobs in $B_2(u)$. This union may consist of a number of disjoint time intervals. Let ℓ be its total length. Clearly, $T(u) \le \ell$.

Let $M \subseteq B_2$ be a minimal cardinality subset of B_2 such that the union of spans of jobs in M equals that of B_2. Note that the minimality property implies no three jobs in M have their spans overlapping at a common time. This implies that we can further partition M into M_1 and M_2 such that within M_1 (resp. M_2), any two jobs have disjoint spans. Now, either M_1 or M_2 has total span of length at least half of that of M. Without loss of generality, suppose that it is M_1. Note that each interval in M_1 corresponds to a span of some job in B_2. Applying Observation 3 to each such interval, it follows that the density of the online schedule is at least $\frac{u}{c}$ for at least $\frac{\delta}{1+2\delta}$ fraction of time during the intervals of M_1. Thus, $L(\frac{u}{c}) \geq \frac{\delta}{2(1+2\delta)} \cdot T(u)$.

Lemma 8. $||B||_O \leq (1 + \frac{2(1+\delta)c}{\delta})||R||_A$.

Proof. Let $\{\phi_1, \phi_2, \ldots, \phi_m\}$ be the set of the adversarial densities of jobs in B_2, where $\phi_i > \phi_{i+1}$ for $i = 1, \ldots, m - 1$. For $i = 1, \ldots, m$, let ℓ_i be the sum over all processors of the length of time that the adversary is running jobs of density ϕ_i on that processor. Similarly, for $i = 1, \ldots, m$, let α_i be the sum over all processors of the length of time that the online schedule on that processor has density in the range $[\phi_i/c, \phi_{i-1}/c)$. Let q_i be the total profit for jobs whose density for the online algorithm is in the range of $[\phi_i/c, \phi_{i-1}/c)$. Then applying Lemma 7

$$||B_2||_O \leq \frac{2(1+2\delta)}{\delta} \sum_{i=1}^{m} \alpha_i \phi_i \leq \frac{2(1+\delta)c}{\delta} \sum_{i=1}^{m} q_i \leq \frac{2(1+\delta)c}{\delta}||R||_A$$

The proof then follows by noting that $||B||_O = ||B_1||_O + ||B_2||_O \leq ||R||_A + ||B_2||_O$.

That the online algorithm is $(1 + \epsilon)$-speed, $O(\frac{1}{\epsilon^3})$-competitive now follows immediately from Lemma 5, Lemma 6 and Lemma 8.

5   Conclusions

We introduced a new type of power management problem into the algorithmic literature, and showed that there is a scalable algorithm for the problem of maximizing profit when you have to buy your energy. It would be interesting to investigate other problems within this general framework.

Acknowledgments

We thank Christoph Durr and Nguyen Kim Thang for several helpful discussions.

References

1. Albers, S., Fujiwara, H.: Energy-efficient algorithms for flow time minimization. ACM Transactions on Algorithms 3(4) (2007)
2. Albers, S., Müller, F., Schmelzer, S.: Speed scaling on parallel processors. In: Proc. ACM Symposium on Parallel Algorithms and Architectures (SPAA), pp. 289–298 (2007)

3. Bansal, N., Chan, H.L., Pruhs, K.: Speed scaling with a solar cell. In: International Conference on Algorithmic Aspects in Information and Management (2008)
4. Bansal, N., Kimbrel, T., Pruhs, K.: Speed scaling to manage energy and temperature. JACM 54(1) (2007)
5. Bansal, N., Bunde, D., Chan, H.L., Pruhs, K.: Average rate speed scaling. In: Latin American Theoretical Informatics Symposium (2008)
6. Bansal, N., Chan, H.-L., Lam, T.W., Lee, L.-K.: Scheduling for speed bounded processors. In: Aceto, L., Damgård, I., Goldberg, L.A., Halldórsson, M.M., Ingólfsdóttir, A., Walukiewicz, I. (eds.) ICALP 2008, Part I. LNCS, vol. 5125, pp. 409–420. Springer, Heidelberg (2008)
7. Bansal, N., Chan, H.-L., Pruhs, K.: Competitive algorithms for due date scheduling. In: Arge, L., Cachin, C., Jurdziński, T., Tarlecki, A. (eds.) ICALP 2007. LNCS, vol. 4596, pp. 28–39. Springer, Heidelberg (2007)
8. Bansal, N., Chan, H.-L., Pruhs, K.: Speed scaling with an arbitrary power function. In: SODA, pp. 693–701 (2009)
9. Bansal, N., Chan, H.-L., Pruhs, K., Katz, D.: Improved bounds for speed scaling in devices obeying the cube-root rule. In: Albers, S., Marchetti-Spaccamela, A., Matias, Y., Nikoletseas, S., Thomas, W. (eds.) ICALP 2009. LNCS, vol. 5556, pp. 144–155. Springer, Heidelberg (2009)
10. Bansal, N., Pruhs, K., Stein, C.: Speed scaling for weighted flow time. In: ACM-SIAM Symposium on Discrete Algorithms, pp. 805–813 (2007)
11. Barroso, L.A.: The price of performance. ACM Queue 3(7), 48–53 (2005)
12. Brooks, D.M., Bose, P., Schuster, S.E., Jacobson, H., Kudva, P.N., Buyuktosunoglu, A., Wellman, J.-D., Zyuban, V., Gupta, M., Cook, P.W.: Power-aware microarchitecture: Design and modeling challenges for next-generation microprocessors. IEEE Micro 20(6), 26–44 (2000)
13. Bunde, D.P.: Power-aware scheduling for makespan and flow. J. Scheduling 12(5), 489–500 (2009)
14. Chan, H.L., Chan, W.-T., Lam, T.-W., Lee, L.-K., Mak, K.-S., Wong, P.W.H.: Energy efficient online deadline scheduling. In: ACM-SIAM Symposium on Discrete Algorithms, pp. 795–804 (2007)
15. Chan, H.-L., Chan, W.-T., Lam, T.W., Lee, L.-K., Mak, K.-S., Wong, P.W.H.: Energy efficient online deadline scheduling. In: SODA, pp. 795–804 (2007)
16. Chan, H.-L., Edmonds, J., Lam, T.W., Lee, L.-K., Marchetti-Spaccamela, A., Pruhs, K.: Nonclairvoyant speed scaling for flow and energy. In: STACS, pp. 255–264 (2009)
17. Chan, H.-L., Edmonds, J., Pruhs, K.: Speed scaling of processes with arbitrary speedup curves on a multiprocessor. In: SPAA, pp. 1–10 (2009)
18. Chan, H.-L., Lam, T.W., To, K.-K.: Nonmigratory online deadline scheduling on multiprocessors. SIAM J. Comput. 34(3), 669–682 (2005)
19. Fan, X., Weber, W.-D., Barroso, L.A.: Power provisioning for a warehouse-sized computer. In: International Symposium on Computer Architecture, pp. 13–23 (2007)
20. Kalyanasundaram, B., Pruhs, K.: Speed is as powerful as clairvoyance. J. ACM 47(4), 617–643 (2000)
21. Kalyanasundaram, B., Pruhs, K.: Eliminating migration in multi-processor scheduling. J. Algorithms 38(1), 2–24 (2001)
22. Li, M., Liu, B.J., Yao, F.F.: Min-energy voltage allocation for tree-structured tasks. Journal of Combinatorial Optimization 11(3), 305–319 (2006)
23. Li, M., Yao, F.F.: An efficient algorithm for computing optimal discrete voltage schedules. SIAM J. on Computing 35, 658–671 (2005)

24. Markoff, J., Lohr, S.: Intel's huge bet turns iffy. New York Times (September 29, 2002)
25. Mudge, T.: Power: A first-class architectural design constraint. Computer 34(4), 52–58 (2001)
26. Phillips, C.A., Stein, C., Torng, E., Wein, J.: Optimal time-critical scheduling via resource augmentation. Algorithmica 32(2), 163–200 (2002)
27. Pruhs, K.: Competitive online scheduling for server systems. SIGMETRICS Performance Evaluation Review 34(4), 52–58 (2007)
28. Pruhs, K., Sgall, J., Torng, E.: Online scheduling. In: Handbook on Scheduling. CRC Press, Boca Raton (2004)
29. Pruhs, K., Uthaisombut, P., Woeginger, G.: Getting the best response for your erg. In: Scandanavian Workshop on Algorithms and Theory (2004)
30. Pruhs, K., van Stee, R., Uthaisombut, P.: Speed scaling of tasks with precedence constraints. Theory Comput. Syst. 43(1), 67–80 (2008)
31. Yao, F., Demers, A., Shenker, S.: A scheduling model for reduced CPU energy. In: Proc. IEEE Symp. Foundations of Computer Science, pp. 374–382 (1995)

Improving Integrality Gaps via Chvátal-Gomory Rounding

Mohit Singh[1] and Kunal Talwar[2]

[1] McGill University, Montreal, Canada
mohit@cs.mcgill.ca
[2] Microsoft Research, Silicon Valley, CA
kunal@microsoft.com

Abstract. In this work, we study the strength of the Chvátal-Gomory cut generating procedure for several hard optimization problems. For hypergraph matching on k-uniform hypergraphs, we show that using Chvátal-Gomory cuts of low rank can reduce the integrality gap significantly even though Sherali-Adams relaxation has a large gap even after linear number of rounds. On the other hand, we show that for other problems such as k-CSP, unique label cover, maximum cut, and vertex cover, the integrality gap remains large even after adding all Chvátal-Gomory cuts of large rank.

1 Introduction

Linear Programming is an enormously useful tool in the study of combinatorial optimization problems, giving exact algorithms for several problems in P, and approximation algorithms for several NP-hard problems. Typically, one writes an integer linear program for the problem at hand, and solves its linear programming relaxation. For a large number of problems of interest, such a relaxation has an optimum value that is within a small multiplicative factor of the optimal. A more powerful tool that sometimes gives better polynomial time approximations is semidefinite programming. In both cases, the approximation factor one gets depends on the integer linear program (or the vector program) that one starts with. For many problems, a natural linear program suggests itself and can be shown to have the best possible gap (e.g. bipartite matching, set cover). In many other cases (e.g. graph matching, sparsest cut), the "natural" linear program for the problem does not suffice and one needs to add carefully designed constraints that force the linear program to reveal information about optimal solutions.

Cut generating procedures are algorithms for adding constraints to the linear relaxation with the property that every integer solution in the polytope satisfies the new constraints. Starting with a polytope P, such a procedure gives a new polytope that is closer to P_I, the convex hull of integer points in P. Thus they provide a generic way to strengthen the linear relaxation of the integer program, without changing the set of integer feasible solutions. They can thus be thought of as an alternative to the addition of the carefully designed constraints that have been used. Indeed for several problems, the ingeniously added constraints can

M. Serna et al. (Eds.): APPROX and RANDOM 2010, LNCS 6302, pp. 366–379, 2010.

in hindsight be shown to be also generated by these cut generating procedures. A number of such procedures have been proposed including Chvátal-Gomory (CG) [11,22,23], Lovász-Schrijver (LS, LS+) [31], Sherali-Adams (SA) [37] and Lassere [29].

For a large class of combinatorial optimization problems, the best known approximation algorithms are matched by hardness of approximation results, ruling out the possibility of better approximations based on smarter LP relaxations (or on other techniques) unless P=NP. Certain interesting problems such as Vertex Cover, Max Cut, Sparsest Cut and Unique Label cover have so far resisted attempts to prove matching upper and lower bounds. For these problems, it is therefore natural and interesting to ask if one can design stronger LP (or SDP) relaxations. A negative answer would rule out a large class of algorithmic approaches, suggesting that computing better approximations may in fact be NP-hard. Arora, Lovász and Bollobas [2] initiated this direction of research, and showed that starting with a natural linear program for vertex cover, and iteratively applying the LS cut generating procedure does not reduce the integrality gap below $(2 - \epsilon)$, even after a linear number of rounds. Similar results have been shown for other problems, and for LS+, SA and Lassere, which strengthen LS.

Somewhat disconcertingly however, such gap results can also be shown for some polynomial-time solvable problems. This is not surprising since despite its generality, linear programming does not capture all algorithmic tools at our disposal, and other tools such as dynamic programming and local search are often useful in cases where natural convex relaxations fail. However, such gap results exist even for problems where good LP relaxations exist. Indeed if one starts with the natural LP for maximum matching, it can be shown that the gap is at least $(1 + \epsilon)$ even after $\frac{1}{\epsilon}$ rounds of SA [32], even though the problem is polynomial time solvable using an (exponentially sized) LP relaxation. Starker gaps exist for hypergraph matching on k-uniform hypergraphs, where the gap stays above $(k - 2)$ even after a linear number of rounds of SA starting with the natural LP. On the other hand, when k is a constant, there is a polynomial sized linear program that has gap at most $\frac{k+1}{2}$ [9]. Thus even for simple combinatorial problems, SA can fail to capture the power of LP based algorithms.

The gap results from these lift-and-project schemes can be interpreted in several different ways. The *guide-the-algorithmicist* viewpoint looks at such result as a strong integrality gap for a family of linear programs. Thus an algorithm designer considering a new strengthened linear program could check whether or not the constraints in her LP are quickly generated by this procedure, and if so, she would conclude that the new LP will not help in the worst case, and thus may be guided towards other constraints to add. With this viewpoint, it is interesting to try to strengthen the integrality gaps to other cut generating procedures that may capture large families of efficient linear programs (even though the cut generating procedure considered in its full generality may not be efficient). A somewhat more controversial viewpoint is the *limits-of-techniques* viewpoint, where one interprets a gap result as suggesting that "LP based approaches" will not be able to give good approximation algorithms. However the above examples of matching and hypergraph matching make such a viewpoint less appealing. Finally, one can view these results as *structural results* which prove the limits of a certain proof system (e.g. SA).

In this work we study Chvátal-Gomory rounding, a popular cut generating procedure that is often used in practice. Buresh-Oppenheim *et al.* [7] previously showed that optimal integrality gaps survive a linear number of rounds of CG for MAX kSAT and MAX kXORSAT, for $k \geq 5$ (see also [3]). For problems such as unique label cover, where known hardness results do not match the best known upper bounds, most of the attention has been diverted to LS and other procedures, and little is known about CG cuts. It is particularly interesting to look at this procedure since it does in fact handle the (graph) matching example above: one round of Chvátal-Gomory suffices to make the matching polytope integral! Further as we show, the polynomial-sized linear program for hypergraph matching from [9] is also captured by a few rounds of C-G. Thus C-G does in fact capture useful and efficient linear programs that SA fails to capture, making it interesting to study C-G gaps from the guide-the-algorithmicist viewpoint. Moreover, C-G is an interesting proof system in its own right.

Chvátal-Gomory rounding is defined as follows. Let P be a polyhedron in $\mathcal{R}^n$, define

$$P' = \{x \in P : a^T x \geq b \text{ whenever } a \in Z^n, b \in Z, \text{ and } \min\{a^T x : x \in P\} > b-1\}$$

to be the polyhedron obtained after doing a single round of Chvátal-Gomory rounding. Trivially $P \cap Z^n \subseteq P'$, define $P^{(0)} = P$ and recursively,

$$P^{(j)} = (P^{(j-1)})'$$

for all positive integers j. Also let P_I denote the convex hull of $P \cap Z^n$. We clearly have $P_I \subseteq P^{(j)} \subseteq P^{(j-1)}$ for each $j \geq 1$. We call $P^{(j)}$ to be the polyhedron obtained after j rounds of CG rounding.

1.1   Our Contributions and Results

In this work, we study the power of Chvátal-Gomory rounding to reduce integrality gaps for various combinatorial optimization problems as compared to lift and project procedures like Sherali-Adams.

Our first result shows an integrality gap separation between C-G and SA which show that C-G cuts can be much stronger than SA hierarchy.

Theorem 1. *For the maximum matching problem in k-uniform hypergraphs, $O(k^2)$ rounds of CG suffice to reduce the integrality gap to $\frac{k+1}{2}$.*

We contrast the above theorem with result from Chan and Lau [9] that the integrality gap remains at least $k - 2$ after $\Omega(n)$ rounds of the SA hierarchy. Thus C-G can generate significantly stronger linear programs than SA can.

Can C-G rounding then lead to better LP relaxations for other problems? Our next set of results show that CG rounding performs as poorly as the Sherali-Adams hierarchy on a number of problems. We show integrality gaps for the max-cut problem, Unique Label Cover problem, k-CSP$_q$ and the vertex cover problem. We prove the following theorems.

Theorem 2. *For any $\epsilon > 0$, there exists a $\gamma > 0$ such that integrality gap of linear programming relaxation for the max-cut problem obtained using all cuts of CG rank at most r is at least $2 - \epsilon$ where $r = n^\gamma$.*

Theorem 3. *For any $\epsilon > 0$, and integer q, there exists a $\gamma > 0$ such that integrality gap of linear programming relaxation for the unique label cover problem on q labels, using all cuts of CG rank at most r is at least $q - \epsilon$ where $r = n^\gamma$.*

Theorem 4. *For any $\epsilon > 0$, integer k and prime q, there exists a $\gamma > 0$ such that integrality gap of linear programming relaxation for the k-CSP$_q$ problem using all cuts of CG rank at most r is at least $\frac{q^k}{kq(q-1)} - \epsilon$ where $r = \gamma n$.*

We note that the integrality gaps above resemble closely the bounds obtained for the Sherali-Adams hierarchy for the corresponding problems [10,39]. Interestingly, the proofs of all the above results follow a similar outline and use the integrality gap instances for the Sherali-Adams hierarchy as a starting point. Using our general technique we also show the following integrality gap for the vertex cover problem.

Theorem 5. *For any $\epsilon > 0$, there exists a $\gamma > 0$ such that integrality gap of relaxation for the vertex cover problem obtained after r rounds of C-G rounding is at least $2 - \epsilon$ where $r = n^\gamma$.*

We believe that our positive result gives strong motivation for studying C-G cuts as an algorithmic technique[1]. The resulting hopes are somewhat dashed by our negative results. In the process we enlarge the class of linear programs that are provably ineffectual for the problems studied. Moreover, our results enhance our understanding of C-G as a proof system.

1.2   Related Work

Gomory [22,23] introduced the Chvátal-Gomory rounding and proved that for every bounded polyhedron P, there exists a non-negative integer j such that $P^{(j)} = P_I$. Chvátal [11] gave an alternate proof of the result. The smallest such integer j is called the *Chvátal* rank of P. There has been a significant work on both lower and upper bounding Chvátal rank of a polyhedron. Although the Chvátal rank can, in general, be very large, Bockmayr *et al.* [6] proved that it is bounded by $O(n^3 \log n)$ when the polytope is contained in the hypercube $[0, 1]^n$. This bound was improved to $O(n^2 \log n)$ by Eisenbrand and Schulz [18]. Chvátal,

---

[1] One important difference between Chvátal-Gomory rounding and other hierarchies such as SA, is that unlike the latter, C-G does not come with a general efficient algorithmic procedure. Indeed optimizing over the Chvátal-Gomory closure is actually NP-hard in general [17]. Nevertheless, these cuts are commonly used by practitioners [13].

Cook and Hartmann [12] proved lower bounds on the Chvátal rank of many combinatorial optimization problems including maximum cut problem, stable set problem and traveling salesman problem. We also note that their results can also be used to show $(1 + \epsilon)$ integrality gaps after $\Omega(\frac{1}{\epsilon})$ rounds for the vertex cover problem and maximum cut problem while our results show much stronger integrality gaps. However, the Chvátal-Gomory closure from a theoretical point of view does not behave very well algorithmically; Eisenbrand [17] proved that optimizing over the polytope resulting from one round of C-G cuts is a NP-hard problem in general. Nevertheless, Bienstock and Zuckerberg [5] show that for a large class of polytopes (e.g. covering problems), one can optimize over (a subset of) the rth iterate of the polytope, up to an arbitrarily small error, for any constant r in polynomial time.

Arora, Lovász and Bollobas [2] initiated the study of integrality gaps of linear programming relaxations obtained via lift and project hierarchies. Since then there has been a series of works [1,24,10,16,36,39,35] showing integrality gaps for linear and semi-definite relaxations for various combinatorial optimization problems. Closely related to our work is the work of Charikar, Makarychev and Makarychev [10] who show integrality gaps for linear programming relaxations obtained via Sherali-Adams hierarchy for the maximum cut, vertex cover and the unique games problem. We also note that the integrality gap for the vertex cover problem obtained in Theorem 5 can also be obtained using the results of Arora et al [2]. Lift and project hierarchies and CG rounding can also be used as proof systems for satisfiability and other problems. There has been a series of works [7,34,33,14,15] which lower bound the size or depth of the proofs obtained using these hierarchies. Buresh-Oppenheim *et al.* [7] show that for MAX k-SAT, and MAX k-XOR SAT, a linear number of rounds of CG are needed to reduce the integrality gap.

2   Maximum Matching in k-Uniform Hypergraphs

The maximum matching problem on a hypergraph $G = (V, E)$ is to find the maximum cardinality subset $F \subseteq E$ of hyperedges such that for any vertex $v \in V$, there is at most one hyperedge in F incident on v. A hypergraph $G = (V, E)$ is said to be k-uniform if $|e| = k$ for every $e \in E$. We study the (unweighted) maximum matching problem in k-uniform hypergraphs. We note that the problem is NP-hard and APX-hard even for $k = 3$ [4]. Hazan, Safra and Schwartz [26] show an $\Omega(k/\log k)$-inapproximability result, while Hurkens and Schrijver [27] give a $(\frac{k}{2} + \epsilon)$-approximation algorithm.

Figure 1 gives the natural linear programming relaxation for the hypergraph matching problem. Here $\delta(v)$ denotes the set of edges incident at vertex $v \in V$. Let P denote the polytope defined by feasible solutions to this linear program. Chan and Lau [9] show that the integrality gap of this linear program remains at least $k - 2$ even after $O(n/k^3)$ rounds of the Sherali-Adams hierarchy. On the other hand, they show a polynomial sized linear program with integrality gap at most $\frac{k+1}{2}$, for any constant k.

$$\max \sum_{e \in E} x_e$$

s.t.
$$\sum_{e \in \delta(v)} x_e \le 1 \qquad \forall v \in V$$
$$x_e \qquad\qquad \ge 0 \qquad \forall \ e \in E$$

Fig. 1. Linear program for the Hypergraph Matching Problem

This latter result is derived in two steps. First, Chan and Lau [9] define a rather large linear program whose gap is shown to be bounded by $\frac{k+1}{2}$. Next they use a result in extremal combinatorics to construct an equivalent linear program with a polynomial number of constraints. We use similar techniques to show that the polytope $P^{(2k^2)}$ satisfies all the constraints defining the polytope considered by Chan and Lau [9].

A set of hyperedges K is said to be an *intersecting family* if every pair of hyperedges in K has a non-empty intersection. Clearly, for any intersecting family in E, a matching can contain at most one hyperedge. Thus one can add to the linear program the constraint $\sum_{e \in K} x_e \le 1$ for any intersecting family K. Chan and Lau [9] show that

Theorem 6 ([9]). *Consider the linear program in Figure 1 above, augmented with the constraints $\sum_{e \in K} x_e \le 1$ for all intersecting families $K \subseteq E$. For a k-uniform hypergraph, the integrality gap of this program is bounded by $\frac{k+1}{2}$.*

Next we define a *Kernel*. Given a subset $S \subseteq V$ and a hyperedge e, we let e_S denote $e \cap S$. For a subset K of hyperedges, we can then define $K_S = \{e_S : e \in K\}$. A subset $S \subseteq V$ is a kernel for an intersecting family K, if the family K_S is intersecting. In other words, S is a kernel for K if every pair of hyperedges in K has an non-empty intersection in S. It can be shown [8] that every intersecting family has a Kernel of size $s(k)$ for some function $s(k)$ independent of $|V|$.

Theorem 7 ([8]). *There exists a function $s(k)$ such that for any k-uniform hypergraph H, and any intersecting family K of hyperedges in H, there is a kernel S containing at most $s(k)$ vertices.*

The best bounds on $s(k)$ are $\Theta(\binom{2k}{k})$ [19,40,41].

Note that if S is a kernel of K, then the constraint $\sum_{e \in K} x_e \le 1$ is equivalent to the constraint $\sum_{f \in K_S} \sum_{e \in K : e_S = f} x_e \le 1$. We next argue that all constraints of the latter form are derived in a small number of rounds of C-G. In the lemma below,

Lemma 1. *Let $P^{(0)} = P$ be the polytope in figure 1, let $P^{(j)} = (P^{(j-1)})'$ and let $l_0 = 2$ and $l_{t+1} = 2l_t - 1$. Then for any S and any intersecting family K_S on S, $P^{(j)}$ satisfies all constraints of the form*

$$\sum_{f \in L} \sum_{e : e_S = f} x_e \le 1,$$

where $L \subseteq K_S$ is arbitrary with $|L| = l_j$.

Proof. The proof is by induction on j. For $j = 0$, the claim follows from the definition of an intersecting family. Indeed, in this case, L contains two hyperedges which intersect in a vertex, and the relevant inequality is implied by the packing constraint for that vertex. Now suppose that the claim holds for $j \leq t$. We prove the claim for $j = t + 1$. Let $L \subseteq K_S$ be arbitrary with $|L| = l_{t+1}$. By the induction hypothesis, the constraint is satisfied for each of the $\binom{l_{t+1}}{l_t}$ subsets of L of size l_t. Adding up these constraints and dividing by $\binom{l_{t+1}-1}{l_t-1}$, we conclude that $P^{(t)}$ satisfies the constraint

$$\sum_{f \in L} \sum_{e:e_S=f} x_e \leq \frac{\binom{l_{t+1}}{l_t}}{\binom{l_{t+1}-1}{l_t-1}} = \frac{l_{t+1}}{l_t}.$$

Thus $P^{(t+1)}$ satisfies the above constraints with the right hand side replaced by its floor. Since the ratio on the right hand side is strictly smaller than two, this completes the induction.   □

It is easy to see that $l_{2t} \geq 2^t$. Moreover, for $|S| < s(k)$, any intersecting family K_S is of size at most $s(k)^k$. It follows that

Theorem 8. *Let $P^{(0)} = P$ be the polytope in figure 1, and let $P^{(j)} = (P^{(j-1)})'$. Then the integrality gap of $P^{(2k \log s(k))}$ is bounded by $\frac{k+1}{2}$.*

Using the bound of $s(k)$ above, we conclude that $O(k^2)$ rounds of C-G suffice to bring down the integrality gap to $\frac{k+1}{2}$.

3   Integrality Gaps for Max-Cut

Let P denote the linear programming relaxation for the max-cut problem given in Figure 2. The variables x_{uv} for an edge $\{u, v\} \in E$ denote whether the edge is in the cut. The variable y_u for each vertex $u \in V$ denotes whether the vertex is on the *left* side of the cut.

$$\max \sum_{\{u,v\} \in E} w_{uv} x_{uv}$$

$$\begin{aligned}
\text{s.t.} \quad & & \\
x_{uv} &\leq y_u + y_v & \forall \{u, v\} \in E \\
x_{uv} &\leq 2 - (y_u + y_v) & \forall \{u, v\} \in E \\
x_{uv} &\geq 0 & \forall \{u, v\} \in E \\
0 \leq y_u &\leq 1 & \forall u \in V
\end{aligned}$$

Fig. 2. Linear program for the Max-Cut Problem

The following lemma characterizes the constraints for $P^{(k)}$ and is crucial in showing integrality gaps.

Lemma 2. *Let $\mathbf{a}^T \mathbf{x} \leq b + \mathbf{c}^T \mathbf{y}$ be a non-trivial facet of $P^{(k)}$ for any k. We can assume without loss of generality that $\mathbf{a}, b$ and $\mathbf{c}$ are integral, $\mathbf{a} \geq 0$.*

Proof. The integrality follows simply from the fact that P is a rational poly-hedron and hence $P^{(k)}$ is rational for each integer k. The non-negativity of $\boldsymbol{a}$ follows since using the constraint $x_{uv} \geq 0$, one can obtain a stronger constraint. □

Proof of Theorem 2: The proof uses the integrality gap example for the Sherali-Adams hierarchy to argue nearly the same integrality gap. We show that the fractional solution which survives the Sherali-Adams hierarchy, with a small scaling, also survives the Chvátal-Gomory hierarchy. Let the norm of a constraint $\boldsymbol{a}^T \boldsymbol{x} \leq b + \boldsymbol{c}^T \boldsymbol{y}$ be defined as the size of the support of $\boldsymbol{a}$. To show that the fractional solution satisfies all the constraints generated by the Chvátal-Gomory rounding, we argue separately for the constraints which have small norm and large norm. Using the properties of the Sherali-Adams hierarchy, one can show that the constraints with small norm are implied by the Sherali-Adams hierarchy and thus the fractional solution to the integrality gap example satisfies these constraints. For the constraints with large norm, we show that in each round of C-G rounding, the constraint is strengthened by at most 1 in the constant term. Since the constraint had large norm, this implies that slight degradation of the original fractional solution satisfies the new tighter constraint. We now expand on the above outline.

We use the following theorem which follows from the integrality gap example given by Charikar, Makarychev and Makarychev [10] for the Sherali-Adams Hierarchy.

Theorem 9 ([10]). *For any $\epsilon > 0$, there exists a $\gamma > 0$ and a graph $G = (V, E)$ such that any integral cut has at most $(\frac{1}{2} + \frac{\epsilon}{8})$ fraction of the edges but the fractional solution $x_{uv}^0 = 1 - \frac{\epsilon}{16}$ for each $\{u, v\} \in E$ and $y_u^0 = \frac{1}{2}$ for each $u \in V$ is in P_{SA}^t for $t = \frac{32n^\gamma}{\epsilon}$. Therefore, for every subset $S \subset V$ of size at most t, there exists a distribution $\mathcal{D}$ of solutions such that (i) expected value of the solutions equals $(\mathbf{x}^0, \mathbf{y}^0)$ and (ii) each of the solution with non-zero probability in $\mathcal{D}$ is integral over S.*

Let G be the graph given by Theorem 9. We prove the following lemma.

Lemma 3. *Let $x_{uv}^k = (1 - \frac{\epsilon}{16} - \frac{2k}{t})$ for each $(u, v) \in E$ and $y_v^k = \frac{1}{2}$ for each $v \in V$ for any nonnegative integer k. Then the fractional solution $(\mathbf{x}^k, \mathbf{y}^k) \in P^{(k)}$ for each $0 \leq k \leq n^\gamma$.*

Before we prove Lemma 3, we complete the proof of Theorem 2. Consider $k = n^\gamma$. Lemma 3 implies that

$$x_{uv}^k = (1 - \frac{\epsilon}{16} - \frac{2n^\gamma}{t}) = 1 - \frac{\epsilon}{8}$$

for each $(u, v) \in E$. Consider the weight vector which is uniformly 1. Then

$$\max\{\boldsymbol{w}^T \boldsymbol{x} : (\boldsymbol{x}, \boldsymbol{y}) \in P_I\} \leq (\frac{1}{2} + \frac{\epsilon}{8})|E|$$

but

$$\max\{\boldsymbol{w}^T\boldsymbol{x} : (\boldsymbol{x},\boldsymbol{y}) \in P^{(k)}\} \geq \mathbf{1}^T\boldsymbol{x}^k \geq \left(1 - \frac{\epsilon}{8}\right)|E|$$

proving Theorem 2.

Now we prove Lemma 3. We show $(\boldsymbol{x}^k, \boldsymbol{y}^k) \in P^{(k)}$ by induction on k. For $k = 0$, the claim is trivially true. Suppose that the claim is true for $k - 1 \geq 0$; we prove that the claim holds for k if $k \leq r = n^\gamma$.

Let $\boldsymbol{a}^T\boldsymbol{x} \leq b + \boldsymbol{c}^T\boldsymbol{y}$ be a non-trivial facet of $P^{(k)}$. First suppose that the size of the support of $\boldsymbol{a}$, $\|\boldsymbol{a}\|_0 \leq \frac{t}{2}$. Let S denote the set of vertices at which some edge in support of $\boldsymbol{a}$ is incident. We have $|S| \leq t$. From Theorem 9, there exists a distribution $\mathcal{D}$ over a set of feasible solutions to P which are integral on S and whose expectation is $(\boldsymbol{x}^0, \boldsymbol{y}^0)$. Modify these integral solutions in the following manner. For each edge not incident at a vertex in S, set $x_e = 0$ and for each vertex v not in S, set $y_v = y_u$ where u is the smallest index vertex in S (or any fixed vertex in S). Thus, we obtain a distribution $\mathcal{D}$ over *integral* feasible solutions. Let $(\boldsymbol{x}^*, \boldsymbol{y}^*)$ denote the expectation of these solutions under distribution $\mathcal{D}$. We have the following properties for $(\boldsymbol{x}^*, \boldsymbol{y}^*)$.

1. $x^*_{uv} = x^0_{uv}$ if both $u, v \in S$.
2. $y^*_v = \frac{1}{2}$ for each $v \in V$.

The second property holds for each vertex $v \in S$ from Theorem 9 and for each vertex $v \notin S$ by construction. Observe that $(\boldsymbol{x}^*, \boldsymbol{y}^*)$ satisfies $\boldsymbol{a}^T\boldsymbol{x}^* \leq b + \boldsymbol{c}^T\boldsymbol{y}^*$ since $(\boldsymbol{x}^*, \boldsymbol{y}^*) \in P_I$. But $y^k_v = y^*_v$ and $x^k_e \leq x^0_e = x^*_e$ for each e with $a_e > 0$. Thus $\boldsymbol{a}^T\boldsymbol{x}^k - \boldsymbol{c}^T\boldsymbol{y}^k \leq \boldsymbol{a}^T\boldsymbol{x}^* - \boldsymbol{c}^T\boldsymbol{y}^*$ thus showing that $(\boldsymbol{x}^k, \boldsymbol{y}^k)$ satisfies the constraint.

Now, suppose that $\|\boldsymbol{a}\|_0 > \frac{t}{2}$. Since $\boldsymbol{a}^T\boldsymbol{x} \leq b + \boldsymbol{c}^T\boldsymbol{y}$ is valid for $P^{(k)}$, we must have $\max\{\boldsymbol{a}^T\boldsymbol{x} - \boldsymbol{c}^T\boldsymbol{y} : (\boldsymbol{x},\boldsymbol{y}) \in P^{(k-1)}\} < b + 1$. But we have $(\boldsymbol{x}^{k-1}, \boldsymbol{y}^{k-1}) \in P^{(k-1)}$. Thus we have

$$\boldsymbol{a}^T\boldsymbol{x}^k - \boldsymbol{c}^T\boldsymbol{y}^k = (\boldsymbol{a}^T\boldsymbol{x}^{k-1} - \boldsymbol{a}^T \cdot (\frac{2}{t}\mathbf{1})) - \boldsymbol{c}^T\boldsymbol{y}^{k-1} \quad \text{(By definition of } \boldsymbol{x}^k, \boldsymbol{y}^k)$$

$$= \boldsymbol{a}^T\boldsymbol{x}^{k-1} - \boldsymbol{c}^T\boldsymbol{y}^{k-1} - \frac{2}{t}\|\boldsymbol{a}\|_1 \quad \text{(Rearranging)}$$

$$\leq \boldsymbol{a}^T\boldsymbol{x}^{k-1} - \boldsymbol{c}^T\boldsymbol{y}^{k-1} - \frac{2\|\boldsymbol{a}\|_0}{t} \quad \text{(For integer vectors, } \|\cdot\|_1 \geq \|\cdot\|_0)$$

$$< b + 1 - 1 \quad \text{(By definition of CG)}$$

$$= b \qquad \qquad \square$$

4   Integrality Gaps for Unique Games

We now prove Theorem 3 and present integrality gap result for the unique games problem. The problem is defined as follows. Given a graph $G = (V, E)$, a set of q labels $\mathcal{L} = \{1, \ldots, q\}$ and permutation $\pi_{uv} : \mathcal{L} \to \mathcal{L}$ for each edge $\{u, v\} \in E$, the task is to assign a label $\Lambda(v)$ to each vertex v of G to maximize the number of *satisfied* edges $\pi_{uv}(\Lambda(u)) = \Lambda(v)$.

Figure 3 is a linear program for the unique label cover problem. Here variable $y(u, i)$ denotes whether the vertex u gets label i. The variable $x(uv, i)$ denotes whether edge $(u, v) \in E$ is violated (value 1) with u getting label i and v not getting label $\pi_{uv}(i)$. Note that the LP here is for maximizing the number of satisfied constraints; the LP for minimizing the number of satisfied constraints can be obtained by changing the objective function to $\sum_{(u,v) \in E} \sum_{i \in \mathcal{L}} x(uv, i)$.

$$\max \sum_{(u,v) \in E} (1 - \sum_{i \in \mathcal{L}} x(uv, i))$$

s.t.

$$
\begin{array}{lll}
x(uv, i) & \geq y(u, i) - y(v, \pi_{uv}(i)) & \forall (u, v) \in E, i \in \mathcal{L} \\
\sum_{i \in \mathcal{L}} y(u, i) & = 1 & \forall\ u \in V \\
\sum_{i=1}^{t} x(u_{i-1} u_i, l_{i-1}) & \geq y(u, l_0) & \forall C, \forall u \in C, \forall l_0 \in B(u, C) \\
x(uv, i) & \geq 0 & \forall\ (u, v) \in E \\
y(u, i) & \geq 0 & \forall\ u \in V
\end{array}
$$

Fig. 3. Linear program for the Unique Label Cover Problem

We will in fact look at a richer LP from [25]. Let C be a simple cycle $u = v_0, v_1, \ldots, v_t = u$ in G containing u. Let l_0 be a label for v_0: for each value of $i \in [1, t]$, inductively define l_i as $l_i = \pi_{v_{i-1} v_i}(l_{i-1})$. I.e., the l_i's are defined so that $l_0, l_1, \ldots, l_i$ are labels that satisfy each of the edges $(v_0, v_1), \ldots, (v_{i-1}, v_i)$. Note that this process also defines another label l_t for $u = v_t$ which may or may not agree with the initial label l_0: indeed, we say that the label l_0 is *bad* for u with respect to C if $l_t \neq l_0$. Let $B_{u,C}$ be the set of labels that are bad for u with respect to C. Note that for any labeling f, if the label $f(u) = l_0$ lies in $B_{u,C}$, there must be at least one position i such that the label $f(v_i) = l_i$ and the next label $f(v_{i+1}) \neq l_{i+1}$; i.e., there must be at least one edge (v_i, v_{i+1}) that is violated. Hence for every such cycle C and every label $l_0 \in B_{u,C}$, we can write a constraint $\sum_{i=1}^{t} x(u_{i-1} u_i, l_{i-1}) \geq y(u, l_0)$.

We use the following gap results for the unique label cover problem shown by Charikar, Makarychev and Makarychev [10].

Theorem 10 ([10]). *For any $\epsilon > 0$, integer q there exists a $\gamma > 0$ and a unique label cover instance on a graph $G = (V, E)$ on n vertices such that a) Any labeling satisfies at most $(1 + \epsilon)/q$ fraction of the constraints, but b) for any set S of $t = n^\gamma$ vertices, there is a distribution $\mathcal{D}$ over assignments Λ_S of labels to these vertices such that (i) the marginal on any vertex is uniform over the labels, i.e. $\Pr_{\Lambda_S \sim \mathcal{D}}[\Lambda_S(v) = l] = \frac{1}{q}$ for any $l \in [q]$, and $v \in S$, and (ii) for any $e = (u, v) \in E$ with $u, v \in S$, $\Pr_{\Lambda_S \sim \mathcal{D}}[\Lambda_S(v) = \pi_{uv}(\Lambda_S(u)) = l] \geq \frac{1-\epsilon}{q}$.*

The result then follows along lines similar to the previous section. We inductively construct feasible solutions for the polytope $P^{(k)}$. Valid constraints involving few x variables are handled by the fact that local distributions Λ_S exist with the right marginals. Valid constraints involving many x variables are satisfied by induction due to the right scaling.

We set $(\boldsymbol{x}^k, \boldsymbol{y}^k)$ as follows: $y^k(u, i)$ is set to $\frac{1}{q}$ for each $u \in V, i \in [q]$. $x^k(uv, i)$ is set to $\frac{\epsilon}{q} + \frac{2(k+1)}{t}$. We will show by induction that $(\boldsymbol{x}^k, \boldsymbol{y}^k)$ lies in $P^{(k)}$.

We first show that any constraint in $P^{(k)}$ has a specific structure.

Lemma 4. *Let $\mathbf{a}^T \mathbf{x} + \mathbf{b}^T \mathbf{y} \geq c$ be a valid non-trivial constraint for $P^{(k)}$. Then the following hold without loss of generality.*

- *$\mathbf{a}, \mathbf{b}$ and c are integral and $a_i \geq 0$ for each i.*
- *Every vector $(\mathbf{x}, \mathbf{y})$ in P_I satisfies $\mathbf{a}^T \mathbf{x} + \mathbf{b}^T \mathbf{y} \geq c$.*
- *$\mathbf{a}^T \mathbf{x} + \mathbf{b}^T \mathbf{y} > c - 1$ for any $(\mathbf{x}, \mathbf{y}) \in P^{(k-1)}$.*

Proof. The first property follows by observing that they hold for the inequalities in P, and are preserved under summation. The last two properties are a consequence of the definition of $P^{(k)}$. □

Lemma 5. *Under the definitions above, $(\mathbf{x}^k, \mathbf{y}^k) \in P^{(k)}$.*

Proof. For the base case, note that $\boldsymbol{x}^0, \boldsymbol{y}^0$ satisfies all equation of the type $x(uv, i) \geq y(u, i) - y(v, \pi_{uv}(i))$ since the right hand side is zero. Also $\sum_i y(u, i)$ is indeed 1. For the cycle constraints, note that any cycle of length greater than $\lceil \frac{t}{2} \rceil$ is satisfied since each $x^0(uv, i)$ is at least $2/t$. For a constraint $\boldsymbol{a}^T \boldsymbol{x} + \boldsymbol{b}^T \boldsymbol{y} \geq c$ corresponding to a shorter cycle C, let $F = \{e \in E : \exists i : a_e^i > 0\}$ denote the set of edges with a positive a, and let $S = \{u \in V : \exists e \in F \cap \delta(u)\}$. Thus $|S| \leq |C| \leq \frac{t}{2}$. Let $\mathcal{D}$ denote the distribution of labelings of S guaranteed by Theorem 10. For a partial labeling Λ_S, let $Comp(\Lambda_S)$ denote a *completion* of Λ_S to all of V giving each vertex the same label as the lexicographically smallest vertex in S, and let $(\boldsymbol{x}^*, \boldsymbol{y}^*)$ denote the expected value of the integer solution defined by $Comp(\Lambda_S)$, when Λ_S is drawn from $\mathcal{D}$. Clearly $(\boldsymbol{x}^*, \boldsymbol{y}^*) \in P_I$ so that $\boldsymbol{a}^T \boldsymbol{x}^* + \boldsymbol{b}^T \boldsymbol{y}^* \geq c$. By Theorem 10, for any $u, v \in S$, $x^*(uv, i) \leq \frac{\epsilon}{q} \leq x^0(uv, i)$. Moreover, for any $u \in S$, $y^*(u, i) = \frac{1}{q} = y^0(u, i)$. Thus $\boldsymbol{a}^T \boldsymbol{x}^0 + \boldsymbol{b}^T \boldsymbol{y}^0 \geq \boldsymbol{a}^T \boldsymbol{x}^* + \boldsymbol{b}^T \boldsymbol{y}^* \geq c$.

Suppose that the claim holds for $k - 1$, i.e. $(\boldsymbol{x}^{k-1}, \boldsymbol{y}^{k-1}) \in P^{(k-1)}$. We argue that the claim holds for k. Now let $\boldsymbol{a}^T \boldsymbol{x} + \boldsymbol{b}^T \boldsymbol{y} \geq c$ be a constraint in $P^{(k)}$. We wish to argue that the solution $(\boldsymbol{x}^k, \boldsymbol{y}^k)$ above satisfies this constraint. Let $F = \{e \in E : \exists i : a_e^i > 0\}$ denote the set of edges with in support of $\boldsymbol{a}$, and let $S = \{u \in V : \exists e \in F \cap \delta(u)\}$. It is easy to see that $|S| \leq 2|F| \leq 2\|\boldsymbol{a}\|_0$.

First suppose that $|S| \leq t$. Let $\mathcal{D}$ denote the distribution of labelings of S guaranteed by theorem 10. For $Comp(\Lambda_S)$ as above, let $(\boldsymbol{x}^*, \boldsymbol{y}^*)$ denote the expected value of the integer solution defined by $Comp(\Lambda_S)$, when Λ_S is drawn from $\mathcal{D}$. Clearly $(\boldsymbol{x}^*, \boldsymbol{y}^*) \in P_I$ so that $\boldsymbol{a}^T \boldsymbol{x}^* + \boldsymbol{b}^T \boldsymbol{y}^* \geq c$. By theorem 10, for any $u, v \in S$, $x^*(uv, i) \leq \frac{\epsilon}{q} \leq x^k(uv, i)$. Moreover, for any $u \in V$, $y^*(u, i) = \frac{1}{q} = y^k(u, i)$. Thus $\boldsymbol{a}^T \boldsymbol{x}^k + \boldsymbol{b}^T \boldsymbol{y}^k \geq \boldsymbol{a}^T \boldsymbol{x}^* + \boldsymbol{b}^T \boldsymbol{y}^* \geq c$.

Now suppose that $|S| > t$. Then $\sum_i a_i \geq \|\boldsymbol{a}\|_0 > \frac{t}{2}$. By the last property is Lemma 4, $\boldsymbol{a}^T \boldsymbol{x}^{k-1} + \boldsymbol{b}^T \boldsymbol{y}^k > c - 1$. Thus

$$\boldsymbol{a}^T \boldsymbol{x}^k + \boldsymbol{b}^T \boldsymbol{y}^k = \boldsymbol{a}^T \boldsymbol{x}^{k-1} + \frac{2}{t} \sum_i a_i + \boldsymbol{b}^T \boldsymbol{y}^{k-1} \geq \boldsymbol{a}^T \boldsymbol{x}^{k-1} + 1 + \boldsymbol{b}^T \boldsymbol{y}^{k-1} \geq c - 1 + 1 = c$$

This completes the induction and the claim follows. □

Proof of Theorem 3 now follows form observing that the solution $(\boldsymbol{x}^k, \boldsymbol{y}^k)$ has an objective value at least $(1 - 2\epsilon)$ times the number of constraints while by Theorem 10 no integral solution satisfies more than $\frac{1+\epsilon}{q}$ fraction of the constraints.

5 Other Results

Our results for k-CSP and vertex cover go along similar lines and are proved in the full version of the paper [38].

6 Open Problems

Our negative results suggest that the connection between SA and C-G integrality gaps may extend to a fairly general class of linear programs. While this class would have to exclude hypergraph matching due to our negative result, it may include other interesting problems such as the sparsest cut. It also seems natural to investigate whether combining the various cut generation procedures improves integrality gaps when they individually do not.

References

1. Arora, S., Bollobás, B., Lovász, L., Tourlakis, I.: Proving Integrality Gaps without Knowing the Linear Program. Theory of Computing 2(1), 19–51 (2006)
2. Arora, S., Bollobás, B., Lovász, L.: Proving Integrality Gaps without Knowing the Linear Program. In: Proceedings of the 43rd Symposium on Foundations of Computer Science (FOCS), pp. 313–322 (2002)
3. Beame, P., Huynh, T., Pitassi, T.: Hardness amplification in proof complexity. In: Proceedings of the Fourty-Second Annual ACM Symposium on Theory of Computing, STOC (2010)
4. Berman, P., Karpinski, M.: Improved Approximation Lower Bounds on Small Occurrence Optimization. In: Electronic Colloquium on Computational Complexity (ECCC), vol. 10, p. 008 (2003)
5. Bienstock, D., Zuckerberg, M.: Approximate fixed-rank closures of covering problems. Math. Program. 105(1), 9–27 (2006)
6. Bockmayr, A., Eisenbrand, F., Hartmann, M.E., Schulz, A.S.: On the Chvtal Rank of Polytopes in the 0/1 Cube. Discrete Applied Mathematics 98(1-2), 21–27 (1999)
7. Buresh-Oppenheim., J., Galesi, N., Hoory, S., Magen, A., Pitassi, T.: Rank Bounds and Integrality Gaps for Cutting Planes Procedures. In: Proceedings of the 44th Annual IEEE Symposium on Foundations of Computer Science, October 11-14 (2003)
8. Calczyńska-Karlowicz, M.: Theorem on Families of Finite Sets. Bulletin de lAcadémie Polonaise des Sciences. Série des Sciences Mathématiques, Astronomiques et Physiques 12, 87–89 (1964)
9. Chan, Y.H., Lau, L.C.: On Linear and Semidefinite Programming Relaxations for Hypergraph Matching. In: Proceedings of ACM-SIAM Symposium on Discrete Algorithms (SODA) (January 2010)

10. Charikar, M., Makarychev, K., Makarychev, Y.: Integrality Gaps for Sherali-Adams Relaxations. In: Proceedings of the 41st Annual ACM Symposium on theory of Computing, STOC 2009, pp. 283–292 (2009)

11. Chvátal, V.: Edmonds Polytopes and a Hierarchy of Combinatorial Problems. Discrete Math. 4, 305–337 (1973)

12. Chvátal, V., Cook, W., Hartmann, M.: On Cutting-Plane Proofs in Combinatorial Optimization. Linear Algebra and its Applications 114/115, 455–499 (1989)

13. Cornuéjols, G.: Revival of the Gomory cuts in the 1990's. Annals of Operations Research 149(1), 63–66 (2007)

14. Dash, S.: On the Matrix Cuts of Lovsz and Schrijver and their use in Integer Programming. PhD thesis, Department of Computer Science, Rice University (March 2001)

15. Dash, S.: An Exponential Lower Bound on the Length of Some Classes of Branch-and-Cut Proofs. In: Cook, W.J., Schulz, A.S. (eds.) IPCO 2002. LNCS, vol. 2337, pp. 145–160. Springer, Heidelberg (2002)

16. de la Vega, W.F., Kenyon-Mathieu, C.: Linear Programming Relaxations of Max-cut. In: ACM-SIAM Symposium on Discrete Algorithms (SODA), January 2007, pp. 53–61 (2007)

17. Eisenbrand, F.: On the Membership Problem for the Elementary Closure of a Polyhedron. Combinatorica 19, 297–300 (1999)

18. Eisenbrand, F., Schulz, A.S.: Bounds on the Chvtal rank of polytopes in the 0/1-cube. Combinatorica 23, 245–261 (2003)

19. Erdös, P., Lovász, L.: Problems and Results on 3- chromatic Hypergraphs and Some Related Questions. In: Hajnal, A., Rado, R., Sós, V.T. (eds.) Proceedings of Colloquia Mathematica Societatis János Bolyai, Infinite and Finite Sets, vol. 10, pp. 609–627. Keszthely, Hungary (1973)

20. Eisenbrand, F., Schulz, A.S.: Bounds on the chvátal rank of polytopes in the 0/1-cube. In: Cornuéjols, G., Burkard, R.E., Woeginger, G.J. (eds.) IPCO 1999. LNCS, vol. 1610, pp. 137–150. Springer, Heidelberg (1999)

21. Goemans, M.X., Tuncel, L.: When does the Positive Semidefiniteness Constraint help in Lifting Procedures. Mathematics of Operations Research 26, 796–815 (2001)

22. Gomory, R.E.: Outline of an Algorithm for Integer Solutions to Linear Programs. Bulletin of the American Mathematical Society 64(5), 275–278 (1958)

23. Gomory, R.E.: Solving Linear Programming Problems in Integers. In: Bellman, R., Hall Jr., M. (eds.) Combinatorial Analysis. Symposia in Applied Mathematics X, pp. 211–215. American Mathematical Society, Providence (1960)

24. Georgiou, K., Magen, A., Pitassi, T., Tourlakis, I.: Integrality gaps of 2 - o(1) for Vertex Cover SDPs in the Lovsz-Schrijver Hierarchy. In: Proceedings of the 50th Annual Symposium on Foundations of Computer Science (FOCS), pp. 702–712 (2007)

25. Gupta, A., Talwar, K.: Approximating Unique Games. In: Proceedings of the Seventeenth Annual ACM-SIAM Symposium on Discrete Algorithm SODA 2006, pp. 99–106. ACM, New York (2006)

26. Hazan, E., Safra, M., Schwartz, O.: On the Complexity of Approximating k-Dimensional Matching. In: Arora, S., Jansen, K., Rolim, J.D.P., Sahai, A. (eds.) RANDOM 2003 and APPROX 2003. LNCS, vol. 2764, pp. 59–70. Springer, Heidelberg (2003)

27. Hurkens, C.A.J., Schrijver, A.: On the Size of Systems of Sets Every t of which have an SDR, with an Application to the Worst-case Ratio of Heuristics for Packing Problems. SIAM Journal on Discrete Mathematics 2, 68–72 (1989)

28. Khot, S., Saket, R.: SDP Integrality Gaps with Local ℓ_1-Embeddability. In: Proceedings of the 50th Annual Symposium on Foundations of Computer Science, FOCS (2009)

29. Lasserre, J.B.: An Explicit Exact SDP Relaxation for Nonlinear 0-1 Programs. In: Aardal, K., Gerards, B. (eds.) IPCO 2001. LNCS, vol. 2081, pp. 293–303. Springer, Heidelberg (2001)

30. Laurent, M.: A comparison of the Sherali-Adams, Lovász-Schrijver and Lasserre relaxations for 0-1 programming. Mathematics of Operations Research 28(3), 470–496 (2003)

31. Lovász, L., Schrijver, A.: Cones of Matrices and Set-functions and 0-1 Optimization. SIAM J. Optimization 1(2), 166–190 (1991)

32. Mathieu, C., Sinclair, A.: Sherali-Adams Relaxations of the Matching Polytope. In: Proceedings of the 41st Annual ACM Symposium on Theory of Computing, STOC 2009, pp. 293–302 (2009)

33. Pitassi, T., Segerlind, N.: Exponential lower bounds and Integrality gaps for Treelike Lovsz-Schrijver Procedures. In: Proceedings of the Twentieth Annual ACM-SIAM Symposium on Discrete Algorithms, New York, January 04-06 (2009)

34. Pudlák, P.: Lower Bounds for Resolution and Cutting Plane Proofs and Monotone Computations. Journal of Symbolic Logic 62(3), 981–998 (1997)

35. Raghavendra, P., Steurer, D.: Integrality Gaps for Strong SDP Relaxations of Unique Games. In: Proceedings of the 50th Annual Symposium on Foundations of Computer Science, FOCS (2009)

36. Schoenebeck, G.: Linear Level Lasserre Lower Bounds for Certain k-CSPs. In: Proceedings of the 50th Annual Symposium on Foundations of Computer Science (FOCS), pp. 593–602 (2008)

37. Sherali, H.D., Adams, W.P.: A Hierarchy of Relaxations between the Continuous and Convex Hull Representations for Zero-One Programming Problems. SIAM J. Discrete Math. 3(3), 411–430 (1990)

38. Singh, M., Talwar, K.: Improving Integrality Gaps via chvátal-Gomory Rounding (2010) (manuscript)

39. Tulsiani, M.: CSP gaps and Reductions in the Lasserre Hierarchy. In: Proceedings of the 41st Annual ACM Symposium on Theory of Computing, STOC 2009, pp. 303–312 (2009)

40. Tuza, Z.: Critical Hypergraphs and Intersecting Set-pair Systems. Journal of Combinatorial Theory (B) 39, 134–145 (1985)

41. Tuza, Z.: On Two Intersecting Set Systems and k- continuous Boolean functions. Discrete Applied Mathematics 16, 183–185 (1987)

Uniform Derandomization from Pathetic Lower Bounds

Eric Allender[1], V. Arvind[2], and Fengming Wang[1]

[1] Department of Computer Science, Rutgers University, Piscataway, NJ 08855, USA
{allender,fengming}@cs.rutgers.edu
[2] The Institute of Mathematical Sciences, C.I.T. Campus Chennai 600 113, India
arvind@imsc.res.in

Abstract. A recurring theme in the literature on derandomization is that probabilistic algorithms can be simulated quickly by deterministic algorithms, if one can obtain *impressive* (i.e., superpolynomial, or even nearly-exponential) circuit size lower bounds for certain problems. In contrast to what is needed for derandomization, existing lower bounds seem rather pathetic (linear-size lower bounds for general circuits [30], nearly cubic lower bounds for formula size [23], nearly $n \log \log n$ size lower bounds for branching programs [12], n^{1+c_d} for depth d threshold circuits [26]). Here, we present two instances where "pathetic" lower bounds of the form $n^{1+\epsilon}$ would suffice to derandomize interesting classes of probabilistic algorithms. We show:

- If the word problem over S_5 requires constant-depth threshold circuits of size $n^{1+\epsilon}$ for some $\epsilon > 0$, then any language accepted by uniform polynomial-size probabilistic threshold circuits can be solved in subexponential time (and more strongly, can be accepted by a uniform family of deterministic constant-depth threshold circuits of subexponential size.)
- If no constant-depth arithmetic circuits of size $n^{1+\epsilon}$ can multiply a sequence of n 3-by-3 matrices, then for every constant d, black-box identity testing for depth-d arithmetic circuits with bounded individual degree can be performed in subexponential time (and even by a uniform family of deterministic constant-depth AC^0 circuits of subexponential size).

Keywords: Derandomization, Circuit Complexity, Polynomial Identity Testing.

1   Introduction

Hardness-based derandomization is one of the success stories of the past quarter century. This line of research dates back to the work of Shamir, Yao, and Blum and Micali [39,46,14], and involves showing that, if given a suitably hard function f, one can construct pseudorandom generators and hitting-set generators. Much of the progress on this front over the years has involved showing how to weaken the hardness assumption on f and still obtain useful derandomizations [10,9,28,29,32,6,5,7,15,33,18,17,27,41,38,42]. In rare instances, it has been possible to obtain *unconditional* derandomizations using this framework; Nisan and Wigderson showed that uniform families of probabilistic AC^0 circuits can be simulated by uniform deterministic AC^0 circuits of size $n^{\log^{O(1)} n}$ [34]. More often, the derandomizations that have been obtained are conditional, and

M. Serna et al. (Eds.): APPROX and RANDOM 2010, LNCS 6302, pp. 380–393, 2010.

rely on the existence of functions f that are hard on average. For certain large complexity classes $\mathcal{C}$ (notably including #P, PSPACE, and exponential time), various types of random self-reducibility and hardness amplification have been employed to show that such hard-on-average functions f exist in $\mathcal{C}$ if and only if there is some problem in $\mathcal{C}$ that requires large Boolean circuits [10,28].

Some more recent work in derandomization studies the implications of *arithmetic* circuit lower bounds for derandomization. Kabanets and Impagliazzo showed that the probabilistic algorithm to test if two arithmetic *formulae* (or more generally, two arithmetic circuits of polynomial degree) are equivalent can be simulated by a quick deterministic algorithm, if the Permanent requires large *arithmetic circuits* [31]. Dvir, Shpilka, and Yehudayoff subsequently adapted those techniques, to show that if one could present a multilinear polynomial (such as the permanent) that requires depth d arithmetic formulae of size 2^{n^ϵ}, then the probabilistic algorithm to test if two arithmetic circuits of depth $d-5$ are equivalent (where in addition, the variables in these circuits have degree at most $\log^{O(1)} n$) can be derandomized to obtain a $2^{\log^{O(1)} n}$ deterministic algorithm for the problem.

Here, we combine these two lines of work with the recent insight that, in some cases, extremely modest-sounding (or even "pathetic") lower bounds can be amplified to obtain superpolynomial bounds [4]. In order to do this, we must identify and exploit some special properties of certain functions in and near NC^1.

- The word problem over S_5 is one of the standard complete problems for NC^1 [11]. Many of the most familiar complete problems for NC^1 have very efficient *strong downward self-reductions* [4]. We show that the word problem over S_5, in addition, is *randomly self-reducible*. (This was observed previously by Goldwasser *et al.* [19].) This enables us to transform a "pathetic" *worst-case* size lower bound of $n^{1+\epsilon}$ on constant-depth threshold circuits, to a superpolynomial size *average-case* lower bound for this class of circuits. In turn, by making some adjustments to the Nisan-Wigderson generator, this average-case hard function can be used to give uniform subexponential derandomizations of probabilistic TC^0 circuits.
- Iterated Multiplication of n three-by-three matrices is a multilinear polynomial that is complete for arithmetic NC^1 [13]. In the Boolean setting, this function is strongly downward self-reducible via TC^0 self-reductions [4]. Here we present an *arithmetic* self-reduction; this enables us to amplify a lower bound of size $n^{1+\epsilon}$ for constant-depth arithmetic circuits, to obtain a superpolynomial lower bound for constant-depth arithmetic circuits. Then, by building on the approach of Dvir *et al.* [16], we obtain subexponential derandomizations of the identity testing problem for a class of constant-depth arithmetic circuits.

The rest of the paper is organized as follows: In Section 2 we give the preliminary definitions and notation. In Section 3 we convert a modest worst-case hardness assumption to a strong average-case hardness separation of NC^1 from TC^0, and in Section 4 we use this to give a uniform derandomization of probabilistic TC^0 circuits. Finally, in Section 5 we prove our derandomization of a special case of polynomial identity testing under a modest hardness assumption.

2 Preliminaries

This paper will mainly discuss NC^1 and its subclass TC^0. The languages in NC^1 are accepted by families of circuits of depth $O(\log n)$ that are built with fan-in two AND and OR gates, and NOT gates of fan-in one. For any function $s(n)$, $TC^0(s(n))$ consists of languages that are decided by constant-depth circuit families of size at most $s(n)$ which contain only unbounded fan-in MAJORITY gates as well as unary NOT gates. $TC^0 = \cup_{k \geq 0} TC^0(n^k)$. $TC^0(\text{SUBEXP}) = \cap_{\delta > 0} TC^0(2^{n^\delta})$. The definitions of $AC^0(s(n))$, AC^0, and $AC^0(\text{SUBEXP})$ are similar, although MAJORITY gates are not allowed, and unbounded fan-in AND and OR gates are used instead.

As is usual in arguments in derandomization based on the hardness of some function f, we require not only that f not have small circuits in order to be considered "hard", but furthermore we require that f needs large circuits at *every* relevant input length. This motivates the following definition.

Definition 2.1. *Let A be a language, and let D_A be the set $\{n : A \cap \Sigma^n \neq \emptyset\}$. We say that $A \in$ io-$TC^0_\infty(s(n))$ if there is an infinite set $I \subseteq D_A$ and a language $B \in TC^0(s(n))$ such that, for all $n \in I$, $A_n = B_n$ (where, for a language C, we let C_n denote the set of all strings of length n in C). Similarly, we define io-TC^0_∞ to be $\cup_{k \geq 0}$io-$TC^0_\infty(n^k)$.*

Thus A requires large threshold circuits on *all* relevant input lengths if $A \notin$ io-TC^0_∞. (A peculiarity of this definition is that if A is a *finite* set, or A^n is empty for infinitely many n, then $A \notin$ io-TC^0_∞. This differs starkly from most notions of "io" circuit complexity that have been considered, but it allows us to consider "complex" sets A that are empty on infinitely many input lengths; the alternative would be to consider artificial variants of the "complex" sets that we construct, having strings of every length.)

Probabilistic circuits take an input divided into two pieces, the actual input and the random coin flips. We say an input x is accepted by such a circuit C if, with respect to the uniform distribution U_R over coin flips, $Pr_{r \sim U_R}[C(x, r) = 1] \geq \frac{2}{3}$ while x is rejected by C if $Pr_{r \sim U_R}[C(x, r) = 1] \leq \frac{1}{3}$.

DLOGTIME-uniformity is the standard uniformity condition for small complexity classes. In order to provide its proper definition, we need to mention the direct connection language associated with a circuit family.

Definition 2.2. *Let $C = (C_n)_{n \in \mathbb{N}}$ be a circuit family. The direct connection language L_{DC} of C is the set of all tuples having either the form $\langle n, p, q, b \rangle$ or $\langle n, p, d \rangle$, where*

- *If $q = \epsilon$, then b is the type of gate p in C_n;*
- *If q is the binary encoding of k, then b is the kth input to p in C_n.*
- *The gate p has fan-in d in C_n.*

The circuit family C is **DLOGTIME**-uniform if there is a deterministic Turing machine that accepts L_{DC} in linear time. For any circuit complexity class C, uC is its uniform counterpart, consisting of languages that are accepted by **DLOGTIME**-uniform circuit families. For more background on circuit complexity, we refer the reader to the textbook by Vollmer [45]. The term "uniform derandomization" in the title refers to the fact that we are presenting uniform circuit families that compute derandomized algorithms; this should not be confused with doing derandomization based on uniform hardness assumptions.

A particularly important complete language for NC^1 is the word problem WP for S_5, where S_5 is the symmetric group over 5 distinct elements [11]. The input to the word problem is a sequence of permutations from S_5 and it is accepted if and only if the product of the sequence evaluates to the identity permutation. The corresponding *search* problem FWP is required to output the exact result of the iterated multiplication. A closely related *balanced* language is BWP, which stands for Balanced Word Problem.

Definition 2.3. *The input to BWP is a pair* $\langle w_1 w_2 .. w_n, S \rangle$, *where* $\forall i \in [1..n]$, $w_i \in S_5$, $S \subseteq S_5$ *and* $|S| = 60$. BWP *contains* $\langle w_1 w_2 .. w_n, S \rangle$ *if and only if* $\Pi_{i=1}^n w_i \in S$.

It is easy to verify that BWP is complete for NC^1 as well.

In the following sections, let FWP_n be the sub-problem of FWP where the domain is restricted to inputs of length n and let BWP_n be BWP $\cap \{\langle \phi, S \rangle \mid \phi \in S_5^n, |\phi| = n, S \subseteq S_5, |S| = 60\}$. Note that BWP_n accepts exactly half of the instances in $\{\langle \phi, S \rangle \mid \phi \in S_5^n, |\phi| = n, S \subseteq S_5, |S| = 60\}$ since $|S_5| = 120$.

The following simplified version of Chernoff's bound is useful in our application.

Lemma 2.4 (Chernoff's bound). *Let* $X_1, .., X_m$ *be i.i.d. 0-1 random variables with* $E[X_i] = p$. *Let* $X = \Sigma_{i=1}^n X_i$. *Then for any* $0 < \delta \le 1$,

$$Pr[X < (1 - \delta)pm] \le e^{-\frac{\delta^2 pm}{2}}.$$

3   The Existence of an Average-Case Hard Language

In this section, we use random self-reducibility to show that, if $NC^1 \neq TC^0$, then there are problems in NC^1 that are hard on average for TC^0. First we recall the definition of hardness on average for decision problems.

Definition 3.1. *Let* U_D *denote the uniform distribution over all inputs in a finite domain* D. *For any Boolean function* $f : D \to \{0, 1\}$, f *is* $(1 - \epsilon)$-*hard for a set of circuits* S, *if, for every* $C \in S$, *we have that* $Pr_{x \sim U_D}[f(x) = C(x)] < 1 - \epsilon$.

We will sometimes abuse notation by identifying a set with its characteristic function. For languages to be considered hard on average, we consider only those input lengths where the language contains some strings.

Definition 3.2. *Let* Σ *be an alphabet. Consider a language* $L = \cup_n L_n$, *where* $L_n = L \cap \Sigma^n$, *and let* $D_L = \{n : L_n \neq \emptyset\}$. *We say that* L *is* $(1-\epsilon)$-*hard for a class of circuit families* $\mathcal{C}$ *if* D_L *is an infinite set and, for any circuit family* $\{C_n\}$ *in* $\mathcal{C}$, *there exists* m_0 *such that for all* $m \in D_L$ *such that* $m \ge m_0$, $Pr_{x \in \Sigma^m}[f(x) = C(x)] < 1 - \epsilon$.

The next theorem shows that BWP is hard on average for TC^0 if FWP $\notin$ io-TC_∞^0.

Theorem 3.3. *There exist constants* $c, \delta > 0$ *and* $0 < \epsilon < 1$ *such that for any constant* $d > 0$, *if* FWP_n *is not computable by* $TC^0(\delta n(s(n) + cn))$ *circuits of depth at most* $d + c$, *then* BWP_n *is* $(1 - \epsilon)$-*hard for* TC^0 *circuits of size* $s(n)$ *and depth* d.

Proof. Let $\epsilon < \frac{1}{4\binom{120}{60}}$. We prove the contrapositive. Assume there is a circuit C of size $s(n)$ and depth d such that $Pr_x[\text{BWP}_n(x) = C(x)] \geq 1 - \epsilon$. We first present a probabilistic algorithm for FWP_n.

Let the input instance for FWP_n be $w_1 w_2 \ldots w_n$. Generate a sequence of $n + 1$ random permutations $u_0, u_1, \ldots, u_n$ in S_5 and a random set $S \subseteq S_5$ of size 60. Let ϕ be the sequence $(u_0 \cdot w_1 \cdot u_1)(u_1^{-1} \cdot w_2 \cdot u_2)..(u_{n-1}^{-1} \cdot w_n \cdot u_n)$. Note that ϕ is a completely random sequence in S_5^n.

Let us say that ϕ is a "good" sequence if $\forall S' \subset S_5$ with $|S'| = 60$, $C(\langle \phi, S' \rangle) = \text{BWP}_n(\langle \phi, S' \rangle)$.

If we have a "good" sequence ϕ (meaning that for *every* set S' of size 60, C gives the "correct" answer $\text{BWP}_n(\phi, S)$ on input (ϕ, S')), then we can easily find the unique value r that is equal to $\Pi_{i=1}^n \phi_i$ where $\phi_i = u_{i-1} w_i u_i$, as follows:

- If $C(\phi, S) = 1$, then it must be the case that $r \in S$. Pick any element $r' \in S_5 \setminus S$ and observe that r is the only element such that $C(\phi, (S \setminus \{r\}) \cup \{r'\}) = 0$.
- If $C(\phi, S) = 0$, then it must be the case that $r \notin S$. Pick any element $r' \in S$ and observe that r is the only element such that $C(\phi, (S \setminus \{r'\}) \cup \{r\}) = 1$.

Thus the correct value r can be found by trying all such r'. Hence, if ϕ is good, we have

$$r = \Pi_{i=1}^n \phi_i = u_0 w_1 u_1 \Pi_{i=2}^n u_{i-1}^{-1} w_i u_i.$$

Produce as output the value $u_0^{-1} r u_n^{-1} = \Pi_{i=1}^n w_i = \text{FWP}_n(w)$.

Since $\epsilon < \frac{1}{4\binom{120}{60}}$, a standard averaging argument shows that at least $\frac{3}{4}$ of the sequences in S_5^n are good. Thus with probability at least $\frac{3}{4}$, the probabilistic algorithm computes FWP_n correctly. The algorithm can be computed by a threshold circuit of depth $d + O(1)$ since the subroutines related to C can be invoked in parallel and moreover, the preparation of ϕ and the aggregation of results of subroutines can be done by constant-depth threshold circuits. Its size is at most $122s(n) + O(n)$ since there are 122 calls to C. Next, we put $10^4 n$ independent copies together in parallel and output the majority vote. Let X_i be the random variable that the outcome of the ith copy is $\Pi_{i=1}^n w_i$. By Lemma 2.4, on every input the new circuit computes FWP_n with probability at least $1 - \frac{120^{-n}}{2}$. Thus there is a random sequence that can be hardwired in to the circuit, with the property that the resulting circuit gives the correct output on *every* input (and in fact, at least half of the random sequences have this property). This yields a deterministic TC^0 circuit computing FWP_n exactly which is of depth at most $d + c$ and of size no more than $(122 * 10^4)n(s(n) + cn)$ for some universal constant c. Choosing $\delta \geq (122 * 10^4)$ completes the proof.

The problem FWP is strongly downward self-reducible [4, Definition , Proposition 7]. Hence, its worst-case hardness against TC^0 circuit families can be amplified as observed by Allender and Koucký [4, Corollary 17].

Theorem 3.4. *[4] If there is a $\gamma > 0$ such that* FWP $\notin$ io-$\text{TC}^0_\infty(n^{1+\gamma})$, *then* FWP $\notin$ io-TC^0_∞.

(Theorem 3.4 is not stated in terms of io-TC^0_∞ in [4], but the proof shows that if there are infinitely many input lengths n where FWP has circuits of of size n^k, then there are infinitely many input lengths m where FWP has circuits of size $m^{1+\gamma}$. The strong downward self-reducibility property allows small circuits for inputs of size m to be constructed by efficiently using circuits for size $n < m$ as subcomponents.)

Since FWP is equivalent to WP via linear-size reductions on the same input length, the following corollary is its easy consequence.

Corollary 3.5. *If there is a $\gamma > 0$ such that* WP $\notin$ io-TC$^0_\infty(n^{1+\gamma})$, *then* FWP $\notin$ io-TC$^0_\infty$.

Combining Corollary 3.5 with Theorem 3.3 yields the average-case hardness of BWP from nearly-linear-size worst-case lower bounds for WP against TC^0 circuit families.

Corollary 3.6. *There exists a constant $\epsilon > 0$ such that if $\exists \gamma > 0$ such that* WP $\notin$ io-TC$^0_\infty(n^{1+\gamma})$, *then for any k and d there exists $n_0 > 0$ such that when $n \geq n_0$,* BWP$_n$ *is $(1 - \epsilon)$-hard for any* TC^0 *circuit of size n^k and depth d.*

Define the following Boolean function WPM$_n$: $S^n \times S^{60} \rightarrow \{0, 1\}$, where WPM$_n$ stands for Word Problem over Multi-set.

Definition 3.7. *The input to* WPM$_n$ *is a pair $\langle w_1 w_2..w_n, v_1 v_2..v_{60} \rangle$, where $\forall i \in [1..n]$, $w_i \in S_5$ and $\forall j \in [1..60]$, $v_i \in S_5$. $\langle w_1 w_2..w_n, v_1 v_2..v_{60} \rangle \in$ WPM if and only if $\exists j \in [1..60]$, $\Pi^n_{i=1} w_i = v_j$.*

BWP is the restriction of WPM$_n$ to the case where all v_is are distinct. Hence, WPM inherits the average-case hardness of BWP, since any circuit that computes WPM$_n$ on a sufficiently large fraction of inputs also approximates BWP well. Formally,

Lemma 3.8. *There is an absolute constant $0 < c < 1$ such that for every $\epsilon > 0$, if* BWP$_n$ *is $(1 - \epsilon)$-hard for* TC^0 *circuits of size n^k and depth d, then* WPM$_n$ *is $(1 - c\epsilon)$-hard for* TC^0 *circuits of size n^k and depth d.*

Proof. Let $c = \frac{\binom{120}{60}}{(120)^{60}}$. Note that c is the probability that a sequence of 60 permutations contains no duplicates and is in sorted order. Suppose there is a circuit C with the property that $Pr_{x \in S^n \times S^{60}}[C(x) \neq \text{WPM}(x)] \leq c\epsilon$. Then the conditional probability that $C(x) \neq \text{WPM}(x)$ given that the last 60 items in x give a list in sorted order with no duplicates is at most ϵ. This yields a circuit having the same size, solving BWP with error at most ϵ, using the uniform distribution over its domain, contrary to our assumption.

Corollary 3.9. *There exists a constant $\epsilon > 0$ such that if $\exists \gamma > 0$ such that* WP $\notin$ io-TC$^0_\infty(n^{1+\gamma})$, *then for any k and d there exists $n_0 > 0$ such that when $n \geq n_0$,* WPM$_n$ *is $(1 - \epsilon)$-hard for* TC^0 *circuits of size n^k and depth d.*

Yao's XOR lemma [46] is a powerful tool to boost average-case hardness. We utilize a specialized version of the XOR lemma for our purpose. Several proofs of this useful result have been published. For instance, see the text by Arora and Barak [8] for a proof that is based on Impagliazzo's hardcore lemma [25]. For our application here, we need a version of the XOR lemma that is slightly different from the statement given by Arora and Barak. In the statement of the lemma as given by them, g is a function of the form $\{0,1\}^n \to \{0,1\}$. However, their proof works for any Boolean function g defined over any finite alphabet, because both the hardcore lemma and its application in the proof of the XOR lemma are insensitive to the encoding of the alphabet. Hence, we state the XOR Lemma in terms of functions over an alphabet set Σ.

For any Boolean function g over some domain Σ^n, define $g^{\oplus m} : \Sigma^{nm} \to \{0,1\}$ by $g^{\oplus m}(x_1, x_2, .., x_m) = g(x_1) \oplus g(x_2) \oplus .. \oplus g(x_m)$ where $\oplus$ is the parity function.

Lemma 3.10. *[46] Let $\frac{1}{2} < \epsilon < 1$, $k \in \mathbb{N}$ and $\theta > 2(1-\epsilon)^k$. There is a constant $c > 1$ that depends only on $|\Sigma|$ such that if g is $(1 - \epsilon)$-hard for TC^0 circuits of size s and depth d, then $g^{\oplus k}$ is $(\frac{1}{2} + \theta)$-hard for TC^0 circuits of size $\frac{\theta^2 s}{cn}$ and depth $d - 1$.*

Let $\Sigma = S_5$. The next corollary follows from Corollary 3.9 and Lemma 3.10.

Corollary 3.11. *If there is a $\gamma > 0$ such that $\mathrm{WP} \notin \mathrm{io\text{-}TC}_\infty^0(n^{1+\gamma})$, then for any k, k' and d there exists $n_0 > 0$ such that when $n \geq n_0$ $(\mathrm{WPM}_n)^{\oplus n}$ is $(\frac{1}{2} + \frac{1}{n^{k'}})$-hard for TC^0 circuits of size n^k and depth d.*

Let $\mathrm{WP}^\otimes = \cup_{n \geq 1} \{x \mid (\mathrm{WPM}_n)^{\oplus n}(x) = 1\}$. Note that it is a language in uNC^1 and, moreover, it is decidable in linear time.

Theorem 3.12. *If there is a $\gamma > 0$ such that $\mathrm{WP} \notin \mathrm{io\text{-}TC}_\infty^0(n^{1+\gamma})$, then for any integer $k > 0$, $\mathrm{WP}^\otimes$ is $(\frac{1}{2} + \frac{1}{n^k})$-hard for TC^0.*

4 Uniform Derandomization

The Nisan-Wigderson generator, the canonical way to build pseudo-random generators based on hard functions, relies on the following definition of combinatorial designs.

Definition 4.1 (Combinatorial Designs). *Fix a universe of size u. An (m, l)-design of size n on $[u]$ is a list of subsets $S_1, S_2, ..., S_n$ satisfying:*

1. *$\forall i \in [1..n]$, $|S_i| = m$;*
2. *$\forall i \neq j \in [1..n]$, $|S_i \cap S_j| \leq l$.*

Nisan and Wigderson [34] invented a general approach to construct combinatorial designs for various ranges of parameters. The proof given by Nisan and Wigderson gives designs where $l = \log n$, and most applications have used that value of l. For our application, l can be considerably smaller, and furthermore, we need the S_i's to be very efficiently computable. For completeness, we present the details here. (Other variants of the Nisan-Wigderson construction have been developed for different settings; we refer the reader to one such construction by Viola [44], as well as to a survey of related work [44, Remark 5.3].)

Lemma 4.2. *[43] For $l > 0$, the polynomial $x^{2 \cdot 3^l} + x^{3^l} + 1$ is irreducible over $\mathbb{F}_2[x]$.*

Lemma 4.3. *[34] For any integer n, any α such that $\log \log n / \log n < \alpha < 1$, let $b = \lceil \alpha^{-1} \rceil$ and $m = \lceil n^\alpha \rceil$, there is a (m, b)-design with $u = O(m^6)$. Furthermore, each S_i can be computed within $O(bm^2)$ time.*

Proof. Fix $q = 2^{2 \cdot 3^l}$ for some l such that $m \leq q \leq m^3$. Let the universe be $\mathbb{F}_q \times \mathbb{F}_q$ and S_i be the graph of the ith univariate polynomial of degree at most b in the standard order. Since $q^b \geq (n^\alpha)^b \geq n$, there are at least n distinct S_is. No two polynomials share more than b points, hence, the second condition is satisfied. The first condition holds because we could simply drop elements without increasing the size of intersections.

The arithmetic operations in $\mathbb{F}_q$ are performed within $\log^{O(1)} q$ time because of the explicitness of the irreducible polynomial by Lemma 4.2. It is evident that for any $i \in [n]$, we are able to enumerate all elements of S_i in time $O(m \cdot b(\log^{O(1)} q)) = O(bm^2)$.

Lemma 4.4. *For any constant $\alpha > 0$ and for any large enough integer n, if g is $(\frac{1}{2} + \frac{1}{n^2})$-hard for TC^0 circuits of size n^2 and depth $d+2$, then any probabilistic TC^0 circuit C of size n and depth d can be simulated by another probabilistic TC^0 circuit of size $O(n^{1+\alpha})$ and depth $d+1$ which is given oracle access to $g_{\lceil n^\alpha \rceil}$ and uses at most $O(n^{6\alpha})$ many random bits.*

The proof is omitted due to space limitations. For details, see [2].

The simulation in Lemma 4.4 is quite uniform, thus, plugging in appropriate segments of $WP^\otimes$ as our candidates for the hard function g, we derive our first main result.

Theorem 4.5. *If* WP *is not infinitely often computed by $TC^0(n^{1+\gamma})$ circuit families for some constant $\gamma > 0$, then any language accepted by polynomial-size probabilistic uniform TC^0 circuit family is in $uTC^0(SUBEXP)$.*

Proof. Fix any small constant $\delta > 0$. Let L be a language accepted by some probabilistic uniform TC^0 circuit family of size at most n^k and of depth at most d for some constants k, d.

Choose m such that $n^{\frac{\delta}{12}} \leq m \leq n^{\frac{\delta}{6}}$, and let α be such that $m = n^\alpha$. By Theorem 3.12, when m is large enough, $WP_m^\otimes$ is $(\frac{1}{2} + \frac{1}{n^{2k}})$-hard for TC^0 circuits of size n^{2k} and depth $d+c$, where c is any constant. Hence, as a consequence of Lemma 4.4, we obtain a probabilistic oracle TC^0 circuit for L_n of depth $d+1$. Since the computation only needs $O(m^6)$ random bits, it can be turned into a deterministic oracle TC^0 circuit of depth $d+2$ and of size at most $O(n^{2k}) * 2^{O(m^6)} \leq 2^{O(n^\delta)}$ (when n is large enough), where we evaluate the previous circuit on every possible random string and add an extra MAJORITY gate at the top. The oracle gates all have fan-in $m \leq n^{\delta/6}$, and thus can be replaced by DNF circuits of size $2^{O(n^\delta)}$, yielding a deterministic TC^0 circuit of size $2^{O(n^\delta)}$ and depth $d+3$.

We need to show that this construction is uniform, so that the direct connection language can be recognized in time $O(n^\delta)$. The analysis consists of three parts.

- The connectivity between the top gate and the output gate of individual copies is obviously computable in time $m^6 \leq n^\delta$.
- The connectivity inside individual copies is **DLOGTIME**-uniform.
- By Lemma 4.3 each S_i is computable in time $O(dm^2)$ which is $O(m^2)$ since d is a constant only depending on δ. Moreover, notice that $WP^\otimes$ is a linear-time decidable language. Therefore, the DNF expression corresponding to each oracle gate can be computed within time $O(m^2) \leq n^\delta$.

In conclusion, the above construction produces a uniform TC^0 circuit of size 2^{n^δ}. Since δ is arbitrarily chosen, our statement holds.

This can be strengthened as follows: Any language accepted by a polynomial-size probabilistic $o(n)$-uniform TC^0 circuit family is in $uTC^0(SUBEXP)$.

5   Consequences of Pathetic Arithmetic Circuit Lower Bounds

In this section we show that a pathetic lower bound assumption for *arithmetic circuits* yields a uniform derandomization of a special case of polynomial identity testing (introduced and studied by Dvir et al [16]).

The explicit polynomial that we consider is $\{IMM_n\}_{n>0}$, where IMM_n is the $(1,1)$ entry of the product of n 3×3 matrices whose entries are all distinct indeterminates. Notice that IMM_n is a degree n multilinear polynomial in $9n$ indeterminates, and IMM_n can be considered as a polynomial over any field $\mathbb{F}$.

Arithmetic circuits computing a polynomial in the ring $\mathbb{F}[x_1, x_2, \ldots, x_n]$ are directed acyclic graphs with the indegree zero nodes (the inputs nodes) labeled by either a variable x_i or a scalar constant. Each internal node is either a $+$ gate or a $\times$ gate, and the circuit *computes* the polynomial that is naturally computed at the output gate. The circuit is a *formula* if the fanout of each gate is 1.

Before going further, we pause to clarify a point of possible confusion. There is another way that an arithmetic circuit C can be said to compute a given polynomial $f(x_1, x_2, \ldots, x_n)$ over a field $\mathbb{F}$; even if C does not compute f in the sense described in the preceding paragraph, it can still be the case that for all scalars $a_i \in \mathbb{F}$ we have $f(a_1, \ldots, a_n) = C(a_1, \ldots, a_n)$. In this case, we say that C *functionally* computes f over $\mathbb{F}$. If the field size is larger than the syntactic degree of circuit C and the degree of f, then the two notions coincide. Assuming that f is not *functionally* computed by a class of circuits is a *stronger* assumption than assuming that f is not computed by a class of circuits (in the usual sense). In our work in this paper, we use the weaker intractability assumption.

An *oracle* arithmetic circuit is one that has *oracle* gates: For a given sequence of polynomials $A = \{A_n\}$ as oracle, an oracle gate of fan-in n in the circuit evaluates the n-variate polynomial A_n on the values carried by its n input wires. An oracle arithmetic circuit is called *pure* (following [4]) if all non-oracle gates are of bounded fan-in. (Note that this use of the term "pure" is unrelated to the "pure" arithmetic circuits defined by Nisan and Wigderson [35].)

Arithmetic NC^1 is the class of polynomials computed by polynomial-size arithmetic formulas; by [13] the polynomial IMM_n is complete for this class. Whether IMM_n has polynomial size *constant-depth* arithmetic circuits is a long-standing open problem in the area of arithmetic circuits [35]. In this context, the known lower bound result is that IMM_n requires exponential size multilinear depth-3 circuits [35].

Very little is known about lower bounds for general constant-depth arithmetic circuits, compared to what is known about constant-depth Boolean circuits. Exponential lower bounds for depth-3 arithmetic circuits over finite fields were shown in [21] and [20]. On the other hand, for depth-3 arithmetic circuits over fields of characteristic zero only quadratic lower bounds are known [40]. However, it is shown in [37] that the determinant and the permanent require exponential size *multilinear* constant-depth arithmetic circuits. More details on the current status of arithmetic circuit lower bounds can be found in Raz's paper [36, Section 1.3].

Definition 5.1. *We say that a sequence of polynomials* $\{p_n\}_{n>0}$ *in* $\mathbb{F}[x_1, x_2, \ldots, x_n]$ *is* $(s(n), m(n), d)$*-downward self-reducible if there is a pure oracle arithmetic circuit* C_n *of depth* $O(d)$ *and size* $O(s(n))$ *that computes the polynomial* p_n *using oracle gates only for* $p_{m'}$, *for* $m' \leq m(n)$.

Analogous to [4, Proposition 7], we can easily observe the following. It is a direct divide and conquer argument using the iterated product structure.

Lemma 5.2. *For each* $1 > \epsilon > 0$ *the polynomial sequence* $\{IMM_n\}$ *is* $(n^{1-\epsilon}, n^\epsilon, 1/\epsilon)$*-downward self-reducible.*

An easy argument, analogous to Theorem 3.4, shows that Lemma 5.2 allows for the amplification of weak lower bounds for $\{IMM_n\}$ against arithmetic circuits of constant depth:

Theorem 5.3. *Suppose there is a constant* $\delta > 0$ *such that for all* d *and every* n, *the polynomial sequence* $\{IMM_n\}$ *requires depth-*d *arithmetic circuits of size at least* $n^{1+\delta}$. *Then, for any constant depth* d *the sequence* $\{IMM_n\}$ *is not computable by depth-*d *arithmetic circuits of size* n^k *for any constant* $k > 0$.

Our goal is to apply Theorem 5.3 to derandomize a special case of polynomial identity testing (first studied in [16]). To this end we restate a result of Dvir et. al [16].

Theorem 5.4 (Theorem 4 in [16]). *Let* n, s, r, m, t, d *be integers such that* $s \geq n$. *Let* $\mathbb{F}$ *be a field which has at least* $2mt$ *elements. Let* $P(x, y) \in \mathbb{F}[x_1, \ldots, x_n, y]$ *be a nonzero polynomial with* $\deg(P) \leq t$ *and* $\deg_y(P) \leq r$ *such that* P *has an arithmetic circuit of size* s *and depth* d *over* $\mathbb{F}$. *Let* $f(x) \in \mathbb{F}[x_1, \ldots, x_n]$ *be a polynomial with* $\deg(f) = m$ *such that* $P(x, f(x)) \equiv 0$. *Then* $f(x)$ *can be computed by a circuit of size* $s' = poly(s, m^r)$ *and depth* $d' = d + O(1)$ *over* $\mathbb{F}$.

Let the underlying field $\mathbb{F}$ be large enough ($\mathbb{Q}$, for instance). The following lemma is a variant of Lemma 4.1 in [16]. For a detailed proof, see [2].

Lemma 5.5 (Variant of Lemma 4.1 in [16]). *Let* n, r, s *be integers and let* $f \in \mathbb{F}[x_1, x_2, \ldots, x_n]$ *be a nonzero polynomial with individual degrees at most* r *that is computed by an arithmetic circuit of size* $s \geq n$ *and depth* d. *Let* $m = n^\alpha$ *be an integer where* $\alpha > 0$ *is an arbitrary constant. Let* $S_1, S_2, \ldots, S_n$ *be the sets of the* (m, b)*-design constructed in Lemma 4.3 where* $b = \lceil \frac{1}{\alpha} \rceil$. *Let* $p \in \mathbb{F}[z_1, \ldots, z_m]$ *be a multilinear polynomial with the property that*

$$F(y) = F(y_1, y_2, \ldots, y_u) \triangleq f(p(y|_{S_1}), \ldots, p(y|_{S_n})) \equiv 0 \tag{1}$$

Then there exists absolute constants a *and* k *such that* $p(z)$ *is computable by an arithmetic circuit over* $\mathbb{F}$ *with size bounded by* $O((sm^r)^a)$ *and having depth* $d + k$.

At this point we describe our deterministic black-box identity testing algorithm for constant-depth arithmetic circuits of polynomial size and bounded individual degree. Let n, m, u, α be the parameters as in Lemma 4.3. Given such a circuit C over variables $\{x_i \mid i \in [n]\}$ of size n^t, depth d and individual degree r, we simply replace x_i with $\text{IMM}(y|S_i)$ where y is a new set of variables $\{y_j \mid j \in [u]\}$. Let $\tilde{C}[y_1, \ldots, y_u]$ denote the polynomial computed by the new circuit.

Notice that the total degree of $\tilde{C}$ is bounded by u^c where c is a constant depending on the combinatorial design and r. Let $R \subseteq \mathbb{F}$ be any set of $u^c + 1$ distinct points. Then by the Schwartz-Zippel Lemma the polynomial computed by $\tilde{C}$ is identically zero if and only if $\tilde{C}(a_1, a_2, \ldots, a_u) = 0$ for all $(a_1, a_2, \ldots, a_u) \in R^u$.

This gives us the claimed algorithm. Its running time is bounded by $O((u^c + 1)^u) = O(2^{7\alpha n^{6\alpha}})$. Since α can be chosen to be arbitrarily small, we have shown that this identity testing problem is in deterministic sub-exponential time. The correctness of the algorithm follows from the next lemma.

Lemma 5.6. *If for every constant $d' > 0$, the polynomial sequence $\{IMM_n\}$ is not computable by depth-d' arithmetic circuits of size n^k for any $k > 0$, then $C[x_1, \ldots, x_n] \equiv 0$ if and only if $\tilde{C}[y_1, \ldots, y_u] \equiv 0$.*

Proof. The only-if part is easy to see. Let us focus on the if part. Suppose it is not the case, which means that $\tilde{C}[y_1, \ldots, y_u] \equiv 0$ but $C[x_1, \ldots, x_n] \not\equiv 0$. Then let the role of $f[x_1, \ldots, x_n]$ in Lemma 5.5 be played by $C[x_1, \ldots, x_n]$ and let $\text{IMM}[z_1, \ldots, z_m]$ take the place of $p[z_1, \ldots, z_m]$. Therefore, $\text{IMM}[z_1, \ldots, z_m]$ is computable by a circuit of depth $d + k$ and size at most $(n^t m^r)^a = m^{O(1)}$, a contradiction.

Putting it together, we get the following result.

Theorem 5.7. *If there exists $\delta > 0$ such that for any constant $e > 0$, IMM requires depth-e arithmetic circuits of size at least $n^{1+\delta}$, then the black-box identity testing problem for constant-depth arithmetic circuits of polynomial size and bounded individual degree is in deterministic sub-exponential time.*

Next, we notice that the above upper bound can be sharpened considerably. The algorithm simply takes the OR over subexponentially-many evaluations of an arithmetic circuit; if any of the evaluations does not evaluate to zero, then we know that the expressions are not equivalent; otherwise they are. Note that evaluating an arithmetic circuit can be accomplished in logspace. (When evaluating a circuit over $\mathbb{Q}$, this is shown in [24, Corollary 6.8]; the argument for other fields is similar, using standard results about the complexity of field arithmetic.) Note also that every language computable in logspace has AC^0 circuits of subexponential size. (This appears to have been observed first by Gutfreund and Viola [22]; see also [3] for a proof.) This yields the following uniform derandomization result.

Theorem 5.8. *If there are no constant-depth arithmetic circuits of size $n^{1+\epsilon}$ for the polynomial sequence $\{IMM_n\}$, then for every constant d, black-box identity testing for depth-d arithmetic circuits with bounded individual degree can be performed by a uniform family of constant-depth AC^0 circuits of subexponential size.*

We call attention to an interesting difference between Theorems 4.5 and 5.8. In Theorem 5.8, in order to solve the identity testing problem with uniform AC^0 circuits of size 2^{n^ϵ} for smaller and smaller ϵ, the depth of the AC^0 circuits increases as ϵ decreases. In contrast, in order to obtain a deterministic threshold circuit of size 2^{n^ϵ} to simulate a given probabilistic TC^0 algorithm, the argument that we present in the proof of Theorem 4.5 gives a circuit whose depth is not affected by the choice of ϵ. We do not know if a similar improvement of Theorem 5.8 is possible, but we observe here that the depth need not depend on ϵ if we use threshold circuits for the identity test.

Theorem 5.9. *If there are no constant-depth arithmetic circuits of size $n^{1+\epsilon}$ for the polynomial sequence $\{IMM_n\}$, then there is a constant c such that, for every constant d and every $\gamma > 0$, black-box identity testing for depth-d arithmetic circuits with bounded individual degree can be performed by a uniform family of depth $d+c$ threshold circuits of size 2^{n^γ}.*

Proof. (Sketch.) Choose $\alpha < \gamma/14$, where α is the constant from the discussion in the paragraph before Lemma 5.6. Thus, our identity testing algorithm will evaluate a depth d arithmetic circuit $C(x_1, \ldots, x_n)$ at fewer than $2^{n^{\gamma/2}}$ points $v = (v_1, \ldots, v_n)$, where each v_i is obtained by computing an instance of IMM_{n^α} consisting of n^α 3-by-3 matrices, whose entries without loss of generality have representations having length at most n^α. Thus these instances of IMM have DNF representations of size $2^{O(n^{2\alpha})}$. These DNF representations are uniform, since the direct connection language can be evaluated by computing, for a given input assignment to IMM_{n^α}, the product of the matrices represented by that assignment, which takes time at most $(n^\alpha)^3 < \log(2^{n^{\gamma/2}})$. Evaluating the circuit C on v can be done in uniform TC^0 [1,24].

Acknowledgments

The possibility of applying random self-reductions to derandomize small classes was suggested to us by Rahul Santhanam. We thank Luke Friedman for many helpful discussions, and we thank Lance Fortnow for some useful suggestions. The first author is supported in part by NSF Grants DMS-0652582, CCF-0830133, and CCF-0832787; some of this work was performed while this author was a visiting scholar at the University of Cape Town. The third author was supported in part by NSF Grants CCF-0830133 and CCF-0832787.

References

1. Agrawal, M., Allender, E., Datta, S.: On TC^0, AC^0, and arithmetic circuits. Journal of Computer and System Sciences 60(2), 395–421 (2000)
2. Allender, E., Arvind, V., Wang, F.: Uniform derandomization from pathetic lower bounds. Technical Report TR10-069, Electronic Colloquium on Computational Complexity, ECCC (2010)
3. Allender, E., Hellerstein, L., McCabe, P., Pitassi, T., Saks, M.E.: Minimizing disjunctive normal form formulas and AC^0 circuits given a truth table. SIAM Journal on Computing 38(1), 63–84 (2008)

4. Allender, E., Koucký, M.: Amplifying lower bounds by means of self-reducibility. Journal of the ACM (to appear); A preliminary version appeared in Proc. IEEE Conference on Computational Complexity (2008)

5. Andreev, A.E., Clementi, A.E.F., Rolim, J.D.P.: A new general derandomization method. Journal of the ACM 45(1), 179–213 (1998)

6. Andreev, A.E., Clementi, A.E.F., Rolim, J.D.P.: Worst-case hardness suffices for derandomization: A new method for hardness-randomness trade-offs. Theoretical Computer Science 221(1-2), 3–18 (1999)

7. Andreev, A.E., Clementi, A.E.F., Rolim, J.D.P., Trevisan, L.: Weak random sources, hitting sets, and BPP simulations. SIAM Journal on Computing 28(6), 2103–2116 (1999)

8. Arora, S., Barak, B.: Computational Complexity, a modern approach. Cambridge University Press, Cambridge (2009)

9. Arvind, V., Köbler, J.: On resource-bounded measure and pseudorandomness. In: Ramesh, S., Sivakumar, G. (eds.) FST TCS 1997. LNCS, vol. 1346, pp. 235–249. Springer, Heidelberg (1997)

10. Babai, L., Fortnow, L., Nisan, N., Wigderson, A.: BPP has subexponential time simulations unless EXPTIME has publishable proofs. Computational Complexity 3, 307–318 (1993)

11. Barrington, D.A.: Bounded-width polynomial-size branching programs recognize exactly those languages in NC^1. Journal of Computer and System Sciences 38(1), 150–164 (1989)

12. Beame, P., Saks, M.E., Sun, X., Vee, E.: Time-space trade-off lower bounds for randomized computation of decision problems. Journal of the ACM 50(2), 154–195 (2003)

13. Ben-Or, M., Cleve, R.: Computing algebraic formulas using a constant number of registers. SIAM Journal on Computing 21(1), 54–58 (1992)

14. Blum, M., Micali, S.: How to generate cryptographically strong sequences of pseudo-random bits. SIAM Journal on Computing 13(4), 850–864 (1984)

15. Buhrman, H., Fortnow, L.: One-sided versus two-sided error in probabilistic computation. In: Meinel, C., Tison, S. (eds.) STACS 1999. LNCS, vol. 1563, pp. 100–109. Springer, Heidelberg (1999)

16. Dvir, Z., Shpilka, A., Yehudayoff, A.: Hardness-randomness tradeoffs for bounded depth circuits. SIAM Journal on Computing 39(4), 1279–1293 (2009)

17. Goldreich, O., Vadhan, S.P., Wigderson, A.: Simplified derandomization of BPP using a hitting set generator. Electronic Colloquium on Computational Complexity 7(4) (2000)

18. Goldreich, O., Wigderson, A.: Improved derandomization of BPP using a hitting set generator. In: Hochbaum, D.S., Jansen, K., Rolim, J.D.P., Sinclair, A. (eds.) RANDOM 1999 and APPROX 1999. LNCS, vol. 1671, pp. 131–137. Springer, Heidelberg (1999)

19. Goldwasser, S., Gutfreund, D., Healy, A., Kaufman, T., Rothblum, G.N.: A (de)constructive approach to program checking. Technical Report TR07-047, Electronic Colloquium on Computational Complexity, ECCC (2007); See also STOC 2008

20. Grigoriev, D., Razborov, A.: Exponential complexity lower bounds for depth 3 arithmetic circuits in algebras of functions over finite fields. Applicable Algebra in Engineering, Communication and Computing 10, 465–487 (2000)

21. Grigoriev, D., Karpinski, M.: An exponential lower bound for depth 3 arithmetic circuits. In: Proc. ACM Symp. on Theory of Computing (STOC), pp. 577–582 (1998)

22. Gutfreund, D., Viola, E.: Fooling parity tests with parity gates. In: APPROX-RANDOM, pp. 381–392 (2004)

23. Håstad, J.: The shrinkage exponent of de morgan formulas is 2. SIAM Journal on Computing 27(1), 48–64 (1998)

24. Hesse, W., Allender, E., Barrington, D.A.M.: Uniform constant-depth threshold circuits for division and iterated multiplication. Journal of Computer and System Sciences 65(4), 695–716 (2002)

25. Impagliazzo, R.: Hard-core distributions for somewhat hard problems. In: Proc. IEEE Symp. on Found. of Comp. Sci. (FOCS), pp. 538–545 (1995)

26. Impagliazzo, R., Paturi, R., Saks, M.E.: Size-depth tradeoffs for threshold circuits. SIAM Journal on Computing 26(3), 693–707 (1997)

27. Impagliazzo, R., Shaltiel, R., Wigderson, A.: Reducing the seed length in the Nisan-Wigderson generator. Combinatorica 26(6), 647–681 (2006)

28. Impagliazzo, R., Wigderson, A.: P = BPP if E requires exponential circuits: Derandomizing the XOR lemma. In: Proc. ACM Symp. on Theory of Computing (STOC), pp. 220–229 (1997)

29. Impagliazzo, R., Wigderson, A.: Randomness vs time: Derandomization under a uniform assumption. Journal of Computer and System Sciences 63(4), 672–688 (2001)

30. Iwama, K., Morizumi, H.: An explicit lower bound of $5n - o(n)$ for boolean circuits. In: Diks, K., Rytter, W. (eds.) MFCS 2002. LNCS, vol. 2420, pp. 353–364. Springer, Heidelberg (2002)

31. Kabanets, V., Impagliazzo, R.: Derandomizing polynomial identity tests means proving circuit lower bounds. Computational Complexity 13(1-2), 1–46 (2004)

32. Klivans, A., van Melkebeek, D.: Graph nonisomorphism has subexponential size proofs unless the polynomial-time hierarchy collapses. SIAM Journal on Computing 31(5), 1501–1526 (2002)

33. Miltersen, P.B., Vinodchandran, N.V.: Derandomizing Arthur-Merlin games using hitting sets. Computational Complexity 14(3), 256–279 (2005)

34. Nisan, N., Wigderson, A.: Hardness vs randomness. Journal of Computer and System Sciences 49(2), 149–167 (1994)

35. Nisan, N., Wigderson, A.: Lower bounds on arithmetic circuits via partial derivatives. Computational Complexity 6(3), 217–234 (1997)

36. Raz, R.: Elusive functions and lower bounds for arithmetic circuits. In: Proc. ACM Symp. on Theory of Computing (STOC), pp. 711–720 (2008)

37. Raz, R., Yehudayoff, A.: Lower bounds and separations for constant depth multilinear circuits. Journal of Computational Complexity 18(2), 171–207 (2009)

38. Shaltiel, R., Umans, C.: Simple extractors for all min-entropies and a new pseudorandom generator. Journal of the ACM 52(2), 172–216 (2005)

39. Shamir, A.: On the generation of cryptographically strong pseudo-random sequences. In: Even, S., Kariv, O. (eds.) ICALP 1981. LNCS, vol. 115, pp. 544–550. Springer, Heidelberg (1981)

40. Shpilka, A., Wigderson, A.: Depth-3 arithmetic circuits over fields of characteristic zero. Computational Complexity 10(1), 1–27 (2001)

41. Sudan, M., Trevisan, L., Vadhan, S.P.: Pseudorandom generators without the XOR lemma. Journal of Computer and System Sciences 62(2), 236–266 (2001)

42. Umans, C.: Pseudo-random generators for all hardnesses. Journal of Computer and System Sciences 67(2), 419–440 (2003)

43. van Lint, J.H.: Introduction to Coding Complexity. Springer, Heidelberg (1999)

44. Viola, E.: The complexity of constructing pseudorandom generators from hard functions. Computational Complexity 13(3-4), 147–188 (2005)

45. Vollmer, H.: Introduction to Circuit Complexity. Springer, Heidelberg (1999)

46. Yao, A.C.-C.: Theory and applications of trapdoor functions (extended abstract). In: Proc. IEEE Symp. on Found. of Comp. Sci. (FOCS), pp. 80–91 (1982)

Testing Boolean Function Isomorphism

Noga Alon[1],[*] and Eric Blais[2]

[1] Schools of Mathematics and Computer Science, Sackler Faculty of Exact Sciences,
Tel Aviv University, Tel Aviv 69978, Israel
nogaa@tau.ac.il
[2] School of Computer Science, Carnegie Mellon University, Pittsburgh 15213, USA
eblais@cs.cmu.edu

Abstract. Two boolean functions $f, g : \{0,1\}^n \to \{0,1\}$ are *isomorphic* if they are identical up to relabeling of the input variables. We consider the problem of *testing* whether two functions are isomorphic or far from being isomorphic with as few queries as possible.

In the setting where one of the functions is known in advance, we show that the non-adaptive query complexity of the isomorphism testing problem is $\tilde{\Theta}(n)$. In fact, we show that the lower bound of $\Omega(n)$ queries for testing isomorphism to g holds for almost all functions g.

In the setting where both functions are unknown to the testing algorithm, we show that the query complexity of the isomorphism testing problem is $\tilde{\Theta}(2^{n/2})$. The bound in this result holds for both adaptive and non-adaptive testing algorithms.

1 Introduction

The field of property testing, originally introduced by Rubinfeld and Sudan [20], considers the following general problem: given a property $\mathcal{P}$, determine the minimum number q of queries required to determine with high probability whether an input has the property $\mathcal{P}$ or whether it is "far" from $\mathcal{P}$. The field has been extremely active over the last few years – see, e.g., the recent surveys [18,19].

In this paper, we concern ourselves with property testing of boolean functions. Despite significant progress in the study of the query complexity of many properties of boolean functions (e.g., monotonicity [7,11,13], juntas [10,5], having concise representations [6], halfspaces [16,17]), our overall understanding of the testability of boolean function properties still lags behind our understanding of the testability of graph properties, whose study was initiated by Goldreich, Goldwasser, and Ron [14].

A notable example that illustrates the gap between our understanding of graph and boolean function properties is *isomorphism*. Two graphs are isomorphic if they are identical up to relabeling of the vertices, while two boolean functions are isomorphic if they are identical up to relabeling of the input variables. There are three main variants to the isomorphism testing problem. (In the following list, an "object" refers to either a graph or a boolean function.)

---

[*] Research supported in part by an ERC Advanced grant, by a USA-Israeli BSF grant and by the Hermann Minkowski Minerva Center for Geometry at Tel Aviv University.

M. Serna et al. (Eds.): APPROX and RANDOM 2010, LNCS 6302, pp. 394–405, 2010.

1. **Testing isomorphism to a given object** $\mathcal{O}$. The query complexity required to test isomorphism in this variant depends on the object $\mathcal{O}$; the goal for this problem is to characterize the query complexity for *every* graph or boolean function.
2. **Testing isomorphism to the hardest known object.** A less fine-grained variant of the first problem asks to determine the maximum query complexity of testing isomorphism to $\mathcal{O}$ over objects of a given size.
3. **Testing isomorphism of two unknown objects.** In this variant, the testing algorithm has query access to two unknown objects $\mathcal{O}_1$ and $\mathcal{O}_2$ and must distinguish between the cases where they are isomorphic to each other or far from isomorphic to each other.

The problem of testing graph isomorphism was first raised by Alon, Fischer, Krivelevich, and Szegedy [1] (see also [8]), who used a lower bound on testing isomorphism of two unknown graphs to give an example of a non-testable first-order graph property of a certain type. Fischer [9] studied the problem of testing isomorphism to a given graph G and characterized the query complexity of the problem in terms of a complexity measure of G. Tight asymptotic bounds on the query complexity of the problem of testing isomorphism to a known graph and testing isomorphism of two unknown graphs were then obtained by Fischer and Matsliah [12]. As a result, all three versions of the graph isomorphism testing problem are well understood.

The picture is much less complete in the setting of boolean functions. Testing isomorphism against a known function f was first studied by Fischer, Kindler, Ron, Safra, and Samorodnitsky [10]. They gave a general upper bound on the problem showing that for every function f that depends on k variables (that is, for every k-junta), the problem of testing isomorphism to f requires $\text{poly}(k/\epsilon)$ queries. Conversely, they showed that when f is a parity function on $k < o(\sqrt{n})$ variables, testing isomorphism to f requires $\widetilde{\Omega}(k)$ queries. No other progress was made on the problem of testing isomorphism on boolean functions until very recently, when Blais and O'Donnell [3] showed that for every function f that "strongly" depends on k variables, testing isomorphism to f requires $\Omega(\log k)$ queries. Taken together, the results in [10,3] give only an incomplete solution to the problem of testing isomorphism to a given boolean function and provide only weak bounds on the other two versions of the isomorphism testing problem.

Our results. We introduce new results for all three variants of the problem of testing isomorphism to boolean functions.

In the problem of testing isomorphism to a given function $g : \{0,1\}^n \to \{0,1\}$, it is easy to show that $O(\frac{n \log n}{\epsilon})$ queries always suffice to ϵ-test isomorphism to any function g. (For completeness, we give the proof of this statement in Section 3.1.) Our main result is a matching lower bound (up to a logarithmic factor) that applies for *almost all* functions g.

Theorem 1.1. *Fix* $0 < \epsilon < \frac{1}{2}$. *For a* $1 - o(1)$ *fraction of the functions* $g : \{0,1\}^n \to \{0,1\}$, *any non-adaptive algorithm for* ϵ-*testing isomorphism to* g *must make at least* $\frac{n}{100}$ *queries.*

We present the proof of Theorem 1.1 in Sections 3.2 and 3.3. The lower bound of the theorem and the aforementioned upper bound immediately give a tight bound on the query complexity of testing isomorphism to a known function:

Corollary 1.2. *The maximum possible query complexity for testing isomorphism to a known function* $\{0,1\}^n \to \{0,1\}$ *non-adaptively is* $\tilde{\Theta}(n)$. *This bound holds for testing algorithms with 1-sided and 2-sided error.*

Finally, we examine the problem of testing two unknown functions for the property of being isomorphic. A simple algorithm can ϵ-test isomorphism in this setting with $\tilde{O}(2^{n/2}/\sqrt{\epsilon})$ queries. We give a matching lower bound establishing that no other algorithm can do better.

Theorem 1.3. *The query complexity for testing isomorphism of two unknown functions in* $\{0,1\}^n \to \{0,1\}$ *is* $\tilde{\Theta}(2^{n/2})$. *This bound holds for all testing algorithms (adaptive or non-adaptive, with 1-sided or 2-sided error).*

We present the proof of Theorem 1.3 in Section 4.

Related work. Recently, Chakraborty, García-Soriano, and Matsliah [4] independently obtained results very similar to Corollary 1.2 and Theorem 1.3. In fact, their version of Corollary 1.2 contains a stronger lower bound that also applies to adaptive testing algorithms.

Furthermore, [4] also show tight bounds on the query complexity for testing isomorphism to the hardest known function within some *restricted* classes of functions. Notably, they show that $O(k \log k)$ queries are sufficient to test isomorphism to any k-juntas and that $\Omega(k)$ queries are required to test isomorphism to some k-juntas.

2   Preliminaries and Notation

Throughout the paper, f and g represent boolean functions $\{0,1\}^n \to \{0,1\}$. The *weight* of an input $x = (x_1, \ldots, x_n) \in \{0,1\}^n$ is $|x| = x_1 + \cdots + x_n$. All big O notation in this paper refers to asymptotic statements as $n \to \infty$ while the other parameters (typically, ϵ) remain constant. Tilde notation is used to hide polylogarithmic factors – for example $f = \tilde{\Theta}(n)$ if there is a positive constant c such that $f \geq \Omega(\frac{n}{\log^c n})$ and $f \leq O(n \log^c n)$.

For a permutation $\pi : [n] \to [n]$ and $x = (x_1, \ldots, x_n) \in \{0,1\}^n$, we write $\pi(x) = (x_{\pi(1)}, \ldots, x_{\pi(n)})$. The function $g_\pi : \{0,1\}^n \to \{0,1\}$ represents the function defined by $g_\pi(x) = g(\pi(x))$ for every $x \in \{0,1\}^n$. Two functions f and g are *isomorphic* if there is a permutation π such that $f = g_\pi$.

Given a set $X \subseteq \{0,1\}^n$ and a permutation π on $[n]$, we write $\pi(X) = \{\pi(x) : x \in X\}$. With some abuse of notation, we also write $f(X) \in \{0,1\}^{|X|}$ to represent the value of f over each $x \in X$, over some ordering of X. In particular, $f(X) = g(X)$ iff $f(x) = g(x)$ for every $x \in X$.

Given two random variables A, B defined on a common discrete sample space Ω, the *total variation* distance between A and B is

$$d_{TV}(A, B) = \frac{1}{2} \sum_{\omega \in \Omega} \left| \Pr[A = \omega] - \Pr[B = \omega] \right|.$$

A *property* $\mathcal{P}$ of boolean functions $\{0, 1\}^n \to \{0, 1\}$ is simply a subset of those functions. The distance of a function f to $\mathcal{P}$ is the minimum distance between f and g over all $g \in \mathcal{P}$, where the distance between two functions is $\mathrm{dist}(f, g) = \Pr_x[f(x) \neq g(x)] = \frac{1}{2^n} \sum_{x \in \{0,1\}^n} \mathbf{1}[f(x) \neq g(x)]$.

A (q, ϵ)-*tester* for the property $\mathcal{P}$ is a randomized algorithm $\mathcal{T}$ that queries an unknown function f on q different inputs in $\{0, 1\}^n$ and then (1) accepts f with probability at least $\frac{2}{3}$ when $f \in \mathcal{P}$, and (2) rejects f with probability at least $\frac{2}{3}$ when f is ϵ-far from $\mathcal{P}$. (If the property deals with a pair of input functions, the algorithm may query both.)

When a tester $\mathcal{T}$ chooses all its queries in advance, it is *non-adaptive*; if it uses the responses to some of its queries to decide what queries to make afterwards, it is *adaptive*. A tester that accepts functions in $\mathcal{P}$ with probability 1 (instead of $\frac{2}{3}$) has *1-sided error*, otherwise it has *2-sided error*.

The *query complexity* of a property $\mathcal{P}$ for a given $\epsilon > 0$ is the minimum value of q for which there is a (q, ϵ)-tester for $\mathcal{P}$.

3 Testing Isomorphism to a Given Function

3.1 Upper Bound

The trivial algorithm $\mathcal{T}$ for testing isomorphism to g queries the unknown function $f : \{0, 1\}^n \to \{0, 1\}$ on a set $Q \subseteq \{0, 1\}^n$ of $\frac{n \ln n}{\epsilon}$ randomly selected inputs. The algorithm accepts f if and only if there is a permutation $\pi \in \mathcal{S}_n$ such that $f(x) = g(\pi(x))$ for every $x \in Q$.

Clearly, the trivial algorithm $\mathcal{T}$ is non-adaptive and accepts functions isomorphic to g with probability 1. The following simple proposition completes the proof of correctness of $\mathcal{T}$ by showing that it rejects functions ϵ-far from isomorphic to g with probability at least $\frac{2}{3}$.

Proposition 3.1. *Fix $\epsilon > 0$. Let $g : \{0, 1\}^n \to \{0, 1\}$ be any boolean function and $f : \{0, 1\}^n \to \{0, 1\}$ be a function ϵ-far from isomorphic to g. Then $\mathcal{T}$ accepts f with probability $o(1)$.*

Proof. For any permutation $\pi \in \mathcal{S}_n$, there are at least $\epsilon 2^n$ values of $x \in \{0, 1\}^n$ for which $f(x) \neq g(\pi(x))$. The probability that none of those inputs are queried by $\mathcal{T}$ is at most $(1 - \epsilon)^{|Q|} \leq e^{-\epsilon(n \ln n / \epsilon)} = n^{-n}$. Thus, by the union bound, the probability that there is a permutation $\pi \in \mathcal{S}_n$ such that $f(x) = g(\pi(x))$ for every $x \in Q$ is at most $n! / n^n = o(1)$. $\qquad \square$

3.2   Lower Bound

We prove Theorem 1.1 in this section. The proof of this theorem combined with the upper bound of the previous section immediately yields Corollary 1.2.

The proof of Theorem 1.1 uses Yao's Minimax Principle [21]. For a fixed function g we introduce two distributions $\mathcal{F}_{\text{yes}}$ and $\mathcal{F}_{\text{no}}$ such that a function $f \sim \mathcal{F}_{\text{yes}}$ is isomorphic to g and a function $f \sim \mathcal{F}_{\text{no}}$ is ϵ-far from isomorphic to g with high probability. We then show that for most choices of g, deterministic non-adaptive testing algorithms cannot distinguish functions drawn from either of these distributions with only $\frac{n}{100}$ queries.

We define $\mathcal{F}_{\text{yes}}$ to be the uniform distribution over functions isomorphic to g. In other words, we draw a function $f \sim \mathcal{F}_{\text{yes}}$ by choosing $\pi \in \mathcal{S}_n$ uniformly at random and setting $f = g_\pi$.

A first idea for $\mathcal{F}_{\text{no}}$ may be to make it the uniform distribution over all boolean functions $\{0,1\}^n \to \{0,1\}$. This idea does not quite work, since, for example, a random function differs from g and all functions isomorphic to it on the all 0 input or the all 1 input with probability at least $3/4$. However, a simple modification of this idea does work: to draw a function $f \sim \mathcal{F}_{\text{no}}$, we choose a permutation $\pi \in \mathcal{S}_n$ uniformly at random and we choose a function f_{rand} uniformly at random from all boolean functions on n variables. We then let f be the function defined by

$$f(x) = \begin{cases} f_{\text{rand}}(x) & \text{if } \frac{n}{3} \leq |x| \leq \frac{2n}{3}, \\ g_\pi(x) & \text{otherwise.} \end{cases}$$

With high probability, a function $f \sim \mathcal{F}_{\text{no}}$ is far from isomorphic to g.

Proposition 3.2. *Fix $0 < \epsilon < \frac{1}{2}$. For any function $g : \{0,1\}^n \to \{0,1\}$, the function $f \sim \mathcal{F}_{\text{no}}$ is ϵ-close to isomorphic to g with probability at most $o(1)$.*

Proof. Fix any permutation $\pi \in \mathcal{S}_n$. Let f_{rand} be the random function generated in the draw of $f \sim \mathcal{F}_{\text{no}}$. By the triangle inequality,

$$\text{dist}(f, g_\pi) \geq \text{dist}(f_{\text{rand}}, g_\pi) - \text{dist}(f, f_{\text{rand}}).$$

Since $\text{dist}(f, f_{\text{rand}}) \leq 2 \sum_{i=0}^{n/3} \binom{n}{i}/2^n \leq o(1)$, to complete the proof it suffices to fix $\epsilon < \epsilon' < \frac{1}{2}$ and show that $\text{dist}(f_{\text{rand}}, g_\pi) > \epsilon'$ with high probability.

Let $\eta = 1 - 2\epsilon'$. For any $x \in \{0,1\}^n$, $f_{\text{rand}}(x) = g_\pi(x)$ with probability $\frac{1}{2}$, so $\mathbb{E}[\text{dist}(f_{\text{rand}}, g_\pi)] = \frac{1}{2}$. By Chernoff's bound (see, e.g., Appendix A in [2]),

$$\Pr[\text{dist}(f_{\text{rand}}, g_\pi) < \epsilon'] = \Pr[\text{dist}(f_{\text{rand}}, g_\pi) < (1 - \eta)\tfrac{1}{2}] \leq e^{-2^n \eta^2/6} \leq o(\tfrac{1}{n!}).$$

Taking the union bound over all choices of $\pi \in \mathcal{S}_n$ completes the proof.   □

Let $\mathcal{T}$ be any deterministic non-adaptive algorithm that attempts to test g-isomorphism with at most $\frac{n}{100}$ queries to an unknown function f. We will show that $\mathcal{T}$ cannot reliably distinguish between the cases where f was drawn from $\mathcal{F}_{\text{yes}}$ or from $\mathcal{F}_{\text{no}}$.

Let $Q \subseteq \{0,1\}^n$ be the set of queries performed by $\mathcal{T}$ on f. We partition the queries in Q in two: the set $Q_b = \{q \in Q : \frac{n}{3} \leq |q| \leq \frac{2n}{3}\}$ of *balanced* queries, and the set $Q_u = Q \setminus Q_b$ of *unbalanced* queries.

When f is drawn from $\mathcal{F}_{\text{yes}}$ or from $\mathcal{F}_{\text{no}}$, the responses to the unbalanced queries Q_u are consistent with some function g_π isomorphic to g. Our next proposition shows that when $\mathcal{T}$ makes only $\frac{n}{100}$ queries to f, then in fact the responses to the unbalanced queries will be consistent with *many* functions isomorphic to g. More precisely, define

$$\Pi_g(f, Q_u) = \{\pi \in \mathcal{S}_n : g_\pi(Q_u) = f(Q_u)\}$$

to be the set of permutations π for which g_π is consistent with the responses to the queries Q_u. The following proposition shows that when the unknown function is drawn from $\mathcal{F}_{\text{yes}}$ or from $\mathcal{F}_{\text{no}}$, then with high probability the set $\Pi_g(f, Q_u)$ is large.

Proposition 3.3. *Let Q_u be any set of unbalanced queries and let f be a function drawn from $\mathcal{F}_{\text{yes}}$ or from $\mathcal{F}_{\text{no}}$. Then for any $0 < t < 1$,*

$$\Pr_f \left[|\Pi_g(f, Q_u)| < t \cdot \frac{n!}{2^{|Q_u|}} \right] \leq t.$$

Proof. When $f \sim \mathcal{F}_{\text{yes}}$ or $f \sim \mathcal{F}_{\text{no}}$, then $f(x) = g_\pi(x)$ for every unbalanced input x, where π is chosen uniformly at random from $\mathcal{S}_n$. So it suffices to show that $\Pr_\pi[|\Pi_g(g_\pi, Q_u)| < t \cdot \frac{n!}{2^{|Q_u|}}] \leq t$.

For every $r \in \{0,1\}^{|Q_u|}$, let $S_r \subseteq \mathcal{S}_n$ be the set of permutations σ for which $g_\sigma(Q_u) = r$. A set S_r is *small* if $|S_r| \leq t \frac{n!}{2^{|Q_u|}}$. The union of all small sets covers at most $2^{|Q_u|} \cdot t \frac{n!}{2^{|Q_u|}} = tn!$ permutations, so the probability that a randomly chosen permutation π belongs to a small set is at most t.   □

The last proposition showed that when f is drawn from $\mathcal{F}_{\text{yes}}$ or from $\mathcal{F}_{\text{no}}$, then with high probability $\Pi_g(f, Q_u)$ is large; the next lemma shows that conditioned on $\Pi_g(f, Q_u)$ being large, the distribution on the responses to the balanced queries is nearly uniform, even when $f \sim \mathcal{F}_{\text{yes}}$. Specifically, given a function g and a set S of permutations, we define the *discrepancy of g on S* to be

$$\Delta_S(g) = \max_{\substack{Q_b : |Q_b| = \frac{n}{100} \\ r \in \{0,1\}^{|Q_b|}}} \left| \Pr_{\pi \in S}[g_\pi(Q_b) = r] - 2^{-\frac{n}{100}} \right|.$$

We then define the *discrepancy* of g to be

$$\Delta(g) = \max_{\substack{Q_u : |Q_u| = \frac{n}{100} \\ \pi : |\Pi_g(g_\pi, Q_u)| \geq n!/2^{n/50}}} \Delta_{\Pi_g(g_\pi, Q_u)}(g).$$

The following lemma shows that $\Delta(g)$ is small for almost all functions g.

Lemma 3.4. *When g is drawn uniformly at random from the set of functions $\{0,1\}^n \to \{0,1\}$,*

$$\Pr_g \left[\Delta(g) > \frac{1}{3} \cdot 2^{-\frac{n}{100}} \right] \leq 2^{-\Omega(2^{n/25})}.$$

We prove Lemma 3.4 in the next section, but first we show how it implies Theorem 1.1.

Proof (Theorem 1.1). By Lemma 3.4, with probability at least $1 - 2^{-\Omega(2^{n/25})} = 1 - o(1)$, the discrepancy of a randomly drawn function $g : \{0,1\}^n \to \{0,1\}$ is $\Delta(g) \leq \frac{1}{3} 2^{-\frac{n}{100}}$. Fix g to be any function that satisfies this condition. We will show that testing isomorphism to g requires at least $\frac{n}{100}$ queries.

As discussed earlier, we complete the proof with Yao's Minimax Principle, with the distributions $\mathcal{F}_{yes}$ and $\mathcal{F}_{no}$ as defined at the beginning of the section. Let $\mathcal{T}$ be any deterministic non-adaptive algorithm that makes at most $\frac{n}{100}$ queries to the input function f, and let $Q = Q_u \cup Q_b$ represent the queries made by $\mathcal{T}$. Without loss of generality, we can assume $|Q_u| = |Q_b| = \frac{n}{100}$. (If $|Q_b| < \frac{n}{100}$, simply add extra balanced queries to Q_b; this can only help $\mathcal{T}$ determine whether f was drawn from $\mathcal{F}_{yes}$ or from $\mathcal{F}_{no}$. Similarly, adding unbalanced queries to Q_u can only help $\mathcal{T}$.)

By Proposition 3.3, the probability that $|\Pi_g(f, Q_u)| < \frac{n!}{2^{n/50}}$ is at most $\frac{1}{2^{n/100}} = o(1)$. Assume, thus, that this event does not happen. Let $\mathcal{R}_{yes}$ and $\mathcal{R}_{no}$ be the distribution of the responses to the balanced queries Q_b. Then the total variation distance between $\mathcal{R}_{yes}$ and $\mathcal{R}_{no}$ is bounded by

$$d_{TV}(\mathcal{R}_{yes}, \mathcal{R}_{no}) = \frac{1}{2} \sum_{r \in \{0,1\}^{\frac{n}{100}}} \left| \Pr_{\pi \in \Pi_g(f, Q_u)} [g_\pi(Q_b) = r] - 2^{-\frac{n}{100}} \right|$$

$$\leq \frac{1}{2} \cdot 2^{\frac{n}{100}} \Delta(g) \leq \frac{1}{6}. \tag{1}$$

Therefore, if $\mathcal{T}$ accepts functions drawn from $\mathcal{F}_{yes}$ with probability at least $\frac{2}{3}$, (1) implies that $\mathcal{T}$ also accepts functions drawn from $\mathcal{F}_{no}$ with probability at least $\frac{2}{3} - \frac{1}{6} = \frac{1}{2}$. But by Proposition 3.2, a function drawn from $\mathcal{F}_{no}$ is ϵ-far from isomorphic to g with probability $1 - o(1)$, so $\mathcal{T}$ can't be a valid ϵ-tester for isomorphism to g.   □

3.3   Proof of Lemma 3.4

The first step in the proof of Lemma 3.4 is to show that for any sufficiently small set Q of balanced queries and sufficiently large set S of permutations, the set $\{\pi(Q)\}_{\pi \in S}$ can be partitioned into a number of large pairwise disjoint sets. The proof of this claim uses the celebrated theorem of Hajnal and Szemerédi [15].

Hajnal-Szemerédi Theorem. Let G be a graph on n vertices with maximum vertex degree $\Delta(G) \leq d$. Then G has a $(d + 1)$-coloring in which all the color classes have size $\lfloor \frac{n}{d+1} \rfloor$ or $\lceil \frac{n}{d+1} \rceil$.

Lemma 3.5. *Let S be a set of at least $\frac{n!}{2^{n/50}}$ permutations on $[n]$, and let Q_b be a set of at most $\frac{n}{100}$ balanced queries. Then there exists a partition $S_1 \dot\cup \cdots \dot\cup S_k$ of the permutations in S such that for $i = 1, 2, \ldots, k$,*

(i) $|S_i| \geq 2^{n/20}$, and
(ii) The sets $\{\pi(Q_b)\}_{\pi \in S_i}$ are pairwise disjoint.

Proof. Construct a graph G on S where two permutations σ, τ are adjacent iff there exist $u, v \in Q_b$ such that $\sigma(u) = \tau(v)$. By this construction, when T is a set of permutations that form an independent set in G, then $\{\pi(Q_b)\}_{\pi \in T}$ are pairwise disjoint.

Consider a fixed permutation $\sigma \in S$. A second permutation τ is adjacent to σ in G iff there are two vectors u, v in Q_b such that the permutation $\tau\sigma^{-1}$ maps the indices where u has value 1 to the indices where v has value 1 as well. There are $\binom{|Q_b|}{2} \leq (\frac{n}{100})^2$ ways to choose $u, v \in Q_b$ and at most $|u|!(n - |u|)!$ ways to satisfy the mapping condition, so the graph has degree at most

$$\max_{\frac{n}{3} \leq k \leq \frac{2n}{3}} \left(\frac{n}{100}\right)^2 \cdot k!\,(n-k)! = \left(\frac{n}{100}\right)^2 \cdot \left(\frac{n}{3}\right)! \left(\frac{2n}{3}\right)! = \left(\frac{n}{100}\right)^2 \cdot \frac{n!}{\binom{n}{n/3}} \leq \frac{n!}{2^{cn}} - 1$$

for a constant $c = 1 - H_2(\frac{1}{3}) - o(1) \geq 0.07$.[1] Therefore, by the Hajnal-Szemerédi Theorem, G can be colored with $n!/2^{0.07n}$ colors, with each color class having size at least $\frac{n!/2^{n/50}}{n!/2^{0.07n}} = 2^{n/20}$. □

Lemma 3.5 is useful because most functions g have low discrepancy on large pairwise disjoint sets.

Lemma 3.6. *Fix Q_b to be a set of $\frac{n}{100}$ balanced queries and fix $r \in \{0, 1\}^{\frac{n}{100}}$. Let S be a fixed set of at least $2^{\frac{n}{20}}$ permutations such that the sets $\{\pi(Q_b)\}_{\pi \in S}$ are pairwise disjoint. Then*

$$\Pr_g\left[\left|\Pr_{\pi \in S}[g_\pi(Q_b) = r] - 2^{-\frac{n}{100}}\right| > \frac{1}{3} \cdot 2^{-\frac{n}{100}}\right] < 2^{-\Omega(2^{n/25})}.$$

Proof. For every function $g : \{0, 1\}^n \to \{0, 1\}$ and every permutation π of $[n]$, define the indicator random variable

$$X_{g,\pi} = \begin{cases} 1 & \text{if } g_\pi(Q_b) = r, \\ 0 & \text{otherwise.} \end{cases}$$

When g is chosen uniformly at random from the set of all boolean functions $\{0, 1\}^n \to \{0, 1\}$, $E_g[X_{g,\pi}] = \Pr_g[g_\pi(Q_b) = r] = 2^{-\frac{n}{100}}$, so

$$E_g\left[\Pr_{\pi \in S}[g_\pi(Q_b) = r]\right] = \frac{1}{|S|}\sum_{\pi \in S} E_g[X_{g,\pi}] = 2^{-\frac{n}{100}}.$$

Furthermore, the pairwise disjointness property of S guarantees that the indicator variables $X_{g,\pi}$ are pairwise independent. Therefore, by Chernoff's bound,

$$\Pr_g\left[\left|\Pr_{\pi \in S}[g_\pi(Q_b) = r] - 2^{-\frac{n}{100}}\right| > \frac{1}{3} \cdot 2^{-\frac{n}{100}}\right] < e^{-\Omega(|S|2^{-n/100})}. \qquad \square$$

---

[1] $H_2(p)$ represents the binary entropy of p. $H_2(\frac{1}{3}) \approx 0.918$.

The proof of Lemma 3.4 can now be completed as follows.

Proof (Lemma 3.4). Fix a permutation π and a set Q_u of $\frac{n}{100}$ unbalanced queries such that $|\Pi_g(g_\pi, Q_u)| \geq \frac{n!}{2^{n/50}}$. Let $S = \Pi_g(g_\pi, Q_u)$, and fix a set Q_b of $\frac{n}{100}$ balanced queries.

By Lemma 3.5, there exists a partition $S_1 \dot\cup \cdots \dot\cup S_k$ of S such that for each part S_i, $|S_i| \geq 2^{n/20}$ and $\{\pi(Q_b)\}_{\pi \in S_i}$ are pairwise disjoint. By Lemma 3.6, for every set S_i in the partition,

$$\Pr_g \left[\left| \Pr_{\pi \in S_i}[g_\pi(Q_b) = r] - 2^{-\frac{n}{100}} \right| > \tfrac{1}{3} \cdot 2^{-\frac{n}{100}} \right] \leq 2^{-\Omega(2^{n/25})}.$$

Taking the union bound over all $k < n!$ sets S_i, we get that

$$\Pr_g \left[\left| \Pr_{\pi \in S}[g_\pi(Q_b) = r] - 2^{-\frac{n}{100}} \right| > \tfrac{1}{3} \cdot 2^{-\frac{n}{100}} \right] < n! \cdot 2^{-\Omega(2^{n/25})}.$$

Applying a union bound once again, this time over all $\binom{2^n}{n/100} < 2^{\frac{n^2}{100}}$ choices of Q_b and $2^{\frac{n}{100}}$ choices for r, we obtain

$$\Pr_g \left[\Delta_S(g) > \tfrac{1}{3} \cdot 2^{-\frac{n}{100}} \right] < 2^{\frac{n^2}{100} + \frac{n}{100}} \cdot n! \cdot 2^{-\Omega(2^{n/25})}.$$

Finally, applying the union bound one last time over the $n!$ choices for π and $\binom{2^n}{n/100} \leq 2^{\frac{n^2}{100}}$ choices for Q_u, we get

$$\Pr_g \left[\Delta(g) > \frac{1}{3} \cdot 2^{-\frac{n}{100}} \right] < 2^{\frac{2n^2}{100} + \frac{n}{100}} \cdot n!^2 \cdot 2^{-\Omega(2^{n/25})} = 2^{-\Omega(2^{n/25})}. \qquad \square$$

4   Testing Isomorphism of Two Unknown Functions

4.1   Upper Bound

ALGORITHM $\mathcal{T}$
1. Generate two sets $Q_f, Q_g \subset \{0,1\}^n$ of $2^{n/2}\sqrt{\frac{n \ln n}{\epsilon}}$ queries independently and uniformly at random.
2. Query $f(x)$ for every $x \in Q_f$.
3. Query $g(x)$ for every $x \in Q_g$.
4. Accept iff there exists $\pi \in \mathcal{S}_n$ such that for every element $x \in Q_f$ where $\pi(x) \in Q_g$, $f(x) = g(\pi(x))$.

The algorithm $\mathcal{T}$ is non-adaptive and makes $\tilde{O}(2^{n/2})$ queries. Clearly, it always accepts when f and g are isomorphic. The following simple argument completes the proof of correctness of the algorithm by showing that it rejects functions that are ϵ-far from isomorphic with high probability.

Proposition 4.1. *Fix $\epsilon > 0$. Let f and g be ϵ-far from isomorphic. Then $\mathcal{T}$ rejects (f,g) with probability $1 - o(1)$.*

Proof. For any permutation $\pi \in \mathcal{S}_n$, there are at least $\epsilon 2^n$ inputs $x \in \{0,1\}^n$ for which $f(x) \neq g(\pi(x))$. It is not too difficult to show that the probability that none of these inputs satisfy $x \in Q_f$ and $\pi(x) \in Q_g$ is at most

$$4 \left(1 - \frac{|Q_f|}{2^n} \cdot \frac{|Q_g|}{2^n} \right)^{\epsilon 2^n} = 4 \left(1 - \frac{n \ln n}{\epsilon 2^n} \right)^{\epsilon 2^n} \leq 4 e^{-\frac{n \ln n}{\epsilon 2^n} \cdot \epsilon 2^n} = 4 n^{-n}.$$

By the union bound, the probability that f and g are accepted by the algorithm is at most $n!/n^n = o(1)$. □

4.2   Lower Bound

The following Lemma, combined with the upper bound in the previous section, implies Theorem 1.3.

Lemma 4.2. *Any algorithm for testing two unknown functions $f, g : \{0,1\}^n \to \{0,1\}$ for the property of being isomorphic must make at least $\Omega(\frac{2^{n/2}}{n^{1/4}})$ queries to the functions.*

Proof. Let $\mathcal{T}$ be an algorithm making $o(\frac{2^{n/2}}{n^{1/4}})$ queries to f and g. We will define two distributions $\mathcal{D}_{\text{yes}}$ and $\mathcal{D}_{\text{no}}$ on pairs of functions (f, g) that are isomorphic and ϵ-far from isomorphic with probability $1 - o(1)$, respectively, and show that $\mathcal{T}$ can not determine with probability greater than $\frac{1}{2} + o(1)$ which distribution generated an input.

Let $T = \{x \in \{0,1\}^n : \frac{n}{2} - \sqrt{n} \leq |x| \leq \frac{n}{2} + \sqrt{n}\}$ consist of the elements in the middle slice of the hypercube and let M be the set of all functions from $\{0,1\}^n$ to $\{0,1\}$ that map each $x \notin T$ to 0. A pair of functions (f, g) from $\mathcal{D}_{\text{yes}}$ is drawn by the following procedure:

1. Pick $\pi \in \mathcal{S}_n$ uniformly at random.
2. Choose $f \in M$ uniformly at random.
3. Let $g = f_\pi$.

A pair of functions (f, g) is drawn from $\mathcal{D}_{\text{no}}$ by independently choosing two functions uniformly at random from M. With probability $1 - o(1)$, f is $\frac{1}{4}$-far from isomorphic to g.

We now introduce two random processes P_{yes} and P_{no} that answer the queries of $\mathcal{T}$ while generating a pair of functions (f, g) from $\mathcal{D}_{\text{yes}}$ or from $\mathcal{D}_{\text{no}}$, respectively. Without loss of generality, we can assume that the tester queries the value of f or of g only on inputs $x \in T$, since functions drawn from $\mathcal{D}_{\text{yes}}$ or from $\mathcal{D}_{\text{no}}$ always take the value 0 on the remaining inputs.

The process P_{yes} starts by choosing a permutation $\pi \in \mathcal{S}_n$ uniformly at random. It then proceeds to answer all the queries of the algorithm $\mathcal{T}$ randomly, with one exception: P_{yes} "quits" if $\mathcal{T}$ queries the value of $f(x)$ after previously having queried $g(\pi(x))$, and similarly P_{yes} quits if $\mathcal{T}$ queries $g(x)$ after having queried $f(\pi^{-1}(x))$.

When P_{yes} quits or reaches the end of the queries, it completes the generation of (f, g) by choosing f uniformly at random from all the functions that are consistent with the previously-answered queries (note: in this step, the value of $f(x)$ for every x where $g(\pi(x))$ was queried is also determined by the value that was returned to the tester) and setting $g = f_\pi$. If there are more queries that have not yet been answered because P_{yes} quit, they are answered as per the generated f and g.

The process P_{no} is defined similarly. First, it chooses a permutation $\pi \in \mathcal{S}_n$ uniformly at random. It then answers the queries of $\mathcal{T}$ randomly, with the same exception as in the P_{yes} case: if $\mathcal{T}$ queries $f(x)$ after having queried $g(\pi(x))$, or if $\mathcal{T}$ queries $g(x)$ after having queried $f(\pi^{-1}(x))$, then P_{no} "quits".

When P_{no} quits or reaches the end of the queries, it completes the definitions of f and of g independently, randomly fixing the value of $f(x)$ and $g(x)$ for every input $x \in T$ that has not been queried by $\mathcal{T}$. If P_{no} quit before answering all the queries, those queries are then answered with the values of f and g that have been fixed.

It is easy to check that P_{yes} and P_{no} generate pairs of functions from $\mathcal{D}_{\text{yes}}$ and $\mathcal{D}_{\text{no}}$, respectively. Furthermore, when P_{yes} and P_{no} do not quit, they induce the same (i.e., uniformly random) distribution on the responses. So to complete the proof of the Lemma, it suffices to show that neither process quits with probability greater than $o(1)$.

The process P_{yes} or P_{no} quits if there is a pair of inputs $x_f, x_g \in T$ such that $f(x_f)$ and $g(x_g)$ are queried by $\mathcal{T}$ and $\pi(x_f) = x_g$. For any such pair, the probability that $\pi(x_f) = x_g$ is at most $O(\frac{\sqrt{n}}{2^n})$. But the answers to the queries yield no information about π to the tester $\mathcal{T}$, so the probability that it causes P_{yes} or P_{no} to quit is at most $o(\frac{2^{n/2}}{n^{1/4}})^2 \cdot O(\frac{\sqrt{n}}{2^n}) = o(1)$. $\qquad\square$

Acknowledgements

E.B. wishes to thank Ryan O'Donnell for much valuable advice throughout the course of this research and Michael Saks for enlightening discussions.

Some of this work was completed while the authors were participating in the 2010 ITCS mini-workshop on property testing. We thank the organizers, Oded Goldreich and Yuexuan Wang, for inviting us and everyone at the Institute for Theoretical Computer Science at Tsinghua University for their wonderful hospitality.

References

1. Alon, N., Fischer, E., Krivelevich, M., Szegedy, M.: Efficient testing of large graphs. Combinatorica 20(4), 451–476 (2000)
2. Alon, N., Spencer, J.H.: The Probabilistic Method, 3rd edn. Wiley, Chichester (2008)
3. Blais, E., O'Donnell, R.: Lower bounds for testing function isomorphism. In: Conference on Computational Complexity (2010)

4. Chakraborty, S., García-Soriano, D., Matsliah, A.: Nearly tight bounds for testing function isomorphism. Manuscript (2010)

5. Chockler, H., Gutfreund, D.: A lower bound for testing juntas. Information Processing Letters 90(6), 301–305 (2004)

6. Diakonikolas, I., Lee, H.K., Matulef, K., Onak, K., Rubinfeld, R., Servedio, R.A., Wan, A.: Testing for concise representations. In: Proc. 48th Symposium on Foundations of Computer Science, pp. 549–558 (2007)

7. Dodis, Y., Goldreich, O., Lehman, E., Raskhodnikova, S., Ron, D., Samorodnitsky, A.: Improved testing algorithms for monotonicity. In: Hochbaum, D.S., Jansen, K., Rolim, J.D.P., Sinclair, A. (eds.) RANDOM 1999 and APPROX 1999. LNCS, vol. 1671, pp. 97–108. Springer, Heidelberg (1999)

8. Fischer, E.: The art of uninformed decisions: a primer to property testing. Bull. Eur. Assoc. for Theoretical Comp. Sci. 75, 97–126 (2001)

9. Fischer, E.: The difficulty of testing for isomorphism against a graph that is given in advance. SIAM J. on Comp. 34(5), 1147–1158 (2005)

10. Fischer, E., Kindler, G., Ron, D., Safra, S., Samorodnitsky, A.: Testing juntas. J. Comput. Syst. Sci. 68(4), 753–787 (2004)

11. Fischer, E., Lehman, E., Newman, I., Raskhodnikova, S., Rubinfeld, R., Samorodnitsky, A.: Monotonicity testing over general poset domains. In: STOC 2002: Proceedings of the Thiry-Fourth Annual ACM Symposium on Theory of Computing, pp. 474–483 (2002)

12. Fischer, E., Matsliah, A.: Testing graph isomorphism. SIAM J. Comput. 38(1), 207–225 (2008)

13. Goldreich, O., Goldwasser, S., Lehman, E., Ron, D., Samorodnitsky, A.: Testing monotonicity. Combinatorica 20(3), 301–337 (2000)

14. Goldreich, O., Goldwasser, S., Ron, D.: Property testing and its connection to learning and approximation. J. ACM 45(4), 653–750 (1998)

15. Hajnal, A., Szemerédi, E.: Proof of a conjecture of Paul Erdős. In: Erdős, P., Rényi, A., Sós, V.T. (eds.) Combinatorial Theory and its Applications, pp. 601–623 (1969)

16. Matulef, K., O'Donnell, R., Rubinfeld, R., Servedio, R.A.: Testing halfspaces. In: SODA 2009: Proceedings of the Nineteenth Annual ACM -SIAM Symposium on Discrete Algorithms, pp. 256–264 (2009)

17. Matulef, K., O'Donnell, R., Rubinfeld, R., Servedio, R.A.: Testing ±1-weight halfspace. In: Dinur, I., Jansen, K., Naor, J., Rolim, J. (eds.) RANDOM '09. LNCS, vol. 5687, pp. 646–657. Springer, Heidelberg (2009)

18. Ron, D.: Property testing: a learning theory perspective. Foundations and Trends in Machine Learning 1(3), 307–402 (2008)

19. Ron, D.: Algorithmic and analysis techniques in property testing. Foundations and Trends in Theoretical Computer Science 5(2), 73–205 (2009)

20. Rubinfeld, R., Sudan, M.: Robust characterizations of polynomials with applications to program testing. SIAM J. Comput. 25(2), 252–271 (1996)

21. Yao, A.C.-C.: Probabilistic computations: Toward a unified measure of complexity. In: Proceedings of the 28th IEEE Symposium on Foundations of Computer Science, pp. 222–227 (1977)

Better Size Estimation for Sparse Matrix Products[*]

Rasmus Resen Amossen, Andrea Campagna, and Rasmus Pagh

IT University of Copenhagen, DK-2300 Copenhagen S, Denmark
{resen,acam,pagh}@itu.dk

Abstract. We consider the problem of doing fast and reliable estimation of the number of non-zero entries in a sparse boolean matrix product.

Let n denote the total number of non-zero entries in the input matrices. We show how to compute a $1 \pm \varepsilon$ approximation (with small probability of error) in expected time $\mathcal{O}(n)$ for any $\varepsilon > 4/\sqrt[4]{n}$. The previously best estimation algorithm, due to Cohen (JCSS 1997), uses time $\mathcal{O}(n/\varepsilon^2)$. We also present a variant using $\mathcal{O}(\text{sort}(n))$ I/Os in expectation in the cache-oblivious model.

We also describe how sampling can be used to maintain (independent) sketches of matrices that allow estimation to be performed in time $o(n)$ if z is sufficiently large. This gives a simpler alternative to the sketching technique of Ganguly et al. (PODS 2005), and matches a space lower bound shown in that paper.

1 Introduction

In this paper we will consider a $d \times d$ boolean matrix as the subset of $[d] \times [d]$ corresponding to the nonzero entries. The product of two matrices R_1 and R_2 contains (i, k) if and only if there exists j such that $(i, j) \in R_1$ and $(j, k) \in R_2$. The matrix product can also be expressed using basic operators of relational algebra: $R_1 \bowtie R_2$ denotes the set of tuples (i, j, k) where $(i, j) \in R_1$ and $(j, k) \in R_2$, and the projection operator π can be used to compute the tuples (i, k) where there exists a tuple of the form $(i, \cdot, k)$ in $R_1 \bowtie R_2$. Since most of our applications are in database systems we will primarily use the notation of relational algebra.

We consider the following question: given relations R_1 and R_2 with schemas (a, b) and (b, c), estimate the number of *distinct* tuples in the relation $Z = \pi_{ac}(R_1 \bowtie R_2)$. This problem has been referred to in the literature as *join-project* or *join-distinct*[1]. We define $n_1 = |R_1|$, $n_2 = |R_2|$, and $n = n_1 + n_2$. As observed

---

[*] This work was supported by the Danish National Research Foundation, as part of the project "Scalable Query Evaluation in Relational Database Systems". A full version of this paper is available on arXiv [2].

[1] Readers familiar with the database literature may notice that we consider projections that return a set, i.e., that projection is duplicate eliminating. We also observe that any equi-join followed by a projection can be reduced to the case above, having two variables in each relation and projecting away the single join attribute. Thus, there is no loss of generality in considering this minimal case.

M. Serna et al. (Eds.): APPROX and RANDOM 2010, LNCS 6302, pp. 406–419, 2010.
© Springer-Verlag Berlin Heidelberg 2010

above, the join-project problem is equivalent to the problem of estimating the number of non-zero entries in the product of two boolean matrices, having n_1 and n_2 non-zero entries, respectively.

In recent years there has been several papers presenting new algorithms for sparse matrix multiplication [3,12,14]. In particular, these algorithms can be used to implement boolean matrix multiplication. However, the proposed algorithms all have substantially superlinear time complexity in the input size n: On worst-case inputs they require time $\omega(n^{4/3})$, even when the output Z only has size $\mathcal{O}(n)$.

In an influential work, Cohen [5] presented an estimation algorithm that, for any constant error probability $\delta > 0$, and any $\varepsilon > 0$, can compute a $1 \pm \varepsilon$ approximation of $z = |Z|$ in time $\mathcal{O}(n/\varepsilon^2)$. Cohen's algorithm applies to the more general problem of computing the size of the transitive closure of a graph.

Our main result is that in the special case of sparse matrix product size estimation, we can improve this to expected time $\mathcal{O}(n)$ for $\varepsilon > 4/\sqrt[4]{n}$. This means that we have a linear time algorithm for relative error where Cohen's algorithm would use time $\mathcal{O}(n^{3/2})$.

Approach. To build intuition on the size estimation question, consider the sets $\mathcal{A}_j = \{i \mid (i,j) \in R_1\}$ and $\mathcal{C}_j = \{k \mid (j,k) \in R_2\}$. By definition, $Z = \bigcup_j \mathcal{A}_j \times \mathcal{C}_j$. The size of Z depends crucially on the extent of overlap among the sets $\{\mathcal{A}_j \times \mathcal{C}_j\}_j$. However, the total size of these sets may be much larger than both input and output (see [3]), so any approach that explicitly processes them is unattractive.

The starting point for our improved estimation algorithm is a well-known algorithm for estimating the number of distinct elements in a data streaming context [4]. (We remark that the idea underlying this algorithm is similar to that of Cohen [5].) Our main insight is that this algorithm can be extended such that a set of the form $\mathcal{A}_j \times \mathcal{C}_j$ can be added to the sketch in expected time $\mathcal{O}(|\mathcal{A}_j| + |\mathcal{C}_j|)$, i.e., without explicitly generating all pairs. The idea is to use a hash function that is particularly well suited for the purpose: sufficiently structured to make hash values easy to handle algorithmically, and sufficiently random to make the analysis of sketching accuracy go through.

1.1   Motivation

Cohen [6] investigated the use of the size estimation technique in sparse matrix computations. In particular, it can be used to find the optimal order of multiplying sparse matrices, and in memory allocation for sparse matrix computations.

In addition, we are motivated by applications in database systems, where size estimation is an important part of query optimization. Our result also has an application in data mining, where it gives a way choosing the support threshold of the Apriori algorithm [1] to obtain a given space usage. Further details can be found in the full version of this paper [2].

1.2   Further Related Work

JD sketch. Ganguly et al. [8] previously considered techniques that compute a data structure (a *sketch*) for R_1 and R_2 (individually), such that the two sketches suffice to compute an approximation of z.

Define $n_a = |\{i \mid \exists j.(i,j) \in R_1\}|$ and $n_c = |\{k \mid \exists j.(j,k) \in R_2\}|$. Ganguly et al. show that for any constant c and any β, a sketching method that returns a c-approximation with probability $\Omega(1)$ whenever $z \geq \beta$ must, on a worst-case input, use expected space

$$\Omega(\min(n_1+n_2, n_a n_c(n_1/n_a+n_2/n_c)/\beta)) = \Omega(\min(n_1+n_2, (n_1 n_c+n_2 n_a)/\beta)) \text{ bits.}$$

The lower bound proof applies to the case where $n_1 = n_2$, $n_a = n_c$, and $z < n_a + n_c$. We note that [8] claims a stronger lower bound, but their proof does not establish a lower bound above $n_1 + n_2$ bits. Ganguly et al. present a sketch whose worst-case space usage matches the lower bound times polylogarithmic factors (while not stated in [8], the trivial sketch that stores the whole input can be used to nearly match the first term in the minimum).

In Section 3 we analyze a simple sketch, previously considered in other contexts by Gibbons [10] and Ganguly and Saha [9]. It similarly matches the above worst-case bound, but the exact space usage is incomparable to that of [8].

The focus of [8] is on space usage, and so the time for updating sketches, and for computing the estimate from two sketches, is not discussed in the paper. Looking at the data structure description we see that the update time grows linearly with the quantity s_1, which is $\Omega(n)$ in the worst case. Also, the sketch uses a number of summary data structures that are accessed in a random fashion, meaning that the worst case number of I/Os is at least $\Omega(n)$ *unless* the sketch fits internal memory. By the above lower bound we see that keeping the sketch in internal memory is not feasible in general. In contrast, the sketch we consider allows collection and combination of sketches to be done efficiently in linear time and I/O.

Distinct elements and distinct paths estimation. Our work is related in terms of techniques to papers on estimating the number of distinct items in a data stream (see [4] and its references). However, our basic estimation algorithm does not work in a general streaming model, since it crucially needs the ability to access all tuples with a particular value on the join attribute together.

Ganguly and Saha [9] consider the problem of estimating the number of distinct vertex pairs connected by a length-2 path in a graph whose edges are given as a data stream of n edges. This corresponds to size estimation for the special case of *squaring* a matrix (or self-join in database terminology). It is shown that space $\sqrt{n}$ is required, and that space roughly $\mathcal{O}(n^{3/4})$ suffices for constant ε (unless there are close to n connected components). The estimation itself is a join-distinct size estimation of a sample of the input having size no smaller than $\mathcal{O}(n^{3/4}/\varepsilon^2)$. Using Cohen's estimation algorithm this would require time $\mathcal{O}(n^{3/4}/\varepsilon^4)$, so this is $\mathcal{O}(n)$ time only for $\varepsilon > 1/\sqrt[16]{n}$.

Distinct sampling. Gibbons [10] considered different samples that can be extracted by a scan over the input, and proposed *distinct samples*, which offer much better guarantees with respect to estimating the number of distinct values in query results. Gibbons shows that this technique applies to single relations, and to foreign key joins where the join result has the same number of tuples as one of the relations. In Section 3 we show that the distinct samples, with suitable settings of parameters, can often be used in our setting to get an accurate estimate of $z = |Z|$.

2   Our Algorithm

The task is to estimate the size z of $Z = \pi_{ac}(R_1 \bowtie R_2)$. We may assume that attribute values are $\mathcal{O}(\log n)$-bits integers, since any domain can be mapped into this one using hashing, without changing the join result size with high probability. When discussing I/O bounds, B is the number of such integers that fits in a disk block. In linear expected time (by hashing) or sort(n) I/Os we can cluster the relations according to the value of the join attribute b. By initially eliminating input tuples that do not have any matching tuples in the other relation we may assume without loss of generality that $z \geq n/2$.

In what follows, k is a positive integer parameter that determines the space usage and accuracy of our method. The technique used is to compute the kth smallest value v of a hash function $h(x, y)$, for $(x, y) \in Z$. Analogously to the result by Bar-Yossef et al. [4] we can then use $\tilde{z} = k/v$ as an estimator for z.

Our main building block is an efficient iteration over all tuples $(x, \cdot, y) \in R_1 \bowtie R_2$ for which $h(x, y)$ is smaller than a carefully chosen threshold p, and is therefore a candidate for being among the k smallest hash values. The essence of our result lies in how the pairs being output by this iteration are computed in expected linear time. We also introduce a new buffering trick to update the sketch in expected amortized $\mathcal{O}(1)$ time per pair. In a nutshell, each time k new elements have been retrieved, they are merged using a linear time selection procedure with the previous k smallest values to produce a new (unordered) list of the k smallest values.

Theorem 1. *Let $R_1(a, b)$ and $R_2(b, c)$ be relations with n tuples in total, and define $z = |\pi_{ac}(R_1 \bowtie R_2)|$. Let ε, $0 < \varepsilon < \frac{1}{2}$ be given. There are algorithms that run in expected $\mathcal{O}(n)$ time on a RAM, and expected $\mathcal{O}(\text{sort}(n))$ I/Os in the cache-oblivious model, and output a number $\tilde{z}$ such that for $k = 9/\varepsilon^2$:*

- $\mathbf{Pr}[(1 - \varepsilon)z < \tilde{z} < (1 + \varepsilon)z] \geq 2/3$ *when* $z > k^2$, *and*
- $\mathbf{Pr}[\tilde{z} < (1 + \varepsilon)k^2] \geq 2/3$ *when* $z \leq k^2$.

Observe that for $\varepsilon > 4/\sqrt[4]{n}$, since $z \geq n/2$ we will be in the first case, and get the desired $1 \pm \varepsilon$ approximation with probability 2/3. The error probability can be reduced from 1/3 to δ by the standard technique of doing $\mathcal{O}(\log(1/\delta))$ runs and taking the median (the analysis follows from a Chernoff bound). We remark that this can be done in such a way that the $\mathcal{O}(\log(1/\delta))$ factor affects only the RAM running time and not the number of I/Os. For constant relative error $\varepsilon > 0$ we have the following result:

Theorem 2. *In the setting of Theorem 1, if ε is constant there are algorithms that run in expected $\mathcal{O}(n)$ time on a RAM, and expected $\mathcal{O}(sort(n))$ I/Os in the cache-oblivious model, that output $\tilde{z}$ such that $\mathbf{Pr}[(1 - \varepsilon)z < \tilde{z} < (1 + \varepsilon)z] = \mathcal{O}(1/\sqrt{n})$.*

The error probability can be reduced to n^{-c} for any desired constant c by running the algorithms $\mathcal{O}(c)$ times, and taking the median as above.

Finding pairs. For $\mathcal{B} = \pi_b(R_1) \cup \pi_b(R_2)$ and each $i \in \mathcal{B}$ let $\mathcal{A}_i = \pi_a(\sigma_{b=i}(R_1))$ and $\mathcal{C}_i = \pi_c(\sigma_{b=i}(R_2))$. We would like to efficiently iterate over all pairs $(x, y) \in \mathcal{A}_i \times \mathcal{C}_i$, $i \in \mathcal{B}$, for which $h(x, y)$ is smaller than a threshold p. This is done as follows (see Algorithm 1 for pseudocode).

For a set U, let $h_1, h_2 : U \to [0; 1]$ be hash functions chosen independently at random from a pairwise independent family, and define $h : U \times U \to [0; 1]$ by[2]

$$h(x, y) = (h_1(x) - h_2(y)) \bmod 1.$$

It is easy to show that h is also a pairwise independent hash function — a property we will utilize later. Now, conceptually arrange the values of $h(x, y)$ in an $|\mathcal{A}_i| \times |\mathcal{C}_i|$ matrix, and order the rows by increasing values of $h_1(x)$, and the columns by increasing values of $h_2(y)$. Then the values of $h(x, y)$ will decrease (modulo 1) from left to right, and increase (modulo 1) from top to bottom.

For each $i \in \mathcal{B}$, we traverse the corresponding $|\mathcal{A}_i| \times |\mathcal{C}_i|$ matrix by visiting the columns from left to right, and in each column t finding the row $\bar{s}$ with the smallest value of $h(x_{\bar{s}}, y_t)$. Values smaller than p in that column will be found in rows subsequent to $\bar{s}$. When all such values have been output, the search proceeds in column $t + 1$. Notice, that if $h(x_{\bar{s}}, y_t)$ was the minimum value in column t, then the minimum value in column $t + 1$ is found by increasing $\bar{s}$ until $h(x_{\bar{s}}, y_{t+1}) < h(x_{(\bar{s}-1) \bmod |\mathcal{A}_i|}, y_{t+1})$. We observe that the algorithm is robust to decreasing the value of the threshold p during execution, in the sense that the algorithm still outputs all pairs with hash value at most p.

Estimating the size. While finding the relevant pairs, we will use a technique that allows us to maintain the k smallest hash values in an unordered buffer instead of using a heap data structure (lines 14–18 in Algorithm 1). In this way we are able to maintain the k smallest hash values in constant amortized time per insertion in the buffer, eliminating the $\log k$ factor implied by the heap data structure.

Let S and F be two unordered sets containing, respectively, the k smallest hash values seen so far (all, of course, smaller than p), and the latest up to k

---

[2] We observe that this is different from the "composable hash functions" used by Ganguly et al. [8].

Algorithm 1. Pseudocode for the size estimator

```
 1: procedure DISITEMS(p, ε)
 2:     k ← ⌈9/ε²⌉
 3:     F ← ∅
 4:     for i ∈ B do
 5:         x ← A_i sorted according to h₁-value
 6:         y ← C_i sorted according to h₂-value
 7:         s̄ ← 1
 8:         for t := 1 to |C_i| do
 9:             while h(x_s̄, y_t) > h(x_(s̄−1) mod |A_i|, y_t) do  ▷ Find s̄ s.t. h(x_s̄, y_t) is min.
10:                 s̄ ← (s̄ + 1) mod |A_i|
11:             end while
12:             s ← s̄
13:             while h(x_s, y_t) < p do                        ▷ Find all s where h(x_s, y_t) < p
14:                 F ← F ∪ {(x_s, y_t)}
15:                 if |F| = k then      ▷ Buffer filled, find smallest hash values in S ∪ F
16:                     (p, S) ← COMBINE(S, F)
17:                     F ← ∅
18:                 end if
19:                 s ← (s + 1) mod |A_i|
20:             end while
21:         end for
22:     end for
23:     (p, S) ← COMBINE(S, F)
24:     if |S| = k then
25:         return "z̃ = k/p and z̃ ∈ [(1 ± ε)z] with probability 2/3"
26:     else
27:         return "z̃ = k², z ≤ k² with probability 2/3"
28:     end if
29: end procedure

30: procedure COMBINE(S, F)
31:     v ← RANK(h(S) ∪ h(F), k)        ▷ RANK(·, k) returns the kth smallest value
32:     S ← {x ∈ S ∪ F | h(x) ≤ v}
33:     return (v, S)
34: end procedure
```

elements seen. We avoid duplicates in S and F (i.e., the sets are kept disjoint) by using a simple hash table to check for membership before insertion. Whenever $|F| = k$ the two sets S and F are combined in order to obtain a new sketch S. This is done by finding the median of $S \cup F$, which takes $\mathcal{O}(k)$ time using either deterministic methods (see [7]) or more practical randomized ones [11].

At each iteration the current kth smallest value in S may be smaller than the initial value p, and we use this as a better substitute for the initial value of p. However, in the analysis below we will upper bound both the running time and the error probability using the initial threshold value p.

2.1   Time Analysis

We split the time analysis into two parts. One part accounts for iterations of the inner while loop in lines 13–20, and the other part accounts for everything else. We first consider the RAM model, and then outline the analysis in the cache-oblivious model.

Inner while loop. Observe that for each iteration, one pair (x_s, y_t) is added to F (if it is not already there). For each $t \in C_i$, $p|A_i|$ elements are expected to be added since each pair (x_s, y_t) is added with probability p. This means that the expected total number of iterations is $O(p|A_i||C_i|)$. Each call to COMBINE costs time $O(k)$, but we notice that there must be at least k iterations between successive calls, since the size of F must go from 0 to k. Inserting a new value into F costs $O(1)$ since the set is not sorted. Hence, the total cost of the inner loop is $O(p|A_i||C_i|)$.

Remaining cost. Consider the processing of a single $i \in B$ in Algorithm 1. The initial sorting of hash values can be done with bucket sort requiring expected time $O(|A_i| + |C_i|)$ since the numbers sorted are pairwise independent (by the same analysis as for hashing with chaining).

For the iteration in lines 9–11 observe that $h(x_{\bar{s}}, y_t)$ is monotone modulo 1, and we have at most a total of $2|A_i|$ increments of $\bar{s}$ among all $t \in C_i$. Thus, the total number of iterations is $O(|A_i|)$, and the total cost for each $i \in B$ is $O(|A_i| + |C_i|)$.

The time for the final call to COMBINE is dominated by the preceding cost of constructing S and F.

I/O efficient variant. As for I/O efficiency, notice that a direct implementation of Algorithm 1 may cause a linear number of cache misses if A_i and C_i do not fit into internal memory. To get an I/O-efficient variant we use a cache-oblivious sorting algorithm, sorting R_1 according to $(b, h_1(a))$, and R_2 according to $(b, h_2(c))$, such that the sorting steps for each $i \in B$ is replaced by one global sorting step.

The rest of the algorithm works directly in a cache-oblivious setting. To see this, notice that it suffices to keep in internal memory the two input blocks that are closest to each of the pointers s, t, and $\bar{s}$. The cache-oblivious model assumes the cache to behave in an optimal fashion, so also in this model there will be $\Omega(B)$ operations between cache misses, and $O(n/B)$ I/Os, expected, in total.

Lemma 1. *Suppose $R_1(a, b)$ and $R_2(b, c)$ are relations with n tuples in total. Let $p > 0$ and $\varepsilon > 0$ be given. Then Algorithm 1 runs in expected $O(n + \sum_i p|A_i||C_i|)$ time and $O(1/\varepsilon^2)$ space on a RAM, and can be modified to use expected $O(\mathrm{sort}(n))$ I/Os in the cache-oblivious model.*

Choice of threshold p. We would like a value of p that ensures the expected processing time is $O(n)$. At the same time p should be large enough that we expect to reach line 25 where an exact estimate is returned (except possibly in the case where z is small).

Lemma 2. *Let $j \in \mathcal{B}$ satisfy $|\mathcal{A}_i||\mathcal{C}_i| \leq |\mathcal{A}_j||\mathcal{C}_j|$ for all $i \in \mathcal{B}$. Then $p = \min(1/k, k/(|\mathcal{A}_j||\mathcal{C}_j|))$ gives an expected $\mathcal{O}(n)$ running time for Algorithm 1.*

Proof. We argue that for each i, $p|\mathcal{A}_i||\mathcal{C}_i| \leq \max(|\mathcal{A}_i|, |\mathcal{C}_i|)$, which by Lemma 1 implies running time $\mathcal{O}(n + \sum_i p|\mathcal{A}_i||\mathcal{C}_i|) = \mathcal{O}(n + \sum_i \max(|\mathcal{A}_i|, |\mathcal{C}_i|)) = \mathcal{O}(n)$. Suppose first that $|\mathcal{A}_i||\mathcal{C}_i| \geq k^2$. Then $p = k/(|\mathcal{A}_j||\mathcal{C}_j|)$ and $p|\mathcal{A}_i||\mathcal{C}_i| \leq k \leq \sqrt{|\mathcal{A}_i||\mathcal{C}_i|} \leq \max(|\mathcal{A}_i|, |\mathcal{C}_i|)$. Otherwise, when $|\mathcal{A}_i||\mathcal{C}_i| < k^2$, we have $p = 1/k$ and $p|\mathcal{A}_i||\mathcal{C}_i| = |\mathcal{A}_i||\mathcal{C}_i|/k \leq \max(|\mathcal{A}_i|, |\mathcal{C}_i|)$. $\qquad\square$

We note that when R_1 and R_2 are sorted according to b, the value of p specified above can be found by a simple scan over both inputs. Our experiments indicate that in practice this initial scan is not needed, see Section 4 for details.

2.2 Error Probability

Theorem 3. *Let h be a pairwise independent hash function. Suppose we are provided with a stream of elements N with $h(x) < v$ for all $x \in N$. Further, let ε, $0 < \varepsilon < \frac{1}{2}$ be given and assume that $p \geq \min\left(\frac{k}{2z}, \frac{1}{k}\right)$, where $k \geq 9/\varepsilon^2$, and z is the number of distinct items in N. Then Algorithm 1 produces an approximation $\tilde{z}$ of z such that*

- $\mathbf{Pr}[(1-\varepsilon)z < \tilde{z} < (1+\varepsilon)z] \geq 2/3$ *for $z > k^2$, and*
- $\mathbf{Pr}[\tilde{z} < (1+\varepsilon)k^2] \geq 2/3$ *for $z \leq k^2$.*

Proof. The error probability proof is similar to the one that can be found in [4], with some differences and extensions. We bound the error probability of three cases: the estimate being smaller/larger than the multiplicative error bound, and the number of obtained samples being too small.

Estimate too large. Let us first consider the case where $\tilde{z} > (1+\varepsilon)z$, i.e. the algorithm overestimates the number of distinct elements. This happens if the stream N contains at least k entries smaller than $k/(1+\varepsilon)z$. For each pair $(a, c) \in Z$ define an indicator random variable $X_{(a,c)}$ as

$$X_{(a,c)} = \begin{cases} 1 & h(a, c) < k/(1+\varepsilon)z \\ 0 & \text{otherwise} \end{cases}$$

That is, we have z such random variables for which the probability of $X_{(a,c)} = 1$ is exactly $k/(1+\varepsilon)z$ and $\mathbf{E}[X_{(a,c)}] = k/(1+\varepsilon)z$. Now define $Y = \sum_{(a,c)\in Z} X_{(a,c)}$ so that $\mathbf{E}[Y] = \mathbf{E}[\sum_{(a,c)\in Z} X_{(a,c)}] = \sum_{(a,c)\in Z} \mathbf{E}[X_{(a,c)}] = k/(1+\varepsilon)$. By the pairwise independence of the $X_{(a,c)}$ we also get $\mathbf{Var}(Y) \leq k/(1+\varepsilon)$. Using Chebyshev's inequality [13] we can bound the probability of having too many pairs reported:

$$\mathbf{Pr}\left[Y > k\right] \leq \mathbf{Pr}\left[|Y - \mathbf{E}[Y]| > k - \frac{k}{1+\varepsilon}\right] \leq \frac{\mathbf{Var}[Y]}{\left(k - \frac{k}{1+\varepsilon}\right)^2} \leq \frac{k/(1+\varepsilon)}{\left(k - \frac{k}{1+\varepsilon}\right)^2} \leq \frac{1}{6}$$

since $k \geq 9/\varepsilon^2$.

Estimate too small. Now, consider the case where $\tilde{z} < (1 - \varepsilon)z$ which happens when at most k hash values are smaller than $k/(1 - \varepsilon)z$ and at least k hash values are smaller than p. Define $X'_{(a,c)}$ as

$$X'_{(a,c)} = \begin{cases} 1 & h(a,c) < k/(1-\varepsilon)z \\ 0 & \text{otherwise} \end{cases}$$

so that $\mathbf{E}[X'_{(a,c)}] = k/(1-\varepsilon)z < (1+\varepsilon)k/z$. Moreover, with $Y' = \sum_{(a,c)\in Z} X'_{(a,c)}$ we have $\mathbf{E}[Y'] = k/(1-\varepsilon)$, and since the indicator random variables defined above are pairwise independent, we also have $\mathbf{Var}[Y'] \leq \mathbf{E}[Y'] < (1+\varepsilon)k$. Chebyshev's inequality gives:

$$\mathbf{Pr}\left[Y' > k\right] \leq \mathbf{Pr}\left[|Y' - \mathbf{E}[Y']| > \tfrac{k}{1-\varepsilon} - k\right] \leq \frac{\mathbf{Var}[Y']}{\left(k - \tfrac{k}{1+\varepsilon}\right)^2} \leq \frac{(1+\varepsilon)k}{\left(\tfrac{k}{1-\varepsilon} - k\right)^2} < \tfrac{1}{9}$$

since $k \geq 9/\varepsilon^2$.

Not enough samples. Consider the case where $|S| < k$ after all pairs have been retrieved. In this case the algorithm returns $\beta = k^2$ as an upper bound on the number of distinct elements in the output, and we have two possible situations: either there is actually less than k^2 distinct pairs in the output, in which case the algorithm is correct, or there are more than k^2 distinct elements in the output, in which case it is incorrect. In the latter case, less than k hash values have been smaller than p and the kth smallest value v is therefore larger than p. Define $X''_{(a,c)}$ as

$$X''_{(a,c)} = \begin{cases} 1 & h(a,c) < p \\ 0 & \text{otherwise} \end{cases}$$

and let again $Y'' = \sum_{(a,c)\in Z} X''_{(a,c)}$. It results that $\mathbf{E}[X''_{(a,c)}] = p$ and $\mathbf{E}[Y''] = zp$, and because of pairwise independancy of $X''_{(a,c)}$, also $\mathbf{Var}[Y''] \leq \mathbf{E}[Y'']$. Using Chebyshev's inequality and remembering that $z > k^2$ in this case we have:

$$\mathbf{Pr}[Y'' < k] \leq \mathbf{Pr}[|Y'' - \mathbf{E}[Y'']| > zp - k] \leq \frac{zp}{(zp - k)^2} \leq \frac{zp}{\left(\tfrac{1}{2}zp\right)^2} \leq 2/k \leq 1/18.$$

using that $k \geq 9/\varepsilon^2 \geq 36$.

In conclusion, the probability that the algorithm fails to output an estimate within the given limits is at most $1/6 + 1/9 + 1/18 = 1/3$. □

For the proof of Theorem 2 we observe that in the above proof, if ε is constant the error probability is $\mathcal{O}(1/k)$. Using $k = \sqrt{n}$ we get linear running time and error probability $\mathcal{O}(1/\sqrt{n})$.

Realization of hash functions. We have used the idealized assumption that hash values were real numbers in $(0; 1)$. Let $m = n^3$. To get an actual implementation we approximate (by rounding down) the real numbers used by rational numbers of the form i/m, for integer i. This changes each hash value by at most $2/m$. Now, because of the way hash values are computed, the probability that we get a different result when comparing two real-valued hash values and two rational ones is bounded by $2/m$. Similarly, the probability that we get a different result when looking up a hash value in the dictionary is bounded by $2k/m$. Thus, the probability that the algorithm makes a different decision based on the approximation, in any of its steps, is $\mathcal{O}(kn/m) = o(1)$. Also, for the final output the error introduced by rounding is negligible.

3   Distinct Sketches

A well-known approach to size estimation in, described in generality by Gibbons [10] and explicitly for join-project operations in [9,3], is to sample random subsets $R_1' \subseteq R_1$ and $R_2' \subseteq R_2$, compute $Z' = \pi_{ac}(R_1' \bowtie R_2')$, and use the size of Z' to derive an estimate for z. This is possible if $R_1' = \sigma_{a \in S_a}(R_1)$, where $S_a \subseteq \pi_a(R_1)$ is a random subset where each element is picked independently with probability p_1, and similarly $R_2' = \sigma_{c \in S_c}(R_2)$, where $S_c \subseteq \pi_c(R_2)$ includes each element independently with probability p_2. Then $z' = |Z'|/(p_1 p_2)$ is an unbiased estimator for z. The samples can be obtained in small space using hash functions whose values determine which elements are picked for S_a and S_c. The value $|Z'|$ can be approximated in linear time using the method described in section 2 if the samples are sorted — otherwise one has to add the cost of sorting. In either case, the estimation algorithm is I/O-efficient.

Below we analyze the variance of the estimator z', to identify the minimum sampling probability that introduces only a small relative error with good probability. The usual technique of repetition can be used to reduce the error probability. Recall that we have two relations with n_1 and n_2 tuples, respectively, and that n_a and n_c denotes the number of distinct values of attributes a and c, respectively. Our method will pick samples R_1' and R_2' of expected size s from each relation, where $s = p_1 n_1 = p_2 n_2$ is a parameter to be specified.

Theorem 4. *Let R_1' and R_2' be samples of size s, obtained as described above. Then $z' = |\pi_{ac}(R_1' \bowtie R_2')|/(p_1 p_2)$ is a $1 \pm \varepsilon$ approximation of $z = |\pi_{ac}(R_1 \bowtie R_2)|$ with probability $5/6$ if $z > \beta$, where $\beta = \frac{14}{\varepsilon^2}\left(\frac{n_c n_1 + n_a n_2}{s}\right)$. If $z \leq \beta$ then $z' < (1 + \varepsilon)\beta$ with probability $5/6$.*

3.1   Analysis of Variance

To arrive at a sufficient condition that z' is a $1 \pm \varepsilon$ approximation of z with good probability, we analyze its variance. To this end define $Z_{i\cdot} = \{j \mid (i,j) \in Z\}$, $Z_{\cdot j} = \{i \mid (i,j) \in Z\}$, and let

$$X_i = \begin{cases} 1 - p_1, & \text{if } i \in S_a \\ -p_1, & \text{otherwise} \end{cases} \qquad Y_j = \begin{cases} 1 - p_2, & \text{if } j \in S_c \\ -p_2, & \text{otherwise} \end{cases}.$$

By definition of S_a, $\mathbf{E}[X_i] = \mathbf{Pr}[i \in S_a](1 - p_1) - \mathbf{Pr}[i \notin S_a]p_1 = 0$. Similarly, $\mathbf{E}[Y_i] = 0$. We have that $(i,j) \in Z'$ if and only if $(i,j) \in Z$ and $(i,j) \in S_a \times S_c$.

This means that $z'p_1p_2 = \sum_{(i,j) \in Z}(X_i + p_1)(Y_j + p_2)$. By linearity of expectation, $\mathbf{E}[(X_i + p_1)(Y_j + p_2)] = p_1 p_2$, and we can write the variance of $z'p_1p_2$, $\mathbf{Var}(z'p_1p_2)$ as

$$\mathbf{E}\left[\left(\sum_{(i,j) \in Z}((X_i + p_1)(Y_j + p_2) - p_1p_2)\right)^2\right].$$

Expanding the product and using linearity of expectation, we get

$$\mathbf{Var}(z'p_1p_2) = \sum_{(i,j) \in Z}\sum_{(i,j') \in Z}\mathbf{E}\left[X_i^2 p_2^2\right] + \sum_{(i,j) \in Z}\sum_{(i',j) \in Z}\mathbf{E}\left[Y_j^2 p_1^2\right] + \sum_{(i,j) \in Z}\mathbf{E}\left[X_i^2 Y_j^2\right]$$

$$= \sum_{i \in A}\sum_{j,j' \in Z_i.} p_2^2\,\mathbf{E}\left[X_i^2\right] + \sum_{j \in C}\sum_{i,i' \in Z.j} p_1^2\,\mathbf{E}\left[Y_i^2\right] + z\,\mathbf{E}\left[X_i^2\right]\mathbf{E}\left[Y_i^2\right]$$

Since $\mathbf{E}\left[X_i^2\right] = p_1(1 - p_1)^2 + (1 - p_1)(-p_1)^2 = p_1 - p_1^2 < p_1$, and similarly $\mathbf{E}\left[Y_j^2\right] < p_2$ we can upper bound $\mathbf{Var}(z')$ as follows:

$$\mathbf{Var}(z') = (p_1 p_2)^{-2}\,\mathbf{Var}(z'p_1p_2)$$

$$< (p_1 p_2)^{-2}\left(\sum_{i \in A}\sum_{j,j' \in Z_i.} p_1 p_2^2 + \sum_{j \in C}\sum_{i,i' \in Z.j} p_1^2 p_2 + z\,p_1 p_2\right)$$

$$\leq (p_1 p_2)^{-2}\left(n_c z\,p_1 p_2^2 + n_a z\,p_1^2 p_2 + z\,p_1 p_2\right)$$

$$= \left(n_c/p_1 + n_a/p_2 + (p_1 p_2)^{-1}\right) z\ .$$

3.2   Sufficient Sample Size

We are ready to derive a bound on the probability that z' deviates significantly from z. Choose $0 < \varepsilon < 1$. Since $z = \mathbf{E}[z']$ Chebyshev's inequality says

$$\mathbf{Pr}[|z' - z| > \varepsilon z] < \frac{\mathbf{Var}(z')}{(\varepsilon z)^2} \leq \left(n_c/p_1 + n_a/p_2 + (p_1 p_2)^{-1}\right)/(\varepsilon^2 z).$$

This can equivalently be expressed in terms of the sample size s, since $p_1 = s/n_1$ and $p_2 = s/n_2$:

$$\mathbf{Pr}[|z' - z| > \varepsilon z] < (n_c n_1 + n_a n_2 + n_1 n_2/s)/(s\varepsilon^2 z).$$

We seek a sufficient condition on s that the above probability is bounded by some constant $\delta < \frac{1}{2}$ (e.g. $\delta = 1/6$). In particular it must be the case that $n_1 n_2/(s^2 \varepsilon^2 z) < \delta$, which implies $s > \sqrt{n_1, n_2/(\delta z)} \geq \sqrt{n_1, n_2/(\delta n_a n_c)}$. Hence, using the arithmetic-geometric inequality:

$$n_1 n_2/s < \sqrt{n_c n_1 n_a n_2 \delta} \leq (n_c n_1 + n_a n_2)/(2\sqrt{\delta}).$$

In other words, it suffices that

$$\frac{(n_c n_1 + n_a n_2)\left(1 + (2\sqrt{\delta})^{-1}\right)}{s\varepsilon^2 z} < \delta \iff s > \left(\frac{n_c n_1 + n_a n_2}{z}\right)\left(\frac{1 + (2\sqrt{\delta})^{-1}}{\varepsilon^2 \delta}\right).$$

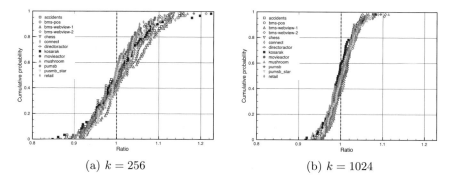

(a) $k = 256$ (b) $k = 1024$

Fig. 1. The cumulative distribution functions for $k = 256$ and $k = 1024$. It is seen that $k = 1024$ yields a more precise estimate than $k = 256$ with 2/3 of the estimates being within 4% and 10% of the exact size, respectively.

One apparent problem is the chicken-egg situation: z is not known in advance. If a lower bound on z is known, this can be used to compute a sufficient sample size. Alternatively, if we allow a larger relative error whenever $z \leq \beta$ we may compute a sufficient value of s based on the assumption $z \geq \beta$. Whenever $z < \beta$ we then get the guarantee that $z' < (1 + \varepsilon)\beta$ with probability $1 - \delta$. Theorem 4 follows by fixing s and solving for β.

Optimality. For constant ε and δ our upper bound matches the lower bound of Ganguly et al. [8] whenever this does not exceed $n_1 + n_2$. It is trivial to achieve a sketch of size $\mathcal{O}((n_1 + n_2)\log(n_1 + n_2))$ bits (simply store hash signatures for the entire relations). We also note that the lower bound proof in [8] uses certain restrictions of parameters ($n_1 = n_2$, $n_a = n_c$, and $z < n_a + n_c$), so it may be possible to do better in some settings.

4   Experiments

We have run our algorithm on most of the datasets from the Frequent Itemset Mining Implementations (FIMI) Repository[3] together with some datasets extracted from the Internet Movie Database (IMDB). Each dataset represents a single relation, and motivated by the Apriori space estimation example in the introduction, we perform the size estimation on self-joins of these relations.

Rather than selecting h_1 and h_2 from an arbitrary pairwise independent family, we store functions that map the attribute values to fully random and independent values of the form $d/2^{64}$, where d is a 64 bit random integer formed by reading 64 random bits from the Marsaglia Random Number CDROM[4].

We have chosen an initial value of $p = 1$ for our tests in order to be certain to always arrive at an estimate. In most cases we observed that p quickly decreases

---

[3] http://fimi.cs.helsinki.fi
[4] http://www.stat.fsu.edu/pub/diehard/

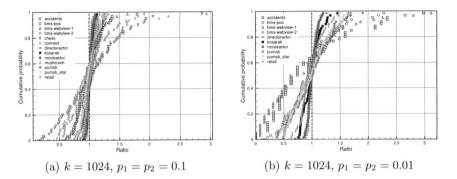

(a) $k = 1024$, $p_1 = p_2 = 0.1$ (b) $k = 1024$, $p_1 = p_2 = 0.01$

Fig. 2. Plots for sampling with probability 10% and 1%. If the sampling probability is too small, no elements at all may reach the sketch and in these cases we are not able to return an estimate. Instances with no estimates have been left out of the graph.

to a value below $1/k$ anyway. But as the sampling probability decreases, the probability that the sketch will never be filled increases, implying that we will not get a linear time complexity with an initial value of $p = 1$. In the cases where the sketch is not filled, we report $|F|/(p_1 p_2)$ as the estimate, where $|F|$ is the number of elements in the buffer.

Tests have been performed for $k = 256$ and $k = 1024$. In each test, 60 independent estimates were made and compared to the exact size of the join-project. By sorting the ratios "estimate"/"exact size" we can draw the cumulative distribution function for each instance that, for each ratio-value on the x-axis, displays on the y-axis the probability that an estimate will have this ratio or less. Figure 1 shows plots for $k = 256$ and $k = 1024$.

In Figure 2 we perform sampling with 10% and 1% probability, as described in Section 3. Again, the samples are chosen using truly random bits. The variance of estimates increase as the probability decreases, but increases more for smaller than for larger instances. If the sampling probability is too small, no elements at all may reach the sketch and in these cases we are not able to return an estimate.

5   Conclusion

We have presented improved algorithms for estimating the size of boolean matrix products, for the first time allowing $o(1)$ relative error to be achieved in linear time. An interesting open problem is if this can be extended to transitive closure in general graphs, and/or to products of more than two matrices.

Acknowledgement. We would like to thank Jelani Nelson for useful discussions, and in particular for introducing us to the idea of buffering to achieve faster data stream algorithms. Also, we thank Sumit Ganguly for clarifying the lower bound proof of [8] to us.

References

1. Agrawal, R., Srikant, R.: Fast algorithms for mining association rules. In: Proceedings of 20th International Conference on Very Large Data Bases (VLDB 1994), pp. 487–499. Morgan Kaufmann Publishers, San Francisco (1994)
2. Amossen, R.R., Campagna, A., Pagh, R.: Better size estimation for sparse matrix products. Technical Report arXiv:1006.4173, arxiv.org (2010)
3. Amossen, R.R., Pagh, R.: Faster join-projects and sparse matrix multiplications. In: Proceedings of the 12th International Conference on Database Theory (ICDT 2009), pp. 121–126. ACM, New York (2009)
4. Bar-Yossef, Z., Jayram, T.S., Kumar, R., Sivakumar, D., Trevisan, L.: Counting distinct elements in a data stream. In: Rolim, J.D.P., Vadhan, S.P. (eds.) RANDOM 2002. LNCS, vol. 2483, pp. 1–10. Springer, Heidelberg (2002)
5. Cohen, E.: Size-estimation framework with applications to transitive closure and reachability. Journal of Computer and System Sciences 55(3), 441–453 (1997)
6. Cohen, E.: Structure prediction and computation of sparse matrix products. J. Comb. Optim. 2(4), 307–332 (1998)
7. Dor, D., Zwick, U.: Selecting the median. In: Proceedings of the 6th annual ACM-SIAM Symposium on Discrete Algorithms (SODA 1995), pp. 28–37. SIAM, Philadelphia (1995)
8. Ganguly, S., Garofalakis, M., Kumar, A., Rastogi, R.: Join-distinct aggregate estimation over update streams. In: Proceedings of the 24th ACM Symposium on Principles of Database Systems (PODS '05), pp. 259–270. ACM, New York (2005)
9. Ganguly, S., Saha, B.: On estimating path aggregates over streaming graphs. In: Asano, T. (ed.) ISAAC 2006. LNCS, vol. 4288, pp. 163–172. Springer, Heidelberg (2006)
10. Gibbons, P.B.: Distinct sampling for highly-accurate answers to distinct values queries and event reports. In: Proceedings of the 27th International Conference on Very Large Data Bases (VLDB 2001), pp. 541–550. Morgan Kaufmann Publishers, San Francisco (2001)
11. Hoare, C.A.R.: Algorithm 65: find. Commun. ACM 4(7), 321–322 (1961)
12. Lingas, A.: A fast output-sensitive algorithm for boolean matrix multiplication. In: Fiat, A., Sanders, P. (eds.) ESA 2009. LNCS, vol. 5757, pp. 408–419. Springer, Heidelberg (2009)
13. Motwani, R., Raghavan, P.: Randomized Algorithms. Cambridge University Press, Cambridge (1995)
14. Yuster, R., Zwick, U.: Fast sparse matrix multiplication. ACM Trans. Algorithms 1(1), 2–13 (2005)

Low Rate Is Insufficient for Local Testability

Eli Ben-Sasson* and Michael Viderman

Computer Science Department
Technion — Israel Institute of Technology
Haifa, 32000, Israel
{eli,viderman}@cs.technion.ac.il

Abstract. Three results are shown regarding locally testable and locally decodable linear codes. All three results rely on the observation that repetition codes have the same local testability and local decodability parameters as the unrepeated base code used to create them.

The first two results deal with families of sparse linear codes, i.e., codes with dimension logarithmic in the code blocklength n. Such codes have been shown by Kaufman and Sudan [8] to be locally testable and decodable as long as all nonzero codewords have Hamming weight $n \cdot \left(\frac{1}{2} \pm n^{-\Omega(1)} \right)$. Our first result shows that certain sparse codes are neither locally testable, nor locally decodable. This refutes a conjecture of Kopparty and Saraf [9] which postulated that all sparse codes are locally testable. Our second result shows that the result of Kaufman and Sudan is surprisingly tight, and for any function $h(n) = o(1)$ there exist families of sparse codes all of whose codewords have weight $n \cdot \left(\frac{1}{2} \pm n^{-h(n)} \right)$ and these codes are neither locally testable, nor locally decodable.

Our third and final result is about the redundancy of locally testable codes. Informally, the redundancy of a locally testable code is the minimal number of redundant tests sampled by a tester, where a test is said to be redundant if is a linear combination of other tests. Ben-Sasson et al. [1] introduced the notion of redundancy and showed that for every linear locally testable code the redundancy is at least linear in the dimension of the code. Our last result shows that redundancy is indeed a function of the code dimension, not blocklength, and that the bound given in [1] is nearly tight.

1   Introduction

This paper deals with locally testable, and locally decodable, linear codes, which have drawn much attention in recent years in theoretical computer science (cf. [12]). We show tightness of various parameters associated with locally testable

* The research leading to these results has received funding from the European Community's Seventh Framework Programme (FP7/2007-2013) under grant agreement number 240258. Research of both authors supported by grant number 2006104 by the US-Israel Binational Science Foundation and by grant number 679/06 by the Israeli Science Foundation.

M. Serna et al. (Eds.): APPROX and RANDOM 2010, LNCS 6302, pp. 420–433, 2010.

(decodable) codes and refute a recently suggested conjecture regarding them. All results in this paper rely on the same simple observation: If C is a code and C' is its t-wise repetition, obtained by repeating each codeword of C a t number of times, then C and C' have essentially the same local-testability and local-decodability parameters. The rate of the repeated code is, however, smaller, and we use this fact in our proofs time and again.

Given a linear code $C \subseteq \mathbf{F}^n$, the dimension of C, denoted by $\dim(C)$, is its dimension as a vector space and its distance, denoted by $\Delta(C)$, is the minimal Hamming distance between two closest different codewords. A family of linear codes $\mathcal{C} = \{C^{(n)} \subset \mathbf{F}_2^n\}_{n \in \mathbf{Z}}$ is called *sparse* if $\dim(C^{(n)}) = O(\log n)$. We say that a code C is ϵ-*biased* if all nonzero codewords of C have relative weight in the range $(\frac{1}{2} - \epsilon, \frac{1}{2} + \epsilon)$, where ϵ may be a function of n. (Notice that an ϵ-biased code has relative distance is at least $\frac{1}{2} - \epsilon$.)

Inverse polynomial bias is necessary for local testability and decodability of sparse codes. Kaufman and Sudan [8] showed that local testability and decodability exists in random sparse linear codes. They showed that for any constant $\gamma > 0$ all sparse linear codes with relative distance $\frac{1}{2} - n^{-\gamma}$ are (strongly) locally testable. They also showed that for any constant $\gamma > 0$ all sparse $n^{-\gamma}$-biased linear codes are locally decodable [1]. In particular, sparse random linear codes have low-bias and hence are locally testable and decodable. This result was later generalized by Kopparty and Saraf [10] to the problem known as "local list-decoding and testing in the high error regime" (see [10] for the definition and discussion of the problem), i.e., they proved that all sparse $n^{-\Omega(1)}$-biased linear codes are locally testable and locally list-decodable even in the high error regime.

One of our results (Theorem 5) shows that the results of [8,10] are surprisingly tight. We show a family of sparse $n^{-o(1)}$-biased linear codes which are non-locally testable/decodable. In plain words, relaxing the bias-requirement of $n^{-\Omega(1)}$ to any slightly larger bias of the form $n^{-o(1)}$ implies that local testability and decodability of sparse codes are no longer guaranteed, even in the easier to obtain, low-error, regime.

Sparse codes are not necessarily locally testable/decodable. Kopparty and Saraf [9] conjectured that all families of sparse linear codes — even those with sub-constant relative distance — are locally testable. In particular, this conjecture suggested that the result of Kaufman and Sudan could be extended to all sparse linear codes.

Another result of this paper (Theorem 4) refutes several plausible relaxations of this conjecture by showing that for any $d(n)$ ranging from $\omega(1)$ to $\Omega(n)$ there exists a family of linear codes $\{C^{(n)} \subset \mathbf{F}_2^n\}_{n \in \mathbf{Z}}$ with linear distance and $\dim(C^{(n)}) = \Theta(d(n))$ which are non-locally testable/decodable.

Tester redundancy is proportional to code dimension and unrelated to code block-length. Ben-Sasson et al. suggested in [1] to study the *redundancy* of a locally

---

[1] In fact, Kaufman and Sudan [8] proved a stronger result. They showed that sparse "low-bias" linear codes are locally self-correctable and thus are locally decodable.

testable code. A tester for a linear code can be viewed as a distribution over words in the dual code that have small weight. The support of a tester is the set of (small-weight) dual words sampled by the tester and the redundancy of the tester is the number of dual words in its support that are linearly dependent on the rest of the support (they are called redundant because they are not required for characterizing the code). The redundancy of a locally testable code is the minimal redundancy of a tester of the code. The main result of [1] was to show that the redundancy of any locally testable code must be linear in the *dimension* of the code. This dependence seems a bit strange and one could imagine the redundancy being linear in the *blocklength* of the code. Our final result (Theorem 6) says that redundancy should indeed be related to code dimension, and not to blocklength, and that the dependence showed in [1] is tight up to a polylogarithmic factor in the dimension.

Organization of the paper. In the following section we provide standard definitions regarding locally testable and locally decodable codes. In Section 3 we state our main results. Section 4 contains our main observation — that repetition does not affect testability/decodability parameters. Finally, in Section 5 we use this observation to prove our main theorems.

2 Definitions

Notation. Let $\mathbf{F}$ be a finite field and $[n]$ be the set $\{1, \ldots, n\}$. Let $C \subseteq \mathbf{F}^n$ be a linear code over $\mathbf{F}$ (In this work, we consider only linear codes). For $w \in \mathbf{F}^n$, let $supp(w) = \{i \in [n] \mid w_i \neq 0\}$ and $|w| = |supp(w)|$. We define the relative distance between two words $x, y \in \mathbf{F}^n$ to be $\delta(x, y) = \frac{\Delta(x,y)}{n}$ and let $\Delta(x, y) = \delta(x, y) \cdot n$. The distance of a code is denoted by $\Delta(C)$ and defined to be the minimal value of $\Delta(x, y)$ for two distinct codewords $x, y \in C$. Similarly, the relative distance of the code is denoted $\delta(C) = \frac{\Delta(C)}{n}$. For $x \in \mathbf{F}^n$ and $C \subseteq \mathbf{F}^n$, let $\delta(x, C) = \delta_C(x) = \min_{y \in C} \{\delta(x, y)\}$ denote the relative distance of x from the code C. We note that $\Delta(C) = \min_{c \in C \setminus \{0\}} \{wt(c)\}$. For two linear codes $C_1, C_2 \subseteq \mathbf{F}^n$ we let $\delta(C_1, C_2) = \min_{c_1 \in C_1 \setminus \{0\}} \{\delta(c_1, C_2)\}$. If $\delta(x, C) \geq \epsilon$, we say that x is ϵ-far from C and otherwise x is ϵ-close to C. Let $dim(C)$ be the dimension of C. For $u = (u_1, u_2, \ldots, u_n), v = (v_1, v_2, \ldots, v_n) \in \mathbf{F}^n$ let $\langle u, v \rangle$ denote the bilinear function from $\mathbf{F}^n \times \mathbf{F}^n$ to $\mathbf{F}$ defined by $\langle u, v \rangle = \sum_{i=1}^{n} u_i v_i$ The dual code $C^\perp$ is defined as $C^\perp = \{u \in \mathbf{F}^n \mid \forall c \in C : \langle u, c \rangle = 0\}$. In a similar way we define $C_{\leq t}^\perp = \{u \in C^\perp \mid |u| \leq t\}$ and $C_t^\perp = \{u \in C^\perp \mid |u| = t\}$. For $w \in \mathbf{F}^n$ and $S = \{j_1, j_2, \ldots, j_m\} \subseteq [n]$ we let $w|_S = (w_{j_1}, w_{j_2}, \ldots, w_{j_m})$, where $j_1 < j_2 < \ldots < j_m$, be the restriction of w to the subset S. Similarly, we let $C|_S = \{c|_S \mid c \in C\}$ denote the projection of the code C onto S. We say that a code C has a q-characterization if $span(C_{\leq q}^\perp) = C^\perp$. For $w \in F^n$ and $t \in \mathbf{N}$ let $w^{(t)} \in \mathbf{F}^{nt}$ be the concatenation of w to itself t times. For $S \subseteq \mathbf{F}^n$, where S is not a vector space, with some abuse of notation we let $dim(S) = dim(span(S))$.

2.1   Locally Testable and Locally Decodable Codes

We use the definition of a (linear) locally testable code provided in [4]. The justification for this definition is given there. The notion of redundancy is defined (and explained) in [1].

Definition 1 ((Locally testable codes (LTCs))). *A linear code $C \subseteq \mathbf{F}^n$ is said to be a (q, ϵ, δ)-LTC if it has a (q, ϵ, δ)-tester, defined next. A (q, ϵ, δ)-tester for C is a distribution $\mathcal{D}$ over subsets $I \subseteq [n]$ such that $|I| \leq q$ and for all $w \in \mathbf{F}^n$ that is δ-far from C*

$$\Pr_{I \sim \mathcal{D}}[w|_I \notin C|_I] \geq \epsilon.$$

(Note that I defines $C|_I$ which is a linear subspace of $\mathbf{F}^I$.)

We shall say that the tester *accepts* w when the set I selected according to $\mathcal{D}$ is such that $w|_I \in C|_I$, otherwise we say the tester *rejects* w.

A family of codes $\{C^{(n)} \mid n \in \mathbf{Z}\}$ is *locally testable* if there exist constants $q, \epsilon, \delta > 0$ such that for infinitely many n it holds that $C^{(n)} \subseteq \mathbf{F}^n$ is a (q, ϵ, δ)-LTC.

Remark. Usually we assume that $\delta \leq 1/3$ (see the discussion in [1] regarding this issue). So later on we shall say that a family of codes is *not* locally testable if for all constants $q, \epsilon > 0$, large enough n and distribution $\mathcal{D}_n$ over subsets $I \subseteq [n]$, such that $|I| \leq q$ there is a word w that is $(1/3)$-far from $C^{(n)}$ and $\Pr_{I \sim \mathcal{D}_n}[w|_I \notin C|_I] < \epsilon$.

Ben-Sasson et al. [1] defined the redundancy of a tester using the following definition. We point out that this definition assumes that a q-query tester is a distribution over dual words of support at most q. While this definition seems to be less general than Definition 1 (for instance, a tester according to the former definition can select I that is the support of two words in the dual code), it turns out that the two definitions are essentially equivalent (see [1, Section 2] for details).

Definition 2 ((Distributions and redundancy)). *The support of the distribution $\mathcal{D}$ over $C^\perp_{\leq q}$ is*

$$\mathcal{D}_S = \{u \in C^\perp_{\leq q} \mid \mathcal{D}(u) > 0\}.$$

The redundancy *of the distribution $\mathcal{D}$ is defined to be* $\mathrm{Redun}(\mathcal{D}) = |\mathcal{D}_S| - \dim(\mathcal{D}_S)$.

Now we define locally decodable codes.

Definition 3 ((Locally decodable codes (LDCs) and decoders)). *Let $C \subseteq \mathbf{F}_2^n$ and $E_C : \mathbf{F}_2^k \to \mathbf{F}_2^n$ be its encoding function, i.e., $C = \{E_C(x) \mid x \in \mathbf{F}_2^k\}$. Then C is a (q, ϵ, δ)-LDC if there exists a randomized decoder (**D**) such that:*

- *In every invocation, **D** makes at most q queries.*
- *For all $x \in \mathbf{F}^k$, $i \in [k]$ and $\hat{c} \in \mathbf{F}^n$ such that $\Delta(E_C(x), \hat{c}) \leq \delta n$ we have*
  $$\mathbf{Pr}\big[\mathbf{D}^{\hat{c}}[i] = x_i\big] \geq \frac{1}{2} + \epsilon,$$ *i.e., with probability at least $\frac{1}{2} + \epsilon$ entry x_i will be recovered correctly.*

Note that the definition implies that $\delta < \delta(C)/2$ and $\epsilon \leq \frac{1}{2}$. We can assume without loss of generality that the decoder for a linear code is non-adaptive [5]. We say that a family of codes $\big\{ C^{(n)} \mid n \in \mathbf{Z} \big\}$ is locally decodable if there exist constants $q, \epsilon, \delta > 0$ such that for infinitely many n it holds that $C^{(n)} \subseteq \mathbf{F}^n$ is a (q, ϵ, δ)-LDC.

Informally, locally correctable codes (LCCs) allow to recover each entry of a codeword with high probability by reading only a few entries of the codeword even if a large fraction of it is adversely corrupted (see [8]). It is well-known that q-query LCCs are also q-query LDCs so if we show that a code C is not q-query LDC then it is not q-query LCC either.

Remark. Notice that a message space M for an LDC can be a strict subspace of $\mathbf{F}^k$, i.e., not every word in $\mathbf{F}^k$ is a message. The linearity of C the message space M must be a linear subspace and the dimension of the code is defined to be $\dim(M)$.

3   Main Results

In this section we present our main results. Theorem 4 refutes several possible formulations of the conjecture of Kopparty and Saraf [9], which says that all sparse linear codes are locally testable. Theorem 5 shows the surprising tightness of the results in [8]. Theorem 6 shows that the redundancy of a tester should be measured as a function of the dimension of the code, not its blocklength.

3.1   Sparse Biased Codes Can Be Non-locally Testable

Our first main result that there exists a family of sparse linear codes with relative distance ≥ 0.49 which are non-locally testable (decodable). One could conjecture that all sparse linear codes with small characterization are locally testable or decodable. So, Theorem 4 also shows that there are sparse linear codes which have a small characterization and linear distance, but are non-locally testable (decodable).

We also show that there are sparse linear codes with dual distance strictly greater than 2 which are non-locally testable (decodable). We believe that this case is interesting since it shows that non-redundant[2] sparse codes can be non-locally testable (decodable). Equivalently, picking a large fraction of columns

---

[2] The term "non-redundant" here means the dual distance of the code is at least 3. Note that if the dual distance of a code is 1 then some of its bits are identically 0, and if the dual distance is 2 then some pairs of bits are equal to each other and hence one of each pair is redundant.

from the Hadamard generating matrix $(G \in F_2^{[n] \times [2^n]})$ may result in a generating matrix of a code which is *non*-locally testable (decodable). This contrasts with the result of [8] which says that taking a large *random* fraction of columns from the Hadamard generating matrix gives, with high probability, a generating matrix of a locally testable (decodable) code.

Theorem 4 ((Low Rate does not imply LTC or LDC)). *For every $q, \epsilon > 0$, function $w(1) \leq d(n) \leq O(n)$ and infinitely many $n > 0$*

1. *There exists $C_\epsilon \subset \mathbf{F}_2^n$ such that $\delta(C) \geq \frac{1}{2} - \epsilon$, $\dim(C_\epsilon) = \Theta(d(n))$ and C_ϵ is non-locally testable with $o(d(n))$ queries and is not-locally decodable with q queries.*
2. *There exists $C \subset \mathbf{F}_2^n$ such that $\dim(C) = \Theta(d(n))$, $\Delta(C) \geq \Omega(n)$, $span(C_{\leq 3}^\perp) = C^\perp$ and C is non-locally testable with $o(d(n))$ queries and is non-locally decodable with q queries.*
3. *There exists $C \subset \mathbf{F}_2^n$ such that $\dim(C) = 1.1 \log(n)$, $\Delta(C) \geq \Omega(n)$, $\Delta(C^\perp) > 2$ and C is non-locally testable with $o(\log(n))$ queries and is non-locally decodable with q queries.*

Remark. Folklore claim 17 states that every linear code C is testable by $\dim(C) + 1$ queries. Intuitively, C is non-trivially testable if it can be testable with $o(\dim(C))$ queries. So, Theorem 4 shows the families of linear codes that cannot be non-trivially testable.

Remark. The construction in the third bullet of Theorem 4 cannot achieve dimension lower than $\log(n)$, since every linear code $C \subseteq \mathbf{F}_2^n$ such that $\dim(C) < \log(n)$ has $\Delta(C^\perp) \leq 2$.

3.2   Inverse Polynomial Bias Is Necessary for Local Testability and Decodability of Sparse Codes

Recall that Kaufman and Sudan showed that sparse linear codes with bias $n^{-\Omega(1)}$ are locally testable and decodable. Our next main result is that the requirement of such small bias is necessary because for every $h(n) = o(1)$ there exists a family of sparse linear codes with bias $n^{-h(n)}$ which are non-locally testable (decodable).

Theorem 5 ((Inverse polynomial bias needed for sparse LTCs (LDCs))). *For every constant $q > 0$, computable function $h(n) = o(1)$ and infinitely many $n > 0$ there exists $C \subset \mathbf{F}_2^n$ such that $\dim(C) = \log(n)$, C is $n^{-h(n)}$-biased and C is non-locally testable (decodable) with q queries.*

3.3   Redundancy Is Proportional to Dimension, Not to Blocklength

[1] showed that if $C \subseteq \mathbf{F}^n$ is a LTC, $\dim(C) \geq w(1)$ and $\mathcal{D}$ is its tester then $\text{Redun}(\mathcal{D}) \geq \Omega(\dim(C))$. However, the only known upper bound on $\text{Redun}(\mathcal{D})$ was $\text{Redun}(\mathcal{D}) = O(n)$.

One could conjecture a stronger lower bound on redundancy, of the form $\text{Redun}(\mathcal{D}) \geq \Omega(n)$. We refute this conjecture in our last main theorem that the bound $\text{Redun}(\mathcal{D}) \geq \Omega(\dim(C))$ is nearly tight. In plain words, this theorem proves that the redundancy of a tester can be close to the dimension of the tested code, up to a multiplicative polylogarithmic factor.

Theorem 6. *For every* $\omega(1) \leq f(n) \leq n/\text{poly}(\log n)$ *there exists a locally testable code* C *with tester* $\mathcal{D}$ *such that* $\text{Redun}(\mathcal{D}) \leq O(f(n))$, *where* $\delta(C) \geq \Omega(1)$ *and* $\dim(C) \geq f(n)/\text{poly}(\log f(n)) \geq \omega(1)$.

4   Repetition Does Not Affect LTCs and LDCs

In this section we show that repeating codewords does not affect testability and decodability. To do this we first define projected testers and decoders and then prove our main propositions — Propositions 8 and 9.

4.1   Projected Testers and Decoders

Definition 7 ((Repetition Code)). *Let* $R \subseteq \mathbf{F}^m$ *be a linear code and* $t > 0$. *We say that* $C \subseteq \mathbf{F}^{(im)}$ *is the* t-*wise repeated code of* R, *or, simply, the* t-*repetition of* R *if every codeword of* C *is a codeword of* R *repeated* t *times. Formally,* $c \in C$ *if and only if* $c = r^{(t)}$ *for some* $r \in R$.

Notice that the linearity of R implies the linearity of its repetition.

Projected testers and decoders. Let $R \subset \mathbf{F}_2^m$ be a linear code and $C \subset \mathbf{F}_2^{(mt)}$ be its t-repetition. Let $n = m \cdot t$. For $I \subseteq [n]$ let $I \bmod m = \{i \bmod m \mid i \in I\} \subseteq [m]$. Given a tester $\mathcal{D}_C$ for C we define its *projected* tester by the distribution obtained from picking tests $I \bmod m$ where $I \sim \mathcal{D}_C$.

The *projected decoder* is defined in a similar way, i.e., invokes the decoder of C and queries all indices modulo m.

Note that if $w \in \mathbf{F}_2^m$ then the tester (decoder) for C will view the same values on the t-wise repetition of w as the projected tester (decoder) views on w. In this way, if tester for C rejects $w^{(t)}$ then the tester for R rejects w and if the decoder for C recovers correctly from $w^{(t)}$ a certain message bit then the decoder for R recovers correctly the same message bit.

4.2   Main Propositions

We prove a number of simple and important propositions which say that repetition does not affect the testability and the decodability of a code. These propositions (Proposition 8 and Proposition 9) are shown for binary codes but can be easily extended to any field.

Proposition 8. *Let* $R \subset \mathbf{F}_2^m$ *be a linear code and* $t > 0$ *be an integer. Let* $C \subset \mathbf{F}_2^{(mt)}$ *be a* (t)-*repetition of* R. *Then,*

- if R is a (q, ϵ, δ)-LTC then C is a $(q, \min\{\epsilon/2, \delta/2\}, 2\delta)$-LTC
- if C is a (q, ϵ, δ)-LTC then R is a (q, ϵ, δ)-LTC.

Proof. Note that $C|_{[m]} = R$. For the first part, assume that R is a (q, ϵ, δ)-LTC and let $\mathcal{D}_R$ be a (q, ϵ, δ)-tester for R. We define the following tester $\mathcal{D}_C$ for C.

- Flip a coin
- **If** "heads,"
    - pick $j \in [m]$ and $\ell_1 \in [t-1]$ independently at random,
    - pick $I = \{j, j + m \cdot \ell_1\}$ (note that $I \subseteq [mt]$);
- **Else** pick $I \sim \mathcal{D}_R$ (note that $I \subseteq [m]$).

We argue that $\mathcal{D}_C$ is a $(q, \min\{\epsilon/2, \delta/2\}, 2\delta)$-tester for C. Let $w \in \mathbf{F}^{(mt)}$ be a word such that $\delta(w, C) \geq 2\delta$. If $\delta(w|_{[m]}, C|_{[m]}) = \delta(w|_{[m]}, R) \geq \delta$ we are done, since

$$\Pr_{I \sim \mathcal{D}_C} [w|_I \notin C|_I] \geq \frac{1}{2} \cdot \Pr_{I \sim \mathcal{D}_R} [(w|_{[m]})|_I \notin R|_I] \geq \frac{\epsilon}{2}.$$

Otherwise we have $\delta(w|_{[m]}, C|_{[m]}) = \delta(w|_{[m]}, R) < \delta$.
But $\delta(w, C) \geq 2\delta$ implies that

$$\mathop{\mathbf{E}}_{j \in [t-1]} \left[\delta(w|_{\{jm+1, \dots, (j+1)m\}}, w|_{[m]})\right] \geq 2\delta - \delta = \delta.$$

Hence $\Pr_{I \sim \mathcal{D}_C} [w|_I \notin C|_I] \geq \frac{1}{2} \Pr_{j \in [m], \ell_1 \in [t-1]} \left[w|_{\{j, j+m \cdot \ell_1\}} \notin C|_{\{j, j+m \cdot \ell_1\}}\right] \geq \delta/2$. This completes the proof of the first part and now we deal with the second part.

Assume that C is a (q, ϵ, δ)-LTC and let $\mathcal{D}_C$ be its (q, ϵ, δ)-tester. Let $\mathcal{D}_R$ be a projected tester of C. Note that $\mathcal{D}_R$ is a distribution over subsets $I \subseteq [m]$ such that $|I| \leq q$.

We argue that $\mathcal{D}_R$ is a (q, ϵ, δ)-tester for R. Let $w \in F_2^m$ be a word such that $\delta(w, R) \geq \delta$. Assume by way of contradiction that $\Pr_{I \sim \mathcal{D}_R} [w|_I \notin R|_I] < \epsilon$. Notice that $\delta(w^{(tm)}, C) = \delta(w, R) \geq \delta$. We have $\Pr_{I \sim \mathcal{D}_C} \left[w^{(tm)}|_I \notin C|_I\right] < \epsilon$ since if for $I \subset [im]$ it holds that $w^{(tm)}|_I \notin C|_I$ then $w|_{I \bmod m} \notin R$. We conclude that $\mathcal{D}_C$ is not a (q, ϵ, δ)-distribution for C. Contradiction.

Proposition 9. *Let $R \subset \mathbf{F}_2^m$ be a code such that the first $\dim(R)$ bits of R are message bits. Let $t > 0$ and C be an (t)-repetition of R. Then,*

- *If R is not (q, ϵ, δ)-LDC then C is not (q, ϵ, δ)-LDC.*
- *If R is a (q, ϵ, δ)-LDC then C is a $(q, \epsilon/2, \frac{\epsilon\delta}{2})$-LDC.*

Proof. Let $k = \dim(R) = \dim(C)$. Assume without loss of generality that the first k bits of C are message bits. For the first part, if R is not a (q, ϵ, δ)-LDC then for every q-query decoder there exists a word $w \in \mathbf{F}_2^m$ and $i \in [k]$ such that $\delta(w, R) \leq \delta$ and the probability that the decoder recovers correctly the ith message bit is less than $\frac{1+\epsilon}{2}$.

Assume by way of contradiction that C is a (q, ϵ, δ)-LDC with decoder $\mathbf{D}_C$. Let $\mathbf{D}_R$ be the projected decoder of $\mathbf{D}_C$. But then there exists $w \in \mathbf{F}^m, \delta(w, R) \leq \delta$

such that the decoder $\mathbf{D}_R$ recovers some message bit i with probability less than $\frac{1+\epsilon}{2}$. But $\mathbf{D}_C$ views the same values on $w^{(n/m)}$ as $\mathbf{D}_R$ views on w and moreover $\delta(w^{(n/m)}, C) = \delta(w, R) \leq \delta$. Thus $\mathbf{D}_C$ is not a (q, ϵ, δ)-decoder for C. Contradiction.

For the second bullet assume R is (q, ϵ, δ)-LDC and let $\mathbf{D}_R$ be its (q, ϵ, δ)-decoder. For $j \in [t]$ and $I = \{i_1, i_2, \ldots, i_q\} \subseteq [m]$ let $jm + I = \{jm + i_1, jm + i_2, \ldots, jm + i_q\}$. Note that $jm + I \subseteq \{mj + 1, \ldots, m(j + 1))\}$. The decoder $\mathbf{D}_C$ for C recovers message bit ℓ from the given word $w \in \mathbf{F}^{im}$ as follows:

- Pick random $j \in [t - 1]$ and select $I = jm + [m]$,
- Return $\mathbf{D}_R^{(w|_I)}[\ell]$, i.e., return the value output by $\mathbf{D}_R[\ell]$ on $w|_I$.

We argue that if $\delta(w, C) \leq \frac{\epsilon\delta}{2}$ then $\mathbf{D}_C$ recovers correctly the ℓth message bit of C with probability at least $\frac{1}{2} + \frac{\epsilon}{2}$. Let $r^{(t)} \in C$ be a codeword of C closest to w, i.e., $\delta(w, r^{(t)}) = \delta(w, C) \leq \frac{\epsilon\delta}{2}$.

For $j \in [t]$ we say that $w|_{(jm+[m])}$ is a j-block of w. We say that j-block is corrupted if $\delta(w|_{jm+[m]}, r) > \delta$. The fraction of corrupted blocks is bounded by $\frac{\epsilon}{2}$, because otherwise we have $\delta(w, r^{(t)}) > \frac{\epsilon\delta}{2}$. Recall that the decoder $\mathbf{D}_C$ for C picks random $j \in [t - 1]$ and invokes the decoder for R on the j-block of w. The probability that $\mathbf{D}_R$ will be invoked on a non-corrupted block and will recover a message bit correctly is at least $(\frac{1}{2} + \epsilon) \cdot (1 - \frac{\epsilon}{2}) \geq \frac{1}{2} + \epsilon - \frac{\epsilon}{2} = \frac{1+\epsilon}{2}$, where the inequality follows since $0 \leq \epsilon \leq \frac{1}{2}$.

Summing up the previous propositions we get:

Corollary 10. *Let $R \subset \mathbf{F}_2^m$ be a linear code and $t > 0$ be an integer. Let $C \subset \mathbf{F}_2^{(mt)}$ be an (t)-repetition code of R. Then,*

- *if R is not (q, ϵ, δ)-LTC then C is not (q, ϵ, δ)-LTC.*
- *if R is not (q, ϵ, δ)-LDC then C is not (q, ϵ, δ)-LDC.*

5   Proof of Main Results

5.1   Sparse Biased Codes Are Not Necessarily Locally Testable/Decodable — Proof of Theorem 4

Proof of the first bullet of Theorem 4. Let $\epsilon > 0$ be a constant and $m = d(n) \geq w(1)$. Let $R_\epsilon \subset \mathbf{F}_2^m$ be a linear code such that $\delta(R_\epsilon) \geq \frac{1}{2} - \epsilon$ and $\delta(R_\epsilon^\perp) \geq \Omega(1)$ and $\dim(R_\epsilon) = \Theta(m)$ (e.g., a random linear code of constant rate will have these properties). Claim 16 implies that R_ϵ is non-locally testable with $o(m)$ queries. Lower bounds on the blocklength of locally decodable codes from [7] imply that R_ϵ is not q-query locally decodable code.

Let $C_\epsilon \subset \mathbf{F}_2^n$ be a (n/m)-repetition code of R_ϵ. We have $\dim(C_\epsilon) = \Theta(m) = \Theta(d(n))$ and $\delta(C_\epsilon) \geq 1/2 - \epsilon$. Furthermore, Corollary 10 implies that C_ϵ is not locally testable with $o(m)$ queries and non-locally decodable with q queries and the proof is complete.

Proof of the second bullet of Theorem 4. Let $m = d(n) \geq w(1)$. Then, for sufficiently large m, Theorem 18 implies the existence of a linear code $R \subset \mathbf{F}_2^m$ such that $\dim(R) = \Theta(d(n))$, $\Delta(R) \geq \Omega(m)$, $span(R_{\leq 3}^\perp) = R^\perp$ and R is non-locally testable with $o(m)$ queries and non-locally decodable with q queries.

Let $C \subset \mathbf{F}_2^n$ be a (n/m)-repetition code of R. Then Proposition 8 implies that C is non-locally testable with $o(m) = o(d(n))$ queries. Proposition 9 implies that C is non-locally decodable with q queries. Notice that $span(C_{\leq 3}^\perp) = C^\perp$. This completes the proof.

Proof of the third bullet of Theorem 4. Given two linear codes $C_1, C_2 \subseteq \mathbf{F}^n$, let $C_1 + C_2 = \{c_1 + c_2 \mid c_1 \in C_1, c_2 \in C_2\}$. Notice that $(C_1 + C_2)$ is a linear code.

We start from a straightforward proposition that will be useful in the next theorem.

Proposition 11. *Let $C_1, C_2 \subseteq \mathbf{F}_2^n$ be two linear binary codes. Then $(C_1 + C_2)^\perp = C_1^\perp \cap C_2^\perp$ and so $(C_1 + C_2)^\perp \subseteq C_1^\perp$ and $(C_1 + C_2)^\perp \subseteq C_2^\perp$.*

Proof. We have $u \in (C_1 + C_2)^\perp$ iff $(u \in C_1^\perp$ and $u \in C_2^\perp)$ iff $u \in (C_1^\perp \cap C_2^\perp)$.

It follows that if C_1 is a repetition code but C_2 is not then $C_1 + C_2$ is not repetition code. Moreover, if there is a "small-size" intersection between low-weight dual words of C_1 and of C_2 then $C_1 + C_2$ will have a small number of low-weight dual words and hence, intuitively will not be a locally testable code.

Claim 12. *Let $C \subseteq \mathbf{F}^n$ be a linear code and a (q, ϵ, δ)-LDC. Assume that $C' \subset C$ is a linear code (subcode of C). Then C' is a (q, ϵ, δ)-LDC.*

Proof. Assume C is associated with the (linear) message space $S \subseteq F^k$ and its decoder is $\mathbf{D}$. Let $S' \subset S$ be a (linear) message space for C'. We argue that C' has the same decoder $\mathbf{D}$. Let w be δ-close to C' (δ-close to the encoding of some message $m \in S'$). Then w is δ-close to C and thus for all $i \in [k]$ the decoder $\mathbf{D}$ recovers correctly the message entry (m_i) with probability at least $\frac{1}{2} + \epsilon$.

Notice that the message space S' of the subcode C' will have smaller dimension than S, i.e., $\dim(S') < \dim(S)$. S' is a linear vector space because for every two messages $x_1, x_2 \in S'$ which are encoded to $c_1, c_2 \in C'$, respectively, we have $(x_1 + x_2) \in S'$ and $(x_1 + x_2)$ is encoded to $c_1 + c_2$. For every $\alpha \in \mathbf{F}$ we also have that $\alpha x_1 \in S'$ and is encoded to αc_1.

Proposition 13. *Let $C_1, C_2 \subseteq \mathbf{F}_2^n$ be linear codes. If C_1 is non-locally decodable with q queries then $C_1 + C_2$ is non-locally decodable with q queries.*

Proof. If $C_1 + C_2$ is locally decodable with q queries then C_1 is locally decodable with q queries by Claim 12, because C_1 is a subcode of $C_1 + C_2$.

We are ready to prove the third bullet of Theorem 4. Let $m = \log(n)$ and $R \subset \mathbf{F}_2^m$ be a linear code with $\Delta(R) \geq m/5$, $\Delta(R^\perp) \geq \Theta(m)$ and $\dim(R) = m/10$ (e.g., a random linear code of constant rate will have these properties). Claim 16 implies that R is non-locally testable with $o(m)$ queries and in particular non-locally testable with q queries.

Let $w \in \mathbf{F}_2^m$ be a word such that $\delta(w, R) \geq \frac{1.1}{3}$ (a random $w \in \mathbf{F}^m$ satisfies this condition with high probability). Notice that for all $u \in R_{\leq o(m)}^{\perp}$ we have $\langle u, w \rangle = 0$ because $R^{\perp}$ has no words of weight less then $o(m)$, i.e., $R_{\leq o(m)}^{\perp} = \emptyset$.

Let $C_1 \subset \mathbf{F}_2^n$ be a (n/m)-repetition code of R. Then $\delta(w^{(n/m)}, C_1) = \delta(w, R) \geq \frac{1.1}{3}$. Notice that by construction for all $u \in C_1{}_{\leq o(m)}^{\perp}$ we have $\langle u, w^{(n/m)} \rangle = 0$. We also have $\delta(C_1) \geq 1/5$ and $\dim(C_1) = \log(n)/10$. Let $C_2 \subseteq \mathbf{F}_2^n$ be the Hadamard code[3] (assume w.l.o.g. that n is a power of 2). Note that $\Delta(C_2^{\perp}) > 2$.

Let $\pi : [n] \mapsto [n]$ be a permutation. With some abuse of notation, for $w = (w_1, w_2, \ldots, w_n) \in \mathbf{F}^n$ let $\pi(w) = (w_{\pi(1)}, w_{\pi(2)}, \ldots, w_{\pi(n)})$ be a π-permuted word. Let $\pi(C_2) = \{\pi(c) \mid c \in C_2\}$ be a set of all permuted codewords of C_2. Note that for every permutation $\pi : [n] \mapsto [n]$ it holds that $\delta(\pi(C_2)) = \delta(C_2)$, $\dim(\pi(C_2)) = \dim(C_2)$ and $\Delta((\pi(C_2))^{\perp}) = \Delta(C_2^{\perp}) > 2$.

Recall that $\delta(C_1, C_2) = \min\limits_{c_1 \in C_1 \setminus \{0\}} \{\delta(c_1, C_2)\}$. We say that a permutation $\pi : [n] \mapsto [n]$ is *good* if $\delta(w^{(n/m)}, C_1 + \pi(C_2)) \geq \frac{1}{3}$ and $\delta(C_1, \pi(C_2)) \geq 1/10$.

We argue that a random permutation $\pi : [n] \mapsto [n]$ is good with probability at least $1 - o(1)$. It is sufficient to show that a random permutation $\pi : [n] \mapsto [n]$ is bad with probability at most $o(1)$.

By the Chernoff inequality the probability that for some $c_1 \in (C_1 \setminus \{0\})$ and $c_2 \in (C_2 \setminus \{0\})$ we get $\delta(w^{(n/m)}, c_1 + \pi(c_2)) < 1/3$ with probability at most $\frac{1}{2^{\Omega(m)}}$. Note that if $c_2 = 0$ then $\delta(w^{(n/m)}, c_1 + \pi(c_2)) = \delta(w^{(n/m)}, c_1) \geq \frac{1.1}{3}$ by construction.

Take a union bound over all $c_1 \in (C_1 \setminus \{0\})$ and $c_2 \in C_2$ to get that $\delta(w^{(n/m)}, C_1 + \pi(C_2)) < 1/3$ with probability at most $\frac{O(n^2)}{2^{\Omega(n)}} = o(1)$. Moreover, the probability that for given $c_1 \in (C_1 \setminus \{0\})$ and $c_2 \in C_2$ we have $\delta(c_1, \pi(c_2)) \leq 1/10$ is bounded by $\frac{1}{2^{\Omega(n)}}$. Take a union bound over all $c_1 \in (C_1 \setminus \{0\})$ and $c_2 \in C_2$ to get that the probability that $\delta(C_1, \pi(C_2)) \leq 1/10$ is bounded by $\frac{O(n^2)}{2^{\Omega(n)}} = o(1)$.

We conclude that a random permutation $\pi : [n] \mapsto [n]$ is bad with probability at most $o(1)$. So, let $\pi : [n] \mapsto [n]$ be a good permutation. Then, we have $\delta(C_1 + \pi(C_2)) = \delta(C_1, \pi(C_2)) \geq 1/10$ and $\delta(w^{(n/m)}, C_1 + C_2) \geq 1/3$. Proposition 11 implies that $w^{(n/m)}$ satisfies all constraints in $(C_1 + C_2)_{\leq o(m)}^{\perp}$ and thus $w^{(n/m)}$ will be accepted with probability 1 by any tester for $C_1 + \pi(C_2)$ with query complexity $\leq o(m)$. We conclude that $(C_1 + \pi(C_2))$ is non-locally testable with $o(m)$ queries. Notice that $\dim(C_1 + \pi(C_2)) = \dim(C_1) + \dim(C_2) = 1.1 \log(n)$.

The proof for local decodability is almost the same. R is not q-query locally decodable by the lower bound on the blocklength of locally decodable codes of [7]. Corollary 10 implies that C_1 is not q-query locally decodable. Proposition 13 implies that $C_1 + \pi(C_2)$ is not q-query locally decodable.

---

[3] Instead of the Hadamard code we could take any binary, sparse code with linear distance and dual distance > 2.

5.2 Inverse Polynomial Bias Necessary for Local Testability/Decodability — Proof of Theorem 5

In this proof we always assume that $n > 2$. Let $h(n) = o(1)$ such that $h(n) > 0$ for all $n > 2$. Without loss of generality we can assume that $h(n) \geq \frac{1}{3\log(\log(n))}$ (otherwise let $h(n) = \frac{1}{3\log(\log(n))}$) because if $h(n) < \frac{1}{3\log(\log(n))}$ then $n^{-h(n)} > n^{-\frac{1}{3\log(\log(n))}}$, so we will prove the Theorem even for a lower bias than $n^{-h(n)}$. Hence we assume that $h(n) \geq \frac{1}{3\log(\log(n))}$.

Let $g(n) = 3h(n)$ and then $g(n) \geq \frac{1}{\log(\log(n))}$ for all $n > 2$. Let $f(n) = \frac{1}{g(n)} = \omega(1)$, note that $f(n) \leq \log(\log(n))$. Let $m = n^{g(n)}$ and $R \subseteq \mathbf{F}_2^m$ be a random linear code such that $\dim(R) = \log(m) \cdot f(n) = \log(n)$. The probability that at least one nonzero codeword of R has relative weight less than $\frac{1}{2} - m^{-1/3}$ or more than $\frac{1}{2} + m^{-1/3}$ is bounded by $\frac{2 \cdot 2^{(\log(m) \cdot f(n))}}{2^{\Omega(m^{1/3})}} = o(1)$, and this follows from the Chernoff inequality and the union bound. Moreover, the probability that $\Delta(R^\perp) < \log(f(n)) = \omega(1)$ is bounded by $\frac{m^{\log(f(n))}}{2^{\dim(R)}} = o(1)$, and this follows from the union bound. So, let R be a $m^{-1/3}$-biased code such that $\dim(R) = \log(m) \cdot f(n)$ and $\Delta(R^\perp) \geq \log(f(n)) = \omega(1)$, i.e., $R^\perp_{\leq \log(f(n))} = R^\perp_{\leq \omega(1)} = \emptyset$. Assume without loss of generality that the first $\dim(R)$ bits of R are message bits.

Notice that $\log(m) \cdot f(n) = \log(n)$ and $m^{-1/3} = n^{-(1/3)g(n)} = n^{-h(n)}$. Claim 16 implies that R is non-locally testable (decodable) with $q = O(1)$ queries.

Let $C \subseteq \mathbf{F}_2^n$ be the (n/m)-repetition code of R. We have $\dim(C) = \dim(R) = \log(n)$ and C has the same bias as R, i.e., C is $n^{-h(n)}$-biased. In particular we have $\delta(C) \geq \frac{1}{2} - n^{-h(n)}$. Furthermore, Corollary 10 implies that C is not locally testable (decodable) with $q = O(1)$ queries. The Theorem follows.

5.3 Repetition Preserves Redundancy of Testers — Proof of Theorem 6

We prove Theorem 6 by showing (in the following proposition) that the repetition of a code preserves the redundancy of the associated tester.

Proposition 14. *Let $t > 1$ be an integer. Let $R \subset \mathbf{F}_2^m$ be a (q, ϵ, δ)-LTC and let $\mathcal{D}$ be its (q, ϵ, δ)-tester. Let $C \subset \mathbf{F}_2^{(mt)}$ be the (t)-repetition of R. Then C has a $(q, \min\{\epsilon/2, \delta/2\}, 2\delta)$-tester $\mathcal{D}'$ such that $\mathrm{Redun}(\mathcal{D}'_S) = \mathrm{Redun}(\mathcal{D}_S)$.*

Proof. The proof of Proposition 14 is omitted due to space limitations.

[3,6] constructed a family of LTCs of inverse poly-logarithmic rate with linear distance and only $O(1)$ queries (see also [11]). Then, [1] showed that a tester for every locally testable code "uses" only $\leq 3n$ tests. Let us record these two results for future reference.

Proposition 15. *There exist constants $q, \epsilon, \delta > 0$, polynomial $p(\cdot)$ and a family of (q, ϵ, δ)-LTCs $\{R_m\}_{m \in \mathbf{Z}}$ with testers $\mathcal{D}_n$, where $R_m \subseteq \mathbf{F}^m$, $\dim(C) \geq m/p(\log m)$ and $\mathcal{D}_m$ is a (q, ϵ, δ)-tester for R_n. Moreover, $\mathrm{Redun}(\mathcal{D}_m) \leq 3m$.*

We are ready to prove Theorem 6.

Proof of Theorem 6. Let $R \subseteq \mathbf{F}^m$ be the LTC described in Proposition 15 and let $\mathcal{D}_m$ be its tester. The code and tester satisfy $\text{Redun}(\mathcal{D}_m) \leq 3m$, $\dim(R) \geq m/\text{poly}(\log m)$ and $\Delta(R) = \Theta(m)$. Let $t > 0$ be an integer such that $f(mt) = \Theta(m)$ and let $n = mt$ and $C \subseteq \mathbf{F}^n$ be the (t)-repetition code of R. Proposition 14 implies that C is a LTC and has tester $\mathcal{D}_n$ such that $\text{Redun}(\mathcal{D}_n) = \text{Redun}(\mathcal{D}_m) \leq 3m = O(f(mt)) = O(f(n))$. Furthermore, $\dim(C) = \dim(R) \geq m/\text{poly}(\log m) = f(n)/\text{poly}(\log f(n))$.

Acknowledgements. We thank Tali Kaufman, Swastik Kopparty and Shubhangi Saraf for helpful discussions. We would like to thank the anonymous referees for valuable comments on an earlier version of this article.

References

1. Ben-Sasson, E., Guruswami, V., Kaufman, T., Sudan, M., Viderman, M.: Locally testable codes require redundant testers. In: IEEE Conference on Computational Complexity, pp. 52–61. IEEE Computer Society, Los Alamitos (2009)
2. Ben-Sasson, E., Harsha, P., Raskhodnikova, S.: Some 3CNF properties are hard to test. SIAM Journal on Computing 35(1), 1–21 (2005)
3. Ben-Sasson, E., Sudan, M.: Short PCPs with polylog query complexity. SIAM J. Comput. 38(2), 551–607 (2008)
4. Ben-Sasson, E., Viderman, M.: Composition of semi-LTCs by two-wise tensor products. In: Dinur, I., Jansen, K., Naor, J., Rolim, J.D.P. (eds.) APPROX-RANDOM. LNCS, vol. 5687, pp. 378–391. Springer, Heidelberg (2009)
5. Deshpande, A., Jain, R., Kavitha, T., Lokam, S.V., Radhakrishnan, J.: Lower bounds for adaptive locally decodable codes. Random Struct. Algorithms 27(3), 358–378 (2005)
6. Dinur, I.: The PCP theorem by gap amplification. Journal of the ACM 54(3), 12:1–12:44 (2007)
7. Katz, J., Trevisan, L.: On the efficiency of local decoding procedures for error-correcting codes. In: STOC, pp. 80–86 (2000)
8. Kaufman, T., Sudan, M.: Sparse random linear codes are locally decodable and testable. In: FOCS, pp. 590–600. IEEE Computer Society, Los Alamitos (2007)
9. Kopparty, S., Saraf, S.: Tolerant linearity testing and locally testable codes. In: Dinur, I., Jansen, K., Naor, J., Rolim, J.D.P. (eds.) APPROX-RANDOM. LNCS, vol. 5687, pp. 601–614. Springer, Heidelberg (2009)
10. Kopparty, S., Saraf, S.: Local list-decoding and testing of random linear codes from high error. In: Mitzenmacher, M., Schulman, L.J. (eds.) STOC, pp. 417–426. ACM, New York (2010)
11. Meir, O.: Combinatorial construction of locally testable codes. In: STOC, pp. 285–294. ACM, New York (2008)
12. Trevisan, L.: Some applications of coding theory in computational complexity (September 23, 2004)

Appendix

Claim 16 states that the small dual distance of the linear code $C \subseteq \mathbf{F}^n$ is necessary for its local testing and local decoding. We explain this claim now.

[2] showed that a q-query tester for a locally testable code is (w.l.o.g.) a distribution over dual codewords of weight at most q. In particular, if $\Delta(C^\perp) \geq q+1$ we conclude that C is non-locally testable with q queries. Now, assume that the first $\dim(C)$ entries of the code C are message entries and $\Delta(C^\perp) \geq q+1$. Then any local decoder which makes only $q-1$ queries always obtains a "local view" that contains no information about the message entries and hence message entries cannot be recovered with non-trivial probability.

Claim 16 (Folklore). *Let $C \subseteq \mathbf{F}^n$ be a linear code such that $\Delta(C^\perp) \geq (q+1)$, where $q \geq 1$. Assume that the first $\dim(C)$ entries of C are message entries. Then C is non-locally testable with q queries and non-locally decodable with $q-1$ queries.*

The other folklore claim (stated e.g. in [1]) says that every linear code is testable with query complexity equal to its dimension plus one.

Claim 17 (Folklore 2). *Every linear code C is testable by $\dim(C)+1$ queries.*

Let us state the central theorem (which we rephrase) from [2]. [2] showed a family of codes $C_m \subset \mathbf{F}_2^m$ which has linear distance, constant rate and was characterized by 3 weight dual words. They proved that this family is non-locally testable with $o(m)$ queries. Note that this family of codes is not local decodable with constant number of queries (q) because of the lower bound on the blocklength of locally decodable codes due to Katz and Trevisan [7].

Theorem 18. *Let $q > 0$ be a constant integer. For infinitely many $m > 0$ there exists a family of codes $C_m \subset \mathbf{F}_2^m$ which has $\delta(C_m) = \Theta(1)$, $\dim(C_m) = \Theta(m)$ and $span((C_m)_3^\perp) = (C_m)^\perp$. Moreover, C_m is non-locally testable with $o(m)$ queries and non-locally decodable with q queries.*

Reconstruction Threshold for the Hardcore Model

Nayantara Bhatnagar, Allan Sly, and Prasad Tetali[*]

Hebrew University, Microsoft Research, Georgia Tech

Abstract. In this paper we consider the reconstruction problem on the tree for the hardcore model. We determine new bounds for the non-reconstruction regime on the k-regular tree showing non-reconstruction when

$$\lambda < \frac{(\ln 2 - o(1)) \ln^2 k}{2 \ln \ln k}$$

improving the previous best bound of $\lambda < e - 1$. This is almost tight as reconstruction is known to hold when $\lambda > (e + o(1)) \ln^2 k$. We discuss the relationship for finding large independent sets in sparse random graphs and to the mixing time of Markov chains for sampling independent sets on trees.

1 Introduction

The reconstruction problem on the tree was originally studied as a problem in statistical physics but has since found many applications including in computational phylogenetic reconstruction [8], the study of the geometry of the space of random constraint satisfaction problems [1,14] and the mixing time of Markov chains [5,17]. For a Markov model on an infinite tree the reconstruction problem asks when do the states at level n provide non-trivial information about the state at the root as n goes to infinity. In general the problem involves determining the existence of solutions of distribution valued equations and as such exact thresholds are known only in a small number of examples [4,10,5,25].

In this paper we analyze the reconstruction problem for the hardcore model on the k-regular tree, where each vertex of the tree has degree k. The hardcore model is a probability distribution over independent sets I weighted proportionally to $\lambda^{|I|}$. Previously Brightwell and Winkler [7] showed that reconstruction is possible when $\lambda > (e + o(1)) \ln^2 k$. Improving on their bound for the non-reconstruction regime, Martin [16] showed that non-reconstruction holds when $\lambda < e - 1$ still leaving a wide gap between the two thresholds. Our main result establishes that the bound of Brightwell and Winkler is tight up to a $\ln \ln k$ multiplicative factor.

Theorem 1. *The hardcore model on the k-regular tree has non-reconstruction when*

$$\lambda < \frac{(\ln 2 - o(1)) \ln^2 k}{2 \ln \ln k}.$$

[*] Research supported in part by NSF grants DMS-0701043 and-CCR 0910584.

M. Serna et al. (Eds.): APPROX and RANDOM 2010, LNCS 6302, pp. 434–447, 2010.

1.1   The Hardcore Model

For a finite graph G the independent sets $I(G)$ are subsets of the vertices containing no adjacent vertices. The hardcore model is a probability measure over $\sigma \in I(G) \subset \{0,1\}^G$ such that

$$\mathbb{P}(\sigma) = \frac{1}{Z} \lambda^{\sum_{v \in G} \sigma_v} \mathbb{1}_{\sigma \in I(G)} \qquad (1)$$

where λ is the *fugacity* parameter and Z is a normalizing constant. The definition of the hardcore model can be extended to infinite graphs by way of the Dobrushin-Lanford-Ruelle condition which essentially says that for every finite set A the configuration on A is given by the Gibbs distribution given by a random boundary generated by the measure outside of A. Such a measure is called a Gibbs measure and there may be one or infinitely many such measures (see e.g. [12] for more details). For every λ, there exists a unique translation invariant Gibbs measure on the k-regular tree and it is this measure which we study.

An alternative equivalent formulation of the hardcore model is as a Markov model on the k-regular tree. An independent set σ is generated by first choosing the root according to the distribution

$$(\pi_1, \pi_0) = \left(\frac{\omega}{1 + 2\omega}, \frac{1 + \omega}{1 + 2\omega} \right)$$

for some $0 < \omega < 1$. The states of the remaining vertices of the graph are generated from their parents' states by taking one step of the Markov transition matrix

$$M = \begin{pmatrix} p_{11} & p_{10} \\ p_{01} & p_{00} \end{pmatrix} = \begin{pmatrix} 0 & 1 \\ \frac{\omega}{1+\omega} & \frac{1}{1+\omega} \end{pmatrix}.$$

It can easily be checked that π is reversible with respect to M and that this generates a translation invariant Gibbs measure on the tree with fugacity

$$\lambda = \omega(1 + \omega)^{k-1}.$$

Restating Theorem 1 in terms of ω we have non-reconstruction when

$$\omega \leq \frac{1}{k}\left[\ln k + \ln \ln k - \ln \ln \ln k - \ln 2 + \ln \ln 2 - o(1) \right] =: \bar{\omega}. \qquad (2)$$

In contrast, from [7], for every fixed $\varepsilon > 0$ and k large enough, reconstruction is known to hold when

$$\omega \geq \frac{1}{k}\left[\ln k + \ln \ln k + 1 + \varepsilon \right]. \qquad (3)$$

We will introduce some further notation which we will make use of in the proof.

$$\pi_{01} \equiv \frac{\pi_0}{\pi_1} = \frac{1 + \omega}{\omega}, \qquad \Delta \equiv \pi_{01} - 1 = \frac{1}{\omega},$$

$$\theta \equiv p_{00} - p_{10} = p_{11} - p_{01} = -\frac{\omega}{1 + \omega}$$

A particularly important role is played by θ, the second eigenvalue of M as is discussed in the following subsection. We denote by $\mathbb{P}_T^1, \mathbb{E}_T^1$ (and resp. $\mathbb{P}_T^0, \mathbb{E}_T^0$ and $\mathbb{P}_T, \mathbb{E}_T$) the probability and expectations with respect to the measure obtained by conditioning on the root ρ of T to be 1 (resp. 0, and stationary). We let $L = L(n)$ denote the set of vertices at depth n and $\sigma(L) = \sigma(L(n))$ denote the configuration on level n. We will write $\Pr_T[\cdot|\sigma(L) = A]$ to denote the measure conditioned on the leaves being in state $A \in \{0, 1\}^{L(n)}$.

1.2   The Reconstruction Problem

The reconstruction problem on the tree essentially asks if we can recover information on the root from the spins deep inside the tree. In particular we say that the model has *non-reconstruction* if

$$\mathbb{P}_T[\sigma_\rho = 1|\sigma(L)] \to \pi_1 \tag{4}$$

in probability as $n \to \infty$, otherwise the model has *reconstruction*. Note that if we do not condition on the configuration at the leaves, the probability above is exactly π_1 by the Markov model formulation of the hardcore measure. Equivalent formulations of non-reconstruction are that the Gibbs measure is extremal or that the tail σ-algebra of the Gibbs measure is trivial [23]. It follows from Proposition 12 of [22] that there exists a λ_R such that reconstruction holds for $\lambda > \lambda_R$ and non-reconstruction holds for $\lambda < \lambda_R$. The reconstruction problem is to determine the threshold λ_R.

1.3   Related Work

A significant body of work has been devoted to the reconstruction problem on the tree by probabilists, computer scientists and physicists. The earliest such result is the Kesten-Stigum bound [15] which states that reconstruction holds whenever $\theta^2(k - 1) > 1$. This bound was shown to be tight in the case of the Ising model [4,10] where it was shown that non-reconstruction holds when $\theta^2(k-1) \leq 1$. Similar results were derived for the Ising model with small external field [2] and the 3-state Potts model [25] which constitute the only models for which exact thresholds are known. On the other hand, at least when k is large, the Kesten-Stigum bound is known not to be tight for the hardcore model [7]. As such, the most one can reasonably ask to show is the asymptotics of the reconstruction threshold $\lambda_R(k)$ for large k.

The Kesten-Stigum bound is known to be the correct bound for robust reconstruction for all Markov models [13]. Robust reconstruction asks whether reconstruction is possible after adding a large amount of noise to the spins in level n. It was shown in [13] that when $\theta^2(k - 1) < 1$ after adding enough noise to the spins at level n, the "information" provided by the modified spins at level n decays exponentially quickly.

In both the coloring model and the hardcore model the reconstruction threshold is far from the Kesten-Stigum bound for large k. In the case of the hardcore model $\theta^2(k-1) = (1 + o(1))\frac{1}{k}\ln^2 k$. As such, given a noisy version of the spins at level n, the information on the root decays rapidly as n grows. In the coloring model close to optimal bounds [3,24] were obtained by first showing that, when n is small, the information on the root is sufficiently small. Then a quantitative version of [13] establishes that the information on the root converges to 0 exponentially quickly. The hardcore model behaves similarly. Indeed, the form of our bound in equation (2) is strikingly similar to the bound for the q-coloring model which states that reconstruction (resp. non-reconstruction) holds when the degree is at least (resp. at most) $q[\ln q + \ln \ln q + O(1)]$.

Our proof then proceeds as follows. We first establish that when ω satisfies (2) then even for a tree of depth 3 there is already significant loss of information of the spin at the root. In particular we show that if the state of the root is 1 then the typical posterior probability that the state of the root is 1 given the spins at level 3 will be less than $\frac{1}{2}$. The result is completed by linearizing the standard tree recursion as in [5,25]. In this part of the proof we closely follow the notation of [5] who analyzed the reconstruction problem for the Ising model with small external field. We do not require the full strength of their analysis as in our case we are far from the Kesten-Stigum bound. We show that a quantity which we refer to as the *magnetization* decays exponentially fast to 0. The magnetization provides a bound on the posterior probabilities and this completes the result.

The $\ln \ln k$ term in our bound on λ is explained as the first point at which there is significant decay of information at level 3 on the tree. In particular the analysis in Proposition 2 part c) is essentially tight. It may be possible to get improved bounds by considering higher depth trees although the description of the posterior distribution necessarily becomes more complex. A sharper analysis of this sort was done in [24] for the coloring model although the method there made crucial use of the symmetry of the states.

Replica Symmetry Breaking and Finding Large Independent Sets. The reconstruction problem plays a deep role in the geometry of the space of solutions of random constraint satisfaction problems. While for problems with few constraints the space of solutions is connected and finding solutions is generally easy, as the number of constraints increases the space may break into exponentially many small clusters. Physicists, using powerful but non-rigorous "replica symmetry breaking" heuristics, predicted that the clustering phase transition exactly coincides with the reconstruction region on the associated tree model [19,14]. This picture was rigorously established (up to first order terms) for the coloring and satisfiability problems [1] and further extended to sparse random graphs by [20]. As solutions are far apart, local search algorithms will in general fail. Indeed for both the coloring and SAT models, no algorithm is known to find solutions in the clustered phase. It has been conjectured to be computationally intractable beyond this phase transition [1].

The associated CSP for the hardcore model corresponds to finding large independent sets in random k-regular graphs. The replica heuristics again predict that the space of large independent sets should be clustered in the reconstruction regime. Specifically this refers to independent sets of size sn where $s > \pi_1(R)$, the density of 1's in the hardcore model at the reconstruction threshold. It is known that the largest independent set is with high probability $\frac{(2-o(1))\ln k}{k}n$ [6]. On the other hand the best known algorithm finds independent sets only of size $\frac{(1+o(1))\ln k}{k}n$ which is equal to $\pi_1(R)n$ [27]. This is consistent with the physics predictions and it would be of interest to determine if the space of independent sets indeed exhibits the same clustering phenomena as colorings and SAT at the reconstruction threshold. Determining the reconstruction threshold more precisely thus has implications for the problem of finding large independent sets in random graphs.

Glauber Dynamics on Trees. The reconstruction threshold plays a key role in the study of the rate of convergence of the Glauber dynamics Markov chain for sampling spin systems on trees. This problem has received considerable attention (see e.g. [2,9,17,18,26]) and in the case of the Ising model, the mixing time is known to undergo a phase transition from $\Theta(n \ln n)$ in the non-reconstruction regime to $n^{1+\Theta(1)}$ in the reconstruction regime [2]. In fact, the mixing time is $n^{1+\Theta(1)}$ for any spin system above the reconstruction threshold. A similar transition was shown to take place for the coloring model [26]. Sharp bounds of this type are not known for the hardcore model, however, it is predicted that the Glauber dynamics should again be $O(n \log n)$ in the non-reconstruction regime.

Phylogenetic Reconstruction. Phylogenetic reconstruction is an important problem in evolutionary biology [11]. The results of [8,21] imply that for binary symmetric channels the amount of data needed for phylogenetic reconstruction is closely related to the corresponding reconstruction threshold on the tree. This sampling efficiency undergoes a phase transition from $O(\log(N))$ in the reconstruction regime to $N^{\Omega(1)}$ in the non-reconstruction regime, where N is the number of leaves (or species) at the bottom of the phylogenetic tree. The results of [5] for non-reconstruction also imply lower bounds on the sample complexity for phylogenetic reconstruction for asymmetric channels.

2   Proof of Theorem 1

For ease of notation we establish our bounds for the k-ary tree (where each vertex has k children) instead of on the k-regular tree. It is not difficult to modify the recursion we will obtain for the k-ary tree to a recursion for the $(k + 1)$-regular tree, showing that non-reconstruction also holds in that case. Finally, we can show that non-reconstruction on the k-regular tree is equivalent to non-reconstruction on the $(k+1)$-regular tree once we note that in equation (2) we have that $\bar{\omega}(k + 1) - \bar{\omega}(k) = o(k)$ so the difference can be absorbed in the error term. Let $\mathcal{T}$ denote the infinite k-ary tree and let $\mathcal{T}_n$ denote the restriction of $\mathcal{T}$ to its first n levels.

Before reading further, it might help the reader to quickly recall the notation from the end of Section 1.1. As in [5] we analyze a random variable X which denotes *weighted magnetization of the root* which is a function of the leaf states of the tree. We define $X = X(n)$ on T_n by

$$X = \pi_0^{-1}[\pi_0 \mathbb{P}(\sigma_\rho = 1|\sigma(L) = A) - \pi_1 \mathbb{P}(\sigma_\rho = 0|\sigma(L) = A)]$$
$$= \frac{1}{\pi_{01}}\left[\frac{\mathbb{P}[\sigma_\rho = 1|\sigma(L) = A]}{\pi_1} - 1\right] \tag{5}$$

Since $\mathbb{E}_T[\mathbb{P}[\sigma_\rho = 1|\sigma(L) = A]] = \mathbb{P}[\sigma_\rho = 1] = \pi_1$, from the above expression, we have that $\mathbb{E}[X] = 0$. Also, $X \le 1$ since $\mathbb{P}[\sigma_\rho = 1|\sigma(L) = A] \le 1$. We will make extensive use of the following second moments of the magnetization.

$$\overline{X} = \mathbb{E}_T[X^2], \quad \overline{X}_1 = \mathbb{E}_T^1[X^2], \quad \overline{X}_0 = \mathbb{E}_T^0[X^2]$$

The following equivalent definition of non-reconstruction is well known and follows from the definition in (4) using (5).

Proposition 1. *Non-reconstruction for the model (T, M) is equivalent to*

$$\lim_{n \to \infty} \overline{X}(n) = 0,$$

where $\overline{X}(n) = \mathbb{E}_{T_n}[X^2]$.

In the remainder of the proof we derive bounds for $\overline{X}$. We begin by showing that already for a 3 level tree, $\overline{X}$ becomes small. Then we establish a recurrence along the lines of [5] that shows that once $\overline{X}$ is sufficiently small, it must converge to 0. As this part of the derivation follows the calculation in [5] we will adopt their notation in places. Non-reconstruction is then a consequence of Proposition 1. In the next lemma we determine some basic properties of X.

Lemma 1. *The following relations hold:*
a) $\mathbb{E}_T[X] = \pi_1 \mathbb{E}_T^1[X] + \pi_0 \mathbb{E}_T^0[X] = 0.$
b) $\overline{X} = \pi_1 \overline{X}_1 + \pi_0 \overline{X}_0.$
c) $\mathbb{E}_T^1[X] = \pi_{01}\overline{X}$ and $\mathbb{E}_T^0[X] = -\overline{X}.$

Proof. Note that for any random variable which depends only on the states at the leaves, $f = f(A)$, we have $\mathbb{E}_T[f] = \pi_1 \mathbb{E}_T^1[f] + \pi_0 \mathbb{E}_T^0[f]$. Parts $a)$ and $b)$ therefore follow since X is a random variable that is a function of the states at the leaves. For part $c)$ we proceed as follows. The first and last equalities below follow from (5).

$$\mathbb{E}_T^1[X] = \pi_{01}^{-1}\sum_A \mathbb{P}_T[\sigma(L) = A|\sigma_\rho = 1]\left(\frac{\mathbb{P}_T[\sigma_\rho = 1|\sigma(L) = A]}{\pi_1} - 1\right)$$

$$= \pi_{01}^{-1}\sum_A \mathbb{P}_T[\sigma(L) = A]\frac{\mathbb{P}_T[\sigma_\rho = 1|\sigma(L) = A]}{\pi_1}\left(\frac{\mathbb{P}_T[\sigma_\rho = 1|\sigma(L) = A]}{\pi_1} - 1\right)$$

$$= \pi_{01}^{-1}\left(\frac{\mathbb{E}_T[(\mathbb{P}_T[\sigma_\rho = 1|\sigma(L) = A'])^2]}{\pi_1^2} - 1\right)$$

$$= \pi_{01}\mathbb{E}[X^2]$$

The second part of c) follows by combining this with a).   □

The following proposition estimates typical posterior probabilities which we will use to bound $\overline{X}$. For a finite k-ary subtree T of the $k+1$-regular tree let T^i be the subtrees rooted at the children of the root u_i.

Proposition 2. *For a finite k-ary subtree T we have that*

a) For any configuration at the leaves $A = (A_1, \cdots, A_k)$,

$$\mathbb{P}_T[\sigma_\rho = 0 | \sigma(L) = A] = \left(1 + \lambda \prod_i \mathbb{P}_{T^i}[\sigma_{u_i} = 0 | \sigma_{L_i} = A_i]\right)^{-1}.$$

b) Let $\mathcal{G}$ be the set of leaf configurations

$$\mathcal{G} = \left\{ \sigma(L) \mid \mathbb{P}[\sigma_\rho = 0 | \sigma(L)] = \frac{1}{2}\left(1 + \frac{1}{1 + 2\lambda}\right) \right\}.$$

Then

$$\frac{\mathbb{P}^0_T[\sigma(L) \in \mathcal{G}]}{\mathbb{P}^1_T[\sigma(L) \in \mathcal{G}]} = \frac{\pi_1}{\pi_0} \frac{1 + \lambda}{\lambda}.$$

c) Let $\beta > \ln 2 - \ln \ln 2$ and $\omega = \frac{1}{k}\left[\ln k + \ln \ln k - \ln \ln \ln k - \beta\right]$. Then in the 3 level k-ary tree T_3 we have that

$$\mathbb{E}^1_{T_3}[\mathbb{P}[\sigma_\rho = 1 | \sigma(L)]] \leq \frac{1}{2}.$$

Proof. Part a) is a consequence of standard tree recursions for Markov models established using Bayes rule.

For part b) first note that

$$\mathbb{P}[\sigma_\rho = 1 \mid \sigma(L) \in \mathcal{G}] = 1 - \mathbb{P}[\sigma_\rho = 0 \mid \sigma(L) \in \mathcal{G}]$$
$$= \frac{1}{2}\left(1 - \frac{1}{1 + 2\lambda}\right) \qquad (6)$$

Now,

$$\mathbb{P}^0_T[\sigma(L) \in \mathcal{G}] = \frac{\mathbb{P}[\sigma_\rho = 0 \mid \sigma(L) \in \mathcal{G}]\mathbb{P}[\sigma(L) \in \mathcal{G}]}{\pi_0}$$
$$= \frac{\pi_1}{\pi_0} \frac{1 + \lambda}{\lambda}\left(\frac{\mathbb{P}[\sigma_\rho = 1 \mid \sigma(L) \in \mathcal{G}]\mathbb{P}[\sigma(L) \in \mathcal{G}]}{\pi_1}\right)$$
$$= \frac{\pi_1}{\pi_0} \frac{1 + \lambda}{\lambda}\mathbb{P}^1_T[\sigma(L) \in \mathcal{G}]$$

where the first and third equations follow by definition of conditional probabilities and the second follows from (6) which establishes b).

For part $c)$, we start by calculating the probability of certain posterior probabilities for trees of small depth. With our assumption on ω we have that

$$\lambda = \omega(1+\omega)^k = \frac{(1+o_k(1))e^{-\beta}\ln^2 k}{\ln \ln k}$$

Since $\sigma(L) \equiv 0$ under $\mathbb{P}^1_{T_1}$, by part $a)$ we have that

$$\mathbb{P}^1_{T_1}[\sigma_\rho = 0|\sigma(L)] = \frac{1}{1+\lambda} \ w.p. \ 1.$$

Also,

$$\mathbb{P}_{T_1}(u_i = 0 \ \forall \ i|\sigma_\rho = 0) = \left(\frac{1}{1+\omega}\right)^k$$

Using the two equations above, we have that

$$\mathbb{P}^0_{T_1}(\sigma_\rho = 0|\sigma(L)) = \begin{cases} 1 & w.p. \ 1 - \left(\frac{1}{1+\omega}\right)^k \\ \frac{1}{1+\lambda} & w.p. \ \left(\frac{1}{1+\omega}\right)^k. \end{cases}$$

The first case above corresponds to leaf configurations of the tree T_1 where at least one of the leaves is 1, while the second case corresponds to the configurations where all the leaves are 0. Next, applying part $a)$ to a tree of depth 2, we have

$$\mathbb{P}^1_{T_2}[\sigma_\rho = 0|\sigma(L)] = \frac{1}{1 + \lambda\prod_i \mathbb{P}^0_{T_1}[\sigma_{u_i} = 0|\sigma(L)]}$$

Using this expression we can write down this conditional probability based on the leaf configurations of the subtrees of the root of depth 1.

$$\mathbb{P}^1_{T_2}[\sigma_\rho = 0|\sigma(L)] = \begin{cases} \frac{1}{1+\lambda} & w.p. \ \left(1 - \left(\frac{1}{1+\omega}\right)^k\right)^k \\ \frac{1}{2}\left(1 + \frac{1}{1+2\lambda}\right) & w.p. \ \left(1 - \left(\frac{1}{1+\omega}\right)^k\right)^{k-1}\left(\frac{1}{1+\omega}\right)^k k \\ > \frac{1}{2}\left(1 + \frac{1}{1+2\lambda}\right) & o.w. \end{cases} \tag{7}$$

The first case above corresponds to the situation when each subtree of the root of depth 1 has a leaf configuration where at least one of the leaves is 1. The second case is when one of the k subtrees has a leaf configuration where all leaves are 0, while the remaining subtrees have leaf configurations where at least one leaf is 1. The third case corresponds to the remaining possibilities.

By part $b)$ with $\mathcal{G}$ as defined, and (7) we have that after substituting the expressions for λ and ω,

$$\mathbb{P}^0_{T_2}[\sigma(L) \in \mathcal{G}] = \frac{\pi_1}{\pi_0}\frac{1+\lambda}{\lambda}\mathbb{P}^1_{T_2}[\sigma(L) \in \mathcal{G}]$$

$$= \frac{\omega(1+\lambda)}{\lambda(1+\omega)}\left(1 - \left(\frac{1}{1+w}\right)^k\right)^{k-1}\left(\frac{1}{1+\omega}\right)^k k$$

$$\geq (1 - o_k(1))\frac{e^\beta \ln \ln k}{k} \tag{8}$$

We can now calculate the values of $P^1_{T_3}[\sigma_\rho = 0 | \sigma(L)]$ as follows. By part $a)$

$$P^1_{T_3}[\sigma_\rho = 0 | \sigma(L)] = \frac{1}{1 + \lambda \prod_i P^0_{T_2}[\sigma_{u_i} = 0 | \sigma(L)]}$$

Denote

$$p = \frac{\omega(1+\lambda)}{\lambda(1+\omega)} \left(1 - \left(\frac{1}{1+w}\right)^k \right)^{k-1} \left(\frac{1}{1+\omega}\right)^k k$$

Thus, p is the probability that if we started with $\sigma_\rho = 0$ in T_2, the configuration at the leaves is from $\mathcal{G}$. If we start with $\sigma_\rho = 1$ in T_3, the number subtrees of the root with leaf configurations in $\mathcal{G}$ is distributed binomially and will be about kp. By Chernoff bounds, and the bound on p from (8),

$$\mathbb{P}(Bin(k, p) < e^\beta \ln \ln k - 2\sqrt{e^\beta \ln \ln k}) < \frac{1}{3}.$$

Finally, by the definition of $\mathcal{G}$,

$$P^0_{T_2}[\sigma_{u_i} = 0 | \sigma(L) \in \mathcal{G}] = \frac{1}{2}\left(1 + \frac{1}{1 + 2\lambda}\right)$$

and hence,

$$\mathbb{E}^1_{T_3}[\mathbb{P}[\sigma_\rho = 1 | \sigma(L)]] = \mathbb{E}^1_{T_3}[1 - \mathbb{P}[\sigma_\rho = 0 | \sigma(L)]]$$

$$\leq \left(1 - \frac{1}{1 + \lambda[2(1 - o_k(1))]^{-(e^\beta \ln \ln k - 2\sqrt{e^\beta \ln \ln k})}} \right) \frac{2}{3} + \frac{1}{3}$$

By taking k large enough above, we conclude that for β and large enough k,

$$\mathbb{E}^1_{T_3}[\mathbb{P}[\sigma_\rho = 1 | \sigma(L)]] \leq \frac{1}{2} \qquad \qquad \square$$

Lemma 2. *Let $\beta > \ln 2 - \ln \ln 2$ and $\omega = \frac{1}{k}\left[\ln k + \ln \ln k - \ln \ln \ln k - \beta \right]$. For k large enough,*

$$\overline{X}(3) \leq \frac{\omega}{2}.$$

Proof. By part $c)$ of Lemma 1, and part $c)$ of Proposition 2,

$$\overline{X}(3) = \frac{1}{\pi^2_{01}} \left(\frac{\mathbb{E}^1_{T_3}[\mathbb{P}[\sigma_\rho = 1 \mid \sigma(L)]]}{\pi_1} - 1 \right)$$

$$\leq \frac{1}{\pi^2_{01}} \left(\frac{1}{2\pi_1} - 1 \right)$$

$$\leq \frac{\omega}{2}$$

Next, we present a recursion for $\overline{X}$ and complete the proof of the main result. The development of the recursion follows the steps in [5] closely so we follow their notation and omit some of the calculations in this short version.

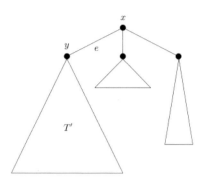

Fig. 1. A finite tree T

Magnetisation of a Child. With T and x as defined previously, let y be a child of x and let T' be the subtree of T rooted at y (see Figure 1). Let A' be the restriction of A to the leaves of T'. Let $Y = Y(A')$ denote the magnetization of y.

Lemma 3. *We have*

a) $\mathbb{E}_T^1[Y] = \theta\mathbb{E}_{T'}^1[Y]$ *and* $\mathbb{E}_T^0[Y] = \theta\mathbb{E}_{T'}^0[Y]$.
b) $\mathbb{E}_T^1[Y^2] = (1-\theta)\mathbb{E}_{T'}[Y^2] + \theta\mathbb{E}_{T'}^1[Y^2]$.
c) $\mathbb{E}_T^0[Y^2] = (1-\theta)\mathbb{E}_{T'}[Y^2] + \theta\mathbb{E}_{T'}^0[Y^2]$.

The proof follows from the first part of Lemma 1 and the Markov property when we condition on x.

Next, we can write the effect on the magnetization of adding an edge to the root and merging roots of two trees as follows. Referring to Figure 2, let T' (resp. T'') be a finite tree rooted at y (resp. z) with the channel on all edges being given M, leaf states A (resp A'') and weighted magnetisation at the root Y (resp. Z). Now add an edge $(\hat{y}, z)$ to T'' to obtain a new tree $\hat{T}$. Then merge $\hat{T}$ with T' by identifying $y = \hat{y}$ to obtain a new tree T. To avoid ambiguities, denote by x the root of T and X the magnetization of the root of T. We let $A = (A', A'')$ be the leaf state of T. Let $\hat{Y}$ be the magnetization of the root of $\hat{T}$.

Note: In the above construction, the vertex y is a vertex "at the same level" as x, and not a child of x as it was in Lemma 3.

Lemma 4. *With the notation above,* $\hat{Y} = \theta Z$.

The proof follows by applying Bayes rule, the Markov property and Lemma 1. These facts also imply that

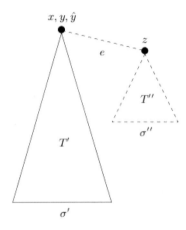

Fig. 2. The tree T after obtained after merging T' and T''. The dashed subtree is $\hat{T}$.

Lemma 5. *For any tree* $\hat{T}$,

$$X = \frac{Y + \hat{Y} + \Delta Y \hat{Y}}{1 + \pi_{01} Y \hat{Y}}.$$

With these lemmas in hand we can use the following relation to derive a recursive upper bound on the second moments. We will use the expansion

$$\frac{1}{1+r} = 1 - r + r^2 \frac{1}{1+r}.$$

Taking $r = \pi_{01} Y \hat{Y}$, by Lemma 5 we have

$$X = (Y + \hat{Y} + \Delta Y \hat{Y}) \left[1 - \pi_{01} Y \hat{Y} + (\pi_{01} Y \hat{Y})^2 \frac{1}{1 + \pi_{01} Y \hat{Y}} \right]$$

$$= Y + \hat{Y} + \Delta Y \hat{Y} - \pi_{01} Y \hat{Y} \left(Y + \hat{Y} + \Delta Y \hat{Y} \right) + (\pi_{01})^2 (Y \hat{Y})^2 X$$

$$\leq Y + \hat{Y} + \Delta Y \hat{Y} - \pi_{01} Y \hat{Y} \left(Y + \hat{Y} + \Delta Y \hat{Y} \right) + (\pi_{01})^2 (Y \hat{Y})^2 \qquad (9)$$

where the last inequality follows since $X \leq 1$ with probability 1.

Let $\rho' = \overline{Y}_1/\overline{Y}$ and $\rho'' = \overline{Z}_1/\overline{Z}$. Below, the moments $\overline{Y}$ etc. are defined according to the appropriate measures over the tree rooted at y (i.e. T') etc.

By applying Lemmas 1, 3 and 4, we have the following relations.

$$\mathbb{E}_T^1[X] = \pi_{01} \overline{X}, \quad \mathbb{E}_T^1[Y] = \pi_{01} \overline{y}, \quad \mathbb{E}_T^1[Y^2] = \overline{Y} \rho'$$
$$\mathbb{E}_T^1[\hat{Y}] = \pi_{01} \theta^2 \overline{Z}, \quad \mathbb{E}_T^1[\hat{Y}^2] = \theta^2 \overline{Z}((1 - \theta) + \theta \rho'') \qquad (10)$$

Applying $(\pi_{01})^{-1}E_T^1[\cdot]$ to both sides of (9), we obtain the following.

$$\overline{X} \leq \overline{Y} + \theta^2\overline{Z} + \Delta\pi_{01}\overline{YZ} - \pi_{01}\theta^2\overline{YZ}\rho' - \pi_{01}\theta^2\overline{YZ}((1-\theta) + \theta\rho'')$$
$$-\Delta\theta^2\overline{YZ}\rho'((1-\theta) + \theta\rho'') + \pi_{01}\theta^2\overline{YZ}\rho'((1-\theta) + \theta\rho'')$$
$$= \overline{Y} + \theta^2\overline{Z} - \pi_{01}\theta^2\overline{YZ}[\mathcal{A} - \Delta\mathcal{B}]$$

where

$$\mathcal{A} = \rho' + (1 - \rho')[(1-\theta) + \theta\rho''],$$
$$\text{and } \mathcal{B} = 1 - (\pi_{01})^{-1}\rho'[(1-\theta) + \theta\rho''] = 1 - \frac{\omega}{1+\omega}\rho'[(1-\theta) + \theta\rho''].$$

If $\mathcal{A} - \Delta\mathcal{B} \geq 0$, this would already give a sufficiently good recursion to show that $\overline{X}(n)$ goes to 0, so we will assume is negative and try to get a good (negative) lower bound. First note that by their definition $\rho', \rho'' \geq 0$. Further since $\overline{Y} = \pi_1\overline{Y}_1 + \pi_0\overline{Y}_0$,

$$\rho' \leq (\pi_1)^{-1} = \frac{1 + 2\omega}{\omega}.$$

Similarly,

$$\rho'' \leq (\pi_1)^{-1} = \frac{1 + 2\omega}{\omega}.$$

Since $E_T^1[\hat{Y}^2]$ and $\overline{Z} \geq 0$, it follows from (10) that $(1 - \theta) + \theta\rho'' \geq 0$. Together with the fact that $\rho' \geq 0$, this implies that $\mathcal{B} \leq 1$.

Since $\mathcal{A}$ is multi-linear in (ρ', ρ''), to minimize it, its sufficient to consider the extreme cases. When $\rho' = 0$, $\mathcal{A}$ is minimized at the upper bound of ρ'' and hence

$$\mathcal{A} \geq 1 - \pi_{01}\frac{\omega}{1+\omega} = 0.$$

When $\rho' = (\pi_1)^{-1}$,

$$\mathcal{A} = (\pi_1)^{-1} + (1 - (\pi_1)^{-1})[1 - \theta(1 - \rho'')] \geq 0.$$

Hence, we have

$$\overline{X} \leq \overline{Y} + \theta^2\overline{Z} + \frac{1}{1+\omega}\overline{YZ}.$$

Applying this recursively to the tree, we obtain the following recursion for the moments.

$$\overline{X} \leq (1+\omega)\theta^2\left[\left(1 + \frac{\overline{Z}}{1+\omega}\right)^k - 1\right]$$

We bound the $(1 + x)^k - 1$ term as,

$$|(1+x)^k - 1| \leq e^{|x|k} - 1 = \int_0^{|x|k} e^s \, ds \leq e^{|x|k}k|x|$$

and this implies the following recursion.

Theorem 2. *If for some n, $\overline{X}(n) \leq \frac{\omega}{2}$, we have that*

$$\overline{X}(n+1) \leq \omega^2 e^{\frac{1}{2}\omega k} k \overline{X}(n).$$

Thus if $\omega^2 e^{\frac{1}{2}\omega k} k < 1$ then it follows from the recursion that

$$\lim_n \overline{X}(n) = 0. \tag{11}$$

When $\omega = \frac{1}{k}\left[\ln k + \ln \ln k - \ln \ln \ln k - \beta\right]$ and $\beta > \ln 2 - \ln \ln 2$, by Lemma 2, for k large enough, $\overline{X}(3) \leq \frac{\omega}{2}$. Hence by equation (11) we have that $\overline{X}(n) \to 0$ and so by Proposition 1 we have non-reconstruction. Since reconstruction is monotone in λ and hence in ω it follows that we have non-reconstruction for $\omega \leq \bar{\omega}$ for large k. This completes the proof of Theorem 1.

References

1. Achlioptas, D., Coja-Oghlan, A.: Algorithmic barriers from phase transitions. In: Proceedings of IEEE FOCS 2008, pp. 793–802 (2008)
2. Berger, N., Kenyon, C., Mossel, E., Peres, Y.: Glauber dynamics on trees and hyperbolic graphs. Probability Theory and Related Fields 131, 311–340 (2005)
3. Bhatnagar, N., Vera, J., Vigoda, E., Weitz, D.: Reconstruction for colorings on trees. To appear in SIAM Journal on Discrete Mathematics
4. Bleher, P.M., Ruiz, J., Zagrebnov, V.A.: On the purity of limiting gibbs state for the Ising model on the Bethe lattice. Journal of Statistical Physics 79, 473–482 (1995)
5. Borgs, C., Chayes, J., Mossel, E., Roch, S.: The Kesten-Stigum reconstruction bound is tight for roughly symmetric binary channels. In: Proceedings of IEEE FOCS 2006, pp. 518–530 (2006)
6. Cooper, C., Frieze, A., Reed, B., Riordan, O.: Random regular graphs of non-Constant degree: Independence and Chromatic Number. Combinatorics, Probability and Computing 11, 323–341 (2002)
7. Brightwell, G., Winkler, P.: A second threshold for the hard-core model on a Bethe lattice. Random structures and algorithms 24, 303–314 (2004)
8. Daskalakis, C., Mossel, E., Roch, S.: Optimal phylogenetic reconstruction. In: Proceedings of the Thirty-Eighth Annual ACM Symposium on Theory of Computing, pp. 159–168 (2006)
9. Ding, J., Lubetzky, E., Peres, Y.: Mixing time of critical Ising model on trees is polynomial in the height. Communications in Mathematical Physics 295, 161–207 (2010)
10. Evans, W., Kenyon, C., Peres, Y., Schulman, L.J.: Broadcasting on trees and the Ising model. Annals of Applied Probabability 10, 410–433 (2000)
11. Felsenstein, J.: Inferring Phylogenies. Sinauer, New York (2004)
12. Georgii, H.O.: Gibbs measures and phase transitions. Walter de Gruyter, Berlin (1988)
13. Janson, S., Mossel, E.: Robust reconstruction on trees is determined by the second eigenvalue. Annals of Probability 32, 2630–2649 (2004)
14. Krzakała, F., Montanari, A., Ricci-Tersenghi, F., Semerjian, G., Zdeborová, L.: Gibbs states and the set of solutions of random constraint satisfaction problems. Proceedings of the National Academy of Sciences 104, 10318 (2007)

15. Kesten, H., Stigum, B.P.: Additional limit theorems for indecomposable multidimensional Galton-Watson processes. Annals of Mathematical Statistics 37, 1463–1481 (1966)
16. Martin, J.: Reconstruction thresholds on regular trees. In: Banderier, C., Krattenthaler, C. (eds.) Discrete Random Walks, DRW 2003. Discrete Mathematics and Theoretical Computer Science Proceedings, pp. 191–204 (2003)
17. Martinelli, F., Sinclair, A., Weitz, D.: Fast mixing for independent sets, colorings, and other models on trees. In: Proceedings of ACM-SIAM SODA, pp. 449–458 (2004)
18. Martinelli, F., Sinclair, A., Weitz, D.: Glauber dynamics on trees: boundary conditions and mixing time. Communications in Mathematical Physics 250, 301–334 (2004)
19. Mézard, M., Montanari, A.: Reconstruction on trees and spin glass transition. Journal of Statistical Physics 124, 1317–1350 (2006)
20. Montanari, A., Restrepo, R., Tetali, P.: Reconstruction and Clustering Thresholds for Random Constraint Satisfaction Problems (2009) (preprint)
21. Mossel, E.: Survey: Information flow on trees. In: Nestril, J., Winkler, P. (eds.) Graphs, Morphisms and Statistical Physics. DIMACS series in discrete mathematics and theoretical computer science, pp. 155–170. Amer. Math. Soc., Providence (2004)
22. Mossel, E.: Reconstruction on trees: beating the second eigenvalue. Annals of Applied Probabability 11, 285–300 (2001)
23. Mossel, E., Peres, Y.: Information flow on trees. Annals of Applied Probabability 13, 817–844 (2003)
24. Sly, A.: Reconstruction of random colourings. Communications of Mathematical Physics 288, 943–961 (2009)
25. Sly, A.: Reconstruction of symmetric Potts Models. In: Proceedings of the 41st ACM Symposium on Theory of Computing, pp. 581–590 (2009)
26. Tetali, P., Vera, J., Vigoda, E., Yang, L.: Phase Transition for the Mixing Time of the Glauber Dynamics for Coloring Regular Trees. In: Proceedings of the Twenty-First Annual ACM-SIAM Symposium on Discrete Algorithms, pp. 1646–1656 (2010)
27. Wormald, N.: Differential equations for random processes and random graphs. Annals of Applied Probability 5, 1217–1235 (1995)

Lower Bounds for Local Monotonicity Reconstruction from Transitive-Closure Spanners

Arnab Bhattacharyya[1,*], Elena Grigorescu[1,*], Madhav Jha[2,**], Kyomin Jung[3], Sofya Raskhodnikova[2,**], and David P. Woodruff[4]

[1] Massachusetts Institute of Technology, USA
{abhatt,elena_g}@mit.edu
[2] Pennsylvania State University, USA
{mxj201,sofya}@cse.psu.edu
[3] Korea Advanced Institute of Science and Technology, Korea
kyomin@kaist.edu
[4] IBM Almaden Research Center, USA
dpwoodru@us.ibm.com

Abstract. Given a directed graph $G = (V, E)$ and an integer $k \geq 1$, a k-*transitive-closure-spanner (k-TC-spanner)* of G is a directed graph $H = (V, E_H)$ that has (1) the same transitive-closure as G and (2) diameter at most k. Transitive-closure spanners are a common abstraction for applications in access control, property testing and data structures.

We show a connection between 2-TC-spanners and *local monotonicity reconstructors*. A local monotonicity reconstructor, introduced by Saks and Seshadhri (SIAM Journal on Computing, 2010), is a randomized algorithm that, given access to an oracle for an almost monotone function $f : [m]^d \rightarrow \mathbb{R}$, can quickly evaluate a related function $g : [m]^d \rightarrow \mathbb{R}$ which is guaranteed to be monotone. Furthermore, the reconstructor can be implemented in a distributed manner. We show that an efficient local monotonicity reconstructor implies a sparse 2-TC-spanner of the directed hypergrid (hypercube), providing a new technique for proving lower bounds for local monotonicity reconstructors. Our connection is, in fact, more general: an efficient local monotonicity reconstructor for functions on any partially ordered set (poset) implies a sparse 2-TC-spanner of the directed acyclic graph corresponding to the poset.

We present tight upper and lower bounds on the size of the sparsest 2-TC-spanners of the directed hypercube and hypergrid. These bounds imply tighter lower bounds for local monotonicity reconstructors that nearly match the known upper bounds.

Keywords: Property Testing, Property Reconstruction, Monotone Functions, Spanners, Hypercube, Hypergrid.

[*] A.B. is supported by a DOE Computational Science Graduate Fellowship and NSF Awards 0514771, 0728645, 0732334. E.G. is supported by NSF award CCR-0829672.
[**] Supported by NSF/CCF award 0729171. S.R. is also supported by NSF/CCF CAREER award 0845701.

M. Serna et al. (Eds.): APPROX and RANDOM 2010, LNCS 6302, pp. 448–461, 2010.

1   Introduction

Graph spanners were introduced in the context of distributed computing [1], and since then have found numerous applications, such as efficient routing [2–6], simulating synchronized protocols in unsynchronized networks [7], parallel and distributed algorithms for approximating shortest paths [8–10], and algorithms for distance oracles [11, 12]. Several variants on graph spanners have been defined. In this work, we focus on *transitive-closure* spanners that were introduced in [13] as a common abstraction for applications in access control, property testing and data structures.

Definition 1.1 (TC-spanner). *Given a directed graph $G = (V, E)$ and an integer $k \geq 1$, a k-**transitive-closure-spanner** (k-**TC-spanner**) of G is a directed graph $H = (V, E_H)$ with the following properties:*

1. *E_H is a subset of the edges in the transitive closure of G.*
2. *For all vertices $u, v \in V$, if $d_G(u, v) < \infty$, then $d_H(u, v) \leq k$.*

Thus, a k-transitive-closure-spanner (or k-TC-spanner) is a graph with small diameter that preserves the connectivity of the original graph. In the applications above, the goal is to find the sparsest k-TC-spanner for a given k and G. The number of edges in the sparsest k-TC-spanner of G is denoted by $S_k(G)$.

Our Contributions. The contributions of this work fall into two categories: (1) We show that an efficient local monotonicity reconstructor implies a sparse 2-TC-spanner of the directed hypergrid (hypercube), providing a new technique for proving lower bounds for local monotonicity reconstructors. (2) We present tight upper and lower bounds on the size of the sparsest 2-TC-spanners of the directed hypercube and hypergrid. These bounds imply tighter lower bounds for local monotonicity reconstructors for these graphs that nearly match the upper bounds given in [14].

1.1   Lower Bounds for Local Monotonicity Reconstruction

Property-preserving data reconstruction was introduced in [15]. In this model, a reconstruction algorithm, called a *filter*, sits between a *client* and a *dataset*. A dataset is viewed as a function $f : \mathcal{D} \to \mathcal{R}$. The client accesses the dataset using *queries* of the form $x \in \mathcal{D}$ to the filter. The filter *looks up* a small number of values in the dataset and outputs $g(x)$, where g must satisfy some fixed *structural* property $\mathcal{P}$. Extending this notion, Saks and Seshadhri [14] defined *local* reconstruction. A filter is *local* if it allows for a local (or distributed) implementation: namely, if the output function g does not depend on the order of the queries.

Definition 1.2 (Local filter). *A local filter for reconstructing property $\mathcal{P}$ is an algorithm A that has oracle access to a function $f : \mathcal{D} \to \mathcal{R}$, and to an auxiliary random string ρ (the "random seed"), and takes as input $x \in \mathcal{D}$. For fixed f and ρ, A runs deterministically on input x to produce an output $A_{f,\rho}(x) \in \mathcal{R}$. (Note that a local filter has no internal state to store previously made queries.) The function $g(x) = A_{f,\rho}(x)$ output by the filter must satisfy the following conditions:*

- *For each f and ρ, the function g must satisfy $\mathcal{P}$.*
- *If f satisfies $\mathcal{P}$, then g must be identical to f with probability at least $1 - \delta$, for some error probability $\delta \leq 1/3$. The probability is taken over ρ.*

In answering query $x \in \mathcal{D}$, the filter A may ask for values of f at domain points of its choice (possibly adaptively) using its oracle access to f. Each such access made to the oracle is called a *lookup* to distinguish it from the client query x. A local filter is *non-adaptive* if the set of domain points that the filter looks up to answer an input query x does not depend on answers given by the oracle.

In [14], the authors also required that g must be sufficiently close to f: *With high probability (over the choice of ρ), $Dist(g, f) \leq B(n) \cdot Dist(f, \mathcal{P})$, where $B(n)$ is called the* error blow-up. *($Dist(g, f)$ is the number of points in the domain on which f and g differ. $Dist(f, \mathcal{P})$ is $\min_{g \in \mathcal{P}} Dist(g, f)$.)* If a local filter along with Definition 1.2 satisfies this condition, we call it *distance-respecting*.

Local Monotonicity Reconstructors. The most studied property in the local reconstruction model is monotonicity of functions [14, 15]. To define monotonicity of functions, consider an n-element poset V_n and let $G_n = (V_n, E)$ be the relation graph, *i.e.*, the Hasse diagram, for V_n. A function $f : V_n \to \mathbb{R}$ is called *monotone* if $f(x) \leq f(y)$ for all $(x, y) \in E$. We particularly focus on posets which have the *directed hypergrid* graph as its relation graph. The *directed hypergrid*, denoted $\mathcal{H}_{m,d}$, has vertex set $\{1, 2, \ldots, m\}^d$ and edge set $\{(x, y) : \exists$ unique $i \in \{1, \ldots, d\}$ such that $y_i - x_i = 1$ and for $j \neq i, y_j = x_j\}$. For the special case $m = 2$, $\mathcal{H}_{2,d}$ is called a *hypercube* and is also denoted by $\mathcal{H}_d$. A monotonicity filter needs to ensure that the output function g is monotone. For instance, if G_n is a directed line, $\mathcal{H}_{n,1}$, the filter needs to ensure that the output sequence specified by g is sorted.

To motivate monotonicity reconstructors for hypergrids, consider the scenario of rolling admissions: An admissions office assigns d scores to each application, such as the applicant's GPA, SAT results, essay quality, etc. Based on these scores, some complicated (third-party) algorithm outputs the probability that a given applicant should be accepted. The admissions office wants to make sure "on the fly" that strictly better applicants are given higher probability, that is, probabilities are *monotone* in scores. A hypergrid monotonicity filter may be used here. A local filter can be implemented in a distributed manner with an additional guarantee that every copy of the filter will correct to the same monotone function of the scores. This can be done by supplying the same random seed to each copy of the filter.

[14] gives a *distance-respecting* local monotonicity filter for the directed hypergrid, $\mathcal{H}_{m,d}$, that makes $(\log m)^{O(d)}$ lookups per query. No non-trivial monotonicity filter for the hypercube $\mathcal{H}_d$ (performing $o(2^d)$ lookups per query) is known. One of the monotonicity filters in [15] is a local filter for the directed line $\mathcal{H}_{m,1}$ with $O(\log m)$ lookups per query (but a worse error blow up than in [14]). As observed in [14], this upper bound is tight. A lower bound of $2^{\alpha d}$, on the number of lookups per query for a *distance-respecting* local monotonicity filter on $\mathcal{H}_d$ with *error blow-up* $2^{\beta d}$, where α, β are sufficiently small constants, appeared in [14]. Notably, all known local monotonicity filters are *non-adaptive*.

We show how to construct sparse 2-TC-spanners from local monotonicity re-constructors with low lookup complexity. These constructions, together with our lower bounds on the size of 2-TC-spanners of the hypergrid and hypercube (Section 1.2), imply lower bounds on lookup complexity of local monotonicity reconstructors for these graphs with arbitrary error blow-up. We state our trans-formations from non-adaptive and adaptive reconstructors separately.

Theorem 1.1 (Transformation from non-adaptive Local Monotonicity Reconstructors to 2-TC-spanners). *Let $G_n = (V_n, E)$ be a poset on n nodes. Suppose there is a* non-adaptive *local monotonicity reconstructor A for G_n that looks up at most $\ell(n)$ values on any query and has* error *probability at most δ. Then there is a 2-TC-Spanner of G_n with $O(n\ell(n) \cdot \lceil \log n / \log(1/\delta) \rceil)$ edges.*

Next theorem applies even to *adaptive* local monotonicity reconstuctors. It takes into account how many lookups on query x are points incomparable to x. In particular, if there are no such lookups, then constructed 2-TC-spanner is of the same size as in Theorem 1.1. (The proof and the implications of Theorem 1.2 are deferred to the full version.)

Theorem 1.2 (Transformation from adaptive Local Monotonicity Reconstructors to 2-TC-spanners). *Let $G_n = (V_n, E)$ be a poset on n nodes. Suppose there is an (adaptive) local monotonicity reconstructor A for G_n that, for any query $x \in V_n$, looks up at most $\ell_1(n)$ vertices comparable to x and at most $\ell_2(n)$ vertices incomparable to x, and has* error *probability at most δ. Then there is a 2-TC-Spanner of G_n with $O(n\ell_1(n) \cdot 2^{\ell_2(n)} \lceil \log n / \log(1/\delta) \rceil)$ edges.*

In Theorem 1.1 and 1.2, when δ is sufficiently small, the bounds on the 2-TC-Spanner size become $O(n\ell(n))$ and $O(n\ell_1(n) \cdot 2^{\ell_2(n)})$, respectively.

As mentioned earlier, all known monotonicity reconstructors are non-adaptive. It is an open question whether it is possible to give a transformation from adaptive local monotonicity reconstructors to 2-TC-spanners without incurring an exponential dependence on the number of lookups made to points incomparable to the query point. We do not know whether this dependence is an artifact of the proof or an indication that lookups to incomparable points might be helpful for adaptive local monotonicity reconstructors.

In Theorems 1.3 and 1.4 (Section 1.2), we present nearly tight bounds on the size of the sparsest 2-TC-spanners of the hypercube and the hypergrid. Theorem 1.1, together with the lower bounds in Theorems 1.3 and 1.4, im-plies the following lower bounds on the lookup complexity of local monotonicity reconstructors for these graphs with arbitrary error blow-up.

Corollary 1.1. *Consider a nonadaptive local monotonicity filter with constant error probability δ. If the filter is for functions $f : \mathcal{H}_{m,d} \to \mathbb{R}$, it must perform $\Omega\left(\frac{\log^{d-1} m}{d^d (2 \log \log m)^{d-1}}\right)$ lookups per query. If the filter is for functions $f : \mathcal{H}_d \to \mathbb{R}$, it must perform $\Omega\left(2^{\alpha d}/d\right)$ lookups per query, where $\alpha \geq 0.1620$.*

Prior to this work, no lower bounds for monotonicity reconstructors on $\mathcal{H}_{m,d}$ with dependence on both m and d were known. Unlike the bound in [14], our lower bounds hold for any error blow-up and for non-distance-respecting filters. Our bounds are tight for non-adaptive reconstructors. Specifically, for the hypergrid $\mathcal{H}_{m,d}$ of constant dimension d, the number of lookups is $(\log m)^{\Theta(d)}$, and for the hypercube $\mathcal{H}_d$, it is $2^{\Theta(d)}$ for any error blow-up.

Testers vs. Reconstructors. [13] obtained monotonicity testers from 2-TC-spanners. Unlike in the application to monotonicity testing, here we use *lower bounds* on the size of 2-TC-spanners to prove *lower bounds* on complexity of local monotonicity reconstuctors. Lower bounds on the size of 2-TC-spanners do not imply corresponding lower bounds on monotonicity testers. *E.g.*, the best monotonicity tester on $\mathcal{H}_d$ runs in $O(d^2)$ time [16, 17], while, as shown in Theorem 1.4, every 2-TC-spanner of $\mathcal{H}_d$ must have size exponential in d.

1.2   Our Results on 2-TC-Spanners of the Hypercube and Hypergrid

Our main theorem gives a set of explicit bounds on $S_2(\mathcal{H}_{m,d})$:

Theorem 1.3 (Hypergrid). *Let $S_2(\mathcal{H}_{m,d})$ denote the number of edges in the sparsest 2-TC-spanner of $\mathcal{H}_{m,d}$. Then[1] for $m \geq 3$,*

$$\Omega\left(\frac{m^d \log^d m}{(2d \log \log m)^{d-1}}\right) \;=\; S_2(\mathcal{H}_{m,d}) \;\leq\; m^d \log^d m.$$

The upper bound in Theorem 1.3 follows from a general construction of k-TC-spanners for graph products for arbitrary $k \geq 2$, presented in the full version. The lower bound is the most technically difficult part of our work. It is proved by a reduction of the 2-TC-spanner construction for $[m]^d$ to that for the $2 \times [m]^{d-1}$ grid and then directly analyzing the number of edges required for a 2-TC-spanner of $2 \times [m]^{d-1}$. We show a tradeoff between the number of edges in the 2-TC-spanner of the $2 \times [m]^{d-1}$ grid that stay within the hyperplanes $\{1\} \times [m]^{d-1}$ and $\{2\} \times [m]^{d-1}$ versus the number of edges that cross from one hyperplane to the other. The proof proceeds in multiple stages. Assuming an upper bound on the number of edges staying within the hyperplanes, each stage is shown to contribute a substantial number of new edges crossing between the hyperplanes. The proof of this tradeoff lemma is already non-trivial for $d = 2$ and is presented in Section 3. The proof for $d > 2$ is deferred to the full version of the paper.

While Theorem 1.3 is most useful when m is large and d is small, in Section 4 we present bounds on $S_2(\mathcal{H}_{m,d})$ which are optimal up to a factor of d^{2m} and, thus, supersede the bounds from Theorem 1.3 when m is small. The general form of these bounds is a somewhat complicated combinatorial expression but they can be estimated numerically. Specifically, $S_2(\mathcal{H}_{m,d}) = 2^{c_m d} \operatorname{poly}(d)$, where $c_2 \approx 1.1620$, $c_3 \approx 2.03$, $c_4 \approx 2.82$ and $c_5 \approx 3.24$, each significantly smaller than the exponents corresponding to the transitive closure sizes for the different m.

As a special case of the above, for $m = 2$ we obtain the following theorem for the hypercube. The proof of this theorem is omitted from this version.

---

[1] Logarithms are always to base 2 unless otherwise indicated.

Theorem 1.4 (Hypercube). *Let $S_2(\mathcal{H}_d)$ be the number of edges in the sparsest 2-TC-spanner of $\mathcal{H}_d$. Then $\Omega(2^{cd}) = S_2(\mathcal{H}_d) = O(d^3 2^{cd})$, where $c \approx 1.1620$.*

As a comparison point for our bounds, note that the obvious bounds on $S_2(\mathcal{H}_d)$ are the number of edges in the d-dimensional hypercube, $2^{d-1}d$, and the number of edges in the transitive closure of $\mathcal{H}_d$, which is $3^d - 2^d$. (An edge in the transitive closure of $\mathcal{H}_d$ has 3 possibilities for each coordinate: both endpoints are 0, both endpoints are 1, or the first endpoint is 0 and the second is 1. This includes self-loops, so we subtract the number of vertices in $\mathcal{H}_d$ to get the desired quantity.) Thus, $2^{d-1}d \leq S_2(\mathcal{H}_d) \leq 3^d - 2^d$. Similarly, the straightforward bounds on the number of edges in a 2-TC-spanner of $\mathcal{H}_{m,d}$ in terms of the number of edges in the directed grid and in its transitive closure are $dm^{d-1}(m-1)$ and $\left(\frac{m^2+m}{2}\right)^d - m^d$, respectively.

1.3   Previous Work on Bounding S_k for Other Families of Graphs

Thorup [18] considered a special case of TC-spanners of graphs G that have at most twice as many edges as G, and conjectured that for all directed graphs G on n nodes there are such k-TC-spanners with k polylogarithmic in n. He proved this for planar graphs [19], but Hesse [20] gave a counterexample for general graphs by constructing a family for which all $n^{\frac{1}{17}}$-TC-spanners need $n^{1+\Omega(1)}$ edges. TC-spanners were studied for directed trees: implicitly in [17, 21–24] and explicitly in [25]. For the directed line, [21] (and later, [22]) expressed $S_k(\mathcal{H}_{n,1})$ in terms of the inverse Ackermann function.

Lemma 1.1 ([13, 21, 22]). *Let $S_k(\mathcal{H}_{n,1})$ denote the number of edges in the sparsest k-TC-spanner of the directed line $\mathcal{H}_{n,1}$. Then $S_2(\mathcal{H}_{n,1}) = \Theta(n \log n)$, $S_3(\mathcal{H}_{n,1}) = \Theta(n \log \log n)$, $S_4(\mathcal{H}_{n,1}) = \Theta(n \log^* n)$ and, more generally, $S_k(\mathcal{H}_{n,1}) = \Theta(n\lambda_k(n))$ where $\lambda_k(n)$ is the inverse Ackermann function.*

The same bound holds for directed trees [21, 23, 25]. An $O(n \log n \cdot \lambda_k(n))$ bound on S_k for H-minor-free graph families (e.g., bounded genus and bounded tree-width graphs) was given in [13].

Notation. For a positive integer m, we denote $\{1, \ldots, m\}$ by $[m]$. For $x \in \{0,1\}^d$, we use $|x|$ to denote the weight of x, that is, the number of non-zero coordinates in x. Level i in a hypercube contains all vertices of weight i. The partial order $\preceq$ on the hypergrid $\mathcal{H}_{m,d}$ is defined as follows: $x \preceq y$ for two vertices $x, y \in [m]^d$ iff $x_i \leq y_i$ for all $i \in [d]$. Similarly, $x \prec y$, if x and y are distinct vertices in $[m]^d$ satisfying $x \preceq y$. Vertices x and y are *comparable* if either y is *above* x (that is, $x \preceq y$) or y is *below* x (that is, $y \preceq x$). We denote a path from v_1 to v_ℓ, consisting of edges $(v_1, v_2), (v_2, v_3), \ldots, (v_{\ell-1}, v_\ell)$ by $(v_1, \ldots, v_\ell)$.

2   From Monotonicity Reconstructors to 2-TC-Spanners

In this section, we prove Theorem 1.1.

Proof (of Theorem 1.1). Let A be a local reconstructor given by the statement of the theorem. Let $\mathcal{F}$ be the set of pairs (x, y) with x, y in V_n such that $x \prec y$. Then, $\mathcal{F}$ is of size at most $\binom{n}{2}$. Given $(x, y) \in \mathcal{F}$, let $\mathsf{cube}(x, y)$ be the set $\{z \in V_n : x \preceq z \preceq y\}$. Define function $f^{(x,y)}(v)$ to be 1 on all $v \succeq x$ and all $v \succeq y$, and 0 everywhere else. Also, define function $f^{(\overline{x},y)}(v)$, which is identical to $f^{(x,y)}(v)$ for all $v \notin \mathsf{cube}(x, y)$ and 0 for $v \in \mathsf{cube}(x, y)$. Both, $f^{(x,y)}$ and $f^{(\overline{x},y)}$, are monotone functions for all $(x, y) \in \mathcal{F}$. Let A_ρ be the deterministic algorithm which runs A with the random seed fixed to ρ. We say a string ρ is *good* for $(x, y) \in \mathcal{F}$ if filter A_ρ on input $f^{(x,y)}$ returns $g = f^{(x,y)}$ *and* on input $f^{(\overline{x},y)}$ returns $g = f^{(\overline{x},y)}$.

Now we show that there exists a set S of size $s \leq \lceil 2 \log n / \log(1/2\delta) \rceil$, consisting of strings used as random seeds by A, such that for every $(x, y) \in \mathcal{F}$ some string $\rho \in S$ is good for (x, y). We choose S by picking strings used as random seeds uniformly and independently at random. Since A has error probability at most δ, we know that for every monotone f, with probability at least $1 - \delta$ (with respect to the choice of ρ), the function $A_{f,\rho}$ is identical to f. Then, for fixed $(x, y) \in \mathcal{F}$ and uniformly random ρ,

$$\Pr[\rho \text{ is not } good \text{ for } (x, y)] \leq \quad \Pr[A_\rho \text{ on input } f^{(x,y)} \text{ fails to output } f^{(x,y)}]$$
$$+ \Pr[A_\rho \text{ on input } f^{(\overline{x},y)} \text{ fails to output } f^{(\overline{x},y)}] \leq 2\delta.$$

Since strings in S are chosen independently, $\Pr[\text{no } \rho \in S \text{ is good for } (x, y)] \leq (2 \cdot \delta)^s$, which, for $s = \lceil 2 \log n / \log(1/2\delta) \rceil$, is at most $1/n^2 < 1/|\mathcal{F}|$. By a union bound over $\mathcal{F}$, $\Pr[\text{for some } (x, y) \in \mathcal{F}, \text{ no } \rho \in S \text{ is good for } (x, y)] < 1$.

Thus, there exists a set S with required properties.

We construct our 2-TC-spanner $H = (V_n, E_H)$ of G_n using set S described above. Let $\mathcal{N}_\rho(x)$ be the set consisting of x and all vertices looked up by A_ρ on query x. (Note that the set $\mathcal{N}_\rho(x)$ is well-defined since algorithm A is assumed to be *non-adaptive*). For each string $\rho \in S$ and each vertex $x \in V_n$, connect x to all comparable vertices in $\mathcal{N}_\rho(x)$ (other than itself) and orient these edges according to their direction in G_n.

We prove H is a 2-TC-Spanner as follows. Suppose not, *i.e.*, there exists $(x, y) \in \mathcal{F}$ with no path of length at most 2 in H from x to y. Consider $\rho \in S$ which is *good* for (x, y). Define function h by setting $h(v) = f^{(x,y)}(v)$ for all $v \notin \mathsf{cube}(x, y)$. Then $h(v) = f^{(\overline{x},y)}(v)$ for all $v \notin \mathsf{cube}(x, y)$, by definition of $f^{(\overline{x},y)}$. For a $v \in \mathsf{cube}(x, y)$, set $h(v)$ to 1 for $v \in \mathcal{N}_\rho(x)$ and to 0 for $v \in \mathcal{N}_\rho(y)$. All unassigned points are set to 0. By the assumption above, $\mathcal{N}_\rho(x) \cap \mathcal{N}_\rho(y)$ does not contain any points in $\mathsf{cube}(x, y)$. Therefore, h is well-defined. Since ρ is *good* for (x, y) and h is identical to $f^{(x,y)}$ for all lookups made on query x, $A_\rho(x) = h(x) = 1$. Similarly, $A_\rho(y) = h(y) = 0$. But $x \prec y$, so $A_{h,\rho}(v)$ is not monotone. Contradiction.

The number of edges in H is at most

$$\sum_{x \in V_n, \rho \in S} |\mathcal{N}_\rho(x)| \leq n \cdot \ell(n) \cdot s \leq n\ell(n) \cdot \lceil 2 \log n / \log(1/2\delta) \rceil. \qquad \square$$

3   2-TC-Spanners for Low-Dimensional Hypergrids

In this section, we describe the proof of Theorem 1.3 which gives explicit bounds on the size of the sparsest 2-TC-spanner for $\mathcal{H}_{m,d}$. The upper bound in Theorem 1.3 follows straightforwardly from a more general statement about TC-spanners of product graphs; details are in the full version. Here, we show the lower bound on $S_2(\mathcal{H}_{m,d})$. Actually, in this extended abstract, we treat only the special case of this lower bound for $d = 2$, since it already contains most of the difficulty of the larger dimensional case. The extension to arbitrary dimension is deferred to the full version due to space constraints.

Theorem 3.1. *Any 2-TC-spanner of the 2-dimensional grid $\mathcal{H}_{m,2}$ must have* $\Omega\left(\frac{m^2 \log^2 m}{\log\log m}\right)$ *edges.*

One way to prove the $\Omega(m \log m)$ lower bound on the size of a 2-TC-spanner for the directed line $\mathcal{H}_{m,1}$, stated in Lemma 1.1, is to observe that at least $\lfloor \frac{m}{2} \rfloor$ edges are cut when the line is halved: namely, at least one per vertex pair $(v, m-v+1)$ for all $v \in \left[\lfloor \frac{m}{2} \rfloor\right]$. Continuing to halve the line recursively, we obtain the desired bound.

A natural extension of this approach to proving a lower for the grid is to recursively halve the grid along both dimensions, hoping that each such operation on an $m \times m$ grid cuts $\Omega(m^2 \log m)$ edges. This would imply that the size $S(m)$ of a 2-TC-spanner of the $m \times m$ grid satisfies the recurrence $S(m) = 4S(m/2) + \Omega(m^2 \log m)$; that is, $S(m) = \Omega(m^2 \log^2 m)$, matching the upper bound in Theorem 1.3.

An immediate problem with this approach is that in some 2-TC-spanners of the grid only $O(m^2)$ edges connect vertices in different quarters. One example of such a 2-TC-spanner is the graph containing the transitive closure of each quarter and only at most $3m^2$ edges crossing from one quarter to another: namely, for each node u and each quarter q with vertices comparable to u, this graph contains an edge (u, v_q), where v_q is the smallest node in q comparable to u.

The TC-spanner in the example above is not optimal because it has too many edges inside the quarters. The first step in our proof of Theorem 3.1 is understanding the tradeoff between the number of edges *crossing* the cut and the number of edges *internal* to the subgrids, resulting from halving the grid along some dimension. The simplest manifestation of this tradeoff occurs when a $2 \times m$ grid is halved into two lines. (In the case of one line, there is no trade off: the $\Omega(m)$ bound on the number of crossing edges holds even if each half-line contains all edges of its transitive closure.) Lemma 3.1 formulates the tradeoff for the two-line case, while taking into account only edges needed to connect comparable vertices on different lines by paths of length at most 2:

Lemma 3.1 (Two-Lines Lemma). *Let U be a graph with vertex set $[2] \times [m]$ that contains a path of length at most 2 from u to v for every $u \in \{1\} \times [m]$ and $v \in \{2\} \times [m]$, where $u \preceq v$. An edge (u, v) in U is called internal if $u_1 = v_1$, and crossing otherwise. If U contains at most $\frac{m \log^2 m}{32}$ internal edges, it must contain at least $\frac{m \log m}{16 \log \log m}$ crossing edges.*

Note that if the number of internal edges is unrestricted, a 2-TC-spanner of $\mathcal{H}_{m,2}$ may have only m crossing edges.

Proof. The proof proceeds in $\frac{\log m}{2 \log \log m}$ stages dealing with pairwise disjoint sets of crossing edges. In each stage, we show that U contains at least $\frac{m}{8}$ crossing edges in the prescribed set.

In the first stage, divide U into $\log^2 m$ blocks, each of length $\frac{m}{\log^2 m}$: namely, a node (v_1, v_2) is in block i if $v_2 \in \left[\frac{(i-1) \cdot m}{\log^2 m} + 1, \frac{i \cdot m}{\log^2 m} \right]$. Call an edge *long* if it starts and ends in different blocks, and *short* otherwise. Assume, for contradiction, that U contains fewer than $\frac{m}{8}$ long crossing edges.

Call a node (v_1, v_2) *low* if $v_1 = 1$ (*high* if $v_1 = 2$), and *left* if $v_2 \in \left[\frac{m}{2} \right]$ (*right* otherwise). Also, call an edge (u, v) *low-internal* if $u_1 = v_1 = 1$ and *high-internal* if $u_1 = v_1 = 2$. Let L be the set of low left nodes that are not incident to long crossing edges. Similarly, let R be the set of high right nodes that are not incident to long crossing edges. Since there are fewer than $\frac{m}{8}$ long crossing edges, $|L| > \frac{m}{4}$ and $|R| > \frac{m}{4}$.

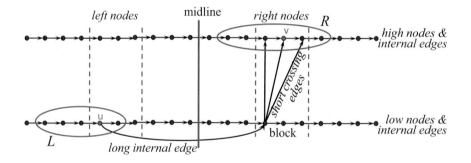

Fig. 1. Illustration of the first stage in the proof of Lemma 3.1

A node $u \in L$ can connect to a node $v \in R$ via a path of length at most 2 only by using a long internal edge. Observe that each long low-internal edge can be used by at most $\frac{m}{\log^2 m}$ such pairs (u, v): one low node u and high nodes v from one block. This is illustrated in Figure 1. Analogously, every long high-internal edge can be used by at most $\frac{m}{\log^2 m}$ such pairs. Since $|L| \cdot |R| > \frac{m^2}{16}$ pairs in $L \times R$ connect via paths of length at most 2, graph U contains more than $\frac{m^2}{16} \cdot \frac{\log^2 m}{m} = \frac{m \log^2 m}{16}$ long internal edges, which is a contradiction.

In each subsequent stage, call blocks used in the previous stage *megablocks*, and denote their length by B. Subdivide each megablock into $\log^2 m$ blocks of equal size. Call an edge *long* if it starts and ends in different blocks, but stays within one megablock. Assume, for contradiction, that U contains fewer than $\frac{m}{8}$ long crossing edges.

Call a node (v_1, v_2) *left* if it is in the left half of its megablock, that is, if $v_2 \leq \frac{\ell+r}{2}$ whenever (v_1, v_2) is in a megablock $[2] \times \{\ell, \ldots, r\}$. (Call it *right* otherwise). Consider megablocks containing fewer than $\frac{B}{4}$ long crossing edges each. By an averaging argument, at least $\frac{m}{2B}$ megablocks are of this type. (Recall that there are $\frac{m}{B}$ megablocks in total). Within each such megablock more than $\frac{B}{4}$ low left nodes and more than $\frac{B}{4}$ high right nodes have no incident long crossing edges. By the argument from the first stage, each such megablock contributes more than $\frac{B^2}{16b}$ long internal edges, where $b = \frac{B}{\log^2 m}$ is the size of the blocks. Hence there must be more than $\frac{B^2}{16b} \cdot \frac{m}{2B} = \frac{m \log^2 m}{32}$ long internal edges, which is a contradiction to the fact that U contains at most $\frac{m \log^2 m}{32}$ internal edges.

We proceed to the next stage until each block is of length 1. Therefore, the number of stages, t, satisfies $\frac{m}{\log^{2t} m} = 1$. That is, $t = \frac{\log m}{2 \log \log m}$, and each stage contributes $\frac{m}{8}$ new crossing edges, as desired. $\square$

Next we generalize Lemma 3.1 to understand the tradeoff between the number of internal edges and crossing edges resulting from halving a 2-TC-spanner of a $2\ell \times m$ grid with the usual partial order.

Lemma 3.2. *Let S be a 2-TC-spanner of the directed $[2\ell] \times [m]$ grid. An edge (u, v) in S is called* internal *if $u_1, v_1 \in [\ell]$ or $u_1, v_1 \in \{\ell + 1, \ldots, 2\ell\}$, and* crossing *otherwise. If S contains at most $\frac{\ell m \log^2 m}{64}$ internal edges, it must contain at least $\frac{\ell m \log m}{32 \log \log m}$ crossing edges.*

Proof. For each $i \in [\ell]$, we match the lines $\{i\} \times [m]$ and $\{2\ell - i + 1\} \times [m]$. Observe that a path of length at most 2 between the matched lines cannot use any edges with both endpoints in $\{i+1, \ldots, 2\ell - i\} \times [m]$. We modify S to ensure that there are no edges with only one endpoint in $\{i+1, \ldots, 2\ell - i\} \times [m]$ for all $i \in [\ell]$, and then apply Lemma 3.1 to the matched pairs of lines.

Call the $[\ell] \times [m]$ subgrid and all vertices and edges it contains *low*, and the remaining $\{\ell+1, \ldots, 2\ell\} \times [m]$ subgrid and its vertices and edges *high*. Transform S into S' as follows: change each low internal edge (u, v) to $(u, (u_1, v_2))$, change each high internal edge (u, v) to $((v_1, u_2), v)$, and finally change each crossing edge $((i_1, j_1), (2\ell - i_2 + 1, j_2))$ to $((i, j_1), (2\ell - i + 1, j_2))$, where $i = \min(i_1, i_2)$. Intuitively, we are projecting the edges in S to be fully contained in one of the matched pairs of lines, while preserving whether the edge is internal or crossing. Crossing edges are projected onto the outer matched pair of lines chosen from the two pairs that contain the endpoints of a given edge.

Clearly, S' contains at most the number of internal (crossing) edges as S. Observe that S' contains a path of length at most 2 from u to v for every comparable pair (u, v) where u is low, v is high, and u and v belong to the same pair of matched lines. Indeed, since S is a 2-TC-spanner, it contains either the edge (u, v) or a path (u, w, v). In the first case, S' also contains (u, v). In the second case, if (u, w) is a crossing edge S' contains $(u, (v_1, w_2), v)$, and if (u, w) is an internal edge S' contains $(u, (u_1, w_2), v)$. As claimed, each edge in S' belongs to one of the matched pairs of lines.

Finally, we apply Lemma 3.1. If S contains at most $\frac{\ell m \log^2 m}{64}$ internal edges, then so does S', and so at least half (i.e., $\frac{\ell}{2}$) of the matched line pairs each contain at most $\frac{m \log^2 m}{32}$ internal edges. By Lemma 3.1, each of these pairs contributes at least $\frac{m \log m}{16 \log \log m}$ crossing edges. Thus S' must contain at least $\frac{\ell m \log m}{32 \log \log m}$ crossing edges. Since S contains as many crossing edges as S', the lemma follows.   □

Now we prove Theorem 3.1 by recursively halving $\mathcal{H}_{m,2}$ along the horizontal dimension. Some resulting $\ell \times m$ subgrids may violate Lemma 3.2, but we can guarantee that the lemma holds for a constant fraction of the recursive steps for which $\ell \geq \sqrt{m}$. This is sufficient for obtaining the lower bound in the theorem.

Proof (of Theorem 3.1). Assume m is a power of 2 for simplicity. For each step $i \in \{1, \ldots, \frac{1}{2} \log m\}$, partition $\mathcal{H}_{m,2}$ into the following 2^{i-1} equal-sized subgrids: $\{1, \ldots, l_i\} \times [m]$, $\{l_i + 1, \ldots, 2l_i\} \times [m]$, $\ldots$, $\{m - l_i + 1, \ldots, m\} \times [m]$ where $l_i = m/2^{i-1}$. For each of these subgrids, define internal and crossing edges as in Lemma 3.2. Now, suppose that there exists a step i such that at least half of the 2^{i-1} subgrids have $> \frac{l_i m \log^2 m}{64}$ internal edges. Since at a fixed i, the subgrids are disjoint, there are $2^{i-1} \Omega(l_i m \log^2 m) = \Omega(m^2 \log^2 m)$ edges in S, proving the theorem. On the other hand, suppose that for every $i \in \{1, \ldots, \frac{1}{2} \log m\}$, at least half of the 2^{i-1} subgrids have $\leq \frac{l_i m \log^2 m}{64}$ internal edges. Then, applying Lemma 3.2, the number of crossing edges in those subgrids is $\geq \frac{l_i m \log m}{32 \log \log m}$. Counting over all steps i and for all appropriate subgrids from those steps, the number of edges in S is bounded by $\Omega \left(m^2 \log m \frac{\log m}{\log \log m} \right) = \Omega \left(m^2 \frac{\log^2 m}{\log \log m} \right)$.   □

In the full version, we extend the above proof to establish lower bounds on $S_2(\mathcal{H}_{m,d})$ for arbitrary $d \geq 2$. The main technical deferred result is a tradeoff lemma between internal and crossing edges with respect to two $(d-1)$-dimensional hyperplanes. An important part of the generalization is the appropriate definition of the notions of blocks and megablocks, so that the iterative argument in the proof of Lemma 3.1 applies in the high-dimensional setting.

4   2-TC-Spanners for High-Dimensional Hypergrids

Theorem 4.1 gives matching upper and lower bounds up to a d^{2m} factor in terms of an expression involving binomial coefficients. This result supersedes the results of the previous section when, for instance, m is constant and d is growing.

Before stating Theorem 4.1, we introduce some notation.

Definition 4.1. *For the hypergrid $\mathcal{H}_{m,d}$, define a* level *to be a set of vertices, indexed by vector $\mathbf{i} \in [d]^m$ with $i_1 + \cdots + i_m = d$, that consists of vertices $x = (x_1, \ldots, x_d) \in [m]^d$ containing i_1 positions of value 1, i_2 positions of value 2, $\ldots$, and i_m positions of value m.*

Notice that the number of vertices in level $\mathbf{i} = (i_1, i_2, \ldots, i_m)$ is the multinomial coefficient

$$\binom{d}{\mathbf{i}} = \binom{d}{i_1, \ldots, i_d} = \binom{d}{i_1} \binom{d - i_1}{i_2} \binom{d - i_1 - i_2}{i_3} \cdots \binom{d - \sum_{l=1}^{m-1} i_l}{i_m}.$$

Indeed, there are $\binom{d}{i_1}$ choices for the coordinates of value 1. For each such choice there are $\binom{d-i_1}{i_2}$ choices for the coordinates of value 2, and repeating this argument one obtains the above expression.

For levels $\mathbf{i}, \mathbf{j} \in [d]^m$, say $\mathbf{j}$ *majorizes* $\mathbf{i}$, denoted $\mathbf{j} \succ \mathbf{i}$, if $\mathbf{j}$ contains a vertex which is above some vertex in $\mathbf{i}$, *i.e.*, , if $\sum_{\ell=t}^{m} j_\ell \geq \sum_{\ell=t}^{m} i_\ell$ for all $t \in \{m, m-1, ..., 1\}$.

For $\mathbf{j} \succ \mathbf{i}$, the number of vertices y at level $\mathbf{i}$ comparable to a fixed vertex x at level $\mathbf{j}$ is $\mathcal{M}(\mathbf{i}, \mathbf{j})$:

$$\binom{j_m}{i_m}\binom{j_m + j_{m-1} - i_m}{i_{m-1}}\binom{j_m + j_{m-1} + j_{m-2} - i_m - i_{m-1}}{i_{m-2}} \cdots \binom{\sum_{l=1}^{m} j_l - \sum_{l=2}^{m} i_l}{i_1}.$$

Indeed, there are $\binom{j_m}{i_m}$ choices for the coordinates of value m in y. For each such choice, there are $\binom{j_m + j_{m-1} - i_m}{i_{m-1}}$ choices for the coordinates of value $m-1$ in y, and one can repeat this argument to obtain the claimed expression.

For $\mathbf{j} \succ \mathbf{i}$, the number of vertices y at level $\mathbf{j}$ comparable to a fixed vertex x at level $\mathbf{i}$ is

$$\mathcal{N}(\mathbf{i}, \mathbf{j}) = \frac{\mathcal{M}(\mathbf{i}, \mathbf{j})\binom{d}{\mathbf{j}}}{\binom{d}{\mathbf{i}}}.$$

Indeed, there are $\mathcal{M}(\mathbf{i}, \mathbf{j})\binom{d}{\mathbf{j}}$ comparable pairs of vertices in levels $\mathbf{i}$ and $\mathbf{j}$, and level $\mathbf{i}$ contains $\binom{d}{\mathbf{i}}$ vertices. Since, by symmetry, each vertex in $\mathbf{i}$ is comparable to the same number of vertices in level $\mathbf{j}$, we get the desired expression.

Theorem 4.1. *Let*

$$B(m, d) = \max_{\mathbf{i},\mathbf{j}:\mathbf{j} \succ \mathbf{i}} \min_{\mathbf{k}:\mathbf{i} \prec \mathbf{k} \prec \mathbf{j}} \frac{\mathcal{M}(\mathbf{i}, \mathbf{j})\binom{d}{\mathbf{j}}}{\mathcal{M}(\mathbf{i}, \mathbf{k})\mathcal{N}(\mathbf{k}, \mathbf{j})} \max\{\mathcal{M}(\mathbf{i}, \mathbf{k}), \mathcal{N}(\mathbf{k}, \mathbf{j})\}.$$

Then the number of edges in the sparsest 2-TC-spanner of the directed hypergrid $\mathcal{H}_{m,d}$ *is* $O\left(d^{2m}B(m, d)\right)$ *and* $\Omega\left(B(m, d)\right)$.

The proof for the upper bound part of Theorem 4.1 appears in the full version. We now prove the lower bound.

Lemma 4.1. *Any 2-TC-spanner of* $\mathcal{H}_{m,d}$ *has at least* $\Omega(B(m, d))$ *many edges, where* $B(m, d)$ *is defined as in Theorem 4.1.*

Proof. Let S be a 2-TC-spanner for $\mathcal{H}_{m,d}$. We count the edges in S that occur on paths connecting two particular levels of $\mathcal{H}_{m,d}$. Let $P_{\mathbf{i},\mathbf{j}} = \{(v_1, v_2) : v_1 \in \mathbf{i}, v_2 \in \mathbf{j}, v_1 \prec v_2\}$. We will lower bound $e^*_{\mathbf{i},\mathbf{j}}$, the number of edges in the paths of length at most 2 in S, that connect the pairs $P_{\mathbf{i},\mathbf{j}}$. Notice that $|P(\mathbf{i}, \mathbf{j})| = \binom{d}{\mathbf{j}}\mathcal{M}(\mathbf{i}, \mathbf{j})$.

Let $e_{\mathbf{k},\boldsymbol{\ell}}$ denote the number of edges in S that connect vertices in level $\mathbf{k}$ to vertices in level $\boldsymbol{\ell}$. Then

$$e^*_{\mathbf{i},\mathbf{j}} = e_{\mathbf{i},\mathbf{j}} + \sum_{\mathbf{i} \prec \mathbf{k} \prec \mathbf{j}} (e_{\mathbf{i},\mathbf{k}} + e_{\mathbf{k},\mathbf{j}}). \qquad (1)$$

We say that a vertex v *covers* a pair of vertices (v_1, v_2) if S contains the edges (v_1, v) and (v, v_2) or, for the special case $v = v_1$, if S contains (v_1, v_2). Let $V_{i,j}^{(k)}$ be the set of vertices in level k that cover pairs in $P_{i,j}$. Let α_k be the fraction of pairs in $P_{i,j}$ that are covered by the vertices in $V_{i,j}^{(k)}$. Since each pair in $P_{i,j}$ must be covered by a vertex in levels k with $i \prec k \prec j$, we must have $\sum_{i \prec k \prec j} \alpha_k \geq 1$.

For any vertex $v \in V_{i,j}^{(k)}$, let in_v be the number of incoming edges from vertices of level i incident to v and let out_v be the number of outgoing edges to vertices of level j incident to v. For each level k with $i \prec k \prec j$, since each vertex $v \in V_{i,j}^{(k)}$ covers $in_v \cdot out_v$ pairs,

$$\sum_{v \in V_{i,j}^{(k)}} in_v \cdot out_v \geq \alpha_k |P_{i,j}| \geq \alpha_k \mathcal{M}(i,j) \binom{d}{j}. \tag{2}$$

We upper bound $\sum_{v \in V_{i,j}^{(k)}} in_v \cdot out_v$ as a function of $e_{i,k} + e_{k,j}$, and then use Equation (2) to lower bound $e_{i,k} + e_{k,j}$. For all k with $i \prec k \prec j$, variables in_v and out_v satisfy the following constraints:

$$\sum_{v \in V_{i,j}^{(k)}} in_v \leq e_{i,k} \leq e_{i,k} + e_{k,j}, \qquad \sum_{v \in V_{i,j}^{(k)}} out_v \leq e_{k,j} \leq e_{i,k} + e_{k,j},$$

$$in_v \leq \mathcal{M}(i,k) \ \forall v \in V_{i,j}^{(k)}, \qquad out_v \leq \mathcal{N}(k,j) \ \forall v \in V_{i,j}^{(k)}.$$

The last two constraints hold because in_v and out_v count the number of edges to a vertex of level k from vertices of level i, and from a vertex of level k to vertices of level j, respectively. Using these bounds we obtain

$$\sum_{v \in V_{i,j}^{(k)}} in_v \cdot out_v \leq \sum_{v \in V_{i,j}^{(k)}} \mathcal{M}(i,k) \cdot out_v = \mathcal{M}(i,k) \cdot \sum_{v \in V_{i,j}^{(k)}} out_v \leq \mathcal{M}(i,k) \cdot (e_{i,k} + e_{k,j}).$$

Similarly, $\sum_{v \in V_{i,j}^{(k)}} in_v \cdot out_v \leq \mathcal{N}(k,j) \cdot (e_{i,k} + e_{k,j})$. Therefore,

$$\sum_{v \in V_{i,j}^{(k)}} in_v \cdot out_v \leq (e_{i,k} + e_{k,j}) \min \{\mathcal{M}(i,k), \mathcal{N}(k,j)\}.$$

From Equation (2), $e_{i,k} + e_{k,j} \geq \alpha_k \mathcal{M}(i,j) \binom{d}{j} \dfrac{1}{\min \{\mathcal{M}(i,k), \mathcal{N}(k,j)\}}$ for all $i \prec k \prec j$. Applying Equation (1) and the fact that $\sum_{i \prec k \prec j} \alpha_k \geq 1$, we get

$$e_{i,j}^* = e_{i,j} + \sum_{i \prec k \prec j} (e_{i,k} + e_{k,j}) \geq \sum_k \alpha_k \frac{1}{\min \{\mathcal{M}(i,k), \mathcal{N}(k,j)\}} \mathcal{M}(i,j) \binom{d}{j}$$

$$\geq \min_k \frac{1}{\min \{\mathcal{M}(i,k), \mathcal{N}(k,j)\}} \mathcal{M}(i,j) \binom{d}{j}$$

$$= \min_k \frac{1}{\mathcal{M}(i,k) \mathcal{N}(k,j)} \mathcal{M}(i,j) \binom{d}{j} \max \{\mathcal{M}(i,k), \mathcal{N}(k,j)\}.$$

Since this holds for arbitrary i and j, the size of the 2-TC-spanner is $|S| \geq \mathcal{B}(m, d)$. $\qquad \square$

References

1. Peleg, D., Schäffer, A.A.: Graph spanners. Journal of Graph Theory 13, 99–116 (1989)
2. Cowen, L.: Compact routing with minimum stretch. J. Algorithms 38, 170–183 (2001)
3. Cowen, L., Wagner, C.G.: Compact roundtrip routing in directed networks. J. Algorithms 50, 79–95 (2004)
4. Peleg, D., Upfal, E.: A trade-off between space and efficiency for routing tables. JACM 36, 510–530 (1989)
5. Roditty, L., Thorup, M., Zwick, U.: Roundtrip spanners and roundtrip routing in directed graphs. In: SODA, pp. 844–851 (2002)
6. Thorup, M., Zwick, U.: Compact routing schemes. In: ACM Symposium on Parallel Algorithms and Architectures, pp. 1–10 (2001)
7. Peleg, D., Ullman, J.D.: An optimal synchronizer for the hypercube. SIAM J. Comput. 18, 740–747 (1989)
8. Cohen, E.: Fast algorithms for constructing t-spanners and paths with stretch t. SIAM J. Comput. 28, 210–236 (1998)
9. Cohen, E.: Polylog-time and near-linear work approximation scheme for undirected shortest paths. JACM 47, 132–166 (2000)
10. Elkin, M.: Computing almost shortest paths. In: PODC, pp. 53–62 (2001)
11. Baswana, S., Sen, S.: Approximate distance oracles for unweighted graphs in expected $\tilde{O}(n^2)$ time. ACM Transactions on Algorithms 2, 557–577 (2006)
12. Thorup, M., Zwick, U.: Approximate distance oracles. JACM 52, 1–24 (2005)
13. Bhattacharyya, A., Grigorescu, E., Jung, K., Raskhodnikova, S., Woodruff, D.P.: Transitive-closure spanners. In: SODA, pp. 932–941 (2009)
14. Saks, M., Seshadhri, C.: Local monotonicity reconstruction. SIAM Journal on Computing 39, 2897–2926 (2010)
15. Ailon, N., Chazelle, B., Comandur, S., Liu, D.: Property-preserving data reconstruction. Algorithmica 51, 160–182 (2008)
16. Goldreich, O., Goldwasser, S., Lehman, E., Ron, D., Samorodnitsky, A.: Testing monotonicity. Combinatorica 20, 301–337 (2000)
17. Dodis, Y., Goldreich, O., Lehman, E., Raskhodnikova, S., Ron, D., Samorodnitsky, A.: Improved testing algorithms for monotonicity. In: Rolim, J.D.P. (ed.) RANDOM 1997. LNCS, vol. 1269, pp. 97–108. Springer, Heidelberg (1997)
18. Thorup, M.: On shortcutting digraphs. In: Mayr, E.W. (ed.) WG 1992. LNCS, vol. 657, pp. 205–211. Springer, Heidelberg (1993)
19. Thorup, M.: Shortcutting planar digraphs. Combinatorics, Probability & Computing 4, 287–315 (1995)
20. Hesse, W.: Directed graphs requiring large numbers of shortcuts. In: SODA, pp. 665–669 (2003)
21. Alon, N., Schieber, B.: Optimal preprocessing for answering on-line product queries. Technical Report 71/87, Tel-Aviv University (1987)
22. Atallah, M.J., Frikken, K.B., Fazio, N., Blanton, M.: Dynamic and efficient key management for access hierarchies. In: ACM Conference on Computer and Communications Security, pp. 190–202 (2005)
23. Chazelle, B.: Computing on a free tree via complexity-preserving mappings. Algorithmica 2, 337–361 (1987)
24. Yao, A.C.C.: Space-time tradeoff for answering range queries (extended abstract). In: STOC, pp. 128–136 (1982)
25. Thorup, M.: Parallel shortcutting of rooted trees. J. Algorithms 23, 139–159 (1997)

Monotonicity Testing and Shortest-Path Routing on the Cube

Jop Briët, Sourav Chakraborty, David García-Soriano, and Arie Matsliah

Centrum Wiskunde & Informatica,
Amsterdam, Netherlands
{jop.briet,sourav.chakraborty,david,ariem}@cwi.nl

Abstract. We study the problem of monotonicity testing over the hypercube. As previously observed in several works, a positive answer to a natural question about routing properties of the hypercube network would imply the existence of efficient monotonicity testers. In particular, if any set of source-sink pairs on the directed hypercube (with all sources and all sinks distinct) can be connected with edge-disjoint paths, then monotonicity of functions $f : \{0,1\}^n \to \mathcal{R}$ can be tested with $O(n/\epsilon)$ queries, for any totally ordered range $\mathcal{R}$. More generally, if at least a $\mu(n)$ fraction of the pairs can always be connected with edge-disjoint paths then the query complexity is $O(n/(\epsilon\mu(n)))$.

We construct a family of instances of $\Omega(2^n)$ pairs in n-dimensional hypercubes such that no more than roughly a $\frac{1}{\sqrt{n}}$ fraction of the pairs can be simultaneously connected with edge-disjoint paths. This answers an open question of Lehman and Ron [LR01], and suggests that the aforementioned appealing combinatorial approach for deriving query-complexity upper bounds from routing properties cannot yield, by itself, query-complexity bounds better than $\approx n^{3/2}$. Additionally, our construction can also be used to obtain a strong counterexample to Szymanski's conjecture about routing on the hypercube. In particular, we show that for any $\delta > 0$, the n-dimensional hypercube is not $n^{\frac{1}{2}-\delta}$-realizable with shortest paths, while previously it was only known that hypercubes are not 1-realizable with shortest paths.

We also prove a lower bound of $\Omega(n/\epsilon)$ queries for one-sided non-adaptive testing of monotonicity over the n-dimensional hypercube, as well as additional bounds for specific classes of functions and testers.

1 Background

Testing monotonicity of functions [DGL+99],[Ras99],[GGL+00],[EKK+00],[Fis04], [FLN+02],[AC06],[Bha08],[HK08] is one of the oldest and most studied problems in Property Testing. The problem is defined as follows: Let $\mathcal{D}$ be a partially ordered set (poset) and let $\mathcal{R} \subseteq \mathbb{Z}$. A function $f : \mathcal{D} \to \mathcal{R}$ is monotone if for every (comparable) pair $x, y \in \mathcal{D}$, $x \leq y$ implies $f(x) \leq f(y)$. A function f is ϵ-far from monotone if it has to be changed on at least an ϵ-fraction of the domain $\mathcal{D}$ to become monotone. A

M. Serna et al. (Eds.): APPROX and RANDOM 2010, LNCS 6302, pp. 462–475, 2010.

(q, ϵ)-monotonicity tester for domain $\mathcal{D}$ and range $\mathcal{R}$ is a probabilistic algorithm that, given oracle access to a function $f : \mathcal{D} \to \mathcal{R}$, satisfies the following: (a) it makes at most q queries to f; (b) it accepts with probability at least $2/3$ if f is monotone; (c) it rejects with probability at least $2/3$ if f is ϵ-far from monotone.

The simplest monotonicity testers are those that specify all their queries in advance (non-adaptively) and reject if and only if they reveal a violation, i.e. if $f(x) > f(y)$ for some comparable pair $x \le y$ of points queried from $\mathcal{D}$. These *non-adaptive* testers with *one-sided error* are the only ones considered in this paper, unless explicitly stated otherwise. We note that nearly all known monotonicity testers are non-adaptive and have one-sided error. Furthermore, it is also known that if $\mathcal{D}$ is totally ordered then non-adaptive testers with one-sided error are as powerful (in terms of query complexity) as general ones [Fis04].

For general domains $\mathcal{D}$, Fischer et al. [FLN+02] proved that testing monotonicity is equivalent to several natural problems, including testing certain graph properties and testing assignments for Boolean formulae. Domains of the form $\{0, 1, \ldots, m\}^n$, however, received most of the attention [DGL+99], [EKK+00], [GGL+00], [Fis04], [Ras99], [Bha08], [BGJ+09a], [BGJ+09b]. Here the order relation $x \le y$ is defined to hold for $x, y \in \{0, \ldots, m\}^n$ when $x_i \le y_i$ for all $i \in [n]$. In this paper we focus on a well-studied subcase of the above, where $m = 1$ and $\mathcal{R} \subseteq \mathbb{Z}$.

1.1 Preliminaries

Every $x \in \{0, 1\}^n$ is identified with the subset $support(x) = \{i \in [n] : x_i = 1\}$ as usual. With a slight abuse of notation, we interpret binary strings as sets (and vice-versa). E.g., we write $x \subseteq y$ (or $x \le y$) for two strings $x, y \in \{0, 1\}^n$ such that $support(x) \subseteq support(y)$.

The *directed n-dimensional hypercube* (or simply n-cube) is a directed graph $H_n = (V_n, E_n)$ with $V_n = \{0, 1\}^n$ and $E_n = \{(x, y) : x \subseteq y \text{ and } |y| = |x| + 1\}$. The h-th *layer* (or *level*) of H_n contains all $x \in V_n$ with $|x| = h$.

Definition 1. *A set $\mathcal{P} \subseteq V_n \times V_n$ of ℓ pairs $\{(s^i, t^i)\}_{i=1}^{\ell}$ is called a* source-sink pairing *(of size ℓ), with sources $s^1, \ldots, s^\ell$ and sinks $t^1, \ldots, t^\ell$, if*

- $s^i \subset t^i$ *for all $i \in [\ell]$ and*
- $s^i \ne s^j$, $s^i \ne t^j$ *and $t^i \ne t^j$ for all $i, j \in [\ell], i \ne j$.*

$\mathcal{P}$ *is* aligned *if in addition $|s^i| = |s^j|$ and $|t^i| = |t^j|$ for all $i, j \in [\ell]$.*

Notice that $\mathcal{P}$ is a source-sink pairing if and only if it forms a (partial) matching in the transitive closure of H_n. Throughout this paper we denote by $\mathcal{P}$ only sets of pairs that form a source-sink pairing, even when it is not explicitly mentioned.

A (directed) path in H_n is called a $\mathcal{P}$-*path* if it connects some source s^i from $\mathcal{P}$ to its sink t^i. A subset $C \subseteq E_n$ is called a $\mathcal{P}$-*cut* if every $\mathcal{P}$-path in H_n uses at least one edge from C. Similarly, a subset $S \subseteq V_n$ is called a $\mathcal{P}$-*vertex-cut* if every $\mathcal{P}$-path uses at least one vertex from S. We write maxflow($\mathcal{P}$) for the size of the largest set of edge-disjoint $\mathcal{P}$-paths, mincut($\mathcal{P}$) for the size of the smallest $\mathcal{P}$-cut and minvertexcut($\mathcal{P}$) for the size of the smallest $\mathcal{P}$-vertex-cut. Clearly mincut($\mathcal{P}$) is an upper bound on both minvertexcut($\mathcal{P}$) and maxflow($\mathcal{P}$). Unlike the case with a single pair in $\mathcal{P}$, these quantities need not coincide.

We define the terms *sparsity* and *meagerness* as in [RL05], [ABY08], [AHJ$^+$06]. The *sparsity* of $\mathcal{P}$ is the ratio mincut($\mathcal{P}$)/$|\mathcal{P}|$, and the *vertex-sparsity* of $\mathcal{P}$ is the ratio minvertexcut($\mathcal{P}$)/$|\mathcal{P}|$. The *sparsity* and the *vertex-sparsity* of H_n are defined as $\min_{\mathcal{P}}\{$mincut($\mathcal{P}$)/$|\mathcal{P}|\}$ and $\min_{\mathcal{P}}\{$minvertexcut($\mathcal{P}$)/$|\mathcal{P}|\}$, respectively. In other words, *sparsity* is the average number of edges per source-sink pair that one has to remove to disconnect every source from its sink, whereas *vertex-sparsity* is the average number of vertices per source-sink pair that one has to remove to disconnect every source from its sink. The definitions of *meagerness* and *vertex-meagerness* are similar, except for the stronger requirement that the corresponding cuts disconnect *all* sources s^i from *all* sinks t^j.

Observe that (1) sparsity $\geq$ vertex-sparsity; (2) meagerness $\geq$ vertex-meagerness; (3) meagerness $\geq$ sparsity and (4) vertex-meagerness $\geq$ vertex-sparsity.

Given $\mathcal{R} \subseteq \mathbb{Z}$ and a function $f : \{0,1\}^n \to \mathcal{R}$, we say that a pair $(x,y) \in V_n \times V_n$ is *violated* by f if $x \leq y$ and $f(x) > f(y)$. If in addition $(x,y) \in E_n$, we call it a *violated edge*. We denote by Viol(f) the set of all pairs (x,y) violated by f, and by EdgeViol(f) the set of all *edges* violated by f. Thus, f is monotone if and only if Viol(f) = EdgeViol(f) = $\emptyset$.

We denote by $\epsilon_M(f) \in [0,1]$ the relative distance of f from being monotone, i.e. the minimum of $\Pr_x[f(x) \neq g(x)]$ taken over all monotone functions $g : \{0,1\}^n \to \mathcal{R}$. Let $\delta_M(f) \in [0,1]$ denote the fraction $|$EdgeViol(f)$|$/$|E_n| = |$EdgeViol(f)$|$/$(n2^{n-1})$ of edges violated by f.

2   Our Results and Related Work

2.1   Monotonicity Testers via Sparsity Lower Bounds

One of the earliest upper bounds on the query complexity of monotonicity testing on the hypercube used an approach based on the concepts of meagerness and sparsity [GGLR98]. In particular, [GGLR98] observed that if the meagerness of H_n is at least 1, then monotonicity of Boolean functions can be tested with $O(n/\epsilon)$ queries. Then they proved that vertex-meagerness (and hence meagerness too) is 1 if the possible pairings $\mathcal{P}$ are restricted to *aligned* sets, satisfying $|s^i| = |s^j|$ and $|t^i| = |t^j|$ for all i, j (see also [LR01] for a detailed proof). This sufficed to derive an upper-bound of $O(n^2)$ queries for any constant $\epsilon > 0$.

While a lower bound on meagerness implies query-complexity upper bounds for Boolean functions, a lower bound on sparsity implies query-complexity upper bounds for functions with general range (see Section 3.1 for details). In particular, if the sparsity of H_n is at least $\mu = \mu(n)$, then monotonicity of functions with any linearly ordered range can be tested with $O(n/(\epsilon\mu))$ queries. In [LR01] the authors ask whether the sparsity of any $\mathcal{P}$ (or even just of the aligned ones) is at least 1, noting that this would imply the existence of efficient monotonicity testers as well as progress on some long-standing questions regarding routing in the hypercube network. We prove that the answer to both of their questions is *no*. The following theorem is proved in Section 3.2:

Theorem 2. *The sparsity of H_n is at most $n^{-\frac{1}{2}+o(1)}$. Furthermore, this upper bound on the sparsity can be demonstrated both with aligned sets and with $\Omega(2^n)$-sized sets:*

- *for any $\delta > 0$ and large enough n there is an aligned set $\mathcal{P}$ in H_n with sparsity at most $n^{-\frac{1}{2}+\delta}$;*
- *for any $\delta > 0$ there is $\epsilon > 0$, such that for large enough n there is a set $\mathcal{P}$ in H_n of size $|\mathcal{P}| \geq \epsilon 2^n$ with sparsity at most $n^{-\frac{1}{2}+\delta}$.*

2.2   Routing in the Hypercube and Szymanski's Conjecture

The hypercube is a natural and well-studied architecture for multi-processor systems and networks. The ability to route arbitrary permutations on it models flow of information in a network of processors. In this context, a doubly-directed version of H_n is usually considered, where each edge in E_n is replaced with a pair of anti-parallel edges. Let us denote the doubly-directed version of H_n by $H_n^{\uparrow\downarrow}$. A permutation π of V_n is 1-*realizable* if there exist pairwise edge-disjoint paths in $H_n^{\uparrow\downarrow}$ that connect every v with $\pi(v)$. A permutation π is k-*realizable* if there exist paths connecting every v with $\pi(v)$ such that each edge is used in at most k paths. Szymanski [Szy89] conjectured that any permutation π of V_n is 1-realizable with *shortest paths*. It was proved that the conjecture holds up to dimension 3, but later Lubiw [Lub90] provided a counterexample in dimension 5 that is not 1-realizable using shortest paths. While it is still unknown whether or not every permutation is 1-realizable *without* requiring shortest paths[1], the fact that any permutation is 2-realizable follows from the classical work of Beneš [Ben65] (see [Lub90] for details). In contrast, we prove that if we insist on the shortest-path condition, there are permutations that are not k-realizable for any k significantly smaller than $\sqrt{n}$. Specifically, the construction in Theorem 2 can be used (see Section 3.3) to prove the following.

Theorem 3. *For any $\delta > 0$ and large enough n, there are permutations on V_n that cannot be $n^{\frac{1}{2}-\delta}$-realized in $H_n^{\uparrow\downarrow}$ with shortest paths.*

Remark 1. Any upper bound $\mu(n)$ on the sparsity of H_n can be used to show that $H_n^{\uparrow\downarrow}$ is not $1/\mu(n)$-realizable with shortest paths. But the opposite is not true; in particular, the counterexample from [Lub90] does not imply that the sparsity of H_5 is less than 1.

---

[1] Since the original conjecture was shown to be false, the weaker version that does not require shortest paths is now called Szymanski's conjecture.

2.3   New Bounds on Testing Monotonicity

At the moment the best known query-complexity bounds for testing monotonicity (non-adaptively with one-sided error) of functions $f : \{0,1\}^n \to \mathcal{R}$ are:

- an upper bound of $O(\frac{n}{\epsilon} \log |\mathcal{R}|)$ for any range $\mathcal{R}$ [DGL+99];
- a lower bound of $\Omega(\sqrt{n}/\epsilon)$ for Boolean ranges (and hence for wider ranges too) [FLN+02].

The tester used in the upper bound of [DGL+99] is perhaps the most natural one: it picks an edge $(x, y) \in E_n$ uniformly at random, and rejects if $f(x) > f(y)$. Let us call this test an *edge-test*. [DGL+99] prove that the probability that a single execution of an edge-test rejects is $\Omega(\frac{\epsilon_M(f)}{n \log |\mathcal{R}|})$, by relating the distance of a function from monotone to the number of edges that it violates.

It is an interesting open question whether the general upper bound of [DGL+99] can be improved into one that is independent of $|\mathcal{R}|$ (or at least has a better dependence on it). Since we can assume without loss of generality that $|\mathcal{R}| \leq 2^n$, any upper bound of $o(n^2/\epsilon)$ queries would be an improvement. We make a small step in this direction. Call a function $f : \{0,1\}^n \to \mathcal{R}$ *dist-k monotone* if $f(y) \geq f(x)$ for every $y > x$ with $|y| > |x| + k$. In this terminology dist-0 monotone is simply monotone. In Section 3.4 we prove that given a dist-3 monotone function f, we can test if f is monotone with $O(n^{3/2}/\epsilon)$ queries. We actually prove the following stronger claim:

Theorem 4. *Let $\epsilon > 0$, $\mathcal{R} \subseteq \mathbb{Z}$ and let $f : \{0,1\}^n \to \mathcal{R}$ be a dist-3 monotone function. If f is ϵ-far from being monotone then $|\mathsf{EdgeViol}(f)| \geq \Omega\left(\frac{2^n}{\epsilon\sqrt{n}}\right)$.*

The upper bound on the query complexity follows using the edge-tests described above.

The reasons for considering dist-3 monotonicity here are twofold. Firstly, it is the first non-trivial case (it is easy to see that both dist-1 and dist-2 monotone functions can be tested in $O(n/\epsilon)$ queries). Secondly, we will see later that non-trivial sparsity upper bounds already exist for pairings in which every source is at distance 3 from its sink.

In Section 3.5 we also extend the lower bound of $\Omega(\sqrt{n}/\epsilon)$ of [FLN+02] to $\Omega(n/\epsilon)$, for large enough $|\mathcal{R}|$. Using the "Range-Reduction Lemma" of [DGL+99], the new bound implies an improved lower bound of $\Omega(n/(\epsilon \log n))$ for the Boolean range, in the special case of pair-testers whose query complexity can be written as $q(n)/\epsilon$ for some function q. (A *pair-tester* picks independent pairs of comparable vertices according to some distribution, and rejects if and only if one of them forms a violation). We note that such testers are not overly restricted: essentially all known query-complexity upper bounds for monotonicity-testing use (or can be easily converted into ones that use) pair-tests of this kind. Furthermore, the new lower-bound almost matches the aforementioned upper-bound of $O(n/\epsilon)$ achieved by edge-tests (a special case of pair-tests).

3   Proofs

3.1   From Sparsity to Monotonicity Testers

The basic combinatorial interpretation of $\epsilon_M(f)$ is given in the following lemma:

Lemma 1. *[DGL+99], [FLN+02], [GGL+00] Let $f : \{0,1\}^n \to \mathcal{R}$ be a function, and define the violation graph of f as the undirected graph $G = (\{0,1\}^n, E)$, where $\{x, y\} \in E$ if either (x, y) or (y, x) is in $\mathsf{Viol}(f)$. Then $\epsilon_M(f)2^n$ is exactly the size of a minimum vertex cover of G. Consequently, there is a matching in G of size at least $\epsilon_M(f)2^{n-1}$.*

An important observation is that since G is a subgraph of the transitive closure of H_n, the matching of violated pairs in Lemma 1 forms a source-sink pairing $\mathcal{P}$ (see Definition 1) of size $\epsilon_M(f)2^{n-1}$.

As we mentioned earlier, the best known upper bounds for testing monotonicity over hypercubes are obtained by a simple edge-tester, which picks a set of edges from H_n uniformly at random, queries f on their endpoints, and rejects if one of them is violated. Recall that $\delta_M(f)$ denotes the fraction of edges in H_n that are violated by f; thus the success probability of the edge-tester is determined by $\delta_M(f)$. Goldreich et al prove the following:

Theorem 5. *[GGLR98], [GGL+00] For any $f : \{0,1\}^n \to \{0,1\}$, $\delta_M(f) \geq \frac{\epsilon_M(f)}{n}$.*

More generally, [DGL+99] use their range-reduction lemma to conclude that for any $f : \{0,1\}^n \to \mathcal{R}$, $\delta_M(f) \geq \frac{\epsilon_M(f)}{n \log |\mathcal{R}|}$. Since without loss of generality $|\mathcal{R}| \leq 2^n$, this gives an upper bound of $O(n^2/\epsilon)$ queries for testing monotonicity of all functions $f : \{0,1\}^n \to \mathcal{R}$.

Clearly, obtaining better lower bounds on $\delta_M(f)$ is sufficient for improving the upper bounds on the query complexity of testing monotonicity. (It may even be the case that Theorem 5 holds for any $\mathcal{R}$). The next lemma states that this can also be done by proving lower bounds on the sparsity of H_n.

Lemma 2. *Let $\mu(n)$ denote the sparsity of H_n. For any $\epsilon > 0$ and $\mathcal{R} \subseteq \mathbb{Z}$, monotonicity of functions $f : \{0,1\}^n \to \mathcal{R}$ can be tested with $O(\frac{n}{\epsilon\mu(n)})$ queries.*

Proof: Let $\epsilon > 0$ and let $f : \{0,1\}^n \to \mathcal{R}$ be ϵ-far from monotone. Let $\mathcal{P}$ be the set of $\epsilon_M(f)2^{n-1} \geq \epsilon 2^{n-1}$ vertex-disjoint violated pairs promised by Lemma 1. By definition, $\mathcal{P}$ is a source-sink pairing. Notice that since every $(s^i, t^i) \in \mathcal{P}$ is violated, we have that every path from s^i to t^i must contain at least one violated edge. It follows that the set $\mathsf{EdgeViol}(f)$ is a $\mathcal{P}$-cut and $|\mathsf{EdgeViol}(f)|/|\mathcal{P}| \geq \mu(n)$. Hence $\delta_M(f) = \frac{|\mathsf{EdgeViol}(f)|}{|E_n|} \geq \frac{\epsilon\mu(n)}{n}$. We can thus conclude that $O(\frac{n}{\epsilon\mu(n)})$ edge queries suffice to find an edge-violation with constant probability.   □

3.2 Proof of Theorem 2

We use a number of properties of the parity-check matrix of Hamming codes, which we now describe. For an integer $k \geq 1$, let the strings $y \in \{0,1\}^k \backslash \{0\}^k$ represent the indices of bit positions of binary strings of length $n = 2^k - 1$. The Hamming code consists of the n-bit strings $x \in \{0,1\}^n$ that, for every $i \in [k]$, have an even number of positions y for which $y_i = 1$ and $x_y = 1$. The columns of its $k \times n$ parity check matrix p are all possible non-zero k-bit vectors y; this matrix represents a linear map $p : \{0,1\}^n \rightarrow \{0,1\}^k$, with arithmetic done modulo 2. Therefore, for any unit vector e_y (i.e., the vector having 1 at position y and 0 elsewhere), $p(e_y) = y$. Consequently, for all x, y, $p(x \oplus e_y) = p(x) \oplus y$.

Codewords of the Hamming code correspond to strings satisfying $p(x) = 0$ (here and in what follows we use 0 to denote the all-zero vector of the appropriate size). The k bit positions of the form 2^i (i.e., $1, 2, 4, \ldots, (n+1)/2$) can be viewed as the parity bits of the code; in a codeword they are determined by the remaining $n - k$ bits.

Warm-up. To showcase the main ideas in the construction, we first show that the sparsity of the hypercube is at most $O(\frac{1}{n^{1/3}})$; better bounds are derived in Section 3.2.

Proposition 6. *Let $k > 0$ be a multiple of three, and $n = 2^k - 1$. There is a pairing $\mathcal{P} \subseteq V_n \times V_n$ in H_n of size $|\mathcal{P}| = \Omega(2^n)$ having a $\mathcal{P}$-cut $C \subseteq E_n$ of size $|C| = O(2^n/n^{1/3})$.*

Proof: For $a \in \{0,1\}^n$, consider the k parity bits $p(a)$ and divide them into three groups of size $k/3$ each, denoted $x(a), y(a)$ and $z(a)$. For convenience, we will write (v_1, v_2, v_3) to denote the concatenation of three vectors $v_1, v_2, v_3 \in \{0,1\}^{k/3}$, and whenever no confusion may arise, we interpret every $v \in \{0,1\}^k$ as an element of $\{0\} \cup [n]$. With this convention, we have $p(a) = (x(a), y(a), z(a))$, and if one of v_1, v_2 or v_3 is non-zero, then $(v_1, v_2, v_3) \in [n]$.

The set S of sources of $\mathcal{P}$ is the set of all $s \in \{0,1\}^n$ that satisfy

$$\Big(x(s) \neq 0 \wedge y(s) \neq 0 \wedge z(s) \neq 0\Big) \wedge \Big(s_{(x(s),y(s),0)} = s_{(x(s),0,z(s))} = s_{(0,y(s),z(s))} = 0\Big).$$

For each source $s \in S$, we define its sink t as

$$t = s \cup \{(x(s), y(s), 0), (x(s), 0, z(s)), (0, y(s), z(s))\}.$$

That is, the three directions leading from s to t are $(x(s), y(s), 0), (x(s), 0, z(s))$ and $(x(s), 0, z(s))$. The first three conditions on a member s of S ensure that all three directions are (1) distinct; (2) proper (i.e. non-zero); and (3) have a k-bit binary representation with Hamming weight strictly greater than one. The last condition ensures that the relevant bits of s are set to zero.

The pairing $\mathcal{P}$ will be given by all pairs (s, t) defined in this way. Clearly $s \subseteq t$ and $|t - s| = 3$. It is easy to verify that $|S| = (2^{k/3} - 1)^3 2^{n-k-3} = \Omega(2^n)$, since none of the directions used corresponds to a parity bit, i.e., none of them is a power of 2.

To prove that $\mathcal{P}$ is a pairing, it remains to show that all sources are distinct, and that no source is also a sink. Because of the properties of map p, after flipping e.g. bit $(x, y, 0)$ from a source s with parity (x, y, z), we reach a vertex with parity $(0, 0, z)$. Thus, we see that the parities of the eight vertices in the cube from s to t are:

- Level 3 (sink): (x, y, z).
- Level 2: $(x, 0, 0), (0, y, 0), (0, 0, z)$.
- Level 1: $(0, 0, z), (0, y, 0), (x, 0, 0)$.
- Level 0 (source): (x, y, z).

Notice that the parities at level 1 are distinct, as are the parities at level 2.

Since the three directions from s to t are determined by $p(s) = (x, y, z) = p(t)$, it follows that the set of sinks is disjoint from the set of sources (these bits already belong to t, so $t \notin S$). Likewise, if two different sources s_1 and s_2 were associated with the same sink t, we would get $p(s_1) = p(t) = p(s_2)$, so the three directions from s_1 to t are the same as from s_2 to t, implying $s_1 = s_2$. Hence $\mathcal{P}$ is indeed a pairing.

Let $Q \subseteq V_n$ be the set of vertices at level 1 or 2 for some pair $(s, t) \in \mathcal{P}$ (that is, lying on a path from s to t and different from s and t). All vertices in Q have parities of one of the forms $(0, 0, z), (0, y, 0), (x, 0, 0)$, hence $|Q| = O(2^n/n^{2/3})$. Now take the set $C \subseteq E_n$ of all edges of H_n with both endpoints in Q; it is clearly a $\mathcal{P}$-cut. Furthermore, each vertex of Q is incident with at most $3 \cdot 2^{k/3} = O(n^{1/3})$ edges from C. This follows from the fact that every $v \in Q$ with parity vector, say, $(x, 0, 0)$, can be incident only with those edges in C that have directions corresponding to vectors of the form $(x, y, 0), (x, 0, z)$ or $(x', 0, 0)$, for various $y, z, x' \in \{0, 1\}^{k/3}$. Therefore, $|C| = O(2^n/n^{1/3})$, concluding the proof. □

Improved Bounds. In the main construction, we divide the length-k strings into m equally-sized parts, we let d be the distance between pairs in the pairing and w be the number of non-zero length-(k/m) parts of the parity strings of the direction vectors. The main tool is the following lemma about certain sets of vectors used to generalize the proof in the warm-up. The reader should keep in mind that an example of such a set of vectors for $m = 3$, $d = 3$, $w = 2$, is $V = \{110, 101, 011\}$, and was used implicitly in the previous proof.

For our purposes, all parameters involved except k and n should be thought of as constants, although the constants hidden in the Big-O notation are absolute.

Lemma 3. *Suppose* $V \subseteq \{0, 1\}^m$, $d = |V|$, *and* $w \in \mathbb{N}$ *are such that:*
1. $2 \leq |v| \leq w$ *for all* $v \in V$,
2. $\bigoplus_{v \in V} v = 0$, *and*
3. *For all* $W \subseteq V$ *of size* $|W| = \lfloor d/2 \rfloor, |\bigoplus_{v \in W} v| \geq \lceil m/2 \rceil$

Let k *be a positive multiple of* m *and* $n = 2^k - 1$. *Then there is a pairing* $\mathcal{P} \subseteq V_n \times V_n$ *of vertices of* H_n *of size* $|\mathcal{P}| = \Omega(2^{n-d})$ *that has a* $\mathcal{P}$-cut $C \subseteq E_n$ *of size* $|C| = O\left(\frac{2^n}{\sqrt{n}} n^{w/m} \sqrt{d 2^d}\right)$ *and with the additional property that each source in* $\mathcal{P}$ *is at distance exactly d from its sink.*

Proof: Divide $[k]$ into m disjoint subsets $G_1, \ldots, G_m \subseteq [k]$ of size k/m; e.g. $G_i = \{(i-1)k/m+1, \ldots, ik/m\}$. For $a \in \{0,1\}^n$, consider the k parity bits $p(a) \in \{0,1\}^k$ of a, and split them into m blocks according to $G_1, \ldots, G_m$ [2]; let us call each of the corresponding k/m-bit substrings $x_1(a), \ldots, x_m(a)$. Thus, $p(a)$ is the concatenation of $x_1(a), x_2(a), \ldots, x_m(a)$.

For a subset $v \subseteq [m]$, let $Z_v = \bigcup_{i \in v} G_i \subseteq [k]$. Given $p \subseteq [k]$, define the *projection* of p on v to be $\Pi_v(p) = p \cap Z_v$, (remember that p and $\Pi_v(x)$ can be interpreted as strings in $\{0,1\}^k$ as well). For example, in the preceding subsection, $\Pi_{110}((x,y,z)) = (x,y,0)$. Consider the set

$$S = \{a \in \{0,1\}^n : \forall_{i \in [m]} \, x_i(a) \neq 0 \text{ and } \forall_{v \in V} \, a_{\Pi_v(p(a))} = 0\}.$$

This will be set of sources in $\mathcal{P}$. Note that the expression $a_{\Pi_v(p(a))}$, referring to bit number $\Pi_v(p(a))$ of a, is well-defined, because the condition $\forall_i x_i(a) \neq 0$, along with $v \neq 0$, implies $\Pi_v(p(a)) \neq 0$.

The set of d directions between a source s and the corresponding sink t will be determined by the parity of s alone, in the following way: for $p \in \{0,1\}^k$, let $D(p) = \bigcup_{v \in V} \{\Pi_v(p)\}$. Condition 1 of the hypothesis of the lemma implies that if $s \in S$, $|D(p(s))| = |V| = d$, and all elements of $D(p(s))$ have weight ≥ 2.

For each source $s \in S$, we define the sink $t = s \cup D(p(s))$; by construction $s \subseteq t$, and $t - s = |D(p(s))| = d$. $\mathcal{P}$ is defined as the union of all such ordered pairs (s,t): $\mathcal{P} = \bigcup_{s \in S} \{(s, s \cup D(p(s)))\}$. Notice that $|\mathcal{P}| = |S| = (2^{k/m} - 1)^m 2^{n-k-d} = \Omega(2^{n-d})$.

We prove now that $\mathcal{P}$ forms a pairing: the set of sinks is disjoint from the set of sources, and no two different sources have the same sink. Because of the afore-mentioned properties of the parity check p, for any source-sink pair (s,t) we have $p(t) = p(s) \oplus \bigoplus_{v \in V} \Pi_v(p(s)) = p(s) \oplus \Pi_{\bigoplus_{v \in V} v}(p(s)) = p(s)$ (where we used the second property of V and simple properties of the projection operator). Since for every $d \in D(p)$, $d \notin s$ but $d \in t$, it follows that no sink is a source too. Likewise, if two sinks t_1 and t_2 (corresponding to sources s_1 and s_2) were the same ($t_1 = t_2$), we would have $p(s_1) = p(s_2)$, which implies $D(p(s_1)) = D(p(s_2))$ and therefore $s_1 = s_2$.

To conclude, we only need to bound the size of a smallest $\mathcal{P}$-cut. Consider the set of vertices halfway between a source and a sink:
$Q = \{x \in \{0,1\}^n : \text{there exists } (s,t) \in \mathcal{P} \text{ such that } s \subseteq x \subseteq t \text{ and } |x - s| = \lfloor d/2 \rfloor\}$

(notice the slightly different definition of Q, compared to that in 3.2).

Due to the third property of V and the definition of $D(p(s))$, it follows that $b \in Q$ implies that at least half of $x_1(b), \ldots, x_m(b)$ are zero. For any $b \in \{0,1\}^n$, if $r(b)$ is the m-bit string such that for all $1 \leq i \leq m$, $x_i(b) = 0$ iff $r(b)_i = 0$, then the set $\{r(b) : b \in Q\}$ has size bounded by $\binom{d}{d/2}$: for all $s \in S$, $r(s)$ is the all-ones string and any for any $b \in Q$, $r(b)$ is $r(s)$ XORed with some $d/2$ vectors in V. So the set $\{p(b) : b \in Q\}$ has size at most $\binom{d}{d/2}(2^{k/m} - 1)^{m/2}$, and does not contain unit vectors; therefore $|Q| \leq \frac{2^n}{n+1} \binom{d}{d/2}(2^{k/m} - 1)^{m/2} = O(\frac{2^n}{\sqrt{n}} \frac{2^d}{\sqrt{d}})$.

---

[2] Actually, in order to do this we first impose an arbitrary ordering on the elements of each G_i.

An edge cut is given by $C = \{(b, c) \in E_n : b \in Q \wedge c - b \in D(p(S))\}$, where $D(p(S)) = \bigcup_{s \in S}\{D(p(s))\}$. Thus, $|C| \leq |Q||D(p(S))|$. The claim follows since $|D(p(S))| \leq d(2^{k/m} - 1)^w$. $\qquad\square$

Proof of Theorem 2: We prove a strengthening of the second part of the theorem that implies the first as well. To be precise, we show that, for every $1 > \delta > 0$, there exist $\epsilon > 0$ and d such that, for all large enough n, there is a pairing $\mathcal{P}$ in H_n of size $|\mathcal{P}| \geq \epsilon 2^n$, sparsity at most $n^{-1/2+\delta}$ and with the additional property that all pairs in $\mathcal{P}$ have distance exactly d. By partitioning the pairs in $\mathcal{P}$ according the level modulo d of their source, and applying a simple averaging argument, we conclude that there must exist an *aligned* pairing in H_n with sparsity at most $n^{-1/2+\delta}$.

First note that, whatever our choice of m, w and d (as long as m and w are constants depending only on δ), we can assume without loss of generality that n is of the form $n = 2^k - 1$ and m divides k. Otherwise, let n' be the largest integer less than n such that n' is of the form $n' = 2^k - 1$ and m divides k. Note that $n' > n/2^{m+1}$. $H_{n'}$ can be embedded into H_n, so if we find a set $\mathcal{P}$ in $H_{n'}$ that satisfies the conclusion of the theorem for n' then the embedding of $\mathcal{P}$ in H_n will also suffice for n with a smaller ϵ'.

Let $w = \lceil 1/\delta \rceil$, $m = w^2$, $d = 2w$. It only remains to show that sets with parameters m, d, w, as in the hypotheses of Lemma 3, exist. The size of $\mathcal{P}$ is $\Omega(2^{n-d})$ and hence the ϵ we get depends on d and hence on δ.

Arrange the w^2 elements of $[m]$ into a square matrix $A \in \{0, 1\}^{w \times w}$. Associate one vector with each row and each column of A ($2w$ vectors in total). The i-th row is associated with the subset (or vector in $\{0, 1\}^w$) $R_i = \{r \in [m] : (i-1)w < r \leq iw\}$; the j-th column will correspond to the subset $C_j = \{r \in [m] : (r-1) \bmod w = j-1\}$. Let $V = \bigcup_{i \in [w]}\{R_i, S_i\}$. Clearly, $|V| = 2w$ and for all $v \in V$, we have $|v| = w > 1$. It is also apparent that $\oplus_{v \in V} v = 0$, because any $k \in [m]$ belongs to exactly two vectors in V, namely R_i and C_j, where $k = (i-1)w + j$ with $i, j \in [w]$.

Finally, we show that, for any $W \subseteq V$ with $|V| = d/2 = w$, $|\oplus_{v \in W} v| \geq \frac{m}{2} = \frac{w^2}{2}$. Suppose W contains a row elements R_i and $w - a$ column elements C_j; then $|\oplus_{v \in W} v| = a^2 + (w-a)^2 \geq \frac{w^2}{2}$ by the QM-AM inequality. $\qquad\square$

3.3   Proof of Theorem 3

Let $\mathcal{P}$ and C be the pairing and the cut constructed in the proof of Theorem 2. Let π be any permutation on V_n that maps each source in $\mathcal{P}$ to its sink. Notice that any shortest path in $H_n^{\uparrow\downarrow}$ that connects a source of $\mathcal{P}$ to its sink must also be a directed path in H_n. Hence, any realization of $\mathcal{P}$ with shortest paths must use some edge in C at least $|\mathcal{P}|/|C| = \Omega(n^{1/2-\delta})$ times. $\qquad\square$

3.4   Proof of Theorem 4

Let $\epsilon > 0$, $\mathcal{R} \subseteq \mathbb{Z}$ and let $f : \{0, 1\}^n \to \mathcal{R}$ be a dist-3 monotone function. If f is ϵ-far from being monotone, then by Lemma 1 there is a set $\mathcal{P}$ of $\epsilon 2^{n-1}$ vertex disjoint pairs in H_n that are violated by f. Furthermore, since f is dist-3 monotone, for every $(s^i, t^i) \in \mathcal{P}$ we have $|t^i| \leq |s^i| + 3$. To prove Theorem 4 we show that the sparsity of such $\mathcal{P}$ must be $\Omega(1/\sqrt{n})$.

Let C be a smallest $\mathcal{P}$-cut, and let us prove that $|C|/|\mathcal{P}| \geq \Omega(1/\sqrt{n})$. First we note that it is possible to assume that C has *no* edges that are incident with any source s^i or sink t^j from $\mathcal{P}$ (and in particular, this will mean that no pair in $\mathcal{P}$ has distance 1 or 2): Let $p > 0$ be the number of edges in C that are incident to some source or sink of a pair in $\mathcal{P}$. If $p \geq |\mathcal{P}|/4$ then we are done, since clearly $|C| \geq p$. Otherwise, removing these p edges from C and the corresponding pairs from $\mathcal{P}$ leaves a set C' of size $|C| - p$ that cuts a subset $\mathcal{P}' \subseteq \mathcal{P}$ of at least $|\mathcal{P}| - 2p$ pairs. This is due to the fact that the pairs in $\mathcal{P}$ are disjoint, and hence each edge can be incident with at most two pairs. Since $p \leq |\mathcal{P}|/4$, we have $\frac{|C|-p}{|\mathcal{P}|-2p} \leq 2\frac{C}{|\mathcal{P}|}$, so it is enough to prove the claim for $C \triangleq C'$ and $\mathcal{P} \triangleq \mathcal{P}'$.

For $0 \leq h \leq n - 3$, let $\mathcal{P}^h \subseteq \mathcal{P}$ be the set of pairs $(s^i, t^i) \in \mathcal{P}$ with $|s^i| = h$ (and $|t^i| = h + 3$). Clearly C is a $\mathcal{P}^h$-cut for every h. Let $C^h \subseteq C$ denote the set of edges in C that lie on some $\mathcal{P}^h$-path. Since C^h has no edges incident to any s^i or t^j, in order to cut $\mathcal{P}^h$ we must use exactly those edges between levels $h + 1$ and $h + 2$ that lie on some $\mathcal{P}^h$-path. So the sets C^h, $0 \leq h \leq n - 3$, are in fact disjoint. Therefore it is sufficient to prove that $C^h/|\mathcal{P}^h| \geq \Omega(1/\sqrt{n})$ for all h.

Fix h, and for clarity let us redefine $\mathcal{P} \triangleq \mathcal{P}^h$ and $C \triangleq C^h$. Each pair $(s^i, t^i) \in \mathcal{P}$ defines a sub-cube of dimension 3, which we will denote by H_3^i, that contains all vertices and edges that belong to one of the six possible paths from s^i to t^i.

Observation 7. *For any two pairs* $(s^i, t^i), (s^j, t^j) \in \mathcal{P}$, $|E(H_3^i) \cap E(H_3^j)| \leq 1$.

Proof: Assume that $|E(H_3^i) \cap E(H_3^j)| \geq 2$ for some $i \neq j$, and let $e = (a, b)$ and $e' = (a', b')$ be two edges in $E(H_3^i) \cap E(H_3^j)$. Since the pairs (s^i, t^i) and (s^j, t^j) are disjoint, both e and e' should lie between layers $h + 1$ and $h + 2$. Therefore, $a = a' = s^i \cup s^j$ and $b = b' = t^i \cap t^j$, contradicting the assumption that $e \neq e'$.   □

Consider the directed graph $G = (V, E)$ with $V = \bigcup_{(s^i, t^i) \in \mathcal{P}} V(H_3^i)$ and $E = \bigcup_{(s^i, t^i) \in \mathcal{P}} E(H_3^i)$. Since every s^i has out-degree 3 in G (and in-degree 0), the number of edges between layers h and $h + 1$ of H_n that belong to G is exactly $3|\mathcal{P}|$. Let $A = a_1, \ldots, a_k$ be the vertices in layer $h + 1$ of H_n that belong to G, let $\alpha_1, \ldots, \alpha_k$ denote their in-degrees and let $\beta_1, \ldots, \beta_k$ denote their out-degrees in G. We have that $\sum_{i \in [k]} \alpha_i = 3|\mathcal{P}|$, and our goal is to prove that $|C| \equiv \sum_{i \in [k]} \beta_i = \Omega(|\mathcal{P}|/\sqrt{n})$.

Consider vertex a_i. For every pair $(s^j, t^j) \in \mathcal{P}$ such that $a_i \in V(H_3^j)$ there are two edges in H_3^j going out of a_i. Since for any two pairs $(s^j, t^j), (s^{j'}, t^{j'}) \in \mathcal{P}$ we have $|E(H_3^j) \cap E(H_3^{j'})| \leq 1$, it follows that $\binom{\beta_i}{2} \geq \alpha_i$. So $\beta_i > \sqrt{\alpha_i}$ for all i and hence $|C| = \sum_{i \in [k]} \beta_i > \sum_{i \in [k]} \sqrt{\alpha_i} = \sum_{i \in [k]} \frac{\alpha_i}{\sqrt{\alpha_i}} \geq \frac{3|\mathcal{P}|}{\sqrt{n}}$, as $\alpha_i \leq n$.

3.5   An $\Omega(n/\epsilon)$ Lower Bound for General Functions

Theorem 8. *Let $\mathcal{R} \subseteq \mathbb{Z}$, $|\mathcal{R}| = \Omega(\sqrt{n})$. Testing monotonicity of functions $f\{0,1\}^n \to \mathcal{R}$ (non-adaptively with one-sided error) requires $\Omega(n/\epsilon)$ queries.*

Proof: We first prove a lower bound of $\Omega(n)$ for some constant ϵ and argue at the end how we can achieve the promised lower bound of $\Omega(n/\epsilon)$.

A non-adaptive q-query monotonicity tester with one-sided error queries f on a set Q of at most q vertices and rejects if and only if one of the comparable pairs in Q is violated. Hence, it is sufficient to show a family $\mathcal{F}_n$ of functions $f : \{0,1\}^n \to \mathcal{R}$ that are ϵ-far from monotone (for a fixed $\epsilon > 0$ and all n) and such that, for any fixed set $Q \subseteq \{0,1\}^n$ of size $o(n)$, a random $f \sim_U \mathcal{F}_n$ induces a violated pair in Q with probability less than $1/3$.

For every n, we will define a family $\mathcal{F}_n = \{f_1, \ldots, f_n\}$ of n functions $f_i : \{0,1\}^n \to \mathcal{R}$ with the following properties:

- every f_i is ϵ-far from monotone, for some absolute constant $\epsilon > 0$;
- for any set $Q \subseteq \{0,1\}^n$, $\Pr_{i \sim_U [n]}[(Q \times Q) \cap \mathsf{Viol}(f_i) \neq \emptyset] \leq \frac{|Q|-1}{n}$.

This implies any tester making fewer than $\frac{2n}{3}$ queries will fail with probability $\geq 1/3$.

Similarly to [FLN+02], each $f_i \in \mathcal{F}_n$ will violate some pairs that differ in the i-th coordinate. But here we will make sure that only the actual *edges* of H_n are violated, making it more difficult to catch violated pairs.

We now formally define $\mathcal{F}_n$. Let $\mathcal{R} = \{0, 1, \ldots, 2\sqrt{n}\}$, and let $h(x) \triangleq |x| - n/2 + \sqrt{n}$ for all $x \in \{0,1\}^n$. For each $i \in [n]$ we define $f_i : \{0,1\}^n \to \mathcal{R}$ as follows:

$$f_i(x) = \begin{cases} 0, & h(x) < 0 \\ 2\sqrt{n}, & h(x) > 2\sqrt{n} \\ h(x), & h(x) \in \mathcal{R} \text{ and } x_i \neq h(x) \bmod 2 \\ h(x) + (-1)^{x_i}, & h(x) \in \mathcal{R} \text{ and } x_i = h(x) \bmod 2 \end{cases}$$

Notice that for all $i \in [n]$, $\mathsf{Viol}(f_i) = \mathsf{EdgeViol}(f_i)$, and the edges in $\mathsf{EdgeViol}(f_i)$ are vertex disjoint. So by Lemma 1, the functions $f_i \in \mathcal{F}_n$ are ϵ-far from monotone (for some fixed $\epsilon > 0$) if $|\mathsf{EdgeViol}(f_i)| \geq \epsilon 2^n$. Indeed, $|\mathsf{EdgeViol}(f_i)|$ equals the number of points $x \in \{0,1\}^n$ such that: $h(x) \in \mathcal{R}$, $h(x) = 0 \pmod 2$ and $x_i = 0$. Notice that for $n > 10$, these constitute roughly a quarter of all points $y \in \{0,1\}^n$ with $h(y) \in \mathcal{R}$. On the other hand, it follows from Chernoff bounds that for some constant $\rho > 0$ and for all $n > 10$, the number of points $y \in \{0,1\}^n$ with $h(y) \in \mathcal{R}$ is at least $\rho 2^n$. Setting $\epsilon = \rho/5$, we conclude that all functions $f_i \in \mathcal{F}_n$ are ϵ-far from monotone.

Now we prove that $\Pr_{i \sim_U [n]}[(Q \times Q) \cap \mathsf{Viol}(f_i) \neq \emptyset] \leq \frac{|Q|-1}{n}$. Fix Q and consider the *undirected* graph $G = (V, E)$, where $V = Q$ and $E = \{\{x, y\} \in Q \times Q : (x, y) \in E_n\}$. In other words, G is the undirected skeleton of the subgraph of H_n induced on Q. For $x, y \in \{0,1\}^n$ we write $x = y^{(j)}$ if x equals y in all coordinates except j. Let $T \subseteq [n]$ be a set of directions spanned by E, namely, $T = \{j : \text{there exists } \{x, y\} \in$

E such that $x = y^{(j)}$}. Clearly, the success probability of the test is bounded by $|T|/n$. To finish the proof, we show that $|T| \le |Q| - 1$.

Consider a minimal subgraph G' of G that spans all directions in T. Then clearly, $|E(G')| = |T|$. Since any cycle in the undirected skeleton of H_n travels in any direction even number of times so G' is acyclic. So $|T| = |E(G')| \le |V(G')| - 1 = |Q| - 1$.

We proved a lower bound of $\Omega(n)$ queries for some constant $\epsilon > 0$. To get a lower bound of $\Omega(n/\epsilon)$ for any $\epsilon = \epsilon(n)$ we need to compose our lower bound with a simple "hiding" procedure. Namely, we define a distribution $\mathcal{F}'_n$ that fools any deterministic tester with $o(n/\epsilon)$ queries as follows: first, partition H_n into disjoint subcubes, each of size $\epsilon 2^n$ (for simplicity we assume that $1/\epsilon$ is a power of 2); then pick a random subcube C in this partition, and value it with a random $f_i \in \mathcal{F}_{n-\log 1/\epsilon}$; value the other subcubes so that there are no violations outside C. Now for any fixed set Q of $o(n/\epsilon)$ queries, the expected number of queries that hit C is $o(n)$, and we know that with $o(n)$ queries it is impossible to find a violation in a random f_i. □

Notice that the range $\mathcal{R}$ of the functions f_i is of size $O(\sqrt{n})$ - much smaller than the 2^n different values a function on the hypercube may have. Consider pair-testers (see Section 2.3) of *Boolean* monotonicity making at most $q(n)/\epsilon$ queries for some function $q : \mathbb{N} \to \mathbb{N}$ and any $\epsilon > 0$; it follows from the range-reduction lemma of [DGL+99] and Theorem 8 that for any such tester, $q(n) = \Omega(n/\log n)$ must hold. This is tight up to the $\log n$ factor.

4   Concluding Remarks

We suggest three open problems related to this line of work:

First, is it true that the best testers for monotonicity over H_n are in fact pair-testers? The question is of interest even just for Boolean-range functions, since a positive answer coupled with our $\Omega(\frac{n}{\epsilon \log n})$ lower bound for pair testers would give an almost-tight lower bound.

Another challenge is to find better upper bounds for the special case of testing monotonicity of dist-k monotone functions, for some $k \ge 3$. As we saw in Section 3.2, non-trivial sparsity upper bounds can be found even if we restrict ourselves to pairings in which all pairs are at distance 3. This seems to indicate, in our opinion, that a better understanding of the small-distance situations will yield new insights that may be applicable in the general case.

Finally, recall from Section 3.4 that for $k \le 3$, dist-k monotonicity can be tested with $O(n^{3/2})$ queries; on the other hand, the construction in Section 3.2 shows that sparsity considerations alone will never yield upper bounds better than this. In view of these results, it is natural to ask whether these two measures need to coincide for larger k; that is, whether the complexity of edge-testers may be better than the values derived from sparsity upper-bounds.

References

[ABY08]   Al-Bashabsheh, A., Yongaçoglu, A.: On the k-pairs problem. CoRR, abs/0805.0050 (2008)

[AC06]   Ailon, N., Chazelle, B.: Information theory in property testing and monotonicity testing in higher dimension. Inf. Comput. 204(11), 1704–1717 (2006)

[AHJ$^+$06] Adler, M., Harvey, N.J.A., Jain, K., Kleinberg, R.D., Lehman, A.R.: On the capacity of information networks. In: SODA, pp. 241–250 (2006)

[Ben65]   Benes, V.E.: Mathematical theory of connecting networks and telephone traffic. Academic Press, New York (1965)

[BGJ$^+$09a] Bhattacharyya, A., Grigorescu, E., Jung, K., Raskhodnikova, S., Woodruff, D.: Transitive-closure spanners. In: SODA 2009: Proceedings of the Nineteenth Annual ACM -SIAM Symposium on Discrete Algorithms, pp. 932–941. Society for Industrial and Applied Mathematics, Philadelphia (2009)

[BGJ$^+$09b] Bhattacharyya, A., Grigorescu, E., Jung, K., Raskhodnikova, S., Woodruff, D.: Transitive-closure spanners of the hypercube and the hypergrid. Electronic Colloquium on Computational Complexity (ECCC) 09(046) (2009)

[Bha08]   Bhattacharyya, A.: A note on the distance to monotonicity of boolean functions. Electronic Colloquium on Computational Complexity (ECCC) 15(012) (2008)

[DGL$^+$99] Dodis, Y., Goldreich, O., Lehman, E., Raskhodnikova, S., Ron, D., Samorodnitsky, A.: Improved testing algorithms for monotonicity. In: Hochbaum, D.S., Jansen, K., Rolim, J.D.P., Sinclair, A. (eds.) RANDOM 1999 and APPROX 1999. LNCS, vol. 1671, pp. 97–108. Springer, Heidelberg (1999)

[EKK$^+$00] Ergün, F., Kannan, S., Kumar, R., Rubinfeld, R., Viswanathan, M.: Spot-checkers. J. Comput. Syst. Sci. 60(3), 717–751 (2000)

[Fis04]   Fischer, E.: On the strength of comparisons in property testing. Inf. Comput. 189(1), 107–116 (2004)

[FLN$^+$02] Fischer, E., Lehman, E., Newman, I., Raskhodnikova, S., Rubinfeld, R., Samorodnitsky, A.: Monotonicity testing over general poset domains. In: STOC, pp. 474–483 (2002)

[GGL$^+$00] Goldreich, O., Goldwasser, S., Lehman, E., Ron, D., Samorodnitsky, A.: Testing monotonicity. Combinatorica 20(3), 301–337 (2000)

[GGLR98]   Goldreich, O., Goldwasser, S., Lehman, E., Ron, D.: Testing monotonicity. In: FOCS, pp. 426–435 (1998)

[HK08]   Halevy, S., Kushilevitz, E.: Testing monotonicity over graph products. Random Struct. Algorithms 33(1), 44–67 (2008)

[LR01]   Lehman, E., Ron, D.: On disjoint chains of subsets. J. Comb. Theory, Ser. A 94(2), 399–404 (2001)

[Lub90]   Lubiw, A.: Counterexample to a conjecture of szymanski on hypercube routing. Inf. Process. Lett. 35(2), 57–61 (1990)

[Ras99]   Raskhodnikova, S.: Monotonicity testing. Master's thesis, Department of Electrical Engineering and Computer Science. MIT, Cambridge, MA (1999)

[RL05]   Rasala-Lehman, A.: Network coding. PhD thesis, Department of Electrical Engineering and Computer Science. MIT, Cambridge, MA (2005)

[Szy89]   Szymanski, T.H.: On the permutation capability of a circuit-switched hypercube. In: ICPP (1), pp. 103–110 (1989)

Better Gap-Hamming Lower Bounds via Better Round Elimination

Joshua Brody[1,*], Amit Chakrabarti[1,**], Oded Regev[2,***],
Thomas Vidick[3,†], and Ronald de Wolf[4,‡]

[1] Department of Computer Science, Dartmouth College, Hanover, NH 03755
[2] Blavatnik School of Computer Science, Tel Aviv University, Tel Aviv 69978, Israel
[3] Department of Computer Science, UC Berkeley
[4] CWI Amsterdam

Abstract. Gap Hamming Distance is a well-studied problem in communication complexity, in which Alice and Bob have to decide whether the Hamming distance between their respective n-bit inputs is less than $n/2 - \sqrt{n}$ or greater than $n/2 + \sqrt{n}$. We show that every k-round bounded-error communication protocol for this problem sends a message of at least $\Omega(n/(k^2 \log k))$ bits. This lower bound has an exponentially better dependence on the number of rounds than the previous best bound, due to Brody and Chakrabarti. Our communication lower bound implies strong space lower bounds on algorithms for a number of data stream computations, such as approximating the number of distinct elements in a stream.

Keywords: Communication Complexity, Gap Hamming Distance, Round Elimination, Measure Concentration.

1 Introduction

1.1 The Communication Complexity of the Gap Hamming Distance Problem

Communication complexity studies the communication requirements of distributed computing. In its simplest and best-studied setting, two players, Alice and Bob, receive inputs x and y, respectively, and are required to compute some function $f(x, y)$. Clearly, for most functions f, the two players need to communicate to solve this problem. The basic question of communication complexity is the *minimal amount* of communication

* Supported in part by NSF Grant CCF-0448277. Part of this work was done while the author was visiting CWI and Tel Aviv University.
** Supported in part by NSF Grants CCF-0448277 and IIS-0916565 and a McLane Family Fellowship.
*** Supported by the Israel Science Foundation, by the European Commission under the Integrated Project QAP funded by the IST directorate as Contract Number 015848, by the Wolfson Family Charitable Trust, and by a European Research Council (ERC) Starting Grant.
† Supported by ARO Grant W911NF-09-1-0440 and NSF Grant CCF-0905626. Part of this work was done while the author was visiting CWI and Tel Aviv University.
‡ Supported by a Vidi grant from Netherlands Organization for Scientific Research (NWO).

M. Serna et al. (Eds.): APPROX and RANDOM 2010, LNCS 6302, pp. 476–489, 2010.

needed. By abstracting away from the resources of local computation time and space, communication complexity gives us a bare-bones but elegant model of distributed computing. It is interesting for its own sake but is also useful as one of our main sources of lower bounds in many other models of computation, including data structures, circuits, Turing machines, VLSI, and streaming algorithms. The basic results are excellently covered in the book of Kushilevitz and Nisan [12], but many additional fundamental results have appeared since its publication in 1997.

One of the few basic problems whose randomized communication complexity is not yet well-understood, is the *Gap Hamming Distance* (GHD) problem, defined as follows.

> GHD: Alice receives input $x \in \{0,1\}^n$ and Bob receives input $y \in \{0,1\}^n$, with the promise that $|\Delta(x,y) - n/2| \geq \sqrt{n}$, where Δ denotes the Hamming distance. Decide whether $\Delta(x,y) < n/2$ or $\Delta(x,y) > n/2$.

Mind the gap between $n/2 - \sqrt{n}$ and $n/2 + \sqrt{n}$, which is what makes this problem interesting and useful. Indeed, the communication complexity of the gapless version, where there is no promise on the inputs, can easily be seen to be linear (for instance by a reduction from disjointness). The gap makes the problem easier, and the question is how it affects the communication complexity: does it remain linear? A gap size of $\Theta(\sqrt{n})$ is the natural choice – a $\Theta(1)$ fraction of the inputs lie inside the promise area for this gap size, and as we'll see below, it is precisely this choice of gap size that has strong implications for streaming algorithms lower bounds. Moreover, understanding the complexity of the $\sqrt{n}$-gap version can be shown to imply a complete understanding of the GHD problem for all gaps.

Randomized protocols for GHD and more general problems can be obtained by sampling. Suppose for instance that it is promised that either $\Delta(x,y) \leq (1/2 - \gamma)n$ or $\Delta(x,y) \geq (1/2 + \gamma)n$. Choosing an index $i \in [n]$ at random, the predicate $[x_i \neq y_i]$ is a coin flip with heads probability $\leq 1/2 - \gamma$ in the first case and $\geq 1/2 + \gamma$ in the second. It is known that flipping such a coin $\Theta(1/\gamma^2)$ times suffices to distinguish these two cases with probability at least $2/3$. Hence if we use shared randomness to choose $\Theta(1/\gamma^2)$ indices, we obtain a one-round bounded-error protocol with communication $\Theta(1/\gamma^2)$ bits. In particular, for GHD (where $\gamma = 1/\sqrt{n}$), the communication is $\Theta(n)$ bits, which is no better than the trivial upper bound of n when Alice just sends x to Bob.

What about lower bounds? Indyk and Woodruff [10] managed to prove a linear lower bound for the case of one-round protocols for GHD, where there is only one message from Alice to Bob (see also [20,11]). However, going beyond one-round bounds turned out to be quite a difficult problem. Recently, Brody and Chakrabarti [5] obtained linear lower bounds for all *constant*-round protocols:

Theorem 1. *[5] Every k-round bounded-error protocol for GHD sends a message of length $\frac{n}{2^{O(k^2)}}$.*

In fact their bound is significant as long as the number of rounds is $k \leq c_0\sqrt{\log n}$, for a universal constant c_0. Regarding lower bounds that hold irrespective of the number of rounds, an easy reduction gives an $\Omega(\sqrt{n})$ lower bound (which is folklore): take an instance of the gapless version of the problem on $x,y \in \{0,1\}^{\sqrt{n}}$ and "repeat" x and y $\sqrt{n}$ times each. This blows up the gap from 1 to $\sqrt{n}$, giving an instance of GHD on

n bits. Solving this n-bit instance of GHD solves the $\sqrt{n}$-bit instance of the gapless problem. Since we have a linear lower bound for the latter, we obtain a general $\Omega(\sqrt{n})$ bound for GHD.[1]

1.2   Our Results

Our main result is an improvement of the bound of Brody and Chakrabarti, with an exponentially better dependence on the number of rounds:

Theorem 2. *Every k-round bounded-error protocol for GHD sends a message of length*
$$\Omega\left(\frac{n}{k^2 \log k}\right).$$

In fact we get a bound for the more general problem of distinguishing distance $\Delta(x,y) \leq (1/2 - \gamma)n$ from $\Delta(x,y) \geq (1/2 + \gamma)n$, as long as $\gamma = \Omega(1/\sqrt{n})$: for this problem every k-round protocol sends a message of $\Omega\left(\frac{1}{k^2 \log k} \frac{1}{\gamma^2}\right)$ bits.

Like the result of [5], our lower bound deteriorates with the number of rounds. Also like their result, our proof is based on *round elimination*, an important framework for proving communication lower bounds. Our proof contains an important insight into this framework that we now explain.

A communication problem usually involves a number of parameters, such as the input size, an error bound, and in our case the gap size. The round elimination framework consists of showing that a k-round protocol solving a communication problem for a class $\mathcal{C}$ of parameters can be turned into a $(k-1)$-round protocol for an easier class $\mathcal{C}'$, provided the message communicated in the first round is short. This fact is then applied repeatedly to obtain a 0-round protocol (say), for some nontrivial class of instances. The resulting contradiction can then be recast as a communication lower bound. Historically, the easier class $\mathcal{C}'$ has contained *smaller input lengths*[2] than those in $\mathcal{C}$.

In contrast to previous applications of round elimination, we manage to *avoid shrinking the input length*: the simplification will instead come from a slight deterioration in the error parameter. Here is how this works. If Alice's first message is short, then there is a specific message and a large set A of inputs on which Alice would have sent that message. Roughly speaking, we can use the largeness of A to show that *almost any* input $\tilde{x}$ for Alice is close to A in Hamming distance. Therefore, Alice can "move" $\tilde{x}$ to its nearest neighbor, x, in A: this makes her first message redundant, as it is constant for all inputs $x \in A$. Since x and $\tilde{x}$ have small Hamming distance, it is likely that both pairs

---

[1] In fact the same proof lower-bounds the *quantum* communication complexity; a linear quantum lower bound for the gapless version follows easily from Razborov's work [18] and the observation that $\Delta(x,y) = |x| + |y| - 2|x \wedge y|$. However, as Brody and Chakrabarti observed, in the quantum case this $\sqrt{n}$ lower bound is essentially tight: there is a bounded-error quantum protocol, based on a well-known quantum algorithm for approximate counting, that communicates $O(\sqrt{n} \log n)$ qubits. This also implies that lower bound techniques which apply to quantum protocols, such as discrepancy, factorization norms [15,13], and the pattern matrix method [19], cannot prove better bounds for classical protocols.

[2] In fact, $\mathcal{C}$ and $\mathcal{C}'$ are often designed such that an instance in $\mathcal{C}$ is a "direct sum" of several independent instances in $\mathcal{C}'$.

$(\tilde{x}, y)$ and (x, y) are on the same side of the gap, i.e. have the same GHD value. Hence the correctness of the new protocol, which is one round shorter, is only mildly affected by the move. Eliminating all k rounds in this manner, while carefully keeping track of the accumulating errors, yields a lower bound of $\Omega(n/(k^4 \log^2 k))$ on the maximum message length of any k-round bounded-error protocol for GHD.

Notice that this lower bound is slightly weaker than the above-stated bound of $\Omega(n/(k^2 \log k))$. To obtain the stronger bound, we leave the purely combinatorial setting and analyze a version of GHD *on the unit sphere*:[3] Alice's input is now a unit vector $x \in \mathbb{R}^n$ and Bob's input is a unit vector $y \in \mathbb{R}^n$, with the promise that either $x \cdot y \geq 1/\sqrt{n}$ or $x \cdot y \leq -1/\sqrt{n}$ (as we show below in Section 2, this version and the Boolean one are essentially equivalent in terms of communication complexity). Alice's input is now close to the large, constant-message set A in *Euclidean distance*. The rest of the proof is as outlined above, but the final bound is stronger than in the combinatorial proof for reasons that are discussed in Section 2.2. Although this proof uses arguments from high-dimensional geometry, such as measure concentration, it arguably remains conceptually simpler than the one in [5].

Related work. The round elimination technique was first formalized in Miltersen et al. [17] and dates back even further, at least to Ajtai's lower bound for predecessor data structures [1]. For us, the most relevant previous use of this technique is in the result by Brody and Chakrabarti [5], where a weaker lower bound is proved on GHD.

Their proof, as ours, identifies a large subset A of inputs on which Alice sends the same message. The "largeness" of A is used to identify a suitable subset of $(n/3)$ coordinates such that Alice can "lift" any $(n/3)$-bit input $\tilde{x}$, defined on these coordinates, to some n-bit input $x \in A$. In the resulting protocol for $(n/3)$-bit inputs, the first message is now constant, hence redundant, and can be eliminated.

The input size thus shrinks from n to $n/3$ in one round elimination step. As a result of this constant-factor shrinkage, the Brody-Chakrabarti final lower bound necessarily decays exponentially with the number of rounds. Our proof crucially avoids this shrinkage of input size by instead considering the *geometry* of the set A, and exploiting the natural invariance of the GHD predicate to small perturbations of the inputs.

Remark. After we obtained our results, a subset of the authors independently proved an optimal $\Omega(n)$ lower bound, independent of the number of rounds [6]. However, the techniques they introduce are completely different, and rather involved. In contrast, our result, through its relatively simple and elegant proof, should be of independent interest to the community.

1.3   Applications to Streaming

The introduction of gapped versions of the Hamming distance problem by Indyk and Woodruff [10] was motivated by the streaming model of computation, in particular the

---

[3] The idea of going to the unit sphere was also used by Jayram et al. [11] for a simplified one-round lower bound. As we will see in Section 2, doing so is perhaps even more natural than working with the combinatorial version; in particular it is then easy to make GHD into a *dimension-independent* problem.

problem of approximating the number of distinct elements in a data stream. For many data stream problems, including the distinct elements problem, the goal is to output a multiplicative approximation of some real-valued quantity. Usually, both *randomization* and *approximation* are required. When both are allowed, there are often remarkably space-efficient solutions.

As Indyk and Woodruff showed, *communication lower bounds* for the Gap Hamming Distance problem imply *space lower bounds* on algorithms that output the number of distinct elements in a data stream up to a multiplicative approximation factor $1 \pm \gamma$. The reduction from GHD works as follows. Alice converts her n-bit string $x = x_1 x_2 \cdots x_n$ into a stream of tuples $\sigma = \langle (1, x_1), (2, x_2), \ldots, (n, x_n) \rangle$. Bob converts y into $\tau = \langle (1, y_1), (2, y_2), \ldots, (n, y_n) \rangle$ in a similar fashion. Using a streaming algorithm for the distinct elements problem, Alice processes σ and sends the memory contents to Bob, who then processes τ starting from where Alice left off. In this way, they estimate the number of distinct elements in $\sigma \circ \tau$. Note that each element in σ is unique, and that elements in τ are distinct from elements in σ precisely when $x_i \neq y_i$. Hence, an accurate approximation ($\gamma = \Omega(1/\sqrt{n})$ is required) for the number of distinct elements in $\sigma \circ \tau$ gives an answer to the original GHD instance. This reduction can be extended to multi-pass streaming algorithms in a natural way: when Bob is finished processing τ, he sends the memory contents back to Alice, who begins processing σ a second time. Generalizing, it is easy to see that a p-pass streaming algorithm gives a $(2p - 1)$-round communication protocol, where each message is the memory contents of the streaming algorithm. Accordingly, a lower bound on the length of the largest message of $(2p - 1)$-round protocols gives a space lower bound for the p-pass streaming algorithm.

Thus, the one-round linear lower bound by Indyk and Woodruff [10] yields the desired $\Omega(1/\gamma^2)$ (one-pass) space lower bound for the streaming problem. Similarly, our new communication lower bounds imply $\Omega(1/(\gamma^2 p^2 \log p))$ space lower bounds for p-pass algorithms for the streaming problem. This improves on previous bounds for all $p = o(n^{1/4}/\sqrt{\log n})$.

Organization of the paper. We start with some preliminaries in Section 2, including a discussion of the key measure concentration results that we will use, both for the sphere and for the Hamming cube, in Section 2.2. In Section 3 we prove our main result, while in Section 4 we give the simple combinatorial proof of the slightly weaker result mentioned above.

2   Preliminaries

Notation. For $x, y \in \mathbb{R}^n$, let $d(x, y) := \|x - y\|$ be the Euclidean distance between x and y, and $x \cdot y$ their inner product. For $z \in \mathbb{R}$, define $\text{sgn}(z) := 0$ if $z \geq 0$, and $\text{sgn}(z) = 1$ otherwise. For a set $S \subseteq \mathbb{R}^n$, let $d(x, S)$ be the infimum over all $y \in S$ of $d(x, y)$. The unique rotationally-invariant probability distribution on the n-dimensional sphere $\mathbb{S}^{n-1}$ is the Haar measure, which we denote by ν. When we say that a vector is taken from the uniform distribution over a measurable subset of the sphere, we will always mean that it is distributed according to the Haar measure, conditioned on being in that subset.

Define the max-cost of a communication protocol to be the length of the longest *single* message sent during an execution of the protocol, for a worst-case input. We use $R_\varepsilon^k(f)$ to denote the minimal max-cost amongst all two-party, k-round, public-coin protocols that compute f with error probability at most ε on every input (here a "round" is one message).

2.1 Problem Definition

We will prove our lower bounds for the problem $\mathcal{GHD}_{d,\gamma}$, where d is an integer and $\gamma > 0$. In this problem Alice receives a d-dimensional unit vector x, and Bob receives a d-dimensional unit vector y, with the promise that $|x \cdot y| \geq \gamma$. Alice and Bob should output $\mathrm{sgn}(x \cdot y)$.

We show that $\mathcal{GHD}_{n,1/\sqrt{n}}$ has essentially the same randomized communication complexity as the problem GHD that we defined in the introduction. Generalizing that definition, for any $g > 0$ define the problem $\mathrm{GHD}_{n,g}$, in which the input is formed of two n-bit strings x and y, with the promise that $|\Delta(x,y) - n/2| \geq g$, where Δ is the Hamming distance. Alice and Bob should output 0 if $\Delta(x,y) < n/2$ and 1 otherwise.

The following proposition shows that for any $\sqrt{n} \leq g \leq n$, the problems $\mathrm{GHD}_{n,g}$ and $\mathcal{GHD}_{d,\gamma}$ are essentially equivalent from the point of view of randomized communication complexity (with shared randomness) as long as $d \geq n$ and $\gamma = \Theta(g/n)$. It also shows that the randomized communication complexity of $\mathcal{GHD}_{d,\gamma}$ is independent of the dimension d of the input, as long as d is large enough with respect to γ.

Proposition 1. *For every $\varepsilon > 0$, there is a constant $C_0 = C_0(\varepsilon)$ such that for all integers $k, d \geq 0$ and $\sqrt{n} \leq g \leq n$, we have*

$$R_{2\varepsilon}^k(\mathcal{GHD}_{d,C_0 g/n}) \leq R_\varepsilon^k(\mathrm{GHD}_{n,g}) \leq R_\varepsilon^k(\mathcal{GHD}_{n,2g/n}).$$

Proof. We begin with the right inequality. The idea is that a $\mathrm{GHD}_{n,g}$ protocol can be obtained by applying a given $\mathcal{GHD}$ protocol to a suitably transformed input. Let $x, y \in \{0,1\}^n$ be two inputs to $\mathrm{GHD}_{n,g}$. Define $\tilde{x} = \frac{1}{\sqrt{n}}((-1)^{x_i})_{i \in [n]}$ and $\tilde{y} = \frac{1}{\sqrt{n}}((-1)^{y_i})_{i \in [n]}$. Then $\tilde{x}, \tilde{y} \in \mathbb{S}^{n-1}$. Moreover, $\tilde{x} \cdot \tilde{y} = 1 - 2\Delta(x,y)/n$. Therefore, if $\Delta(x,y) \geq n/2 + g$ then $\tilde{x} \cdot \tilde{y} \leq -2g/n$, and if $\Delta(x,y) \leq n/2 - g$ then $\tilde{x} \cdot \tilde{y} \geq 2g/n$. This proves $R_\varepsilon^k(\mathrm{GHD}_{n,g}) \leq R_\varepsilon^k(\mathcal{GHD}_{n,2g/n})$.

For the left inequality, let x and y be two unit vectors (in any dimension) such that $|x \cdot y| \geq \gamma$, where $\gamma = C_0 g/n$. Note that since $g \geq \sqrt{n}$, we have $n = \Omega(\gamma^{-2})$. Using shared randomness, Alice and Bob pick a sequence of vectors $w_1, \ldots, w_n$, each independently and uniformly drawn from the unit sphere. Define two n-bit strings $\tilde{x} = (\mathrm{sgn}(x \cdot w_i))_{i \in [n]}$ and $\tilde{y} = (\mathrm{sgn}(y \cdot w_i))_{i \in [n]}$. Let $\alpha = \cos^{-1}(x \cdot y)$ be the angle between x and y. Then a simple argument (used, e.g., by Goemans and Williamson [8]) shows that the probability that a random unit vector w is such that $\mathrm{sgn}(x \cdot w) \neq \mathrm{sgn}(y \cdot w)$ is exactly α/π. This means that for each i, the bits $\tilde{x}_i$ and $\tilde{y}_i$ differ with probability $\frac{1}{\pi}\cos^{-1}(x \cdot y)$, independently of the other bits of $\tilde{x}$ and $\tilde{y}$. The first few terms in the Taylor series expansion of $\cos^{-1}$ are $\cos^{-1}(z) = \frac{\pi}{2} - z - \frac{z^3}{6} + O(z^5)$. Hence, for each i, $\Pr_{w_i}(\tilde{x}_i \neq \tilde{y}_i) = 1/2 - \Theta(x \cdot y)$, and these events are independent for different i. Choosing C_0 sufficiently large, with probability at least $1 - \varepsilon$, the Hamming distance between $\tilde{x}$ and $\tilde{y}$ is at most $n/2 - g$ if $x \cdot y \geq \gamma$, and it is at least $n/2 + g$ if $x \cdot y \leq -\gamma$.

2.2 Concentration of Measure

It is well known that the Haar measure ν on a high-dimensional sphere is tightly concentrated around the equator — around *any* equator, which makes it a fairly counter-intuitive phenomenon. The original phrasing of this phenomenon, usually attributed to P. Lévy [14], goes by showing that among all subsets of the sphere, the one with the smallest "boundary" is the spherical cap $S_\gamma^x = \{y \in \mathbb{S}^{n-1} : x \cdot y \geq \gamma\}$. The following standard volume estimate will prove useful (see, e.g., [2], Lemma 2.2).

Fact 3. *Let $x \in \mathbb{S}^{n-1}$ and $\gamma > 0$. Then $\nu(S_\gamma^x) \leq e^{-\gamma^2 n/2}$.*

Given a measurable set A, define its *t-boundary* $A_t := \{x \in \mathbb{S}^{n-1} : d(x, A) \leq t\}$, for any $t > 0$. At the core of our results will be the standard fact that, for any not-too-small set A, the set A_t contains almost all the sphere, even for moderately small values of t.

Fact 4 (Concentration of measure on the sphere). *For any measurable $A \subseteq \mathbb{S}^{n-1}$ and any $t > 0$,*

$$\Pr(x \in A)\Pr(x \notin A_t) \leq 4\,e^{-t^2 n/4}, \tag{1}$$

where the probabilities are taken according to the Haar measure on the sphere.

Proof. The usual measure concentration inequality for the sphere (Theorem 14.1.1 in [16]) says that for any set $B \subseteq \mathbb{S}^{n-1}$ of measure at least $1/2$ and any $t' > 0$,

$$\Pr(x \notin B_{t'}) \leq 2\,e^{-(t')^2 n/2}.$$

This suffices to prove the fact if $\Pr(x \in A) \geq 1/2$, so assume that $\Pr(x \in A) < 1/2$. Let t_0 be such that A_{t_0} has measure $1/2$; such a t_0 exists by continuity. Applying measure concentration to $B = A_{t_0}$ gives

$$\Pr(x \notin A_{t'+t_0}) \leq 2\,e^{-(t')^2 n/2}, \tag{2}$$

for all $t' > 0$, while applying it to $B = \overline{A_{t_0}}$ yields

$$\Pr(x \in A_{t_0-t''}) \leq \Pr(x \notin B_{t''}) \leq 2\,e^{-(t'')^2 n/2} \tag{3}$$

for all $t'' \leq t_0$, since $A_{t_0-t''}$ is included in the complement of $(\overline{A_{t_0}})_{t''}$. Taking $t'' = t_0$ gives us $\Pr(x \in A) \leq 2\,e^{-t_0^2 n/2}$. If $t \leq t_0$ then this suffices to prove the inequality. Otherwise, set $t' := t - t_0$ in (2) and $t'' := t_0$ in (3) and multiply the two inequalities to obtain the required bound, by using that $t_0^2 + (t - t_0)^2 \geq t^2/2$ (which holds since $2t_0^2 + t^2/2 - 2t\,t_0 = (\sqrt{2}t_0 - t/\sqrt{2})^2 \geq 0$).

Why the sphere? In Section 4 we give a proof of a slightly weaker lower bound than the one in our main result by using measure concentration facts on the Hamming cube only. We present those useful facts now, together with a brief discussion of the differences, in terms of concentration of measure phenomenon, between the Haar measure on the sphere and the uniform distribution over the hypercube. These differences point to the reasons why the proof of Section 4 gives an inferior bound.

On the Hamming cube, the analogous notion of spherical cap is the Hamming ball: let $T_c^x = \{y \in \{0,1\}^n : \Delta(x,y) \leq n/2 - c\sqrt{n}\}$ be the Hamming ball of radius $n/2 - c\sqrt{n}$ centered at x. The analogue of Fact 3 is given by the Chernoff bound:

Fact 5. *For all $c > 0$, we have $2^{-n}|T_c^x| \leq e^{-2c^2}$.*

A result similar to Lévy's, attributed to Harper [9], states that among all subsets (of the Hamming cube) of a given size, the ball is the one with the smallest boundary. Following a similar proof as for Fact 4, one can get the following statement for the Hamming cube (see e.g. Corollary 4.4 in [3]):

Fact 6 (Concentration of measure on the Hamming cube). *Let $A \subseteq \{0,1\}^n$ be any set, and define $A_c = \{x \in \{0,1\}^n : \exists y \in A, \Delta(x,y) \leq c\sqrt{n}\}$. Then*

$$\Pr(x \in A)\Pr(x \notin A_c) \leq e^{-c^2}, \tag{4}$$

where the probabilities are taken according to the uniform distribution on the Hamming cube.

To compare these two statements, embed the Hamming cube in the sphere by mapping $x \in \{0,1\}^n$ to the vector $v_x = \frac{1}{\sqrt{n}}((-1)^{x_i})_{i\in[n]}$, so that two strings of Hamming distance $c\sqrt{n}$ are mapped to vectors with Euclidean distance $\sqrt{2c}/n^{1/4}$. While on the sphere inequality (1) indicates that most points are at distance roughly $1/\sqrt{n}$ from any set of measure half, if we are restricted to the Hamming cube then very few points are at a corresponding Hamming distance of 1 from, say, the set of all strings with fewer than $n/2$ 1s, which has measure roughly $1/2$ in the cube. This difference is crucial: it indicates that the n-dimensional cube is too rough an approximation of the n-dimensional sphere for our purposes, perhaps explaining why our combinatorial bound in Section 4 yields a somewhat weaker dependence on the number of rounds.

3   Main Result

Our main result is the following.

Theorem 7. *Let $0 \leq \varepsilon \leq 1/50$. There exist constants C, C' depending only on ε such that the following holds for any $\gamma > 0$ and any integers $n \geq \varepsilon^2/(4\gamma^2)$ and $k \leq C'/(\gamma\ln(1/\gamma))$: if P is a randomized ε-error k-round communication protocol for $\mathcal{GHD}_{n,\gamma}$ then some message has length at least $\frac{C}{k^2\ln k} \cdot \frac{1}{\gamma^2}$ bits.*

Using Proposition 1 we immediately get a lower bound for the Hamming cube version $GHD = GHD_{n,\sqrt{n}}$:

Corollary 1. *Any ε-error k-round randomized protocol for GHD communicates at least $\Omega(n/(k^2 \ln k))$ bits.*

This follows from Theorem 7 when $k = o(\sqrt{n}/\log n)$. If k is larger, then the bound stated in the Corollary is in fact weaker than the general $\Omega(\sqrt{n})$ lower bound which we sketched in the introduction.

3.1   Proof Outline

We now turn to the proof of Theorem 7. Let ε, γ and n be as in the statement of the theorem. Since lowering n only makes the $\mathcal{GHD}_{n,\gamma}$ problem easier, for the rest of this section we assume that $n := \varepsilon^2/(4\gamma^2)$ is fixed, and for simplicity of notation we write $\mathcal{GHD}_\gamma$ for $\mathcal{GHD}_{n,\gamma}$.

Measurability. Before proceeding with the proof, we first need to handle a small technicality arising from the continuous nature of the input space: namely, that the distributional protocol might make decisions based on subsets of the input space that are not measurable. To make sure that this does not happen, set $\delta = \gamma/6$ and consider players Alice and Bob who first round their inputs to the closest vector in a fixed δ-net, and then proceed with an ε-error protocol for $\mathcal{GHD}_{\gamma/2}$. Since by definition rounding to the δ-net moves any vector a distance at most δ, the rounding will affect the inner product $x \cdot y$ by at most $2\delta + \delta^2 \leq \gamma/2$. As a result, Alice and Bob will succeed with probability $1 - \varepsilon$ provided they are given valid inputs to $\mathcal{GHD}_\gamma$. Hence any randomized ε-error protocol for $\mathcal{GHD}_{\gamma/2}$ can be transformed into a randomized ε-error protocol for $\mathcal{GHD}_\gamma$ with the same communication, but which initially rounds its inputs to a discrete set. We prove a lower bound on the latter type of protocol. This will ensure that all sets encountered in the proof are measurable.

Distributional complexity. By Yao's principle it suffices to lower-bound the *distributional complexity*, i.e., to analyze *deterministic* protocols that are correct with probability $1 - \varepsilon$ under some input distribution. As our input distribution for $\mathcal{GHD}_\gamma$ we take the distribution that is uniform over the inputs satisfying the promise $|x \cdot y| \geq \gamma$. Given our choice of n, Claim 8 below guarantees that the $\nu \times \nu$-measure of non-promise inputs is at most ε. Hence it will suffice to lower-bound the distributional complexity of protocols making error at most 2ε under the distribution $\nu \times \nu$. We define an ε-*protocol* to be a deterministic communication protocol for $\mathcal{GHD}_{n,\gamma}$ whose error under the distribution $\nu \times \nu$ is at most ε, where we say that a protocol P makes an error if $P(x, y) \neq \mathrm{sgn}(x, y)$.

We prove a lower bound on the maximum length of a message sent by any ε-protocol, via round elimination. The main reduction step is given by the following technical lemma:

Lemma 1 (Round Elimination on the sphere). *Let $\varepsilon, \gamma > 0$, $n = \varepsilon^2/(4\gamma^2)$, and $1 \leq \kappa \leq k$. Assume there is a κ-round ε-protocol P such that the first message has length bounded as $c_1 \leq C_1 \frac{n}{k^2 \ln k} - 7\ln(2k)$ where C_1 is a universal constant. Then there is a $(\kappa - 1)$-round ε'-protocol Q (obtained by eliminating the first message of P), where $\varepsilon' \leq \left(1 + \frac{1}{k}\right)\varepsilon + \frac{1}{16k}$.*

Before proving this lemma in Section 3.2, we show how it implies Theorem 7.

Proof (of Theorem 7). We will show that in any k-round (2ε)-protocol, there is a message sent of length at least $C_1 n/(k^2 \ln k) - 7 \ln(2k)$. The discussion in the "Distributional complexity" paragraph above shows this suffices to prove the theorem, by setting $C = C_1 \varepsilon^2/8$, and choosing C' small enough so that the bound on k in the statement of the theorem implies that $7 \ln(2k) < C_1 n/(2k^2 \ln k)$.

Let P be a k-round (2ε)-protocol, and assume for contradiction that each round of communication uses at most $C_1 n/(k^2 \ln k) - 7 \ln(2k)$ bits. Solving the recurrence $\varepsilon_\kappa = (1 + 1/k)\varepsilon_{\kappa-1} + 1/(16k)$, $\varepsilon_0 = 2\varepsilon$ gives $\varepsilon_\kappa = (1 + 1/k)^\kappa (2\varepsilon + 1/16) - 1/16$, so that applying Lemma 1 k times leads to a *0-round* protocol for $\mathcal{GHD}_\gamma$ that errs with probability at most $\varepsilon' \leq e\,(2\varepsilon + 1/16) - 1/16 \leq 1/4$ over the input distribution $\nu \times \nu$. We have reached a contradiction: such a protocol needs communication and hence cannot be 0-round. Hence P must send a message of length at least $C_1 n/(k^2 \ln k) - 7 \ln(2k)$.

3.2   The Main Reduction Step

Proof (of Lemma 1). Let $P(x, y)$ denote the output of the protocol on input x, y. Define $x \in \mathbb{S}^{n-1}$ to be *good* if $\Pr_{\nu \times \nu}(P(x, y) \text{ errs } |x) \leq (1 + 1/k)\varepsilon$. By Markov's inequality, at least a $1/(k+1)$-fraction of x (distributed according to ν) are good. For a given message m, let A_m be the set of all good x on which Alice sends m as her first message. The sets A_m, over all messages $m \in \{0, 1\}^{c_1}$, form a partition of the set of good x. Define $m_1 := \mathrm{argmax}_m \nu(A_m)$ and let $A := A_{m_1}$. We then have $\nu(A) \geq \frac{1}{k+1} 2^{-c_1} \geq e^{-c_1 - \ln(k+1)}$.

We now define protocol Q. Alice receives an input $\tilde{x}$, Bob receives $\tilde{y}$, both distributed according to ν. Alice computes the point $x \in A$ that is closest to $\tilde{x}$, and Bob sets $y := \tilde{y}$. They run protocol $P(x, y)$ without Alice sending the first message, so Bob starts and proceeds as if he received m_1 from Alice.

To prove the lemma, it suffices to bound the error probability ε' of Q with input $\tilde{x}, \tilde{y}$ distributed according to $\nu \times \nu$. Define $d_1 = 2\sqrt{\frac{c_1 + 6\ln(2k) + 2}{n}}$. We consider the following bad events:

- $\mathrm{BAD}_1 : d(\tilde{x}, A) > d_1$,
- $\mathrm{BAD}_2 : P(x, y) \neq \mathrm{sgn}(x \cdot y)$,
- $\mathrm{BAD}_3 : d(\tilde{x}, A) \leq d_1$ but $\mathrm{sgn}(x \cdot y) \neq \mathrm{sgn}(\tilde{x} \cdot \tilde{y})$.

If none of those events occurs, then protocol P outputs the correct answer. We bound each of them separately, and will conclude by upper bounding ε' with a union bound.

The first bad event can be easily bounded using the measure concentration inequality from Fact 4. Since $\tilde{x}$ is uniformly distributed in $\mathbb{S}^{n-1}$ and $\Pr(A) \geq e^{-c_1 - \ln(k+1)}$, we get

$$\Pr(\mathrm{BAD}_1) \leq 4\,e^{-d_1^2 n/4 + c_1 + \ln(k+1)} \leq 4\,e^{-5\ln(2k) - 2} \leq \frac{1}{32k}.$$

The second bad event has probability bounded by $(1 + 1/k)\,\varepsilon$ by the goodness of x. Now consider event BAD_3. Without loss of generality, we may assume that $\tilde{x} \cdot \tilde{y} = \tilde{x} \cdot y > 0$

but $x \cdot y < 0$ (the other case is treated symmetrically). In order to bound BAD_3, we will use two claims. The first shows that the probability that $\tilde{x} \cdot y$ is close to 0 for a random $\tilde{x}$ and y is small. The second uses measure concentration to show that, if $\tilde{x} \cdot y$ is not too close to 0, then moving $\tilde{x}$ to the nearby x is unlikely to change the sign of the inner product.

Claim 8. *Let x, y be distributed according to ν. For any real $\alpha \geq 0$, we have*

$$\Pr(0 \leq x \cdot y \leq \alpha) \leq \alpha\sqrt{n}.$$

Proof. With ω_n the volume of the n-dimensional Euclidean unit ball, we write (see e.g. [4], Lemma 5.1)

$$\Pr(0 \leq x \cdot y \leq \alpha) = \frac{(n-1)\,\omega_{n-1}}{n\,\omega_n} \int_0^\alpha (1 - t^2)^{\frac{n-3}{2}}\,\mathrm{dt} \leq \alpha\sqrt{n},$$

where we used $\frac{\omega_{n-1}}{\omega_n} < \sqrt{\frac{n+1}{2\pi}} < \sqrt{n}$.

Claim 9. *Let $x, \tilde{x}$ be two fixed unit vectors at distance $\|x - \tilde{x}\| = d \in [0, d_1]$, and $0 < \alpha \leq 1/(4\sqrt{n})$. Let y be taken according to ν. Then*

$$\Pr(\tilde{x} \cdot y \geq \alpha \wedge x \cdot y < 0) \leq e^{-\alpha^2 n/(8d_1^2)}.$$

Proof. Note that $x \cdot \tilde{x} = 1 - \|x - \tilde{x}\|^2/2 = 1 - d^2/2$. Since the statement of the lemma is rotationally-invariant, we may assume without loss of generality that

$$\tilde{x} = (1, 0, 0 \ldots, 0),$$
$$x = (1 - d^2/2, -\sqrt{d^2 - d^4/4}, 0, \ldots, 0),$$
$$y = (y_1, y_2, y_3, \ldots, y_n).$$

Therefore, $y_1 \geq \alpha$ when $\tilde{x} \cdot y \geq \alpha$. Note that

$$x \cdot y = x_1 y_1 + x_2 y_2 \geq (1 - d^2/2)\alpha - \sqrt{d^2 - d^4/4}\, y_2.$$

Hence the event $\tilde{x} \cdot y \geq \alpha \wedge x \cdot y < 0$ implies

$$y_2 > \frac{(1 - d^2/2)\,\alpha}{\sqrt{d^2 - d^4/4}} \geq \frac{\alpha}{2d},$$

where we used the fact that $d \leq d_1 \leq 1$, given our assumption on c_1. By Fact 3, the probability that, when y is sampled from ν, y_2 is larger than $\alpha/(2d)$ is at most $e^{-\alpha^2 n/(8d^2)}$. Hence the probability that both $\tilde{x} \cdot y \geq \alpha$ and $x \cdot y < 0$ happen is at most as much.

Setting $\alpha = 1/(128k\sqrt{n})$, by Claim 8 we find that the probability that $0 \leq \tilde{x} \cdot y \leq \alpha$ is at most $1/(128k)$. Furthermore, the probability that $\tilde{x} \cdot y \geq \alpha$ and $x \cdot y < 0$ is at most $\exp\left(-\frac{n}{2^{19}k^2(c_1 + 6\ln(2k) + 2)}\right)$ by Claim 9. This bound is less than $1/(128k)$ given our assumption on c_1, provided C_1 is a small enough constant. Putting both bounds together, we see that

$$\Pr(\tilde{x} \cdot y \geq 0 \wedge x \cdot y < 0) < 1/(64k).$$

The event that $\tilde{x} \cdot y < 0$ but $x \cdot y \geq 0$ is bounded by $1/(64k)$ in a similar manner. Hence, $\Pr(\text{BAD}_3) < 1/(32k)$. Taking the union bound over all three bad events concludes the proof of the lemma.

4   A Simple Combinatorial Proof

In this section we present a combinatorial proof of the following:

Theorem 10. *Let* $0 \leq \varepsilon \leq 1/50$. *There exists a constant* C'' *depending on* ε *only, such that the following holds for any* $g \leq C'' \sqrt{n}$ *and* $k \leq n^{1/4}/(1024 \log n)$: *if* P *is a randomized* ε-*error* k-*round communication protocol for* $\text{GHD}_{n,g}$ *then some message has length at least* $\frac{n}{(512k)^4 \log^2 k}$ *bits.*

Even though this is a weaker result than Theorem 7, its proof is simpler and is based on concentration of measure in the Hamming cube rather than on the sphere (we refer to Section 2.2 for a high-level comparison of the two proofs). Interestingly, the dependence on the number of rounds that we obtain is quadratically worse than that of the proof using concentration on the sphere. We do not know if this can be improved using the same technique.

We proceed as in Section 3.1, observing that it suffices to lower-bound the distributional complexity of $\text{GHD}_{n,g}$ under a distribution uniform over the inputs satisfying the promise $|\Delta(x, y) - n/2| \geq g$. In fact, as we did before, by taking C'' small enough we can guarantee that the number of non-promise inputs is at most $\varepsilon \, 2^n$. Hence it will suffice to lower-bound the distributional complexity of protocols making error at most 2ε under the uniform input distribution. We define an ε-*protocol* to be a deterministic communication protocol for GHD whose distributional error under the uniform distribution is at most ε. The following is the analogue of Lemma 1, from which the proof of Theorem 10 follows as in Section 3.1.

Lemma 2 (Round Elimination on the Hamming cube). *Let* $\varepsilon > 0$ *and* κ, k *be two integers such that* $k \geq 128$ *and* $1 \leq \kappa \leq k \leq n^{1/4}/(1024 \log n)$. *Assume that there is a* κ-*round* ε-*protocol* P *such that the first message has length bounded by* $c_1 \leq n/((512k)^4 \log^2 k)$. *Then there exists a* $(\kappa - 1)$-*round* ε'-*protocol* Q *(obtained by eliminating the first message of* P*) where* $\varepsilon' \leq \left(1 + \frac{1}{k}\right) \varepsilon + \frac{1}{16k}$.

Proof. Define $x \in \{0, 1\}^n$ to be *good* if $\Pr(P(x, y) \text{ errs } | x) \leq (1 + 1/k)\varepsilon$. By Markov's inequality, at least a $1/(k + 1)$-fraction of $x \in \{0, 1\}^n$ are good. For a given message m, let $A_m := \{\text{good } x : \text{Alice sends } m \text{ given } x\}$. The sets A_m, over all messages $m \in \{0, 1\}^{c_1}$, together form a partition of the set of good x. Define $m_1 := \text{argmax}_m |A_m|$, and let $A := A_{m_1}$. By the pigeonhole principle, we have $|A| \geq \frac{1}{k+1} 2^{n-c_1}$.

We now define protocol Q. Alice receives an input $\tilde{x}$, Bob receives $\tilde{y}$, uniformly distributed. Alice computes the string $x \in A$ that is closest to $\tilde{x}$ in Hamming distance, and Bob sets $y := \tilde{y}$. They run protocol $P(x, y)$ without Alice sending the first message, so Bob starts and proceeds as if he received the fixed message m_1 from Alice.

To prove the lemma, it suffices to bound the error probability ε' of Q under the uniform distribution. Define $d_1 := 9\sqrt{n}/((1024k)^2 \log k)$. As in the proof of Lemma 1, we consider the following bad events:

- BAD$_1$: $\Delta(x, \tilde{x}) > d_1\sqrt{n}$,
- BAD$_2$: $P(x, y) \neq \mathrm{GHD}(x, y)$,
- BAD$_3$: $\Delta(x, \tilde{x}) \leq d_1\sqrt{n}$ but $\mathrm{GHD}(\tilde{x}, y) \neq \mathrm{GHD}(x, y)$.

If none of those events occurs, then protocol P outputs the correct answer. We bound each of them separately, and will conclude by a union bound. BAD$_1$ is easily bounded using Fact 6, which implies

$$\Pr(\tilde{x} \notin A_{d_1}) \leq e^{-81n/((1024k)^4 \log^2 k)} 2^{c_1 + \log(k+1)} \leq \frac{2}{k^2} \leq \frac{1}{32k},$$

given our assumptions on c_1 and k. The second bad event is bounded by $(1 + 1/k)\varepsilon$, by definition of A.

We now turn to BAD$_3$. The event that $\mathrm{GHD}(\tilde{x}, y) \neq \mathrm{GHD}(x, y)$ only depends on the relative distances between x, $\tilde{x}$, and y, so we may apply a shift to assume that $x = (0, \ldots, 0)$. Without loss of generality, we assume that $\Delta(\tilde{x}, y) > n/2$ and $|y| < n/2$ (the error bound when $\Delta(\tilde{x}, y) < n/2$ and $|y| > n/2$ is proved in a symmetric manner). Note that, since y is uniformly random (subject to $|y| < n/2$), by a standard head estimate for the binomial distribution with probability at least $1 - 1/(128k)$ we have $|y| \leq n/2 - \sqrt{n}/(128k)$ (this is analoguous to the estimate from Claim 8 that we used in the continuous setting). Hence we may assume that this holds with an additive loss of at most $1/(128k)$ in the error. Now

$$\Delta(\tilde{x}, y) > n/2 \iff |\tilde{x}| + |y| - 2|\tilde{x} \cap y| > n/2 \iff |\tilde{x} \cap y| < \frac{|\tilde{x}| + |y| - n/2}{2}.$$

It is clear that the worst case in this statement is for $|y| = n/2 - \sqrt{n}/(128k)$ and $|\tilde{x}| = \Delta(x, \tilde{x}) = d_1\sqrt{n}$. By symmetry, the probability that this event happens is the same as if we fix any y of the correct weight, and $\tilde{x}$ is a random string of weight $d_1\sqrt{n}$. The expected intersection size is $|y||\tilde{x}|/n = |\tilde{x}|/2 - d_1/(128k)$, and so by Hoeffding's inequality (see e.g. the bound on the tail of the hypergeometric distribution given in [7]), for $a = \sqrt{n}/(256k) - d_1/(128k)$, we have

$$\Pr\left(|\tilde{x} \cap y| \leq \frac{|\tilde{x}| + |y| - n/2}{2}\right) = \Pr\left(|\tilde{x} \cap y| \leq \mathbb{E}[|\tilde{x} \cap y|] - a\right) \leq e^{-2a^2/(d_1\sqrt{n})}.$$

Given our choice of d_1 we have $a \geq 3\sqrt{n}/(4 \cdot 256k)$, and hence the upper bound is at most $1/k^2 \leq 1/(128k)$, given our assumption on k. Applying the union bound over all bad events then yields the lemma.

Acknowledgments. We thank Ishay Haviv for discussions during the early stages of this work.

References

1. Ajtai, M.: A lower bound for finding predecessors in Yao's cell probe model. Combinatorica 8, 235–247 (1988)
2. Ball, K.: An elementary introduction to modern convex geometry. Flavors of Geometry 31 (1997)
3. Barvinok, A.: Lecture notes on measure concentration (2005), http://www.math.lsa.umich.edu/~barvinok/total710.pdf
4. Brieden, A., Gritzmann, P., Kannan, R., Klee, V., Lovász, L., Simonovits, M.: Approximation of diameters: Randomization doesn't help. In: Proceedings of 39th IEEE Symposium on Foundations of Computer Science (FOCS 1998), pp. 244–251 (1998)
5. Brody, J., Chakrabarti, A.: A multi-round communication lower bound for Gap Hamming and some consequences. In: Proceedings of 24th IEEE Conference on Computational Complexity (CCC 2009), pp. 358–368 (2009)
6. Chakrabarti, A., Regev, O.: Tight lower bound for the Gap Hamming problem. Personal Communication (2009)
7. Chvátal, V.: The tail of the hypergeometric distribution. Discrete Mathematics 25(3), 285–287 (1979)
8. Goemans, M., Williamson, D.: Improved approximation algorithms for maximum cut and satisfiability problems using semidefinite programming. Journal of the ACM 42, 1115–1145 (1995)
9. Harper, L.: Optimal numbering and isoperimetric problems on graphs. Journal of Combinatorial Theory 1, 385–393 (1966)
10. Indyk, P., Woodruff, D.: Tight lower bounds for the distinct elements problem. In: Proceedings of 44th IEEE Symposium on Foundations of Computer Science (FOCS 2003), pp. 283–289 (2003)
11. Jayram, T.S., Kumar, R., Sivakumar, D.: The one-way communication complexity of Hamming distance. Theory of Computing 4(1), 129–135 (2008)
12. Kushilevitz, E., Nisan, N.: Communication Complexity. Cambridge University Press, Cambridge (1997)
13. Lee, T.,, S.: Disjointness is hard in the multi-party number-on-the-forehead model. In: Proceedings of 23rd IEEE Conference on Computational Complexity (CCC 2008), pp. 81–91 (2008)
14. Lévy, P.: Problèmes concrets d'analyse fonctionnelle. Gauthier-Villars (1951)
15. Linial, N., Shraibman, A.: Lower bounds in communication complexity based on factorization norms. In: Proceedings of 39th ACM Symposium on the Theory of Computing (STOC 2007), pp. 699–708 (2007)
16. Matoušek, J.: Lectures on Discrete Geometry. Springer, Heidelberg (2002)
17. Miltersen, P., Nisan, N., Safra, S., Wigderson, A.: On data structures and asymmetric communication complexity. J. Comput. Syst. Sci. 57(1), 37–49 (1998); preliminary version in Proceedings of 27th ACM Symposium on the Theory of Computing (STOC 1995), pp. 103–111 (1995)
18. Razborov, A.: Quantum communication complexity of symmetric predicates. Izvestiya of the Russian Academy of Science, Mathematics 67, 0204025 (2002)
19. Sherstov, A.: The pattern matrix method for lower bounds on quantum communication. In: Proceedings of 40th ACM Symposium on the Theory of Computing (STOC 2008), pp. 85–94 (2008)
20. Woodruff, D.: Optimal space lower bounds for all frequency moments. In: Proceedings of 15th ACM-SIAM Symposium on Discrete Algorithms (SODA 2004), pp. 167–175 (2004)

Propagation Connectivity
of Random Hypergraphs

Amin Coja-Oghlan[1,*], Mikael Onsjö[2,**], and Osamu Watanabe[2,**]

[1] Mathematics and Computer Science
University of Warwick, Zeeman building, Coventry CV4 7AL, UK
a.coja-oghlan@warwick.ac.uk
[2] Department of Mathematical and Computing Sciences
Tokyo Institute of Technology, Meguro-ku Ookayama 2-12-1 W8-25
mikael@is.titech.ac.jp, watanabe@is.titech.ac.jp

Abstract. We study the concept of *propagation connectivity* on random 3-uniform hypergraphs. This concept is defined for investigating the performance of a simple algorithm for solving instances of certain constraint satisfaction problems. We derive upper and lower bounds for edge probability of random 3-uniform hypergraphs such that the propagation connectivity holds. Based on our analysis, we also show the way to implement the simple algorithm so that it runs in linear time on average.

1   Introduction and Results

There are several natural ways to define connectivity for 3-uniform hypergraphs $H = (V, E)$ (recall that in a 3-uniform hypergraph each edge is a set of three vertices). For instance, a standard concept is to consider H connected if the graph obtained by replacing each edge e by a triangle is connected. In this paper we study a different concept that we call *propagation connectivity*.

Definition 1. *Let $H = (V, E)$ be a 3-uniform hypergraph on $n = |V|$ vertices. We call a sequence $e_1, \ldots, e_{n-2} \in E$ a* propagation sequence *if for any $1 \leq l < n - 2$ we have $|e_{l+1} \cap \bigcup_{i=1}^{l} e_l| = 2$. If H has a propagation sequence, then we say that H is* propagation connected.

This definition is motivated by a simple algorithm for a certain kind of constraint satisfaction problem. For the time being, let us focus on the concrete example of a system of linear equations over a finite field with three variables per equation. We can associate a hypergraph H with this system by thinking of the variables as vertices and of the equations as hyperedges. If we are given a propagation sequence $e_1, \ldots, e_{n-2}$ for H, then we can find a solution to the system of equations in linear time (if there is one). Namely, suppose that the

---

* Supported by EPSRC grant EP/G039070/2 and DIMAP.
** Supported in part by the "CompView" Global COE of JSPS, and Grants-in-Aid for Scientific Research on the Priority Areas "Dex-SMI" and No. 22300003 of MEXT.

M. Serna et al. (Eds.): APPROX and RANDOM 2010, LNCS 6302, pp. 490–503, 2010.

variables of e_1 are x, y, z. We can easily 'guess' the correct values of x, y (i.e., we can try all possible assignments because the field is finite). Then the value of z is implied. Now, assume inductively that we have obtained the values of the variables occurring in the first l edges/equations $e_1, \ldots, e_l$ already. Then e_{l+1} contains precisely one additional variable (by the definition of propagation sequence), whose value we can thus infer directly. Thus, after passing through the entire sequence $e_1, \ldots, e_{n-2}$, we have determined the values of all n variables. If this solves the linear system, we are done. Conversely, if we find that no assignment to the first two variables x, y leads to a solution, then it is safe to conclude that no solution exists.

The contribution of this paper is close upper and lower bounds on the edge probability that the propagation connectivity holds in random hypergraphs. More precisely, we consider the following random hypergraph model $\mathcal{H}(n, p)$: the vertex set of the random hypergraph is $V = [n] = \{1, \ldots, n\}$, and each of the $\binom{n}{3}$ possible edges is present with probability $0 \le p \le 1$ independently. We write $H : \mathcal{H}(n, p)$ to indicate that H is a random hypergraph chosen from this distribution. Moreover, we say that the random hypergraph has some property *with high probability* (w.h.p.) if the probability that the property holds converges to one as $n \to \infty$.

Theorem 1. *Suppose that $p = c/(n \ln n)$ for a constant $c > 0$.*

(1) If $c < 0.16$, then $H : \mathcal{H}(n, p)$ fails to be propagation connected w.h.p.
(2) If $c > 0.25$, then $H : \mathcal{H}(n, p)$ is propagation connected w.h.p.

Determining the threshold for 'standard' connectivity (where each hyperedge is replaced by a triangle) is easy. The result is a hardly surprising $p \sim 2n^{-2} \ln n$, and the proof is via a simple coupon collecting argument. By contrast, analyzing propagation connectivity is quite non-trivial. Our proof is based on a kind of large deviations analysis of a time-dependent random walk. A precise solution of this problem might close the gap left by Theorem 1.

For a propagation connected hypergraph H one can determine a propagation sequence in polynomial time via a generalized breadth first search procedure. However, the running time of this algorithm is superlinear (in contrast to BFS on graphs). Based on our analysis, we derive a simple algorithm with linear expected running time.

Theorem 2. *There is a randomized algorithm A that satisfies the following. For any given hypergraph, A finds a propagation sequence if it exists. For any parameter p, the expected running time of A applied to $H : \mathcal{H}(n, p)$ is linear in the number of edges of H.*

As an application, we show how Theorem 2 yields an algorithm for deciding a class of random constraint satisfaction problems. A *CSP instance with domain* $[k] = \{1, \ldots, k\}$ consists of a 3-uniform hypergraph $H = (V, E)$ with $V = [n]$ and a family $(f_e)_{e \in E}$ of maps $f_e : [k] \times [k] \times [k] \to \{0, 1\}$. Moreover, a *solution* is a map $\sigma : V \to [k]$ such that for any triple $1 \le x < y < z \le n$ of vertices with $e = \{x, y, z\} \in E$ we have $f_e(\sigma(x), \sigma(y), \sigma(z)) = 1$. Thus, intuitively the hypergraph H describes the interactions of the variables V, and for any edge e the map f_e characterizes the values that can be assigned to the variables in e so as to satisfy the constraint that e represents.

Furthermore, we say that a CSP instance is *uniquely extendable* if for any $x, y \in [k]$, any $i \in \{1, 2, 3\}$, and any edge $e \in E$ there is precisely one value $z_i \in [k]$ such that $f_e(z_1, x, y) = f_e(x, z_2, y) = f_e(x, y, z_3) = 1$. In other words, once we assign two variable in a constraint e, there is precisely one way to assign the third variable so as to satisfy e. Clearly, systems of linear equations over a finite field provide an example of uniquely extendable problems, but there are many others.

By combining Theorem 2 with the simple propagation procedure outlined after Definition 1, we obtain the following result.

Corollary 1. *Fix $c > 0.25$ and $k \geq 2$ and let $p = c/(n \ln n)$. Moreover, assume that P is a probability distribution over uniquely extendable CSP instance with domain $[k]$ such that the distribution of the random hypergraph underlying the problem instance coincides with the distribution $\mathcal{H}(n, p)$. There is an algorithm with linear expected running time that decides whether a random CSP instance chosen from the distribution P has a solution w.h.p.*

There are a variety of probability distribution over CSPs that satisfy the assumptions of Corollary 1. Examples include uniformly random systems of linear equations, which at the density assumed in Corollary 1 do not have solutions w.h.p. Thus, for these problems running the algorithm in Corollary 1 will provide a *succinct proof* that no solution exists w.h.p. On the other hand, distributions that do admit solutions w.h.p. include systems of linear equations with a 'planted' solution, for which the algorithm will find a solution w.h.p.

Related Work

The 'standard' concept of random hypergraph connectivity (where edges are replaced by triangles) has been studied, e.g., in [BCK07, CMV07], particularly with respect to the emergence and size of the giant component. These results generalize what was known for random graphs (see [JLR00] for a comprehensive summary). A further related random hypergraph concept is that of a core. This concept is related to local search algorithms such as the 'pure literal rule' for the satisfiability. Contributions on these subjects include [DN05, Mol05].

Berke and Onsjö [BO09] approached the propagation connectivity threshold for random 3-uniform hypergraphs. They established a lower bound of $p = \Omega(1/n(\log n)^2)$ and an upper bound of $p = O(1/n(\log n)^{0.4})$. As Theorem 1 shows, the correct order of magnitude is $p = \Theta(1/(n \ln n))$.

With respect to the application to random constraint satisfaction problems, it is clear that the case of linear equations over finite fields can be solved in polynomial (albeit superlinear) time by Gaussian elimination. However, if the underlying hypergraph comes with a propagation sequence, then the problem can be solved in linear time as indicated. While linear equations provide an example of uniquely extendable constraint satisfaction problems, there are more; in fact, some of them are NP-hard [CM04].

2   The Propagation Process

In this section we show how the propagation connectivity problem can be modeled by a stochastic process, which we call the *propagation process*. We start out by describing this process for a fixed hypergraph $H = (V, E)$ with vertex set $V = \{1, \ldots, n\}$. Let (v_1, v_2) be a pair of distinct vertices, which we refer to as the *initial pair*. In the course of the prcoess, vertices are either *active*, *neutral*, or *dead*. Initially v_1 is dead, v_2 is active, and all other vertices are neutral; formally, we let

$$\mathcal{D}_0^{(v_1, v_2)}[H] = \{v_1\}, \quad \mathcal{A}_0^{(v_1, v_2)}[H] = \{v_2\}.$$

Once there is no active vertex left, the process stops. Otherwise at each time $t \geq 1$, the least active vertex u is chosen (recall that $V = [n]$ is an ordered set). All neutral vertices v for which there is a dead vertex w such that $\{u, v, w\} \in E$ are declared active, and then u is declared dead. In symbols, we let $u = \min \mathcal{A}_{t-1}^{(v_1, v_2)}[H]$ and

$$\mathcal{D}_t^{(v_1, v_2)}[H] = \mathcal{D}_{t-1}^{(v_1, v_2)}[H] \cup \{u\},$$
$$\mathcal{A}_t^{(v_1, v_2)}[H] = \left(\mathcal{A}_{t-1}^{(v_1, v_2)}[H] \setminus \{u\}\right)$$
$$\cup \left\{v \notin \mathcal{D}_{t-1}^{(v_1, v_2)}[H] : \exists w \in \mathcal{D}_{t-1}^{(v_1, v_2)}[H] : \{u, v, w\} \in E\right\}.$$

Thus, at time t the total number of dead vertices equals $t + 1$. Let $T^{(v_1, v_2)}[H]$ be the time when the process stops. To avoid case distinctions, we consider vertices dead (or active, or neutral) at times $t > T^{(v_1, v_2)}[H]$ if they had the corresponding predicate at time $T^{(v_1, v_2)}[H]$. Observe that for a fixed hypergraph H, the process is entirely deterministic.

The process is related to the propagation connectivity problem as follows. Assume that vertex v was declared active at time $t \geq 2$. Then H has an edge e_t that contains v and two vertices from $\mathcal{D}_t^{(v_1, v_2)}[H]$. Proceeding inductively, we obtain a sequence $e_2, \ldots, e_t$ such that $v_1, v_2 \in e_2$ and $|e_{l+1} \cap \bigcup_{i=2}^{l} e_l| \geq 2$ for all $2 \leq l < t$. Hence, if all vertices are declared dead eventually, i.e., if $T^{(v_1, v_2)}[H] = n - 1$, then we obtain a propagation sequence. Conversely, if there is a propagation sequence $e_2, \ldots, e_{n-1}$ such that $v_1, v_2 \in e_2$, then the propagation process will not stop before time $n - 1$. Thus, we have the following.

Fact 1. *H is propagation connected iff there is a pair (v_1, v_2) such that $T^{(v_1, v_2)}[H] = n - 1$.*

To prove Theorem 1, we are going to study the propagation process on a random hypergraph $H : \mathcal{H}(n, p)$. In this case we omit the reference to H, i.e., we just write $\mathcal{D}_t^{(v_1, v_2)}$ etc. It will be convenient to use the terminology of stochastic processes. In particular, for $t \geq 0$ we let $\mathcal{F}_t^{(v_1, v_2)}$ signify the coarsest σ-algebra on $\mathcal{H}(n, p)$ in which all events $\{v \in \mathcal{D}_s^{(v_1, v_2)}\}$ and $\{v \in \mathcal{A}_s^{(v_1, v_2)}\}$ for $s \leq t$ and $v \in V$ are measurable. Then $(\mathcal{F}_t^{(v_1, v_2)})_{t \geq 0}$ is a filtration. We will also use the concept of conditional probabilities with respect to the filtration $(\mathcal{F}_t)_{t \geq 0}$ (see [D05]). To remind the reader, for an event A and a (fixed) hypergraph H_0 the conditional probability is

$$\Pr\left[A|\mathcal{F}_t^{(v_1,v_2)}\right](H_0) =$$

$$\frac{\Pr\left[A \text{ occurs and } \mathcal{D}_s^{(v_1,v_2)} = \mathcal{D}_s^{(v_1,v_2)}[H_0], \, \mathcal{A}_s^{(v_1,v_2)} = \mathcal{A}^{(v_1,v_2)}[H_0] \text{ for all } s \leq t\right]}{\Pr\left[\mathcal{D}_s^{(v_1,v_2)} = \mathcal{D}_s^{(v_1,v_2)}[H_0], \, \mathcal{A}_s^{(v_1,v_2)} = \mathcal{A}^{(v_1,v_2)}[H_0] \text{ for all } s \leq t\right]}.$$

In words, $\Pr\left[A|\mathcal{F}_t^{(v_1,v_2)}\right](H_0)$ is the probability of the event A in a random hypergraph $H : \mathcal{H}(n,p)$ given that the first t steps of the propagation process on H work out the same as in H_0. When the argument H_0 is omitted, it is understood that the corresponding statement holds for all H_0.

For any $t \geq 1$ the first t steps of the propagation process on the random hypergraph $H : \mathcal{H}(n,p)$ *only* depend on the presence (or absence) of edges that contain at least two vertices that have been declared dead by time t, i.e., from the set $\mathcal{D}_t^{(v_1,v_2)}$. This means that the presence of edges e with $|e \cap \mathcal{D}_t^{(v_1,v_2)}| < 2$ is stochastically independent of the first t steps.

Fact 2. *Given $\mathcal{F}_t$, for all triples $e = \{u,v,w\}$ such that $\left|e \cap \mathcal{D}_t^{(v_1,v_2)}\right| < 2$, the edge e is present in $H : \mathcal{H}(n,p)$ with probability p independently. In symbols, for any set*

$$\mathcal{E} \subset \left\{e \in \binom{V}{3} : \left|e \cap \mathcal{D}_t^{(v_1,v_2)}\right| < 2\right\}$$

we have $\Pr\left[\mathcal{E} \subset E(H)|\mathcal{F}_t^{(v_1,v_2)}\right] = p^{|\mathcal{E}|}$.

The above propagation process is similar in spirit to the branching process approach for the giant component problem in random graphs/digraphs [Kar90]. The difference between our proofs and the standard argument is that we need to investigate whether *there exists* a pair (v_1, v_2) such that $T^{(v_1,v_2)} \geq n - 1$ (cf. Fact 1). Since there are a total of $\binom{n}{2}$ initial pairs to choose from, this means that we need to study *unlikely* trajectories of the propagation process (that occur with probability merely about $1/\binom{n}{2}$).

By contrast, for the giant component problem the corresponding process has to be studied only from a *random* start vertex, a problem which relatively easily reduces to the typical behavior of a standard Galton-Watson branching process. Alternatively, the problem can be tackled via a whole arsenal of different techniques, ranging from differential equations to random walks. Unfortunately, the fact that here we need to study an 'exceptional' event puts these standard arguments out of business.

To get started, we point out that the hypergraph distribution $H : \mathcal{H}(n,p)$ is invariant w.r.t. permutations of the vertices. Therefore, the distribution of the propagation process is the same for any initial pair. For the sake of concreteness we will refer to $(v_1, v_2) = (1, 2)$. For this initial pair we will omit the superscript (v_1, v_2) from the notation. Moreover, we let $A_t = |\mathcal{A}_t|$ be the number of active vertices at time t (from the initial pair $(1, 2)$). Then $A_0 = 1$ by construction. For any $t \geq 1$, we define a further random variable X_t via

$$X_t = A_t - A_{t-1} + 1. \tag{1}$$

That is, X_t is the number of vertices that got declared active at time t.

Fact 3. *If $1 \leq t \leq T$, then given $\mathcal{F}_{t-1}$, the random variable X_t is binomially distributed* $\mathrm{Bin}(n - t - A_{t-1}, 1 - (1-p)^t)$.

Proof. The number of neutral vertices at time $t - 1$ equals $n - A_{t-1} - |\mathcal{D}_{t-1}| = n - A_{t-1} - t$. Suppose that v is neutral at time $t - 1$ and let $u = \min \mathcal{A}_{t-1}$. Then v becomes active at time t iff there is $w \in \mathcal{D}_{t-1}$ such that $\{u, v, w\} \in E$. By Fact 2 each of these t edges is present in H with probability p independently. Hence, the probability that all of them are absent is $1 - (1-p)^t$. □

To outline the proof of Theorem 1, let us interpret the propagation process in terms of a time-dependent random walk. The process continues up to time t iff $A_s > 0$ for all $1 \leq s \leq t$. Due to (1), this is true iff $\sum_{q=1}^{s}(X_q - 1) \geq 0$ for all $1 \leq s \leq t$. Thus, if we think of the random variables $X_s - 1$ as the steps of a random walk, then the propagation process continues to time t iff the random walk stays non-negative at all times $s \leq t$. As Fact 3 shows, this random walk is time-dependent.

In the regime $p = \Theta(1/(n \ln n))$ that we are interested in, and for times $s \ll \ln n$, the random walk has a negative drift. More precisely, for $s \ll \ln n$ Fact 3 implies that the expectation of $X_s - 1$ is $(1 + o(1))nps - 1 < 0$. Therefore, standard results on random walks show that the probability that the random walk will continue to time, say, $\ln n$ is $o(1)$. If, however, the process happens to survive up to time $t = (1 + \epsilon)/(np) = \Theta(\ln n)$ for a fixed $\epsilon > 0$, then Fact 3 shows that the 'drift' of $X_t - 1$ becomes positive and thus the process is likely to continue up to time $n - 1$.

The previous paragraph shows that the probability that one specific initial pair leads to a propagation sequence is $o(1)$. But this does *not* imply that the random hypergraph $H : \mathcal{H}(n, p)$ is not propagation connected w.h.p., because there is a total $\binom{n}{2}$ initial pairs to choose from. This observation suggests that in order find the threshold for propagation connectivity we need to determine for what p the random walk continues to time $1/(np)$ with probability $1/\binom{n}{2}$. In Section 3 we will derive a lower bound on this value of p. The more challenging problem is to obtain an upper bound, which we address in Section 4.

3   The Lower Bound

In this section we prove the first part of Theorem 1, i.e., we show that the random hypergraph $H : \mathcal{H}(n, p)$ is *not* propagation connected w.h.p. if $p < 0.16/(n \ln n)$.

(Throughout this section, we fix $p = c/(n \ln n)$ and let $c > 0$ denote this constant.) To this end, we will derive that the probability that the initial pair $(1, 2)$ leads to a propagation sequence is $o(n^{-2})$. By symmetry and the union bound, this implies that w.h.p. *no* initial pair (v_1, v_2) does. We start by reducing the problem of estimating the probability that $(1, 2)$ yields a propagation sequence to an exercise in calculus. The proof employs the following Chernoff bound on the tails of a binomially distributed random variable X with mean μ (e.g., [JLR00, p. 21]): letting $\varphi(x) = (1 + x) \ln(1 + x) - x$, we have for any $t > 0$

$$
\begin{aligned}
\Pr[X \leq \mu - t] &\leq \exp(-\mu \cdot \varphi(-t/\mu)), \quad \text{and} \\
\Pr[X \geq \mu + t] &\leq \exp(-\mu \cdot \varphi(t/\mu)),
\end{aligned}
\tag{2}
$$

Lemma 1. *For the constant c, assume that d satisfies $0 < d \leq 2/c$ and $d(cd/2 + \ln(2/cd) - 1) > 2$. Let $t_0 = d \ln n$. Then $\Pr[T > t_0] = o(n^{-2})$.*

Proof. Let $\{\widetilde{X}_t\}_{t \geq 1}$ be a family of mutually independent random variables such that $\widetilde{X}_t$ has distribution $\mathrm{Bin}(nt, p)$. Let $t \geq 1$. By construction, for each vertex $v \in A_t \setminus A_{t-1}$ that becomes active at time t, there is an edge $\{u, v, w\}$ in $H : \mathcal{H}(n, p)$ such that $u = \min A_{t-1}$ and $w \in \mathcal{D}_{t-1}$. In particular, the number X_t of newly active vertices v is bounded by the number of such edges $\{u, v, w\}$. By Fact 2, given $\mathcal{F}_{t-1}$, each such edge is present in H with probability p independently. As $|\mathcal{D}_{t-1}| = t$ and because the number of neutral vertices v to choose from is bounded by n, this shows that $X_t|\mathcal{F}_{t-1}$ is stochastically dominated by the binomial variable $\widetilde{X}_t = \mathrm{Bin}(nt, p)$.

If the stopping time T exceeds some specific time t_0, then $A_t \geq 1$ for all $t \in [t_0]$. Hence, (1) implies $\sum_{1 \leq t \leq t_0} X_t \geq t_0$. Because each X_t is dominated by $\widetilde{X}_t$, we can bound the probability of this event by

$$
\Pr[T \geq t_0] \leq \Pr\left[\sum_{1 \leq t \leq t_0} \widetilde{X}_t \geq t_0\right] = \Pr\left[\mathrm{Bin}\left(n \cdot \frac{t_0(t_0 + 1)}{2}, p\right) \geq t_0\right]. \tag{3}
$$

Let μ_0 denote the expectation of this last binomial distribution. Then

$$
\mu_0 = n \cdot \frac{t_0(t_0 + 1)}{2} \cdot p = \frac{cd^2}{2}\left(1 + \frac{2}{d \ln n}\right) \ln n \sim \frac{cd^2}{2} \ln n. \tag{4}
$$

We are going to verify that our assumption on c, d implies that the r.h.s. of (3) is $o(n^{-2})$. Since we assume $d \leq 2/c$, we have $\frac{cd^2}{2} \ln n = \mu_0 \leq t_0 = d \ln n$. Therefore, we can bound the probability (3) via Chernoff (2) as follows:

$$
\Pr\left[\mathrm{Bin}\left(n \cdot \frac{t_0(t_0 + 1)}{2}, p\right) \geq t_0\right] \leq \mathrm{e}^{-\mu_0 \cdot \varphi\left(\frac{t_0}{\mu_0} - 1\right)} = n^{-\frac{\mu_0}{\ln n} \cdot \varphi\left(\frac{t_0}{\mu_0} - 1\right)}.
$$

Thus, we just need to verify that

$$
\frac{\mu_0}{\ln n} \cdot \varphi\left(\frac{t_0}{\mu_0} - 1\right) > 2. \tag{5}
$$

Using the approximation (4), we obtain

$$\frac{\mu_0}{\ln n} \cdot \varphi\left(\frac{t_0}{\mu_0} - 1\right) \sim \frac{cd^2}{2}\left(\frac{2}{cd}\ln\frac{2}{cd} - \frac{2}{cd} + 1\right) = d\left(\frac{cd}{2} + \ln\frac{2}{cd} - 1\right).$$

Thus, our assumption on c, d implies (5). □

Proof of Theorem 1, part (1). Let $c = 0.16$ and $f(d) = d(cd/2 + \ln(2/cd) - 1)$. Then we see that $\max_{0 < d < 2/c} f(d) > 2$. Hence, Lemma 1 entails that for $c < 0.16$, we have $\Pr[T > t_0] = o(n^{-2})$ for a certain $t_0 = O(\ln n)$. By the union bound, this implies that w.h.p. there is no pair (v_1, v_2) such that $T^{(v_1,v_2)} = n-1$, whence $H : \mathcal{H}(n, p)$ is not propagation connected w.h.p. by Fact 1. □

4   The Upper Bound

In this section we sketch the proof of part (2) of our main theorem, that is, an upper bound for p such that $H : \mathcal{H}(n, p)$ is propagation connected w.h.p. Due to space limitation, some of the proofs will be omitted, which can be found in [COW10].

Here again we fix p and c such that $p = c/(n \ln n)$. As we saw in Section 2, the propagation process can be viewed as a time-dependent random walk. At first, the drift of this random walk is negative, but after a certain time the drift turns positive. The following proposition reflects this fact by showing that once the process has survived up to a certain time, it will likely continue to time $n - 1$. In the following, we use $\nu = \lceil \ln^3 n \rceil$.

Proposition 1. *W.h.p. there is no pair (u, v) such that $\nu \leq T^{(u,v)} < n - 1$.*

In the light of Proposition 1, we call a pair of vertices (u, v) *good* if $T^{(u,v)} \geq \nu$. Let N be the number of good pairs of $H : \mathcal{H}(n, p)$. Then by Proposition 1 in order to prove that $H : \mathcal{H}(n, p)$ is propagation connected w.h.p., we just need to establish that $N > 0$ w.h.p. We first estimate the *expected* number of good pairs.

Proposition 2. *For any fixed $c > 0.25$ there is a number $\delta = \delta(c) > 0$ such that $\mathrm{E}[N] \geq \Omega(n^\delta)$ holds.*

Then by the following proposition, we relate the above result on the expectation of N to showing that $N > 0$ w.h.p. The proof of this proposition is based on a second moment argument; see [COW10] for the proof.

Proposition 3. *Assume that $\delta > 0$ is a constant such that $\mathrm{E}[N] \geq \Omega(n^\delta)$ holds for $c > 0$ and $p = c/(n \ln n)$. Then in fact $N \geq \Omega(n^\delta) > 0$ w.h.p.*

The second part of Theorem 1 is a direct consequence of Propositions 1–3.

In the rest of this section we sketch the proof of Proposition 2. As indicated in Section 2, this basically means that we need to analyze the probability that the random walk described by the variables $X_t = A_t - A_{t-1} + 1$ remains positive.

From now on, we fix a number $c > 0.25$ and let $p = c/(n \ln n)$ for n sufficiently large. We will keep the notation from Section 2.

For a time t and a number $g \geq 1$, we let $\mathrm{AT}(t, g)$ denote the event that $A_t \geq g$. That is, the process does not stop before time t, and at this time there are at least g active vertices. As we saw in Section 2, the 'drift' of the time-dependent random walk described by the variables X_t is negative for small $t \ll \ln n$. The following lemma will help us get over the first few steps of the process. Intuitively, it shows that with a decent probability the process will not only survive up to time $\gamma \ln n$, but also amass a small excess of $\gamma \ln n$ active vertices for a small $\gamma > 0$.

Lemma 2. *For any $\delta > 0$, there is $\gamma_0 = \gamma_0(c, \delta) > 0$ such that for all $0 < \gamma < \gamma_0$, the event $\mathrm{AT}(\lceil \gamma \ln n \rceil, \lceil \gamma \ln n \rceil)$ holds with probability at least $n^{-\delta}$.*

Proof. As $\lim_{\gamma \to 0} 2\gamma \ln(c) - c\gamma^2/2 + 2\gamma \ln(\gamma/2) = 0$, for any $\delta > 0$, there is $\gamma_0 > 0$ such that for all $0 < \gamma < \gamma_0$, we have $2\gamma \ln(c) - c\gamma^2/2 + 2\gamma \ln(\gamma/2) > -\delta$. Assume that γ, $0 < \gamma < \gamma_0$, is sufficiently small so that this is the case. Let $t_1 = \lceil \gamma \ln n \rceil$ and $t_0 = \lfloor t_1/2 \rfloor$. Then

$$\Pr\left[\mathrm{AT}(\lceil \gamma \ln n \rceil, \lceil \gamma \ln n \rceil)\right] \geq \Pr\left[\bigwedge_{1 \leq t \leq t_0} X_t = 1 \wedge \bigwedge_{t_0 < t \leq t_1} X_t = 3\right].$$

(For if $X_t > 0$ for all $t \in [t_1]$, then the process won't stop before time t_1, i.e., $T \geq t_1$. Moreover, the number of active vertices at time t_1 equals $\sum_{t=1}^{t_1}(X_t - 1) = 2(t_1 - t_0) \geq \gamma \ln n$.) For $0 \leq t \leq t_1$, we let $\mathcal{E}_t$ signify the event that $X_s = 1$ for all $1 \leq s \leq \min\{t, t_0\}$ and $X_s = 3$ for all $t_0 < s \leq t$. Then our objective is to lower bound $\Pr[\mathcal{E}_{t_1}]$.

If we condition on the event $\mathcal{E}_{t-1}$ for some $t \in [t_1]$, then the number of neutral vertices at time t works out to be $n - (t+1) - A_t \geq n - 2t_1 - 2 = n - O(\ln n)$. Furthermore, Fact 3 entails that X_t given $\mathcal{E}_{t-1}$ is binomially distributed $\mathrm{Bin}(n - t - A_{t-1}, 1 - (1-p)^t)$. Consequently,

$$\Pr[X_t = 1 | \mathcal{E}_{t-1}] \geq (n - O(\ln n))(1 - (1-p)^t)(1-p)^{tn}$$

$$\sim \frac{ct}{\ln n} \cdot \exp(-ct/\ln n), \quad \text{and}$$

$$\Pr[X_t = 3 | \mathcal{E}_{t-1}] \geq \binom{n - O(\ln n)}{3}(1 - (1-p)^t)^3 (1-p)^{tn}$$

$$\sim \frac{(ct)^3}{(\ln n)^3} \cdot \exp(-ct/\ln n).$$

Therefore,

$$\Pr[\mathcal{E}_{t_1}] = \prod_{1 \leq t \leq t_0} \Pr[X_t = 1 | \mathcal{E}_{t-1}] \prod_{t_0 < t \leq t_1} \Pr[X_t = 3 | \mathcal{E}_{t-1}] \geq c^{3t_1 - 2t_0} n^{-c\gamma^2/2} \cdot$$

$$\left(\frac{\gamma}{2}\right)^{3\gamma \ln(n)/2} \cdot \frac{\exp\left(\sum_{t=1}^{t_0} \ln t\right)}{(\ln n)^{t_0}} \geq \Omega\left(n^{2\gamma \ln(c) - c\gamma^2/2 + 2\gamma \ln(\gamma/2)}\right).$$

Since we have chosen γ so that $2\gamma \ln(c) - c\gamma^2/2 + 2\gamma \ln(\gamma/2) > -\delta$, the assertion follows. $\qquad\square$

Lemma 2 shows that with a decent probability the first few steps of the process will yield a good number of active vertices. The following lemma studies the continuation of the process up to the time $c^{-1}\ln n$ where the 'drift' of the random walk turns positive.

Lemma 3. *There exists $\delta > 0$ such that $\Pr[T \geq \lceil(c^{-1} + \delta)\ln n\rceil] \geq n^{\delta-2}$.*

Proof. Since $c > 0.25$, we can choose $\delta > 0$ so that $4c(1 - \delta) > 1$. Let γ_0 be the number promised by Lemma 2. Moreover, choose $0 < \gamma < \gamma_0$ sufficiently small so that $1 + 4c\gamma - \ln(1 - c\gamma) < 4c(1 - \delta)$. We may also assume that $\lceil(c^{-1}+\delta)\ln n\rceil \leq \lceil\gamma \ln n\rceil \cdot (\lfloor(c\gamma)^{-1}\rfloor + 1)$.

Let $g = \lceil\gamma \ln n\rceil$ and $s_0 = \lfloor(c\gamma)^{-1}\rfloor$. Then our goal is to estimate the probability that the propagation process lasts at least $(s_0 + 1)g$ steps. To this end, we partition this period into $s_0 + 1$ chunks of size g. That is, for each $s \in [s_0]$, we define $Y_s = \sum_{sg < t \leq (s+1)g} X_t$. We are going to lower bound the probability of the event

$$\mathrm{AT}(g, g) \wedge (Y_1 \geq g) \wedge \cdots \wedge (Y_{s_0} \geq g). \tag{6}$$

If this event occurs, then $T \geq g(s_0 + 1)$. To see this, we show by induction that for each $1 \leq s \leq s_0$ at time $t = sg$ there are at least g active vertices. For $s = 1$ this follows directly from the definition for $\mathrm{AT}(g, g)$. Proceeding inductively, we note that the following period up to time $(s + 1)g$ will generate g new active vertices, because $Y_{s+1} \geq g$. This ensures that at time $(s + 1)g$ there are at least g active vertices as well.

Thus, in order to establish the proposition, we just need to prove that the event (6) holds with probability $n^{\delta-2}$. Lemma 2 shows that $\Pr[\mathrm{AT}(g, g)] \geq n^{-\delta}$. In addition, we are going to estimate probability that $Y_s \geq g$ given $\mathrm{AT}(g, g) \wedge (Y_1 \geq g) \wedge \cdots \wedge (Y_{s-1} \geq g)$ for any $s \in [s_0]$. In doing so we may assume that $A_{sg} \leq 2c^{-1}\ln n$, because otherwise the process will continue to time $2c^{-1}\ln n > (c^{-1} + \delta)\ln n$ with certainty. Hence, we may assume that there are always more than $n' = (n - 2c^{-1}\ln n) = n(1 - o(1))$ neutral vertices. On the other hand, at times $sg < t \leq (s+1)g$ there are at least sg dead vertices. Thus, Fact 3 implies that

$$\Pr\left[A_{sg} \geq 2c^{-1}\ln n \vee Y_s \geq g \mid \mathrm{AT}(g, g) \wedge (Y_1 \geq g) \wedge \cdots \wedge (Y_{s-1} \geq g)\right]$$

$$\geq \Pr\left[\sum_{sg < t \leq (s+1)g} X_t \geq g \,\middle|\, (A_{sg} < 2c^{-1}\ln n) \wedge \mathrm{AT}(g, g) \wedge (Y_1 \geq g) \wedge \cdots\right]$$

$$\geq \Pr\left[\mathrm{Bin}(gn', 1 - (1 - p)^{sg}) \geq g\right] \geq \binom{g^2 n' s}{g} p^g (1 - p)^{g^2 n' s - g}.$$

Let $\mu_s = g^2 n' sp$ and $x_s = g/\mu_s$. Applying Stirling's formula, we obtain

$$\binom{g^2 n' s}{g} p^g (1 - p)^{g^2 n' s - g} \geq \exp\left(-g \ln x_s + g - \mu_s - O(\ln \ln n)\right).$$

Hence,

$$\Pr[\,(6)\,] = \Pr[\,\mathrm{AT}(g,g)\,] \cdot \Pr[\,(Y_1 \geq g) \wedge \cdots \wedge (Y_{s_0} \geq g) \,|\, \mathrm{AT}(g,g)\,]$$

$$\geq n^{-\delta} \cdot \prod_{1 \leq s \leq s_0} \exp\left(-g \ln x_s + g - \mu_s - c'' \ln \ln n\right)$$

$$= n^{-\delta} \cdot \exp\left(\sum_{1 \leq s \leq s_0} (-g \ln x_s + g - \mu_s - c'' \ln \ln n)\right) \qquad (7)$$

Approximating the sum in the exponent by an integral, we see that

$$\sum_{1 \leq s \leq s_0} (-g \ln x_s + g - \mu_s - c'' \ln \ln n) \geq -\frac{\ln n}{2c} \cdot \left(1 + o(1) + 3c\gamma - \ln(1 - c\gamma)\right)$$

$$> -2 \ln n + 2\delta \ln n,$$

where the last step is due to our choice of γ and δ. Finally, combining this estimate with (7) yields $\Pr[\,(6)\,] \geq n^{-\delta} \cdot n^{-1/2c_{\mathrm{pos}} + 2\delta} = n^{\delta-2}$, as desired. $\qquad \square$

The basic idea in the above proof was to study the behavior of the random walk by partitioning the time up to about $c^{-1} \ln n$ in short periods of length $g = \lceil \gamma \ln n \rceil$ with a small $\gamma > 0$. What we estimated was the probability that for each of these periods the *total* number of newly generated active vertices is at least g, without taking into account how these g vertices are distributed over the period. Alternatively, one could lower bound the probability that the process survives up to time $c^{-1} \ln n$ by the probability that the process generates at least one active vertex *at each individual step*. However, this argument gives a significantly weaker result. Intuitively, this means that typically the process will generate a little bit of 'leeway' for itself by aggregating a certain excess of active vertices.

Once the process 'survives' up to time $c^{-1} \ln n$, we are on firm ground, because then the 'drift' of the underlying random walk becomes positive. This observation yields the following corollary to Lemma 3, which in turn implies Proposition 2; again see [COW10] for the details.

Corollary 2. *There is $\delta > 0$ such that $\Pr[(1,2)$ is good$] = \Omega(n^{\delta-2})$.*

5 Computing a Propagation Sequence

An algorithm A with the properties claimed in Theorem 2 is outlined in Figure 1. This algorithm behaves as stated for any probability parameter $p = c(n)/(n \ln n)$; for convenience we will assume $c(n) \geq \ln n/n$.

The algorithm is divided into two steps. At the first step (i.e., step (1) of Figure 1) a random hash table $\mathtt{HashTable}$ is constructed. For any given pair of vertices x and y, the hash table yields a list (e.g. by reference to a linked list) of edges in E that contain both x and y. By using a standard pairwise

---

algorithm A (for computing a propagation sequence);
given $H = (V, E)$ following $\mathcal{H}(n, p)$, where $V = [n]$ and $E = \{e_1, \ldots, e_m\}$;
(1) prepare a random hash table `HashTable` with $O(m)$ entries such that one may
      search for edges containing any pair of vertices using the pair as a key;
(2) **for each** initial edge **e** $\in E$ that has not been examined **do** {
          let the candidate propagation sequence be (**e**);
          let u, v, w be the vertices of edge **e**;
          `D` $\leftarrow \{u\}$;   `A` $\leftarrow \{v\}$;   `N` $\leftarrow V - $ `D` $\cup$ `A`;
          while `A` $\neq \emptyset$ and `N` $\neq \emptyset$ **do** { //try propagation from **e**
               `x` $\leftarrow$ any one element of `A`;
    (a)   **for each** `y` $\in$ `D` **do** {
               use `HashTable` to search for edges containing `x` and `y`;
      (b)   **for each** such edge with `x` and `y` **do** {
                  `z` $\leftarrow$ the third vertex of the edge;
                  if `z` $\in$ `N` **then** {
                     append this edge to the candidate propagation sequence;
                     `A` $\leftarrow$ `A` $\cup \{$`z`$\}$;   `N` $\leftarrow$ `N` $- \{$`z`$\}$;
          } } }
             `A` $\leftarrow$ `A` $- \{$`x`$\}$;   `D` $\leftarrow$ `D` $\cup \{$`x`$\}$;
          }
          if `N` $= \emptyset$ (i.e., the process succeeds) **then**
             output the candidate propagation sequence and terminate;
    } (if for loop completes) **output** failure;

Fig. 1. Outline of Algorithm A

independent random hash function family (see, e.g., [MU05, Theorem 13.11]), we can construct a 'perfect' such hash table with $O(m)$ entries in $O(m)$ time *on average*.

The second step (i.e., step (2) of of Figure 1) is based on the propagation process we introduced in section 2. In this step the algorithm searches for an initial pair, that is called *good*, such the propagation process *succeeds*, i.e., such that $n - 1$ edges are obtained in order with each (after the first) covering exactly one previously uncovered vertex. Pairs that are not good are called *bad*. Let `D`, `A`, and `N` be variables for the current set of dead, active, and neutral vertices respectively. Notice that for the propagation process of step (2) to reach some i vertices, the worst case time is trivially $O(i^3)$ as each vertex pair is used at most once and each time at most i new vertices are processed. Furthermore the worst case time of the entire algorithm is $O(n^6)$ as there are less than n^2 initial vertex pairs to try.

From the proof of Proposition 1, it is easy to show that the probability there is some bad initial vertex pair from which we reach $(\ln n)^3$ vertices, is smaller than $O(n^{-6})$. Since the worst case time of algorithm A is trivially $O(n^6)$ we may therefore assume without further remark that any initial pair that propagates to $(\ln n)^3$ vertices will propagate to the entire graph; the total expected time for the other case is at most a constant and therefore negligible.

Before proceeding, let us point out that for the sake of brevity and readability, the exposition of algorithm A in Figure 1 is slightly wrong on one point and should be refined, namely: If an initial edge in step (2) is good, the process will eventually deplete the set, N, of neutral vertices; but when N becomes small the number of pairs x, y the algorithm must try to find an appropriate z, may be too large.

Modification of the algorithm: Let $n' = n - |N|$ be the number of active and dead vertices at some point in stage (2) of the algorithm. If $n' > (\ln n)^3$, then the way to search for a new vertex z is changed as follows: Fix any active vertex x and scan the edges containing x for an edge $\{x, y, z\}$ where $z \in N$ and $y \in D$.

Proposition 4. *If an initial vertex pair is good, the time to complete step (2) in algorithm A with the modification described, is $O(m)$ on average.*

Proof. No vertex is investigated more than once as it is moved from active to dead after. Since an edge contains exactly three vertices, no edge is investigated more than three times. Thus the total time is $O((\ln n)^9 + 3m) = O(m)$. □

We now come to the key points of our analysis of the expected running time for step (2). First we consider a parameter range in which we know the graph is almost surely propagation connected.

Proposition 5. *Let $c(n) > 0.3$. The time spent of step (2) in algorithm is $O(m)$ on average.*

Proof. By the proof of Proposition 3 we have that with probability $> 1 - n^{-\epsilon}$ the graph has at least n^δ good pairs for some positive constants δ and ϵ. This is true for $c(n) > 0.3$. If there are indeed more than n^δ good pairs, and the pairs are investigated in random order, we expect to hit on a good pair in time $O(n^{2-\delta})$. Since the propagation from bad pairs must finish before reaching $(\ln n)^3$ vertices the expected time is trivially $O((\ln n)^9 n^{2-\delta}) = O(m)$. With probability less than $n^{-\epsilon}$ it happens that there is not a sufficient number of good pairs in the graph. But then the expected time spent on bad pairs is still $O((\ln n)^9 n^{2-\epsilon}) = O(m)$. □

Finally we have the case where there may or may not exist a propagation sequence, but the graph at any rate has not too many edges. In this region, any given vertex pair is in fact *very bad* with high probability, as will be seen.

Proposition 6. *Let $c(n) \leq 0.3$. The time spent of step (2) in algorithm is $O(m)$ on average.*

Proof. Consider the argument for the lower bound in Section 3, but this time choose a constant stopping time, $t_0 = 20$. The probability that a given vertex pair will propagate past the corresponding number of vertices is therefore bounded as

$$\Pr\left[\text{Bin}\left(n\frac{t_0(t_0+1)}{2}, p \right) \geq t_0 \right] \leq e^{-\mu_0 \varphi(t_0/\mu_0 - 1)} \leq e^{-\frac{t_0}{2} \ln \ln n} = (\ln n)^{-10}$$

for sufficiently large n. We have the following three cases, each of which has expected linear time.

1. The process stops short of 20 vertices: $O(20^3 m) = O(m)$.
2. It stops between 20 and $(\ln n)^3$ vertices: $O(m(\ln n)^{9-10}) = O(m)$.
3. It propagates to the entire graph: $O(m)$ as given by Proposition 4.   □

Proof of Theorem 2. The Propositions 5 and 6 together with the previous argument about step (1), show that the expected time of the algorithm is linear in m. If the algorithm fails it must have tried all possible initial pairs and found them bad, hence there is no propagation sequence. On the other hand any sequence outputted by the algorithm is trivially a propagation sequence.   □

References

[BCK07]   Behrisch, M., Coja-Oghlan, A., Kang, M.: Local limit theorems for the giant component of random hypergraphs. In: Charikar, M., Jansen, K., Reingold, O., Rolim, J.D.P. (eds.) RANDOM 2007 and APPROX 2007. LNCS, vol. 4627, pp. 341–352. Springer, Heidelberg (2007)

[BO09]   Berke, R., Onsjö, M.: Propagation connectivity of random hyptergraphs. In: Watanabe, O., Zeugmann, T. (eds.) SAGA 2009. LNCS, vol. 5792, pp. 117–126. Springer, Heidelberg (2009)

[CMV07]   Coja-Oghlan, A., Moore, C., Sanwalani, V.: Counting connected graphs and hypergraphs via the probabilistic method. Random Structure and Algorithms 31, 288–329 (2007)

[CM04]   Connamacher, H., Molloy, M.: The exact satisfiability threshold for a potentially intractable random constraint satisfaction problem. In: Proc. 45th Annual Symposium on Foundations of Computer Science (FOCS 2004), pp. 590–599. IEEE, Los Alamitos (2004)

[COW10]   Coja-Oghlan, A., Onsjö, M., Watanabe, O.: Propagation connectivity of random hypergraphs, Research Report C-271, Dept. Math. Comput. Sci., Tokyo Inst. of Tech. (2010)

[D05]   Durrett, R.: Probability and examples, 3rd edn. (2005)

[DN05]   Darling, R.W.R., Norris, J.R.: Structure of large random hypergraphs. Ann. App. Probability 15(1A), 125–152 (2005)

[Fe50]   Feller, W.: An introduction to probability theory and its applications. Wiley, Chichester (1950)

[JLR00]   Janson, S., Łuczak, T., Ruciński, A.: Random Graphs. Wiley, Chichester (2000)

[MU05]   Mitzenmacher, M., Upfal, E.: Probability and Computing, Randomized Algorithms and Probabilistic Analysis. Cambridge Univ. Press, Cambridge (2005)

[Kar90]   Karp, R.M.: The transitive closure of a random digraph. Random Structures and Algorithms 1, 73–93 (1990)

[Mol05]   Molloy, M.: Cores in random hypergraphs and Boolean formulas. Random Structures and Algorithms 27(1), 124–135 (2005)

Improved Pseudorandom Generators for Depth 2 Circuits

Anindya De[1,*], Omid Etesami[1,**], Luca Trevisan[2,***], and Madhur Tulsiani[3,†]

[1] University of California at Berkeley
{anindya,etesami}@cs.berkeley.edu
[2] University of California at Berkeley and Stanford University
luca@cs.berkeley.edu
[3] Institute for Advanced Study, Princeton
madhurt@math.ias.edu

Abstract. We prove the existence of a $poly(n, m)$-time computable pseudorandom generator which "$1/poly(n, m)$-fools" DNFs with n variables and m terms, and has seed length $O(\log^2 nm \cdot \log\log nm)$. Previously, the best pseudorandom generator for depth-2 circuits had seed length $O(\log^3 nm)$, and was due to Bazzi (FOCS 2007).

It follows from our proof that a $1/m^{\tilde{O}(\log mn)}$-biased distribution $1/poly(nm)$-fools DNFs with m terms and n variables. For inverse polynomial distinguishing probability this is nearly tight because we show that for every m, δ there is a $1/m^{\Omega(\log 1/\delta)}$-biased distribution X and a DNF ϕ with m terms such that ϕ is not δ-fooled by X.

For the case of *read-once* DNFs, we show that seed length $O(\log mn \cdot \log 1/\delta)$ suffices, which is an improvement for large δ.

It also follows from our proof that a $1/m^{O(\log 1/\delta)}$-biased distribution δ-fools all read-once DNF with m terms. We show that this result too is nearly tight, by constructing a $1/m^{\tilde{\Omega}(\log 1/\delta)}$-biased distribution that does not δ-fool a certain m-term read-once DNF.

Keywords: DNF, pseudorandom generators, small bias spaces.

1 Introduction

One of the main open questions in *unconditional* pseudorandomness and derandomization is to construct logarithmic-seed pseudorandom generators that "fool"

---

* Supported by the "Berkeley fellowship for Graduate Study" and by the BSF under grant 2006060.
** This material is based upon work supported by the National Science Foundation under grant No. CCF-0729137 and by the BSF under grant 2006060.
*** This material is based upon work supported by the National Science Foundation under grant No. CCF-0729137 and by the BSF under grant 2006060.
† This material is based upon work supported by the National Science Foundation under grant No. CCF-0832797 and IAS Sub-contract no. 00001583. Work done partly when the author was a graduate student at UC Berkeley.

M. Serna et al. (Eds.): APPROX and RANDOM 2010, LNCS 6302, pp. 504–517, 2010.
© Springer-Verlag Berlin Heidelberg 2010

bounded-depth circuits.[1] Ajtai and Wigderson [1] first considered the problem of pseudorandomness against bounded-depth circuits, and constructed a pseudorandom generator against AC^0 with a seed of length $O(n^\epsilon)$ for any $\epsilon > 0$. This was substantially improved by Nisan [17], who used the hardness of parity against AC^0 [8] to construct a pseudorandom generator against depth d circuits with a seed of length $O(\log^{2d+6} n)$. This remains the best known result for AC^0.

Even for depth-2 circuits, the construction of optimal pseudorandom generators remains a challenging open question. A depth-2 circuit is either a CNF or a DNF formula, and a pseudorandom generator that fools DNFs must also fool CNFs with the same distinguishing probability, so from now on we will focus without loss of generality on DNFs, and denote by n the number of variables and m the number of terms.

Nisan's result quoted above gives a pseudorandom generator for DNFs with seed length $O(\log^{10} nm)$. Luby, Velickovic and Wigderson [13] reduced the seed length to $O(\log^4 nm)$ via various optimizations. For the simpler task of approximating the number of satisfying assignments to a DNF formula, Luby and Velickovic [12] provide a deterministic algorithm running in time $(m \log n)^{\exp(O(\sqrt{\log \log m}))}$.

The current best pseudorandom generator for DNFs is due to Bazzi [5]. In 1990, Linial and Nisan [11] conjectured that depth-d circuits are fooled by every distribution that is $(\log mn)^{O_d(1)}$-wise independent. Bazzi proved the depth-2 case of the Linial-Nisan conjecture, and showed that every $O(\log^2(m/\delta))$-wise independent distribution δ-fools DNFs. This result gives two approaches to constructing a pseudorandom generator for DNFs of seed $O(\log n \cdot \log^2(m/\delta))$, which is $O(\log^3 nm)$ when $\delta = 1/poly(n, m)$. One is to use one of the known constructions of k-wise independent generators of seed length $O(k \log n)$. The other is to use a result of Alon, Goldreich and Mansour [3] showing that every ϵ-biased distribution, in the sense of Naor and Naor [16], over n bits is ϵn^k-close to a k-wise independent distribution. This means that, because of Bazzi's theorem, every $exp(-O(\log n \cdot \log^2(m/\delta)))$-biased distribution fools DNFs; Naor and Naor [16] prove that an ϵ-biased distribution over n bits can be sampled using a seed of $O(\log(n/\epsilon))$ random bits, and so a $exp(-O(\log n \cdot \log^2(m/\delta)))$-biased distribution can be sampled using $O(\log n \cdot \log^2(m/\delta))$ random bits.

Razborov [19] considerably simplified Bazzi's proof (retaining the same quantitative bounds). In a recent breakthrough, building on Razborov's argument, Braverman [6] has proved the full Linian-Nisan conjecture.

For width-w DNF formulas[2], better bounds are known for small w. Luby and Velickovic [12] prove the existence of a generator with seed length $O(\log n + w2^w \log 1/\delta)$ which δ-fools all width-w DNFs. It follows from their proof that

---

[1] We say that a random variable X, ranging over $\{0,1\}^n$, "δ-fools" a function $f : \{0,1\}^n \to \mathbb{R}$ if

$$|\mathbb{E}f(X) - \mathbb{E}f(U_n)| \leq \delta,$$

where U_n is uniformly distributed over $\{0,1\}^n$. If $\mathcal{C}$ is a class of functions, then we say that X δ-fools $\mathcal{C}$ if X δ-fools every function $f \in \mathcal{C}$.

[2] Each term involves at most w variables.

	DNF Family	Seed length
[17]	general DNFs	$O(\log^{10}(mn/\delta))$
[13]	general DNFs	$O(\log^4(mn/\delta))$
[5]	general DNFs	$O(\log n \cdot \log^2(m/\delta))$
This work	general DNFs	$O(\log n + \log^2(m/\delta) \cdot \log\log(m/\delta))$
[12]	width-w DNFs	$O(\log n + w2^w \cdot \log(1/\delta))$
This work	width-w DNFs	$O(\log n + w\log w \cdot \log(m/\delta))$
[4]	read-once DNFs	$O(\log n \cdot \log m \cdot \log(1/\delta))$
This work	read-once DNFs	$O(\log n + \log m \cdot \log(1/\delta))$

Fig. 1. Pseudorandom generators to δ-fool DNFs with m terms and n variables

every $exp(-O(w2^w \log 1/\delta))$-biased distribution δ-fools width-w DNFs. One may always assume without loss of generality that $w = O(\log(m/\delta))$, and so if the Luby-Velickovic result could be improved to a seed length of $O(w + \log(n/\delta))$, the result would be a generator of optimal seed length $O(\log(mn/\delta))$.

For read-once DNFs, Bazzi proves that every $O(\log m \cdot \log 1/\delta)$-wise independent distribution δ-fools every read-once DNF, and hence every $exp(-O(\log n \cdot \log m \cdot \log 1/\delta))$-biased distribution δ-fools read-once DNFs. This gives a generator of seed length $O(\log n \cdot \log m \cdot \log 1/\delta)$, which is $O(\log^2 nm)$ for constant δ.

Our Results. We prove that every width-w DNF is δ-fooled by every $exp(-O(\log n + w\log w(\log m + \log 1/\delta)))$-biased distribution. This gives a pseudorandom generator of seed length $O(\log^2 mn \cdot \log\log mn)$ for general DNFs and $\delta = 1/poly(n, m)$.

Regarding read-once DNFs, we show that they are δ-fooled by every $exp(-O(\log m \cdot \log 1/\delta))$-biased distribution, leading to a generator with seed length $O(\log n + \log m \cdot \log 1/\delta)$, which is $O(\log nm)$ for constant δ.

We prove that our quantitative connections between small bias and DNF derandomization are nearly tight. Specifically, we construct an m-term DNF that is not δ-fooled by a certain $1/m^{\Omega(\log 1/\delta)}$-biased distribution, which means that seed length $\Omega(\log n + \log m \cdot \log 1/\delta)$ is necessary if one wants to δ-fool DNFs using a generic small bias distribution. This matches our positive result up to a $\log\log nm$ term when $\delta = 1/poly(n, m)$. It remains open whether seed length $O(\log nm)$ is achievable for constant δ.

We also construct an m-term *read-once* DNF that is not δ-fooled by a certain $1/m^{\tilde{\Omega}(\log 1/\delta)}$-biased distribution (where the $\tilde{\Omega}$ notation hides a $1/\log\log 1/\delta$ term). This means that seed length $\Omega(\log^2 nm/\log\log nm)$ is necessary if one wants to $1/poly(nm)$-fool read-once DNFs using a generic small bias distribution.

Due to lack of space, we defer the first example to the full version.

Our Techniques. Our positive results for DNFs and read-once DNFs are based on techniques similar to the ones developed by Bazzi [5] and simplified by Razborov [19].

Bazzi shows that a sufficient (and necessary) condition for a function g to be δ-fooled fooled by a k-wise independent distribution is that the function be "sandwiched" between two real-valued functions f_ℓ, f_u which are degree-k polynomials and such that $f_\ell(x) \leq g(x) \leq f_u(x)$ holds for every x, and $\mathbb{E}_{x \in U_n}[f_u(x) - f_\ell(x)] \leq \delta$. We provide a similar sufficient (and necessary) condition for a function g to be δ-fooled by an ϵ-biased distribution in terms of g being sandwiched between functions *whose Fourier transform has small ℓ_1 norm*.

Bazzi and Razborov then proceed to show how to construct the sandwiching functions for every DNF by showing that it suffices to find just one low-degree function that approximates the DNF in the ℓ_2 norm, and such a function is provided by a result of Linial, Mansour and Nisan [10] on the Fourier spectrum of DNFs. Our goal, instead, is to find a function of small ℓ_1 Fourier norm which approximates the given DNF well in the ℓ_2 norm. The existence of such a function follows from the result in [10]. However, we use the better quantitative bounds obtained by Mansour[14].

For the case of read-once DNFs we explicitly construct the sandwiching functions with bounded Fourier ℓ_1 norm, using the inclusion-exclusion formula for the DNF. To analyze the error in the truncated inclusion-exclusion formula, we apply an argument which is similar to the one appearing in a paper by Even *et al.* [7] on the related subject of pseudorandomness for combinatorial rectangles. The technical difference between our argument and the one in [7] is that while they use the kth-truncations of the inclusion-exclusion series to directly show that k-wise independence fools combinatorial rectangles, we use these to compute functions with low ℓ_1 norm sandwiching the given DNF.

Our negative example for read-once DNFs goes as follows. We start from a "tribe" function, a read-once DNF with m terms each with $\log m$ literals, and we show how to construct a $1/m^{\tilde{\Omega}(\log 1/\delta)}$-biased distribution that does not δ-fool the tribe function. We show that for every parameter d we can construct a distribution X that is roughly $1/m^d$-biased, and is such that the distinguishing probability of the tribe between X and the uniform distribution is the same as the error of the d-th term of the inclusion-exclusion formula in approximating the tribe. The latter error is roughly $1/d!$, so we get our result by setting $d = (\log 1/\delta)/(\log \log 1/\delta)$.

Subsequent Work. Klivans *et al.* [9] obtained better bounds on the ℓ_1 norm of sandwiching functions for the case of *random* and *read-k* DNF formulas. They thus get pseudorandom generators with seed length $O(\log n + \log m \cdot \log(1/\delta))$ for *random* and *read-k* DNF formulas. For the case of *read-k* DNF formulas, the implicit constant in the seed length is exponentially dependent on k and it is open to improve this dependence.

2   Preliminaries

An arbitrary function $f : \{0,1\}^n \to \mathbb{R}$ can be expressed as a linear combination of the character functions $\chi_S(x) = \prod_{i \in S}(-1)^{x_i}$, as $f(x) = \sum_S \hat{f}(S)\chi_S(x)$. The coefficient $\hat{f}(S)$ is referred to as the Fourier coefficient of f corresponding to the set S. We use the following notation for the Fourier ℓ_1 norm of f and a minor variant of it as below:

$$\|f\|_1 := \sum_S \left|\hat{f}(S)\right| \quad \text{and} \quad \|f\|_1^{\neq\emptyset} := \sum_{S\neq\emptyset} \left|\hat{f}(S)\right|$$

For a probability distribution X over $\{0,1\}^n$, we say that X ϵ-fools a real function $f : \{0,1\}^n \to \mathbb{R}$ if

$$|\mathbb{E}[f(X)] - \mathbb{E}[f(U_n)]| \le \epsilon$$

where U_n denotes the uniform distribution over $\{0,1\}^n$. We say a probability distribution X over $\{0,1\}^n$ is ϵ-biased if it ϵ-fools the character functions χ_S.

Proposition 1 (Efficient construction of ϵ-biased sets [16,2]). *A subset $B \subseteq \{0,1\}^n$ is called an ϵ-biased set if the uniform distribution with support B is ϵ-biased. There exist ϵ-biased sets of size $O(n^2/\epsilon^2)$ such that a random element from the set can be sampled using a seed of length $O(\log(n/\epsilon))$, in time $\mathrm{poly}(n, \log(1/\epsilon))$.*

A DNF formula ϕ is of the form $\phi = \bigvee_{i=1}^m C_i$ where each term C_i is an AND of literals (variables or negations). ϕ is said to be of **width** w if every term C_i involves at most w distinct variables. A DNF is said to be **read-once** if every variable appears in at most one of the terms. A DNF is said to be **read-k** if every variable appears in at most k of the terms.

2.1   Sandwich Bound

The following claims give a a characterization of functions that can be fooled well by ϵ-biased probability distributions, similar to one derived by Bazzi [5]. The proofs of these propositions can be found in the full version.

Proposition 2 (Sandwich bound). *Suppose $f, f_\ell, f_u : \{0,1\}^n \to \mathbb{R}$ are three functions such that for every $x \in \{0,1\}^n$ we have $f_\ell(x) \le f(x) \le f_u(x)$. Furthermore, assume $\mathbb{E}[f(U_n)] - \mathbb{E}[f_\ell(U_n)] \le \delta$ and $\mathbb{E}[f_u(U_n)] - \mathbb{E}[f(U_n)] \le \delta$. Let $l = \max(\|f_\ell(x)\|_1^{\neq\emptyset}, \|f_u(x)\|_1^{\neq\emptyset})$. Then any ϵ-biased probability distribution $(\delta + \epsilon l)$-fools f.*

The following result shows that the condition of Proposition 2 is not only a sufficient condition for being fooled by ϵ-biased distributions but also a necessary condition.

Proposition 3 (Inverse of the sandwich bound). *Suppose $f : \{0,1\}^n \to \mathbb{R}$ is ϵ'-fooled by any ϵ-biased set. Then there exist functions $f_\ell, f_u : \{0,1\}^n \to \mathbb{R}$ and $\delta, l \in \mathbb{R} \ge 0$ with the following properties:*

- *For every $x \in \{0,1\}^n$ we have $f_\ell(x) \le f(x) \le f_u(x)$.*
- $\mathbb{E}[f(x)] - \mathbb{E}[f_\ell(x)] \le \delta$ *and* $\mathbb{E}[f_u(x)] - \mathbb{E}[f(x)] \le \delta$,
- $\|f_\ell(x)\|_1^{\neq \emptyset} \le l$, $\|f_u(x)\|_1^{\neq \emptyset} \le l$, *and* $\delta + \epsilon l \le \epsilon'$.

It is easy to check the following properties of ℓ_1 norm of functions over the fourier domain.

Observation 1. *If $f, g : \{0,1\}^n \to \mathbb{R}$, then $\|f+g\|_1 \le \|f\|_1 + \|g\|_1$ and $\|fg\|_1 \le \|f\|_1 \|g\|_1$.*

Observation 2. *If $\phi : \{0,1\}^n \to \{0,1\}$ is an AND of some subset of literals (i.e., variables or their negations), then $\|\phi\|_1 = 1$.*

3   Fooling Read-Once DNF Formulas

The following result shows that ϵ-biased sets can fool read-once DNFs.

Theorem 3. *Let ϕ be a read-once DNF formula with m terms. For $1 \le k \le m$, ϵ-biased distributions $O(2^{-\Omega(k)} + \epsilon m^k)$-fool ϕ. In particular, we can δ-fool ϕ by an ϵ-biased distribution, for $\epsilon = m^{-O(\log(1/\delta))}$.*

We first recall the inclusion-exclusion principle. Let $A_1, \ldots, A_m$ be m arbitrary events in a probability space. The principle of inclusion and exclusion asserts that

$$\Pr[A_1 \cup \cdots \cup A_m] = \sum_{j=1}^{m} (-1)^{j-1} T_j,$$

where

$$T_j = \sum_{S \subseteq [m], |S|=j} \Pr\left[\bigcap_{i \in S} A_i\right].$$

Moreover, the partial sum $\sum_{j=1}^{r} (-1)^{j-1} T_j$ is an upper bound for $\Pr[A_1 \cup \cdots \cup A_m]$ for odd values of r, and a lower bound for $\Pr[A_1 \cup \cdots \cup A_m]$ for even values of r.

We now return to the proof of Theorem 3. The proof follows that of Theorem 2 in [7].

Proof of Theorem 3:   Let $\phi = C_1 \vee \cdots \vee C_m$ be the read-once formula. For $1 \le i \le m$, let A_i denote the event that term C_i is satisfied. We divide the analysis into two cases depending on whether $\sum_{i=1}^{m} \Pr[A_i] \le k/(2e)$ or not.

<u>Case 1</u>: $\sum_{i=1}^{m} \Pr[A_i] \le k/(2e)$.
   Let T_k denote the kth term of the inclusion-exclusion formula. Since the terms are disjoint, we have

$$T_k = \sum_{S \subseteq [m], |S|=k} \prod_{i \in S} \Pr[A_i].$$

We now observe that $T_k \leq 2^{-k}$. Indeed, subject to the restriction $\sum_{i=1}^{m} \Pr[A_i] = \alpha$ and $\Pr[A_i] \geq 0$, a convexity based argument implies that T_k is maximized when all the $\Pr[A_i]$'s are equal implying that $T_k \leq \binom{m}{k}(2em/k)^{-k} \leq 2^{-k}$.

Consider the rth approximation to ϕ, obtained by inclusion-exclusion:

$$\phi_r(x) = \sum_{j=1}^{r}(-1)^{j-1} \sum_{S \subseteq [m], |S|=j} \bigwedge_{l \in S} C_l(x),$$

where $\bigwedge$ is the AND function. The functions ϕ_{k-1} and ϕ_k sandwich ϕ and we shall use them in applying Proposition 2. To verify the conditions, we note that the function $\bigwedge_{l \in S} C_l(x)$ is an AND of AND terms, therefore $\|\bigwedge_{l \in S} C_l(x)\|_1^{\neq \emptyset} = O(1)$, and hence $\|\phi_r\|_1^{\neq \emptyset} = O(m^r)$. We also have $|\mathbb{E}[f_k(U_n)] - \mathbb{E}[f_{k-1}(U_n)]| = T_k \leq 2^{-k}$. and hence, by Proposition 2, ϕ is $O(2^{-k} + \epsilon m^k)$-fooled by ϵ-biased distributions.

<u>Case 2</u>: $\sum_{i=1}^{m} \Pr[A_i] > k/(2e)$.

Consider the first m' where $\sum_{i=1}^{m'} \Pr[A_i] \geq k/(2e)$. Define $\phi' = C_1 \vee \cdots \vee C_{m'}$. Observe that the DNF ϕ' is satisfied with probability $1 - 2^{-\Omega(k)}$, for it is not satisfied with probability $\prod_{i=1}^{m'}(1 - \Pr[A_i]) \leq (1 - k/(2em'))^{m'} \leq 2^{-\Omega(k)}$. (Again by a convexity argument, $\prod_i(1 - \Pr[A_i])$ is maximized when $\Pr[A_i]$s are equal.)

Let $\phi_r'(x)$ denote the rth approximation to ϕ'. Also, (without loss of generality) let k be even so that $\phi_k' \leq \phi' \leq \phi$. Note that while ϕ_{k-1}' is a an upper bound on ϕ', it is *not* an upper bound on ϕ. We shall use ϕ_k' and identically 1 function respectively as lower and upper bounds for applying Proposition 2 to ϕ.

From argument above, we know that $\mathbb{E}[1 - \phi] \leq \mathbb{E}[1 - \phi'] \leq 2^{-\Omega(k)}$. To bound $\mathbb{E}[\phi - \phi_k']$, we note that

$$\mathbb{E}\left[\phi - \phi_k'\right] = \mathbb{E}\left[\phi - \phi'\right] + \mathbb{E}\left[\phi' - \phi_k'\right] \leq \mathbb{E}\left[1 - \phi'\right] + \mathbb{E}\left[\phi_{k-1}' - \phi_k'\right] \leq 2^{-\Omega(k)}$$

where in the last inequality we used that $\mathbb{E}[\phi_{k-1}' - \phi_k']$ as in the previous case, since $\sum_{i=1}^{m'} \Pr[A_i] < k/(2e)+1$. The bound on the $\|\phi_k'\|_1^{\neq \emptyset}$ is as before. Applying Proposition 2, we then get that ϵ-biased sets $O(2^{-\Omega(k)} + \epsilon m'^k)$-fool ϕ. ∎

If we plug in the construction from Proposition 1 in Theorem 3, we get a pseudorandom generator which δ-fools a read-once DNF with n variables and m terms and has seed length $O(\log n + \log m \cdot \log(1/\delta))$.

4   Fooling General DNF Formulas

In this section, we show that small biased distributions fool general DNFs. While the seed length will not be as good as in the previous section, the result will be more general. Also, this section will involve use of more analytic tools. Our proof shall be along the lines of Razborov's simplified proof of Bazzi's theorem [19]. The following two theorems will be the main theorems of this section.

Theorem 4. *Let ϕ be a width w-DNF formula with m terms. Then, ϕ is δ-fooled by every ϵ-biased distribution where $\epsilon = w^{-O(w \log(m/\delta))}$.*

Theorem 5. *Let ϕ be a DNF formula with m terms. Then, ϕ is δ-fooled by every ϵ-biased distribution where $\epsilon = (\log(m/\delta))^{O(-\log^2(m/\delta))}$.*

Plugging in the pseudorandom generator construction from Proposition 1 in Theorem 4, we get a pseudorandom generator which δ-fools width-w DNFs with m terms over n variables and has a seed of length $O(\log n + w \log w \log(m/\delta))$. Doing the same for Theorem 5, we get a pseudorandom generator which δ-fools DNFs with m terms over n variables and has a seed of length $O(\log n + \log^2(m/\delta) \log \log(m/\delta))$. Theorem 5 follows by a reduction to Theorem 4, by deleting the terms with large width, as we describe later. For most of this section, we will be concerned with DNFs of a bounded width. To prove Theorem 4, we will be interested in finding sandwiching functions f_l and f_u to apply Proposition 2.

Using an argument similar to [5], we reduce this to the problem of finding a function g such that $\|\phi - g\|_2$ and $\|g\|_1$ are small, and $\phi(x) = 0 \implies g(x) = 0$. We then show how to remove the last condition and then find an appropriate g using a Fourier concentration result of Mansour [15]. More formally, we prove the following three lemmas.

Lemma 1. *Let $\phi : \{0,1\}^n \to \{0,1\}$ be a DNF with m terms and $g : \{0,1\}^n \to \mathbb{R}$ be such that: $\|g\|_1 \le l$, $\|\phi - g\|_2 \le \epsilon_1$ and $g(x) = 0$ whenever $\phi(x) = 0$. Then, we can get $f_\ell, f_u : \{0,1\}^n \to \mathbb{R}$ such that*

- $\forall\, x,\ f_\ell(x) \le \phi(x) \le f_u(x)$
- $\mathbb{E}_{x \in U_n}[f_u(x) - \phi(x)] \le m\epsilon_1^2$ and $\mathbb{E}_{x \in U_n}[\phi(x) - f_\ell(x)] \le m\epsilon_1^2$.
- $\|f_\ell\|_1, \|f_u\|_1 \le (m+1)(l+1)^2 + 1$

Lemma 2. *Let $\phi : \{0,1\}^n \to \{0,1\}$ be a width-w DNF with m terms. Suppose for every width-w DNF ϕ_1, there is a function $g_1 : \{0,1\}^n \to \mathbb{R}$ such that: $\|g_1\|_1 \le l_1$ and $\|\phi_1 - g_1\|_2 \le \epsilon_2$. Then, we can get $g : \{0,1\}^n \to \mathbb{R}$ such that $\|g\|_1 \le m(l_1 + 1)$, $\|\phi - g\|_2 \le m\epsilon_2$ and $g(x) = 0$ whenever $\phi(x) = 0$.*

Lemma 3. *Let $\phi : \{0,1\}^n \to \{0,1\}$ be a width w DNF and $\epsilon_2 > 0$. Then there is a function $g_1 : \{0,1\}^n \to \mathbb{R}$ such that $\|\phi - g_1\|_2 \le \epsilon_2$ and $\|g_1\|_1 = w^{O(w \log(1/\epsilon_2))}$*

Before, we prove these lemmas, we show how they imply Theorem 4.

Proof (of Theorem 4). Set $\epsilon_2 = \sqrt{\delta/2m^3}$ and $\epsilon_1 = \sqrt{\delta/2m}$. By applying Lemma 3, for every width-w DNF ϕ_1, we can get a function $g_1 : \{0,1\}^n \to \mathbb{R}$ such that

- $\|\phi_1 - g_1\|_2 \le \epsilon_2 = \sqrt{\delta/2m^3}$
- $\|g_1\|_1 = w^{O(w \log(1/\epsilon_2))} = w^{O(w \log(m/\delta))}$

Now, we apply Lemma 2 with $l_1 = w^{O(w \log(m/\delta))}$ and $\epsilon_2 = \sqrt{\delta/2m^3}$. Then, for the given DNF ϕ, we get a function g such that $\|g\|_1 = w^{O(w \log(m/\delta))}$ and $\|g - \phi\|_2 \le m\epsilon_2 = \epsilon_1 = \sqrt{\delta/2m}$. Finally, we apply Lemma 1 with g and ϵ_1 as defined and $l = w^{O(w \log(m/\delta))}$ to get f_ℓ and f_u such that ϕ is sandwiched by f_ℓ and f_u, $\|f_\ell\|_1, \|f_u\|_1 \leq w^{O(w \log(m/\delta))}$ and

$$\mathop{\mathbb{E}}_{x \in U_n}[f_u(x) - \phi(x)] \leq \frac{\delta}{2} \quad \text{and} \quad \mathop{\mathbb{E}}_{x \in U_n}[\phi(x) - f_\ell(x)] \leq \frac{\delta}{2}$$

By applying Proposition 2, we get that an $\epsilon = w^{-O(w \log(m/\delta))}$ (for an appropriately large constant inside $O(\cdot)$) biased set fools ϕ by $\delta/2 + \epsilon l \leq \delta$.

We first sketch the proof of Lemma 1.

Proof Sketch of Lemma 1: Let $\phi = \bigvee_{i=1}^m A_i$ where A_i are the terms. We define f_ℓ and f_u as follows:

- $f_\ell = 1 - (1 - g)^2$
- $f_u = 1 - (1 - \sum_{i=1}^m A_i)(1 - g)^2$

We note that this is the same construction of functions as in Lemma 3.3 in [5]. In particular, the first two claims about f_ℓ and f_u are already proven in [5]. The third claim regarding f_ℓ and f_u (about their ℓ_1 norm) follows easily by repeated applications of Observation 1. $\qquad\square$

We next give the proof of Lemma 2 which follows the proof by Razborov [19].

Proof of Lemma 2: We first observe as in [19] (attributed to Avi Wigderson) that if $\phi = \bigvee_{i=1}^m A_i$ where $A_i \in \{0, 1\}$ are the individual terms, then ϕ can be rewritten as $\sum_{i=1}^m A_i(1 - \bigvee_{j=1}^{i-1} A_j)$. Let ϕ_i denote $\bigvee_{j=1}^{i-1} A_j$ ($\phi_i = 0$ if $i = 1$). Then, we can say that $\phi = \sum_{i=1}^m A_i(1 - \phi_i)$. Note that each of the ϕ_i is a width w-DNF and we can apply the hypothesis to get functions $g_1, \ldots, g_m : \{0, 1\}^n \to \mathbb{R}$ such that for all i, $\|g_i\|_1 \leq l_1$ and $\|g_i - \phi_i\|_2 \leq \epsilon_2$. Let $g : \{0, 1\}^n \to \mathbb{R}$ be defined as

$$g = \sum_{i=1}^m A_i(1 - g_i)$$

We observe that if $\phi(x) = 0$ for some x, then $\forall i$, $A_i(x) = 0$ which implies that $g(x) = 0$. Applying Observation 1 and using that A_i's are terms and hence $\|A_i\|_1 = 1$, we also get that $\|g\|_1 \leq m(l_1 + 1)$. So, the only thing that remains to be proven is that $\|\phi - g\|_2 \leq m\epsilon_2$. Though this is done in [19], we do it here for the sake of completeness.

$$\|g - \phi\|_2^2 = \mathop{\mathbb{E}}_{x \in U_n}\left[\left(\sum_{i=1}^m A_i(\phi_i - g_i)(x)\right)^2\right] \leq m \mathop{\mathbb{E}}_{x \in U_n}\left[\sum_{i=1}^m (A_i \cdot (\phi_i - g_i)(x))^2\right]$$

$$= m \sum_{i=1}^m \mathop{\mathbb{E}}_{x \in U_n}\left[(A_i \cdot (\phi_i - g_i)(x))^2\right] \leq m \sum_{i=1}^m \mathop{\mathbb{E}}_{x \in U_n}\left[(\phi_i - g_i)(x)^2\right]$$

$$= m \sum_{i=1}^m \|\phi_i - g_i\|_2^2 \leq m^2 \epsilon_2^2$$

This proves that $\|\phi - g\|_2 \leq m\epsilon_2$ which finishes the proof. $\blacksquare$

We now come to the proof of Lemma 3. The proof is dependent upon the following well-known concentration result by Mansour [15] (or see Ryan O'Donnell's lecture notes on fourier analysis [18]).

Theorem 6. *[15] Let $\phi : \{0,1\}^n \to \{0,1\}$ be a width w-DNF with m terms and $\epsilon_2 > 0$. Let $\sum_{S \subset [n]} \hat{\phi}(S) \chi_S$ be the fourier expansion of ϕ. Then there is a subset $\Gamma \subset 2^{[n]}$ of size $w^{O(w \log(1/\epsilon_2))}$ such that g defined as $g_1 = \sum_{S \in \Gamma} \hat{\phi}(S) \chi_S$ is such that $\|\phi - g_1\|_2 \leq \epsilon_2$.*

Proof of Lemma 3:   For the given ϕ and ϵ_2, let g_1 be the function given by Theorem 6. Clearly, it satisfies $\|\phi - g_1\|_2 \leq \epsilon_2$. To bound $\|g_1\|_1$, note that $\|g_1\|_1 = \sum_{S \in \Gamma} |\hat{\phi}(S)|$ where $|\Gamma| = w^{O(w \log(1/\epsilon_2))}$. Note that $\sum_{S \in \Gamma} |\hat{\phi}(S)|^2 = \alpha$ for some $\alpha \in [0,1]$ (by Parseval-Plancherel identity and the fact that ϕ lies in $[0,1]$). Now, we have

$$\left(\sum_{S \in \Gamma} |\hat{\phi}(S)| \right)^2 \leq |\Gamma| \left(\sum_{S \in \Gamma} |\hat{\phi}(S)|^2 \right) \leq |\Gamma| \qquad \text{(By Jensen's inequality)}$$

Hence, this gives us $\sum_{S \in \Gamma} |\hat{\phi}(S)| \leq \sqrt{|\Gamma|} = w^{O(w \log(1/\epsilon_2))}$ which proves the lemma. ∎

To prove Theorem 5, we prove the following proposition (proof is deferred to the full version).

Proposition 4. *Let ϕ be any m-term DNF and ϕ' be the DNF obtained from ϕ by removing all the terms which have more than w distinct variables. Also, let D be an ϵ'-biased distribution which δ'-fools ϕ'. Then ϕ is $\delta' + m2^{-w} + m\epsilon'$ fooled by D.*

To get Theorem 5, we plug in $w = \log(m/2\delta)$, $\epsilon' = w^{-O(w \log(m/\delta))}$ and $\delta' = \delta/4$ in the above proposition and use Theorem 4.

5   Almost Tight Lower Bounds for Fooling Read-Once DNFs

We now investigate the question of how small a bias is required for fooling DNFs using arbitrary small bias distributions. In section 3, we showed that to δ-fool an m-term read-once DNF using an ϵ biased distribution, it suffices to have $\epsilon = m^{-O(\log(1/\delta))}$. Below we give a construction of a specific ϵ-biased distribution which shows that to δ-fool the "tribes" DNF (described below), one must have $\epsilon = m^{-\Omega(\log(1/\delta)/\log\log(1/\delta))}$.

The following theorem states the existence of the required read-once DNF and small bias distribution. The lower bound of $\epsilon = m^{-\Omega(\log(1/\delta)/\log\log(1/\delta))}$ follows as an easy corollary.

Theorem 7. *For every sufficiently large integer n of the form $n = m \log m$ for m which is power of 2 and for every integer $d \geq 1$, there is an $(m/2)^{-d}$-biased distribution D over $\{0,1\}^n$ and a read-once DNF ϕ with m terms such that ϕ distinguishes D from uniform by at least $1/(2d+3)!$.*

Proof. We first describe the DNF. The DNF is defined by splitting the n variables into m chunks of size $\log m$. Let the variables in the ith chunk be $x_{i,1}, \ldots, x_{i,\log m}$. The DNF is

$$\phi(x) = \bigvee_{i=1}^{m} C_i \quad \text{where } C_i \equiv \bigwedge_{j=1}^{\log m} x_{i,j}$$

The following two claims, describe the required distribution D.

Claim 8. *There is a distribution $Y = Y_1 \circ \ldots \circ Y_m$ over $\{0,1\}^m$ with the following properties*

- *for every $1 \leq i \leq m$, $\Pr[Y_i = 1] = 1/m$.*
- *$Y_1, \ldots, Y_m$ are d-wise independent;*
- *For every $y \in Supp(Y)$, $y_1 + \ldots + y_m \leq d$.*

We can now describe the distribution D in terms of the random variables $Y_1, \ldots, Y_m$. Given values $y_1, \ldots, y_m$, we choose $x_{i,1}, \ldots, x_{i,\log m}$ to be all 1, if $y_i = 1$ and uniformly from $\{0,1\}^{\log m} \setminus 1^{\log m}$ if $y_i = 0$. In particular, this ensures that $\bigwedge_{j=1}^{\log m} x_{i,j} = y_i$ and hence C_i is satisfied if and only if $y_i = 1$. We claim that the distribution has a small bias.

Claim 9. *The distribution D defined above has bias at most $(m/2)^{-d}$.*

Before proving these two claims, lets see why they suffice to construct the counterexample. First, observe that by Claim 8, term C_i being satisfied is equivalent to $y_i = 1$. By inclusion-exclusion principle, the probability that $x \in_r D$ satisfies ϕ is

$$\Pr_{x \in D}[\phi \text{ is satisfied}] = \sum_{S \in [m], |S| > 0} (-1)^{|S|-1} \Pr[\forall i \in S, \, C_i \text{ is satisfied}]$$

$$= \sum_{S \in [m], |S| > 0} (-1)^{|S|-1} \Pr[\forall i \in S, y_i = 1]$$

$$= \sum_{S \in [m], d \geq |S| > 0} (-1)^{|S|} \Pr[\forall i \in S, \, y_i = 1] \quad \left(\text{Using } \sum_i^m y_i \leq d\right)$$

$$= \sum_{t=1}^{d} (-1)^{t-1} \binom{m}{t} \frac{1}{m^t}$$

The last equality uses that y_i's are d-wise independent and $\Pr[y_i = 1] = 1/m$. To estimate the above probability for the uniform distribution, we can obtain upper and lower bounds on it by truncating the inclusion-exclusion respectively at $d + 1$ and $d + 2$ when d is even (the upper and lower bounds are switched when d is odd). Thus ϕ distinguishes D from uniform with probability at least

$$\binom{m}{d+1}\frac{1}{m^{d+1}} - \binom{m}{d+2}\frac{1}{m^{d+2}}$$

$$= \frac{m!}{m^{d+1}(d+1)!(m-d-2)!}\left(\frac{1}{m-d-1} - \frac{1}{m(d+2)}\right)$$

$$\geq \frac{m!}{m^{d+1}(d+1)!(m-d-2)!}\frac{1}{2m}$$

$$\geq \frac{1}{2(d+1)!}\prod_{i=1}^{d+1}\left(1 - \frac{i}{m}\right)$$

$$= \frac{1}{2(2d+2)!}\prod_{i=1}^{d+1}\left((d+1+i)\left(1 - \frac{i}{m}\right)\right) \geq \frac{1}{(2d+3)!}$$

The last inequality uses that $(d+1+i)(1-i/m) \geq 1$. Hence, we need to prove Claims 8 and 9. We start with Claim 8.

Proof of Claim 8:   Let $p_0,\ldots,p_d \geq 0$ such that $\sum p_i = 1$ (We will non-constructively describe p_i's later). The distribution Y is chosen as following. Pick i, $0 \leq i \leq d$ with probability p_i. Choose a uniformly random subset $S \subset [m]$ of size i and set $y_i = 1$ if $i \in S$ and $y_i = 0$ if $i \notin S$. By construction, trivially the third property is satisfied. We need to set $p_0,\ldots,p_d$ such that the first and the second properties are satisfied. Note that to ensure that Y_i's are d-wise independent, it suffices to show that for every $0 \leq i \leq d$ and $1 \leq j_1 < \ldots < j_i \leq m$, we have $\mathbb{E}[y_{j_1} \cdot \ldots \cdot y_{j_i}] = \mathbb{E}[y_{j_1}] \cdot \ldots \cdot \mathbb{E}[y_{j_i}] = 1/m^i$ (because each variable y_k takes only two possible values.) By symmetry of the construction, it suffices to ensure these properties when $\{j_1,\ldots,j_i\} = \{1,\ldots,i\}$ for every $0 \leq i \leq d$. Thus we only need to select $p_0,\ldots,p_d$ such that for every $0 \leq i \leq d$,

$$\mathbb{E}[y_1 \cdot \ldots \cdot y_i] = \sum_{t=i}^{d}\frac{\binom{m-i}{t-i}}{\binom{m}{t}}p_t = 1/m^i.$$

This is a triangular system of $d+1$ linear equations which has a unique solution $p_0,\ldots,p_d$. However, we must make sure that the values of the solution $p_0,\ldots,p_d$ are nonnegative. We use descent on i to show $p_i \geq 0$. We have $p_d = \binom{m}{d}/m^d \geq 0$. For $i < d$, we have:

$$p_i = \binom{m}{i}\left[\frac{1}{m^i} - \sum_{t=i+1}^{d}\frac{\binom{m-i}{t-i}}{\binom{m}{t}}p_t\right] \geq \binom{m}{i}\left[\frac{1}{m^i} - \sum_{t=i+1}^{d}\frac{\binom{m-i-1}{t-i-1}}{\binom{m}{t}}mp_t\right]$$

$$= m\binom{m}{i}\left[\frac{1}{m^{i+1}} - \sum_{t=i+1}^{d}\frac{\binom{m-i-1}{t-i-1}}{\binom{m}{t}}p_t\right] = 0$$

∎

Proof of Claim 9: To compute the bias of the distribution D, consider any character χ_S where $S \subset [m \log m]$ is non-empty. For any $i \in [m]$, let us define $S_i = S \cap \{(i-1)\log m + 1, \ldots, i \log m\}$. Note that

$$\mathop{\mathbb{E}}_{x \in D} [\chi_S(x)] = \mathop{\mathbb{E}}_{x \in D} \left[\prod_{i: S_i \neq \phi} \chi_{S_i}(x) \right]$$

Our proof will only depend on the number of non-empty sets S_i. Without loss of generality, we can assume that the non-empty sets are $S_1, \ldots, S_t$ for some $t > 0$. We denote the set of variables $x_{i,1}, \ldots, x_{i,\log m}$ by x_i. To compute the bias, we then need to calculate

$$\mathbb{E}_{x \in D} \left[\prod_{i=1}^{t} \chi_{S_i}(x_i) \right] = \mathbb{E}_Y \left[\prod_{i=1}^{t} \mathbb{E}_{x_i} [\chi_{S_i}(x_i)|y_i] \right]$$

as the variables $x_1, \ldots x_m$ are independent given Y. We now note that

$$\mathbb{E}_{x_i} [\chi_{S_i}(x_i)|y_i = 1] = (-1)^{|S_i|} \quad \text{and} \quad \mathbb{E}_{x_i} [\chi_{S_i}(x_i)|y_i = 0] = -\frac{(-1)^{|S_i|}}{m-1}.$$

If $t \leq d$, then $y_1, \ldots, y_t$ are independent and the bias simply becomes 0 as below.

$$\mathbb{E}_Y \left[\prod_{i=1}^{t} \mathbb{E}_{x_i} [\chi_{S_i}(x_i)|y_i] \right] = \prod_{i=1}^{t} \mathbb{E}_{x_i, y_i} [\chi_{S_i}(x_i)]$$

$$= \prod_{i=1}^{t} \left(\frac{1}{m} \cdot (-1)^{|S_i|} - \left(1 - \frac{1}{m}\right) \cdot \frac{(-1)^{|S_i|}}{m-1} \right) = 0$$

If $t > d$, we can bound the bias as

$$\mathbb{E}_Y \left[\prod_{i=1}^{t} \mathbb{E}_{x_i} [\chi_{S_i}(x_i)|y_i] \right] \leq \mathbb{E}_Y \left[\prod_{i=1}^{t} |\mathbb{E}_{x_i} [\chi_{S_i}(x_i)|y_i]| \right]$$

$$\leq \mathbb{E}_Y \left[\prod_{i=1}^{d} |\mathbb{E}_{x_i} [\chi_{S_i}(x_i)|y_i]| \right]$$

$$= \prod_{i=1}^{d} \left(\frac{1}{m} + \left(1 - \frac{1}{m}\right) \cdot \frac{1}{m-1} \right) = \left(\frac{2}{m}\right)^d$$

which proves the claim.   ∎

By plugging $d = \log(1/\delta)/\log\log(1/\delta)$ in the above theorem, we get the following corollary.

Corollary 1. *For m which is a power of 2 and $\delta > 0$, there is a read-once DNF ϕ over $n = m \log m$ variables and a distribution D over $\{0,1\}^n$ which has bias $m^{-O(\log(1/\delta)/\log\log(1/\delta))}$ and ϕ distinguishes D from uniform by δ.*

Acknowledgements

We thank Salil Vadhan for help with the negative examples.

References

1. Ajtai, M., Wigderson, A.: Deterministic simulation of probabilistic constand-depth circuits. Advances in Computing Research - Randomness and Computation 5, 199–223 (1989); Preliminary version in Proc. of FOCS 1985
2. Alon, N., Goldreich, O., Håstad, J., Peralta, R.: Simple constructions of almost k-wise independent random variables. Random Structures and Algorithms 3(3), 289–304 (1992)
3. Alon, N., Goldreich, O., Mansour, Y.: Almost k-wise independence versus k-wise independence. Information Processing Letters 88(3), 107–110 (2003)
4. Bazzi, L.: Minimum Distance of Error Correcting Codes versus Encoding Complexity, Symmetry, and Pseudorandomness. PhD thesis, MIT (2003)
5. Bazzi, L.: Polylogarithmic independence can fool DNF formulas. In: Proceedings of the 48th IEEE Symposium on Foundations of Computer Science, pp. 63–73 (2007)
6. Braverman, M.: Poly-logarithmic independence fools AC^0 circuits. In: Proceedings of the 24th IEEE Conference on Computational Complexity, pp. 3–8 (2009)
7. Even, G., Goldreich, O., Luby, M., Nisan, N., Velickovic, B.: Approximations of general independent distributions. In: Proceedings of the 24th ACM Symposium on Theory of Computing, pp. 10–16 (1992)
8. Håstad, J.: Almost optimal lower bounds for small depth circuits. In: Proceedings of the 18th ACM Symposium on Theory of Computing, pp. 6–20 (1986)
9. Klivans, A., Lee, H., Wan, A.: Mansour's conjecture is true for random DNF formulas. Technical Report TR10-023, Electronic Colloquium on Computational Complexity (2010)
10. Linial, N., Mansour, Y., Nisan, N.: Constant depth circuits, fourier transform and learnability. Journal of the ACM 40(3), 607–620 (1993)
11. Linial, N., Nisan, N.: Approximate inclusion-exclusion. Combinatorica 10(4), 349–365 (1990)
12. Luby, M., Velickovic, B.: On deterministic approximation of DNF. Algorithmica 16(4/5), 415–433 (1996)
13. Luby, M., Velickovic, B., Wigderson, A.: Deterministic approximate counting of depth-2 circuits. In: Proceedings of the 2nd ISTCS, pp. 18–24 (1993)
14. Mak, L.: Parallelism always helps. Manuscript (1993)
15. Mansour, Y.: An $o(n^{\log \log n})$ learning algorithm for DNF under the uniform distribution. Journal of Computer and System Sciences 50(3), 543–550 (1995)
16. Naor, J., Naor, M.: Small-bias probability spaces: efficient constructions and applications. SIAM Journal on Computing 22(4), 838–856 (1993)
17. Nisan, N.: Pseudorandom bits for constant depth circuits. Combinatorica 12(4), 63–70 (1991)
18. O'Donnell, R.: Lecture notes for analysis of boolean functions (2007), http://www.cs.cmu.edu/~odonnell/boolean-analysis
19. Razborov, A.: A Simple Proof of Bazzi's Theorem. ACM Trans. Comput. Theory 1(1), 1–5 (2009)
20. Viola, E., Wigderson, A.: Norms, XOR lemmas, and lower bounds for polynomials and protocols. Theory of Computing 4(1), 137–168 (2008)

The Structure of Winning Strategies in Parallel Repetition Games

Irit Dinur[1],[*] and Elazar Goldenberg[2]

[1] Weizmann Institute of Science,
irit.dinur@weizmann.ac.il
[2] Weizmann Institute of Science,
elazar.goldenberg@weizmann.ac.il

Abstract. Given a function $f : X \to \Sigma$, its ℓ-wise direct product is the function $F = f^\ell : X^\ell \to \Sigma^\ell$ defined by $F(x_1, \ldots, x_\ell) = (f(x_1), \ldots, f(x_\ell))$. A two prover game G is a game that involves 3 participants: $V, \mathcal{A},$ and $\mathcal{B}$. V picks a random pair (x, y) and sends x to $\mathcal{A}$, and y to $\mathcal{B}$. $\mathcal{A}$ responds with $f(x)$, $\mathcal{B}$ with $g(y)$. $\mathcal{A}, \mathcal{B}$ win if $V(x, y, f(x), g(y)) = 1$. The repeated game G^ℓ is the game where $\mathcal{A}, \mathcal{B}$ get ℓ questions in a single round and each of them responds with an ℓ symbol string (this is also called the parallel repetition of the game). $\mathcal{A}, \mathcal{B}$ win if they win each of the questions.

In this work we analyze the structure of the provers that win the repeated game with non negligible probability. We would like to deduce that in such a case $\mathcal{A}, \mathcal{B}$ must have a global structure, and in particular they are close to some direct product encoding.

A similar question was studied by the authors and by Impagliazzo et. al. in the context of testing Direct Product. Their result can be be interpreted as follows: For a specific game G, if $\mathcal{A}, \mathcal{B}$ win G^ℓ with non negligible probability, then $\mathcal{A}, \mathcal{B}$ must be close to be a direct product encoding. We would like to generalize these results for any 2-prover game.

In this work we prove two main results: In the first part of the work we show that for a certain type of games, there exist $\mathcal{A}, \mathcal{B}$ that win the repeated game with non negligible probability yet are still very far from any Direct Product encoding. In contrast, in the second part of the work we show that for a certain type of games, called "miss match" games, we have the following behavior. Whenever $\mathcal{A}, \mathcal{B}$ win non negligibly then they are both close to a Direct Product strategy.

1   Introduction

Given a function $f : S \to \Sigma$ its ℓ-wise direct product is the function $f^\ell : S^\ell \to \Sigma^\ell$ defined by: $f^\ell(s_1, \ldots, s_\ell) = (f(s_1), \ldots, f(s_\ell))$. The Direct Product Testing Theorem by [DG08] and [IKW09] asserts that there exists a two query test T such that, whenever a function $F : S^\ell \to \Sigma^\ell$ passes T with non negligible probability, then F is somewhat close to an ℓ-wise direct product for some global function $f : S \to \Sigma$.

[*] Work supported by ISF grant 1179/09, BSF grant 2008293, and ERC starting grant 239985.

M. Serna et al. (Eds.): APPROX and RANDOM 2010, LNCS 6302, pp. 518–530, 2010.

Let us describe the 2-query direct product test T. The test picks a random tuple $\mathbf{x} \in S^\ell$ and then picks another tuple $\mathbf{x}'$ as follows: For each such coordinate i with probability α $\mathbf{x}'_i = \mathbf{x}_i$, otherwise, $\mathbf{x}'_i$ is drawn uniformly at random from S. The test queries $F(\mathbf{x}), F(\mathbf{x}')$ and accepts if and only if $F(\mathbf{x}), F(\mathbf{x}')$ are consistent among the common values of $\mathbf{x}$ and $\mathbf{x}'$.

The test can be viewed as a repeated 2-prover equality game in the following way: The original game, EQ, is the game in which with probability α $\mathcal{A}, \mathcal{B}$ get the same question x and with probability $1 - \alpha$ they get two independent questions x and x'. $\mathcal{A}$ responds with $a \in \Sigma$ and $\mathcal{B}$ with $b \in \Sigma$. If $\mathcal{A}, \mathcal{B}$ get the same question the verifier checks that $a = b$ otherwise it always accepts. The repeated game EQ^ℓ- the game where the verifier picks ℓ independent pairs of questions and sends them in a single round- is exactly the test T described above. The Direct Product Testing Theorem asserts that for this specific game: Whenever the provers win with non negligible probability, then the provers' strategy has a global structure: They have a global agreement with some direct product function.

The Parallel Repetition Theorem by [Raz98] asserts that, for any 2-prover game, the value of the repeated game decreases exponentially with the number of repetitions. Thus, if the provers win the repeated game with probability above 1%, then the value of the original game is almost 1. The Parallel Repetition Theorem concludes nothing about the structure of the provers' strategy assuming they win with probability above 1%. Furthermore, it is easy to see that the value of the EQ game is 1. Therefore, the Parallel Repetition Theorem, unlike the Direct Product Testing Theorem, tells us nothing about EQ^ℓ.

This work is a bridge between the Parallel Repetition Theorem and the Direct Product Testing Theorem showing that for every 2-prover game, if $\mathcal{A}, \mathcal{B}$ win with non negligible probability, then $\mathcal{A}, \mathcal{B}$ have global structure, namely $\mathcal{A}, \mathcal{B}$ are close to a direct product encoding.

Let us introduce some of our notations: A two-prover game G is defined by a distribution $\mathcal{D}$ on questions (X, Y) and a verifier V. The verifier V picks a questions pair $(x, y) \in (X, Y)$ according to $\mathcal{D}$. Then, the verifier sends the question x to prover $\mathcal{A}$ and the question y to prover $\mathcal{B}$. The provers $\mathcal{A}, \mathcal{B}$ are not allowed to communicate with each other during the game, and $\mathcal{A}$ responds with $f(x)$, while $\mathcal{B}$ responds with $g(y)$. The players win if $V(x, y, f(x), g(y)) = 1$. The value of the game G, denoted $val(G)$ is the maximum success probability of the players.

For functions $f : X \to \Sigma_A$ and $g : Y \to \Sigma_B$ we denote by $val(G, f, g)$ the value of the game if A plays according to f and B according to g, i.e. $val(G, f, g) = \mathbf{E}_{(x,y) \sim \mathcal{D}} V(x, y, f(x), g(y))$. We call the pair (f, g) a perfect strategy if $val(G, f, g) = 1$.

The repeated game G^ℓ is the the game where V samples ℓ independent questions: $(x_1, y_1), \ldots, (x_\ell, y_\ell)$ each is distributed according to $\mathcal{D}$. The verifier sends $\mathbf{x} = (x_1, \ldots, x_\ell)$ to $\mathcal{A}$ and $\mathbf{y} = (y_1, \ldots, y_\ell)$ to $\mathcal{B}$. Each prover responds with ℓ answers. The provers win if they win each of the ℓ coordinates. A *projection game* is a game in which the predicate V has a special structure- every pair (x, y) defines a function $\Pi_{x,y} : \Sigma_A \rightarrow \Sigma_B$, and $V(x, y, a, b)$ is satisfied iff $\Pi_{x,y}(f(x)) = g(y)$.

As mentioned earlier, the Parallel Repetition Theorem by [Raz98] bounds the value of the repeated game. Roughly speaking, it says that for every game G, if $val(G) < 1 - \varepsilon$, then $val(G^{\ell}) < (1 - \varepsilon')^{\ell}$ (where ε' depends on ε and on the length of the answer in G).

How would honest verifiers $\mathcal{A}, \mathcal{B}$ play in order to win the repeated game? They choose a pair of perfect strategies (f, g). $\mathcal{A}$, upon receiving $(x_1, \ldots, x_\ell)$, answers with $(f(x_1), \ldots, f(x_\ell))$ while $\mathcal{B}$ answers with $(g(y_1), \ldots, g(y_\ell))$. In fact, $\mathcal{A}, \mathcal{B}$ can choose ℓ pairs of perfect strategies $(f_1, g_1), \ldots, (f_\ell, g_\ell)$ and $\mathcal{A}$ answers with $(f_1(x_1), \ldots, f_\ell(x_\ell))$ while $\mathcal{B}$ answers with $(g_1(y_1), \ldots, g_\ell(y_\ell))$ and still win with probability 1. We call such strategies $\mathcal{A}, \mathcal{B}$ direct product strategies and denote them by $\prod f_i$ and $\prod g_i$.

In this work, we consider the case where the provers win the repeated game with non negligible probability. We would like to deduce a **structure** for the provers' strategies. Ideally, such strategies are approximately direct product strategies, in other words, global structure. Let us call this the **Global Structure Hypothesis**.

Without loss of generality we focus only on non trivial games, i.e. games in which for every questions pair (x, y) there exists a pair of answers (a, b) such that $V(x, y, a, b) = 0$. Otherwise, if the verifier always accepts, then it is trivial that we cannot expect of $\mathcal{A}, \mathcal{B}$ being structured, since every $\mathcal{A}, \mathcal{B}$ win with probability 1.

Results. Our first result is that the Global Structure Hypothesis does not hold in general even for non trivial games. We show games, for which there exists a strategy for $\mathcal{A}, \mathcal{B}$ that is extremely far from any direct product strategy (i.e. has no global structure) while attaining constant winning probability . We conclude, (perhaps surprisingly), that high success probability does not imply global structure.

Our main negative result shows that the Global Structure Hypothesis fails for any constant degree game [1] that has a large number of perfect strategies that are pairwise far apart:

Theorem 1 (Anti Structural Theorem- Informal Statement). *There exists a non-trivial constant degree game G, and constant α such that for every ℓ: There exist strategies $\mathcal{A}, \mathcal{B}$ such that the maximal agreement between $\mathcal{A}$ and $\prod f_i$ for any $\prod f_i$ is at most $2^{-\omega(\ell)}$, and similarly for $\mathcal{B}$. Yet, $\mathcal{A}, \mathcal{B}$ win with probability α.*

We extend Theorem 1 for games with unbounded degree, and also for the so-called "permuting" verifiers that permute the questions (these were called "clever" in [FK95]). For details see Section 3.

In the second part of the work we show, as our second result, that in contrast to Theorem 1 the Global Structure Hypothesis is true for a certain type of games called "miss-match" games. These games were first studied in [FK94].

---

[1] The degree is the maximal number of neighbors of a certain question.

Given a 2-prover game G its repeated "miss-match" game, denoted by $G^{m,\ell}$, $0 < m < \ell$, is as follows: The verifier chooses m coordinates, on each such coordinate it samples a pair (x, y) according to the distribution of G, these are called the *match* coordinates. As for the rest of the coordinates, the so called *miss* coordinates, the verifier picks $x \in X$ and $y \in Y$ independently uniformly at random. Then the verifier performs a random shuffle on the coordinates. The provers answer with an ℓ symbols string and they win the game if they win each of the match coordinates.[2] The random shuffle of the coordinates means that the only direct product strategies that will succeed are roughly of the form f^ℓ and not $\prod f_i$ for distinct f_i. See also Remark 1.

We first show that for every projection game G, if the provers win $G^{m,\ell}$ with non negligible probability ε, then $\mathcal{B}$ plays according to a direct product strategy: We show that there exists a small $(poly(1/\varepsilon))$ list of functions $g_1, \ldots, g_t : Y \to \Sigma_B$ such that $\mathcal{B}$ agrees non-negligibly with $(g_i)^\ell$ for each i. Furthermore, we show that essentially the only way $\mathcal{A}, \mathcal{B}$ win is whenever $\mathcal{B}(\mathbf{y}) \approx (g_i)^\ell(\mathbf{y})$ where g_i is some function from the list.

Theorem 2 (Informal Statement). *Let G be a projection game. Assume $\mathcal{A}, \mathcal{B}$ win $G^{m,\ell}$ with probability $\varepsilon > \ell^{-\Omega(1)}$, then there exists a small list of t functions $g_1, \ldots g_t : Y \to \Sigma_A$ such that:*

- *For each $i \in [t]$: $\Pr_{\mathbf{y}}[\mathcal{B}(\mathbf{y}) \approx g_i^\ell(\mathbf{y})] > \varepsilon'$, where $\varepsilon' = poly(\varepsilon)$.*
- *$\Pr[\exists i \text{ s.t. } \mathcal{B}(\mathbf{y}) \approx g_i^\ell(\mathbf{y}) | \mathcal{A}, \mathcal{B} \text{ win}] \geq 1 - o(1)$.*

The proof resembles [DG08] and [IKW09] and appears in Section 4.1.

Note that Theorem 2 only discusses $\mathcal{B}$'s strategy. It turns out that deducing a similar result for $\mathcal{A}$ is more subtle, and is only true if G is **smooth** enough. This smoothness parameter, first defined by[HK04], is as follows:

Definition 1. *A projection game G is called α-smooth if for every $x \in X$ and distinct answers $a, a' \in \Sigma_A$, we have: $\Pr_y[\Pi_{x,y}(a) = \Pi_{x,y}(a')] < 1 - \alpha$, where y is a random neighbor of x.*

Assuming the game is sufficiently smooth, we show an analog of Theorem 2, namely: we show that whenever $\mathcal{A}, \mathcal{B}$ win $G^{m,\ell}$ with non negligible probability, then there exists a short list of functions pairs $(f_1, g_1), \ldots, (f_s, g_s)$ such that: $\mathcal{A}, \mathcal{B}$ agree with f_i^ℓ, g_i^ℓ non-negligibly, and $val(G, f_i, g_i)$ is close to 1. We also prove that if $\mathcal{B}$ plays on $\mathbf{y}$ according to g_i^ℓ, and $\mathcal{A}$ does not play according to f_i^ℓ, or vice versus, then with high probability $\mathcal{A}, \mathcal{B}$ lose. Combining with Theorem 2 we get that there exists a small list of functions pairs (f_i, g_i), such that the only way to win the repeated game is whenever $\mathcal{A}$ plays according to direct product of f_i^ℓ while $\mathcal{B}$ plays according to g_i^ℓ. Thus, we fully explain the high winning probability of the provers through a direct product structure of their strategies.

---

[2] Alternatively, we can define "miss-match" as follows: Given a game G, we define $mm - G$ as the game that with probability $\alpha = (m/\ell)$ the verifier plays the original game G, and with probability $1 - \alpha$ it picks two independent questions and always accept. The repeated game $(mm - G)^\ell$ is very similar to $G^{m,\ell}$.

Theorem 3 (Informal Statement). *Let G be a an α smooth projection game (where α is a constant). Assume $\mathcal{A}, \mathcal{B}$ win $G^{m,\ell}$ with probability $\varepsilon > \ell^{-\Omega(1)}$, then there exists a small list of s pairs of functions $(f_1, g_1), \ldots (f_s, g_s)$ such that:*

- *$f_i : X \to \Sigma_A, \ g_i : Y \to \Sigma_B$ and: $val(G, f_i, g_i) > 1 - o(1)$.*
- *Let (A, B) be a random pair of questions, then:*

$$\Pr[\exists i \ s.t. \ \mathcal{A}(\mathbf{x}) \approx f_i^\ell(\mathbf{x}) \ and \ \mathcal{B}(\mathbf{y}) \not\approx g_i^\ell(\mathbf{y}) | \mathcal{A}, \mathcal{B} \ win \ ] > 1 - o(1).$$

The smoothness property is essential for Theorem3. Theorem 4 shows a game that is not smooth enough, for which there exist strategies $\mathcal{A}, \mathcal{B}$ that win the game with probability 1, yet $\mathcal{A}$ is unstructured.

Theorem 4 (Informal Statement). *There exists a projection game G, such that for every ℓ there exist strategies $\mathcal{A}, \mathcal{B}$ such that the maximal agreement between $\mathcal{A}$ and f^ℓ for any f is at most $2^{-\omega(\ell)}$. Yet, $\mathcal{A}, \mathcal{B}$ win $G^{m,\ell}$ with probability 1.*

Additional Motivation and Context. The study of structure of winning strategies, aside from being an interesting generalization of the direct product testing question, has also some additional motivation coming from PCP constructions.

In recent years, stronger variants of PCPs called PCPPs [BSGH+06] or assignment testers [DR06] and more recently dPCPs [DH09] have been introduced. These are constructs that are similar to PCPs but are stronger, and much more useful in composition. Without getting into the details, let us say that the main difference between these objects and regular PCPs lies in the soundness criterion. The difference is closely related to the difference between just knowing that the soundness error of repeated games is small (this only gives a PCP), and between being able to say that strategies that have non-negligible winning probability must be structured as direct products (such a result will give you the stronger object, i.e., a dPCP or a PCPP). Whereas the former is already given by the parallel repetition theorem of [Raz98], the later is the content of this work.

In fact, our structure result (Theorem 3) can be used in order to show that a parallel repetition of a dPCP is a dPCP with amplified soundness. However, since this has already been done (with better parameter setting) in [DM10], we do not work out the details here.

Future Work. In this work we deal with several types of games and repetitions. We show that for part of them, such as $G^{m,\ell}$ the Global Structure Hypothesis holds. Contrary, we show hat for other types of games, such as constant degree games with many perfect strategies, the hypothesis fails. It would be interesting to characterize the types of games and repetitions for which the hypothesis holds.

Organization of the Paper. Subsection 2.1 shows the Direct Product Lemma which is the basis for our approach. In section 3 prove Theorem 1. Finally, in Section 4 we prove Theorem 2 and Theorem 3.

2   Preliminaries

In this work we deal with several kinds of repetitions: Repetition where the provers gets ordered tuples, sets and multisets.

When the provers get ordered tuples, then we see them as **tuple oracles**: $\mathcal{A}$ gets a tuple $\mathbf{x} = (x_1, \ldots, x_\ell) \in X^\ell$ and responds with $\mathcal{A}(\mathbf{x}) \in \Sigma_A^\ell$, and $\mathcal{B}$ gets a tuple $\mathbf{y} = (y_1, \ldots, y_\ell) \in Y^\ell$ and responds with $\mathcal{B}(\mathbf{y}) \in \Sigma_B^\ell$. Let us define the product encoding of functions $(f_1, \ldots, f_\ell)$, $f_i : S \to \Sigma$, to be a tuple oracle, $\prod f_i$, assigning for every tuple $(s_1, \ldots, s_\ell)$ the value $(f_1(s_1), \ldots, f_\ell(s_\ell))$. In the case where $f_1 = \ldots = f_\ell = f$ we denote $\prod f_i$ by f^ℓ.

When the provers get multi-sets, then we see them as **multi-set oracles**: $\mathcal{A}$ gets a multi-set $A = \{x_1, \ldots, x_\ell\}$ and responds with $\mathcal{A}(A)$ which is a function $A \to \Sigma_A$. $\mathcal{B}$ gets a multi-set $B = \{y_1, \ldots, y_\ell\}$ and responds with $\mathcal{B}(B)$ which is a function $B \to \Sigma_B$. Let us define the ℓ multi-set direct product encoding of a function $f : S \to \Sigma$ to be a multi-set oracle, f^ℓ, assigning for every $T \subset S$ of cardinality ℓ the restriction of f to T.

When the provers get sets, then see them as a **set oracles**: The definitions are identical to multi-sets oracles besides that in this case the provers gets sets rather than multi-sets.

For a function $f : S \to \Sigma$, and $T \subset S$ we denote by f_T the restriction of f to T. The definition of the support of f is important in our discussion:

Definition 2. *For two vectors* $\mathbf{v}, \mathbf{w}$ *in some alphabet* Σ^ℓ *we write* $\mathbf{v} \overset{\rho}{\approx} \mathbf{w}$ *to denote* $\Pr_{i \in [\ell]}[\mathbf{v}_i = \mathbf{w}_i] \geq 1 - \rho$ *and* $\mathbf{v} \overset{\rho}{\not\approx} \mathbf{w}$ *to denote* $\Pr_{i \in [\ell]}[\mathbf{v}_i \neq \mathbf{w}_i] \geq \rho$.

For two function $f, g : T \to \Sigma$ *we write:* $f \overset{\rho}{\approx} g$ *to denote* $\Pr_{t \in T}[f(t) = g(t)] \geq 1 - \rho$ *and* $f \overset{\rho}{\not\approx} g$ *to denote* $\Pr_{t \in T}[f(t) \neq g(t)] \geq \rho$.

For a tuple oracle F *and* $f : S \to \Sigma$ *the* $\rho-$*support denoted by* $\mathrm{supp}_\rho^F(f)$ *defined as follows:* $\mathrm{supp}_\rho^F(f) = \{\mathbf{s} \in S^\ell \mid F(\mathbf{s}) \overset{\rho}{\approx} f^\ell(\mathbf{s})\}$.

For a multi-set oracle F *and* $f : S \to \Sigma$ *the* $\rho-$*support denoted by* $\mathrm{supp}_\rho^F(f)$ *defined as follows:* $\mathrm{supp}_\rho^F(f) = \{A \subset S \mid |A| = \ell \text{ and } F(A) \overset{\rho}{\approx} f^\ell(A)\}$.

Now we would like to introduce "miss match" games in these settings:

Definition 3. *"Miss-Match" Games:* Let G be a game, let ℓ, m be integers $0 < m < \ell$, then we define the miss-match, $G^{m,\ell}$, as follows:

1. *The verifier picks* m *pairs* (x_i, y_i) *where each pair is selected independently according to* $\mathcal{D}$. *The verifier defines a multiset* $A' = \{x_1, \ldots x_m\}$ *and* $B' = \{y_1, \ldots, y_m\}$. *These are the match elements, each pair* (x_i, y_i) *is called a match pair and* A', B' *are called the match questions.*
2. *The verifier picks* $\ell - m$ *additional pairs* (x_j, y_j), *where* x_j, y_j *are chosen independently at random from* X, Y *(respectively). The verifier defines multisets* $A'' = \{x_{m+1}, \ldots, x_\ell\}$ *and* $B'' = \{y_{m+1}, \ldots, y_\ell\}$. *These are the confuse elements.*
3. V *sends* $A = A' \cup A''$ *to* $\mathcal{A}$, *and* $B = B' \cup B''$ *to* $\mathcal{B}$.

4. $\mathcal{A}$ *responds with* $\mathcal{A}(A) : A \to \Sigma_A$, *and* $\mathcal{B}$ *responds with* $\mathcal{B}(B) : B \to \Sigma_B$ *($\mathcal{A}, \mathcal{B}$ are multiset-oracles). The provers win* G^ℓ *if they win each of the match elements, i.e. for every match pair (x, y) we have:*

$$V(x, y, \mathcal{A}(A)_x, \mathcal{B}(B)_y) = 1.$$

Remark 1. Note that in our definition of miss-match games we assumed that the provers are set-oracles and not tuple oracles, as done also in [IKW09]. This simplifying assumption allows us to consider only direct products of the form f^ℓ rather than $\prod f_i$.

One reason for this assumption is the fact that it is implicitly "enforced" if the verifier randomly shuffles the coordinates. Indeed in such a case the provers' answers must not depend on the order of the questions too much. A similar situation was analyzed in [DG08].

If the verifier does not shuffle the coordinates, our results should generalize appropriately to $\prod f_i$ instead of f^ℓ, but we did not check the details.

2.1   Testing Direct Product

We now turn to describe the Direct Product Testing Lemma as in [DG08] and in [IKW09]. Let F be a ℓ set oracle that works over a set X. The goal is to test whether F is close to a direct product encoding- i.e. whether there exists f such that F is the direct product encoding of f. A two queries test that resemble the "miss match" game is used. The test chooses a random subset A and a random subset B as follows: A and B share m elements in common. As for the rest elements of B the test picks $\ell - m$ random elements from X. Then the test checks for consistency among $F(A)$ and $F(B)$ i.e. for each common element x it verifies that $F(A)_x = F(B)_x$.

The following definition is quoted from [DG08].

Definition 4. *Let $\mathcal{B}$ a ℓ set oracle that works over a set Y. Let $B' \subset Y$ of cardinality m. We call B' ε-alive if there exists $b' : B' \to \Sigma_B$ such that:*

$$\Pr_{B \supset B'}[\mathcal{B}(B)_{B'} = b'] \geq \varepsilon$$

Such an answer b' is called a live answer for B'.

Now we are ready to state Theorem 3.14 from [IKW09]. This is a local to global Lemma that claims that as long as there exist many live sets (the local property), then this implies an existence of a direct product function with a large support (the global property).

Theorem 5 (Direct Product Testing:). *There exists $\ell_0 \in \mathbb{N}$ and $c > 0$ such that for every $\ell > \ell_0$: Let $\mathcal{B}$ be a ℓ set-oracle such that*

$$\Pr_{B' \subset Y || B'| = \sqrt{\ell}}[B' \text{ is } \varepsilon/2\text{-alive}] \geq \varepsilon/2,$$

where $\varepsilon \geq 1/\sqrt{\ell}$. Then, there exists a function $g : Y \to \Sigma_B$ such that $\mathcal{B}(B) \overset{\rho}{\approx} g^\ell(B)$ for at least $\Omega(\varepsilon^6)$ of the $B \in \binom{Y}{\ell}$, where $\rho \leq \ell^{-c}$.

3 Negative Results

In this section we prove Theorem 1 showing that, for any constant degree game G with many perfect strategies, $\mathcal{A}, \mathcal{B}$ can win G^ℓ with constant probability and still be very far from any direct product strategy. We extend Theorem 1 for games of of non-constant degree in Theorem 6. Theorem 7 extends Theorem 1 to handle "Permuting Verifiers".

For a game G we define a bipartite weighted graph, where $L = X$, $R = Y$ and $w_{x,y} = \Pr_D[y|x]$. The game is called d regular if the degree of every left node is d, the degree of every right node is $d|X|/|Y|$, and $w_{x,y} = 1/d$ for every adjacent x and y. d is called the degree of the game. Another property that we take into consideration is the the the rate between the cardinalities X and Y. We denote by r the ratio $|X|/|Y|$, and without loss of generality we assume $r > 1$.

3.1 Proof of Theorem 1

In this section we prove Theorem 1. We define $P_A = \{\prod f_i | f_i : X \to \Sigma_A\}$, and $P_B = \{\prod g_i | g_i : Y \to \Sigma_B\}$. For two functions $F, G : S^\ell \to \Sigma^\ell$ we define their relaxed Hamming distance with parameter γ as: $dist_\gamma(F, G) = \Pr_{\mathbf{s} \in S^\ell}[F(\mathbf{s}) \overset{\gamma}{\not\approx} G(\mathbf{s})]$. Let us first state Theorem 1 formally:

Theorem 1 (Formal Statement). *For every constants $d > 1$ and $0 < \gamma < 1/20$ there exists a non-trivial constant degree d game G, and tuples-oracles strategies $\mathcal{A}, \mathcal{B}$ such that $dist_\gamma(\mathcal{A}, P_A) \geq 1 - (1/|Y| + 2^{-\omega(\ell)})$, and $dist_\gamma(\mathcal{B}, P_B) \geq 1 - (1/|Y| + 2^{-\omega(\ell)})$. Yet, $\mathcal{A}, \mathcal{B}$ win G^ℓ with probability at least $1/d$.*

The theorem holds for any constant degree d game G, for which there exists a large list of $t = |Y|$ pairs of perfect strategies $(f_1, g_1), \ldots, (f_t, g_t)$ that satisfy: For $i \neq j : dist(f_i, f_j)$, and $dist(g_i, g_j)$ are both greater than 10γ.

The requirement for the distance between the pairs prevents the case where all the perfect strategies have a small relative distance. In such a case all of the above functions pairs (f_i, g_i) could be clustered into a single function pair (f, g) for which: $\mathcal{A}(\mathbf{x}) \approx f^\ell$ and $\mathcal{B}(\mathbf{y}) \approx g^\ell$. Such a behavior can still be viewed as a direct product structure for $\mathcal{A}, \mathcal{B}$.

Proof. The strategies of $\mathcal{A}, \mathcal{B}$ are based on the following combinatorial claim:

Claim. Let $G = (V, E)$ be a bipartite (c, d) regular graph (the left degree is c, and the right degree is d), and assume wlog $c \leq d$. Then there exists a subgraph $G' = (V, E')$ such that G' is $(1, d/c)$ regular.

Due to space limitations the proof is omitted and can found in the full version of the paper.

Let us present the strategies $\mathcal{A}, \mathcal{B}$: As a first step $\mathcal{A}, \mathcal{B}$ match for every $y \in Y$ a pair (f_i, g_i) from the list, so we associate the strategies list with the set Y and we write (f_y, g_y). Then they choose a subgraph G' as in claim 3.1.

$\mathcal{B}$ decides according to value of the first coordinate $\mathbf{y}_1$- i.e. given $\mathbf{y} = (\mathbf{y}_1, \ldots, \mathbf{y}_\ell)$, $\mathcal{B}(\mathbf{y}) = g_{\mathbf{y}_1}^\ell(\mathbf{y})$.

$\mathcal{A}$ strategy is similar, it is also based just on the value of the first coordinate $\mathbf{x}_1$: Given $\mathbf{x} = (\mathbf{x}_1, \ldots, \mathbf{x}_\ell)$, $\mathcal{A}(\mathbf{x}) = f_{N(\mathbf{x}_1)}^\ell(\mathbf{x})$ where $N(\mathbf{x}_1)$ is the vertex y such that $(\mathbf{x}_1, y) \in E'$.

We now turn to prove the success probability of the proves, and the distance between $\mathcal{A}, \mathcal{B}$ and any direct product strategy.

Note that if $\mathbf{y}_1 = N(\mathbf{x}_1)$, then $\mathcal{A}, \mathcal{B}$ win the game, since they are playing according to $f_{\mathbf{y}_1}, g_{\mathbf{y}_1}$, which is a perfect strategy.

What is the probability that indeed $\mathbf{y}_1 = N(\mathbf{x}_1)$? Note that we care only about the values of the first coordinate. Once $\mathbf{x}_1$ is fixed, the probability that $\mathbf{y}_1 = N(\mathbf{x}_1)$ is exactly $1/d$. Therefore, the winning probability is $1/d$.

What is the distance between $\mathcal{B}$ and any product $\prod g_i$?

Let $\prod g_i$ be a product strategy, we divide the proof into cases: The case where for every $y \in Y$ it holds that $dist(g_i, g_y) > 5\gamma$ for at least $1/4$ fraction of the g_i, and the case where there exists $y \in Y$ such that $dist(g_i, g_y) \leq 5\gamma$ for at least $3/4$ fraction of the g_i. Note, that since $dist(g_y, g_{y'}) > 10\gamma$ for $y \neq y' \in Y$, then every function g agrees with at most a single function g_y on more than $1 - 5\gamma$ fraction of the domain, and in particular for every i there can be only a single y with $dist(g_i, g_y) \leq 5\gamma$.

Assume we are in the first case:

$$\Pr_{\mathbf{y}}[\mathcal{B}(\mathbf{y}) \overset{\gamma}{\approx} \prod g_i(\mathbf{y})] = \Pr[\Pr_{y_1}[\mathcal{B}(\mathbf{y}) \overset{\gamma}{\approx} \prod g_i(\mathbf{y})]] = \Pr[\Pr_{y_1 \, y_2 \ldots, y_\ell}[g_{y_1}^\ell(\mathbf{y}) \overset{\gamma}{\approx} \prod g_i(\mathbf{y})]]$$

Now, we can use Chernoff inequality to deduce that $\Pr_{y_2 \ldots, y_\ell}[g_{y_1}^\ell(\mathbf{y}) \overset{\gamma}{\approx} \prod g_i(\mathbf{y})] < 2^{-\omega(\ell)}$ (the expected number of coordinates on which there is an inequality is at least $5\gamma/4$), so we get that in the first case: $dist(\mathcal{B}, \prod g_i) > 1 - 2^{-\omega(\ell)}$.

As for the second case, where we assume that $\prod g_i$ is close for some function g_y, then:

$$\Pr_{\mathbf{y}}[\mathcal{B}(\mathbf{y}) \overset{\gamma}{\approx} \prod g_i(\mathbf{y})] = \Pr[y_1 = y] \Pr_{y_2 \ldots, y_\ell}[g_y^\ell(\mathbf{y}) \overset{\gamma}{\approx} \prod g_i(\mathbf{y})]$$

$$+ \Pr[y_1 \neq y] \Pr_{y_2 \ldots, y_\ell}[g_{y_1}^\ell(\mathbf{y}) \overset{\gamma}{\approx} \prod g_i(\mathbf{y})]$$

$$\leq 1/|Y| + 2^{-\omega(\ell)}$$

We get that in this case $dist_\gamma(\mathcal{B}, \prod g_i) \geq 1 - (1/|Y| + 2^{\omega(\ell)})$, and we are done. The analysis for $\mathcal{A}$ is similar.

One may think that Theorem 1 is true just for constant degree game. However, in Theorem 6 we extend Theorem 1 for a certain non-constant game:

Theorem 6. *For every constant $d > 1, 0 < \gamma < 1/8$ there exists a non-trivial non-constant degree $\tilde{d}$ game $\tilde{G}$, and tuple-oracles strategies $\mathcal{A}, \mathcal{B}$ such that $dist_\gamma(\mathcal{A}, P_\mathcal{A}) \geq 1 - (\frac{\tilde{d}}{d|Y|} + 2^{-\omega(\ell)})$, and $dist_\gamma(\mathcal{B}, P_\mathcal{B}) \geq 1 - (\frac{\tilde{d}}{d|Y|} + 2^{-\omega(\ell)})$. Yet, $\mathcal{A}, \mathcal{B}$ win $\tilde{G}^\ell$ with probability $1/d$.*

The proof of Theorem 6 can be found in the full version of the paper.

Our next negative result is Theorem 7 that extends for "Permuting Verifiers". In this case we would like to view the provers as multi-sets oracles. [FK95] studied this type of verifiers and called them "Clever Verifiers". Let us first introduce them:

Definition 5 (Permuting Verifiers:). *The verifier selects ℓ pairs of questions $(x_1, y_1), \ldots, (x_\ell, y_\ell)$ Each is pair is drawn independently according to the distribution of G. V sends $A = \{x_1, \ldots, x_\ell\}$ to $\mathcal{A}$ (note that A is a multi-set), and $B = \{y_1, \ldots, y_\ell\}$ to $\mathcal{B}$. $\mathcal{A}$ answers with $\mathcal{A}(A)$, and $\mathcal{B}$ with $\mathcal{B}(B)$. The verifier accepts if for every i: $V(x_i, y_i, \mathcal{A}(A)_{x_i}, \mathcal{B}(B)_{y_i}) = 1$.*

We define $DP_A = \{f^\ell | f : X \to \Sigma_A\}$, and $DP_B = \{g^\ell | g : Y \to \Sigma_B\}$. Now let us state Theorem 7 formally:

Theorem 7. *For every constants $d > 1$ and $0 < \gamma < 1/8$, there exists a nontrivial constant degree d game G a constant c, and multiset-oracles strategies $\mathcal{A}, \mathcal{B}$ such that $dist_\gamma(\mathcal{A}, DP_A) \geq 1 - O(\ell/|Y| + 2^{-\omega(\ell)})$, and $dist_\gamma(\mathcal{B}, DP_B) \geq 1 - O(\ell/|Y| + 2^{-\omega(\ell)})$ Yet, $\mathcal{A}, \mathcal{B}$ win G^ℓ against "Permuting Verifier" with probability at least c/d.*

The proof of Theorem 7 can be found in the full version of the paper.

Now we would like to extend Theorem 7 to "miss-match" games. The result for "miss-match" game is weaker: $\mathcal{A}, \mathcal{B}$ can be unstructured and win the the game only with probability $\Omega(\frac{m}{d\ell})$ (and not $1/d$ as before). We address here that if $\mathcal{A}, \mathcal{B}$ win the game with probability $\gg m/\ell$ then we can prove that such a behavior is impossible, see section 4 for details.

Claim. For every constants $d > 1$ and $0 < \gamma < 1/8$, there exists a constant degree d game G, a constant c, and multiset-oracle strategies $\mathcal{A}, \mathcal{B}$ such that $dist_\gamma(\mathcal{A}, DP_A) \geq 1 - O(\ell/|Y| + 2^{-\omega(\ell)})$, and $dist_\gamma(\mathcal{B}, DP_B) \geq 1 - O(\ell/|Y| + 2^{-\omega(\ell)})$. Yet, $\mathcal{A}, \mathcal{B}$ win $G^{m,\ell}$ with probability at least $\frac{cm}{d\ell}$.

The proof of Claim 3.1 can be found in the full version of the paper.

4   Positive Results: "Miss-Match" Games

In this section we show that, unlike general games, "miss match" games have the following property: If $\mathcal{A}, \mathcal{B}$ win "miss match" games with non negligible probability, then there exists a small list of pairs $(f_1, g_1), (f_2, g_2), \ldots$ such that $Val(G, f_i, g_i) \approx 1$ and: If $\mathcal{A}, \mathcal{B}$ win then $\mathcal{A}(A) \approx f_i^\ell(A)$ and $\mathcal{B}(B) \approx g_i^\ell(B)$ for some pair from the list, except with negligible probability.

We first prove Theorem 2. The theorem asserts the above only for $\mathcal{B}$, namely: if the provers win $G^{m,\ell}$ with non negligible probability ε, then $\mathcal{B}$ plays according to a direct product strategy.

It turns out that deducing a similar result for $\mathcal{A}$ is more subtle, and depends on the **smoothness** of the game (see Definition 1). Assuming the game is sufficiently smooth, we obtain in Theorem 3 the desired result claimed above.

We also address the question of whether smoothness is essential for direct product behavior. In subsection 4.3 we show that it is essential. In Theorem 4 we show a game that is not smooth such that $G^{m,\ell}$ can be won with probability 1 and still $\mathcal{A}$ is far from being a direct product strategy.

4.1   Direct Product Structure for $\mathcal{B}$

In this section we prove Theorem 2, let us state it formally:

Theorem 2   (Formal Statement). *There exists $\ell_0 \in \mathbb{N}$ and $c > 0$ such that for every $\ell > \ell_0$ the following holds. Let G be a projection game, and let $m = \sqrt{\ell}$, $\varepsilon_0 = 2\sqrt{m/\ell}$ and $\delta = \sqrt{\varepsilon_0}$.*

Assume $\mathcal{A}, \mathcal{B}$ win $G^{m,\ell}$ with probability $\varepsilon > \sqrt{\varepsilon_0}$, then there exists a list of $t = O(1/(\delta \cdot \varepsilon)^6)$ functions $g_1, \ldots g_t : Y \to \Sigma_A$ such that:

- *For each $i \in [t]$: $\Pr_B[\mathcal{B}(B) \overset{\rho}{\approx} g_i^\ell(B)] > \Omega((\delta \cdot \varepsilon)^6)$, where $\rho = \ell^{-c}$.*
- *$\Pr[\exists i \text{ s.t. } \mathcal{B}(B) \overset{\rho}{\approx} g_i^\ell(B) | \mathcal{A}, \mathcal{B} \text{ win }] \geq 1 - \delta$*

Before we proceed with the proof, let us make a few remarks:

- The theorem concludes that on many Bs, $\mathcal{B}(B) \overset{\rho}{\approx} g^\ell(B)$ rather than $\mathcal{B}(B) = g^\ell(B)$. This weaker conclusion is inherent as seen by the following example. Take $\mathcal{B} = g^\ell$ and then change each $\mathcal{B}(B)$ arbitrarily in fewer than ℓ/m of the coordinates. With high probability the verifier would not notice the difference between $\mathcal{B}$ and g^ℓ, yet $\mathcal{B}$ is only close to g^ℓ in the above sense.
- We would like to address the relation between m and ℓ and the value of ε in Theorem 2. We have already proved Claim 3.1 that asserts that $\mathcal{A}, \mathcal{B}$ can be far away from direct product encoding and still win $G^{m,\ell}$ with probability $\Omega(\frac{m}{d\ell})$. This enforces two constraints regarding our choice of parameters: First, we need that the winning probability ε would be greater than m/ℓ. Indeed, we prove our theorem for values of ε that are bigger than $\sqrt[4]{m/\ell}$. Second, we must choose $m \ll \ell$, and in this work we focus on $m = \sqrt{\ell}$. We leave the study of the entire range of m, ℓ for future work (We mention that this is an open question even in the Direct Product Testing settings see [GS00], [DR06], [DG08] and [IKW09]).
- We work in the settings where $\ell \ll |Y|$ and in particular $\ell < \sqrt[6]{|Y|}$. This enables us an easy transition between sets and multi-sets.

In order to prove Theorem 2, we first show if $\mathcal{A}, \mathcal{B}$ win then there exists at least one function $g : Y \to \Sigma_B$ such that $\mathcal{B}(B) \approx g^\ell(B)$ on a non negligible part of the domain.

Lemma 1. *There exist $\ell_0 \in \mathbb{N}$, and $c > 0$ such that for every $\ell > \ell_0$ the following holds. Let G be a projection game, and let $\varepsilon_0 = 2\sqrt{\frac{m}{\ell}}$.*

Assume $\mathcal{A}, \mathcal{B}$ win $G^{m,\ell}$ where $m = \sqrt{\ell}$, with probability $\varepsilon > \varepsilon_0$, then there exists a function $g : Y \to \Sigma_B$ such that for at least $\Omega(\varepsilon^6)$ of the ℓ multi-sets B, we have $\mathcal{B}(B) \overset{\rho}{\approx} g^\ell(B)$, where $\rho = \ell^{-c}$.

The proof of Theorem 2 and Lemma 1 can be found in the full version of the paper.

4.2 Direct Product Structure for $\mathcal{A}$

In Section 4.1 we show that for every projection game G, whenever $\mathcal{A}, \mathcal{B}$ win $G^{m,\ell}$ with non-negligible probability, then $\mathcal{B}$'s strategy has a direct product structure. However, we have not involved $\mathcal{A}$ strategy at all. In this section we deduce a similar behavior for $\mathcal{A}$ for smooth games. Let us state Theorem 3 formally:

Theorem 3 (Formal Statement). *There exist $\ell_0 \in \mathbb{N}$, $0 < \alpha < 1$ and $c > 0$ such that for every $\ell > \ell_0$ the following holds. Let G be an α-smooth projection game, and let $\rho = \ell^{-c}, \varepsilon_0 = 2\sqrt{\frac{m}{\ell}}$ and $\delta = \sqrt{\varepsilon_0}$.*

Assume $\mathcal{A}, \mathcal{B}$ win $G^{m,\ell}$, with $m = \sqrt{\ell}$, with probability $\varepsilon > \sqrt{\varepsilon_0}$, then there exists a list of $s = O(1/(\delta\varepsilon)^6)$ pairs of functions $(f_1, g_1), \ldots (f_s, g_s)$ such that:

- *$f_i : X \to \Sigma_A$, $g_i : Y \to \Sigma_B$ and: $val(G, f_i, g_i) > 1 - 10\rho/\alpha$.*
- *Let (A, B) be a random pair of questions. Define the following events:*
  - *$B_1 := B \notin \cup_{i \in [s]}\mathrm{supp}_\rho(g_i)$*
  - *$B_2 := \exists i \in [s]$ s.t. $B \in \mathrm{supp}_\rho(g_i)$ while $A \notin \mathrm{supp}_{6\rho/\alpha}(f_i))$.*
  - *$B_3 := \exists i \in [s]$ s.t. $A \in \mathrm{supp}_{6\rho/\alpha}(f_i))$ while $B \notin \mathrm{supp}_{40\rho/\alpha}(g_i)$.*

  Then:
$$\Pr[\mathcal{A}, \mathcal{B} \text{ win } | B_1 \text{ or } B_2 \text{ or } B_3] < \delta + O(\exp^{-\Omega(\rho^2 m)}).$$

In order to prove Theorem 3 we use the following three lemmas:

Lemma 2. *There exist $\ell_0 \in \mathbb{N}$, $0 < \alpha < 1$ and $c > 0$ such that for every $\ell > \ell_0$ the following holds. Let G be an α-smooth projection game, $g : Y \to \Sigma_B$ and $\rho = \ell^{-c}$. Let $f : X \to \Sigma_A$ be a function that maximizes $val(G, f, g)$, then: If $B \in \mathrm{supp}_\rho(g)$ while $A \notin \mathrm{supp}_{6\rho/\alpha}(f)$. Then $\mathcal{A}, \mathcal{B}$ win with probability at most $3\exp^{-\Omega(\rho^2 m)}$.*

Lemma 3. *There exist $\ell_0 \in \mathbb{N}$, $0 < \alpha < 1$ and $c > 0$ such that for every $\ell > \ell_0$ the following holds. Let G be an α-smooth projection game, $f : X \to \Sigma_A$ and $\rho = \ell^{-c}$. Let $g : Y \to \Sigma_B$ be a function such that $val(G, f, g) > 1 - 10\rho/\alpha$, then: If $A \in \mathrm{supp}_{6\rho/\alpha}(f)$ while $B \notin \mathrm{supp}_{40\rho/\alpha}(g)$. Then $\mathcal{A}, \mathcal{B}$ win with probability at most $4\exp^{-\Omega(\rho^2 m)}$.*

Lemma 4. *There exist $\ell_0 \in \mathbb{N}$, $0 < \alpha < 1$ and $c > 0$ such that for every $\ell > \ell_0$ the following holds. Let G be an α-smooth projection game, $g : Y \to \Sigma_B$ and $\rho = \ell^{-c}$. Let $f : X \to \Sigma_A$ be a function that maximizes $val(G, f, g)$, then: If $val(G, f, g) < 1 - 10\rho/\alpha$, and assuming $B \in \mathrm{supp}_\rho(g)$ and $A \in \mathrm{supp}_{6\rho/\alpha}(f)$ then $\mathcal{A}, \mathcal{B}$ win with probability at most $3\exp^{-\Omega(\rho^2 m)}$.*

The proofs of Theorem 3, Lemma 2, Lemma 3 and Lemma 4 can be found in the full version of the paper.

4.3   The Smoothness Is Essential

In this section we show that the smoothness property is crucial. We show the existence of a game G that is not smooth, such that $G^{m,\ell}$ has perfect strategies $\mathcal{A}, \mathcal{B}$ and $\mathcal{A}$ is far from being a direct product strategy. Let us state Theorem 4 formally:

Theorem 4 (Formal Statement). *There exists a projection game G, such that for every ℓ and $0 < m < \ell$: There exist multiset oracles $\mathcal{A}, \mathcal{B}$ such that for every $f : X \to \Sigma_A$: $dist_{1/2}(A, f^\ell) > 1 - 2^{-\omega(\ell)}$. Yet, $\mathcal{A}, \mathcal{B}$ win $G^{m,\ell}$ with probability 1.*

The proof of Theorem 4 can be found in the full version of the paper.

References

[BSGH+06]   Ben-Sasson, E., Goldreich, O., Harsha, P., Sudan, M., Vadhan, S.P.: Robust pcps of proximity, shorter pcps, and applications to coding. SIAM J. Comput. 36(4), 889–974 (2006)

[DG08]   Dinur, I., Goldenberg, E.: Locally testing direct product in the low error range. In: FOCS 2008: Proceedings of the 2008 49th Annual IEEE Symposium on Foundations of Computer Science, Washington, DC, USA, pp. 613–622. IEEE Computer Society, Los Alamitos (2008)

[DH09]   Dinur, I., Harsha, P.: Composition of low-error 2-query pcps using decodable pcps. In: FOCS, pp. 472–481 (2009)

[DM10]   Dinur, I., Meir, O.: Derandomized parallel repetition of structured pcps, CoRR abs/1002.1606 (2010)

[DR06]   Dinur, I., Reingold, O.: Assignment testers: Towards a combinatorial proof of the pcp theorem. SIAM J. Comput. 36(4), 975–1024 (2006)

[FK94]   Feige, U., Kilian, J.: Two prover protocols: low error at affordable rates. In: STOC, pp. 172–183 (1994)

[FK95]   Feige, U., Kilian, J.: Impossibility results for recycling random bits in two-prover proof systems. In: STOC, pp. 457–468 (1995)

[GS00]   Goldreich, O., Safra, S.: A combinatorial consistency lemma with application to proving the pcp theorem. SIAM J. Comput. 29(4), 1132–1154 (2000)

[HK04]   Holmerin, J., Khot, S.: A new pcp outer verifier with applications to homogeneous linear equations and max-bisection. In: STOC, pp. 11–20 (2004)

[IKW09]   Impagliazzo, R., Kabanets, V., Wigderson, A.: New direct-product testers and 2-query pcps. In: STOC 2009: Proceedings of the 41st Annual ACM Symposium on Theory of Computing, pp. 131–140. ACM, New York (2009)

[Raz98]   Raz, R.: A parallel repetition theorem. SIAM Journal on Computing 27, 763–803 (1998)

Distribution-Free Testing Algorithms for Monomials with a Sublinear Number of Queries

Elya Dolev and Dana Ron*

School of Electrical Engineering, Tel Aviv University, Tel Aviv 69978, Israel
elyadolev@msn.com,
danar@eng.tau.ac.il

Abstract. We consider the problem of distribution-free testing of the class of monotone monomials and the class of monomials over n variables. While there are very efficient algorithms for testing a variety of functions classes when the underlying distribution is uniform, designing distribution-free algorithms (which must work under any arbitrary and unknown distribution), tends to be a more challenging task. When the underlying distribution is uniform, Parnas et al. (*SIAM Journal on Discrete Math, 2002*) give an algorithm for testing (monotone) monomials whose query complexity does not depends on n, and whose dependence on the distance parameter is (inverse) linear. In contrast, Glasner and Servedio (in *Proceedings of RANDOM, 2007*) prove that every distribution-free testing algorithm for monotone monomials as well as for general monomials must have query complexity $\tilde{\Omega}(n^{1/5})$ (for a constant distance parameter ϵ).

In this paper we present distribution-free testing algorithms for these classes where the query complexity of the algorithms is $\tilde{O}(n^{1/2}/\epsilon)$. We note that as opposed to previous results for distribution-free testing, our algorithms do not build on the algorithms that work under the uniform distribution. Rather, we define and exploit certain structural properties of monomials (and functions that differ from them in a non-negligible manner), which were not used in previous work on property testing.

1   Introduction

Testers (for properties of functions) are algorithms that decide whether a given function has a prespecified property or is "far" from having the property with respect to some fixed distance measure. In most works on property testing, distance is measured with respect to the uniform distribution over the function domain. While in many contexts this distance is appropriate, as it corresponds to assigning equal "importance" (weight) to each point in the domain, there are scenarios in which we may want to deal with an underlying weight distribution that is not uniform, and furthermore, is not known to the algorithm. We refer to the latter model as *distribution-free* property testing, while testing under the

---

* Research supported by a grant from the Israel Science Foundation (grant No. 246/08).

M. Serna et al. (Eds.): APPROX and RANDOM 2010, LNCS 6302, pp. 531–544, 2010.

uniform distribution is considered to be the *standard* model. In both models the algorithm is given query access to the tested function and in the distribution-free model the algorithm is also given access to examples distributed according to the unknown underlying distribution.

Indeed, the notion of distribution-free testing is inspired by the distribution-free (Probably Approximately Correct (PAC)) learning model [20] and understanding the relation between testing and learning is one of the motivations for property testing. As observed in [9], the complexity of testing a function class $\mathcal{F}$ (that is, testing the property of membership in $\mathcal{F}$), is not higher than (proper) learning the class $\mathcal{F}$ (under the same conditions, e.g., with respect to the uniform distribution or distribution-free). In view of this, a natural question is for what classes of functions is the complexity of testing *strictly* lower than that of learning. Here, when we refer to the complexity of the algorithm, our main focus is on its query complexity (where in this complexity we count all queries: both on arbitrary points selected by the algorithm, and on points sampled according to the underlying distribution). Note that, as opposed to learning, if we have a testing algorithm for (membership in) a class of functions $\mathcal{F}$, this does not imply that we have a testing algorithm (with similar complexity) for all subclasses $\mathcal{F}'$ of $\mathcal{F}$.

There is quite a large variety of function classes for which the complexity of testing is strictly lower than that of learning when the underlying distribution is uniform (e.g., linear functions [1], low-degree polynomials [18], singletons, monomials [17] and small monotone DNF [17], monotone functions (e.g., [5,3]), small juntas [6], small decision lists, decision trees and (general) DNF [2] linear threshold functions [15], and more). In contrast, there are relatively few such positive results for distribution-free testing [10,11,13], and, in general, designing distribution-free testing algorithms tends to be more challenging.

One of the main positive results for distribution-free testing [13] is that every function class that has a standard tester and can be efficiently *self-corrected* [1], has a distribution-free tester whose complexity is similar to that of the standard tester. In particular this implies that there are efficient distribution-free testers for linear functions and more generally, for low-degree polynomials [13]. However, there are function classes of interest (in particular from the point of view of learning theory), which have efficient standard testers, but for which self-correctors do not exist (or are not known to exist). Several such classes (of Boolean functions over $\{0,1\}^n$) were studied by Glasner and Servedio [7]. Specifically, they consider monotone monomials, general monomials, decisions lists, and linear threshold functions. They prove that for these classes, in contrast to standard testing, where the query complexity does not depend on the number of variables n, every distribution-free testing algorithm must make $\Omega((n/\log n)^{1/5})$ queries (for a constant distance parameter ϵ). While these negative results establish that a strong dependence on n is unavoidable for these functions classes in the distribution-free case, it still leaves open the question of whether some *sublinear* dependence on n can be obtained (where distribution-free learning (with queries) requires at least linear complexity [19]).

OUR RESULTS. In this work we prove that both for monotone monomials and for general monomials, a sublinear dependence on n can be obtained for distribution-free testing. Specifically, we describe distribution-free testing algorithms for these families whose query complexity is $O(\sqrt{n}\log n/\epsilon)$. Thus we advance our knowledge concerning efficient distribution-free testing for two basic function classes. Furthermore, while previous distribution-free testing algorithms are based on, and are similar to the corresponding standard testing algorithms, this is not the case for our algorithms. Rather, we define and exploit certain structural properties of monomials (and functions that differ from them in a non-negligible manner), which were not used in previous work on property testing in the standard model. In what follows we give some intuition concerning the difficulty encountered when trying to extend standard testing of (monotone) monomials to distribution-free testing and then shortly discuss the ideas behind our algorithms.

STANDARD VS. DISTRIBUTION-FREE TESTING OF MONOMIALS. The first simple observation concerning testing monomials under the uniform distribution is the following. If f is a k-monomial (that is, a conjunction of k literals), then $\Pr[f(x) = 1] = 2^{-k}$ (where the probability is over a uniformly selected x). This implies that we can effectively consider only relatively small monomials, that is, k-monomials for which $k = \log(O(1/\epsilon))$, and it allows the testing algorithm to have an exponential dependence on k (since this translates to a linear dependence on $1/\epsilon$). This is not in general the case when the underlying distribution is arbitrary. In particular, the functions considered in the lower bound proof of [7] (some of which are monomials, and some of which are far from being monomials), depend on $\Omega(n)$ variables. Thus, for these functions, considering uniformly selected points, essentially gives no information (since the function assigns value 0 to all but a tiny fraction of the points). Furthermore, the support of the distribution D defined in [7] is such that the following holds. If one takes a sample (distributed according to D) of size smaller than the square-root of the support size of D, (where there are roughly $n^{2/5}$ points in the support), and performs queries on the sampled points, then it is not possible to distinguish between the monomials and the functions that are far from being monomials (with respect to D). Thus, by sampling according to D, we essentially get no information unless the size of the sample is above a (fairly high) threshold. On the other hand, if we perform queries outside the support of D, then intuitively (and this is formalized in [7]), violations (with respect to being a monomial) are hard to find.

Before continuing with a high level description of our algorithms, we note that if we restrict the task of testing to distribution-free testing of (monotone) k-monomials, where k is fixed, then there is an algorithm whose query complexity grows exponentially with k. This follows by combining two results: (1) The aforementioned result of Halevy and Kushilevitz [13] concerning the use of "self-correction" in transforming standard testing algorithm to distribution-free testing algorithms; (2) The result of Parnas et al. [17] for testing (monotone) monomials, which has a self-corrector (with complexity 2^k) as a building block. Hence, for small k (i.e., k that is strictly smaller than $\log n$) we have an algorithm with complexity that is sublinear in n, and the question is what can be done when it is not assumed that k is small.

OUR ALGORITHMS: IDEAS AND TECHNIQUES. In what follows we discuss the algorithm for testing monotone monomials over $\{0,1\}^n$. The algorithm for testing general monomials has the same high-level structure, and can be viewed as a generalization of the algorithm for testing monotone monomials.

We start by introducing the notion of a *violation* hypergraph for a function f. The vertex-set of this hypergraph is $\{0,1\}^n$, and its edge-set corresponds to subsets that contain evidence that f is not a monotone monomial. Each (hyper)edge includes a single point y^0 such that $f(y^0) = 0$, and for each additional point y^j in the edge, $f(y^j) = 1$. For each such subset there is no monotone monomial that is consistent with f on the subset. For example, we may have $y^0 = 010$, $y^1 = 011$ and $y^2 = 110$ (since y^0 "forces" either x_1 or x_3 to be in the monomial whereas y^1 and y^2 "disallow" these possibilities). Thus, the edge-set of the hypergraph may be exponentially large in n, and edges may have large size (e.g., $\Omega(n)$). Clearly, if f is a monotone monomial, then the hypergraph has no edges. On the other hand, we prove that if f is far from being a monotone monomial (with respect to the underlying distribution, D), then every vertex cover of the (edges of the) hypergraph must have relatively large weight (with respect to D).

Assuming from this point on that f is far from being a monotone monomial (and hence its violation hypergraph has a relatively large weight minimum vertex cover), our algorithm tries to find a small edge in the hypergraph. To this end we do the following. First we take a random sample T of $\Theta(\sqrt{n}/\epsilon)$ points (generated according to D), and consider all points in the sample that are labeled 0 by f. Observe that if f were a monotone monomial, then for each such sample point $y \in f^{-1}(0)$, there must exist at least one index, j, such that x_j is a variable in the monomial and $y_j = 0$. But then $f(1^{j-1}01^{n-j})$ must be 0. In view of this, for each $y \in T \cap f^{-1}(0)$, we search for such an index j (satisfying $y_j = 0$ and $f(1^{j-1}01^{n-j}) = 0$). The search is initiated with the candidate set $\{j : y_j = 0\}$, and the set is cut in half in each iteration (by performing two queries). Thus, if f were a monotone monomial, then such a search must always succeed. On the other hand, if the search fails for any $y \in f^{-1}(0)$, then we obtain evidence that f is not a monotone monomial, and the algorithm may reject. This evidence is an edge (of size 3) in the violation hypergraph (e.g., $(0^t 1^{n-t}, 0^{t/2} 1^{n-t/2}, 1^{t/2} 0^{t/2} 1^{n-t})$ where the first point is in $f^{-1}(0)$ and the latter two are in $f^{-1}(1)$).

Assuming no search fails, the algorithm has a set J of "representative indices". These indices are such that if f were a monotone monomial, then for each $j \in J$, the variable x_j would be among the variables in f. This means that for every $w \in f^{-1}(1)$ and $j \in J$, it would hold that $w_j = 1$. In the second stage of the algorithm we take an additional sample of $\Theta(\sqrt{n}/\epsilon)$ points and consider all sample points in $f^{-1}(1)$. If for any such sample point w we have that $w_j = 0$ for some $j \in J$, then the algorithm has evidence that f is not a monotone monomial (an edge of size 2 in the violation hypergraph) and it rejects. The crux of the proof is showing that if the probability that the algorithm does not find evidence (in both stages) is small, then it is possible to construct a small-weight vertex cover in the violation hypergraph (implying that f is close to being a monotone monomial).

OTHER RELATED WORK. In addition to the results mentioned previously, Halevy and Kushilevitz [10,13] study distribution-free testing of monotonicity for functions $f : \Sigma^n \to R$ (where Σ and R are fully ordered). Building on the (one-dimensional) standard testing algorithm in [5] they give a distribution-free testing algorithm whose query complexity is $O((2\log|\Sigma|)^n/\epsilon)$. Thus, the dependence on the dimension, n is exponential, in contrast to some of the standard testing algorithms for monotonicity [8,3] where the dependence on n is linear.[1] In follow-up work [12,13], Halevy and Kushilevitz showed that the exponential dependence on n is unavoidable for distribution-free testing even in the case of Boolean functions over the Boolean hypercube (that is, $|\Sigma| = |R| = 2$).

Halevy and Kushilevitz [11] also study distribution-free testing of graph properties in sparse graphs, and give an algorithm for distribution-free testing of connectivity, with similar complexity to the standard testing algorithm for this property.

We note that for some properties that have efficient standard testers, the algorithms can be extended to work under more general families of distributions such as product distributions (e.g., [6,2]). In recent work, Kopparty and Saraf [14] consider tolerant testing [16] of linearity under non-uniform distributions (that have certain properties).

FURTHER RESEARCH. Perhaps the first question that comes to mind is what is the exact complexity of distribution-free testing of (monotone) monomials given the gap between our upper bound and the lower bound of [7]. It will also be interesting to design sublinear algorithms for testing the other function classes studied in [7]. Another direction is to study testing of monomials and other basic function classes under known distributions (other than the uniform distribution).

ORGANIZATION. We start by introducing some notation and definitions in Section 2. In Section 3 we describe and analyze the distribution-free testing algorithm for monotone monomials, and in Section 4 we explain how to extend it to general monomials. All missing proofs can be found in the full version of this paper [4].

2   Preliminaries

For an integer k we let $[k] \overset{\text{def}}{=} \{1, \ldots, k\}$. In all that follows we consider Boolean functions f whose domain is $\{0,1\}^n$.

---

[1] To be precise, the complexity of the algorithm in [3] is $O(n \log|\Sigma| \log|R|/\epsilon)$, where $|R|$ is the effective size of the range of the function, that is, the number of distinct values of the function.

Definition 1 (Monomials). *A function* $f : \{0,1\}^n \to \{0,1\}$ *is a monomial if it is a conjunction ("and") of a subset of the literals* $\{x_1, \bar{x}_1, \ldots, x_n, \bar{x}_n\}$. *It is a* monotone monomial *if it is a conjunction only of variables (and no negations of variables). We denote the class of monomials by* $\mathcal{M}$ *and the class of monotone monomials by* $\mathcal{M}_m$.

We note that we allow the special case that the subset of literals (variables) is empty, in which case f is the all-1 function. the full version of this paper [4] we discuss how to augment our tests so that they work for the case that the subset of literals (variables) must be non-empty.

Definition 2 (Distance). *For two functions* $f, g : \{0,1\}^n \to \{0,1\}$ *and a distribution* D *over* $\{0,1\}^n$, *we let* $\mathrm{dist}_D(f,g) \overset{\text{def}}{=} \Pr_{x \sim D}[f(x) \neq g(x)]$ *denote the distance between* f *and* g *with respect to* D. *For a class of Boolean functions* $\mathcal{F}$ *over* $\{0,1\}^n$ *and a function* $f : \{0,1\}^n \to \{0,1\}$, *we let* $\mathrm{dist}_D(f,\mathcal{F}) \overset{\text{def}}{=} \min_{g \in \mathcal{F}} \{\mathrm{dist}_D(f,g)\}$ *denote the distance between* f *and the class of functions* $\mathcal{F}$.

Definition 3 (Distribution-Free Testing). *Let* $\mathcal{F}$ *be a class of Boolean functions over* $\{0,1\}^n$. *A* distribution-free testing *algorithm for (membership in)* $\mathcal{F}$ *is given access to examples that are distributed according to an unknown distribution* D *and is given query access to* f. *The algorithm is also given a distance parameter* $0 < \epsilon < 1$, *and is required to behave as follows.*

- *If* $f \in \mathcal{F}$, *then the algorithm should output* accept *with probability at least* 2/3.
- *If* $\mathrm{dist}_D(f,\mathcal{F}) > \epsilon$, *then the algorithm should output* reject *with probability at least* 2/3.

If the algorithm accepts every $f \in \mathcal{F}$ *with probability* 1, *then it is a* one-sided error *algorithm.*

In all that follows f always denotes the (unknown) tested function, and D denotes the (unknown) underlying distribution with respect to which the testing algorithm should work. For a point $y \in \{0,1\}^n$ let $D(y)$ denote the probability assigned to y by D, and for a subset $S \subseteq \{0,1\}^n$ let $D(S) = \sum_{y \in S} D(y)$ denote the weight that D assigns to the subset S.

We assume without loss of generality that $\epsilon \geq 2^{-n}$, or else, by performing a number of queries that is linear in $1/\epsilon$ (that is, querying f on all domain elements) it is possible to determine whether f is a monotone monomial.

3   Distribution-Free Testing of Monotone Monomials

We start by introducing the notion of a *violation hypergraph* of a function and establishing its relation to (the distance to) monotone monomials.

3.1   The Violation Hypergraph

Before defining the violation hypergraph, we introduce some notation. For each point $y \in \{0,1\}^n$ let $Z(y) \overset{\text{def}}{=} \{i : y_i = 0\}$. We use 1^n to denote the all-1 vector (point).

Let g be a Boolean function over $\{0,1\}^n$ and let $\{y^0, y^1, \ldots, y^t\} \subseteq \{0,1\}^n$ be a subset of points such that $g(y^0) = 0$ and $g(y^j) = 1$ for all $1 \le j \le t$. A simple but useful observation is that if g is a monotone monomial, then $Z(y^0)$ must include at least one index i such that $i \notin \bigcup_{j=1}^{t} Z(y^j)$. This observation motivates the next definition.

Definition 4 (Violation Hypergraph). *Let* $H_f = (V(H_f)), E(H_f))$ *be the hypergraph whose vertex set,* $V(H_f)$ *is* $\{0,1\}^n$, *and whose edge set,* $E(H_f)$, *contains all subsets* $\{y^0, y^1, \ldots, y^t\} \subseteq \{0,1\}^n$ *of the following form:*

- *$f(y^0) = 0$ and $f(y^j) = 1$ for all $1 \le j \le t$.*
- *$Z(y^0) \subseteq \bigcup_{j=1}^{t} Z(y^j)$.*

For example, if $f(0011) = 0$, $f(0110) = 1$ and $f(1011) = 1$, then $Z(0011) = \{1,2\}$, $Z(0110) = \{1,4\}$, and $Z(1011) = \{2\}$, and so $\{0011, 0110, 1011\}$ is an edge in H_f. Note that if $f(1^n) = 0$, then $E(H_f)$ contains the edge $\{y^0 = 1^n\}$ (because $Z(y^0) = \emptyset$ and $\bigcup_{j=1}^{t} Z(y^j)$ is trivially empty as well).

By the observation preceding Definition 4, if f is a monotone monomial, then $E(H_f) = \emptyset$. We next claim that the reverse implication holds as well, so that we obtain a characterization of monotone monomials that is based on H_f.

Lemma 1. *If $E(H_f) = \emptyset$, then f is a monotone monomial.*

Lemma 1 is proved by defining a monotone monomial h, which is the conjunction of all variables x_i such that $y_i = 1$ for all $y \in f^{-1}(1)$, and showing that it agrees with f on the whole domain.

Recall that a vertex cover of a hypergraph is a subset of the vertices that intersects every edge in the hypergraph. We next establish that if f is far from being a monotone monomial (with respect to D), then every vertex cover of H_f must have large weight (with respect to D). This lemma strengthens Lemma 1 in the following sense. Lemma 1 is equivalent to saying that if f is not a monotone monomial, then $E(H_f) \ne \emptyset$. In particular this implies that if f is not a monotone monomial, then every vertex cover of H_f is non-empty. Lemma 2 can be viewed as quantifying this statement (and taking into account the underlying distribution D). Lemma 2 follows by extending the proof of Lemma 1.

Lemma 2. *If $\text{dist}_D(f, \mathcal{M}_m) > \epsilon$, then for every vertex cover C of H_f we have $D(C) > \epsilon$.*

By Lemmas 1 and 2, if f is a monotone monomial, then $E(H_f) = \emptyset$, so that trivially every minimum vertex cover of H_f is empty, while if $\text{dist}_D(f, \mathcal{M}_m) > \epsilon$, then every vertex cover of H_f has weight greater than ϵ with respect to D. We would like to show that this implies that if $\text{dist}_D(f, \mathcal{M}_m) > \epsilon$, then we can actually find (with high probability) an edge in H_f, which provides evidence to the fact that f is not a monotone monomial.

3.2   The Testing Algorithm

We first introduce a few more notation. Let $\bar{e}^i = 1^{i-1}01^{n-i}$. For any subset $Z \subseteq [n]$, let $y(Z)$ be the point in $\{0,1\}^n$ such that for every $i \in Z$ its ith coordinate is 1, and for every $i \notin Z$ its ith coordinate is 0. For any subset $S \subseteq \{0,1\}^n$, let $S_{f,0} = \{y \in S : f(y) = 0\}$ and $S_{f,1} = \{y \in S : f(y) = 1\}$.

The first observation on which our algorithm is based is that for every point $y \in f^{-1}(0)$, there must be at least one index $i \in Z(y)$ for which $f(\bar{e}^i) = 0$, or else we have evidence that f is not a monotone monomial. In fact, we don't need to verify that $f(\bar{e}^i) \neq 0$ for *every* $i \in Z(y)$ in order to obtain evidence that f is not a monotone monomial. Rather, if we search for such an index (in a manner described momentarily), and this search fails, then we already have evidence that f is not a monotone monomial.

The search procedure (which performs a binary search), receives as input a point $y \in f^{-1}(0)$ and searches for an index $j \in Z(y)$ such that $f(\bar{e}_j) = 0$. This is done by repeatedly partitioning a set of indices, Z, starting with $Z = Z(y)$, into two parts Z_1 and Z_2 of (almost) equal size, and continuing the search with a part Z_i, $i \in \{1,2\}$ for which $f(y(Z_i)) = 0$. (If both parts satisfy the condition, then we continue with Z_1.) Note that if both $f(y(Z_1)) = 1$ and $f(y(Z_2)) = 1$, then we have evidence that f is not a monotone monomial because $f(y(Z_1 \cup Z_2)) = 0$ (so that $\{y(Z_1 \cup Z_2), y(Z_1), y(Z_2)\}$ is an edge in H_f). The search also fails (from the start) if $Z(y) = \emptyset$ (that is, $y = 1^n$). For the precise pseudo-code of the procedure, see Fig. 1.

Algorithm 1: Binary Search (Input: $y \in \{0,1\}^n$)

  1. $Z \leftarrow Z(y)$.
  2. if $|Z| = 0$, then output fail *and halt.*
  3. While ($|Z| \geq 2$) do
      (a) Let (Z_1, Z_2) be a fixed partition of Z where $||Z_1| - |Z_2|| \leq 1$.
          Specifically, Z_1 *is the set of the first* $\lfloor |Z|/2 \rfloor$ *indices in* Z.
          − If $f(y(Z_1)) = 0$, then $Z \leftarrow Z_1$;
          − else if $f(y(Z_2)) = 0$, then $Z \leftarrow Z_2$;
          − else output fail *and halt.*
  4. *Output the single index that remains in* Z.

Fig. 1. The binary search procedure for monotone monomials

The testing algorithm starts by obtaining a sample of $\Theta(\sqrt{n}/\epsilon)$ points, where each point is generated independently according to D. (Since the points are generated independently, repetitions may occur.) For each point in the sample that belongs to $f^{-1}(0)$, the algorithm calls the binary search procedure. If any search fails, then the algorithm rejects f (recall that in such a case the algorithm has evidence that f is not a monotone monomial). Otherwise, the algorithm has a collection of indices J such that $f(\bar{e}_j) = 0$ for every $j \in J$. The algorithm then takes an additional sample, also of size $\Theta(\sqrt{n}/\epsilon)$, and checks whether there exists a point y in the sample such that $f(y) = 1$ and $Z(y)$ contains some $j \in J$. In such a case the algorithm has evidence that f is not a monotone monomial (specifically, $\{\bar{e}_j, y\}$ is an edge in H_f), and it rejects. For the precise pseudo-code of the algorithm, see Fig. 2.

We shall use Lemma 2 to show that if $\text{dist}_D(f, \mathcal{M}_m)$ is relatively large, then either the first sample will contain a point on which the binary search procedure fails (with high probability over the choice of the first sample), or the second sample will contain a point y such that $f(y) = 1$ and $Z(y) \cap J \neq \emptyset$ (with high probability over the choice of both samples).

Algorithm 2: Monotone Monomials Test

1. *Take a sample T of $\Theta(\sqrt{n}/\epsilon)$ points, generated independently according to D.*
2. *For each point $y \in T_{f,0}$ run the binary search procedure (Algorithm 1) on y.*
3. *If the binary search fails for any of the points, then output* reject *and halt. Otherwise, for each $y \in T_{f,0}$ let $j(y)$ be the index returned for y, and let $J(T_{f,0}) = \bigcup_{y \in T_{f,0}} j(y)$.*
4. *Take a sample T' of size $\Theta(\sqrt{n}/\epsilon)$ (generated independently according to D).*
5. *If there is a point $y \in T'_{f,1}$ such that $Z(y) \cap J(T_{f,0}) \neq \emptyset$, then output* reject, *otherwise output* accept.

Fig. 2. The distribution-free testing algorithm for monotone monomials

3.3 The Analysis of the Testing Algorithm for Monotone Monomials

The next definition will serve us in the analysis of the algorithm.

Definition 5 (Empty points and representative indices). *For a point $y \in f^{-1}(0)$, we say that y is* empty *(with respect to f) if the binary search procedure (Algorithm 1) fails on y. We denote the set of empty points (with respect to f) by $Y_\emptyset(f)$. If y is not empty, then we let $j(y) \in Z(y)$ denote the index that the binary search procedure returns. We refer to this index as the* representative *index for y. If $y \in Y_\emptyset(f)$, then $j(y)$ is defined to be 0.*

Note that since the binary search procedure is deterministic, the index $j(y)$ is uniquely defined for each $y \notin Y_\emptyset(f)$.

As in Algorithm 2, for a sample T and $T_{f,0} = T \cap f^{-1}(0)$, we let $J(T_{f,0}) = \{j(y) : y \in T_{f,0} \setminus Y_\emptyset(f)\}$ denote the set of representative indices for the sample. For any subset $J \subseteq [n]$, let $Y_{f,1}(J)$ denote the set of all points $y \in f^{-1}(1)$ for which $Z(y) \cap J \neq \emptyset$. In particular, if we set $J = J(T_{f,0})$, then each point $y \in Y_{f,1}(J)$, together with any index j in its intersection with J, provide evidence that f is not a monotone monomial (i.e., $\{\bar{e}_j, y\} \in E(H_f)\}$). We next state our main lemma.

Lemma 3. *Suppose that* $\text{dist}_D(f, \mathcal{M}_m) > \epsilon$ *and consider a sample* T *of* $c_1\sqrt{n}/\epsilon$ *points generated independently according to* D. *For a sufficiently large constant* c_1, *with probability at least* $5/6$ *over the choice of* T, *either* $T_{f,0}$ *contains an empty point (with respect to* f) *or* $D(Y_{f,1}(J(T_{f,0}))) \geq \frac{\epsilon}{4\sqrt{n}}$.

Lemma 3 is established by proving the contrapositive statement. Namely, that if the probability (over the choice of T) that $T_{f,0}$ does not contain an empty point (with respect to f) and $D(Y_{f,1}(J(T_{f,0}))) < \frac{\epsilon}{4\sqrt{n}}$ is at least $1/6$, then $\text{dist}_D(f, \mathcal{M}_m) \leq \epsilon$. This is done by applying a probabilistic argument to construct a vertex cover C of H_f such that $D(C) < \epsilon$ (assuming the counter-assumption holds).

Theorem 4. *Algorithm 2 is a distribution-free 1-sided-error testing algorithm for (membership in)* $\mathcal{M}_m$. *Its query complexity is* $O(\sqrt{n}\log n/\epsilon)$.

Proof. Consider first the case that f is a monotone monomial. Observe that the algorithm rejects only if it finds evidence that f is not a monotone monomial. This evidence is either in the form of two (disjoint) subsets of indices, Z_1 and Z_2 such that $f(y(Z_1)) = f(y(Z_2)) = 1$ while $f(y(Z_1 \cup Z_2))) = 0$ (found by the binary search procedure), or it is of the form of an index j and a point $y \in f^{-1}(1)$, such that $f(\bar{e}_j) = 0$ and $j \in Z(y)$. Therefore, the algorithm never rejects a monotone monomial.

Consider next the case that $\text{dist}_D(f, \mathcal{M}_m) > \epsilon$. By Lemma 3, for a sufficiently large constant c_1 in the $\Theta(\cdot)$ notation for T (the first sample), with probability at least $5/6$ over the choice of T, either there is an empty point in $T_{f,0} \subseteq T$, or $D(Y_{f,1}(J(T_{f,0}))) \geq \frac{\epsilon}{4\sqrt{n}}$. If there is an empty point in $T_{f,0}$, then the binary search will fail on that point and the algorithm will reject. On the other hand, if $D(Y_{f,1}(J(T_{f,0}))) \geq \frac{\epsilon}{4\sqrt{n}}$, then, since the size of the second sample, T', is $c_1'\sqrt{n}/\epsilon$, the probability that no point $y \in Y_{f,1}(J(T_{f,0}))$ is selected in T' is at most $(1 - \frac{\epsilon}{4\sqrt{n}})^{c_1'\sqrt{n}/\epsilon}$, which is upper bounded by $1/6$ for $c_1' \geq 8$. But if such a point is selected, then the algorithm rejects.[2] Therefore, the probability that the algorithm rejects a function f for which $\text{dist}_D(f, \mathcal{M}_m) > \epsilon$ is at least $2/3$.

Finally, the number of points sampled is $O(\sqrt{n}/\epsilon)$ since the algorithm obtains two samples of this size. Since for each point in the first sample that belongs to $f^{-1}(0)$ the algorithm performs a binary search, the query complexity of the algorithm is $O(\sqrt{n}\log n/\epsilon)$.

---

[2] We note that the analysis doesn't explicitly address the case that $T_{f,0} = \emptyset$, where the algorithm accepts (for every T') simply because $J(T_{f,0})$ is empty. What the analysis implies (implicitly) is that the probability that such an event occurs when $\text{dist}_D(f, \mathcal{M}_m) > \epsilon$ is at most a small constant. It is possible to argue this directly (since if $\text{dist}_D(f, \mathcal{M}_m) > \epsilon$, then in particular, f is ϵ-far with respect to D from the all-1 function, so that the probability that no point in $f^{-1}(0)$ is selected in the first sample is very small).

4   Distribution-Free Testing of (General) Monomials

The high-level structure of the algorithm for testing general monomials is similar to the algorithm for testing monotone monomials, but several modifications have to be made (and hence the algorithm and the notions it is based on are seemingly more complex). In this section we explain what the modifications are.

Recall that for a point $y \in \{0,1\}^n$ we let $Z(y) = \{i : y_i = 0\}$. Analogously, we let $O(y) = \{i : y_i = 1\}$. For a non-empty set of points $Y \subseteq \{0,1\}^n$, let $Z(Y) = \bigcap_{y \in Y} Z(y)$ and $O(Y) = \bigcap_{y \in Y} O(y)$. We shall use the convention that $Z(\emptyset) = O(\emptyset) = [n]$.

4.1   The Violation Hypergraph (for General Monomials)

A basic observation concerning (general) monomials is that if g is a monomial, then for every subset $\{y^0, y^1, \ldots, y^t\}$ such that $g(y^0) = 0$ and $g(y^j) = 1$ for each $1 \le j \le t$, the following holds: There must be at least one index $i \in [n]$ such that either $i \in Z(\{y^1, \ldots, y^t\})$ and $y_i^0 = 1$ or $i \in O(\{y^1, \ldots, y^t\})$ and $y_i^0 = 0$. This implies that if we have a subset $\{y^0, y^1, \ldots, y^t\}$ such that $g(y^0) = 0$ and $g(y^j) = 1$ for each $1 \le j \le t$, and such that $Z(\{y^1, \ldots, y^t\}) \subseteq Z(y^0)$ and $O(\{y^1, \ldots, y^t\}) \subseteq O(y^0)$, then we have evidence that g is not a monomial. This motivates the next (modified) definition of a violation hypergraph.

Definition 6 (Violation hypergraph (w.r.t. general monomials)). *Let $H_f = (V(H_f)), E(H_f))$ be the hypergraph whose vertex set, $V(H_f)$ is $\{0,1\}^n$, and whose edge set, $E(H_f)$, contains all subsets $\{y^0, y^1, \ldots, y^t\} \subseteq \{0,1\}^n$ of the following form:*

- *$f(y^0) = 0$ and $f(y^j) = 1$ for all $1 \le j \le t$.*
- *$Z(\{y^1, ..., y^t\}) \subseteq Z(y^0)$ and $O(\{y^1, ..., y^t\}) \subseteq O(y^0)$.*

Observe that the second item in Definition 4 (of the violation hypergraph for monotone monomials), which requires that $Z(y^0) \subseteq \bigcup_{j=1}^{t} Z(y^j)$, is equivalent to $O(\{y^1, ..., y^t\}) \subseteq O(y^0)$. Therefore, the difference between Definition 4 and Definition 6 is in the additional requirement that $Z(\{y^1, ..., y^t\}) \subseteq Z(y^0)$.

Similarly to Lemma 1 here we have the next lemma (whose proof is very similar to the proof of Lemma 1).

Lemma 5. *If $E(H_f) = \emptyset$ and $f^{-1}(1) \ne \emptyset$, then f is a monomial.*

Note that slightly differently from Lemma 1, in Lemma 5 we explicitly added the condition that $f^{-1}(1) \ne \emptyset$. The reason is that while in the case of monotone monomials, the fact that $E(H_f) = \emptyset$ implies that $f^{-1}(1) \ne \emptyset$ (because if $f(1^n) = 0$, then $E(H_f)$ is not empty since it contains 1^n), in the case of general monomials this implication does not hold.

The next lemma is analogous to Lemma 2.

Lemma 6. *If $\mathrm{dist}_D(f, \mathcal{M}) > \epsilon$ and $f^{-1}(1) \ne \emptyset$, then for every vertex cover C of H_f we have $D(C) > \epsilon$.*

4.2   The Algorithm for Testing General Monomials

For a vector $y \in \{0,1\}^n$, and for an index $i \in [n]$, let $y^{\neg i}$ be the same as y except that the ith coordinate in y is flipped. That is, $y_\ell^{\neg i} = y_\ell$ for all $\ell \neq i$ and $y_i^{\neg i} = \bar{y}_i$. For a subset $I \subseteq [n]$ let $y^{\neg I}$ be the vector y with each coordinate $i \in I$ flipped. That is, $y_\ell^{\neg I} = y_\ell$ for all $\ell \notin I$ and $y_\ell^{\neg i} = \bar{y}_\ell$ for all $\ell \in I$. Let $\Delta(y,w) \subseteq [n]$ be the subset of indices i such that $y_i \neq w_i$, and note that $y = w^{\neg \Delta}$ for $\Delta = \Delta(y,w)$.

Algorithm 3: Binary Search for General Monomials (Input: $y \in f^{-1}(1)$, $w \in f^{-1}(0)$)

1. $\Delta \leftarrow \Delta(y,w)$;
2. *While ($|\Delta| \geq 2$) do*
   (a) *Let (Δ_1, Δ_2) be a fixed partition of Δ where $||\Delta_1| - |\Delta_2|| \leq 1$. Specifically, Δ_1 is the set of the first $\lfloor |\Delta|/2 \rfloor$ indices in Δ.*
   (b) *If $f(w^{\neg \Delta_1}) = 0$, then $\Delta \leftarrow \Delta_1$;*
   (c) *else if $f(w^{\neg \Delta_2}) = 0$, then $\Delta \leftarrow \Delta_2$;*
   (d) *else output* fail *and halt.*
3. *Output the single index $j \in \Delta$;*

Fig. 3. The binary search procedure for general monomials

We start by describing the binary search procedure (for general monomials). Its pseudo-code is given in Algorithm 3 (see Fig. 3). The procedure receives as input two points $w, y \in \{0,1\}^n$ such that $f(w) = 1$ and $f(y) = 0$ and outputs an index $j \in [n]$ such that $y_j \neq w_j$ and such that $f(w^{\neg j}) = 0$. If f is a monomial, then at least one such index must exist. Note that if $w = 1^n$, then the output of the search is as specified by the binary search procedure for monotone monomials (Algorithm 1). In fact, Algorithm 1 itself (and not only its output specification) is essentially the same as Algorithm 3 for the special case of $w = 1^n$. (Since $f(1^n)$ must equal 1 if f is a monotone monomial, we can think of the binary search procedure for monotone monomials as implicitly working under this assumption.)

The search is performed by repeatedly partitioning a set of indices Δ, starting with $\Delta = \Delta(y,w)$, into two parts Δ_1 and Δ_2 of (almost) equal size, and querying f on the two points, $w^{\neg \Delta_1}$ and $w^{\neg \Delta_2}$. If f returns 1 for both, then the search fails. Otherwise, the search continues with Δ_i for which $f(w^{\neg \Delta_i}) = 0$, unless $|\Delta_i| = 1$, in which case the desired index is found. If the search fails, then we have evidence that f is not a monomial. Namely, we have three points, $w^{\neg \Delta_1}$, $w^{\neg \Delta_2}$ and $w^{\neg \Delta}$, where $\Delta = \Delta_1 \cup \Delta_2$, such that $f(w^{\neg \Delta}) = 0$ and $f(w^{\neg \Delta_1}) = f(w^{\neg \Delta_2}) = 1$. Since $w^{\neg \Delta_1}$ and $w^{\neg \Delta_2}$ disagree on all coordinates in Δ, and all three points agree on all coordinates in $[n] \setminus \Delta$, we have that $Z(\{w^{\neg \Delta_1}, w^{\neg \Delta_2}\}) \subseteq Z(w^{\neg \Delta})$ and $O(\{w^{\neg \Delta_1}, w^{\neg \Delta_2}\}) \subseteq O(w^{\neg \Delta})$, so that the three points constitute an edge in H_f.

The testing algorithm for general monomials starts by obtaining a sample of $\Theta(1/\epsilon)$ points, each generated independently according to D. The algorithm arbitrarily selects a point w in this sample that belongs to $f^{-1}(1)$. If no such point exists, then the algorithm simply accepts f (and halts). Otherwise, this

Algorithm 4 : General Monomials Test

1. *Take a sample S of $\Theta(1/\epsilon)$ points, generated independently according to D.*
2. *If $S_{f,1} = \emptyset$, then output* accept *and halt. Otherwise, arbitrarily select a point $w \in S_{f,1}$.*
3. *Take a sample T of $\Theta(\sqrt{n}/\epsilon)$ points (generated independently according to D).*
4. *For each point $y \in T_{f,0}$ run the binary search procedure (Algorithm 3) on w, y.*
5. *If the binary search fails for any of the points, then output* reject *and halt. Otherwise, for each $y \in T_{f,0}$ let $j^w(y)$ be the index returned for y, and let $J^w(T_{f,0}) = \{j^w(y) : y \in T_{f,0}\}$.*
6. *Take a sample T' of size $\Theta(\sqrt{n}/\epsilon)$ (generated independently according to D).*
7. *If there is a point $y \in T'_{f,1}$ and an index $j \in J^w(T_{f,0})$ such that $y_j \neq w_j$, then output* reject, *otherwise output* accept.

Fig. 4. The testing algorithm for general monomials

point serves as as a kind of *reference point*. As in the case of the binary search procedure, the testing algorithm for monotone monomials (Algorithm 2) is essentially the same as the testing algorithm for general monomials (Algorithm 4) with w (implicitly) set to be 1^n.

Next, the algorithm obtains a sample of $\Theta(\sqrt{n}/\epsilon)$ points (each generated independently according to D). For each point y in the sample that belongs to $f^{-1}(0)$, the algorithm performs a binary search on the pair w, y. If any search fails, then the algorithm rejects (recall that in such a case it has evidence that f is not a monomial). Otherwise, for each point y in the sample that belongs to $f^{-1}(0)$, the algorithm has an index, $j^w(y) \in \Delta(y, w)$, such that $f(w^{-j^w(y)}) = 0$. Let the subset of all these indices be denoted by J. Note that by the construction of J, if f is a monomial, then for every $j \in J$, if $w_j = 1$, then the variable x_j must belong to the conjunction defining f and if $w_j = 0$, then $\bar{x}_j$ must belong to the conjunction.

The algorithm then takes an additional sample, also of size $\Theta(\sqrt{n}/\epsilon)$, and checks whether there exists a point y in the sample that belongs to $f^{-1}(1)$ and an index $j \in J$ such that $y_j \neq w_j$. In such a case the algorithm has evidence that f is not a monomial. Viewing this in terms of the hypergraph H_f, we have that $f(w^{-j}) = 0$, $f(y) = f(w) = 1$, and both $Z(\{y, w\}) \subseteq Z(w^{-j})$ and $O(\{y, w\}) \subseteq O(w^{-j})$, so that $\{w^{-j}, y, w\} \in E(H_f)$. The pseudo-code of the algorithm is given in Fig. 4, and our main result in this section is stated next and its proof can be found in [4].

Theorem 7. *Algorithm 4 is a distribution-free 1-sided-error testing algorithm for (membership in) $\mathcal{M}$. Its query complexity is $O(\sqrt{n}\log n/\epsilon)$.*

Acknowledgements. We would like to thank an anonymous reviewer of STOC 2010 for suggesting to simplify the proof of Lemma 2, and to thank the anonymous reviewers of RANDOM 2010 for helpful comments.

References

1. Blum, M., Luby, M., Rubinfeld, R.: Self-testing/correcting with applications to numerical problems. JACM 47, 549–595 (1993)
2. Diakonikolas, I., Lee, H.K., Matulef, K., Onak, K., Rubinfeld, R., Servedio, R.A., Wan, A.: Testing for concise representations. In: Proceedings of the 48th FOCS, pp. 549–557 (2007)
3. Dodis, Y., Goldreich, O., Lehman, E., Raskhodnikova, S., Ron, D., Samorodnitsky, A.: Improved testing algorithms for monotonocity. In: Hochbaum, D.S., Jansen, K., Rolim, J.D.P., Sinclair, A. (eds.) RANDOM 1999 and APPROX 1999. LNCS, vol. 1671, pp. 97–108. Springer, Heidelberg (1999)
4. Dolev, E., Ron, D.: Distribution-free testing algorithms for monomials with a sub-linear number of queries (2010), http://www.eng.tau.ac.il/~danar
5. Ergun, F., Kannan, S., Kumar, S.R., Rubinfeld, R., Viswanathan, M.: Spot-checkers. JCSS 60(3), 717–751 (2000)
6. Fischer, E., Kindler, G., Ron, D., Safra, S., Samorodnitsky, S.: Testing juntas. JCSS 68(4), 753–787 (2004)
7. Glasner, D., Servedio, R.A.: Distribution-free testing lower bounds for basic Boolean functions. In: Charikar, M., Jansen, K., Reingold, O., Rolim, J.D.P. (eds.) RANDOM 2007 and APPROX 2007. LNCS, vol. 4627, pp. 494–508. Springer, Heidelberg (2007)
8. Goldreich, O., Goldwasser, S., Lehman, E., Ron, D., Samordinsky, A.: Testing monotonicity. Combinatorica 20(3), 301–337 (2000)
9. Goldreich, O., Goldwasser, S., Ron, D.: Property testing and its connection to learning and approximation. JACM 45(4), 653–750 (1998)
10. Halevy, S., Kushilevitz, E.: Distribution-free property testing. In: Arora, S., Jansen, K., Rolim, J.D.P., Sahai, A. (eds.) RANDOM 2003 and APPROX 2003. LNCS, vol. 2764, pp. 341–353. Springer, Heidelberg (2003)
11. Halevy, S., Kushilevitz, E.: Distribution-free connectivity testing. In: Jansen, K., Khanna, S., Rolim, J.D.P., Ron, D. (eds.) RANDOM 2004 and APPROX 2004. LNCS, vol. 3122, pp. 393–404. Springer, Heidelberg (2004)
12. Halevy, S., Kushilevitz, E.: A lower bound for distribution-free monotonicity testing. In: Chekuri, C., Jansen, K., Rolim, J.D.P., Trevisan, L. (eds.) APPROX 2005 and RANDOM 2005. LNCS, vol. 3624, pp. 330–341. Springer, Heidelberg (2005)
13. Halevy, S., Kushilevitz, E.: Distribution-free property testing. SICOMP 37(4), 1107–1138 (2007)
14. Kopparty, S., Saraf, S.: Tolerant linearity testing and locally testable codes. In: Dinur, I., Jansen, K., Naor, J., Rolim, J. (eds.) RANDOM 2009. LNCS, vol. 5687, pp. 601–614. Springer, Heidelberg (2009)
15. Matulef, K., O'Donnell, R., Rubinfed, R., Servedio, R.A.: Testing halfspaces. In: Proceedings of the 20th SODA, pp. 256–264 (2009)
16. Parnas, M., Ron, D., Rubinfeld, R.: Tolerant property testing and distance approximation. JCSS 72(6), 1012–1042 (2006)
17. Parnas, M., Ron, D., Samorodnitsky, A.: Testing basic boolean formulae. SIDMA 16(1), 20–46 (2002)
18. Rubinfeld, R., Sudan, M.: Robust characterization of polynomials with applications to program testing. SICOMP 25(2), 252–271 (1996)
19. Turán, G.: Lower bounds for PAC learning with queries. In: Proceedings of the 6th COLT, pp. 384–391 (1993)
20. Valiant, L.G.: A theory of the learnable. CACM 27(11), 1134–1142 (1984)

Periodicity in Streams

Funda Ergun, Hossein Jowhari, and Mert Sağlam

Simon Fraser University
{funda,hjowhari,msa99}@cs.sfu.ca

Abstract. In this work we study sublinear space algorithms for detecting periodicity over data streams. A sequence of length n is said to be periodic if it consists of repetitions of a block of length p for some $p \leq \frac{n}{2}$. In the first part of this paper, we give a 1-pass randomized streaming algorithm that uses $O(\log^2 n)$ space and reports the shortest period if the given stream is periodic. At the heart of this result is a 1-pass $O(\log n \log m)$ space streaming pattern matching algorithm. This algorithm uses similar ideas to Porat and Porat's algorithm in FOCS 2009 but it does not need an offline pre-processing stage and is simpler.

In the second part, we study distance to p-periodicity under the Hamming metric, where we estimate the minimum number of character substitutions needed to make a given sequence p-periodic. In streaming terminology, this problem can be described as computing the cascaded aggregate $L_1 \circ F_1^{res(1)}$ over a matrix $A_{p \times \lfloor \frac{n}{p} \rfloor}$ given in column ordering. For this problem, we present a randomized streaming algorithm with approximation factor $2 + \epsilon$ that takes $\tilde{O}(\frac{1}{\epsilon^2})$ space. We also show a $1 + \epsilon$ randomized streaming algorithm which uses $\tilde{O}(\frac{1}{\epsilon^{5.5}} p^{1/2})$ space.

1 Introduction

A sequence, informally speaking, is said to be *periodic* if it consists of repetitions of the same block of characters. In this work we study detecting periodicity over a sequence given as a stream. We present 1-pass randomized algorithms for discovering periodic properties of a given stream that use sublinear (in most cases polylogarithmic) space and per-character running time.

The study of periodic sequences and patterns has been important in many fields such as algorithms, data mining, and computational biology. Applications involving weather patterns, stock market data mining, intrusion detection, etc. (eg see [11]) aim to identify self-similar trends in large data in almost real time. The search for efficient algorithms for periodicity has also generated fundamental algorithmic tools for solving problems on sequences/strings [19,8,2].

Formally, a sequence s of length n is said to be p-periodic if $s[i] = s[i+p]$ for all $i = 1, \ldots, |s| - p$. The *smallest* $p > 0$ for which s is p-periodic is referred to as *the period* of s. By convention, if the length of the period of s is at most $n/2$, then s is said to be *periodic*, otherwise it is *aperiodic*.

Given the intimate relationship between periodicity and pattern matching, we first investigate sublinear space solutions for finding patterns. Recently Porat and

M. Serna et al. (Eds.): APPROX and RANDOM 2010, LNCS 6302, pp. 545–559, 2010.

Porat, in a breakthrough result, presented a polylogarithmic space randomized algorithm for pattern matching that does not require the storage of the entire pattern [25]. Briefly, given a pattern u of length m, in an off-line step, they pre-process u and build $O(\log n)$-size sketches of $\log m$ prefixes of u and use them to find occurrences of the pattern in the stream. In order to find the period, we first develop a simple and more streaming-friendly algorithm for pattern match-ing which does not require any offline preprocessing. While our solution utilizes ideas similar in essence to those used by [25], we achieve this by computing only the Rabin-Karp fingerprints of the prefixes $u_1, \ldots, u_{\log d}$ where $u_i = u[1, 2^i]$, and we manage to get the same $O(\log n \log m)$ bit space bound. In a high level, our algorithm consists of $\log m$ layers. In the ith layer, we recursively match u_i using the information from the previous layers.

We then use our pattern matching algorithm to compute the period of s in a single pass over s using $O(\log^2 n)$ space when s is periodic. (otherwise we re-port that s is aperiodic). The limitation in computing the period for aperiodic sequences turns out to be necessary as we later show a lower bound that comput-ing the period in 1-pass for these sequences requires linear space. On the other hand we show that an additional pass will give us a $O(\log^2 n)$ space solution for periods of any length.

In addition to periodicity, our pattern matching algorithm enables us to get sublinear solutions for frequency moments defined over substrings.

In real-world applications, periodic trends might be hidden or infected with noise; thus, where exact periodicity is hard to come by, one is likely to encounter instances where a stream is close to periodic. As result, measures that capture approximate periodicity are a natural course to investigate. In this direction we study distance to periodicity under Hamming distance: we define the distance of s to p-periodicity as the minimum number of character substitutions required to make s p-periodic.

$$D_p(s) = \min_{x \text{ is } p\text{-periodic}} \{\mathcal{H}(s, x)\}.$$

It turns out that $D_p(s)$ can be expressed as a product-sum of a certain func-tion defined over rows of a matrix $A_{p \times d}$ where $n/p = d$. The problem then is to compute $L_1 \circ F_1^{res(1)}(A) = \sum_{i=1}^{p} F_1^{res(1)}(A_i)$ where A_i is the ith row of A and $F_1^{res(1)}(s)$, known as the *residual tail* of sequence s, equals $|s| - F_\infty(s)$. In general $F_k^{res(r)}(s) = \sum_{i>r}^{m} f_i^k$, where $f_1, \ldots, f_m$ are the character frequencies in decreasing order. Note that when $r = 0$ this is the same as F_k, the kth frequency moment of s. While there are space efficient algorithms for approximating $F_1^{res(1)}$ and $F_2^{res(r)}$ [7,13,4], aggregate computation of $F_1^{res(1)}$ over multiple streams has a different nature and is a new challenge. In fact, this problem can be viewed as a generalization of the Hamming distance to multiple vectors (when $d = 2$, we get the classical Hamming distance), and thus might be of independent in-terest. For this problem, we present two 1-pass randomized algorithms. The first algorithm approximates $L_1 \circ F_1^{res(1)}$ within $2 + \epsilon$ factor and uses $O(\frac{1}{\epsilon^2} \log \frac{1}{\epsilon})$ words of space. This algorithm uses a straightforward reduction to computing L_0

difference of two vectors that are generated on the fly. Using this algorithm and a combination of sampling and exact sparse recovery, we get a $1 + \epsilon$ approximation solution that uses $O(\frac{1}{\epsilon^{5.5}}(p \log p)^{1/2} \log n)$ words of space. For constant alphabet size, the space bound is $O(\frac{1}{\epsilon^3}(p \log p)^{1/2})$.

Related Work. The streaming model is well studied; see [24] for a recent survey. Aside from the implicit implications of [25], to our knowledge, our paper is the first to investigate the space complexity of computing the period in the streaming model. In a related direction, Ergun et al. [12] gave an $O(\sqrt{n})$ tester for distinguishing periodic strings from highly aperiodic ones under the Hamming distance in the property testing model. Subsequently Lachish and Newman [20] showed a lower bound of $\Omega(\sqrt{n})$ for testing periodicity in the query model. With a focus on time complexity, Czumaj and Gasieniec [10] presented an average case analysis for computing the exact period. Numerous studies have been done in the data mining community for detecting periodicity in *time-series databases* and online data (e.g. see [11]), typically with quite different space considerations than in our model. Streaming complexity of cascaded norms $L_k \circ L_p$ over matrices is investigated in depth by Jayram and Woodruff in [16]; also see [9,23].

2   Preliminaries

Throughout this paper $[n]$ denotes the set of integers $\{1, \ldots, n\}$. We assume the input stream is a sequence of length n over the alphabet $\Sigma = \{0, 1, \ldots, L\}$. We represent the length of a string s with $|s|$, the ith element of s with $s[i]$, and the substring of s between locations i and j (inclusive) with $s[i, j]$. A d-substring is a substring of length d. The concatenation of two sequences (or vectors) u, v is written as $u \circ v$ and u^i represents the concatenation of i instances of u.

The smallest $p > 0$ for which s is p-periodic, i.e., $s[i] = s[i + p]$ for all $i = 1, \ldots, |s| - p$, is called *the* period of s and is denoted per(s). We use $M_s(t)$ to denote the set of all positions in s where an exact occurrence of string t starts; i.e., $M_s(t) = \{i \mid s[i, i + |t| - 1] = t\}$. The following lemma shows the relation between per(t) and $M_s(t)$. (This and other missing proofs are left to the full version of the paper.)

Lemma 1. *Let $i \in M_s(t)$ and let $U = M_s(t) \cap [i, i + |t| - 1]$. The following are true.*

i. *Let $j \in U$ where $j > i$ and there is no $k \in U$ such that $i < k < j$. If $|i - j| \le |t|/2$ then $|i - j| = $ per(t).*
ii. *There is at most one $j \in U$ such that $|i - j|$ is not a multiple of per(t). Moreover if $|i - j|$ is not a multiple of per(t), then $j = \max(U)$.*

Fingerprints. In Section 3 we use Rabin-Karp fingerprints [18], a standard sketching tool which allows us to compare strings of arbitrary length in constant time. Fix an integer alphabet Σ. Let $q > |\Sigma|$ be a prime and $r \in \mathbb{Z}_q^*$ be arbitrary. The Rabin-Karp fingerprint of a string $s \in \Sigma^*$ is defined as

$\Phi_{q,r}(s) = \sum_{i=1}^{|s|} s[i] \cdot r^{i-1} \pmod{q}$. The following facts are well-known and the reader is referred to [18,25] for the proofs.

(P1) $\Phi_{q,r}(s)$ can be computed in one pass over s using $O(\log q)$ bits of space.

(P2) Let $s \neq t$ be two strings and $l = \max(|s|, |t|)$. $\Pr_r[\Phi_{q,r}(s) = \Phi_{q,r}(t)] \leq \frac{l}{q-1}$.

(P3) Given $\Phi_{q,r}(s)$ and $\Phi_{q,r}(t)$, we can obtain $\Phi_{q,r}(s \circ t)$ by constant arithmetic operations in $\mathbb{Z}_q$.

(P4) Given $\Phi_{q,r}(s \circ t)$ and $\Phi_{q,r}(s)$, we can obtain $\Phi_{q,r}(t)$ by constant arithmetic operations in $\mathbb{Z}_q$.

Henceforth we set $q = \Theta(n^4)$ and assume that r is chosen uniformly at random from $\mathbb{Z}_q^*$ at the beginning of the respective algorithm. We also omit the subscripts and denote the fingerprint of s by $\Phi(s)$.

3   Periodicity and Pattern Matching

In this section first we show a streaming algorithm for pattern matching and then we present our results for periodicity and frequency moments over substrings.

3.1   The Pattern Matching Algorithm

We assume the input stream $S = u \circ s$ is the concatenation of the pattern u of length m and the text s of length n. Here we present a 1-pass streaming algorithm that *generates* the starting positions of the matches of u in s (equivalently, $M_s(u)$), on the fly using logarithmic space and per-item time. Strictly speaking, if $s[i - m + 1, i] = u$, after receiving $s[i]$ our algorithm reports a match with high probability. Also, the probability that our algorithm reports a match where there is no occurrence of u is bounded by n^{-1}.

While it is easy to generate $M_s(u)$ when u is small, the problem is non-trivial for large u. The following lemma implies that given a streaming algorithm that finds length-m patterns, by taking advantage of the Rabin-Karp fingerprints, we can obtain a streaming algorithm for length-cm patterns using only $O(c \log n)$ extra space.

Lemma 2. *Let k be an integer greater than m. Let $\mathcal{A}$ be a 1-pass algorithm that generates $M_s(u)$ using $O(g)$ bits space. Given $\mathcal{A}$ and $\Phi(u)$, there is a 1-pass algorithm that outputs $\Phi(s[i, i+k])$ at position $i+k$ for all $i \in M_s(u)$ using space $O(g + \frac{k}{m} \log n)$ bits.*

Proof. The algorithm partitions the sequence of positions in $M_s(u)$ (as generated by $\mathcal{A}$) into maximal contiguous subsequences where in each subsequence the distance between consecutive positions is at most $\frac{m}{2}$. To do this we only need to keep track of the last position in $M_s(u)$. If the next position is more than $\frac{m}{2}$ characters apart then we start a new maximal subsequence, otherwise the new position is appended to the last subsequence.

Now let $a_1, a_2, \ldots, a_h \in M_s(u)$ be a maximal sequence of consecutive positions in $M_s(u)$ where $|a_{l+1} - a_l| \leq \frac{1}{2}m$ for all $l \in [h-1]$. We claim that for this sequence we need to maintain at most four fingerprints to generate $\Phi(s[a_l, a_l+k])$ for all $l \in [h]$. To do this, first we launch an individual process to generate $\Phi(s[a_1, a_1 + k])$ and $\Phi(s[a_2, a_2 + k])$. By Property (P3) from Section 2, this can be done by adding $\Phi(s[a_1, a_1 + m - 1])$ and $\Phi(s[a_1 + m, a_1 + k])$. Now if $h < 3$, our claim is proved. So suppose $h \geq 3$.

First we note that by Lemma 1, we should have $|a_{l+1} - a_l| = \mathrm{per}(u)$ for all $l \in [h-1]$. As a result, when we reach the position $a_2 + m - 1$, we have obtained the value of $\mathrm{per}(u)$. Now let $x = u[1, \mathrm{per}(u)]$. We show that it is possible to compute $\Phi(x)$ when we reach $a_3 + m - 1$. To this end, when we are in $a_1 + m - 1$, starting from the next character we build a fingerprint until we reach $a_2 + m - 1$. This gives us $\Phi(s[a_1 + m, a_2 + m - 1])$. Note that if $\mathrm{per}(u)$ divides m, then $s[a_1 + m, a_2 + m - 1] = x$ and we are done. Otherwise $s[a_1 + m, a_2 + m - 1]$ is x shifted r times to the left (cyclic shift), where $r = m \pmod{\mathrm{per}(u)}$. Therefore

$$s[a_1 + m, a_2 + m - 1] = x[r + 1, \mathrm{per}(u)] \circ x[1, r].$$

Likewise, we have $s[a_2 + m, a_3 + m - 1] = x[r + 1, \mathrm{per}(u)] \circ x[1, r]$. Therefore, at location $a_2 + m$, we know the value of r and $\mathrm{per}(u)$, and consequently using this information, we can build the fingerprints $\Phi(x[r+1, \mathrm{per}(u)])$ and $\Phi(x[1, r])$ when we go over $s[a_2 + m, a_3 + m - 1]$. Note that here we have used the properties (P3) and (P4) from Section 2. It follows that we are able to construct $\Phi(x)$ when we get to $a_3 + m - 1$.

Now observe that $s[a_l, a_l + k]$ is equivalent to the substring $s[a_{l-1}, a_{l-1} + k]$ after removing a block of length $\mathrm{per}(u)$ from the left-end of it and adding $s[a_{l-1} + k, a_l - 1]$ to the right-end. Therefore we can generate $\Phi(s[a_l, a_l + k])$ by having $\Phi(s[a_{l-1}, a_{l-1} + k])$, $\Phi(s[a_{l-1} + k, a_l - 1])$, and $\Phi(x)$. This proves our claim.

It should be clear that at each point in time, we run at most $\frac{4k}{m}$ parallel fingerprint computations. Each fingerprint takes $O(\log n)$ space. This finishes the proof of the lemma. $\qquad\square$

Our pattern matching algorithm is the result of a recursive application of Lemma 2. First as we go over u, we build $\Phi(u[1, 2^i])$ for all $i \in [\log m]$. By Property (P1) this can be done in 1-pass and using $O(\log m \log n)$ bits of space. Let $\mathcal{A}_i$ be an algorithm that generates $M_s(u[1, 2^i])$ in space g_i. When $i < c$ where c is a small constant, we can use the naive solution of storing the entire pattern which gives $g_i = O(\log n)$. By Lemma 2, we get an algorithm $\mathcal{A}_{i+1}$ for $M_s(u[1, 2^{i+1}])$ in space $O(g_i + \log n)$ by fingerprint comparisons. Applying this $O(\log |u|)$ times we obtain an algorithm for $M_s(u)$ using space $O(\log |u| \log n)$ bits. The success probability is at least $1 - \log m / n^2$ and this is due to the Property (P2) in Section 2 and the observation that we make at most $O(n \log |u|)$ fingerprint comparisons.

Theorem 1. *There is a 1-pass streaming algorithm that generates $M_s(u)$ in $O(\log |u| \log n)$ bits of space and $O(\log |u|)$ per-item processing time. The error probability is bounded by n^{-1}.*

Remark. Since our pattern matching algorithm only requires the fingerprints of a small set of prefixes of the pattern, it can be used to generate $M_s(s[1, m])$ (where the pattern itself is a prefix of the text) in one pass and in space $O(\log m \log n)$ bits. This property of our algorithm will be essential in Section 3.3. Furthermore, in addition to $M_s(u)$, our algorithm generates $M_s(u[1, 2^i])$ for each $i = 1, \ldots, \log m$, which leads to further space economy in our algorithms in the next section.

3.2   Finding the Period

Testing whether the steuqence s is periodic or not is equivalent to testing if there is a suffix of s of length at least $\frac{n}{2}$ that matches a prefix of s. Hence for finding the period of s, we just need to check the positions that match a certain prefix of s. Basically our algorithms for testing periodicity has two stages. In the first stage (search), it finds the positions where they match the first half of s. This is done by using the pattern matching algorithm we described above. Then, in the second stage (verfication), we check if the detected position can be the start of a suffix that matches a prefix of s. However these stages are performed in parallel as the search and verification of different positions might overlap. In the following, to demonstrate the idea, first we present a weaker bound and then we handle the general case.

Let $T = M_s(s[1, n/2])$.[1] By definition, s is periodic if there exists $i \in T$ where $s[i + 1, n] = s[1, n - i]$. Now if $i \leq n/4$, we can build both $\Phi(s[i + 1, n])$ and $\Phi(s[1, n - i])$ in one pass over s and thus we can test whether $\mathrm{per}(s) \leq n/4$ or not as follows.

Run the pattern matching algorithm to find $i = \min (T \cap [1, n/4])$. Build $\Phi(s[i + 1, n])$ and $\Phi(s[1, n - i])$. If $\Phi(s[i + 1, n]) = \Phi(s[1, n - i])$ then $\mathrm{per}(s) = i$ otherwise output that $\mathrm{per}(s) > n/4$.

The reason that we only perform the test for $\min (T \cap [1, n/4])$ is a consequence of Lemma 1. We do not need to check whether $s[i + 1, n] = s[1, n - i]$ for $i = c \min (T)$ when c is an integer greater than 1 as, in this case, $s[1, i]$ would be of the form $u \circ \ldots \circ u$ (a cyclic string) and thus can not be the period of s. From these observations we get the following lemma.

Lemma 3. *There is a 1-pass streaming algorithm that decides whether* $\mathrm{per}(s) \leq n/4$ *or not in space* $O(\log^2 n)$ *bits. The algorithm also outputs the exact period if* $\mathrm{per}(s) \leq n/4$.

For $i > n/4$, checking whether $s[i+1, n] = s[1, n-i]$ is not straightforward. This is because when we find out $i \in T$, we have already crossed the point $n-i$ and lost the opportunity to build $\Phi(s[1, n - i])$. To solve this problem we conservatively maintain a superset of T and prune it as we learn more about the input stream. First observe that, for $i \in T$, since $s[1, n - i] = s[1, n/2] \circ s[n/2 + 1, n - i]$, it is enough to build $\Phi(s[n/2+1, n-i])$. Now for $i \in [1, n/2]$, let $s_i = s[n/2+1, n-i]$. Roughly speaking, at each point in time, we maintain a dynamic set of positions

---

[1] To make the presentation simpler, we assume n is a power of 2.

R that will contain T and for each $i \in R$ we collect enough information to be able to construct $\Phi(s_i)$. Also in parallel we run a pattern matching process to generate T. Finally for each position in $\{i \in R \cap T \mid i \neq c \min (T) \text{ for } c \in \mathbb{N}\}$ we check whether $\Phi(s[i+1, n]) = \Phi(s[1, n-i])$. If $\Phi(s[i+1, n]) = \Phi(s[1, n-i])$ holds in one case, then we declare s to be periodic, otherwise it is reported aperiodic.

The dynamic set R. Let $I_k = [n/2 - 2^k + 1, n/2 - 2^{k-1}]$ and let $H = H_1 \cup H_2 \cup \ldots \cup H_{\log(n/4)}$ where $H_k = M_s(s[1, 2^k]) \cap I_k$. In other words, H_k is the positions of all occurrences of $s[1, 2^k]$ that start within the interval I_k. Clearly $T \subseteq H$. In what follows, for a fixed k we show how to compute $R_k \subseteq H_k$ and, more importantly, how to maintain $\Phi(s_i)$ for each $i \in R_k$. Also we guarantee that every member of T will be added to $R = R_1 \cup \ldots \cup R_{\log(n/4)}$ at some point. Initially all R_k are empty. First we distinguish two main cases. In both cases, we use the pattern matching algorithm described in Section 3 to get the sequence of positions in H. Also, when we detect $i \in H_k$, we add it to R_k. However, we might prune R_k and remove some unnecessary elements. In the following let $p = \mathrm{per}(s[1, 2^k])$.

The case $p > \frac{1}{4}2^k$. By Lemma 1, we get $|H_k| < 4$. Moreover, we detect $i \in H_k$ before reaching the end of s_i, and thus, we can build $\Phi(s_i)$ at the right time. In this case we let $R_k = H_k$. Clearly we can maintain R and the associated fingerprints in $O(\log n)$ space.

The case $p \leq \frac{1}{4}2^k$. Here things get a bit complicated. In this case H_k could be large and if we maintain $\Phi(s_i)$ for each $i \in H_k$ individually, this might take linear space. To solve this problem, first we note that, by Lemma 1, the positions in H_k have a succinct representation as the distance between consecutive positions is exactly p. As result, we can encode R_k using $O(\log n)$ space. Further, we take advantage of the periodic structure of $s[1, 2^k]$ and possibly the substring $s[2^k + 1, 2^{k+1}]$. Consider that for $i \in H_k$, s_i is a substring of $s[i, i + 2^{k+1} - 1]$. Now (informally) if the substrings $\{s_i\}$ fall in a periodic region, we can maintain all $\Phi(s_i)$ by saving a constant number of fingerprints. On the other hand, if the substring $s[i, i + 2^{k+1} - 1]$ is not periodic then we use the period information of $s[1, 2^{k+1}]$ to prune R_k. To do this, we collect the following information when we process the first half of the stream.

- Using the tester from Lemma 3, we compute p. If it is reported that $p > \frac{1}{4}2^k$, then I_k falls into the previous case. We also compute $\Phi(s[1, p])$ and $\Phi(s[2^k - p + 1, 2^k])$.
- Let $u_1 \circ u_2 \circ \ldots \circ u_t \circ u'$ be a decomposition of $s[2^k + 1, 2^{k+1}]$ into consecutive blocks of length p except possibly for the last block. Let x to be the maximum j such that $s[1, 2^k] \circ u_1 \circ \ldots \circ u_j$ is p-periodic. We compute x.

Now let $b_1, b_2, \ldots, b_r$ be the elements of H_k in increasing order. Since $|I_k| \leq \frac{1}{2}2^k$, we have $|b_{i+1} - b_i| = p$ for all $i \in [r-1]$. Let $v_1 \circ v_2 \circ \ldots \circ v_l \circ v'$ be a decomposition

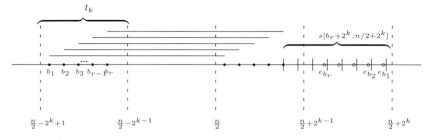

Fig. 1. A sample run of the algorithm in Section 3.2

of the substring $s[b_r + 2^k, n/2 + 2^k]$ into consecutive blocks of length p except possibly the last block (see Figure 1 for a pictorial presentation of the substrings). Now let y be the maximum j such that $s[b_r, b_r + 2^k - 1] \circ v_1 \circ \ldots \circ v_j$ is p-periodic. We consider two cases. If $y = l$ then $\{s_i \mid i \in H_k\}$ are substrings of a periodic interval. Let e_{b_1} be the right endpoint of s_{b_1}, i.e. $e_{b_1} = n - b_1$. Note that we have $e_{b_1} > e_{b_2} > \ldots > e_{b_r}$. In this case, all the following substrings (except possibly the last one) are equal: $s[e_{b_r} + 1, e_{b_{r-1}}], s[e_{b_{r-1}} + 1, e_{b_{r-2}}], \ldots, s[e_{b_2} + 1, e_{b_1}]$. Therefore to compute $\Phi(s_{b_j})$, we just need to maintain $\Phi(s_{b_1})$ and $\Phi(s[e_{b_2} + 1, e_{b_1}])$. We compute $\Phi(s_{b_1})$ individually. So in this case $R_k = H_k$. In the other case, we have $y < l$. We make the following claim.

Claim. If $y < l$ and $|r - j| + y \neq x$ then $b_j \notin T$.

By Claim 3.2, $|H_k \cap T| \leq 1$. Consequently it is enough to maintain $\Phi(s_{b_j})$ where $|j - r| + y = x$ and $\Phi(s_{b_1})$. So in this case $|R_k| \leq 2$.

It remains to state how to compute x and y. To compute x, we need to know p and $\Phi(s[2^k - p + 1, 2^k])$. This information can be obtained in one pass (see the observations before Lemma 3). Computation of y is similar to x. Finally, given the above discussion, for each $k \in \{1, 2, \ldots, \log(n/4)\}$, we need to keep $O(1)$ number of fingerprints to maintain R_k and its associated fingerprints which makes the total space $O(\log^2 n)$ bits. Hence, we get the following result.

Theorem 2. *There is a 1-pass randomized streaming algorithm that given $s \in \Sigma^n$ outputs per(s) if s is periodic, otherwise it reports that s is aperiodic. The algorithm uses $O(\log^2 n)$ bits of space and has $O(\log n)$ per-item running time. The error probability is at most $O(n^{-1})$.*

In general finding the period in one pass requires linear space. With one additional pass, however, the period of an arbitrary string can be found in $O(\log^2 n)$ space. The proof of the following theorem is left to the full version.

Theorem 3. *Every 1-pass exact algorithm for per(s) requires $\Omega(n)$ space.*

3.3   Frequency Moments over Substrings

Let s be a string of length n, and $k \geq 0$, $d \leq n$ be integers. We define the kth frequency moment of d-substrings of s as

$$F_{k,d}(s) = \sum_{u \in \Sigma^d} |M_s(u)|^k.$$

To approximate $F_{k,d}$, one can create a fingerprint for each d-substring and feed this stream of fingerprints to a standard F_k algorithm. Thus, using the algorithms of [15,5,14,17] one can $(1 + \epsilon)$-approximate $F_{k,d}$ with $\tilde{O}(d + n^{1-2/k})$ space and $\tilde{O}(1)$ per item processing time for any $k \geq 0$. It is not possible to obtain a $o(d)$ algorithm however, if we insist on constructing a fingerprint for each d-substring[2]. We note that by replacing the reservoir sampling procedure of [1] with the pattern matching algorithm above, one can $(1 + \epsilon)$-approximate $F_{k,d}$ using space $\tilde{O}(\frac{1}{\epsilon^2} n^{1-1/k})$, in particular independent of d.

Unfortunately, the estimator of [1] does not give a bound for $F_{0,d}$ which is perhaps the most commonly used moment for substrings, also known as the q-gram measure. Here we present an $\tilde{O}(\frac{1}{\epsilon}\sqrt{n})$ space randomized algorithm that $(1 + \epsilon)$-approximates $F_{0,d}$.

Theorem 4. *There exists a 1-pass streaming algorithm that $(1+\epsilon)$-approximates $F_{0,d}$ using $\tilde{O}(\frac{1}{\epsilon}\sqrt{n})$ space.*

Proof. Let $s \in \Sigma^n$ be the stream. Let K be the set of all d-substrings of s and $n' = n - d + 1$. Our basic estimator X is defined as follows. Let i be random position between 1 and n'. We set $X = 0$ if there exists a $j > i$ such that $s[i, i+d-1] = s[j, j+d-1]$, we set $X = n'$ otherwise. We have $\mathbb{E}[X] = \frac{1}{n'} \sum_{w \in K} n' = F_{0,d}$. Also, $\text{Var}(X) \leq \mathbb{E}[X^2] = \frac{1}{n'} \sum_{w \in K} n'^2 \leq n \cdot F_{0,d}$. Let Y be the average of $\frac{3}{\epsilon}\sqrt{n}$ repetitions of X. By Chebyshev's inequality,

$$\Pr[|Y - F_{0,d}| \geq \epsilon F_{0,d}] \leq \frac{\text{Var}(Y)}{\epsilon^2 F_{0,d}^2} \leq \frac{\sqrt{n}}{3\epsilon F_{0,d}}.$$

Right hand side is smaller than $1/3$ when $F_{0,d} \geq \frac{1}{\epsilon}\sqrt{n}$. Note that we can compute each X in $O(\log n \log d)$ space in one pass using the pattern matching algorithm of Section 3.1. It can be shown that one can compute $F_{0,d}$ exactly using space $\tilde{O}(F_{0,d})$. Therefore we compute $\frac{3}{\epsilon}\sqrt{n}$ estimates for X, while we run the exact algorithm in parallel. If at any point in the stream the exact algorithm detects that $F_{0,d} \geq \frac{1}{\epsilon}\sqrt{n}$ we terminate it and output the sampling estimate, otherwise we output the value computed by the exact algorithm. $\square$

4   Approximating the Distance to Periodicity

Recall that $D_p(s)$ is the minimum number of character changes on $s \in \Sigma^n$ to make it p-periodic. Assume WLOG that p divides n where $n = dp$, and

---

[2] An easy information theoretic observation shows that sliding a fingerprint for d-substrings that preserves equality with high probability requires $\Omega(d)$ space.

view s as a $p \times d$ matrix A where $A(i, j) = s[(i - 1)p + j]$. If p does not divide n, s can be represented by two matrices. Then, $D_p(s)$ is the the minimum number of substitutions in A to make every row consist of d repetitions of the same character. Also, $D_p(s) = L_1 \circ F_1^{res(1)}(A) = \sum_{i=1}^{p} F_1^{res(1)}(A_i)$. It is challenging to compute this quantity since we receive A in the column order: $A(1, 1), \ldots, A(p, 1), A(1, 2), \ldots, A(p, 2), \ldots$ To compute $L_1 \circ F_1^{res(1)}(A)$ exactly, one can compute the residual tail of each row in parallel using independent counters, in $O(|\Sigma|p)$ words of space. On the other hand, one can estimate $F_i^{res(1)}(A_i)$ within $1 - \epsilon$ factor in $O(1/\epsilon)$ words of space in several ways. For instance, using the Heavy Hitters algorithms in [22,6] we can approximate $F_\infty(A_i)$ with additive error $\epsilon F_1^{res(1)}(A_i)$, giving the following bound.

Theorem 5. *There is a deterministic streaming algorithm that approximates* $L_1 \circ F_1^{res(1)}(A)$ *within* $1 - \epsilon$ *factor using* $O(\frac{p}{\epsilon})$ *words of space.*

Now we turn our attention to randomized algorithms. In the following, let $F(A_i) = F_1^{res(1)}(A_i)$ and $F(A) = L_1 \circ F_1^{res(1)}(A)$.

4.1   A $(2 + \epsilon)$ Algorithm

The idea of this algorithm is to reduce $F(A)$ to L_0 of a vector where each item in s represents a set of updates to this vector. Let $f_i(a)$ be the number of occurrences of $a \in [m]$ in A_i. We first observe the following.

Fact 6. $F_1^{res(1)}(A_i) \geq \frac{1}{d} \sum_{a<b} f_i(a) f_i(b) \geq \frac{1}{2} F_1^{res(1)}(A_i)$.

Proof. Notice that $\frac{1}{d} \sum_{a<b} f_i(a) f_i(b) = \frac{1}{2}(d - \frac{1}{d} \sum_a f_i^2(a))$. Clearly $\frac{1}{d} \sum_a f_i^2(a) \leq \max\{f_i(a)\}$. This proves the right hand side inequality. To prove the left inequality, we need to show $d \geq 2 \max\{f_i(a)\} - \frac{1}{d} \sum_a f_i^2(a)$. This is true because the RHS is maximized when $\max\{f_i(a)\} = d$. □

One way to produce $\sum_{a<b} f_i(a) f_i(b)$ is to compare each location of A_i with all other locations and sum up the mismatches. To express this in terms of L_0, let v_i be an all zero vector of length d^2 with a coordinate for each $(j, k) \in [d] \times [d]$. Given $A_i(j) = l$, add l to $v_i(j, k)$ and subtract l from $v_i(k, j)$ for all $k \in [d]$. Then, $L_0(v_i) = 2 \sum_{a<b} f_i(a) f_i(b)$. We generate the updates to vector $v = v_1 \circ \ldots \circ v_p$ as we go over A and estimate L_0 using the following result by Kane et al. [17].

Theorem 7. *[17] Let* $x = (x_1, \ldots, x_n)$ *be an initially zero vector. Let the input stream be a sequence of* t *updates to the coordinates of* x *of the form* (i, u) *where* $u \in \{-M, \ldots, M\}$ *for an integer* M *and* i *is an index. There is a 1-pass streaming algorithm for* $(1+\epsilon)$-*approximating* $L_0(x)$ *using space* $O(1/\epsilon^2 \log n(\log(1/\epsilon) + \log\log(tM)))$, *with success probability* $7/8$, *and with* $O(1)$ *per-item processing time.*

By Theorem 7 and Fact 6, we get a $2 + \epsilon$ approximation for $F(A)$ in space $O(1/\epsilon^2 \log(1/\epsilon) \log(n))$ bits. However, per-item processing time is $\Omega(d)$. To overcome this, we pick a random subset S from $[d]$ of size $O(\frac{1}{\epsilon^2} \log p)$ and, for $j \in S$, we compare $A_i(j)$ with all the coordinates of A_i. Now this gives us a vector v_i' with dimension $d|S|$. Fix an i and consider random variable $L_0(v_i')$. Let Y_j be an indicator random variable which is 1 iff $j \in S$. We have $\mathbb{E}[L_0(v_i')] = \sum_{j=1}^{d} \mathbb{E}[Y_j] \sum_{k=1}^{d} \mathcal{H}(A_i(j), A_i(k)) = \frac{2|S|}{d} \sum_{a<b} f_i(a) f_i(b)$. Since $\{Y_j\}$ are independent, using Chernoff bounds,

$$\Pr\left[|L_0(v_i') - \mathbb{E}[L_0(v_i')]| > \epsilon \mathbb{E}[L_0(v_i')]\right] \leq \frac{1}{8p}.$$

By the union bound, the probability that $\frac{L_0(v')}{2|S|}$ is away from $\frac{1}{d}\sum_{i=1}^{p} \sum_{a<b} f_i(a) f_i(b)$ by a factor of ϵ is at most $1/8$. Given this and the fact that the underlying L_0 estimation itself gives a $1 + \epsilon$ approximation we get a $(1 + \epsilon)^2 = 1 + \theta(\epsilon)$ approximation using polylogarithmic space and $O(1/\epsilon^2 \log p)$ per-item processing time.

Theorem 8. *Let $\epsilon > 0$. There is a 1-pass randomized streaming algorithm that approximates $L_1 \circ F_1^{res(1)}(A)$ within $2 + \epsilon$ factor using $O(1/\epsilon^2 \log(1/\epsilon))$ words of space. The error probability is at most $1/4$.*

4.2   A $(1 + \epsilon)$ Algorithm

To find a better estimate for $F(A)$ we use a combination of naive sampling and exact sparse recovery. If $F(A)$ is high, naive sampling gives us a good estimate. If $F(A)$ is low, then A has few non-uniform rows (we call A_i non-uniform if $F_1^{res(1)}(A_i) > 0$) and in space roughly proportional to the number of non-uniform rows, we can use sparse recovery to find all non-uniform rows with high probability. In the latter case, we obtain $F(A)$ exactly, or with a large alphabet, to within $1 + \epsilon$ factor. A generic implementation of this gives a $\tilde{O}(n^{1/2})$ space solution, where $n = dp$. Below we describe a $\tilde{O}(p^{1/2})$ space algorithm which is in line with this this approach but uses a combination of sampling, exact sparse recovery, and the $2 + \epsilon$ algorithm described earlier.

Let $F'(A_i) = 1/d \sum_{a<b} f_i(a) f_i(b)$. Recall that in the previous algorithm we used $F'(A_i)$ as an approximation for $F_1^{res(1)}(A_i)$. The worst case for this approximation happens when $F_1^{res(1)}(A_i)$ is maximized, i.e., $F_\infty(A_i) = d/F_0(A_i)$. On the other hand, when $F_1^{res(1)}(A_i)$ is low, the above quantity gives us a good estimate. This is because $F'(A_i)$ is lowerbounded by $\frac{1}{d}(d - F_\infty(A_i))F_\infty(A_i)$ which implies the following.

Fact 9. *Let $\epsilon \geq 0$. Suppose $F_1^{res(1)}(A_i) \leq \epsilon d$. We have $F'(A_i) \geq (1-\epsilon)F_1^{res(1)}(A_i)$.*

Define $F'(A) = \sum_{i=1}^{p} F'(A_i)$. From the definitions, we get

$$F'(A) + \frac{1}{2d} \sum_{i=1}^{p} ((F_1^{res(1)}(A_i))^2 + F_2^{res(1)}(A_i)) = F(A). \tag{1}$$

Now let $F''(A_i) = \frac{1}{2d}((F_1^{res(1)}(A_i))^2 + F_2^{res(1)}(A_i))$. From (1) it follows that if we are given $F''(A) = \sum_{i=1}^{p} F''(A_i)$, by using the algorithm from the previous section, we get a $1+\epsilon$ approximation for $F(A)$. On the other hand, Fact 9 tells us that we only need to compute $F''(A_i)$ for rows with high contribution. For $t \leq d$ define H_t to be the set $\{j \mid F_1^{res(1)}(A_j) \geq t\}$. The following is a consequence of Fact 9 and (1).

$$F(A) \geq F'(A) + \sum_{i \in H_{ed}} F''(A_i) \geq (1 - \epsilon)F(A). \tag{2}$$

In our algorithm we do not compute $F''(A_i)$ for H_{ed} but approximate them with error proportional to $F(A_i)$. This is achieved by sampling a few columns from A and using a sparse recovery procedure to find non-uniform rows in the sampled matrix. For our sparse recovery procedure, we use the following result from [21].

Theorem 10. *[21] Let $x, y \in \Sigma^n$. There is a randomized 1-pass streaming algorithm that, given the coordinates of x and y in arbitrary order, can check if $\mathcal{H}(x, y) > r$ or not using $O(r(\log n + \log |\Sigma|))$ bits of space and $O(\log n)$ per-item time. Moreover in case $\mathcal{H}(x, y) \leq r$, the algorithm finds all pairs $(x[i], y[i])$ where $x[i] \neq y[i]$. The probability of error is at most n^{-1}.*

Now we are ready to describe our algorithm. Let ϵ be an arbitrary constant smaller than 1. Let $\delta < \epsilon$ (we determine the value of δ later) and $k \geq \frac{8 \log n}{\delta^2}$. For $r \leq p$, denote by $\mathrm{SR}(r)$ the exact sparse recovery algorithm from Theorem 10. We run the following three threads in parallel.

T1 Run the $2+\epsilon$-approximation algorithm from Section 4.1. Let t_1 be the output.
T2 Let $K \geq 8k(\frac{p \log p}{\epsilon})^{1/2}$. Let B be K sampled rows of A (picked uniformly and independently). Compute a $1 - \epsilon$ approximation of $F(B)$ using the algorithm from Theorem 5. Let t' be the answer. Let $t_2 = \frac{pt'}{K}$.
T3 Let $r_0 > 4 \log p$ be an odd integer. Run the following r_0 times in parallel. In run j, let $r = \frac{8}{\epsilon^{1.5}}(\frac{p}{\log p})^{1/2}$. Sample k columns of A uniformly and independently, obtaining matrix C. Run $\mathrm{SR}(r)$ over consecutive columns in C. If more than r non-uniform rows are detected, abort the run. Otherwise for each non-uniform row C_i do the following. Let $f_{C_i}(a)$ be the frequency of a in C_i. Use $f_i'(a) = \frac{d}{k} f_{C_i}(a)$ to estimate $f_i(a)$. Let A_i' be a sequence corresponding to the frequency vector f_i'. Compute $X_i = \frac{d^2}{2\binom{k}{2}} \sum_a f_{C_i}(a)(f_{C_i}(a) - 1)$ and let $Y_i = \frac{1}{2d}(F_1^{res(1)}(A_i'))^2 + \frac{1}{2d}(X_i - F_\infty^2(A_i'))$. At the end we let $G_j = \{(i, Y_i) \mid F_1^{res}(A_i') \geq \epsilon d\}$.

In the end, if the majority of the runs have aborted, the algorithm outputs t_2. Otherwise, WLOG, assume the first $l > 2 \log n$ runs have survived. Let G be the set of pairs $(i, g(A_i))$, where i appears in all $G_1, \ldots, G_l$ and $g(A_i)$ is the median of Y_i's produced by the surviving runs. Then we output $t_3 = t_1 + \sum_{i \in G} g(A_i)$.

Lemma 4. *Assuming p is greater than a large enough constant, the above algorithm gives a $1 \pm 3\epsilon$ approximation for $F(A)$ with probability is at least $3/4$.*

Proof. We first consider the case when we ignore T3 and take the answer of T2. For each aborting run we have $F(C) > r$. Based on the observation that $\mathbb{E}[\frac{d}{k}F(C)] \leq F(A)$, and by Markov inequality, we have $\Pr[\frac{d}{k}F(C) > 8F(A)] < 1/8$. By Chernoff bound, if $F(A) < \frac{rd}{8k}$, the probability that more than half of the runs abort is at most $1/p^2$. So in this case, with high probability, we have $F(A) \geq \frac{rd}{8k}$. On the other hand, by Chebyshev's bound, we have $\Pr[|t_2 - F(A)| \geq 2\epsilon F(A)] < pd/(\epsilon^2 F(A)K)$. Plugging the values of r and K, we get that the probability is bounded by $1/8 + 1/p^2$.

Now consider the case where the output is t_3. In this case, we need to analyze the quality of the approximation of $F''(A_i)$ produced by a fixed run. The below claim follows by Chernoff bounds.

Claim. For $a \in \Sigma$, with probability at least $1 - \frac{1}{8n^2}$, $|f_i'(a) - f_i(a)| \leq \delta d$.

From Claim 4.2 it follows that, with probability at least $1 - 1/(8np)$, the error of the first term in Y_i, i.e., $\frac{1}{2d}(F_1^{res(1)}(A_i'))^2$, is bounded by $2\delta d$. To bound the error of the second term in Y_i, we use Chebyshev bound and the variance analysis of [3] (cf. Lemma 5.3) to estimate F_2. From [3], we have $\mathbb{E}[X_i] = F_2(A_i)$ and $\mathrm{Var}(X_i) \leq \frac{d}{k}(F_2(A_i))^{3/2}$. Using Chebyshev's inequality, we get

$$\Pr[|X_i - F_2(A_i)| > \delta d^2] \leq \frac{(F_2(A_i))^{3/2}}{\delta^2 k d^3}.$$

Given that $k > \frac{8}{\delta^2}\log n$, this probability is bounded by $1/(8\log n)$. Therefore, with probability at least $1 - 1/(8\log n)$, the second term of Y_i has error at most $3\delta d$. Since we took the median of at least $2\log p$ outcomes, with probability at least $1 - 1/(p^2\log n)$, for $i \in G$, we have $|g(A_i) - F''(A_i)| < 5\delta d$. Also with probability at least $1 - (\log p)/(2n)$, we have

$$H_{(\epsilon+\delta)d} \subseteq G, \quad ([p] \setminus H_{(\epsilon-\delta)d}) \cap G = \emptyset \tag{3}$$

Now we choose δ so that $5\delta d \leq \epsilon(\epsilon - \delta)d$. This gives us $\delta = O(\epsilon^2)$ and now we guarantee that, for all $i \in G$, $g(A_i)$ is away from $F''(A_i)$ by at most $\epsilon F_1^{res}(A_i)$. Putting these observations and (1),(2), and (3) together we get $|t_3 - F(A)| \leq 3\epsilon F(A)$. This proves our lemma. □

Threads T2 and T3 dominate our space complexity. The sampling algorithm in T2 takes $O(\frac{1}{\epsilon}K)$ space with $O(1)$ time per item. The runs in T3 take $O(r_0 rk)$ space in total. However since the decoding time of the sparse recovery is $O(r\log n)$, this makes the worst-case per-item time $O(r_0 r\log^2 n)$. Since $\delta = O(\epsilon^2)$, our final space bound becomes $O(1/\epsilon^{5.5}(p\log p)^{1/2}\log n)$. Note that with a consant alphabet, eliminating the repetitions in T3 and choosing parameters differently, we can get $O(1/\epsilon^3(p\log p)^{1/2})$ space.

Theorem 11. *There is a randomized 1-pass streaming algorithm that outputs a $1 \pm \epsilon$ approximation of $L_1 \circ F_1^{res}(A_{p\times d})$ with probability at least $3/4$ using $O(1/\epsilon^{5.5}(p\log p)^{1/2}\log n)$ words of space.*

References

1. Alon, N., Matias, Y., Szegedy, M.: Space complexity of approximating the frequency moments. In: STOC 1996 (1996)
2. Amir, A., Lewenstein, M., Porat, E.: Faster algorithms for string matching with k mismatches. In: SODA 2000 (2000)
3. Bar-Yossef, Z., Kumar, R., Sivakumar, D.: Sampling algorithms: lower bounds and applications. In: CCC 2002 (2002)
4. Berinde, R., Cormode, G., Indyk, P., Strauss, M.: Space-optimal heavy hitters with strong error bounds. In: PODS 2009 (2009)
5. Bhuvanagiri, L., Ganguly, S., Kesh, D., Saha, C.: Simpler algorithm for estimating frequency moments of data streams. In: SODA 2006 (2006)
6. Bose, P., Kranakis, E., Morin, P., Tang, Y.: Bounds for frequency estimation of packet streams. In: Proceedings of the 10th International Colloquium on Structural Information and Communication Complexity (2003)
7. Charikar, M., Chen, K., Farach-Colton, M.: Finding frequent items in data streams. Theor. Comput. Sci. 312(1), 3–15 (2004)
8. Cole, R., Hariharan, R.: Approximate String Matching: A Simpler Faster Algorithm. In: SODA 1998 (1998)
9. Cormode, G., Muthukrishnan, S.: Space efficient mining of multigraph streams. In: PODS 2005, pp. 271–282 (2005)
10. Czumaj, A., Gasieniec, L.: On the complexity of determining the period of a string. In: Giancarlo, R., Sankoff, D. (eds.) CPM 2000. LNCS, vol. 1848, pp. 412–422. Springer, Heidelberg (2000)
11. Elfeky, M.G., Aref, W.G., Elmagarmid, A.K.: STAGGER: periodicity mining of data streams using expanding sliding windows. In: ICDM 2006 (2006)
12. Ergun, F., Muthukrishnan, S., Sahinalp, C.: Sublinear methods for detecting periodic trends in data streams. In: Farach-Colton, M. (ed.) LATIN 2004. LNCS, vol. 2976, pp. 16–28. Springer, Heidelberg (2004)
13. Ganguly, S., Kesh, D., Saha, C.: Practical algorithms for tracking database join sizes. In: Sarukkai, S., Sen, S. (eds.) FSTTCS 2005. LNCS, vol. 3821, pp. 297–309. Springer, Heidelberg (2005)
14. Indyk, P.: Stable distributions, pseudorandom generators, embeddings, and data stream computation. J. ACM 53(3), 307–323 (2006)
15. Indyk, P., Woodruff, D.: Optimal approximations of the frequency moments of data streams. In: STOC 2005 (2005)
16. Jayram, T.S., Woodruff, D.: The data stream space complexity of cascaded norms. In: FOCS 2009 (2009)
17. Kane, D.M., Nelson, J., Woodruff, D.: An optimal algorithm for the distinct elements problem. In: PODS 2010 (2010)
18. Karp, R.M., Rabin, M.O.: Efficient randomized pattern matching algorithms. IBM Journal of Res. and Dev. 249, 260 (1987)
19. Knuth, D.E., Morris, J.H., Pratt, V.R.: Fast pattern matching in strings. SIAM J. Comp. 6, 323–350 (1977)
20. Lachish, O., Newman, I.: Testing periodicity. In: Chekuri, C., Jansen, K., Rolim, J.D.P., Trevisan, L. (eds.) APPROX 2005 and RANDOM 2005. LNCS, vol. 3624, pp. 366–377. Springer, Heidelberg (2005)
21. Lipsky, O., Porat, E.: Improved sketching of hamming distance with error correcting. In: Ma, B., Zhang, K. (eds.) CPM 2007. LNCS, vol. 4580, pp. 173–182. Springer, Heidelberg (2007)

22. Misra, J., Gries, D.: Finding repeated elements. Technical Report, Cornell University (1982)
23. Monemizadeh, M., Woodruff, D.: 1-Pass relative-error Lp-sampling with applications. In: SODA 2010 (2010)
24. Muthukrishnan, S.: Data stream algorithms. In: The Barbados Workshop on Computational Complexity (2009)
25. Porat, B., Porat, E.: Exact and approximate pattern matching in the streaming model. In: FOCS 2009 (2009)

Rumor Spreading on Random Regular Graphs and Expanders

Nikolaos Fountoulakis and Konstantinos Panagiotou*

Max-Planck-Institute for Informatics
Saarbrücken, Germany

Abstract. Broadcasting algorithms are important building blocks of distributed systems. In this work we investigate further the performance of the classical and well-studied *push model*. Assume that initially one node in a given network holds some piece of information. In each round, every one of the informed nodes chooses independently a neighbor uniformly at random and transmits the message to it.

In this paper, we consider random networks where each vertex has degree $d \geq 3$, i.e., the underlying graph is drawn uniformly at random from the set of all d-regular graphs with n vertices. We show that with probability $1 - o(1)$ the push model broadcasts the message to all nodes in $C_d \ln n + \xi$ rounds, where $|\xi| = O((\ln \ln n)^2)$ and

$$C_d = \frac{1}{\ln(2(1 - \frac{1}{d}))} - \frac{1}{d \ln(1 - \frac{1}{d})}.$$

In particular, we determine precisely the effect of the node degree to the typical broadcast time of the push model. Moreover, we consider pseudo-random regular networks, where we assume that the degree of each node depends on n. There we show that the broadcast time is $(1 + o(1))C \ln n$ with probability $1 - o(1)$, where $C = \lim_{d \to \infty} C_d = \frac{1}{\ln 2} + 1$.

Keywords: Rumor Spreading, Random Regular Graphs.

1 Introduction

1.1 Rumor Spreading and the Push Model

In this work we consider the classical and well-studied *push model* (or *push protocol*) for disseminating information in networks. Initially, one of the nodes has some piece of information. In each succeeding round, every node who has the information passes it to another node, which is chosen independently and uniformly at random among its neighbors. The important question is: how many rounds are typically needed until all nodes are informed?

The main advantage of randomized broadcasting is its inherent robustness against several kinds of failures and sudden changes in the network topology. Thus, the push model has been the topic of many theoretical works, and its

---

* This author was supported by the Humboldt Foundation.

M. Serna et al. (Eds.): APPROX and RANDOM 2010, LNCS 6302, pp. 560–573, 2010.

performance was evaluated on several types of networks. In the case where the underlying network is the complete graph, Frieze and Grimmett [17] proved that with high probability (whp.) (i.e., with probability $1 - o(1)$) the broadcasting is completed within $(1 + o(1))(\log_2 n + \ln n)$ rounds, where n denotes the total number of nodes. Recently, this result was extended by the two authors and Huber [14] to the classical Erdős-Rényi graph $G_{n,p}$, which is obtained by including each of the possible $\binom{n}{2}$ edges with probability p, independently of all other edges. The main result of [14] is that if $p = \omega(\frac{\ln n}{n})$, then the typical broadcast time essentially coincides with the broadcast time on the complete graph. In other words, as long as the average degree of the underlying graph is significantly larger than $\ln n$, the number of rounds needed is not affected. Prior to this work, there has been no result describing precisely the performance of the push model on significantly sparser networks.

The typical broadcast time of the push model was also investigated for other types of networks. Feige et al. [13] derived rough upper bounds that hold for arbitrary graphs. Moreover, they proved a logarithmic upper bound for the number of rounds needed to broadcast the information if the underlying network is a hypercube. This result was generalized by Elsässer and Sauerwald [11], who determined similar bounds for several classes of Cayley graphs, and in [12], where, among other results, lower bounds for d-regular and general graphs were shown. Bradonjic et al. [3] considered random geometric graphs as underlying networks, and proved that whp. the broadcast time is essentially proportional to the diameter of these graphs. Finally, the effect of the conductance of the underlying network on the broadcast time of the push protocol was considered by Chierichetti, Lattanzi and Panconesi in [4].

1.2   Our Contribution

The main contribution of this paper is the precise analysis of the push model on sparse random networks. Note that in this context the study of the $G_{n,p}$ distribution is not appropriate, as we would have to set $p = c/n$ for some constant $c > 0$. However, for such p the random graph $G_{n,p}$ is typically not connected. In fact, if we took any $p = o\left(\frac{\ln n}{n}\right)$, we would face the same problem, as such a p is below the connectivity threshold for $G_{n,p}$ (see for example [20]).

A candidate class of random graphs that combines the feature of constant average degree with that of connectivity is the class of random d-regular graphs $\mathbb{G}(n,d)$ for $d \geq 3$. It is well-known that a random d-regular graph on n vertices is connected with probability $1 - o(1)$. Thus, a typical member of this class of graphs is suitable for the analysis of the push protocol as far as the effect of density is concerned. Let $T = T(\mathbb{G}(n,d))$ denote the broadcast time of the push model on $\mathbb{G}(n,d)$. Note that in this case the choice of the vertex where the information is placed initially does not matter. Moreover, note that getting a crude bound of $O(\ln n)$ for the number of rounds is easy, as the diameter of a random d-regular graph is $O(\ln n)$ with high probability. However, our theorem determines precisely the effect of the density to the probable broadcast time.

Theorem 1. *Let $d \geq 3$. With probability $1 - o(1)$*

$$|T(\mathbb{G}(n, d)) - C_d \ln n| = O((\ln \ln n)^2),$$

where $C_d = \frac{1}{\ln(2(1-\frac{1}{d}))} - \frac{1}{d \ln(1-\frac{1}{d})}$.

The above theorem is interpreted as follows: for almost all d-regular graphs on n vertices, with probability $1 - o(1)$ the push protocol broadcasts the information within the claimed number of rounds. It is easy to see that as d grows C_d converges to $\frac{1}{\ln 2} + 1$, which is the constant factor of the broadcast time of the push protocol on the complete graph, as shown by Frieze and Grimmett [17]. Thus, our result reveals the essential insensitivity of the performance of the push protocol regarding the density of the underlying network as d grows and shows that the crucial factor is the "uniformity" of its structure. In Subsection 1.3 below we give an informal description of the evolution of the push protocol, thus explaining also how do the two summands involved in C_d come up.

A lower bound that holds for all d-regular graphs was shown in [12]. More precisely, let G be any d-regular graph with n vertices. Then, in [12] it is proved that $T(G) \geq C'_d \ln n - o(\ln n)$ with high probability, where $C'_d = \frac{1}{\ln(2-\frac{1}{d})} - \frac{1}{d \ln(1-\frac{1}{d})}$. So, our result demonstrates that this bound is not tight for almost all d-regular graphs.

One question that remains open is the performance of the push protocol in the case $d = d(n)$. We explore further this aspect and we consider regular graphs whose structural characteristics resemble those of a random regular graph. In particular, we consider expanding graphs whose "geometry" is determined by the spectrum of their adjacency matrix.

Regular Expanding Graphs. Expanding graphs have found numerous applications in modern theoretical computer science as well as in pure mathematics. Their properties together with the theory of finite Markov chains have led to the solution of central problems such as the approximation of the volume of a convex body, approximate counting or the approximate uniform sampling from a class of combinatorial objects. The latter applications have had further impact outside computer science such as in the field of statistical physics. We refer the reader to the excellent survey of Hoory et al. [18] for a detailed exploration of the properties and the numerous applications of expanding graphs.

The main feature of an expanding graph is that every set of vertices is connected to the rest of the graph by a large number of edges. This key property makes random walks on such graphs rapidly mixing and has led to the above mentioned applications. Moreover, this property makes expanding graphs an attractive candidate for communication networks. Intuitively, the high expansion of a graph implies that information that is initially located on a small part of the graph can be spread quickly to the rest of the graph. This becomes possible as the high expansion of a graph ensures the lack of "bottlenecks", that is, local obstructions on which a broadcasting protocol would need a significant amount of time in order to bypass them.

We focus on a spectral characterization of expanding graphs, which is related to the spectral gap of their adjacency matrix. Let $G = (V, E)$ be a connected d-regular graph and let A be its adjacency matrix. The Perron-Frobenius Theorem implies (see Proposition 2.10 in [21]) that the largest eigenvalue of A equals d and that the corresponding eigenvector is proportional to the all-ones vector $[1, \ldots, 1]^T$. Let $\lambda_1, \ldots, \lambda_n$ be the eigenvalues of A ordered according to their value (note that since A is symmetric, these are all real). Set $\lambda := \lambda(A) := \max_{2 \leq i \leq n} |\lambda_i|$. If G has n vertices we say that G is an (n, d, λ) *graph*. One can show (see for example p. 19 in [21]) that $\lambda = \Omega(\sqrt{d})$. In particular, Alon and Boppana (see Nilli [24]) and Friedman [16] have shown that for every d-regular graph on n vertices we have $\lambda_2 \geq 2\sqrt{d - 1}(1 - o(1))$.

We are interested in the class of d-regular graphs for which λ almost attains this lower bound. In particular, we are concerned with the broadcast time of the push protocol on expanding d-regular graphs on n vertices with $\lambda = O(\sqrt{d})$. Such graphs can be explicitly constructed through number-theoretic or group-theoretic methods (see the survey of Krivelevich and Sudakov [21] where numerous examples are presented). Informally, we show that if $d = \omega(\sqrt{n})$, then the broadcast time is that of the complete graph.

Theorem 2. *Let G be a connected (n, d, λ) graph with $\lambda \leq C\sqrt{d}$ and $d \geq 2C\sqrt{n}\ln^{1/9} n$. Then for any $v \in V$, with probability $1 - o(1)$*

$$|T(G, v) - (\log_2 n + \ln n)| = o(\ln n).$$

Again, this theorem shows the insensitivity of the broadcast time regarding the density of the underlying network. In fact, the assumption that $\lambda = O(\sqrt{d})$ does not merely yield the high expansion of the graph, but it also implies that the edges of the graph are distributed in a uniform way among each subset of vertices. This is an important fact exploited in the proof of Theorem 2, as the assumption implies that the structure of the graph is not very different from that of a random graph on n vertices and edge probability equal to d/n. For example, the number of edges between a subset S and its complement is close to $\frac{d}{n}|S|(n - |S|)$, which is the expected value in the random graph with edge probability d/n. In this sense, such graphs are *pseudorandom*. This notion was introduced by Thomason [27] and was explored further by Chung, Graham and Wilson [5], especially regarding its spectral characterization.

1.3   The Evolution of the Randomized Protocol in a Nutshell

Roughly speaking, the evolution of the protocol consists of three phases, which have different characteristics regarding the rate in which the information is spread.

Let us consider the first phase, which ends when there are at least εn informed vertices, for some very small $\varepsilon > 0$. Let us denote by $\mathcal{I}_t$ the set of informed vertices (i.e., those who possess the information), and by $\mathcal{U}_t$ the set of uninformed vertices at the beginning of round $t + 1$ of the push model. Moreover, let e be some edge that is incident to a vertex in $\mathcal{I}_t$ that *has not been used up to now* to transmit a message, and let $\mathcal{E}_t$ be the set of such edges. Then we show that the subgraph of $\mathbb{G}(n,d)$ induced by $\mathcal{I}_t$ is essentially a *tree*, and moreover, that $\mathcal{E}_t$ contains approximately $2^t(1 - \frac{1}{d})^t$ edges. To see this, note that as every vertex informs some specific neighbor with probability $1/d$, the expected number of edges from $\mathcal{E}_t$ that are going to be used is $|\mathcal{E}_t|/d$. This means that approximately $|\mathcal{E}_t|/d$ new vertices are going to be informed (as the set of informed vertices induces a tree), implying that $|\mathcal{E}_{t+1}| \approx |\mathcal{E}_t| - |\mathcal{E}_t|/d + (d-1)|\mathcal{E}_t|/d$, as for every vertex that becomes informed in this round the number of edges counted in $\mathcal{E}_t$ increases by $d - 1$. So, $|\mathcal{E}_{t+1}| \approx 2(1 - \frac{1}{d})|\mathcal{E}_t|$. Note that in this calculation we worked only with expected values. In the actual proof we have to show that all the relevant quantities are sharply concentrated around their expectations. To this end, we use a variant of *Talagrand's* inequality by McDiarmid [23] (Theorem 4), which has not been used very frequently in the analysis of distributed algorithms. We believe that it could be widely applicable to the analysis of existing or future randomized protocols with several different degrees of dependency.

As soon as the number of informed vertices exceeds εn, then after very few rounds the number of informed vertices is already $(1 - \varepsilon)n$. Here, it is essentially the expansion properties of $\mathbb{G}(n,d)$, which guarantee that every large set of vertices has linearly many neighbors and, thus, with high probability a certain fraction of those become informed in each round.

During the final phase, the number of remaining uninformed vertices shrinks by a factor of $(1 - \frac{1}{d})^d$. Indeed, suppose that there are $o(n)$ uninformed vertices. Then we expect that almost all of them have the property that the number of their neighbors in $\mathcal{I}_t$ is d, implying that the probability that any one of the remains uninformed is precisely $(1 - \frac{1}{d})^d$. An easy calculation shows that a "typical" subset of $\mathbb{G}(n,d)$ has this property. However, the set of uninformed vertices might not be typical at all, implying that we need additional effort to guarantee the desired properties.

2   Concentration Inequalities

In this section we will state two concentration inequalities that will serve as the backbone of our proofs. The first one is a Chernoff-type bound for sums of negatively correlated random variables, see e.g. [7].

Theorem 3. *Let $I_1, \ldots, I_n$ be a family of indicator random variables on a common probability space, which are identically distributed and negatively correlated, i.e., $\mathbb{E}(I_i I_j) \le \mathbb{E}(I_i)\mathbb{E}(I_j)$ for all $1 \le i, j \le n$. Let $X := \sum_{i=1}^{n} I_i$. Then, for any $t > 0$*

$$\mathbb{P}\left(|X - \mathbb{E}(X)| > t\right) < 2\exp\left(-\frac{t^2}{2\left(\mathbb{E}(X) + t/3\right)}\right).$$

The next concentration inequality that we will need is due to McDiarmid [23], and it is based on the work of Talagrand [26]. We give first a few necessary definitions. Let B be a finite set and let $Sym(B)$ be the set of all permutations on B. Assume that π is an element of $Sym(B)$, drawn uniformly at random. Also, let $\mathbf{X} = (X_1, \ldots, X_n)$ be a finite family of independent random variables, where X_j takes values in a set Ω_j. Finally, set $\Omega = Sym(B) \times \prod_{j=1}^{n} \Omega_j$.

Theorem 4. *Let c and r be positive constants. Suppose that $h : \Omega \to \mathbb{R}_+$ satisfies the following conditions. For each $(\sigma, \mathbf{x}) \in \Omega$ we have*

- *if $\mathbf{x}'$ differs from $\mathbf{x}$ in only one coordinate, then $|h(\sigma, \mathbf{x}) - h(\sigma, \mathbf{x}')| \leq 2c$;*
- *if σ' can be obtained from σ by swapping two elements, then $|h(\sigma, \mathbf{x}) - h(\sigma', \mathbf{x})| \leq c$;*
- *if $h(\sigma, \mathbf{x}) = s$, then there is a set of at most rs coordinates such that $h(\sigma', \mathbf{x}') \geq s$ for any $(\sigma', \mathbf{x}') \in \Omega$ that agrees with $(\sigma, \mathbf{x})$ on these coordinates.*

Let $Z = h(\pi, \mathbf{X})$ and let m be the median of Z. Then, for any $t > 0$

$$\mathbb{P}\left(|Z - m| > t\right) \leq 4 \exp\left(-\frac{t^2}{16rc^2(m+t)}\right).$$

3 Properties of Random Regular Graphs and the Configuration Model

3.1 The Configuration Model

We perform the analysis of the randomized protocol using the *configuration model* introduced by Bender and Canfield [1] and independently by Bollobás [2]. For $n \geq 1$ let $V_n := \{1, \ldots, n\}$ and $[d] = \{1, \ldots, d\}$. Also for those n for which dn is even, we let $P := V_n \times [d]$. We call the elements of P *clones*. A *configuration* is a perfect matching on P. If we project a configuration onto V_n, then we obtain a d-regular multigraph on V_n. Let $\widetilde{\mathbb{G}}(n, d)$ denote the multigraph that is obtained by choosing the configuration on P uniformly at random. It can be shown (see e.g. [20, p. 236]) that if we condition on $\widetilde{\mathbb{G}}(n, d)$ being simple (i.e. it does not have loops or multiple edges), then this is distributed uniformly among all d-regular graphs on V_n. In other words, $\widetilde{\mathbb{G}}(n, d)$ conditional on being simple has the same distribution as $\mathbb{G}(n, d)$. Moreover, Corollary 9.7 in [20] guarantees that

$$\liminf_{n \to \infty} \mathbb{P}(\widetilde{\mathbb{G}}(n, d) \text{ is simple}) > 0. \tag{1}$$

(Of course the above limit is taken over those n for which dn is even.) Let A_n be a subset of the set of d-regular multigraphs on V_n. Altogether the above facts imply that if $\mathbb{P}(\widetilde{\mathbb{G}}(n, d) \in A_n) \to 0$ as $n \to \infty$ then also $\mathbb{P}(\mathbb{G}(n, d) \in A_n) \to 0$. This allows us to work with $\widetilde{\mathbb{G}}(n, d)$ instead of $\mathbb{G}(n, d)$ itself.

3.2 Some Useful Facts

We continue by introducing some notation. Let G be a graph, and let S, S' be subsets of its vertices. Then we denote by $e_G(S)$ the number of edges in G joining vertices only in S, and by $e_G(S, S')$ the number of edges in G joining a vertex in S to a vertex in S'. Moreover, we denote by $\Gamma_G(v)$ the set of neighbors of a vertex v in G.

Lemma 1. *Let $\mathcal{A}, \mathcal{B} \subseteq V_n \times [d]$ be two disjoint sets of clones, and let $\mathcal{C} \subseteq V_n$ be a set of vertices such that $(\mathcal{C} \times [d]) \cap (\mathcal{A} \cup \mathcal{B}) = \emptyset$. Let M be a matching drawn uniformly at random from the set of perfect matchings on the union of the clones in $\mathcal{A}, \mathcal{B}$ and $\mathcal{C} \times [d]$, and set $N := |\mathcal{A}| + |\mathcal{B}| + d|\mathcal{C}| - 1$. Then*

$$\mathbb{E}(e_M(\mathcal{A})) = \binom{|\mathcal{A}|}{2} \frac{1}{N}, \quad \mathbb{E}(e_M(\mathcal{A}, \mathcal{B})) = |\mathcal{A}||\mathcal{B}| \frac{1}{N}, \quad and \quad \mathbb{E}(e_M(\mathcal{A}, \mathcal{C})) = d|\mathcal{A}||\mathcal{C}| \frac{1}{N}.$$
$$(2)$$

Moreover, let H_ℓ denote the number of vertices in $\mathcal{C}$ that are adjacent to exactly ℓ clones in $\mathcal{A}$ in M, where $0 \leq \ell \leq d$. Then, if $|\mathcal{B}| \geq |\mathcal{A}| = \omega(\ln n)$

$$\mathbb{E}(H_\ell) = \left(1 + o\left(\frac{1}{\ln n}\right)\right) \cdot |\mathcal{C}| \binom{d}{\ell} \left(\frac{|\mathcal{A}|}{N}\right)^\ell \left(1 - \frac{|\mathcal{A}|}{N}\right)^{d-\ell}.$$
$$(3)$$

Finally, let $Q = \sum_{\ell \geq 2} H_\ell$. Then, if $N \geq 4$

$$\mathbb{E}(Q) \leq d^2 |\mathcal{A}|^2 |\mathcal{C}| N^{-2}.$$
$$(4)$$

Let X be any of $e_M(\mathcal{A}), e_M(\mathcal{A}, \mathcal{B}), e_M(\mathcal{A}, \mathcal{C})$ or H_ℓ, and let $\mu = \mathbb{E}(X)$. Then, if $\mu = \omega(\ln^2 n)$, for any $\varepsilon = \omega(\mu^{-1/2})$

$$\mathbb{P}(|X - \mu| \geq \varepsilon \mu) \leq 4 e^{-\frac{\varepsilon^2}{64 d(1+\varepsilon)} \mu}.$$
$$(5)$$

The proof of this lemma, and all other omitted proofs can be found in the full version [15] due to space limitations.

4   Analysis of the Randomized Broadcasting Algorithm

4.1   The Preliminary Phase

Let T_0 be the first round in which the number of informed vertices exceeds $\ln^7 n$. We will show the following statement; it is not best possible, but it suffices for our purposes.

Lemma 2. *With probability $1 - o(1)$ we have that $T_0 = O(\ln \ln n)$. Moreover, for sufficiently large n the subgraph induced by the vertices in $\mathcal{I}_{T_0}$ is with probability $1 - o(1)$ a tree.*

4.2   The Exposure Strategy

In this section we will describe our general strategy for determining the probable broadcast time of the randomized rumor spreading protocol. We will denote by $\mathcal{I}_t$ the set of informed vertices and by $\mathcal{U}_t$ the set consisting of the uninformed vertices, i.e., $\mathcal{U}_t = [n] \setminus \mathcal{I}_t$, at the beginning of round t. We have that $\mathcal{I}_1 = \{1\}$. We can simulate the execution of the rumor spreading protocol as follows in two steps. First, we choose one of the clones of vertex 1 uniformly at random, say c_1. Then, we expose the edge in the random matching whose one endpoint is c_1, and pass the message to the other endpoint, say c_2. Note that this is equivalent to selecting uniformly at random a clone c' different from c_1, and joining c_1 and c' by an edge. Clearly, c_2 is a clone that corresponds to some vertex in the original graph, which now becomes informed. This completes the first round, and $\mathcal{I}_2$ consists of vertex 1 and the vertex corresponding to c_2.

This gradual exposure of the graph can be generalized to any other round in the following manner. Suppose that we are in the beginning of round $t + 1 \geq 2$. We will simulate the execution of the protocol as follows in two steps.

1. For each $v \in \mathcal{I}_t$ we choose one of its clones uniformly at random, independently for every such vertex. We shall denote the selected clone by $c_v = c_v(t)$.

2. Set $\mathcal{I}_{t+1} = \mathcal{I}_t$ and let $v \in \mathcal{I}_t$. If c_v belongs to an edge in the random matching that was exposed in one of the previous rounds, do nothing. Otherwise, choose uniformly at random one of the remaining unmatched clones, say c, and connect it to c_v by an edge. Add the vertex corresponding to c to $\mathcal{I}_{t+1}$, if it is not already contained in $\mathcal{I}_{t+1}$.

If a clone of a vertex in $\mathcal{U}_t$ is matched to c_v, for some $v \in \mathcal{I}_t$, then that vertex becomes *informed* – we denote by $\mathcal{N}_{t+1}$ the set of those vertices. In short, $\mathcal{N}_{t+1}$ is the set of *newly informed* vertices in the $t + 1$st round. Let us introduce some further notation regarding the two exposure steps. At the beginning of round $t + 1$, we denote by $\mathcal{P}_t$ the set of clones of the vertices in $\mathcal{I}_t$ whose neighbors have not been exposed yet (i.e., in none of the previous rounds the edges in the matching containing those clones were exposed). Among those, during Step 1 we choose a set $\mathcal{A}_{t+1} \subseteq \mathcal{P}_t$ of clones, which are those clones chosen in *Step 1* and whose neighbors have not been exposed in previous rounds. Thus, $\mathcal{A}_{t+1}$ contains the clones through which new vertices might get informed. Finally, we write $N_{t+1} = |\mathcal{N}_{t+1}|$, $A_{t+1} = |\mathcal{A}_{t+1}|$ and $P_t = |\mathcal{P}_t|$, and note that $\mathcal{P}_0$ consists of the d clones of vertex 1.

The two steps of our exposure strategy can be also viewed as follows. In the first step we choose according to the rule described above a random subset $\mathcal{A}_{t+1}$ of $\mathcal{P}_t$. Then, in *Step 2*, the clones in $\mathcal{A}_{t+1}$ are matched to the union of the clones in $\mathcal{P}_t$ and the clones corresponding to the vertices in $\mathcal{U}_t$ (as, per definition, all other clones are already matched). In other words, we consider a random perfect matching $\mathcal{M}_{t+1}$ on the set of clones in $\mathcal{P}_t$ and $\mathcal{U}_t$, and we will study its combinatorial properties. In particular, the following claim relates the random quantities in question.

Proposition 1. *Let $H_{i,t+1}$ denote the number of vertices in $\mathcal{U}_t$ that were informed i times in round $t + 1$, i.e., a vertex v is counted in $H_{i,t+1}$, if there are i clones in $\mathcal{A}_{t+1}$ that are matched to the clones of v in $\mathcal{M}_{t+1}$. Then*

$$I_{t+1} = I_t + N_{t+1} \quad \text{and} \quad U_{t+1} = U_t - N_{t+1}, \tag{6}$$

$$N_{t+1} = \sum_{i=1}^{d} H_{i,t+1} \le e_{\mathcal{M}_{t+1}}(\mathcal{A}_{t+1}, \mathcal{U}_t), \tag{7}$$

$$P_{t+1} = P_t - A_{t+1} - e_{\mathcal{M}_{t+1}}(\mathcal{A}_{t+1}, \mathcal{P}_t \setminus \mathcal{A}_{t+1}) + \sum_{i=1}^{d}(d-i)H_{i,t+1}. \tag{8}$$

Proof. The first two equations are easy to see. To deduce (8), note first that all clones in $\mathcal{A}_{t+1}$ are excluded from $\mathcal{P}_{t+1}$, as they are matched to other clones in $\mathcal{P}_t$ or $\mathcal{U}_t$; this accounts for the "$-A_{t+1}$" term. Moreover, all clones in $\mathcal{P}_t \setminus \mathcal{A}_{t+1}$ that are contained in edges of $\mathcal{M}_{t+1}$ with the other endpoint in $\mathcal{A}_{t+1}$ are excluded from $\mathcal{P}_{t+1}$ as well, as the edge including them was exposed; this accounts for the the "$-e_{\mathcal{M}_{t+1}}(\mathcal{A}_t, \mathcal{P}_t \setminus \mathcal{A}_{t+1})$" term. Finally, for each newly informed vertex counted in $H_{i,t+1}$, i.e., which was informed i times in round t, the number of clones counted in $\mathcal{P}_t$ increases by $d-i$. □

For future reference we prove already here a lemma that addresses the concentration properties of A_{t+1}.

Lemma 3. *For any $t \ge 1$ and $n \ge 5$*

$$\mathbb{P}\left(\left|A_t - \frac{P_t}{d}\right| \ge \frac{P_t}{d \ln^2 n} \,\Big|\, P_t\right) \le 2e^{-\frac{P_t}{3d \ln^4 n}}.$$

Proof. For each clone $c \in \mathcal{P}_t$ let I_c be the indicator variable for the event that c is selected in the first step of the tth round, i.e., "$I_c = 1$" iff the random decisions in *Step 1* are such that $c \in \mathcal{A}_{t+1}$. Since each clone has probability $1/d$ to be selected we have $\mathbb{E}(I_c) = 1/d$. Moreover, for two distinct clones c, c' we have that

$$\mathbb{E}(I_c I_{c'}) = \begin{cases} 0 & \text{, if } c, c' \text{ are clones of the same } v \in V_n \\ 1/d^2 & \text{, otherwise} \end{cases} \le \frac{1}{d^2} = \mathbb{E}(I_c)\mathbb{E}(I_{c'}),$$

i.e., the I_c's are negatively correlated. We infer that $\mu := \mathbb{E}(A_{t+1} \mid P_t) = \frac{P_t}{d}$, and Theorem 3 implies that the sought probability is at most

$$\mathbb{P}(|A_{t+1} - \mu| \ge \mu / \ln^2 n \mid P_t) \le 2\exp\left(-\frac{\mu^2 \ln^{-4} n}{2(\mu + \mu/(3\ln^2 n))}\right) \le 2\exp\left(-\frac{\mu}{3\ln^4 n}\right).$$

□

4.3   The Middle Phases

Let T_1 be the first round where the number of informed vertices is at least $n - \ln^7 n$, or equivalently, where $U_{T_1} \le \ln^7 n$. The main accomplishment of this section is the proof of the following lemma, which describes the likely evolution of the number of (un)informed vertices and of P_t until $t = T_1$.

Lemma 4. *Suppose that $P_t, U_t \geq \ln^7 n$. Abbreviate $F_t = 1 - \frac{P_t}{d(P_t+dU_t)}$. Then, uniformly with probability at least $1 - o(\frac{1}{\ln n})$,*

$$P_{t+1} = \left(1 - o\left(\frac{1}{\ln n}\right)\right) \cdot \left(\left(1 - \frac{1}{d}\right) F_t \cdot P_t + dU_t(F_t - F_t^d)\right), \quad (9)$$

$$U_{t+1} = \left(1 - o\left(\frac{1}{\ln n}\right)\right) \cdot F_t^d \cdot U_t. \quad (10)$$

Lemma 4 allows us now to derive probable bounds for T_1.

Corollary 1. *With probability $1-o(1)$ we have that $T_1 - T_0 = C_d \ln n + O(\ln \ln n)$, where*

$$C_d = \frac{1}{\ln(2(1 - \frac{1}{d}))} - \frac{1}{d \ln(1 - \frac{1}{d})}.$$

Proof. By applying Lemma 2 we infer that at round T_0 with high probability there are for the first time at least $\ln^7 n$ informed vertices, and the set of informed vertices induces a tree. Hence, we may assume that

$$\ln^7 n \leq I_{T_0} \leq 2 \ln^7 n \quad \text{and} \quad (d - 1)I_{T_0} \leq P_{T_0} \leq dI_{T_0}.$$

We will use those facts in the sequel without further reference.

Let p_t and u_t be given by the recursions

$$p_{t+1} = \left(1 - \frac{1}{d}\right) f_t p_t + d u_t(f_t - f_t^d) \quad \text{and} \quad u_{t+1} = f_t^d u_t,$$

where $f_t = 1 - \frac{p_t}{d(p_t+du_t)}$, and $p_{T_0} = P_{T_0}, u_{T_0} = n - I_{T_0}$. As we are interested in the probable values of P_t and U_t for $t = O(\ln n)$ we infer by applying Lemma 4 that $p_t = (1 + o(1))P_t$ and $u_t = (1 + o(1))U_t$ for all such t, provided that $U_t, P_t \geq \ln^7 n$. In what follows, we shall therefore consider only the evolution of p_t and u_t.

Let $q := 2\left(1 - \frac{1}{d}\right)$, $\varepsilon = 0.01$ and t_1 be the maximal t such that $q^{t-T_0} \leq \frac{\varepsilon n}{\ln^7 n}$. We will first show that for all $T_0 \leq t \leq t_1$

$$p_t \leq P_{T_0} \cdot q^{t-T_0} \quad \text{and} \quad p_t \geq P_{T_0} \cdot q^{t-T_0} - 3P_{T_0}^2 \cdot q^{2(t-T_0)}/n, \quad (11)$$

and

$$u_t = n - I_{T_0} - P_{T_0} \frac{q^{t-T_0} - 1}{d(q - 1)} \pm 9 \cdot P_{T_0}^2 q^{2(t-T_0)}/n. \quad (12)$$

We proceed by induction on t. Note that for $t = T_0$ the statement trivially holds. In order to perform the induction step $(t \rightarrow t+1)$ we will need some facts. First, let $x = 1 - f_t$ and note that

$$f_t - f_t^d = (1 - x) - (1 - x)^d \leq (d - 1)x = (d - 1)\frac{p_t}{d(p_t + du_t)} \leq \frac{d - 1}{d^2} \frac{p_t}{u_t}.$$

So, we readily obtain the upper bound for p_t in (11) by using the the recursion for p_t as follows.

$$p_{t+1} \leq \left(1 - \frac{1}{d}\right) f_t p_t + d u_t \cdot \frac{d-1}{d^2} \frac{p_t}{u_t} \leq 2\left(1 - \frac{1}{d}\right) p_t = q p_t \Rightarrow p_{t+1} \leq P_{T_0} \cdot q^{t+1-T_0}.$$

To see the lower bound for p_t, note that $\frac{1}{1+\frac{p_t}{d u_t}} \geq 1 - \frac{p_t}{d u_t}$. Also, a similar calculation as above and by using the fact $(1-x)^d \leq 1 - dx + \binom{d}{2}x^2$ for $x \geq 0$ reveals that

$$f_t - f_t^d \geq (d-1)x - \binom{d}{2}x^2 \geq \frac{d-1}{d} \frac{p_t}{p_t + d u_t} - \frac{d^2}{2} \frac{p_t^2}{d^2(p_t + d u_t)^2}$$

$$\geq \frac{d-1}{d^2} \frac{p_t}{u_t(1+\frac{p_t}{d u_t})} - \frac{p_t^2}{2d^2 u_t^2} \geq \frac{d-1}{d^2} \frac{p_t}{u_t} - \frac{3p_t^2}{2d^2 u_t^2}.$$

By using again the recursion for p_t we infer that

$$p_{t+1} \geq \left(1 - \frac{1}{d}\right) f_t p_t + d u_t \cdot \left(\frac{d-1}{d^2} \frac{p_t}{u_t} - \frac{3p_t^2}{2d^2 u_t^2}\right) \geq q p_t - \frac{2}{d} \frac{p_t^2}{u_t}.$$

Note that the induction hypothesis and the fact $q^{t-T_0} \leq \frac{\varepsilon n}{\ln^7 n}$ imply that $u_t \geq n/2$. So,

$$p_{t+1} \geq q p_t - \frac{4}{dn} p_t^2 \geq P_{T_0} q^{t+1-T_0} - \frac{3P_{T_0}^2 q^{2(t-T_0)+1}}{n} - \frac{4}{dn}\left(P_{T_0} q^{t-T_0}\right)^2$$

$$= P_{T_0} q^{t+1-T_0} - \frac{P_{T_0}^2 q^{2(t-T_0+1)}}{n}\left(\frac{3}{q} + \frac{4}{dq^2}\right) \geq P_{T_0} q^{t+1-T_0} - 3\frac{P_{T_0}^2 q^{2(t-T_0+1)}}{n}.$$

This proves the lower bound for p_t in (11). Next we prove the bounds for u_{t+1}. Note that

$$\frac{u_{t+1}}{u_t} = \left(1 - \frac{p_t}{d(p_t + d u_t)}\right)^d \geq 1 - \frac{p_t}{p_t + d u_t} \geq 1 - \frac{p_t}{d u_t} \quad \Rightarrow \quad u_{t+1} \geq u_t - \frac{p_t}{d}.$$

A similar calculation using the fact $(1-x)^d \leq 1 - dx + \binom{d}{2}x^2$ for $x \geq 0$ reveals that

$$\frac{u_{t+1}}{u_t} \leq 1 - \frac{p_t}{p_t + d u_t} + \binom{d}{2}\frac{p_t^2}{d^2(p_t + d u_t)^2} \leq 1 - \frac{p_t}{d u_t} + \frac{3}{4}\frac{p_t^2}{u_t^2}.$$

Recall that the induction hypothesis guarantees $u_t \geq n/2$. The above facts together with the bounds for p_t imply after a straightforward but lengthy calculation (12). We omit the details.

The above discussion settles the growth of p_t and u_t up to the time t_1. Note that $t_1 = \ln(2(1-\frac{1}{d}))^{-1} \ln n + \Theta(\ln \ln n)$. In order to deal with $t > t_1$ let us first make two important observations. First, note that at t_1 we have that

$$\frac{p_{t_1}}{u_{t_1}} = \Omega(1). \tag{13}$$

Let us next consider the ratio $r_t := p_t/u_t$. Note that $f_t = 1 - \frac{p_t}{d(p_t+du_t)} = 1 - \frac{1}{d(1+d/r_t)}$. The recursions for p_t and u_t imply that

$$r_{t+1} = \left(1 - \frac{1}{d}\right) f_t^{-d+1} r_t + d(f_t^{-d+1} - 1) \Rightarrow \frac{r_{t+1}}{r_t} = \left(1 - \frac{1}{d}\right) f_t^{-d+1} + \frac{d}{r_t}(f_t^{-d+1} - 1).$$

Consider the function

$$g(x) = \left(1 - \frac{1}{d} + \frac{d}{x}\right)\left(1 - \frac{1}{d(1+d/x)}\right)^{-d+1} - \frac{d}{x},$$

and   note   that $\frac{r_{t+1}}{r_t}$
$= g(r_t)$. A straightforward calculation shows that $\lim_{x\to 0} g(x) = 2(1 - \frac{1}{d})$. In the sequel we will argue that g is monotone increasing. This implies $\frac{r_{t+1}}{r_t} \geq g(0) \geq \frac{4}{3}$, and so we have for any $t' > 0$

$$r_{t+t'} \geq r_t \left(\frac{4}{3}\right)^{t'} \Rightarrow p_{t+t'} \geq \left(\frac{4}{3}\right)^{t'} u_{t+t'}. \tag{14}$$

This fact will become very useful later on. To see why g is increasing, note that

$$g'(x) = \frac{-T(1 + d^2/x) + d + d^2/x}{x^2 + xd}, \quad \text{where } T = \left(1 - \frac{1}{d(1+d/x)}\right)^{-d+1}.$$

To show that this is positive for $x \geq 0$, it suffices to show that

$$1 + \frac{d^2}{x} \leq d\left(1 + \frac{d}{x}\right)\left(1 - \frac{1}{d(1+d/x)}\right)^{d-1}.$$

But

$$d\left(1 + \frac{d}{x}\right)\left(1 - \frac{1}{d(1+d/x)}\right)^{d-1} \geq d\left(1 + \frac{d}{x}\right)\left(1 - \frac{d-1}{d(1+d/x)}\right)$$
$$\geq d\left(1 + \frac{d}{x}\right)\left(\frac{d + d^2/x - d + 1}{d(1+d/x)}\right) = 1 + \frac{d^2}{x},$$

which concludes the proof of the monotonicity of g.

Let t_2 be the minimal t such that $p_{t_2} \geq u_{t_2} \ln^2 n$. The Equations (13) and (14) guarantee that $t_2 = t_1 + O(\ln \ln n)$, and moreover that for any $t > t_2$ such that $u_t > 0$ we have $p_t \geq u_t \ln^2 n \geq 1$. Under these conditions note that

$$f_t^d = \left(1 - \frac{p_t}{d(p_t + du_t)}\right)^d = (1 + O(\ln^{-2} n))\left(1 - \frac{1}{d}\right)^d.$$

Thus, for any t such that $t = t_2 + O(\ln n)$ we have that

$$u_t = (1 + o(1)) \cdot \left(1 - \frac{1}{d}\right)^{d(t-t_2)} u_{t_2}.$$

Recall that T_1 is the first t such that $U_{T_1} \leq \ln^7 n$. As $u_{t_2} \leq n$, we readily obtain that $T_1 \leq t_1 + O(\ln \ln n) - \frac{1}{d \ln((1-\frac{1}{d}))} \ln n = C_d \ln n + O(\ln \ln n)$. To see the corresponding lower bound for T_1, note that as long as $p_t \geq 1$ we always have

$$u_{t+1} \geq \left(1 - \frac{1}{d}\right)^d u_t.$$

The proof completes with the fact $u_{t_1} = \Theta(n)$. $\quad\square$

4.4   The Final Phase

Let T_1 be the first time such that the number of uninformed vertices drops below $\ln^7 n$. In the previous section we argued that $T_1 = C_d \ln n + O(\ln \ln n)$, where C_d is given in Corollary 1. We conclude the proof of Theorem 1 with the following.

Lemma 5. *With probability* $1 - o(1)$ *we have* $T - T_1 = O((\ln \ln n)^2)$.

Acknowledgment. We would like to thank Colin McDiarmid for suggesting the use of his concentration inequality (Theorem 4), which greatly facilitated our proofs. We would also like to thank Carola Winzen for a careful reading of our manuscript.

References

1. Bender, E.A., Canfield, E.R.: The asymptotic number of labelled graphs with given degree sequences. J. Combin. Theory Ser. A 24, 296–307 (1978)
2. Bollobás, B.: A probabilistic proof of an asymptotic formula for the number of labelled regular graphs. Europ. J. Combin. 1, 311–316 (1980)
3. Bradonjic, M., Elsässer, R., Friedrich, T., Sauerwald, T., Stauffer, A.: Efficient broadcast on random geometric graphs. In: SODA 2010, pp. 1412–1421 (2010)
4. Chierichetti, F., Lattanzi, S., Panconesi, A.: Almost Tight Bounds for Rumour Spreading with Conductance. In: STOC 2010, pp. 399–408 (2010)
5. Chung, F.R.K., Graham, R., Wilson, R.M.: Quasi-random graphs. Combinatorica 9, 345–362 (1989)
6. Demers, A., Greene, D., Hauser, C., Irish, W., Larson, J., Shenker, S., Sturgis, H., Swinehart, D., Terry, D.: Epidemic algorithms for replicated database maintenance. In: PODC 1987, pp. 1–12 (1987)
7. Dubhashi, D., Panconesi, A.: Concentration of Measure for the Analysis of Randomized Algorithms. Cambridge University Press, Cambridge (2009)
8. Durrett, R.: Random Graph Dynamics. Cambridge University Press, New York (2007)
9. Elsässer, R.: On randomized broadcasting in power law networks. In: Dolev, S. (ed.) DISC 2006. LNCS, vol. 4167, pp. 370–384. Springer, Heidelberg (2006)
10. Elsässer, R., Gasieniec, L., Sauerwald, T.: On radio broadcasting in random geometric graphs. In: Taubenfeld, G. (ed.) DISC 2008. LNCS, vol. 5218, pp. 212–226. Springer, Heidelberg (2008)

11. Elsässer, R., Sauerwald, T.: On Broadcasting vs. mixing and information dissemination on Caley graphs. In: Thomas, W., Weil, P. (eds.) STACS 2007. LNCS, vol. 4393, pp. 163–174. Springer, Heidelberg (2007)
12. Elsässer, R., Sauerwald, T.: On the runtime and robustness of randomized broadcasting. Theoretical Computer Science 410, 3414–3427 (2009)
13. Feige, U., Peleg, D., Raghavan, P., Upfal, E.: Randomized broadcast in networks. Random Structures and Algorithms 1(4), 447–460 (1990)
14. Fountoulakis, N., Huber, A., Panagiotou, K.: Reliable broadcasting and the effect of density. In: IEEE INFOCOM 2010, pp. TS58 (2010)
15. Fountoulakis, N., Panagiotou, K.: Rumor spreading on random regular graphs and expanders, full version, http://arxiv.org/abs/1002.3518
16. Friedman, J.: Some geometric aspects of graphs and their eigenfunctions. Duke Math. J. 69, 487–525 (1993)
17. Frieze, A.M., Grimmett, G.R.: The shortest-path problem for graphs with random arc-lengths. Discrete Appl. Math. 10, 57–77 (1985)
18. Hoory, S., Linial, N., Wigderson, A.: Expander graphs and their applications. Bull. AMS 43, 439–561 (2006)
19. Jagannathan, S., Pandrurangan, G., Srinivasan, S.: Query protocols for highly resilient peer-to-peer networks. In: ISCA PDCS 2006, pp. 247–252 (2006)
20. Janson, S., Łuczak, T., Ruciński, A.: Random Graphs. Wiley, Chichester (2000)
21. Krivelevich, M., Sudakov, B.: Pseudo-random graphs. In: Proceedings of the Conference on Finite and Infinite Sets. Bolyai Society Mathematical Studies, vol. 15, pp. 199–262 (2006)
22. Law, C., Siu, K.-Y.: Distributed construction of random expander networks. In: IEEE INFOCOM 2003, pp. 2133–2143 (2003)
23. McDiarmid, C.: Concentration for independent permutations. Combinatorics, Probability and Computing 11, 163–178 (2002)
24. Nilli, A.: On the second eigenvalue of a graph. Discrete Math. 91, 207–210 (1991)
25. Pandurangan, G., Raghavan, P., Upfal, E.: Building low-diameter peer-to-peer networks. IEEE Journal on Selected Areas in Communications 21, 995–1002 (2003)
26. Talagrand, M.: Concentration of measure and isoperimetric inequalities in product spaces. Inst. Hautes Études Sci. Publ. Math. 81, 73–205 (1995)
27. Thomason, A.: Pseudo-random graphs. In: Proceedings of Random Graphs, pp. 307–331 (1987)

On Testing Computability by Small Width OBDDs

Oded Goldreich

Department of Computer Science, Weizmann Institute of Science, Rehovot, Israel
`oded.goldreich@weizmann.ac.il`

Abstract. We take another step in the study of the testability of small-width OBDDs, initiated by Ron and Tsur (Random'09). That is, we consider algorithms that, given oracle access to a function $f : \{0,1\}^n \to \{0,1\}$, need to determine whether f can be implemented by some restricted class of OBDDs or is far from any such function.

Ron and Tsur showed that testing whether a function $f : \{0,1\}^n \to \{0,1\}$ is implementable by a width-2 OBDD has query complexity $\Theta(\log n)$. Thus, testing width-2 OBDD functions is significantly easier than learning such functions (which requires $\Omega(n)$ queries). We show that such exponential gaps do not hold for several related classes. Specifically:

1. Testing whether $f : \{0,1\}^n \to \{0,1\}$ is implementable by a width-4 OBDD requires $\Omega(\sqrt{n})$ queries.
2. Testing whether $f : \mathrm{GF}(3)^n \to \mathrm{GF}(3)$ is a linear function with 0-1 coefficients requires $\Omega(\sqrt{n})$ queries. Note that this class of functions is a subset of the class of all linear functions over $\mathrm{GF}(3)$, and that each such linear function can be implemented by a width-3 OBDD.
3. There exists a subclass $\mathcal{C}$ of the linear functions from $\mathrm{GF}(2)^n$ to $\mathrm{GF}(2)$ such that testing membership in $\mathcal{C}$ has query complexity $\Theta(n)$. Note that each linear function over $\mathrm{GF}(2)$ can be implemented by a width-2 OBDD.

Recall that each of these classes has a proper learning algorithm of query complexity $O(n)$.

Keywords: Property Testing, Small Width OBDDs.

1 Introduction

In the last couple of decades, the area of property testing has attracted much attention (see, e.g., a couple of recent surveys [18,19]). Loosely speaking, property testing typically refers to super-fast probabilistic algorithms for deciding whether a given object has a predetermined property or is far from any object having this property. Such algorithms, called testers, obtain local views of the object by making suitable queries. The current work belongs to the study of property testing, but pursues what we perceive as somewhat different themes than the standard ones.

M. Serna et al. (Eds.): APPROX and RANDOM 2010, LNCS 6302, pp. 574–587, 2010.

1.1   Testing Membership in Complexity Classes

In the foregoing description, objects are viewed as functions, and so properties are sets of functions. Given this perspective, it is most natural to ask whether various *traditional complexity classes* are testable. Arguably, this question was not addressed till [20].[1] Instead, whenever (before [20]) standard computational devices were referred to in the context of property testing, the perspective was that each fixed *computational device* defines a set of strings and the testing problem studied was of membership of the input string in this set (cf. [2,16,14]). In contrast, following Ron and Tsur [20], we fix a *complexity class* and study the testing problem that refers to whether the input function is in this class.

To illustrate the difference recall that Alon *et al.* [2] fix any regular set, and study the problem of testing whether a given (input) string is in the set. In contrast, Ron and Tsur [20] consider the complexity class of width-2 OBDDs,[2] and study the problem of testing whether a given (input) function belongs to this complexity class.

The main result of [20] is that testing width-2 OBDD has query complexity $\Theta(\log n)$, where n denotes the length of the argument to the function being tested (i.e., the question is whether $f : \{0,1\}^n \to \{0,1\}$ can be implemented by a width-2 OBDD). This should be compared to the query complexity of *learning* this very class, which is $\Theta(n)$. Thus, testing this complexity class is significantly easier than learning this class. Two natural questions arise:

1. What about width-w OBDDs, for any fixed $w > 2$?

   That is, is testing width-w OBDDs significantly easier (i.e., (poly) logarithmically easier) than learning width-w OBDDs? (Recall that learning width-w OBDDs requires $\Omega(n)$ queries, whereas proper learning is possible with $O(n)$ queries.)

2. What about testing subclasses of width-w OBDDs, for any fixed $w \geq 2$ (i.e., testing whether a given function belongs to a fixed subclass of width-w OBDDs)? Specifically, is every subclass of width-2 OBDDs testable in query complexity $O(\log n)$ or poly$(\log n)$?[3]

---

[1] Indeed, this is a controversial statement, which relies on not viewing the classes of dictatorship functions, juntas, monomials, and constant-term DNFs as traditional complexity classes. The testability of these classes was studied in various works; see, for example [17,9,6]. Some readers have expressed strong disagreement with our views, claiming that the foregoing classes are not that different from constant-width OBDDs. We remain unconvinced by their objections, and argue that traditional complexity classes refer to natural computing devices (ruling out polynomials) and furthermore to computing devices that at the very least can scan their entire input (ruling out constant-size decision trees, etc).

[2] OBDDs are ordered binary decision diagrams, which are a restricted type of read-once branching programs in which the variables are read in a fixed order (across all possible computation paths). See definition in Section 1.5.

[3] Note that the query complexity of testing such a subclass need not be smaller that the query complexity of testing the class.

We provide rather gloomy answers to both questions: We prove that even at low computational complexity levels such as constant-width OBDDs, testing may not be significantly easier than learning; that is, the complexities of these two tasks are polynomially related rather than being exponentially related. Specifically:

Theorem 1 (see Theorem 4.2): *Testing width-4 OBDD requires* $\Omega(\sqrt{n})$ *queries.*

We conjecture that the actual query complexity is $\Theta(n)$.

Theorem 2 (see Theorem 2.1): *There exists a subclass of width-2 OBDDs such that testing this subclass requires* $\Omega(n)$ *queries. Furthermore, this subclass is a class of linear functions* (over GF(2)).

1.2   Subclasses of Linear and Quadratic Functions

A different perspective on our results is best illustrated by a question of Shafi Goldwasser, who asked whether there is more to algebraic property testing than testing low degree. (Needless to say, this was a rhetorical question; she meant to advocate such studies.) We mention that a clear example of such a study was provided by Rubinfeld [22] in the mid 1990s, and that various properties of polynomials (e.g., dictatorship functions [17], juntas [9,4], sparse polynomials [6,7]) were studied in the last decade (although these studies were not viewed from this perspective).

In any case, taking this perspective, we view Theorem 2 as saying that a certain property of linear functions (from GF(2)^n to GF(2)) cannot be tested significantly faster than learning (i.e., cannot be tested with $o(n)$ queries). More generally, we present a full hierarchy of properties (or classes) of linear functions arranged by their query complexity:

Theorem 3 (see Theorem 2.3): *For every function* $t : \mathsf{N} \rightarrow \mathsf{N}$ *that is at most linear, there exists a property of linear functions* (over GF(2)) *such that testing this property has query complexity* $\Theta(t + \epsilon^{-1})$. *Furthermore, learning each of the corresponding concept classes requires* $\Omega(n)$ *queries.*

This leads to the question of how natural are these properties, which build on the property used in the proof of Theorem 2. Since the property is not very natural, we also prove the following.

Theorem 4 (see Theorem 2.7 in our technical report [10]): *Testing the set of linear functions from* GF(2)^n *to* GF(2) *with at most* $n/2$ *influential variables requires* $\Omega(\sqrt{n})$ *queries.*

Here too, we conjecture that the actual query complexity is $\Theta(n)$. Another natural property of linear functions is the subject of the following result.

Theorem 5 (see Theorem 3.2): *Testing the class of linear functions from* GF(3)^n *to* GF(3) *that have 0-1 coefficients requires* $\Omega(\sqrt{n})$ *queries.*

Again, we conjecture that the actual query complexity is $\Theta(n)$. (Note that the foregoing class is implemented by width-3 OBDDs.) Lastly, we mention that the proof of Theorem 1 actually establishes also the following.

Theorem 6 (see end of Section 4): *Testing the class of linear functions from* $GF(2)^n$ *to* $GF(2)$ *that have no consecutive influential variables requires* $\Omega(\sqrt{n})$ *queries.*

And, again, we conjecture that the actual query complexity is $\Theta(n)$.

1.3   Techniques

The proofs of all the foregoing lower bounds, with the exception of Theorem 2, follow a common theme and cope with a similar difficulty. The common theme is that in all these cases the analysis reduces to upper-bounding the ability of query-bounded observers to distinguish two specific distributions of linear functions. In each case, these two distributions are very natural, and the difficulty is in analyzing the corresponding answer distributions (i.e., the distributions of the sequence of answers obtained by querying each function distribution).

To illustrate the difficulty, consider the set of linear functions from $GF(3)^n$ to $GF(3)$, denoted $\mathcal{L}_3$. It is well known that if f is uniformly distributed in $\mathcal{L}_3$, then its values on a sequence of t linearly independent vectors are uniformly distributed over $GF(3)^t$. But it is less clear what happens when f is uniformly distributed in some natural subset $\mathcal{L}_3' \subset \mathcal{L}_3$. In particular, what happens when $\mathcal{L}_3'$ is the set of all linear functions with 0-1 coefficients? Furthermore, what if these t strings are selected adaptively?

Our proofs deal with these types of problems. For example, in the case of the set of $GF(3)$-linear functions with either no or a single 2-valued coefficient, we prove that *the deviation of the answers to t (adaptively chosen) queries is at most* $O(t^2/n)$ (cf. Theorem 3.2).

1.4   Discussion

In response to comments of some anonymous reviewers, we further articulate what we perceive to be the main conceptual messages of this work.

As stated in Section 1.1, most works in property testing that mention standard notions of computational complexity refer to the complexity of the properties being tested (i.e., the complexity of determining whether a given object has the said property). In contrast, following Ron and Tsur [20], we consider the complexity of evaluating (or implementing) single functions that have the tested property. We ask how simple may such functions be as to form a class that is relatively hard to test in the sense that testing membership in the class has almost the same query complexity as learning functions in the class.

We note that the hardness result of [11,12] can be interpreted as addressing this question. For example, one may obtain a class of functions such that each function can be evaluated by a polynomial-size circuit, while testing membership in this class requires essentially as many queries as learning functions in this class. A closer look at these constructions reveals that the functions can be implemented by a poly(ℓ)-sized circuit, where ℓ is logarithmic in the query complexity of testing.

The results of this paper indicate that such hardness (of testing) results may hold for classes of functions that are implementable by computing devices of very low complexity. We mention that this assertion holds in two different senses. The first (and weaker) sense is that there exist hard-to-test properties that *consist of* functions that are all implementable by computing devices of very low complexity (i.e., width-2 OBDDs). The second (and stronger) sense is that there exists a natural low complexity class (i.e., width-4 OBDDs) such that the property of *belonging to that class* is hard to test.

The gap between the two aforementioned senses is demonstrated by contrasting the tester of the class of width-2 OBDDs obtained in [20] with the lower bound stated in Theorem 2. We also note that, while it seems that almost every natural class of functions has a subclass that is hard to test, our results regarding the hardness of testing subclasses of linear functions refer to *natural subclasses* (i.e., natural properties).

1.5   Preliminaries: OBDDs and Property Testing

In this section we review the quite standard definitions used in this paper. We merely stress that when we talk of OBDDs, we assume (as in [20]) that the order of the variables is fixed (and known).

OBDDs: Ordered Binary Decision Diagrams. Several different definitions of this notion appear in the literature, and we adopt the one that calls for a fixed ordering of the variables (knows as "strict"). That is, an ordered binary decision diagram (OBDD) is a read-once branching program in which the order in which the variables are read is fixed for all computing devices in the model. Specifically, we shall assume, without loss of generality, that the ith variable is always read at the ith level. This yields the following definition.

Definition 7. *An* OBDD *is a directed acyclic graph with labeled edges and marked sinks that satisfies the following conditions:*

1. *The graph contains a single source vertex.*
2. *Each sink vertex in the graph is marked either 0 or 1.*
3. *Each non-sink vertex has two out-going edges (which may be parallel) one labeled 0 and the other labeled 1.*
4. *The graph edges connect vertices in consecutive levels, where the* level *of a vertex is its distance from the source.*
5. *All sinks have the same level, called the graph* length.

The width *of an OBDD is the maximum number of vertices that have the same level. An OBDD of length n computes the function $f : \{0,1\}^n \to \{0,1\}$ such that, for every $x \in \{0,1\}^n$ it holds that the sink that is reached from the source by following the path with edge labels x is marked $f(x)$.*

Indeed, we may view $x = x_1 \cdots x_n$ as a sequence of variables, and observe that in the ith step (i.e., when moving from the $i - 1^{\text{st}}$ level to the ith level) the OBDD branches according to the value of x_i.

We mention that in a subsequent work, Ron and Tsur [21] considered OBDDs with a variable ordering of the variables. Indeed, in such a case, one should specify the ordering, and in more general models that allow different variables to be queried along different computation paths it is necessary to specify the variable queried at each non-sink vertex (by marking the non-sink vertices with variable names).

Property Testing. We merely recall the standard definition.

Definition 8. *Let $\Pi = \bigcup_{n \in \mathbb{N}} \Pi_n$, where Π_n contains functions defined over the domain D_n (and range R_n). A* tester for a property *Π is a probabilistic oracle machine T that satisfies the following two conditions:*

1. *The tester accepts each $f \in \Pi$ with probability at least $2/3$; that is, for every $n \in \mathbb{N}$ and $f \in \Pi_n$ (and every $\epsilon > 0$), it holds that $\Pr[T^f(n, \epsilon) = 1] \geq 2/3$.*
2. *Given $\epsilon > 0$ and oracle access to any f that is ϵ-far from Π, the tester rejects with probability at least $2/3$; that is, for every $\epsilon > 0$ and $n \in \mathbb{N}$, if $f : D_n \to R_n$ is ϵ-far from Π_n, then $\Pr[T^f(n, \epsilon) = 0] \geq 2/3$, where f is ϵ-far from Π_n if, for every $g \in \Pi_n$, it holds that $|\{e \in D_n : f(e) \neq g(e)\}| > \epsilon \cdot |D_n|$.*

If the tester accepts every function in Π with probability 1, then we say that it has one-sided error*; that is, T has one-sided error if for every $f \in \Pi$ and every $\epsilon > 0$, it holds that $\Pr[T^f(n, \epsilon) = 1] = 1$. A tester is called* non-adaptive *if it determines all its queries based solely on its internal coin tosses (and the parameters n and ϵ); otherwise it is called* adaptive*.*

Almost all our results are lower bounds on the query complexity of property testing tasks, and they are obtained for fixed values of the proximity parameter ϵ (i.e., $\epsilon = 1/16$ will do in all). In these cases we omit mention of the proximity parameter.

2   Testing Subclasses of Width 2 OBDDs

We consider various subclasses of linear functions over $GF(2)$, which in particular are realizable by width-2 OBDDs. For a set of strings $S \subseteq \{0, 1\}^n$ we denote by $\mathcal{L}_S$ the set of linear functions $\{f_v : v \in S\}$, where $f_v : \{0, 1\}^n \to \{0, 1\}$ satisfies $f_v(x) \overset{\text{def}}{=} \langle v, x \rangle = \sum_{i=1}^{n} v_i x_i \bmod 2$.

We present a hierarchy of properties of linear functions arranged according to the query complexity of testing them. Our starting point is a property of linear functions having maximal query complexity, and the hierarchy can be derived using any such property. (This is indeed reminiscent of [12].) After establishing the said hierarchy (and since it refers to somewhat unnatural properties), we also consider the natural property of linear function having a bounded number of influential variables.

2.1   A Hierarchy of Classes of Linear Functions

We start by presenting a class of linear functions that is hard to test, and then exhibit the full hierarchy by combining any such class with the class of all linear functions.

Linear functions with coefficients from a small-bias space. Let $S \subset \{0,1\}^n$ be a small bias sample space [15,1], say, of size $2^{0.99n}$ and bias $2^{-0.3n}$. Then, testing $\mathcal{L}_S$ requires $\Omega(n)$ queries, even if we allow two-sided error and adaptive testers. More generally, we have the following.

Theorem 2.1 (Theorem 2, restated): *Let $S \subset \{0,1\}^n$ be a δ-bias sample space; that is, for every $c \in \{0,1\}^n \setminus \{0^n\}$, it holds that $|\mathrm{Pr}_{v \in S}[\langle c, v \rangle = 1] - 0.5| \leq \delta$. Then, testing $\mathcal{L}_S$ requires $\log_2((1 - \rho)/3\delta)$ queries, where $\rho = |S|/2^n$.*

The proof of Theorem 2.1 can be found in our technical report [10]. Typically (e.g., in the following example), ρ is small (i.e., $\rho \leq 1/2$), and so the lower bound simplifies to $\log_2(1/6\delta)$. An appealing example consists of the set of all n-bit long strings having a number of 1 that is a multiple of 3 (i.e., $S = \{v \in \{0,1\}^n : \sum_{i=1}^{n} v_i \equiv 0 \pmod 3\}$), which has exponentially small bias and density $\approx 1/3$ (see Proposition A.1 in our technical report [10]). Thus, we get

Corollary 2.2. *Let S be the set of all n-bit strings having a number of 1-entries that is divisible by three. Then, testing $\mathcal{L}_S$ requires $\Omega(n)$ queries.*

The Hierarchy. The following hierarchy theorem follows by combining any set of hard-to-test linear functions (from $\mathrm{GF}(2)^t$ to $\mathrm{GF}(2)$) with the class of all linear functions (from $\mathrm{GF}(2)^{n-t}$ to $\mathrm{GF}(2)$).

Theorem 2.3 (Theorem 3, restated): *For every function $t : \mathsf{N} \to \mathsf{N}$ that is at most linear, there exist sets $S \subseteq \{0,1\}^n$ such that testing $\mathcal{L}_S$ has query complexity $\Theta(t + \epsilon^{-1})$. Furthermore, learning $\mathcal{L}_S$ requires $\Omega(n)$ queries.*

The proof of Theorem 2.3 can be found in our technical report [10]. Here we only mention the fact that a straightforward tester for the aforementioned class has query complexity $\widetilde{O}(t)$, but a slightly more sophisticated tester obtains the $O(t)$ bound.

Linear functions in a fixed linear space. Recall that the standard linearity property (i.e., the set of all linear functions over $\mathrm{GF}(2)$) is testable by $O(1/\epsilon)$ non-adaptive queries. Here we point out that this is not the only property of linear functions having $\Theta(1/\epsilon)$ testing complexity, but is merely a special case of a larger class of properties. Specifically, we consider arbitrary classes $\mathcal{L}_S$ such that S is a linear space. That is, let $S = \{Gs : s \in \{0,1\}^k\}$, where G is an k-dimensional generator matrix. Thus, for every $s \in \{0,1\}^k$, we define the function $g_s \in \mathcal{L}_S$ as $g_s(x) = f_{Gs}(x) = \langle Gs, x \rangle$, and note that $\langle Gs, x \rangle = \langle s, G^\top x \rangle$.

Theorem 2.4. *Let $S \subseteq \{0,1\}^n$ be a linear space, and $\mathcal{L}_S = \{f_v : v \in S\}$. Then, $\mathcal{L}_S$ can be tested with $O(1/\epsilon)$ non-adaptive queries.*

The proof of Theorem 2.4 can be found in our technical report [10].

2.2   Linear Functions with at Most ρn Influential Variables

For any constant $\rho > 0$, let W_ρ denote the class of linear functions with at most ρn influential variables. That is, $W_\rho = \mathcal{L}_S$ for $S = \{v : \mathrm{wt}(v) \le \rho n\}$, where $\mathrm{wt}(v) = |\{i : v_i = 1\}|$.

Conjecture 2.5. *Testing $W_{0.5}$ requires $\Omega(n)$ queries, even when allowing adaptive testers of two-sided error.*

If true, then (by using techniques as in the proof of Theorem 2.3) it will follow that, for any function $\rho : \mathsf{N} \to [0,1]$, testing W_ρ requires $\Omega(\rho(n) \cdot n)$ queries. In our technical report [10] we present two partial results that support Conjecture 2.5: the first is an $\Omega(n)$ lower bound for non-adaptive testers and the second is an $\Omega(\sqrt{n})$ lower bound for general (adaptive) testers. In particular, this establishes Theorem 4.

3   Hardness of Testing a Subclass of Width 3 OBDDs

We shall consider the class of linear functions over $\mathrm{GF}(3)$, consisting of all such functions that have binary coefficients. That is, for every $v \in \{0,1\}^n$, we consider the function $f_v : \mathrm{GF}(3)^n \to \mathrm{GF}(3)$ defined by $f_v(x) = \sum_{i=1}^n v_i x_i$, where the arithmetic is modulo 3. Let $\mathcal{BL}_3 = \{f_v : v \in \{0,1\}^n\}$.

Conjecture 3.1. *Testing $\mathcal{BL}_3$ requires $\Omega(n)$ queries, even when allowing adaptive testers of two-sided error.*

Theorem 3.2 (Theorem 5, restated): *Testing $\mathcal{BL}_3$ requires $\Omega(\sqrt{n})$ queries, even when allowing adaptive testers of two-sided error.*

Proof: We consider the class $\mathrm{BAD} = \{b_{j_0,v} : j_0 \in [n], v \in \{0,1\}^n\}$ such that $b_{j_0,v}(x) \stackrel{\mathrm{def}}{=} f_v(x) + x_{j_0}$. Note that all functions in BAD are linear and that exactly half of BAD is not in $\mathcal{BL}_3$ (since $b_{j_0,v} \in \mathcal{BL}_3$ if and only if $v_{j_0} = 0$). Hence, with probability $1/2$, a uniformly selected function in BAD is $2/3$-far from $\mathcal{BL}_3$. Our goal is to prove that distinguishing a uniformly selected function in $\mathcal{BL}_3$ from a uniformly selected function in BAD requires $\Omega(\sqrt{n})$ queries.

   Recall that an element in either sets is selected by specifying an index $j_0 \in [n]$ and an n-bit string. Fixing any sequence of queries $\bar{q} = (q^{(1)}, ..., q^{(t)})$, we shall show that if this sequence has a certain feature with respect to j_0, then the answers are distributed almost identically in the two distributions. This feature is defined next, where w is an integer (i.e., we shall use $w = \sqrt{n}$).

Definition 3.2.1. *An index $j \in [n]$ is called w-special with respect to a sequence of queries $\bar{q} = (q^{(1)}, ..., q^{(t)})$ if there exists a linear combination of these queries that yields an n-bit string q such that $j \in \mathrm{supp}(q)$ and $|\mathrm{supp}(q)| \leq w$, where $\mathrm{supp}(q) \overset{\mathrm{def}}{=} \{i : q_i \neq 0\}$.*

It will be convenient to use matrix notation in our analysis. Presenting $\bar{q}$ as a matrix, denoted Q, such that the ith row of Q equals $q^{(i)}$, the foregoing condition asserts that there exists a t-vector c such that $\mathrm{supp}(cQ)$ contains j as well as at most $w - 1$ other indices. Thus, we get:

Claim 3.2.2. *For any sequence of t queries, $\bar{q}$, there exists at most $w \cdot t$ indices that are w-special with respect to $\bar{q}$.*

Proof: Let S denote the set of w-special indices with respect to $\bar{q}$. For every $j \in S$, there exists a t-vector $c^{(j)}$ such that $\mathrm{supp}(c^{(j)}Q)$ contains j as well as at most $w - 1$ other elements of S. Using a greedy strategy, we can obtain a set I of at least $|S|/w$ elements of S such that for every $j \in I$ it holds that $\mathrm{supp}(c^{(j)}Q) \cap I = \{j\}$. Thus, the rank of Q is lower bounded by $|S|/w$, and the claim follows.   □

Claim 3.2.3. *Suppose that j_0 is not w-special with respect to $\bar{q} = (q^{(1)}, ..., q^{(t)})$. Then, for every $\alpha \in \{0, 1, 2\}^t$, when $v = (v_1, ..., v_n)$ is selected uniformly in $\{0, 1\}^n$, it holds that*

$$\mathrm{Pr}_v[(f_v(q^{(1)}), ..., f_v(q^{(t)})) = \alpha] = \mathrm{Pr}_v[(b_{j_0,v}(q^{(1)}), ..., b_{j_0,v}(q^{(t)})) = \alpha] \pm 2^{-(w-1)}. \tag{1}$$

Proof: For every $\alpha \in \{0, 1, 2\}^n$, we denote by $D_{j_0,\bar{q}}(\alpha)$ the difference between the two probabilities in Eq. (1); that is,

$$D_{j_0,\bar{q}}(\alpha) \overset{\mathrm{def}}{=} \mathrm{Pr}_v[(f_v(q^{(1)}), ..., f_v(q^{(t)})) = \alpha] - \mathrm{Pr}_v[(b_{j_0,v}(q^{(1)}), ..., b_{j_0,v}(q^{(t)})) = \alpha]. \tag{2}$$

Our aim is to prove that the max-norm of $D_{j_0,\bar{q}}(\cdot)$ is at most $2^{-(w-1)}$. By using the relation between bases (cf. Lemma A.5 in our technical report [10]).[4] it suffices to show that for every $c \in \{0, 1, 2\}^t$ it holds that

$$\sum_{\tau \in \{0,1,2\}} \left| \sum_{\alpha \in S_{c,\tau}} D_{j_0,\bar{q}}(\alpha) \right| \leq 2^{-(w-1)}, \tag{3}$$

where $S_{c,\tau} \overset{\mathrm{def}}{=} \{\alpha \in \{0, 1\}^t : \sum_{i=1}^t c_i \alpha_i = \tau\}$ denotes the set of all t-bit vectors that have $3k + \tau$ non-zero entries (for some k). The l.h.s of Eq. (3) equals

$$\sum_{\tau \in \{0,1,2\}} \left| \mathrm{Pr}_v \left[\sum_{i=1}^t c_i f_v(q^{(i)}) = \tau \right] - \mathrm{Pr}_v \left[\sum_{i=1}^t c_i b_{j_0,v}(q^{(i)}) = \tau \right] \right| \tag{4}$$

---

[4] Specifically, letting ω denote the third root of unity, it suffices to upper-bound $|\sum_{\tau \in \mathrm{GF}(3)} \omega^\tau \sum_{\alpha \in S_{c,\tau}} D_{j_0,\bar{q}}(\alpha)|$, where $S_{c,\tau} = \{\alpha : \sum_i c_i \alpha_i = \tau\}$. Instead, we upper-bound each of the three terms of the outer summation (and use $|\omega| = 1$).

Using the linearity of both functions, and moving to matrix notation, each term in Eq. (4) equals

$$\Pr_v[f_v(cQ) = \tau] \; - \; \Pr_v[b_{j_0,v}(cQ) = \tau], \tag{5}$$

which equals $\Pr_v[cQv = \tau] - \Pr_v[cQ(v + u^{j_0}) = \tau]$, where $u^{j_0} = 0^{j_0-1}10^{n-j_0}$ is the j_0^{th} unit vector. Thus, Eq. (4) equals

$$\sum_{\tau \in \{0,1,2\}} \left| \Pr_v \left[cQv = \tau \right] \; - \; \Pr_v[cQv + cQu^{j_0} = \tau] \right|. \tag{6}$$

To upper-bound Eq. (6), we consider two cases (regarding the value of cQu^{j_0}). If $cQu^{j_0} = 0$, then Eq. (6) equals zero. On the other hand, if $cQu^{j_0} \neq 0$, then $\text{supp}(cQ)$ contains j_0, and it follows that $|\text{supp}(cQ)| > w$ (because otherwise j_0 would have been w-special w.r.t $\bar{q}$). But in this case, it follows that $\sum_{\tau \in \{0,1,2\}} |\Pr_v[cQv = \tau] - \frac{1}{3}| < 2^{-w}$ and the same holds for $\Pr_v[cQv = \tau - cQu^{j_0}]$. Thus, Eq. (6) is upper-bounded by $2 \cdot 2^{-w}$, and the claim follows. □

Armed with Claims 3.2.2 and 3.2.3, we prove the theorem by considering the sequence of queries in the order they were issued. Setting $w = \sqrt{n}$, we evaluate the situation after each additional query. Using Claim 3.2.3, we note that as long as j_0 is not special with respect to the queries made, the answers are almost oblivious of whether the function is uniformly selected in BAD or in $\mathcal{BL}_3$ in the sense that the probabilistic deviation on each possible sequence of answers (i.e., α) is at most $2^{-(w-1)}$. Recalling that the functions in $\mathcal{BL}_3$ are oblivious of j_0, it follows that the answers obtained from a random function in BAD are also almost oblivious of j_0 (as long as j_0 is not special with respect to the queries made). Noting that the answers determine the next query, we infer that this query is also almost oblivious of the currently non-special value of j_0, and so the probability that j_0 is special with respect to the augmented sequence of queries can be bounded using Claim 3.2.2. Details follow.

We may assume, (as usual and) without loss of generality, that the tester is deterministic, and so the query sequence is determined adaptively by the previous answers. Thus, we consider the 3^{t-1} possible t-query sequences that arise from each possible sequence of t answers. For each such sequence, we first dispose of the case that j_0 is special with respect to it, which by Claim 3.2.2 happens with probability at most tw/n. Assuming that j_0 is not special with respect to that sequence, we conclude (by Claim 3.2.3) that the corresponding sequence of answers occurs with about the same probability in both distributions. Over all, the statistical distance between the observed answers is at most $(tw/n) + 3^{t-1} \cdot 2^{-(w-1)}$, and the theorem follows. ∎

4   Hardness of Testing the Class of Width 4 OBDDs

In this section we establish Theorems 1 and 6.

Conjecture 4.1. *Testing the class of functions that are implementable by width-4 OBDDs requires $\Omega(n)$ queries, even when allowing adaptive testers of two-sided error.*

Theorem 4.2 (Theorem 1, restated): *Testing the class of functions that are implementable by width-4 OBDDs requires $\Omega(\sqrt{n})$ queries, even when allowing adaptive testers of two-sided error.*

Proof: We consider Boolean functions of $4n$-bit long strings, which are quadratic polynomials over $GF(2)$. Specifically, these functions are linear combinations of n quadratic expressions, where each quadratic expression refers to a distinct block of four variables. A generic block, containing the variables x_1, x_2, x_3, x_4, will contribute a linear combination of x_1x_3 and x_2x_4, where the combination $x_1x_3 + x_2x_4$ is considered bad because the expression $x_0 + x_1x_3 + x_2x_4$ cannot be computed by a width-4 OBDDs. Specifically, letting $f_0(x_1, x_2, x_3, x_4) = 0$, $f_1(x_1, x_2, x_3, x_4) = x_1x_3$, and $f_2(x_1, x_2, x_3, x_4) = x_2x_4$, we will consider the class GOOD that consists of functions of the form $g_{\sigma_1,...,\sigma_n}$ such that

$$g_{\sigma_1,...,\sigma_n}(x_1, ..., x_{4n}) = \sum_{j \in [n]} f_{\sigma_j}(x_{4(j-1)+1}, ..., x_{4(j-1)+4}), \qquad (7)$$

where $\sigma_1, ..., \sigma_n \in \{0, 1, 2\}$. Note that each such function can be computed by a width-4 OBDD, which uses one "bit" to store the accumulated sum and another "bit" to compute the value of the current block. In contrast, the class BAD consists of functions of the form $b_{j_0,\sigma_1,...,\sigma_n}$ such that

$$b_{j_0,\sigma_1,...,\sigma_n}(x_1, ..., x_{4n}) = \sum_{j \in [n] \setminus \{j_0\}} f_{\sigma_j}(x_{4(j-1)+1}, ..., x_{4(j-1)+4})$$
$$+ x_{4(j_0-1)+1}x_{4(j_0-1)+3} + x_{4(j_0-1)+2}x_{4(j_0-1)+4} \qquad (8)$$

Since, except when $\sigma_1 \cdots \sigma_{j_0-1} = 0^{j_0-1}$, the j_0^{th} block can not be computed by a width-4 OBDD (while maintaining the accumulated sum), it follows that such functions are $1/16$-far from the set of functions that are computable by width-4 OBDDs (see [10, Lem. A.6], which is a simple version of Yao's XOR Lemma for OBDDs, which is also an over-kill).

Our goal is to prove that a random function in GOOD is hard to distinguish from a random function in BAD, where "random" does not necessarily refer to the uniform distribution over the corresponding set (but rather any two distributions will do). Specifically, we consider a distribution over GOOD, in which each σ_i is set to 0 with probability $1/2$ and is uniformly distributed in $\{1, 2\}$ otherwise. (This random selection process determines a function $g_{\sigma_1,...,\sigma_n} \in$ GOOD.) We consider a related distribution over GOOD $\cup$ BAD, where $\sigma_1, ..., \sigma_n$ are selected as above, the index j_0 is selected uniformly in $[n]$, and the function being determined is $g_{\sigma_1,...,\sigma_n} + a_{j_0}$, where $a_{j_0}(x_1, ..., x_{4n}) = x_{4(j_0-1)+1}x_{4(j_0-1)+3} + x_{4(j_0-1)+2}x_{4(j_0-1)+4}$. Note that the resulting function is in BAD if and only if both $\sigma_1 \cdots \sigma_{j_0-1} \neq 0^{j_0-1}$ and $\sigma_{j_0} = 0$, which means that it is in BAD with probability $\frac{1}{2} - o(1)$.

Our analysis reduces to analyzing related families of linear functions defined over variables $y_1, ..., y_{2n}$ such that $y_{2(j-1)+1} = x_{4(j-1)+1}x_{4(j-1)+3}$ and $y_{2(j-1)+2} = x_{4(j-1)+2}x_{4(j-1)+4}$. Specifically, we first show that distinguishing the foregoing two distributions (of quadratic functions) leads to distinguishing the two corresponding distributions of linear functions, where in both the latter distributions $\sigma_1, ..., \sigma_n$ and j_0 are selected as above (i.e., j_0 is distributed uniformly in $[n]$ and each σ_i is set to 0 with probability $1/2$ and is uniformly distributed in $\{1, 2\}$ otherwise). Letting $f_0'(y_1, y_2) = 0$, $f_1'(y_1, y_2) = y_1$, and $f_2'(y_1, y_2) = y_2$, the linear functions in these two distributions are:

$$g_{\sigma_1,...,\sigma_n}'(y_1, ..., y_{2n}) = \sum_{j \in [n]} f_{\sigma_j}'(y_{2(j-1)+1}, y_{2(j-1)+2}) \tag{9}$$

$$b_{j_0,\sigma_1,...,\sigma_n}'(y_1, ..., y_{2n}) = g_{\sigma_1,...,\sigma_n}'(y_1, ..., y_{2n}) + y_{2(j_0-1)+1} + y_{2(j_0-1)+2} \tag{10}$$

The reduction between these distinguishing problems is quite straightforward: Given a distinguisher D for the original distinguishing problem (i.e., regarding quadratic functions), we obtain a distinguisher D' for the distinguishing problem regarding linear functions. The new distinguisher (i.e., D') invokes D and serves each query $q = (q_1, ..., q_{4n})$ that it issues (to its quadratic oracle) by forwarding the query $q' = (q_1', ..., q_{2n}')$ to the actual (linear function) oracle, where $q_{2(j-1)+1}' = q_{4(j-1)+1}q_{4(j-1)+3}$ and $q_{2(j-1)+2}' = q_{4(j-1)+2}q_{4(j-1)+4}$ for every $j \in [n]$. Thus, when given oracle access to $g_{\sigma_1,...,\sigma_n}'$, we emulate an execution of D with $g_{\sigma_1,...,\sigma_n}$, whereas when given oracle access to $b_{j_0,\sigma_1,...,\sigma_n}'$, we emulate an execution of D with $b_{j_0,\sigma_1,...,\sigma_n}$.

We now turn to prove that distinguishing the two aforementioned distributions on linear functions requires $\Omega(\sqrt{n})$ queries. Our proof follows the structure of the proof of Theorem 3.2. Specifically, in analogy to Definition 3.2.1, we say that $j \in [n]$ is w-special with respect to a sequence of queries $\bar{q}$ if there exists a linear combination of these queries that yields a $2n$-bit string q such that $\{2j - 1, 2j\} \cap \text{supp}(q) \neq \emptyset$ and $|\text{supp}(q)| \leq w$. Analogously to Claim 3.2.2, the number of w-special indices with respect to a sequence of t queries is bounded by $w \cdot t$. Next, analogously to Claim 3.2.3 we upper-bound the deviation of the answers whenever j_0 is not w-special with respect to the sequence of queries.

Claim 4.2.1. *Suppose that j_0 is not w-special with respect to $\bar{q} = (q^{(1)}, ..., q^{(t)}) \in (\{0, 1\}^{2n})^t$. Then, for every $\alpha \in \{0, 1\}^t$, when $\sigma = (\sigma_1, ..., \sigma_n)$ is selected as above, it holds that*

$$\Pr[(g_\sigma'(q^{(1)}), ..., g_\sigma'(q^{(t)})) = \alpha] = \Pr[(b_{j_0,\sigma}'(q^{(1)}), ..., b_{j_0,\sigma}'(q^{(t)})) = \alpha] \pm 2^{-\Omega(w)}.$$

Proof: Like in the proof of Claim 3.2.3, it suffices to show that, for every $c \in \{0, 1\}^t$,

$$\left| \Pr_\sigma [g_\sigma'(cQ) = 1] - \Pr_\sigma [b_{j_0,\sigma}'(cQ) = 1] \right| \leq 2^{-\Omega(w)}, \tag{11}$$

where Q is the matrix with the $q^{(i)}$'s as rows. Let $q = cQ$ and recall that $b_{j_0,\sigma}'(q) = g_\sigma'(q) + q_{2j_0-1} + q_{2j_0}$. We consider two cases. If $q_{2j_0-1} = q_{2j_0} = 0$, then the l.h.s of Eq. (11) equals zero. Otherwise (i.e., $\{2j_0-1, 2j_0\} \cap \text{supp}(q) \neq \emptyset$), since j_0 is not w-special, it holds that $|\text{supp}(q) \setminus \{2j_0 - 1, 2j_0\}| \geq w - 1$. Hence, there exists at least $(w - 1)/2$ indices j in $[n] \setminus \{j_0\}$ such that $(q_{2j-1}, q_{2j}) \neq (0,0)$, which means that for each such j the value of $f'_{\sigma_j}(q_{2(j-1)+1}, q_{2(j-1)+2})$ is not fixed when σ_j is random as above. Specifically, for each such j (i.e., j such that $(q_{2j-1}, q_{2j}) \neq (0,0)$), it holds that

$$\Pr_{\sigma_j}\left[f'_{\sigma_j}(q_{2(j-1)+1}, q_{2(j-1)+2}) = 1\right] = \begin{cases} \frac{1}{4} & \text{if } q_{2(j-1)+1} + q_{2(j-1)+2} = 1 \\ \frac{1}{2} & \text{if } q_{2(j-1)+1} = q_{2(j-1)+2} = 1 \end{cases}$$

and these events, which refer to different j's, are independent. Recalling Eq. (9)&(10), we conclude that each of the two probabilities in the l.h.s of Eq. (11) is $\frac{1}{2} \pm 2^{-\Omega(w)}$, and the claim follows.   □

The rest of the analysis mimics the proof of Theorem 3.2.   ∎

Establishing Theorem 6. In the course of the proof of Theorem 4.2 we actually established a lower bound on the complexity of testing the set of linear functions defined in Eq. (9). Letting $g''_\sigma(z_1, ..., z_{3n})$ equal $g'_\sigma(z_1, z_2, z_4, z_5, ..., z_{3n-2}, z_{3n-1})$ we obtain a set of linear functions in which there are no consecutive influential variables. Theorem 6 follows by observing that the argument establishing the hardness of testing the former property also establishes the hardness of testing the latter property.

Acknowledgments

Part of this work is based on joint research with Dana Ron, who refused to co-author it. The research was partially supported by the Israel Science Foundation (grants No. 1041/08).

References

1. Alon, N., Goldreich, O., Håstad, J., Peralta, R.: Simple Constructions of Almost k-wise Independent Random Variables. Journal of Random Structures and Algorithms 3(3), 289–304 (1992)
2. Alon, N., Krivelevich, M., Newman, I., Szegedy, M.: Regular languages are testable with a constant number of queries. In: SICOMP, pp. 1842–1862 (2001)
3. Bellare, M., Coppersmith, D., Håstad, J., Kiwi, M., Sudan, M.: Linearity testing in characteristic two. In: 36th FOCS, pp. 432–441 (1995)
4. Blais, E.: Testing juntas almost optimally. In: 41st STOC, pp. 151–158 (2009)
5. Blum, M., Luby, M., Rubinfeld, R.: Self-Testing/Correcting with Applications to Numerical Problems. JCSS 47(3), 549–595 (1993)
6. Diakonikolas, I., Lee, H.K., Matulef, K., Onak, K., Rubinfeld, R., Servedio, R.A., Wan, A.: Testing for concise representations. In: 48th FOCS, pp. 549–557 (2007)
7. Diakonikolas, I., Lee, H.K., Matulef, K., Servedio, R.A., Wan, A.: Efficient testing of sparse GF(2) polynomials. In: Aceto, L., Damgård, I., Goldberg, L.A., Halldórsson, M.M., Ingólfsdóttir, A., Walukiewicz, I. (eds.) ICALP 2008, Part I. LNCS, vol. 5125, pp. 502–514. Springer, Heidelberg (2008)

8. Even, G.: Construction of Small Probabilistic Spaces for Deterministic Simulation. M.Sc. Thesis, Computer Science Dept., Technion – Israel Institute of Technology (August 1991) (in Hebrew, abstract in English)

9. Fischer, E., Kindler, G., Ron, D., Safra, S., Samorodnitsky, S.: Testing Juntas. JCSS 68(4), 753–787 (2004)

10. Goldreich, O.: On Testing Computability by Small Width OBDDs. ECCC, TR10-061 (2010)

11. Goldreich, O., Goldwasser, S., Ron, D.: Property testing and its connection to learning and approximation. JACM 1996, 653–750 (1998)

12. Goldreich, O., Krivelevich, M., Newman, I., Rozenberg, E.: Hierarchy Theorems for Property Testing. ECCC, TR08-097 (2008)

13. Goldreich, O., Ron, D.: On Proximity Oblivious Testing. ECCC, TR08-041 (2008); Extended abstract in the proceedings of the 41st STOC (2009)

14. Lachish, O., Newman, I., Shapira, A.: Space Complexity vs. Query Complexity. Computational Complexity 17, 70–93 (2008)

15. Naor, J., Naor, M.: Small-bias Probability Spaces: Efficient Constructions and Applications. SICOMP 22, 838–856 (1993)

16. Newman, I.: Testing membership in languages that have small width branching programs. SIAM Journal on Computing 31(5), 1557–1570 (2002)

17. Parnas, M., Ron, D., Samorodnitsky, A.: Testing basic boolean formulae. SIDMA 16(1), 20–46 (2002)

18. Ron, D.: Property Testing: A Learning Theory Perspective. Foundations and Trends in Machine Learning 1(3), 307–402 (2008)

19. Ron, D.: Algorithmic and Analysis Techniques in Property Testing. In: Foundations and Trends in TCS (to appear)

20. Ron, D., Tsur, G.: Testing Computability by Width Two OBDDs. In: Dinur, I., Jansen, K., Naor, J., Rolim, J. (eds.) RANDOM 2009. LNCS, vol. 5687, pp. 686–699. Springer, Heidelberg (2009)

21. Ron, D., Tsur, G.: Testing Computability by Width Two OBDDs where the Variable Order is Unknown. In: 7th CIAC (to appear)

22. Rubinfeld, R.: On the Robustness of Functional Equations. SIAM Journal on Computing 28(6), 1972–1997 (1999)

23. Rubinfeld, R., Sudan, M.: Robust characterization of polynomials with applications to program testing. SIAM Journal on Computing 25(2), 252–271 (1996)

Learning and Lower Bounds for AC^0 with Threshold Gates

Parikshit Gopalan[1] and Rocco A. Servedio[2],[⋆]

[1] Microsoft Research Silicon Valley, Mountain View, CA, 94203, U.S.A.
parik@microsoft.com
[2] Columbia University, New York, NY, 10027, U.S.A.
rocco@cs.columbia.edu

Abstract. In 2002 Jackson et al. [JKS02] asked whether AC^0 circuits augmented with a threshold gate at the output can be efficiently learned from uniform random examples. We answer this question affirmatively by showing that such circuits have fairly strong Fourier concentration; hence the low-degree algorithm of Linial, Mansour and Nisan [LMN93] learns such circuits in sub-exponential time. Under a conjecture of Gotsman and Linial [GL94] which upper bounds the total influence of low-degree polynomial threshold functions, the running time is quasi-polynomial. Our results extend to AC^0 circuits augmented with a small super-constant number of threshold gates at arbitrary locations in the circuit. We also establish some new structural properties of AC^0 circuits augmented with threshold gates, which allow us to prove a range of separation results and lower bounds.

Keywords: Computational learning theory, AC^0, Fourier concentration, threshold gates, polynomial threshold functions.

1 Introduction

The seminal result of Linial, Mansour and Nisan [LMN93] showed how to learn the class AC^0 of constant depth circuits in quasi-polynomial time under the uniform distribution with random examples. Their work introduced the *Low-Degree Algorithm* which can learn functions where the Fourier spectrum is concentrated on low-degree coefficients; this algorithm and its extensions have since found many applications in learning, see e.g. [FJS91, BT96] for some early work and [JKS02, KOS04, MOS04, OS07, BOW08, KKMS08, KOS08] for more recent results.

In the two decades since their work, despite much effort, there has been limited progress in designing learning algorithms for more expressive circuit classes. Circuit classes like AC^0 with parity gates ($\mathsf{AC}^0[2]$) and depth-2 circuits of arbitrary threshold gates remain beyond the reach of currently known algorithms. One obstacle is that there are no lower bounds known for some of these classes, such

[⋆] Supported by NSF grants CCF-0347282, CCF-0523664 and CNS-0716245, and by DARPA award HR0011-08-1-0069.

M. Serna et al. (Eds.): APPROX and RANDOM 2010, LNCS 6302, pp. 588–601, 2010.
© Springer-Verlag Berlin Heidelberg 2010

as depth-2 TC^0, and the existence of lower bounds seems to be a pre-requisite for any learning algorithm (see [FK09]). Devising learning algorithms and lower bound techniques that can handle more powerful classes of circuits is a central open problem at the intersection of computational learning theory and circuit lower bounds.

Jackson et al. made some progress on learning circuits more expressive than AC^0 in [JKS02]. They gave a quasipolynomial-time algorithm that can learn Majority-of-AC^0 circuits – polynomial-size, constant-depth circuits augmented with a single Majority gate at the output – under the uniform distribution. Using a result of [Bei94], this yields a quasipolynomial-time algorithm that can learn AC^0 circuits augmented with $\mathrm{polylog}(n)$ many Majority gates at arbitrary locations in the circuit. The algorithm of Jackson et al. uses the low-degree algorithm as a weak learner and combines it with boosting. [JKS02] posed as an open question whether any efficient algorithm can learn Threshold-of-AC^0 circuits, in which the the topmost gate is a threshold gate (i.e. a weighted majority in which the weights may be arbitrary). It is observed in [JKS02] via an explicit counterexample that the analysis of their boosting-based algorithm breaks down for Threshold-of-AC^0. In this work, we take a significant step towards answering the question of [JKS02].

AC^0 circuits augmented with a few threshold gates have been well studied in the complexity theory literature, see e.g. [ABFR94, Bei94, GHR92, Gol97, Han07]. This is a natural class of circuits lying between the classes AC^0 (which we understand well) and TC^0 (for which we do not know lower bounds). One focus of this work has been on understanding the difference in power between unweighted threshold gates (i.e. majorities) versus threshold gates with arbitrary weights. Aspnes et al. [ABFR94] prove that any AC^0 circuit with a single threshold gate at the top cannot compute (or even approximate) parity, and more recently Hansen [Han07] has established $n^{\Omega(\log n)}$ size bounds on AC^0 circuits augmented with up to $\epsilon \log^2 n$ threshold gates. In contrast, when we restrict ourselves to Majority gates, an elegant result of Beigel [Bei94] alluded to above shows that any polynomial-size AC^0 circuit with $\mathrm{polylog}(n)$ Majority gates is equivalent to a quasi-polynomial size AC^0 circuit with a single majority gate at the top, and lower bounds for such circuits follow from [ABFR94].

1.1   Our Results

We show that AC^0 circuits augmented with a few threshold gates with arbitrary weights can be learned in subexponential time under the uniform distribution. In doing this we establish some new structural properties of such circuits, which allow us to prove new lower bounds and separations for such circuits.

Learning AC^0 with threshold gates. Our first main result is a Fourier concentration bound for Threshold-of-AC^0 circuits: roughly speaking, this bound says that any size-M, constant-depth Threshold-of-AC^0 circuit C must satisfy

$$\sum_{|\alpha|>t} \widehat{C}(\alpha)^2 \leq \epsilon \qquad \text{for} \qquad t = \frac{(\log M)^{\Theta(d)} 2^{\Theta((\log M)^{2/3})}}{\epsilon^{(\log M)^{1/3}}}.$$

This can be viewed as a natural extension of the [ABFR94] result showing that Threshold-of-AC^0 cannot compute parity; we show that such circuits in fact exhibit strong Fourier concentration. (Thus, roughly speaking, our result is to [ABFR94] as the [LMN93] Fourier concentration bound for AC^0 is to the earlier AC^0 lower bounds of Håstad [Hås86].) We note that Fourier concentration bounds of the sort we establish were not known even for Majority-of-AC^0 prior to this work; the [JKS02] algorithm requires boosting and its analysis does not establish Fourier concentration.

With our Fourier concentration bound for Threshold-of-AC^0 in hand, applying the Low-Degree Algorithm of [LMN93] we get the first subexponential-time learning result for this class: any size-M, constant-depth Threshold-of-AC^0 can be learned to any constant accuracy ϵ in time $n^{2^{\Theta((\log M)^{2/3})}}$.

An important ingredient in our proof is a recent $2^{O(d)}n^{1-1/O(d)}$ upper bound on the total influence of degree-d polynomial threshold functions over n Boolean variables, proved recently by [HKM09] and [DRST09]. In 1994 Gotsman and Linial [GL94] conjectured a stronger bound, that every degree-d PTF has total influence $O(d\sqrt{n})$. We show that under the [GL94] conjecture our results become significantly stronger: every size-M depth-d Threshold-of-AC^0 circuit C has Fourier concentration

$$\sum_{|\alpha|>t} \widehat{C}(\alpha)^2 \leq \epsilon \qquad \text{for} \qquad t = \frac{2^{O(d)}(\log M)^d}{\epsilon^2}$$

and consequently such circuits can be learned to constant accuracy in time $n^{2^{O(d)}(\log M)^d}$.

We extend the above results by giving Fourier concentration and learning results for AC^0 circuits with r threshold gates in arbitrary locations in the circuit. We unconditionally learn such circuits with $r = O((\log M)^{1/3})$ many threshold gates, to any constant accuracy, in time $n^{2^{\Theta((\log M)^{2/3})}}$. Assuming the [GL94] conjecture, we learn such circuits with $r = O(\log \log M)$ to any constant accuracy in time $n^{2^{O(d)}(\log M)^{O(d)}}$. These results are achieved building on our results for Threshold-of-AC^0.

Lower bounds and separation results. To complement the positive (learning) results described above, in Section 6 we establish new lower bounds and separation results for AC^0 circuits augmented with threshold gates. These results separate the classes Majority-of-AC^0 and Threshold-of-AC^0 and highlight some interesting contrasts between them.

1. Since Majority-of-AC^0 is already known to be learnable in quasi-polynomial time, our learning results are only of interest if Threshold-of-AC^0 is actually a broader class than Majority-of-AC^0. We show that this is indeed the case, by exhibiting a single threshold gate for which any equivalent depth-d Majority-of-AC^0 circuit must have size $2^{\Omega(n^{1/(d-1)})}$. (See Section 6.1.)

2. Beigel [Bei94] showed that any size-s, depth-d circuit that contains m Majority gates is computed by a size-$2^{m(O(\log s))^{2d+1}}$, depth-$(d+2)$ circuit with

a single Majority gate at the root. We show that this size bound cannot be improved to polynomial, by showing that a simple AND of two Majority gates requires any constant-depth circuit with a single Majority gate at the top (or even an arbitrary Threshold gate at the top) to have $n^{\Omega_d(\log n)}$ size. (See Section 6.2.)

3. A natural question is whether Beigel's result can be extended from Majority gates to arbitrary Threshold gates. Perhaps every AC^0 circuit which contains polylog(n) many Threshold gates is equivalent to a quasipoly(n)-size Threshold-of-AC^0? In fact the answer is no: we show that no analogue of Beigel's result is possible for Threshold gates, by showing that any Threshold-of-AC^0 circuit that computes the AND of two (high-weight) Threshold gates must have exponential size. (See Section 6.3.)

4. We also give lower bounds for AC^0 circuits with relatively many Threshold gates. We prove that any AC^0 circuit with $\epsilon \log n$ Threshold gates cannot compute parity, for a small constant $\epsilon > 0$. Previously, Aspnes et al. [ABFR94] proved this claim for AC^0 with a single threshold gate at the top. Beigel [Bei94] showed that any AC^0 circuit must be augmented with $n^{\Omega(1)}$ many Majority gates in order to compute parity. Our bound allows for a smaller number of gates augmenting the basic AC^0 circuit, but the gates (Threshold instead of Majority) are more powerful. (See Section 6.4.)

We note that the previous lower bounds on Threshold-of-AC^0 due to [ABFR94] apply to functions which have high PTF degree. This approach cannot be used for results (1) and (2) above, where we are proving lower bounds against functions which have low PTF degree. As mentioned above, using different techniques Hansen [Han07] has established $n^{\Omega(\log n)}$ size bounds on AC^0 circuits augmented with up to $\epsilon \log^2 n$ threshold gates. Very recently V. Podolskii [Pod10] has shown that any AC^0 circuit augmented with $O(\log n)$ threshold gates that approximates the parity function to high accuracy must have exponential size.

2   Preliminaries

2.1 MAC^0 and TAC^0 and $\mathsf{TAC}^0[r]$

Recall that a *threshold function*, or halfspace, over n variables is a Boolean function $h : \{-1, 1\}^n \to \{-1, 1\}$, $h(x) = \text{sign}\left(\sum_{i=1}^n w_i x_i - \theta\right)$, where $w_1, \dots, w_n, \theta$ may be arbitrary real values. We will sometimes write Thr to denote a single threshold gate and Maj to denote a single Majority gate, where the Majority function is the threshold function for which each w_i equals 1 and the threshold θ equals 0.

A *Threshold-of-AC^0 circuit*, or TAC^0, is a circuit consisting of a threshold function (with arbitrary weights and fanin) as the output gate and AC^0 circuits feeding into it. A depth-d TAC^0 is one in which each of the AC^0 circuits feeding into the output threshold gate has depth at most $d - 1$. The size of a TAC^0 is the total number of gates (so in particular, in a size-M TAC^0 each of the AC^0 circuits is of size at most M).

A *Majority-of*-AC^0 *circuit*, or MAC^0, is a TAC^0 in which the top threshold function is a majority gate.

Finally, we will also consider AC^0 circuits that have r arbitrary Thr gates buried at arbitrary locations in the circuit; we refer to such a circuit as a "Threshold-of-r-AC^0s", or $\mathsf{TAC}^0[r]$.

We give standard definitions of polynomial threshold functions, influence of variables on Boolean functions, noise sensitivity, and the basics of Fourier analysis in the full version of the paper [GS10].

2.2   Random Restrictions and AC^0

We write "$\rho \sim \mathcal{R}_p$" to indicate that ρ is a *random restriction with parameter p*. Such a restriction ρ is chosen by independently fixing each variable to $+1$ or -1 each with probability $\frac{1-p}{2}$, and leaving the variable unfixed with probability p. We write f_ρ to denote the function that results from applying ρ to f.

We will use several facts about the behavior of AC^0 circuits under random restrictions. The first of these facts is Håstad's Switching Lemma:

Lemma 1 ([Hås86]). *Let C be a depth-2 circuit (i.e. a DNF or a CNF) of bottom fan-in s. Then $\Pr_\rho[C_\rho$ cannot be written as a depth-t decision tree] $\leq (5ps)^t$, where ρ is a random restriction with parameter p.*

(The above statement is implicit in [Hås86] and is made explicit in e.g. [Hås01].) Repeated applications of the Switching Lemma can be used to prove the following in a rather straightforward way:

Lemma 2 ([LMN93], Lemma 2). *If C is a size-M depth-d AC^0 circuit, then for any $t \geq 0$ we have $\Pr_\rho[C_\rho$ cannot be written as a depth-t decision tree] $\leq M2^{-t}$, where ρ is a random restriction with parameter $p = \frac{1}{10^d t^{d-1}}$.*

([LMN93] actually state a slightly weaker form in which the LHS is replaced by "$\Pr_\rho[\deg(C_\rho) > t]$." It is easy to check that using Lemma 1, the [LMN93] proof directly yields Lemma 2 as stated above.)

2.3   Sketch of the Random Restriction Argument

The high-level idea of our proof is quite simple, and is similar to that of [LMN93]. We show that when a TAC^0 is hit with a random restriction, with high probability it collapses into a "much simpler function," specifically a low-degree PTF. Recent results on the Fourier concentration of low-degree PTFs due to [DRST09, HKM09] let us infer that the original TAC^0 must also have had good Fourier concentration. In the rest of this section we elaborate on this argument.

We begin by recalling the basic outline of [LMN93]'s Fourier concentration bound for AC^0 circuits. It will be useful for us to view the [LMN93] argument as proceeding in two stages:

1. The first stage analyzes what happens to a size-M, depth-d AC^0 circuit C when it is hit with a random restriction with parameter $p \approx \frac{1}{(\log M)^{d-1}}$ (recall that p is the probability that a variable "survives" the restriction, i.e. is left unfixed). [LMN93] show that with high probability such a restriction causes C_ρ to collapse down to a $(\log M)$-depth decision tree.
2. The second stage is the observation that a $(\log M)$-depth decision tree T, being a degree $\log M$ polynomial, has extremely strong Fourier concentration: $\sum_{|\alpha|>\log M} \hat{T}(\alpha)^2 = 0$. Linial et al. then use the Fourier concentration of C_ρ to argue that the original AC^0 function computed by C must have had most of its Fourier weight at levels $\le (\log M)^d$.

Our argument for TAC^0 has a similar high-level structure, but with some significant differences in both stages. Let C now denote a size-M, depth-d TAC^0 circuit.

1'. In the first stage, we consider hitting C with a "stronger" random restriction with a smaller value of p (so fewer variables survive the restriction). We show that with high probability such a restriction causes C_ρ to collapse down to a "low-degree" PTF of degree $k \ll \log M$. The stronger restriction is necessary since the results of [DRST09, HKM09] are non-trivial only when the degree of the PTF is $o(\sqrt{\log n})$.
2'. The results of [DRST09, HKM09] imply that C_ρ must have some nontrivial Fourier concentration. The Fourier concentration for C_ρ is much weaker than what one gets for decision trees, but one can adapt the original [LMN93] argument to show that the original circuit C itself must have had some Fourier concentration.

The conjecture of Gotsman & Linial significantly strengthens the bounds on total influence and noise sensitivity of low-degree PTFs that are currently known; it implies non-trivial bounds as long as the degree is $o(\sqrt{n})$. This in turn strengthens the Fourier concentration that we get for C_ρ in Stage 2', and hence also for C. We present each of the stages of the above argument in as self-contained a way as possible in Section 3. Section 4 puts the pieces together to prove the main results.

3   Random Restrictions of TAC^0

3.1   Stage 1: Collapsing TAC^0 to a Low-Degree PTF

In this section we prove the following:

Lemma 3. *Let C be a size-M, depth-d TAC^0. Let ρ be a random restriction with parameter p (specified below) and let $k \ge 1$. Then for any $0 < p' < 1$, with failure probability at most δ the function C_ρ is a degree-k PTF, where*

$$\delta = M^{-2} + M^5 \left(\frac{4e \log(M)p'}{k} \right)^k \quad and \quad p = \frac{1}{10^{d-1}(4 \log M)^{d-2}} \cdot p'.$$

Proof. The proof is conceptually quite simple. Let $C = \mathsf{Thr}(C_1, \ldots, C_\ell)$ where Thr is the topmost threshold gate, $\ell \leq M$ is its fan-in, and each C_i is an AC^0 circuit of depth at most $d-1$ and size M_i, where $M_1, \ldots, M_\ell \leq M$. We view the restriction ρ as being obtained in two steps. The first step collapses each C_i to a decision tree of depth $O(\log M)$. The second step significantly reduces the depth of each decision tree, down to k. After these two steps, with high probability each C_i has collapsed down to $(C_i)_\rho$ which is a degree-k polynomial. Thus C_ρ is a PTF of degree k.

In the first step we take a random restriction ρ_1 with parameter $p_1 = \frac{1}{10^{d-1}(4 \log M)^{d-2}}$. For a given i, Lemma 2 gives that with failure probability at most $M_i \cdot M^{-4}$, the function $(C_i)_{\rho_1}$ is equivalent to a decision tree T_i of depth $4 \log M$. Summing failure probabilities over all $i = 1, \ldots, \ell$, this occurs for every C_i with overall failure probability at most $(M_1 + \cdots + M_\ell)M^{-4} \leq M^{-2}$.

In the second step, we take a random restriction with parameter p' (thus the overall probability that a variable survives the combined restriction is $p = p_1 p'$ as desired). The following simple lemma analyzes the effect of a random restriction on a depth-t decision tree:

Lemma 4. *Let T be a depth-t decision tree and ρ be a random restriction with parameter p'. Then for $k \geq 1$, we have $\Pr[T_\rho$ cannot be written as a depth-k decision tree$] \leq 2^t \left((etp')/k\right)^k$.*

Proof. Suppose that under ρ at most k variables survive in each root-to-leaf path in T. Then it is clear that T_ρ can be written as a decision tree of depth at most k. So fix any given path of length at most t in T; wlog the variables appearing on this path are $x_1, \ldots, x_t$. The probability that at least k of these variables survive ρ is at most

$$\binom{t}{k}(p')^k \leq \left(\frac{et}{k}\right)^k (p')^k = \left(\frac{etp'}{k}\right)^k.$$

A union bound over all (at most 2^t) paths in T finishes the proof.

We apply this lemma to each of the $\ell \leq M$ decision trees T_i from step 1, taking $t = 4 \log M$. A union bound gives that the probability that any T_i fails to have its depth reduced to k is at most $M \cdot 2^t \cdot (etp'/k)^k$. Any decision tree of depth k is exactly computed by a Fourier polynomial of degree at most k; the top-level Thr gate takes the sign of a weighted sum of these polynomials, and we obtain Lemma 3.

3.2   Stage 2: From Fourier Concentration of C_ρ to Fourier Concentration of C

We will use the following recent bound on the noise sensitivity of degree-k PTFs due to Diakonikolas et al. [DRST09] and Harsha et al. [HKM09]:

Theorem 1. *For any degree-k PTF f over $\{-1,1\}^n$ and any $0 \leq \epsilon \leq 1$, we have $\mathrm{ns}_\epsilon(f) \leq 2^{O(k)} \cdot \epsilon^{\frac{1}{O(k)}}$.*

The following simple result (Corollary 17 of [KOS04]) converts noise sensitivity upper bounds to Fourier concentration bounds:

Lemma 5. *Let* $f : \{-1,1\}^n \to \{-1,1\}$ *be any Boolean function and let* $\kappa : [0,1/2] \to \mathbf{R}^+$ *be an increasing function such that* $\mathrm{ns}_\epsilon(f) \leq \kappa(\epsilon)$. *Then*

$$\sum_{|\alpha| \geq m} \widehat{f}(\alpha)^2 \leq \epsilon \quad for \quad m = \frac{1}{\kappa^{-1}(\epsilon/2.32)}.$$

Plugging in Theorem 1 gives the following Fourier concentration bound:

Corollary 1. *For any degree-k PTF f over* $\{-1,1\}^n$ *and any* $0 \leq \epsilon \leq 1$, *we have*

$$\sum_{|\alpha| \geq m(\epsilon)} \widehat{f}(\alpha)^2 \leq \epsilon \quad where \quad m(\epsilon) = \frac{2^{\Theta(k^2)}}{\epsilon^{\Theta(k)}}.$$

We now show that if f_ρ has good Fourier concentration (w.h.p. over the choice of random restriction ρ), then f itself has good Fourier concentration. This is done by the following lemma, adapting arguments from [LMN93]; the proof is in the full version [GS10].

Lemma 6. *Let* $f : \{-1,1\}^n \to \{-1,1\}$ *and let* t,p *be parameters such that* $pt > 8$. *Then*

$$\sum_{|\alpha| > t} \widehat{f}(\alpha)^2 \leq 2\mathbf{E}_\rho[\sum_{|\beta| > pt/2} \widehat{f_\rho}(\beta)^2],$$

where ρ *is a random restriction with parameter p.*

As an easy corollary of Lemma 6 we have the following:

Corollary 2. *Let* $f : \{-1,1\}^n \to \{-1,1\}$ *and let* t,p *be parameters such that* $tp > 8$. *Suppose that with probability at least* $1 - \delta$ *(over the choice of a random restriction ρ with parameter p) the function f_ρ has Fourier concentration* $\sum_{|\beta|>pt/2} \widehat{f_\rho}(\beta)^2 \leq \epsilon$. *Then we have* $\sum_{|\alpha|>t} \widehat{f}(\alpha)^2 \leq 2\epsilon + 2\delta$.

(This follows from the lemma because $\widehat{f_\rho}$ is a Boolean function and consequently always has total Fourier weight at most 1.)

4   Proof of the Fourier Concentration Results for TAC^0

Throughout this section C is a size-M, depth-d TAC^0. The regime we are most interested in is when the circuit size M is poly(n) and the error parameter ϵ is something like a small constant; in particular, we are most interested in situations where $\epsilon > M^{-1}$. (We note that even the Majority function has $\widehat{f}([n])^2 = \Theta(1/n)$, so Fourier concentration bounds for TAC^0 must certainly be vacuous for $\epsilon < 1/n$.)

4.1   The Unconditional Result

Putting together all the pieces, we have established a Fourier concentration bound for TAC^0:

Theorem 2. *Let C be a size-M, depth-d TAC^0. Let $\epsilon \geq 2M^{-2}$. Then C has Fourier concentration*

$$\sum_{|\alpha|>t} \widehat{C}(\alpha)^2 \leq 4\epsilon \qquad \text{for} \qquad t = \frac{(\log M)^{\Theta(d)} \cdot 2^{\Theta((\log M)^{2/3})}}{\epsilon^{\Theta((\log M)^{1/3})}}. \tag{1}$$

Proof. In Stage 1 we shall take (with foresight) $k = (\log M)^{1/3}$ and $p' = \frac{k}{4eM^{7/k}\log M}$, so consequently $p = \frac{k}{(40\log M)^{d-1}\cdot e \cdot M^{7/k}}$. This choice of parameters gives failure probability at most $\delta = 2M^{-2}$ in Lemma 3, so with this failure probability we have that C_ρ is a degree-k PTF which satisfies

$$\sum_{\alpha \geq m} \widehat{f}(\alpha)^2 \leq \epsilon \text{ where } m = \frac{2^{\Theta((\log M)^{2/3})}}{\epsilon^{\Theta((\log M)^{1/3})}}.$$

In Step 3, we take $t = 2m/p$ so $tp/2 = m$ which is at least 8. Corollary 2 thus gives us

$$\sum_{|\alpha|>t} \widehat{C}(\alpha)^2 \leq 2\epsilon + 2\delta \text{ where } t = \frac{(\log M)^{\Theta(d)} \cdot 2^{\Theta((\log M)^{2/3})}}{\epsilon^{\Theta((\log M)^{1/3})}}.$$

Applying the well-known [LMN93] machinery for uniform distribution learning of Boolean functions with good Fourier concentration, we get the following:

Corollary 3. *Size-M depth-d TAC^0 circuits can be learned to accuracy ϵ in time n^t where*

$$t = \frac{(\log M)^{\Theta(d)} \cdot 2^{\Theta((\log M)^{2/3})}}{\epsilon^{\Theta((\log M)^{1/3})}}.$$

Thus as long as $\epsilon \geq 1/2^{O((\log M)^{1/3})}$ and $d \leq O((\log M)^{2/3}/(\log\log M))$ this gives an algorithm to learn size-M depth-d TAC^0 in time $n^{2^{\Theta((\log M)^{2/3})}}$, i.e. subexponential time $(2^{n^{o(1)}})$ for any $M = \text{poly}(n)$.

4.2   The Gotsman-Linial Conjecture and Its Consequences

In 1994 Gotsman and Linial [GL94] asked the question of what is the maximum total influence of any degree-k PTF over n variables. They conjectured that the symmetric function which changes sign on the k middle layers of the Boolean hypercube has the highest total influence of any degree-k PTF (it is easy to see that this function is indeed a degree-k PTF). Since each layer of edges in the Boolean hypercube contains at most $\sqrt{n}2^{n-1}$ edges, a direct consequence of their conjecture (which is nearly equivalent to it for $k = o(\sqrt{n})$) is the following:

Conjecture 1 ([GL94]). Every degree-k PTF f over n variables has $\text{Inf}(f) \leq k\sqrt{n}$.

We show that using our approach, Conjecture 1 yields significantly improved Fourier concentration (and significantly more efficient learnability) for TAC^0. The noise sensitivity bounds of [DRST09] and [HKM09] follow from a bound of $2^{O(k)}n^{1-1/O(k)}$ on the average sensitivity of degree-k PTFs. This bound becomes trivial for $k = \Omega(\sqrt{\log n})$, and hence we needed to use a very strong random restriction in order to reduce our initial TAC^0 to a PTF of degree $o(\sqrt{\log n})$. Conjecture 1 implies that a weaker random restriction will suffice. We use the following noise sensitivity and Fourier concentration consequences of the Gotsman-Linial conjecture:

Corollary 4. *If Conjecture 1 holds, then for any degree k PTF f over $\{-1,1\}^n$ and any $0 \leq \epsilon \leq 1$,*

$$\text{ns}_\epsilon(f) \leq 2k\sqrt{\epsilon} \qquad and \qquad \sum_{|\alpha| \geq m} \widehat{f}(\alpha)^2 \leq \epsilon \quad where \quad m = \frac{24k^2}{\epsilon^2}.$$

The first inequality follows from the reduction from total influence to noise sensitivity for PTFs given in [DRST09] (see Section 7), and the second inequality then follows from Lemma 5. We thus obtain:

Theorem 3. *Let C be a size-M, depth-d TAC^0. Let $\epsilon \geq 2M^{-2}$. If Conjecture 1 is true, then we have*

$$\sum_{|\alpha| > t} \widehat{C}(\alpha)^2 \leq 4\epsilon \qquad for \qquad t = \frac{2^{O(d)}(\log M)^d}{\epsilon^2}.$$

Proof. In Stage 1 we shall take $k = \log M$ and $p' = 10^{-4}$, so $p = \frac{1}{10^{d+3}(4 \log M)^{d-2}}$. Lemma 3 gives that with probability at least $1 - 2M^{-2}$, the function C_ρ is a degree-k PTF, in which case we have, for $m = 24k^2/\epsilon^2$, $\sum_{|\alpha| > m} \widehat{C_\rho}(\alpha)^2 \leq \epsilon$. For Stage 3, in Corollary 2 we take $t = \frac{2m}{p} = \frac{(c_1 \log M)^d}{\epsilon^2}$ for some absolute constant c_1. Corollary 2 thus gives us $\sum_{|\alpha| > t} \widehat{C}(\alpha)^2 \leq 2\epsilon + 2\delta$.

Similar to before, the [LMN93] low-degree algorithm gives us:

Corollary 5. *If Conjecture 1 is true, then size-M, depth-d TAC^0 can be learned to accuracy ϵ in time*

$$n^{\frac{2^{O(d)}(\log M)^d}{\epsilon^2}}.$$

This gives quasi-polynomial time learning for $M = \text{poly}(n)$-size TAC^0 for any constant (or even $1/\text{polylog}(n)$) accuracy ϵ.

5 Learning $\mathsf{TAC}^0[r]$

Our learning results can be extended from TAC^0 circuits to $\mathsf{TAC}^0[r]$ circuits for small (but superconstant) values of r. The high-level approach is as follows: We first prove a general result showing that if a class $\mathcal{C}$ has Fourier concentration, then any R-junta-of-functions-from-$\mathcal{C}$ must also have fairly good Fourier concentration provided that R is not too large. We then argue that any $\mathsf{TAC}^0[r]$ is equivalent to a R-junta-of-TAC^0 for $R = (r+1)2^r$. This lemma and the arguments used in its proof are similar to arguments found in [BRS95]. Combining the above two ingredients with the Fourier concentration bounds for TAC^0 which we obtained in Section 4, we get Fourier concentration bounds for $\mathsf{TAC}^0[r]$.

Because of space limits here we only state the results and defer full proofs to [GS10]. We show unconditionally that $\mathsf{TAC}^0[O((\log M)^{1/3})]$ circuits can be learned in essentially the same time bound that we achieved for unconditionally learning TAC^0 circuits:

Theorem 4. *The class of* $\mathsf{TAC}^0[O((\log M)^{1/3})]$ *circuits of size M and depth d can be learned to accuracy ϵ (for $\epsilon > 2M^{-2}$) in time n^t, where*

$$t = (\log M)^{\Theta(d)} \cdot 2^{\Theta((\log M)^{2/3})} \epsilon^{-\Theta((\log M)^{1/3})}.$$

Assuming the Gotsman-Linial conjecture, we obtain

Theorem 5. *If Conjecture 1 is true, then the class of* $\mathsf{TAC}^0[r]$ *circuits of size M and depth d can be learned to accuracy ϵ (for $\epsilon > 2M^{-2}$) in time n^t, where*

$$t = 2^{O(d+r)} \cdot (\log M)^{3d} \epsilon^{-3}.$$

For constant d this gives quasi-polynomial time learning for r as large as $O(\log \log M)$.

6 Lower Bounds

6.1 MAC^0 Cannot Compute TAC^0

In this section we prove that there are TAC^0 circuits that have no small equivalent MAC^0 circuit.

Theorem 6. *There is a threshold function over $N = O(n^2)$ variables such that any equivalent MAC^0 circuit of depth $d \geq 2, d = \Theta(1)$ must have size $2^{\Omega(n^{1/(d-1)})}$.*

The desired function is the function $U_{n,4n}(x)$ defined by Goldmann *et al.* in Section 4 of [GHR92] (all variables below take values ± 1):

$$U_{n,4n}(x) = \mathrm{sign}(2r_{n,4n}(x) + 1), \qquad r_{n,4n}(x) = \sum_{i=0}^{n-1} \sum_{j=0}^{4n-1} 2^i x_{ij}. \tag{2}$$

It is clear that $U_{n,4n}$ is a TAC^0 circuit (of depth 1), consisting of a single threshold gate over $N = 4n^2$ input variables. It remains to show that any depth-d MAC^0 circuit for $U_{n,4n}(x)$ must be large. We do this in two steps as follows. Suppose that C is a depth-d, size-M MAC^0 circuit that computes $U_{n,4n}(x)$. If $M = 2^{\Omega(n^{1/(d-1)})}$ then there is nothing to show, so we assume $M = 2^{O(n^{1/(d-1)})}$. We shall consider the effect of applying a random restriction with parameter $r = \frac{1}{10^{d-1} s^{d-2}}$ to C, where we select $s = 3 \log M$. We establish the following two lemmas in [GS10]:

Lemma 7. *With probability at least $1 - M^{-1}$ over the random choice of ρ, the function $(U_{n,4n})_\rho$ is a polynomial threshold function of total weight at most M^7.*

Lemma 8. *With probability at least $1 - 2n^{-2}$ over the random choice of ρ, the function $(U_{n,4n})_\rho$ has a sub-function (obtained by possibly fixing some additional variables in $(U_{n,4n})_\rho$) that is equivalent, up to renaming variables, to $U_{m,4m}$ where $m = \Omega(n/(\log M)^{d-2})$.*

Fix a restriction ρ that satisfies both Lemmas (such a ρ must exist since each of the two events has probability greater than $1/2$). The function $U_{m,4m}$ is a restriction of the function $(U_{n,4n})_\rho$ from Lemma 7, and thus $(U_{m,4m})_\rho$ must have a polynomial threshold function of weight at most M^7. However, the discussion following Corollary 8 of [GHR92] shows that the total weight of any PTF for $U_{m,4m}$ must be at least $\Omega(2^{m/2}/\sqrt{m})$. Since $m = \Omega(n/(\log M)^{d-2})$, straightforward manipulation yields the desired lower bound $M = 2^{\Omega(n^{1/(d-1)})}$.

6.2   Lower Bounds on MAC^0

Beigel [Bei94] showed that any size-s, depth-d circuit that contains m Maj gates is computed by a size-$2^{m(O(\log s))^{2d+1}}$, depth-$(d+2)$ circuit with a single Maj gate at the root. It is natural to ask whether this simulation can be improved to a polynomial-size (rather than quasi-polynomial) Maj of AC^0. In this section we observe that no such strengthened version of Beigel's theorem can exist, by proving that there is no polynomial-size MAC^0 (or even TAC^0) for an AND of two Maj gates:

Theorem 7. *For any constant d, any TAC^0 circuit of depth d that computes $f(x,y) = \mathsf{Maj}(x_1, \ldots, x_n) \wedge \mathsf{Maj}(y_1, \ldots, y_n)$ must have size $n^{\Omega_d(\log n)}$.*

Proof. The proof is by contradiction. Let $M = n^{o(\log n)}$ and let C be a depth-d TAC^0 of size M that computes $f(x,y)$. We analyze the effect of hitting C with a very strong random restriction ρ, one which has parameter $p = n^{-0.1}$. It is easy to see that with extremely high probability – much more than $1/2$ – f_ρ turns into some function of the form

$$f_\rho(x,y) = \mathrm{sign}(\sum_{i \in S_1} x_i + C_1) \wedge \mathrm{sign}(\sum_{j \in S_2} y_j + C_2),$$

where $|S_1|, |S_2| \geq n^{0.8}$ and $|C_1|, |C_2| \leq n^{0.51}$. For any such ρ, by fixing at most $2n^{.51}$ additional variables, we get $\mathsf{Maj}(x') \wedge \mathsf{Maj}(y')$ where x', y' are $\Omega(n^{0.8})$-bit strings. By the recent result of Sherstov [She09], any PTF for this function must have degree at least $c_1 \log n$ for some absolute constant $c_1 > 0$.

On the other hand, let us consider what happens to the TAC^0 C under such a strong random restriction using Lemma 3. Since $p = n^{-0.1}$, we have $p' = n^{-0.1} \cdot 10^{d-1}(4 \log M)^{d-2} < n^{-0.09}$ for n sufficiently large. Taking $k = (c_1/2) \log n$, Lemma 3 gives us that C_ρ has a PTF of degree at most $(c_1/2) \log n$ with failure probability at most

$$M^{-2} + M^5(4e \log(M)p'/k)^k = M^{-2} + M^5 n^{-\Omega(\log n)} < 1/2$$

since $M = n^{o(\log n)}$. Thus, there must be some restriction ρ such that f_ρ has PTF degree at least $c_1 \log n$, but C_ρ has PTF degree at most $(c_1/2) \log n$. This contradiction proves the theorem.

Aspnes et al. [ABFR94] prove lower bounds on the size of TAC^0 circuits that compute various functions such as parity. The method of [ABFR94] is useful for functions that have high weak PTF degree (such as parity). In contrast, our argument above gives us a TAC^0 lower bound for the function $\mathsf{Maj}(x) \wedge \mathsf{Maj}(y)$, which is known [BRS95] to have PTF degree only $O(\log n)$.

6.3   Lower Bounds on TAC^0

We prove that no analogue of Beigel's theorem [Bei94] is possible for Thr gates: even an AND of two Thr gates may require a TAC^0 of more than quasi-polynomial size. The proof (see [GS10]) is similar to that of Theorem 7, it uses a recent result of Sherstov [She09] showing that the function $f(x, y) = U_{n,4n}(x) \wedge U_{n,4n}(y)$ (see Section 6.1) has PTF degree $\Omega(n)$.

Theorem 8. *Fix any absolute constant d. Any TAC^0 circuit of depth d that computes $f(x, y) = U_{n,4n}(x) \wedge U_{n,4n}(y)$ must have size $2^{\Omega(n^{1/(d-1)})}$.*

6.4   Lower Bounds on $\mathsf{TAC}^0[t(n)]$

Inspection of the proof of Theorem 4 is easily seen to imply that the parity function cannot be computed by a $\mathsf{TAC}^0[(\log n)^{2/3}]$ circuit. In the full version [GS10] we give an improved bound that allows up to $O(\log n)$ threshold gates.

Theorem 9. *Fix any absolute constant d. Any poly(n)-size, depth-d $\mathsf{TAC}^0[t(n)]$ circuit that computes the parity function must have $t(n) = \Omega(\log n)$.*

References

[ABFR94]   Aspnes, J., Beigel, R., Furst, M., Rudich, S.: The expressive power of voting polynomials. Combinatorica 14(2), 1–14 (1994)

[Bei94]   Beigel, R.: When do extra majority gates help? polylog(n) majority gates are equivalent to one. Computational Complexity 4, 314–324 (1994)

[BOW08]   Blais, E., O'Donnell, R., Wimmer, K.: Polynomial regression under arbitrary product distributions. In: COLT, pp. 193–204 (2008)

[BRS95]   Beigel, R., Reingold, N., Spielman, D.: PP is closed under intersection. Journal of Computer & System Sciences 50(2), 191–202 (1995)

[BT96]   Bshouty, N., Tamon, C.: On the Fourier spectrum of monotone functions. Journal of the ACM 43(4), 747–770 (1996)

[DLM+07]   Diakonikolas, I., Lee, H., Matulef, K., Onak, K., Rubinfeld, R., Servedio, R., Wan, A.: Testing for concise representations. In: FOCS, pp. 549–558 (2007)

[DRST09]   Diakonikolas, I., Raghavendra, P., Servedio, R., Tan, L.-Y.: Average sensitivity and noise sensitivity of polynomial threshold functions (2009), http://arxiv.org/abs/0909.5011

[FJS91]   Furst, M., Jackson, J., Smith, S.: Improved learning of AC^0 functions. In: COLT, pp. 317–325 (1991)

[FK09]   Fortnow, L., Klivans, A.: Efficient learning algorithms yield circuit lower bounds. Journal of Computer & System Sciences 75(1), 27–36 (2009)

[GHR92]   Goldmann, M., Håstad, J., Razborov, A.: Majority gates vs. general weighted threshold gates. Computational Complexity 2, 277–300 (1992)

[GL94]   Gotsman, C., Linial, N.: Spectral properties of threshold functions. Combinatorica 14(1), 35–50 (1994)

[Gol97]   Goldmann, M.: On the power of a threshold gate at the top. Information Processing Letters 63(6), 287–293 (1997)

[GS10]   Gopalan, P., Servedio, R.A.: Learning and Lower Bounds for AC^0 with Threshold Gates (2010), http://eccc.hpi-web.de/report/2010/074/

[Han07]   Hansen, K.: Computing symmetric Boolean functions by circuits with few exact threshold gates. In: COCOON, pp. 448–458 (2007)

[Hås86]   Håstad, J.: Computational Limitations for Small Depth Circuits. MIT Press, Cambridge (1986)

[Hås01]   Håstad, J.: A slight sharpening of LMN. Journal of Computer and System Sciences 63(3), 498–508 (2001)

[HKM09]   Harsha, P., Klivans, A., Meka, R.: Bounding the sensitivity of polynomial threshold functions (2009), http://arxiv.org/abs/0909.5175

[JKS02]   Jackson, J., Klivans, A., Servedio, R.: Learnability beyond AC^0. In: STOC, pp. 776–784 (2002)

[KKMS08]   Kalai, A., Klivans, A., Mansour, Y., Servedio, R.: Agnostically learning halfspaces. SIAM Journal on Computing 37(6), 1777–1805 (2008)

[KOS04]   Klivans, A., O'Donnell, R., Servedio, R.: Learning intersections and thresholds of halfspaces. Journal of Computer & System Sciences 68(4), 808–840 (2004)

[KOS08]   Klivans, A., O'Donnell, R., Servedio, R.: Learning geometric concepts via Gaussian surface area. In: FOCS, pp. 541–550 (2008)

[LMN93]   Linial, N., Mansour, Y., Nisan, N.: Constant depth circuits, Fourier transform and learnability. Journal of the ACM 40(3), 607–620 (1993)

[MOS04]   Mossel, E., O'Donnell, R., Servedio, R.: Learning functions of k relevant variables. Journal of Computer & System Sciences 69(3), 421–434 (2004)

[OS07]   O'Donnell, R., Servedio, R.: Learning monotone decision trees in polynomial time. SIAM J. Comput. 37(3), 827–844 (2007)

[Pod10]   Podolskii, V.: Personal communication (2010)

[She09]   Sherstov, A.: The intersection of two halfspaces has high threshold degree. In: FOCS, pp. 343–362 (2009)

Liftings of Tree-Structured Markov Chains

(Extended Abstract)

Thomas P. Hayes[1] and Alistair Sinclair[2],[*]

[1] Department of Computer Science, University of New Mexico
[2] Computer Science Division, University of California at Berkeley
hayes@cs.unm.edu,
sinclair@cs.berkeley.edu

Abstract. A "lifting" of a Markov chain is a larger chain obtained by replacing each state of the original chain by a set of states, with transition probabilities defined in such a way that the lifted chain projects down exactly to the original one. It is well known that lifting can potentially speed up the mixing time substantially. Essentially all known examples of efficiently implementable liftings have required a high degree of symmetry in the original chain. Addressing an open question of Chen, Lovász and Pak, we present the first example of a successful lifting for a complex Markov chain that has been used in sampling algorithms. This chain, first introduced by Sinclair and Jerrum, samples a leaf uniformly at random in a large tree, given approximate information about the number of leaves in any subtree, and has applications to the theory of approximate counting and to importance sampling in Statistics. Our lifted version of the chain (which, unlike the original one, is non-reversible) gives a significant speedup over the original version whenever the error in the leaf counting estimates is $o(1)$. Our lifting construction, based on flows, is systematic, and we conjecture that it may be applicable to other Markov chains used in sampling algorithms.

1 Introduction

1.1 Background and Motivation

As the field of Markov chain Monte Carlo (MCMC) algorithms matures, attention is turning to refinements of these algorithms with improved running times. A general framework for speeding up MCMC algorithms, known as "lifting," was introduced ten years ago by Chen, Lovász and Pak [2]. A *lifting* of a Markov chain $\mathcal{M}$ is a larger chain $\mathcal{M}'$ obtained by replacing each state of $\mathcal{M}$ by a set of states; the lifting is required to preserve the structure of $\mathcal{M}$ in the sense that the obvious projection obtained by merging appropriate states of $\mathcal{M}'$ gets us back to $\mathcal{M}$ itself. (See Section 2 for a precise definition.) The intriguing fact, first observed by Diaconis, Holmes and Neal [4] and explored further by Chen, Lovász and Pak [2], is that lifting can in certain cases reduce the mixing time of the chain substantially, and hence potentially improve the running time of algorithms in which it is used.

---

[*] Supported in part by NSF grant CCF-0635153 and by a UC Berkeley Chancellor's Professorship.

M. Serna et al. (Eds.): APPROX and RANDOM 2010, LNCS 6302, pp. 602–616, 2010.

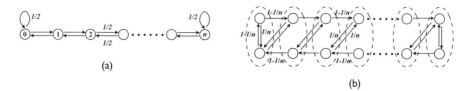

Fig. 1. (a) Simple random walk on a path of length n. (b) The lifted walk; dotted ovals indicate nodes that project to the same node in the original walk.

The simplest example of lifting, due to [4], is for simple random walk on the path of length n, with uniform stationary distribution. The mixing time of this chain is well known to be $\Theta(n^2)$. This chain can be lifted by replacing each node by a pair of nodes, with the two sets of copies connected in two *directed* paths with opposite senses, and bidirected crossing edges between the paths. (See Fig. 1.) If the crossing probabilities are chosen appropriately (of order $1/n$), then the mixing time drops to $\Theta(n)$. The lifting achieves this speedup by almost eliminating the diffusive behavior of the original symmetric walk, and instead giving the walking particle "momentum" in its current direction of travel. In particular, after $t < n$ steps the lifted walk will typically be at distance $\Theta(t)$ from its starting point, in contrast to $\Theta(\sqrt{t})$ for the original walk.

This idea was extended by Chen et al. [2] to random walks on Cayley graphs. The strategy, roughly, is to lift the state space Ω to $\Omega \times \{1, \ldots, r\}$ where each $i \in \{1, \ldots, r\}$ is associated with a generator, and then to give the walk momentum around a carefully chosen cycle through the generators. The authors give several examples of significant speedups using this construction.

Chen et al. also give a general lifting construction that applies to arbitrary Markov chains, and achieves a mixing time of $O(\rho)$, where ρ is a multicommodity flow parameter (in the original chain) that is almost the inverse of the more familiar "conductance" (or sparsest cut); they also show that this is essentially best possible. Unfortunately, however, this construction is in general not feasible to implement, as simulating even one step of the lifted chain may be as hard as sampling from the stationary distribution π. Chen et al. pose the open question whether lifting can be used to speed up actual sampling algorithms.

In this paper we prove what is apparently the first result in this direction. We revisit a Markov chain introduced by Sinclair and Jerrum [26] which samples a leaf of a tree uniformly at random given crude estimates of the number of leaves in each subtree. This Markov chain was used in [26] to prove that approximate counting for all self-reducible problems in $\#P$ is robust, in the sense that such problems either have a fully-polynomial randomized approximation scheme, or cannot be approximated in polynomial time within *any* polynomial factor (even, say, n^{100}). It was also used in the same paper to give a polynomial time algorithm for uniformly generating random graphs with specified vertex degrees, based on analytic estimates for the number of such graphs [22].

To describe the setting more precisely, let T be a binary[1] tree, all of whose leaves are at the same depth d. Our goal is to sample a leaf of T uniformly at random, in time

---

[1] We make this assumption for simplicity of presentation only; T may in fact have an arbitrary branching factor.

polynomial in d. This fundamental problem goes back at least to Knuth [17]. Suppose we are given partial information about T in the form of an estimate $\widetilde{N}_v$ of the number of leaves N_v in the subtree rooted at each node v. This estimate is guaranteed to be within ratio $1 + \delta$, i.e., $(1 + \delta)^{-1} N_v \le \widetilde{N}_v \le (1 + \delta) N_v$. (Such estimates may be available, e.g., from a crude approximate counting algorithm as in the abstract framework of [26], from analytic approximations as for graphs with given degrees in [22], or from the solution of idealized approximations as in the derivative pricing framework discussed in [3]. In the Statistics literature, the use of such estimates is often referred to as "importance sampling.") If $\delta = O(\frac{1}{d})$ then we can solve the problem rather easily by choosing a random path from the root to a leaf, branching left or right at each node with probabilities proportional to the counting estimates at its two children. Because of the bound on δ, we will accumulate at most a constant bias at the leaves, which can be eliminated by "rejection sampling" with a constant number (in expectation) of repeated trials. (In rejection sampling, if a leaf ℓ is sampled with probability p_ℓ then we output the leaf with probability p^*/p_ℓ, where p^* is a lower bound on p_ℓ for all ℓ, and start again otherwise. The reader is referred to the full version of the paper for a detailed discussion of rejection sampling, including a comparison with the Markov chain approach.)

For larger values of δ the above approach breaks down. To overcome this obstacle, Sinclair and Jerrum [26] introduced a more involved sampling algorithm that runs in polynomial time provided δ is bounded by any constant (or indeed, by any polynomial in d). This algorithm works by simulating a Markov chain on T whose transition probabilities are proportional to the edge weights $\widetilde{N}_v$ (where we think of $\widetilde{N}_v$ as being associated with the edge whose lower endpoint is v). Note that transitions from a node to its parent are allowed, so backtracking occurs. The stationary distribution of this Markov chain is easily seen to be uniform over leaves, and to put a constant fraction of its weight on the leaves.[2] Perhaps surprisingly, the mixing time for $\delta = O(1)$ was shown in [26] to be $\widetilde{O}(d^2)$, implying that the algorithm outputs a uniformly random leaf with bias ε in expected time $\widetilde{O}(d^2 \log \varepsilon^{-1})$. The intuition for the effectiveness of this algorithm is that an overestimate $\widetilde{N}_v$, which leads the chain to choose a downward edge to v with too large probability, also acts to increase the probability of backtracking from v; thus the process is "self-correcting." We note also that the $\Omega(d^2)$ dependence on d is unavoidable as, even in the case of perfect estimates ($\delta = 0$), the process reduces to symmetric random walk on the levels $[0, d]$.

1.2   Results

In this paper we consider lifting the above Markov chain in the regime $\delta \in [\frac{1}{d}, 1]$. (Recall that the problem is trivial for $\delta = O(\frac{1}{d})$.) Our main result is a (non-reversible) lifting that speeds up the mixing time to $O(\delta d^2)$ throughout this range. Thus our lifted chain interpolates smoothly between a trivial linear time rejection sampling algorithm when $\delta = O(\frac{1}{d})$ and the Sinclair-Jerrum quadratic time algorithm when $\delta = \Omega(1)$. In particular, for all $\delta = o(1)$ the lifted chain overcomes the $\Omega(d^2)$ diffusion lower bound

---

[2] The original chain in [26] puts weight $O(1/d)$ on the leaves; a simple modification, which we provide, improves this to a constant with the same bound on mixing time.

on the mixing time of the original chain. (For example, when $\delta = O(\frac{1}{\sqrt{d}})$, we are able to sample leaves in time $O(d^{3/2})$.) We leave open the question of whether a fast lifting exists for larger values of δ.

We believe that the main interest value of this result is as the first application of lifting to a complex Markov chain used in random sampling.[3] However, we briefly mention as an example one potential concrete application. Let $\mathbf{g} = (g_1, \ldots, g_n)$ be a graphical degree sequence on n vertices, and suppose we wish to sample a random graph in which vertex i has degree g_i for each i. We can construct such graphs edge-by-edge, giving rise to a "self-reducibility tree" in which each node corresponds to a partial graph (of edges previously chosen) and a residual degree sequence; the leaves of the tree are precisely the desired graphs (see [26] for details). Note that the depth of this tree is $d = |E(\mathbf{g})|$, the number of edges in the graphs. Classical work of McKay [22] (see also [10,23]) provides analytical estimates for the number of graphs with given vertex degrees that are within ratio $1 + O(\frac{g_{\max}^4}{|E(\mathbf{g})|})$, where $g_{\max} = \max_i g_i$. In [26] the Sinclair-Jerrum Markov chain was used with these estimates to sample graphs from sequences in which $g_{\max} = O(|E(\mathbf{g})|^{1/4})$. The lifting in the present paper would potentially improve the mixing time of this Markov chain from $\widetilde{O}(|E(\mathbf{g})|^2)$ to $O(g_{\max}^4 |E(\mathbf{g})|)$, which is significantly less when $g_{\max} \ll |E(\mathbf{g})|^{1/4}$.

Since our construction is the main contribution of the paper, we say a few words about it here. We stress that the construction is purely local and can be implemented efficiently, unlike the optimal liftings discussed in [2]. Our lifting creates two copies of the tree, having "upward" and "downward" momentum respectively. To eliminate diffusive behavior, we need to arrange for small crossing probabilities between the two copies; this we achieve using a "flow cancellation" idea that is facilitated by our view throughout the paper of Markov chains as flows. Another key ingredient is smoothing of the holding time distribution at some nodes; we achieve this by lifting certain self-loops in the original chain to two-state "traps." This smoothing makes possible our analysis of the mixing time via a non-Markovian coupling argument.

While some of the above features can be identified with hindsight in the efficient liftings of [4,2], our construction is considerably more general and systematic. In particular, we do not exploit strong symmetries in the original Markov chain which make the liftings in those papers rather simpler to construct and to analyze. Indeed, in our case the original Markov chain is not at all symmetrical, as the tree may have arbitrary structure and its edge weights may vary arbitrarily within their respective ranges. For the same reason, the tree is also very far from the one-dimensional processes analyzed in [4,7,8]. We conjecture that our flow-based approach may lead in future to a systematic framework for constructing liftings in a larger class of Markov chains where it is possible to identify generalized "directions" along which momentum can be defined.

In the full version of the paper, we discuss alternative approaches to the leaf-sampling problem for $\delta \in (0, 1]$ based on rejection sampling combined with Markov chain Monte Carlo.

---

[3] We mention that, in hindsight, the "hit-and-run" Markov chain [19] used for sampling points in a convex body has the flavor of a "lifting" of the more classical "ball walk" [20]. We return to this point in Section 5.

1.3   Related Work

The first authors to implicitly discuss lifting of Markov chains to speed up mixing were Diaconis, Holmes and Neal [4], who observed that the mixing time of simple random walk on a path of length n can be improved from $\Theta(n^2)$ to $\Theta(n)$. They also proposed an extension to more general one-dimensional chains with non-uniform stationary distribution, but did not provide bounds on the mixing time. Such an extension was subsequently analyzed by Hildebrand [7,8], who showed that a similar acceleration to $\Theta(n)$ occurs when the stationary distribution is log-concave.

Chen, Lovász and Pak [2] studied lifting in a more general framework. In addition to giving several examples of liftings for random walks on Cayley graphs, they also proved general results on the scope and limitations of lifting. For example, they show that the best possible lifting of any given Markov chain has mixing time (suitably defined) $\Theta(\rho)$, where ρ is the flow parameter mentioned earlier. Since the mixing time is always $\widetilde{O}(\rho^2)$, the optimal speedup via lifting is at most roughly a square root. Chen et al. also give a theoretical construction that achieves this optimal lifting (up to a constant factor) for an arbitrary Markov chain; however, as mentioned earlier, this construction is in general not efficiently implementable. Moreover, they show that if the lifted Markov chain is reversible then the speedup obtainable is (relatively) negligible; hence any useful lifting needs to be non-reversible (as are all the liftings mentioned in this paper).

Jung, Shah and Shin [14] build on the work of Chen et al. by considering the problem of minimizing the size of the lifted Markov chain while still achieving a similar speedup. This measure has applications to distributed algorithms for computing averages in networks, which the same authors discuss in [15].

We mention that all of the above lifting constructions, like our own, seek to eliminate or reduce diffusive behavior in the Markov chain. This is also the idea behind other, more classical techniques for speeding up Markov chain Monte Carlo algorithms, notably Hybrid Monte Carlo [5] and Horowitz's method [9] (see also [27] for more recent work in this direction). However, to the best of our knowledge, these methods lack rigorous analysis in non-trivial examples.

The problem of sampling leaves of a tree can be traced back at least to Knuth [17] in his work on estimating the efficiency of branching programs. Knuth sampled leaves by branching uniformly to children regardless of the number of leaves in the corresponding subtree, which yields a non-uniform distribution $\{p_\ell\}$ over leaves ℓ; he then used the quantity p_ℓ^{-1} as an unbiased estimator of the number of leaves in the tree. This can be seen as the origin of the rejection sampling approach mentioned earlier. A paper by Rosenbaum [24] provides some further analysis and refinement of Knuth's scheme.

The Markov chain approach to leaf sampling appeared in the work of Sinclair and Jerrum [26], where the main application was to show robustness of approximate counting for self-reducible problems. The version of the Sinclair-Jerrum chain presented here is slightly more efficient than the original one. The same paper also applied this Markov chain to give the first polynomial time sampling algorithm for subgraphs of a given graph that have specified vertex degrees, under certain constraints on the maximum degree, using the fact that analytic approximations exist for the number of such graphs (see, e.g., [22]). For subsequent developments on this problem, see [11,16,1].

2   Preliminaries

2.1   Markov Chains, Liftings and Mixing Times

Markov chains. Let Ω be a finite state space. We shall specify Markov chains on Ω using the following weighted graph framework.

A reversible chain is specified by an undirected graph $G = (\Omega, E)$ (possibly with self-loops) with a positive weight Q_e on each edge $e \in E$. Transitions from any vertex $u \in \Omega$ are made with probabilities proportional to the edge weights: i.e., the transition probability from u to v is $P(u, v) = \frac{Q_{(u,v)}}{W_u}$, where $W_u = \sum_{e \ni u} Q_e$ is the sum of the edge weights incident at u.

This Markov chain is easily seen to be reversible with respect to the distribution $\pi(u) = \frac{W_u}{W}$, where $W := \sum_u W_u$ (i.e., π is proportional to the weighted vertex degrees). As is well known, if G is connected and not bipartite (e.g., a single self-loop suffices) then it is ergodic and converges to π from any initial state. Note that the edge weights Q_e are, up to scaling by W, the *ergodic flows* in the stationary distribution; i.e., $Q_{(u,v)} = W\pi(u)P(u, v) = W\pi(v)P(v, u) = Q_{(v,u)}$.

The above framework can be extended to general, non-reversible Markov chains by making G directed and requiring that the edge weights Q_e satisfy the flow condition $\sum_{u:(u,v)\in E} Q_{(u,v)} = \sum_{u:(v,u)\in E} Q_{(v,u)} =: W_v$ for all $v \in \Omega$. If G is strongly connected and aperiodic (again, a single self-loop suffices) then it again converges to the unique stationary distribution $\pi(v) \propto W_v$. Again Q_e is proportional to the ergodic flow along (directed) edge e.

Mixing times. For an ergodic Markov chain $(X_t)_{t\geq0}$ on Ω with stationary distribution π, any $x \in \Omega$ and any $\varepsilon \in (0, 1]$, we define

$$\tau_x(\varepsilon) = \min\{t : \|\eta_{x,t} - \pi\| \leq \varepsilon\},$$

where $\eta_{x,t}$ denotes the distribution of X_t (the state at time t) starting from initial state $X_0 = x$, and $\|\cdot\|$ is total variation distance. We will refer to $\tau_x(\varepsilon)$ as the *mixing time starting from state x*. The *mixing time*, $\tau(\varepsilon)$, is defined as the maximum over $x \in \Omega$ of $\tau_x(\varepsilon)$. We shall sometimes abuse terminology by dropping the dependence on ε from the mixing time.

In this paper we will bound the mixing time using *couplings*. By a *coupling* of a Markov chain, we mean a joint distribution $(X_t, Y_t)_{t\geq0}$ such that the two random processes $(X_t)_{t\geq0}$ and $(Y_t)_{t\geq0}$, considered separately, each obey the transition rule for the given chain. In addition, if $X_t = Y_t$ then we require $X_{t'} = Y_{t'}$ for all $t' \geq t$. One way of defining such a coupling is to specify a suitable transition matrix indexed by the product space $\Omega \times \Omega$, thereby defining a Markov chain with this state space. As long as the two marginal transition probabilities agree with the original Markov chain, this defines a coupling, often referred to as a "Markovian coupling." However, in general, couplings are not required to be Markovian, and in fact, even conditioned on the previous states X_{t-1}, Y_{t-1}, it is perfectly possible for the state X_t to be correlated non-trivially with the sequence of states $Y_0, \ldots, Y_{t-2}$ (as will be the case for the coupling we define in Section 4).

When we speak of couplings in the present paper, we will always mean that a class of couplings has been defined, one for each possible initial pair of states (X_0, Y_0). We say that the coupling has *coalesced by time t* if the event $\{X_t = Y_t\}$ occurs. The following theorem relates the mixing time to the worst-case time until coalescence, and dates back to work of Doeblin in the 1930's (see [18]).

Theorem 2.1 (Coupling Theorem). *Let $(X_t, Y_t)_{t\geq 0}$ be any coupling of a Markov chain on state space Ω, and define*

$$\tau_{\text{couple}}(\varepsilon) = \max_{(X_0, Y_0)\in\Omega\times\Omega} \min\{t\colon \Pr[X_t \neq Y_t] \leq \varepsilon\}.$$

Then, for every $\varepsilon > 0$, $\tau(\varepsilon) \leq \tau_{\text{couple}}(\varepsilon)$.

Liftings. Let $\mathcal{M}$ and $\widehat{\mathcal{M}}$ be Markov chains on finite state spaces Ω, $\widehat{\Omega}$ respectively. We use Q, π to denote the flows and stationary distribution of $\mathcal{M}$, and $\widehat{Q}, \widehat{\pi}$ for the same quantities in $\widehat{\mathcal{M}}$.

We say that $\widehat{\mathcal{M}}$ is a *lifting* of $\mathcal{M}$ if there is a function $f : \widehat{\Omega} \to \Omega$ such that

$$Q_{(u,v)} = \sum_{x\in f^{-1}(u),\ y\in f^{-1}(v)} \widehat{Q}_{(x,y)} \qquad \text{for all } u, v \in \Omega. \tag{1}$$

Informally, if we "collapse" $\widehat{\mathcal{M}}$ by merging into a single state all states that have the same image under f, and aggregate the flows between these merged states, then we obtain precisely the chain $\mathcal{M}$. Note that equation (1) can be viewed as a homomorphism between flows. An immediate consequence of (1) is that $\pi(v) = \sum_{x\in f^{-1}(v)} \widehat{\pi}(x)$ for all $v \in \Omega$. The reader may wish to verify that the construction in Fig. 1 is indeed a valid lifting. We observe that our definition of lifting based on flows makes it particularly easy to design liftings for a given Markov chain (cf. the equivalent definition given in [2]).

Note that $\widehat{\mathcal{M}}$ may be non-reversible even when $\mathcal{M}$ is reversible. Indeed, as Chen et al. [2] show, to substantially speed up a reversible chain one must consider non-reversible liftings. (Note that the lifting in Fig. 1(b) is non-reversible.)

2.2   Approximate Counting and Leaf Sampling

Framework. Let $T = (V, E)$ be a binary[4] tree with root r, all of whose leaves are at the same depth d. As discussed in the Introduction, our goal is to sample a leaf of T u.a.r. We think of T as being very large, so we want an algorithm that is polynomial in the *depth* d of T.

For each node v, let N_v denote the number of leaves in the subtree rooted at v. (Thus $N := N_r$ is the total number of leaves of T, and $N_v = 1$ for each leaf v.) We are given an estimate $\widetilde{N}_v$ of each N_v satisfying

$$(1 + \delta)^{-1} N_v \leq \widetilde{N}_v \leq (1 + \delta) N_v, \tag{2}$$

and $\widetilde{N}_v = N_v = 1$ for leaves v.[5]

---

[4] The assumption that the tree is binary is made for simplicity of presentation only.
[5] Note that it is not necessary to know the structure of T a priori: since (2) implies that $N_v = 0$ (the subtree below v is empty) iff $\widetilde{N}_v = 0$, we can actually infer the structure of T locally from the estimates $\widetilde{N}_v$ for all vertices v.

Throughout the paper, unless otherwise stated, we will assume that δ lies in the range $[\frac{1}{d}, 1]$. The case when $\delta = O(1/d)$ is of little interest, since in this case, as noted in the Introduction, there is a simple linear time sampling algorithm based on rejection sampling. On the other hand, for larger values, $\delta = \Omega(1)$, our lifting construction cannot offer more than a constant factor speedup over the original Sinclair-Jerrum Markov chain, which we now describe.

The Sinclair-Jerrum chain. Sinclair and Jerrum [26] proposed a reversible Markov chain for sampling leaves from a uniform distribution in polynomial time, even when δ is an arbitrarily large constant (or indeed polynomially large in d). We specify the chain by giving the flows Q_e on each edge of T. We set $Q_e = \tilde{N}_v$, where v is the lower endpoint of e. Additionally we introduce at each non-leaf node a self-loop of weight $Q_{(v,v)}$ equal to the total weight of the other edges incident at v, and at each leaf v a self-loop of weight $Q_{(v,v)} = 4d - 1$. (Thus the self-loop probabilities are $\frac{1}{2}$ for non-leaves and $1 - \frac{1}{4d}$ for leaves.) The self-loops of $\frac{1}{2}$ are a standard device to make the chain aperiodic (the resulting chain is usually called "lazy"). The large self-loops at the leaves are included to ensure that the stationary distribution puts large weight on the leaves[6].

As discussed above, the stationary distribution is given by $\pi(v) \propto W_v$, where $W_v := \sum_{e \ni v} Q_e$ is the sum of the edge weights incident at v. Now for any non-leaf node v, since $W_v = 2(\tilde{N}_v + \sum_{u \text{ a child of } v} \tilde{N}_u)$ we have $W_v \in [4(1+\delta)^{-1}N_v, 4(1+\delta)N_v]$. And for any leaf v we have $W_v = 4d$. This implies the following properties of the stationary distribution π:

1. π is uniform over the leaves.
2. $\sum_{v \text{ a leaf}} \pi(v) \geq \frac{1}{2+\delta}$. [To see this, note that the sum of W_v over all nodes v in any level above the leaves is at most $4(1+\delta)\sum_v N_v = 4(1+\delta)N$, while the sum of W_v over leaves is $4dN$.]

Therefore, we can sample leaves as follows. Simulate the Markov chain, starting from the root, until the distribution is close to π. If the final node is a leaf then output it, else fail and repeat. This gives us an almost uniformly distributed leaf (within any desired variation distance ε) in expected time $O(\tau_r(\varepsilon))$, where $\tau_r(\varepsilon)$ is the mixing time starting from the root r. The following theorem, which is a slightly improved version of the original result of Sinclair and Jerrum [26], bounds the mixing time. A proof is given in the full version of the paper.

Theorem 2.2. *For any $\delta \geq 0$, the mixing time of the Sinclair-Jerrum chain starting from the root satisfies $\tau_r(\varepsilon) = O(d^2(1+\delta)^2 \log(d\varepsilon^{-1}))$.*

Thus, for δ bounded by a constant (which is our range of interest in this paper), the mixing time is $\tilde{O}(d^2)$. (The Theorem actually also shows that the mixing time remains polynomial for any $\delta \leq \text{poly}(d)$.)

We note that a lower bound of $\Omega(d^2)$ follows easily, even in the case where the counting estimates are all exact (i.e., $\tilde{N}_v = N_v \; \forall v$), since the height of the walking

---

[6] The construction in [26] did not include these large self-loops; this simple modification actually leads to greater efficiency, since without it a leaf is sampled only with probability $O(\frac{1}{d})$, leading to a factor $O(d)$ overhead in the time to output a leaf.

particle then behaves like symmetric random walk on $[0, d]$. Our main goal in this paper is to give a lifting that improves the above mixing time to $O(\delta d^2)$, thus beating the $\Omega(d^2)$ lower bound for all $\delta = o(1)$.

3   The Lifted Chain

We will define a non-reversible lifted Markov chain having exactly two states for every node of the tree, with the exception of the root which will only have one lifted state. Roughly speaking, one set of these nodes correspond to "particles with downward momentum," and the others to "particles with upward momentum." The root and the leaves are exceptions. In the case of the root there is no need for the "upward" copy, so we retain just a single root node. In the case of a leaf we correspondingly have no need for a "downward" copy; however, we do need a second copy to act as a "trap" node, whose purpose will be to give the distribution of the departure time from the leaf a heavier tail than that provided by the self-loop in the original chain. We describe our construction in three steps:

Step 1: Lazy edges become 4-cycles. Let e be any non-loop edge in the original tree, joining nodes v, w, and with bidirectional flow Q_e through it. In the lifted chain, the original node v corresponds to two nodes, v^+ and v^-, and likewise for w. Suppose v is the parent of w. The new chain has a directed 4-cycle, (v^+, w^+, w^-, v^-), with each of the four directed edges carrying flow Q_e. Under the "projection" sending $v^+, v^- \mapsto v$ and $w^+, w^- \mapsto w$, this directed 4-cycle maps down to the original bidirectional flow Q_e on edge e, plus self-loops at v and w, each also of flow Q_e. Note that Q_e is exactly the contribution of edge e to the lazy self-loops at v, w in the original chain. (See Fig. 2(a).)

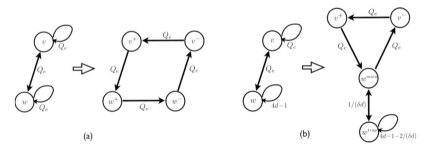

Fig. 2. (a) Lifting of an internal edge $\{v, w\}$. (b) Lifting of a leaf node w.

Applying the above construction to every non-loop edge e in the original tree yields a directed flow which exactly projects back onto the original undirected flow, with the sole exception of the large self-loops on the leaves.

Before proceeding, we first modify the above construction slightly. In the case when w is a leaf and $e = \{v, w\}$ is the edge joining it to its parent, our lifted flow looks slightly different. In this case, the self-loop of flow Q_e at v, and the bidirectional flow Q_e on e lift to a directed 3-cycle, $(v^+, w^{\mathrm{main}}, v^-)$, with each of the three edges carrying flow Q_e. Similarly, in the case when $v = r$ is the root and $e = \{v, w\}$ is the edge joining it to one of its children, the self-loops of flow Q_e at v and w, and the bidirected flow Q_e

on e lift to a directed 3-cycle, (r, w^+, w^-), plus a self-loop at r, with each of these four edges carrying flow Q_e.

Step 2: Set traps at the leaves. Let w be any leaf of the original tree. In the lifted chain, there will be two nodes, w^{main} and w^{trap}, corresponding to w. We next describe the lifted version of the self-loop of flow $4d - 1$ at w. This consists of a self-loop of flow $4d - 1 - \frac{2}{\delta d}$ at w^{trap}, together with a bidirectional flow of $1/(\delta d)$ between w^{main} and w^{trap}. (See Fig. 2(b).)

Step 3: Cancel the crossing edges. After the above two steps we have a lifted flow which projects down onto the original flow. However, in order to avoid the diffusive behavior of the original Markov chain, we need to reduce the "crossing flows" between nodes v^+ and v^-. We do this in a systematic way which preserves the projection onto the original flow. Let v be a non-leaf node in the tree, with flow Q_{up} to its parent and aggregated flow Q_{down} to its children. Then, as described in Step 1, we have crossing flows of value Q_{up} from v^+ to v^-, and Q_{down} from v^- to v^+.

Let $Q_{\min} = \min\{Q_{\mathrm{up}}, Q_{\mathrm{down}}\}$. We now cancel $Q_{\min}$ of the crossing flow in each direction, replacing it with self-loops at v^+ and v^-, each of flow $Q_{\min}$. This leaves us with crossing flow in just one direction, of value $|Q_{\mathrm{up}} - Q_{\mathrm{down}}|$. Note that this modification does not violate the flow condition, nor does it change the projection onto the original Markov chain. (See Fig. 3.)

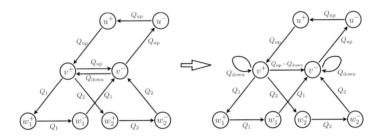

Fig. 3. Cancelling the crossing edges. Here $Q_{\mathrm{down}} = Q_1 + Q_2$, where Q_1, Q_2 are the flows between node v and its two children. The diagram assumes that $Q_{\mathrm{up}} \geq Q_{\mathrm{down}}$.

Since each step of the above construction preserves the lifting condition (1), the resulting chain is indeed a lifting of the Sinclair-Jerrum chain defined in Section 2.2. An immediate consequence is that the stationary distribution $\hat{\pi}$ is the pull-back of π along the projection, and hence has the desired properties on the lifted copies of the leaves, namely that it projects down to a uniform distribution over the leaves having probability mass at least $1/(2 + \delta)$. Thus, in order to use our lifted chain as an improved sampler of leaves, all that remains is to prove that its mixing time is faster than the original undirected chain.

4   Analysis of the Lifted Chain

In this section we prove our main result, which is the following bound on the mixing time of the lifted chain of the previous section. As we have noted earlier, the mixing time overcomes the $\Omega(d^2)$ diffusion lower bound for the original chain for all $\delta = o(1)$, and interpolates smoothly between the trivial $O(d)$ rejection sampling algorithm for $\delta = O(\frac{1}{d})$ and the original $\widetilde{O}(d^2)$ Sinclair-Jerrum algorithm for $\delta = \Omega(1)$.

Theorem 4.1. *For any $\delta \in [\frac{1}{d}, 1]$, the mixing time of the lifted Markov chain defined in Section 3 satisfies $\tau(\varepsilon) = O(\delta d^2 \log(1/\varepsilon))$.*

Proof. We proceed by constructing a non-Markovian coupling for the lifted chain. Let $X_0 \neq Y_0$ be arbitrary states of this lifted chain. We will define a coupled joint evolution $(X_t, Y_t)_{t \geq 0}$ in such a way that each of (X_t) and (Y_t), considered separately, obeys the law of our lifted Markov chain. We will do this in three asynchronous stages. First, let (X_t), (Y_t) each run independently until reaching the root, r, at times ρ_X, ρ_Y, respectively.

Subsequently, let both (X_t) and (Y_t) follow the same trajectory until they reach a "leaf trap" node, at respective times σ_X and $\sigma_Y = \sigma_X + (\rho_Y - \rho_X)$. Since ρ_X may not equal ρ_Y, this portion of the coupling is non-Markovian.

The third stage is empty for whichever chain had reached the root later, and lasts for $|\rho_X - \rho_Y|$ steps for the chain that reached the root earlier. This means that at the end of the third stage, the same (random) number of time steps will have elapsed for both chains. Also note that, since both chains begin stage 3 at the same leaf trap node, there is at least a probability of

$$(1 - 1/((\delta d)(4d - 1 - \delta d)))^{|\rho_X - \rho_Y|} = \exp(-O(|\rho_X - \rho_Y|/(\delta d^2))) \qquad (3)$$

that both chains remain at this node throughout stage 3, and have therefore coalesced by the end. If not, we can simply start over again with the first stage.

Our analysis of this coupling rests on two lemmas.

Lemma 4.2. *There exists an absolute constant C such that, from any initial node $X_0 = v$, the expected hitting time from v to the root is $\leq C\delta d^2$.*

Proof. We split the proof into three cases, according to whether v is a downward node, a leaf node, or an upward node. (In the case when v is the root, the hitting time is 0.)

Case 1: $v = w^+$ **is a downward node.** Since v is a downward node, every non-self-loop move either increases the depth by 1, or crosses to a rootward-oriented node. Hence, since the self-loop probability at v is at most $1/2$, in expected time at most $2d = O(\delta d^2)$ we will reach one of the other two cases. Thus, it suffices to handle cases 2 and 3.

Case 2: $v = w^{\text{main}}$ **or** $v = w^{\text{trap}}$ **is a leaf node.** Let u denote the parent of w in the original tree. Now, in our lifted chain, starting from $X_0 = v$, the first node reached by X_t that is not in $\{w^{\text{main}}, w^{\text{trap}}\}$ must be u^-. What is the hitting time to u^-? Solving a system of two linear equations in two unknowns, we find that this hitting time is $4d$ when starting from w^{main}, and $(4\delta d^2 - \delta d - 1 + 4d)$ when starting from w^{trap}. Since in both cases this is $O(\delta d^2)$, and u^- is an upward node, it thus suffices to handle case 3.

Case 3: $v = w^-$ **is an upward node.** Let u be the parent of w in the original tree. As in case 2, note that, starting from $X_0 = w^-$, the first node that will be reached by X_t that does not project into the subtree rooted at w must be u^-.

Let $Q_{up} = \tilde{N}_w$ denote the flow up from w, and Q_{down} the aggregated flow down from w to its children. When $Q_{up} \geq Q_{down}$ (as in Fig. 3), the only edges out from w^- are a self-loop and the edge (w^-, u^-), so it is easy to calculate that the expected hitting time from w^- to u^- equals $1 + Q_{down}/Q_{up}$, which is at most $1 + (1 + \delta)^2 = O(1)$.

Claim. Suppose $Q_{up} < Q_{down}$. Let H denote the hitting time from w^- to u^-. Then $\mathbb{E}(H) = O(\delta d)$.

Assuming the Claim is true, we have shown that the expected hitting time from w^- to u^- is always $O(\delta d)$. It follows by induction that the hitting time from w^- to r is $O(\delta d^2)$, since the depth of w is at most d, which completes our analysis of case 3 and the proof of the lemma.

All that remains is to prove the Claim. To see this, consider what happens to our lifted walk if we re-route the flow on the edge (w^-, u^-) to instead go along the edge (w^-, w^+). In this case, starting from w^-, we can never leave the subtree rooted at w, and in fact the random walk is exactly the same as would be produced by our lifting construction applied just to the subtree rooted at w, except that the transition probabilities at the leaves are still based on d rather than on the height of the subtree below w. Let us compute the stationary probability of w^- in this modified chain.

Using the well-known fact that the stationary probability at any node is the reciprocal of the expected return time to that node, it follows that

$$\frac{1}{\tilde{\pi}(w^-)} = 1 + \frac{Q_{down}}{Q_{down} + Q_{up}} \mathbb{E}(H'), \tag{4}$$

where H' is the hitting time from w^+ to w^-, and $\tilde{\pi}$ is the stationary distribution for the modified lifted chain rooted at w. A straightforward calculation yields $\tilde{\pi}(w^-) \geq 1/((1 + \delta)^2(2d + i)) \geq 1/(4d)$, where i is the height of node w, whence by (4) it follows that $\mathbb{E}(H') \leq 2(4d - 1)$.

Returning now to the full lifted chain, since from w^- the flows out are Q_{up} to u^-, Q_{up} in a self-loop, and $Q_{down} - Q_{up}$ to w^+, it follows that

$$\mathbb{E}(H) = 1 + \frac{Q_{up}}{Q_{down} + Q_{up}} \mathbb{E}(H) + \frac{Q_{down} - Q_{up}}{Q_{down} + Q_{up}} \mathbb{E}(H'),$$

which implies

$$\mathbb{E}(H) = 1 + \frac{Q_{up}}{Q_{down}} + \left(\frac{Q_{down} - Q_{up}}{Q_{down}}\right) \mathbb{E}(H') \leq 2 + 2P_{cross}(w^-)\mathbb{E}(H'),$$

where $P_{cross}(w^-) = O(\delta)$ is the transition probability from w^- to w^+ in the lifted chain. Since we already know that $\mathbb{E}(H') = O(d)$, it follows that $\mathbb{E}(H) \leq 2 + O(\delta d) = O(\delta d)$. This concludes the proof of the Claim, and of Lemma 4.2. $\qquad \square$

Lemma 4.3. *There exists an absolute constant C' such that the expected hitting time from the root to the set of "leaf trap" nodes is $\leq C' \delta d^2$.*

Proof. Consider an infinite run of the Markov chain, and partition the positive integers into epochs, where the even epochs end at the first time (after they start) that the root is reached, and the odd epochs end at the first time (after they start) that a leaf trap is reached. Let us denote by L the set of all leaf trap nodes. Since no leaf traps are visited during the odd epochs, the fraction of time in even epochs is at least $\pi(L)$. But the average length of an even epoch is at most $C\delta d^2$, by Lemma 4.2. Hence the average length of an odd epoch must be at most $C\delta d^2/\pi(L)$, which is $O(\delta d^2)$ since we arranged for $\pi(L) = \Theta(1)$. This concludes the proof, as the average length of an odd epoch equals the expected hitting time from the root to the set of leaf trap nodes. □

We now continue with the proof of Theorem 4.1. By Lemmas 4.2 and 4.3, the expected total length of stages 1 and 2 combined is $O(\delta d^2)$. Hence, by Markov's inequality, with probability at least $7/8$, the total length is at most eight times the expectation, which is $O(\delta d^2)$. An application of the triangle inequality implies that therefore $\mathbb{E}(|\rho_X - \rho_Y|) = O(\delta d^2)$ (where the O hides an explicit constant of moderate size). By Markov's inequality, it follows that with probability $\Omega(1)$, $|\rho_X - \rho_Y| = O(\delta d^2)$. By (3), the coupling has coalesced by the end of the third stage with probability $\Omega(1)$. Thus the chain coalesces within $O(\delta d^2 \log(1/\varepsilon))$ time steps with probability at least $1 - \varepsilon$. The corresponding bound on the mixing time follows from Theorem 2.1. □

5   Conclusions and Future Work

We have shown that non-reversible liftings can be used to speed up MCMC sampling (and hence also approximate counting) algorithms, even without the high degree of symmetry present in previous examples. Although it is still highly specialized, the class of Markov chains we consider, being random walks on trees with an approximation oracle for the number of leaves, is nevertheless natural in the context of computation, and encompasses many combinatorial problems with interesting and complex structure.

The first open question is whether our construction can be improved to reduce the mixing time of the lifted chain down to the asymptotically optimal value $O(\rho)$, where $\rho = O(d(1 + \delta)^2)$ is the flow parameter for the original chain, while retaining the local character of the current construction which makes it a practical tool for sampling. In the case of large bias, $\delta = \Omega(1)$, our current lifting exhibits (potentially) nearly as much diffusive behavior as the unlifted chain; intuitively this happens because "excursions" upward or downward may typically be of length $1/\delta = O(1)$, as is the case for symmetric random walk. However, at least in the special case when the tree is a path, we have developed a more complex (yet still local) construction that eliminates this diffusive behavior to a large extent; the idea is to keep track of multiple momentum values (rather than just "up" and "down"). This will be discussed in the full version of the paper.

A second natural question is whether our techniques can be profitably applied to other Markov chains used in sampling algorithms. Prime candidates here are Markov chains for matchings [12,13] and for sampling points in a convex body [6,21]. The latter example seems particularly intriguing as there is a well-defined notion of "direction" along which momentum can be preserved. Indeed, we note that lifting ideas have already appeared, albeit not explicitly, in this example: the "hit-and-run" Markov chain [19], which at each step moves to a random point on a randomly chosen chord of the body through the current point, has the flavor of a "lifting" of the more local "ball walk"[20],[7] which moves to a random point within a ball centered at the current point.

---

[7] See the full version for a more precise discussion of this point.

We conjecture that understanding this connection more formally within a lifting framework may illuminate previous work on random walks on convex bodies, and perhaps even lead to further algorithmic improvements.

References

1. Bayati, M., Kim, J.-H., Saberi, A.: A sequential algorithm for generating random graphs. In: Charikar, M., Jansen, K., Reingold, O., Rolim, J.D.P. (eds.) RANDOM 2007 and APPROX 2007. LNCS, vol. 4627, pp. 326–340. Springer, Heidelberg (2007)
2. Chen, F., Lovász, L., Pak, I.: Lifting Markov chains to speed up mixing. In: Proc. 17th Annual ACM Symposium on Theory of Computing, pp. 275–281 (1999)
3. Das, S.R., Sinclair, A.: A Markov chain Monte Carlo method for derivative pricing and risk assessment. J. Investment Management 3, 29–44 (2005)
4. Diaconis, P., Holmes, S., Neal, R.: Analysis of a nonreversible Markov chain sampler. Annals of Applied Probability 10, 726–752 (2000)
5. Duane, S., Kennedy, A., Pendleton, B., Roweth, D.: Hybrid Monte Carlo. Physics Letters B 195, 216–222 (1987)
6. Dyer, M., Frieze, A., Kannan, R.: A random polynomial-time algorithm for approximating the volume of convex bodies. JACM 38, 1–17 (1991)
7. Hildebrand, M.: Rates of convergence of the Diaconis-Holmes-Neal Markov chain sampler with a V-shaped stationary probability. Markov Proc. Rel. Fields 10, 687–704 (2004)
8. Hildebrand, M.: Analysis of the Diaconis-Holmes-Neal Markov chain sampler for log-concave probabilities (2002) (preprint),
   http://nyjm.albany.edu:8000/~martinhi/preprints.html
9. Horowitz, A.M.: A generalized guided Monte Carlo algorithm. Physics Letters B 268, 247–252 (1991)
10. Janson, S.: The probability that a random multigraph is simple. Combinatorics, Probability and Computing 18, 205–225 (2009)
11. Jerrum, M., Sinclair, A.: Fast uniform generation of regular graphs. Theoretical Computer Science 73, 91–100 (1990)
12. Jerrum, M., Sinclair, A.: Approximating the permanent. SIAM Journal on Computing 18, 1149–1178 (1989)
13. Jerrum, M., Sinclair, A., Vigoda, E.: A polynomial-time approximation algorithm for the permanent of a matrix with non-negative entries. JACM 51, 671–697 (2004)
14. Jung, K., Shah, D., Shin, J.: Distributed averaging via lifted Markov chains (August 2009) (preprint), arxiv.org/pdf/0908.4073v1
15. Jung, K., Shah, D., Shin, J.: Fast and slim lifted Markov chains. In: Allerton Conference on Communication, Control and Computing (2007)
16. Kim, J.-H., Vu, V.: Generating random regular graphs. In: Proc. 21st Annual ACM Symposium on Theory of Computing, pp. 213–222 (2003)
17. Knuth, D.: Estimating the efficiency of backtrack programs. Mathematics of Computation 29, 121–136 (1975)
18. Lindvall, T.: Lectures on the coupling method. Dover, Mineola (2002)
19. Lovász, L.: Hit-and-run mixes fast. Mathematical Programming 86, 443–461 (1998)
20. Lovász, L., Simonovits, M.: Random walks in a convex body and an improved volume algorithm. Random Structures & Algorithms 4, 359–412 (1993)
21. Lovász, L., Vempala, S.: Simulated annealing in convex bodies and an $O^*(n^4)$ volume algorithm. J. Computer and System Sciences 72, 392–417 (2006)
22. McKay, B.: Asymptotics for symmetric 0-1 matrices with prescribed row sums. Ars Combinatorica 19A, 15–25 (1985)

23. McKay, B.D., Wormald, N.: Asymptotic enumeration by degree sequence of graphs with degrees $o(n^{1/2})$. Combinatorica 11, 369–382 (1991)
24. Rosenbaum, P.: Sampling the leaves of a tree with equal probabilities. Journal of the American Statistical Association 88, 1455–1457 (1993)
25. Sinclair, A.: Improved bounds for mixing rates of Markov chains and multicommodity flow. Combinatorics, Probability and Computing 1, 351–370 (1992)
26. Sinclair, A., Jerrum, M.: Approximate counting, uniform generation and rapidly mixing Markov chains. Information & Computation 82, 93–133 (1989)
27. Turitsyn, K., Chertkov, M., Vucelja, M.: Irreversible Monte Carlo algorithms for efficient sampling (2008) (preprint), arxiv.org/pdf/0809.0916v2

Constructive Proofs of Concentration Bounds

Russell Impagliazzo[1,⋆] and Valentine Kabanets[2,⋆⋆]

[1] Institute for Advanced Study & University of California, San Diego
russell@cs.ucsd.edu
[2] Institute for Advanced Study & Simon Fraser University
kabanets@cs.sfu.ca

Abstract. We give a combinatorial proof of the Chernoff-Hoeffding concentration bound [9,16], which says that the sum of independent $\{0,1\}$-valued random variables is highly concentrated around the expected value. Unlike the standard proofs, our proof does not use the method of higher moments, but rather uses a simple and intuitive counting argument. In addition, our proof is constructive in the following sense: if the sum of the given random variables is not concentrated around the expectation, then we can efficiently find (with high probability) a subset of the random variables that are statistically dependent. As simple corollaries, we also get the concentration bounds for $[0,1]$-valued random variables and Azuma's inequality for martingales [4].

We interpret the Chernoff-Hoeffding bound as a statement about Direct Product Theorems. Informally, a Direct Product Theorem says that the complexity of solving all k instances of a hard problem increases exponentially with k; a Threshold Direct Product Theorem says that it is exponentially hard in k to solve even a significant fraction of the given k instances of a hard problem. We show the equivalence between optimal Direct Product Theorems and optimal Threshold Direct Product Theorems. As an application of this connection, we get the Chernoff bound for expander walks [12] from the (simpler to prove) hitting property [2], as well as an optimal (in a certain range of parameters) Threshold Direct Product Theorem for weakly verifiable puzzles from the optimal Direct Product Theorem [8]. We also get a simple constructive proof of Unger's result [38] saying that XOR Lemmas imply Threshold Direct Product Theorems.

1 Introduction

Randomized algorithms and random constructions have become common objects of study in modern computer science. Equally ubiquitous are the basic tools of probability theory used for their analysis. Some of the most widely used such tools are various *concentration bounds*. Informally, these are statements saying

---

⋆ Supported by NSF grants DMS-0835373, CNS-0716790, CCF-0832797, DMS-0635607, Simonyi Foundation, the Bell Company Fellowship, and Fund for Math.
⋆⋆ Supported by an NSERC Discovery grant and NSF grant DMS-0635607.

M. Serna et al. (Eds.): APPROX and RANDOM 2010, LNCS 6302, pp. 617–631, 2010.
© Springer-Verlag Berlin Heidelberg 2010

that the outcome of a random experiment is likely to be close to what is expected (concentrated near the expectation). The well-known Chernoff bound [9] is a prime example, and is probably one of the most-often used such concentration bounds. Basically, it says that repeating a random experiment many times independently and taking the average of the outcomes results in a value that is extremely likely to be very close to the expected outcome of the experiment, with the probability of deviation diminishing exponentially fast with the number of repetitions.

A computational analogue of concentration bounds in complexity are *Direct Product* Theorems. Informally, these are statements saying that solving a somewhat hard problem on many independent random instances becomes extremely hard, with the hardness growing at an exponential rate with the number of repetitions. The main application of direct product theorems is to hardness amplification: taking a problem that is somewhat hard-on-average to solve, and turning it into a problem that is extremely hard-on-average to solve. Such hardness amplification is important for cryptography and complexity; for example, in cryptography, the increased hardness of a function translates into the increased security of a cryptographic protocol.

In this paper, we show a close connection between probability-theoretic and complexity-theoretic concentration bounds. We give a new, constructive proof of the Chernoff bound, and use this proof to establish an equivalence between two versions of direct product theorems: the standard Direct Product Theorem and the *Threshold* Direct Product. In the standard direct product, we want to upperbound the probability of efficiently solving *all* given instances of a somewhat hard problem, whereas in the threshold direct product, we want to upperbound the probability of solving more than a certain *fraction* of the instances.

To motivate the need for Threshold Direct Product Theorems, we give an example of its typical use in cryptography. CAPTCHAs [1] are now widely used to distinguish human users from artificially intelligent "bots". Here a user is issued a random puzzle, say distorted text, and is asked to decipher the text. Say that a legitimate user succeeds with probability $c \leqslant 1$, whereas an attacker succeeds with probability at most $s < c$. To boost our confidence that we are dealing with a legitimate user, we will issue k random puzzles in parallel, and see how many of them get answered correctly. If $c = 1$, then we know that the legitimate user will answer all k instances correctly. A standard Direct Product Theorem for CAPTCHAs [5,8] could then be used to argue that it's very unlikely that an attacker will answer *all* k instances. In reality, however, even a legitimate user can make an occasional mistake, and so $c < 1$. Thus we can't distinguish between legitimate users and attackers by checking if all k instances are answered correctly. Intuitively, though, we still expect that a legitimate user should answer almost all instances (close to c fraction), whereas the attacker can't answer significantly more than s fraction of them. This intuition is formalized in the Threshold Direct Product Theorem for CAPTCHAs [22], which thus allows us to make CAPTCHAs reliably easy for humans but reliably hard for "bots".

The probability-theoretic analogue of a Direct Product Theorem is the statement that if a random experiment succeeds with probability at most p, then the probability that it succeeds in k independent trials is at most p^k. The analogue of a Threshold Direct Product is the Chernoff bound saying that the probability of getting significantly more than the expected pk successes is exponentially small in k. We give a *constructive* proof of the equivalence between these two probability-theoretic statements. Namely, we show that if the probability of getting more than pk successes is *noticeably larger* than it should be (by the Chernoff bound), then we can *efficiently* find a subset S of the k trials such that the random experiment succeeds in all trials $i \in S$ with probability *noticeably larger* than $p^{|S|}$.

In the language of direct products, this means that there is an equivalence between standard direct product theorems and threshold direct product theorems. Moreover, the constructive nature of the proof of this equivalence means that it applies to the *uniform* setting of computation, where the hardness (security) is measured with respect to uniform algorithms (rather than non-uniform circuits). In particular, we get that for a wide variety of classes of cryptographic protocols, there is a Direct Product Theorem for the class iff there is a Threshold Direct Product theorem.

The formalized equivalence between standard and threshold direct products also allows us to quantify the information-theoretic limitations of simple reductions between the two. We then show how to overcome this limitation with slightly more complicated reductions (using conditioning).

1.1   Chernoff-Hoeffding Bounds, Martingales and Expander Walks

The well-known Chernoff-Hoeffding bound [9,16] states that the sum of independent $\{0, 1\}$-valued random variables is highly concentrated around the expected value. Numerous variants of this concentration bound have been proved, with Bernstein's inequalities from 1920's and 1930's being probably the earliest [7]. The known proofs of these bounds rely on the idea of Bernstein to use the moment-generating function of the given sum of independent random variables $X_1 + \cdots + X_n$; recall that the moment-generating function of a random variables X is $M_X(t) = \mathbf{Exp}[e^{t \cdot X}]$, where $\mathbf{Exp}[\cdot]$ denotes the expectation.

While not difficult technically, the standard proof, in our opinion, does not provide intuition why concentration is likely. One of the main results of our paper is a different proof of the Chernoff bound, using a simple combinatorial argument (and, in particular, avoiding any use of the moment-generating functions). We actually prove a generalization of the Chernoff bound, originally due to Panconesi and Srinivasan [30] (who also used the standard method of moment-generating functions in their proof). In this generalization, the assumption of independence of the variables $X_1, \ldots, X_n$ is replaced with the following weaker assumption: There exists some $\delta > 0$ such that, for all subsets $S \subseteq [n]$ of indices, $\mathbf{Pr}[\wedge_{i \in S} X_i = 1] \leqslant \delta^{|S|}$. Observe that if the variables X_i's are independent, with each $\mathbf{Exp}[X_i] \leqslant \delta$, then, for all $S \subseteq [n]$, $\mathbf{Pr}[\wedge_{i \in S} X_i = 1] \leqslant \delta^{|S|}$.

Theorem 1 (Generalized Chernoff bound [30]). *Let $X_1, \ldots, X_n$ be Boolean random variables such that, for some $0 \leqslant \delta \leqslant 1$, we have that, for every subset $S \subseteq [n]$, $\mathbf{Pr}[\wedge_{i \in S} X_i = 1] \leqslant \delta^{|S|}$. Then, for any $0 \leqslant \delta \leqslant \gamma \leqslant 1$, $\mathbf{Pr}\left[\sum_{i=1}^n X_i \geqslant \gamma n\right] \leqslant e^{-nD(\gamma \| \delta)}$, where $D(\cdot \| \cdot)$ is the relative entropy function (defined in Section 2 below), satisfying $D(\gamma \| \delta) \geqslant 2(\gamma - \delta)^2$.*

We now sketch our proof of Theorem 1. Imagine sampling a random subset $S \subseteq [n]$ where each index $i \in [n]$ is put in S independently with some probability q (to be optimally chosen). We compute, in two ways, $\mathbf{Pr}[\wedge_{i \in S} X_i = 1]$, where the probability is over S and $X_1, \ldots, X_n$.

On one hand, since $\mathbf{Pr}[\wedge_{i \in S} X_i = 1] \leqslant \delta^{|S|}$ for *all* $S \subseteq [n]$, the probability of choosing $S \subseteq [n]$ with $\wedge_{i \in S} X_i = 1$ is *small*. On the other hand, if $p = \mathbf{Pr}[\sum_{i=1}^n X_i \geqslant \gamma n]$ is relatively large, we are likely to sample a n-tuple $X_1, \ldots, X_n$ with very many (at least γn) 1's. Given such a tuple, we are then likely to sample a subset $S \subseteq [n]$ with $\wedge_{i \in S} X_i = 1$. Thus the overall probability of choosing $S \subseteq [n]$ with $\wedge_{i \in S} X_i = 1$ is relatively *large*. The resulting contradiction shows that p must be small. (The complete proof is given in Section 3.1.)

We also get several other concentration bounds as simple corollaries of Theorem 1. First, we get a version of Theorem 1 in the setting of real-valued random variables that take their values in the interval $[0, 1]$, the Hoeffding bound [16] (Theorem 6). Then we prove a concentration bound for martingales, known as Azuma's inequality [4] (Theorem 7). In another application of our Theorem 1, we obtain a Chernoff-type concentration bound for random walks on expander graphs (Theorem 11), almost matching the parameters of [12,15].

1.2 Applications to Direct Product Theorems

We interpret Theorem 1 as giving an equivalence between certain versions of Direct Product Theorems (DPTs), which are statements of the form "k-wise parallel repetition increases the complexity of a problem at an exponential rate in the number of repetitions k". Such theorems are known for a variety of models: Boolean circuits [40,13], 2-prover games [34], decision trees [29], communication complexity [31], polynomials [39], puzzles [5], and quantum XOR games [11], just to mention a few. However, there are also examples where a direct product statement is false (see, e.g., [5,32,36]).

More formally, for a function $F: U \to R$, its k-wise direct product is the function $F^k: U^k \to R^k$, where $F^k(x_1, \ldots, x_k) = (F(x_1), \ldots, F(x_k))$. The main application of this construction is to *hardness amplification*. Intuitively, if $F(x)$ is easy to compute on at most p fraction of inputs x (by a certain resource-bounded class of algorithms), then we expect $F^k(x_1, \ldots, x_k)$ to be easy on at most (close to) p^k fraction of k-tuples $(x_1, \ldots, x_k)$ (for a related class of algorithms).

A DPT may be viewed as a computational analogue of the following (obvious) probabilistic statement: Given k random independent Boolean variables $X_1, \ldots, X_k$, where each $X_i = 1$ with probability at most p, we have $\mathbf{Pr}[\wedge_{i=1}^k X_i = 1] \leqslant p^k$. The Chernoff bound says that with all but exponentially small probability at most about pk of the random variables $X_1, \ldots, X_k$ will be 1. The computational analogue of this concentration bound is often called a *Threshold Direct Product Theorem (TDPT)*, saying that if a function F is easy to compute on at most p fraction of inputs (by a certain class of algorithms), then computing $F^k(x_1, \ldots, x_k)$ correctly in significantly more than pk positions $1 \leqslant i \leqslant k$ is possible for at most a (negligibly more than) exponentially small fraction of k-tuples $(x_1, \ldots, x_k)$ (for a related class of algorithms). TDPTs are also known for a number of models, e.g., Boolean circuits (follows from [20,17]), 2-prover games [33], puzzles [22], and quantum XOR games [11].

Observe that Theorem 1 says that the Chernoff concentration bound for random variables $X_1, \ldots, X_n$ follows from the assumption that $\mathbf{Pr}[\wedge_{i \in S} X_i = 1] \leqslant p^{|S|}$ for all subsets S of $[n]$. In the language of direct products, this means that Threshold Direct Product Theorems follow from Direct Product Theorems. We explain this connection in more detail next.

Equivalence between DPTs and TDPTs. Let us call a DPT *optimal* if has perfect exponential increase in complexity: A function F that is computable on at most p fraction of inputs gives rise to the function F^k that is computable on at most p^k fraction of inputs. Similarly, we call a TDPT optimal, if its parameters match exactly its probabilistic analogue, the Chernoff-Hoeffding bound.

As an immediate application of Theorem 1, we get that an optimal DPT implies an optimal TDPT. We illustrate it for the case of the DPT for Boolean circuits. Suppose F is a Boolean function that can be computed on at most p fraction of inputs (by circuits of certain size s). The optimal DPT for circuits (provable, e.g., using [20,17]) says that for any k, the function F^k is computable on at most p^k fraction of inputs (by any circuit of appropriate size $s' < s$).

Towards a contradiction, suppose there is an algorithm A that computes $F^k(x_1, \ldots, x_k)$ in significantly more than pk positions $1 \leqslant i \leqslant k$, for more than the exponentially small fraction of inputs $(x_1, \ldots, x_k)$. Define Boolean random variables $X_1, \ldots, X_k$, dependent on F, A, and a random k-tuple $(x_1, \ldots, x_k)$, so that $X_i = 1$ iff $A(x_1, \ldots, x_k)_i = F(x_i)$. By our assumption, these variables $X_1, \ldots, X_k$ fail the Chernoff concentration bound. Hence, by Theorem 1, there is a subset $S \subseteq \{1, \ldots, k\}$ such that $\mathbf{Pr}[\wedge_{i \in S} X_i = 1] > p^{|S|}$. But the latter means that our algorithm A, restricted to the positions $i \in S$, computes $F^{|S|}$ with probability greater than $p^{|S|}$, contradicting the optimal DPT.

In an analogous way, we get an optimal TDPT for every non-uniform model where an optimal DPT is known: e.g., decision trees [29] and quantum XOR games [11]; for the latter model, an optimal TDPT was already proved in [11].

A constructive version of Theorem 1. For non-uniform models (as in the example of Boolean circuits considered above), it suffices to use Theorem 1 which only says that if the random variables $X_1, \ldots, X_n$ fail to satisfy the concentration bound, then there *must exist* a subset S of them such that $\wedge_{i \in S} X_i = 1$ with large probability. To obtain the Direct Product Theorems in the uniform model of computation, it is important that such a subset S be efficiently computable by a *uniform* algorithm.

Our combinatorial proof of Theorem 1 immediately yields such an algorithm. Namely, we just randomly sample a subset S by including each index i, $1 \leqslant i \leqslant n$, into S with probability q, where q is chosen as a function of how far the variables $X_1, \ldots, X_n$ are from satisfying the concentration bound. We then output S if $\wedge_{i \in S} X_i = 1$ has "high" probability; otherwise we sample another set S. Here we assume that our algorithm has a way to sample from the distribution $X_1, \ldots, X_n$. This reasoning yields the following.

Theorem 2. *There is a randomized algorithm $\mathcal{A}$ such that the following holds. Let $X_1, \ldots, X_n$ be 0-1-valued random variables. Let $0 < \delta < \gamma \leqslant 1$ be such that $\mathbf{Pr}[\sum_{i=1}^{n} X_i \geqslant \gamma n] = p > 2\alpha$, for some $\alpha \geqslant e^{-nD(\gamma \| \delta)}$. Then, on inputs $n, \gamma, \delta, \alpha$, the algorithm $\mathcal{A}$, using oracle access to the distribution $X_1, \ldots, X_n$, runs in time $\mathrm{poly}(\alpha^{-1/((\gamma - \delta)\delta)}, n)$ and outputs a set $S \subseteq [n]$ such that, with probability at least $1 - o(1)$, $\mathbf{Pr}[\wedge_{i \in S} X_i = 1] > \delta^{|S|} + \Omega(\alpha^{4/((\gamma - \delta)\delta)})$.*

Using this constructive version, we prove an optimal TDPT also for uniform models. In particular, we get such a result for the case of CAPTCHA-like puzzles, called weakly verifiable puzzles [8] (see Theorem 13).[1] DPTs for puzzles are known [5,8], with [8] giving an optimal DPT. Also TDPTs are known [22,27], but they are not optimal. Here we immediately get an optimal TDPT for puzzles, using the optimal DPT of [8], when the success probabilities of the legitimate user and the attacker are constant.

We also show that the limitation on the success probabilities being constant is *unavoidable* for the *naive* reductions between DPTs and TDPTs, as those in Theorem 2. Namely, we give an example of a distribution $X_1, \ldots, X_n$ where the dependence on $\gamma - \delta$ in the exponent of α (stated in Theorem 2) is necessary.

Lemma 1. *There are Boolean random variables $X_1, \ldots, X_n$, and parameters $0 < \delta < \gamma < 1$ such that $\mathbf{Pr}[\sum_{i=1}^{n} X_i \geqslant \gamma n] = p/2 > 2\alpha$, for $\alpha \geqslant e^{-nD(\gamma \| \delta)}$, but, for every subset $S \subseteq [n]$, $\mathbf{Pr}[\wedge_{i \in S} X_i = 1] - \delta^{|S|} \leqslant (4\alpha)^{\delta(\ln 1/\delta)/(\gamma - \delta)}$.*

We also show that this limitation can be overcome by *conditioned* reductions which are allowed to use conditioning (albeit the concentration bound we get in this case is not as tight as before).

Theorem 3. *There is a randomized algorithm $\mathcal{A}$ satisfying the following. Let $X_1, \ldots, X_n$ be Boolean-valued random variables, and let $0 \leqslant \delta < \gamma \leqslant 1$. Suppose that $\mathbf{Pr}\left[\frac{1}{n} \sum_{i=1}^{n} X_i \geqslant \gamma\right] > \alpha$, where $\alpha > (32/(\gamma - \delta)) \cdot e^{-(\gamma - \delta)^2 n/64}$. Then, the algorithm $\mathcal{A}$ on inputs $n, \gamma, \delta, \alpha$, using oracle access to the conditional distribution $(X_1, \ldots, X_n \mid \sum_{j \in S} X_j \geqslant \gamma n/2)$, runs in time $\mathrm{poly}(n, 1/\alpha, 1/\gamma, 1/\delta)$ and outputs a subset $S \subset [n]$ (of size $n/2$) and an index $i_0 \in \bar{S}$ (where $\bar{S} = [n] - S$) such that, with probability at least $1 - o(1)$, $\mathbf{Pr}\left[X_{i_0} = 1 \mid \sum_{j \in S} X_j \geqslant \gamma n/2\right] > \delta + (\gamma - \delta)/16$.*

---

[1] Unger [38] claims to get a TDPT for puzzles, but in fact only proves a TDPT for circuits from Yao's XOR Lemma. Actually, no XOR Lemma for puzzles is known, and so Unger's methods don't apply.

Remark 1. Naive reductions between DPT and TDPT (as in Theorem 2) are applicable in any setting, whereas conditioned reductions (as in Theorem 3) need an additional assumption (sampleability from a conditional distribution). The universality of naive reductions, however, comes at an unavoidable cost, as witnessed by Lemma 1. Together with Theorem 3, this shows that there is an actual quantitative difference between naive and conditioned reductions. In particular, while, surprisingly, naive reductions are optimal in terms of quantitative hardness amplification, they are suboptimal in terms of preserving the adversary's advantage, which is only polynomailly preserved if $\gamma - \delta = \Omega(1)$. In contrast, conditioned reductions can preserve this advantage linearly.

Finally, our Theorem 1 implies some TDPT even when we only have a weak (suboptimal) DPT for the model. For example, we can get some version of a TDPT for 2-prover games, using the best available DPT for such games [34,18,33];[2] however, a better TDPT for 2-prover games is known [33]. Also, as shown by Haitner [14], for a wide class of cryptographic protocols (interactive arguments), even if the original protocol doesn't satisfy any DPT, there is a slight modification of the protocol satisfying some weak DPT. Then, our results imply that these modified protocols also satisfy some weak TDPT.

Direct Product Theorems vs. XOR Lemmas. A close relative of DPTs is an XOR Theorem. For a Boolean function $F \colon \{0,1\}^n \to \{0,1\}$, its k-wise XOR function is $F^{\oplus k} \colon (\{0,1\}^n)^k \to \{0,1\}$, where $F^{\oplus k}(x_1, \ldots, x_k) = \oplus_{i=1}^k F(x_i)$. Intuitively, taking XOR of the k independent copies of a function F, where F can be computed on at most p fraction of inputs, is similar to taking the XOR of k independent random Boolean variables $X_1, \ldots, X_k$, where each $X_i = 1$ with probability at most p. In the latter case, it is easy to compute that $\mathbf{Pr}[\oplus_{i=1}^k X_i = 1] \leqslant 1/2 + (2p-1)^k/2$, i.e., the k-wise XOR approaches a fair coin flip exponentially fast in k. In the computational setting, one would like to argue that $F^{\oplus k}$ becomes essentially unpredictable. Such XOR results are also known, the most famous being Yao's XOR Lemma for Boolean circuits [40,28,13] (many proofs of this lemma have been given over the years, see, e.g., [23] for the most recent proof, and the references).

   We call an XOR lemma *optimal* if its parameters exactly match the probabilistic analogue given above. Recently, Unger [38] essentially showed that an optimal XOR result implies an optimal TDPT (and hence also an optimal DPT). More precisely, he proved the following generalization of the Chernoff-Hoeffding bound: Let $X_1, \ldots, X_k$ be Boolean random variables such that for some $-1 \leqslant \beta \leqslant 1$, we have that, for every subset $S \subseteq \{1, \ldots, k\}$, $\mathbf{Pr}[\oplus_{i \in S} X_i = 1] \leqslant 1/2 + \beta^{|S|}/2$. Then for any $\beta \leqslant \rho \leqslant 1$, $\mathbf{Pr}[\sum_{i=1}^k X_i \geqslant (1/2 + \rho/2)k] \leqslant e^{-kD(1/2+\rho/2\|1/2+\beta/2)}$.

   Unger's original proof uses the method of moment-generating functions and some basic tools from Fourier analysis. In contrast, we give a simple reduction showing that the assumption in Unger's theorem implies the assumption in

---

[2] In fact, for 2-prover games, it is impossible to achieve the "optimal" decrease in the success probability from p to p^k, for k parallel repetitions of the game [35].

Theorem 1, and thus we immediately get an alternative (and simpler) proof of Unger's result. For a random variable $X \in \{0, 1\}$, we define $bias(X) = \mathbf{Pr}[X = 0] - \mathbf{Pr}[X = 1]$. We show the following.

Theorem 4. *Let $X_1, \ldots, X_n$ be 0-1-valued random variables. Suppose that there is $-1 \leqslant \beta \leqslant 1$ such that, for every $S \subseteq [n]$, $bias(\oplus_{i \in S} X_i) \leqslant \beta^{|S|}$. Then, for every $S \subseteq [n]$, $\mathbf{Pr}[\wedge_{i \in S}(X_i = 0)] \leqslant (1/2 + \beta/2)^{|S|}$.*

Moreover, the reduction in the proof of Theorem 4 is constructive. Combining it with the constructive version of Theorem 1, we get a *constructive* version of Unger's result: if the variables $X_1, \ldots, X_n$ fail to satisfy the concentration bound, then we can efficiently find (using a randomized algorithm) a subset S of indices such that $\oplus_{i \in S} X_i$ has "large" bias. Such a constructive version is not implied by the original proof of [38].

1.3   Related Work

Chernoff bounds for negatively correlated random variables. The assumption on the random variables $X_1, \ldots, X_n$ used in Theorem 1 is similar to the assumption that the X_i's are *negatively correlated*; the latter means that for every subset $S \subseteq [n]$, $\mathbf{Pr}[\wedge_{i \in S} X_i = 1] \leqslant \prod_{i \in S} \mathbf{Pr}[X_i = 1]$. The only difference between the negative correlation assumption and the assumption in Theorem 1 is that the latter upperbounds $\mathbf{Pr}[\wedge_{i \in S} X_i = 1]$ by some $\delta^{|S|}$, where δ is an upper bound on $\mathbf{Pr}[X_i = 1]$. Panconesi and Srinivasan [30] observed that the Chernoff-Hoeffding bound continues to hold for the case of random variables that satisfy this generalized version of negative correlation. The proof in [30] follows the standard, Bernstein-style, proof of the Chernoff-Hoeffding bound.

TDPTs from DPTs, and DPTs from XOR lemmas. A simple idea for converting DPTs into TDPTs by randomly sampling a subset of a given n-tuple of instances was also suggested by Ben-Aroya et al. [6, Theorem 10], but their reduction doesn't give the optimal parameters. In the setting of interactive protocols, Chung and Liu [10] show how to obtain an almost-optimal TDPT from an optimal DPT, also using a very similar sampling-based argument. The fact that XOR Lemma implies DPT was also shown by Viola and Wigderson [39, Proposition 1.4]. Our proof of Theorem 4 (showing that optimal XOR Lemma implies optimal DPT) is a very similar argument.

While the idea of using sampling to get weak versions of TDPTs from DPTs has been used in earlier works, the difference in our paper is to use it in the *abstract setting* of probability-theoretic concentration bounds, and achieve *tight parameters*. It is actually surprising that such a simple idea is powerful enough to yield tight concentration bounds. The advantage of the abstract framework is also that it suggests applications in settings where one doesn't usually think in terms of standard direct products and threshold direct products. For example, we use our Theorem 1 to prove the Chernoff concentration bound for expander walks [12] from the hitting property of [2]. We also show the *information-theoretic limitations* of simple reductions between DPTs and TDPTs, and suggest a way to overcome these limitations with stronger reductions.

We consider the new proof of Chernoff-type concentration bounds more revealing and intuitive than the standard Bernstein-style proofs, and hope that its constructiveness will have other applications in computer science.

2   Preliminaries

For a natural number n, we denote by $[n]$ the set $\{1, 2, \ldots, n\}$. For $0 \leqslant \rho, \sigma \leqslant 1$, let $D(\rho \parallel \sigma)$ be the binary relative entropy defined as $D(\rho \parallel \sigma) = \rho \ln \frac{\rho}{\sigma} + (1 - \rho) \ln \frac{1-\rho}{1-\sigma}$, with $0 \ln 0 = 0$. We shall also use the following simple estimate: $D(\sigma + \epsilon \parallel \sigma) \geqslant 2\epsilon^2$ (obtained by considering the Taylor expansion of the function $g(x) = D(p + x \parallel p)$ up to the second derivative).

For parameters $0 \leqslant \delta \leqslant \gamma \leqslant 1$, we define the function $f_{\delta, \gamma}(q) = \frac{1 - q(1-\delta)}{(1-q)^{1-\gamma}}$; we shall be interested in the case where $0 \leqslant q < 1$. When δ, γ are clear from the context, we drop the subscripts and simply write $f(q)$. Taking the derivative of the function $f(q)$, we get that $f(q)$ achieves its minimum at $q^* = \frac{\gamma - \delta}{\gamma(1-\delta)}$. It is easy to see that $f(q^*) = \left(\frac{\delta}{\gamma}\right)^{\gamma} \left(\frac{1-\delta}{1-\gamma}\right)^{1-\gamma} = e^{-D(\gamma \parallel \delta)}$.

For parameters $n \in \mathbb{N}$ and $0 \leqslant q \leqslant 1$, we denote by $Bin(n, q)$ the *binomial distribution* on sets $S \subseteq [n]$, where a set S is obtained by picking each index $1 \leqslant i \leqslant n$, independently, with probability q. We will denote by $S \sim Bin(n, q)$ the random choice of $S \subseteq [n]$ according to $Bin(n, q)$.

We use the following "mean is median" result of Jogdeo and Samuels [26] for general binomial distributions (where the probabilities of choosing an index i may be different for different i's).

Lemma 2 ([26]). *For every n-tuple of real numbers $p_1, \ldots, p_n$, $0 \leqslant p_i \leqslant 1$ for all $1 \leqslant i \leqslant n$, and for the Boolean random variables $X_1, \ldots, X_n$ where each $X_i = 1$ with probability p_i, and $X_i = 0$ with probability $1 - p_i$, let $S = \sum_{i=1}^{n} X_i$ and let $\mu = \sum_{i=1}^{n} p_i$. Then the median of the distribution S is either $\lfloor \mu \rfloor$ or $\lceil \mu \rceil$ (and is equal to μ if μ is an integer). In particular, we have $\mathbf{Pr}\left[S \geqslant \lfloor \mu \rfloor\right] \geqslant 1/2$.*

3   Concentration Bounds

3.1   Boolean Random Variables

Theorem 1 is the special case of the following theorem (when $\delta_1 = \cdots = \delta_n$).

Theorem 5. *Let $X_1, \ldots, X_n$ be 0-1-valued random variables. Suppose that there are $0 \leqslant \delta_i \leqslant 1$, for $1 \leqslant i \leqslant n$, such that, for every set $S \subseteq [n]$, $\mathbf{Pr}\left[\wedge_{i \in S} X_i = 1\right] \leqslant \prod_{i \in S} \delta_i$. Let $\delta = (1/n) \sum_{i=1}^{n} \delta_i$. Then, for any γ such that $\delta \leqslant \gamma \leqslant 1$, we have $\mathbf{Pr}\left[\sum_{i=1}^{n} X_i \geqslant \gamma n\right] \leqslant e^{-n D(\gamma \parallel \delta)}$.*

Proof. For a parameter $0 \leqslant q \leqslant 1$ to be chosen later, consider the following random experiment. Pick a random n-tuple $(x_1, \ldots, x_n)$ from the given distribution $X_1, \ldots, X_n$. Pick a set $S \sim Bin(n, q)$ (i.e., each position $1 \leqslant i \leqslant n$, independently, is in S with probability q).

Let $\mathcal{E}$ be the event that $\sum_{j=1}^{n} X_j \geqslant \gamma n$, and let $p = \mathbf{Pr}[\mathcal{E}]$. By conditioning,

$$\mathbf{Exp}[\wedge_{i \in S} X_i = 1] \geqslant \mathbf{Exp}[\wedge_{i \in S} X_i = 1 \mid \mathcal{E}] \cdot p, \tag{1}$$

where the expectations are over random choices of $S \sim Bin(n, q)$ and $X_1, \ldots, X_n$.

For every $S \subseteq [n]$, we have $\mathbf{Pr}[\wedge_{i \in S} X_i = 1] \leqslant \prod_{i \in S} \delta_i$. Hence,

$$\mathbf{Exp}[\wedge_{i \in S} X_i = 1] \leqslant \sum_{S \subseteq [n]} \left[q^{|S|}(1 - q)^{n-|S|} \prod_{i \in S} \delta_i \right]. \tag{2}$$

Let us denote by $(z_1, \ldots, z_n) \in \{0, 1\}^n$ the characteristic vector of a set S chosen in the random experiment above. That is, each z_i is 1 with probability q, and 0 with probability $1 - q$; all z_i's are independent. In this new notation, the expression in (2) equals $\mathbf{Exp}_{z_1, \ldots, z_n}[\prod_{i=1}^{n} \delta_i^{z_i}] = \prod_{i=1}^{n} \mathbf{Exp}_{z_i}[\delta_i^{z_i}] = \prod_{i=1}^{n}(q\delta_i + 1 - q)$, where the first equality is by the independence of the z_i's. By convexity, $(1/n) \sum_{i=1}^{n} \ln(q\delta_i + 1 - q) \leqslant \ln(q\delta + 1 - q)$, and hence $\prod_{i=1}^{n}(q\delta_i + 1 - q) \leqslant (q\delta + 1 - q)^n$. (When $\delta_1 = \cdots = \delta_n$, the same upper bound on the r.h.s. of (2) follows immediately from the binomial formula.)

On the other hand, $\mathbf{Exp}[\wedge_{i \in S} X_i = 1 \mid \mathcal{E}]$ is the probability that a random $S \sim Bin(n, q)$ misses all the 0 positions in the chosen sample from $X_1, \ldots, X_n$, conditioned on $\mathcal{E}$. Since there are at most $n - \gamma n$ such 0 positions, we get $\mathbf{Exp}[\wedge_{i \in S} X_i = 1 \mid \mathcal{E}] \geqslant (1 - q)^{n-\gamma n}$. Combining this with Eqs. (1)–(2), we get $p \leqslant \left(\frac{q\delta + 1 - q}{(1-q)^{(1-\gamma)}} \right)^n = (f(q))^n$, where $f(q)$ is the function defined in Sect. 2 above. Choosing $q = q^*$ to minimize $f(q)$ (see Sect. 2), we get $p \leqslant e^{-nD(\gamma \| \delta)}$.

Remark 2. For $\gamma = 1$, Theorem 5 is tight, as $e^{-nD(1\|\delta)} = \delta^n$.

3.2   Real-Valued Random Variables, and Martingales

We prove a version of Theorem 1 for the case of real-valued random variables.

Theorem 6. *Let $X_1, \ldots, X_n \in [0, 1]$ be real-valued random variables. Suppose that there is a $0 \leqslant \delta \leqslant 1$ such that, for every set $S \subseteq [n]$, $\mathbf{Exp}\left[\prod_{i \in S} X_i\right] \leqslant \delta^{|S|}$. Then, for any γ such that $\delta \leqslant \gamma \leqslant 1$, $\mathbf{Pr}\left[\sum_{i=1}^{n} X_i \geqslant \lceil \gamma n \rceil\right] \leqslant 2 \cdot e^{-nD(\gamma\|\delta)}$.*

Proof. Let $p = \mathbf{Pr}\left[\sum_{i=1}^{n} X_i \geqslant \lceil \gamma n \rceil\right]$. Suppose that $p > 2 \cdot \exp(-nD(\gamma \| \delta))$. Our proof is by a reduction to the Boolean case. Consider Boolean random variables $Y_1, \ldots, Y_n$, where $\mathbf{Pr}[Y_i = 1] = X_i$, for all $1 \leqslant i \leqslant n$; that is, we think of the real value X_i as the probability that a Boolean variable Y_i is 1. Suppose we sample $x_1, \ldots, x_n$ from the distribution $X_1, \ldots, X_n$. Conditioned on $\sum_{i=1}^{n} x_i \geqslant \lceil \gamma n \rceil$, we have by Lemma 2 that $\mathbf{Pr}[\sum_{i=1}^{n} Y_i \geqslant \lceil \gamma n \rceil] \geqslant 1/2$. Lifting the conditioning (and using the assumed lower bound on the probability p), we get $\mathbf{Pr}\left[\sum_{i=1}^{n} Y_i \geqslant \lceil\gamma n\rceil\right] \geqslant p/2 > e^{-nD(\gamma\|\delta)}$, where the probability is over X_i's and Y_i's.

By Theorem 1, we have that there is a subset $S \subseteq [n]$ such that $\mathbf{Pr}[\wedge_{i\in S}Y_i = 1] > \delta^{|S|}$. Denote $\boldsymbol{X} = (X_1, \ldots, X_n)$, and similarly for $\boldsymbol{Y}$. We can equivalently write $\mathbf{Pr}[\wedge_{i\in S}Y_i = 1] = \mathbf{Exp}_{\boldsymbol{X}}\left[\mathbf{Exp}_{\boldsymbol{Y}}\left[\prod_{i\in S} Y_i\right]\right] = \mathbf{Exp}_{\boldsymbol{X}}\left[\prod_{i\in S} \mathbf{Exp}_{\boldsymbol{Y}}[Y_i]\right] = \mathbf{Exp}_{\boldsymbol{X}}\left[\prod_{i\in S} X_i\right]$, where the second equality is by the independence of Y_i's (given any fixing of X_i's), and the last equality by the definition of Y_i's. Thus, $\mathbf{Exp}[\prod_{i\in S} X_i] > \delta^{|S|}$, which is a contradiction.

A sequence of random variables $X_0, \ldots, X_n$ is a *martingale* if $\mathbf{Exp}[X_{i+1} \mid X_i, X_{i-1}, \ldots, X_0] = X_i$, for all $0 \leqslant i < n$. Suppose that $X_0 = 0$. The concentration bound for martingales (Azuma's inequality [4]) says that if $|X_{i+1} - X_i| \leqslant 1$ for all $1 \leqslant i \leqslant n$, then X_n is unlikely to deviate from 0 by more than $\sqrt{n}$. More precisely, for any $\lambda > 0$, $\mathbf{Pr}[X_n \geqslant \lambda\sqrt{n}] \leqslant \exp(-\lambda^2/2)$.

Theorem 7. *Let $0 = X_0, X_1, \ldots, X_n$ be a martingale such that $|X_{i+1} - X_i| \leqslant 1$ for all $0 \leqslant i < n$. Then, for any $\lambda > 0$, $\mathbf{Pr}[X_n \geqslant \lceil\lambda\sqrt{n}\rceil] \leqslant 2 \cdot \exp(-\lambda^2/2)$.*

Proof. Define new random variables $Y_i = X_i - X_{i-1}$, for all $1 \leqslant i \leqslant n$; the sequence $Y_1, \ldots, Y_n$ is a martingale difference sequence. Note that each $Y_i \in [-1, 1]$. Clearly, $\mathbf{Exp}[Y_{i+1} \mid Y_i, Y_{i-1}, \ldots, Y_1] = \mathbf{Exp}[Y_{i+1} \mid X_i, X_{i-1}, \ldots, X_0] = 0$. Let us also define the random variables $Z_i = (1+Y_i)/2$, for $1 \leqslant i \leqslant n$. Observe that each $Z_i \in [0, 1]$. We want to apply Theorem 6 to the Z_i's. To this end, we show that, for every subset $S \subseteq [n]$, $\mathbf{Exp}[\prod_{i\in S} Z_i] = (1/2)^{|S|}$. The proof of this is by induction on $|S|$, and using the martingale property of Y_i's.

Applying Theorem 6 to the Z_i's (with $\delta = 1/2$ and $\gamma = 1/2 + \epsilon$), we get that, for every $0 \leqslant \epsilon \leqslant 1/2$, $\mathbf{Pr}[\sum_{i=1}^{n} Z_i \geqslant \lceil(1/2 + \epsilon)n\rceil] \leqslant 2 \cdot \exp(-nD(1/2 + \epsilon \| 1/2)) \leqslant 2\cdot\exp(-2\epsilon^2 n)$. Since $\sum_{i=1}^{n} Z_i = n/2+(\sum_{i=1}^{n} Y_i)/2$, we get $\mathbf{Pr}[\sum_{i=1}^{n} Y_i \geqslant \lceil 2\epsilon n\rceil] \leqslant 2 \cdot \exp(-2\epsilon^2 n)$. Using the fact that $\sum_{i=1}^{n} Y_i = X_n$ and choosing ϵ so that $\lambda = 2\epsilon\sqrt{n}$, we conclude that $\mathbf{Pr}[X_n \geqslant \lceil\lambda\sqrt{n}\rceil] \leqslant 2 \cdot \exp(-\lambda^2/2)$.

3.3 Expander Walks

We recall some basic definitions (for more details on expanders, see the excellent survey [19]). For a d-regular undirected graph $G = (V, E)$ on n vertices, let $A = (a_{i,j})$ be its normalized adjacency matrix (where each entry of the adjacency matrix is divided by d). All eigenvalues of A are between -1 and 1, with the largest eigenvalue being equal to 1. Order all eigenvalues according to their absolute values. For $0 \leqslant \lambda \leqslant 1$, we call G a λ-*expander* if the second largest (in absolute value) eigenvalue of A is at most λ.

Expanders have numerous applications in computer science and mathematics (cf. [19]), in particular, due to the following sampling properties. The *hitting* property of expanders, first shown by Ajtai, Komlos, and Szemeredi [2], and later improved by Alon et al. [3], is the following.

Theorem 8 (Hitting property of expander walks [2,3]). *Let $G = (V, E)$ be a λ-expander, and let $W \subset V$ be any vertex subset of measure μ, with $\mu \geqslant 6\lambda$.*

Then the probability that a $(t-1)$-step random walk started from a uniformly random vertex stays inside W is at most $\mu(\mu+2\lambda)^{t-1}$. Moreover, for any subset $S \subseteq [t]$, the probability that, in each of the time steps $i \in S$, the random walk hits a vertex in W is at most $(\mu+2\lambda)^{|S|}$.

The second sampling property, originally proved by Gillman [12], is similar to the Chernoff-Hoeffding concentration bound, and is sometimes called the *Chernoff bound for expander walks*.

Theorem 9 (Chernoff bound for expander walks [12]). *Let $G = (V, E)$ be a λ-expander, and let $W \subset V$ be any vertex subset of measure μ. Then the probability that a $(t-1)$-step random walk started from a uniformly random vertex contains at least $(\mu+\epsilon)t$ vertices from W is at most $e^{-\epsilon^2(1-\lambda)t/4}$.*

The hitting property of Theorem 8 is fairly easy to prove, using basic linear algebra. In contrast, the original proof of Theorem 9 relied on some tools from perturbation theory and complex analysis. Subsequently, the proof was significantly simplified by Healy [15], who used only basic linear algebra.

We first observe the following.

Theorem 10. *Let $G = (V, E)$ be a λ-expander, and let $W \subset V$ be of measure μ, where $\mu \geqslant 6\lambda$. Let $1 > \epsilon > 2\lambda$. Then the probability that $(t-1)$-step random walk started from a uniformly random vertex contains at least $(\mu+\epsilon)t$ vertices from W is at most $e^{-tD(\mu+\epsilon\|\mu+2\lambda)} \leqslant e^{-2(\epsilon-2\lambda)^2 t}$.*

Proof. Define the 0-1-valued random variables $X_1, \ldots, X_t$ where $X_i = 1$ if the ith step of a random walk in G lands in W, and $X_i = 0$ otherwise. By Theorem 8, we have that for every subset $S \subseteq [t]$, $\mathbf{Pr}[\wedge_{i \in S} X_i = 1] \leqslant (\mu+2\lambda)^{|S|}$. By Theorem 1, the probability that a random walk in G contains at least $(\mu+\epsilon)t$ vertices from W is at most $e^{-tD(\mu+\epsilon\|\mu+2\lambda)}$. Using $D(\sigma + \rho \,\|\, \sigma) \geqslant 2\rho^2$, we can upperbound this probability by $e^{-2(\epsilon-2\lambda)^2 t}$.

We can lift the assumption of Theorem 10 that $\epsilon > 2\lambda$, thereby getting

Theorem 11. *Let $G = (V, E)$ be a λ-expander, and let $W \subset V$ be of measure μ. Then the probability that a $(t-1)$-step random walk started from a uniformly random vertex contains at least $(\mu+\epsilon)t$ vertices from W (where $\epsilon \leqslant (2/3)\mu$) is at most $e^{-\epsilon^2(1-\lambda)t/(2\ln 4/\epsilon)}$.*

Proof (sketch). The idea is to view random t-vertex walks in the graph G also as t/c-vertex walks in the graph G^c (the cth power of the graph G), for a suitably chosen integer c. The second largest eigenvalue of G^c is at most λ^c. By choosing c so that $\lambda^c < \epsilon/2$, we will satisfy the assumptions of Theorem 10, for walks of length t/c, thus getting an exponentially small upper bound on the fraction of imbalanced walks in G. Since this probability is computed based on walks of length t/c rather than t, we lose an extra factor (namely, $(1-\lambda)/(\ln 1/\epsilon)$) in the exponent.

4   Application: Uniform TDPTs for CAPTCHAs

CAPTCHAs are a special case of weakly verifiable puzzles defined by [8]. A *weakly verifiable puzzle* has two components: *(1)* a polynomial-time sampleable distribution ensemble $D = \{D_n\}_{n \geqslant 1}$ on pairs (x, α), where x is called the puzzle and α the check string (n is the security parameter); and *(2)* a polynomial-time computable relation $R((x, \alpha), y)$, where y is a string of a fixed polynomially-related length. Here we think of α as a uniform random string used to generate the puzzle x. The *k-wise direct product puzzle* P^k is defined in the obvious way.

A puzzle P is called δ-*hard* (for some $0 \leqslant \delta \leqslant 1$) if, for every randomized polynomial-time algorithm A, there is a negligible function *negl* so that the success probability of A on a random P-instance is at most $(1 - \delta) + negl$.

Theorem 12 ([8]). *If a puzzle P is $(1 - \rho)$-hard, for some $0 \leqslant \rho \leqslant 1$, then P^k is $(1 - \rho^k)$-hard.*

We show the following optimal threshold direct-product result for P^k.

Theorem 13. *Suppose a puzzle P is $(1 - \rho)$-hard, for a constant $0 \leqslant \rho \leqslant 1$. Let $\gamma = \rho + \nu \leqslant 1$, for any constant $0 \leqslant \nu \leqslant 1$. Then, for every randomized polynomial-time algorithm A, there is a negligible function negl such that the following holds: The fraction of k-tuples $\boldsymbol{x} = (x_1, \ldots, x_k)$ of instances of P^k where A solves correctly at least γk of the x_i's, is at most $e^{-kD(\gamma \| \rho)} + negl$.*

Proof. Suppose A is a randomized polynomial-time algorithm that violates the conclusion of the theorem. For random strings $\alpha_1, \ldots, \alpha_k$, define the 0-1-valued random variables $Z_1, \ldots, Z_k$ so that, for each $1 \leqslant i \leqslant k$, $Z_i = 1$ iff the algorithm $A(x_1, \ldots, x_k)$ is correct on x_i, where $x_1, \ldots, x_k$ are the puzzles determined by the random strings $\alpha_1, \ldots, \alpha_k$. Note that the distribution of $Z_1, \ldots, Z_k$ is efficiently sampleable since A is efficient (and since the puzzle P is defined for a polynomial-time sampleable distribution D).

By assumption, there is some nonnegligible function $\eta \geqslant e^{-kD(\gamma \| \rho)}$ so that $\mathbf{Pr}[\sum_{i=1}^k Z_i \geqslant \gamma k] \geqslant e^{-kD(\gamma \| \rho)} + 2\eta$. By Theorem 2, we can efficiently find a subset $S \subseteq [k]$ such that $\mathbf{Pr}[\wedge_{i \in S} Z_i = 1] > \rho^{|S|} + \eta'$, where $\eta' = \Omega(\eta^{4/(\nu \rho)})$ is nonnegligible. Thus we have an efficient algorithm that solves $P^{|S|}$ with success probability noticeably higher than $\rho^{|S|}$, contradicting Theorem 12.

Remark 3. The proof argument of Theorem 13 applies to *any* cryptographic interactive protocol as long as the protocol can be *efficiently simulated* (so that the corresponding distribution $Z_1, \ldots, Z_k$ is efficiently sampleable). Hence, for every class of protocols that can be efficiently simulated, *there is an optimal DPT for the class iff there is an optimal TDPT*; here the hardness parameters (as ρ and ν in Theorem 13) are assumed to be constants.

Theorem 13 provides an optimal concentration bound, but under the assumption that the probabilities γ and ρ are constant; the same assumption is also needed for the similar result of [10]. The earlier bounds of [22,27] do not make such an assumption, but they are not optimal. Using conditioning in the reductions, we can remove the said limitation on γ and δ, albeit at the expense of losing the tightness of the probability bound.

5 Summary

Let $X_1, \ldots, X_n$ be Boolean random variables such that, for some $0 \leqslant \delta \leqslant 1$, $\mathbf{Pr}[X_i = 0] \leqslant \delta$, for $1 \leqslant i \leqslant n$. Let $bias(X_i) = \mathbf{Pr}[X_i = 0] - \mathbf{Pr}[X_i = 1] \leqslant \beta = 2\delta - 1$, for $1 \leqslant i \leqslant n$. Consider the following statements.

1. $X_1, \ldots, X_n$ are independent.
2. $\forall S \subseteq [n], \, bias(\oplus_{i \in S} X_i) \leqslant \beta^{|S|}$.
3. $\forall S \subseteq [n], \, \mathbf{Pr}[\wedge_{i \in S}(X_i = 0)] \leqslant \delta^{|S|}$.
4. $\forall S \subseteq [n], \, \forall 0 \leqslant \delta \leqslant \gamma \leqslant 1, \, \mathbf{Pr}[\{X_i\}_{i \in S} \text{ has } \geqslant \gamma|S| \text{ zeros }] \leqslant e^{-|S| \cdot D(\gamma \| \delta)}$.

Theorem 14. $(1) \Rightarrow (2) \Rightarrow (3) \Leftrightarrow (4)$.

Proof. $(1) \Rightarrow (2)$ is trivial. For $(2) \Rightarrow (3)$, see Theorem 4. For $(3) \Rightarrow (4)$, see Theorem 5 (the implication $(4) \Rightarrow (3)$ is trivial).

The analogous statement for direct product theorems is: optimal XOR Theorems $\Rightarrow$ optimal DPTs $\Leftrightarrow$ optimal TDPTs. Moreover, the implications have *constructive* proofs.

References

1. von Ahn, L., Blum, M., Hopper, N.J., Langford, J.: CAPTCHA: Using hard AI problems for security. In: Biham, E. (ed.) EUROCRYPT 2003. LNCS, vol. 2656, pp. 294–311. Springer, Heidelberg (2003)
2. Ajtai, M., Komlos, J., Szemeredy, E.: Deterministic simulation in LOGSPACE. In: STOC, pp. 132–140 (1987)
3. Alon, N., Feige, U., Wigderson, A., Zuckerman, D.: Derandomized graph products. Comp. Compl. 5(1), 60–75 (1995)
4. Azuma, K.: Weighted sums of certain dependent random variables. Tohoku Math. J. 19, 357–367 (1967)
5. Bellare, M., Impagliazzo, R., Naor, M.: Does parallel repetition lower the error in computationally sound protocols? In: FOCS, pp. 374–383 (1997)
6. Ben-Aroya, A., Regev, O., de Wolf, R.: A hypercontractive inequality for matrix-valued functions with applications to quantum computing and LDCs. In: FOCS, pp. 477–486 (2008)
7. Bernstein, S.N.: Collected works, vol. 4. Nauka, Moscow (1964) (in Russian)
8. Canetti, R., Halevi, S., Steiner, M.: Hardness amplification of weakly verifiable puzzles. In: Kilian, J. (ed.) TCC 2005. LNCS, vol. 3378, pp. 17–33. Springer, Heidelberg (2005)
9. Chernoff, H.: A measure of asymptotic efficiency for tests of a hypothesis based on the sum of observations. Ann. Math. Stat. 23, 493–509 (1952)
10. Chung, K.M., Liu, F.H.: Tight parallel repetition theorems for public-coin arguments. In: Micciancio, D. (ed.) Theory of Cryptography. LNCS, vol. 5978, pp. 19–36. Springer, Heidelberg (2010)
11. Cleve, R., Slofstra, W., Unger, F., Upadhyay, S.: Perfect parallel repetition theorem for quantum XOR proof systems. In: CCC, pp. 109–114 (2007)
12. Gillman, D.: A Chernoff bound for random walks on expander graphs. SICOMP 27(4), 1203–1220 (1998)
13. Goldreich, O., Nisan, N., Wigderson, A.: On Yao's XOR-Lemma. In: ECCC (1995)
14. Haitner, I.: A parallel repetition theorem for any interactive argument. In: FOCS, pp. 241–250 (2009)

15. Healy, A.: Randomness-efficient sampling within NC^1. Comp. Compl. 17(1), 3–37 (2008)
16. Hoeffding, W.: Probability inequalities for sums of bounded random variables. American Stat. J., 13–30 (1963)
17. Holenstein, T.: Key agreement from weak bit agreement. In: STOC (2005)
18. Holenstein, T.: Parallel repetition: Simplifications and the no-signaling case. In: STOC, pp. 411–419 (2007)
19. Hoory, S., Linial, N., Wigderson, A.: Expander graphs and their applications. Bull. AMS 43(4), 439–561 (2006)
20. Impagliazzo, R.: Hard-core distributions for somewhat hard problems. In: FOCS (1995)
21. Impagliazzo, R., Jaiswal, R., Kabanets, V.: Approximately list-decoding direct product codes and uniform hardness amplification. SICOMP 39(2), 564–605 (2009)
22. Impagliazzo, R., Jaiswal, R., Kabanets, V.: Chernoff-type direct product theorems. J. Cryptology 22(1), 75–92 (2009)
23. Impagliazzo, R., Jaiswal, R., Kabanets, V., Wigderson, A.: Uniform direct-product theorems: Simplified, optimized, and derandomized. SICOMP 39(4), 1637–1665 (2010)
24. Impagliazzo, R., Wigderson, A.: P=BPP if E requires exponential circuits: Derandomizing the XOR Lemma. In: STOC, pp. 220–229 (1997)
25. Janson, S., Luczak, T., Rucinski, A.: Random Graphs (2000)
26. Jogdeo, K., Samuels, S.: Monotone convergence of binomial probabilities and a generalization of Ramanujan's equation. Ann. Math. Stat. 39, 1191–1195 (1968)
27. Jutla, C.S.: Almost optimal bounds for direct product threshold theorem. In: Micciancio, D. (ed.) TCC 2010. LNCS, vol. 5978, pp. 37–51. Springer, Heidelberg (2010)
28. Levin, L.A.: One-way functions and pseudorandom generators. Combinatorica 7(4), 357–363 (1987)
29. Nisan, N., Rudich, S., Saks, M.: Products and help bits in decision trees. In: FOCS, pp. 318–329 (1994)
30. Panconesi, A., Srinivasan, A.: Randomized distributed edge coloring via an extension of the Chernoff-Hoeffding bounds. SICOMP 26(2), 350–368 (1997)
31. Parnafes, I., Raz, R., Wigderson, A.: Direct product results and the GCD problem, in old and new communication models. In: STOC, pp. 363–372 (1997)
32. Pietrzak, K., Wikström, D.: Parallel repetition of computationally sound protocols revisited. In: Vadhan, S.P. (ed.) TCC 2007. LNCS, vol. 4392, pp. 86–102. Springer, Heidelberg (2007)
33. Rao, A.: Parallel repetition in projection games and a concentration bound. In: STOC, pp. 1–10 (2008)
34. Raz, R.: A parallel repetition theorem. SICOMP 27(3), 763–803 (1998)
35. Raz, R.: A counterexample to strong parallel repetition. In: FOCS (2008)
36. Shaltiel, R.: Towards proving strong direct product theorems. Comp. Compl. 12(1-2), 1–22 (2003)
37. Siegel, A.: Median bounds and their application. J. Algo. 38, 184–236 (2001)
38. Unger, F.: A probabilistic inequality with applications to threshold direct-product theorems. In: FOCS, pp. 221–229 (2009)
39. Viola, E., Wigderson, A.: Norms, XOR lemmas, and lower bounds for polynomials and protocols. Theory Comp. 4(1), 137–168 (2008)
40. Yao, A.C.: Theory and applications of trapdoor functions. In: FOCS (1982)

Almost-Euclidean Subspaces of ℓ_1^N via Tensor Products: A Simple Approach to Randomness Reduction

Piotr Indyk[1,*] and Stanislaw Szarek[2,**]

[1] MIT
indyk@mit.edu
[2] CWRU & Paris 6
szarek@math.jussieu.fr

Abstract. It has been known since 1970's that the N-dimensional ℓ_1-space contains almost Euclidean subspaces whose dimension is $\Omega(N)$. However, proofs of existence of such subspaces were probabilistic, hence non-constructive, which made the results not-quite-suitable for subsequently discovered applications to high-dimensional nearest neighbor search, error-correcting codes over the reals, compressive sensing and other computational problems. In this paper we present a "low-tech" scheme which, for any $\gamma > 0$, allows us to exhibit almost Euclidean $\Omega(N)$-dimensional subspaces of ℓ_1^N while using only N^γ random bits. Our results extend and complement (particularly) recent work by Guruswami-Lee-Wigderson. Characteristic features of our approach include (1) simplicity (we use only tensor products) and (2) yielding almost Euclidean subspaces with arbitrarily small distortions.

1 Introduction

It is a well-known fact that for any vector $x \in \mathbb{R}^N$, its ℓ_2 and ℓ_1 norms are related by the (optimal) inequality $\|x\|_2 \le \|x\|_1 \le \sqrt{N}\|x\|_2$. However, classical results in geometric functional analysis show that for a "substantial fraction" of vectors , the relation between its 1-norm and 2-norm can be made much tighter. Specifically, [FLM77, Kas77, GG84] show that there exists a subspace $E \subset \mathbb{R}^N$ of dimension $m = \alpha N$, and a scaling constant S such that for all $x \in E$

$$1/D \cdot \sqrt{N}\|x\|_2 \le S\|x\|_1 \le \sqrt{N}\|x\|_2 \tag{1}$$

where $\alpha \in (0,1)$ and $D = D(\alpha)$, called the *distortion* of E, are absolute (notably dimension-free) constants. Over the last few years, such "almost-Euclidean" subspaces of ℓ_1^N have found numerous applications, to high-dimensional nearest

---

* This research has been supported in part by David and Lucille Packard Fellowship, MADALGO (Center for Massive Data Algorithmics, funded by the Danish National Research Association) and NSF grant CCF-0728645.

** Supported in part by grants from the National Science Foundation (U.S.A.) and the U.S.-Israel BSF.

M. Serna et al. (Eds.): APPROX and RANDOM 2010, LNCS 6302, pp. 632–641, 2010.
© Springer-Verlag Berlin Heidelberg 2010

neighbor search [Ind00], error-correcting codes over reals and compressive sensing [GLR08, GLW08], vector quantization [LV06], oblivious dimensionality reduction and ϵ-samples for high-dimensional half-spaces [KRS09], and to other problems.

For the above applications, it is convenient and sometimes crucial that the subspace E is defined in an explicit manner[1]. However, the aforementioned results do not provide much guidance in this regard, since they use the *probabilistic method*. Specifically, either the vectors spanning E, or the vectors spanning the space dual to E, are i.i.d. random variables from some distribution. As a result, the constructions require $\Omega(N^2)$ independent random variables as starting point. Until recently, the largest *explicitly* constructible almost-Euclidean subspace of ℓ_1^N, due to Rudin [Rud60] (cf. [LLR94]), had only a dimension of $\Theta(\sqrt{N})$.

During the last few years, there has been a renewed interest in the problem [AM06, Sza06, Ind07, LS07, GLR08, GLW08], with researchers using ideas gained from the study of expanders, extractors and error-correcting codes to obtain several explicit constructions. The work progressed on two fronts, focusing on (a) fully explicit constructions of subspaces attempting to maximize the dimension and minimize the distortion [Ind07, GLR08], as well as (b) constructions using limited randomness, with dimension and distortion matching (at least qualitatively) the existential dimension and distortion bounds [Ind00, AM06, LS07, GLW08]. The parameters of the constructions are depicted in Figure 1. Qualitatively, they show that in the fully explicit case, one can achieve either arbitrarily low distortion or arbitrarily high subspace dimension, but not (yet?) both. In the low-randomness case, one can achieve arbitrarily high subspace dimension and constant distortion while using randomness that is sub-linear in N; achieving arbitrarily low distortion was possible as well, albeit at a price of (super)-linear randomness.

Reference	Distortion	Subspace dimension	Randomness
[Ind07]	$1 + \epsilon$	$N^{1-o_\epsilon(1)}$	explicit
[GLR08]	$(\log N)^{O_\eta(\log \log \log N)}$	$(1 - \eta)N$	explicit
[Ind00]	$1 + \epsilon$	$\Omega(\epsilon^2 / \log(1/\epsilon))N$	$O(N \log^2 N)$
[AM06, LS07]	$O_\eta(1)$	$(1 - \eta)N$	$O(N)$
[GLW08]	$2^{O_\eta(1/\gamma)}$	$(1 - \eta)N$	$O(N^\gamma)$
This paper	$1 + \epsilon$	$(\gamma\epsilon)^{O(1/\gamma)}N$	$O(N^\gamma)$

Fig. 1. The best known results for constructing almost-Euclidean subspaces of ℓ_1^N. The parameters $\epsilon, \eta, \gamma \in (0, 1)$ are assumed to be constants, although we explicitly point out when the dependence on them is subsumed by the big-Oh notation.

Our result. In this paper we show that, using sub-linear randomness, one can construct a subspace with arbitrarily small distortion while keeping its dimension proportional to N. More precisely, we have:

---

[1] For the purpose of this paper "explicit" means "the basis of E can be generated by a deterministic algorithm with running time polynomial in N." However, the individual constructions can be even "more explicit" than that.

Theorem 1 *Let $\epsilon, \gamma \in (0, 1)$. Given $N \in \mathbb{N}$, assume that we have at our disposal a sequence of random bits of length $\max\{N^\gamma, C(\epsilon, \gamma)\} \log(N/(\epsilon\gamma))$. Then, in deterministic polynomial (in N) time, we can generate numbers $M > 0$, $m \geq c(\epsilon, \gamma)N$ and an m-dimensional subspace of ℓ_1^N E, for which we have*

$$\forall x \in E, \quad (1 - \epsilon)M\|x\|_2 \leq \|x\|_1 \leq (1 + \epsilon)M\|x\|_2$$

with probability greater than 98%.

In a sense, this complements the result of [GLW08], optimizing the distortion of the subspace at the expense of its dimension. Our approach also allows to retrieve – using a simpler and low-tech approach – the results of [GLW08] (see the comments at the end of the Introduction).

Overview of techniques. The ideas behind many of the prior constructions as well as this work can be viewed as variants of the related developments in the context of error-correcting codes. Specifically, the construction of [Ind07] resembles the approach of amplifying minimum distance of a code using expanders developed in [ABN+92], while the constructions of [GLR08, GLW08] were inspired by low-density parity check codes. The reason for this state of affairs is that a vector whose ℓ_1 norms and ℓ_2 norms are very different must be "well-spread", i.e., a small subset of its coordinates cannot contain most of its ℓ_2 mass (cf. [Ind07, GLR08]). This is akin to a property required from a good error-correcting code, where the weight (a.k.a. the ℓ_0 norm) of each codeword cannot be concentrated on a small subset of its coordinates.

In this vein, our construction utilizes a tool frequently used for (linear) error-correcting codes, namely the *tensor product*. Recall that, for two linear codes $C_1 \subset \{0, 1\}^{n_1}$ and $C_2 \subset \{0, 1\}^{n_2}$, their tensor product is a code $C \subset \{0, 1\}^{n_1 n_2}$, such that for any codeword $c \in C$ (viewed as an $n_1 \times n_2$ matrix), each column of c belongs to C_1 and each row of c belongs to C_2. It is known that the dimension of C is a product of the dimensions of C_1 and C_2, and that the same holds for the minimum distance. This enables constructing a code of "large" block-length N^k by starting from a code of "small" block-length N and tensoring it k times. Here, we roughly show that the tensor product of two subspaces yields a subspace whose distortion is a product of the distortions of the subspaces. Thus, we can randomly choose an initial small low-distortion subspace, and tensor it with itself to yield the desired dimension.

However, tensoring alone does not seem sufficient to give a subspace with distortion arbitrarily close to 1. This is because we can only analyze the distortion of the product space for the case when the scaling factor S in Equation 1 is equal to 1 (technically, we only prove the left inequality, and rely on the general relation between the ℓ_2 and ℓ_1 for the upper bound). For $S = 1$, however, the best achievable distortion is strictly greater than 1, and tensoring can make it only larger. To avoid this problem, instead of the ℓ_1^N norm we use the $\ell_1^{N/B}(\ell_2^B)$ norm, for a "small" value of B. The latter norm (say, denoted by $\| \cdot \|$) treats the vector as a sequence of N/B "blocks" of length B, and returns the sum of the ℓ_2 norms of the blocks. We show that there exist subspaces $E \subset \ell_1^{N/B}(\ell_2^B)$ such that for any $x \in E$ we have

$$1/D \cdot \sqrt{N/B}\|x\|_2 \leq \|x\| \leq \sqrt{N/B}\|x\|_2$$

for D that is arbitrarily close to 1. Thus, we can construct almost-Euclidean subspaces of $\ell_1(\ell_2)$ of desired dimensions using tensoring, and get rid of the "inner" ℓ_2 norm at the end of the process.

We point out that if we do not insist on distortion arbitrarily close to 1, the "blocks" are not needed and the argument simplifies substantially. In particular, to retrieve the results of [GLW08], it is enough to combine the scalar-valued version of Proposition 1 below with "off-the-shelf" random constructions [Kas77, GG84] yielding – in the notation of Equation 1 – a subspace E, for which the parameter α is close to 1.

2   Tensoring Subspaces of L_1

We start by defining some basic notions and notation used in this section.

Norms and distortion. In this section we adopt the "continuous" notation for vectors and norms. Specifically, consider a real Hilbert space $\mathcal{H}$ and a probability measure μ over $[0, 1]$. For $p \in [1, \infty]$ consider the space $L_p(\mathcal{H})$ of $\mathcal{H}$-valued p-integrable functions f endowed with the norm

$$\|f\|_p = \|f\|_{L_p(\mathcal{H})} = \left(\int \|f(x)\|_{\mathcal{H}}^p \, d\mu(x) \right)^{1/p}$$

In what follows we will omit μ from the formulae since the measure will be clear from the context (and largely irrelevant). As our main result concerns finite dimensional spaces, it suffices to focus on the case where μ is simply the normalized counting measure over the discrete set $\{0, 1/n, \ldots (n-1)/n\}$ for some fixed $n \in \mathbb{N}$ (although the statements hold in full generality). In this setting, the functions f from $L_p(\mathcal{H})$ are equivalent to n-dimensional vectors with coordinates in $\mathcal{H}$.[2] The advantage of using the L_p norms as opposed to the ℓ_p norms that the relation between the 1-norm and the 2-norm does not involve scaling factors that depend on dimension, i.e., we have $\|f\|_2 \geq \|f\|_1$ for all $f \in L_2(\mathcal{H})$ (note that, for the ℓ_p norms, the "trivial" inequality goes in the other direction than for the ℓ_p norms). This simplifies the notation considerably.

We will be interested in lialmost subspaces $E \subset L_2(\mathcal{H})$ on which the 1-norm and 2-norm uniformly agree, i.e., for some $c \in (0, 1]$,

$$\|f\|_2 \geq \|f\|_1 \geq c\|f\|_2 \tag{2}$$

for all $f \in E$. The best (the largest) constant c that works in (2) will be denoted $\Lambda_1(E)$. For completeness, we also define $\Lambda_1(E) = 0$ if no $c > 0$ works.

---

[2] The values from $\mathcal{H}$ roughly correspond to the finite-dimensional "blocks" in the construction sketched in the introduction. Note that $\mathcal{H}$ can be discretized similarly as the L_p-spaces; alternatively, functions that are constant on intervals of the type $((k-1)/N, k/N)$ can be considered in lieu of discrete measures.

Tensor products. If $\mathcal{H}, \mathcal{K}$ are Hilbert spaces, $\mathcal{H} \otimes_2 \mathcal{K}$ is their (Hilbertian) tensor product, which may be (for example) described by the following property: if (e_j) is an orthonormal sequence in $\mathcal{H}$ and (f_k) is an orthonormal sequence in $\mathcal{K}$, then $(e_j \otimes f_k)$ is an orthonormal sequence in $\mathcal{H} \otimes_2 \mathcal{K}$ (a basis if (e_j) and (f_k) were bases). Next, any element of $L_2(\mathcal{H}) \otimes \mathcal{K}$ is canonically identified with a function in the space $L_2(\mathcal{H} \otimes_2 \mathcal{K})$; note that such functions are $\mathcal{H} \otimes \mathcal{K}$-valued, but are defined on the same probability space as their counterparts from $L_2(\mathcal{H})$. If $E \subset L_2(\mathcal{H})$ is a linear subspace, $E \otimes \mathcal{K}$ is – under this identification – a linear subspace of $L_2(\mathcal{H} \otimes_2 \mathcal{K})$.

As hinted in the Introduction, our argument depends (roughly) on the fact that the property expressed by (1) or (2) "passes" to tensor products of subspaces, and that it "survives" replacing scalar-valued functions by ones that have values in a Hilbert space. Statements to similar effect of various degrees of generality and precision are widely available in the mathematical literature, see for example [MZ39, Bec75, And80, FJ80]. However, we are not aware of a reference that subsumes all the facts needed here and so we present an elementary self-contained proof.

We start with two preliminary lemmas.

Lemma 1. *If* $g_1, g_2, \ldots \in E \subset L_2(\mathcal{H})$, *then*

$$\int \Big(\sum_k \|g_k(x)\|_{\mathcal{H}}^2 \Big)^{1/2} dx \geq \Lambda_1(E) \Big(\int \sum_k \|g_k(x)\|_{\mathcal{H}}^2 \, dx \Big)^{1/2}.$$

Proof Let $\mathcal{K}$ be an auxiliary Hilbert space and (e_k) an orthonormal sequence (O.N.S.) in $\mathcal{K}$. We will apply Minkowski inequality – a continuous version of the triangle inequality, which says that for vector valued functions $\| \int h \| \leq \int \|h\|$ – to the $\mathcal{K}$-valued function $h(x) = \sum_k \|g_k(x)\|_{\mathcal{H}} e_k$. As is easily seen, $\| \int h \|_{\mathcal{K}} = \| \sum_k \big(\int \|g_k(x)\|_{\mathcal{H}} dx \big) e_k \|_{\mathcal{K}} = \big(\sum_k \|g_k\|_{L_1(\mathcal{H})}^2 \big)^{1/2}$. Given that $g_k \in E$, $\|g_k\|_{L_1(\mathcal{H})} \geq \Lambda_1(E) \|g_k\|_{L_2(\mathcal{H})}$ and so

$$\Big\| \int h \Big\|_{\mathcal{K}} \geq \Lambda_1(E) \Big(\int \sum_k \|g_k(x)\|_{\mathcal{H}}^2 \, dx \Big)^{1/2}$$

On the other hand, the left hand side of the inequality in Lemma 1 is exactly $\int \|h\|_{\mathcal{K}}$, so the Minkowski inequality yields the required estimate.

We are now ready to state the next lemma. Recall that E is a linear subspace of $L_2(\mathcal{H})$, and $\mathcal{K}$ is a Hilbert space.

Lemma 2. $\Lambda_1(E \otimes \mathcal{K}) = \Lambda_1(E)$

If $E \subset L_2 = L_2(\mathbb{R})$, the lemma says that any estimate of type (2) for scalar functions $f \in E$ carries over to their linear combinations with vector coefficients, namely to functions of the type $\sum_j v_j f_j$, $f_j \in E, v_j \in \mathcal{K}$. In the general case, any estimate for $\mathcal{H}$-valued functions $f \in E \subset L_2(\mathcal{H})$ carries over to functions of the form $\sum_j f_j \otimes v_j \in L_2(\mathcal{H} \otimes_2 \mathcal{K})$, with $f_j \in E, v_j \in \mathcal{K}$.

Proof of Lemma 2. Let (e_k) be an orthonormal basis of $\mathcal{K}$. In fact w.l.o.g. we may assume that $\mathcal{K} = \ell_2$ and that (e_k) is the canonical orthonormal basis. Consider $g = \sum_j f_j \otimes v_j$, where $f_j \in E$ and $v_j \in \mathcal{K}$. Then also $g = \sum_k g_k \otimes e_k$ for some $g_k \in E$ and hence (pointwise) $\|g(x)\|_{\mathcal{H} \otimes_2 \mathcal{K}} = \left(\sum_k \|g_k(x)\|_{\mathcal{H}}^2 \right)^{1/2}$. Accordingly, $\|g\|_{L_2(\mathcal{H} \otimes_2 \mathcal{K})} = \left(\int \sum_k \|g_k(x)\|_{\mathcal{H}}^2 \, dx \right)^{1/2}$, while $\|g\|_{L_1(\mathcal{H} \otimes_2 \mathcal{K})} = \int \left(\sum_k \|g_k(x)\|_{\mathcal{H}}^2 \right)^{1/2} dx$. Comparing such quantities is exactly the object of Lemma 1, which implies that $\|g\|_{L_1(\mathcal{H} \otimes_2 \mathcal{K})} \geq \Lambda_1(E) \|g\|_{L_2(\mathcal{H} \otimes_2 \mathcal{K})}$. Since $g \in E \otimes \mathcal{K}$ was arbitrary, it follows that $\Lambda_1(E \otimes \mathcal{K}) \geq \Lambda_1(E)$. The reverse inequality is automatic (except in the trivial case $\dim \mathcal{K} = 0$, which we will ignore).

If $E \subset L_2(\mathcal{H})$ and $F \subset L_2(\mathcal{K})$ are subspaces, $E \otimes F$ is the subspace of $L_2(\mathcal{H} \otimes_2 \mathcal{K})$ spanned by $f \otimes g$ with $f \in E, g \in F$. (For clarity, $f \otimes g$ is a function on the *product* of the underlying probability spaces and is defined by $(x, y) \rightarrow f(x) \otimes g(y) \in \mathcal{H} \otimes \mathcal{K}$.)

The next proposition shows the key property of tensoring almost-Euclidean spaces.

Proposition 1. $\Lambda_1(E \otimes F) \geq \Lambda_1(E)\Lambda_1(F)$

Proof Let (φ_j) and (ψ_k) be orthonormal bases of respectively E and F and let $g = \sum_{j,k} t_{jk} \varphi_j \otimes \psi_k$. We need to show that $\|g\|_{L_1(\mathcal{H} \otimes_2 \mathcal{K})} \geq \Lambda_1(E)\Lambda_1(F)\|g\|_{L_2(\mathcal{H} \otimes_2 \mathcal{K})}$, where the p-norms refer to the product probability space, for example

$$\|g\|_{L_1(\mathcal{H} \otimes_2 \mathcal{K})} = \int \int \Big\| \sum_{j,k} t_{jk} \varphi_j(x) \otimes \psi_k(y) \Big\|_{\mathcal{H} \otimes_2 \mathcal{K}} dx \, dy.$$

Rewriting the expression under the sum and subsequently applying Lemma 2 to the inner integral for fixed y gives

$$\int \Big\| \sum_{j,k} t_{jk} \varphi_j(x) \otimes \psi_k(y) \Big\|_{\mathcal{H} \otimes_2 \mathcal{K}} dx = \int \Big\| \sum_j \varphi_j(x) \otimes \Big(\sum_k t_{jk} \psi_k(y) \Big) \Big\|_{\mathcal{H} \otimes_2 \mathcal{K}} dx$$

$$\geq \Lambda_1(E) \Big(\int \Big\| \sum_j \varphi_j(x) \otimes \Big(\sum_k t_{jk} \psi_k(y) \Big) \Big\|_{\mathcal{H} \otimes_2 \mathcal{K}}^2 dx \Big)^{1/2}$$

$$= \Lambda_1(E) \Big(\sum_j \Big\| \sum_k t_{jk} \psi_k(y) \Big\|_{\mathcal{K}}^2 \Big)^{1/2}$$

In turn, $\sum_k t_{jk} \psi_k \in F$ (for all j) and so, by Lemma 1,

$$\int \Big(\sum_j \Big\| \sum_k t_{jk} \psi_k(y) \Big\|_{\mathcal{K}}^2 \Big)^{1/2} dy \geq \Lambda_1(F) \Big(\int \sum_j \Big\| \sum_k t_{jk} \psi_k(y) \Big\|_{\mathcal{K}}^2 dy \Big)^{1/2}$$

$$= \Lambda_1(F) \, \|g\|_{L_2(\mathcal{H} \otimes_2 \mathcal{K})}.$$

Combining the above formulae yields the conclusion of the Proposition.

3   The Construction

In this section we describe our low-randomness construction. We start from a recap of the probabilistic construction, since we use it as a building block.

3.1   Dvoretzky's Theorem, and Its "Tangible" Version

For general normed spaces, the following is one possible statement of the well-known Dvoretzky's theorem:

Given $m \in \mathbb{N}$ and $\varepsilon > 0$ there is $N = N(m, \varepsilon)$ such that, for any norm on $\mathbb{R}^N$ there is an m-dimensional subspace on which the ratio of ℓ_1 and ℓ_2 norms is (approximately) constant, up to a multiplicative factor $1 + \varepsilon$.

For specific norms this statement can be made more precise, both in describing the dependence $N = N(m, \varepsilon)$ and in identifying the constant of (approximate) proportionality of norms. The following version is (essentially) due to Milman [Mil71].

Dvoretzky's theorem. (Tangible version)   *Consider the N-dimensional Euclidean space (real or complex) endowed with the Euclidean norm $\|\cdot\|_2$ and some other norm $\|\cdot\|$ such that, for some $b > 0$, $\|\cdot\| \leq b\|\cdot\|_2$. Let $M = \mathbb{E}\|X\|$, where X is a random variable uniformly distributed on the unit Euclidean sphere. Then there exists a computable universal constant $c > 0$, so that if $0 < \varepsilon < 1$ and $m \leq c\varepsilon^2 (M/b)^2 N$, then for more than 99% (with respect to the Haar measure) m-dimensional subspaces E we have*

$$\forall x \in E, \quad (1 - \varepsilon)M\|x\|_2 \leq \|x\| \leq (1 + \varepsilon)M\|x\|_2. \tag{3}$$

Alternative good expositions of the theorem are in, e.g., [FLM77], [MS86] and [Pis89]. We point out that standard and most elementary proofs yield $m \leq c\varepsilon^2/\log(1/\varepsilon)(M/b)^2 N$; the dependence on ε of order ε^2 was obtained in the important papers [Gor85, Sch89].

3.2   The Case of $\ell_1^n(\ell_2^B)$

Our objective now is to apply Dvoretzky's theorem and subsequently Proposition 1 to spaces of the form $\ell_1^n(\ell_2^B)$ for some $n, B \in \mathbb{N}$, so from now on we set $\|\cdot\| := \|\cdot\|_{\ell_1^n(\ell_2^B)}$ To that end, we need to determine the values of the parameter M that appears in the theorem. (The optimal value of b is clearly $\sqrt{n}$, as in the scalar case, i.e., when $B = 1$.) We have the following standard (cf. [Bal97], Lecture 9)

Lemma 3

$$M(n, B) := \mathbb{E}_{x \in S^{nB-1}} \|x\| = \frac{\Gamma(\frac{B+1}{2})}{\Gamma(\frac{B}{2})} \frac{\Gamma(\frac{nB}{2})}{\Gamma(\frac{nB+1}{2})} n.$$

In particular, $\sqrt{1 + \frac{1}{n-1}}\sqrt{\frac{2}{\pi}}\sqrt{n} > M(n, 1) > \sqrt{\frac{2}{\pi}}\sqrt{n}$ for all $n \in \mathbb{N}$ (the scalar case) and $M(n, B) > \sqrt{1 - \frac{1}{B}}\sqrt{n}$ for all $n, B \in \mathbb{N}$.

The equality is shown by relating (via passing to polar coordinates) spherical averages of norms to Gaussian means: if X is a random variable uniformly distributed on the Euclidean sphere S^{N-1} and Y has the standard Gaussian distribution on $\mathbb{R}^N$, then, for any norm $\|\cdot\|$,

$$\mathbb{E}\|Y\| \;=\; \frac{\sqrt{2}\,\Gamma(\frac{N+1}{2})}{\Gamma(\frac{N}{2})}\,\mathbb{E}\|X\|$$

The inequalities follow from the estimates $\sqrt{x-\frac{1}{2}} < \frac{\Gamma(x+\frac{1}{2})}{\Gamma(x)} < \sqrt{x}$ (for $x \geq \frac{1}{2}$), which in turn are consequences of log-convexity of Γ and its functional equation $\Gamma(y+1) = y\Gamma(y)$. (Alternatively, Stirling's formula may be used to arrive at a similar conclusion.)

Combining Dvoretzky's theorem with Lemma 3 yields

Corollary 1. *If $0 < \varepsilon < 1$ and $m \leq c_1\varepsilon^2 n$, then for more than 99% of the m-dimensional subspaces $E \subset \ell_1^n$ we have*

$$\forall x \in E \quad (1-\varepsilon)\sqrt{\frac{2}{\pi}}\,\sqrt{n}\|x\|_2 \leq \|x\|_1 \leq (1+\varepsilon)\sqrt{1+\frac{1}{n-1}}\sqrt{\frac{2}{\pi}}\,\sqrt{n}\|x\|_2 \quad (4)$$

Similarly, if $B > 1$ and $m \leq c_2\varepsilon^2 nB$, then for more than 99% of the m-dimensional subspaces $E \subset \ell_1^n(\ell_2^B)$ we have

$$\forall x \in E \quad (1-\varepsilon)\sqrt{1-\frac{1}{B}}\,\sqrt{n}\|x\|_2 \leq \|x\| \leq \sqrt{n}\|x\|_2 \quad (5)$$

We point out that the upper estimate on $\|x\|$ in the second inequality is valid for all $x \in \ell_1^n(\ell_2^B)$ and, like the estimate $M(n,B) \leq \sqrt{n}$, follows just from the Cauchy-Schwarz inequality.

Since a random subspace chosen uniformly according to the Haar measure on the manifold of m-dimensional subspaces of $\mathbb{R}^N$ (or $\mathbb{C}^N$) can be constructed from an $N \times m$ random Gaussian matrix, we may apply standard discretization techniques to obtain the following

Corollary 2. *There is a deterministic algorithm that, given ε, B, m, n as in Corollary 1 and a sequence of $O(mn\log(mn/\epsilon))$ random bits, generates subspaces E as in Corollary 1 with probability greater than 98%, in time polynomial in $1/\varepsilon + B + m + n$.*

We point out that in the literature on the "randomness-reduction", one typically uses Bernoulli matrices in lieu of Gaussian ones. This enables avoiding the discretization issue, since the problem is phrased directly in terms of random bits. Still, since proofs of Dvoretzky type theorems for Bernoulli matrices are often much harder than for their Gaussian counterparts, we prefer to appeal instead to a simple discretization of Gaussian random variables. We note, however, that the early approach of [Kas77] was based on Bernoulli matrices.

We are now ready to conclude the proof of Theorem 1. Given $\varepsilon \in (0,1)$ and $n \in \mathbb{N}$, choose $B = \lceil \varepsilon^{-1} \rceil$ and $m = \lfloor c\varepsilon^2(1 - \frac{1}{B})nB \rfloor \geq c_0\varepsilon^2 nB$. Corollary 2 (Equation 5) and repeated application of Proposition 1 give us a subspace $F \subset \ell_1^\nu(\ell_2^\beta)$ (where $\nu = n^k$ and $\beta = B^k$) of dimension $m^k \geq (c_0\varepsilon^2)^k\nu\beta$ such that

$$\forall x \in F \quad (1 - \varepsilon)^{3k/2}n^{k/2}\|x\|_2 \leq \|x\| \leq n^{k/2}\|x\|_2.$$

Moreover, $F = E \otimes E \otimes \ldots \otimes E$, where $E \subset \ell_1^n(\ell_2^B)$ is a typical m-dimensional subspace. Thus in order to produce E, hence F, we only need to generate a "typical" $m \approx c_0\varepsilon^2(\nu\beta))^{1/k}$ subspace of the $nB = (\nu\beta))^{1/k}$-dimensional space $\ell_1^n(\ell_2^B)$. Note that for fixed ε and $k > 1$, nB and m are asymptotically (substantially) smaller than $\dim F$. Further, in order to efficiently represent F as a subspace of an ℓ_1-space, we only need to find a good embedding of ℓ_2^β into ℓ_1. This can be done using Corollary 2 (Equation 4); note that β depends only on ε and k. Thus we reduced the problem of finding "large" almost Euclidean subspaces of ℓ_1^N to similar problems for much smaller dimensions.

Theorem 1 now follows from the above discussion. The argument gives, e.g., $c(\varepsilon, \gamma) = (c\varepsilon\gamma)^{3/\gamma}$ and $C(\varepsilon, \gamma) = c(\varepsilon, \gamma)^{-1}$.

References

[ABN+92]   Alon, N., Bruck, J., Naor, J., Naor, M., Roth, R.: Construction of asymptotically good low-rate error-correcting codes through pseudo-random graphs. IEEE Transactions on Information Theory 38, 509–516 (1992)

[And80]   Andersen, K.F.: Inequalities for Scalar-Valued Linear Operators That Extend to Their Vector-Valued Analogues. J. Math. Anal. Appl. 77, 264–269 (1980)

[AM06]   Artstein-Avidan, S., Milman, V.D.: Logarithmic reduction of the level of randomness in some probabilistic geometric constructions. Journal of Functional Analysis 235, 297–329 (2006)

[Bal97]   Ball, K.: An elementary introduction to modern convex geometry. In: Levy, S. (ed.) Flavors of geometry. Math. Sci. Res. Inst. Publ., vol. 31, pp. 1–58. Cambridge Univ. Press, Cambridge (1997)

[Bec75]   Beckner, W.: Inequalities in Fourier analysis. Annals of Math. 102, 159–182 (1975)

[DS01]   Davidson, K.R., Szarek, S.J.: Local operator theory, random matrices and Banach spaces. In: Johnson, W.B., Lindenstrauss, J. (eds.) Handbook of the geometry of Banach spaces, vol. 1, pp. 317–366. North-Holland, Amsterdam (2001); Vol. 2, pp. 1819–1820

[FJ80]   Figiel, T., Johnson, W.B.: Large subspaces of ℓ_∞^n and estimates of the Gordon-Lewis constant. Israel J. Math. 37, 92–112 (1980)

[FLM77]   Figiel, T., Lindenstrauss, J., Milman, V.D.: The dimension of almost spherical sections of convex bodies. Acta Math. 139(1-2), 53–94 (1977)

[GG84]   Garnaev, A., Gluskin, E.: The widths of a Euclidean ball. Soviet Math. Dokl. 30, 200–204 (1984) (English translation)

[Gor85]   Gordon, Y.: Some inequalities for Gaussian processes and applications. Israel J. Math. 50, 265–289 (1985)

[GLR08]   Guruswami, V., Lee, J., Razborov, A.: Almost euclidean subspaces of l1 via expander codes. In: SODA (2008)

[GLW08]   Guruswami, V., Lee, J., Wigderson, A.: Euclidean sections with sublinear randomness and error-correction over the reals. In: Goel, A., Jansen, K., Rolim, J.D.P., Rubinfeld, R. (eds.) APPROX and RANDOM 2008. LNCS, vol. 5171, pp. 444–454. Springer, Heidelberg (2008)

[Ind00]   Indyk, P.: Dimensionality reduction techniques for proximity problems. In: Proceedings of the Ninth ACM-SIAM Symposium on Discrete Algorithms (2000)

[Ind07]   Indyk, P.: Uncertainty principles, extractors and explicit embedding of l2 into l1. In: STOC (2007)

[Kas77]   Kashin, B.S.: The widths of certain finite-dimensional sets and classes of smooth functions. Izv. Akad. Nauk SSSR Ser. Mat. 41(2), 334–351 (1977)

[KRS09]   Karnin, Z., Rabani, Y., Shpilka, A.: Explicit Dimension Reduction and Its Applications. ECCC TR09-121 (2009)

[LLR94]   Linial, N., London, E., Rabinovich, Y.: The geometry of graphs and some of its algorithmic applications. In: FOCS, pp. 577–591 (1994)

[LPRTV05]   Litvak, A., Pajor, A., Rudelson, M., Tomczak-Jaegermann, N., Vershynin, R.: Euclidean embeddings in spaces of finite volume ratio via random matrices. J. Reine Angew. Math. 589, 1–19 (2005)

[LS07]   Lovett, S., Sodin, S.: Almost euclidean sections of the n-dimensional cross-polytope using $O(n)$ random bits. ECCC Report TR07-012 (2007)

[LV06]   Lyubarskii, Y., Vershynin, R.: Uncertainty principles and vector quantization. Arxiv.org eprint math.NA/0611343

[MZ39]   Marcinkiewicz, J., Zygmund, A.: Quelques inégalités pour les opérations linéaires. Fund. Math. 32, 113–121 (1939)

[Mil71]   Milman, V.: A new proof of the theorem of A. Dvoretzky on sections of convex bodies. Funct. Anal. Appl. 5, 28–37 (1971) (English translation)

[Mil00]   Milman, V.: Topics in asymptotic geometric analysis. In: Visions in mathematics. Towards 2000 (Tel Aviv, 1999); Geom. Funct. Anal., Special Volume, Part II, pp. 792–815 (2000)

[MS86]   Milman, V.D., Schechtman, G.: Asymptotic theory of finite-dimensional normed spaces. With an appendix by M. Gromov. Lecture Notes in Math., vol. 1200. Springer, Berlin (1986)

[Pis89]   Pisier, G.: The volume of convex bodies and Banach space geometry. Cambridge Tracts in Mathematics, vol. 94. Cambridge University Press, Cambridge (1989)

[Rud60]   Rudin, W.: Trigonometric series with gaps. J. Math. Mech. 9, 203–227 (1960)

[Sch89]   Schechtman, G.: A remark concerning the dependence on ϵ in Dvoretzky's theorem. In: Geometric aspects of functional analysis (1987-88). Lecture Notes in Math., vol. 1376, pp. 274–277. Springer, Berlin (1989)

[Sza06]   Szarek, S.: Convexity, complexity and high dimensions. In: Proceedings of the International Congress of Mathematicians (Madrid, 2006), vol. II, pp. 1599–1621. European Math. Soc. (2006), icm2006.org

[TJ89]   Tomczak-Jaegermann, N.: Banach-Mazur distances and finite-dimensional operator ideals. Longman Scientific & Technical, Harlow (1989)

Testing Outerplanarity of Bounded Degree Graphs

Yuichi Yoshida[1] and Hiro Ito[2]

[1] School of Informatics, Kyoto University, Kyoto 606-8501, Japan
yyoshida@lab2.kuis.kyoto-u.ac.jp
[2] School of Informatics, Kyoto University, Kyoto 606-8501, Japan
itohiro@kuis.kyoto-u.ac.jp

Abstract. We present an efficient algorithm for testing outerplanarity of graphs in the bounded degree model. In this model, given a graph G with n vertices and degree bound d, we should distinguish with high probability the case that G is outerplanar from the case that modifying at least an ϵ-fraction of the edge set of G is necessary to make G outerplanar.

Our algorithm runs in $\tilde{O}\left(\frac{1}{\epsilon^{13}d^6} + \frac{d}{\epsilon^2}\right)$ time, which is independent of the size of graphs. This is the first algorithm for a non-trivial minor-closed property whose time complexity is polynomial in $\frac{1}{\epsilon}$ and d. To achieve the time complexity, we exploit the tree-like structure inherent to an outerplanar graph using the microtree/macrotree decomposition of a tree.

As a corollary, we also show an algorithm that tests whether a given graph is a cactus with time complexity $\tilde{O}\left(\frac{1}{\epsilon^{13}d^6} + \frac{d}{\epsilon^2}\right)$.

1   Introduction

Property testing [1] is a relaxation of decision. The objective of property testing is to distinguish between the case that an object (e.g., a graph or a function) has a predetermined property and the case that it differs significantly from any such object. It is known that various properties are testable in constant time, i.e., independent of the size of objects. See [2, 3] for excellent surveys.

In this paper, we study testing outerplanarity of graphs in the bounded degree model. We only consider simple graphs, i.e., graphs without loops and multi-edges. A graph is called *outerplanar* if it has an embedding in the plane such that there is some face that includes every vertex. The *bounded degree model* [4] is a model for sparse graphs, under which the maximum degree of an input graph is bounded by a constant d. We obtain information of an input graph $G = (V, E)$ through an oracle by asking what the ith neighbor $(1 \le i \le d)$ of a vertex $v \in V$ is. The efficiency of a testing algorithm is measured by the number of queries to the oracle, which is called *query complexity*. A graph G with n vertices and a degree bound d is called ϵ-far from a property $\mathcal{P}$ if we must add or remove at least $\frac{\epsilon dn}{2}$ edges to make G satisfy $\mathcal{P}$, preserving the degree bound. An algorithm is called a *testing algorithm* for a property $\mathcal{P}$ if it accepts graphs satisfying $\mathcal{P}$

M. Serna et al. (Eds.): APPROX and RANDOM 2010, LNCS 6302, pp. 642–655, 2010.
© Springer-Verlag Berlin Heidelberg 2010

with probability of at least $\frac{2}{3}$ and rejects graphs ϵ-far from $\mathcal{P}$ with probability of at least $\frac{2}{3}$.

We state our main result.

Theorem 1 (main). *There is a testing algorithm for outerplanarity in the bounded degree model with query complexity* $\tilde{O}\left(\frac{1}{\epsilon^{13}d^6} + \frac{d}{\epsilon^2}\right)$.

A property $\mathcal{P}$ of graphs is called a *minor-closed property* if every minor of a graph satisfying $\mathcal{P}$ also satisfies $\mathcal{P}$. It is well-known that outerplanarity is a minor-closed property, and it is characterized in such a way that a graph is outerplanar iff the graph is K_4-minor free and $K_{2,3}$-minor free. Testing minor-closed properties is a well-studied problem in the bounded degree model. The first result is achieved by [5], in which it is shown that every minor-closed property is testable with query complexity $2^{2^{2^{\text{poly}(1/\epsilon)}}}$. Recently, the query complexity is reduced to $2^{\text{poly}(\frac{1}{\epsilon})}$ [6]. It is an important challenge to construct an algorithm for minor-closed properties whose running time is polynomial in $\frac{1}{\epsilon}$. Cycle-freeness, which is a minor-closed property, can be tested in $O\left(\frac{1}{\epsilon^3} + \frac{d}{\epsilon^2}\right)$ time [4]. Our work extends the algorithm for cycle-freeness to outerplanarity.

Some other properties are known to be testable with constant queries, such as k-edge-connectivity [4] and k-vertex-connectivity [7]. Bipartiteness [8] and being an α-expander [9] are known to be testable with $\tilde{\Theta}(\sqrt{n})$ queries.

We explain why we use the bounded degree model. Another well-studied model for graphs might be the *adjacency matrix model* [10]. The adjacency matrix model mainly concerns dense graphs, and a graph G with n vertices is called ϵ-far from a property $\mathcal{P}$ if we must add or remove at least $\frac{\epsilon n^2}{2}$ edges to make G satisfy $\mathcal{P}$. However, any outerplanar graph cannot have more than $2n - 3$ edges [11], and any graph ϵ-far from outerplanarity must have at least $\frac{\epsilon n^2}{2}$ edges. Since it is easy to estimate the number of edges up to, say, $\frac{\epsilon n^2}{100}$ in the adjacency matrix model, testing outerplanarity in the model is rather trivial. Thus, we use the bounded degree model, under which the number of edges is linear in the number of vertices.

Sketch of our algorithm: We describe a brief sketch of our algorithm. If many (linear or larger number of) constant-size K_4-minors or $K_{2,3}$-minors exist, it is easy to detect them from a vertex in them. An issue is how to distinguish outerplanar graphs from graphs ϵ-far from outerplanarity with a small number of constant-size K_4-minors and $K_{2,3}$-minors since we cannot detect them by locally searching around a constant number of randomly selected vertices. The crucial observation to resolve this issue is that an outerplanar graph cannot be an expander [12–14]. Thus, it is hoped that almost all cycles reside in local parts of an outerplanar graph. In particular, we concentrate on base cycles of the cycle space of G since the number of base cycles (or the rank of the cycle space) can be well-approximated. On the other hand, let us consider a graph ϵ-far from outerplanarity with no constant-size K_4-minor and $K_{2,3}$-minor. Such a graph must have a lot of cycles since it is ϵ-far from outerplanarity. Also, most of the cycles cannot reside in the local part of the graph since it otherwise leads to the existence of constant-size K_4-minors or $K_{2,3}$-minors. Thus, it is hoped that a significant number of base cycles cannot be found in the local part of the graph, and we will show that this is the case. Therefore, by comparing the estimated rank of the cycle space and the number of found base cycles by local search, we can decide that the graph is outerplanar or ϵ-far from outerplanarity.

The length of base cycles we try to find by local search is $r = O\left(\frac{1}{\epsilon d}\right)$. To achieve polynomial query complexity, we cannot enumerate such base cycles by naively performing a BFS with radius r since it already makes $d^r = d^{O\left(\frac{1}{\epsilon d}\right)}$ queries. Thus, we must selectively explore the place where base cycles exist. To do so, we exploit the tree-like structure inherent to an outerplanar graph. Also, we decompose the tree-like structure with the technique called *microtree/macrotree decomposition* [15], which was originally designed for the marked ancestor problem. It decomposes a tree into one macrotree and microtrees in such a way that the number of vertices with degree at least 3 in the macrotree is small and the number of leaves in a microtree is also small. Using this structure, even if we ignore base cycles such that the number of vertices around them are exponential to the radius, we can detect almost all base cycles of an outerplanar graph.

Organization: The organization of this paper is as follows. In Sect. 2, we give definitions used in this paper. We give our testing algorithm for outerplanarity in Sect. 3. In Sect. 4, we describe how many base cycles will be detected by the algorithm, and the proof of the main theorem is shown in Sect. 6. In Sect. 7, we describe possible future work.

2   Definitions

Throughout this paper, we let $\delta < \frac{1}{100}$ be a small constant. Let $\mathcal{G}_{n,d}$ be the set of graphs with n vertices and a degree bound d. Let $G \in \mathcal{G}_{n,d}$ be a graph. For a set of vertices X, let $G[X]$ denote the subgraph of G induced by X.

The *cycle space* of G is the linear space generated by all simple cycles of G where the addition is defined as a symmetric difference on edges. It is well-known that the rank of the cycle space $\rho(G)$ is equal to $m - n + c$ where m is the number of edges and c is the number of connected components in G. The number of edges in a cycle C is denoted by $|C|$. The *minimum cycle basis* $\mathcal{B}^*$ is a basis of the cycle space such that $\sum_{B \in \mathcal{B}^*} |B|$ is minimum. Though there may be many minimum cycle bases, we only consider the one obtained in the following way. First, we introduce a total ordering on cycles in G such that a cycle C is regarded as smaller than C' if $|C| < |C'|$. When $|C| = |C'|$, we break ties arbitrarily. We write $C < C'$ when C is smaller than C' in this ordering. We start with an empty set of cycles. Seeing cycles from the smaller one, we add a cycle to the set if it is independent of the set. We let $\mathcal{B}^*$ be the resulting set. Since a linear space is a matroid, $\mathcal{B}^*$ is indeed a minimum cycle basis.

We say C is a *base cycle* if $C \in \mathcal{B}^*$ and a *non-base cycle* otherwise. A *chord* of a cycle C is an edge connecting two vertices not adjacent in C. A cycle C is called *chordless* if C has no chord. Note that a cycle in an outerplanar graph is a base cycle iff the cycle is chordless. A vertex or a pair of adjacent vertices is called a *hinge* if the number of connected components increases by removing it. Also, an edge is called a *bridge* if the number of connected components increases by removing it.

Let $\mathcal{B}_v^* \subseteq \mathcal{B}^*$ be the set of base cycles containing a vertex v and $\rho_v(G) = \sum_{B \in \mathcal{B}_v^*} \frac{1}{|B|}$. Note that $\rho(G) = \sum_{v \in V} \rho_v(G)$.

Next, we briefly introduce the basic notions in the area of graph minors used in this paper. For more details, see [16, 17]. A graph H is called a *minor* of graph G if H can be obtained from G by iteratively performing edge removals, vertex removals and edge contractions. If H is not a minor of G, then G is called *H-minor free*. *Subdividing* an edge e of a graph G is an operation that inserts a new vertex into the interior, thereby splitting e into two edges. For a graph H, an *H-subdivision* is a graph obtained by iteratively subdividing edges of H. If G has no H-subdivision as a subgraph, G is called *H-subdivision free*. It is easily seen that if G is H-minor free, then G is H-subdivision free. The other direction does not hold in general. However, for outerplanarity, we can characterize it by means of subdivisions as well as minors. The following proposition is well-known (e.g., [16]).

Proposition 1. *A graph is outerplanar if and only if the graph is K_4-minor free and $K_{2,3}$-minor free. Also, a graph is outerplanar if and only if the graph is K_4-subdivision free and $K_{2,3}$-subdivision free.* $\square$

3   Algorithm Description

3.1   Outline

Let $G \in \mathcal{G}_{n,d}$ be a graph with m edges and c connected components. As described in the introduction, our testing algorithm compares the (estimated) number of base cycles $\rho(G) = m - n + c$[1] and the (estimated) number of locally residing base cycles. The next lemma asserts that $\rho(G)$ can be easily approximated.

Lemma 1. *There is an algorithm with query complexity $\tilde{O}(\frac{d}{\epsilon^2})$ that calculates $\tilde{\rho}(G)$ such that $|\tilde{\rho}(G) - \rho(G)| \le \frac{\delta \epsilon d n}{2}$ with probability of at least $1 - \delta$.*

Proof. Let m and c be the number of edges and connected components, respectively. Sampling $O(\frac{1}{\epsilon^2})$ vertices and seeing degrees of them, we can obtain $\tilde{m}$ such that $|\tilde{m} - m| \le \frac{\delta \epsilon d n}{4}$ with probability of at least $1 - \frac{\delta}{2}$ and query complexity $O\left(\frac{d}{\epsilon^2}\right)$. Also, there is an algorithm that outputs $\tilde{c}$ such that $|\tilde{c} - c| \le \frac{\delta \epsilon d n}{4}$ with probability of at least $1 - \frac{\delta}{2}$ and query complexity $\tilde{O}\left(\frac{1}{\epsilon^2 d}\right)$ [18]. By union bound, $\tilde{\rho}(G) = \tilde{m} - n + \tilde{c}$ satisfies $|\tilde{\rho}(G) - \rho(G)| \le \frac{\delta \epsilon d n}{2}$ with probability of at least $1 - \delta$. $\square$

Our algorithm is composed of three parts, i.e., BASE-CYCLE-FINDER, BASE-CYCLE-CLEANSER, and OUTERPLANARITY-TESTER. We describe the role of these three parts first and go into detail later.

---

[1] We assume that n and d are known in advance.

Algorithm 1. OUTERPLANARITY-TESTER:

OUTERPLANARITY-TESTER:
Let $\tilde{\rho}(G)$ be the output of the algorithm in Lemma 1.
Let $w = 0$ and S be a set of $s = O(\frac{1}{\epsilon^2})$ vertices chosen uniformly at random from V.
for $v \in S$ **do**
    let $\varphi_{v,\epsilon}(G)$ be the returned value of BASE-CYCLE-CLEANSER(v).
    Set $w = w + \varphi_{v,\epsilon}(G)$.
$\tilde{\varphi}_\epsilon(G) = \frac{wn}{s}$.
Accept if $\tilde{\varphi}_\epsilon(G) \geq \tilde{\rho}(G) - \frac{\epsilon dn}{8}$, reject otherwise.

Given a vertex v, BASE-CYCLE-FINDER locally searches around v and tries to enumerate chordless cycles containing v with length of at most $r = O(\frac{1}{\epsilon d})$. Note that chordless cycles are base cycles for outerplanar graphs. If BASE-CYCLE-FINDER finds a non-outerplanar subgraph in the process, it immediately rejects the graph. This can be simply done by performing a BFS with radius r from v. However, it may take d^r queries if we do it naively. To achieve polynomial query complexity, we exploit the tree-like structure inherent to outerplanar graphs. Using the structure, we can show that, for outerplanar graphs, almost all chordless cycles reside in places where a BFS with radius r visits only a polynomial number of vertices after removing at most one hinge. Thus, by removing at most one hinge and performing a BFS with radius r from v, BASE-CYCLE-FINDER enumerates almost all base cycles containing v. Since we do not know where the hinge is, we try all possibilities. A problem that occurs by this modification is that we can no longer guarantee that the subgraph induced by vertices with distance at most r from v is outerplanar. Nevertheless, for graphs ϵ-far from outerplanarity, we can show that the number of chordless cycles detected by BASE-CYCLE-FINDER is still small.

BASE-CYCLE-CLEANSER "cleanses" cycles returned by BASE-CYCLE-FINDER. Let C be one of the cycles returned by BASE-CYCLE-FINDER(v). It might occur that BASE-CYCLE-FINDER(v') for a vertex $v' \in C, v' \neq v$ does not return C. This phenomenon makes our analysis harder. Therefore, we only adopt cycles such that BASE-CYCLE-FINDER returns them for every vertex in them. We can show that the number of base cycles ignored by this cleansing is sufficiently small. Let $\mathcal{B}_v$ be a set of remaining cycles. Then, BASE-CYCLE-CLEANSER returns $\sum_{B \in \mathcal{B}_v} \frac{1}{|B|}$ as an approximation to $\rho_v(G)$.

Finally, OUTERPLANARITY-TESTER decides whether a given graph is outerplanar using BASE-CYCLE-CLEANSER as a subroutine. Since OUTERPLANARITY-TESTER is a simple algorithm, we describe the entire process here in Algorithm 1.

First, it calculates $\tilde{\rho}(G)$ using Lemma 1. Then, it invokes BASE-CYCLE-CLEANSER for a sufficiently large constant number of vertices chosen uniformly at random and let w be the sum of returned values. It immediately rejects the graph if an invoked BASE-CYCLE-CLEANSER rejects the graph (i.e., it finds evidence of non-outerplanarity). From w, it calculates an approximation to $\rho(G)$, which is denoted by $\tilde{\varphi}_\epsilon(G)$. It accepts the graph if $\tilde{\varphi}_\epsilon(G) \geq \tilde{\rho}(G) - \frac{\epsilon dn}{8}$ and rejects otherwise.

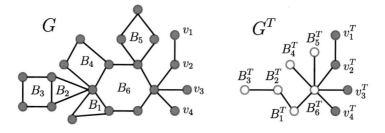

Fig. 1. Left: an outerplanar graph G. Right: the tree representation G^T of G. White vertices indicate cycle vertices and black vertices indicate basic vertices.

3.2   Tree Representation

Next, we go into the details of BASE-CYCLE-FINDER. To utilize the tree-like structure of outerplanar graphs, we introduce the *tree representation* G^T of an outerplanar graph G. Here we assume that G is connected. The tree representation of a disconnected graph is just a union of the tree representation of each connected component. An example of a tree representation is depicted in Fig. 1. Each base cycle B in G has a corresponding vertex B^T in G^T, which is called a *cycle vertex*. Each vertex v not contained in any cycle in G has a corresponding vertex v^T in G^T, which is called a *basic vertex*. We use *T-vertices* as a general name for cycle vertices and basic vertices.

First, for each pair of cycle vertices B_1^T and B_2^T such that B_1 and B_2 share an edge (not just a vertex), we connect them. Note that the end vertices of the edge should form a hinge. After this process, we have connected components of T-vertices. We call these connected components *T-components*. Note that each T-component forms a tree from the outerplanarity of G.

We construct a graph G^T starting with an arbitrary T-component. Then, we add other T-components one by one as follows. Through this process, we keep G^T connected. Let C^T be a T-component not added to G^T yet. There are two cases that C^T will be added to the current G^T.

- There exist T-vertices $w_1^T \in C^T$ and $w_2^T \in G^T$ such that w_1 and w_2 are connected by a bridge.
- There exist cycle vertices $B_1^T \in C^T$ and $B_2^T \in G^T$ such that B_1 and B_2 share a vertex.

From the outerplanarity of G, only one of the two cases can occur. We note that w_2^T is unique in the former case and the shared vertex is unique in the latter case. In the former case, we make an edge between w_1^T and w_2^T and add C^T to G^T. In the latter case, though there may be many possibilities for B_2^T, we just choose any of them and make an edge between B_1^T and B_2^T and add C^T to G^T. The order in which we add T-components is unimportant.

Since G^T is connected and the number of edges is exactly the number of vertices minus one (note that each T-component is a tree), G^T forms a tree. We can think that each edge in G^T corresponds to a hinge or a bridge in G. For two cycle vertices B_1^T, B_2^T such that B_1 and B_2 share an edge (u, v), the hinge (u, v) corresponds to the edge (B_1^T, B_2^T). For two T-vertices w_1^T, w_2^T connected by a bridge (u, v) in G, the bridge (u, v) corresponds to the edge (w_1^T, w_2^T). Finally, for two cycle vertices B_1^T, B_2^T such that B_1 and B_2 share a vertex v in G, the hinge v corresponds to the edge (B_1^T, B_2^T). For each edge (w_1^T, w_2^T) in G^T, w_1 and w_2 become disconnected in G by removing the corresponding hinge or bridge.

Our main tool for analysis is the *microtree/macrotree decomposition* of a tree [15]. Given a parameter l, the microtree/macrotree decomposition is defined as follows (we arbitrarily select a root and regard the tree as rooted).

- For every maximally high vertex whose subtree contains no more than l leaves, we designate the subtree as a *microtree*[2].
- Vertices not in any microtree form a *macrotree*.

In [15], l is always set to be $O(\log n)$. However, for our applications, l must be a parameter. It is easy to see that the following propositions hold.

Proposition 2. *Let n be the number of vertices in a tree T and l be the parameter for the microtree/macrotree representation. The macrotree of T has at most $\frac{n}{l}$ vertices with degree at least 3. (Note that degrees are calculated in the macrotree after removing microtrees.)* $\square$

Proposition 3. *Let T be a tree and S be any microtree in the microtree/macrotree decomposition of T with the parameter l. Suppose that we perform a BFS with radius r from $v \in S$ on T after removing the edge between the root of S and its parent. Then, the BFS reaches at most rl vertices.* $\square$

Let G be an outerplanar graph and G^T be its tree representation. Let v be a vertex in G and suppose that a base cycle B contains v. What we want to do is to find B given v. Assume that the cycle vertex B^T corresponding to B is in a microtree S of the microtree/macrotree decomposition of G^T with parameter l. If we can simulate the removal of the edge between the root of S and its parent, we can efficiently find B by performing a BFS with radius $|B|$ on G. The number of seen vertices in G^T will be $|B|l$ from Proposition 3. Since one step of a BFS inside a cycle vertex in G^T will see at most two vertices in G, the number of seen vertices in G will be at most $2|B|l$. If the upper bound on $|B|$ and l are set to be polynomial in $\frac{1}{\epsilon}$ and d, the total query complexity also becomes polynomial. Also, for base cycles whose corresponding cycle vertices are in the macrotree, we will show that almost all of them can be found in polynomial time (see Sect. 4).

3.3   BASE-CYCLE-FINDER

We formalize the intuition described so far. We explained earlier that a removal of an edge in G^T can be simulated by a removal of a hinge or a bridge. Because of a

---

[2] The original definition is described in terms of the number of vertices with degree at least 3 instead of the number of leaves. This is essentially equivalent to our definition.

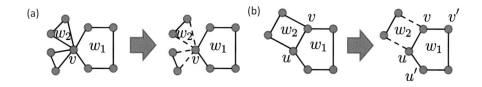

Fig. 2. Expected behavior of invalidating temporary hinges. Dashed lines indicate removed edges. (a) invalidating a temporary hinge v, which separates a cycle w_1 from a cycle w_2. (b) invalidating a temporary hinge (u, v), which separates a cycle w_1 from a cycle w_2.

technical reason, we simulate it by removing edges incident to a hinge. Any vertex or any pair of adjacent vertices is called a *temporary hinge*. Since we actually do not know where hinges are, we regard each temporary hinge as a hinge and check whether it behaves like a hinge. Now, we define *invalidating* a temporary hinge h. Let w_1^T and w_2^T be T-vertices, and then we design the invalidation so that it generates (at least) two connected components, one of which contains w_1 and the other one of which contains w_2 if we put vertices shared by w_1 and w_2 in either of the two components. If h is a vertex v, invalidating h is an operation that removes edges incident to v except at least one and at most two edges (see Fig. 2 (a)). We have at most $O(d^2)$ ways of invalidating v. If h is a pair of adjacent vertices (u, v), invalidating h is an operation that removes edges incident to u or v except (precisely) three edges $(u, v), (u, u')$, and (v, v'), where u' and v' are any adjacent vertices of u and v, respectively (see Fig. 2 (b)). Also, we have at most $O(d^2)$ ways of invalidating (u, v).

In the rest of the paper, we set $r_\epsilon = \frac{2}{\delta \epsilon d} + 2$, $l_\epsilon = \frac{16 r_\epsilon}{\epsilon d}$ and $p_\epsilon = \frac{8 r_\epsilon}{\epsilon d}$. Here r_ϵ means an upper bound on the length of a base cycle we consider, l_ϵ means an upper bound on the number of leaves in a microtree, and p_ϵ means an upper bound on the number of microtrees we see in one BFS. The description of BASE-CYCLE-FINDER is shown in Algorithm 2. Given a vertex v, BASE-CYCLE-FINDER tries to enumerate base cycles containing v with length of at most r_ϵ. First, we perform a BFS to get a set of $2 l_\epsilon p_\epsilon r_\epsilon$ vertices. And after invalidating a temporary hinge, we perform a BFS with radius $2 r_\epsilon$ again. Let Y be the reached vertices in the BFS. We reject the graph immediately if $G[Y]$ is not outerplanar. If $G[Y]$ is outerplanar, we adopt chordless cycles containing v in $G[Y]$.

3.4   BASE-CYCLE-CLEANSER

Algorithm 3 shows the description of BASE-CYCLE-CLEANSER. As described earlier, BASE-CYCLE-CLEANSER(v) only adopt chordless cycles B such that BASE-CYCLE-FINDER(v') returns them for every $v' \in B$. Let $\varphi_{v,\epsilon}(G)$ be the returned value of BASE-CYCLE-CLEANSER(v). Note that $\varphi_{v,\epsilon}(G)$ is not a random variable. We use $\varphi_\epsilon(G) = \sum_v \varphi_{v,\epsilon}(G)$ as the expectation of $\tilde{\varphi}_\epsilon(G)$ calculated by OUTERPLANARITY-TESTER.

Algorithm 2. BASE-CYCLE-FINDER(v):

Let $r_\epsilon = O(\frac{1}{\epsilon d}), l_\epsilon = O(\frac{r_\epsilon}{\epsilon^2 d^2}), p_\epsilon = O(\frac{r_\epsilon}{\epsilon d})$ and $\mathcal{B}_v = \emptyset$.
Perform a BFS until it reaches $2l_\epsilon p_\epsilon r_\epsilon$ vertices and let X be the set of reached vertices.
for every way of invalidating at most one temporary hinge in $G[X]$ (possibly no hinge is invalidated) **do**
    Execute a BFS with radius $2r_\epsilon$ without using invalidated edges until it reaches $2l_\epsilon p_\epsilon \cdot 2r_\epsilon$ vertices and let Y be the set of reached vertices.
    if $G[Y]$ is not outerplanar **then**
        reject the graph.
    else if the BFS reaches all the vertices whose distance from v is at most $2r_\epsilon$ (the distance is measured without using invalidated edges) and the BFS does not reach end vertices of invalidated edges except the current temporary hinge **then**
        Let $\mathcal{C}$ be the set of chordless cycles with length of at most r_ϵ containing v and the invalidated hinge. Then add $\mathcal{C}$ to $\mathcal{B}_v$.
return $\mathcal{B}_v$.

Algorithm 3. BASE-CYCLE-CLEANSER(v):

Let $\mathcal{B}_v = \emptyset$ and $\mathcal{B}'_v$ be the set of cycles returned by BASE-CYCLE-FINDER given v.
for $B \in \mathcal{B}'_v$ **do**
    if B is returned by BASE-CYCLE-FINDER(v') for every $v' \in B$ **then**
        Add B to $\mathcal{B}_v$.
return $\sum_{B \in \mathcal{B}_v} \frac{1}{|B|}$.

We call a vertex v r_ϵ-*locally outerplanar* if BASE-CYCLE-FINDER(v) does not reject the graph. Also, we call a graph r_ϵ-*locally outerplanar* if every vertex in the graph is r_ϵ-locally outerplanar. We note that, even if a graph is r_ϵ-locally outerplanar, its subgraph may not be r_ϵ-locally outerplanar. We have the following.

Lemma 2. *Let G be a graph in $\mathcal{G}_{n,d}$. Any chordless cycle C in G returned by* BASE-CYCLE-CLEANSER *is a base cycle in G.*

The proof is omitted due to the space. Thus, any cycle that BASE-CYCLE-CLEANSER returns is indeed a base cycle and we call B *detectable* if BASE-CYCLE-CLEANSER(v) returns B for $v \in B$.

4   The Number of Detectable Base Cycles in Outerplanar Graphs

In this section, we show the following lemma.

Lemma 3. *Let $G \in \mathcal{G}_{n,d}$ be an outerplanar graph, then $\varphi_\epsilon(G) \geq \rho(G) - \frac{3\delta\epsilon dn}{2}$.*

We classify base cycles by their length. First, we show that the number of base cycles with length of more than r_ϵ (that means undetectable) is small. The next proposition is elementary.

Proposition 4. *Let $G \in \mathcal{G}_{n,d}$ be an outerplanar graph and $r \geq 3$ be an integer. There are at most $\frac{n}{r-2}$ base cycles with length of more than r.*   □

Proof (of Lemma 3). Let z be the number of base cycles with length of more than $r_\epsilon = \frac{2}{\delta\epsilon d} + 2$. From Proposition 4, we have $z \leq \frac{\delta\epsilon dn}{2}$.

We describe which base cycle with length of at most r_ϵ is detectable by BASE-CYCLE-CLEANSER. First, a base cycle B with $|B| \leq r_\epsilon$ is detectable if B^T is in a microtree. If a BFS with radius $2r_\epsilon$ does not reach vertices in the macrotree, then the BFS can reach at most $2l_\epsilon \cdot 2r_\epsilon$ vertices, since the number of leaves in a microtree is bounded by l_ϵ. Here, the coefficient 2 comes from the fact that the BFS may run on cycles. Also, even if the BFS reaches vertices in the macrotree, the same argument follows by invalidating one hinge between the microtree and the macrotree from Proposition 3.

The number of undetectable base cycles in the macrotree is also bounded. The degree of a vertex in the macrotree is calculated in the tree after removing microtrees. Let B be a base cycle such that B^T is in the macrotree and the degree of B^T is 2. If the distance between B^T and the nearest vertex in the macrotree with degree at least 3 is at most $2r_\epsilon$, B^T may be undetectable. The number of vertices with degree at least 3 in the macrotree is at most $\frac{n}{l_\epsilon}$ from Proposition 2. It follows that the number of vertices in the macrotree with degree 1 or at least 3 is at most $\frac{2n}{l_\epsilon}$. Thus, the number of B^T that are undetectable by such reason is at most $\frac{8r_\epsilon n}{l_\epsilon}$.

Suppose that the distance between B^T and the nearest vertex in G^T with degree at least 3 is more than $2r_\epsilon$. Then, the number of vertices in the macrotree that a BFS of BASE-CYCLE-FINDER sees is at most $4r_\epsilon$. When the total number of microtrees incident to those vertices in the macrotree is more than p_ϵ, B may be undetectable. However, since the number of microtrees is bounded by n, the number of such B is at most $\frac{4r_\epsilon n}{p_\epsilon}$.

When neither of these holds, a BFS with radius $2r_\epsilon$ from a vertex in B can reach at most $2l_\epsilon p_\epsilon \cdot 2r_\epsilon$ vertices. Thus, B is detectable. We have $\varphi_\epsilon(G) \geq \rho(G) - z - \frac{8r_\epsilon n}{l_\epsilon} - \frac{4r_\epsilon n}{p_\epsilon} \geq \rho(G) - \frac{3\delta\epsilon dn}{2}$.   □

5   The Number of Detectable Base Cycles in Graphs ϵ-Far from Outerplanarity

In this section, we prove the following lemma.

Lemma 4. *Let $G \in \mathcal{G}_{n,d}$ be a graph ϵ-far from outerplanarity. Then, at least $\frac{\delta\epsilon n}{2}$ vertices in G are not r_ϵ-locally outerplanar or $\varphi_\epsilon(G) \leq \rho(G) - \frac{1-5\delta}{2} \cdot \frac{\epsilon dn}{2}$.*

Let $G = (V, E_G)$ be a graph and $H = (V, E_H)$ be a supergraph of G. We introduce an algorithm BASE-CYCLE-FINDER$_H(v)$, which is a variant of BASE-CYCLE-FINDER(v). BASE-CYCLE-FINDER$_H(v)$ on G basically runs in the same manner as BASE-CYCLE-FINDER(v) on G. The only difference is that it uses E_H instead of E_G when performing BFS. That is, the sets of seen vertices by

BASE-CYCLE-FINDER$_H(v)$ and BASE-CYCLE-FINDER(v) are the same. BASE-CYCLE-FINDER$_H(v)$ checks outerplanarity using E_G and returns cycles existing in E_G. Obviously, we cannot perform such procedure when we do not know H. The algorithm is introduced to define notions below. We call a vertex v of G (H, r_ϵ)-*locally outerplanar* if BASE-CYCLE-FINDER$_H(v)$ does not reject the graph. Also, we call a graph (H, r_ϵ)-*locally outerplanar* if every vertex in the graph is (H, r_ϵ)-locally outerplanar. The important fact is that, if a graph G is (H, r_ϵ)-locally outerplanar, any subgraph of G is also (H, r_ϵ)-locally outerplanar. Similarly, if a vertex v in G is (H, r_ϵ)-locally outerplanar, v is also (H, r_ϵ)-locally outerplanar w.r.t. any subgraph of G.

We introduce BASE-CYCLE-CLEANSER$_H$, which is a variant of BASE-CYCLE-CLEANSER. The only difference between them is that BASE-CYCLE-CLEANSER$_H$ uses BASE-CYCLE-FINDER$_H(v)$ instead of BASE-CYCLE-FINDER(v). A base cycle B is called H-*detectable* if BASE-CYCLE-CLEANSER$_H(v)$ returns B for $v \in B$. Let $\varphi_{v,\epsilon}^H(G)$ be the returned value of BASE-CYCLE-CLEANSER$_H(v)$. We define $\varphi_\epsilon^H(G) = \sum_v \varphi_{v,\epsilon}^H(G)$.

We introduce the *contracted representation* G^U of an (H, r_ϵ)-locally outerplanar graph G such that any edge in H-undetectable base cycles is also contained in some other H-detectable base cycle. The contracted representation is similar to the tree representation of an outerplanar graph. The difference is that it can be applied for a general graph if the graph satisfies the conditions above. We assume that G is connected. The contracted representation of a disconnected graph is just a union of the contracted representation of each connected component. Each H-detectable base cycle B in G has a corresponding vertex B^U in G^U, which is called a *cycle vertex*. Also, each vertex v not contained in any cycle in G has a corresponding vertex v^U in G^U, which is called a *basic vertex*. We use U-*vertices* as a general name for cycle vertices and basic vertices.

First, for each pair of cycle vertices B_1^U and B_2^U such that B_1 and B_2 share an edge (not just a vertex), we connect them. After this process, we have connected components of U-vertices. We call these connected components U-*components*.

We construct a graph G^U starting with an arbitrary U-component. Then, we add other U-components one by one as follows. Through this process, we keep G^U connected. Let C^U be an U-component not added to G^U yet. There are two cases that C^U will be added to the current G^U.

1. There exist U-vertices $w_1^U \in C^U$ and $w_2^U \in G^U$ such that w_1 and w_2 are connected by an edge not used in any H-detectable base cycles.
2. There exist cycle vertices $B_1^U \in C^U$ and $B_2^U \in G^U$ such that B_1 and B_2 share a vertex.

We note that w_2^U is unique in the former case. If otherwise, the edge between w_1 and w_2 must be in a cycle. It follows that the edge is in a H-detectable cycle, contradicting the assumption. From the same reason, only one of the two cases can occur. In the latter case, there may exist many shared vertices, and for each shared vertex v, there may exist many B_2 such that $v \in B_2$.

In the former case, we make an edge between w_1^U and w_2^U and add C^U to G^U. In the latter case, for each shared vertex $v \in B_1$, we make an edge. Though there may be many possibilities for B_2^U, we just choose any of them and make an edge between B_1^U and B_2^U. Finally, we add C^U to G^U. We do not care about the order in which we add U-components.

The following proposition is simple but useful.

Proposition 5. *At most two (H-)detectable base cycles can share an edge.*

Proof. Suppose that there are at least three detectable base cycles sharing a common edge. Then, those base cycles themselves form a non-outerplanar subgraph, and it contradicts the detectability of them.   □

We need the following lemmas, the proofs of which appear in the full version.

Lemma 5. *Let $G \in \mathcal{G}_{n,d}$ be an (H, r_ϵ)-locally outerplanar graph such that any edge in H-undetectable base cycles is contained in a H-detectable base cycle. Then, $\rho(G^U) = \rho(G) - \varphi_\epsilon^H(G)$.*

Lemma 6. *Let $G \in \mathcal{G}_{n,d}$ be an (H, r_ϵ)-locally outerplanar graph with $\varphi_\epsilon^H(G) = \rho(G)$. Then, there exists neither K_4-subdivision nor $K_{2,3}$-subdivision in G.*

The next lemma states that there are a small number of detectable base cycles in a graph that is locally outerplanar but far from outerplanarity.

Lemma 7. *Let $G \in \mathcal{G}_{n,d}$ be an (H, r_ϵ)-locally outerplanar graph $(1 - \delta)\epsilon$-far from outerplanarity. Then, $\varphi_\epsilon^H(G) \leq \rho(G) - \frac{1-\delta}{2} \cdot \frac{\epsilon dn}{2}$.*

Proof. Suppose that $\varphi_\epsilon^H(G) > \rho(G) - \frac{1-\delta}{2} \cdot \frac{\epsilon dn}{2}$. While G contains an H-undetectable base cycle with an edge not contained in any other H-detectable base cycles, we remove the edge. Let K be the resulting graph and $\frac{\gamma \epsilon dn}{2}$ denote the number of removed edges where $0 \leq \gamma \leq \frac{1-\delta}{2}$. Note that $\rho(K) = \rho(G) - \frac{\gamma \epsilon dn}{2}$, and $\varphi_\epsilon^H(K) = \varphi_\epsilon^H(G)$.

Next, we create the contracted representation K^U of K. From Lemma 5, $\rho(K^U) = \rho(K) - \varphi_\epsilon^H(K) < (\frac{1-\delta}{2} - \gamma)\frac{\epsilon dn}{2}$. We take any spanning forest of K^U. This process is done by removing $\rho(K^U)$ edges from K^U. Each removal of an edge in K^U can be simulated by removing at most two edges in K (in a way similar to invalidating a hinge). The resulting graph K' is (H, r_ϵ)-locally outerplanar and has no H-undetectable base cycles, i.e., $\varphi_\epsilon^H(K') = \rho(K')$. Thus, from Lemma 6, K' is an outerplanar graph.

The number of removed edges from G is less than $\frac{\gamma \epsilon dn}{2} + \frac{(1-\delta-2\gamma)\epsilon dn}{2} = \frac{(1-\delta-\gamma)\epsilon dn}{2}$. This contradicts the $(1 - \delta)\epsilon$-farness of G.

Proof (of Lemma 4). Suppose that the former statement does not hold, i.e., the number of r_ϵ-locally outerplanar vertices is less than $\frac{\delta \epsilon n}{2}$. Also, suppose that $\varphi_\epsilon(G) > \rho(G) - \frac{1-5\delta}{2} \cdot \frac{\epsilon dn}{2}$ holds.

We remove every edge incident to vertices that are not r_ϵ-locally outerplanar and let G' be the resulting graph. One edge removal decreases φ_ϵ by at most two since at most two $(G$-$)$detectable base cycles share the edge in common from Proposition 5. Thus, $\varphi_\epsilon^G(G') \geq \varphi_\epsilon(G) - \delta \epsilon dn > \rho(G) - \frac{1-\delta}{2} \cdot \frac{\epsilon dn}{2}$ holds. On the other hand, G' is (G, r_ϵ)-locally outerplanar and $(1-\delta)\epsilon$-far from outerplanarity. From Lemma 7, we have $\varphi_\epsilon^G(G') \leq \rho(G') - \frac{1-\delta}{2} \cdot \frac{\epsilon dn}{2} \leq \rho(G) - \frac{1-\delta}{2} \cdot \frac{\epsilon dn}{2}$, which is a contradiction.   □

6   Proof of the Main Result

Proof (of Theorem 1). First, we show the query complexity of OUTERPLANARITY-TESTER. The query complexity of calculating $\tilde{\rho}(G)$ is $\tilde{O}\left(\frac{d}{\epsilon^2}\right)$ from Lemma 1.

The query complexity of BASE-CYCLE-FINDER is as follows. Since the size of X is $O(l_\epsilon p_\epsilon r_\epsilon)$, there are $O(d^3 l_\epsilon p_\epsilon r_\epsilon)$ ways of invalidating hinges in X. For each way of invalidating hinges, we perform a BFS, which requires $O(dl_\epsilon p_\epsilon r_\epsilon)$ queries. Thus, the total query complexity of BASE-CYCLE-FINDER is $O(d^4 l_\epsilon^2 p_\epsilon^2 r_\epsilon^2)$. For each cycle BASE-CYCLE-FINDER returns, BASE-CYCLE-CLEANSER invokes BASE-CYCLE-FINDER for each vertex in the cycle. Since the number of cycles returned by BASE-CYCLE-FINDER is at most d and the number of vertices in one cycle is at most r_ϵ, the total query complexity of BASE-CYCLE-CLEANSER is $O(dr_\epsilon d^4 l_\epsilon^2 p_\epsilon^2 r_\epsilon^2) = \frac{1}{\epsilon^{11}d^6}$. OUTERPLANARITY-TESTER selects $s = O\left(\frac{1}{\epsilon^2}\right)$ vertices and performs BASE-CYCLE-CLEANSER for each selected vertex. Thus, the total query complexity of OUTERPLANARITY-TESTER is $\tilde{O}\left(\frac{1}{\epsilon^{13}d^6} + \frac{d}{\epsilon^2}\right)$.

Next, we show that OUTERPLANARITY-TESTER actually tests outerplanarity. From Lemma 1, $|\tilde{\rho}(G) - \rho(G)| \leq \frac{\delta \epsilon dn}{2}$ holds with probability of at least $1-\delta$. Also, from Hoeffding's inequality [19], $|\tilde{\varphi}_\epsilon(G) - \varphi_\epsilon(G)| < \frac{\delta \epsilon dn}{2}$ holds with probability of at least $1 - \delta$ by choosing sufficiently large s. From the union bound, both of them simultaneously hold with probability of at least $1 - 2\delta$.

Suppose that the given graph is outerplanar. Then, from Lemma 3, $\varphi_\epsilon(G) \geq \rho(G) - \frac{3\delta \epsilon dn}{2}$. Thus, $\tilde{\varphi}_\epsilon(G) > \tilde{\rho}(G) - \frac{\epsilon dn}{8}$ holds with probability of at least $1 - 2\delta$. Hence, OUTERPLANARITY-TESTER w.h.p. accepts the graph.

Suppose that the given graph is ϵ-far from outerplanarity. Then from Lemma 4, either at least $\frac{\delta \epsilon n}{2}$ vertices are not r_ϵ-locally outerplanar or $\varphi_\epsilon(G) \leq \rho(G) - \frac{1-5\delta}{2}\frac{\epsilon dn}{2}$ holds. In the former case, OUTERPLANARITY-TESTER w.h.p. rejects G since we choose $s = O\left(\frac{1}{\epsilon^2}\right)$. In the latter case, $\tilde{\varphi}_\epsilon(G) < \tilde{\rho}(G) - \frac{\epsilon dn}{8}$ holds with probability of at least $1 - 2\delta$. Hence, OUTERPLANARITY-TESTER w.h.p. rejects the graph.   □

7   Conclusions

In this paper, we showed a testing algorithm for outerplanarity in the bounded degree model. The query complexity of the algorithm is $\tilde{O}\left(\frac{1}{\epsilon^{13}d^6} + \frac{d}{\epsilon^2}\right)$.

A graph is called a *cactus* if every edge resides on at most one cycle. Our algorithm can be easily modified to test whether a given graph is a cactus. The detail is omitted due to the space. We just state the result here.

Theorem 2. *There is a testing algorithm for being a cactus in the bounded degree model with query complexity* $\tilde{O}\left(\frac{1}{\epsilon^{13}d^6} + \frac{d}{\epsilon^2}\right)$.   □

A natural extension of this work is developing polynomial time algorithms for testing wider minor-closed properties such as planar graphs. Our observation that the number of detectable base cycles is much different between outerplanar graphs and graphs far from it might be helpful. To achieve polynomial running time, we utilize the microtree/macrotree decomposition. This technique might be applicable to testing algorithms for other classes of graphs, in particular series-parallel graphs and graphs with a bounded tree-width.

References

1. Rubinfeld, R., Sudan, M.: Robust characterizations of polynomials with applications to program testing. SIAM J. Comput. 25(2), 252–271 (1996)
2. Fischer, E.: The art of uninformed decisions: A primer to property testing. Bulletin of the European Association for Theoretical Computer Science 75 (2001)
3. Goldreich, O.: Combinatorial property testing (a survey). DIMACS Series in Discrete Mathematics and Theoretical Computer Science 43, 45–59 (1999)
4. Goldreich, O., Ron, D.: Property testing in bounded degree graphs. Algorithmica 32(2), 302–343 (2008)
5. Benjamini, I., Schramm, O., Shapira, A.: Every minor-closed property of sparse graphs is testable. In: Proc. of STOC 2008, pp. 393–402 (2008)
6. Hassidim, A., Kelner, J.A., Nguyen, H.N., Onak, K.: Local graph partitions for approximation and testing. In: Proc. of FOCS 2009 (2009) (to appear)
7. Yoshida, Y., Ito, H.: Property testing on k-vertex-connectivity of graphs. In: Aceto, L., Damgård, I., Goldberg, L.A., Halldórsson, M.M., Ingólfsdóttir, A., Walukiewicz, I. (eds.) ICALP 2008, Part I. LNCS, vol. 5125, pp. 539–550. Springer, Heidelberg (2008)
8. Goldreich, O., Ron, D.: A sublinear bipartiteness tester for bounded degree graphs. In: Proc. of STOC 1998, pp. 289–298. ACM, New York (1998)
9. Czumaj, A., Sohler, C.: Testing expansion in bounded-degree graphs. In: Proc. of FOCS 2007, pp. 570–578 (2007)
10. Goldreich, O., Goldwasser, S., Ron, D.: Property testing and its connection to learning and approximation. J. ACM 45(4), 653–750 (1998)
11. Mitchell, S.L.: Linear algorithms to recognize outerplanar and maximal outerplanar graphs. Information Processing Letters 9(5), 229–232 (1979)
12. Alon, N., Seymour, P., Thomas, R.: A separator theorem for graphs with an excluded minor and its applications. In: Proc. of STOC 1990, pp. 293–299 (1990)
13. Kleinberg, J., Rubinfeld, R.: Short paths in expander graphs. In: Proc. of FOCS 1996, p. 86 (1996)
14. Plotkin, S., Rao, S., Smith, W.D.: Shallow excluded minors and improved graph decompositions. In: Proc. of SODA 1994, pp. 462–470 (1994)
15. Alstrup, S., Husfeldt, T., Rauhe, T.: Marked ancestor problems. In: Proc. of FOCS 1998, pp. 534–543 (1998)
16. Diestel, R.: Graph Theory, 3rd edn. Graduate Texts in Mathematics, vol. 173. Springer, Heidelberg (2005)
17. Lovász, L.: Graph minor theory. Bull. Amer. Math. Soc. 43(1), 75–86 (2006)
18. Chazelle, B., Rubinfeld, R., Trevisan, L.: Approximating the minimum spanning tree weight in sublinear time. In: Orejas, F., Spirakis, P.G., van Leeuwen, J. (eds.) ICALP 2001. LNCS, vol. 2076, pp. 190–200. Springer, Heidelberg (2001)
19. Hoeffding, W.: Probability inequalities for sums of bounded random variables. Amer. Statistical Assoc. J. 58(301), 13–30 (1963)
20. El-Mallah, E., Colbourn, C.J.: The complexity of some edge deletion problems. IEEE Transactions on Circuits and Systems 35(3), 354–362 (1988)

Two-Source Extractors Secure against Quantum Adversaries

Roy Kasher[*] and Julia Kempe[**]

School of Computer Science, Tel Aviv University

Abstract. We initiate the study of multi-source extractors in the quantum world. In this setting, our goal is to extract random bits from two independent weak random sources, on which two quantum adversaries store a bounded amount of information. Our main result is a two-source extractor secure against quantum adversaries, with parameters closely matching the classical case and tight in several instances. Moreover, the extractor is secure even if the adversaries share entanglement. The construction is the Chor-Goldreich [5] two-source inner product extractor and its multibit variant by Dodis et al. [9]. Previously, research in this area focused on the construction of seeded extractors secure against quantum adversaries; the multi-source setting poses new challenges, among which is the presence of entanglement that could potentially break the independence of the sources.

Keywords: Extractors, Quantum Information.

1 Introduction and Results

Randomness extractors are fundamental in many areas of computer science, with numerous applications to derandomization, error-correcting codes, expanders, combinatorics and cryptography, to name just a few. Randomness extractors generate almost uniform randomness from imperfect sources, as they appear either in nature, or in various applications. Typically, the imperfect source is modelled as a distribution over n-bit strings whose *min-entropy* is at least k, i.e., a distribution in which no string occurs with probability greater than 2^{-k} [26,5,34]. Such sources are known as *weak sources*. One way to arrive at a weak source is to imagine that an adversary (or some process in nature), when in contact with a uniform source, *stores* $n - k$ bits of information about the string (which are later used to break the security of the extractor, i.e. to distinguish its output from uniform). Then, from the adversary's point of view, the source essentially has min-entropy k.

---

[*] Supported by JK's ERC Starting Grant QUCO.

[**] Supported by the European Commission under the Integrated Project Qubit Applications (QAP) funded by the IST directorate as Contract Number 015848, by an Alon Fellowship of the Israeli Higher Council of Academic Research, by an Individual Research Grant of the Israeli Science Foundation, by a European Research Council (ERC) Starting Grant and by the Wolfson Family Charitable Trust.

M. Serna et al. (Eds.): APPROX and RANDOM 2010, LNCS 6302, pp. 656–669, 2010.

Ideally, we would like to extract randomness from a weak source. However, it is easy to see that no deterministic function can extract even one bit of randomness from all such sources, even for min-entropies as high as $n - 1$ (see e.g. [26]). One main approach to circumvent this problem is to use a short truly random *seed* for extraction from the weak source (*seeded extractors*) (see, e.g., [27]). The other main approach, which is the focus of the current work, is to use several independent weak sources (*seedless extractors*) (e.g. [5,31,9,4,23] and many more).

With the advent of quantum computation, we must now deal with the possibility of quantum adversaries (or quantum physical processes) interfering with the sources used for randomness extraction. For instance, one could imagine that a quantum adversary now stores $n - k$ *qubits* of information about the string sampled from the source. This scenario of a *bounded storage quantum adversary* arises in several applications, in particular in cryptography.

Some constructions of *seeded* extractors were shown to be secure in the presence of quantum adversaries: König, Maurer, and Renner [25,19,24] proved that the pairwise independent extractor of [16] is also good against quantum adversaries, and with the same parameters. König and Terhal [20] showed that any one-bit output extractor is also good against quantum adversaries, with roughly the same parameters. In light of this, it was tempting to conjecture that *any* extractor is also secure against quantum storage. Somewhat surprisingly, Gavinsky et al. [12] gave an example of a seeded extractor that is secure against classical storage but becomes insecure even against very small quantum storage. This example has initiated a series of recent ground-breaking work that examined which seeded extractors stay secure against bounded storage quantum adversaries. Ta-Shma [28] gave an extractor with a short (polylogarithmic) seed extracting a polynomial fraction of the min-entropy. His result was improved by De and Vidick [8] extracting almost all of the min-entropy. Both constructions are based on Trevisan's extractor [30].

However, the question of whether *seedless* multi-source extractors can remain secure against quantum adversaries has remained wide open. The multi-source scenario corresponds to several independent adversaries, each tampering with one of the sources, and then jointly trying to distinguish the extractor's output from uniform. In the classical setting this leads to several independent weak sources. In the quantum world, measuring the adversaries' stored information might break the independence of the sources, thus jeopardizing the performance of the extractor.[1] Moreover, the multi-source setting offers a completely new aspect of the problem: the adversaries could potentially share *entanglement* prior to tampering with the sources. Entanglement between several parties is known to yield several astonishing effects with no counterpart in the classical world, e.g., non-local correlations [1] and superdense coding [3].

We note that the example of Gavinsky et al. can also be viewed as an example in the two-source model; we can imagine that the seed comes from a

---

[1] Such an effect appears also in *strong seeded* extractors and has been discussed in more detail in [20].

second source (of full entropy in this case, just like any seeded extractor can be artificially viewed as a two-source extractor). And obviously, in the same way, recent work on quantum secure seeded extractors artificially gives secure two-source extractors, albeit for a limited range of parameters and without allowing for entanglement. However, no one has as of yet explored how more realistic multi-source extractors fare against quantum adversaries, and in particular how entanglement might change the picture. We ask: Are there any good multi-source extractors secure against quantum bounded storage? And does this remain true when considering entanglement?

Our results: In this paper we answer all these questions in the positive. We focus on the inner-product based two-source extractor of Dodis et al. [9] (DEOR-extractor). Given two independent weak sources X and Y with the same length n and min-entropies k_1 and k_2 satisfying $k_1 + k_2 \gtrsim n$, this extractor gives m close to uniform random bits, where $m \approx \max(k_1, k_2) + k_1 + k_2 - n$. In recent years several two-source extractors with better parameters have been presented; however, the DEOR-construction stands out through its elegance and simplicity and its parameters still fare very well in comparison with recent work (e.g., [4,23]).

A first conceptual step in this paper is to define the model of quantum adversaries and of security in the two-source scenario (see Defs. 2 and 3): Each adversary gets access to an independent weak source X (resp. Y), and is allowed to store a *short* arbitrary quantum state. In the entangled setting, the two adversaries may share arbitrary prior entanglement, and hence their final joint stored state is the possibly entangled state ρ_{XY}. In the non-entangled case their joint state is of the form $\rho_{XY} = \rho_X \otimes \rho_Y$. In both cases, the security of the extractor is defined with respect to the joint state they store.

Definition 1. *[Two-source extractor against (entangled) quantum storage (informal):] A function $E : \{0,1\}^n \times \{0,1\}^n \to \{0,1\}^m$ is a (k_1, k_2, ε) extractor against (b_1, b_2) (entangled) quantum storage if for any sources X, Y with min-entropies k_1, k_2, and any joint stored quantum state ρ_{XY} prepared as above, with X-register of b_1 qubits and Y-register of b_2 qubits, the distribution $E(X,Y)$ is ε-close to uniform even when given access to ρ_{XY}.*

Depending on the type of adversaries, we will say E is secure against *entangled* or *non-entangled* storage. Note again that entanglement between the adversaries is specific to the multi-source scenario and does not arise in the case of seeded extractors.

Having set the framework, we show that the construction of Dodis et al. [9] is secure, first in the case of non-entangled adversaries.

Theorem 1. *The DEOR-construction is a (k_1, k_2, ε) extractor against (b_1, b_2) non-entangled storage with $m = (1 - o(1)) \max(k_1 - \frac{b_1}{2}, k_2 - \frac{b_2}{2}) + \frac{1}{2}(k_1 - b_1 + k_2 - b_2 - n) - 9\log\varepsilon^{-1} - O(1)$ output bits, provided $k_1 + k_2 - \max(b_1, b_2) > n + \Omega(\log^3(n/\varepsilon))$.*

As we show next the extractor remains secure even in the case of entangled adversaries. Notice the loss of essentially a factor of 2 in the allowed storage; this is related to the fact that superdense coding allows to store n bits using only $n/2$ entangled qubit pairs.

Theorem 2. *The DEOR-construction is a (k_1, k_2, ε) extractor against (b_1, b_2) entangled storage with $m = (1 - o(1)) \max(k_1 - b_2, k_2 - b_1) + \frac{1}{2}(k_1 - 2b_1 + k_2 - 2b_2 - n) - 9 \log \varepsilon^{-1} - O(1)$ output bits, provided $k_1 + k_2 - 2 \max(b_1, b_2) > n + \Omega(\log^3(n/\varepsilon))$.*

Note that in both cases, when the storage is linear in the source entropy we can output $\Omega(n)$ bits with exponentially small error. To compare to the performance of the DEOR-extractor in the classical case, note that a source with min-entropy k and *classical* storage of size b roughly corresponds to a source of min-entropy $k - b$ (see, e.g., [28] Lem. 3.1). Using this correspondence, the extractor of [9] gives $m = \max(k_1, k_2) + k_1 - b_1 + k_2 - b_2 - n - 6 \log \varepsilon^{-1} - O(1)$ output bits against classical storage, whenever $k_1 + k_2 - \max(b_1, b_2) > n + \Omega(\log n \cdot (\log^2 n + \log \varepsilon^{-1}))$. Hence the conditions under which one can extract randomness are essentially the same for DEOR and for our Thm. 1. The amount of random bits we can extract is somewhat less than in the classical case, even when disregarding storage.

In the non-entangled case, we are able to generalize our result to the stronger notion of guessing entropy adversaries or so called *quantum knowledge* (see discussion below and the full version [18] of this paper for details). We show that the DEOR-extractor remains secure even in this case, albeit with slightly weaker parameters.

Theorem 3. *The DEOR-construction is a (k_1, k_2, ε) extractor against quantum knowledge with $m = (1 - o(1)) \max(k_1, k_2) + \frac{1}{6}(k_1 + k_2 - n) - 9 \log \varepsilon^{-1} - O(1)$ output bits, provided $k_1 + k_2 > n + \Omega(\log^3(n/\varepsilon))$.*

For the proof of Thm. 3 refer to the full version of the paper [18].

Strong extractors: The extractor in Thms. 1, 2 and 3 is a so called *weak* extractor, meaning that when trying to break the extractor, no full access to any of the sources is given (which is natural in the multi-source setting). We also obtain several results in the so called *strong* case (see Cor. 2 and Lem. 5). A *strong* extractor has the additional property that the output remains secure even if the adversaries later gain full access to any one (but obviously not both) of the sources.[2] See Sec. 2 for details and a discussion of the subtleties in defining a strong extractor in the entangled case, and Secs. 3, 4 for our results in the strong case.

Tightness: In the one-bit output case, we show that our results are *tight*, both in the entangled and non-entangled setting.

---

[2] In [9], this is called a *strong blender*.

Proof ideas and tools: To show both of our results, we first focus on the simplest case of one-bit outputs. In this case the DEOR extractor [9] simply computes the inner product $E(x,y) = x \cdot y \pmod 2$ of the n-bit strings x and y coming from the two sources. Assume that the two adversaries are allowed quantum storage of b qubits each. Given their stored information they jointly wish to distinguish $E(x,y)$ from uniform, or, in other words, to predict $x \cdot y$. We start by observing that this setting corresponds to the well known simultaneous message passing (SMP) model in communication complexity,[3] where two parties, Alice and Bob, have access to an input each (which is unknown to the other). They each send a message of length b to a referee, who, upon reception of both messages, is to compute a function $E(x,y)$ of the two inputs. When E is hard to compute, it is a good extractor. Moreover, the entangled adversaries case corresponds to the case of SMP with entanglement between Alice and Bob, a model that has been studied in recent work (see e.g. [13,14]).

Before we proceed, let us remark, that there are cases, where entanglement is known to add tremendous power to the SMP model. Namely, Gavinsky et al. [13] showed an exponential saving in communication in the entangled SMP model, compared to the non-entangled case.[4] This points to the possibility that some extractors can be secure against a large amount of storage in the non-entangled case, but be insecure against drastically smaller amounts of entangled storage. Our results show that this is not the case for the DEOR extractor, i.e., that this construction is secure against the potentially harmful effects of entanglement.

In the one-bit output DEOR case we can tap into known results on the quantum communication complexity of the inner product problem (IP). Cleve et al. [6] and Nayak and Salzman [21] have given tight lower bounds in the one-way and two-way communication model, with and without entanglement (which also gives bounds in the SMP model). For instance, in the non-entangled case, to compute IP exactly in the one-way model, n qubits of communication are needed, and in the SMP model, n qubits of communication are needed from Alice and from Bob, just like in the classical case. Note that whereas in the communication setting typically worst case problems are studied, extractors correspond to *average case* (w.r.t. to weak randomness) problems. With some extra work we can adapt the communication lower bounds to weak sources and to the average bias which is needed for the extractor result. In fact, the results we obtain hold in the strong case (where later one of the sources is completely exposed), which corresponds to one-way communication complexity.

Tightness of our results comes from matching upper bounds on the one-way and SMP model communication complexity of the inner product. Adapting the

---

[3] The connection between extractors and communication complexity has been long known, see, e.g., [31].

[4] This result has been shown for a relation, not a function. It is tempting to conjecture that this result can be turned into an exponential separation for an extractor with entangled vs. non-entangled adversaries. It is, however, not immediate how to turn a worst case relation lower bound into an average case function bound, as needed in the extractor setting, so we leave this problem open.

work of [5] we can obtain tight bounds for any bias ε. Somewhat surprisingly, it seems no one has looked at tight upper bounds for IP in the *entangled SMP model*, where [6] give an $n/2$ lower bound for the message length for Alice and Bob. It turns out this bound is tight,[5] which essentially leads to the factor 2 separation in our results for the entangled vs. non-entangled case (see Sec. 3).

To show our results for the case of multi-bit extractors, we use the nice properties of the DEOR construction (and its precursors [31,10]). The extractor outputs bits of the form $Ax \cdot y$. Vazirani's XOR-Lemma allows to reduce the multi-bit to the one-bit case by relating the distance from uniform of the multi-bit extractor to the sum of biases of XOR's of subsets of its bits. Each such XOR, in turn, is just a (linearly transformed) inner product, for which we already know how to bound the bias. Our main technical challenge is to adapt the XOR lemma to the case of *quantum* side-information (see Sec. 2). This way we obtain results for multi-bit extractors, which even hold in the case of strong extractors. Following [9], we further improve the parameters in the *weak* extractor setting by combining our strong two-source extractor with a good seeded extractor (in our case with the construction of [7]) to extract even more bits. See Sec. 4 for details.

Guessing entropy: One can weaken the requirement of bounded storage, and instead only place a lower bound on the *guessing entropy* of the source given the adversary's storage, leading to the more general definition of extractors secure against guessing entropy. Informally, a guessing entropy of at least k means that the adversary's probability of correctly guessing the source is at most 2^{-k} (or equivalently, that given the adversary's state, the source has essentially min-entropy at least k). Working with guessing entropy has the advantage that we no longer have to worry about two parameters (min-entropy and storage) instead only working with one parameter (guessing entropy), and that the resulting extractors are stronger (assuming all other parameters are the same).

In the classical world, a guessing entropy of k is more or less equivalent to a source with k min-entropy; in the quantum world, however, things become less trivial. In the case of seeded extractors, this more general model has been successfully introduced and studied in [24,20,11,7,29], where several constructions secure against bounded guessing entropy were shown.

In the case of *non-entangled* two-source extractors, we can show (based on [20]) that any classical *one-bit* output two-source extractor remains secure against bounded guessing entropy adversaries, albeit with slightly worse parameters. Moreover, our XOR-Lemma allow us to prove security of the DEOR-extractor against guessing entropy adversaries even in the multi-bit case (Thm. 3, see the full version [18] for the details).[6]

In the *entangled* adversaries case, one natural way to define the model is to require the guessing entropy of each source given the corresponding adversary's

---

[5] We thank Ronald de Wolf [32] for generously allowing us to adapt his upper bound to our setting.

[6] We are grateful to Thomas Vidick for pointing out that our XOR-Lemma allows us to obtain results also in this setting.

storage to be high. This definition, however, is too strong: it is easy to see that no extractor can be secure against such adversaries. This follows from the observation that by sharing a random string $r_1 r_2$ (which is a special case of shared entanglement) and having the first adversary store $r_1 \oplus x, r_2$ and the other store $r_1, r_2 \oplus y$, we keep the guessing entropy of X (resp. Y) relative to the adversary's storage unchanged yet we can recover x and y completely from the combined storage.

Hence we are naturally lead to consider the weaker requirement that the guessing entropy of each source given the combined storage of *both* adversaries is high. We now observe that already the DEOR one-bit extractor (where the output is simply the inner product) is not secure under this definition, indicating that this definition is still too strong. To see this, consider uniform n-bit sources X, Y, and say Alice stores $x \oplus r$, and Bob stores $y \oplus r$, where r is a shared random string. Obviously, their joint state does not help in guessing X (or Y), hence the guessing entropy of the sources is still n; but their joint state does give $x \oplus y$. If, in addition, Alice also stores the Hamming weight $|x| \bmod 4$ and Bob $|y| \bmod 4$, the guessing entropy is barely affected, and indeed one can easily show it is $n - O(1)$. However, their information now suffices to compute $x \cdot y$ exactly, since $x \cdot y = \frac{1}{2}((|x| + |y| - |x \oplus y|) \bmod 4)$. Hence inner product is insecure in this model even for very high guessing entropies, even though it is secure against a fair amount of bounded storage.

In light of this, it is not clear if and how entangled guessing entropy sources can be incorporated into the model, and hence we only consider bounded storage adversaries in the entangled case.

Related work: We are the first to consider two-source extractors in the quantum world, especially against entanglement. As mentioned, previous work on seeded extractors against quantum adversaries [25,19,24,20,28,8,7,2] gives rise to trivial two-source extractors where one of the sources is not touched by the adversaries. However, the only previous work that allows to derive results in the genuine two-source scenario is the work by König and Terhal [20]. Using what is implicit in their work, and with some extra effort, it is possible to obtain results in the one-bit output non-entangled two-source scenario (which hold against guessing entropy adversaries, but with worse performance than our results for the inner product extractor), and we give this result in detail in the full version of the paper [18]. Moreover, [20] show that any classical multi-bit extractor is secure against bounded storage adversaries, albeit with an exponential decay in the error parameter. This easily extends to the non-entangled two-source scenario, to give results in the spirit of Thm. 1. We have worked out the details and comparison to Thm. 1 in the full version of the paper [18]. Note, however, that to our knowledge no previous work gives results in the entangled scenario.

Discussion and Open Problems: We have, for the first time, studied two-source extractors in the quantum world. Previously, only seeded extractors have been studied in the quantum setting. In the two-source scenario a new phenomenon appears: entanglement between the (otherwise independent) sources. We have formalized what we believe the strongest possible notion of quantum adversaries in this setting and shown that one of the best performing extractors, the DEOR-construction, remains secure. We also show that our results are tight in the one-bit output case.

Our results for the multi-bit output DEOR-construction allow to extract slightly less bits compared to what is possible classically. An interesting open question is whether it is possible to obtain matching parameters in the (non-entangled) quantum case. One might have to refine the analysis and not rely solely on communication complexity lower bounds. Alternatively, our quantum XOR-Lemma currently incurs a penalty exponential in either the length of the output or the length of the storage. Any improvement here also immediately improves all three main theorems. In particular, by removing the penalty entirely, Thm. 1 can be made essentially optimal (with respect to the classical case).

We have shown that inner-product-based constructions are necessarily insecure in two reasonable models of entangled guessing entropy adversaries (and hence that bounded storage adversaries are the more appropriate model in the entangled case). It should be noted that it is possible that other extractor constructions (not based on inner product) could remain secure in this setting, and this subject warrants further exploration.

As pointed out, it is conceivable that entanglement could break the security of two-source extractors. Evidence for this is provided by the communication complexity separation in the entangled vs. non-entangled SMP-model, given in [13]. A fascinating open problem is to turn this relational separation into an extractor that is secure against non-entangled quantum adversaries but completely broken when entanglement is present.

Our work leaves several other open questions. It would be interesting to see if other multi-source extractors remain secure against entangled adversaries, in particular the recent breakthrough construction by Bourgain [4] which works for two sources with min-entropy $(1/2 - \alpha)n$ each for some small constant α, or the construction of Raz [23], where one source is allowed to have logarithmic min-entropy while the other has min-entropy slightly larger than $n/2$. Both extractors output $\Omega(n)$ almost uniform bits.

And lastly, it would be interesting to see other application of secure multi-source extractors in the quantum world. One possible scenario is multi-party computation. Classically, Kalai et al. [17] show that sufficiently strong two-source extractors allow to perform multi-party communication with weak sources when at least two parties are honest. Perhaps similar results hold in the quantum setting.

Structure of the paper: In Sec. 2 we introduce our basic notation and definitions, and describe the DEOR construction. Here we also present one of our tools, the "quantum" XOR-Lemma. Sec. 3 is dedicated to the one-bit output case and the connection to communication complexity and gives our tightness results. In Sec. 4 we deal with the multi-bit output case and prove our main result, Thms. 1 and 2. The details of several of the proofs and an in depth treatment of guessing entropy are relegated to the full version [18].

2   Preliminaries and Tools

In this section we provide the necessary notation, formalize Def. 1, describe the DEOR-extractor and present and prove our quantum XOR-Lemma. For background on quantum information see e.g. [22].

Notation: Given a classical random variable Z and a set of density matrices $\{\rho_z\}_{z \in Z}$ we denote by $Z\rho_Z$ the cq-state $\sum_{z \in Z} \Pr[Z = z]|z\rangle\langle z| \otimes \rho_z$. When the distribution is clear from the context we write $p(z)$ instead of $\Pr[Z = z]$. For any random variable Z' on the domain of Z, we define $\rho_{Z'} := \sum_{z \in Z'} \Pr[Z' = z]\rho_z$. For any random variable Y, let $Y\rho_Z := \sum_{y \in Y} \Pr[Y = y]|y\rangle\langle y| \otimes \rho_{Z|Y=y}$. We denote by U_m the uniform distribution on m bits. For matrix norms, we define $|A|_{\mathrm{tr}} = \frac{1}{2}\|A\|_1 = \frac{1}{2}Tr(\sqrt{A^\dagger A})$ and $\|A\|_2 = \sqrt{Tr(A^\dagger A)}$.

Extractors against quantum storage: We first formalize the different types of quantum storage.

Definition 2. *For two random variables X, Y we say ρ_{XY} is a (b_1, b_2) entangled storage if it is generated by two non-communicating parties, Alice and Bob, in the following way. Alice and Bob initially share an arbitrary entangled state. Alice receives $x \in X$, Bob receives $y \in Y$. They each apply an arbitrary quantum operation on their qubits. Alice then stores b_1 of her qubits (and discards the rest), and Bob stores b_2 of his qubits, giving the state ρ_{xy}.*

We denote by ρ^A_{XY} the state obtained when Alice stores her entire state, whereas Bob stores only b_2 qubits of his, and similarly for ρ^B_{XY}.

We say ρ_{XY} is (b_1, b_2) non-entangled storage if $\rho_{xy} = \rho_x \otimes \rho_y$ for all $x \in X, y \in Y$.

The security of the extractor is defined relative to the storage.

Definition 3. *A (k_1, k_2, ε) two-source extractor against (b_1, b_2) (entangled) quantum storage is a function $E : \{0,1\}^n \times \{0,1\}^n \to \{0,1\}^m$ such that for any independent n-bit weak sources X, Y with respective min-entropies k_1, k_2, and any (b_1, b_2) (entangled) storage ρ_{XY}, $|E(X,Y)\rho_{XY} - U_m\rho_{XY}|_{\mathrm{tr}} \leq \varepsilon$.*

The extractor is called X-strong if $|E(X,Y)\rho_{XY}X - U_m\rho_{XY}X|_{\mathrm{tr}} \leq \varepsilon$, X-superstrong when ρ_{XY} is replaced by ρ^A_{XY}, and similarly for Y. It is called (super)strong if it is both X- and Y- (super)strong.

A note on the definition: A strong extractor is secure even if at the distinguishing stage one of the sources is completely exposed. A superstrong extractor is secure even if, in addition, the matching party's entire state is also given. Without entanglement, the two are equivalent, as the state can be completely reconstructed from the source. Equivalently, in terms of storage, we can say an extractor is X-strong if it is secure against (k_1, b_2) storage, and X-superstrong if secure against (∞, b_2) storage. Further note that in the communication complexity setting the model of strong extractors corresponds to the SMP model where the referee also gets access to one of the inputs, whereas the model of superstrong extractors corresponds to the one-way model, where one party also has access to its share of the entangled state.

To prove E is an extractor, it suffices to show that it is either X-strong or Y-strong. All our proofs follow this route.

Flat sources: It is well known that any source with min-entropy k is a convex combination of flat sources (i.e., sources that are uniformly distributed over their support) with min-entropy k. In what follows we will therefore only consider such sources in our analysis of extractors, as one can easily verify that for every sources X, Y and quantum storage ρ_{XY},

$$|E(X, Y)\rho_{XY} - U_m\rho_{XY}|_{\mathrm{tr}} \leq \max_{i,j} |E(X_i, Y_j)\rho_{X_iY_j} - U_m\rho_{X_iY_j}|_{\mathrm{tr}},$$

where $X = \sum \alpha_i X_i$ and $Y = \sum \beta_j Y_j$ are convex combinations of flat sources.

The DEOR construction: The following (strong) extractor construction is due to Dodis et al. [9]. Every output bit is a linearly transformed inner product, namely $A_i x \cdot y$ for some full rank matrix A_i, where x and y are the n-bit input vectors. Here $x \cdot y := \sum_{j=1}^n x_j y_j \pmod 2$. The matrices A_i have the additional property that every subset sum is also of full rank. This ensures that any XOR of some bits of the output is itself a linearly transformed inner product.

Lemma 1 ([9]). *For all $n > 0$, there exist an efficiently computable set of $n \times n$ matrices $A_1, A_2, \ldots, A_n$ over GF(2) such that for any non-empty set $S \subseteq [n]$, $A_S := \sum_{i \in S} A_i$ has full rank.*

Definition 4 (strong blender of [9]). *Let $n \geq m > 0$, and let $\{A_i\}_{i=1}^m$ be a set as above. The DEOR-extractor $E_D : \{0,1\}^n \times \{0,1\}^n \to \{0,1\}^m$ is given by $E_D(x, y) = A_1 x \cdot y, A_2 x \cdot y, \ldots, A_m x \cdot y$.*

The XOR-Lemma: Vazirani's XOR-Lemma [31] relates the non-uniformity of a distribution to the non-uniformity of the characters of the distribution, i.e., the XOR of certain bit positions. For the DEOR-extractor it allows to reduce the multi-bit output case to the binary output case.

Lemma 2 (Classical XOR-Lemma [31,15]). *For every m-bit random variable Z*

$$|Z - U_m|_1^2 \leq \sum_{0 \neq S \in \{0,1\}^m} |(S \cdot Z) - U_1|_1^2.$$

This lemma is not immediately applicable in our scenario, as we need to take into account *quantum* side information. For this, we need a slightly more general XOR-Lemma.

Lemma 3 (Classical-Quantum XOR-Lemma).[7] *Let $Z\rho_Z$ be an arbitrary cq-state, where Z is an m-bit classical random variable and ρ_Z is of dimension 2^d. Then*

---

[7] We thank Thomas Vidick for pointing out that we can also have a bound in terms of m and not only d.

$$|Z\rho_Z - U_m\rho_Z|_{\mathrm{tr}}^2 \leq 2^{\min(d,m)} \cdot \sum_{0 \neq S \in \{0,1\}^m} |(S \cdot Z)\rho_Z - U_1\rho_Z|_{\mathrm{tr}}^2 .$$

Following the proof of the classical XOR-Lemma in [15], we first relate $\|Z\rho_Z - U_m\rho_Z\|_1$ to $\|Z\rho_Z - U_m\rho_Z\|_2$, and then view $Z\rho_Z - U_m\rho_Z$ in the Hadamard (or Fourier) basis, giving us the desired result. The detailed proof can be found in the full version of the paper [18].

3   Communication Complexity and One-Bit Extractors

3.1   Average Case Lower Bound for Inner Product

Cleve et al. [6] give a lower bound for the worst case one-way quantum communication complexity of inner product with arbitrary prior entanglement. It is achieved by first reducing the problem of computing the inner product to that of transmitting one input over a quantum channel, and then using an extended Holevo bound. Nayak and Salzman [21] obtained an optimal lower bound by replacing Holevo with a more "mission-specific" bound.

Revisiting Cleve et al.'s reduction, we now show how to adapt it to flat sources, to the average case error and to the linearly transformed inner product. The main challenge is to carefully treat the error terms so as to not cancel out the (small) amplitude of the correct state.

Lemma 4. *Let X, Y be flat sources over n bits with min-entropies k_1, k_2, and A, B full rank n by n matrices over $GF(2)$. Let P be a b qubit one-way protocol for $(AX) \cdot (BY)$ with success probability $\frac{1}{2} + \varepsilon$. Then*

(a) $\varepsilon \leq 2^{-(k_1+k_2-2b-n+2)/2}$, *if the parties share prior entanglement and*
(b) $\varepsilon \leq 2^{-(k_1+k_2-b-n+2)/2}$ *otherwise.*

The proof can be found in the full version of the paper [18].

3.2   One Bit Extractor

When the extractor's output is binary, distinguishing it from uniform is equivalent to computing the output on average. This was shown by Yao [33] when the storage is classical and is trivially extended to the quantum setting. With this observation, reformulating Lem. 4 in the language of trace distance yields a one bit extractor.

Corollary 1. *The function $E_{IP}(x, y) = x \cdot y$ is a (k_1, k_2, ε) extractor against (b_1, b_2) (entangled) quantum storage provided*

(a) *(entangled)* $k_1 + k_2 - 2\min(b_1, b_2) \geq n - 2 + 2\log\varepsilon^{-1}$,
(b) *(non-entangled)* $k_1 + k_2 - \min(b_1, b_2) \geq n - 2 + 2\log\varepsilon^{-1}$.

Proof. With Yao's equivalence, Lem. 4.(a) immediately gives

$$|(AX \cdot Y)\rho_{XY}X - U\rho_{XY}X|_{\mathrm{tr}} \leq 2^{-(k_1+k_2-2b_2-n+2)/2} \tag{1}$$

$$|(AX \cdot Y)\rho_{XY}Y - U\rho_{XY}Y|_{\mathrm{tr}} \leq 2^{-(k_1+k_2-2b_1-n+2)/2} \tag{2}$$

for any full rank matrix A, and specifically for $A = I$. By the assumption on ε, E_{IP} is either Y-strong or X-strong. Repeating this argument with Lem. 4.(b) gives the non-entangled case.

Recall (see Def. 3 and discussion thereafter) that one-way communication corresponds to the model of *superstrong* extractors. It is not surprising then that Lem. 4 actually implies a superstrong extractor. By choosing ε in the above proof of Cor. 1 such that both inequalities (1) and (2) are satisfied, where we replace ρ_{xY} by ρ_{xY}^A to include Alice's complete state as well as Bob's entangled qubits and similarly for ρ_{Xy}^B, we obtain:

Corollary 2. *The function $E_{IP}(x, y) = x \cdot y$ is a (k_1, k_2, ε) superstrong extractor against (b_1, b_2) (entangled) quantum storage provided*

  (a) (entangled) $k_1 + k_2 - 2\max(b_1, b_2) \geq n - 2 + 2\log\varepsilon^{-1}$,
  (b) (non-entangled) $k_1 + k_2 - \max(b_1, b_2) \geq n - 2 + 2\log\varepsilon^{-1}$.

We show that the parameters of all our extractors are *tight* in the full version of the paper [18].

4   Many Bit Extractors

Here we prove our main theorems, Thms. 1 and 2. First, using our quantum XOR-Lemma, Lem. 3, we obtain results in the *strong* case.

Lemma 5. *E_D is a (k_1, k_2, ε) X-strong extractor against (b_1, b_2) (entangled) quantum storage provided*

  (a) (entangled) $k_1 + k_2 - 2b_2 \geq 2m + n - 2 + 2\log\varepsilon^{-1}$,
  (b) (non-entangled) $k_1 + k_2 - b_2 \geq 2m + n - 2 + 2\log\varepsilon^{-1}$.

Proof. Recall that $E_D(x, y) = A_1x \cdot y, A_2x \cdot y, \ldots, A_mx \cdot y$ (see Def. 4). For $0 \neq S \in \{0, 1\}^m$, let $A_S = \sum_{i:S_i=1} A_i$ and note that $S \cdot E(x, y) = A_Sx \cdot y$. By the XOR-Lemma 3,

$$|E(X, Y)\rho_{XY}X - U_m\rho_{XY}X|_{\mathrm{tr}} \leq \sqrt{2^m \sum_{S \neq 0} |(A_SX \cdot Y)\rho_{XY}X - U_1\rho_{XY}X|_{\mathrm{tr}}^2}.$$

The result then follows by Ineq. (1) in the proof of Cor. 1 and its non-entangled analogue.

In a similar way, we also obtain a *Y-strong* extractor with analogous parameters. Following [9], we now apply a seeded extractor against quantum storage to the output of an X-strong (Y-strong) extractor to obtain a two-source extractor with more output bits.

Lemma 6. *Let $E_B : \{0,1\}^n \times \{0,1\}^n \to \{0,1\}^d$ be a (k_1, k_2, ε) X-strong extractor against (b_1, b_2) (entangled) quantum storage, let $E_S : \{0,1\}^n \times \{0,1\}^d \to \{0,1\}^m$ be some function and define $E(x,y) = E_S(x, E_B(x,y))$.*

  (a) (entangled) If E_S is a (k_1, ε) seeded extractor against $b_1 + b_2$ quantum storage then E is a $(k_1, k_2, 2\varepsilon)$ extractor against (b_1, b_2) entangled quantum storage.
  (b) (non-entangled) If E_S is a (k_1, ε) seeded extractor against b_1 quantum storage then E is a $(k_1, k_2, 2\varepsilon)$ extractor against (b_1, b_2) non-entangled quantum storage.

The proof of this lemma follows from a simple application of the triangle inequality (see the full version [18] for details). Thms. 1 and 2 now follow by composing the strong two-source extractor of Lem. 5 with the seeded extractor of [7] as in Lem. 6. See the full version [18] for all details.

Acknowledgments

The authors would like to thank Nir Bitansky, Ashwin Nayak, Oded Regev, Amnon Ta-Shma, Thomas Vidick and Ronald de Wolf for valuable discussions. We are especially indebted to Ronald de Wolf for allowing us to use his exact protocol for IP in the SMP model with entanglement, and to Thomas Vidick for pointing out how to replace 2^d with 2^m in our XOR-Lemma, which allowed us to prove Thm. 3.

References

1. Bell, J.S.: On the Einstein-Podolsky-Rosen paradox. Physics 1, 195–200 (1964)
2. Ben-Aroya, A., Ta-Shma, A.: Better short-seed extractors against quantum knowledge. CoRR abs/1004.3737 (2010)
3. Bennett, C.H., Wiesner, S.J.: Communication via one- and two-particle operators on Einstein-Podolsky-Rosen states. Phys. Rev. Lett. 69(20), 2881–2884 (1992)
4. Bourgain, J.: More on the sum-product phenomenon in prime fields and its applications. IJNT 1(1), 1–32 (2005)
5. Chor, B., Goldreich, O.: Unbiased bits from sources of weak randomness and probabilistic communication complexity. SIAM J. Comput. 17(2), 230–261 (1988)
6. Cleve, R., van Dam, W., Nielsen, M., Tapp, A.: Quantum entanglement and the communication complexity of the inner product function. In: Williams, C.P. (ed.) QCQC 1998. LNCS, vol. 1509, pp. 61–74. Springer, Heidelberg (1999)
7. De, A., Portmann, C., Vidick, T., Renner, R.: Trevisan's extractor in the presence of quantum side information. CoRR abs/0912.5514 (2009)
8. De, A., Vidick, T.: Near-optimal extractors against quantum storage. In: Proc. of STOC (2010) (to appear)
9. Dodis, Y., Elbaz, A., Oliveira, R., Raz, R.: Improved randomness extraction from two independent sources. In: Jansen, K., Khanna, S., Rolim, J.D.P., Ron, D. (eds.) RANDOM 2004 and APPROX 2004. LNCS, vol. 3122, pp. 334–344. Springer, Heidelberg (2004)
10. Dodis, Y., Oliveira, R.: On extracting private randomness over a public channel. In: Arora, S., Jansen, K., Rolim, J.D.P., Sahai, A. (eds.) RANDOM 2003 and APPROX 2003. LNCS, vol. 2764, pp. 252–263. Springer, Heidelberg (2003)

11. Fehr, S., Schaffner, C.: Randomness extraction via delta-biased masking in the presence of a quantum attacker. In: Canetti, R. (ed.) TCC 2008. LNCS, vol. 4948, pp. 465–481. Springer, Heidelberg (2008)
12. Gavinsky, D., Kempe, J., Kerenidis, I., Raz, R., de Wolf, R.: Exponential separation for one-way quantum communication complexity, with applications to cryptography. SIAM J. Comput. 38(5), 1695–1708 (2008)
13. Gavinsky, D., Kempe, J., Regev, O., de Wolf, R.: Bounded-error quantum state identification and exponential separations in communication complexity. SIAM J. Comput. 39(1), 1–24 (2009)
14. Gavinsky, D., Kempe, J., de Wolf, R.: Strengths and weaknesses of quantum fingerprinting. In: Proc. of CCC, pp. 288–298 (2006)
15. Goldreich, O.: Three xor-lemmas - an exposition. ECCC 2(56) (1995)
16. Impagliazzo, R., Levin, L.A., Luby, M.: Pseudo-random generation from one-way functions (extended abstracts). In: Proc. of STOC, pp. 12–24 (1989)
17. Kalai, Y.T., Li, X., Rao, A.: 2-source extractors under computational assumptions and cryptography with defective randomness. In: Proc. of FOCS, pp. 617–626 (2009)
18. Kasher, R., Kempe, J.: Two-source extractors secure against quantum adversaries. CoRR abs/1005.0512 (2009)
19. König, R., Maurer, U.M., Renner, R.: On the power of quantum memory. IEEE Trans. Inform. Theory 51(7), 2391–2401 (2005)
20. König, R.T., Terhal, B.M.: The bounded-storage model in the presence of a quantum adversary. IEEE Trans. Inform. Theory 54(2), 749–762 (2008)
21. Nayak, A., Salzman, J.: Limits on the ability of quantum states to convey classical messages. Journal of the ACM 53(1), 184–206 (2006)
22. Nielsen, M.A., Chuang, I.L.: Quantum Computation and Quantum Information, 1st edn. Cambridge University Press, Cambridge (2000)
23. Raz, R.: Extractors with weak random seeds. In: Proc. of STOC, pp. 11–20 (2005)
24. Renner, R.: Security of Quantum Key Distribution. Ph.D. thesis, ETH Zurich (September 2005), http://arxiv.org/abs/quant-ph/0512258
25. Renner, R., König, R.: Universally composable privacy amplification against quantum adversaries. In: Kilian, J. (ed.) TCC 2005. LNCS, vol. 3378, pp. 407–425. Springer, Heidelberg (2005)
26. Santha, M., Vazirani, U.V.: Generating quasi-random sequences from slightly-random sources (extended abstract). In: Proc. of FOCS, pp. 434–440 (1984)
27. Shaltiel, R.: Recent developments in explicit constructions of extractors. Bulletin of the EATCS 77, 67–95 (2002)
28. Ta-Shma, A.: Short seed extractors against quantum storage. In: Proc. of STOC, pp. 401–408 (2009)
29. Tomamichel, M., Schaffner, C., Smith, A., Renner, R.: Leftover hashing against quantum side information. In: Proc. of ISIT (to appear, 2010)
30. Trevisan, L.: Extractors and pseudorandom generators. Journal of the ACM 48(4), 860–879 (2001)
31. Vazirani, U.V.: Strong communication complexity or generating quasirandom sequences form two communicating semi-random sources. Combinatorica 7(4), 375–392 (1987)
32. de Wolf, R.: Personal communication (2010)
33. Yao, A.C.C.: Theory and applications of trapdoor functions (extended abstract). In: Proc. of FOCS, pp. 80–91 (1982)
34. Zuckerman, D.: General weak random sources. In: Proc. of FOCS, pp. 534–543 (1990)

Locally Testable vs. Locally Decodable Codes

Tali Kaufman[1,*] and Michael Viderman[2,**]

[1] Department of Computer Science and Applied Math.
Weizmann Institute, Rehovot, Israel
kaufmant@mit.edu
[2] Computer Science Department
Technion, Haifa, Israel
viderman@cs.technion.ac.il

Abstract. We study the relation between locally testable and locally decodable codes. Locally testable codes (LTCs) are error-correcting codes for which membership of a given word in the code can be tested probabilistically by examining it in very few locations. Locally decodable codes (LDCs) allow to recover each message entry with high probability by reading only a few entries of a slightly corrupted codeword. A linear code $\mathcal{C} \subseteq \mathbf{F}_2^n$ is called sparse if $n \geq 2^{\Omega(\dim(\mathcal{C}))}$.

It is well-known that LTCs do not imply LDCs and that there is an intersection between these two families. E.g. the Hadamard code is both LDC and LTC. However, it was not known whether LDC implies LTC. We show the following results.

- Two-transitive codes with a local constraint imply LDCs, while they do not imply LTCs.
- Every non-sparse LDC contains a large subcode which is not LTC, while every subcode of an LDC remains LDC. Hence, every non-sparse LDC contains a subcode that is LDC but is not LTC.

The above results demonstrate inherent differences between LDCs and LTCs, in particular, they imply that LDCs do not imply LTCs.

1 Introduction

A linear code over a finite field $\mathbf{F}$ is a linear subspace $\mathcal{C} \subseteq \mathbf{F}^n$. The dimension of $\mathcal{C}$ is its dimension as a vector space, and its rate is the ratio of its dimension to n. The distance of $\mathcal{C}$ is the minimal Hamming distance between two different codewords. Typically, we are interested in codes whose distance is a linear to the block length n, i.e., $\Omega(n)$.

Locally testable codes (LTCs) are error-correcting codes for which membership of a given word in the code can be tested probabilistically by examining it in very few locations. More precisely such a code has a *tester*, which is a randomized algorithm with oracle access to the received word x. The tester reads

* This research was partially supported by the Koshland Fellowship.
** This research was partially supported by the Israeli Science Foundation (grant No. 679/06).

M. Serna et al. (Eds.): APPROX and RANDOM 2010, LNCS 6302, pp. 670–682, 2010.

at most q symbols from x and based on this local view decides if $x \in C$ or not. It should accept codewords with probability one, and reject words that are far (in Hamming distance) from the code with noticeable probability.

In recent years, starting with the work of Goldreich and Sudan [12], several surprising constructions of LTCs have been given (see [10] for an extensive survey of some of these constructions). The principal challenge is to understand the largest asymptotic rate possible for LTCs, and to construct LTCs approaching this limit. We now know constructions of LTCs of dimension $n/\log^{O(1)} n$ which can be tested with only three queries [5,7], [21]. The main open question in the subject is whether there are asymptotically good LTCs, i.e., LTCs that have dimension $\Omega(n)$ and distance $\Omega(n)$.

The only negative results on LTCs concern binary codes testable with just 2-queries [2,15] (which is a severe restriction), random LDPC codes [4], cyclic codes [1][1], Solvable codes [19] and codes with small redundancy in the small weight dual words [3].

On the other hand, locally decodable codes (LDCs) allow to recover each message entry with high probability by reading only a few entries of the codeword even if a constant fraction of it is adversely corrupted.

The best construction of LDCs was initiated by the breakthrough results of Yekhanin [26] who showed a (conditional) subexponential construction of 3-query LDCs. Later Efremenko [9] showed unconditional subexponential construction of LDCs. Gopalan showed that these codes can be considered as a sub-family of Reed-Muller codes [13].

Katz and Trevisan [16] were first who defined formally LDCs and showed that LDCs have superlinear blocklength. Goldreich et al. [11] showed that linear 2-query LDCs have exponential blocklength. This result was generalized by Dvir and Shpilka [8] for all arbitrarily large fields. Obata [22] and then Shiowattana and Lokam [23] showed asymptotically tight (exponential) lower bounds on the blocklength of 2-query LDCs. Kerenidis and de Wolf [20] showed exponential lower bounds for 2-query LDC and improved superlinear lower bound for q-query LDCs, where $q \geq 3$. Then Woodruff [25] improved this result for odd q and showed that q-query LDCs ($q \geq 3$) with k message bits and blocklength n have $n \geq \Omega(k^{1+\frac{1}{\lceil q/2-1 \rceil}})/\log(k)$ and for 3-query linear LDCs showed that $n \geq k^2/\log\log(k)$. The known lower bounds for q-query LDCs for $q \geq 3$ seems to be very far from tight.

LDCs are related to private information retrieval protocols, initiated by [6], while LTCs are related to PCPs [12]. Both these families of error correcting codes are explicitly studied, for survey see e.g. [24]. In spite of the fact, the distinction between the two families of the codes was not made. Namely, it is well-known that there is an intersection between the two families of codes, e.g. the famous Hadamard code is 3-query LTC and 2-query LDC. Moreover, it is well-known

---

[1] The last result rules out asymptotically good *cyclic* LTCs; the existence of asymptotically good cyclic codes has been a longstanding open problem, and the result shows the "intersection" of these questions concerning LTCs and cyclic codes has a negative answer.

that LTCs do not imply LDCs, i.e., there are LTCs which are not LDCs. This follows simply by comparing the upper and lower bounds on the blocklength of these families of codes. If $\mathcal{C} \subseteq \mathbf{F}^n$ is a q-query LDCs then $n \geq \Omega(\dim(C)^{q/(q-1)})$ (by Katz and Trevisan [16]), while there exist (best known) LTCs s.t. $n \leq O(\dim(\mathcal{C}) \cdot \text{poly}(\log \dim(\mathcal{C})))$ [5,7], [21]. However, the other direction, i.e., whether LDCs imply LTCs, was not known.

1.1   Our Results

We show that LDC does not imply LTC, and in fact there are inherent differences between LDCs and LTCs. Specifically we show the following results.

- In Theorem 5 we show that codes invariant under two-transitive groups that obey a local constraint are LDCs, while they are not necessarily LTCs. This provides a general proof to the local decodability of polynomial codes such as Hadamard code, Reed-Muller codes and dual-BCH codes. Combining this with a recent result of [14], we obtain an explicit family of linear codes which is locally decodable but is not locally testable.
- In Theorem 9 we show that every non-sparse code contains a large subcode which is not LTC, while every subcode of an LDC remains LDC (Corollary 11). Hence, every non-sparse LDC contains a subcode that is LDC but is not LTC. Moreover, we show (Theorem 13) that if we consider uniform-LTCs (for which a tester picks every possible local constraint with the same probability) then, in fact, *every* non-sparse LDC has *many* large subcodes which are not uniform-LTCs (but still LDCs).

1.2   On Sparse Codes vs. Non-sparse Codes

Recall that a code $\mathcal{C} \subseteq \mathbf{F}_2^n$ is called sparse if $n \geq 2^{\Omega(\dim(\mathcal{C}))}$, otherwise the code is non-sparse. A sparse code $\mathcal{C}$ is called unbiased if all nonzero codewords $c \in \mathcal{C}$ have relative weight ranging in $(\frac{1}{2} - n^{-\gamma}, \frac{1}{2} + n^{-\gamma})$ for some constant $\gamma > 0$. Kaufman and Sudan [17] showed that *all* sparse unbiased codes are LTCs and LDCs. Since every subcode of a sparse unbiased code is a sparse unbiased code we conclude that it is an LTC and LDC. However, sparse codes have exponential blocklength.

Our Theorem 9 shows that *every* non-sparse LDC contains a large subcode which not LTC. Hence, every non-sparse LDC contains a subcode that is LDC but is not LTC. This demonstrates an inherent difference between sparse and non-sparse codes. In sparse codes local testability is preserved in subcodes, while in non-sparse codes local testability is not preserved in subcodes. In contrast to local testability, local decodability of all codes is always preserved in their subcodes (Corollary 11).

1.3   Paper Organization

We start with some definitions in Section 2. In Section 3 we provide the proof to our first theorem. In Section 4 we prove our second theorem, moreover we also prove a stronger version of our second theorem that applies only for a special type of LTCs known as uniform-LTCs.

2   Preliminaries

Let $\mathbf{F}$ be a finite field and $[n]$ be the set $\{1, \ldots, n\}$. In this work, we consider only linear codes. We start with a few definitions.

Let $\mathcal{C} \subseteq \mathbf{F}^n$ be a linear code over $\mathbf{F}$. For $w \in \mathbf{F}^n$, let $supp(w) = \{i \in [n] \mid w_i \neq 0\}$ and $|w| = |supp(w)|$. We define the *distance* between two words $x, y \in \mathbf{F}^n$ to be $\Delta(x, y) = |\{i \mid x_i \neq y_i\}|$ and the relative distance to be $\delta(x, y) = \frac{\Delta(x,y)}{n}$. The distance of a code is denoted by $\Delta(\mathcal{C})$ and defined to be the minimal value of $\Delta(x, y)$ for two distinct codewords $x, y \in \mathcal{C}$. Similarly, the relative distance of the code is denoted $\delta(\mathcal{C}) = \frac{\Delta(\mathcal{C})}{n}$. For $x \in \mathbf{F}^n$ and $\mathcal{C} \subseteq \mathbf{F}^n$, let $\delta(x, \mathcal{C}) = \min_{y \in \mathcal{C}} \{\delta(x, y)\}$ denote the relative distance of x from the code $\mathcal{C}$. We note that $\Delta(\mathcal{C}) = \min_{c \in \mathcal{C} \setminus \{0\}} \{|c|\}$. If $\delta(x, \mathcal{C}) \geq \epsilon$, we say that x is ϵ-far from $\mathcal{C}$ and otherwise x is ϵ-close to $\mathcal{C}$. Let $\dim(\mathcal{C})$ be the dimension of $\mathcal{C}$. The vector inner product between u_1 and u_2 is denoted by $\langle u_1, u_2 \rangle$. The dual code $\mathcal{C}^\perp$ is defined as $\mathcal{C}^\perp = \{u \in \mathbf{F}^n \mid \forall c \in \mathcal{C} : \langle u, c \rangle = 0\}$. In a similar way we define $\mathcal{C}^\perp_{\leq t} = \{u \in \mathcal{C}^\perp \mid |u| \leq t\}$ and $\mathcal{C}^\perp_t = \{u \in \mathcal{C}^\perp \mid |u| = t\}$.

For $w \in F^n$ and $S = \{j_1, j_2, \ldots, j_m\} \subseteq [n]$, where $j_1 < j_2 < \ldots < j_m$, let $w|_S = (w_{j_1}, w_{j_2}, \ldots, w_{j_m})$ be the *restriction* of w to the subset S. Let $\mathcal{C}|_S = \{c|_S \mid c \in \mathcal{C}\}$ denote the restriction of the code $\mathcal{C}$ to the subset S. For $T \subseteq \mathbf{F}^n$ and $w \in \mathbf{F}^n$ we say that $w \perp T$ if for all $t \in T$ we have $\langle w, t \rangle = 0$.

2.1   Codes Invariant under Groups

Let G be a group of permutations over $[n]$. For $\pi \in G$ and $w = (w_1, w_2, \ldots, w_n) \in F^n$ with some abuse of notation we let $\pi(w) = (w_{\pi^{-1}(1)}, \ldots, w_{\pi^{-1}(n)})$ be a π-permuted word. Note that since G is a group and $\pi \in G$ we have $\pi^{-1} \in G$. A linear code $\mathcal{C}$ is invariant under G if for every $\pi \in G$ and $c \in \mathcal{C}$ we have $\pi(c) \in \mathcal{C}$. Note that if $\mathcal{C}$ is invariant under G then also $\mathcal{C}^\perp$ is invariant under G. G is called 2-transitive if for all $i \neq j \in [n]$ and $i' \neq j' \in [n]$ we have $\pi \in G$ such that $\pi(i) = i'$ and $\pi(j) = j'$. A linear code $\mathcal{C}$ is 2-transitive if it is invariant under some 2-transitive permutation group G.

2.2   LTCs, LDCs and LCCs

In this section, we define LTCs, LDCs and LCCs formally and recall a few concepts that will be used later in this paper. We define LTCs following [3].

Definition 1 ((LTCs and Testers)). Let $\mathcal{C} \in \mathbf{F}^n$ be a linear code. Given a distribution $\mathcal{D}$ over set $\mathcal{C}^\perp$, we define the support of $\mathcal{D}$ over $\mathcal{C}^\perp$ as $\mathcal{D}_S = \{u \in \mathcal{C}^\perp \mid \mathcal{D}(u) > 0\}$. We say that $\mathcal{D}$ is a (q, ϵ, δ)-distribution for the code $\mathcal{C}$, if the following conditions are satisfied:

- $\mathcal{D}_S \subseteq \mathcal{C}_{\leq q}^{\perp}$.
- For all $x \in \mathbf{F}^n$ such that $\delta(x, \mathcal{C}) \geq \delta$ it holds that $\displaystyle \Pr_{u \sim \mathcal{D}}[\langle u, x \rangle \neq 0] \geq \epsilon$.

We say that $C \subseteq \mathbf{F}^n$ is a (q, ϵ, δ)-LTC if it has a (q, ϵ, δ)-distribution $\mathcal{D}$. If $\mathcal{D}$ is uniform over $\mathcal{C}_{\leq q}^{\perp}$ we say that $\mathcal{C}$ is a (q, ϵ, δ)-uniform LTC.

The parameter q is known as query complexity, ϵ is the rejection probability and δ is the distance threshold.

Note that if $\mathcal{C}$ is a (q, ϵ, δ)-LTC then $\mathcal{C}$ is also a (q, ϵ, δ')-LTC for all $\delta' \geq \delta$.

We say that a family of codes $\{\mathcal{C}^{(n)} \mid n \in \mathbf{Z}\}$ is locally testable if there exist constants $q, \epsilon, \delta > 0$ such that for infinitely many n it holds that $\mathcal{C}^{(n)} \subseteq \mathbf{F}^n$ is a (q, ϵ, δ)-LTC, where $\delta \leq \delta(\mathcal{C}^{(n)})/3$.

Note that every perfect code $\mathcal{C}$ is $(0, 1, \delta > \delta(\mathcal{C})/2)$-LTC, i.e., the code is locally testable with 0 queries and highest possible rejection probability when the distance threshold is $\delta > \delta(\mathcal{C})/2$ since there are no words which are δ-far from the code. Hence, to avoid trivial cases we must require the distance threshold δ to be at most $\delta(\mathcal{C})/2$. Moreover, in the area of locally testable codes we usually require even less distance threshold, at most $\delta \leq \delta(\mathcal{C})/3$. E.g., all known constructions of LTCs satisfy this requirement (see e.g., [12,17,18,21,7]). From the other side, if for all constants $q, \epsilon > 0$ the code $\mathcal{C}$ is not $(q, \epsilon, \delta(\mathcal{C})/3)$-LTC we say that $\mathcal{C}$ is not locally testable (see e.g., [1,3,14]).

Remark. Our proofs will follow also if we define "uniform LTC" as LTC with a uniform distribution over $\mathcal{C}_q^{\perp}$.

We note that sometimes (e.g. [3]) uniform LTCs mean that the associated distribution is uniform over its support and not over all $\mathcal{C}_{\leq q}^{\perp}$, which is a less restrictive assumption.

Now we define Locally Decodable Codes (LDCs).

Definition 2 ((LDCs)). Let $\mathcal{C} \subseteq \mathbf{F}^n$ be a linear code of dimension k. Let $E_{\mathcal{C}}$ be the encoding function, i.e., $\mathcal{C} = \{E_{\mathcal{C}}(x) \mid x \in \mathbf{F}^k\}$. Then $\mathcal{C}$ is a (q, ϵ, δ)-LDC if there exists a randomized decoder $(\mathbf{D})$ that reads at most q entries and the following condition holds:

- For all $x \in \mathbf{F}^k$, $i \in [k]$ and $\hat{c} \in \mathbf{F}^n$ such that $\Delta(E_{\mathcal{C}}(x), \hat{c}) \leq \delta n$ we have
$$\Pr[\mathbf{D}^{\hat{c}}[i] = x_i] \geq \frac{1}{|\mathbf{F}|} + \epsilon,$$ i.e., with probability at least $\frac{1}{|\mathbf{F}|} + \epsilon$ entry x_i will be recovered correctly.

Note that definition implies that $\delta < \delta(\mathcal{C})/2$. We say that a family of codes $\{C^{(n)} \mid n \in \mathbf{Z}\}$ is locally decodable if there exist constants $q, \epsilon, \delta > 0$ such that for infinitely many n it holds that $C^{(n)} \subseteq \mathbf{F}^n$ is a (q, ϵ, δ)-LDC.

Remark. Notice that a message space for LDC can be a linear subspace $M \subset \mathbf{F}^k$, i.e., the messages are of the length k but not every word in $\mathbf{F}^k$ is a (legal) message. In this case the dimension of the code is $\dim(M)$. Note that because of the linearity of $\mathcal{C}$ the message space M must be a linear subspace.

Now we define locally self-correctable codes (LCCs).

Definition 3 ((LCCs)). Let $\mathcal{C} \subseteq \mathbf{F}^n$ be a linear code of dimension k. Then $\mathcal{C}$ is a (q, ϵ, δ)-LCC if there exists a self-corrector (**SC**) that reads at most q entries and the following condition holds:

– For all $c \in \mathcal{C}$, $i \in [n]$ and $\hat{c} \in F^n$ such that $\Delta(c, \hat{c}) \leq \delta n$ we have

$$\mathbf{Pr}\Big[\mathbf{SC}^{\hat{c}}[i] = c_i\Big] \geq \frac{1}{|\mathbf{F}|} + \epsilon,$$

i.e., with probability at least $\frac{1}{|\mathbf{F}|} + \epsilon$ entry c_i will be recovered correctly.

We say that a code $\mathcal{C}$ is locally self-correctable when $q, \epsilon, \delta > 0$ are constants. Note that the definition implies that $\delta < \delta(\mathcal{C})/2$.

The following folklore claim says that LCCs imply LDCs with the same parameters.

Claim 4. *If $\mathcal{C} \subseteq \mathbf{F}^n$ is a (q, ϵ, δ)-LCC then $\mathcal{C}$ is a (q, ϵ, δ)-LDC.*

Proof. Let $k = \dim(\mathcal{C})$. We pick a generator matrix $G \in \mathbf{F}^{n \times k}$ for $\mathcal{C}$, i.e., $\mathcal{C} = \{Gm \mid m \in \mathbf{F}^k\}$ such that the first k rows of G form identity matrix[2]. Hence the first k symbols of the code are message symbols, i.e., for all $m \in \mathbf{F}^k$ we have $(Gm)|_{[k]} = m$.

Let **SC** be a self-corrector for a code $\mathcal{C}$ that for every $i \in [n]$ reads at most q symbols and recovers the symbol i with probability at least $\frac{1}{|\mathbf{F}|} + \epsilon$ even if at most δ-fraction of the symbols was adversely corrupted. In particular, **SC** recovers with probability at least $\frac{1}{|\mathbf{F}|} + \epsilon$ every coordinate $i \in [k]$, i.e., every message symbol. We conclude that $\mathcal{C}$ is a (q, ϵ, δ)-LDC.

We stress that LDCs do not imply LCCs. To see this let $\mathcal{C} \subseteq \mathbf{F}^n$ be a LDC, append to it one entry (with coordinate $(n + 1)$) obtaining $\mathcal{C}' \subseteq \mathbf{F}^{(n+1)}$, such that this entry will not be involved in too much low-weight constraints of $\mathcal{C}'$ and thus could not be recovered with constant query complexity after the codeword will be corrupted, however the code remains LDC.

3   Two Transitivity with a Local Constraint Implies Local Correction

In this section we show (Theorem 5) that 2-transitive codes with local constraints imply LCCs and hence also LDCs. However, there exists a family of two-transitive codes with local constraints which is not locally testable, due to [14]. We conclude in Corollary 6 that a family of codes $\{\mathcal{C}^{(n)}\}_{n \in \mathbf{Z}}$ (explicitly) shown in [14] is LCC (and LDC) but is not LTC.

---

[2] $\mathcal{C}$ need not be systematic but it can be easily converted into one as was stated.

Theorem 5 ((2-transitivity implies LCCs)). *If $\mathcal{C} \subseteq \mathbf{F}^n$ is a 2-transitive code such that $\mathcal{C}_q^\perp \neq \emptyset$ then $\mathcal{C}$ is a $(q-1, \frac{1}{6}, \frac{1}{3q})$-LCC (LDC).*

Moreover, there exists a family of 2-transitive codes $\left\{\mathcal{C}^{(n)}\right\}_{n\in\mathbf{Z}}$, where $\mathcal{C}^{(n)} \subseteq \mathbf{F}^n$ and $(\mathcal{C}^{(n)})_8^\perp \neq \emptyset$, which is not $(q', \epsilon', 1/7)$-LTC for all constants $q', \epsilon' > 0$.

The following corollary follows immediately from Theorem 5.

Corollary 6. *There exists a family of linear codes $\{\mathcal{C}_n\}_{n\in\mathbf{N}}$, where $\mathcal{C}_n \subseteq \mathbf{F}^n$, which is a $(7, \frac{1}{6}, \frac{1}{24})$-LCC (LDC) but is not $(q', \epsilon', 1/7)$-LTC for all constants $q', \epsilon' > 0$.*

Since by Claim 4 (q, ϵ, δ)-LCC is also a (q, ϵ, δ)-LDC then Theorem 5 and the lower bound on the blocklength of LDCs by Kerenidis and de Wolf [20] imply the next corollary.

Corollary 7. *Let $\mathcal{C} \subseteq \mathbf{F}^n$ be a 2-transitive linear code and $k = \dim(\mathcal{C})$. If $\mathcal{C}_q^\perp \neq \emptyset$ then $n \geq \Omega(k/\log(k))^{1+1/(\lceil \frac{q}{2}-1\rceil)}$.*

Notice that under the famous conjecture that LDCs have superpolynomial blocklength we have that 2-transitive codes with constant weight duals have superpolynomial blocklength.

Proof of Theorem 5. Assume $\mathcal{C}$ (and thus $\mathcal{C}^\perp$) is invariant under a 2-transitive permutations group G (note that $G \neq \emptyset$, e.g., G contains the identity permutation). Let $u \in \mathcal{C}_q^\perp$ (note $\mathcal{C}_q^\perp \neq \emptyset$) and let $supp(u) = \{i_1, i_2, \ldots, i_q\}$. Hence for every $i \in [n]$ there exists $u' \in \mathcal{C}_q^\perp$ such that $i \in supp(u')$, e.g. pick $g \in G$ s.t. $g(i_1) = i$ and let $u' = g(u)$.

We define the self-corrector of entry $i \in [n]$ ($\mathbf{SC}_i$) which on word w

- picks random $g \in G$ such that $g(i) = i$
- queries all entries of $w|_{supp(g(u'))\setminus\{i\}}$
- and recovers the entry $w|_{(i)}$ by $\dfrac{-\sum_{j\in(supp(g(u')))\setminus\{i\}} w|_j \cdot g(u')|_j}{u'_{(i)}}$.

This self-corrector queries only $q - 1$ entries and has perfect completeness, i.e., for all $c = (c_1, \ldots, c_n) \in \mathcal{C}$ and $i \in [n]$ it holds that $\mathbf{SC}_i[c]$ returns c_i. Assume the self-corrector $\mathbf{SC}_i$ is given a word w such that for some $c \in C$ we have $\delta(w, c) \leq \frac{1}{3q}$. Let $I = supp(w - c)$ and note that $|I| \leq \frac{n}{3q}$. Think of I as a set of corrupted coordinates. Notice that if $\mathbf{SC}_i$ picks $g \in G$, $g(i) = i$ such that $(supp(g(u')) \setminus \{i\}) \cap I = \emptyset$ then $\mathbf{SC}_i$ recovers correctly the entry w_i, i.e., $SC_i[w] = c_i$. This is true because

$$\frac{\sum_{j\in(supp(g(u')))\setminus\{i\}} w|_j \cdot g(u')|_j}{u'_{(i)}} = \frac{\sum_{j\in(supp(g(u')))\setminus\{i\}} c|_j \cdot g(u')|_j}{u'_{(i)}} = c_i,$$

where the last equality follows because $\langle c, g(u') \rangle = 0$. In other words, whenever all the coordinates of $g(u')$ are correct but may be the i's coordinate, $\mathbf{SC}_i$ recovers correctly the entry w_i. Proposition 8 implies that for $j \neq i$ and random

$g \in G$ such that $g(i) = i$ we have that $g(j)$ is uniformly distributed in $[n] \setminus \{i\}$. We conclude that

$$\Pr_{g \in G, g(i) = i}[(supp(g(u')) \cap I) \setminus \{i\} \neq \emptyset] \leq \frac{(q-1)|I|}{n-1} \leq \frac{q|I|}{n}.$$

It follows that the probability that $\mathbf{SC}_i$ picks $g \in G$, $g(i) = i$ such that $supp(g(u')) \cap I \subseteq \{i\}$ is at least $1 - \frac{q}{3q} = \frac{2}{3}$. So, with probability at least $2/3$ the self-corrector $\mathbf{SC}_i$ picks $g \in G$ such that $g(i) = i$ and $|supp(g(u')) \setminus \{i\} \cap I| = \emptyset$ and the correction succeeds.

To see that two transitivity does not imply local testability we recall the main result of [14] (Theorem 16) that shows a family of 2-transitive codes $\{\mathcal{C}^{(n)}\}_n$ such that $\mathcal{C}_8^\perp \neq \emptyset$ which is not $(q', \epsilon', 1/7)$-LTC for all constants $q', \epsilon' > 0$. The proof of Proposition 8 is inspired by [1, Section 7].

Proposition 8. *Let G be a 2-transitive group and $G_{(i)} = \{g \in G \mid g(i) = i\}$. Then $G_{(i)}$ is a group of permutations s.t. for all $i' \neq i$ and $j' \neq i$ there exists $g \in G_{(i)}$ s.t. $g(i') = j'$. Furthermore, for any $i' \neq i$ and $j' \neq i$ we have*

$$\Pr_{g \in G_{(i)}}[g(i') = j'] = \frac{1}{n-1}.$$

Proof. Let $id \in G$ be the identity permutation, i.e., for all $j \in [n]$ we have $id(j) = j$. We know that $id \in G_{(i)}$, for every $g \in G_{(i)}$ there exists $g^{-1} \in G_{(i)}$, and if $h_1, h_2 \in G_{(i)}$ then also $h_1 \circ h_2 \in G_{(i)}$. We conclude that $G_{(i)}$ is a group of permutations.

For any $i' \neq i$ and $j' \neq i$ there exists $g \in G$ s.t. $g(i) = i$ and $g(i') = j'$ because G is 2-transitive, moreover, $g \in G_{(i)}$.

We argue that for any $i' \neq i$ and $j' \neq i$ we have $\Pr_{g \in G_{(i)}}[g(i') = j'] = \frac{1}{n-1}$. It is sufficient to show that for any $i', j'_1, j'_2 \neq i$ we have $\Pr_{g \in G_{(i)}}[g(i') = j'_1] = \Pr_{g \in G_{(i)}}[g(i') = j'_2]$.

Assume by a way of contradiction that $\Pr_{g \in G_{(i)}}[g(i') = j'_1] > \Pr_{g \in G_{(i)}}[g(i') = j'_2]$. Let $h \in G_{(i)}$ s.t. $h(j'_2) = j'_1$. Since $G_{(i)}$ is a group then random g is distributed in $G_{(i)}$ exactly as hg is distributed in $G_{(i)}$ and thus

$$\Pr_{g \in G_{(i)}}[g(i') = j'_1] > \Pr_{g \in G_{(i)}}[g(i') = j'_2] = \Pr_{g \in G_{(i)}}[h(g(i')) = j'_1] = \Pr_{g \in G_{(i)}}[g(i') = j'_1].$$

Contradiction.

4 Non-sparse LDCs Contain Subcodes That Are Not LTCs

In this section we show (Theorem 9) that non-sparse LDCs contain subcodes that are not LTCs. This demonstrates an important difference between LTCs and LDCs.

It turns out that reducing dimension of LDCs remains LDCs (Corollary 11), however LTCs are not stable to the dimension reduction. This leads to an interesting observation that every non-sparse LDC has a large subcode which is not LTC (but still LDC). Notice that non-sparse LDCs include Reed-Muller codes of low degree as well as subexponential LDCs that were recently discovered by Yekhanin [26] and Efremenko [9].

Theorem 9 ((Non-sparse LDCs contain non-LTCs as subcodes)). *Let* $q > 0$ *and* $0 < \epsilon, \delta < 1$ *be constants. Then for every linear code* $\mathcal{C} \subseteq \mathbf{F}^n$ *that is a* (q, ϵ, δ)*-LDC with* $\dim(\mathcal{C}) \geq \omega(\log(n))$ *and any constants* $q', \epsilon' > 0$ *there exists a linear subcode* $\mathcal{C}' \subset \mathcal{C}$ *such that* $\dim(\mathcal{C}') \geq \omega(\log(n))$, $\mathcal{C}'$ *is a* (q, ϵ, δ)*-LDC but* $\mathcal{C}'$ *is not* $(q', \epsilon', \delta(\mathcal{C}))$*-LTC.*

In the following we show that every subcode of an LDC remains an LDC.

Claim 10. *Let* $\mathcal{C} \subseteq \mathbf{F}^n$ *be a linear code and a* (q, ϵ, δ)*-LDC. Assume that* $\mathcal{C}' \subset \mathcal{C}$ *is a linear subcode of* $\mathcal{C}$. *Then* $\mathcal{C}'$ *is a* (q, ϵ, δ)*-LDC.*

Proof. Assume $\mathcal{C}$ has the (linear) message space $S \subseteq \mathbf{F}^k$ and has the decoder D. Let $S' \subset S$ be a (linear) message space for $\mathcal{C}'$. We argue that $\mathcal{C}'$ has the same decoder D. Let w be δ-close to $\mathcal{C}'$ (δ-close to the encoding of some message $m \in S'$) then w is δ-close to $\mathcal{C}$ and thus for all $i \in [k]$ the decoder D recovers correctly the message entry (m_i) with probability at least $\frac{1}{2} + \epsilon$.

Notice that $\dim(S') < \dim(S)$, i.e., the message space of the linear subcode $\mathcal{C}'$ has smaller dimension than the message space of $\mathcal{C}$. S' is a linear vector space because for every two messages $x_1, x_2 \in S'$ which encoded to c_1 and c_2 of $\mathcal{C}'$, respectively; $(x_1 + x_2) \in S'$ and $(x_1 + x_2)$ is encoded to $c_1 + c_2$.

Remark. The special case of reducing dimension is a removing of columns from the generator matrix. E.g., given a code $\mathcal{C} = \{Gx \mid x \in \mathbf{F}^k\}$, where $G \in \mathbf{F}^{n \times k}$ is a generator matrix for $\mathcal{C}$. Let $G' \in \mathbf{F}^{n \times (k-1)}$ be obtained by removing the last column of G. Then $\mathcal{C}' = \{G'x \mid x \in \mathbf{F}^{k-1}\}$ is a linear subcode of $\mathcal{C}$ and $\dim(\mathcal{C}') = \dim(\mathcal{C}) - 1$. In this case message space of $\mathcal{C}'$ is $\mathbf{F}^{k-1}$, while the message space of $\mathcal{C}$ is $\mathbf{F}^k$.

Corollary 11 ((LDCs are stable for dimension reduction)). *Let* $\mathcal{C} \subseteq \mathbf{F}^n$ *be a linear code and a* (q, ϵ, δ)*-LDC. Take any sequence of linear subcodes:* $\mathcal{C}_1 \subset \mathcal{C}_2 \subset \ldots \subset \mathcal{C}_f = \mathcal{C}$. *Then for all* $i \in [f]$ *it holds that* $\mathcal{C}_i$ *is a* (q, ϵ, δ)*-LDC.*

Now we show the auxiliary claim and then prove Theorem 9.

Claim 12. *Let* $\mathcal{C} \subseteq \mathbf{F}^n$ *be a linear code such that* $\dim(\mathcal{C}) = \omega(\log(n))$. *Then there exists* $w \in \mathbf{F}^n$ *such that* $\Delta(w, \mathcal{C}^\perp) \geq \omega(1)$.

Proof. For integer R let $V(n, R) = \sum_{i=0}^{R} \binom{n}{i} \cdot (|\mathbf{F}| - 1)^i$ be the volume of a sphere in $\mathbf{F}^n$ of radius R. Let $k = \dim(\mathcal{C}) \geq \omega(\log(n))$ and $S = \mathcal{C}^\perp$. Then $\dim(S) = n - k$ and $|S| = |\mathbf{F}|^{n-k} = |\mathbf{F}|^n / |\mathbf{F}|^k$. Recall that a covering radius of a code S is $R_S = \max_{w \in \mathbf{F}^n} \Delta(w, S)$, i.e., the largest Hamming distance of any word in $\mathbf{F}^n$ from S. Note that if R_S is constant then $V(n, R_S)$ is polynomial in n and vice versa, if $V(n, R_S)$ is super-polynomial in n then R_S goes to infinity with n. Assume by a way of contradiction that there exists a constant $t > 0$ such that for all $w \in \mathbf{F}^n$ we have $\Delta(w, S) \leq t$, i.e., $R_S \leq t = O(1)$.

The covering radius bound[3] states that

$$|S| \cdot V(n, R_S) \geq |\mathbf{F}|^n.$$

But then $V(n, R_S) \geq |\mathbf{F}|^k$, where $k \geq \omega(\log(n))$. Hence $V(n, R_S)$ must be super-polynomial in n, and $R_S \geq \omega(1)$. Contradiction.

Proof of Theorem 9. Claim 12 implies that there exists $u \in \mathbf{F}^n$ such that $\Delta(u, \mathcal{C}^\perp) \geq \omega(1) > q'$. Let $S = span(\mathcal{C}^\perp \cup \{u\})$. Note that for all $u' \in S$ if $|u'| \leq q'$ then $u' \in \mathcal{C}^\perp$. Let $\mathcal{C}' = S^\perp$ and then $\mathcal{C}'^\perp = S$. We have $\mathcal{C}^\perp \subset \mathcal{C}'^\perp$, $\mathcal{C}' \subset \mathcal{C}$ and in particular, $\dim(\mathcal{C}') = \dim(\mathcal{C}) - 1$. We argue that $\mathcal{C}'$ is not $(q', \epsilon', \delta(\mathcal{C}))$-LTC. We have $c \in \mathcal{C} \setminus \mathcal{C}'$ since $\mathcal{C}' \subset \mathcal{C}$. However $c \perp (\mathcal{C}')^\perp_{\leq q'}$ because $(\mathcal{C}')^\perp_{\leq q'} \subseteq \mathcal{C}^\perp$ by construction. Hence c is $\delta(\mathcal{C})$-far from $\mathcal{C}'$ but will be accepted with probability 1 by any q'-query tester of $\mathcal{C}'$.

We conclude that $\mathcal{C}' \subset \mathcal{C}$ and $\dim(\mathcal{C}') = \dim(\mathcal{C}) - 1$ but $\mathcal{C}'$ is not $(q', \epsilon', \delta(\mathcal{C}))$-LTC. Claim 10 guarantees that $\mathcal{C}'$ is a (q, ϵ, δ)-LDC since $\mathcal{C}'$ is a linear subcode of $\mathcal{C}$. The Theorem follows.

4.1   Non-sparse LDCs Contain Many Subcodes Which Are Not Uniform-LTCs

Theorem 9 shows that every non-sparse LDC $\mathcal{C}$ contains a *single* sequence of linear subcodes $\mathcal{C}_1 \subset \mathcal{C}_2 \subset \mathcal{C}_3 \subset \cdots \subset \mathcal{C}$ which are all not LTCs. In the following (Theorem 13) we show that every non-sparse LDC and *every* long enough sequence of linear subcodes $\mathcal{C}_1 \subset \mathcal{C}_2 \subset \mathcal{C}_3 \subset \cdots \subset \mathcal{C}_\ell = \mathcal{C}$ contains at least one subcode $\mathcal{C}_i$ which is not uniform LTC.

Theorem 13. *Let $q, q', \epsilon, \epsilon', \delta > 0$ be constants. Let $\mathcal{C} \subseteq \mathbf{F}^n$ be a linear code such that $\dim(\mathcal{C}) \geq \omega(\log(n))$. Then every sequence of ℓ linear subcodes $\mathcal{C}_1 \subset \mathcal{C}_2 \subset \mathcal{C}_3 \subset \cdots \subset \mathcal{C}_\ell = \mathcal{C}$, where $\ell \geq (q' \log(n))/\epsilon'$, contains at least one code $\mathcal{C}_i$ which is not $(q', \epsilon', \delta(\mathcal{C})/2)$-uniform LTC. Moreover, if $\mathcal{C}$ is a (q, ϵ, δ)-LDC then all linear subcodes $\mathcal{C}_i$ in the sequence are (q, ϵ, δ)-LDCs.*

Note that if $\dim(\mathcal{C}) \geq \omega(\log(n))$ then $\mathcal{C}$ contains sequences of subcodes of length $\omega(\log(n))$. Now we prove two simple claims that will be useful in the proof of Theorem 13.

Claim 14. *Let $\mathcal{C} \subseteq \mathbf{F}^n$ be a linear code. Moreover, let a linear code $\mathcal{C}' \subset \mathcal{C}$ be a $(q, \epsilon, \delta(\mathcal{C}))$-uniform LTC. Then $|\mathcal{C}'^\perp_{\leq q}| \leq (1 - \epsilon)|\mathcal{C}'^\perp_{\leq q}|$.*

---

[3] For any code $\mathcal{C} \subseteq \mathbf{F}^n$ (whether linear or not) the covering bound states that the covering radius R of $\mathcal{C}$ relates to n and $|\mathcal{C}|$ by $|\mathcal{C}| \cdot V(n, R) \geq |\mathbf{F}|^n$.

Proof. Let $\mathcal{D}$ be the uniform distribution over $\mathcal{C}'^{\perp}_{\leq q}$. We know that $\mathcal{C}^{\perp}_{\leq q} \subset \mathcal{C}'^{\perp}_{\leq q}$. Consider any $w \in \mathcal{C} \setminus \mathcal{C}'$ (note that $\mathcal{C} \setminus \mathcal{C}' \neq \emptyset$). Since $\mathcal{C}'$ is a $(q, \epsilon, \delta(\mathcal{C}))$-uniform LTC and $\delta(w, \mathcal{C}') \geq \delta(\mathcal{C})$ it holds that $\mathbf{Pr}_{u \sim \mathcal{D}}[\langle u, w \rangle \neq 0] \geq \epsilon$. Notice that if for $u \in \mathcal{C}'^{\perp}$ it holds that $\langle u, w \rangle \neq 0$ then $u \notin \mathcal{C}^{\perp}$. So, there are at least $\epsilon |\mathcal{C}'^{\perp}_{\leq q}|$ words in $\mathcal{C}'^{\perp}_{\leq q}$ that are not in $\mathcal{C}^{\perp}_{\leq q}$. Thus we have $|\mathcal{C}^{\perp}_{\leq q}| \leq (1 - \epsilon)|\mathcal{C}'^{\perp}_{\leq q}|$.

Claim 15. *Let $q', \epsilon' > 0$ be constants. Let ℓ be the minimal integer such that $\ell \geq \frac{q' \log n}{\epsilon'}$ and $\mathcal{C} \subseteq \mathbf{F}^n$ be a linear code such that $\dim(\mathcal{C}) > \frac{q' \log n}{\epsilon'}$. Then at least one of the codes in the sequence of the linear subcodes $\mathcal{C}_1 \subset \mathcal{C}_2 \subset \cdots \subset \mathcal{C}_\ell = \mathcal{C}$ is not $(q', \epsilon', \delta(\mathcal{C})/2)$-uniform LTC.*

Proof. Note that for all $i \in [\ell - 1]$ we have $\dim(\mathcal{C}_i) < \dim(\mathcal{C}_{i+1})$. Assume that for all $i \in [\ell]$, $\mathcal{C}_i$ is a $(q', \epsilon', \delta(\mathcal{C})/2)$-uniform LTC. If $(\mathcal{C}_\ell)^{\perp}_{\leq q} = \emptyset$ then for any word $w \in \mathbf{F}^n$ such that $|supp(w)| = \delta(\mathcal{C})/2$ (i.e., $\delta(w, \mathcal{C}_\ell) \geq \overline{\delta}(\mathcal{C})/2$) it holds that $w \perp (\mathcal{C}_\ell)^{\perp}_{\leq q}$. Contradiction. We conclude that $|(\mathcal{C}_\ell)^{\perp}_{\leq q'}| \geq 1$.

Claim 14 implies that for all $i \in [\ell - 1]$ we have that $|(\mathcal{C}_{i+1})^{\perp}_{\leq q'}| \leq (1 - \epsilon') \cdot |(\mathcal{C}_i)^{\perp}_{\leq q}|$. Then it holds that

$$|(\mathcal{C}_\ell)^{\perp}_{\leq q'}| \leq (1 - \epsilon')^\ell \cdot |(\mathcal{C}_1)^{\perp}_{\leq q'}| < e^{-\epsilon'\ell} \cdot n^{q'} \leq 1,$$

where $\ell \geq \frac{q' \log n}{\epsilon'}$. We conclude that $(\mathcal{C}_\ell)^{\perp}_{\leq q'} = \emptyset$. Contradiction.

Proof of Theorem 13. Assume $\mathcal{C}_1 \subset \mathcal{C}_2 \subset \ldots \subset \mathcal{C}_\ell = \mathcal{C}$, where $\ell \geq (q' \log(n))/\epsilon'$. Claim 15 says that at least one of the codes in the sequence is not $(q', \epsilon', \delta(\mathcal{C})/2)$-uniform LTC. Corollary 11 implies that for all $i \in [\ell]$ the code $\mathcal{C}_i$ is a (q, ϵ, δ)-LDC.

Acknowledgements. We thank Eli Ben-Sasson, who refused to co-author this paper for many valuable discussions and ideas. We also thank Anna Gal for asking the question about the relation between LTCs and LDCs. We would like to thank Ronny Roth and Partha Mukhopadhyay for helpful discussions. We thank the anonymous referees for valuable comments on an earlier version of this article.

References

1. Babai, L., Shpilka, A., Stefankovic, D.: Locally testable cyclic codes. In: IEEE (ed.) Proceedings: 44th Annual IEEE Symposium on Foundations of Computer Science, FOCS 2003, Cambridge, Massachusetts, October 11-14, pp. 116–125. IEEE Computer Society Press, Los Alamitos (2003)
2. Ben-Sasson, E., Goldreich, O., Sudan, M.: Bounds on 2-query codeword testing. In: Arora, S., Jansen, K., Rolim, J.D.P., Sahai, A. (eds.) RANDOM 2003 and APPROX 2003. LNCS, vol. 2764, pp. 216–227. Springer, Heidelberg (2003)
3. Ben-Sasson, E., Guruswami, V., Kaufman, T., Sudan, M., Viderman, M.: Locally testable codes require redundant testers. In: IEEE Conference on Computational Complexity, pp. 52–61. IEEE Computer Society, Los Alamitos (2009)

4. Ben-Sasson, E., Harsha, P., Raskhodnikova, S.: Some 3CNF properties are hard to test. SIAM Journal on Computing 35(1), 1–21 (2005)
5. Ben-Sasson, E., Sudan, M.: Simple PCPs with poly-log rate and query complexity. In: STOC, pp. 266–275. ACM, New York (2005)
6. Chor, B., Goldreich, O., Kushilevitz, E., Sudan, M.: Private information retrieval. JACM: Journal of the ACM 45 (1998)
7. Dinur, I.: The PCP theorem by gap amplification. Journal of the ACM 54(3), 12:1–12:44 (2007)
8. Dvir, Z., Shpilka, A.: Locally decodable codes with two queries and polynomial identity testing for depth 3 circuits. SIAM J. Comput. 36(5), 1404–1434 (2007)
9. Efremenko, K.: 3-query locally decodable codes of subexponential length. In: Mitzenmacher, M. (ed.) STOC, pp. 39–44. ACM, New York (2009)
10. Goldreich, O.: Short locally testable codes and proofs (survey). Electronic Colloquium on Computational Complexity (ECCC) (014) (2005)
11. Goldreich, O., Karloff, H.J., Schulman, L.J., Trevisan, L.: Lower bounds for linear locally decodable codes and private information retrieval. Computational Complexity 15(3), 263–296 (2006)
12. Goldreich, O., Sudan, M.: Locally testable codes and PCPs of almost-linear length. Journal of the ACM 53(4), 558–655 (2006)
13. Gopalan, P.: A note on efremenko's locally decodable codes. Electronic Colloquium on Computational Complexity (ECCC) (069) (2009)
14. Grigorescu, E., Kaufman, T., Sudan, M.: 2-transitivity is insufficient for local testability. In: IEEE Conference on Computational Complexity, pp. 259–267. IEEE Computer Society, Los Alamitos (2008)
15. Guruswami, V.: On 2-query codeword testing with near-perfect completeness. In: Asano, T. (ed.) ISAAC 2006. LNCS, vol. 4288, pp. 267–276. Springer, Heidelberg (2006)
16. Katz, J., Trevisan, L.: On the efficiency of local decoding procedures for error-correcting codes. In: STOC, pp. 80–86 (2000)
17. Kaufman, T., Sudan, M.: Sparse random linear codes are locally decodable and testable. In: FOCS, pp. 590–600. IEEE Computer Society, Los Alamitos (2007)
18. Kaufman, T., Sudan, M.: Algebraic property testing: the role of invariance. In: STOC, pp. 403–412. ACM, New York (2008)
19. Kaufman, T., Wigderson, A.: Symmetric ldpc and local testing. In: ICS (2010)
20. Kerenidis, I., de Wolf, R.: Exponential lower bound for 2-query locally decodable codes. Electronic Colloquium on Computational Complexity (ECCC) (059) (2002)
21. Meir, O.: Combinatorial construction of locally testable codes. In: STOC, pp. 285–294. ACM, New York (2008)
22. Obata, K.: Optimal lower bounds for 2-query locally decodable linear codes. In: Rolim, J.D.P., Vadhan, S.P. (eds.) RANDOM 2002. LNCS, vol. 2483, pp. 39–50. Springer, Heidelberg (2002)
23. Shiowattana, Lokam: An optimal lower bound for 2-query locally decodable linear codes. IPL: Information Processing Letters 97 (2006)
24. Trevisan, L.: Some applications of coding theory in computational complexity, September 23 (2004)
25. Woodruff: New lower bounds for general locally decodable codes. ECCC: Electronic Colloquium on Computational Complexity, technical reports (2007)
26. Yekhanin, S.: Towards 3-query locally decodable codes of subexponential length. J. ACM 55(1) (2008)

Appendix

For the sake of completeness we state the main result of Grigorescu et al. [14], who showed a 2-transitive code with dual codewords of weight 8 which is not LTC. For additional details see [14].

Let Tr be a trace function from $\mathbf{F}_{2^s}$ to $\mathbf{F}_2$. For positive integers $k < s$ let

$$\mathcal{F}^*_{k,s} = \left\{ f : \mathbf{F}_{2^s} \mapsto \mathbf{F}_2 \mid \exists \beta, \beta_0, \ldots, \beta_k \in \mathbf{F}_{2^n} \text{ s.t. } f(x) = \mathrm{Tr}\left(\beta + \beta_0 x + \sum_{i=1}^{k} \beta_i x^{2^i+1}\right) \right\}.$$

Theorem 16 [14]. *Assume* $k = \omega(1)$ *and* $s > 2k + 1$. *Let* $\mathcal{C} = \mathcal{F}^*_{k,s}$ *be a linear code. Then* $\mathcal{C}$ *is 2-transitive and* $\mathcal{C}_8^\perp \neq \emptyset$, *but* $\mathcal{C}$ *is not* $(q, \epsilon, 1/7)$*-LTC for all constants* $q, \epsilon > 0$.

Differential Privacy and the Fat-Shattering Dimension of Linear Queries

Aaron Roth[*]

Computer Science Department
Carnegie Mellon University
and
Microsoft Research New England

Abstract. In this paper, we consider the task of answering linear queries under the constraint of differential privacy. This is a general and well-studied class of queries that captures other commonly studied classes, including predicate queries and histogram queries. We show that the accuracy to which a set of linear queries can be answered is closely related to its *fat-shattering dimension*, a property that characterizes the learnability of real-valued functions in the agnostic-learning setting.

1 Introduction

The administrator of a database consisting of sensitive, but valuable information faces two conflicting objectives. Because the data is valuable, she would like to make statistical information about it available to the public. However, because the data is sensitive, she must take care not to release information that exposes the data of any particular individual in the data set. The central question in the field of *private data analysis* is how these two objectives can be traded off, and more specifically, how many queries of what type can be answered to given degrees of accuracy, while still preserving privacy.

Recent work on *differential privacy* provides a mathematical framework to reason about such questions. Informally, a probabilistic function f from a database D to some range $\mathcal{R}$ is α-differentially private if adding or removing a single individual from the dataset does not change the probability that $f(D) = r$ for any outcome $r \in \mathcal{R}$ by more than an e^α factor. The intuition behind this definition is that an individual's privacy should not be considered to have been violated by some event r, if r would have been almost as likely to occur even without the individual's data.

In this paper, we consider databases D which are real valued vectors, and the class of queries that we consider correspond to linear combinations of the entries of D. Formally, we consider databases $D \in \mathbb{R}_+^n$, and queries of the form $q \in [0, 1]^n$. The answer to query q on database D is simply the dot-product of the two vectors: $q(D) = q \cdot D$. This model has previously been considered ([DN03, DMT07, DY08, HT10]), and generalizes the class of *count queries* or *predicate*

---

[*] This work has been supported in part by an NSF Graduate Research Fellowship.

M. Serna et al. (Eds.): APPROX and RANDOM 2010, LNCS 6302, pp. 683–695, 2010.

queries, which has also been well studied ([DMNS06, BLR08, DNRRV09, RR10, UV10]).

The *fat-shattering dimension* (FSD) of a class of real-valued functions C over some domain is a generalization of the Vapnik-Chervonenkis dimension, and characterizes a distribution-free convergence property of the mean value of each $f \in C$ to its expectation. The fat-shattering dimension of a class of functions C is known to characterize the sample complexity necessary to PAC learn C in the agnostic framework [ABCH97, BLW94]: that is, ignoring computation, the sample complexity that is both necessary and sufficient to learn C in the agnostic framework is polynomially related to the fat-shattering dimension of C.

Our main result is a similar information theoretic characterization of the magnitude of the noise that must be added to the answer to each query in some class C in terms of the fat-shattering dimension of C, FSD(C). We show polynomially related information theoretic upper and lower bounds on the noise that must be added to each query in C in terms of FSD(C). This generalizes the results of [BLR08] to linear queries, and to our knowledge gives the first analysis of generic linear queries using some parameter other than their cardinality. This yields the first mechanism capable of answering a possibly infinite set of generic linear queries, and the first non-trivial lower bound for infinite classes of non-boolean linear queries. As a consequence, we extend results of Kasiviswanathan et al. and Blum et al. [KLNRS08, BLR08] relating the sample complexity necessary for agnostic PAC learning and private agnostic PAC learning from classes of boolean valued functions to classes of real valued functions.

1.1   Related Work and Our Results

Dinur and Nissim studied the special case of linear queries for which both the database and the query are elements of the boolean hypercube $\{0,1\}^n$ [DN03]. Even in this special case, they showed that there cannot be any private mechanism that answers n queries with error $o(\sqrt{n})$, because an adversary could use any such mechanism to reconstruct a $1 - o(1)$ fraction of the original database, a condition which they called *blatant non-privacy*. This result was strengthened by several subsequent papers [DMT07, DY08, KRSU10].

Beimel et al. consider the class of basis vectors on the boolean hypercube, and show that even though this class has a constant VC-dimension (and hence fat-shattering dimension), it requires a superconstant number of samples for useful private release [BKN10]. Specifically, they show that the $\log n$ factor which appears in the upper bound in this paper and in [BLR08], but not in the lower bound, is in fact necessary in some cases.

Dwork et al. gave the original definition of differential privacy, as well as the Laplace mechanism, which is capable of answering any k "low sensitivity" queries (including linear queries) up to error $O(k)$. A more refined analysis of the relationship between the laplace mechanism and function sensitivity was later given by [NRS07].

In a different setting, Blum Ligett and Roth considered the question of answering *predicate queries* over a database drawn from some domain X [BLR08]. This can be viewed as a special case of linear queries in which the queries are restricted to lie on the boolean hypercube, and the database must be integer valued: $D \in \mathbb{Z}_+^n$. They give a mechanism for answering every query in some class C with noise that depends linearly on the VC-dimension of the class of queries. This is a quantity that is at most $\log |C|$ for finite classes C, and can be finite even for *infinite* classes. Roth and Roughgarden later gave a mechanism which achieved similar bounds in the online model, in which the mechanism does not know the set of queries that must be answered ahead of time, and instead must answer them as they arrive [RR10]. We generalize the technique of [BLR08, RR10] to apply to general linear queries. VC-dimension is no longer an appropriate measure of query complexity in this setting, but we show that a quantity known as Fat-Shattering dimension plays an analogous role.

Dwork et al. [DNRRV09] also gave upper and lower bounds for predicate queries, which are incomparable to the bounds of [BLR08, RR10] (and those presented in this paper). The upper bounds of [DNRRV09] are for an approximate form of differential privacy, and have a better dependence on α, but a worse dependence on k. Their lower bounds are computational, whereas the lower bounds presented in this paper are information theoretic.

Hardt and Talwar [HT10] give matching upper and lower bounds on the noise that must be added for α-differential privacy when answering $k \leq n$ linear queries of roughly $\Theta(\frac{\sqrt{k}\log(n/k)}{\alpha})$. In contrast, we prove bounds in terms of different parameters, and can handle arbitrarily (even infinitely) large values of k. For finite sets of k queries, our mechanism adds noise roughly $O\left(||D||_1^{2/3} \cdot \left(\frac{\log k \log n}{\alpha}\right)^{1/3}\right)$. Note that for some settings of the parameters, this is significantly less noise than the bounds of [HT10]: specifically, for $k \geq \Omega(||D||_1^{4/3})$. To achieve low *relative* error η (i.e. error $\epsilon = \eta ||D||_1$), our mechanism requires only that $||D||_1$ be polylogarithmic in k, rather than polynomial in k. For infinite classes of queries $|C|$, the $\log k$ in our bound can be replaced with the fat shattering dimension of the class C. We also show a lower bound in terms of the fat shattering dimension of the class C, which is the first non-trivial lower bound for infinite classes of non-boolean linear queries.

2   Preliminaries

A database is some vector $D \in \mathbb{R}_+^n$, and a query is some vector $q \in [0,1]^n$. We write that the evaluation of q on D is $q(D) = q \cdot D$. We write $||D||_1 = \sum_{i=1}^n D_i$ to denote the ℓ_1 norm of D, and note that for any query q, $q(D) \in [0, ||D||_1]$. We let C denote a (possibly infinite) class of queries. We are interested in mechanisms that are able to provide answers a_i for each $q_i \in C$ so that the maximum error, defined to be $\max_{i \in C} |q_i(D) - a_i|$ is as small as possible. Without loss of generality, we restrict our attention to mechanisms which actually output some synthetic database: mechanisms with range $\mathcal{R} = \mathbb{R}^n_+$. That is, if our mechanism outputs some synthetic database D', we take a_i to be $q_i(D')$ for each i.[1]

We formalize our notion of utility and relative utility for a randomized mechanism M:

Definition 1 (Usefulness and Relative Usefulness). *A mechanism* M : $\mathbb{R}^n_+ \to \mathbb{R}^n_+$ *is* (ϵ, δ)-*useful with respect to a class of queries* C *if with probability at least* $1 - \delta$ *(over the internal coins of the mechanism), it outputs a synthetic database* D' *such that:*

$$\sup_{q_i \in C} |q_i(D) - q_i(D')| \leq \epsilon$$

For $0 < \eta \leq 1$, M *is* (η, δ)-*relatively useful with respect to* C *for databases of size* s *if it is* $(\eta ||D||_1, \delta)$-*useful with respect to* C *for all input databases* D *with* $||D||_1 \geq s$.

That is, useful mechanisms should have low error for each query in C. We now define differential privacy:

Definition 2 (Differential Privacy [DMNS06]). *A mechanism* $M : \mathbb{R}^n_+ \to \mathbb{R}^n_+$ *is* α-*differentially private, if for any two databases* D_1, D_2 *such that* $||D_1 - D_2||_1 \leq 1$, *and for any* $S \subseteq \mathbb{R}^n_+$:

$$\Pr[M(D_1) \in S] \leq e^\alpha \Pr[M(D_2) \in S]$$

The standard notion of differential privacy need only hold for mechanisms defined over integer valued databases $D_1, D_2 \in \mathbb{N}^n$, which is a weaker condition. Our upper bounds will hold for the stronger notion of differential privacy, and our lower bounds for the weaker notion. A useful observation is that arbitrary (database independent) functions of differentially private mechanisms are also differentially private:

Fact 1. *If* $M : \mathbb{R}^n_+ \to \mathbb{R}^n_+$ *is* α-*differentially private, and if* $f : \mathbb{R}^n_+ \to \mathbb{R}^n_+$ *is a (possibly randomized) function, then* $f(M)$ *is* α-*differentially private.*

2.1   Fat Shattering Dimension

Fat-shattering-dimension is a combinatorial property describing classes of functions of the form $f : X \to [0, 1]$ for some domain X. It is a generalization of the Vapnik-Chervonenkis-dimension, which is a property only of classes of boolean valued functions of the form $f : X \to \{0, 1\}$. In this section, we generalize these concepts slightly to classes of linear queries, where we view our linear queries as linear combinations of functions $f : X \to [0, 1]$, where we let X be the set of standard basis vectors of $\mathbb{R}^n$.

Let $B = \{e_i\}^n_{i=1}$ denote the set of n standard basis vectors of $\mathbb{R}^n$ (e_i is the vector with a 1 in the i'th coordinate, and a 0 in all other coordinates). For any

---

[1] This is without loss of generality, because given a different representation for each answer a_i to error ϵ, it is possible to compute a synthetic database D' with error at most 2ϵ using the linear program of [DNRRV09].

$S \subseteq B$ of size $|S| = d$, we say that S is γ-shattered by C if there exists a vector $r \in [0,1]^d$ such that for every $b \in \{0,1\}^d$, there exists a query $q_b \in C$ such that for each $e_i \in S$:

$$q_b(e_i) \begin{cases} \geq r_i + \gamma, \text{ if } b_i = 1; \\ \leq r_i - \gamma, \text{ if } b_i = 0. \end{cases}$$

Note that since the range of each query is $[0,1]$, γ can range from 0 to $1/2$.

Definition 3 (Fat Shattering Dimension [BLW94, KS94]). *The γ-fat-shattering dimension of a class of linear queries C is:*

$$FSD_\gamma(C) = \max\{d \in \mathbb{N} : C \ \gamma - shatters \ some \ S \subseteq B \ with \ |S| = d\}$$

In the special case when $\gamma = r_i = 1/2$ for all i, note that the fat shattering dimension of a class of boolean valued functions is equal to its VC-dimension.

For finite classes C, we will let $k = |C|$ denote the cardinality of C. The following observation follows immediately from the definition of fat-shattering dimension:

Observation 1. *For finite classes C, $FSD_\gamma(C) \leq \log k$ for all $\gamma > 0$, where $k = |C|$.*

3   Lower Bound

In this section, we show that any α-differentially private mechanism that answers every linear query in some class C must add noise at least linear in the fat-shattering dimension of C at any scale. The bound that we prove in this section is in terms of the privacy parameter α and the fat shattering dimension of the class. It differs from the upper bound proved in the next section by several important parameters, which include a $\log n$ term and a term depending on the size of the database. Beimel et al. [BKN10] have shown that the $\log n$ term in the upper bound is necessary in some contexts. The database that we construct in our lower bound is of size $O(\gamma \cdot FSD_\gamma(C))$. Therefore, in order to prove a nontrivial lower bound on the *relative* error achievable by a private mechanism, it would be necessary to remove a factor of γ from our current bound. This is possible in the context of VC-dimension, and we conjecture that it should also be possible for a bound in terms of fat-shattering dimension, and is merely a limitation of our techniques as present. The problem of proving a tight lower bound encapsulating all of the relevant parameters remains an interesting open question. We now proceed with the lower bound:

Theorem 2. *For any δ bounded away from 1 by a constant, let M be a mechanism M that is (ϵ, δ) useful with with respect to some class of linear queries C. If M preserves α-differential privacy, then*

$$\epsilon \geq \Omega \left(\sup_{0 < \gamma \leq 1/2} \frac{\gamma^2 \cdot FSD_\gamma(C)}{e^\alpha} \right)$$

We begin with some preliminaries which allow us to prove some useful lemmas:

Given some class of linear queries C and any $\gamma > 0$, let $S \subseteq B$ be a collection of basis-vectors of size $\mathrm{FSD}_\gamma(C)$ that are γ-shattered by C, and let $r \in [0,1]^{\mathrm{FSD}_\gamma(C)}$ be the corresponding vector as in the definition of fat-shattering dimension. We now partition S into $1/\gamma$ pieces. For each $j \in \{1,\ldots,1/\gamma\}$, let:

$$S^j = \{e_i \in S : (j-1) \cdot \gamma < r_i \le j \cdot \gamma\}$$

Since the sets $\{S_j\}$ partition S, By the pigeon-hole principle, there exists some j^* such that $|S^{j^*}| \ge \gamma \cdot |S| = \gamma \cdot \mathrm{FSD}_\gamma(C)$. Let $d = |S^{j^*}|$.

We consider subsets $T \subset S^{j^*}$ of size $|T| = d/2$. For each such subset, we consider the database $D_T = \sum_{e_i \in T} e_i$. Let $b^T \in \{0,1\}^d$ be the vector guaranteed by the definition of fat shattering dimension such that:

$$b_i^T = \begin{cases} 1, \ e_i \in T; \\ 0, \ \text{otherwise.} \end{cases}$$

Let $q_T \in C$ be the query that corresponds to b^T as in the definition of fat shattering dimension, and let $C_{S^{j^*}} = \{q_T : T \subseteq S^{j^*}, |T| = d/2\}$.

We first show that each function q_T takes its highest value on D_T and cannot take large values on databases $D_{T'}$ for sets T' that differ significantly from T.

Lemma 1. *For all $q_T \in C_{S^{j^*}}$ and for all $T' \subseteq S^{j^*}$ with $|T'| = d/2$:*

$$q_T(D_T) - q_T(D_{T'}) \ge \frac{\gamma}{2} \cdot |T \triangle T'|$$

Proof.

$$q_T(D_T) - q_T(D_{T'}) = \sum_{e_i \in T} q_T(e_i) - \sum_{e_i \in T'} q_T(e_i)$$

$$= \left(\sum_{e_i \in T \cap T'} q_T(e_i) - q_T(e_i) \right) + \sum_{e_i \in T \setminus T'} q_T(e_i)$$
$$- \sum_{e_i \in T' \setminus T} q_T(e_i)$$

$$\ge \left(\sum_{e_i \in T \setminus T'} r_i + \gamma \right) - \left(\sum_{e_i \in T' \setminus T} r_i - \gamma \right)$$

$$\ge 2\gamma \cdot |T \setminus T'| - \left(\max_{i \in T' \setminus T} r_i - \min_{i \in T \setminus T'} r_i \right) \cdot |T \setminus T'|$$

$$\ge \gamma \cdot |T \setminus T'|$$

where the last inequality follows from the fact that $T, T' \subset S^{j^*}$ which was constructed such that:

$$\left(\max_{i \in S^{j^*}} r_i - \min_{i \in S^{j^*}} r_i \right) \le \gamma$$

holds. Observing that $|T \triangle T'| = 2|T \setminus T'|$ completes the proof.

With this lemma, we are ready to prove the main technical lemma for our lower bound:

Lemma 2. *For any δ bounded away from 1 by a constant, let M be an (ϵ, δ)-useful mechanism with respect to class C. Given as input $M(D_T)$, where D_T is an unknown private database for some $T \subseteq S^{j^*}$ with $|T| = d/2$, with constant probability $1 - \delta$, there is a procedure to reconstruct a new database D_{T^*} such that $|T \triangle T^*| \leq \frac{4\epsilon}{\gamma}$.*

Proof. Suppose that mechanism M is (ϵ, δ) useful with respect to C for some constant δ bounded away from 1. Then by definition, with constant probability, given input D_T, it outputs some database D' such that for all $q_i \in C$, $|q_i(D_T) - q_i(D')| \leq \epsilon$. For each $T' \subseteq S^{j^*}$ with $|T'| = d/2$ let:

$$v(T') = q_{T'}(D_{T'}) - q_{T'}(D')$$

and let $T^* = \text{argmin}_{T'} v(T')$. Therefore, we have:

$$v(T^*) \leq v(T) = q_T(D_T) - q_T(D') \leq \epsilon \tag{1}$$

where the last inequality follows from the usefulness of the mechanism. We also have:

$$
\begin{aligned}
v(T^*) &= q_{T^*}(D_{T^*}) - q_{T^*}(D') \\
&\geq q_{T^*}(D_{T^*}) - q_{T^*}(D_T) - \epsilon \\
&\geq \frac{\gamma}{2} \cdot |T \triangle T^*| - \epsilon
\end{aligned}
$$

where the first inequality follows from the usefulness of the mechanism, and the second inequality follows from lemma 1. Combining this with equation 1, we get:

$$|T \triangle T^*| \leq \frac{4\epsilon}{\gamma}$$

We are now ready to prove the lower bound:

Proof (Proof of Theorem). Let $T \subset S^{j^*}$ with $|T| = d/2$ be some randomly selected subset. Let $D_T = \sum_{e_i \in T} e_i$ be the corresponding database. By lemma 2, given $M(D_T, \epsilon)$, with probability $1 - \delta$ there is a procedure P to reconstruct a database D_{T^*} such that $|T \triangle T^*| \leq 4\epsilon/\gamma$. Throughout the rest of the argument, we assume that this event occurs. Let $x \in T$ be an element selected from T uniformly at random, and let $y \in S \setminus T$ be an element selected from $S \setminus T$ uniformly at random. Let $T' = T \setminus x \cup \{y\}$. Observe that:

$$\Pr[x \in P(M(D_T, \epsilon))] \geq \frac{d/2 - 2\epsilon/\gamma}{d/2} = 1 - \frac{4\epsilon}{\gamma \cdot d}$$

$$\Pr[x \in P(M(D_{T'}, \epsilon))] \leq \frac{2\epsilon/\gamma}{d/2} = \frac{4\epsilon}{\gamma \cdot d}$$

Since $||D_T - D_{T'}||_1 \leq 2$, we have by the definition of α-differential privacy and fact 1:

$$e^\alpha \geq \frac{\Pr[x \in P(M(D_T, \epsilon))]}{\Pr[x \in P(M(D_{T'}, \epsilon))]}$$

$$\geq \frac{1 - \frac{4\epsilon}{\gamma \cdot d}}{\frac{4\epsilon}{\gamma \cdot d}}$$

$$= \frac{\gamma \cdot d}{4\epsilon} - 1$$

Solving for ϵ, we find that:

$$\epsilon \geq \Omega\left(\frac{\gamma \cdot d}{e^\alpha}\right)$$

Since this holds for all choices of γ, the claim follows from the fact that $d \geq \gamma \mathrm{FSD}_\gamma(C)$.

4   Upper Bound

We now show that (ignoring the other important parameters), it is sufficient to add noise linear in the fat shattering dimension of C to simultaneously guarantee usefulness with respect to C and differential privacy. Unlike our lower bound which was not quite strong enough to state in terms of relative error, our upper bound is most naturally stated as a bound on relative error.

We make use of a theorem of Bartlett and Long [BL95] (improving a bound of Alon et al. [ABCH97]) concerning the rate of convergence of uniform Glivenko-Cantelli classes with respect to their fat-shattering dimension.

Theorem 3 ([BL95] Theorem 9). *Let C be a class of functions from some domain X into $[0, 1]$. Then for all distributions $\mathbb{P}$ over X and for all $\eta, \delta \geq 0$:*

$$\Pr\left[\sup_{f \in C}\left|\frac{1}{m}\sum_{i=1}^{m} f(x_i) - \mathbb{E}_{x \sim \mathbb{P}}[f(x)]\right| \geq \eta\right] \leq \delta$$

where $\{x_i\}_{i=1}^m$ are m independent draws from $\mathbb{P}$ and

$$m = O\left(\frac{1}{\eta^2}\left(d_{\eta/5} \ln^2 \frac{1}{\eta} + \ln \frac{1}{\delta}\right)\right)$$

where $d_{\eta/5} = FSD_{\eta/5}(C)$.

We use this theorem to prove the following useful corollary:

Corollary 1. *Let C be a class of linear functions with coefficients in $[0, 1]$ from $\mathbb{R}_+^n$ to $\mathbb{R}$. For any database $D \in \mathbb{R}_+^n$, there is a database $D' \in \mathbb{N}^n$ with*

$$||D'||_1 = O\left(\frac{d_{\eta/5}}{\eta^2} \cdot \log^2\left(\frac{1}{\eta}\right)\right)$$

such that for each $q \in C$,

$$\left| q(D) - \frac{||D||_1}{||D'||_1} q(D') \right| \leq \eta ||D||_1$$

where $d_{\eta/5} = FSD_{\eta/5}(C)$.

Proof. Let $B = \{e_i\}_{i=1}^n$ denote the set of n standard basis vectors over $\mathbb{R}^n$. Let $\mathbb{P}_D$ be the probability distribution over B that places probability $D_i/||D||_1$ on e_i. Note that for any $q \in C$:

$$\mathrm{E}_{e_i \sim \mathbb{P}_D}[q(e_i)] = \sum_{i=1}^n \frac{D_i}{||D||_1} q(e_i) = \frac{1}{||D||_1} \sum_{i=1}^n q(D_i e_i) = \frac{q(D)}{||D||_1}$$

Let $x_1, \ldots, x_m$ be $m = O\left(\frac{1}{\eta^2}\left(d_{\eta/5} \ln^2 \frac{1}{\eta} + \ln 2\right)\right)$ independent draws from $\mathbb{P}_D$, and let $D' = \sum_{i=1}^m x_i$. Then:

$$q(D') = \sum_{i=1}^n q(D'_i e_i) = \sum_{i=1}^m q(x_i)$$

By lemma 3, we have:

$$\Pr\left[\left| \frac{q(D')}{m} - \frac{q(D)}{||D||_1} \right| \geq \eta\right] = \Pr\left[\left| \frac{1}{m} \sum_{i=1}^m q(x_i) - \mathrm{E}_{e_i \sim \mathbb{P}_D}[q(e_i)] \right| \geq \eta\right]$$
$$\leq \frac{1}{2}$$

In particular, there exists some $D' \in \mathbb{N}^n$ with $||D'||_1 = m$ that satisfies $\left| \frac{q(D')}{||D'||_1} - \frac{q(D)}{||D||_1} \right| \leq \eta$. Multiplying through by $||D||_1$ gives the desired bound.

Armed with Corollary 1, we may now proceed to instantiate the exponential mechanism over a sparse domain, analogously to the instantiation of the exponential mechanism in [BLR08].

Definition 4 (The Exponential Mechanism [MT07]). *Let $\mathcal{D}$ be some domain, and let $s : \mathbb{R}_+^n \times \mathcal{D} \to \mathbb{R}$ be some* quality score *mapping database/domain-element pairs to some real value. Let*

$$\Delta_s \geq \max_{r \in \mathcal{D}} \sup_{D_1, D_2 \in \mathbb{R}_n^+ : ||D_1 - D_2||_1 \leq 1} |s(D_1, r) - s(D_2, r)|$$

be an upper bound on the ℓ_1 sensitivity of s. The exponential mechanism defined with respect to domain $\mathcal{D}$ and score s is the probability distribution (parameterized by the private database D) which outputs each $r \in \mathcal{D}$ with probability proportional to:

$$r \sim \exp\left(\frac{s(D, r) \cdot \alpha}{2\Delta_s}\right)$$

Theorem 4 (McSherry and Talwar [MT07]). *The exponential mechanism preserves α-differential privacy.*

We let $m = O\left(\frac{d_{\eta/5}}{\eta^2} \cdot \log^2\left(\frac{1}{\eta}\right)\right)$, and define the domain of our instantiation of the exponential mechanism to be:

$$\mathcal{D} = \{D' \in \mathbb{N}^n : ||D'||_1 = m\}$$

We note that $|\mathcal{D}| = n^m$. Finally, we sample each $D' \in \mathcal{D}$ with probability proportional to:

$$D' \sim \exp\left(-\frac{\sup_{q \in C}\left|q(D) - \frac{||D||_1}{||D'||_1} \cdot q(D')\right|\alpha}{4}\right) \tag{2}$$

and output the database $D_{\text{out}} \equiv \frac{||D||_1}{||D'||_1} \cdot D'^2$. Observe that for any two databases D_1, D_2 such that $||D_1 - D_2||_1 \le 1$ we have:

$$\sup_{q \in C}|q(D_1) - \frac{||D_1||_1}{||D'||_1} \cdot q(D')| - \sup_{q \in C}|q(D_2) - \frac{||D_2||_1}{||D'||_1} \cdot q(D')| \le$$

$$||D_1 - D_2||_1 + \frac{|||D_1||_1 - ||D_2||_1|}{m} \le$$

$$1 + \frac{1}{m}$$

Therefore, the distribution defined in equation 2 is a valid instantiation of the exponential mechanism, and by [MT07] preserves α-differential privacy. It remains to show that the above instantiation of the exponential mechanism yields a useful mechanism with low error. In particular, it gives us a *relatively useful* mechanisms with respect to classes C for databases that have size linear in the fat shattering dimension of C, or only logarithmic in $|C|$ for finite classes C. This is in contrast to the bounds of [HT10] that require databases to be of size polynomial in $|C|$ before giving relatively-useful mechanisms.

Theorem 5. *For any constant δ and any query class C, there is an (η, δ)-relatively useful mechanism that preserves α-differential privacy for any database of size at least:*

$$||D||_1 \ge \tilde{\Omega}\left(\frac{FSD_{2\eta/5}(C)\log n}{\alpha\eta^3}\right)$$

Proof (Proof of Theorem). Recall that the domain $\mathcal{D}$ of our instantiation of the exponential mechanism consists of all databases $D' \in \mathbb{N}^n$ with $||D'||_1 = m$ with $m = O\left(\frac{d_{\eta/5}}{\eta^2} \cdot \log^2\left(\frac{1}{\eta}\right)\right)\}$ In particular, by corollary 1, there exists a $D^* \in \mathcal{D}$ such that:

$$\left|q(D) - \frac{||D^*||_1}{||D'||_1}q(D^*)\right| \le \eta||D||_1$$

---

[2] If $||D||_1$ is not public knowledge, it can be estimated to small constant error using the Laplace mechanism [DMNS06], losing only additive constants in the approximation parameter ϵ and privacy parameter α. This does not affect our results.

By the definition of our mechanism, such a D^* is output with probability proportional to at least:

$$D^* \sim \exp(-\frac{\eta||D||_1\alpha}{4})$$

Similarly, any $D^B \in \mathcal{D}$ such that $\left| q(D) - \frac{||D^B||_1}{||D'||_1}q(D^*) \right| \geq 2\eta||D||_1$ is output with probability proportional to at most:

$$D^B \sim \exp(-\frac{\eta||D||_1\alpha}{2})$$

Let $\mathcal{D}_B$ denote the set of all such D^B. Because $|\mathcal{D}| = n^m$, we have that:

$$\frac{\Pr[D' = D^*]}{\Pr[D' \in \mathcal{D}_B]} \geq \frac{\exp(-\frac{\eta||D||_1\alpha}{4})}{n^m \cdot \exp(-\frac{\eta||D||_1\alpha}{2})} = n^{-m} \cdot \exp\left(\frac{\eta||D||_1\alpha}{2}\right)$$

Rearranging terms, we have:

$$\Pr[D' \in \mathcal{D}_B] \leq n^m \exp\left(-\frac{\eta||D||_1\alpha}{2}\right)$$

Solving, we find that this bad event occurs with probability at most δ for any database D with:

$$||D||_1 \geq \Omega\left(\frac{m\log n}{\eta\alpha} + \log\frac{1}{\delta}\right)$$
$$= \tilde{\Omega}\left(\frac{FSD_{2\eta/5}(C)\log n}{\alpha\eta^3}\right)$$

We remark that the above mechanism is the analogue of the general release mechanism of [BLR08], and answers linear queries in the *offline* setting, when all queries C are known to the mechanism in advance. This is not necessary, however. In the same way as above, corollary 1 can also be used to generalize the Median Mechanism of Roth and Roughgarden [RR10], to achieve roughly the same bounds, but in the *online* setting, in which queries arrive online, and the mechanism must privately answer queries as they arrive, without knowledge of future queries. This results in the following theorem:

Theorem 6. *There exists a mechanism such that for every sequence of adaptively chosen queries $q_1, q_2, \ldots$ arriving online, chosen from some (possibly infinite) set C (unknown to the mechanism), the mechanism is (η, δ) useful with respect to C and preserves (α, τ)-differential privacy[3], where τ is a negligible function of n, for any database D with size at least:*

$$||D||_1 \geq \tilde{\Omega}\left(\frac{FSD_{2\eta/5}(C)\log n}{\alpha\eta^3}\right)$$

---

[3] This is an approximate form of differential privacy. Specifically, a mechanism $M : \mathbb{R}_+^n \to \mathbb{R}_+^n$ is (α, τ)-differentially private, if for any two databases D_1, D_2 such that $||D_1 - D_2||_1 \leq 1$, and for any $S \subseteq \mathbb{R}_+^n$:

$$\Pr[M(D_1) \in S] \leq e^\alpha \Pr[M(D_2) \in S] + \tau$$

Remark 1. Notice that for finite classes of linear queries, we may replace the fat shattering dimension in the bounds of both theorems 5 and 6 with $\log |C|$ if we so choose.

5   Conclusion

In this paper, we have generalized the techniques used by Blum Ligett and Roth, [BLR08] and Roth and Roughgarden [RR10] from the class of predicate queries to the more general class of *linear* queries. This gives the first mechanism for answering every linear query from some class C with noise that is bounded by a parameter other than the cardinality of C; in particular, we have given the first mechanism for answering all of the linear queries in certain *infinite* classes of queries beyond predicate queries. We have shown that the relevant parameter is the Fat-Shattering dimension of the class, which is a generalization of VC-dimension to non-boolean valued queries. In particular (ignoring other parameters), it is necessary and sufficient to add noise proportional to the fat shattering dimension of C. Our results show, among other things, that the sample complexity needed to privately agnostically learn real valued functions is polynomially related to the sample complexity needed to non-privately agnostically learn real valued functions.

At a high level, the same technique can be applied for any class of queries, all of the answers to which can be summarized by some 'small' object. It is then sufficient to instantiate the exponential mechanism only over this much smaller set of objects (rather than the set of all databases) to obtain a useful mechanism. In the case of linear queries, we have shown that the answers to many queries can be summarized by integer valued databases with small ℓ_1 norm. An interesting future direction is to determine what types of nonlinear (but low sensitivity) queries have similar small summarizes from which useful mechanisms can be derived.

Acknowledgements

The author would like to thank Avrim Blum for many insightful discussions, and the anonymous reviewers for extremely detailed and helpful comments.

References

[ABCH97]   Alon, N., Ben David, S., Cesa Bianchi, N., Haussler, D.: Scale-sensitive dimensions, uniform convergence, and learnability. Journal of the ACM (JACM) 44(4), 615–631 (1997)

[BKN10]   Beimel, A., Kasiviswanathan, S., Nissim, K.: Bounds on the sample complexity for private learning and private data release. In: Micciancio, D. (ed.) Theory of Cryptography. LNCS, vol. 5978, pp. 437–454. Springer, Heidelberg (2010)

[BL95]   Bartlett, P.L., Long, P.M.: More theorems about scale-sensitive dimensions and learning. In: Proceedings of the eighth annual conference on Computational learning theory, pp. 392–401. ACM, New York (1995)

[BLR08]   Blum, A., Ligett, K., Roth, A.: A learning theory approach to non-
            interactive database privacy. In: Proceedings of the 40th annual ACM
            symposium on Theory of computing, pp. 609–618. ACM, New York
            (2008)
[BLW94]   Bartlett, P.L., Long, P.M., Williamson, R.C.: Fat-shattering and the
            learnability of real-valued functions. In: Proceedings of the seventh an-
            nual conference on Computational learning theory, pp. 299–310. ACM,
            New York (1994)
[DMNS06]   Dwork, C., McSherry, F., Nissim, K., Smith, A.: Calibrating noise to
            sensitivity in private data analysis. In: Halevi, S., Rabin, T. (eds.) TCC
            2006. LNCS, vol. 3876, pp. 265–284. Springer, Heidelberg (2006)
[DMT07]   Dwork, C., McSherry, F., Talwar, K.: The price of privacy and the limits
            of LP decoding. In: Proceedings of the thirty-ninth annual ACM Sym-
            posium on Theory of Computing, p. 94. ACM, New York (2007)
[DN03]   Dinur, I., Nissim, K.: Revealing information while preserving privacy.
            In: 22nd ACM SIGACT-SIGMOD-SIGART Symposium on Principles
            of Database Systems (PODS), pp. 202–210 (2003)
[DNRRV09]   Dwork, C., Naor, M., Reingold, O., Rothblum, G.N., Vadhan, S.: On the
            complexity of differentially private data release: efficient algorithms and
            hardness results. In: Proceedings of the 41st annual ACM symposium
            on Symposium on theory of computing, pp. 381–390. ACM, New York
            (2009)
[DY08]   Dwork, C., Yekhanin, S.: New efficient attacks on statistical disclo-
            sure control mechanisms. In: Wagner, D. (ed.) CRYPTO 2008. LNCS,
            vol. 5157, pp. 469–480. Springer, Heidelberg (2008)
[HT10]   Hardt, M., Talwar, K.: On the Geometry of Differential Privacy. In: The
            42nd ACM Symposium on the Theory of Computing, STOC 2010 (2010)
[KLNRS08]   Kasiviswanathan, S.P., Lee, H.K., Nissim, K., Raskhodnikova, S., Smith,
            A.: What Can We Learn Privately? In: IEEE 49th Annual IEEE Sym-
            posium on Foundations of Computer Science, FOCS 2008, pp. 531–540
            (2008)
[KRSU10]   Kasiviswanathan, S., Rudelson, M., Smith, A., Ullman, J.: The Price
            of Privately Releasing Contingency Tables and the Spectra of Random
            Matrices with Correlated Rows. In: The 42nd ACM Symposium on the
            Theory of Computing, STOC 2010 (2010)
[KS94]   Kearns, M.J., Schapire, R.E.: Efficient distribution-free learning of prob-
            abilistic concepts*. Journal of Computer and System Sciences 48(3),
            464–497 (1994)
[MT07]   McSherry, F., Talwar, K.: Mechanism design via differential privacy. In:
            Proceedings of the 48th Annual Symposium on Foundations of Com-
            puter Science (2007)
[NRS07]   Nissim, K., Raskhodnikova, S., Smith, A.: Smooth sensitivity and sam-
            pling in private data analysis. In: Annual ACM Symposium on Theory
            of Computing: Proceedings of the thirty-ninth annual ACM symposium
            on Theory of computing. Association for Computing Machinery, Inc.,
            New York (2007)
[RR10]   Roth, A., Roughgarden, T.: Interactive Privacy via the Median Mech-
            anism. In: The 42nd ACM Symposium on the Theory of Computing,
            STOC 2010 (2010)
[UV10]   Ullman, J., Vadhan, S.: PCPs and the Hardness of Generating Synthetic
            Data (manuscript) (2010)

Two Theorems on List Decoding[*]
(Extended Abstract)

Atri Rudra and Steve Uurtamo

Dept. of Computer Sc. & Engg., University at Buffalo, SUNY, Buffalo, NY, 14620

Abstract. We prove the following results concerning the list decoding of error-correcting codes:

1. We show that for *any* code with a relative distance of δ (over a large enough alphabet), the following result holds for *random errors*: With high probability, for a $\rho \leq \delta - \varepsilon$ fraction of random errors (for any $\varepsilon > 0$), the received word will have only the transmitted codeword in a Hamming ball of radius ρ around it. Thus, for random errors, one can correct *twice* the number of errors uniquely correctable from worst-case errors for any code. A variant of our result also gives a simple algorithm to decode Reed-Solomon codes from random errors that, to the best of our knowledge, runs faster than known algorithms for certain ranges of parameters.

2. We show that concatenated codes can achieve the list decoding capacity for *erasures*. A similar result for worst-case errors was proven by Guruswami and Rudra (SODA 08), although their result does not directly imply our result. Our results show that a subset of the random ensemble of codes considered by Guruswami and Rudra also achieve the list decoding capacity for erasures. We also show that the exponential list size bound in our result with outer random linear codes cannot be improved using the recent techniques of Guruswami, Håstad and Kopparty that achieved similar improvements for errors.

Our proofs employ simple counting and probabilistic arguments.

1 Introduction

List decoding is a relaxation of the traditional unique decoding paradigm, in which one is allowed to output a list of codewords that are close to the received word. This relaxation allows for designing list decoding algorithms that can recover from scenarios where almost all of the redundancy could have been corrupted [22,10,17,3]. In particular, one can design binary codes from which one can recover from a $1/2 - \varepsilon$ fraction of worst-case errors. This fact has lead to many surprising applications in complexity theory– see e.g. the survey by Sudan [23] and Guruswami's thesis [6, Chap. 12].

The results mentioned above mostly deal with worst-case errors, where the channel is considered to be an adversary that can corrupt any arbitrary fraction

---

[*] Research supported by NSF CAREER Award CCF-0844796.

M. Serna et al. (Eds.): APPROX and RANDOM 2010, LNCS 6302, pp. 696–709, 2010.

of symbols (with an upper bound on the maximum fraction of such errors). In this work, we deal with both random and erasure noise models, which are weaker than the worst-case errors model, and which also have interesting applications in complexity theory (though less frequently than worst-case errors).

1.1 Random Errors

It is well-known that for worst-case errors, one cannot uniquely recover the transmitted codeword if the total number of errors exceeds half the distance. (We refer the reader to Section 2 for definitions related to codes.) List decoding circumvents this by allowing the decoder to output multiple nearby codewords. In situations where the decoder has access to some side information, one can prune the output list to obtain the transmitted codeword. In fact, most of the applications of list decoding in complexity theory crucially use side information. However, a natural question to ask is what one can do in situations where there is no side information (this is not an uncommon assumption in the traditional point-to-point communication model).

In such a scenario, it makes sense to look at a weaker random noise model and try to argue that the pathological cases that prevent us from decoding a code with relative distance δ from more than $\delta/2$ fraction of errors are rarely encountered.

Before we move on, we digress a bit to establish our notion of random errors. In our somewhat non-standard model, we assume that the adversary can pick the location of the ρ fraction of error *positions* but that the errors themselves are random. For the binary case, this model coincides with worst-case errors, so in this work, we consider alphabet size $q \geq 3$. We believe that this is a nice intermediary to the worst-case noise model and the more popular models of random noise, where errors are *independent* across different symbols. Indeed, a result with high probability in our random noise model (for roughly ρ errors) immediately implies a similar result for a more benign random noise model such as the q-ary symmetric noise channel with cross-over probability ρ. For the rest of the paper, when we say random errors, we will be referring to the stronger random noise model above.

Related Work. The intuition that pathological worst-case errors are rare has been formalized for certain families of codes. For example, McEliece showed that for Reed-Solomon codes with distance δ, with high probability, for a fraction $\rho \leq \delta - \varepsilon$ of random errors, the output list size is one [16].[1] Further, for *most* codes of rate $1 - H_q(\rho) - \varepsilon$, with high probability, for a ρ fraction of random errors, the output list size is one. (This follows from Shannon's famous result on the capacity of the q-ary symmetric channel: for a proof, see e.g. [19].) It is also known that most codes of rate $1 - H_q(\rho) - \varepsilon$ have relative distance at least ρ. Further, for $q \geq 2^{\Omega(1/\varepsilon)}$, it is known that such a code cannot have distance more

---

[1] The actual result is slightly weaker: see Section 3 for more details.

than $\rho + \varepsilon$: this follows from the Singleton bound and the fact that for such an alphabet size, $1 - H_q(\rho) \geq 1 - \rho - \varepsilon$ (cf. [18, Sec 2.2.2]).

Our Results. In our first main result, we show that the phenomenon above is universal, that is, for *every* q-ary code, with $q \geq 2^{\Omega(1/\varepsilon)}$, the following property holds: if the code has relative distance δ, then for any $\rho \leq \delta - \varepsilon$ fraction of random errors, with high probability, the Hamming ball of fractional radius ρ around the received word will only have the transmitted codeword in it. We would like to point out three related points. First, our result implies that if we relax the worst-case error model to a random error model, then combinatorially one can always correct *twice* the number of errors. Second, one cannot hope to correct more than a δ fraction of random errors: it is easy to see that, for instance, for Reed-Solomon codes, *any* error pattern of relative Hamming weight $\rho > \delta$ will give rise to a list size greater than one. Finally, the proof of our result follows from a fairly straightforward counting argument.

A natural follow-up question to our result is whether the lower bound of $2^{\Omega(1/\varepsilon)}$ on q can be relaxed. We show that if q is $2^{o(1/\varepsilon)}$, then the result above is *not true*. This negative result follows from the following two observations/results. First, it is known that for *any* code with rate $1 - H_q(\rho) + \varepsilon$, the average list size, over all possible received words, is exponential. Second, it is known that Algebraic-Geometric (AG) codes over alphabets of size at least 49 can have relative distance *strictly* bigger than $1 - H_q(\rho)$ (cf. [11]). However, these two results do not immediately imply the negative result for the random error case. In particular, what we need to show is that there is at least one codeword $\mathbf{c}$ such that for most error patterns $\mathbf{e}$ of relative Hamming weight ρ, the received word $\mathbf{c} + \mathbf{e}$ has at least one codeword other than $\mathbf{c}$ within a relative Hamming distance of ρ from it. To show that this can indeed be true for AG codes, we use a generalization of an "Inverse Markov argument" from Dumer et al. [1].

A Cryptographic Application. In addition to being a natural noise model to study, list decoding in the random error model has applications in cryptography. In particular, Kiayias and Yung have proposed cryptosystems based on the hardness of decoding Reed-Solomon codes [12]. However, if for Reed-Solomon codes (of rate R), one can list decode ρ fraction of *random* errors then the cryptosystem from [12] can be broken for the corresponding parameter settings. Since Guruswami-Sudan can solve this problem for $\rho \leq 1 - \sqrt{R}$ for *worst-case* errors [10], Kiayias and Yung set the parameter $\rho > 1 - \sqrt{R}$. Beyond the $1 - \sqrt{R}$ bound, to the best of our knowledge, the only known algorithms to decode Reed-Solomon codes are the following trivial ones: (i) Go through all possible q^k codewords and output all the codewords with Hamming distance of ρ from the received word; and (ii) Go through all possible $\binom{n}{\rho n}$ error locations and output the codeword, if any, that agrees in the $(1 - \rho)n$ "non-error" locations.

It is interesting to note that each of the three algorithms mentioned above work in the stronger model of worst-case errors. However, since we only care about decoding from random errors, one might hope to design better algorithms that make use of the fact that the errors are random. In this paper, we show that (essentially) the proof of our first main result implies a related result that in turn implies a modest improvement in the running time of algorithms to decode Reed-Solomon codes from $\rho > 1 - \sqrt{R}$ fraction of random errors. The related result states the following: for any code with relative distance δ (over a large enough alphabet) with high probability, for a ρ fraction of random errors, Hamming balls of fractional radius $\delta - \varepsilon$ around the received word only have the transmitted codeword in them.[2] Note that unlike the statement of our result mentioned earlier, we are considering Hamming balls of radius *larger* than the fraction of errors. This allows us to improve the second trivial algorithm in the paragraph above so that one needs to verify fewer "error patterns." This leads to an asymptotic improvement in the running time over both of the trivial algorithms for certain setting of parameters, though the running time is still exponential and thus, too expensive to break the Kiayias-Yung cryptosystem.

Other Applications. In addition to the application in cryptography our results on random errors have found the following two applications. The authors have presented a data stream algorithm to perform tolerant testing for Reed-Solomon codes under random errors [20]. Our result in this paper on the list size being one with high probability is one crucial component of the analysis of the performance of the data stream algorithm. Further, Kulhandjian, Rudra and Pados have observed that our result implies that for the Gilbert-Elliott channel, the Reed-Solomon code with high probability has a list size of one [13], which shows the viability of list decoding as an error recovery paradigm for communication on the Gilbert-Elliott channel. (The Gilbert-Elliott channel is one of the well studied stochastic channel that models bursty errors.) We point out that the observation in [13] crucially uses the fact that the error locations can be arbitrary and thus, does *not* follow from the result of McEliece [16].

1.2   Erasures

In the second part of the paper, we consider the erasure noise model, where the decoder knows the locations of the errors. (However, the error locations are still chosen by the adversary.) Intuitively, this noise model is weaker than the general worst-case noise model as the decoder knows for sure which locations are uncorrupted. This intuition can also be formalized. E.g., it is known that for a ρ fraction of worst-case errors, the list decoding capacity is $1 - H_q(\rho)$, whereas for a ρ fraction of erasures, the list decoding capacity is $1 - \rho$ (cf. [6, Chapter 10]). Note that the capacity for erasures is *independent* of the alphabet size. As another example, for a linear code, a combinatorial guarantee on list decodability from erasures gives a polynomial time list decoding algorithm. By contrast, such a result is not known for worst-case errors.

As is often the case, the capacity result is proven by random coding arguments. A natural quest then is to design explicit linear codes that achieve the list decoding capacity for erasures, and is an important milestone in the program of

---

[2] A similar result was shown for Reed-Solomon codes by McEliece [16].

designing explicit codes that achieve list decoding capacity for worst-case errors. This goal is the primary motivation for our second main result.

Our Result and Related Work. For large enough alphabets, explicit linear codes that achieve list decoding capacity for erasures are not hard to find: e.g., Reed-Solomon codes achieve the capacity. For smaller alphabets, the situation is much different. For binary codes, Guruswami presented explicit linear codes that can handle $\rho = 1 - \varepsilon$ fraction of erasures with rate $\Omega \left(\frac{\varepsilon^2}{\log(1/\varepsilon)} \right)$ [5]. For alphabets of size 2^t, a $1 - \varepsilon$ fraction of erasures can be list decoded with explicit linear codes of rate $\Omega \left(\frac{\varepsilon^{1+1/t}}{t^2 \log(1/\varepsilon)} \right)$ [6, Chapter 10]. Thus, especially for binary codes, an explicit code with a capacity of $1 - \rho$ is still a lofty goal. (In fact, breaking the ε^2 rate barrier for *polynomially* small ε would imply explicit construction of certain bipartite Ramsey graphs, solving an open question [5].)

To gain a better understanding about codes that achieve the list decoding capacity for erasures, a natural question is to ask whether *concatenated codes* can achieve the list decoding capacity for erasures. Concatenated codes are the preeminent method to construct good list decodable codes over small alphabets. In fact, the best explicit list decodable binary codes (for both erasures [5] and worst-case errors [9]) are concatenated codes. Briefly, in code concatenation, an "outer" code over a large alphabet is first used to encode the message. Then "inner" codes over the smaller alphabet are used to encode each of the symbols in the outer codeword. These inner codes typically have a much smaller block length than the outer code, which allows one to use brute-force type algorithms to search for "good" inner codes. Also note that the rate of the concatenated code is the product of the rate of the outer and inner codes.

Given that concatenated codes have such a rigid structure, it seems plausible that such codes would not be able to achieve list decoding capacity. For the worst-case error model, Guruswami and Rudra showed that there do exist concatenated codes that achieve list decoding capacity [8]. However, for erasures there is an additional potential complication that does not arise for the worst-case error case. In particular, consider erasure patterns in which a ρ fraction of the outer symbols are completely erased. It is clear by this example that the outer code needs to have rate very close to $1 - \rho$. However, note that to approach the list decoding capacity for erasures, the concatenated code needs to have a rate of $1 - \rho - \varepsilon$. This means that the inner codes need to have a rate very close to 1. By contrast, even though the result of [8] has some restrictions on the rate of the inner codes, it is not nearly as stringent as the requirement above. (The restriction in [8] seems to be an artifact of the proof, whereas for erasures, the restriction is unavoidable.) Further, this restriction on the inner rate arises just by looking at a specific class of erasure patterns. It is reasonable to wonder if when taking into account all possible erasure patterns, we can rule out the possibility of concatenated codes achieving the list decoding capacity for erasures.

In our second main result, we show that concatenated codes *can* achieve the list decoding capacity for erasures. In fact, we show that choosing the outer code to be a Folded Reed-Solomon code ([3]) and picking the inner codes to be random independent linear codes with rate 1, will with high probability, result in a linear code that achieves the list decoding capacity for erasures. We show a similar result (but with better bounds on the list size) when the outer code is also chosen to be a random linear code. Both of these ensembles were shown to achieve the list decoding capacity for errors in [8], although, as mentioned earlier, the result for errors holds for a superset of concatenated codes (as the inner codes could have rates strictly less than 1). The proof of our result is similar to the proof structure in [8]. Because we are dealing with the more benign erasure noise model, some of the calculations in our proofs are much simpler than the corresponding ones in [8].

Our result for concatenated codes where the outer code is a random linear code shows that most concatenated linear codes with rate $1 - \rho - \varepsilon$ can be list decoded from (a constant) ρ fraction of erasures with a list size of $2^{O(1/\varepsilon)}$. A similar result holds for general random linear codes [5,6]. By contrast, for general random codes for the same setting of parameters, the list size for erasures is known to be $O(1/\varepsilon)$. Until recently such an exponential gap was also true for the error case. Guruswami, Håstad and Kopparty have shown that for random linear codes of rate $1 - H(\rho) - \varepsilon$, one can correct ρ fraction of errors with a list size of $O(1/\varepsilon)$. (A similar result for general random codes has been known since the 1980s.) It is then natural to wonder if the techniques of [7] can be used to improve the list size for the erasure case or if the exponential gap is inherent for erasures. (The gap is known to be necessary when $\rho = 1 - \varepsilon$ [5]– nothing is known for constant ρ.) We provide partial evidence for the latter by showing that a random linear code of rate $1 - p - \varepsilon$ that can handle ρ fraction of errors w.h.p. needs a list size of $2^{\Omega(1/\varepsilon)}$. Our proof structure follows that of an analogous result of Rudra for errors [19]. The proof, however is a bit more complicated. The main idea in [19] was to (a) look at only a possible subset of "bad events" and (b) make all such bad events disjoint. Unfortunately, there is no natural notion of a large enough subset of bad events for the erasure case. Thus, in our proof we pick a large subset of bad events that are nearly disjoint. This fact makes our calculations slightly more complicated.

Approximating NP Witnesses. We conclude this section by pointing out that binary codes that are list decodable from erasures have an application in the problem of approximating NP-witnesses [2,14]. For any NP-language L, we have a polynomial-time decidable relation $R_L(\cdot, \cdot)$ such that $x \in L$ if and only if there exists a polynomially sized witness w such that $R_L(x, w)$ accepts. Thus, for an NP-complete language we do not expect to be able to compute the witness w in polynomial time given x. A natural notion of approximation is the following: given an ε fraction of the bits in a a correct witness w, can we verify if $x \in L$ in polynomial time? The results in [2,14] show that such an approximation is not possible unless P=NP.

Guruswami and Sudan ([4]) have shown that the reductions for witnesses of size N in [2,14] can be made to work with $\varepsilon = N^{-1/2+\gamma}$ for the Kumar and Sivakumar problem and with $\varepsilon = N^{-1/4+\gamma}$ for the Gál et al. problem (for any constant $\gamma > 0$). An explicit linear code that meets the list decoding capacity for erasures will improve the value of ε above to $N^{-1+\gamma}$ and $N^{-1/2+\gamma}$, respectively.

Organization of the Paper. We begin with some preliminaries in Section 2. We present our first main result on random codes in Section 3 and our results on erasures in Section 4. Due to lack of space, omitted proofs are deferred to the full version of the paper [21].

2   Preliminaries

For an integer $m \geq 1$, we will use $[m]$ to denote the set $\{1, \ldots, m\}$.

Basic Coding Definitions. A code C of *dimension* k and *block length* n over an alphabet Σ is a subset of Σ^n of size $|\Sigma|^k$. The *rate* of such a code equals k/n. Each n-tuple in C is called a codeword. Let $\mathbb{F}_q$ denote the field with q elements. A code C over $\mathbb{F}_q$ is called a linear code if C is a subspace of $\mathbb{F}_q^n$. In this case the dimension of the code coincides with the dimension of C as a vector space over $\mathbb{F}_q$. By abuse of notation we can also think of a linear code C as a map from an element in $\mathbb{F}_q^k$ to its corresponding codeword in $\mathbb{F}_q^n$, mapping a row vector $\mathbf{x} \in \mathbb{F}_q^k$ to a vector $\mathbf{x}\mathbf{G} \in \mathbb{F}_q^n$ via a $k \times n$ matrix $\mathbf{G}$ over $\mathbb{F}_q$ which is referred to as the generator matrix.

The Hamming distance between two vectors in $\mathbf{x}, \mathbf{y} \in \Sigma^n$, denoted by $\Delta(\mathbf{x}, \mathbf{y})$, is the number of places they differ in. The (minimum) distance of a code C is the minimum Hamming distance between any two distinct codewords from C. The relative distance is the ratio of the distance to the block length.

We will need the following notions of the weight of a vector. Given a vector $\mathbf{v} \in \{0, 1, \ldots, q-1\}^n$, its Hamming weight, which is the number of non-zero entries in the vector, is denoted by $\mathrm{WT}(\mathbf{v})$. Given a vector $\mathbf{y} = (y_1, \ldots, y_n) \in \{0, \ldots, q-1\}^n$ and a subset $S \subseteq [n]$, $\mathbf{y}_S$ will denote the subvector $(y_i)_{i \in S}$, and $\mathrm{WT}_S(\mathbf{y})$ will denote the Hamming weight of $\mathbf{y}_S$.

Code Concatenation. Concatenated codes are constructed from two different types of codes that are defined over alphabets of different sizes. If we are interested in a concatenated code over $\mathbb{F}_q$, then the *outer code* C_{out} is defined over $\mathbb{F}_Q$, where $Q = q^k$ for some positive integer k, and has block length N. The second type of codes, called the *inner codes*, and which are denoted by $C_{\mathrm{in}}^1, \ldots, C_{\mathrm{in}}^N$, are defined over $\mathbb{F}_q$ and are each of dimension k (note that the message space of C_{in}^i for all i and the alphabet of C_{out} have the same size). The concatenated code, denoted by $C = C_{\mathrm{out}} \circ (C_{\mathrm{in}}^1, \ldots, C_{\mathrm{in}}^N)$, is defined as follows: Let the rate of C_{out} be R and let the block lengths of C_{in}^i be n (for $1 \leq i \leq N$). Define $K = RN$ and $r = k/n$. The input to C is a vector $\mathbf{m} = \langle m_1, \ldots, m_K \rangle \in (\mathbb{F}_q^k)^K$. Let $C_{\mathrm{out}}(\mathbf{m}) = \langle x_1, \ldots, x_N \rangle$. The codeword in C corresponding to $\mathbf{m}$ is defined as follows

$$C(\mathbf{m}) = \langle C_{\mathrm{in}}^1(x_1), C_{\mathrm{in}}^2(x_2), \ldots, C_{\mathrm{in}}^N(x_N) \rangle.$$

The outer code C_{out} in this paper will either be a random linear code over $\mathbb{F}_Q$ or the folded Reed-Solomon code from [3]. In the case when C_{out} is random linear, we will pick C_{out} by selecting $K = RN$ vectors uniformly at random from $\mathbb{F}_Q^N$ to form the rows of the generator matrix. For every position $1 \leq i \leq N$, we will choose an inner code C_{in}^i to be a random linear code over $\mathbb{F}_q$ of block length n and rate $r = k/n$. In particular, we will work with the corresponding generator matrices $\mathbf{G}_i$, where every $\mathbf{G}_i$ is a random $k \times n$ matrix over $\mathbb{F}_q$. All the generator matrices $\mathbf{G}_i$ (as well as the generator matrix for C_{out}, when we choose a random C_{out}) are chosen independently. This fact will be used crucially in our proofs.

List Decoding. We define some terms related to list decoding.

Definition 1 (List decodable code for errors). *For* $0 < \rho < 1$ *and an integer* $L \geq 1$, *a code* $C \subseteq \Sigma^n$ *is said to be* (ρ, L)-*list decodable if for every* $\mathbf{y} \in \Sigma^n$, *the number of codewords in* C *that are within Hamming distance* ρn *from* $\mathbf{y}$ *is at most* L.

Given a vector $\mathbf{c} = (c_1, \ldots, c_n) \in \Sigma^n$ and an erased received word $\mathbf{y} = (y_1, \ldots, y_n) \in (\Sigma \cup \{?\})^n,$[3] we will use $\mathbf{c} \simeq \mathbf{y}$ to denote the fact that for every $i \in [n]$ such that $y_i \neq ?$, $c_i = y_i$. With this definition, we are ready to define the notion of list decodability for erasures. Further, for an erased received word, we will use $\text{WT}(\mathbf{y})$ to denote the number of erased positions.

Definition 2 (List decodable code for erasures). *For* $0 < \rho < 1$ *and an integer* $L \geq 1$, *a code* $C \subseteq \Sigma^n$ *is said to be* $(\rho, L)_{led}$-*list decodable if for every* $\mathbf{y} \in (\Sigma \cup \{?\})^n$ *with* $\text{WT}(\mathbf{y}) \leq \rho n$, *the number of codewords* $\mathbf{c} \in C$ *such that* $\mathbf{c} \simeq \mathbf{y}$ *is at most* L.

Reed-Solomon and Related Codes. The classical family of Reed-Solomon (RS) codes over a field $\mathbb{F}$ are defined to be the evaluations of low-degree polynomials at a sequence of distinct points of $\mathbb{F}$. Folded Reed-Solomon codes are obtained by viewing the RS code as a code over a larger alphabet $\mathbb{F}^s$ by bundling together s consecutive symbols for some folding parameter s.

3   Random Errors

In this section we consider the random noise model mentioned in the introduction: the error locations are adversarial but the errors themselves are random. Our main result is the following.

Theorem 1. *Let* $0 < \varepsilon, \delta < 1$ *be reals and let* q *and* $n \geq \Omega(1/\varepsilon)$ *be positive integers. Let* $\Sigma = \{0, 1, \ldots, q-1\}$.[4] *Let* $0 < \rho \leq \delta - \varepsilon$ *be a real. Let* C *be a code over* Σ *of block length* n *and relative distance* δ. *Let* $S \subseteq [n]$ *with* $|S| = (1 - \rho)n$. *Then the following hold:*

---

[3] ? denotes an erasure.

[4] We will assume that Σ is equipped with a monoid structure, i.e. for any $a, b \in \Sigma$, $a + b \in \Sigma$ and 0 is the identity element.

(a) If $q \geq 2^{\Omega(1/\varepsilon)}$, then for every codeword $\mathbf{c}$ and all but a $q^{-\Omega(\varepsilon n)}$ fraction of error patterns $\mathbf{e} \in \Sigma^n$ with $\mathrm{WT}(\mathbf{e}) = \rho n$ and $\mathrm{WT}_S(\mathbf{e}) = 0$, the only codeword within the Hamming ball of radius ρn around the received word $\mathbf{c} + \mathbf{e}$ is $\mathbf{c}$.

(b) Let $\gamma > 0$. If $q > \max\left(n, \left(\frac{e}{1-\delta+\varepsilon}\right)^{\lceil\frac{1}{\gamma}\rceil}\right)$, then for every codeword $\mathbf{c}$ and all but a $(q-1)^{-((1-\gamma)\varepsilon/2-(1-\delta)\gamma)n}$ fraction of error patterns $\mathbf{e} \in \Sigma^n$ with $\mathrm{WT}(\mathbf{e}) = \rho n$ and $\mathrm{WT}_S(\mathbf{e}) = 0$, the only codeword within the Hamming ball of radius $(\delta - \varepsilon)n$ around the received word $\mathbf{c} + \mathbf{e}$ is $\mathbf{c}$.

A weaker version of Theorem 1 was previously known for RS codes [16]. (Though the bounds for part (b) are better in [16].) In particular, McEliece showed Theorem 1 for RS codes but over *all* error patterns of Hamming weight ρn. In other words, Theorem 1 implies the result in [16] if we average our result over all subsets $S \subseteq [n]$ with $|S| = \rho n$.

Part (a) of Theorem 1 implies that for $e \leq (\delta - \varepsilon)n$ random errors, with high probability, the Hamming ball of radius e has one codeword in it. Note that this is *twice* the number of errors for which an analogous result can be shown for worst-case errors. Part (b) of Theorem 1 implies the following property of Reed-Solomon codes (where we pick $\varepsilon = 4R$ and $\gamma = 1/2$).

Corollary 1. *Let $k \leq n < q$ be integers such that $q > \left(\frac{n}{k}\right)^2$. Then the following property holds for Reed-Solomon codes of dimension k and block length n over $\mathbb{F}_q$. For at least $1 - q^{-\Omega(k)}$ fraction of error patterns $\mathbf{e}$ of Hamming weight at most $n - 4k$ and any codeword $\mathbf{c}$, the only codeword that agrees in at least $4k$ positions with $\mathbf{c} + \mathbf{e}$ is $\mathbf{c}$.*

We would like to point out that in Corollary 1, the radius of the Hamming ball can be *larger* than the number of errors. This can be used to slightly improve upon the best known algorithms to decode RS codes from random errors beyond the Johnson bound for super-polynomially large q. See Section 3.1 for more details.

A natural question is whether the lower bound of $q \geq 2^{\Omega(1/\varepsilon)}$ in part (a) of Theorem 1 can be improved. In Section 3.2 we show that this is not possible.

Proof of Theorem 1. Let $\mathbf{c} \in C$ be the transmitted codeword. For an $\alpha \geq 1 - \delta + \varepsilon$, we call an error pattern $\mathbf{e}$ (with $\mathrm{WT}(\mathbf{e}) = \rho n$ and $\mathrm{WT}_S(\mathbf{e}) = 0$) α-*bad* if there exits a codeword $\mathbf{c}' \neq \mathbf{c} \in C$ such that $\Delta(\mathbf{c} + \mathbf{e}, \mathbf{c}') = (1 - \alpha)n$ (and every other codeword has a larger Hamming distance from $\mathbf{c} + \mathbf{e}$). We will show that the number of α-bad error patterns (over all $\alpha \geq 1 - \delta + \varepsilon$) is an exponentially small fraction of error patterns $\mathbf{e}$ with $\mathrm{WT}(\mathbf{e}) = \rho n$ and $\mathrm{WT}_S(\mathbf{e}) = 0$, which will prove the theorem.

Fix $\alpha \geq 1 - \delta + \varepsilon$. Associate every α-bad error pattern $\mathbf{e}$ with the lexicographically first codeword $\mathbf{c}' \neq \mathbf{c} \in C$ such that $\Delta(\mathbf{c} + \mathbf{e}, \mathbf{c}') = (1 - \alpha)n$. Let $A \subseteq [n]$ be the set of positions where $\mathbf{c}'$ and $\mathbf{c} + \mathbf{e}$ agree. Further, define $S_0 = S \cap A$, $S_1 = A \cap ([n] \setminus S)$ and $\beta = |S_0|/n$. Thus, for every α-bad error pattern $\mathbf{e}$, we can associate such a pair of subsets $(S_0, S_1) \subseteq S \times ([n] \setminus S)$. Hence, to count the number of α-bad error patterns it suffices to count for each possible pair (S_0, S_1), with $|S_0| = \beta n$ and $|S_1| = (\alpha - \beta)n$ for some $\alpha - \rho \leq \beta \leq \alpha$, the number of α-bad patterns that can be associated with it. (The lower and upper bounds on β follow from the fact that $S_1 \subseteq [n] \setminus S$ and $S_0 \subseteq A$, respectively.)

Fix sets $S_0 \subseteq S$ and $S_1 \subseteq [n] \setminus S$ with $|S_0| = \beta n$ and $|S_1| = (\alpha - \beta)n$ for some $\alpha - \rho \leq \beta \leq \alpha$. To upper bound the number of α-bad error patterns that are associated with (S_0, S_1), first note that such error patterns take all the $(q-1)^{(\rho - \alpha + \beta)n}$ possible values at the positions in $[n] \setminus (S \cup S_1)$. Fix a vector $\mathbf{x}$ of length $n - |S| - |S_1|$ and consider all the α-bad error patterns $\mathbf{e}$ such that $\mathbf{e}_{[n] \setminus (S \cup S_1)} = \mathbf{x}$. Recall that each error pattern is associated with a codeword $\mathbf{c}' \neq \mathbf{c}$ such that $\mathbf{c}'$ and $\mathbf{c} + \mathbf{e}$ agree exactly in the positions $S_0 \cup S_1$. Further, such a codeword $\mathbf{c}'$ is associated with exactly one α-bad error pattern $\mathbf{e}$, where $\mathbf{e}_{[n] \setminus (S \cup S_1)} = \mathbf{x}$. (This is because fixing $\mathbf{c}'$ fixes $\mathbf{e}_{S_1}$ and $\mathbf{e}_S$ is already fixed by the definition of S.) Thus, to upper bound the number of α-bad error patterns associated with (S_0, S_1), where $\mathbf{e}_{[n] \setminus (S \cup S_1)} = \mathbf{x}$ (call this number $N_{\alpha, S_0, S_1, \mathbf{x}}$), we will upper bound the number of such codewords $\mathbf{c}'$. Note that as C has relative distance δn, once any $(1 - \delta)n + 1$ positions are fixed, there is at most one codeword that agrees with the fixed positions (if there is no such codeword then the corresponding "error pattern" does not exist). Thus, there is at most one possible $\mathbf{c}'$ once we fix (say) the "first" $(1 - \delta)n + 1 - |S_0|$ values of $\mathbf{e}_{S_1}$ (recall that $\mathbf{c}'_{S_0} = \mathbf{c}_{S_0}$). This implies that

$$N_{\alpha, S_0, S_1, \mathbf{x}} \leq (q-1)^{(1 - \delta - \beta)n + 1}.$$

Let M_α be the number of choices for (S_0, S_1), which is just the number of choices for A. As the number of choices for $\mathbf{x}$ is $(q-1)^{(\rho - \alpha + \beta)n}$, the number of α-bad error patterns is at most

$$M_\alpha \cdot (q-1)^{(\rho - \alpha + \beta)n} \cdot (q-1)^{(1 - \delta - \beta)n + 1} = M_\alpha \cdot (q-1)^{(1 - \delta - \alpha)n + 1} \cdot (q-1)^{\rho n}. \quad (1)$$

Proof of part(a). Note that the number of α-bad patterns for any $\alpha \geq 1 - \delta + \varepsilon$ is upper bounded by

$$M_\alpha \cdot (q-1)^{-\varepsilon n + 1} \cdot (q-1)^{\rho n}.$$

We trivially upper bound M_α by 2^n. Recalling that there are $(q-1)^{\rho n}$ error patterns $\mathbf{e}$ with $\mathrm{WT}(\mathbf{e}) = \rho n$ and $\mathrm{WT}_S(\mathbf{e}) = 0$ and that α can take at most n values, the fraction of α-bad patterns (over all $\alpha \geq 1 - \rho \geq 1 - \delta + \varepsilon$) is at most

$$n 2^n (q-1)^{-\varepsilon n + 1} \leq (q-1)^{\left(-\varepsilon + \frac{2}{\log(q-1)} + \frac{1}{n}\right)n} \leq (q-1)^{-\varepsilon n / 3} \leq q^{-\varepsilon n / 6},$$

where the first inequality follows from the fact that $n \leq 2^n$, the second inequality is true for $n \geq 3/\varepsilon$ and $q \geq 2^{6/\varepsilon}$ and the last inequality follows from the inequality $(q-1) \geq \sqrt{q}$ (which in turn is true for $q \geq 3$).

Proof of part (b). The proof is similar to that of part (b), except we use the bound $M_\alpha = \binom{n}{\alpha n} \leq (e/\alpha)^{\alpha n}$. ∎

3.1   An Implication of Corollary 1

To the best of our knowledge, for $e > n - \sqrt{kn}$, the only known algorithms to decode Reed-Solomon (RS) codes from e random errors are the trivial ones: (i) Go through all possible codewords and output the closest codeword– this takes $2^{O(k \log q)} \cdot n$ time and (ii) Go through all possible $\binom{n}{e}$ error locations and check that the received word outside the purported error locations is indeed a RS codeword– this takes $2^{O((n-e) \log(n/(n-e)))} \cdot O(n^2)$ time.

If $e \le n - 4k$, then by Corollary 1, we can go through all the $\binom{n}{4k}$ choices of subsets of size $4k$ and check if the received word projected down to the subset lies in the corresponding projected down RS code. This algorithm takes $2^{O(k \log(n/k))} \cdot O(n^2)$ time, which is better than trivial algorithm (ii) mentioned above when e is of size $n - \omega(k)$. Further, this algorithm is better than trivial algorithm (i) when q is super-polynomially large in n.

3.2   On the Alphabet Size in Theorem 1

It is well-known that *any* code that is (ρ, L)-list decodable that also has rate at least $1 - H_q(\rho) + \varepsilon$ needs to satisfy $L = q^{\Omega(\varepsilon n)}$ (cf. [6]). A natural way to try to show that part (a) of Theorem 1 is false for $q \le 2^{o(1/\varepsilon)}$ is to look at codes whose relative distance is strictly larger than $1 - H_q(\rho)$. Algebraic-geometric (AG) codes are a natural candidate since they can beat the Gilbert-Varshamov bound for an alphabet size of at least 49 (cf. [11]). The only catch is that the lower bound on L follows from an average case argument and we need to show that over *most* error patterns, the list size is more than one. For this we need an "Inverse Markov argument," like one in [1].

(The argument above was suggested to us by Venkat Guruswami.)

We begin with the more general statement of the "Inverse Markov argument" from [1]. (We thank Madhu Sudan for the statement and its proof.)

Lemma 1. *Let $G = (L, R, E)$ be a bipartite graph with $|L| = n_L$ and $|R| = n_R$. Let the average left degree of G be denoted by $\bar{d}_L$. Note that the average right degree is $\bar{d}_R = \frac{n_L \cdot d_L}{n_R}$. Then the following statements are true:*

(i) If we pick an edge $e = (u, v)$ uniformly at random from E, then the probability that[5] $d(v) \le \varepsilon \bar{d}_R$ is at most ε.

(ii) If G is d-left regular then consider the following process: Uniformly at random pick a vertex $u \in L$. Then uniformly at random pick a vertex $v \in R$ in u's neighborhood. Then the probability that $d(v) \le \varepsilon \frac{dn_L}{n_R}$ is at most ε.

Proof. We first note that (ii) follows from (i) as the random process in (ii) ends up picking edges uniformly at random from E.

To conclude, we prove part (i). Consider the set $R' \subseteq R$ such that $v \in R'$ satisfies $d(v) \le \varepsilon \bar{d}_R$. Note that that the maximum number of edges that have an end-point in R' is at most $\varepsilon \bar{d}_R \cdot n_R = \varepsilon |E|$. Thus, the probability that a uniformly random edge in E has an end point in R' is upper bounded by $\varepsilon |E|/|E| = \varepsilon$, as desired.

---

[5] For any vertex v, we denote its degree by $d(v)$.

The following is an easy consequence of Lemma 1 and the standard probabilistic method used to prove the lower bound for list decoding capacity.

Lemma 2. *Let $q \geq 2$ and $0 \leq \rho < 1 - 1/q$. Then the following holds for large enough n. Let $C \subseteq \{0, \ldots, q-1\}^n$ be a code with rate $1 - H_q(\rho) + \gamma$. Then there exists a codeword $\mathbf{c} \in C$ such that for at least a $1 - q^{-\Omega(\gamma n)}$ fraction of error patterns $\mathbf{e}$ of Hamming weight at most ρn, it is true that the Hamming ball of radius ρn around $\mathbf{c} + \mathbf{e}$ has at least two codewords from C in it.*

Proof. Define the bipartite graph $G_{C,\rho} = (C, \{0, \ldots, q-1\}^n, E)$ as follows. For every $\mathbf{c} \in C$, add $(\mathbf{c}, \mathbf{y}) \in E$ such that $\Delta(\mathbf{c}, \mathbf{y}) \leq \rho n$. Note that $G_{C,\rho}$ is a $\mathrm{Vol}_q(\rho n)$-left regular bipartite graph, where $\mathrm{Vol}_q(r)$ is the volume of the q-ary Hamming ball with radius r. Note that the graph has an average right degree of

$$\bar{d}_R = \frac{\mathrm{Vol}_q(\rho n) \cdot q^{(1 - H_q(\rho) + \gamma)n}}{q^n} \geq q^{\gamma n - o(n)},$$

where in the above we have used the following well known inequality (cf. [15]):

$$\mathrm{Vol}_q(\rho n) \geq q^{H_q(\rho)n - o(n)}.$$

Thus, by part (b) of Lemma 1 (with $\varepsilon = (\bar{d}_R)^{-1} \leq q^{-\gamma n + o(n)}$), we have

$$\Pr_{\substack{\mathbf{c} \in C}} \Pr_{\substack{\mathbf{e} \in \{0, \ldots, q-1\}^n \\ \mathrm{WT}(\mathbf{e}) \leq \rho n}} [\mathbf{c} + \mathbf{e} \text{ has only } \mathbf{c} \text{ within Hamming distance } \rho n] \leq q^{-\gamma n + o(n)}.$$

Thus, there must exist at least one codeword $\mathbf{c} \in C$ with the required property.

Thus, given Lemma 2, we can prove that part (a) of Theorem 1 is not true for a certain value of q if there exists a code $C \subseteq \{0, \ldots, q-1\}^n$ with relative distance δ such that it has rate at least $1 - H_q(\delta - \varepsilon) + \gamma$ for some $\gamma > 0$. Now it is known that for fixed $\alpha > 0$, $H_q(\alpha) \geq \alpha + \Omega\left(\frac{1}{\log q}\right)$ (cf. [24, Lecture 7]). Thus, we would be done if we could find a code with relative distance δ and rate at least

$$1 - \delta + \varepsilon + \gamma - O(1/\log q).$$

For $q \leq 2^{o(1/\varepsilon)}$, the bound above for small enough ε is upper bounded by $1 - \delta - \varepsilon - \frac{1}{\sqrt{q}-1}$ (assuming that $\gamma = \Theta(\varepsilon)$). It is known that AG codes over alphabets of size ≥ 49 with relative distance δ exist that achieve a rate of $1 - \delta - \frac{1}{\sqrt{q}-1}$. Thus, for $49 \leq q \leq 2^{o(1/\varepsilon)}$, AG codes over alphabets of size q are the required codes.

4   Results on Erasures

We first show that with folded Reed-Solomon codes and independently chosen small random linear inner codes, the resulting concatenated code can achieve erasure capacity in a list decoding setting.

Theorem 2. *Let q be a prime power and let $0 < R \leq 1$ be an arbitrary rational number. Let $n, K, N \geq 1$ be large enough integers such that $K = RN$. Let C_{out} be a folded Reed-Solomon code over $\mathbb{F}_{q^n}$ of block length N and rate R. Let $C_{in}^1, \ldots, C_{in}^N$ be random linear codes over $\mathbb{F}_q$, where C_{in}^i is generated by a random $n \times n$ matrix $\mathbf{G}_i$ over $\mathbb{F}_q$ and the random choices for $\mathbf{G}_1, \ldots, \mathbf{G}_N$ are all independent.[6] Then the concatenated code $C^* = C_{out} \circ (C_{in}^1, \ldots, C_{in}^N)$ is a $\left(1 - R - \varepsilon, \left(\frac{N}{\varepsilon^2}\right)^{O\left(\varepsilon^{-2} \log(1/R)\right)}\right)_{led}$-list decodable code with probability at least $1 - q^{-\Omega(nN)}$ over the choices of $\mathbf{G}_1, \ldots, \mathbf{G}_N$. Further, C^* has rate R w.h.p.*

A similar result holds when the outer code is a random linear code:

Theorem 3. *Let q be a prime power and let $0 < R \leq 1$ be an arbitrary rational. Let $n, K, N \geq 1$ be large enough integers such that $K = RN$. Let C_{out} be a random linear code over $\mathbb{F}_{q^n}$ that is generated by a random $K \times N$ matrix over $\mathbb{F}_{q^n}$. Let $C_{in}^1, \ldots, C_{in}^N$ be random linear codes over $\mathbb{F}_q$, where C_{in}^i is generated by a random $n \times n$ matrix $\mathbf{G}_i$ and the random choices for $C_{out}, \mathbf{G}_1, \ldots, \mathbf{G}_N$ are all independent. Then the concatenated code $C^* = C_{out} \circ (C_{in}^1, \ldots, C_{in}^N)$ is a $\left(1 - R - \varepsilon, 2^{O(1/\varepsilon)}\right)_{led}$-list decodable code with probability at least $1 - q^{-\Omega(nN)}$ over the choices of $C_{out}, \mathbf{G}_1, \ldots, \mathbf{G}_N$. Further, with high probability, C^* has rate R.*

Finally, we give some evidence that the list size of $2^{O(1/\varepsilon)}$ in the result above is unavoidable.

Theorem 4. *Let $\gamma > 0$ and $\sqrt{\gamma} < \rho < 1 - 2\sqrt{\gamma}$ be reals and q be a prime power. Then for $0 < \varepsilon \leq \gamma/4$ the following result holds. With high probability a random q-ary linear code of rate $1 - \rho - \varepsilon$ that is $(\rho, L)_{led}$-list decodable needs to satisfy $L \geq 2^{\Omega(\gamma^2/\varepsilon^2)}$.*

Acknowledgments. We thank Venkat Guruswami and Parikshit Gopalan for helpful discussions. Thanks again to Madhu Sudan for kindly allowing us to include Lemma 1 in this paper.

References

1. Dumer, I., Micciancio, D., Sudan, M.: Hardness of approximating the minimum distance of a linear code. IEEE Transactions on Information Theory 49(1), 22–37 (2003)
2. Gál, A., Halevi, S., Lipton, R.J., Petrank, E.: Computing from partial solutions. In: Proceedings of the 14th Annual IEEE Conference on Computation Complexity, pp. 34–45 (1999)
3. Guruswami, V., Rudra, A.: Explicit codes achieving list decoding capacity: Error-correction up to the Singleton bound. IEEE Transactions on Information Theory 54(1), 135–150 (2008)
4. Guruswami, V., Sudan, M.: List decoding algorithms for certain concatenated codes. In: Proceedings of the 32nd Annual ACM Symposium on Theory of Computing (STOC), pp. 181–190 (2000)

---

[6] We stress that we do *not* require that the $\mathbf{G}_i$'s have rank n.

5. Guruswami, V.: List decoding from erasures: bounds and code constructions. IEEE Transactions on Information Theory 49(11), 2826–2833 (2003)
6. Guruswami, V. (ed.): List Decoding of Error-Correcting Codes. LNCS, vol. 3282. Springer, Heidelberg (2004)
7. Guruswami, V., Håstad, J., Kopparty, S.: On the list-decodability of random linear codes. In: Proceedings of the 42nd ACM Symposium on Theory of Computing, STOC (to appear, 2010)
8. Guruswami, V., Rudra, A.: Concatenated codes can achieve list-decoding capacity. In: Proceedings of the Nineteenth Annual ACM-SIAM Symposium on Discrete Algorithms (SODA), pp. 258–267 (2008)
9. Guruswami, V., Rudra, A.: Better binary list decodable codes via multilevel concatenation. IEEE Transactions on Information Theory 55(1), 19–26 (2009)
10. Guruswami, V., Sudan, M.: Improved decoding of Reed-Solomon and algebraic-geometric codes. IEEE Transactions on Information Theory 45, 1757–1767 (1999)
11. Høholdt, T., van Lint, J.H., Pellikaan, R.: Algebraic Geometry Codes. In: Pless, V.S., Huffamn, W.C., Brualdi, R.A. (eds.) Handbook of Coding Theory. Elsevier, Amsterdam (1998)
12. Kiayias, A., Yung, M.: Cryptographic hardness based on the decoding of Reed-Solomon codes. IEEE Transactions on Information Theory 54(6), 2752–2769 (2008)
13. Kulhandjian, M., Rudra, A., Pados, D.: Improved soft decoding of Reed-Solomon codes on Gilbert-Elliott channels (March 2010) (manuscript)
14. Kumar, S.R., Sivakumar, D.: Proofs, codes, and polynomial-time reducibilities. In: Proceedings of the 14th Annual IEEE Conference on Computation Complexity (1999)
15. MacWilliams, F.J., Sloane, N.J.A.: The Theory of Error-Correcting Codes. Elsevier/North-Holland, Amsterdam (1981)
16. McEliece, R.J.: On the average list size for the Guruswami-Sudan decoder. In: 7th International Symposium on Communications Theory and Applications (ISCTA) (July 2003)
17. Parvaresh, F., Vardy, A.: Correcting errors beyond the Guruswami-Sudan radius in polynomial time. In: Proceedings of the 46th Annual IEEE Symposium on Foundations of Computer Science, pp. 285–294 (2005)
18. Rudra, A.: List Decoding and Property Testing of Error Correcting Codes. Ph.D. thesis, University of Washington (2007)
19. Rudra, A.: Limits to list decoding random codes. In: Ngo, H.Q. (ed.) COCOON 2009. LNCS, vol. 5609, pp. 27–36. Springer, Heidelberg (2009)
20. Rudra, A., Uurtamo, S.: Datastream algorithms for codeword testing. In: Proceedings of the 37th International Colloquium on Automata, Languages and Programming, ICALP (to appear, 2010)
21. Rudra, A., Uurtamo, S.: Two theorems in list decoding. Technical Report TR10-007, Electronic Colloquium on Computational Complexity (January 2010)
22. Sudan, M.: Decoding of Reed-Solomon codes beyond the error-correction bound. Journal of Complexity 13(1), 180–193 (1997)
23. Sudan, M.: List decoding: Algorithms and applications. SIGACT News 31, 16–27 (2000)
24. Sudan, M.: Algorithmic introduction to coding theory (2001), lecture Notes available at http://people.csail.mit.edu/madhu/FT01/

Delaying Satisfiability for Random 2SAT

Alistair Sinclair[1,*] and Dan Vilenchik[2,**]

[1] Computer Science Division, University of California, Berkeley CA 94720–1776
sinclair@cs.berkeley.edu
[2] Department of Mathematics, University of California, Los Angeles, CA 90095
vilenchik@math.ucla.edu

Abstract. Let $(C_1, C_1'), (C_2, C_2'), \ldots, (C_m, C_m')$ be a sequence of ordered pairs of 2CNF clauses chosen uniformly at random (with replacement) from the set of all $4\binom{n}{2}$ clauses on n variables. Choosing exactly one clause from each pair defines a probability distribution over 2CNF formulas. The choice at each step must be made on-line, without backtracking, but may depend on the clauses chosen previously. We show that there exists an on-line choice algorithm in the above process which results *whp* in a satisfiable 2CNF formula as long as $m/n \leq (1000/999)^{1/4}$. This contrasts with the well-known fact that a random m-clause formula constructed without the choice of two clauses at each step is unsatisfiable *whp* whenever $m/n > 1$. Thus the choice algorithm is able to *delay* satisfiability of a random 2CNF formula beyond the classical satisfiability threshold. Choice processes of this kind in random structures are known as "Achlioptas processes." This paper joins a series of previous results studying Achlioptas processes in different settings, such as delaying the appearance of a giant component or a Hamilton cycle in a random graph. In addition to the on-line setting above, we also consider an off-line version in which all m clause-pairs are presented in advance, and the algorithm chooses one clause from each pair with knowledge of all pairs. For the off-line setting, we show that the two-choice satisfiability threshold for k-SAT for any fixed k coincides with the standard satisfiability threshold for random $2k$-SAT.

1 Introduction

The random graph process, introduced by Erdős and Rényi in the 1960's, begins with an empty graph on n vertices and adds a single new edge to the graph in each round $i = 1, \ldots, m$. Each new edge is chosen uniformly at random from all unchosen edges. The resulting distribution over graphs is commonly denoted $G_{n,m}$. One of the most fundamental random graph properties to be studied is the emergence of a giant component: at what density m/n does a connected component of size $\Omega(n)$ first appear? A classical result by Erdős and Rényi [9]

* Supported in part by NSF grant CCF-0635153 and by a UC Berkeley Chancellor's Professorship.
** This work was done while the author was a postdoctoral researcher at UC Berkeley, supported by NSF grants CCF-0635153 and DMS-0528488.

M. Serna et al. (Eds.): APPROX and RANDOM 2010, LNCS 6302, pp. 710–723, 2010.
© Springer-Verlag Berlin Heidelberg 2010

asserts that for $m/n = c$, $c > 1/2$ a constant, a random graph with m edges will have a unique giant component whp[1].

Inspired by the celebrated "power of two choices" phenomenon for balls-and-bins [2] (n balls are randomly thrown into n bins, each ball inspecting two random bins and choosing the less heavily loaded of the two, resulting in a significant decrease in the maximum bin load), Dimitris Achlioptas posed the following question for the random graph process. Suppose that edges arrive in pairs, i.e., in round i the pair (e_i, e_i') appears, and one of these edges is chosen for inclusion in the graph. The decision is to be made on-line, possibly based on the history of the process, but with no backtracking. Does there exist an algorithm A that delays the appearance of the giant component? Frieze and Bohman answered this question positively [3], describing an on-line algorithm A whose greedy rule postpones the appearance of the giant component until $m/n \geq 0.53$. Spencer and Wormald [18] improved this result to $m/n \leq 0.83$. An upper bound of $m/n = 0.964$ for every on-line algorithm was proved in [4].

Quite a few subsequent papers have addressed various other facets of the above model, including speeding up the appearance of the giant component, delaying the appearance of certain fixed subgraphs, speeding up the appearance of a Hamilton cycle, and so on (see, e.g., [4,10,6,16,17]).

Another class of random structures that has been widely studied is that of random 2CNF formulas. To generate a random 2CNF formula with m clauses over n variables, choose uniformly at random m clauses out of all $4\binom{n}{2}$ possible ones. We call the resulting distribution $F_{n,m}$. Goerdt [13], and independently Chvátal and Reed [8], showed that whenever $m/n < 1$ a random $F_{n,m}$ formula is whp satisfiable, while if $m/n > 1$ it is whp unsatisfiable. In this paper we consider an Achlioptas process for random 2CNF formulas. Specifically, we answer the question whether one can delay the sat/unsat threshold if at each step two random clauses are available to choose from. As far as we are aware, this is the first time that an Achlioptas process for random formulas has been studied.

Formally, we examine the following process: at each round $i = 1, \ldots, m$, generate two random clauses independently and uniformly at random (with replacement) out of all $4\binom{n}{2}$ possible clauses; then choose one of the two to be included in the formula. The decision is to be made on-line, without backtracking, but possibly dependent on the clauses seen so far. We note that, to avoid technical complications, our distribution is slightly different from the Achlioptas process for $F_{n,m}$ because in our distribution some clauses may appear twice. However, a simple calculation shows that the number of pairs of identical clauses among the $2m = \Theta(n)$ clauses appearing in the process is $o(n)$. Disregarding steps involving such clauses we get exactly the Achlioptas process for $F_{n,m-o(n)}$. But removing $o(n)$ clauses from the formula doesn't change our result as our advantage over the threshold will be of order $\Theta(n)$.

A random $F_{n,m}$ formula can be viewed in an obvious way as a random $G_{2n,m}$ graph on the set of $2n$ literals (i.e., variables and their negations), in which a

---

[1] Throughout, we shall take the phrase whp (with high probability) to mean "with probability tending to 1 as $n \to \infty$."

clause $(\ell \vee \ell')$ is translated into an edge between ℓ and ℓ'. With this picture, one might naively think that an application of the Bohman-Frieze greedy rule [3] for delaying the giant component suffices to delay the sat/unsat threshold for 2SAT. After all, until the emergence of the giant component the connected components of $F_{n,m}$ are simple (mostly trees, plus a few unicyclic components), and hence one may think that the formula is likely to be satisfiable.

However, this intuition turns out to be false. A recent result of Kravitz [15] implies that, if one only tampers with the degrees of literals in the random formula (leaving the parities of the variables uniformly random), then once the average degree exceeds 1 the formula will be unsatisfiable *whp*. The underlying reason, of course, is that the satisfiability of a 2CNF formula is determined not by the literal graph above, but by the *implication graph*. The vertices of the implication graph are again the $2n$ literals; however, for every clause $(\ell \vee \ell')$ in the formula, two *directed* edges are added to the implication graph: $\bar{\ell} \rightarrow \ell'$ and $\bar{\ell'} \rightarrow \ell$. As is well known [1], a 2CNF formula is satisfiable iff in its implication graph there is no variable x such that x and $\bar{x}$ belong to the same strongly connected component. Thus the fact that the literal graph has a simple structure does not exclude the possibility of contradictory cycles in the implication graph. This tells us that any rule we use to determine which clause to choose at each step must take into account the parities of the variables, and thus the Achlioptas process for 2SAT will depart from the realm of pure random graph structure that has been explored in previous such results.

Before stating our result, let us mention that alongside the on-line version, an *off-line* version of the Achlioptas process has also been studied. In the off-line version (formulated for random 2SAT), m random pairs of clauses are generated; then an algorithm chooses one clause from each pair, given full information about all the pairs. For the analogous off-line version of the random graph process, Bohman and Kim [5] prove an exact threshold for avoiding the giant component, whose value is roughly $m = 0.97677n$. As we shall see shortly (Theorem 1), we are able to obtain the exact threshold for the off-line k-SAT process.

1.1 Our Contributions

Let us first state our threshold result for the off-line version. Observe that it is not *a priori* clear that such a process will have a threshold, in the sense of [11]. However, we show that it does have a threshold, and that this threshold coincides with that of random $2k$-SAT. In what follows, we denote by d_k the satisfiability threshold for random k-SAT. (This threshold exists by virtue of [11]; note that d_k may depend on n as well as on k).

Theorem 1. *Given m pairs $(C_1, C_1'), (C_2, C_2'), \ldots, (C_m, C_m')$ of random k-SAT clauses over n variables, if $m/n < d_{2k}$ there exists whp an off-line choice of one clause per pair so that the resulting formula is satisfiable. If $m/n > d_{2k}$ whp every such choice will result in an unsatisfiable formula.*

The proof uses a somewhat similar idea to that used in [4] in the context of avoiding the giant component in the random graph process, and can be found in Section 2.

We turn now to the on-line case, which is rather more challenging. Of course, the off-line threshold provides an upper bound for the on-line setting, so we immediately deduce from Theorem 1 that no on-line choice algorithm can delay satisfiability beyond $m/n = d_{2k}$. In particular, for 2SAT this upper bound is d_4, which is predicted experimentally to be about 9.25 [14]. A rigorous upper bound on d_4 is obtained by plugging $k = 4$ into the first moment bound $2^k \ln 2$, giving 11.09. Thus Theorem 1 proves that no on-line choice algorithm can delay satisfiability for 2SAT beyond $m/n = 11.09$.

What about the more interesting question of a lower bound? Is it possible to delay satisfiability beyond the threshold at all? We are able to answer this question affirmatively for the case $k = 2$, and this is the main technical contribution of our paper.

Before presenting our on-line algorithm we give a few definitions. We denote the set of variables by $x_1, x_2, \ldots, x_n$. Throughout we use ℓ to denote a literal (i.e., $\ell = x_i$ or $\ell = \bar{x}_i$), and $\bar{\ell}$ its negation.

Definition 1. *A clause $C = (\ell \vee \ell')$ is **bad** with respect to a set F of clauses if either $\bar{\ell}$ appears in F or $\bar{\ell}'$ appears in F. A clause is **good** if it is not bad.*

The procedure in Figure 1 specifies our on-line choice rule.

For each round $i = 1, \ldots, m$ do:

1. Pick two clauses C_1, C_2, with replacement, independently at random out of all $4\binom{n}{2}$ possible clauses.

2. Set $F_i = \{D_1, D_2, \ldots, D_{i-1}\}$, where D_j is the clause chosen in round j.

3. If C_1 is good with respect to F_i, choose it, otherwise choose C_2.

Fig. 1. Generating a random 2CNF instance

Our main result is formally stated in the following theorem, which says that the above choice rule succeeds in delaying the sat/unsat phase transition for random 2CNF formulas by a constant factor:

Theorem 2. *Let F be a random 2CNF formula generated by the procedure in Figure 1. If $m/n \leq (1000/999)^{1/4}$ then F is whp satisfiable.*

Experimental results predict that the right critical value of m/n when using the above algorithm is approximately 1.2. However, to keep the analysis clear we did not try to optimize the constant. (We do not claim that simply optimizing over the constants in our proof will yield the value 1.2.)

One may also try other greedy rules, similar in flavor to the one we use, and get different threshold values. The best experimental threshold we achieved with a simple rule was approximately 1.5; this is discussed in more detail in Section 7.

Another way of extending the result is the following: suppose that in each round one is allowed to choose from T clauses rather than just from two. Our analysis easily implies that, using the same rule (but now choose C_T only if $C_1, \ldots, C_{T-1}$ are all bad), the sat/unsat threshold scales as β^T for some fixed $\beta > 1$. We omit the details.

We also remark that the techniques we develop here to prove Theorem 2 may be applicable in other settings. One such setting is delaying the threshold for the pure-literal procedure in random 3SAT formulas. Broder et al [7] showed a tight threshold of 1.63 for this model. We conjecture that using our techniques it is possible to show that, given a choice of two clauses in each round, one can delay the threshold for the pure literal procedure in 3SAT beyond this point. More details are given in Section 7.

Finally, let us state a result concerning k-SAT for $k = \omega(\log n)$.

Theorem 3. *Given a choice of two clauses in the Achlioptas process for random k-SAT with $k = \omega(\log n)$, there exists an on-line algorithm that delays the satisfiability threshold by a factor of $0.99/\ln 2$.*

Note that the sat/unsat threshold for k-SAT is (for any k) at most $m/n = 2^k \ln 2$ (this follows from a simple first moment calculation). Actually, for $k = \omega(\log n)$ this upper bound is tight [12]. The proof of Theorem 3 is self-contained and short, and uses a different (even simpler) choice rule from that in Figure 1.

The remainder of the paper is organized as follows. In the next section, we give the short proof of the off-line threshold, Theorem 1. Then we turn to the proof of our main result, Theorem 2: in Section 3 we give an outline of the proof, then in Section 4 we establish some useful properties of the distribution induced by the algorithm, and finally we use these properties to prove Theorem 2 in Section 5. Section 6 gives the short proof of Theorem 3. We conclude with a brief discussion in Section 7.

2   The Off-Line Setting: Proof of Theorem 1

Consider m pairs of random k-SAT clauses $(C_1, C'_1), (C_2, C'_2), \ldots, (C_m, C'_m)$, and from each pair generate a $2k$-SAT clause by setting $D_i = C_i \vee C'_i$. Set $F^* = D_1 \wedge D_2 \wedge \cdots \wedge D_m$. Observe that the $2k$-SAT formula F^* may contain some "illegal" clauses in which some variable repeats. As we shall see, this is a technicality that is readily overcome. Hence, in what follows we allow such clauses.

The following is a general lemma that does not assume any randomness in the choice of clauses.

Lemma 1. *F^* is satisfiable iff there exists a choice of a satisfiable formula in $(C_1, C'_1), (C_2, C'_2), \ldots, (C_m, C'_m)$.*

Proof. If there exists a choice of satisfiable formula, concatenating the unchosen clause in every pair obviously lifts this to a satisfiable $2k$-SAT formula. So F^* is satisfiable. Conversely, if F^* is satisfiable, let φ be a satisfying assignment and consider the following choice rule: evaluate $D_i = C_i \vee C_i'$ under φ, and choose the clause (C_i or C_i') that contains at least one true literal under φ (breaking ties arbitrarily). Since φ satisfies F^*, every D_i contains such a choice. Clearly, φ satisfies the k-SAT formula that we chose.   □

As mentioned above, the distribution of F^* does not coincide exactly with that of $F_{2k,n,m}$, random $2k$-SAT. The reason is that, for some i, C_i and C_i' may share a variable, so D_i will be an illegal $2k$-SAT clause. However, the following lemma asserts that the satisfiability threshold for the two distributions is the same (both these distributions have a threshold by [11]).

Lemma 2. *Let F^* be distributed as above. Let d_{2k}^* be the satisfiability threshold for that distribution, and let d_{2k} be the threshold for $F_{n,m,2k}$. Then $d_{2k}^* = d_{2k}$ for any fixed k.*

Proof. First let us estimate the probability of a shared variable in a pair (C_i, C_i'). This probability is easily seen to be $O(k^2/n)$. Let T be a random variable that counts the number of such pairs. Since the regime that is relevant for us is $m = O(2^{2k}n)$, we have $E[T] = O(k^2 2^{2k}) = O(1)$ (as k is fixed). Note that T is binomially distributed, and so with constant probability $T = 0$. Hence if for example we assume that $d_{2k}^* < d_{2k}$, then for $d_{2k}^* < m/n < d_{2k}$ we get that with some constant probability, the resulting random formula F^* is a random $F_{n,m,2k}$ formula, and hence satisfiable *whp*. Thus in turn, with constant probability F^* is satisfiable above the threshold d_{2k}^*, which contradicts the definition of a threshold. The same argument shows that $d_{2k}^* > d_{2k}$ cannot occur.   □

Theorem 1 follows immediately from the above two lemmas.

3   The On-Line Setting: Proof Outline

As we have already mentioned, the satisfiability of a 2CNF formula F is determined by certain structures in its implication graph $G(F)$: namely, directed paths from x to $\bar{x}$ and from $\bar{x}$ to x. The formula is unsatisfiable iff for some x paths of both these types exist.

We may view a simple directed path p from x to $\bar{x}$ in $G(F)$ as a sequence of clauses $C_1, C_2, \ldots, C_t$, where $C_1 = (\bar{x} \vee \ell_1)$, $C_i = (\bar{\ell}_{i-1} \vee \ell_i)$ for $2 \leq i \leq t-1$, and $C_t = (\bar{\ell}_{t-1} \vee \bar{x})$. Observe that every variable in p appears twice, and except for the variable x, each variable appears both positively and negatively.

Our proof is a first moment calculation, estimating the number of pairs of simple paths ($x \rightsquigarrow \bar{x}$, $\bar{x} \rightsquigarrow x$) in the implication graph. We will show that this number is $o(1)$ for our choice of m/n, thus proving Theorem 2.

The main challenge is to estimate $\Pr[C_1, C_2, \ldots, C_t]$, the probability of occurrence of a sequence of clauses as above. Note that in the standard random 2SAT model $F_{n,m}$ this is straightforward: $\Pr[C_1, C_2, \ldots, C_t] \sim \left(m/4\binom{n}{2}\right)^t$. However, under the new distribution we will need to do much better in order to achieve a first moment of $o(1)$ for values of m with $m/n > 1$. In fact, the key point is that our choice rule for clauses "punishes" paths in the implication graph, thus enabling us to delay the sat/unsat threshold.

The key observation in the analysis is the following: given that a clause $(\ell' \vee \ell)$ has been included already, then if $(\bar{\ell} \vee \ell'')$ is to be included later, it will have to be as the *second* clause (C_2 in the procedure in Figure 1): it cannot be included as the first clause because it is "bad" by our definition. This in turn means that the first clause, C_1, must itself be bad in order to allow us to choose C_2. This fact reduces the probability of some clauses in the path (at least half of them, as we shall see), which allows us to achieve satisfiability even when $m/n > 1$.

One challenge in the analysis is the fact that the choice of a new clause depends on the history of clauses chosen so far. Thus instead of dealing with a standard "product" space[2] like $F_{n,m}$, we have to analyze a more complicated conditioned probability space.

4   Properties of the Distribution

In this section we establish some properties of the distribution over formulas induced by our choice rule. We will use these in our proof of Theorem 2 in the next section.

Definition 2. *Two clauses C and C' **threaten** each other if there exists a variable x that appears positively in one and negatively in the other.*

Proposition 1. *Consider a simple path of length t in the implication graph $G(F)$. In every ordering π of the clauses on the path, at least $(t/2) - 1$ clauses are threatened by clauses that appear before them in π.*

Proof. Fix an arbitrary ordering π of the clauses; let T be the set of clauses that are threatened by some clause before them in π, and N the other clauses of the path. Every clause, except possibly for the first and last clauses of the path, threatens exactly two other clauses, and every clause in the path is threatened by at most two clauses. Also, no clause in N can threaten another clause in N, since if so the one appearing first in π would threaten the second clause and the latter would be in T ("threatening" is a symmetric relation). To conclude, the clauses in N threaten at least $2(|N| - 2)$ clauses in T (we assume the worst case for us: both the first and last clauses are in N), each of which was counted at most twice. All in all, $|T| \geq 2(|N| - 2)/2 = |N| - 2$. Further, $|T| + |N| = t$, and therefore $|T| \geq (t/2) - 1$ as desired. □

Consider now a possible (simple) path in $G(F)$, corresponding to the clauses $C_1, \ldots, C_t$. Our next task is to derive an upper bound on the probability that

---

[2] Technically $F_{n,m}$ is not a product space, but standard methods allow it to be viewed as one.

this set of clauses is chosen by the algorithm. This probability depends on the order in which the clauses are chosen, which motivates the following definition. Here π is an arbitrary permutation of the clause labels $1, \ldots, t$.

Definition 3. *The clauses $C_1, \ldots, C_t$ are **chosen according to** π if the algorithm described in Figure 1 chooses these clauses in the order specified by the permutation π. The clauses are chosen according to π **in rounds** $k_1, \ldots, k_t$ if in addition the algorithm chooses clause C_i in round k_i. (Note that the k_i must **respect** π in the sense that $k_i < k_j$ iff $\pi(i) < \pi(j)$.)*

Let us fix a simple path in $G(F)$ and a corresponding set of clauses $C_1, \ldots, C_t$, as well as an associated permutation π and a set of rounds $k_1, \ldots, k_t$ respecting π. Let $\mathcal{A}_i^{(k)}$ denote the event that clause C_i is chosen in round k, and $\mathcal{A}_i^{(<k)}$ the event that clause C_i is chosen in some round $k' < k$. Then we have

$$\Pr[C_1, C_2, \ldots, C_t \text{ are chosen according to } \pi \text{ in rounds } k_1, \ldots, k_t] \quad (1)$$

$$= \prod_{i=1}^{t} \Pr\left[\mathcal{A}_i^{(k_i)} \ \bigg| \ \bigcap_{j:\pi(j)<\pi(i)} \mathcal{A}_j^{(<k_i)}\right].$$

To analyze the conditional probabilities appearing in (1), we partition the clauses into three categories as follows:

1. the first and last clauses of the path (as they appear in the implication graph, not in π);
2. the inner clauses which are threatened by a clause that precedes them in π;
3. the inner clauses which are not threatened by a clause that precedes them in π.

We proceed to bound the conditional probability for a clause in each category.

Proposition 2. *For any fixed round $k \leq m$,*

$$\Pr\left[\mathcal{A}_i^{(k)} \ \bigg| \ \bigcap_{j:\pi(j)<\pi(i)} \mathcal{A}_j^{(<k)}\right] \leq \frac{2}{4\binom{n}{2}}.$$

The proof is immediate: since we are conditioning only on the past, each candidate clause in round k is uniformly distributed over the set of all $4\binom{n}{2}$ clauses, and therefore $2/4\binom{n}{2}$ is an upper bound on C_i even being a candidate in round k.

In what follows, we use $|\pi|$ to denote the number of clauses ordered by π. (Thus $|\pi| = t$ for a sequence of clauses $C_1, \ldots, C_t$.)

Proposition 3. *For a fixed round $k \leq m$, $|\pi| = o(n)$, and $\alpha^* = 969/970$,*

$$\Pr\left[\mathcal{A}_i^{(k)} \ \bigg| \ \bigcap_{j:\pi(j)<\pi(i)} \mathcal{A}_j^{(<k)} \cap (C_i \text{ is threatened by some preceding clause in } \pi)\right]$$

$$\leq \frac{\alpha^*}{4\binom{n}{2}}.$$

Proof. In order for C_i to be chosen in round k, given that all clauses that precede it in π have been previously chosen and at least one of these threatens C_i, two events need to occur: (a) the first clause in round k is bad; and (b) C_i appears as the second clause in round k. Since the conditioning is only on the past, these two events are independent. The probability of the second event is trivially $1/4\binom{n}{2}$. As for the first event, with probability $1 - o(1)$ the two literals of the first candidate clause will not occur in the set of clauses we condition upon (because, by assumption, they span only $o(n)$ variables). So we may assume this is the case. We now derive a lower bound on the probability of the first clause being good. Suppose w.l.o.g. that this clause is $C = (x \vee y)$. Let us calculate the probability that either x or y appears more than four times up to round k. The expected number of appearances of each such variable up to round k is at most $2k/n \leq 2m/n$. (Here we are using the fact that the distribution induced by our choice rule is symmetric for all variables that do not appear in the clauses conditioned upon.) It is also easy to see that variable appearances are negatively correlated (conditioning on x already appearing in a chosen clause, if x is to appear again then it will be with at most the probability of the first appearance, as now one or both of its parities are "punished").

Using the Chernoff bound (which we can do due to negative correlation), the probability that a variable appears four times is smaller than 0.467, and therefore with probability at least $1 - 2 \times 0.467$, both x and y appear at most three times. Let us assume this is the case. The worst case for us is that both actually appear three times. Next observe that the configurations x, x, x and $\bar{x}, \bar{x}, \bar{x}$ are at least as likely as any other (again, by our choice rule). Therefore, with probability at least $2/2^3 = 1/4$, the appearances of x are either all negative or all positive (that is, x appears in pure form). With probability at least $1/16$ this is true for both x and y (conditioning on x's configuration will not make y's non-pure configurations more likely than its pure ones). To conclude, with probability at least $(1 - 2 \times 0.467)/16$ both x and y appear in pure form. In that case, with probability $1/4$ the clause C is good (x and y appear in C with the correct parities). Overall, then, C is good with probability at least $(1 - 2 \cdot 0.467)/(16 \cdot 4) > 1/970$. To conclude, for sufficiently large n,

$$\Pr\left[\mathcal{A}_i^{(k)} \Big| \bigcap_{j:\pi(j)<\pi(i)} \mathcal{A}_j^{(<k)} \cap (C_i \text{ is threatened by some preceding clause in } \pi)\right]$$

$$\leq \frac{969}{970} \cdot \frac{1}{4\binom{n}{2}}. \qquad \square$$

Proposition 4. *For a fixed round $k \leq m$ and $|\pi| \leq r$,*

$$\Pr\left[\mathcal{A}_i^{(k)} \Big| \bigcap_{j:\pi(j)<\pi(i)} \mathcal{A}_j^{(<k)} \cap (C_i \text{ is an inner clause not threatened}\right.$$
$$\left. \text{by any preceding clause in } \pi)\right] \leq \frac{1 + O(r/n)}{4\binom{n}{2}}.$$

Proof. Since C_i is not threatened by any clause we condition upon, it can be chosen as either the first or the second clause in its pair. Let $\mathcal{N}$ be the event "C_i

is *not* threatened by any preceding clause in π"; then (suppressing throughout the proof the conditioning on $\bigcap_{j:\pi(j)<\pi(i)} \mathcal{A}_j^{(<k)}$), we have

$$\Pr[\mathcal{A}_i^{(k)}|\mathcal{N}] = \Pr[C_i \text{ is chosen as first clause } |\mathcal{N}]$$
$$+ \Pr[C_i \text{ is chosen as second clause } |\mathcal{N}].$$

Say w.l.o.g. $C_i = (\ell_1 \vee \ell_2)$. For C_i to be chosen as the first clause in round k, it must be the case that $\bar{\ell}_1$ and $\bar{\ell}_2$ have not appeared in a chosen clause before round k (otherwise C_i is bad). Therefore,

$$\Pr[C_i \text{ chosen as first clause } |\mathcal{N}] = \frac{\Pr[\bar{\ell}_1, \bar{\ell}_2 \text{ don't appear before round } k \ |\mathcal{N}]}{4\binom{n}{2}}.$$
$$(2)$$

The denominator, $4\binom{n}{2}$, accounts for the probability of the clause $(\ell_1 \vee \ell_2)$ actually appearing. Observe that in this case ℓ_1 and ℓ_2 do not appear in any of the clauses we condition upon (otherwise C_i is threatened by a clause in that set as C_i is an inner clause).

For C_i to be chosen as the second clause in round k, it must be the case that the first clause is bad and C_i appears as the second clause. Again, these two events are independent (as the conditioning is only on the past, and at each round the two candidate clauses are chosen independently). With probability $O(r/n)$, some variable of the first clause appears in the set of clauses conditioned upon. (This is because there are r such clauses by assumption, involving at most $2r$ variables.) Assume this is not the case, and let ℓ, ℓ' be the two literals in the first candidate clause. Then the probability that the clause is bad is

$$\Pr[\bar{\ell} \text{ or } \bar{\ell}' \text{ appears before round } k \ |\mathcal{N}]$$
$$= 1 - \Pr[\bar{\ell}, \bar{\ell}' \text{ don't appear before round } k \ |\mathcal{N}].$$

Observe that

$$p^* \equiv \Pr[\bar{\ell}, \bar{\ell}' \text{ don't appear before round } k \ |\mathcal{N}]$$

is exactly the numerator in Equation (2): in both cases we ask for the probability that two literals (whose variables do not appear among the clauses conditioned upon) have not appeared before round k. By symmetry, the identity of the literals doesn't matter, and therefore the two expressions are equal. To conclude,

$$\Pr[\mathcal{A}_i^{(k)}|\mathcal{N}] \leq \frac{p^*}{4\binom{n}{2}} + \frac{O(r/n)+1-p^*}{4\binom{n}{2}} = \frac{1+O(r/n)}{4\binom{n}{2}}. \qquad \square$$

5   Proof of Theorem 2

Recall that, to prove Theorem 2, it suffices to exclude the existence of a directed path from x to $\bar{x}$ and from $\bar{x}$ to x (for any x) in the implication graph $G(F)$ of the formula F constructed by our algorithm. We follow the same approach as that in [8] for proving the threshold for random 2SAT; the main challenge in our case is to use the "power of two choices" to get a tighter bound on the appearance of a fixed path, using the bounds we derived in the previous section.

We branch into two cases, depending on the length of the path. We will start with the case where one of the paths $x \rightsquigarrow \bar{x}$ or $\bar{x} \rightsquigarrow x$ is of length, say, at least $\log^2 n$. Take a (simple) prefix of length $t = \log^2 n$ of some such path, consisting of clauses $C_1, C_2, \ldots, C_t$. Let us bound the probability of such a prefix occurring:

$$\Pr[\text{prefix of length } t] \tag{3}$$

$$\leq \binom{n}{t+1} \cdot 2^{t+1} \cdot (t+1)! \cdot \sum_\pi \Pr[C_1, C_2, \ldots, C_t \text{ are chosen according to } \pi].$$

The first factor is the number of ways to choose the $t+1$ variables, the second counts their parities, and the last factor accounts for the probability the clauses are actually chosen, summing over all possible orderings in which they are chosen.

We may bound this final summation over π by $\binom{m}{t}$ (for the number of ways of choosing the rounds) times an upper bound on

$$\Pr[C_1, C_2, \ldots, C_t \text{ are chosen according to } \pi \text{ in rounds } k_1, \ldots, k_t]$$

over all choices of $k_1, \ldots, k_t$. This we obtain via Equation (1), using the bounds on the conditional probabilities derived in Propositions 2, 3 and 4. Note first that, by Proposition 1, at least $(t/2) - 1$ clauses are threatened in every ordering π. To bound the conditional probabilities for these clauses we use Proposition 3. Otherwise, except for the first and last clauses of the prefix, every clause is an inner clause whose probability we bound using Proposition 4. For the first and last clauses we use Proposition 2. Putting all this together yields

$$\sum_\pi \Pr[C_1, C_2, \ldots, C_t \text{ are chosen according to } \pi]$$

$$\leq t! \binom{m}{t} \left(\frac{\alpha^*}{4\binom{n}{2}} \right)^{(t/2)-1} \cdot \left(\frac{1 + O(t/n)}{4\binom{n}{2}} \right)^{(t/2)-1} \cdot \left(\frac{2}{4\binom{n}{2}} \right)^2.$$

Plugging this into Equation (3) and simplifying yields

$$\Pr[\text{prefix of length } t]$$

$$\leq m^t n^{t+1} \cdot 2^t \cdot \left(\frac{1}{4\binom{n}{2}} \right)^t \cdot (\alpha^*)^{(t/2)-1} \cdot O(1) \leq O(n) \cdot \left(\frac{m}{n} \right)^t \cdot (\alpha^*)^{t/2-1}.$$

By our choice of $m/n \leq (1000/999)^{1/4}$, and the fact that $\alpha^* = 969/970$, for sufficiently large t this last expression is at most $O(n)\beta^t$ for some fixed $\beta < 1$. Since $t \geq \log^2 n$, $\Pr[\text{prefix of length } t] = n^{-\Omega(\log n)}$. Finally, there are at most n ways to choose t (the length of the simple path), so the probability that any path of length greater than $\log^2 n$ exists in the implication graph is $o(1)$.

Let us now move to the case of short paths. We shall bound the probability that, for any variable x, there exists a short path from x to $\bar{x}$ and from $\bar{x}$ to x (recall that we only consider simple paths). Denote the paths from x to $\bar{x}$ and from $\bar{x}$ to x by p_1 and p_2 respectively, and let their respective lengths be t_1 and t_2. Suppose w.l.o.g. that $t_1 \geq t_2$. The path p_2 may contain clauses from p_1: say, s segments of total length r from p_1. The number of variables in p_1 is t_1, and p_2 further introduces $t_2 - 1 - r - s$ new variables. (We do the variable counting in p_2 as follows: for every clause in p_2, we count the variable that appears first along the path. However, some variables were counted already in p_1 and need to be subtracted. We subtract one for the first clause of p_2 (x was already counted), then we subtract one for every shared clause, and one for every clause after a segment ends since that variable was counted in p_1.) As for choosing the segments, there are at most t_1^{2s} ways to choose the shared segments in p_1 (starting and ending points) and at most t_2^s ways to choose their starting points in p_2. Once the segments and starting points are fixed, there are $(t_2 - 1 - r - s)!$ ways to arrange the new variables in p_2, and this completely determines p_2.

Call such a path a (t_1, t_2, s, r)-*path*. We now bound the probability of any such path occurring, for all possible choices of x. In light of the above observations, we have

$$\Pr[\ (t_1, t_2, s, r)\text{-path}\ ] \leq$$

$$\binom{n}{t_1} \cdot 2^{t_1} \cdot t_1! \cdot \binom{n}{t_2 - 1 - r - s} \cdot t_1^{2s} \cdot t_2^s \cdot (t_2 - 1 - r - s)! \cdot 2^{t_2 - r - s}$$

$$\times \sum_\pi \Pr[C_1, C_2, \dots, C_{t_1 + t_2 - r} \text{ are chosen according to } \pi].$$

We may bound the summation over π in analogous fashion to the case of long paths above, with a factor $\binom{m}{t_1 + t_2 - r}$ to choose the rounds in which the clauses along the paths are chosen. By Proposition 1, at least $(t_1/2) - 1$ clauses in any ordering of p_1 are threatened by a clause that precedes them, and this of course remains true if we order a superset of those clauses. We can bound the conditional probabilities for these clauses using Proposition 3. For the four clauses containing x and $\bar{x}$ we use Proposition 2. For the rest of the clauses we use Proposition 4 (some of these may be threatened as well, but we only want an upper bound). The calculation now proceeds similarly to the single (long) path case, giving

$$\Pr[\ (t_1, t_2, s, r)\text{-path}\ ] \leq O(1) \cdot \left(\frac{m}{n}\right)^{t_1 + t_2 - r} \cdot (\alpha^*)^{(t_1/2) - 1} \cdot \left(\frac{t_1^2 t_2}{n}\right)^s \cdot \frac{1}{n}.$$

Observe that $t_1^2 t_2 \leq \log^6 n$, so $\left(\frac{t_1^2 t_2}{n}\right)^s = O(1)$. Summing over the at most $\log^4 n$ ways to choose s and r gives

$$O\left(\frac{\log^4 n}{n}\right) \cdot \left(\frac{m}{n}\right)^{t_1 + t_2} \cdot (\alpha^*)^{(t_1/2) - 1}.$$

Summing again over t_1, t_2 such that $t_2 \leq t_1 \leq \log^2 n$, we can bound the probability of a cycle consisting of short paths through any variable and its negation by

$$O\left(\frac{\log^6 n}{n}\right) \cdot \sum_{t_1 \le \log^2 n} \left(\frac{m}{n}\right)^{2t_1} \cdot (\alpha^*)^{(t_1/2)-1}.$$

Now since $\left(\frac{m}{n}\right)^4 \alpha^* \le \left(\frac{1000}{999}\right)\left(\frac{969}{970}\right) < 1$, this final summation is a decreasing geometric series and hence bounded by a constant. Hence the probability of any such cycle is $O\left(\frac{\log^6 n}{n}\right) = o(1)$. This completes the proof of the theorem. □

6   Proof of Theorem 3

We employ the following greedy rule: Fix some assignment ψ to the variables, say the all-TRUE assignment. In each round k, if just one of the two clauses is satisfied by ψ, choose that clause; otherwise (if neither or both are satisfied) choose one of the clauses arbitrarily.

Let us estimate the probability that in round k both clauses are not satisfied by ψ. W.l.o.g., consider the first clause. Regardless of which variables appear in it, only one of the 2^k possible parities of the variables results in a clause that is not satisfied by ψ. Thus the probability of an unsatisfied clause is 2^{-k}. The probability of both clauses being unsatisfied by ψ is thus 2^{-2k}.

Now suppose $m = 0.99n \cdot 2^k$. The expected number of unsatisfied clauses is

$$2^{-2k} \cdot 0.99n \cdot 2^k = 0.99n \cdot 2^{-k}.$$

When $k = \omega(\log n)$ this quantity is $o(1)$, and the result follows from Markov's inequality together with the fact that the k-SAT threshold is at most $m/n = 2^k \ln 2$. □

7   Discussion

In this paper we considered the Achlioptas process for random 2CNF instances, and provided a simple greedy algorithm that provably delays the sat/unsat threshold for random 2SAT in this model.

As we mentioned in the introduction, there are several natural variations of the greedy rule we used to prove this result. For example, one can define a clause to be bad only if both literals appear already in negated form. Intuitively, this rule "punishes" the path structure in the implication graph even more severely than our rule does, and indeed experiments predict the threshold to be $m/n \approx 1.5$ using this rule. However, this rule is a little harder to analyze than ours, and we expect it to yield similar numerical bounds using our methods. We conjecture that the techniques we developed to prove Theorem 2 can be used in other settings as well. As an example, consider the pure literal heuristic for 3CNF instances. It is known that for random 3SAT, whenever $m/n \le 1.63$ the pure literal procedure ends *whp* with all clauses satisfied [7], and this value is tight. The main idea of the proof is that if the peeling process of layers of pure literals gets stuck before the formula is satisfied, then every variable in the remaining formula appears both positively and negatively. This combinatorial structure is similar to the paths that we are "punishing" in the 2SAT case. Therefore one might expect our on-line algorithm (applied to 3CNF formulas) to move the threshold above 1.63, given a choice of two clauses at each step.

Acknowledgments. We thank Benny Sudakov for introducing us to the Achlioptas process, Alan Frieze for a pointer to reference [15], and Uri Feige for useful discussions.

References

1. Apswall, B., Plass, M., Tarjan, R.: A linear-time algorithm for testing the truth of certain quantified Boolean formulas. Inf. Proc. Letters 8, 121–123 (1979)
2. Azar, Y., Broder, A., Karlin, A., Upfal, E.: Balanced allocations. SIAM Journal on Computing 29(1), 180–200 (1999)
3. Bohman, T., Frieze, A.: Avoiding a giant component. Random Structures & Algorithms 19(1), 75–85 (2001)
4. Bohman, T., Frieze, A., Wormald, N.: Avoidance of a giant component in half the edge set of a random graph. Random Structures & Algorithms 25(4), 432–449 (2004)
5. Bohman, T., Kim, J.H.: A phase transition for avoiding a giant component. Random Structures & Algorithms 28(2), 195–214 (2006)
6. Bohman, T., Kravitz, D.: Creating a giant component. Combinatorics, Probability and Computing 15(4), 489–511 (2006)
7. Broder, A., Frieze, A., Upfal, E.: On the satisfiability and maximum satisfiability of random 3CNF formulas. In: Proc. 4th ACM-SIAM Symposium on Discrete Algorithms (SODA), pp. 322–330 (1993)
8. Chvátal, V., Reed, B.: Mick gets some (the odds are on his side). In: Proc. 33rd IEEE Symposium on Foundations of Computer Science (FOCS), pp. 620–627 (1992)
9. Erdös, P., Rényi, A.: On the evolution of random graphs. Publ. Math. Inst. Hung. Acad. Sci. 5, 17–61 (1960)
10. Flaxman, A., Gamarnik, D., Sorkin, G.: Embracing the giant component. Random Structures & Algorithms 27(3), 277–289 (2005)
11. Friedgut, E.: Sharp thresholds of graph properties and the k-SAT problem. Journal of the American Mathematical Society 12(4), 1017–1054 (1998)
12. Frieze, A., Wormald, N.: Random k-Sat: A tight threshold for moderately growing k. Combinatorica 25(3), 297–305 (2005)
13. Goerdt, A.: A threshold for unsatisfiability. Journal of Computer and System Sciences 53(3), 469–486 (1996)
14. Kirkpatrick, S., Selman, B.: Critical behavior in the satisfiability of random Boolean expressions. Science 264, 1297–1301 (1994)
15. Kravitz, D.: Random 2SAT does not depend on a giant. SIAM Journal on Discrete Mathematics 21(2), 408–422 (2007)
16. Krivelevich, M., Loh, P., Sudakov, B.: Avoiding small subgraphs in Achlioptas processes. Random Structures & Algorithms 34(1), 165–195 (2009)
17. Krivelevich, M., Lubetzky, E., Sudakov, B.: Hamiltonicity thresholds in Achlioptas processes. In: Random Structures & Algorithms (to appear)
18. Spencer, J., Wormald, N.: Birth control for giants. Combinatorica 27(5), 587–628 (2007)

Improved Rounding for Parallel Repeated Unique Games

David Steurer[*]

Princeton University, Princeton NJ 08544, USA

Abstract. We show a tight relation between the behavior of unique games under parallel repetition and their semidefinite value. Let G be a unique game with alphabet size k. Suppose the semidefinite value of G, denoted $\mathrm{sdp}(G)$, is at least $1 - \varepsilon$. Then, we show that the optimal value $\mathrm{opt}(G^\ell)$ of the ℓ-fold repetition of G is at least $1 - O(\sqrt{\ell\varepsilon \log k})$. This bound confirms a conjecture of Barak et al. (2008), who showed a lower bound that was worse by $\sqrt{\ell\varepsilon \log(1/\varepsilon)}$. A consequence of our bound is the following tight relation between the semidefinite value of G and the amortized value $\overline{\mathrm{opt}}(G) := \sup_{\ell \in \mathbb{N}} \mathrm{opt}(G^\ell)^{1/\ell}$,

$$\mathrm{sdp}(G)^{O(\log k)} \le \overline{\mathrm{opt}}(G) \le \mathrm{sdp}(G).$$

The proof closely follows the approach of Barak et al. (2008). Our technical contribution is a natural orthogonalization procedure for nonnegative functions. The procedure has the property that it preserves distances up to an absolute constant factor. In particular, our orthogonalization avoids the additive increase in distances caused by the truncation step of Barak et al. (2008).

1 Introduction

A *unique game* G with *vertex set* V and *alphabet* Σ consists of a list of constraints encoded by triples (u, v, π), where $u, v \in V$ are vertices and $\pi \in \mathbb{S}_\Sigma$ is a permutation of Σ. An *assignment* $x \in \Sigma^V$ satisfies a constraint (u, v, π) if $x_v = \pi(x_u)$. The *(optimal) value* of G, denote $\mathrm{opt}(G)$, is defined as the maximum fraction of constraints of G that can be satisfied simultaneously, that is, $\mathrm{opt}(G) := \max_{x \in \Sigma^V} \mathbf{P}_{(u,v,\pi)\sim G} \{x_v = \pi(x_u)\}$. (Here, $(u, v, \pi) \sim G$ means that (u, v, π) is a random constraint of G.)

Khot's Unique Games Conjecture [9] asserts that it is NP-hard to approximate the value of a unique game in a certain regime. (According to this conjecture, for every constant $\varepsilon > 0$, there exists $k \in \mathbb{N}$ such that given a unique game G with alphabet size at most k, it is NP-hard to distinguish the cases $\mathrm{opt}(G) \ge 1 - \varepsilon$ and $\mathrm{opt}(G) \le \varepsilon$.)

A sequence of recent works showed that this conjecture implies (often optimal) hardness results for many basic combinatorial optimization problems [9,11,10,14,12,4,1,15,13,7]. Most strikingly, Raghavendra [15] showed that the

---

[*] Supported by NSF Grants CCF-0832797, 0830673, and 0528414.

M. Serna et al. (Eds.): APPROX and RANDOM 2010, LNCS 6302, pp. 724–737, 2010.

Unique Games Conjecture, if true, implies that for every constraint satisfaction problem, it is NP-hard to achieve a strictly better approximation guarantee than the one obtained by a simple generic semidefinite programming relaxation.

Due to its potential consequences, the Unique Games Conjecture is one of the central open questions about approximation algorithms and hardness of approximation.

In contrast to the numerous implications of a positive resolution, only few consequence of a refutation of the conjecture are known (some consequence for the approximability of graph expansion in a certain regime were recently shown [16]). The most compelling question in this context is whether an algorithm refuting the Unique Games Conjecture would lead to better approximation algorithms for MAX CUT. (Or equivalently: Would the optimality of current algorithms for MAX CUT, imply the Unique Games Conjecture?)

A natural candidate for such a reduction from MAX CUT to UNIQUE GAMES (the problem of computing the value of a unique game) is *parallel repetition*. (For simplicity, we can identify MAX CUT with the problem of computing the value of a unique game with alphabet size 2.) In general, ℓ-fold parallel repetition takes a unique game G with vertex V and alphabet Σ and outputs a unique game, denoted G^ℓ, with vertex set V^ℓ and alphabet Σ^ℓ. For every ℓ-tuple $(u_1, v_1, \pi_1), \ldots, (u_\ell, v_\ell, \pi_\ell)$ of constraints in G, the game G^ℓ contains a constraint $(\underline{u}, \underline{v}, \underline{\pi})$, where $\underline{u} = (u_1, \ldots, u_\ell)$, $\underline{v} = (v_1, \ldots, v_\ell)$, and $\underline{\pi}$ is the permutation of Σ^ℓ obtained by applying π_i to the ith coordinate. By construction, $\text{opt}(G^\ell) \geq \text{opt}(G)^\ell$. However, this lower bound is not always tight. Raz [18] showed the first strong upper bound on the value of parallel-repeated games: If $\text{opt}(G) = 1 - \varepsilon$, then $\text{opt}(G^\ell) \leq (1 - \varepsilon^c)^{\Omega(\ell/s)}$ for some constant c and $s = O(\log k)$. Subsequently, Holenstein [8] simplified this proof and showed an improved parallel repetition bound with $c = 3$. Rao [17] further improved this result and showed a parallel repetition bound with $c = 2$ and $s = O(1)$. We remark that these bounds also hold for *projection games*, where π need not be a permutation but can be an arbitrary function. (In fact, the bounds of Raz and Holenstein hold in an even more general setting.)

Feige et al. [5] noted that a parallel repetition bound with $c < 2$ would imply a reduction from MAX CUT to UNIQUE GAMES that achieves the goal above (an algorithm refuting the Unique Games Conjecture would lead to a better approximation algorithm for MAX CUT). Raz [19] ruled out this possibility and showed that for a simple family of unique games (odd-cycle games) Rao's parallel repetition bound is optimal.

A related, more general question is whether parallel repeated games could be hard instances for UNIQUE GAMES. Concretely, we could consider the following strengthening of the Unique Games Conjecture: For every constant $\varepsilon > 0$, there exists $\ell \in \mathbb{N}$ such that given a unique game G with alphabet size 2 it is NP-hard to distinguish the cases $\text{opt}(G^\ell) \geq 1 - \varepsilon$ and $\text{opt}(G^\ell) \leq \varepsilon$. We remark that the analogous conjecture for projection games is known to be true (a consequence of the PCP Theorem and Raz's parallel repetition bound). Extending the techniques of Raz's analysis of odd-cycle games, Barak et al. [2] showed that this variant of the Unique Games Conjecture is false (unless P = NP). The authors give a polynomial time algorithm that given a unique game G with alphabet size 2 and $\mathrm{opt}(G^\ell) \geq 1 - \varepsilon$, computes an assignment for G^ℓ of value $1 - O(\sqrt{\varepsilon})$. (In particular, for every small enough $\varepsilon > 0$, this algorithm can distinguish between $\mathrm{opt}(G^\ell) \geq 1 - \varepsilon$ and $\mathrm{opt}(G^\ell) \leq \varepsilon$.)

The algorithm of Barak et al. is based on a semidefinite programming relaxation for $\mathrm{opt}(G)$. They show that the behavior of $\mathrm{opt}(G^\ell)$ is closely characterized by the optimal value of this semidefinite relaxation, denoted $\mathrm{sdp}(G)$. For the case that G has alphabet size 2, the bound of Barak et al. is tight up to constant factors. For the case of larger alphabets, the authors show bounds that are tight up to logarithmic factors. In this work, we improve the analysis of Barak et al. and show a relation between $\mathrm{sdp}(G)$ and the behavior of $\mathrm{opt}(G^\ell)$, which is tight up to constant factors even for larger alphabets.

1.1   Results

Let G be a unique game with alphabet size k. We are interested in the behavior of the value of G under parallel repetition. Barak et al. [2] show that if the semidefinite value of G is at least $1 - \varepsilon$, then the optimal value of G^ℓ is at least $1 - O(\sqrt{s\ell\varepsilon})$, where $s = \log k + \log(1/\varepsilon)$. It was conjectured that the $\log(1/\varepsilon)$ term in this bound is not necessary. (The $\log k$ term is known to be necessary.) We confirm this conjecture and show the following lower bound on $\mathrm{opt}(G^\ell)$ in terms of $\mathrm{sdp}(G)$.

Theorem 1. *For every unique game G with alphabet size k and $\mathrm{sdp}(G) \geq 1 - \varepsilon$,*

$$\mathrm{opt}(G^\ell) \geq 1 - O(\sqrt{\ell\varepsilon \log k})\,.$$

As a consequence of this theorem and results of Feige and Lovász [6] and Charikar, Makarychev, and Makarychev [3], we obtain the following tight relation between the *amortized value* $\overline{\mathrm{opt}}(G) := \sup_{\ell \in \mathbb{N}} \mathrm{opt}(G^\ell)^{1/\ell}$ and the semidefinite value of G. (See Section A.1 for a proof of this theorem.)

Theorem 2. *For every unique game G with alphabet size k,*

$$\mathrm{sdp}(G)^{O(\log k)} \leq \overline{\mathrm{opt}}(G) \leq \mathrm{sdp}(G)\,.$$

(We remark that for every $k \in \mathbb{N}$, there exist unique games G that achieve the above lower bound on $\overline{\mathrm{opt}}(G)$. See [2] for details.) The approach of [2] for proving lower bounds on the value of repeated games G^ℓ involves an intermediate relaxation, denoted here $\mathrm{sdp}_+(G)$, which is "sandwiched" between $\mathrm{opt}(G)$ and $\mathrm{sdp}(G)$. (This approach is also implicit in Raz's counterexample to strong parallel repetition [19].) Using Holenstein's correlated sampling technique [8], it is straightforward to derive lower bounds on $\mathrm{opt}(G^\ell)$ in terms of $\mathrm{sdp}_+(G)$ (see Section 2.2 for the definition of $\mathrm{sdp}_+(G)$ and [2] for more discussion about the correlated sampling technique).

The key result in [2] is a lower bound on $\mathrm{sdp}_+(G)$ in terms of the semidefinite value of G. We prove the following improved bound (which is optimal up to the constants hidden in the $O(\cdot)$-notation).

Theorem 3. *For every unique G with alphabet size k and $\mathrm{sdp}(G) \geq 1 - \varepsilon$,*

$$\mathrm{sdp}_+(G) \geq 1 - O(\varepsilon \log k).$$

Assuming this theorem and some properties of the intermediate relaxation $\mathrm{sdp}_+(G)$ (which are presented in Section 2.2), we can prove Theorem 1.

Proof (Theorem 1). Let G be a unique game with alphabet size k and $\mathrm{sdp}(G) \geq 1 - \varepsilon$. Theorem 3 shows that $\mathrm{sdp}_+(G) \geq 1 - O(\varepsilon \log k)$. The intermediate relaxation satisfies $\mathrm{sdp}_+(G^\ell) \geq \mathrm{sdp}_+(G)^\ell$ (Lemma 9). Hence, $\mathrm{sdp}_+(G^\ell) \geq 1 - O(\ell\varepsilon \log k)$. On the other hand, Theorem 8 implies that $\mathrm{opt}(G^\ell) \geq 1 - O(\sqrt{\eta})$ if $\mathrm{sdp}_+(G^\ell) \geq 1 - \eta$. Thus, we can conclude that $\mathrm{opt}(G^\ell) \geq 1 - O(\sqrt{\ell\varepsilon \log k})$.

In the next section (Section 1.2), we give a sketch of the proof of Theorem 3 and compare it to the proof of the previous bounds by [2]. We present a detailed proof of Theorem 3 in Section 3.

We record another consequence of Theorem 3, which shows that the semidefinite value of parallel repeated unique games is a good approximation of the optimal value (assuming that the alphabet size of the underlying unique game is not too large). (We omit the proof.)

Theorem 4. *Let $H = G^\ell$ be a parallel repetition of a unique game G with alphabet size k. Suppose $\mathrm{sdp}(H) \geq 1 - \varepsilon$. Then, $\mathrm{opt}(H) \geq 1 - O(\sqrt{\varepsilon \log k})$.*

We remark that [3] show the same bound with k being replaced by the alphabet size of H. In our setting, H has alphabet size k^ℓ. (The bound of [3] is for general unique games H and does not assume that H is obtained by parallel repetition.) For $\ell = 1$, the two bounds agree. For more repetitions, our bound is strictly stronger. In this sense, parallel repeated unique games are easier to approximate than general unique games.

1.2   Proof Overview and Techniques

Our proof of Theorem 3 closely follows the approach of [2]. It is illustrative to start with an outline of this approach. (This discussion assumes that the reader is somewhat familiar with unique games and the relaxation $\mathrm{sdp}(G)$. See Section 2.1 and Section 2.2 for formal definitions.) Let G be a unique game with vertex set $V = [n]$ and alphabet $\Sigma = [k]$. The unique game G is represented as a list of triples (u, v, π), where $u, v \in V$ and π is a permutation of Σ. The triple (u, v, π) encodes the constraint that the labels of u and v satisfy the relation π. In other words, an assignment $L \in \Sigma^V$ satisfies the constraint (u, v, π) if $L_v = \pi(L_u)$.

In the relaxation $\mathrm{sdp}(G)$, we consider vector-valued assignments instead of the usual assignments. More precisely, we assign orthogonal vectors $\boldsymbol{u}_1, \dots, \boldsymbol{u}_k$ to every vertex $u \in V$. The vectors are normalized such that $\sum_i \|\boldsymbol{u}_i\|^2 = 1$ for every $u \in V$. Such a vector-valued assignment satisfies a constraint (u, v, π) if $\boldsymbol{u}_i$ is close to $\boldsymbol{v}_{\pi(i)}$ for most labels $i \in \Sigma$. Formally, the violation of the constraint (u, v, π) is measured by $\sum_i \|\boldsymbol{u}_i - \boldsymbol{v}_{\pi(i)}\|^2$.

To prove Theorem 3, we need to transform an optimal solution for $\mathrm{sdp}(G)$ to a good solution for $\mathrm{sdp}_+(G)$. As the notation suggests, the relaxation $\mathrm{sdp}_+(G)$ is quite similar to $\mathrm{sdp}(G)$. Instead of assigning arbitrary orthogonal vectors to a vertex $u \in V$, we ask for orthogonal vectors with only nonnegative coordinates. Note that nonnegative vectors are orthogonal only if they are supported on disjoint sets of coordinates. It is notationally more convenient to talk about nonnegative functions instead of vectors with only nonnegative coordinates.

To summarize, the relaxation $\mathrm{sdp}_+(G)$ asks us to assign nonnegative functions $f_{u,1}, \ldots, f_{u,k}$ with disjoint supports to every vertex $u \in V$. As before, the functions are normalized such that $\sum_i \|f_{u,i}\|^2 = 1$ for every $u \in V$, and the violation of a constraint (u, v, π) is measured by $\sum_i \|f_{u,i} - f_{v,\pi(i)}\|^2$.

Let $\{\boldsymbol{u}_i\}_{u \in V, i \in \Sigma} \subseteq \mathbb{R}^d$ be an optimal solution for $\mathrm{sdp}(G)$. To avoid some technical details, let us assume that all vectors have the same length so that $\|\boldsymbol{u}_i\|^2 = 1/k$ for every $u \in V$ and $i \in \Sigma$. To construct a solution for $\mathrm{sdp}_+(G)$, Barak et al. consider the distributions $N(\bar{\boldsymbol{u}}_i, \sigma^2 I)$ on $\mathbb{R}^d$. (Here, $N(\bar{\boldsymbol{u}}_i, \sigma^2 I)$ denotes the standard d-dimensional Gaussian distribution centered at the unit vector in direction $\boldsymbol{u}_i$, with standard deviation σ in each coordinate.) A tentative solution for $\mathrm{sdp}_+(G)$ is constructed by letting $f_{u,i}$ be the square root of the density function of $N(\boldsymbol{u}_i, \sigma^2 I)$ (suitably normalized). It turns out that the new violations $\sum_i \|f_{u,i} - f_{v,\pi(i)}\|^2$ exceed the original violations $\sum_i \|\boldsymbol{u}_i - \boldsymbol{v}_{\pi(i)}\|^2$ by at most a factor $O(\sigma^{-2})$. However, the supports of the functions $f_{u,1}, \ldots, f_{u,k}$ are far from disjoint (in fact, all of them have the same support, namely $\mathbb{R}^d$). Hence, the idea is to massage the functions such that their supports become disjoint. The approach taken by Barak et al. is to restrict $f_{u,i}$ to the Voronoi cell of the vector $\boldsymbol{u}_i$ (the set of points of $\mathbb{R}^d$ that are closer to $\boldsymbol{u}_i$ than to any other vector $\boldsymbol{u}_j$). Since the vectors $\boldsymbol{u}_1, \ldots, \boldsymbol{u}_k$ are pairwise orthogonal, only a small portion of the L_2-mass of $f_{u,i}$ is outside of the Voronoi cell of $\boldsymbol{u}_i$. Concretely, if $f'_{u,i}$ is the restriction of $f_{u,i}$ to the Voronoi cell of $\boldsymbol{u}_i$, the truncated L_2-mass is bounded by $\sum_i \|f_{u,i} - f'_{u,i}\|^2 \le k e^{-\Omega(1/\sigma^2)}$. By choosing σ appropriately, we can balance the additional violation due to this truncation and the initial multiplicative increase of the violations. The reason why this approach falls short of proving Theorem 3 is that the truncation causes an additive increase in the violations, whereas for Theorem 3 we can only afford a multiplicative increase of the violation. (We remark that this additive increase is not an artifact of the analysis but a property of the construction.)

Our contribution is a construction that avoids this additive increase of the violations. We refer to this construction as smooth orthogonalization for nonnegative functions. We construct the functions $f'_{u,i}$ in the following way: As before, $f'_{u,i}$ is identically 0 outside of the Voronoi cell of $\boldsymbol{u}_i$. Inside the Voronoi cell of $\boldsymbol{u}_i$, we define $f'_{u,i}(\boldsymbol{x}) = f_{u,i}(\boldsymbol{x}) - f_{u,j}(\boldsymbol{x})$, where j is the label of the vector $\boldsymbol{u}_j$ that is second-smallest distance to $\boldsymbol{x}$. (In other words, we consider a refinement of the Voronoi partition according to second nearest neighbors.) With this construction, we can write $f'_{u,i}(\boldsymbol{x})$ as a piecewise-linear function of the values $f_{u,1}(\boldsymbol{x}), \ldots, f_{u,k}(\boldsymbol{x})$, i.e.,

$$f'_{u,i}(\boldsymbol{x}) = \max\left\{ f_{u,i}(\boldsymbol{x}) - f_{u,1}(\boldsymbol{x}), \ldots, f_{u,i}(\boldsymbol{x}) - f_{u,k}(\boldsymbol{x}) \right\}.$$

Since such piecewise-linear functions are Lipschitz, it follows that there exists a number L such that $\sum_i \|f'_{u,i} - f'_{v,\pi(i)}\|^2 \le L\sum_i \|f_{u,i} - f_{v,\pi(i)}\|^2$ for all $u, v \in V$ and permutations π of Σ. It follows that this construction causes only a multiplicative increase of the violations. The remaining problem is to bound the Lipschitz constant L. A priori, L could grow with k. In our case, the Lipschitz constant L is bounded by an absolute constant independent of k. The reason is roughly that at every point $\boldsymbol{x} \in \mathbb{R}^d$, at most four of the values $f_{u,1}(\boldsymbol{x}), \ldots, f_{u,k}(\boldsymbol{x}), f_{v,1}(\boldsymbol{x}), \ldots, f_{v,k}(\boldsymbol{x})$ contribute to the distance $\sum_i \|f'_{u,i} - f'_{v,\pi(i)}\|^2$. See Section 3.1 for details.

2   Preliminaries

2.1   Unique Games and Parallel Repetition

Let V and Σ be two (finite) sets. A *unique game* G with *vertex set* V and *alphabet* Σ is defined by a distribution over triples (u, v, π), where $u, v \in V$ are vertices and $\pi\colon \Sigma \to \Sigma$ is a permutation of the alphabet Σ. We refer to the ℓ-fold product distribution G^ℓ as the ℓ-*fold repetition* of G. Note that G^ℓ is a unique game with vertex set V^ℓ and alphabet Σ^ℓ.

The *(optimal) value of G* is defined by

$$\mathrm{opt}(G) \stackrel{\mathrm{def}}{=} \max_{L \in \Sigma^V} \mathop{\mathbf{P}}_{(u,v,\pi)\sim G}\left\{ L_v = \pi(L_u) \right\}. \tag{2.1}$$

The *amortized value of G* is defined by

$$\overline{\mathrm{opt}}(G) \stackrel{\mathrm{def}}{=} \sup_{\ell \in \mathbb{N}} \mathrm{opt}(G^\ell)^{1/\ell}. \tag{2.2}$$

Theorem 5 ([17]). *If G is a unique game with $\mathrm{opt}(G) \le 1 - \eta$, then $\overline{\mathrm{opt}}(G) \le 1 - \Omega(\eta^2)$.*

2.2   Semidefinite and Nonnegative Relaxation

The *semidefinite value* of a unique game G with vertex set V and alphabet Σ is defined by

$$\mathrm{sdp}(G) \stackrel{\mathrm{def}}{=} \max \mathop{\mathbf{E}}_{(u,v,\pi)\sim G} \sum_{i\in\Sigma} \langle \boldsymbol{u}_i, \boldsymbol{v}_{\pi(i)} \rangle, \tag{2.3}$$

where we maximize over all collections $\{\boldsymbol{u}_i\}_{u\in V, i\in\Sigma}$ of vectors that satisfy

$$\sum_{i\in\Sigma} \|\boldsymbol{u}_i\|^2 = 1 \qquad (u \in V), \tag{2.4}$$

$$\langle \boldsymbol{u}_i, \boldsymbol{u}_j \rangle = 0 \qquad (u \in V,\ i \ne j \in \Sigma). \tag{2.5}$$

The optimization problem $\mathrm{sdp}(G)$ is a relaxation of $\mathrm{opt}(G)$. Hence, $\mathrm{sdp}(G) \geq \mathrm{opt}(G)$ for every unique game G.

Theorem 6 ([3]). *Suppose G is a unique game with alphabet size k and $\mathrm{sdp}(G) \geq 1 - \varepsilon$. Then, $\mathrm{opt}(G) \geq 1 - O(\sqrt{\varepsilon \log k})$.*

Theorem 7 ([6]). *For every unique game G and number $\ell \in \mathbb{N}$, we have $\mathrm{sdp}(G^\ell) = \mathrm{sdp}(G)^\ell$.*

Theorem 7 implies that $\mathrm{sdp}(G) \geq \overline{\mathrm{opt}}(G)$, i.e., $\mathrm{sdp}(G)$ is also a relaxation for $\overline{\mathrm{opt}}(G)$. Combining this fact with Theorem 6, we get that $\overline{\mathrm{opt}}(G) \leq 1 - \Omega(\eta^2 / \log k)$ if G is a unique game with alphabet size k and $\mathrm{opt}(G) \leq 1 - \eta$. Note that Theorem 5 shows that the $\log k$ factor in this bound is not necessary.

The *Hellinger*[1] *value* of G is defined as

$$\mathrm{sdp}_+(G) \overset{\mathrm{def}}{=} \max_{(u,v,\pi) \sim G} \mathbf{E} \sum_{i \in \Sigma} \langle f_{u,i}, f_{v,\pi(i)} \rangle, \qquad (2.6)$$

where we maximize over all collections $\{f_{u,i}\}_{u \in V, i \in \Sigma}$ of nonnegative functions on Ω such that

$$\sum_{i \in \Sigma} \|f_{u,i}\|^2 = 1 \qquad (u \in V), \qquad (2.7)$$

$$\mathrm{supp}(f_{u,i}) \cap \mathrm{supp}(f_{u,j}) = \emptyset \qquad (u \in V, \ i \neq j \in \Sigma). \qquad (2.8)$$

Here, (Ω, μ) is some probability space and the norms and inner products for functions $f, g \colon \Omega \to \mathbb{R}$ are defined as $\langle f, g \rangle := \int_\Omega fg \, d\mu$ and $\|f\| := \langle f, f \rangle^{1/2}$. Without loss of generality, we could assume $\Omega = [0, 1]$ and that μ is the usual Lebesgue measure.

Notice that (2.8) is equivalent to the constraint that $f_{u,i}$ and $f_{u,j}$ are orthogonal for all $i \neq j$ and $u \in V$ (at least if Ω is finite and every atom has positive probability mass). Hence, the optimization problem $\mathrm{sdp}_+(G)$ is equivalent to $\mathrm{sdp}(G)$ except for the constraint that the value of the functions (coordinates of the vectors) are nonnegative.

Our result relies on the following theorem of [2].

Theorem 8 ([2]). *If G is a unique game with $\mathrm{sdp}_+(G) \geq 1 - \varepsilon$, then $\mathrm{opt}(G) \geq 1 - 2\sqrt{2\varepsilon}$.*

Furthermore, we can lower bound $\mathrm{sdp}_+(G^\ell)$ in terms of $\mathrm{sdp}_+(G)$ as expected. This lemma follows from the fact that we can construct a solution for $\mathrm{sdp}_+(G^\ell)$ by taking appropriate tensor products of the functions $f_{u,i}$ that form an optimal solution for $\mathrm{sdp}_+(G)$.

Lemma 9. For every unique game G, we have $\mathrm{sdp}_+(G^\ell) \geq \mathrm{sdp}_+(G)^\ell$.

---

[1] The relaxation $\mathrm{sdp}_+(G)$ has an alternative description in terms minimizing squared Hellinger distances of jointly distributed random variables subject to a certain set of constraints (called distributional strategies in [2]). See [2] for details about this alternative description. The two formulations of $\mathrm{sdp}_+(G)$ are equivalent.

2.3   Mapping Vectors to Nonnegative Functions

Let $N(0, \sigma^2)^d$ be the Gaussian measure on $\mathbb{R}^d$ with mean 0 and covariance $\sigma^2 I$ (each coordinate is independent Gaussian with mean 0 and standard deviation σ). Let $\phi_\sigma \colon \mathbb{R}^d \to \mathbb{R}_+$ be the density of the measure $N(0, \sigma^2)^d$ with respect to the usual Lebesgue measure λ^d on $\mathbb{R}^d$,

$$\phi_\sigma(\boldsymbol{x}) \overset{\text{def}}{=} \frac{1}{\left(\sigma\sqrt{2\pi}\right)^d} e^{-\|\boldsymbol{x}\|^2/2\sigma^2} \,.$$

Let $L_2(\mathbb{R}^d)$ be the (Hilbert) space of functions $f \colon \mathbb{R}^d \to \mathbb{R}$ such that $\int_{\mathbb{R}^d} f^2 \, \mathrm{d}\lambda^d$ is bounded. The inner product in $L_2(\mathbb{R}^d)$ is given $\langle f, g \rangle = \int fg \, \mathrm{d}\lambda^d$.

Let $T_{\boldsymbol{u}}$ be the translation operator on $L_2(\mathbb{R}^d)$, so that $T_{\boldsymbol{u}} f(\boldsymbol{x}) = f(\boldsymbol{x} - \boldsymbol{u})$. Barak et al. [2] consider the following mapping M_σ from $\mathbb{R}^d$ to nonnegative functions in $L_2(\mathbb{R}^d)$,

$$M_\sigma(\boldsymbol{u}) \overset{\text{def}}{=} \|\boldsymbol{u}\| \sqrt{T_{\bar{\boldsymbol{u}}} \phi_\sigma} \,. \tag{2.9}$$

Here, $\bar{\boldsymbol{u}}$ denotes the unit vector in the direction of $\boldsymbol{u}$. The mapping M_σ preserves norms, that is, $\|M_\sigma(\boldsymbol{u})\| = \|\boldsymbol{u}\|$ for every vector $\boldsymbol{u} \in \mathbb{R}^n$. We need the following additional properties of M_σ. (See Section A.2 for a proof of this lemma.)

Lemma 10 ([2]).   For any two vectors $\boldsymbol{u}, \boldsymbol{v} \in \mathbb{R}^d$,

$$\|M_\sigma(\boldsymbol{u}) - M_\sigma(\boldsymbol{v})\|^2 \le O(\sigma^{-2}) \cdot \|\boldsymbol{u} - \boldsymbol{v}\|^2 \,.$$

Furthermore, $\langle M_\sigma(\boldsymbol{u}), M_\sigma(\boldsymbol{v}) \rangle = \|\boldsymbol{u}\| \, \|\boldsymbol{v}\| \cdot e^{-1/4\sigma^2}$ if $\boldsymbol{u}$ and $\boldsymbol{v}$ are orthogonal.

3   Improved Rounding of Repeated Unique Games

In this section, we will prove the following theorem — our main result.

Theorem (Restatement of Theorem 3).   For every unique G with alphabet size k and $\mathrm{sdp}(G) \ge 1 - \varepsilon$,

$$\mathrm{sdp}_+(G) \ge 1 - O(\varepsilon \log k) \,.$$

The key ingredients of the proof of Theorem 3 are the mapping M_σ from vectors to nonnegative functions (Section 2.3) and the following smooth orthogonalization procedure for nonnegative functions (or vectors).

Lemma 11 (Smooth Nonnegative Orthogonalization).   There exists a mapping $Q \colon L_2(\mathbb{R}^d)^k \to L_2(\mathbb{R}^d)^k$ with the following properties: Let $f_1, \ldots, f_k$ and $g_1, \ldots, g_k$ be nonnegative functions in $L_2(\mathbb{R}^d)$ such that $\sum_i \|f_i\|^2 = \sum_i \|g_i\|^2 = 1$ and $\sum_{i \ne j} \langle f_i, f_j \rangle + \sum_{i \ne j} \langle g_i, g_j \rangle \le \gamma$. Suppose $(f_1', \ldots, f_k') = Q(f_1, \ldots, f_k)$ and $(g_1', \ldots, g_k') = Q(g_1, \ldots, g_k)$. Then, for every permutation π of $[k]$,

$$\sum_i \|f_i' - g_{\pi(i)}'\|^2 \le \tfrac{32}{1-4\gamma} \sum_i \|f_i - g_{\pi(i)}\|^2 \,.$$

Furthermore, $\sum_i \|f_i'\|^2 = \sum_i \|g_i'\|^2 = 1$ and $\mathrm{supp}(f_i') \cap \mathrm{supp}(f_j') = \mathrm{supp}(g_i') \cap \mathrm{supp}(g_j') = \emptyset$ for all $i \ne j$.

We remark that Lemma 11 does not depend in any way on the fact that the functions f_i and g_i are defined on $\mathbb{R}^d$. (In fact, we could formulate the same lemma for finite dimensional vectors with nonnegative coordinates. We choose to state and prove the lemma in this form because the proof of Theorem 3 naturally leads to nonnegative functions defined on $\mathbb{R}^d$.) We prove Lemma 11 at the end of this section. Assuming this lemma, we can prove Theorem 3 as follows:

Proof (Theorem 3). Let G be a unique game with vertex set V and alphabet $\Sigma = [k]$. Suppose $\mathrm{sdp}(G) \geq 1 - \varepsilon$. For a parameter $\sigma > 0$, which we determine later, let $f_{u,i} := M_\sigma(u_i)$ be the nonnegative functions in $L_2(\mathbb{R}^d)$ obtained by applying M_σ to a collection of vectors $\{u_i\}_{u \in V, i \in \Sigma}$ corresponding to an optimal solution for $\mathrm{sdp}(G)$. Since M_σ preserves norms, we have for every $u \in V$,

$$\sum\nolimits_{i \in \Sigma} \|f_{u,i}\|^2 = \sum\nolimits_{i \in \Sigma} \|u_i\|^2 = 1 . \tag{3.1}$$

Since $\langle u_i, u_j \rangle = 0$ for $i \neq j$, Lemma 10 also shows that for every $u \in V$,

$$\sum_{i \neq j} \langle f_{u,i}, f_{u,j} \rangle \leq \sum_{i \neq j} \|u_i\| \, \|u_j\| \cdot e^{-1/4\sigma^2}$$

$$\leq e^{-1/4\sigma^2} \Big(\sum_{i \in \Sigma} \|u_i\| \Big)^2 \leq k \cdot e^{-1/4\sigma^2} \sum_{i \in \Sigma} \|u_i\|^2 \leq k \cdot e^{-1/4\sigma^2} . \tag{3.2}$$

Hence, for $\sigma^2 = 1/(4 \log(k/\gamma))$, we can make sure that $\sum_{i \neq j} \langle f_{u,i}, f_{u,j} \rangle \leq \gamma$ for every $u \in V$. By Lemma 10, we have

$$\mathop{\mathbf{E}}_{(u,v,\pi) \sim G} \|f_{u,i} - f_{v,\pi(i)}\|^2 \leq O(\sigma^{-2}) \mathop{\mathbf{E}}_{(u,v,\pi) \sim G} \|u_i - v_{\pi(i)}\|^2 \leq O(\sigma^{-2})\varepsilon . \tag{3.3}$$

For the last inequality, we used the assumption $\mathrm{sdp}(G) \geq 1 - \varepsilon$ of Theorem 3 and the fact that the vectors $\{u_i\}_{u \in V, i \in \Sigma}$ form an optimal solution for $\mathrm{sdp}(G)$. The functions $\{f_{u,i}\}_{u \in V, i \in \Sigma}$ form an approximate solution for $\mathrm{sdp}_+(G)$, in the sense of (3.1)–(3.3). Using the smooth nonnegative orthogonalization $Q \colon L_2(\mathbb{R}^d)^k \to L_2(\mathbb{R}^d)^k$ from Lemma 11, we obtain nonnegative functions $f'_{u,i} = Q(f_{u,1}, \ldots, f_{u,k})_i$ that form a feasible solution for $\mathrm{sdp}_+(G)$. Combining (3.3) and Lemma 11 also shows that for small enough γ (say $\gamma = 1/8$),

$$\mathop{\mathbf{E}}_{(u,v,\pi) \sim G} \|f'_{u,i} - f'_{v,\pi(i)}\|^2 \leq O(1) \mathop{\mathbf{E}}_{(u,v,\pi) \sim G} \|f_{u,i} - f_{v,\pi(i)}\|^2 \leq O(\sigma^{-2})\varepsilon .$$

Since we chose $\sigma^2 = \Omega(1/\log k)$, it follows that $\mathrm{sdp}_+(G) \geq 1 - O(\varepsilon \log k)$.

3.1   Proof of Lemma 11 (Smooth Nonnegative Orthogonalization)

First, we consider the following orthogonalization step,

$$Q^{(1)} \colon L_2(\mathbb{R}^d)^k \to L_2(\mathbb{R}^d)^k, \quad (f_1, \ldots, f_k) \mapsto (f'_1, \ldots, f'_k), \tag{3.4}$$

$$f'_i(x) = \begin{cases} f_i(x) - \max_{j \neq i} f_j(x) & \text{if } f_i(x) > \max_{j \neq i} f_j(x), \\ 0 & \text{otherwise.} \end{cases} \tag{3.5}$$

Next, we consider the following renormalization step,

$$Q^{(2)} \colon L_2(\mathrm{I\!R}^d)^k \to L_2(\mathrm{I\!R}^d)^k \,, \tag{3.6}$$

$$\left(f_1, \ldots, f_k\right) \mapsto \left(\tfrac{1}{\lambda} f_1, \ldots, \tfrac{1}{\lambda} f_k\right) \,, \tag{3.7}$$

$$\text{where } \lambda^2 = \textstyle\sum_i \|f_i\|^2 \,. \tag{3.8}$$

To prove Lemma 11, we choose Q as the composition of $Q^{(1)}$ and $Q^{(2)}$. Let $f_1, \ldots, f_k$ and $g_1, \ldots, g_k$ be nonnegative functions in $L_2(\mathrm{I\!R}^d)$ as in Lemma 11, i.e., $\sum_{i \neq j} \langle f_i, f_j \rangle + \sum_{i \neq} \langle g_i, g_j \rangle \leq \gamma$ and $\sum_i \|f_i\|^2 = \sum_i \|g_i\|^2 = 1$.

Let $(f_1^{(1)}, \ldots, f_k^{(1)}) = Q^{(1)}(f_1, \ldots, f_k)$ and $(f_1^{(2)}, \ldots, f_k^{(2)}) = Q^{(2)}(f_1^{(1)}, \ldots, f_k^{(1)})$. Similarly, let $(g_1^{(1)}, \ldots, g_k^{(1)}) = Q^{(1)}(g_1, \ldots, g_k)$ and $(g_1^{(2)}, \ldots, g_k^{(2)}) = Q^{(2)}(g_1^{(1)}, \ldots, g_k^{(1)})$. In the rest of the section, we first establish several properties of these functions (see Claim 12 and Claim 13) and then use these properties to prove Lemma 11.

Claim 12 (Properties of $Q^{(1)}$).

1. For every permutation π of $[k]$,

$$\sum_i \|f_i^{(1)} - g_{\pi(i)}^{(1)}\|^2 \leq 8 \sum_i \|f_i - g_{\pi(i)}\|^2 \,.$$

2. For all $i \neq j$,

$$\operatorname{supp}(f_i^{(1)}) \cap \operatorname{supp}(f_j^{(1)}) = \emptyset \,.$$

3.

$$1 \geq \sum_i \|f_i^{(1)}\|^2 \geq 1 - 2\gamma \,.$$

Proof. Item 2 holds, since by construction $\operatorname{supp}(f_i^{(1)}) = \{x \mid f_i(x) > \max_{j \neq i} f_j(x)\}$. To prove Item 3, we observe that $f_i^{(1)}(x)^2 > f_i(x)^2 - 2\sum_{j \neq i} f_i(x) f_j(x)$ and therefore as desired

$$\sum_i \|f_i^{(1)}\|^2 \geq \sum_i \|f_i\|^2 - 2 \sum_{i \neq j} \langle f_i, f_j \rangle \geq 1 - 2\gamma \,.$$

To prove Item 1, we will show that for every $x \in \mathrm{I\!R}^d$,

$$\sum_i \left(f_i^{(1)}(x) - g_{\pi(i)}^{(1)}(x)\right)^2 \leq 8 \sum_i \left(f_i(x) - g_{\pi(i)}(x)\right)^2 \,. \tag{3.9}$$

Since $Q^{(1)}$ is invariant under permutation of its inputs, we may assume π is the identity permutation. At this point, we can verify (3.9) by an exhaustive case distinction. Fix $x \in \mathrm{I\!R}^d$. Let i_f be the index i that maximizes $f_i(x)$. (We may assume the maximizer is unique.) Let j_f be the index such that $f_{j_f}(x) = \max_{j \neq i_f} f_j(x)$. Similarly, define i_g and j_g such that $g_{i_g}(x) = \max_i g_i(x)$ and $g_{j_g}(x) = \max_{j \neq i_g} g_j(x)$. We may assume that $i_f = 1$ and $j_f = 2$. Furthermore, we may assume $i_g, j_g \in \{1, 2, 3, 4\}$. Notice that the sum on the left-hand side of

(3.9) has at most two non-zero terms (corresponding to the indices $i \in \{i_f, i_g\} \subseteq \{1, \ldots, 4\}$). Hence, to verify (3.9), it is enough to show

$$\max_{i \in \{1, \ldots, 4\}} \left| f_i^{(1)}(\boldsymbol{x}) - g_i^{(1)}(\boldsymbol{x}) \right| \leq 4 \max_{i \in \{1, \ldots, 4\}} \left| f_i(\boldsymbol{x}) - g_i(\boldsymbol{x}) \right|. \tag{3.10}$$

Put $\varepsilon = \max_{i \in \{1, \ldots, 4\}} |f_i(x) - g_i(x)|$. Let $q_i(a_1, \ldots, a_4) := \max\{a_i - \max_{j \neq i} a_j, 0\}$. Note that $f_i^{(1)}(\boldsymbol{x}) = q_i(f_1(\boldsymbol{x}), \ldots, f_4(\boldsymbol{x}))$ and $g_i^{(1)}(\boldsymbol{x}) = q_i(g_1(\boldsymbol{x}), \ldots, g_4(\boldsymbol{x}))$. The functions q_i are 1-Lipschitz in each of their four inputs. It follows as desired that for every $i \in \{1, \ldots, 4\}$,

$$\left| f_i^{(1)}(\boldsymbol{x}) - g_i^{(1)}(\boldsymbol{x}) \right| = \left| q_i(f_1(\boldsymbol{x}), \ldots, f_4(\boldsymbol{x})) - q_i(g_1(\boldsymbol{x}), \ldots, g_4(\boldsymbol{x})) \right| \leq 4\varepsilon.$$

Claim 13 (Properties of $Q^{(2)}$).

1. For every permutation π of $[k]$,

$$\sum_i \| f_i^{(2)} - g_{\pi(i)}^{(2)} \|^2 \leq \tfrac{4}{1 - 4\gamma} \sum_i \| f_i^{(1)} - g_{\pi(i)}^{(1)} \|^2.$$

2. For all $i \in \Sigma$,

$$\mathrm{supp}(f_i^{(2)}) = \mathrm{supp}(f_i^{(1)}).$$

3.

$$\sum_i \| f_i^{(2)} \|^2 = 1.$$

Proof. Again Item 2 and Item 3 follow immediately by definition of the mapping $Q^{(2)}$. To prove Item 1, let $\lambda_f, \lambda_g > 0$ be the multipliers such that $f_i^{(2)} = f_i^{(1)}/\lambda_f$ and $g_i^{(2)} = g_i^{(1)}/\lambda_g$ for all $i \in [k]$. Item 1 of Claim 12 shows that λ_f^2 and λ_g^2 lie in the interval $[1 - 2\gamma, 1]$. We estimate the distances between $f_i^{(2)}$ and $g_{\pi(i)}^{(2)}$ as follows,

$$\sum_i \left\| f_i^{(2)} - g_{\pi(i)}^{(2)} \right\|^2 = \sum_i \left\| \tfrac{1}{\lambda_f} \left(f_i^{(1)} - g_{\pi(i)}^{(1)} \right) + \left(\tfrac{1}{\lambda_f} - \tfrac{1}{\lambda_g} \right) g_i^{(1)} \right\|^2$$

$$\leq \tfrac{2}{\lambda_f^2} \sum_i \left\| f_i^{(1)} - g_{\pi(i)}^{(1)} \right\|^2 + 2 \left(\tfrac{1}{\lambda_f} - \tfrac{1}{\lambda_g} \right)^2 \sum_i \left\| g_i^{(1)} \right\|^2 \quad \text{(since } \|a + b\|^2 \leq 2\|a\|^2 + 2\|b\|^2 \text{)}$$

$$\leq \tfrac{2}{1 - 2\gamma} \sum_i \left\| f_i^{(1)} - g_{\pi(i)}^{(1)} \right\|^2 + 2 \left(\tfrac{1}{\lambda_f} - \tfrac{1}{\lambda_g} \right)^2 \quad \text{(using } \sum_i \| g_i^{(1)} \|^2 \leq 1 \text{)}.$$

It remains to upper bound the second term on the right-hand side, $(1/\lambda_f - 1/\lambda_g)^2$. Since the function $x \mapsto 1/x$ is $1/a^2$-Lipschitz on an interval of the form $[a, \infty)$, we have

$$\left| \tfrac{1}{\lambda_f} - \tfrac{1}{\lambda_g} \right| \leq \tfrac{1}{1 - 2\gamma} |\lambda_f - \lambda_g|$$

$$= \tfrac{1}{1 - 2\gamma} \left| \left(\textstyle\sum_i \| f_i^{(1)} \|^2 \right)^{1/2} - \left(\textstyle\sum_i \| g_{\pi(i)}^{(1)} \|^2 \right)^{1/2} \right|$$

$$\leq \tfrac{1}{1 - 2\gamma} \left(\textstyle\sum_i \left(\| f_i^{(1)} \| - \| g_{\pi(i)}^{(1)} \| \right)^2 \right)^{1/2} \quad \text{(using triangle inequality)}$$

$$\leq \tfrac{1}{1 - 2\gamma} \left(\textstyle\sum_i \| f_i^{(1)} - g_{\pi(i)}^{(1)} \|^2 \right)^{1/2} \quad \text{(using triangle inequality)}.$$

Combining the previous two estimates, we get as desired

$$\sum_i \left\| f_i^{(2)} - g_{\pi(i)}^{(2)} \right\|^2 \le \left(\tfrac{2}{1-2\gamma} + \tfrac{2}{(1-2\gamma)^2} \right) \sum_i \left\| f_i^{(1)} - g_{\pi(i)}^{(1)} \right\|^2 .$$

Combining Claim 12 and Claim 13 yields Lemma 11.

Lemma (Restatement of Lemma 11). There exists a mapping $Q\colon L_2(\mathbb{R}^d)^k \to L_2(\mathbb{R}^d)^k$ with the following properties: Let $f_1,\ldots,f_k$ and $g_1,\ldots,g_k$ be nonnegative functions in $L_2(\mathbb{R}^d)$ such that $\sum_i \|f_i\|^2 = \sum_i \|g_i\|^2 = 1$ and $\sum_{i\neq j}\langle f_i, f_j\rangle + \sum_{i\neq j}\langle g_i, g_j\rangle \le \gamma$. Suppose $(f_1',\ldots,f_k') = Q(f_1,\ldots,f_k)$ and $(g_1',\ldots,g_k') = Q(g_1,\ldots,g_k)$. Then, for every permutation π of $[k]$,

$$\sum_i \|f_i' - g_{\pi(i)}'\|^2 \le \tfrac{32}{1-4\gamma} \sum_i \|f_i - g_{\pi(i)}\|^2 .$$

Furthermore, $\sum_i \|f_i'\|^2 = \sum_i \|g_i'\|^2 = 1$ and $\operatorname{supp}(f_i') \cap \operatorname{supp}(f_j') = \operatorname{supp}(g_i') \cap \operatorname{supp}(g_j') = \emptyset$ for all $i \neq j$.

Proof. We choose $Q = Q^{(2)} \circ Q^{(1)}$ as the composition of $Q^{(1)}$ and $Q^{(2)}$. In this case, $f_i' = f_i^{(2)}$ and $g_i' = g_i^{(2)}$, where the functions $f_1^{(2)},\ldots,f_k^{(2)}$ and $g_1^{(2)},\ldots,g_k^{(2)}$ are constructed as in the beginning of Section 3.1. Combining Item 1 of Claim 12 and Item 1 of Claim 13 gives the desired upper bound on $\sum_i \|f_i' - g_{\pi(i)}'\|^2$. Similarly, the remaining properties desired of f_i' and g_i' also follow by combining Claim 12 and Claim 13. $\square$

Acknowledgments

Thanks to Oded Regev for discussions and helpful comments on an earlier version of this manuscript. Thanks to Sanjeev Arora, Rajat Mittal, and Mario Szegedy for valuable discussions.

References

1. Austrin, P.: Towards sharp inapproximability for any 2-CSP. In: FOCS, pp. 307–317 (2007)
2. Barak, B., Hardt, M., Haviv, I., Rao, A., Regev, O., Steurer, D.: Rounding parallel repetitions of unique games. In: FOCS, pp. 374–383 (2008)
3. Charikar, M., Makarychev, K., Makarychev, Y.: Near-optimal algorithms for unique games. In: STOC, pp. 205–214 (2006)
4. Chawla, S., Krauthgamer, R., Kumar, R., Rabani, Y., Sivakumar, D.: On the hardness of approximating multicut and sparsest-cut. Computational Complexity 15(2), 94–114 (2006)
5. Feige, U., Kindler, G., O'Donnell, R.: Understanding parallel repetition requires understanding foams. In: IEEE Conference on Computational Complexity, pp. 179–192 (2007)
6. Feige, U., Lovász, L.: Two-prover one-round proof systems: Their power and their problems (extended abstract). In: STOC, pp. 733–744 (1992)

7. Guruswami, V., Manokaran, R., Raghavendra, P.: Beating the random ordering is hard: Inapproximability of maximum acyclic subgraph. In: FOCS, pp. 573–582 (2008)
8. Holenstein, T.: Parallel repetition: Simplification and the no-signaling case. Theory of Computing 5(1), 141–172 (2009)
9. Khot, S.: On the power of unique 2-prover 1-round games. In: STOC, pp. 767–775 (2002)
10. Khot, S., Kindler, G., Mossel, E., O'Donnell, R.: Optimal inapproximability results for MAX-CUT and other 2-variable CSPs? SIAM J. Comput. 37(1), 319–357 (2007)
11. Khot, S., Regev, O.: Vertex cover might be hard to approximate to within $2 - \varepsilon$. J. Comput. Syst. Sci. 74(3), 335–349 (2008)
12. Khot, S., Vishnoi, N.K.: The unique games conjecture, integrality gap for cut problems and embeddability of negative type metrics into ℓ_1. In: FOCS, pp. 53–62 (2005)
13. Manokaran, R., Naor, J., Raghavendra, P., Schwartz, R.: SDP gaps and UGC-hardness for multiway cut, 0-extension, and metric labeling. In: STOC, pp. 11–20 (2008)
14. Mossel, E., O'Donnell, R., Oleszkiewicz, K.: Noise stability of functions with low influences invariance and optimality. In: FOCS, pp. 21–30 (2005)
15. Raghavendra, P.: Optimal algorithms and inapproximability results for every CSP? In: STOC, pp. 245–254 (2008)
16. Raghavendra, P., Steurer, D.: Graph Expansion and the Unique Games Conjecture. In: STOC (to appear, 2010)
17. Rao, A.: Parallel repetition in projection games and a concentration bound. In: STOC, pp. 1–10 (2008)
18. Raz, R.: A parallel repetition theorem. SIAM J. Comput. 27(3), 763–803 (1998)
19. Raz, R.: A counterexample to strong parallel repetition. In: FOCS, pp. 369–373 (2008)

A   Further Proofs

A.1   Relation of Amortized Value and Semidefinite Value

Theorem (Restatement of Theorem 2). For every unique game G with alphabet size k,
$$\mathrm{sdp}(G)^{O(\log k)} \leq \overline{\mathrm{opt}}(G) \leq \mathrm{sdp}(G).$$

Proof. Feige and Lovász [6] show the upper bound on $\overline{\mathrm{opt}}(G)$. (In particular, they show $\mathrm{opt}(G^\ell) \leq \mathrm{sdp}(G^\ell)$ and $\mathrm{sdp}(G^\ell) = \mathrm{sdp}(G)^\ell$.) Charikar, Makarychev, and Makarychev [3] show that $\mathrm{opt}(G) \geq \mathrm{sdp}(G)^{-C \log k}/C$ for some absolute constant $C \geq 1$. (This constant C is necessarily larger than 1.) Hence, we are done if $\mathrm{sdp}(G)^{C' \log k} \leq 1/C$ for some absolute constant $C' \geq 1$. (In this case, we would have $\mathrm{opt}(G) \geq \mathrm{sdp}(G)^{-(C+C') \log k}$.)

On the other hand, if $\mathrm{sdp}(G) \geq 1 - \varepsilon$, then Theorem 1 shows that $\mathrm{opt}(G^\ell) \geq 1 - C''\sqrt{\ell \varepsilon \log k}$ for some absolute constant $C'' \geq 1$. Hence, if $\varepsilon \log k \leq 1/(2C'')^2$, then we can find a natural number $\ell = \Omega(1/(\varepsilon \log k))$ such that $\mathrm{opt}(G^\ell) \geq 1/2$, which implies $\overline{\mathrm{opt}}(G) \geq 2^{-1/\ell} \geq (1 - \varepsilon)^{O(\log k)}$. It is straight-forward to check that there exists an absolute constant $C' \geq 1$ (depending on C and C'') such that $\mathrm{sdp}(G)^{C' \log k} > 1/C$ implies that $\varepsilon \log k \leq 1/(2C'')^2$. We conclude that in all cases $\overline{\mathrm{opt}}(G) \geq \mathrm{sdp}(G)^{O(\log k)}$.

A.2   Mapping Unit Vectors to Nonnegative Functions

Recall the definition of the mapping $M_\sigma \colon \mathbb{R}^d \to L_2(\mathbb{R}^d)$, $M_\sigma(u) := \|u\| \sqrt{T_{\bar{u}} \phi_\sigma}$.
Here, T_u is the translation operator on $L_2(\mathbb{R}^d)$, so that $T_u f(x) = f(x - u)$,
and ϕ_σ is the density of the Gaussian measure $N(0, \sigma^2)^d$ with respect to the
usual Lebesgue measure λ^d on $\mathbb{R}^d$, $\phi_\sigma(x) := (\sigma\sqrt{2\pi})^{-d} e^{-\|x\|^2/2\sigma^2}$. From the
definition of M_σ, it follows that the mapping preserves norms, so that $\|M_\sigma(u)\| = \|u\|$ for every $u \in \mathbb{R}^d$. The following fact about the (Hellinger) affinity of
translated Gaussians shows that the mapping M_σ also preserves angles (at least
approximately).

Lemma 14 ([2]). Let u and v be two unit vectors in $\mathbb{R}^d$. Then

$$\int_{\mathbb{R}^d} \sqrt{T_u \phi_\sigma \cdot T_v \phi_\sigma} \, \mathrm{d}\lambda^d = e^{-\|u-v\|^2/8\sigma^2} \, .$$

Proof. Immediate from the identity $\sqrt{T_u \phi_\sigma \cdot T_v \phi_\sigma} = e^{-\|u-v\|^2/8\sigma^2} T_{\frac{1}{2}(u+v)} \phi_\sigma$.

The following technical fact shows that in order for a mapping to preserve distances it is enough to preserve lengths and distances of unit vectors (angles).

Fact 15. For any two vectors $u, v \in \mathbb{R}^n$, we have

$$\|u - v\|^2 = (\|u\| - \|v\|)^2 + \|u\| \, \|v\| \cdot \|\bar{u} - \bar{v}\|^2 \, .$$

Combining Fact 15 and Lemma 14 yields Lemma 10.

Lemma (Restatement of Lemma 10). For any two vectors $u, v \in \mathbb{R}^d$,

$$\|M_\sigma(u) - M_\sigma(v)\|^2 \le O(\sigma^{-2}) \cdot \|u - v\|^2 \, .$$

Furthermore, $\langle M_\sigma(u), M_\sigma(v) \rangle = \|u\| \, \|v\| \cdot e^{-1/4\sigma^2}$ if u and v are orthogonal.

Proof. If u and v are orthogonal, then Lemma 14 shows that $\langle M_\sigma(u), M_\sigma(v) \rangle = \|u\| \, \|v\| e^{-1/4\sigma^2}$, because $\|\bar{u} - \bar{v}\|^2 = 2$ for any two orthogonal vectors u and v.
It remains to show the upper bound on $\|M_\sigma(u) - M_\sigma(v)\|^2$. By construction,
$M_\sigma(u) = \|u\| M_\sigma(\bar{u})$ and $M_\sigma(v) = \|v\| M_\sigma(\bar{v})$. By Fact 15,

$$\|M_\sigma(u) - M_\sigma(v)\|^2 = (\|u\| - \|v\|)^2 + \|u\| \, \|v\| \cdot \|M_\sigma(\bar{u}) - M_\sigma(\bar{v})\|^2 \, .$$

On the other, Lemma 14 implies that

$$\tfrac{1}{2} \|M_\sigma(\bar{u}) - M_\sigma(\bar{v})\|^2 = 1 - e^{-\|\bar{u}-\bar{v}\|^2/8\sigma^2} \le \|\bar{u} - \bar{v}\|^2/8\sigma^2 \, .$$

(Here, we used the approximation $e^{-x} \ge 1 - x$.) Combining these bounds, yields
as desired

$$\|M_\sigma(u) - M_\sigma(v)\|^2 \le (\|u\| - \|v\|)^2 + \|u\| \, \|v\| \cdot \|\bar{u} - \bar{v}\|^2/4\sigma^2 \le \tfrac{1}{4\sigma^2} \|u - v\|^2 \, .$$

(The last inequality follows from Fact 15.)

A Query Efficient Non-adaptive Long Code Test with Perfect Completeness

Suguru Tamaki and Yuichi Yoshida

Kyoto University, Yoshida Honmachi, Sakyo-ku, Kyoto 606-8501, Japan
{tamak,yyoshida}@kuis.kyoto-u.ac.jp

Abstract. Long Code testing is a fundamental problem in the area of property testing and hardness of approximation. Long Code is a function of the form $f(x) = x_i$ for some index i. In the Long Code testing, the problem is, given oracle access to a collection of Boolean functions, to decide whether all the functions are the same Long Code, or cross-influences of any two functions are small. In this paper, we study the following problem: How small the soundness s of the Long Code test with perfect completeness can be by using non-adaptive q queries? We give a Long Code test with $s = (2q+3)/2^q$, where q is of the form $2^k - 1$ for any integer $k > 2$. Our test is a "noisy" version of Samorodnitsky-Trevisan's Hyper Graph linearity test with suitably chosen noise distribution. To bound the soundness, we use Invariance-Principle style analysis in the spirit of O'Donnell and Wu (STOC 2009).

Previously, Håstad and Khot (Theory of Computing, 2005) had shown $s = 2^{4\sqrt{q}}/2^q$ for infinitely many q. Chen (RANDOM 2009) improved this to $s = q^3/2^q$ for infinitely many q with "adaptive" queries. As for the Long Code test with "almost" perfect completeness, Samorodnitsky and Trevisan (SICOMP, 2009) have shown $s = 2q/2^q$ (or even $(q + 1)/2^q$ for infinitely many q). Austrin and Mossel (Computational Complexity, 2009) have improved this to $s = (q + o(q))/2^q$ (or even $(q+4)/2^q$ assuming the famous Hadamard Conjecture) for any q.

Keywords: Long Code test, dictatorship test, influence of variables, amortized query complexity.

1 Introduction

Testing basic properties of Boolean functions is a well studied subject in the area of property testing, complexity theory and learning theory [18,37]. Examples of such properties include: Linear functions, singleton functions, juntas, low-degree polynomials, and several concise representation [12,8,2,35,3,16,10,11]. In property test setting, our goal is, given oracle access to Boolean function(s) of the form $f : \{-1, 1\}^n \to \{-1, 1\}$, to decide whether the functions satisfy a certain property, or they are *far* from the property, where the definition of farness varies according to the property of interest and intended applications of the test. There are several parameters that characterize testers: *Completeness c* is the probability such that the tester accepts functions that satisfy the property

M. Serna et al. (Eds.): APPROX and RANDOM 2010, LNCS 6302, pp. 738–751, 2010.

with probability at least c. *Soundness* s is the probability such that the tester accepts functions that are far from the property with probability at most s. *Query complexity* q is the total number of oracle accesses in one execution of the tester. An *adaptive* tester means that the tester can make queries that depend on the answers of previous queries. Otherwise the tester is said to be *non-adaptive*.

The property of our interest in this paper is the Long Code. *Long Code* is a function of the form $f(x) = x_i$ for some index i. Multi-function Long Code testing problem is, given oracle access to a collection of Boolean functions, to decide whether they are all the same Long Code for some i, or any two functions are not close to some Long Code at the same time. The notion of Long Code was introduced by Bellare et al. [9] in the context of Probabilistically Checkable Proofs (PCPs) (also studied by Parnas et al. [35] in name of "dictatorship function"). We quickly review the use of (multi-function) Long Code test in PCPs and hardness of approximation. From now on, when we say just PCP, it means a PCP for some NP-complete language. Recall that most efficient PCP constructions today such as the celebrated work of Håstad [22] follow the paradigm of composing "outer verifier" and "inner verifier".

An outer verifier is a certain kind of two prover one round games (or equivalently label cover problems); e.g., the combination of the PCP theorem [6,4] and Raz's parallel repetition theorem [36] (we often call the combination Raz verifier), or Unique Games or d-to-1 Games [26]. An inner verifier is essentially a multi-function Long Code tester. By composing outer and inner verifiers, we obtain a PCP, which often has the same parameters of the underlying Long Code tester. Therefore improving the quality of Long Code testers yields better constructions of PCPs. Here we note that criterion for soundness used in a Long Code tester/inner verifier may restrict the type of an outer verifier to ensure that they can be composed. We elaborate on this issue later.

The use of PCP constructions in hardness of approximation is quite standard. The acceptance criteria of Long Code testers/PCPs correspond to specific optimization problems and the ratio of completeness and soundness s/c gives the inapproximability ratio of the corresponding problems. For more details on the relation between Long Code tests, PCPs and hardness of approximation, see [27] for the general framework and [33] for constraint satisfaction problems (CSPs).

With the connection between Long Code testers and PCPs in mind, we sometimes compare the parameters of them at the same time without distinction. There are several interesting trade-offs between parameters as follows:

Criteria for soundness. The definition of farness is quite an important issue in the composition paradigm. Several possible criteria for rejection are; distance from any Long Codes or any juntas, quasi-randomness, and low-degree influences. Which criterion is used in Long Code test poses some restriction on an outer verifier, e.g., if we use quasi-randomness, the tester can be composed with Raz verifier, but if we use low-degree influences, it is hard to combine the tester with Raz verifier though it may be composed with Unique Games. It may also affect the lowest possible value of the soundness with other parameters fixed.

Perfect completeness vs. almost perfect completeness. It is known that the soundness of PCPs with perfect completeness ($c = 1$) can be worse than that of PCPs with almost perfect completeness ($c = 1 - \varepsilon$). Håstad [22] designed non-adaptive 3-query PCPs with almost perfect completeness and $s = 1/2$. Zwick [43] showed that any non-adaptive 3-query PCPs with perfect completeness must have soundness $s \geq 5/8$. He conjectured $s = 5/8$ is optimal. Håstad [22] showed $s = 3/4$ is possible and Khot and Saket [28] improved this to $s = 20/27$. Finally, O'Donnell and Wu [34] showed $s = 5/8$ under Khot's d-to-1 Games conjecture. We remark that perfect completeness is important for several reasons: First, it is natural to require that testers should always accept correct objects. Second, it is required to show the inapproximability results for certain problems such as satisfiable CSPs and approximate coloring problems. Third, it is robust when we compose testers with other protocols.

Adaptive vs. non-adaptive queries. It is also known that adaptive queries are more powerful than non-adaptive queries. Guruswami et al. [19] showed adaptive 3-query PCPs with $c = 1, s = 1/2$. As mentioned previously, Zwick showed non-adaptive 3-query PCPs with $c = 1$ have $s \geq 5/8$. Non-adaptivity is crucial in applications for inapproximability; we need queries to be non-adaptive to relate the parameters of PCPs to the inapproximability ratio of optimization problems.

Query complexity vs. soundness. Trade-offs between query complexity and soundness is also formalized as amortized query complexity, which is defined as $\frac{q}{\log_2(1/s)}$. Observe that repeating the tester t times independently, we can reduce the soundness to s^t while query complexity grows as $t \cdot q$. In this case, the amortized query complexity of the repeated tester remains the same as that of the original one. Hence this can be thought as a good and non-trivial measure for such trade-offs. Thus, constructions of testers that achieve better amortized query complexity is an interesting problem.

In this paper, we study the trade-offs between query complexity and soundness in the Long Code testing problem. Our focus is on non-adaptive testers with perfect completeness ($c = 1$).

1.1   Previous Work

The study of amortized query complexity in PCPs was initiated by Trevisan [41] and many works followed it [38,39,25,24,17]. One of the most notable earlier work is due to Samorodnitsky and Trevisan [39]: They constructed q-query non-adaptive PCPs with almost perfect completeness and $s = 2^{2\sqrt{q}}/2^q$ for infinitely many q. Later Engebretsen and Holmerin [17] improved this to $s = 2^{\sqrt{2q}}/2^q$. Håstad and Khot constructed PCPs with perfect completeness and $s = 2^{4\sqrt{q}}/2^q$. All these constructions use the distance from juntas in soundness condition, which allows them to compose Raz verifier as an outer verifier. Hence the resulting PCPs are unconditional.

Samorodnitsky and Trevisan [40] observed that the trade-offs achieved by these works are essentially tight if the soundness criterion is with respect to the distance from juntas. (Similar limitations were shown by [14,29].) To break the $s = 2^{\Theta(\sqrt{q})}/2^q$ barrier, they reworked the problem under the relaxed soundness criterion, i.e., low-degree influences. They constructed so called the Hyper Graph Long Code test, that achieves almost perfect completeness and $s = 2q/2^q$ (or even $(q+1)/2^q$ for infinitely many q), and combined it with Unique Games to construct conditional PCPs with the same parameters. They use a novel technique from additive combinatorics, namely, Gowers uniformity norm, in the analysis. Later Austrin and Mossel [5] improved the soundness to $s = (q + o(q))/2^q$ (or even $(q + 4)/2^q$ assuming the famous Hadamard Conjecture) for any q. Their technique is completely different from [40] and based on the Invariance Principle, which was developed by Mossel et al. [30,31]. Similar constructions were also given in [23,1,20] where the emphasis is on the hardness of CSPs over "random predicates" or predicates of non-Boolean domains. However, all these works are affected by the loss of perfect completeness. Recently, Chen [15], building on the work of Samorodnitsky and Trevisan [40] and Håstad and Khot [24], showed q-query "adaptive" Long Code testers with perfect completeness and $s = q^3/2^q$ for infinitely many q.

1.2   Our Contribution

In this paper, we prove the following theorem:

Theorem 1. *For every $k > 2$, there exists a non-adaptive Long Code tester that makes $q = 2^k - 1$ queries, has completeness 1 and soundness $(2q + 3)/2^q$.*

1.3   Our Method

Our tester and its analysis is based on a natural but non-trivial extension of O'Donnell and Wu's analysis for hardness of satisfiable 3-CSPs over Not-Two (NTW) predicates [34]. Most technical part of O'Donnell-Wu's work is to bound the term of the form $\mathbf{E}[f(\mathbf{x})g(\mathbf{y})g(\mathbf{z})]$ where $\mathbf{x}, \mathbf{y}, \mathbf{z}$ are drawn from "NTW"-distribution. We extend this to the case of the product of large number of functions with more complex distribution.

First we need to define suitable test distributions to make queries. To achieve good query complexity-soundness trade-offs, we adopt the Hyper Graph test introduced by Samorodnitsky and Trevisan [40]. The Hyper Graph test can only test linearity; for testing Long Code, we need to perturb the Hyper Graph test with some "noise" distribution.

Most often, the noise distribution is the uniform distribution. A mixture of the distribution for linearity test and uniform distribution with sufficiently small probability often yields a desired Long Code test. However, this forces the loss of perfect completeness because possible query-answers supports all over the domain, and "yes-instance" can be rejected with small probability. To avoid this, Chen [15] introduced a noise distribution based on Håstad and Khot's query-efficient PCPs with perfect completeness [24]. His analysis is based on Fourier analysis and Gowers norm as Samorodnitsky and Trevisan did. Although his test achieves good query-soundness trade-offs, one drawback is its adaptivity. In this paper, we use a different noise distribution, modification of the All-Equal noise. Previously, the All-Equal noise was used by O'donnell and Wu [33] combined with 3-bit linearity test to achieve $c = 1, s = 5/8$, non-adaptive Long Code test based on NTW predicate. The analysis is based on Fourier analysis and hypercontractivity inequality. Later the same authors [34] managed to construct PCPs with the same parameters $c = 1$ and $s = 5/8$ under Khot's d-to-1 Games conjecture [26]. The analysis is based on Invariance Principle style argument [30,31]. The crucial point is bounding the correlation of probability spaces.

We follow the approach of [34]. Most of our analysis becomes complicated to bound terms of the form $\mathbf{E}[f_1 f_2 \cdots f_q]$ rather than a "cubic"-term, but some analysis becomes easier because we only work with functions over the same domain (or functions with "unique" constraints) rather than functions with d-to-1 constraints as in the case of [34].

1.4   Organization of Paper

In Section 2, we introduce notations and definitions needed throughout this paper. In Section 3, we define our multi-function Long Code testing problem formally, then we describe our test distribution and Long Code test. In Section 4, we present completeness and soundness analysis of our tester. Finally, we notice some future work in Section 5.

2   Preliminaries

Let $[k] = \{1, 2, \ldots, k\}$, $2^{[k]}_{\geq 1} = \{S \subseteq [k] : |S| \geq 1\}$ and $2^{[k]}_{\geq 2} = \{S \subseteq [k] : |S| \geq 2\}$. For a vector $x \in \{-1, 1\}^n$, x_i denotes the ith element of x. Write an unit vector as $\mathbf{1} = (1, 1, \ldots, 1)$ and denote by $\mathbf{1}_{-i}$ a vector obtained by flipping the ith coordinate of $\mathbf{1}$ to -1. We denote by $|\mathbf{S}|$ the cardinality of a set $\mathbf{S}$. A set $\{i\}$ is sometimes written as i. A *Boolean* function is a function $f : \{-1, 1\}^n \to \{-1, 1\}$.

2.1   Fourier Expansion

Definition 1 (Fourier expansion). *The Fourier expansion of a function $f : \{-1, 1\}^n \to \mathbb{R}$ is*

$$f(x) = \sum_{S \subseteq [n]} \widehat{f}(S) \chi_S(x),$$

where

$$\chi_S(x) = \prod_{i \in S} x_i, \quad \widehat{f}(S) = \mathbf{E}_x [f(x)\chi_S(x)],$$

and expectation is with respect to uniform distribution.

It is easy to see that for any $S, T \subseteq [n]$, $\mathbf{E}_x[\chi_S(x)\chi_T(x)] = 1$ if $S = T$ and 0 otherwise. For more on Fourier expansion and Fourier analysis see, e.g., [32].

2.2   Influences

In this section we recall basic notions from Fourier analysis, influence and the Bonami-Beckner operator. We first define a notion of the influence of a coordinate on a function f:

Definition 2. *For a function $f : \{-1,1\}^n \to \mathbb{R}$, we define the influence of $i \in [n]$ on f to be*

$$\operatorname{Inf}_i(f) = \sum_{S \ni i} \widehat{f}(S)^2.$$

We also define the degree-w influence of $i \in [n]$ on f to be

$$\operatorname{Inf}_i^{\leq w}(f) = \sum_{S \ni i : |S| \leq w} \widehat{f}(S)^2.$$

We next recall the Bonami-Beckner operator T_ρ acting on Boolean functions:

Definition 3. *Let $0 \leq \rho \leq 1$. The Bonami-Beckner operator T_ρ is a linear operator mapping functions $f : \{-1,1\}^n \to \mathbb{R}$ into functions $T_\rho f : \{-1,1\}^n \to \mathbb{R}$ via*

$$(T_\rho f)(x) = \mathbf{E}[f(\mathbf{y})],$$

where in the expectation, $\mathbf{y}$ is formed from x by setting $\mathbf{y}_i = x_i$ with probability ρ and setting $\mathbf{y}_i$ to be a uniformly random bit with probability $1 - \rho$.

The operator T_ρ can alternately be defined by the following formula:

Proposition 1

$$T_\rho f = \sum_{S \subseteq [n]} \rho^{|S|} \widehat{f}(S) \chi_S.$$

Definition 4. *Let $f_1, \ldots, f_t$ be a collection of Boolean functions. Then the cross-influence of i for them is defined as*

$$\operatorname{XInf}_i(f_1, \ldots, f_t) = \max_{j \neq k} \min(\operatorname{Inf}_i(f_j), \operatorname{Inf}_i(f_k)).$$

Similarly,

$$\operatorname{XInf}_i^{\leq w}(f_1, \ldots, f_t) = \max_{j \neq k} \min(\operatorname{Inf}_i^{\leq w}(f_j), \operatorname{Inf}_i^{\leq w}(f_k)).$$

Proposition 2. *Let f be a Boolean function and $\gamma > 0$. Suppose $\operatorname{Inf}_i(T_{1-\gamma/2}f) \geq \tau$, then there exists w such that $\operatorname{Inf}_i^{\leq w}(f) \geq \frac{\tau}{2}$.*

Proof. Set w such that $(1 - \frac{\gamma}{2})^w \leq \frac{\tau}{2}$, then

$$\tau \leq \operatorname{Inf}_i(T_{1-\gamma/2}f) = \sum_{S \in [n]} (1 - \frac{\gamma}{2})^{|S|} \widehat{f}(S)^2$$

$$= \sum_{S \in [n], |S| \leq w} (1 - \frac{\gamma}{2})^{|S|} \widehat{f}(S)^2 + \sum_{S \in [n], |S| > w} (1 - \frac{\gamma}{2})^{|S|} \widehat{f}(S)^2$$

$$\leq \sum_{S \in [n], |S| \leq w} \widehat{f}(S)^2 + (1 - \frac{\gamma}{2})^w = \operatorname{Inf}_i^{\leq w}(f) + \frac{\tau}{2}.$$

$\square$

2.3   Correlations

We now recall the definition of correlation for correlated probability spaces, as introduced by Mossel [31].

Definition 5. *Let (Ω, Ψ, μ) be a (finite) correlated probability space, meaning that μ is a distribution on the finite product set $\Omega \times \Psi$ and that the marginals of μ on Ω and Ψ have full support. Define the "correlation" between Ω and Ψ to be*

$$\rho(\Omega, \Psi; \mu) = \max \left\{ \mathop{\mathbf{Cov}}_{(\omega, \psi) \sim \mu} [f(\omega) g(\psi)] \mid f : \Omega \to \mathbb{R}, g : \Psi \to \mathbb{R}, \right.$$

$$\left. \mathop{\mathbf{Var}}_{(\omega, \psi) \sim \mu} [f(\omega)] = \mathop{\mathbf{Var}}_{(\omega, \psi) \sim \mu} [g(\psi)] = 1 \right\}.$$

It is clear that in the definition of $\rho(\Omega, \Psi; \mu)$, we can equivalently maximize $|\mathbf{E}[fg]|$ over f restricted to have $\mathbf{E}[f] = 0, \mathbf{E}[f^2] \le 1$ under μ's marginal on Ω; or, over similarly restricted g (or both).

3   Long Code Test

Definition 6. *For $i \in [n]$, the ith Long Code is the function $f(x) = x_i$*

Now let us define a t-function Long Code Test. Suppose we are given oracle access to a collection of Boolean functions $f_1, f_2, \ldots, f_t$. We want to make as few queries as possible into these functions to decide if all the functions are the same Long Code, or no two functions have some common structure. More precisely, we have the following definition:

Definition 7. *We say that a test $\mathbf{T} = \mathbf{T}^{f_1, f_2, \ldots, f_t}$ is a t-functions Long Code test with completeness c and soundness s if $\mathbf{T}$ is given oracle access to a family of t functions $f_1, f_2, \ldots, f_t : \{-1, 1\}^n \to \{-1, 1\}$, such that*

- *if there exists some $i \in [n]$ such that for all $a \in [t], f_a(x) = x_i$, then $\mathbf{T}$ accepts with probability at least c, and*
- *for every $\varepsilon > 0$, there exist a constant $\tau > 0$ and a fixed positive integer w such that if $\mathbf{T}$ accepts with probability at least $s + \varepsilon$, then there exists some $i \in [n]$ such that $\mathrm{XInf}_i^{\le w}(f_1, \ldots, f_t) \ge \tau$.*

3.1   Folding

As introduced in [9], we assume that the functions are folded. We do so by requiring our Long Code test to make queries in a special manner. Suppose the test wants to query f at point $x \in \{-1, 1\}^n$. If $x_1 = 1$, then the test queries $f(x)$ as usual. If $x_1 = -1$, then the test queries at the point $-x$ and use $-f(-x)$ as the evaluation of x. Folding ensures that $f(-x) = -f(x)$ and $\widehat{f}(\emptyset) = 0$.

3.2   The Test Distribution

We now view $f_S : \{-1,1\}^n \to \{-1,1\}$ as a function over an n-fold product set $\mathcal{X}_S^1 \times \mathcal{X}_S^2 \times \cdots \times \mathcal{X}_S^n$ where each $\mathcal{X}_S^i = \{-1,1\}^{\{i\}}$.

For each integer $k > 2$, we construct $\mathcal{T}_k$ as a product probability distribution over the n-fold product set

$$\prod_{i=1}^{n} \prod_{S \in 2_{\geq 1}^{[k]}} \mathcal{X}_S^i = \prod_{i=1}^{n} \mathcal{X}_1^i \times \prod_{i=1}^{n} \mathcal{X}_2^i \times \cdots \times \prod_{i=1}^{n} \mathcal{X}_n^i \times \prod_{i=1}^{n} \mathcal{X}_{\{1,2\}}^i \times \cdots \times \prod_{i=1}^{n} \mathcal{X}_{\{1,2,\ldots,k\}}^i.$$

For each i we have a distribution $\mathcal{T}_k^i$ on $\prod_{S \in 2_{\geq 1}^{[k]}} \mathcal{X}_S^i$ and we think of this as a "correlated space".

The first distribution is the "Hyper Graph Test" distribution which is a slight modification of the one used by Samorodnitsky and Trevisan [40].

Definition 8. *Define distribution $\mathcal{H}_k$ generating $(\mathbf{x}_1, \mathbf{x}_2, \ldots, \mathbf{x}_{\{1,2,\ldots,k\}})$ $\in \prod_{S \in 2_{\geq 1}^{[k]}} \mathcal{X}_S^i$ as follows: The bits $\{\mathbf{x}_j\}_{j \in [k]}$ are independent and uniformly at random; then for each $S \in 2_{\geq 2}^{[k]}$, $\mathbf{x}_S$ is set to be $-\prod_{j \in S} \mathbf{x}_j$.*

Note that under $\mathcal{H}_k$, the marginal distribution on $\mathbf{x}_S$ for each $S \in 2_{\geq 2}^{[k]}$ is also uniformly random. The second distribution is the "Noise" distribution.

Definition 9. *Let $p, q, r > 0$ be such that $p + q + (2^k - 1)r = 1, p + r = q + (2^k - 2)r = 1/2, q = r$ ($p = (2^k - 2)/(2^{k+1} - 2), q = r = 1/(2^{k+1} - 2)$). Define distribution $\mathcal{N}_k$ generating $\mathbf{x} = (\mathbf{x}_1, \mathbf{x}_2, \ldots, \mathbf{x}_{\{1,2,\ldots,k\}}) \in \prod_{S \in 2_{\geq 1}^{[k]}} \mathcal{X}_S^i$ as follows: (i) With probability p, set $\mathbf{x} = \mathbf{1}$, (ii) with probability q, set $\mathbf{x} = -\mathbf{1}$, and (iii) with probability $(2^k - 1)r$, first pick $S \in 2_{\geq 1}^{[k]}$ uniformly at random, then set $\mathbf{x} = -\mathbf{1}_{-S}$. Additionally, for $0 < \delta < 1$, define distribution $\mathcal{H}_{k,\delta}$ to be the mixture distribution $\mathcal{H}_{k,\delta} = (1 - \delta)\mathcal{H}_k + \delta\mathcal{N}_k$; i.e., one draws from $\mathcal{H}_k$ with probability $1 - \delta$ and from $\mathcal{N}_k$ with probability δ.*

Again note that under $\mathcal{N}_k$ (also $\mathcal{H}_{k,\delta}$), the marginal distribution on $\mathbf{x}_S$ for each $S \in 2_{\geq 1}^{[k]}$ is uniformly random.

We are now ready to define the test distribution $\mathcal{T}_k^i$:

Definition 10. *For each $i \in [n]$ we define $\mathcal{T}_k^i$ to be $\mathcal{H}_{k,\delta}$, with $\delta = (\frac{\varepsilon}{9 \cdot |2_{\geq 1}^{[k]}|})^2$, where the domain of $\mathcal{H}_{k,\delta}$ is appropriately identified with the domain $\prod_{S \in 2_{\geq 1}^{[k]}} \mathcal{X}_S^i$ of $\mathcal{T}_k^i$. $\mathcal{T}_k$ is the product distribution of these distributions $\mathcal{T}_k = \otimes_{i=1}^{n} \mathcal{T}_k^i$.*

Definition 11. *For any positive integer k and $S_0 \in 2_{\geq 1}^{[k]}$, we define distribution $\mathcal{I}_{k,S_0}$ generating $(\mathbf{x}_1, \mathbf{x}_2, \ldots, \mathbf{x}_{\{1,2,\ldots,k\}}) \in \prod_{S \in 2_{\geq 1}^{[k]}} \mathcal{X}_S^i$ as follows: First draw from $\mathcal{H}_k$; then uniformly rerandomize $\mathbf{x}_{S_0}$.*

Definition 12. *For $0 < \delta < 1$, define distribution $\mathcal{I}_{k,S_0,\delta}$ to be the mixture distribution $\mathcal{I}_{k,S_0,\delta} = (1 - \delta)\mathcal{I}_{k,S_0} + \delta\mathcal{N}_k$.*

3.3   A Query Efficient Long Code Test

Let $k > 2$ be an integer and $\{f_S\}_{S \in 2^{[k]}_{\geq 1}}$ be a collection of Boolean functions. We define the following Long Code test based on distribution $\mathcal{T}_k$:

Long Code Test $\mathbf{T}_k$: with oracle access to $\{f_S\}_{S \in 2^{[k]}_{\geq 1}}$

1. Pick $(\mathbf{x}_1, \mathbf{x}_2, \ldots, \mathbf{x}_{\{1,2,\ldots,k\}}) \in \prod_{S \in 2^{[k]}_{\geq 1}} \prod_{i=1}^{n} \mathcal{X}_S^i$ from $\mathcal{T}_k$.
2. For each $S \in 2^{[k]}_{\geq 1}$, query $f_S(\mathbf{x}_S)$.
3. Accept iff either of the followings holds:
   (a) (Hyper Graph) For every $S \in 2^{[k]}_{\geq 2}$, $\prod_{s \in S} f_s(\mathbf{x}_s) \neq f_S(\mathbf{x}_S)$, or
   (b) (Noise) $(f_S(\mathbf{x}_1), f_S(\mathbf{x}_2), \ldots, f_{\{1,2,\ldots,k\}}(\mathbf{x}_{\{1,2,\ldots,k\}}))$ is either $\mathbf{1}$ or $-\mathbf{1}$ or $-\mathbf{1}_{-S}$ for some $S \in 2^{[k]}_{\geq 1}$.

Note that the acceptance conditions for the test are mutually exclusive.

Theorem 2 (main theorem restated). *For every integer $k > 2$, there exists a non-adaptive Long Code test $\mathbf{T}_k$ that makes $q = t = 2^k - 1$ queries, has completeness 1 and soundness $(2q + 3)/2^q$.*

4   Proof of Theorem 2

Due to the space limitation, we give the sketch of the proof here. The full proof can be found in [42].

Completeness Analysis: The completeness analysis is entirely standard.

Soundness Analysis: We show that if the accepting probability is at least

$$\frac{2q + 3}{2^q} + \epsilon = \frac{2|2^{[k]}_{\geq 1}| + 3}{2^{|2^{[k]}_{\geq 1}|}} + \epsilon$$

for some $\epsilon > 0$, then there exists w, τ, i such that $\mathrm{XInf}_i^{\leq w}(f_1, \ldots, f_{[k]}) \geq \tau$. We arithmetize the probability that $\mathbf{T}_k$ accepts. Let $\mathbf{Z}_S = -f_S(\mathbf{x}_S) \prod_{s \in S} f_s(\mathbf{x}_s)$. Then

$$\mathbf{Pr}[\mathbf{T}_k \text{ accepts}] = \mathop{\mathbf{E}}_{\mathcal{H}^n_{k,\delta}} \left[\prod_{S \in 2^{[k]}_{\geq 2}} \left(\frac{1 + \mathbf{Z}_S}{2} \right) + \prod_{S \in 2^{[k]}_{\geq 1}} \left(\frac{1 - f_S(\mathbf{x}_S)}{2} \right) \right.$$

$$\left. + \prod_{S \in 2^{[k]}_{\geq 1}} \left(\frac{1 + f_S(\mathbf{x}_S)}{2} \right) + \sum_{S' \in 2^{[k]}_{\geq 1}} \left(\frac{1 - f_{S'}(\mathbf{x}_{S'})}{2} \right) \prod_{S \in 2^{[k]}_{\geq 1} \setminus \{S'\}} \left(\frac{1 + f_S(\mathbf{x}_S)}{2} \right) \right].$$

This can be written as

$$
= \mathop{\mathbf{E}}_{\mathcal{H}^n_{k,\delta}} \left[\frac{1}{2^{|2^{[k]}_{\geq 2}|}} \sum_{\mathbf{S} \subseteq 2^{[k]}_{\geq 2}} \prod_{S \in \mathbf{S}} \mathbf{Z}_S + \frac{1}{2^{|2^{[k]}_{\geq 1}|}} \sum_{\mathbf{S} \subseteq 2^{[k]}_{\geq 1}} (-1)^{|\mathbf{S}|} \prod_{S \in \mathbf{S}} f_S(\mathbf{x}_S) \right.
$$

$$
\left. + \frac{1}{2^{|2^{[k]}_{\geq 1}|}} \sum_{\mathbf{S} \subseteq 2^{[k]}_{\geq 1}} \prod_{S \in \mathbf{S}} f_S(\mathbf{x}_S) + \sum_{S' \in 2^{[k]}_{\geq 1}} \left(\frac{1 - f_{S'}(\mathbf{x}_{S'})}{2^{|2^{[k]}_{\geq 1}|}} \right) \sum_{\mathbf{S} \subseteq 2^{[k]}_{\geq 1} \setminus \{S'\}} \prod_{S \in \mathbf{S}} f_S(\mathbf{x}_S) \right]
$$

$$
\leq \frac{1}{2^{|2^{[k]}_{\geq 2}|}} \sum_{\mathbf{S} \subseteq 2^{[k]}_{\geq 2}} \mathop{\mathbf{E}}_{\mathcal{H}^n_{k,\delta}} \left[\prod_{S \in \mathbf{S}} \mathbf{Z}_S \right] + \frac{2 + |2^{[k]}_{\geq 1}|}{2^{|2^{[k]}_{\geq 1}|}} \sum_{\mathbf{S} \subseteq 2^{[k]}_{\geq 1}} \left| \mathop{\mathbf{E}}_{\mathcal{H}^n_{k,\delta}} \left[\prod_{S \in \mathbf{S}} f_S(\mathbf{x}_S) \right] \right|.
$$

We would like to prove

$$
\frac{1}{2^{|2^{[k]}_{\geq 2}|}} \sum_{\mathbf{S} \subseteq 2^{[k]}_{\geq 2}} \mathop{\mathbf{E}}_{\mathcal{H}^n_{k,\delta}} \left[\prod_{S \in \mathbf{S}} \mathbf{Z}_S \right] = \frac{1}{2^{|2^{[k]}_{\geq 2}|}} + \frac{1}{2^{|2^{[k]}_{\geq 2}|}} \sum_{\mathbf{S} \subseteq 2^{[k]}_{\geq 2}: \mathbf{S} \neq \emptyset} \mathop{\mathbf{E}}_{\mathcal{H}^n_{k,\delta}} \left[\prod_{S \in \mathbf{S}} \mathbf{Z}_S \right]
$$

$$
\leq \frac{1}{2^{|2^{[k]}_{\geq 2}|}} + \frac{\varepsilon}{3}
$$

and

$$
\frac{2 + |2^{[k]}_{\geq 1}|}{2^{|2^{[k]}_{\geq 1}|}} \sum_{\mathbf{S} \subseteq 2^{[k]}_{\geq 1}} \left| \mathop{\mathbf{E}}_{\mathcal{H}^n_{k,\delta}} \left[\prod_{S \in \mathbf{S}} f_S(\mathbf{x}_S) \right] \right| = \frac{2 + |2^{[k]}_{\geq 1}|}{2^{|2^{[k]}_{\geq 1}|}}
$$

$$
+ \frac{2 + |2^{[k]}_{\geq 1}|}{2^{|2^{[k]}_{\geq 1}|}} \sum_{\mathbf{S} \subseteq 2^{[k]}_{\geq 1}: \mathbf{S} \neq \emptyset} \left| \mathop{\mathbf{E}}_{\mathcal{H}^n_{k,\delta}} \left[\prod_{S \in \mathbf{S}} f_S(\mathbf{x}_S) \right] \right| \leq \frac{2 + |2^{[k]}_{\geq 1}|}{2^{|2^{[k]}_{\geq 1}|}} + \frac{2\varepsilon}{3}
$$

for $\varepsilon > 0$. To do so, the following lemma is enough.

Lemma 1. *For any $\varepsilon > 0$, there exists τ, δ, w such that if $\mathrm{XInf}_i^{\leq w}(f_1, \ldots, f_{[k]}) < \tau$ for any i then, for any $\mathbf{S} \subseteq 2^{[k]}_{\geq 1}$*

$$
\left| \mathop{\mathbf{E}}_{\mathcal{H}^n_{k,\delta}} \left[\prod_{S \in \mathbf{S}} f_S(\mathbf{x}_S) \right] \right| \leq \frac{\varepsilon}{3 \cdot |2^{[k]}_{\geq 1}|}.
$$

Note that the term $\mathbf{E}_{\mathcal{H}^n_{k,\delta}} \left[\prod_{S \in \mathbf{S}} \mathbf{Z}_S \right]$ is also represented as $\mathbf{E}_{\mathcal{H}^n_{k,\delta}} \left[\prod_{S \in \mathbf{S}} f_S(\mathbf{x}_S) \right]$ since each f_S is a Boolean function. $\qquad \square$

4.1 Proof of Lemma 1

First note that for any $\mathbf{S} \subseteq 2^{[k]}_{\geq 1}, |\mathbf{S}| \leq 2$, the LHS become zero since the marginal distribution of $\mathcal{H}^n_{k,\delta}$ on $\prod_{S \in \mathbf{S}} \prod_{i=1}^n \mathcal{X}^i_S$ is uniformly random and functions are folded. We need to prove Lemma 1 for $\mathbf{S} \subseteq 2^{[k]}_{\geq 1}, |\mathbf{S}| \geq 3$ and actually we only prove Lemma 1 for $\mathbf{S} = 2^{[k]}_{\geq 1}$; any choice of $\mathbf{S}$ can be handled in almost the same way. It is easy to see that following lemmas with the triangle inequality concludes the proof of Lemma 1.

Lemma 2. *By taking* $\gamma > 0$ *small enough as a function of* $\delta, k,$

$$\left| \mathop{\mathbf{E}}_{\mathcal{H}^n_{k,\delta}} \left[\prod_{S \in \mathbf{S}} f_S(\mathbf{x}_S) - \prod_{S \in \mathbf{S}} T_{1-\gamma} f_S(\mathbf{x}_S) \right] \right| \leq \sqrt{\delta}.$$

Lemma 3. *For any* $S_0 \in \mathbf{S}, |\mathbf{S}| \geq 3, \gamma, \delta,$ *there exist* τ, w *such that if* $\mathrm{XInf}_i^{\leq w}(f_1, \ldots, f_{[k]}) < \tau$ *for any* i *then,*

$$\left| \mathop{\mathbf{E}}_{\mathcal{H}^n_{k,\delta}} \left[\prod_{S \in \mathbf{S}} T_{1-\gamma} f_S(\mathbf{x}_S) \right] - \mathop{\mathbf{E}}_{\mathcal{I}^n_{k,S_0,\delta}} \left[\prod_{S \in \mathbf{S}} T_{1-\gamma} f_S(\mathbf{x}_S) \right] \right| \leq \sqrt{\delta}.$$

Lemma 4. *For any* $S_0 \in \mathbf{S}, \gamma, \delta,$ *it holds that*

$$\left| \mathop{\mathbf{E}}_{\mathcal{I}^n_{k,S_0,\delta}} \left[\prod_{S \in \mathbf{S}} T_{1-\gamma} f_S(\mathbf{x}_S) \right] \right| \leq \sqrt{\delta}.$$

The proof of Lemma 4 is given in [42]. $\qquad \square$

4.2   Proof of Lemma 2

We define a lexicographical order $<_{\mathrm{lex}}$ between elements of $2^{[k]}_{\geq 1}$ as

$$1 <_{\mathrm{lex}} 2 <_{\mathrm{lex}} \cdots <_{\mathrm{lex}} n <_{\mathrm{lex}} \{1,2\} <_{\mathrm{lex}} \{1,3\} <_{\mathrm{lex}} \cdots <_{\mathrm{lex}} \{1,2,\ldots,k\},$$

and let $\mathbf{S}_l = \{S \in 2^{[k]}_{\geq 1} : S$ appears within lth by $<_{\mathrm{lex}}\}$. It is easy to see that the following lemma with the triangle inequality completes the proof of Lemma 2.

Lemma 5. *By taking* $\gamma > 0$ *small enough as a function of* $\delta, k,$ *we ensure, for any* $l \in [|2^{[k]}_{\geq 1}|],$

$$\left| \mathop{\mathbf{E}}_{\mathcal{H}^n_{k,\delta}} \left[\left(\prod_{S \in \mathbf{S}_l} f_S(\mathbf{x}_S) \right) \left(\prod_{S \in \mathbf{S} \setminus \mathbf{S}_l} T_{1-\gamma} f_S(\mathbf{x}_S) \right) - \right. \right.$$
$$\left. \left. \left(\prod_{S \in \mathbf{S}_{l-1}} f_S(\mathbf{x}_S) \right) \left(\prod_{S \in \mathbf{S} \setminus \mathbf{S}_{l-1}} T_{1-\gamma} f_S(\mathbf{x}_S) \right) \right] \right| \leq \sqrt{\delta}/2^{|2^{[k]}_{\geq 1}|}.$$

For the proof of Lemma 5, see [42]. $\qquad \square$

4.3 Proof of Lemma 3

From Proposition 2, by setting w as a function of γ and τ, we can assume $\mathrm{XInf}_i(T_{1-\gamma/2}f_1,\ldots,T_{1-\gamma/2}f_{[k]}) < 2\tau$ for all i where γ is the value chosen appropriately (see [42]). To prove Lemma 3 for the case $|\mathbf{S}| \geq 3$, we use the inductive proof. We replace the distribution from $\mathcal{H}^n_{k,\delta} = \otimes^n_{i=1}\mathcal{H}_{k,\delta}$ to $\mathcal{I}^n_{k,\delta} = \otimes^n_{i=1}\mathcal{I}_{k,S_0,\delta}$ one component at a time. We will show the following lemma.

Lemma 6. *For any $l \in [n]$,*

$$
\left|
\underset{\otimes^{l-1}_{i=1}\mathcal{I}_{k,S_0,\delta}\times\otimes^n_{i=l}\mathcal{H}_{k,\delta}}{\mathbf{E}}
\left[\prod_{S\in\mathbf{S}} T_{1-\gamma}f_S(\mathbf{x}_S)\right] -
\right.
$$
$$
\left.
\underset{\otimes^l_{i=1}\mathcal{I}_{k,S_0,\delta}\times\otimes^n_{i=l+1}\mathcal{H}_{k,\delta}}{\mathbf{E}}
\left[\prod_{S\in\mathbf{S}} T_{1-\gamma}f_S(\mathbf{x}_S)\right]
\right| \leq \Delta_l,
$$

where

$$
\Delta_l = \tau^{\gamma/2|\mathbf{S}|}2^{|\mathbf{S}|}\sum_{S\in\mathbf{S}}\mathrm{Inf}_l(T_{1-\gamma/2}f_S).
$$

For the proof of Lemma 6, see [42].

Proof (of Lemma 3). From Lemma 6,

$$
\left|
\underset{\mathcal{H}^n_{k,\delta}}{\mathbf{E}}\left[\prod_{S\in\mathbf{S}} T_{1-\gamma}f_S(\mathbf{x}_S)\right] -
\underset{\mathcal{I}^n_{k,\delta}}{\mathbf{E}}\left[\prod_{S\in\mathbf{S}} T_{1-\gamma}f_S(\mathbf{x}_S)\right]
\right|
$$
$$
\leq (2\tau)^{\gamma/2|\mathbf{S}|}2^{|\mathbf{S}|}\sum_{i\in[n]}\sum_{S\in\mathbf{S}}\mathrm{Inf}_i(T_{1-\gamma/2}f_S) \leq |\mathbf{S}|(2\tau)^{\gamma/2|\mathbf{S}|}2^{|\mathbf{S}|}/\gamma.
$$

By choosing $\tau = O\left((\gamma\sqrt{\delta}/2|\mathbf{S}|)^{\gamma/|\mathbf{S}|}\right)$, the lemma follows. $\square$

5 Discussion

Unfortunately, our Long Code test does not immediately imply a new PCP characterization of NP. A Long Code test without consistency checks is most easily composed with Unique Games [26] as an outer verifier in PCP constructions. However, since Unique Games cannot have perfect completeness, an obvious approach that combines our tester and Unique Games does not imply a new PCP construction. A variant of Unique Games, which is called d-to-1 Games [26], is conjectured to have perfect completeness. Though it seems hard to combine our tester and d-to-1 Games, we hope that it is possible and we obtain a new conditional PCP construction.

Acknowledgement

The authors thank anonymous reviewers for many suggestions on how to improve the presentation.

References

1. Austrin, P., Håstad, J.: Randomly supported independence and resistance. In: Proc. of STOC 2009, pp. 483–492 (2009)
2. Aumann, Y., Håstad, J., Rabin, M.O., Sudan, M.: Linear-Consistency Testing. J. Comput. Syst. Sci. 62(4), 589–607 (2001)
3. Alon, N., Kaufman, T., Krivelevich, M., Litsyn, S., Ron, D.: Testing low-degree polynomials over GF(2). In: Arora, S., Jansen, K., Rolim, J.D.P., Sahai, A. (eds.) RANDOM 2003 and APPROX 2003. LNCS, vol. 2764, pp. 188–199. Springer, Heidelberg (2003)
4. Arora, S., Lund, C., Motwani, R., Sudan, M., Szegedy, M.: Proof Verification and the Hardness of Approximation Problems. J. ACM 45(3), 501–555 (1998)
5. Austrin, P., Mossel, E.: Approximation Resistant Predicates from Pairwise Independence. Computational Complexity 18(2), 249–271 (2009)
6. Arora, S., Safra, S.: Probabilistic Checking of Proofs: A New Characterization of NP. J. ACM 45(1), 70–122 (1998)
7. Beckner, W.: Inequalities in Fourier analysis. Annals of Mathematics 102(1), 159–182 (1975)
8. Bellare, M., Coppersmith, D., Håstad, J., Kiwi, M.A., Sudan, M.: Linearity testing in characteristic two. IEEE Transactions on Information Theory 42(6), 1781–1795 (1996)
9. Bellare, M., Goldreich, O., Sudan, M.: Free Bits, PCPs, and Nonapproximability-Towards Tight Results. SIAM J. Comput. 27(3), 804–915 (1998)
10. Blais, E.: Improved Bounds for Testing Juntas. In: Goel, A., Jansen, K., Rolim, J.D.P., Rubinfeld, R. (eds.) APPROX and RANDOM 2008. LNCS, vol. 5171, pp. 317–330. Springer, Heidelberg (2008)
11. Blais, E.: Testing juntas nearly optimally. In: Proc. of STOC 2009, pp. 151–158 (2009)
12. Blum, M., Luby, M., Rubinfeld, R.: Self-Testing/Correcting with Applications to Numerical Problems. J. Comput. Syst. Sci. 47(3), 549–595 (1993)
13. Bonami, A.: Étude des coefficients de Fourier des fonctions de $L^p(G)$. Annales de l'Institut Fourier 20(2), 335–402 (1970)
14. Ben-Sasson, E., Harsha, P., Raskhodnikova, S.: Some 3CNF Properties Are Hard to Test. SIAM J. Comput. 35(1), 1–21 (2005)
15. Chen, V.: A hypergraph dictatorship test with perfect completeness. In: Dinur, I., Jansen, K., Naor, J., Rolim, J. (eds.) APPROX-RANDOM 2009. LNCS, vol. 5687, pp. 448–461. Springer, Heidelberg (2009)
16. Diakonikolas, I., Lee, H.K., Matulef, K., Onak, K., Rubinfeld, R., Servedio, R.A., Wan, A.: Testing for Concise Representations. In: Proc. of FOCS 2007, pp. 549–558 (2007)
17. Engebretsen, L., Holmerin, J.: More efficient queries in PCPs for NP and improved approximation hardness of maximum CSP. Random Struct. Algorithms 33(4), 497–514 (2008)
18. Fischer, E.: The art of uninformed decisions: A primer to property testing. The Bulletin of the European Association for Theoretical Computer Science 75, 97–126 (2001)
19. Guruswami, V., Lewin, D., Sudan, M., Trevisan, L.: A Tight Characterization of NP with 3 Query PCPs. In: Proc. of FOCS 1998, pp. 8–17 (1998)
20. Guruswami, V., Raghavendra, P.: Constraint Satisfaction over a Non-Boolean Domain: Approximation Algorithms and Unique-Games Hardness. In: Goel, A., Jansen, K., Rolim, J.D.P., Rubinfeld, R. (eds.) APPROX and RANDOM 2008. LNCS, vol. 5171, pp. 77–90. Springer, Heidelberg (2008)

21. Gross, L.: Logarithmic Sobolev inequalities. Amer. J. Math. 97, 1061–1083 (1975)
22. Håstad, J.: Some optimal inapproximability results. J. ACM 48(4), 798–859 (2001)
23. Håstad, J.: On the Approximation Resistance of a Random Predicate. In: Charikar, M., Jansen, K., Reingold, O., Rolim, J.D.P. (eds.) RANDOM 2007 and APPROX 2007. LNCS, vol. 4627, pp. 149–163. Springer, Heidelberg (2007)
24. Håstad, J., Khot, S.: Query Efficient PCPs with Perfect Completeness. Theory of Computing 1, 119–148 (2005)
25. Håstad, J., Wigderson, A.: Simple analysis of graph tests for linearity and PCP. Random Struct. Algorithms 22(2), 139–160 (2003)
26. Khot, S.: On the power of unique 2-prover 1-round games. In: Proc. of STOC 2002, pp. 767–775 (2002)
27. Khot, S.: Inapproximability results via Long Code based PCPs. SIGACT News 36(2), 25–42 (2005)
28. Khot, S., Saket, R.: A 3-Query Non-Adaptive PCP with Perfect Completeness. In: Proc. of IEEE Conference on Computational Complexity 2006, pp. 159–169 (2006)
29. Lovett, S.: Lower bounds for adaptive linearity tests. In: Proc. of STACS 2008, pp. 515–526 (2008)
30. Mossel, E., O'Donnell, R., Oleszkiewicz, K.: Noise stability of functions with low influences: invariance and optimality. Ann. Math. (to appear)
31. Mossel, E.: Gaussian Bounds for Noise Correlation of Functions. In: GAFA (to appear)
32. O'Donnell, R.: Analysis of Boolean functions lecture notes (2007), http://www.cs.cmu.edu/~odonnell/boolean-analysis/
33. O'Donnell, R., Wu, Y.: 3-bit dictator testing: 1 vs. 5/8. In: Proc. of SODA 2009, pp. 365–373 (2009)
34. O'Donnell, R., Wu, Y.: Conditional hardness for satisfiable 3-CSPs. In: Proc. of STOC 2009, pp. 493–502 (2009)
35. Parnas, M., Ron, D., Samorodnitsky, A.: Testing Basic Boolean Formulae. SIAM J. Discrete Math. 16(1), 20–46 (2002)
36. Raz, R.: A Parallel Repetition Theorem. SIAM J. Comput. 27(3), 763–803 (1998)
37. Ron, D.: Property Testing: A Learning Theory Perspective, 112 pages. Now Publishers Inc. (2008)
38. Sudan, M., Trevisan, L.: Probabilistically Checkable Proofs with Low Amortized Query Complexity. In: Proc. of FOCS 1998, pp. 18–27 (1998)
39. Samorodnitsky, A., Trevisan, L.: A PCP characterization of NP with optimal amortized query complexity. In: Proc. of STOC 2000, pp. 191–199 (2000)
40. Samorodnitsky, A., Trevisan, L.: Gowers Uniformity, Influence of Variables, and PCPs. SIAM J. Comput. 39(1), 323–360 (2009)
41. Trevisan, L.: Recycling Queries in PCPs and in Linearity Tests (Extended Abstract). In: Proc. of STOC 1998, pp. 299–308 (1998)
42. Tamaki, S., Yoshida, Y.: A Query Efficient Non-Adaptive Long Code Test with Perfect Completeness. Electronic Colloquium on Computational Complexity, TR09-074 (2009)
43. Zwick, U.: Approximation Algorithms for Constraint Satisfaction Problems Involving at Most Three Variables per Constraint. In: Proc. of SODA 1998, pp. 201–210 (1998)

Relativized Worlds without Worst-Case to Average-Case Reductions for NP

Thomas Watson[*]

Computer Science Division
University of California, Berkeley, CA, USA
tom@cs.berkeley.edu

Abstract. We prove that relative to an oracle, there is no worst-case to average-case reduction for NP. We also handle classes that are somewhat larger than NP, as well as worst-case to *errorless*-average-case reductions. In fact, we prove that relative to an oracle, there is no worst-case to errorless-average-case reduction from NP to BPP_{path}. The latter class contains $\text{P}_{\parallel}^{\text{NP}}$ and captures the power of randomized computations conditioned on efficiently testable events. We also handle reductions from NP to the polynomial-time hierarchy and beyond, under restrictions on the number of queries the reductions can make.

Keywords: Average-case complexity, oracles.

1 Introduction

The study of average-case complexity concerns the power of algorithms that are allowed to make mistakes on a small fraction of inputs. Of particular importance is the relationship between worst-case complexity and average-case complexity. For example, cryptographic applications require average-case hard problems, and it would be desirable to base the existence of such problems on minimal, worst-case complexity assumptions.

For the class PSPACE, it is known that worst-case hardness and average-case hardness are equivalent [1]. That is, if PSPACE is worst-case hard then it is also average-case hard. For the class NP, the situation is not well-understood. A central open problem in average-case complexity is to prove that if NP is worst-case hard then it is also average-case hard. Considering the lack of progress toward proving this proposition, a natural goal is to exhibit barriers to proving it, by ruling out certain general proof techniques. Bogdanov and Trevisan [3] considered the possibility of a *proof by reduction*. Building on [5], they showed that the proposition cannot be proven by a nonadaptive reduction unless the polynomial-time hierarchy collapses; it remains open to provide evidence against the existence of adaptive reductions. Another possibility that has been considered is a *relativizing proof*. In 1995, Impagliazzo and Rudich claimed [9] that they

---

[*] Supported by a National Science Foundation Graduate Research Fellowship.

M. Serna et al. (Eds.): APPROX and RANDOM 2010, LNCS 6302, pp. 752–765, 2010.
© Springer-Verlag Berlin Heidelberg 2010

had constructed a relativized heuristica, which is a world in which NP is worst-case hard but average-case easy, thus ruling out this possibility. However, they have since retracted their claim. We make progress toward obtaining relativized heuristica, by ruling out the possibility of a *relativizing proof by reduction*. Our barrier holds even for adaptive reductions. More formally, we prove that there exists an oracle relative to which there is no reduction of type

$$\big(\text{NP}, \text{PSAMP}\big) \subseteq \text{HeurBPP} \ \Rightarrow \ \text{NP} \subseteq \text{BPP}$$

where $\big(\text{NP}, \text{PSAMP}\big)$ is the class of distributional NP problems under polynomial-time samplable distributions, and HeurBPP is the class of distributional problems with polynomial-time average-case randomized algorithms.

We also generalize this result in various ways. The proposition that if NP is worst-case hard then it is also average-case hard concerns average-case algorithms that may output the wrong answer on a small fraction of inputs. In light of the aforementioned barriers, it is natural to consider the following proposition, which is potentially easier to prove: If NP is worst-case hard then it is also hard for *errorless* average-case algorithms, which may output "don't know" on a small fraction of inputs but must never output the wrong answer.[1] Our result generalizes to rule out relativizing proofs by reduction of this proposition. Further, we show how to rule out relativizing proofs by reduction that if NP is worst-case hard then certain classes larger than NP are errorless-average-case hard.

Independently of our work, Impagliazzo [10] has succeeded in constructing a relativized heuristica, even for errorless average-case algorithms, which subsumes our result for NP. However, this does not subsume our results for classes higher than NP, although Impagliazzo conjectures that this may be possible using his techniques.

1.1   Notions of Reductions and Relationship to Previous Work

Various models of worst-case to average-case reductions for NP have been considered in the literature, and they can be informally taxonomized as follows.

For the moment let us gloss over the issue of which distribution on inputs an average-case algorithm is judged with respect to. A worst-case to average-case reduction for NP must show that for every $L_1 \in \text{NP}$ there exists an $L_2 \in \text{NP}$ such that if L_2 has a polynomial-time average-case algorithm then L_1 has a polynomial-time worst-case algorithm. The worst-case algorithm for L_1 depends on the hypothesized average-case algorithm for L_2 in some way, which we call the *decoding*. There are the following four natural types of dependence, in decreasing order of strength.

(1) Black-box dependence means that the worst-case algorithm for L_1 has oracle access to the average-case algorithm for L_2, and it must solve L_1 on all

---

[1] An equivalent notion of an errorless average-case algorithm is one that always outputs the correct answer but whose running time is only "polynomial-on-average" [14].

inputs for *every* oracle that solves L_2 on most inputs, regardless of whether the oracle represents an efficient algorithm.

(2) The worst-case algorithm for L_1 might have oracle access to the average-case algorithm for L_2 but only be guaranteed to solve L_1 when the oracle is, in fact, an efficient average-case algorithm for L_2.

(3) The worst-case algorithm for L_1 might require the code of an efficient average-case algorithm for L_2.

(4) The dependence can be arbitrary, meaning that if L_2 has an efficient average-case algorithm then L_1 has an efficient worst-case algorithm. This type of dependence allows for arbitrary proofs that if NP is worst-case hard then it is also average-case hard.

For the first three types, the algorithm that solves L_1 with the aid of a hypothesized average-case algorithm for L_2 is called the reduction itself. In this paper we consider type (1) decoding. Note that since our results are about relativization, the reductions we consider have access to two oracles: the reduction oracle (representing the hypothesized average-case algorithm) and the relativization oracle.

Bogdanov and Trevisan [3] also considered type (1) decoding. They showed that such a reduction cannot exist unless the polynomial-time hierarchy collapses, provided the reduction is nonadaptive in its oracle access to the hypothesized average-case algorithm. Compared to the Bogdanov-Trevisan barrier, our barrier has the advantages that it is unconditional and it applies to adaptive reductions, but has the disadvantage that it only applies to reductions that relativize.

Gutfreund et al. [6] showed a *positive* result, namely that there is a worst-case to average-case reduction for NP with type (2) decoding, under a distribution on inputs that is samplable in slightly-superpolynomial time. Building on this result, Gutfreund and Ta-Shma [7] showed that under a certain weak derandomization hypothesis, there is a worst-case to average-case reduction from NP to nondeterministic slightly-superpolynomial time with type (2) decoding, under the uniform distribution on inputs. Moreover, the results of [6,7] relativize.

Another aspect of worst-case to average-case reductions is the *encoding*, which refers to the way in which L_2 depends on L_1. Black-box encoding means that the algorithm that defines L_2 has oracle access to L_1, and for *every* language L_1 (not just those in NP), if the corresponding L_2 has an efficient average-case algorithm then L_1 has an efficient worst-case algorithm (via one of the above four types of decoding).

Viola [16,17] proved two results about worst-case to average-case reductions with black-box encoders implementable in the polynomial-time hierarchy. In [16] he proved unconditionally that such a reduction with type (1) decoding does not exist. In [17] he proved that if such a reduction with type (4) decoding exists then PH is average-case hard, and thus basing the average-case hardness of PH on the worst-case hardness of PH in this way is no easier than unconditionally proving the average-case hardness of PH.

1.2 Results

Our first result concerns the class BPP_{path}, which was introduced by Han et al. [8], who also showed that relative to every oracle, $P_{\parallel}^{NP} \subseteq BPP_{path} \subseteq BPP^{NP}$. The class of distributional problems with polynomial-time errorless average-case randomized algorithms is denoted by AvgZPP.

Theorem 1. *There exists an oracle relative to which there is no reduction of type*

$$\left(BPP_{path}, PSAMP\right) \subseteq AvgZPP \;\Rightarrow\; UP \subseteq BPP.$$

Note that the type of reduction considered in Theorem 1 is weaker than a worst-case to average-case reduction for NP, because BPP_{path} is larger than NP, AvgZPP is smaller than HeurBPP, and UP is smaller than NP. Ruling out weaker reductions yields a stronger result.

We also prove a similar result for $BPP_{\parallel,\, o(n/\log n)}^{NP}$, which denotes the class BPP^{NP} restricted to have $o(n/\log n)$ rounds of adaptivity in the NP oracle access but any number of queries within each round. In the current state of knowledge, $BPP_{\parallel,\, o(n/\log n)}^{NP}$ is incomparable to BPP_{path}.

Theorem 2. *There exists an oracle relative to which there is no reduction of type*

$$\left(BPP_{\parallel,\, o(n/\log n)}^{NP}, PSAMP\right) \subseteq AvgZPP \;\Rightarrow\; UP \subseteq BPP.$$

If we restrict our attention to reductions that use a limited number of queries, then we can handle classes even larger than BPP_{path} and $BPP_{\parallel,\, o(n/\log n)}^{NP}$.

Theorem 3. *For every polynomial q there exists an oracle relative to which there is no q-query reduction of type*

$$\left(PH, PSAMP\right) \subseteq AvgZPP \;\Rightarrow\; UP \subseteq BPP.$$

Since $BPP_{path} \subseteq PH$ and $BPP_{\parallel,\, o(n/\log n)}^{NP} \subseteq PH$ relative to every oracle, it may appear at first glance that Theorem 3 subsumes Theorem 1 and Theorem 2. The reason it does not is because of the order of the quantifiers. In Theorem 3, the reduction may not make as many queries as it likes; it may only make a fixed polynomial q number of queries even though its running time may be an arbitrarily high degree polynomial.

If we are willing to sacrifice all but two queries, then we can go quite a bit further than PH.

Theorem 4. *For every uniform complexity class of languages C there exists an oracle relative to which there is no 2-query reduction of type*

$$\left(C, PSAMP\right) \subseteq AvgZPP \;\Rightarrow\; UP \subseteq BPP.$$

The term "uniform complexity class of languages" has a somewhat technical meaning, which is explained in the full version of this paper, but it encompasses all "ordinary" complexity classes such as PSPACE and EXP^{EXP}.

2   Preliminaries

We refer the reader to the survey paper [4] for background on average-case complexity. In this section we provide preliminaries that are not completely standard.

2.1   Complexity Classes

For any randomized algorithm M, we let M_r denote M using internal randomness r.

Definition 1. BPP_{path} *denotes the class of languages L such that for some polynomial-time randomized algorithm M that outputs two bits, and for all x,*

- $\Pr_r \left[M_r(x)_2 = 1 \right] > 0$ *and*
- $\Pr_r \left[M_r(x)_1 = L(x) \mid M_r(x)_2 = 1 \right] \geq 2/3$.

The above definition of BPP_{path} is not the same as the original one given by Han et al. [8], but it is equivalent relative to every oracle, and it is more convenient for our purposes. Intuitively, BPP_{path} captures the power of polynomial-time randomized computations after conditioning on efficiently testable events.

We now define the average-case complexity classes we need. Recall that in average-case complexity, we study distributional problems (L, D) where L is a language and $D = (D_1, D_2, \ldots)$ is an ensemble of probability distributions, where D_n is distributed over $\{0, 1\}^n$. Recall that PSAMP denotes the class of polynomial-time samplable ensembles, and $\mathcal{U}$ denotes the class consisting of only the uniform ensemble U. If $\mathcal{C}$ is a class of languages and $\mathcal{D}$ is a class of ensembles then $(\mathcal{C}, \mathcal{D}) = \left\{ (L, D) \ : \ L \in \mathcal{C} \text{ and } D \in \mathcal{D} \right\}$.

Definition 2. HeurBPP *denotes the class of distributional problems (L, D) that have a polynomial-time heuristic scheme, that is, a randomized algorithm M that takes as input x and $\delta > 0$, runs in time polynomial in $|x|$ and $1/\delta$, and for all n and all $\delta > 0$ satisfies*

$$\Pr_{x \sim D_n, r} \left[M_r(x, \delta) \neq L(x) \right] \ \leq \ \delta.$$

Definition 3. AvgZPP *denotes the class of distributional problems (L, D) that have a polynomial-time errorless heuristic scheme, that is, a randomized algorithm M that takes as input x and $\delta > 0$, runs in time polynomial in $|x|$ and $1/\delta$, always outputs $L(x)$ or $\bot$, and for all n and all $\delta > 0$ satisfies*

$$\Pr_{x \sim D_n, r} \left[M_r(x, \delta) = \bot \right] \ \leq \ \delta.$$

2.2   Reductions

In this section we informally explain what we mean when we say there exists a reduction of type

$$\mathcal{C}_2' \subseteq \mathcal{C}_2 \ \Rightarrow \ \mathcal{C}_1' \subseteq \mathcal{C}_1$$

where $\mathcal{C}_2', \mathcal{C}_2, \mathcal{C}_1', \mathcal{C}_1$ are four complexity classes. In Section 2.3 below we give formal definitions for the specific classes to which our theorems apply.

A complexity class is a set of computational problems, such as languages or distributional problems. We assume for concreteness that each of $\mathcal{C}_1$ and $\mathcal{C}_2$ is defined in the following way. By an *input-output relationship* we mean a randomized function. There is a set of algorithms, each of which induces an input-output relationship. That is, each algorithm takes an input and produces an output sampled from some distribution depending on the input. There is a predicate that indicates for each input-output relationship and each computational problem whether the input-output relationship *solves* the problem. There is a notion of computational resources used by the algorithms, and an algorithm is said to be *efficient* if it satisfies certain resource constraints. The class is defined as the set of problems solved by efficient algorithms. This type of definition encompasses classes defined in terms of (uniform or nonuniform) deterministic, randomized, or quantum algorithms, but it could be generalized to handle other models as well.

We also assume that for $\mathcal{C}_1$ there is an analogous set of algorithms that can make queries to a *reduction oracle*, which represents an input-output relationship.[2] We assume that plugging any algorithm from $\mathcal{C}_2$'s set into the reduction oracle yields an algorithm from $\mathcal{C}_1$'s set.

Now suppose P_1 is a computational problem of the appropriate kind for $\mathcal{C}_1$ and P_2 is a computational problem of the appropriate kind for $\mathcal{C}_2$.

Definition 4. *A reduction of type*

$$P_2 \in \mathcal{C}_2 \;\Rightarrow\; P_1 \in \mathcal{C}_1$$

is an algorithm from $\mathcal{C}_1$'s set of reduction oracle algorithms, such that for every reduction oracle that solves P_2 according to $\mathcal{C}_2$, the reduction solves P_1 according to $\mathcal{C}_1$ and it satisfies $\mathcal{C}_1$'s resource constraints if we pretend each query to the reduction oracle uses any amount of resources allowed by $\mathcal{C}_2$'s resource constraints.

Note that if we plug an actual, efficient algorithm for P_2 (according to $\mathcal{C}_2$) into the reduction oracle of such a reduction, then the reduction becomes an efficient algorithm for P_1 (according to $\mathcal{C}_1$). Thus if there exists a reduction satisfying Definition 4 then $P_2 \in \mathcal{C}_2$ implies $P_1 \in \mathcal{C}_1$. But the reduction must work even when the reduction oracle is an input-output relationship that is not efficiently implementable.

As an example, suppose $\mathcal{C}_2 = \mathrm{BPTIME}(2^{n^\epsilon})$. Then the reduction must solve P_1 according to $\mathcal{C}_1$ when the reduction oracle is any randomized function from $\{0,1\}^*$ to $\{0,1\}$ that, on input w, returns $P_2(w)$ with probability $\geq 2/3$.[3]

---

[2] In particular, the reduction oracle is not like a relativization oracle, which just answers queries to a language.

[3] One might wonder about reductions that can also choose the randomness used by the reduction oracle. While this would be more general in one sense, it would be more restrictive in the sense that it would limit the randomness complexity of the reduction oracle. In this paper, queries are always just inputs to an input-output relationship as defined above.

Further, the reduction must satisfy the resource constraints of $\mathcal{C}_1$ when we pretend each query of length n to the reduction oracle takes time $O(2^{n^\epsilon})$.

Definition 5. *We say there exists a reduction of type*

$$\mathcal{C}_2' \subseteq \mathcal{C}_2 \;\Rightarrow\; \mathcal{C}_1' \subseteq \mathcal{C}_1$$

if for every $P_1 \in \mathcal{C}_1'$ there exists a $P_2 \in \mathcal{C}_2'$ and a reduction of type

$$P_2 \in \mathcal{C}_2 \;\Rightarrow\; P_1 \in \mathcal{C}_1.$$

We make a few remarks about Definition 5.

- When $\mathcal{C}_1'$ has an appropriately complete problem P_1, this is equivalent to saying there exists a $P_2 \in \mathcal{C}_2'$ and a reduction of the above type, for the fixed problem P_1.
- Note that we do not require that the reduction is uniform in the sense of there being a fixed algorithm R that computes the reduction for every $P_1 \in \mathcal{C}_1'$ given the code for a $\mathcal{C}_1'$-type algorithm for P_1.
- Note that when we say there is a reduction of the above type, this assertion gets weaker as $\mathcal{C}_2'$ and $\mathcal{C}_1$ get larger and $\mathcal{C}_2$ and $\mathcal{C}_1'$ get smaller.

2.3   Relativization

When we relativize to an oracle language A, every computation gets unrestricted oracle access to A. This includes samplers and reductions. Thus reductions have access to two oracles: the reduction oracle and the relativization oracle. When we write $R^{B,A}$ we mean B is the reduction oracle and A is the relativization oracle for reduction R.

To illustrate the formal framework set up so far, we give the precise statement of Theorem 1. There exists a language A and a language $L_1 \in \mathrm{UP}^A$ such that for all languages $L_2 \in \mathrm{BPP}^A_{\mathrm{path}}$, all ensembles $D \in \mathrm{PSAMP}^A$, and all polynomial-time randomized reductions $R^{\circ,\circ}$, $R^{\circ,A}$ is not of type

$$(L_2, D) \in \mathrm{AvgZPP}^A \;\Rightarrow\; L_1 \in \mathrm{BPP}^A.$$

The latter means that there exists an $x \in \{0,1\}^*$ and a randomized function $B : \{0,1\}^* \times \mathbb{R}_{>0} \to \{0,1,\bot\}$ which is a valid AvgZPP oracle for (L_2, D), such that

$$\Pr_{r,B}\left[R_r^{B,A}(x) = L_1(x) \right] < 2/3$$

where the probability is over both the internal randomness of R and the randomness of B (each query is answered with fresh independent randomness). When we say B is a valid AvgZPP oracle for (L_2, D) we mean that $B(w, \delta)$ always returns $L_2(w)$ or $\bot$, and for all n and all $\delta > 0$,

$$\Pr_{w \sim D_n, B}\left[B(w, \delta) = \bot \right] \le \delta.$$

When we say $R^{\circ,\circ}$ runs in polynomial time, this includes the fact that each query $B(w, \delta)$ to the reduction oracle is charged time polynomial in $|w|$ and $1/\delta$. In other words, δ must always be at least inverse polynomial. Throughout the paper we tacitly assume that "polynomial-time reductions" have this restriction, since $\mathcal{C}_2$ is always AvgZPP. We clarify that $D \in \text{PSAMP}^A$ means that for some randomized algorithm S°, $S^A(n)$ runs in time polynomial in n and outputs a sample distributed according to D_n.

Regarding Theorem 3 and Theorem 4, there is one further issue to consider. For reductions that are allowed an unlimited number of queries (like in Theorem 1 and Theorem 2), the error probability of $1/3$ in the definition of BPP is unimportant since it can be amplified from $1/2 - 1/\text{poly}(n)$ to $1/2^{\text{poly}(n)}$. However, amplification increases the number of queries, so the error probability is not arbitrary for Theorem 3 and Theorem 4. For example, the existence of a q-query $\big(1/2 - 1/\text{poly}(n)\big)$-error reduction of type

$$\big(\text{PH}, \text{PSAMP}\big) \subseteq \text{AvgZPP} \;\Rightarrow\; \text{UP} \subseteq \text{BPP}$$

does not seem to imply the existence of a q-query $1/3$-error reduction of the same type, but it still does imply that if $\big(\text{PH}, \text{PSAMP}\big) \subseteq \text{AvgZPP}$ then $\text{UP} \subseteq \text{BPP}$. For this reason, we allow an error probability of $1/2 - 1/\text{poly}(n)$ (for arbitrarily high degree polynomials) in Theorem 3 and Theorem 4.

3   Intuition

In Section 3.1 we give the intuition behind the proof of Theorem 1. The intuition for Theorem 2 is omitted here, but it uses the same technique. In Section 3.2 we give very brief intuition for Theorem 3 and Theorem 4. The full version of this paper contains detailed intuition and formal proofs for all four theorems.

3.1   Intuition for Theorem 1

We start by informally describing how to construct an oracle relative to which there is no reduction of type

$$\big(\text{NP}, \mathcal{U}\big) \subseteq \text{HeurBPP} \;\Rightarrow\; \text{UP} \subseteq \text{BPP}.$$

To obtain Theorem 1, we must strengthen HeurBPP to AvgZPP[4], strengthen $\mathcal{U}$ to PSAMP, and strengthen NP to BPP_{path}. Strengthening HeurBPP to AvgZPP and $\mathcal{U}$ to PSAMP are not difficult, so we omit the intuition here.

Fix an arbitrary NP-type algorithm M and an arbitrary polynomial-time randomized reduction R, and fix a sufficiently large n. We explain how to diagonalize against the pair M, R. For simplicity we assume that on inputs of length n, R only queries the reduction oracle on inputs of length n^d and only with $\delta = 1/n^d$

---

[4] Usually AvgZPP is thought of as being a weaker class than HeurBPP (since $\text{AvgZPP} \subseteq \text{HeurBPP}$), but it is stronger in our situation.

for some positive integer d; thus we can omit the δ. We consider relativization oracles of the form $A : \{0,1\}^n \times \{0,1\}^n \to \{0,1\}$, which we think of as $2^n \times 2^n$ tables. Let $L_1^A : \{0,1\}^n \to \{0,1\}$ be defined by $L_1^A(x) = \bigvee_y A(xy)$. That is, L_1^A is the language of strings x such that there exists a 1 in the xth row of A. Let $L_2^A : \{0,1\}^{n^d} \to \{0,1\}$ denote the language computed by M^A. We only consider A, L_1^A, L_2^A at these input lengths since all other input lengths are irrelevant.

We wish to construct an A such that for some $x \in \{0,1\}^n$ and some deterministic[5] reduction oracle $B : \{0,1\}^{n^d} \to \{0,1\}$, B agrees with L_2^A on at least a $1 - 1/n^d$ fraction of inputs and $R^{B,A}(x)$ outputs $L_1^A(x)$ with probability $< 2/3$. This will show that R fails to be a reduction of type

$$\left(L_2^A, U\right) \in \text{HeurBPP}^A \;\Rightarrow\; L_1^A \in \text{BPP}^A.$$

We also need to ensure that there is at most one 1 in each row of A so that $L_1^A \in \text{UP}^A$, but this will fall right out of the construction. We construct A through an iterative process, and we use a potential function argument to show that this process makes steady progress toward our goal. The process iteratively modifies the relativization oracle, and we use A to denote the relativization oracle throughout the whole process.[6] Thus the table denoted by A changes many times throughout our argument, and the languages L_1^A and L_2^A change accordingly. Initially A is all 0's.

Let us consider the computation of R on some input x. It is trying to figure out whether there is a 1 in the xth row of A, in other words, compute $L_1^A(x)$. It has two sources of information about $L_1^A(x)$: the relativization oracle A itself, and the reduction oracle B. If R did not have access to B, then we could diagonalize in a standard way: Observe how R behaves given that the xth row of A is all 0's. If R outputs 1 with high probability, then we are done. If R outputs 1 with low probability, then we find a bit in the xth row that R queries with only tiny probability and flip that bit (such a bit must exist because R does not have enough time to keep an eye on the entire row); then R still outputs 1 with low probability, but now $x \in L_1^A$. Thus R must rely on the reduction oracle B for help.

Our construction has two stages. The goal of stage 1 is to gain the upper hand by rendering B useless to R. Then in stage 2 we deliver the coup de grâce with the standard diagonalization argument. We cannot guarantee that B is useless for every x, but we only need it to be useless for some x. Specifically, suppose we could set up A in such a way that there exists an x such that

---

[5] B will be deterministic here even though randomness is allowed; this makes the result stronger.

[6] More formally, we could say we define a sequence of relativization oracles $A_0, A_1, A_2, \ldots$ that leads to some final version $A_k = A$. We omit the subscripts throughout the argument and simply refer to A with the understanding that this means the "current" version.

(1) the xth row of A is all 0's, and
(2) for all y, flipping $A(xy)$ would cause $L_2^A(w)$ to change for at most a $1/n^d$ fraction of w's.

Then declaring B to be L_2^A for the particular A we have set up, we know that we can leave A alone or we can flip any bit in the xth row, and for all these possibilities B is a valid HeurBPP oracle for the new L_2^A. Then we can observe the behavior of R on input x, using this fixed B for the reduction oracle, and diagonalize against R in the standard way with the assurance that whatever happens to A during this second stage, B will remain valid.

How do we set up A so that such an x exists? We do this iteratively. In each iteration, we find a certain x whose row is currently all 0's, which is our "best guess" for the good x. If condition (2) is satisfied for this x, then we are done. Otherwise, there is some column y that violates condition (2). Then we flip the bit $A(xy)$ to 1 and continue with the next iteration. We just need to show that there are $< 2^n$ iterations before we succeed. For this, we define a potential function Φ^A that assigns an energy value to A. The key is to show that if y violates condition (2) for our best guess x, then flipping $A(xy)$ must cause a significant decrease in potential. Since Φ^A must remain bounded, there cannot be too many iterations before M is beaten into submission and our best guess x works.

Let us hold off on the definition of Φ^A and focus on finding a best guess x. Our ultimate goal is to ensure that if we flip any bit in the xth row, most of the inputs to L_2^A "don't notice". There is an asymmetry between inputs that are accepted by M^A and those that are rejected. If $w \in \{0,1\}^{n^d}$ is such that $M^A(w)$ rejects, then if any of the exponentially many computation paths "notices" a change in A, the whole computation could become accepting. However, if $M^A(w)$ accepts, then we can pick an arbitrary accepting computation path of $M^A(w)$ to be the "designated" one. Only polynomially many bits of A are queried by M on this path, and as long as none of these bits is flipped, w "won't notice" any change to A because $M^A(w)$ will still accept. In particular, there are only polynomially many x's such that $M^A(w)$ queries some bit in the xth row on the designated path. Thus for every w with $L_2^A(w) = 1$, the vast majority of x have the property that flipping any bit in the xth row does not cause $L_2^A(w)$ to change to 0. By an averaging argument, most x have the property that for most $w \in \{0,1\}^{n^d}$, flipping any bit in the xth row does not cause $L_2^A(w)$ to change from 1 to 0. For the current A, there must exist an x with the latter property and such that the xth row is all 0's, since (by induction) we know there are not very many x's with a 1 in their row currently. This is our best guess x.

We know that flipping any bit in the xth row causes only a small fraction of all $w \in \{0,1\}^{n^d}$ to change from 1 to 0 under L_2^A. This is good, but it is only half the story. We would also like that flipping any bit in the xth row causes only a small fraction of w's to change from 0 to 1. Suppose we budget a $1/2n^d$ fraction of w's to change from 1 to 0, and a $1/2n^d$ fraction to change from 0 to 1. Now if some y violates condition (2), then it must be the case that flipping $A(xy)$ causes at least a $1/2n^d$ fraction of w's to change from 0 to 1. We want to define the potential function so that having w's change from 0 to 1 under L_2^A causes a decrease in potential. A natural choice is

$$\Phi^A = \Pr_{w \sim U_{n^d}} \left[L_2^A(w) = 0 \right].$$

Flipping $A(xy)$ causes at least a $1/2n^d$ probability mass to leave the event $L_2^A(w) = 0$. However, as much as a $1/2n^d$ probability mass could enter the event due to w's that change from 1 to 0, which could essentially cancel out the drop in potential from the w's that changed from 0 to 1! The solution is to change our budgeting. If we budget a $1/3n^d$ fraction of w's to change from 1 to 0 and a $2/3n^d$ fraction to change from 0 to 1, then flipping $A(xy)$, where y violates condition (2), causes at least a $2/3n^d$ probability mass to leave the event, while at most a $1/3n^d$ probability mass enters the event. Thus Φ^A goes down by at least $1/3n^d$, and there are at most $3n^d < 2^n$ iterations before our best guess x works. This concludes the argument.

Very roughly, the big picture is as follows. For an input that is accepted by M^A, it is easy to ensure that the answer under L_2^A does not change when we make modifications to A. For an input that is rejected by M^A, we cannot ensure that the answer does not change, but the point is that if it does change, then *we can ensure that it does not change again*, since the input is now accepted.

Intuition for Strengthening NP to BPP$_{\text{path}}$. The difference from the above proof is in the definition of the potential function Φ^A, the choice of our best guess x, and the argument that if some y violates condition (2) for our best guess x, then flipping $A(xy)$ causes a significant decrease in potential.

Now instead of an NP-type algorithm we have a BPP$_{\text{path}}$-type algorithm M. Let us hold off on how to define Φ^A and how to choose our best guess x. Consider an arbitrary iteration of stage 1, let A denote the current relativization oracle, and suppose we have somehow picked a certain x such that the xth row of A is all 0's. Suppose there is a y such that flipping $A(xy)$ causes $L_2^A(w)$ to change for a significant fraction of w's. We want it to be the case that flipping $A(xy)$ also causes a significant decrease in potential. Let A' denote A with $A(xy)$ flipped to 1.

Consider a w such that $L_2^{A'}(w) \neq L_2^A(w)$. Let us make the bold assumption that for all choices of M's internal randomness r such that $M_r^A(w)_2 = 1$, we have $M_r^{A'}(w) = M_r^A(w)$ (that is, both output bits match). Then by the definition of BPP$_{\text{path}}$ we have

$$\Pr_r \left[M_r^{A'}(w)_2 = 1 \right] \geq 3 \cdot \Pr_r \left[M_r^{A'}(w)_1 = L_2^A(w) \text{ and } M_r^{A'}(w)_2 = 1 \right]$$

$$\geq 3 \cdot \Pr_r \left[M_r^A(w)_1 = L_2^A(w) \text{ and } M_r^A(w)_2 = 1 \right]$$

$$\geq 3 \cdot \left(\Pr_r \left[M_r^A(w)_2 = 1 \right] \cdot 2/3 \right)$$

$$= 2 \cdot \Pr_r \left[M_r^A(w)_2 = 1 \right]$$

where the second line follows because the event in the second line is a subset of the event on the right side of the first line. In other words, switching from A to A' forces the conditioning event to at least double in size, in order to reduce the probability of outputting $L_2^A(w)$ in the first bit (conditioned on that event) from $\geq 2/3$ to $\leq 1/3$. Thus

$$-\log_2 \Pr_r \left[M_r^{A'}(w)_2 = 1 \right] \leq -\log_2 \Pr_r \left[M_r^A(w)_2 = 1 \right] - 1.$$

This suggests using

$$\Phi^A = \mathop{\mathrm{E}}_{w \in \{0,1\}^{nd}} \left[-\log_2 \Pr_r \left[M_r^A(w)_2 = 1 \right] \right]$$

where w is chosen uniformly at random, because then when we flip $A(xy)$, a significant fraction of w's each contribute a significant negative amount to the potential difference $\Phi^{A'} - \Phi^A$. There are three issues.

(1) We need to make sure the potential is not too large to begin with.
(2) We made an unjustified assumption about the behavior of M.
(3) We also need to make sure that the contribution of bad w's to the potential difference does not cancel out the negative contribution of good w's.

Issue (1) is not problematic: Since we may assume r is chosen uniformly from $\{0,1\}^{\mathrm{poly}(n)}$, for every w and every A we must have

$$\Pr_r \left[M_r^A(w)_2 = 1 \right] \geq 2^{-\mathrm{poly}(n)}$$

since otherwise the conditioning event would be empty and M^A would fail to define a language in $\mathrm{BPP}_{\mathrm{path}}^A$ (for the violating A), which would suffice to diagonalize against the pair M, R.

For issue (2), first note that if we relax our assumption to be that for *almost all* r such that $M_r^A(w)_2 = 1$, we have $M_r^{A'}(w) = M_r^A(w)$, then flipping $A(xy)$ still causes the probability of the conditioning event to go up by at least a constant factor (say $3/2$) assuming $L_2^{A'}(w) \neq L_2^A(w)$. Now we use our ability to choose x. Since M runs in polynomial time, it can be shown that most x are *useful*, in the sense that for the vast majority of w's it is the case that for almost all r such that $M_r^A(w)_2 = 1$, $M_r^A(w)$ does not query any bit in the xth row. Thus we can pick our best guess x so that x is useful and the xth row of A is all 0's. Then for our fixed x and y, we know that the vast majority of w's have the property that for almost all r such that $M_r^A(w)_2 = 1$, we have $M_r^{A'}(w) = M_r^A(w)$. Call the remaining w's *horrible*. Call w *good* if $L_2^{A'}(w) \neq L_2^A(w)$ and w is not horrible. Call w *bad* if it is not good. By a union bound we know that a significant fraction of w's are good, and each good w contributes a significant negative amount to the potential difference $\Phi^{A'} - \Phi^A$.

Finally we consider issue (3). We consider the horrible w's and the bad-but-not-horrible w's separately. The contribution of each horrible w to $\Phi^{A'} - \Phi^A$ could be as large as $\mathrm{poly}(n)$ (inside the expectation), but only a tiny fraction of w's are horrible so this only puts a small dent in the negative contribution from the good w's. Almost all of the w's could be bad-but-not-horrible, but the contribution of each such w to $\Phi^{A'} - \Phi^A$ can be at most a tiny positive amount, since of the r's with $M_r^A(w)_2 = 1$, almost all of them are such that $M_r^{A'}(w)_2 = M_r^A(w)_2 = 1$. Thus the bad-but-not-horrible w's only put a small dent in the negative contribution from the good w's.

3.2   Intuition for Theorem 3 and Theorem 4

It is well-known that error-correcting codes with efficient encoders and decoders can be used to construct worst-case to average-case reductions, at least for large complexity classes such as PSPACE [1,15]. Our strategy for proving Theorem 3 and Theorem 4 is to set up the relativization oracle in such a way that error-correcting codes are in some sense the *only* way to construct worst-case to average-case reductions of the appropriate types, and then argue that the efficiency of the resulting encoders and decoders is too good to be true. For Theorem 3, we use a result due to Viola [16] which states that good error-correcting codes cannot be encoded by small constant-depth circuits. For Theorem 4, we use a lower bound due to Kerenidis and de Wolf [13] on the length of 2-query locally decodable codes.

In a nutshell, here is the intuition. The reduction R cannot possibly hope to solve L_1^A by looking for the answers in A, because A is so vast. It must rely on M (the algorithm solving L_2^A) for help. Intuitively, there are only two ways M could help: by telling R the answers directly, or by telling R where to find the answers in A. If M tries to tell R the answers directly, then (roughly) we have an impossibly good error-correcting code where L_1^A is the information word, L_2^A is the code word, M is the encoder, and R is the decoder. If M tries to tell R where to find the answers in A, then R *cannot make enough queries to the reduction oracle to retrieve this information*. See the full version of this paper for details on how to formalize this intuition. We can handle arbitrary complexity classes in Theorem 4 since the lower bound of [13] holds regardless of the complexity of the encoder.

Acknowledgments. First and foremost, I thank Luca Trevisan for suggesting the research topic. I thank Luca Trevisan, Dieter van Melkebeek, and anonymous reviewers for helpful comments.

References

1. Babai, L., Fortnow, L., Nisan, N., Wigderson, A.: BPP Has Subexponential Time Simulations Unless EXPTIME has Publishable Proofs. Computational Complexity 3, 307–318 (1993)
2. Bennett, C., Bernstein, E., Brassard, G., Vazirani, U.: Strengths and Weaknesses of Quantum Computing. SIAM Journal on Computing 26(5), 1510–1523 (1997)
3. Bogdanov, A., Trevisan, L.: On Worst-Case to Average-Case Reductions for NP Problems. SIAM Journal on Computing 36(4), 1119–1159 (2006)

4. Bogdanov, A., Trevisan, L.: Average-Case Complexity. Foundations and Trends in Theoretical Computer Science 2(1) (2006)
5. Feigenbaum, J., Fortnow, L.: Random-Self-Reducibility of Complete Sets. SIAM Journal on Computing 22(5), 994–1005 (1993)
6. Gutfreund, D., Shaltiel, R., Ta-Shma, A.: If NP Languages are Hard on the Worst-Case, Then It Is Easy to Find Their Hard Instances. Computational Complexity 16(4), 412–441 (2007)
7. Gutfreund, D., Ta-Shma, A.: Worst-Case to Average-Case Reductions Revisited. In: Proceedings of the 11th International Workshop on Randomization and Computation, pp. 569–583 (2007)
8. Han, Y., Hemaspaandra, L., Thierauf, T.: Threshold Computation and Cryptographic Security. SIAM Journal on Computing 26(1), 59–78 (1997)
9. Impagliazzo, R.: A Personal View of Average-Case Complexity. In: Proceedings of the 10th IEEE Conference on Structure in Complexity Theory, pp. 134–147 (1995)
10. Impagliazzo, R.: Relativized Separations of Worst-Case and Average-Case Complexities for NP. Manuscript (2010)
11. Impagliazzo, R., Levin, L.: No Better Ways to Generate Hard NP Instances than Picking Uniformly at Random. In: Proceedings of the 31st IEEE Symposium on Foundations of Computer Science, pp. 812–821 (1990)
12. Impagliazzo, R., Luby, M.: One-Way Functions are Essential for Complexity Based Cryptography. In: Proceedings of the 30th IEEE Symposium on Foundations of Computer Science, pp. 230–235 (1989)
13. Kerenidis, I., de Wolf, R.: Exponential Lower Bound for 2-Query Locally Decodable Codes via a Quantum Argument. Journal of Computer and System Sciences 69(3), 395–420 (2004)
14. Levin, L.: Average Case Complete Problems. SIAM Journal on Computing 15(1), 285–286 (1986)
15. Sudan, M., Trevisan, L., Vadhan, S.: Pseudorandom Generators Without the XOR Lemma. Journal of Computer and System Sciences 62(2), 236–266 (2001)
16. Viola, E.: The Complexity of Constructing Pseudorandom Generators from Hard Functions. Computational Complexity 13(3-4), 147–188 (2004)
17. Viola, E.: On Constructing Parallel Pseudorandom Generators from One-Way Functions. In: Proceedings of the 20th IEEE Conference on Computational Complexity, pp. 183–197 (2005)

A Quadratic Lower Bound for Three-Query Linear Locally Decodable Codes over Any Field

David P. Woodruff

IBM Almaden
dpwoodru@us.ibm.com

Abstract. A linear $(q, \delta, \epsilon, m(n))$-locally decodable code (LDC) $C : \mathbb{F}^n \to \mathbb{F}^{m(n)}$ is a linear transformation from the vector space $\mathbb{F}^n$ to the space $\mathbb{F}^{m(n)}$ for which each message symbol x_i can be recovered with probability at least $\frac{1}{|\mathbb{F}|} + \epsilon$ from $C(x)$ by a randomized algorithm that queries only q positions of $C(x)$, even if up to $\delta m(n)$ positions of $C(x)$ are corrupted. In a recent work of Dvir, the author shows that lower bounds for linear LDCs can imply lower bounds for arithmetic circuits. He suggests that proving lower bounds for LDCs over the complex or real field is a good starting point for approaching one of his conjectures.

Our main result is an $m(n) = \Omega(n^2)$ lower bound for linear 3-query LDCs over any, possibly infinite, field. The constant in the $\Omega(\cdot)$ depends only on ε and δ. This is the first lower bound better than the trivial $m(n) = \Omega(n)$ for arbitrary fields and more than two queries.

Keywords: Error-Correcting Codes, Complexity Theory.

1 Introduction

Classical error-correcting codes allow one to encode an n-bit message x into a codeword $C(x)$ such that even if a constant fraction of the bits in $C(x)$ are corrupted, x can still be recovered. It is known how to construct such codes of length $O(n)$ that can tolerate a constant fraction of errors, even in such a way that allows decoding in linear time [1]. However, if one is only interested in recovering a few bits of the message, then these codes have the disadvantage that they require reading most of the codeword.

A locally decodable code (LDC) $C : \mathbb{F}^n \to \mathbb{F}^{m(n)}$ is an encoding from the vector space $\mathbb{F}^n$ to the space $\mathbb{F}^{m(n)}$ such that each message symbol x_i can be recovered with probability at least $\frac{1}{|\mathbb{F}|} + \epsilon$ from $C(x)$ by a randomized algorithm that reads only q positions of $C(x)$, even if up to $\delta m(n)$ positions in $C(x)$ are corrupted (here $\frac{1}{|\mathbb{F}|}$ is zero if $\mathbb{F}$ is infinite). If C is a linear transformation, then the LDC is said to be linear. LDCs in their full generality were formally defined by Katz and Trevisan [2]. Linear LDCs were first considered in work by Goldreich et al [3]. There is a vast body of work on LDCs; we refer the reader to Trevisan's survey [4] or to Yekhanin's thesis [5].

M. Serna et al. (Eds.): APPROX and RANDOM 2010, LNCS 6302, pp. 766–779, 2010.

While in general an LDC need not be linear, there is good motivation for studying this case. On the practical front, it is easy to encode a message and update a codeword given the generator matrix for a linear code. In applications of error-correcting codes to compressed sensing [6–8], the encoding is defined to be linear because of the physics of an optical lens. In large data streams, sketches are linear because they can be updated efficiently. On the theoretical front, lower bounds for linear 2-query LDCs are useful for polynomial identity testing [9]. These applications consider fields $\mathbb{F}$ of large or infinite size, e.g., in compressed sensing and streaming one has $\mathbb{F} = \mathbb{R}$.

In a surprising recent development, Dvir [10] shows that lower bounds for linear locally self-correctable codes and linear locally decodable codes imply lower bounds on the rigidity of a matrix, which in turn imply size/depth tradeoffs for arithmetic circuits [11]. In Section 5.1 of [10], the author suggests that proving lower bounds on linear locally correctable or linear locally decodable codes over the complex or real field is a good starting point for approaching one of his conjectures.

1.1   Results

Our main result is that for any (possibly infinite) field $\mathbb{F}$, any 3-query linear LDC requires $m(n) = \Omega(n^2)$, where the constant in the $\Omega(\cdot)$ notation depends only on ε and δ.

The first reason previous work does not give a non-trivial lower bound over arbitrary fields is that it uses a generic reduction from an adaptive decoder to a non-adaptive decoder, which effectively reduces ε to $\varepsilon/|\mathbb{F}|^{q-1}$. For constant q, if $\mathbb{F}$ is of polynomial size, one cannot beat the trivial $m(n) = \Omega(n)$ bound this way. We give a better reduction to a non-adaptive decoder.

Given our reduction, it then seems possible to obtain a field-independent $\Omega(n^{3/2})$ bound by turning the birthday paradox argument of Katz and Trevisan [2] into a rank argument. This is still weaker than our bound by a factor of $\sqrt{n}$.

Also, by using a technique of Kerenidis and de Wolf [12], it seems possible to obtain a bound of $\Omega(n^2/(|\mathbb{F}|^2 \log^2 n))$. This bound becomes trivial when $\mathbb{F} = \mathbb{R}$ or $\mathbb{F} = \mathbb{C}$, or even $|\mathbb{F}| = \text{poly}(n)$. Note that if taking $|\mathbb{F}| = \text{poly}(n)$ were to imply 3-query linear LDCs of linear size, then the encoding would need only a linear number of machine words. Our result rules out this possibility.

While the parameters of the LDCs considered by Dvir [10] over $\mathbb{R}$ or $\mathbb{C}$ are in a different regime than those considered here, e.g., he needs a bound for $q = \log^{2+\Omega(1)}(n)$ queries, our result provides the first progress on this problem for LDCs for more than two queries. We note that our results are not possible for non-linear codes, as one can encode n real numbers into a single real number.

An earlier technical report [13] by the author contains some of the ideas used here. That version of this paper has a weaker $m(n) = \Omega(n^2/\log\log n)$ bound for 3-query linear LDCs over any field. It also shows an $\Omega(n^2/\log n)$ bound for non-linear 3-query LDCs over $\mathbb{F}_2$ using a similar argument to that given here in Section 3.1. It contains polylogarithmic improvements over [12] for any odd

$q \geq 3$ number of queries. We do not know if for constant-sized fields, an $\Omega(n^2)$ bound holds for non-linear codes.

1.2   Techniques

In this section we give an overview of the techniques we use for our lower bound.

Let $C : \mathbb{F}^n \to \mathbb{F}^m$ be a linear 3-query LDC. Then each of its output coordinates $C_i(x)$ equals $\langle v_i, x \rangle$, for a vector $v_i \in \mathbb{F}^n$. As observed by Katz and Trevisan [2] for finite fields, since C can tolerate a large fraction of errors and is locally decodable, for each $i \in [n] \stackrel{\text{def}}{=} \{1, 2, \ldots, n\}$, there is a large matching (i.e., collection of disjoint sets) M_i of triples $\{v_a, v_b, v_c\}$ for which u_i, the i-th standard unit vector, is in $\text{span}\{v_a, v_b, v_c\}$. We generalize this to infinite fields, which requires some care since the matching sizes of Katz and Trevisan (and subsequent work of [3] and [12]) degrade with the field size for general adaptive decoders. For constant ε and δ (the setting we consider here), we show that for any field, $|M_i| = \Omega(m)$.

Given the matchings, we work in the 3-uniform multi-hypergraph G on vertex set $\{v_1, \ldots, v_m\}$ whose 3-edgeset is $\cup_{j=1}^n M_j$. The average degree of a vertex in G is $\Omega(n)$, and by standard arguments (iteratively remove the minimum degree vertex in the hypergraph and stop once the minimum degree is larger than the original average degree), we can find an induced sub-multi-hypergraph G' with minimum degree βn for a constant $\beta > 0$. In particular, it is easy to show that we can find a set T of αn linearly independent vertices of G' collectively incident to $\Omega(n^2)$ distinct 3-edges, where α is a constant satisfying $0 < \alpha < \beta$.

We now provide a new way to project 3-query LDCs down to 2-query LDCs. Suppose we extend T to a basis $T \cup U$ of $\mathbb{F}^n$ by greedily adding a set U of standard unit vectors. Consider the linear projection P for which T is in the kernel, but P restricted to U is the identity map. Suppose we apply P to every vertex in G'. Let $N(T)$ denote the set of vertices incident to T via a 3-edge $\{a, b, c\}$ in G', i.e., the neighborhood of T. Suppose $\{a, b, c\} \in M_i$. The key point is that after application of P, either the projection of $a, b,$ or c is equal to 0, since one of these vertices is in the kernel of P. But if $u_i \in U$, then $P(u_i) = u_i$. Hence, either $u_i \in \text{span}(P(a), P(b))$, $u_i \in \text{span}(P(a), P(c))$, or $u_i \in \text{span}(P(b), P(c))$. We can thus obtain large matchings of edges (as opposed to 3-edges), for which a standard unit vector is in the span of the endpoints. Notice that since $|U| \geq n - \alpha n$, whereas the minimum degree of each vertex in T is $\beta n > \alpha n$, each vertex is still incident to at least $(\beta - \alpha)n$ edges for different $i \in U$, which is already enough to prove an $\Omega(n^2 / \log n)$ lower bound by now resorting to known techniques for lower bounding 2-query LDCs [9].

The next and harder part is improving the bound to a clean $\Omega(n^2)$. Our lower bound comes from bounding the cardinality of the neighborhood $N(T)$ of T. Suppose this cardinality really were $\Theta(n^2 / \log n)$. Then there are $\Omega(n^2)$ hyperedges from T to its neighborhood. This means that the average degree of a vertex in $N(T)$ using the edges from T to $N(T)$ is $\Omega(\log n)$. By standard arguments we can find a set A of $\alpha' n$ vertices in $N(T)$ incident to a set B of $\Omega(n \log n)$ vertices in $N(T)$ via the edges from T to $N(T)$. Now if we augment the kernel of our projection to additionally include the vertices in A, as well as more standard unit vectors, we can put most of B into the kernel of our projection. We could not do this a priori, since putting a set B of more than n vertices in the kernel of a projection could make the projection equal to zero. Here, though, it is important that a constant fraction of standard unit vectors are preserved under projection.

We assumed that $N(T) = \Theta(n^2/\log n)$, when it could have been anywhere from $\omega(n^2/\log n)$ to $o(n^2)$. However, we can iteratively apply the above procedure, gradually enlarging the kernel while preserving a large number of standard unit vectors under projection. After $O(\log \log n)$ iterations, we show that the neighborhood of our resulting kernel has size $\Omega(n^2 \log n)$. We can then use lower bound techniques developed in the 2-query setting to deduce that $m = \Omega(n^2)$.

1.3   Related Work

Katz and Trevisan [2] show that 1-query LDCs do not exist.

For linear 2-query LDCs, Dvir and Shpilka [9] show that $m(n) \geq \exp(n)$ for[1] any field $\mathbb{F}$, and the Hadamard code shows this is optimal (see also [3], [14], [15]). We note that for non-linear 2-query LDCs, if the field $\mathbb{F}$ has constant size, then $m(n) \geq \exp(n)$ is also known to hold [12].

For more than 2 queries, there is a large gap between upper and lower bounds. This may, in part, be explained by the recent connections of Dvir [10]. The upper bounds for q-query LDCs are linear and have the form $m(n) = \exp(\exp(\log^{c/\log q} n \log^{1-c/\log q} \log n))$ for an absolute constant $c > 0$ ([16], [17], [18]). While the initial constructions were over finite fields, recently it was shown that similar upper bounds hold also over the real or complex numbers ([19], [20]).

The lower bounds are the aforementioned bounds of Katz and Trevisan [2] and of Kerenidis and de Wolf [12].

2   Preliminaries

Definition 2.1. *([2]) Let $\delta, \epsilon \in (0, 1)$, q an integer, and $\mathbb{F}$ a field. A linear transformation $C : \mathbb{F}^n \to \mathbb{F}^m$ is a linear (q, δ, ϵ)-locally decodable code (LDC for short) if there is a probabilistic oracle machine A such that:*

- *For every $x \in \mathbb{F}^n$, for every $y \in \mathbb{F}^m$ with $\Delta(y, C(x)) \leq \delta m$, and for every $i \in [n]$, $\Pr[A^y(i) = x_i] \geq \frac{1}{|\mathbb{F}|} + \epsilon$, where the probability is taken over the internal coin tosses of A. Here $\Delta(C(x), y)$ refers to the number of positions in $C(x)$ and y that differ.*
- *In every invocation, A makes at most q queries (possibly adaptively).*

In Section 4, we prove the following.

---

[1] Here $\exp(n)$ denotes $2^{\Theta(n)}$.

Theorem 2.1. *Let $C : \mathbb{F}^n \rightarrow \mathbb{F}^m$ be a linear $(3, \delta, \epsilon)$-LDC. Then C is also a linear $(3, \delta/9, 2/3 - 1/|\mathbb{F}|)$-LDC with a non-adaptive decoder.*

This improves known reductions to non-adaptive codes since it holds for any $\mathbb{F}$. Thus, we may assume that we have a non-adaptive decoder by changing δ and ϵ by constant factors.

By known results described in Appendix A, for every $i \in [n]$ there is a matching M_i of $\{v_1, \ldots, v_m\}$ of size $\Omega(m)$ (where the constant depends on ε, δ, and q) such that, if $e \in M_i$, then $u_i \in \text{span}(v \mid v \in e)$, where u_i denotes the unit vector in direction i. Consider the multi-hypergraph G with vertex set $\{v_1, \ldots, v_m\}$ and hyperedge set $\uplus_{i=1}^n M_i$, that is, a hyperedge e occurs in G once for each M_i that it occurs in. For readability, we use the term hypergraph to refer to a multi-hypergraph, that is, a hypergraph which may have repeated hyperedges (which we sometimes just refer to as edges).

In Appendix A, we show there is a non-empty hypergraph $G' \subseteq G$ with minimum degree βn, where β is such that the number of hyperedges in G is at least βmn.

3   Lower Bounds for 3-Queries over Any Field

3.1   The Basic Projection

Assume we have a linear $(3, \delta, \epsilon)$-LDC $C : \mathbb{F}^n \rightarrow \mathbb{F}^m$ for an arbitrary (possibly infinite) field $\mathbb{F}$. Throughout this section we shall use the term edge to denote a 3-edge (i.e., there are 3 endpoints) for ease of notation.

Let G be the hypergraph on vertex set $\{v_1, \ldots, v_m\}$ and G' the non-empty sub-hypergraph of G with minimum degree βn defined in Section 2. Let v be an arbitrary vertex in G', and let $T = \{v\} \cup N(v)$, where $N(v)$ denotes the set of neighbors of v in G' (i.e., the vertices in a 3-edge containing v). Remove vertices from T so that we are left with a set T of exactly αn linearly independent vectors, where $\alpha < \beta$ is a small enough constant specified by the analysis below. This is always possible because $\{v\} \cup N(v)$ spans βn linearly independent vectors.

We may assume, by increasing m by a factor of at most 3, that every edge in M_i has size exactly 3, and moreover, for every such edge $\{v_{j_1}, v_{j_2}, v_{j_3}\} \in M_i$, we have $u_i = \gamma_1 v_{j_1} + \gamma_2 v_{j_2} + \gamma_3 v_{j_3}$, where $\gamma_1, \gamma_2, \gamma_3$ are non-zero elements of $\mathbb{F}$. Indeed, we may append $2m$ constant functions which always output 0 to the end of C. Then, if an edge in M_i either has size less than 3 or has size 3 and has the form $\{v_{j_1}, v_{j_2}, v_{j_3}\}$, but satisfies $u_i = \gamma_1 v_{j_1} + \gamma_2 v_{j_2} + \gamma_3 v_{j_3}$ for some $\gamma_k = 0$, we can replace the γ_k with 1 and replace j_k with an index corresponding to one of the zero functions.

Let $v_1, \ldots, v_T$ denote the vectors in T. Extend $\{v_1, \ldots, v_T\}$ to a basis of $\mathbb{F}^n$ by adding a set U of $n - \alpha n$ standard unit vectors. Define a linear projection L as follows:

$$L(v) = 0 \text{ for all } v \in T \text{ and } L(v) = v \text{ for all } v \in U.$$

Since L is specified on a basis, it is specified on all of $\mathbb{F}^n$.

Let M'_i denote the collection of edges in M_i that are incident to some vertex in T. Let $e = \{v_{j_1}, v_{j_2}, v_{j_3}\}$ be an edge in some M'_i. Then there are non-zero $\gamma_1, \gamma_2, \gamma_3 \in \mathbb{F}$ for which $\gamma_1 v_{j_1} + \gamma_2 v_{j_2} + \gamma_3 v_{j_3} = u_i$. By linearity, $L(u_i) = L(\gamma_1 v_{j_1} + \gamma_2 v_{j_2} + \gamma_3 v_{j_3}) = \gamma_1 L(v_{j_1}) + \gamma_2 L(v_{j_2}) + \gamma_3 L(v_{j_3})$. By definition of M'_i, $|\{v_{j_1}, v_{j_2}, v_{j_3}\} \cap T| > 0$, so one of the following must be true: $L(u_i) \in \text{span}(L(v_{j_1}), L(v_{j_2}))$, $L(u_i) \in \text{span}(L(v_{j_1}), L(v_{j_3}))$, or $L(u_i) \in \text{span}(L(v_{j_2}), L(v_{j_3}))$.

Thus, for each such edge $e = \{v_{j_1}, v_{j_2}, v_{j_3}\}$, by removing exactly one vector $v_{j_\ell} \in \{v_{j_1}, v_{j_2}, v_{j_3}\}$ for which $L(v_{j_\ell}) = 0$, we may define matchings W_i of disjoint pairs $\{v_j, v_k\}$ of $\{v_1, \ldots, v_m\}$ such that if $\{v_j, v_k\} \in W_i$, then $L(u_i) \in \text{span}(L(v_j), L(v_k))$. Moreover, $\sum_{i=1}^n |W_i| = \sum_{i=1}^n |M'_i|$.

Say an index $i \in [n]$ *survives* if $L(u_i) = u_i$, and say an edge e *survives* if $e \in M'_i$ for an i that survives. If i survives, then $u_i \in U$, as otherwise we would have $u_i = \sum_{v \in T} \gamma_v v + \sum_{u \in U} \gamma_u u$ for some coefficients $\gamma_v, \gamma_u \in \mathbb{F}$. Applying L to both sides we would obtain $u_i = L(u_i) = \sum_{u \in U} \gamma_u L(u) = \sum_{u \in U} \gamma_u u$, which is impossible unless $u_i \in U$.

Recall that each of the αn vertices v in T has degree at least βn in G'. For any such $v \in T$, there are at least $\beta n - \alpha n$ edges e in the disjoint union of the M'_i for the i the survive. Thus, since each edge that survives can be incident to at most 3 elements of T, and since $\alpha < \beta$,

$$\sum_{i \text{ that survive}} |W_i| \geq \alpha n(\beta - \alpha)n/3 = \Omega(n^2).$$

For i that do not survive, we set $W_i = \emptyset$. We need a theorem due to Dvir and Shpilka [9].

Theorem 3.1. *([9]) Let $\mathbb{F}$ be any field, and let $a_1, \ldots, a_m \in \mathbb{F}^n$. For every $i \in [n]$, let M_i be a set of disjoint pairs $\{a_{j_1}, a_{j_2}\}$ such that $u_i \in \text{span}(a_{j_1}, a_{j_2})$. Then, $\sum_{i=1}^n |M_i| \leq m \log m + m$.*

Applying Theorem 3.1 to our setting, we have m vectors $L(v_j) \in \mathbb{F}^n$ and matchings W_i with $\sum_i |W_i| = \Omega(n^2)$. We conclude that,

Theorem 3.2. *For $\delta, \epsilon \in (0, 1)$, if $C : \mathbb{F}^n \to \mathbb{F}^m$ is a linear $(3, \delta, \epsilon)$-locally decodable code, then $m = \Omega_{\delta, \epsilon}(n^2/\log n)$, independent of the field $\mathbb{F}$.*

3.2 Recursing to Get the $\Omega(n^2)$ Bound

We assume that $\beta > 2\alpha$ and w.l.o.g., that $(\beta - 2\alpha)n$ is a power of 2 and αn is an integer. For a set $A \subseteq \mathbb{F}^n$, let $ex(A)$ denote a maximal linearly independent subset of A.

Base Case: As before, let G' be the hypergraph on 3-edges with minimum degree βn, and let $T_1 = T$ be the set of αn linearly independent vertices defined in Section 3.1. We extend T_1 to a basis of $\mathbb{F}^n$ by greedily adding a set U of $n - \alpha n$ standard unit vectors to T_1. Set $B_1 = U$. Since each vertex in T_1 has degree at least βn, since $|T_1| = \alpha n$, and since each matching edge can be counted at most 3 times, the set E of 3-edges labeled by a $u \in B_1$ and incident to T_1 has size at least $\alpha n(\beta - \alpha)n/3$.

For each $u \in B_1$, let f_u denote the number of edges in E labeled by u, i.e., in the matching M_u. Order the unit vectors so that $f_{u_1} \geq f_{u_2} \geq \cdots \geq f_{u_{|B_1|}}$, and let $E_1 \subset E$ be the subset of edges incident to T_1 labeled by a unit vector in the set U_1 of the first $\frac{(\beta - 2\alpha)n}{2}$ unit vectors. Set $V_1 = T_1$.

Inductive Step: We construct sets T_i, B_i, U_i, E_i, and V_i, $i \geq 2$, as follows. The proof works provided i satisfies $i \leq \min(\lfloor \log_2(\alpha n/2^{i-1}) \rfloor, \log_2(\beta - 2\alpha)n)$, which holds for $i = O(\log n)$. The intuition for the sets is as follows:

- T_i is the set of vertices that are projected to zero by the i-th projection L_i that we construct.
- B_i is a maximal set of standard unit vectors that have not been projected to zero by the projection L_i that we construct.
- U_i is a subset of B_i of the most frequent standard unit vectors, that is, many of the 3-edges incident to a vertex in T_i are labeled by a vector in U_i.
- E_i is a subset of 3-edges incident to T_i that are labeled by a vector in U_i.
- V_i is a small set of vertices that when projected to zero, project T_i to zero.

Let $N(T_{i-1})$ be the neighborhood of vertices of T_{i-1}, that are not themselves in T_{i-1} (so $N(T_{i-1})$ and T_{i-1} are disjoint). We define a multigraph G_{i-1} on vertex set $N(T_{i-1})$ where we connect two vertices by a 2-edge if and only if they are included in a 3-edge in E_{i-1}. Let $r[i-1]$ be the number of connected components of G_{i-1}. Let $C_{i-1,1}, \ldots, C_{i-1,r[i-1]}$ be the connected components of G_{i-1}, where $|C_{i-1,1}| \geq |C_{i-1,2}| \geq \cdots \geq |C_{i-1,r[i-1]}|$. For each connected component $C_{i-1,j}$, arbitrarily choose a vertex $v_{i-1,j} \in C_{i-1,j}$.

Let $T_i = \cup_{j=1}^{\lfloor \alpha n/2^{i-1} \rfloor} C_{i-1,j}$, where $C_{i-1,j} = \emptyset$ if $j > r[i-1]$, and let

$$V_i = V_{i-1} \cup \{v_{i-1,1}, \ldots, v_{i-1,\lfloor \alpha n/2^{i-1} \rfloor}\} \quad \text{(recall that } V_1 = T_1\text{)}.$$

Extend $ex(V_i \cup (\cup_{j=1}^{i-1} U_j))$ to a basis of $\mathbb{F}^n$ by greedily adding a subset B_i of unit vectors in B_{i-1}. Let E be the set of 3-edges incident to some vertex in T_i, labeled by a $u \in B_i$. We will inductively have that $|U_j| = (\beta - 2\alpha)n/2^j$ for all $j \leq i - 1$. Notice that this holds for our above definition of U_1. Notice that

$$|B_i| \geq n - |V_i| - |\cup_{j=1}^{i-1} U_j| \geq n - \sum_{j=1}^{i} \left\lfloor \frac{\alpha n}{2^{j-1}} \right\rfloor - \sum_{j=1}^{i-1} \frac{(\beta - 2\alpha)n}{2^j}$$

$$\geq n - \alpha n - \sum_{j=1}^{i-1} \frac{\alpha n}{2^j} - \sum_{j=1}^{i-1} \frac{(\beta - 2\alpha)n}{2^j}$$

$$= n - \alpha n - \sum_{j=1}^{i-1} \frac{\beta n - \alpha n}{2^j}$$

$$= n - \alpha n - \beta n + \alpha n + \frac{(\beta - \alpha)n}{2^{i-1}}$$

$$= n - \beta n + \frac{(\beta - \alpha)n}{2^{i-1}}$$

Each vertex in T_i has degree at least βn, since all vertices in G' have degree at least βn. It follows that each vertex in T_i is incident to at least $\beta n - (n - |B_i|) \geq \frac{(\beta - \alpha)n}{2^{i-1}}$ edges in E, since a vertex cannot be incident to two different edges of the same label. Since an edge can be counted at most 3 times, $|E| \geq |T_i| \cdot \frac{(\beta - \alpha)n}{3 \cdot 2^{i-1}}$. For each $u \in B_i$, let f_u denote the number of edges in E labeled by u, and order the unit vectors so $f_{u_1} \geq \cdots \geq f_{u_{|B_i|}}$. Let $E_i \subset E$ be the subset of edges incident to T_i labeled by a unit vector in the set U_i of the first $\frac{(\beta - 2\alpha)n}{2^i}$ unit vectors. Notice that our earlier assumption that $|U_j| = (\beta - 2\alpha)n/2^j$ for all $j \leq i - 1$ holds by this definition of U_i.

Recursive projection: $|T_1| = \alpha n$, and for $i > 1$, $|T_i| = \sum_{j=1}^{\lfloor \alpha n/2^{i-1} \rfloor} |C_{i-1,j}|$. Also, for all $i \geq 1$, $|U_i| = (\beta - 2\alpha)n/2^i$. We turn to bounding $|E_i|$. Since we chose the $(\beta - 2\alpha)n/2^i$ most frequent unit vectors (in terms of the number of their occurrences in E) to include in the set U_i, and since E_i is the set of edges in E labeled by a unit vector in U_i, we have that $|E_i|$ must be at least a $(\beta - 2\alpha)/2^i$ fraction of $|E|$ (there are only n possible unit vectors). That is, we have

$$|E_i| \geq \frac{(\beta - 2\alpha)}{2^i} \cdot |E| \geq \frac{(\beta - 2\alpha)}{2^i} \cdot |T_i| \cdot \frac{(\beta - \alpha)n}{3 \cdot 2^{i-1}} = \left[\frac{2(\beta - 2\alpha)(\beta - \alpha)}{3} \right] \cdot \frac{|T_i|n}{4^i}.$$

We define a sequence of linear projections L_i for $i \geq 1$ as follows. We set $L_i(ex(V_i \cup (\cup_{j=1}^{i-1} U_j))) = 0$, and $L_i(u) = u$ for all $u \in B_i$.

Claim. For any $i \geq 2$, if $j \leq \lfloor \alpha n/2^{i-1} \rfloor$, then all vertices $b \in C_{i-1,j}$ satisfy $L_i(b) = 0$.

Proof. We prove this by induction on $i \geq 2$. For the base case $i = 2$, consider any vertex b in $C_{1,j}$, and let $v_{1,j} = a_0, a_1, a_2, \ldots, a_k = b$ be a path from $v_{1,j}$ to b in $C_{1,j}$. Since $\{a_0, a_1\}$ is an edge in $C_{1,j}$, we have $a_0, a_1 \in N(T_1)$ and so there is a 3-edge $e = \{w, a_0, a_1\} \in E_1$ with $w \in T_1$ and labeled by a $u_j \in U_1$. But then $L_2(w) = 0$ since $w \in T_1 = V_1$. Moreover, $L_2(u_j) = 0$ since $u_j \in U_1$. But, for non-zero $\gamma_1, \gamma_2, \gamma_3 \in \mathbb{F}$, $\gamma_1 w + \gamma_2 a_0 + \gamma_3 a_1 = u_j$. These conditions imply that $\gamma_2 L_2(a_0) + \gamma_3 L_2(a_1) = 0$. Now, notice that $v_{1,j} \in V_2$ since $j \leq \lfloor \alpha n/2^{i-1} \rfloor$, and so $L_2(v_{1,j}) = L_2(a_0) = 0$. It follows that $L_2(a_1) = 0$. By repeated application on the path from $v_{1,j}$ to $a_k = b$, we get $L_2(b) = 0$.

Inductively, suppose it is true for all values from 2 up to $i - 1$. We prove it for i. Consider any vertex b in $C_{i-1,j}$ and let $v_{1,j} = a_0, a_1, \ldots, a_k = b$ be a path from $v_{1,j}$ to b in $C_{i-1,j}$. Since $\{a_0, a_1\}$ is an edge in $C_{i-1,j}$, we have $a_0, a_1 \in N(T_{i-1})$ and so there is a 3-edge $e = \{w, a_0, a_1\} \in E_{i-1}$ with $w \in T_{i-1}$ and labeled by a $u_j \in U_{i-1}$. But then $L_i(w) = 0$ since $w \in T_{i-1}$ and so $w \in C_{i-2,j}$ for some $j \leq \lfloor \alpha n/2^{i-2} \rfloor$, which by the inductive hypothesis means $L_{i-1}(w) = 0$, and the kernel of L_{i-1} is contained in the kernel of L_i. Now also $L_i(u_j) = 0$ since $u_j \in U_{i-1}$. For non-zero $\gamma_1, \gamma_2, \gamma_3 \in \mathbb{F}$, we have $\gamma_1 w + \gamma_2 a_0 + \gamma_3 a_1 = u_j$, and so $\gamma_2 L_i(a_0) + \gamma_3 L_i(a_1) = 0$. Notice that $v_{1,j} \in V_i$ since $j \leq \lfloor \alpha n/2^{i-1} \rfloor$, and so $L_i(v_{1,j}) = L_i(a_0) = 0$. Hence, $L_i(a_1) = 0$, and by repeated application on the path from $v_{1,j}$ to $a_k = b$, we get $L_i(b) = 0$. This completes the induction.

For each component $C_{i-1,j}$ for any i and j, let $c_{i-1,j}$ denote $|C_{i-1,j}|$ for notational convenience.

Lemma 3.1. *For any $i \geq 2$, if $j \leq \lfloor \alpha n / 2^{i-1} \rfloor$, then the number of edges in $C_{i-1,j}$ is at most $c_{i-1,j} \log c_{i-1,j} + c_{i-1,j}$.*

Proof. Let $\{a, b\}$ be an edge in $C_{i-1,j}$. Then there is an edge $e = \{a, b, c\} \in E_{i-1}$ with $c \in T_{i-1}$. Then $\gamma_1 a + \gamma_2 b + \gamma_3 c = u_k$ for some $u_k \in U_{i-1}$, for non-zero $\gamma_1, \gamma_2, \gamma_3$ in $\mathbb{F}$. Since $e \in E_{i-1}$, we have $u_k \in U_{i-1} \subseteq B_{i-1}$, and so we have $L_{i-1}(u_k) = u_k$. Now, $c \in T_{i-1}$, and by Claim 3.2, L_{i-1} vanishes on all of T_{i-1} In particular, $L_{i-1}(c) = 0$. By linearity, $\gamma_1 L_{i-1}(a) + \gamma_2 L_{i-1}(b) = u_k$. Moreover, for each $k' \in [n]$, each vertex in $C_{i-1,j}$ can occur in at most one 3-edge labeled by $u_{k'}$ (by definition of the matchings in G'), so we obtain matchings $W_{k'}$, where an edge $\{a, b\}$ in $C_{i-1,j}$ is in $W_{k'}$ iff there is an $e \in E_{i-1}$ labeled by $u_{k'}$. By Theorem 3.1, $\sum_{k'} |W_{k'}| \leq c_{i-1,j} \log c_{i-1,j} + c_{i-1,j}$. But the number of edges in $C_{i-1,j}$ is at most the sum of matching sizes $|W_{k'}|$ for $u_{k'} \in U_{i-1}$.

Define the constant $\gamma = 2(\beta - 2\alpha)(\beta - \alpha)/3$. It follows that for all i, we have the constraints

1. $\frac{\gamma |T_{i-1}| n}{4^{i-1}} \leq |E_{i-1}| \leq \sum_{j=1}^{r[i-1]} (c_{i-1,j} \log c_{i-1,j} + c_{i-1,j})$
2. $|T_i| = \sum_{j=1}^{\lfloor \alpha n / 2^{i-1} \rfloor} c_{i-1,j}$

Lemma 3.2. *Suppose for $i = 1, 2, \ldots, \Theta(\log \log n)$, we have $|T_i| > 8|T_{i-1}|$. Then $m = \Omega(n^2)$.*

Proof. By induction, $|T_i| > 8^{i-1} |T_1| = 8^{i-1} \alpha n$ for $i = 1, 2, \ldots, \Theta(\log \log n)$. We thus have,

$$|E_i| \geq \gamma \cdot \frac{|T_i| n}{4^i} \geq \frac{\gamma \alpha}{8} \cdot 2^i n^2.$$

Hence, for $i = \Theta(\log \log n)$, we have $|E_{i-1}| = \Omega(n^2 \log n)$. Using that $\Omega(n^2 \log n) = |E_{i-1}| \leq \sum_{j=1}^{r[i-1]} (c_{i-1,j} \log c_{i-1,j} + c_{i-1,j})$, we have

$$m \geq \sum_{j=1}^{r[i-1]} c_{i-1,j} = \Omega(n^2 \log n / \log n) = \Omega(n^2),$$

where we have used that $c_{i-1,j} \leq n^2$ for all i and j, as otherwise $m \geq c_{i-1,j} = n^2$ for some i and j, and we would already be done. Hence, we can use $\log c_{i-1,j} = O(\log n)$.

Lemma 3.3. *Suppose for a value $i = O(\log \log n)$, $c_{i-1,1} = \Omega(n^2 / \log n)$. Then $m = \Omega(n^2)$.*

Proof. Notice that $|T_i| \geq c_{i-1,1} = \Omega(n^2 / \log n)$, and also, $|E_i| = \Omega(|T_i| n / 4^i) = \Omega(n^3 / \text{polylog}(n)) = \Omega(n^2 \log n)$. Using the constraint that $m \geq \sum_{j=1}^{r[i-1]} c_{i-1,j} = \Omega(|E_i| / \log n)$, it follows that $m = \Omega(n^2)$. Here we have again upper bounded $\log c_{i-1,j}$ by $O(\log n)$, justified as in the proof of Lemma 3.2.

Lemma 3.4. *Suppose for a value* $i = O(\log\log n)$, $|T_i| \leq 8|T_{i-1}|$. *Then* $m = \Omega(n^2)$.

Proof. Let i^* be the smallest integer i for which $|T_i| \leq 8|T_{i-1}|$. It follows that $|T_{i^*-1}| \geq 8^{i^*-2}|T_1| = 8^{i^*-2}\alpha n$. Note that $|E_{i^*-1}| = \Omega(|T_{i^*-1}|n/4^{i^*-1}) = \Omega(n^2 2^{i^*})$. We attempt to maximize the RHS of constraint 1 defined above, namely

$$\sum_{j=1}^{r[i^*-1]} (c_{i^*-1,j} \log c_{i^*-1,j} + c_{i^*-1,j}), \tag{1}$$

subject to a fixed value of $|T_{i^*}|$, where recall $|T_{i^*}| = \sum_{j=1}^{\lfloor \alpha n/2^{i^*-1}\rfloor} c_{i^*-1,j}$. We can assume that

$$c_{i^*-1,1} \geq c_{i^*-1,2} = c_{i^*-1,3} = \cdots = c_{i^*-1,\lfloor \alpha n/2^{i^*-1}\rfloor},$$

as otherwise we could increase $c_{i^*-1,1}$ while replacing the other values with $c_{i^*-1,\lfloor \alpha n/2^{i^*-1}\rfloor}$, which would preserve the value of $|T_{i^*}|$ and only make constraint 1 defined above easier to satisfy (notice that since $|T_{i^*}|$ is fixed, the LHS of constraint 1 remains fixed, as well as both sides of constraint 2). Moreover, constraint 1 is only easier to satisfy if we make

$$c_{i^*-1,\lfloor \alpha n/2^{i^*-1}\rfloor} = c_{i^*-1,\lfloor \alpha n/2^{i^*-1}\rfloor+1} = \cdots = c_{i^*-1,r[i^*-1]}.$$

We can assume that $c_{i^*-1,1} = o(n^2/\log n)$, as otherwise Lemma 3.3 immediately shows that $m = \Omega(n^2)$. In this case, though, $c_{i^*-1,1}$ does not contribute asymptotically to sum (1) since $|E_{i^*-1}| = \Omega(n^2 2^{i^*})$ and so sum 1 must be at least this large. It follows that we can replace constraint 1 with

$$\Omega(|T_{i^*-1}|n/4^{i^*}) \leq rA(\log A + 1), \tag{2}$$

where A is the common value $c_{i^*-1,x}$, where $r = r[i^* - 1]$, and where $x \in \{2,\ldots,r\}$. Using that $i = O(\log\log n)$, so we can ignore the floor operation in constraint 2, constraint 2 becomes $An/2^{i^*} = \Theta(|T_{i^*}|)$, or equivalently, $A = \Theta(|T_{i^*}|2^{i^*}/n)$.

Using that $|T_{i^*}| \leq 8|T_{i^*-1}|$, it follows that $A = O(|T_{i^*-1}|2^{i^*}/n)$. Combining this with our reformulation of constraint 1 in (2), we have

$$r(\log A + 1) = \Omega(n^2/8^{i^*}),$$

or equivalently, $r = \Omega(n^2/(8^{i^*}(\log A + 1)))$. Now,

$$m = \Omega(Ar) = \Omega\left(\frac{n|T_{i^*-1}|}{4^{i^*}(\log(|T_{i^*-1}|2^{i^*}/n) + 1)}\right).$$

This is minimized when $|T_{i^*-1}|$ is as small as possible, but $|T_{i^*-1}| \geq 8^{i^*-2}\alpha n$. Hence, $m = \Omega\left(\frac{n^2 2^{i^*}}{\log 16^{i^*}}\right)$, which is minimized for $i^* = \Theta(1)$, in which case $m = \Omega(n^2)$, as desired.

Combining Lemma 3.2 and Lemma 3.4, we conclude,

Theorem 3.3. *For $\delta, \epsilon \in (0, 1)$, if $C : \mathbb{F}^n \to \mathbb{F}^m$ is a linear $(3, \delta, \epsilon)$-locally decodable code, then $m = \Omega_{\delta, \epsilon}(n^2)$, independent of the field $\mathbb{F}$.*

4   From Adaptive Decoders to Non-adaptive Decoders

Theorem 4.1. *For given $\delta, \varepsilon \in (0, 1)$, if $C : \mathbb{F}^n \to \mathbb{F}^m$ is a linear $(3, \delta, \epsilon)$-LDC, then C is a linear $(3, \delta/9, 2/3 - 1/|\mathbb{F}|)$-LDC with a non-adaptive decoder.*

Proof. Since C is a linear code, each of its coordinates can be identified with a vector $v_j \in \mathbb{F}^n$, with the function for that coordinate computing $\langle v_j, x \rangle$, where the inner product is over $\mathbb{F}$. Define the ordered list of vectors $B = v_1, \ldots, v_m$.

Fix some $i \in [n]$, and let $\mathcal{C}_i$ be the collection of all non-empty sets $S \subseteq \{v_1, \ldots, v_m\}$, with $|S| \leq 3$, for which $u_i \in \text{span}(v_j \mid v_j \in S)$, where u_i denotes the unit vector in direction i. Let $D_i \subseteq \{v_1, \ldots, v_m\}$ be a smallest dominating set of $\mathcal{C}_i$, that is, a set for which for all $S \in \mathcal{C}_i$, $|S \cap D_i| > 0$.

Claim. $|D_i| > \delta m$.

Proof. Suppose not. Consider the following adversarial strategy: given a codeword $C(x)$, replace all coordinates $C(x)_j$ for which $v_j \in D_i$ with 0. Denote the new string $\tilde{C}(x)$. The coordinates of $\tilde{C}(x)$ compute the functions $\langle \tilde{v}_j, x \rangle$, where $\tilde{v}_j = v_j$ if $v_j \notin D_i$, and $\tilde{v}_j = 0$ otherwise. Let $\tilde{B}$ be the ordered list of vectors $\tilde{v}_1, \ldots, \tilde{v}_m$.

Define 3-span($\tilde{B}$) to be the (possibly infinite) list of all vectors in the span of each subset of $\tilde{B}$ of size at most 3. We claim that $u_i \notin$ 3-span($\tilde{B}$). Indeed, if not, then let $S \subseteq \{\tilde{v}_1, \ldots, \tilde{v}_m\}$ be a smallest set for which $u_i \in \text{span}(S)$. Then $|S| \leq 3$. This is not possible if $|S| = 0$. It follows that $S \cap D_i \neq \emptyset$. This implies that 0 is a non-trivial linear combination of vectors in S. Indeed, there is an ℓ for which $\tilde{v}_\ell \in S$ and $v_\ell \in D_i$, implying $\tilde{v}_\ell = 0$. Hence, $u_i \in \text{span}(S \setminus \tilde{v}_\ell)$. But $|S \setminus \{\tilde{v}_\ell\}| < |S|$, which contradicts that S was smallest.

Let A be the decoder of C, where A computes $A^y(i, r)$ on input index $i \in [n]$ and random string r. Here, for any $x \in \mathbb{F}^n$, we let the string $y = y(x)$ be defined by the adversarial strategy given above. For any $x \in \mathbb{F}^n$, $A^y(i, r)$ first probes coordinate j_1 of y, learning the value $\langle \tilde{v}_{j_1}, x \rangle$. Next, depending on the answer it receives, it probes coordinate j_2, learning the value $\langle \tilde{v}_{j_2} x \rangle$. Finally, depending on the answer it receives, it probes coordinate j_3, learning the value $\langle \tilde{v}_{j_3} x \rangle$. Consider the affine subspace V of dimension $d \geq n - 2$ of all $x \in \mathbb{F}^n$ which cause $A^y(i, r)$ to read positions j_1, j_2, and j_3. Let V_0 be the affine subspace of V of all x for which $A^y(i, r)$ outputs x_i. Since the output of $A^y(i, r)$ is fixed given that it reads positions j_1, j_2, and j_3, and since $u_i \notin \text{span}(\tilde{v}_{j_1}, \tilde{v}_{j_2}, \tilde{v}_{j_3})$, it follows that the dimension of V_0 is at most $d - 1$.

Suppose first that $\mathbb{F}$ is a finite field. Then for any fixed r, the above implies $A^y(i, r)$ is correct on at most a $\frac{1}{|\mathbb{F}|}$ fraction of $x \in \mathbb{F}^n$ since $\frac{|V_0|}{|V|} \leq \frac{1}{|\mathbb{F}|}$ for any set of three indices j_1, j_2, and j_3 that A can read. Thus, by averaging, there exists an $x \in \mathbb{F}^n$ for which $\Pr[A^y(i) = x_i] \leq \frac{1}{|\mathbb{F}|}$, where the probability is over the random coins r of A. This contradicts the correctness of A.

Now suppose that $\mathbb{F}$ is an infinite field. We will show that there exists an $x \in \mathbb{F}^n$ for which $\Pr[A^y(i) = x_i] = 0$, contradicting the correctness of the decoder.

For each random string r, there is a finite non-empty set G_r of linear constraints over $\mathbb{F}$ that any $x \in \mathbb{F}^n$ must satisfy in order for $A^y(i, r) = x_i$. Consider the union $\cup_r G_r$ of all such linear constraints. Since the number of different r is finite, this union contains a finite number of linear constraints.

Since $\mathbb{F}$ is infinite, we claim that we can find an $x \in \mathbb{F}^n$ which violates all constraints in $\cup_r G_r$. We prove this by induction on n. If $n = 1$, then the constraints have the form $x_1 = c_1, x_1 = c_2, \ldots, x_1 = c_s$ for some finite s. Thus, by choosing $x_1 \notin \{c_1, c_2, \ldots, c_s\}$, we are done. Suppose, inductively, that our claim is true for $n - 1$. Now consider $\mathbb{F}^n$. Consider all constraints in $\cup_r G_r$ that have the form $x_1 = c$ for some $c \in \mathbb{F}$. There are a finite number of such constraints, and we can just choose x_1 not to equal any of these values c, since $\mathbb{F}$ is infinite. Now, substituting this value of x_1 into the remaining constraints, we obtain constraints (each depending on at least one variable) on $n - 1$ variables $x_2, \ldots, x_n$. By induction, we can choose the values to these $n - 1$ variables so that all constraints are violated. Since we haven't changed x_1, the constraints of the form $x_1 = c$ are still violated. This completes the proof.

It follows that since $|D_i| > \delta m$ and D_i is a smallest dominating set of C_i, we can greedily construct a matching M_i of $\delta m/3$ disjoint triples $\{v_{j_1}, v_{j_2}, v_{j_3}\}$ of $\{v_1, \ldots, v_m\}$ for which $u_i \in \text{span}(v_{j_1}, v_{j_2}, v_{j_3})$.

Consider the new behavior of the decoder: on input $i \in [n]$, choose a random triple $\{v_{j_1}, v_{j_2}, v_{j_3}\} \in M_i$, and compute x_i as $\gamma_1 \langle v_{j_1}, x \rangle + \gamma_2 \langle v_{j_2}, x \rangle + \gamma_3 \langle v_{j_3}, x \rangle$, where $u_i = \gamma_1 v_{j_1} + \gamma_2 v_{j_2} + \gamma_3 v_{j_3}$. Since the adversary can now corrupt at most $\delta m/9$ positions, it follows that with probability at least $2/3$, the positions queried by the decoder are not corrupt and it outputs x_i. Note that the new decoder also makes at most 3 queries.

This can be extended straightforwardly to any constant $q > 3$ number of queries:

Theorem 4.2. *For given $\delta, \varepsilon \in (0,1)$, if $C : \mathbb{F}^n \to \mathbb{F}^m$ is a linear (q, δ, ϵ)-LDC, then C is a linear $(q, \delta/(3q), 2/3 - 1/|\mathbb{F}|)$-LDC with a non-adaptive decoder.*

Acknowledgment. The author thanks Johnny Chen, Anna Gal, Piotr Indyk, Swastik Kopparty, Alex Samorodnitsky, C. Seshadhri, Ronald de Wolf, Sergey Yekhanin, and the anonymous referees for many helpful comments.

References

1. Sipser, M., Spielman, D.A.: Expander codes. IEEE Trans. Inform. Theory 42, 1710–1722 (1996)
2. Katz, J., Trevisan, L.: On the efficiency of local decoding procedures for error-correcting codes. In: STOC (2000)

3. Goldreich, O., Karloff, H.J., Schulman, L.J., Trevisan, L.: Lower bounds for linear locally decodable codes and private information retrieval. In: CCC (2002)
4. Trevisan, L.: Some applications of coding theory in computational complexity. Quaderni di Matematica 13, 347–424 (2004)
5. Yekhanin, S.: Locally Decodable Codes and Private Information Retrieval Schemes. PhD thesis, MIT (2007)
6. Candès, E.J., Romberg, J.K., Tao, T.: Robust uncertainty principles: exact signal reconstruction from highly incomplete frequency information. IEEE Transactions on Information Theory 52, 489–509 (2006)
7. Donoho, D.L.: Compressed sensing. IEEE Transactions on Information Theory 52, 1289–1306 (2006)
8. Duarte, M., Davenport, M., Takhar, D., Laska, J., Sun, T., Kelly, K., Baraniuk, R.: Single-pixel imaging via compressing sensing. IEEE Signal Processing Magazine (2008)
9. Dvir, Z., Shpilka, A.: Locally decodable codes with 2 queries and polynomial identity testing for depth 3 circuits. In: Symposium on the Theory of Computing, STOC (2005)
10. Dvir, Z.: On matrix rigidity and locally self-correctable codes. In: IEEE Conference on Computational Complexity, CCC (2010)
11. Valiant, L.G.: Graph-theoretic arguments in low-level complexity. In: MFCS, pp. 162–176 (1977)
12. Kerenidis, I., de Wolf, R.: Exponential lower bound for 2-query locally decodable codes. In: STOC (2003)
13. Woodruff, D.P.: New lower bounds for general locally decodable codes. Electronic Colloquium on Computational Complexity (ECCC) 14 (2007)
14. Obata, K.: Optimal lower bounds for 2-query locally decodable linear codes. In: Rolim, J.D.P., Vadhan, S.P. (eds.) RANDOM 2002. LNCS, vol. 2483, pp. 39–50. Springer, Heidelberg (2002)
15. Woodruff, D.P.: Corruption and recovery-efficient locally decodable codes. In: Goel, A., Jansen, K., Rolim, J.D.P., Rubinfeld, R. (eds.) APPROX and RANDOM 2008. LNCS, vol. 5171, pp. 584–595. Springer, Heidelberg (2008)
16. Efremenko, K.: 3-query locally decodable codes of subexponential length. In: STOC (2009)
17. Itoh, T., Suzuki, Y.: New constructions for query-efficient locally decodable codes of subexponential length (manuscript) (2009)
18. Yekhanin, S.: Towards 3-query locally decodable codes of subexponential length. J. ACM 55 (2008)
19. Dvir, Z., Gopalan, P., Yekhanin, S.: Matching vector codes. Electronic Colloquium on Computational Complexity, ECCC (2010)
20. Gopalan, P.: A note on Efremenko's locally decodable codes. Electronic Colloquium on Computational Complexity, ECCC (2009)
21. Diestel, R.: Graph theory. Graduate Texts in Mathematics. Springer, Heidelberg (2005)

A   Basic Reductions

Intuitively, a local-decoding algorithm A cannot query any particular location of the (corrupted) codeword too often, as otherwise an adversary could ruin the success probability of A by corrupting only a few positions. This motivates the definition of a *smooth code*.

Definition A.1. *([2]) For fixed c, ϵ, and integer q, a linear transformation C : $\mathbb{F}^n \to \mathbb{F}^m$ is a linear (q, c, ϵ)-smooth code if there exists a probabilistic oracle machine A such that for every $x \in \mathbb{F}^n$,*

- *For every $i \in [n]$ and $j \in [m]$, $\Pr[A^{C(x)}(i)$ reads index $j] \leq \frac{c}{m}$.*
- *For every $i \in [n]$, $\Pr[A^{C(x)}(i) = x_i] \geq \frac{1}{|\mathbb{F}|} + \epsilon$.*
- *In every invocation A makes at most q queries.*

The probabilities are taken over the coin tosses of A. An algorithm A satisfying the above is called a (q, c, ϵ)-smooth decoding algorithm for C (a decoder for short).

Unlike a local-decoding algorithm, a smooth decoding algorithm is required to work only when given access to a valid codeword, rather than a possibly corrupt one. The following reduction from LDCs to smooth codes was observed by Katz and Trevisan.

Theorem A.1. *([2]) Let $C : \mathbb{F}^n \to \mathbb{F}^m$ be a linear (q, δ, ϵ)-LDC that makes non-adaptive queries. Then C is also a linear $(q, q/\delta, \epsilon)$-smooth code.*

We use a graph-theoretic interpretation of smooth codes given in [3] and [2]. Let $C : \mathbb{F}^n \to \mathbb{F}^m$ be a linear (q, c, ϵ)-smooth code, and let algorithm A be a (q, c, ϵ)-smooth decoding algorithm for C. Since C is linear, each of the m positions of C computes $\langle v_i, x \rangle$ for a vector $v_i \in \mathbb{F}^n$. We say that a given invocation of A *reads* a set $e \subseteq \{v_1, \ldots, v_m\}$ if the set of inner prodcuts that A reads in that invocation equals $\{\langle v_i, x \rangle \mid v_i \in e\}$. Since A is restricted to read at most q entries, $|e| \leq q$.

We say that e is *good* for i if $\Pr[A^{C(x)}(i) = x_i \mid A$ reads $e] \geq \frac{1}{|\mathbb{F}|} + \frac{\epsilon}{2}$, where the probability is over the internal coin tosses of A. It follows that if e is good for i, then the i-th standard unit vector u_i is in the span of the $|e|$ vectors. Indeed, otherwise, one can find two different inputs x which agree on the inner products that are read but differ in coordinate i.

Definition A.2. *([2]) Fixing a smooth code $C : \mathbb{F}^n \to \mathbb{F}^m$ and a q-query recovery algorithm A, the recovery hypergraphs for $i \in [n]$, denoted G_i, consist of the vertex set $\{v_1, \ldots, v_m\}$ and the hyperedge set $C_i = \{e \subseteq \{v_1, \ldots, v_m\} \mid u_i \in span(e)\}$.*

Lemma A.1. *([2]) Let C be a (q, c, ϵ)-smooth code that is good on average, and let $\{G_i\}_{i=1}^n$ be the set of recovery hypergraphs. Then, for every i, the hypergraph $G_i = (\{v_1, \ldots, v_m\}, C_i)$ has a matching M_i of sets of size q with $|M_i| \geq \frac{\epsilon m}{cq}$.*

Consider the multi-hypergraph G with vertex set $\{v_1, \ldots, v_m\}$ and hyperedge set $\biguplus_{i=1}^n M_i$, that is, a hyperedge occurs in G once for each M_i that it occurs in. For readability, we use the term hypergraph to refer to a multi-hypergraph, that is, a hypergraph which may have repeated hyperedges (which we sometimes just refer to as edges). We claim that we can find a non-empty induced sub-hypergraph G' of G with minimum degree βn for a constant $\beta > 0$. The proof is a straightforward generalization of Proposition 1.2.2 in [21] to hypergraphs. For a proof, see Lemma 27 in Appendix 6 of [13] (omitted here due to space constraints).

Author Index

Printing: Mercedes-Druck, Berlin
Binding: Stein+Lehmann, Berlin